Zoology

For Degree Students

Chordata | Physiology I | Biochemistry I

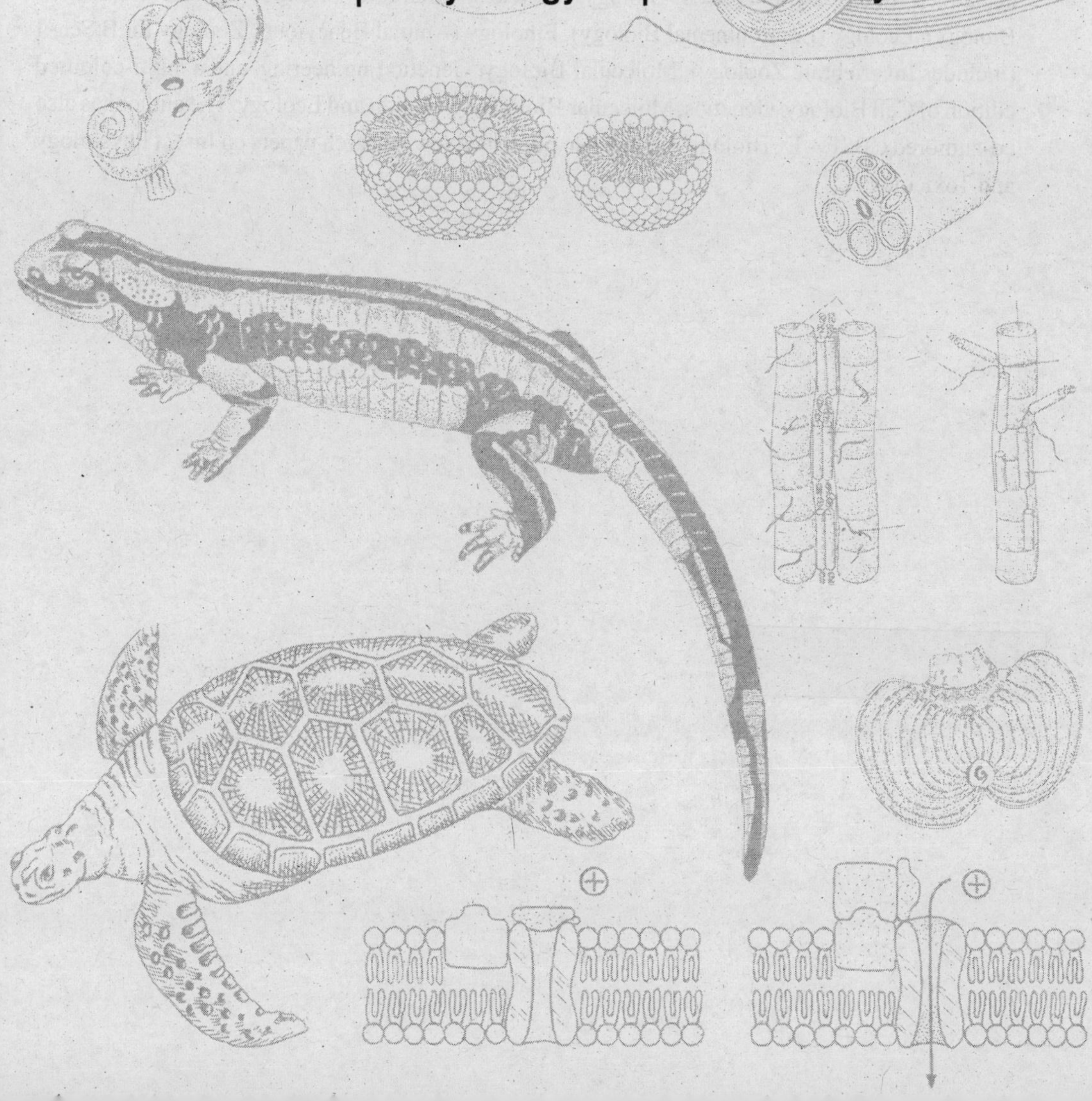

About the Author

Dr V K Agarwal, M.Sc., Ph.D., has rich teaching experience of more than four decades at Meerut College, Meerut. He has taught subjects such as Cell Biology, Genetics, Molecular Biology, Chordate Embryology (Developmental Biology), Invertebrate Zoology, Chordate Zoology, Evolution, Ecology, Animal Behaviour (Ethology) to undergraduate and postgraduate students.

Dr Agarwal has to his credit the authorship of several widely acclaimed books such as Cell Biology, Genetics, Chordate Embryology (Developmental Biology), Ecology (Environmental Biology), Ethology (Animal Behaviour), Zoology for B.Sc. - I (includes Invertebrate Zoology), Molecular Biology, Genetic Engineering, and a multi-coloured edition of Cell Biology, Genetics, Molecular Biology, Evolution and Ecology. The author has also co-authored CBSE – IX Biology. He has also published few research papers on Insect Physiology and Toxicology.

Zoology

For Degree Students

Chordata | Physiology I | Biochemistry I

B.Sc. (Honours) Semester - III

As per new UGC Model Curriculum
Choice Based Credit System (CBCS)

V K Agarwal
MSc, PhD
Former Head of the Zoology Department
Meerut College, Meerut, Uttar Pradesh

S. CHAND
PUBLISHING

S Chand And Company Limited
(ISO 9001 Certified Company)

S. Chand And Company Limited

(ISO 9001 Certified Company)

Head Office: D-92, Sector–2, Noida – 201301, U.P. (India), Ph. 91-120-4682700

Registered Office: A-27, 2nd Floor, Mohan Co-operative Industrial Estate, New Delhi – 110 044, Phone: 011-49731800

www.**schandpublishing.com;** e-mail: **info@schandpublishing.com**

Marketing Offices:

Chennai	:	Ph: 23632120; chennai@schandpublishing.com
Guwahati	:	Ph: 2738811, 2735640; guwahati@schandpublishing.com
Hyderabad	:	Ph: 40186018; hyderabad@schandpublishing.com
Jalandhar	:	Ph: 4645630; jalandhar@schandpublishing.com
Kolkata	:	Ph: 23357458, 23353914; kolkata@schandpublishing.com
Lucknow	:	Ph: 4003633; lucknow@schandpublishing.com
Mumbai	:	Ph: 25000297; mumbai@schandpublishing.com
Patna	:	Ph: 2260011; patna@schandpublishing.com

First Edition 2016

Reprints 2017, 2018 (Twice), 2019, 2020 (Twice), 2021, 2022 (Twice)

Reprint 2023

ISBN: 978-93-525-3311-4 **Product Code:** H6ZOO68ZOOL10ENAA16O

PRINTED IN INDIA

By Vikas Publishing House Private Limited, Plot 20/4, Site-IV, Industrial Area Sahibabad, Ghaziabad – 201 010 and Published by S. Chand And Company Limited, A-27, 2nd Floor, Mohan Co-operative Industrial Estate, New Delhi – 110 044.

Preface

It gives me immense pleasure to present this textbook, **Zoology for Degree Students – Semester III**, covering subjects Chordata, Physiology and Biochemistry. This comprehensively written textbook provides up-to-date information on the subject in an informative and systematic manner.

It has been designed as per the new UGC Model Curriculum – Choice Based Credit System (CBCS). Undergraduate students will find this book immensely useful as it lucidly explains the concepts with the help of well-illustrated images and diagrams. As per the CBCS curriculum, following three core courses of B.Sc. (Hons.) - Semester III have been included in this textbook:

1. **CC–V** **Diversity of Chordata** (19 Chapters)
2. **CC–VI** **Physiology I: Controlling and Coordinating Systems** (9 Chapters)
3. **CC–VII** **Fundamentals of Biochemistry I** (6 Chapters)

Some peculiar features of this book are:

- every topic of the syllabus has been fully justified to its importance
- text has been presented in a simple and lucid language
- subject matter has been arranged in a systematic and well-knit manner
- textbook is profusely illustrated with well-drawn and labelled diagrams not only to supplement the descriptions but also to clearly grasp the text
- running boxed items throughout the text providing information on intricate terms or topics
- engaging and exhaustive question bank at the end of each chapter

This book helps develop an appreciation and understanding of the scientific concepts that unite Zoology and reflect on "how" science works; as **John A. Moore** (1988) aptly puts it "science is a way of knowing."

To friends and family, I remain grateful and thank them for their support during writing of this textbook. Special thanks to my daughter Dr. Surabhi Gupta for helping me in developing the chapter on Endocrine Glands.

Many thanks to the editorial team at S. Chand, especially, Surabhi Khare, Shalini Negi, Rupesh Gupta, Priyanka Sharan, Himani Ramola and Rajesh Singh for timely release of the book.

Suggestions and feedback from the readers will be gratefully acknowledged.

V K Agarwal
Villa–20,
Green Wood City
Haridwar Bypass Road
Meerut - 250 005
Uttar Pradesh

Syllabus

B.Sc. (Hons.) Zoology – Semester III

CORE COURSE V–DIVERSITY OF CHORDATA

Unit 1: Introduction to Chordates 2

General characteristics and outline classification

Unit 2: Protochordata 8

General characteristics of Hemichordata, Urochordata and Cephalochordata; Study of larval forms in protochordates; Retrogressive metamorphosis in Urochordata

Unit 3: Origin of Chordata 3

Dipleurula concept and the Echinoderm theory of origin of chordates, Advanced features of vertebrates over Protochordata

Unit 4: Agnatha 2

General characteristics and classification of cyclostomes up to class

Unit 5: Pisces 8

General characteristics of Chondrichthyes and Osteichthyes, classification up to order Migration, Osmoregulation and Parental care in fishes

Unit 6: Amphibia 6

Origin of Tetrapoda (Evolution of terrestrial ectotherms); General characteristics and classification up to order; Parental care in Amphibians

Unit 7: Reptilia 7

General characteristics and classification up to order; Affinities of Sphenodon; Poison apparatus and Biting mechanism in snakes

Unit 8: Aves 8

General characteristics and classification up to order Archaeopteryx-- a connecting link; Principles and aerodynamics of flight, Flight adaptations and Migration in birds

Unit 9: Mammals 8

General characters and classification up to order; Affinities of Prototheria; Adaptive radiation with reference to locomotory appendages

Unit 10: Zoogeography 8

Zoogeographical realms, Theories pertaining to distribution of animals, Plate tectonic and Continental drift theory, distribution of vertebrates in different realms

CORE COURSE VI–ANIMAL PHYSIOLOGY: CONTROLLING AND COORDINATING SYSTEMS

Unit 1: Tissues 6

Structure, location, classification and functions of epithelial tissue, connective tissue, muscular tissue and nervous tissue

Unit 2: Bone and Cartilage 4

Structure and types of bones and cartilages, Ossification, bone growth and resorption

Unit 3: Nervous System 10

Structure of neuron, resting membrane potential, Origin of action potential and its propagation across the myelinated and unmyelinated nerve fibers; Types of synapse,

Synaptic transmission and, Neuromuscular junction; Reflex action and its types - reflex arc; Physiology of hearing and vision.

Unit 4: Muscle **12**

Histology of different types of muscle; Ultra structure of skeletal muscle; Molecular and chemical basis of muscle contraction; Characteristics of muscle twitch; Motor unit, summation and tetanus

Unit 5: Reproductive System **10**

Histology of testis and ovary; Physiology of male and female reproduction; Puberty, Methods of contraception in male and female

Unit 6: Endocrine System **18**

Histology of endocrine glands - pineal, pituitary, thyroid, parathyroid, pancreas, adrenal; hormones secreted by them and their mechanism of action; Classification of hormones; Regulation of their secretion; Mode of hormone action, Signal transduction pathways for steroidal and non-steroidal hormones; Hypothalamus (neuroendocrine gland) - principal nuclei involved in neuroendocrine control of anterior pituitary and endocrine system; Placental hormones

CORE COURSE VII–FUNDAMENTALS OF BIOCHEMISTRY

Unit 1: Carbohydrates **8**

Structure and Biological importance: Monosaccharides, Disaccharides, Polysaccharides and Glycoconjugates

Unit 2: Lipids **8**

Structure and Significance: Physiologically important saturated and unsaturated fatty acids, Tri-acylglycerols, Phospholipids, Glycolipids, Steroids

Unit 3: Proteins **14**

Amino acids: Structure, Classification and General properties of α-amino acids; Physiological importance of essential and non-essential α-amino acids **Proteins:** Bonds stabilizing protein structure; Levels of organization in proteins; Denaturation; Introduction to simple and conjugate proteins **Immunoglobulins:** Basic Structure, Classes and Function, Antigenic Determinants

Unit 4: Nucleic Acids **12**

Structure: Purines and pyrimidines, Nucleosides, Nucleotides, Nucleic acids Cot Curves: Base pairing, Denaturation and Renaturation of DNA, Types of DNA and RNA, Complementarity of DNA, Hpyo-Hyperchromaticity of DNA

Unit 5: Enzymes **18**

Nomenclature and classification; Cofactors; Specificity of enzyme action; Isozymes; Mechanism of enzyme action; Enzyme kinetics; Factors affecting rate of enzyme-catalyzed reactions; Derivation of Michaelis-Menten equation, Concept of Km and Vmax, Lineweaver-Burk plot; Multi-substrate reactions; Enzyme inhibition; Allosteric enzymes and their kinetics; Regulation of enzyme action

Contents

DIVERSITY OF CHORDATA

FUNDAMENTALS OF BIOCHEMISTRY

Diversity of Chordata
(CC-V)

1

Phylum Chordata

The animal kingdom is divided into two groups: Nonchordata (or Invertebrata) and Chordata, on the basis of a structure, called **notochord** (*Gr., noton* = back; *L. chorda* = *a* cord). The nonchordates lack a notochord and are grouped into 26 phyla. Their major phyla are Porifera, Cnidaria, Platyhelminthes, Nematoda, Annelida, Arthropoda, Mollusca, Echinodermata, Lophophorate phyla and Hemichordata. The chordates contain a notochord and form a single phylum **Chordata**.

Chordata is the largest phylum of deuterostome and includes a majority of living and extinct animals such as tunicates, lancelets, lampreys, fishes, amphibians, reptiles, birds and mammals.

While nonchordates may be either protostomia (in which blastopore becomes mouth) or deuterostomia (in which blastopore becomes anus), chordates are always deuterostomia. Lastly, chordates are most familiar, most adaptable, most successful and most widely distributed animals, showing diversity in form, habitat and habits to an amazing degree. Only the arthropods among the nonchordates rival them in this regard. Indeed, phylum Chordata marks the climax of evolution.

The chordates range from Cambrian to recent period. There are more than 52022 extant (currently living) species of chordates.

1.1 GENERAL CHARACTERS OF CHORDATES

1. Bilateral symmetry; segmented body; three germ-layers (triploblastic); well-developed coelom (coelomate).
2. **Notochord** or skeletal rod present at some stage in the life cycle.
3. **Single, dorsal, tubular nerve cord;** anterior end of the cord usually enlarged to form brain.
4. **Pharyngeal pouches** present at some stage in the life cycle; in aquatic chordates these develop into gill slits.
5. **Postanal tail**, usually projecting beyond the anus at some stage but may or may not persist.
6. **Segmented muscles** in an unsegmented trunk.
7. **Ventral heart**, with dorsal and ventral blood vessels; closed blood system (or circulatory system or vascular system).
8. Complete digestive system (having mouth and anus).
9. A living cartilaginous or bony **endoskeleton** present in the majority of higher chordates (vertebrates).

1.2 FOUR CHORDATE HALLMARKS

Phylum Chordata includes bilaterally symmetrical, metamerically segmented, triploblastic, coelomate metazoa having highly complex organization. All chordates possess the following four characters, called **primary chordate characters** or **hallmarks of chordata** (Fig. 1.1) at least at some stage of their life cycle:

1. Notochord or chorda dorsalis. All chordates possess a dorsal, solid, unsegmented and flexible (elastic), axial rod, the **notochord**, extending along the whole length of the body. It runs mid-dorsally immediately above the alimentary canal and underneath the dorsal nerve cord. Notochord is formed of large vacuolated parenchymatous cells enclosed in an **inner elastic sheath** and an outer **chordal sheath** of dense fibrous connective tissue (Fig. 1.2). It is derived from a mid-dorsal strip of embryonic (archentric) mesoderm that provides structural and locomotory support in the body of larva or adult chordates (**Brusca** and **Brusca** 2003). The notochord serves as a primitive internal skeleton and supports the central nervous system and the segmental muscles.

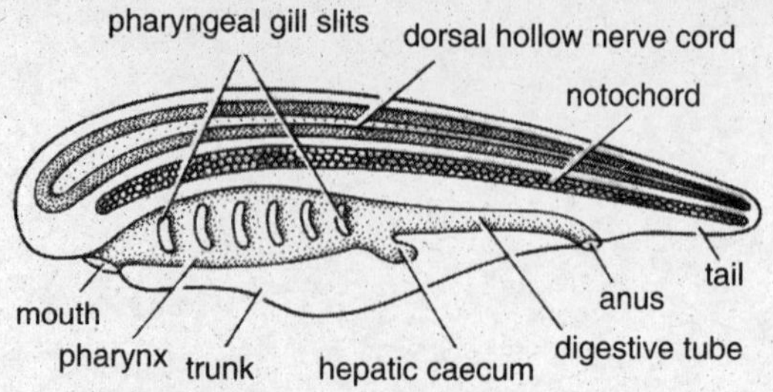

Fig. 1.1. Diagrammatic side view of a chordate showing four chordate hallmarks.

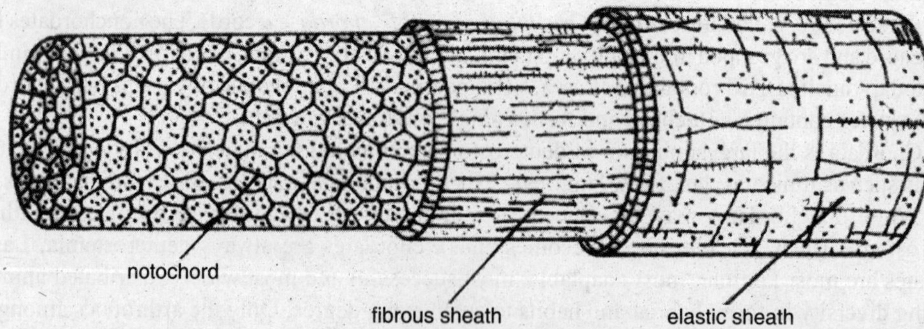

Fig. 1.2. Structure of the notochord and its surrounding sheaths. Cells of the notochord proper are thick walled, pressed together closely, and filled with semifluid. Stiffness is caused mainly by turgidity of fluid-filled cells and surrounding connective tissue sheaths. This primitive type of endoskeleton provides longitudinally stiffening of the main body axis, a base for trunk muscle, and an axis around which the vertebral column develops (after Hickman *et al.*, 1996).

In **Cephalochordata** (*Branchiostoma*) the notochord extends from one end of the body to the other and persists throughout life. In **Urochordata** (*Herdmania*) the notochord is restricted to the caudal region and is present only in larval stage. **Cyclostomata** (lampreys and hagfishes) possess a turgid and rigid notochord persisting throughout life. In *Scoliodon*, the notochord is constricted by a chain of vertebrae formed of cartilage. It is only represented by a notochordal filament in their center. In vertebrates the notochord is completely replaced by the vertebral column.

Notochord is one of key synapomorphies defining the chordates. *Synapomorphy* is a shared derived or new character or evolutionary novelty (**Brusca** and **Brusca** 2003).

2. Dorsal tubular nerve cord. The nervous system of chordates is in the form of a hollow tube which is situated mid-dorsally above the notochord and immediately beneath the body wall. It develops by process of invagination of the neural ectoderm in the mid-dorsal line, so that a medullary (neural) groove is formed. Later on, the two edges (folds) of the groove fuse together to form the medullary tube which sinks down below the surface and forms the **neural tube**. Neural tube develops in nerve cord of the adult. The cavity of the nerve cord is called **neurocoel**. The nerve cord persists throughout the life in most chordates, becoming differentiated into an anterior short, wide **brain** and a posterior long, slender **spinal cord**. In some lower chordates (urochordates), nerve cord degenerates in the adult.

3. Branchial or pharyngeal clefts. The **branchial** or **pharyngeal** clefts are found universally in all the chordates at some stage of their life cycle. These occur as paired series of perforations in the lateral wall of pharynx and lead to the exterior. These develop as a series of ectodermal impushings from the exterior and fuse with the similar outpushings of the endoderm of the pharynx. Such gill clefts appear during development of every chordates.

Duration of gill clefts varies considerably. They persist throughout life in lower chordates (*Herdmania*, *Branchiostoma*), fishes and some tailed amphibians; develop only in the larval stage in tailless amphibians; and appear as nonfunctional, externally closed traces, the **pharyngeal pouches** in reptiles, birds and mammals. In lower chordates the gill clefts do not bear gills and aid mainly in filter feeding and to some extent in respiration. In others, they bear vascular lamellae or filaments, the **gills** and only have a respiratory role.

4. Postanal tail. The tail is the part of the body behind the cloacal or anal aperture. It contains notochord or vertebral column, nerve cord, blood vessels, nerves and muscles, but lacks coelom and viscera. Tail is present in all the chordates (even human embryo contains a short tail). In aquatic forms (fishes), the tail serves as the chief locomotory organ, while in the terrestrial ones, it assumes various shapes and functions. Tail is reduced or absent in the adults of some chordates such as frogs and toads.

Some Additional but Unique Chordate Characters

1. Endostyle. Endostyle is a ciliated, glandular groove in the mid-ventral surface of the pharynx (throat region) in the adult urochordates and cephalochordates, and in the larva (ammocoete) of certain agnathans such as lampreys. It secretes mucus which traps tiny food particles for transport toward the oesophagus. The endostyle is also able to take up **iodine**, a capacity it retains after it is transformed into the **thyroid gland** of the adult lamprey on the completion of metamorphosis. The endostyle is analogous to the thyroid gland of gnathostome. The thyroid functions as an endocrine gland that regulates metabolism. An extract of endostyle stimulates the action of the thyroid gland.

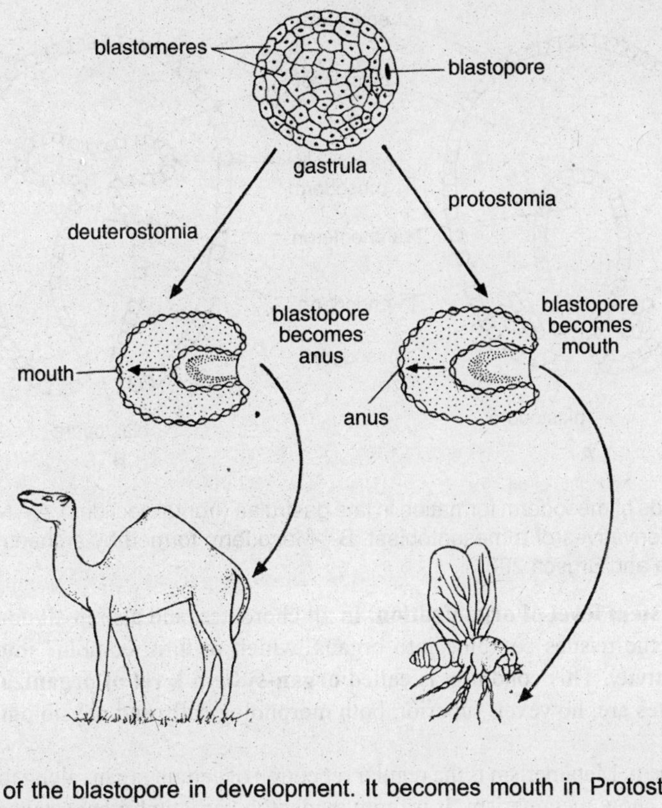

Fig. 1.3. Fate of the blastopore in development. It becomes mouth in Protostomia and anus in Deuterostomia.

2. Photosensitive cells (including optic retina, when present) develop from the ectoderm of the central nervous system (not from the epidermal ectoderm as in invertebrate phyla).

1.3 CHARACTERS SHARED BY CHORDATES WITH NONCHORDATE PHYLA

1. Bilateral symmetry. In bilateral symmetry, anterior and posterior ends, and dorsal and ventral surfaces are distinguishable, right and left sides of the body are mirror images of each other, and only sagittal (median longitudinal vertical) plane divides the body into two similar halves. This is the symmetry of body of all chordates and majority of nonchordates.

Bilateral symmetry gives **polarity** to the animal's body so that it moves with the anterior end forward and dorsal end upward. However, the anterior end of the chordates does not correspond to that of nonchordates. In the nonchordates (except the echinoderms, lophophorates and hemichordates) **mouth** is derived from the blastopore of the embryo (*i.e.*, gastruala), called the **protostomous condition** (Fig. 1.4A). In contrast, in chordates (and also in enchinoderms, lophophorates and hemichordates), **anus** appears near the blastopore and the mouth develops as an invagination at the opposite end (of blastopore), called the **deuterostome condition** (Fig. 1.4B).

2. Triploblastic nature. All chordates, and also all nonchordates above the coelenterates, develop from three germ layers: ectoderm, mesoderm and endoderm, and are said to be **triploblastic.** The origin of mesoderm is, however, different. In chordates, mesoderm arises from the dorsolateral regions of the wall of archenteron, whereas in nonchordates it mostly develops as solid outgrowths from the junction of ectoderm and endoderm (Fig. 1.4A). In echinoderms and chordates, the mesoderm arises from the wall of the archenteron itself (*i.e.*, a preformed endoderm), either as a solid sheet or as pouches (Fig. 1.4B).

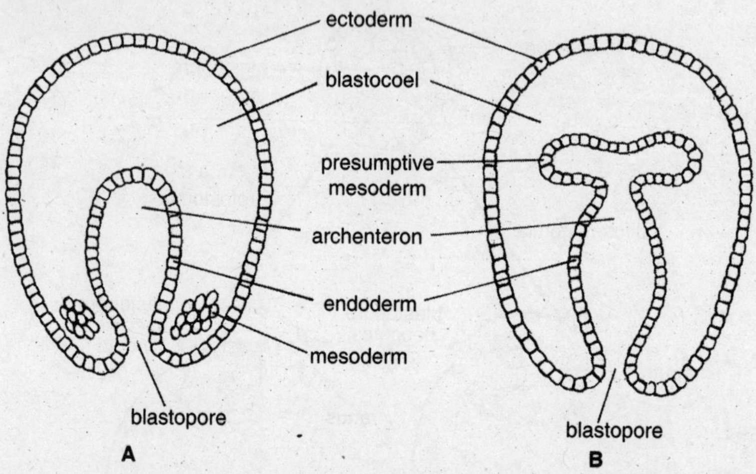

Fig. 1.4. Methods of mesoderm formation in late gastrulae (frontal sections). A—Mesoderm formed from derivatives of a mesentoblast. B—Mesoderm formed by archentric pouches (after Brusca and Brusca 2003).

3. Organ-system level of organization. In all chordates, and also in all nonchordates above the coelenterates, the tissues combine into organs, which in turn, combine into organ-systems, each for a vital activity. This condition is called **organ-system level of organization**. The organ-systems of chordates are, however, superior, both morphologically and physiologically, to those of nonchordates.

4. Metamerism. Metamerism is the regular repetition of certain organs along the anteroposterior axis. All chordates show metamerism. It is quite evident in early embryonic stages. In the adults, it gets masked externally, but persists internally in the arrangement of trunk muscles, paired nerves and blood vessels, ribs and vertebrae. The nonchordate phyla lack metamerism, except Annelida and Arthropoda, in which metamerism, though present, is of different type.

5. Cephalization. Cephalization means the formation of head by concentration of some anterior segments, nervous tissue and sense organs at the anterior end. The head contains brain, mouth and principal sense organs (*e.g.*, nares or olfactory sense organs, eyes, ears, gustatory sense organs). Cephalization dominates the body and directs its activities, for example, head being at the anterior end, it first encounters environmental changes during progression. In chordates, the size and specialization of the head steadily increase from the cyclostomes to the mammals. The phenomenon of cephalization also occurs in many nonchordates.

6. Appendages. The chordates have only two pairs of paired appendages (fins or limbs). One pair or both pairs may be secondarily lost. Mud-eel and whale have lost hindlimbs and snakes have lost both the pairs of limbs. The nonchordates have more than two pairs of paired appendages, if these are present.

7. Endoskeleton. A living endoskeleton of cartilage or/and bone is characteristic of most chordates. It does not interfere with the animal's growth, for it can increase in size with rest of the body. There is, thus, no need to shed it. The skeleton of nonchordates, if present, is always non-living. It may be horny, calcareous or chitinous. The rigid exoskeleton is periodically moulted to allow growth of the body as in arthropods.

8. Coelom. Coelom is the perivisceral space lined by mesodermal epithelium. It is present in most chordates and many nonchordates. The animals with coelom are described as **coelomates**. In nonchordates, the coelom mostly arises by a split in the mesoderm and is said to be **schizocoelic** (Fig. 1.5). For example, in schizocoely of annelids, bilaterally paired packets of mesoderm gradually enlarge and hollow, eventually becoming thin-walled coelomic spaces (Fig. 1.5A, B). The number of such paired coeloms varies among different animals and is frequently associated with segmentation, as it is in annelid worm (Fig. 1.5C).

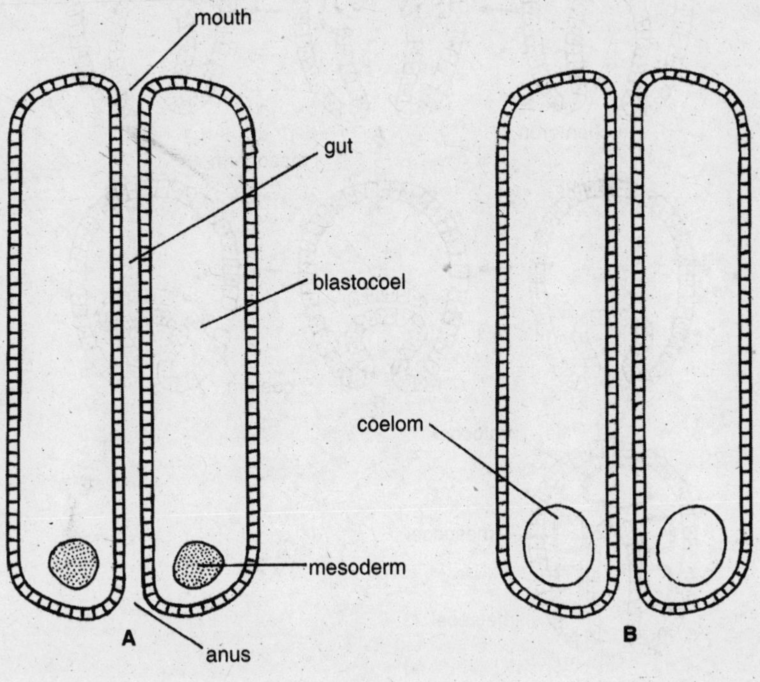

Figure Continue...

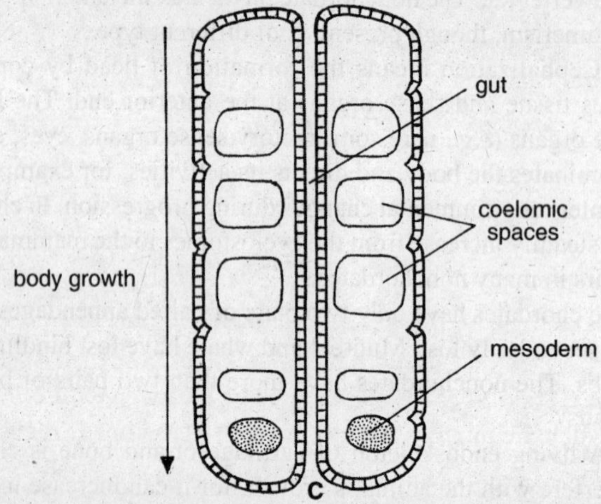

Fig. 1.5. Coelom formation by schizocoely (frontal sections). A—Precoelomic conditions with paired packets of mesoderm. B—Hollowing of the mesodermal packets to produce a pair of coelomic spaces. C—Progressive proliferation of serially arranged pairs of coelomic spaces. This process occurs in metameric annelids (after Brusca and Brusca 2003).

In lower chordates (cephalochordates) the coelom develops as linear series of paired hollow outgrowths from the archenteron and is described as **enterocoelic**. The higher chordates possess a modified form of enterocoelic coelom (Fig. 1.6; Box 1.1).

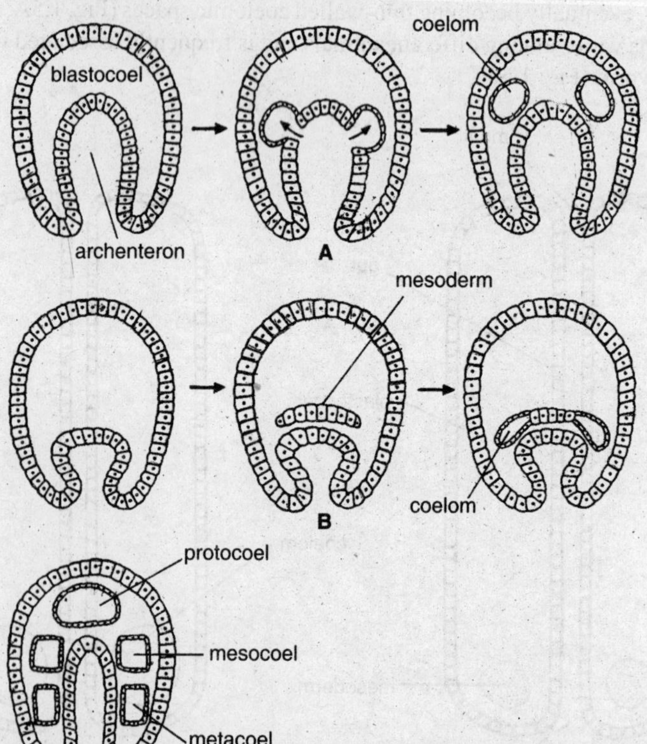

Fig. 1.6. Coelom formation by enterocoely (frontal section). A—Archentric pouching. B—Proliferation and subsequent hollowing of a plate of mesoderm from the archenteron. C—The typical tripartite arrangement of coelom in a deutrosome embryo (after Brusca and Brusca 2003).

Box 1.1.

Chordates have two types of enterocoely. In most direct sort of enterocoely, mesoderm production and coelom formation are one and the same process (Fig. 1.6A). It is called **archentric pouching.** A pouch or pouches form in the gut as a complete coelomic compartment. The walls of these pouches are defined as mesoderm. In some cases the mesoderm arises from the wall of archenteron as a solid sheet or plate that later becomes bilayered and hollow (Fig. 1.6B). Some authors consider this process to be a form of schizocoel (because of the "splitting" of the mesodermal plate), but it is in fact a modified form of enterocoely (**Brusca** and **Brusca** 2003). Enterocoely frequently results in a tripartite arrangement of the body cavities, which are disignated **procoel**, **mesocoel** and **metacoel** (Fig. 1.6C).

9. Digestive tract. Most chordates have a **complete digestive tract**. Such an alimentary canal always has a mouth for ingestion of food and an anus or cloacal aperture for egestion of faeces. Nonchordate phyla above platyhelminthes also possess similar complete digestive tract.

10. Closed circulatory system. In chordates the blood flows in a continuous system of blood vessels (arteries, capillaries and veins), all with definite walls. Such a circulatory system is described as **closed**. In this case blood does not come in direct contact with body tissues. Some nonchordates, *e.g.*, Annelida, also have closed circulatory system. Many nonchordates (arthropods, molluscs), however, possess an **open circulatory system** in which blood flows in sinuses, *i.e.*, blood channels having no walls. In open circulatory system, the blood is in direct contact with the tissues of the body.

11. Special respiratory and excretory organs. For some primitive chordates, respiration and excretion by body surface diffusion is adequate. However, in higher chordates, special organs for respiration (gills or lungs) and for excretion (kidneys) have developed. Special respiratory and excretory organs also occur in many nonchordates, *e.g.*, gills and nephridia in Annelida; tracheae and Malpighian tubules in Insecta; and ctenidia and kidneys in Mollusca.

12. Endocrine glands. The glands that pour their secretions (*i.e.*, the hormones) into the blood for transport to the target organs are called **endocrine** or **ductless glands**. Such endocrine glands are found in chordates as well as nonchordates, but they are better developed in chordates than in nonchordates.

13. Single pair of gonads with gonoducts. The chordates are generally unisexual and bear a single pair of gonads each of which is provided with a gonoduct for the discharge of the sex cells. A similar condition is found in many nonchordates.

14. Mesentries. In chordates, the visceral organs are interconnected and suspended from the dorsal body wall by double-fold of peritoneum, called **mesenteries**. Mesentries are absent in most nonchordates. However, some nonchordates such as *Neries* and *Asterias*, have small mesenteries in connection with the alimentary canal.

15. Energy compound. Chordates and some nonchordates such as echinoderms, use **phosphocreatine** in the energy cycle of muscle contraction. Most nonchordates use phosphoarginine for the same purpose.

16. Cleavage. Chordate eggs undergo **equipotential** and **radial cleavage**. In equipotential cleavage, each early blastomere on separation from others may give rise to a complete embryo (such cleavage is called **indeterminate cleavage**). Most nonchordates have **spiral cleavage** which is **determinate type**, *i.e.*, each early blastomere has a fixed fate.

First two divisions are alike in both radial and spiral cleavages. However, in radial cleavage third and subsequent division planes are symmetrical to the axis of polarity. Therefore in radial cleavage, in eight cell embryo, cells of each tier lie directly above and in line with those of the lower tier (Fig. 1.7). In case of spiral cleavage, subsequent divisions of 4-cell embryo result in the displacement of blastomeres in such a way that they lie in the furrows between one another (Box 1.2).

Box 1.2.

In spiral cleavage, the division from four to eight cells involves a displacement of the cells near the animal pole in a clockwise (**dextrotropic**) direction (viewed from the animal pole). The next division, from eight to sixteen cells, occurs with a displacement in counter-clockwise (**levotropic**) direction; the next is clockwise, and so on (see **Brusca** and **Brusca** 2003).

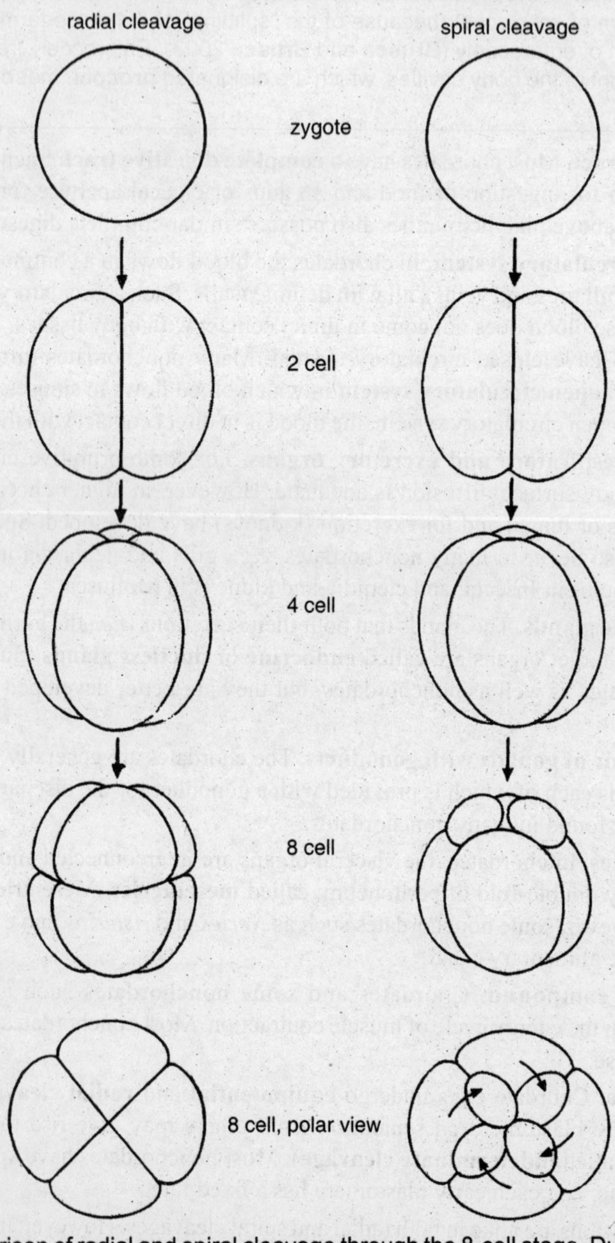

Fig. 1.7. Comparison of radial and spiral cleavage through the 8-cell stage. During radial cleavage, the cleavage planes all pass either perpendicular or parallel to the animal-vegetal axis of the embryo. Spiral cleavage involves a tilting of the mitotic spindles, commencing with the division from 4 to 8 cells. The resulting cleavage planes are neither perpendicular nor parallel to the axis. The polar view of the resulting 8-cell stages illustrate the differences in blastomere orientation (after Brusca and Brusca 2003).

1.4 DIFFERENCES BETWEEN CHORDATES AND NONCHORDATES

1. The chordates possess a hard supporting skeleton of body framework lying inside the body. It is called **endoskeleton**. The nonchordates either lack a skeleton but if present it is always outside the body (called **exoskeleton**) and is formed of nonliving horny, calcareous or chitinous substance.
2. A true tail is present in chordates. The tail is postanal prolongation of the body along the main axis of the body which contains the extension of muscles, nerve cord, notochord or vertebral column and blood vessels but lacks the visceral organs. True tail is absent in nonchordates.
3. In chordates heart is a highly muscular propulsive organ placed below the alimentary canal. In nonchordates the heart is always dorsal to the alimentary canal.
4. The chordates possess a closed type of blood vascular system which contains a definite system of closed blood vessels and their ultimate branches are known as capillaries. In them blood never comes in direct contact with the body tissues. In contrast, majority of nonchordates such as arthropods and molluscs have an open type of blood vascular system because the blood vessels in them open into blood sinuses, devoid of definite walls, and the body organs directly bathe in the blood.
5. In chordates, the food-laden blood from the alimentary canal is collected by the capillaries into bigger veins and is ultimately brought to the liver, where the hepatic portal vein bringing blood, ramifies into capillaries. In this way, the blood from the alimentary canal is not carried to the heart but to the liver. Such hepatic portal system is not found in the nonchordates.
6. In chordates, the respiratory pigment is usually haemoglobin, which is contained in the RBCs (red blood cells). In nonchordates, the haemoglobin if present, is found dissolved in plasma of the blood. (*Note*. The blood in phoronids (lophophorates) possesses nucleated red blood cells that contain haemoglobin for carrying oxygen; **Hickman** *et al.*, 1996).
7. In nonchordates the blood in the dorsal vessel flows from posterior to the anterior direction whereas in chordates blood always flows in anterior direction.
8. In nonchordates the **coelom** occurs as a single wide perivisceral cavity around the visceral organs. In chordates the perivisceral cavity (coelom) is divided into several regions and each has a specific function. These are dorsal **myocoel**, middle **nephrocoel**, lateral and ventral **splanchnocoel**.
9. In the nerve cord, the nerve cells are arranged deep around the central canal in chordates and superficially on the periphery in nonchordates.
10. In nonchordates the nerves arise mostly from the segmental ganglia of the nerve cord, whereas in chordates these arise from the dorsal and ventral nerve roots.
11. The paired appendages in nonchordates vary in number and may even correspond to the number of body segments. In chordates, however, limbs are invariably two pairs (tetrapods).
12. In higher chordates (*i.e.*, vertebrates) the notochord is replaced by the vertebral column. Vertebral column is not found in nonchordates and lower vertebrates.

1.5 CLASSIFICATION

Chordates form a large and heterogenous group. Phylum Chordata is divided into following three subphyla (**Brusca** and **Brusca** 2003):

1. Subphylum Urochordata (= Tunicata). The tunicates. Adult body form varies, but usually lacking obvious trimeric organizations; body covered by thick or thin (test) of a cellulose-like polysaccharide; without bony tissue; notochord restricted to tail and usually found only in larval stage (and in adult appendicularians); gut U-shaped, pharynx (branchial chamber) typically with numerous gill slits (stigmata); coelom not developed; dorsal nerve cord present in larval stages; all marine. It includes four classes:

Class 1. Ascidiacea, *e.g.*, *Herdmania*, *Ascidia*, *Botrylus*, *Ciona*, *Molgula*, etc.

Class 2. Thaliacea, *e.g.*, *Doliolum*, *Pyrosoma*, *Salpa*, etc.

Class 3. Appendicularia (= Larvacea), *e.g.*, *Oikopleura*.

Class 4. Sorberacea, *e.g.*, *Octacnemus*.

2. Subphylum Cephalochordata (= Acrania). Lancelets (amphioxus). Small (to 7 cm), fish-like chordates with notochord, gill slits, dorsal nerve cord and postanal tail present in adults, but without vertebral column or cranial skeleton; gonads numerous (25–38) and serially arranged. Marine and brackish water, usually associated with clean sand or gravel segments in which they burrow, *e.g.*, *Asymmetron*, *Branchiostoma*, *Epigonichthyes*.

3. Subphylum Vertebrata. Vertebrates. Chordates usually possessing a vertebral column that forms the axis of the body skeleton; most with paired appendages, a brain case and (except for members of the class Agnatha) jaws. Several classes are generally recognized, although not all are strictly monophyletic: Class **Myxini** and **Cephalaspidomorphi** are the hagfishes and lampreys, respectively (united as the jawless or agnathan fishes); Class **Chondrichthyes** are the sharks, skates, and rays; class **Osteichthyes** are the bony fishes (*e.g.*, trout, tuna, perch); Class **Amphibia** includes the salamanders, frogs, toads, caecilians; Class **Reptilia** traditionally included the turtles, snakes, lizards and crocodilians, but modern classification place birds (Class **Aves**) and reptiles together as **Reptilomorpha** (or **Sauropsida**). Class **Mammalia** comprises the mammals (see **Brusca** and **Brusca** 2003).

1.6 ANCESTRY AND EVOLUTION

1. Time of origin. The earliest known truely vertebrate animals were freshwater forms abundant during the late Silurian and middle Devonian periods. Their fossils are collectively known as Ostracoderms. Their body was covered by a bony armour forming a shield on head. These jawless bottom-dwelling lamprey eel-like fishes were adapted for filter feeding (Fig. 1.8). The oldest fossils belonging to Ostracoderms occur in late **Cambrian** and middle **Ordovician** periods.

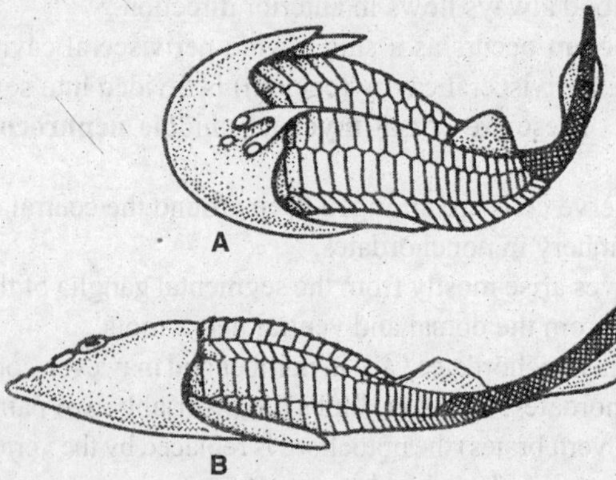

Fig. 1.8. A ostracoderm (*Cephalaspis*) in dorsal (A) and lateral (B) view.

Until recently, it was thought that the earliest known vertebrates were Ordovician jawless "fishes". However, it now appears that fossils long known from the Cambrian, tooth-like structure, called **conodonts**, belong to the earliest known vertebrates. Teeth are made of cellular bone (**Sansom** *et al.*, 1992). Conodont animals (Fig. 1.9) were small and eel-shaped; they lacked vertebrae, but had a notochord. In these respects, they resemble living hag-fishes, to which they may be related.

2. Place of origin. American geologist, **Chamberlain** (1900), gave the idea of **freshwater origin of chordates**. It was supported by **Romer** and **Smith**. However, **marine origin of chordates** was also considered as the protochordates and deuterosome invertebrate phyla are exclusively marine forms.

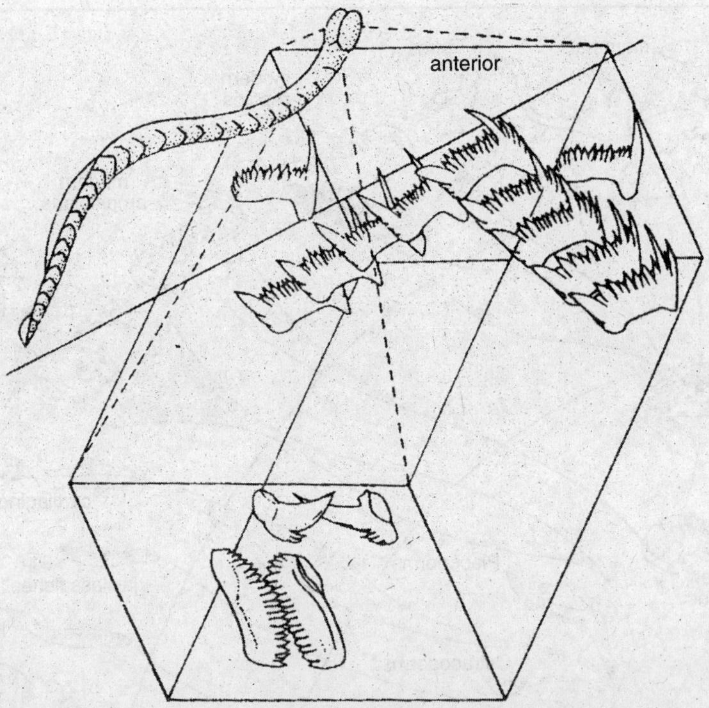

Fig. 1.9. A Cambrian conodont animal showing tooth-like structure.

3. Hypothetical vertebrate ancestor (Prevertebrate). There are several lines of anatomical, developmental and molecular evidences suggesting that somewhat later, at the base of the Cambrian period some 570 million years ago, the first distinctive chordates appeared, evolved from an early echinoderm, early hemichordate, or a common ancestor of both (Fig. 1.10; **Hickman** *et al.*, 1996).

Although modern echinoderms look nothing at all like—modern chordates, there is one curious group of fossil echinoderms, the "**Calcichordata**" which have lived during the Ordovician period (450 million years ago). Calcichordates had gill slits and possibly other chordate attributes. These small nonsymmetrical forms had a head resembling a long-toed medieval boot, a series of gill slits covered with flaps much like the gill openings of sharks, a flexible "arm" that could be interpreted as a postanal tail, and structures that are tentatively interpreted as a notochord and muscle blocks. They apparently used their gill slits for suspension feeding, as do the primitive chordates today. Although the calcichordates apparently had some of the right chordate characters based on soft anatomy, there is no convincing similarity between the hard skeleton of calcichordates (which was calcium carbonate) and that of vertebrates (which is composed of a hydrated complex of calcium and phosphates called **hydroxyapatite**). While such speculation bring us closer to an understanding of chordate origins, we are not yet in a position to know the precise characteristics of the long-sought chordate ancestor. However, we do know a great deal about the two living protochordate groups that descended from it, *i.e.*, Urochordata (Tunicata) and Cephalochordata (see **Hickman** *et al.*, 1996).

4. Theories of Chordate Ancestry.

***(i)* Echinoderm ancestry.** On the basis of anatomical, embryological, palaeontological, biochemical and serological evidences, various workers have tried to establish that chordates probably originated from some primitive echinoderm or echinoderm larva. The hemichordate larva (tornaria) is strikingly similar to the bipinnaria or dipleurula larva of echinoderms. But this view is not accepted these days.

***(ii)* Hemichordate ancestry.** Hemichordates are sedentary and have pharyngeal gill slits and a hollow dorsal nerve cord. But the presence of a true notochord is doubtfull and their adult body plan is different from chordates. Hence they have been put under the separate phylum.

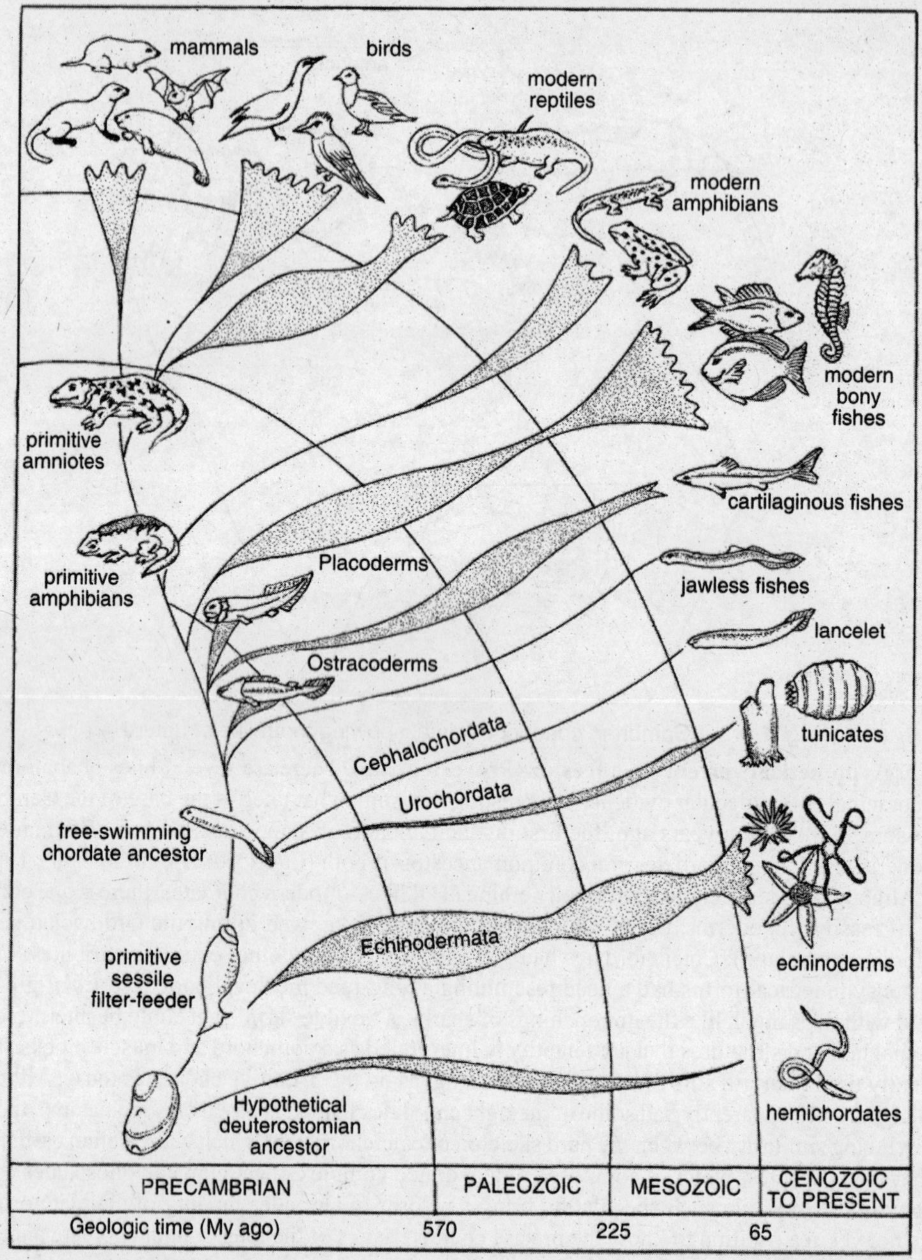

Fig. 1.10. Phylogenetic tree of the chordates. The relative abundance in numbers of species of each group through geological time, as indicated by the fossil record, is suggested by the bulging and thinning of that group's line of descent (after Hickman *et al.*, 1996).

***(iii)* Urochordate ancestry.** The urochordate theory of chordate origin was proposed by **W. Garstang** (1982). The adult tunicates reflect the primitive sessile, marine and filter feeding condition of ancestral vertebrates. But their body plans are divergent from that of chordates. This theory is also rejected.

***(iv)* Cephalochordate ancestry.** *Branchiostoma* (amphioxus) is an interesting animal possessing all the three chordate characters, and hence can act as a model prevertebrate but it can not be taken as ancestral vertebrate. The excretory system of cephalochordates consisting of flame cells, called **solenocytes**, is altogether different from that of vertebrate kidneys.

Further, lack of well marked cephalization and sense organs and the unique forward extension of notochord indicate that the cephalochordates have the identical ancestral body plan of vertebrates but are not themselves ancestral. Both may represent divergent paths of evolution from a common ancestor (see **Hickman** *et al.*, 1996).

***(v)* Barrington's hypothesis.** The most plausible hypothesis regarding the chordate origin was given by **Barrington** in 1965. It is based on the deuterosome line of chordate evolution. The common echinoderm-chordate ancestor was in all probability a small, sessile or semisessile, lophophorate (Box 1.3) or arm feeding creature. This gave rise to early stalked echinoderms. The next step was the derivation of sessile filter feeder or stem chordate. **Pharyngotremy**, *i.e.*, perforated pharynx resulted in the evolution of free living hemichordates and sessile urochordates. Some ancestral tunicates, instead of producing ciliated larva formed tadpole larva with typical somatic features of the chordates. Later on, this larva increased in size, by paedogenesis suppressed the sessile adult stage, developed gonads and become the ancestor of cephalochordates and vertebrates.

Box 1.3.

The lophophorate animals are included into three phyla: Phylum Phoronida, Phylum Ectoprocta (Bryozoa) and Phylum Brachiopoda. These coelomates belong to the deuterosome branch of bilateral animals, but they have some characterstics typical of protostomes. They possess a ciliary feeding and respiratory device called as **lophophore** around their mouth (which is specialized for sedentary filter feeding) (see **Hickman** *et al.*, 1996).

***(vi)* Inversion hypothesis.** One of the most intriguing new views of animal evolution comes from orthologous dorsiventral patterning genes, which suggest that the ventral nerve cord (plus brain) of insects may be homologous to the dorsal spinal cord (plus brain) of chordates (**Nubler-Jung** and **Arendt** 1994, 1996; **Nielsen** 2001). This "**inversion hypothesis**" suggests the possibility that chordates might be dorsiventral inverted with respect to all other animals. The vertebrate gene *Bmp-4*, a ventral determinant in the embryo, has been identified as a counterpart and probable homologue of the insect gene *dpp*, a dorsal determinant. Likewise, *Chordin* gene in vertebrates is a dorsal determinant, while its apparent insect counterpart, *sog*, is a ventral one. Thus, the dorsal surface of vertebrates—and coincidentally, the central nervous system that forms from it—and the ventral surface of insects (and by implication that of all deuterostomes) may be homologous tissue domains. A *sog*, *Chordin* precursor gene may have determined the dorsoventral axis in some ancient wormlike precursor to the proteostome-deuterostome split. This being the case, changes in the early development of chrodates might have somehow flipped the axis, inverting the expression pattern of genes (see **Brusca** and **Brusca** 2003).

QUESTIONS

Long Answer Questions

1. What are the main characteristics of chordates? Justify the inclusion of *Herdmania* and amphioxus in this phylum. *(C.C.S.U. 1999)*
2. What are protochordates? Give their distinguishing features and classify them upto orders giving suitable examples. *(Rohilkhand 2002; Madras 2011)*
3. Describe chordate characters.
4. Describe different hypotheses/theories of origin of chordates. Add a note on Calcichordata.

Short Answer Questions

1. Describe the notochord.
2. How do chordates and nonchordates differ in their mesoderm, coelom and circulatory system?
3. Give an outline classification of chordates upto subphyla only.
4. Describe the general characters of Phylum Chordata. (*Madras 2011*)

Very Short Answer Questions

1. Define the notochord. (*C.C.S.U. 2009*)
2. Give three distinguishing chordate characters. (*C.C.S.U. 2010*)
3. What type of circulatory system is present in chordates?
4. What is the number of paired appendages in chordates?
5. When did origin of chordates took place?
6. In which group of animals, the notochord is replaced by the vertebral column?
7. Name the respiratory pigment of chordates?
8. Cite one example of deuterostomes and one of proteostomes.
9. Name the animal which has the schizocoelic coelom.

Fill in the Blanks

1. Notochord persists throughout life in ________.
2. Notochord acts as a ________ rod.
3. Chordate nerve cord is ________ in position.
4. By ________ pharynx communicates with the exterior.
5. Tail is the ________ or ________ part of body.
6. Chordate heart is ________ in position.
7. Respiratory pigment is present in the ________ of blood in the chordates.
8. Cleavage in chordate egg is ________ and ________.
9. ________ have endoskeleton of living tissues.

Multiple Choice Questions

1. Chordates are different from nonchordates in having
   (*a*) a notochord (*b*) a schizocoel
   (*c*) a ventral solid nerve cord (*d*) all the above
2. Which one of the following is found in the chordates but not in nonchordates
   (*a*) chitinous exoskeleton (*b*) gills
   (*c*) postanal tail (*d*) spiracles
3. How many pairs of appendages are present in vertebrates
   (*a*) one pair (*b*) two pairs
   (*c*) four fairs (*d*) variable from species to species
4. Chordates with a backbone are called
   (*a*) protochordates (*b*) agnatha
   (*c*) gnathostomates (*d*) vertebrates
5. The group amniota includes
   (*a*) birds and mammals (*b*) birds and reptiles
   (*c*) reptiles and mammals (*d*) reptiles, birds and mammals
6. Vertebral column of vertebrates is a modified
   (*a*) nerve cord (*b*) notochord
   (*c*) umbilical cord (*d*) spermatic cord

7. In vertebrates the skin is covered by
   (*a*) scales (*b*) hairs
   (*c*) feathers (*d*) any one of the above
8. Vertebrates respire by
   (*a*) lungs (*b*) gills
   (*c*) skin (*d*) all of the above
9. The amniote egg first evolved in
   (*a*) bony fish (*b*) amphibian
   (*c*) reptiles (*d*) birds
10. Which group of the invertebrates are the closest relatives of the chordates?
    (*a*) Annelida (*b*) Mollusca
    (*c*) Arthropoda (*d*) Echinodermata

ANSWERS

Very Short Answer Questions

1. Notochord is an elongated cellular cord, enclosed in a sheath, which forms the primitive axial skeleton of chordate embryos and adult cephalochordates.
2. (*a*) presence of notochord, (*b*) presence of dorsal tubular central nervous system and (*c*) presence of branchial clefts.
3. Closed circulatory system.
4. Two pairs (hence called tetrapods).
5. Late Cambrian and middle Ordovician.
6. Vertebrates.
7. Haemoglobin.
8. Chordates, nonchordates except Echinodermata and Hemichordata.
9. Annelids.

Fill in the Blanks

1. Cephalochordates. **2.** Skeletal. **3.** Middorsal. **4.** Pharynx. **5.** Postanal, postcloacae aperture. **6.** Ventral. **7.** Erythrocytes. **8.** Radial, equipotential. **9.** Chordates.

Multiple Choice Questions

1. (*a*)	**2.** (*c*)	**3.** (*b*)	**4.** (*d*)	**5.** (*d*)	**6.** (*b*)
7. (*d*)	**8.** (*d*)	**9.** (*c*)	**10.** (*d*)		

2

Phylum Hemichordata (General Characters, Classification and Types)

The **hemichordata** (Gr., *hemi* = half; *chorde* cord) is a small group of worm-like marine invertebrates. They are called hemichordata since notochord in them occurs in the anterior end of the body. Recently, notochord of hemichordates is considered as **buccal diverticulum**. Hemichordata is a group of **deuterostomes** such as the Echinodermata and Chordata. It was once considered a subphylum of the chordates (as non-vertebrate chordates). Hemichordates form an evolutionary link between the echinoderms and chordates.

2.1 DEFINITION

Hemichordates are marine, soft-bodied, bilaterally symmetrical, coelomate, worm-like animals that contain diffused nervous system in the epidermis, a buccal diverticulum and paired gill-slits and no definite nephridia. Their body is divided into proboscis (protosome), collar (mesosome) and trunk (metasome).

2.2 OUTLINE CLASSIFICATION

Phylum Hemichordata

Classes (2)

- **1. Enteropneusta**
  Examples. *Balanoglossus, Saccoglossus*
- **2. Pterobranchia**
  Orders (2)
  - **1. Rhabdopleurida**
    Example. *Rhabdopleura*
  - **2. Cephalodiscida**
    Examples. *Cephalodiscus, Atubaria*

2.3 GENERAL CHARACTERS

1. Simple, solitary or colonial sedentary or free-living and marine animals, found exclusively in sea-water.

2. They are deposit or suspension feeders.
3. They are soft-bodied, bilaterally symmetrical, coelomate and worm-like.
4. Most are burrowers and some live beneath the stones.
5. The body is divided into an anterior muscular **proboscis (protosome)**, a short **collar (mesosome)** and a long **trunk (metasome)**. Proboscis is separated by a narrow stalk from the collar.
6. Mouth opens infront of collar.
7. Epidermis is ciliated and richly glandular. Cuticle is absent.
8. Coelom is **enterocoelous** and divisible into protocoel, mesocoel and metacoel.
9. A buccal diverticulum, formerly regarded as notochord, present in the proboscis.
10. Alimentary canal is a straight tube having a terminal anus.
11. Gill slits are numerous, paired, present in the anterior part of the trunk.
12. Blood vascular system is simple; blood flows forward in dorsal blood vessel.
13. Excretion through a single **glomerulus** situated in the proboscis.
14. Nervous system consists of an inter-epidermal nerve plexus and a dorsal hollow neural tube.
15. Movement occurs by peristaltic contraction of the proboscis.
16. Sexes are separate, fertilization is external and development may be **indirect** (having **tornaria larva**) or direct.

2.4 DETAILED CLASSIFICATION

The phylum Hemichordata is divided into two classes:

Class 1. Enteropneusta

1. Enteropneusta, which means **gut-breather**, includes about 70 species which are known as **acorn worms**, because of the typical shape of their proboscis.
2. They live in shallow sea-water. Most are burrowers (Fig. 2.1). Acorn worms are very **sluggish**, most live in tubes and eat sand or mud; others are free living and ciliary mucus-feeders.

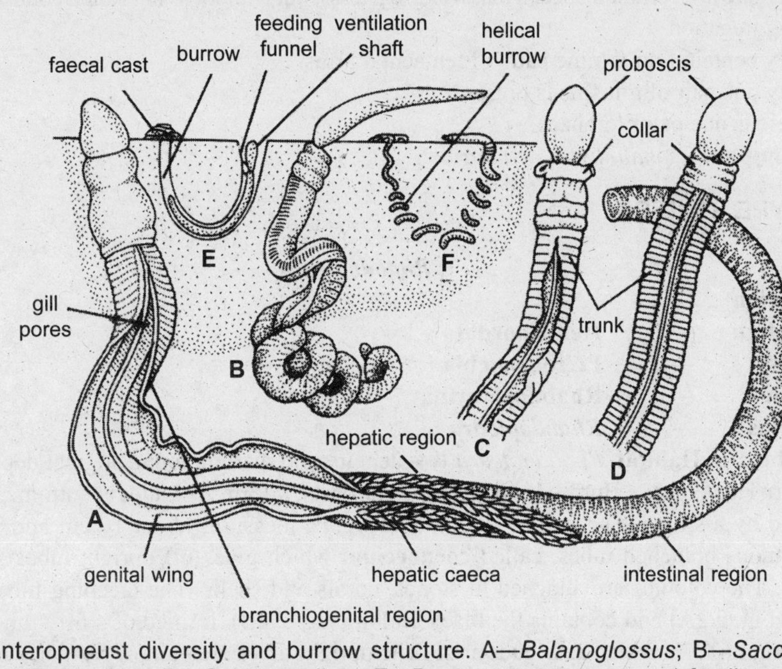

Fig. 2.1. Enteropneust diversity and burrow structure. A—*Balanoglossus*; B—*Saccoglossus*; C—*Ptychodera*; D—*Schizocardium*; E—Burrow arrangement and feeding position of *Balanoglossus*; F—Burrow and feeding position of *Saccoglossus*. Its burrow is helical and trunk is thrown into a permanent twist.

3. Most acorn worms are moderate in size, upto 30 cm or so long, but *Balanoglossus gigas* may reach a length of 2.5 m.
4. They are slimy and very soft animals, *i.e.*, they are fragile. Many smell like iodoform.
5. Many pharyngeal gill slits are present in row on trunk.
6. Alimentary canal is straight.
7. Dioecious. Gonads are located in the trunk coelom just behind the gill slits.

Examples. *Balanoglossus*, *Saccoglossus*, *Protoglossus*, *Planetosphaera*, *Ptychodera*.

Class 2. Pterobranchia

1. Pterobranchia, which means **wing gills**, includes three genera and fifteen species.
2. They are colonial, sedentary and tubicolous (*i.e.*, living in tubular **zoecium**) hemichordates.
3. Collar bears two curved hollow **arms** which bear ciliated and hollow **tentacles**.
4. Proboscis is flattened like a shield and secretes the tube.
5. Alimentary canal is U-shaped, anus is dorsal, close to the mouth.
6. They are with or without pharyngeal slits.
7. Dioecious. Development includes a ciliated larva which is different from tornaria.

Class Pterobranchia is divided into **two** orders:

Order 1. Rhabdopleurida

1. Colonial hemichordata, each living in tubular zoecium secreted by the epidermis and connected to other members by living stolon.
2. They contain two tentacular arms.
3. No gill slits are present.
4. Single gonad is present.

Example. *Rhabdopleura* (It is the only genus).

Order 2. Cephalodiscida (or Cephalodisca)

1. They live in a common zoecial mass (aggregation), but without living stolon connecting members of aggregation.
2. They contain four to nine pairs of tentacular arms.
3. Only one pair of gill slits is present.
4. There is one pair of gonads.

Examples. *Cephalodiscus*, *Atubaria*.

2.5 TYPE STUDIES

1. *Rhabdopleura*

Classification

Phylum	—	**Hemichordata**
Class	—	**Pterobranchia**
Order	—	**Rhabdopleurida**
Genus	—	***Rhabdopleura***

Habit and Habitat. *Rhabdopleura* is sedentary and marine, found at great depths in the seas of Southern Hemisphere, Burmuda and European Coasts. It forms colonies at bottom. The individual is attached by a stolon. The colony of *Rhabodopleura* measures about 10 cm and is enclosed in a membranous branched tubes, called **coenoecium** which gives off upright tubes for the zooids (Fig. 2.2). The colonies are attached to stones, corals and shells. The creeping tube is somewhat irregular (Fig. 2.2A) and contains the black **stolon** (Fig. 2.2B), formed of a firm tube containing a cellular column from which new buds arise. The upright tubes are built of regular, successive rings (Fig. 2.2D).

Structure. A zooid measures 1 mm in length. Body colour is brown. The body is divisible into **proboscis (oral shield** or **cephalic shield)**, **collar**, **trunk** and **stalk**. **Proboscis** is shield-like, contains a red pigment band and secrete tubes in which animal lives. Peristaltic movement of proboscis helps in locomotion. The collar bears two lophophore arms, each bearing two rows of fine ciliated **tentacles**. Both the arms and tentacles are hollow. They help in capturing the food. Collar also bears mouth on the ventral side. *Rhabdopleura* lacks gill slits, so it deflects excess water by means of the fold-like oral lamella near the mouth.

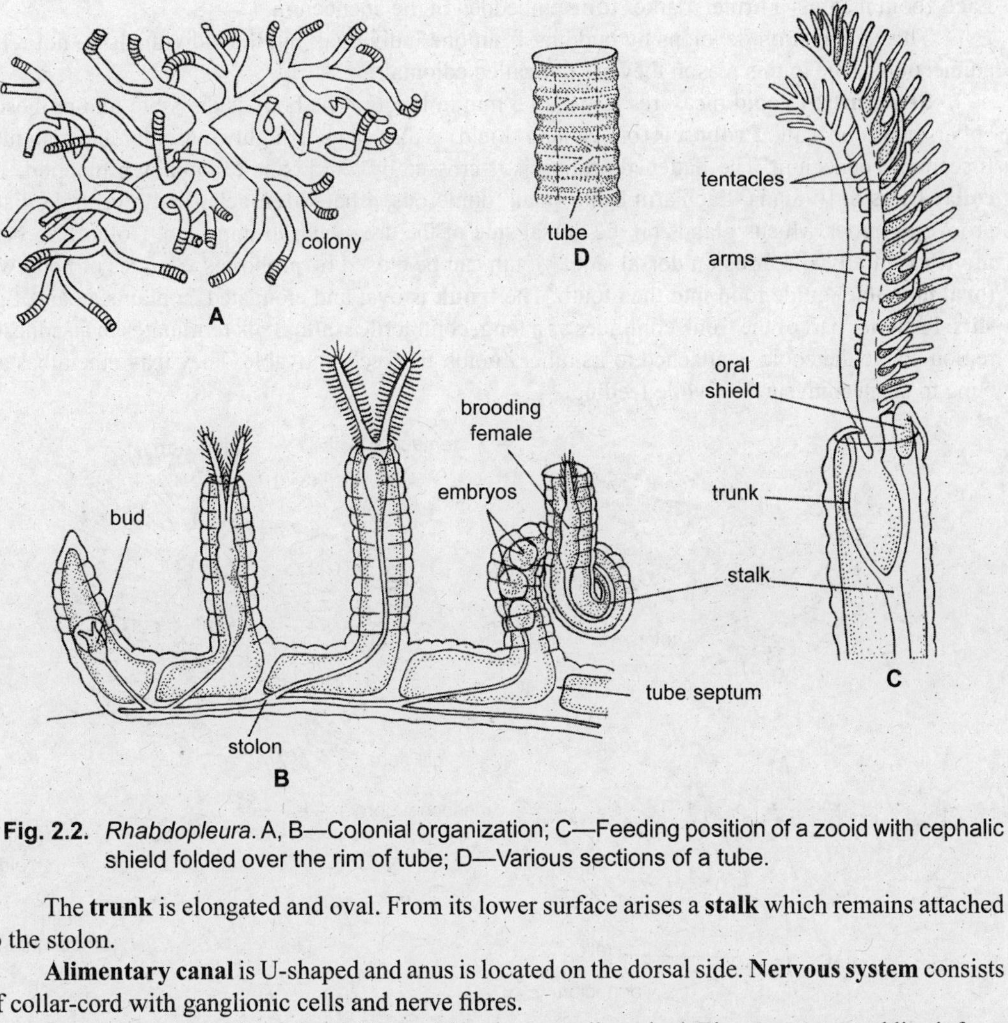

Fig. 2.2. *Rhabdopleura*. A, B—Colonial organization; C—Feeding position of a zooid with cephalic shield folded over the rim of tube; D—Various sections of a tube.

The **trunk** is elongated and oval. From its lower surface arises a **stalk** which remains attached to the stolon.

Alimentary canal is U-shaped and anus is located on the dorsal side. **Nervous system** consists of collar-cord with ganglionic cells and nerve fibres.

Sexes are separate but **sexual dimorphism** is not well marked. The saccate gonad lies infront of the stomach. Short gonoducts open behind the collar-trunk boundary. **Fertilization** of large yolky eggs is external. **Development** is **indirect**. The larva is ciliated and becomes adult-like by the metamorphosis. A colony is formed by budding of the stalk (or common stolon).

2. *Cephalodiscus*

Classification

Phylum — **Hemichordata**
Class — **Pterobranchia**
Order — **Cephalodiscida**
Genus – ***Cephalodiscus***

Habit and Habitat. *Cephalodiscus* is marine and sedentary. It is found at great sea depths of Southern Hemisphere. It is tubicolous and forms **aggregations** which are found attached to corals, shells and stones. *Cephalodiscus* secretes a gelatinous tube to live. Several individuals, called **zooids**, live in association in a common branched **gelatinous investment**, called **coenecium**. Coenecium may be stiffened by firmer spines or have filamentous projections, giving a plant-like appearance. Each zooid inhabits a firmer, darker **tube** embedded in the coenecium.

This aggregation develops by budding from one individual, but the individuals do not retain connections. Due to this reason they are not called colonial.

Structure. A zooid measures less than 5 mm in length. The body is divisible into proboscis, collar, trunk and stalk. **Proboscis** (or cephalic shield) is disc shaped, highly muscular and glandular. It secretes coenecium. The flattened proboscis is crossed by a red band of unknown function. The **collar** bears 8–16 **arms**. Each arm bears small, numerous, ciliated **tentacles** and a ventral ciliated groove or gutter. Mucus glands on the dorsal side of the arms help to trap food. Collar also bears **mouth** ventrally and **anus** on dorsal side. Mouth can be closed by proboscis. Flanges of body wall **(oral lamellae)** guide food into the mouth. The **trunk** is oval and elongated, contains a pair of **gill slits**. Posterior part of the trunk continues as a long, contractile **stalk**. Stalk terminates in an adhesive region where the zooid is attached to its tube. Zooids are freely movable. They leave the tubes and cling to coenecium surface while feeding.

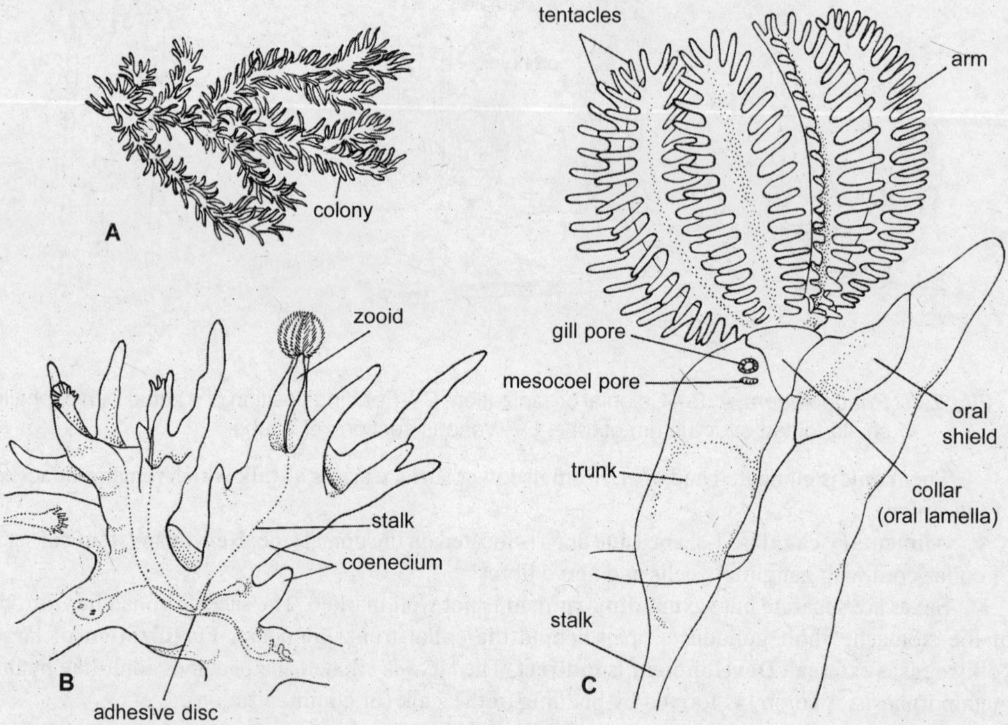

Fig. 2.3. *Cephalodiscus*. A, B—Colonial organization; zooids emerge from openings in the coenecium and perch on branch tips while feeding (B); C—Feeding posture.

Nervous system is epidermal. This consists of a solid collar **ganglion** that join mid-dorsal and latero-dorsal thickenings of the nerve plexus in the cephalic shield.

Gonads are found in the trunk cavities and the development is direct. It also reproduces asexually by budding.

QUESTIONS

Long Answer Questions

1. Give the characters and classification of the phylum Hemichordata.
2. Give classification, habits and habitat and structures of following:
   (*i*) *Rabdopleura*. (*ii*) *Cephalodiscus*.
3. Give classification of Hemichordata with characters and examples. (*CCSU Meerut 2013*)

Short Answer Questions

1. Write short note on hemichordata.

Very Short Answer Questions

1. Name the excretory organ of hemichordates.
2. Cite two examples of hemichordates bearing tentacles on the collar.
3. Name the three coeloms found in hemichordates.
4. Which are primitive, enteropneusts or pterobranchs?
5. In which groups of hemichordates, the epidermal cells are monociliated?
6. Name a hemichordate that lacks tentacles on the collar.
7. Which hemichordates have a dorsal hollow nerve cord?
8. Name the coeloms that do not open out in *Balanoglossus*.

Fill in the Blanks

1. Phylum Hemichordata is divided into two classes: ________ and ________.
2. Acorn worms belongs to the class ________.
3. ________ are colonial hemichordates.
4. ________ is a pterobranch.
5. ________ are tricoelomate animals.
6. There is no peritoneum in ________.
7. U-shaped digestive tract is found in ________.
8. A hemichordate structure considered earlier homologous to the notochord of chordates is named ________.
9. Nervous system is largely intraepidermal in ________.
10. Collar with tentaculate arms is characteristic of ________.

Multiple Choice Questions

1. The acorn worms belong to phylum
   (*a*) Chordata (*b*) Hemichordata
   (*c*) Urochordata (*d*) Cephalochordata
2. A non-contractile heart occurs in
   (*a*) earth worm (*b*) sea star
   (*c*) acorn worm (*d*) apple snail
3. Gastrulation in hemichordates takes place by
   (*a*) invagination (*b*) ingression
   (*c*) delamination (*d*) all these means
4. Stomochord is present in
   (*a*) *Branchiostoma* (*b*) *Herdmania*
   (*c*) *Balanoglossus* (*d*) none of these
5. *Atubaria* belongs to
   (*a*) Rhabdopleurida (*b*) Grapholita
   (*c*) Cephalodiscida (*d*) none of these

6. Larva of Hemichordata is
   (*a*) tornaria (*b*) glochidium
   (*c*) bipinnaria (*d*) nauplius

ANSWERS

Very Short Answer Questions

1. Glomerulus.
2. *Rhabdopleura* and *Cephalodiscus*.
3. Protocoel, a pair of mesocoels, a pair of metacoels.
4. Pterobranchs.
5. Pterobranchia.
6. *Balanoglossus*.
7. Enteropneusts.
8. Metacoels.

Fill in the Blanks

1. Enteropneusta, Pterobranchia. **2.** Enteropneusta. **3.** Pterobranch. **4.** *Rhabdopleura*. **5.** Hemichordates. **6.** Hemichordates. **7.** Pterobranchia. **8.** Stomochord. **9.** Hemichordata. **10.** Pterobranchs.

Multiple Choice Questions

1. (*b*) **2.** (*c*) **3.** (*d*) **4.** (*c*) **5.** (*c*) **6.** (*a*)

3

Balanoglossus (Phylum Hemichordata)

3.1 HABIT, HABITAT AND EXTERNAL FEATURES

Classification

Phylum	—	**Hemichordata**
Class	—	**Enteropneusta**
Genus	—	***Balanoglossus***

Habit and Habitat

Balanoglossus (Gr., *balanus* = acorn, fruit of oak + *glossa* = tongue) is a marine, soft bodied, sluggish, burrowing worm-like animal. It is commonly called **acorn** or **tongue worm**. It lives in U-shaped burrow in shallow water or sand (Fig. 3.1). Burrow is 50–75 cm deep, with a funnel-shaped opening at the anterior end and a round posterior opening. Anterior opening of burrow includes a **feeding funnel** and **ventilation shaft**. The burrow is plastered by mucus. The faecal matter (piles of coiled rope-like castings) lies close to round posterior opening of burrow.

Balanoglossus produces an offensive smell resembling to that of decomposing faecal matter. It is a suspension feeder. It exhibits bioluminescence, emitting bright greenish light.

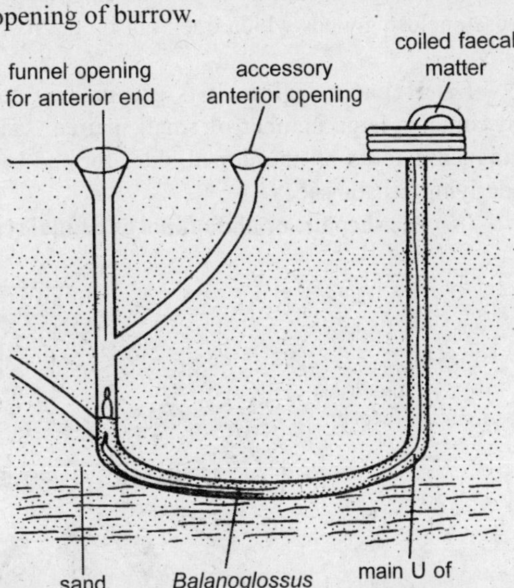

Fig. 3.1. *Balanoglossus.* U-shaped burrow and faecal cast.

External Features

Size, Shape and Colour. Most of species of *Balanoglossus* measures 2–3 cm to 2 metres or more. Body is cylindrical and worm-like. Entire body is covered with cilia. Colour of body is bright with reddish or orange tint. The various body parts are differently coloured.

Division of the Body. The body is unsegmented, bilaterally symmetrical and divided by two constrictions into three parts—(1) proboscis (**protosome**), (2) collar (**mesosome**) and (3) trunk (**metasome**) (Fig. 3.2).

1. Proboscis. It is an anterior club-shaped, highly muscular, tongue-like part. Inside the proboscis, the space is called proboscis coelom or **protocoel** which opens by **proboscis pore**, situated near its base (Fig. 3.3). The posterior narrow part of proboscis is called **stalk** which connect it to the collar. Below the stalk is a U-shaped ciliated **preoral ciliary organ** as a depression, to test the quality of food.

2. Collar. It is short, cylindrical middle part of the body. Its anterior part is flap-like or funnel-like called **collarette**. It bears a wide mouth on the ventral surface. The posterior end projects backward and surrounds a part of trunk. The thick collar wall encloses buccal cavity and paired cavities of **collar coelom (mesocoel**; Fig. 3.4). Mesocoel opens into the first pair of gill pouch through **collar pores**.

3. Trunk. It is long, cylindrical with flattened posterior part and forms major part of the body. It bears mid-dorsal and mid-ventral longitudinal ridge. These ridges extend upto posterior end and the nerves and blood vessels run through these ridges. Trunk contains the **trunk coelom** or **metacoel**. Trunk is divided into following three parts:

(*i*) **Branchiogenital region.** It is the anterior part of trunk. It bears a row of **gill pores** on the dorsal ridge. The number of the **gill pores** range from a few to 100 or more pairs. Its side wall is produced into a pair of thin, flat, lateral longitudinal flaps or genital wings (Fig. 3.5). These enclose gonads which open out by microscopic gonopores.

(*ii*) **Hepatic region.** It is greenish in colour and contains large number of small, paired, sac-like **transverse folds** on the dorsal side, formed by the underlying hepatic caeca.

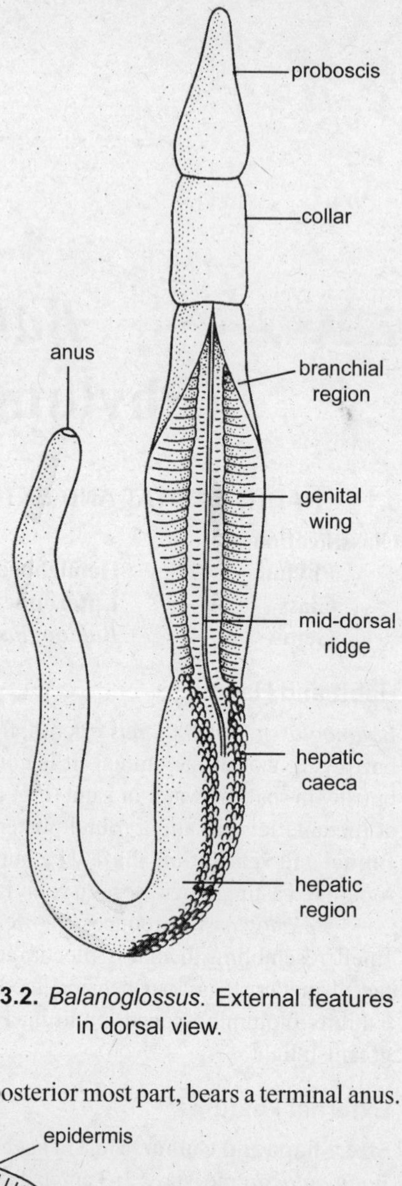

Fig. 3.2. *Balanoglossus*. External features in dorsal view.

(*iii*) **Post-hepatic or abdominal or caudal region.** It is posterior most part, bears a terminal anus.

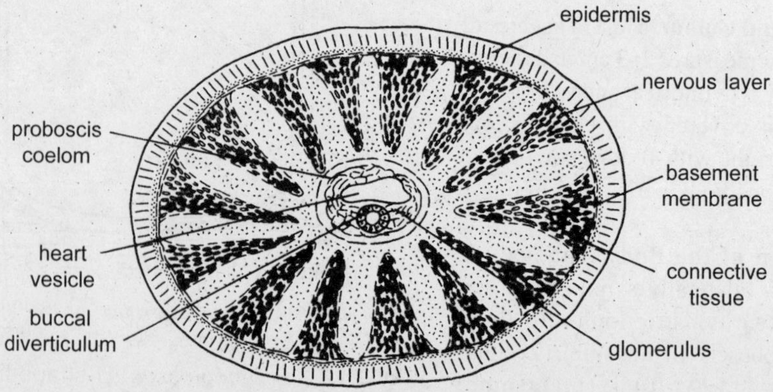

Fig. 3.3. *Balanoglossus*. T.S. of proboscis.

3.2 INTERNAL STRUCTURE

Body Wall

Body wall of *Balanoglossus* is soft, slippery and formed of (*i*) epidermis; (*ii*) muscular layer and (*iii*) peritoneum (Fig. 3.6). It lacks a cuticle.

1. Epidermis. Epidermis is ciliated and richly glandular. It consists of a single layer of tall, slender, columnar cells. They are of three different types:

(*i*) **Ciliated columnar cells.** They are multiciliated and form bulk of epidermis.

(*ii*) **Gland cells.** They are most abundant in collar region. They secrete mucus and are of three types:

(*a*) **Goblet gland cells.** They are flask shaped and have finely granular cytoplasm.

(*b*) **Coarsely granular gland cells.**

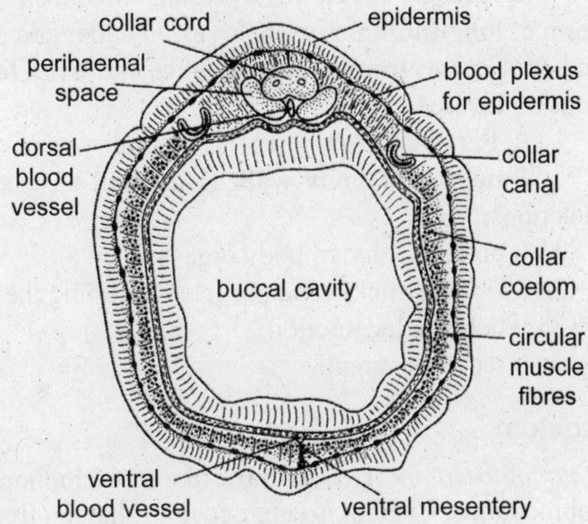

Fig. 3.4. *Balanoglossus*. T.S. of collar.

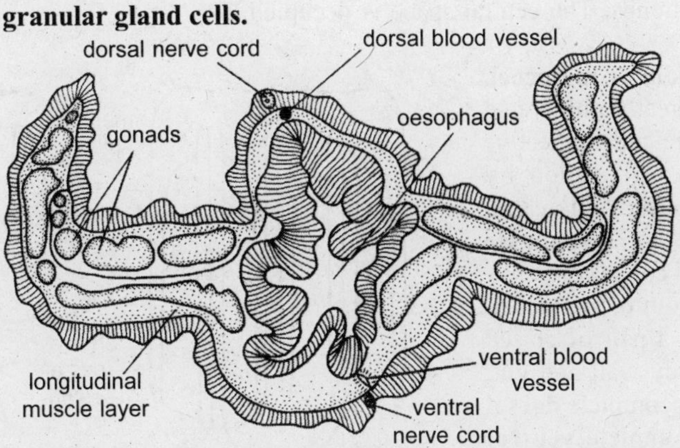

Fig. 3.5. *Balanoglossus*. C.S. through genital wings

(*c*) **Reticulate gland cells.** Globet cells produce bromine-containing compounds that impart strong medicinal odour to acorn worms and may protect them from bacterial infection or animal predation.

A **basal lamina** and a connective tissue **dermis** lie beneath the epidermis, but the dermal fibres are well developed. This fact perhaps accounts for the fragility of body.

The collar is marked by five transverse bands of alternating glandular and non-glandular epithelium.

(*iii*) **Neurosensory cells.** A thick **nervous layer** lies at the base of epidermis immediately above the tough basement membrane. It is formed of a network of **nerve cells** and **nerve fibres**.

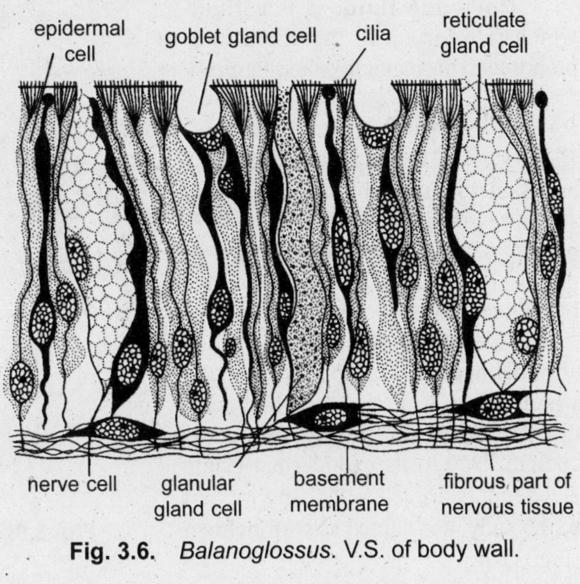

Fig. 3.6. *Balanoglossus*. V.S. of body wall.

2. Muscle layer. It is present in trunk region in the form of **longitudinal muscle layer**. In collar and proboscis, it includes outer **circular muscle layer** and inner **longitudinal muscle layer**.

3. Peritoneum. It lines the coelom.

Function of body wall. Body wall has the following functions:

1. It protects internal body organs.
2. Its mucus sticks sand particles for lining the burrow.
3. It helps in locomotion.
4. It receives stimuli.

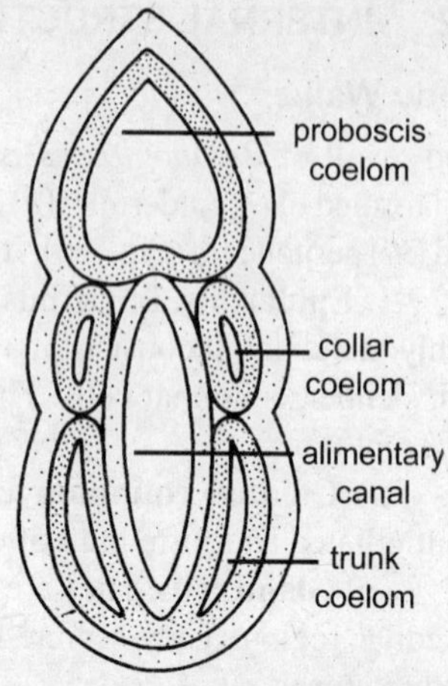

Fig. 3.7. *Balanoglossus.* Diagram of tripartite embryo showing coelomic cavities.

Coelom

Balanoglossus is **tricoelomate** like most lophophorate and echinoderms. Coelom is enterocoelous, having five cavities:

1. Proboscis coelom or protocoel. It is unpaired cavity, opens by **proboscis pore**. It is greatly obliterated by connective tissue and muscle strands. The central space is occupied by **proboscis complex**.

2. Collar coelom or mesocoel. It is paired and is greatly obliterated by collar musculature and connective tissue. It is also partitioned by incomplete middorsal and mid-ventral mesenteries. It opens into first gill sac by a collar canal and collar pore.

3. Trunk coelom or metacoel. It is represented by a pair of closed cavities between body wall and gut. It is partitioned by incomplete **dorsal mesentery** and a complete **ventral mesentery**. This coelom is obliterated by trunk musculature.

Coelomic fluid. It is a fluid which is filled up in the collar and proboscis. This maintains their turgidity.

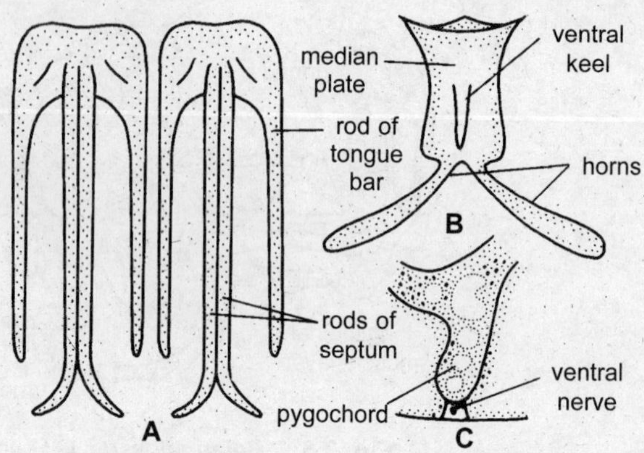

Fig. 3.8. *Balanoglossus.* Skeletal elements. A—Bra-nchial skeleton; B—Proboscis skeleton; C—Pygochord.

Skeletal System

Balanoglossus lacks a definite endoskeleton. But it has following four structures of a supporting nature: (1) buccal diverticulum; (2) skeleton of proboscis; (3) branchial skeleton and (4) pygochord (Fig. 3.8).

1. Buccal diverticulum or stomochord. It is a hollow structure which arises from the roof of buccal cavity and extends upto proboscis coelom. Due to its endodermal origin and being restricted to the anterior part of the body, it is named as **stomochord** (Fig. 3.9).

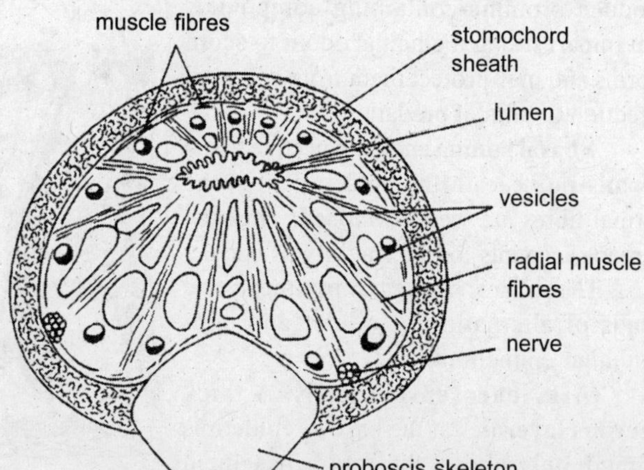

Fig. 3.9. *Balanoglossus.* T.S. stomochord.

2. Proboscis skeleton or nuchal skeleton. It lies ventrally close to the buccal diverticulum. It is a Y-shaped chitinous structure. Its median plate has a median-ventral keel and lies beneath the buccal diverticulum and two arms extend along the margin of the buccal cavity in the collar region.

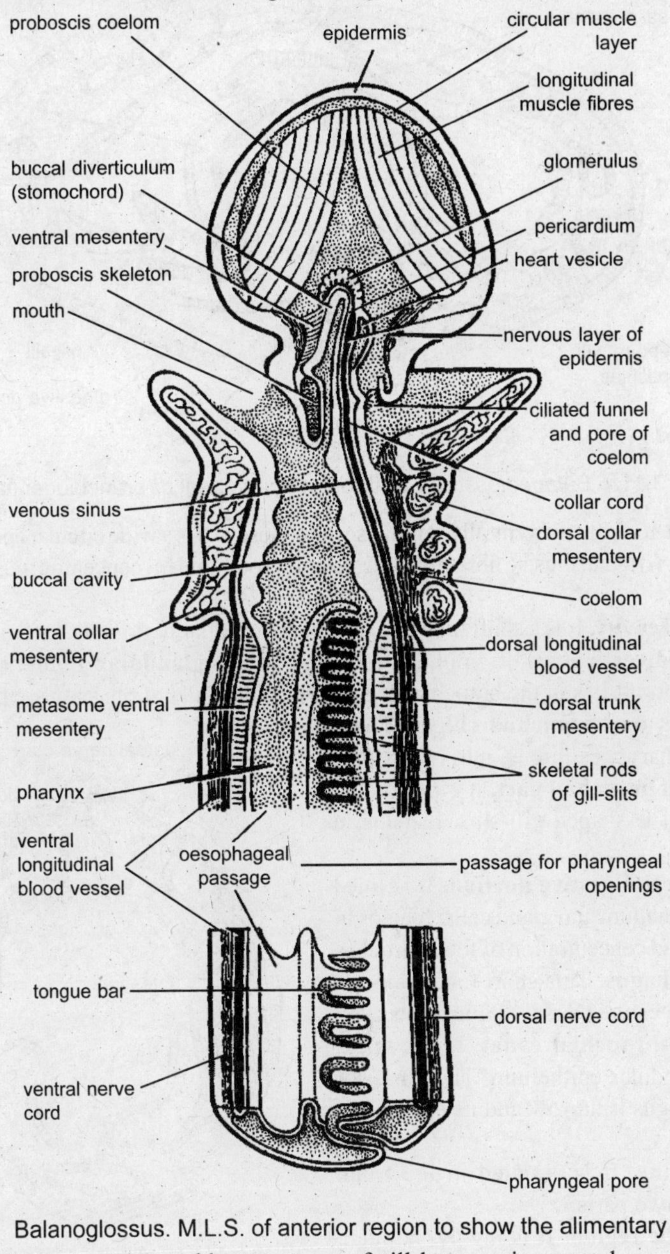

Fig. 3.10. Balanoglossus. M.L.S. of anterior region to show the alimentary canal.

3. Gill bars. They are supported by two types of gill-bars—primary and secondary gill bars. They are formed due to thickening of the basement membrane.

4. Pygochord. It is hard longitudinal rod-like structure which extends from the ventral side of the intestine to the body wall. Its cells are vacuolated. Pygochord may support the soft abdominal region.

3.3 DIGESTIVE SYSTEM

Alimentary canal is a straight tube supported by dorsal and ventral mesenteries and consists of the following structures: mouth, buccal cavity, pharynx, oesophagus, intestine and anus.

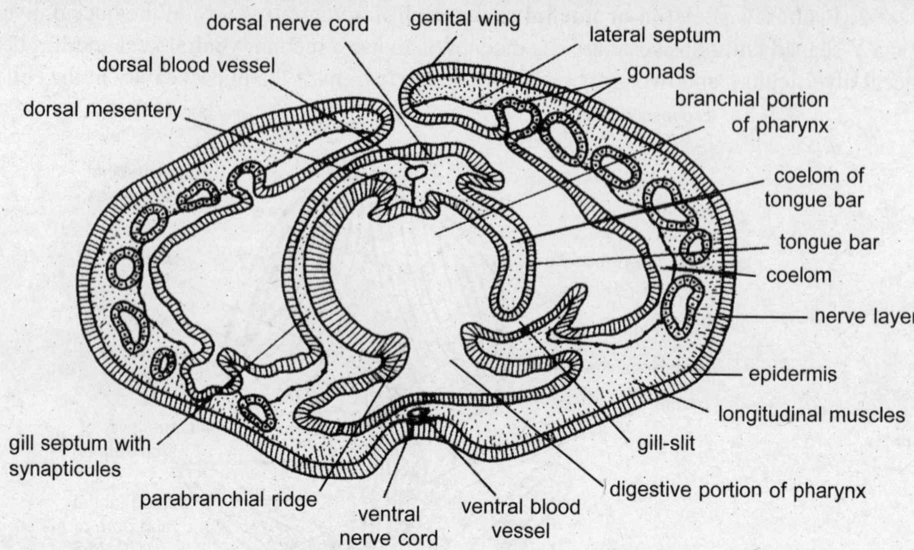

Fig. 3.11. *Balanoglossus*. T.S. through pharyngeal or branchiogenital region.

1. Mouth. It is situated ventrally at the base of proboscis. It is a wide circular aperture. It has two sets of muscle fibres: (*i*) radial muscle fibres for widening the mouth and (*ii*) concentric muscle fibres for closing it.

2. Buccal cavity. It is a short tube that occupies the collar region. Its roof gives out a short, stiff hollow **buccal diverticulum**. Its epithelium lining contains **glandular globlet cells**.

3. Pharynx. It lies in the branchial region. The side walls of pharynx are produced into a pair of ridges called **parabranchial ridges**. These ridges divide pharyngeal lumen into two parts:

(*i*) **Dorsal branchial part.** It is perforated by two rows of U-shaped gill-slits. It helps in respiration.

(*ii*) **Ventral digestive portion.** It is lined by cilia-ted epithelium and gland cells. It helps in the collection and concentration of food particles.

4. Oesophagus. This short tube connects pharynx with intestine Dorsal region of pharynx forms the **post-branchial canal**. It has thick, folded and glandular epithelium. The posterior part of oesophagus is narrow and its epithelium has deep furrows.

5. Intestine. It is situated in the trunk region and has two parts:

(*i*) **Hepatic region.** It is highly vascular. Numerous hepatic caeca are produced from its dorsal wall. Their cells are dark green or dark brown.

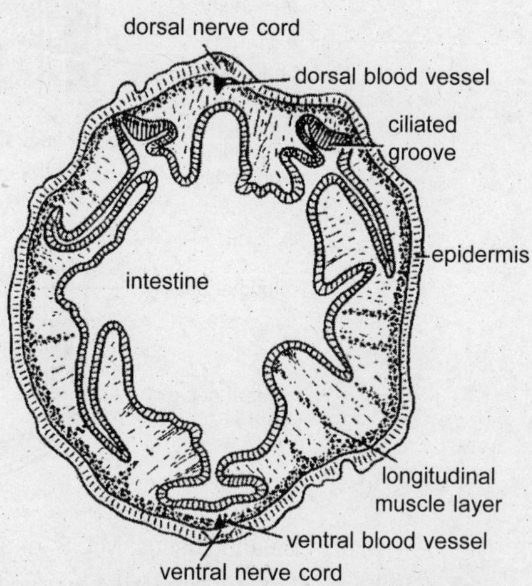

Fig. 3.12. *Balanoglossus*. T.S. through post-hepatic intestine.

(*ii*) **Post hepatic region.** It is straight, thin walled tube and bears a pair of dorsolateral grooves. It is lined with ciliated epithelial cells.

6. Anus. It is situated at the tip of trunk. It is a circular aperture, surrounded by sphincter muscles.

Food and feeding. *Balanoglossus* is a **ciliary feeder**. It feeds on microscopic, dead and decaying organic matter present in water. Due to movement of lateral cilia of gill slits food laden water currents are directed backward which enters mouth, buccal tube, pharynx, gill slits and branchial sacs and leaves through gill-slits (Fig. 3.13). Food particles are entangled in mucus, forming mucous string. Mucus is secreted by gland cells (*i.e.*, coarsely granular gland cells) of proboscis and contains **amylase enzyme**.

Digestion. Digestion of food is brought about in the alimentary canal by the help of enzymes such as **amylase**, **maltase**, **lipase** and **protease**. They are secreted by pharynx, oesophagus and hepatic caeca of intestine.

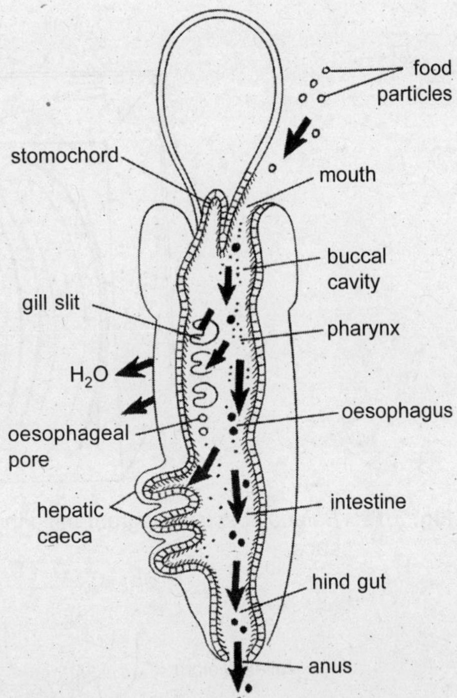

Fig. 3.13. *Balanoglossus.* Diagrammatic path of food and removal of water via gill-slits, in the pharynx.

3.4 RESPIRATORY SYSTEM

In *Balanoglossus* respiratory organs include:

1. Branchial portion of pharynx and gill pouches.
2. Branchial sacs.

1. Branchial portion. The dorsal wall of branchial portion of pharynx is perforated by a longitudinal series of U-shaped **gill slits**. Their number increases with increase in the size of animal.

The two lines of the U are separated by a **tongue bar**. The tongue bar is **hollow** and encloses coelomic cavity. Each gill slit opens into a **gill pouch** or **branchial sac** (Fig. 3.15). The wall of gill pouch is supported by three chitinous rods and form M-shaped rods. These are (*i*) **Median primary bar** which lies in gill septa. It forms middle arms of M and its lower end. (*ii*) **Lateral secondary bar** which are two lateral arms of M and lie in tongue bars. Both of the bars are connected by transverse connection called **synapticulae**, supported by chitinous rod. The gill pouches are richly supplied with blood and gill slits are lined with cilia.

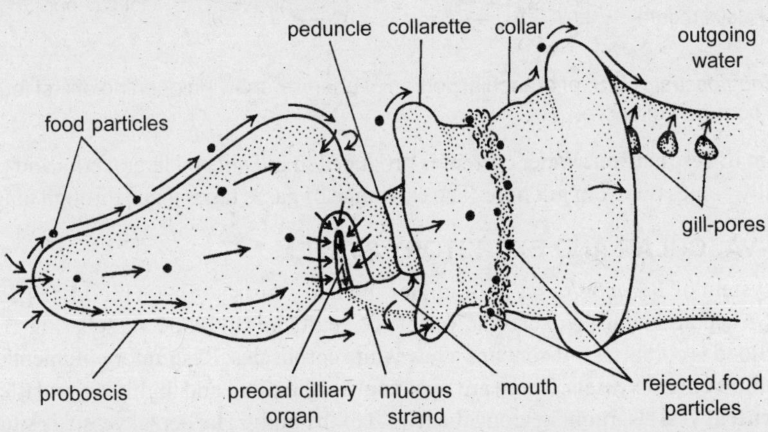

Fig. 3.14. *Balanoglossus.* Feeding current and mode of food testing by preoral ciliary organ.

2. Branchial sac. Each gill slit opens into a **branchial sac** which opens to exterior by a gill pore. Each sac lies between body wall and pharynx.

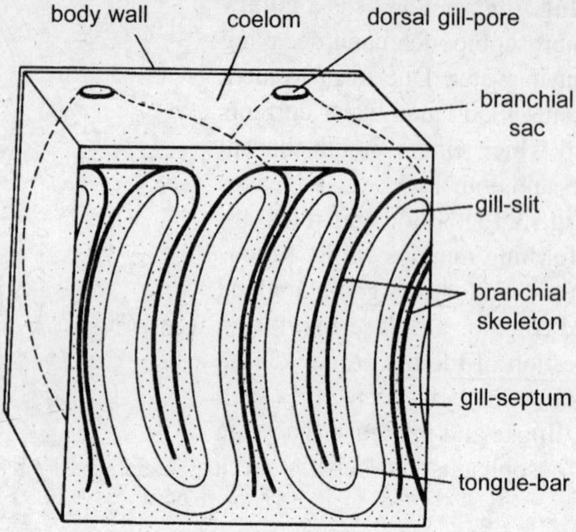

Fig. 3.15. *Balanoglossus*. Diagrammatic three-dimensional view of two gill-slits and two branchial sacs.

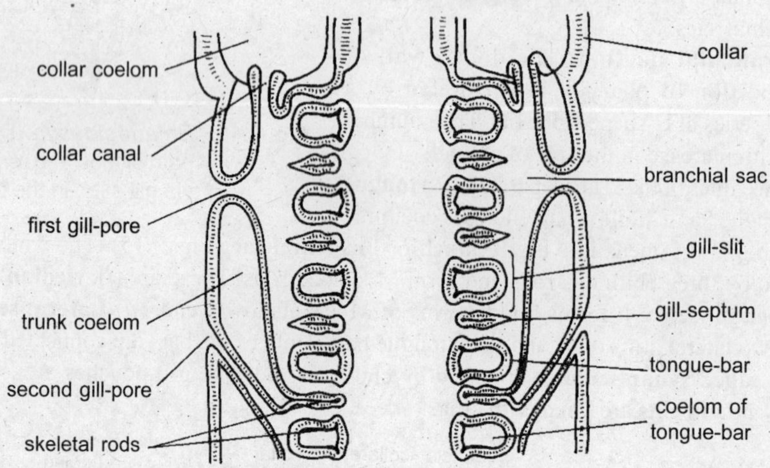

Fig. 3.16. *Balanoglossus*. H.L.S. of branchial portion of pharynx showing opening of first four gill-slits by a common gill-pore.

Mechanism of respiration. Water current is produced in gill slits by lateral cilia and enters the pharynx and finally comes out from gill pore. The exchange of gases takes place through tongue bars.

3.5 BLOOD VASCULAR AND EXCRETORY SYSTEM

Blood vascular system of *Balanoglossus* is of **open type** and consists of—(*i*) blood; (*ii*) dorsal and ventral vessels; (*iii*) central sinus; (*iv*) cardiac vesicle or heart and (*v*) blood vessels (Fig. 3.17).

1. Blood. Blood is colourless. It contains a few white corpuscles. Respiratory pigment is absent.

2. Dorsal and ventral vessels. There are two main longitudinal and highly contractile vessels, the **dorsal** and **ventral vessels**, running along the length of the body. The dorsal vessel is situated just below the dorsal nerve cord and above the alimentary canal and runs through the dorsal mesentery. The blood flows anteriorly through the dorsal vessel. The **ventral vessel** is located in the ventral mesentery and blood flows posteriorly in it.

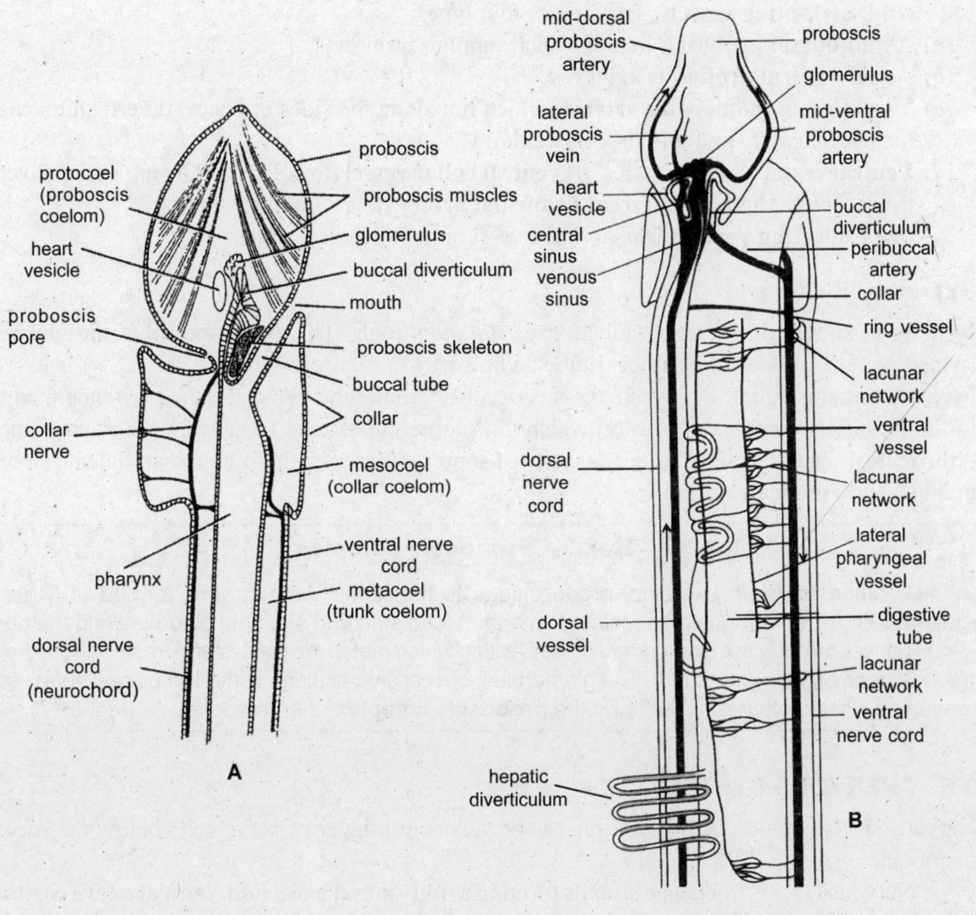

Fig. 3.17. *Balanoglossus.* A—Front end in sagittal section; B—Blood vascular system in anterior end in lateral view.

3. Central sinus. The dorsal vessel is dilated just infront of collar forming a **venous sinus** which passes anteriorly into a **central sinus**. Central sinus receives blood from the collecting vessels at its posterior end. Anteriorly it pumps blood into afferent vessels.

4. Cardiac vesicle or heart. It is situated above the central sinus. It is a closed triangular sac with muscular wall. It produces contraction in central sinus to pump the blood.

5. Blood vessels. These include (*i*) afferent vessels; (*ii*) distributing vessels and (*iii*) collecting vessels.

(*i*) Afferent vessels. They arise from the anterior end of heart (= central glomerulus sinus). These form plexus in **glomerulus**. Here the nitrogenous wastes are removed from the blood.

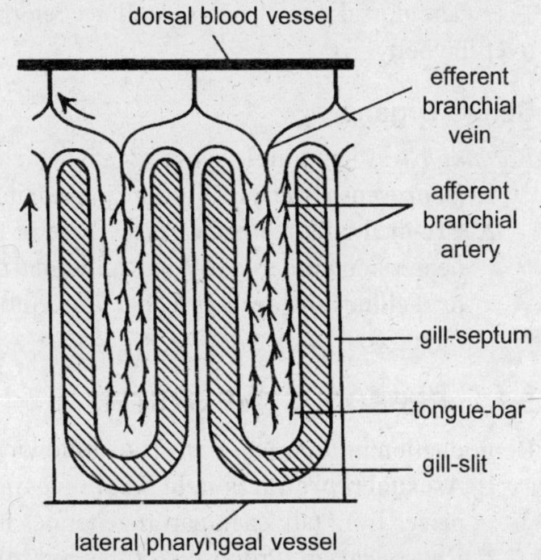

Fig. 3.18. *Balanoglossus.* Branchial vessels in portion of the branchial wall.

***(ii)* Distributing vessels.** They are of four types:

(*a*) A **mid-dorsal proboscis artery** which supplies proboscis.

(*b*) A **mid-ventral proboscis artery**.

(*c*) Two **efferent glomerular arteries** which run along the sides of buccal diverticulum encircle the buccal cavity and join to form ventral vessel.

(*d*) **Ventral vessel** which includes, (*i*) **ventral collar vessel** (to collar), (*ii*) **a ring vessel** (to collar-trunk septum) and (*iii*) **afferent branchial artery** (to each gill septum).

***(iii)* Collecting vessels.** Dorsal blood vessel is a collecting vessel.

Excretory System

Nothing is known of excretory function of *Balanoglossus*. The assumption that the glomerulus (which is a mass of blind peritoneal tube and a part of proboscis complex; Box 3.1) is excretory is purely circumstantial. Brownish crystals occur in glomerular cells, and the peritoneum adjacent to the proboscis complex is also brownish. So glomerulus seems to have an athrocytic function. **Athrocytosis** can be defined as the capacity of some cells to pick up foreign material and store it in granular form in the cytoplasm.

Box 3.1 Proboscis complex

Gloomerulus is an excretory organ unique to the hemichordates; it is formed of finger-like outpocketings of peritoneum associated with the blood sinuses. All of the blood leaving the central sinus passes through the glomerular sinuses, within which metabolic wastes are presumably extracted (see **Brusca** and **Brusca** 2003). The glomerulus, buccal diverticulum, central sinus and heart vesicle compose what is often referred to as the **proboscis complex** of enternopneusts (Fig. 3.17).

3.6 NERVOUS SYSTEM

Nervous system of *Balanoglossus* consists of plexus or a layer of nerve cells below the basement membrane.

Nervous layer is thickened strands forming a **mid-dorsal** and a **mid-ventral nerve cord**. They run the entire length of trunk. In collar region they are connected by a circular **nerve ring**. The dorsal nerve cord is connected by another circular strand called **anterior nerve ring**.

The dorsal cord traverse collar coelom and forms nervous centre. They transport impulses over the body.

Sense Organs

Balanoglossus has simple sense organs which are of following two types:

1. **Neurosensory cells.** These are present in the epidermis of proboscis and in anterior part of collar.
2. **Pre-oral ciliary organ.** At the base of the proboscis, on the ventral side, there is a U-shaped depression bordered by long epidermal cells bearing long cilia. This structure is called the **pre-oral ciliary organ**. It is a chemoreceptor and it tests the quality of food and water entering the mouth.

3.7 REPRODUCTIVE SYSTEM

Reproduction in *Balanoglossus* is of following three methods:

1. **Asexual reproduction.** In *Balanoglossus capensis*, during summer young ones if cut into small pieces from tail, each part regenerates into a complete adult.
2. **Regeneration.** Proboscis, collar and trunk pieces can regenerate into new part.
3. **Sexual reproduction.** Sexes are separate; sexual dimorphism is not well marked. Gonads may or may not be branched, found in longitudinal rows in the genital wings. The testes and ovaries are similar.

Fertilization. Fertilization is external. Mature ova and sperm are released in surrounding sea water to fertilize each other.

3.8 DEVELOPMENT

In *Balanoglossus*, development is **indirect**, *i.e.*, embryonic development of fertilized egg forms a free swimming, ciliated and microphagous **tornaria larva** which undergoes **metamorphosis** to form an adult-like individual (Fig. 3.19).

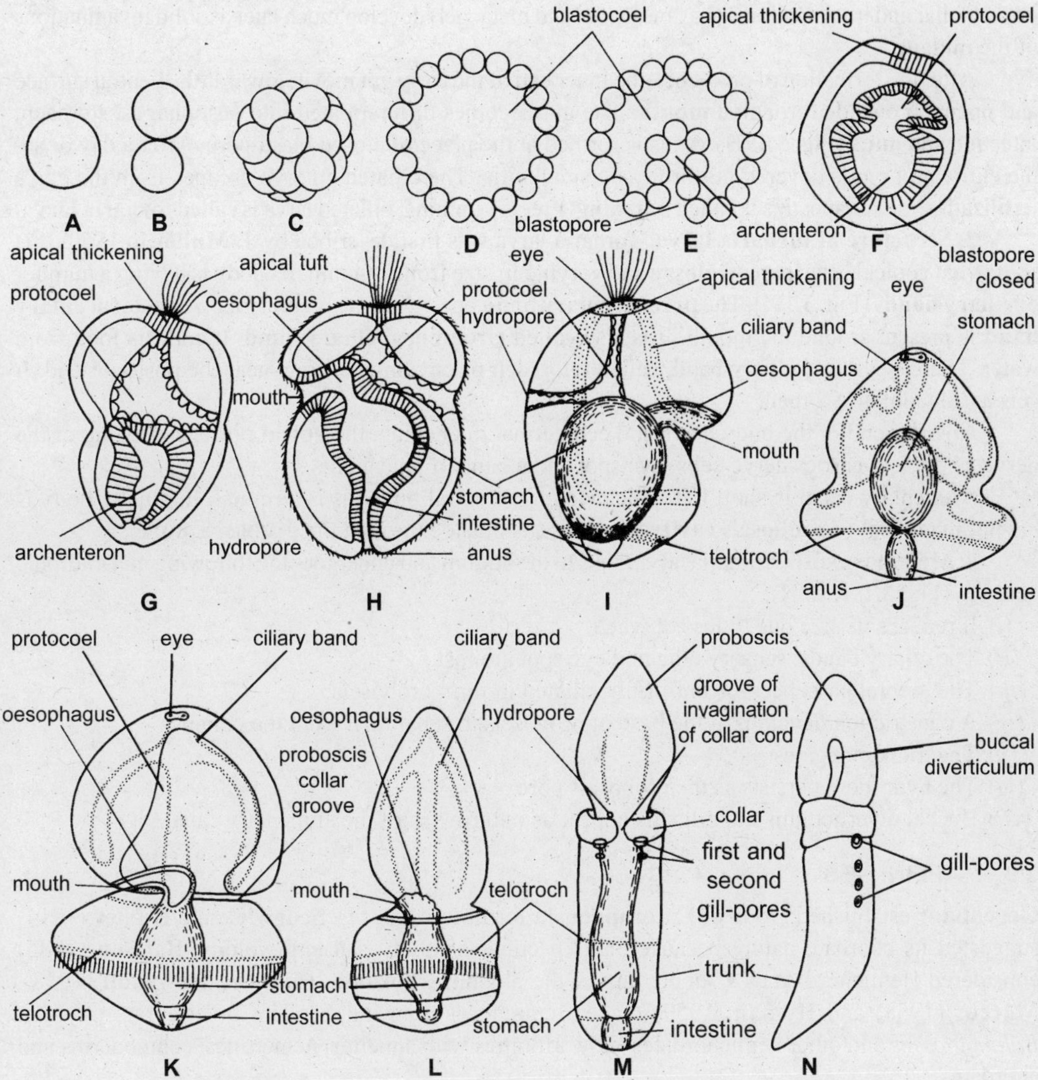

Fig. 3.19. *Balanoglossus*. Stages of development. A—Zygote; B—Two-cell stage; C—Morula; D—Coeloblastula; E—Early gastrula in L.S.; F—Gastrula; G—Early tornaria; H—Young tornaria larva in section; I—Young tornaria; J—Fully formed tornaria; K to N—Metamorphosis of tornaria into a young *Balanoglossus*.

1. Cleavage and gastrulation. Cleavage of zygote is **holoblastic** (equal and radial) and results in a sphere of blastomere called **morula**. By developing a central fluid-filled cavity, called **blastocoel**, morula develops into a single-layered **coeloblastula**.

Soon blastula undergoes an **invagination** of cells of vegetal pole to form an **archenteron** which opens outside through the **blastopore**. When ectoderm and endoderm separates from each other and blastopore is closed, the embryo is called **gastrula**.

2. Formation of larva and hatching. The gastrula elongates along the antero-posterior axis. The anterior tip of archenteron is differentiated as a coelomic vesicle, called **protocoel**. Thus, origin of coelom is **enterocoelic**. The remaining posterior part of the archenteron develops into alimentary canal. The protocoel becomes triangular in shape and opens outside on dorsal side by **hydropore**. Later on, protocoel develops into **proboscis coelom** and hydropore becomes **proboscis pore**. Collar and trunk coeloms (*i.e.*, mesocoel and metacoel) develop much later as solid invaginations of the midgut.

After the formation of protocoel, the inner end of the early gut moves towards the ventral surface and opens to outside through a **mouth**. The gut becomes differentiated into oesophagus, stomach, intestine and **anus** (Fig. 3.19H). Anus is formed at the place of closed blastopore. After a day or so, the entire embryo is covered uniformly by small cilia. The ciliated embryo escapes from the egg's fertilization membrane; this is called **hatching**. Free-swimming, ciliated larva is called **tornaria larva**.

3. Structure of tornaria larva. Tornaria larva was first described by **J. Muller** in 1850. It is somewhat conical, transparent, glossy and varying in size from 1–4 mm. Its body is having a number of **ciliary bands** (Fig. 3.19I). The **preoral ciliary band** encircling the preoral lobe. A **postoral ciliary band** is present around the mouth. It is also called **circumoral ciliary band**. It collects food from water. There is another ciliary band, called **telotroch** or circum-anal ring, near the posterior end. It acts as a locomotory organ.

Anterior end of the body bears and ectodermal thickening, the **apical plate**, consisting of the nerve centre and carries nerve cells, eye spot and a band of apical cilia.

Alimentary canal is short but wide. Mouth is ventral but anus is terminal. Internally body is having **protocoel** or **proboscis cavity** which opens outside through the **proboscis pore**.

4. Metamorphosis. Tornaria larva sinks to the bottom and undergoes the following metamorphic changes (Fig. 3.19K to N):

(*i*) It reduces its size due to loss of water.
(*ii*) The ciliary bands, sensory cilia and eye spot are lost.
(*iii*) The preoral parts become uniformly ciliated to form proboscis.
(*iv*) A constriction develops at the base of proboscis to separate it from the collar.
(*v*) The trunk elongates.
(*vi*) The hydropore persists as the proboscis pore.
(*vii*) Buccal diverticulum and gill slits appear as outgrowths of the alimentary canal.

3.9 AFFINITIES

Gegenbaur established group of **Enteropneusta** for *Balanoglossus*. **Sedgwick** and **Huxley** (1814) suggested its chordate nature because of its peculiar anatomy and embryology. **Bateson** (1885) considered Hemichordata as a subphylum of the phylum **Chordata**. However, **Dawydoff** (1948), **Marcus** (1958) and **Hyman** (1959) made Hemichordata an independent invertebrate phylum. *Balanoglossus* and other hemichordates show affinities with annelids, nemertines, echinoderms and chordates.

I. Affinities with Annelida: Similarities:

1. Both have cylindrical worm-like body.
2. Both are of burrowing habitat and tubicolous nature.
3. Both have similar mode of feeding and casting.
4. Circulatory system is similar. Heart is dorsal to gut as in Annelida.
5. Blood vessels are arranged in same manner in both.
6. In hemichordates nervous system is epidermal (partially) but in annelids, it is totally epidermal.

7. Toranaria larva resembles trochophore larva of annelids.
8. The pharyngeal pockets of hemichordates are comparable to the oesophageal pockets present in *Polygordius*.

Differences:

1. Annelids do not have pharyngeal gill slits.
2. Presence of dorsal nerve in *Balanoglossus*, but it is absent in annelids.
3. Nervous system is primitive and intra-epidermal in Hemichordata.
4. The nephridium is absent in tornaria larva but it is present in **trochopore larva** of annelids.
5. Preoral proboscis coelom is present in tornaria but absent in trochophore.
6. The **blastopore** becomes **mouth** in trochophore while it forms **anus** in tornaria.
7. Process of embryonic development is entirely different in both groups.

II. Affinities with Echinodermata: Similarities in adults:

1. In both coelom has enterocoelic origin.
2. Heart vesicle and glomerulus of hemichordates are homologous to dorsal sac and axial gland of Echinodermata.
3. In both nervous system is poorly developed and form epidermal nerve plexus.
4. Both are similar in habitat and ecological niches.
5. In both similar types of proteins and phosphogens are present.

Similarities in larva:

1. Both are pelagic, small, transparent and oval in shape.
2. Longitudinal ciliary band is present in tornaria and in bipinnaria larva.
3. Both contain tripartite division of body.
4. In both development of coelom is similar (*i.e.*, enterocoelic origin).
5. Opening of proboscis by proboscis pore is similar to the hydrocoel of echinoderm dipleurula and shows the beginning of water vascular system.

Differences:

1. Apical plate, eye-spot and sensory hair are present in tornaria larva but absent in echinoderms.
2. Transverse ciliary band is present in tornaria while absent in echinoderms.
3. Protocoel is single in tornaria while paired in echinoderm larva.

III. Affinities with Nemertines (Rhynchocephalia): Similarities:

1. Feeding and burrowing habit of both are similar.
2. Both are elongated, vermiform without external segmentation.
3. Anus is terminal in both.
4. In both, the skin is smooth containing unicellular glands.
5. Metamerically arranged simple gonads.

Differences:

1. Proboscis is protrusible in nemertines and non-protrusible in *Balanoglossus*.
2. The dorsal nerve is absent in nemertines and present in *Balanoglossus*.
3. The lateral nerve is present in nemertines and absent in *Balanoglossus*.

IV. Affinities with Phoronida: Similarities:

1. Tornaria larva resembles the actinotrocha larva of Phoronida.
2. In *Phoronis*, the paired gastric diverticula are present but in *Balanoglossus*, single oesophageal diverticulum is present.
3. The nervous system is same in both.
4. Both possess power of regeneration.

Differences:

1. The gill-slits and stomochord (= buccal diverticulum) is present in *Balanoglossus* but absent in *Phoronis*.

V. Affinities with Pogonophora: Similarities:

1. Body and coelom is divided into three regions.
2. Enterocoelous mode of origin of coelom.
3. In both, nervous system intraepidermal.
4. Mesosoma and metasoma are divided by a septum.
5. Pericardial sac is present in some pogonophores.
6. Gonads found in trunk.

Differences:

1. Protocoelic nephridial coelomoducts present in pogonophores but not in Hemichordata.
2. Alimentary canal is absent in pogonophores but well developed in hemichordates.
3. Type of nervous system is protosoma in pogonophores but mesosoma in hemichordates.

VI. Affinities with Chordates: Similarities:

1. Presence of a notochord (*i.e.*, stomochord).
2. Presence of pharyngeal gill slits.
3. Presence of a dorsal tubular nerve cord.
4. Structure and function of pharynx and branchial apparatus are similar to those of Urochordata and Cephalochordata.
5. In both, origin of coelom is enterocoelous.

Differences:

1. Notochord:

(*i*) Notochord extends upto proboscis, while in chordates notochord extends upto full length.
(*ii*) In *Balanoglossus* notochord lies below dorsal blood vessels.
(*iii*) In *Balanoglossus* notochord is without any sheath.
(*iv*) Notochord is hollow and lined with epithelial cells in Hemichordata but solid rod-like in chordates.
(*v*) Notochord is supporting in chordates and not in *Balanoglossus*.

2. Dorsal nerve cord:

(*i*) Nervous system of hemichordates is distinctly of invertebrate type.
(*ii*) It is intra-epidermal and consists of dorsal and a ventral nerve cord and a circumenteric nerve ring.

3. Gill slits:

(*i*) Gill slits are dorsal in *Balanoglossus* but lateral in Amphioxus and other chordates.
(*ii*) Differences exist in the arrangement of skeletal gill-bars.
(*iii*) Differences exist in the mode of blood supply.

Conclusion

By considering the affinities, it can be concluded that echinoderms, hemichordates and chordates have arisen from a common ancestral stock, the dipleura larva. Further, the echinoderms deviated greatly from the ancestral stock and formed blind branch in main line of evolution. The main line of evolution continued to give rise to hemichordates and chordates.

Thus, it is most logical to place hemichordates in the invertebrates as an independent phylum. Hemichordata has arisen from an ancestral stock that has given rise, on the one hand, to echinoderms and on the other hand, to hemichordates and chordates.

QUESTIONS

Long Answer Questions

1. Describe habit, habitat and external features of *Balanoglossus*.
2. Describe the morphology of *Balanoglossus*.

3. Describe the habitat and habit of *Balanoglossus*. Draw only a neat and labelled diagram of the blood vascular system of this animal and show the course of blood circulation only by a graphic representation.
4. Draw a neat and labelled diagram of median longitudinal section (M.L.S.) of anterior region of *Balanoglossus*. Describe its internal structure.
5. Describe the development of *Balanoglossus*.
6. Describe the development of *Balanoglossus* upto the formation of tornaria larva.
7. Give an account of development of *Balanoglossus*.
8. Give an account of structure of tornaria larva of *Balanoglossus* and describe its metamorphosis.
9. Discuss the affinities of *Balanoglossus* or Hemichordata. What is its latest position in the scheme of classification?
10. Give general characters of Hemichordata. Discuss affinities of *Balanoglossus*.
11. Give an account of external morphology of *Balanoglossus* in brief. Discuss its affinities in detail.

Short Answer Questions

1. Discuss coelom in *Balanoglossus*.
2. Write a note on tornaria larva.
3. Write short notes on the following:
   (*i*) *Balanoglossus*
   (*ii*) Pharynx of *Balanoglossus*
   (*iii*) Tongue bar
   (*iv*) Tornaria larva
   (*v*) Chordate affinities of *Balanoglossus*
   (*vi*) Stomochord
4. Draw neat and labelled diagram of following (No description is needed).
   (*i*) L.S. anterior part of *Balanoglossus* or V.S. or M.L.S. *Balanoglossus* through proboscis, collar and trunk part of body.
   (*ii*) T.S. *Balanoglossus* through proboscis.
   (*iii*) T.S. *Balanoglossus* through collar.
   (*iv*) T.S. *Balanoglossus* through branchial region.
   (*v*) T.S. *Balanoglossus* passing through branchio-genital region.
   (*vi*) V.S. body wall of *Balanoglossus*.
   (*vii*) Tornaria larva of *Balanoglossus*.
   (*viii*) *Balanoglossus*.

Very Short Answer Questions

1. Where is *Balanoglossus* found.
2. What is a collarette?
3. Name the five coeloms of acorn worm.
4. Which coelom of *Balanoglossus* do not open out?
5. Give the role of oesophagus of acorn worm.
6. Which part of the gut of *Balanoglossus* has gill slits?
7. Name the site of digestion in acorn worm.
8. What is the function of preoral ciliary organ?
9. What are synapticula?
10. Heart of *Balanoglossus* is non-contractile. Is it so?
11. What type of cleavage does the egg of acorn worm undergo?
12. Name the larva of *Balanoglossus*.

Fill in the Blanks

1. *Balanoglossus* is the familiar example of the phylum ________.
2. *Balanoglossus* lives in ________ shaped, burrows.
3. Three regions of body of *Balanoglossus* are ________, mesosome and ________.
4. ________ is tricoelomate.
5. *Balanoglossus* has a pair of mesocoels in the ________ region.
6. Stomochord is a forward diverticulum from the ________.
7. Gill slits of acorn worm are ________ shaped.
8. Blood flows ________ in the dorsal vessel in *Balanoglossus*.
9. Glomerulus of *Balanoglossus* functions as the ________ organ in the acorn worms.
10. Sexes are ________ in *Balanoglossus*.

Multiple Choice Questions

1. *Balanoglossus* belongs to class ....
   (*a*) Enteropneusta (*b*) Plerobranchia
   (*c*) Graplita (*d*) Planctosphaeroidea
2. The body of *Balanoglossus* is
   (*a*) radially symmetrical (*b*) biradially symmetrical
   (*c*) bilaterally symmetrical (*d*) laterally symmetrical
3. The coelom of *Balanoglossus* is
   (*a*) enterocoelous (*b*) haemocoelous
   (*c*) schizocoelous (*d*) homocoelous
4. *Balanoglossus* is
   (*a*) ciliary feeder (*b*) active feeder
   (*c*) filter feeder (*d*) all of the above
5. The excretory organ of *Balanoglossus*
   (*a*) malpighian tubules (*b*) flame cells
   (*c*) nephridia (*d*) proboscis gland
6. In *Balanoglossus*, the function of preoral ciliary organ is
   (*a*) photoreceptor (*b*) chemoreceptor
   (*c*) reoreceptor (*d*) tangoreceptor
7. Respiration in *Balanoglossus* is
   (*a*) cutaneous (*b*) branchial
   (*c*) tracheal (*d*) pulmonary
8. Acorn worm has a plexus of blood sinuses in
   (*a*) gill septa (*b*) tongue bars
   (*c*) both (*a*) and (*b*) (*d*) highly pressurized
9. Circulatory system of *Balanoglossus* is
   (*a*) open (*b*) closed
   (*c*) lacking (*d*) highly pressurized
10. Hollow nerve cord occurs in acorn worms in the
    (*a*) proboscis (*b*) collar
    (*c*) trunk (*d*) entire body
11. The larva of *Balanoglossus* is
    (*a*) tornaria (*b*) miracidium
    (*c*) brachiolaria (*d*) none of these

ANSWERS

Very Short Answer Questions

1. It is tubiculous and found in shallow waters between the tide marks along the coast of warm and temperate oceans.
2. Anterior part of collor around the neck.
3. Protocoel, a pair of mesocoels, a pair of metacoels.
4. Metacoels.
5. Moulding of food particles and mucus into a cord.
6. Respiratory chamber of pharynx.
7. Hepatic region of intestine.
8. To test the water and food particles entering the mouth.
9. Connections between gill septa and tongue bars.
10. Yes.
11. Holoblastic, complete and radial.
12. Tornaria.

Fill in the Blanks

1. Hemichordata. **2.** U. **3.** Protosome, metasome. **4.** Balanoglossus. **5.** Collar. **6.** Buccal cavity. **7.** U. **8.** Forward. **9.** Excretory. **10.** Separate

Multiple Choice Questions

1. (*a*)	**2.** (*c*)	**3.** (*a*)	**4.** (*a*)	**5.** (*d*)	**6.** (*b*)
7. (*b*)	**8.** (*c*)	**9.** (*a*)	**10.** (*b*)	**11.** (*a*)	

4

Herdmania

Herdmania is a simple sedentary tunicate or urochordate which is commonly called **sea potato** or **sea squirt**. Its 12 species are found in all the seas except arctic (*i.e.*, mainly in Atlantic and Pacific Oceans). Four species of *Herdmania* occur in Indian Ocean: *H. pallida*, *H. ceylonica*, *H. ennurensis* and *H. mauritiana*. *Herdmania pallida* is described here.

4.1 SYSTEMATIC POSITION

Phylum	—	**Chordata**
Subphylum	—	**Urochordata or Tunicata**
Class	—	**Ascidiacea**
Order	—	**Pleurogona**
Family	—	**Pyuridae**
Genus	—	*Herdmania*
Species	—	*pallida*

4.2 HABITAT AND HABITS

Habitat. *Herdmania pallida* inhabits the rocky beds in the coastal waters (littoral habitat) of the Indian seas.

Habits. *Herdmania* is a sedentary solitary marine suspension feeder. It is attached to rocky substratum by a flat **base** or anchored in sandy floor by a large **foot**. If *Herdmania* is accidentaly detached, the foot acts as a balancer to keep it erect. If tilted, it can correct its position. *Herdmania* is often gregarious. Many individuals live in close proximity, even on the test of another individual. Sometimes it grows on the shell of living gastropod, *e.g.*, *Xancus* (Shankh). *Herdmania* is carried by the gastropod from place to place, which provides it more food and oxygen, and brings about dispersal of its progeny. The gastropod is, in return, protected from the predators by indedible test of *Herdmania* that covers it. However, the two animals can live equally well separately. Such a mutually beneficial ecological association is called **protocooperation**.

Herdmania is a ciliary feeder and takes suspended microorganisms (diatoms, algae and ciliates) of the plankton. Thus, animal feeds on small organisms wafted into the mouth along with the current of water maintained by the action of cilia in the oral and pharyngeal region. *Herdmania* breathes water (containing dissolved oxygen). When disturbed, *Herdmania* suddenly contracts and emits two jets of sea water with considerable force through its branchial and atrial apertures. [*Note*. According to **Ayyar** and **Ananthakrishnan** (2008) disturbed ascidian squirts out jets of water only through its atrial aperture]. This accounts for its popular name "sea squirt". Sexes are united. Fertilization is external. Development is indirect. Life history of *Herdmania* includes free swimming, bilaterally symmetrical, tailed larva, called **tadpole**. Notochord containing tadpole larva of *Herdmania* undergoes **retrogressive metamorphosis** to become an adult.

4.3 EXTERNAL CHARACTERS

Form, Size and Colour. *Herdmania* has purse-like or potato-like body shape; it is slightly narrower at the fixed end than at the free end. It varies from 6 to 13 cm in length and 4 to 7 cm in width. Living animals (or fresh specimens) are pinkish in colour with red patches at certain places. The body colour of specimen of *Herdmania* changes to yellowish-brown on preservation.

Division of Body. The body is divided into two parts: the body proper and the foot.

1. Body proper. Anterior end of animal's body is called the **body proper**. At its free end the body bears two short projections, the **branchial** and **atrial siphons**. The branchial siphon is a little shorter and has at its tip **branchial** or **incurrent aperture** or **mouth** directed away from the other siphon. The atrial siphon is longer and has at its tip **atrial** or **excurrent** or **cloacal aperture** directed upward (Fig. 4.1). The branchial aperture is wider than the atrial aperture when expanded. Both the apertures are bounded by four lobes, called **lips**, of the test.

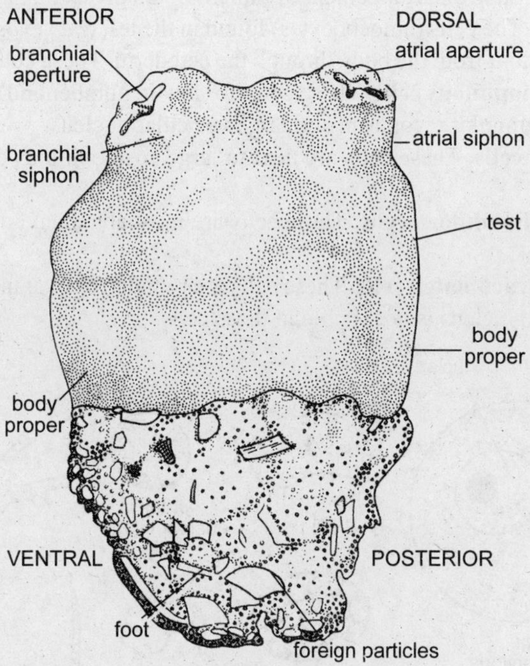

Fig. 4.1. *Herdmania pallida*. External features.

2. Base or Foot. The base of the animal is called **foot**. It is formed by the test alone. Foot looks dirty because variety of foreign objects (sand, shell pieces, spicules, algae, diatoms, hydroids and sea anemones, etc.) remain embedded in it. The animal remains attached to the substratum through its foot.

Orientation

Orientation means a position in relation to something else. Orientation of *Herdmania* is peculiar. The branchial siphon marks the **anterior** end and the side diagonally opposite to it is the **posterior** end. Atrial aperture marks the **dorsal** surface and the side diagonally opposite to it marks the ventral surface. The flat surfaces represent lateral sides.

4.4 TEST OR TUNIC

Only the test or the tunic is seen from outside. Real animal body lies inside it. The test is a thick, leathery, translucent covering of the body. The ascidians are called **tunicates** due to the presence of the test. The test or tunic is regarded as **exoskeleton** which provides support and protection to animal body (**Brusca** and **Brusca** 2003). Test is secreted by the epidermis (ectoderm) of the body wall. It, thus, corresponds to the cuticle of other animals. Its inner surface is smooth and shining whereas outer surface is rough and dull. The test wears off from the surface and is continuously replaced from inner side by epidermis. Test consists of a clear, gelatinous matrix composed of proteins, carbohydrates and water and contain fibers, wandering cells or corpuscles, blood vessels and spicules.

1. Matrix. It is composed of water, various carbohydrates, proteins and a type of cellulose, called **tunicin** (**Welsh** 1984). The production of cellulose-like tunicin is a unique feature of ascidians.

Tunicin occurs in the form of **fibers** which are parallel and occur in thin and successive sheets. In each sheet, the fibers are oriented at a different angle from that in the oriented at a different angle from that in the outer one. Such an arrangement of tunicin fibers gives toughness and strength to the test.

2. Corpuscles. The cells (amoebocytes) found in the test (*i.e.*, exoskeleton) are of mesodermal origin and migrate into it from the body through the ectoderm. These cells are of following six types:

(*i*) Large eosinophilous cells. These cells are few in number and spherical or ovoid in shape. These cells contain granular cytoplasm and large vesicular nucleus.

(*ii*) Amoeboid cells. These cells are next to large eosinophilous cells in size. These cells are very few in number.

(*iii*) Small eosinophilous cells. These cells are most abundant and contain excentric nucleus (at the periphery).

(*iv*) Spherical vacuolated cells. These cells are rounded or oval in shape. They contain a large number of vacuoles. Nucleus is not visible in these cells.

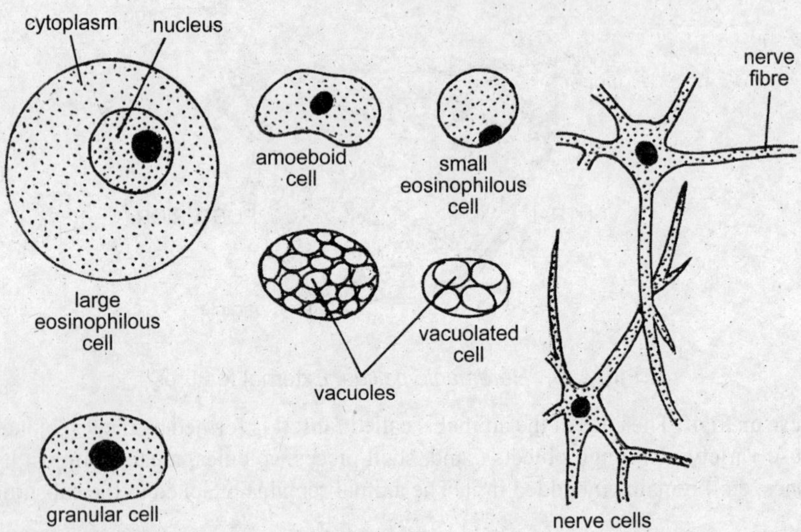

Fig. 4.2. *Herdmania.* Various types of cells found in the test.

(*v*) Granular or receptor cells. These cells are rounded or oval in shape and fewer in number. Their nucleus is large and cytoplasm is granular. They occur near the surface of the test.

(*vi*) Nerve cells. These cells are quite small cells. Each cell is irregular in shape and contains two or six fine branched processes.

3. Interlacing fibrils. Fine fibrils run criss-cross all through the matrx. Some of these resemble the smooth muscle fibrils of the mantle (body wall), others look like nerve fibrils. The nerve fibrils are more abundant round the receptor cells.

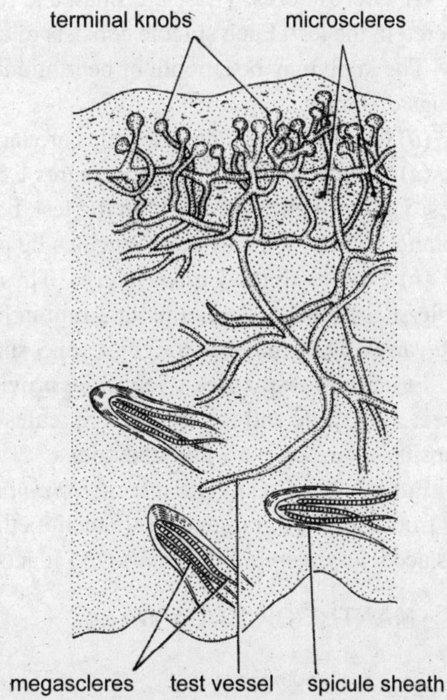

Fig. 4.3. *Herdmania*. V.S. of test showing blood vessels, ampullae and spicules.

4. Blood vessels. The blood vessels found in the test of *Herdmania* extended into it from the connective tissue of the mantle. They bring food to test. Blood vessels form a network throughout the test, and send toward the surface numerous branches which end in rounded ovoid or pear-shaped **vascular ampullae** or **terminal knobs** (Fig. 4.3). These ampullae form red patches on the surface. Each ampulla is covered by a single layer of flat, many-sided cells with red pigment granules. The ampullae are also richly supplied with nerve fibrils and form tactile center (or receptors). Since these ampullae bring the blood very close to the surface, they also facilitate respiratory exchange of gases.

The presence of blood vessels indicates that the tunic is a living tissue. Moreover, the tunic grows as the animal's body enlarges. The tunic is not moulted.

5. Spicules. Calcareous spicules (Fig. 4.4) are found embedded in the matrix of the test, body wall and various other parts of the body. All types of spicules bear equidistant rings of minute **spines** all pointing in one direction. Spicules are of two types:

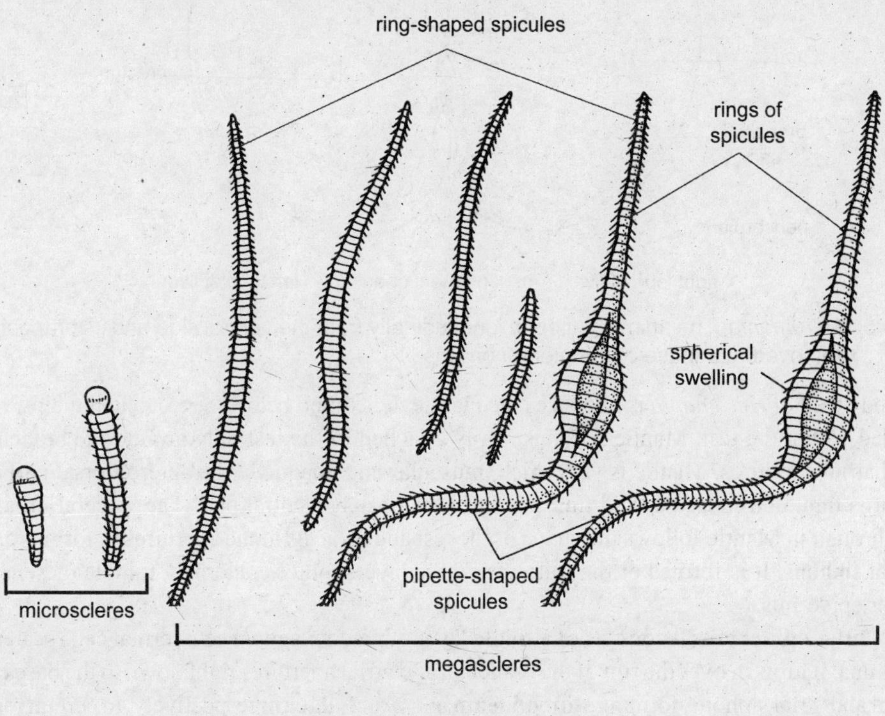

Fig. 4.4. *Herdmania*. Different types of spicules.

***(i)* Microscleres.** These are small and rod-like spicules (50 to 80 μm length). These are found scattered in the test. Each spicule consists of a rounded **knob** or **head** and an elongated and tapering **body**. The knob may be smooth or bear minute spines. The body of each microscleres contains 5 to 20 rings of spines.

***(ii)* Megascleres.** These spicules are larger than the microscleres and are of two types:

***(a)* Spindle shaped.** These measures 1.5 to 2.5 mm in length and contain many rings of minute spines. These spicules are found in the test, body wall, gut, gonads, and around the blood vessels of the branched sac. Some of these spicules lie partly in the test and partly in the mantle.

***(b)* Pipette-shaped megascleres.** These spicules are pipette-shaped and are larger than the spindle-shaped megascleres (about 3 mm in length). Their middle part is swollen like a bulb. Their shape varies. They may be straight, U or S shaped. These are found in mantle, gonads and liver.

Functions of spicules. (*i*) Spicules provide regidity to the test and other body parts. (*ii*) Spicules connect the mantle and the test. (*iii*) Spicules make the blood vessels tough and prevent them from collapsing. They keep off the predators.

Functions of test. The test of *Herdmania* serves the following functions: (*i*) It protects the body from mechanical injury. (*ii*) It anchors the animal by forming a foot. (*iii*) Its vascular ampullae act as accessory respiratory organs. (*iv*) It acts as a tactile receptor.

4.5 MANTLE OR BODY WALL

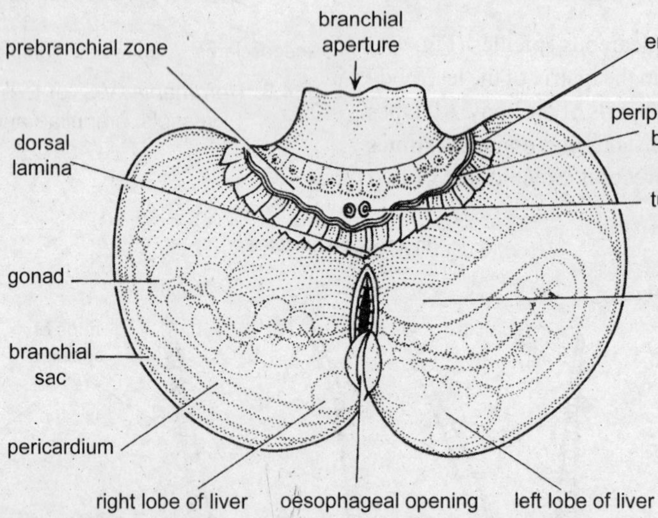

Fig. 4.5. *Herdmania*. The animal cut open longitudinally from the ventral side and its branchial sac removed to expose other internal organs.

Body wall of *Herdmania* is known as the mantle. It is found below the test and remains loosely suspended within the test. Mantle remains firmly attached to the test only around the branchial or oral and atrial apertures. Mantle is very thick, muscular and opaque in the antero-dorsal part of the body but is thin, non-muscular and transparent in the postero-ventral part. The visceral organs are visible through it. Mantle follows the shape of the test and projects at the apertures to form branchial and atrial siphons. It is formed of the following three layers: outer epidermis, middle mesenchyme and inner epidermis.

1. Outer epidermis. It consists of a single layer of flat, hexagonal ectodermal cells seated on a basal lamina. It turns in over the rim of the branchial and atrial apertures right down to the bases of the branchial and atrial siphons, forming **stomodaeum** and **proctodaeum** respectively. In certain regions, *e.g.*, muscular parts of the mantle, bases of the siphons and intersiphonal area, the outer epidermis is bright red due to the presence of red pigment granules in its cells. The outer epidermis is interrupted at places where spindle-shaped spicules and blood vessels pass from the mantle into the test.

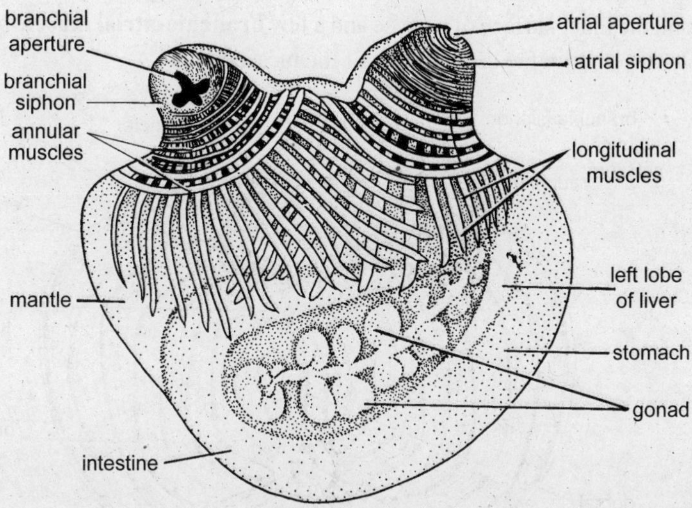

Fig. 4.6. *Herdmania*. The animal as seen from the left side after removal of the test.

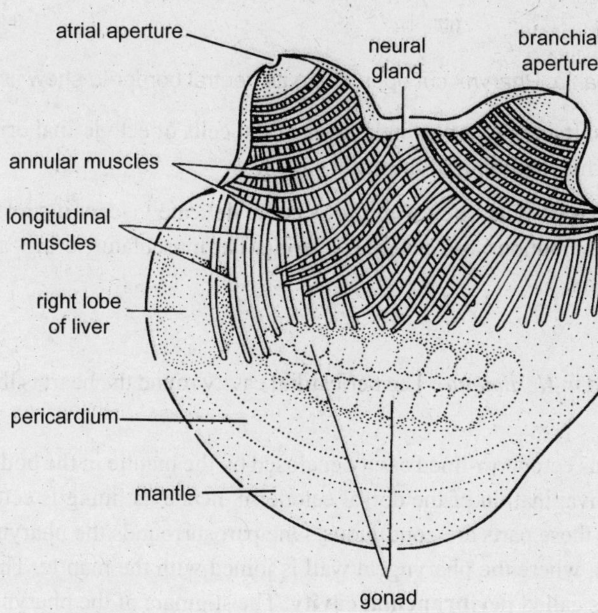

Fig. 4.7. *Herdmania*. The animal as seen from the right side after removal of the test.

2. Mesenchyme. The mesenchyme develops from the mesoderm. It consists of a gelatinous connective tissue containing blood sinuses, muscle fibers, nerve fibers and cells. The connective tissue cells of the mesenchyme are chiefly of amoeboid and vacuolated types. The muscle fibers are long, flat, ribbon-shaped cells, each with a large nucleus. They are unstriated and best developed in the anterodorsal region of the body. They are divided into two sets: **annular** and **longitudinal muscles** (Fig. 4.6 and 4.7). The annular muscles surround the siphons as rings. They act as sphincters that regulate the openings (the apertures). The longitudinal muscles lie inside the annular muscles and extend fanwise from the branchial and atrial apertures on to about the middle of each side of the body. Contraction of longitudinal muscles withdraws the siphons. Some longitudinal muscles cross in the intersiphonal region. Besides these two types of muscles, each siphon has a strong circular **sphincter muscle** around the inner surface of its base and a few **branchio-atrial muscle bands** extending from one siphon to the other in the deeper parts of the mesenchyme.

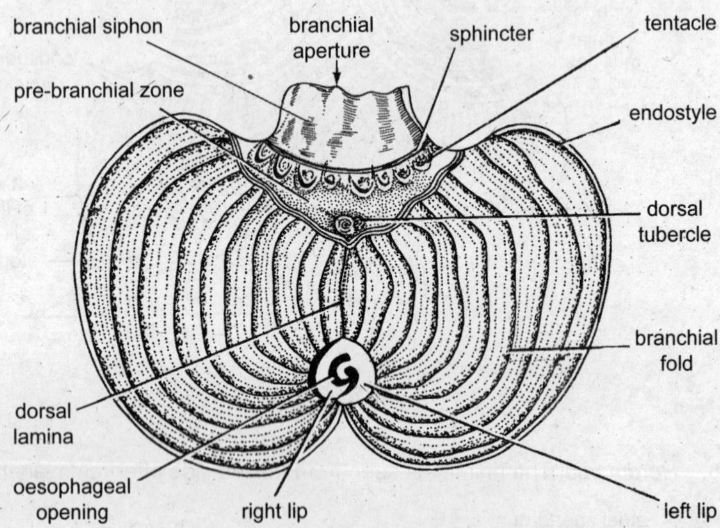

Fig. 4.8. *Herdmania*. Pharynx cut open along its ventral border to show its interior.

3. Inner epidermis. It consists of a single layer of flat cells of ectodermal origin like those of the outer epidermis. It forms the lining of the atrial cavity.

Functions of body wall. The body wall protects the delicate viscera it surrounds. The outer epidermis secretes the test, a mutipurpose external covering. The musculature brings about contraction of the body and the siphons.

Coelom and Atrium

Coelom is greatly reduced in *Herdmania*. The pericardial cavity round the heart is the only remanent of the coelom.

Atrium is a spacious ectoderm-lined cavity enclosed by the mantle in the body of *Herdmania*. It arises in the larva by invagination of the dorsal ectoderm, hence its lining is ectodermal. Atrium shows two parts. Both of these parts are continuous. One part surrounds the pharynx, except on the anterior and ventral sides, where the pharyngeal wall is joined with the mantle. This part of atrium is relatively narrow and is called **peribranchial cavity**. The stigmata of the pharynx opens into this cavity. It is crossed by vascular **trabeculae** stretching between the pharynx and the mantle. Other part of the atrium exists dorsal to the pharynx. It is very wide and termed the **cloaca**. The rectum and the gonoducts open into this part, discharging faeces and gametes into it. The cloaca leads to the exterior by way of the **atrial siphon** and **atrial aperture**. The atrial siphon bears at its base a ring of slightly serrated processes, the **atrial tentacles**.

4.6 DIGESTIVE SYSTEM

Digestive system of *Herdmania* (Fig. 4.9) consists of alimentary canal and digestive glands.

I. Alimentary Canal

Alimentary canal is complete and comprises of mouth or branchial aperture, buccal cavity, pharynx, oesophagus, stomach, intestine, rectum and anus.

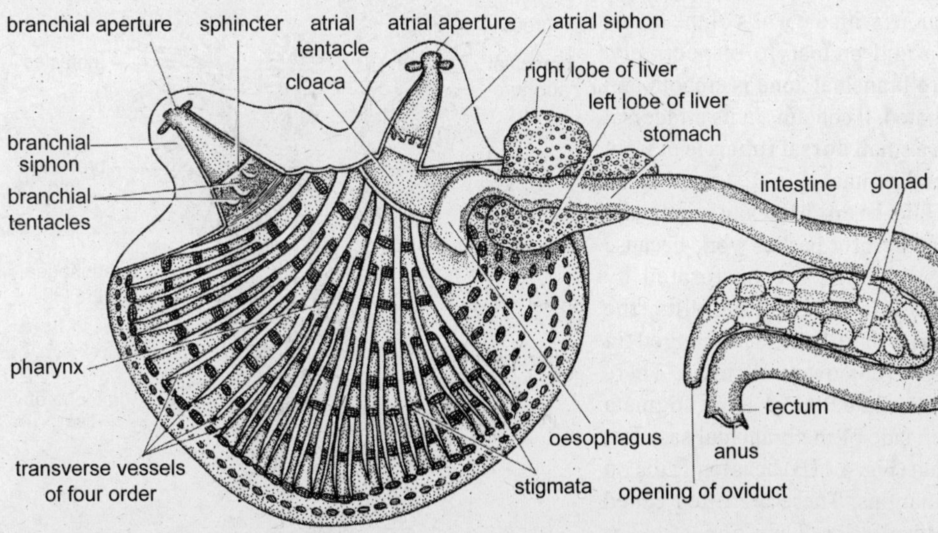

Fig. 4.9. *Herdmania*. Alimentary canal, displaced to one side. Liver is also seen.

The buccal cavity is lined with ectoderm and represents the **stomodaeum**. All other parts are lined with endoderm and constitute the **mesenteron**.

1. Mouth. Branchial aperture represents **mouth**. It lies at the tip of the branchial siphon. Mouth is bordered by four lips and is directed anteriorly. It leads into the buccal cavity.

2. Buccal cavity. It is a short, laterally compressed cavity in the branchial siphon. It is lined with epidermis covered by test, both having turned in from the outer surface of the body over the rim of the mouth. The buccal cavity has at its base a strong **branchial sphincter**, which can close it when required. Close to the sphincter is a ring of sickle-shaped, tripinnate processes called **branchial tentacles** (Box 4.1).

Box 4.1

The tentacles are of four sizes: 8–10 about 5 mm length; 8–10 about 2 to 3 mm in length; 16 of less than 1.5 mm in length and 32 of 0.5 mm in length. These tentacles alternate each other. Each tentacle bears a large number of paired lateral branches (so are called **pinnate**).

During life, the tentacles stand out at right angles to the wall of the siphon and meet in the center, forming a delicate sieve that allows only microscopic food particles to enter the pharynx along with the current of sea water, and keeps out large particles. Rich nerve supply of the tentacles indicates that they also act as chemoreceptors. The buccal cavity is followed by pharynx.

3. Pharynx. The pharynx (Fig. 4.8) is the largest part of the alimentary canal occupies the greater part of the body. It is a laterally compressed sac with a complicated structure. The pharynx is divided into two very unequal regions: a small anterior **prebranchial zone** just behind the branchial sphincter, and a large posterior **branchial sac**, by two circular ridges, the **peripharyngeal bands**, arising from the inner surface. The bands run parallel and close to each other and enclose a narrow, ciliated **peripharyngeal groove** between them (Fig. 4.10). The cilia present in the groove are shorter than those over the bands. The aterior band is a complete ring but the posterior band is interrupted in the middorsal and midventral lines, where it becomes continuous with the walls of the dorsal lamina and endostyle respectively. The pre-branchial zone is smooth and nonciliated. It contains in its middorsal region a small **dorsal tubercle** in front of dorsal lamina.

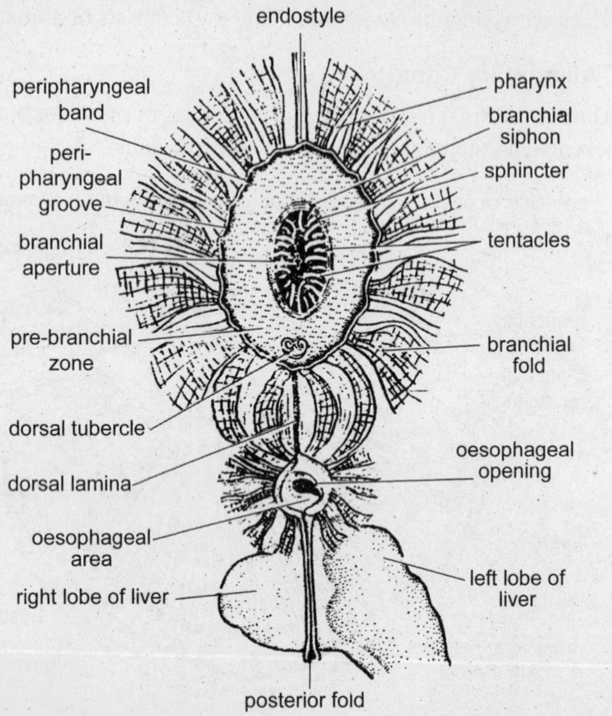

Fig. 4.10. *Herdmania*. Schematic diagram showing the relation of the branchial folds, the peripharyngeal bands, the dorsal lamina, the oesophageal area and the endostyle.

The **branchial sac** gives the appearance of a basket-work because its lateral walls are perforated by innumerable longitudinal slits, the **stigmata** or **gill slits**, arranged in transvers (dorsoventral) rows. There are about 2,00,000 (2 lacs) stigmata on each side of the branchial sac. The stigmata (Fig. 4.11B) bear long cilia on their margins. These are often called the **lateral cilia**. They almost cover the stigmata to prevent the escape of food particles. The lateral walls of the branchial sac have at intervals broad **bars**, transverse and longitudinal, which contain blood vessels and divide the walls into squarish or rectangular **stigmatic areas**. Each stigmatic area has five or six stigmata. Besides the broad bars, thin longitudinal **interstigmatic bars** occur between the adjacent stigmata and thin transverse **intrastigmatic bars** cross each row of stigmata. These bars contain finer blood vessels.

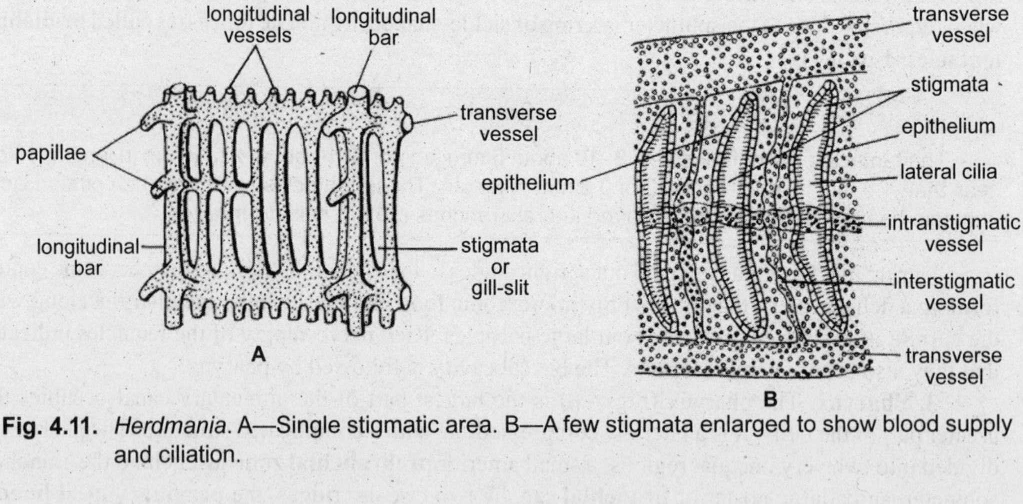

Fig. 4.11. *Herdmania*. A—Single stigmatic area. B—A few stigmata enlarged to show blood supply and ciliation.

Inner surface of each lateral wall of the branchial sac is raised into nine or ten prominent, crescentic **longitudinal folds** that start just behind the posterior peripharyngeal band and converge posteriorly at an area surrounding the oesophageal opening (Fig. 4.8). These folds of branchial sac increase the surface to come in contact with sea water entering the sac. The entire inner surface of the branchial sac including the longitudinal folds bears the cilia. These cilia are shorter than those bordering the stigmata, and are often termed **frontal cilia**, because they face the main water current.

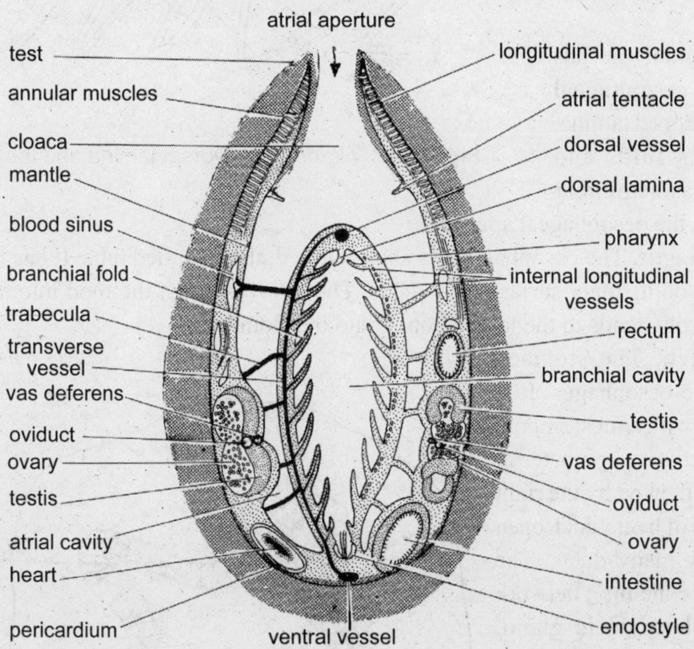

Fig. 4.12. *Herdmania.* Diagrammatic transverse section of the animal through the atrial siphon.

Endostyle. A narrow groove, called the **endostyle** or **hypopharyngeal groove**, run along the anterior and ventral borders of the branchial sac (Fig. 4.12). Anteriorly, the walls of the endostylar groove are continuous with the posterior peripharyngeal band and groove itself merges with the peripharyngeal groove. Posteriorly, the endostylar groove ends a little distance from the oesophageal opening, but its walls extend upto the oesophageal opening as the **retropharyngeal bands**. The endostyle is lined with five longitudinal tracts of ciliated cells alternating with four longitudinal tracts of mucus-secreting gland cells (Fig. 4.13). The cilia of the median tract are much longer than those of the lateral tracts.

Dorsal lamina. A thin flap, called the **hyper-pharyngeal band** or **dorsal lamina**, hangs from the short middorsal border of the branchial sac (Fig. 4.10). Anteriorly, this flap joins the posterior peripharyngeal band and extends posteriorly upto the right lip of the oesophageal opening. The dorsal lamina bears a row of 20 to 30 closely set

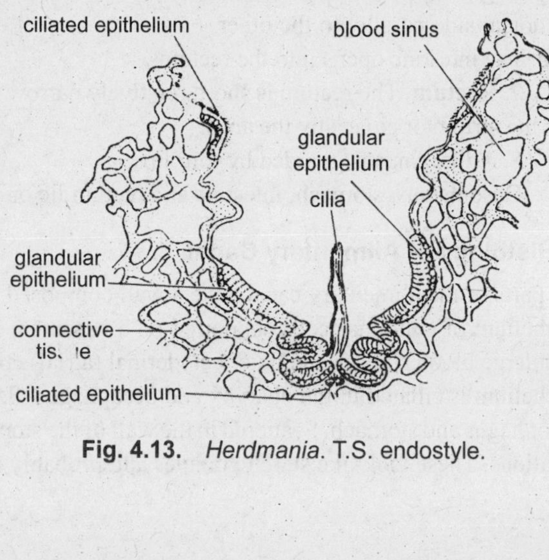

Fig. 4.13. *Herdmania.* T.S. endostyle.

ciliated tongue-like processes, the **languets**, that hang into the cavity of the branchial sac (Fig. 4.14). In a living animal, the languets remain curved to the right, forming a short of gutter.

The branchial sac has a small circular **oesophageal area** behind the dorsal lamina. This area lacks folds and stigmata, but has semilunar lips that enclose the oesophageal aperture.

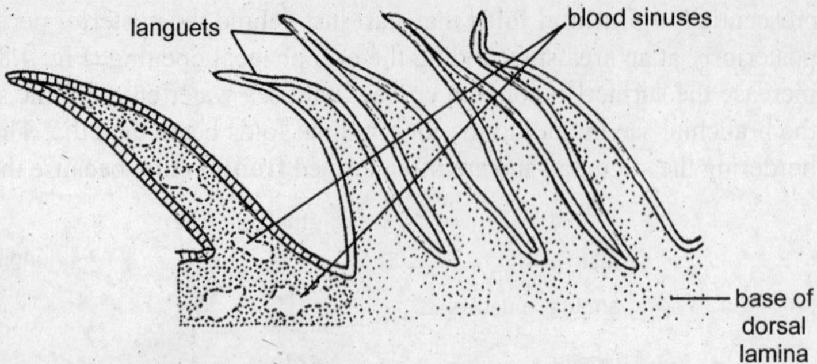

Fig. 4.14. *Herdmania*. Dorsal lamina and languets.

4. Oesophagus. The oesophagus is a short, curved, thick-walled tube. It has four longitudinal ciliated grooves on its inner surface (Fig. 4.15). The grooves direct the food into the stomach. The oesophagus sharply bends to the left and open into the stomach.

5. Stomach. The stomach is wider than the oesophagus. It is thin-walled tube with almost smooth inner surface. It has a sphincter at each end, and is flanked by the right and the left lobes of liver which open into stomach by many ducts. The wall of stomach contains a network of tubules of the pyloric gland. Stomach curves toward the ventral side and unnoticeably passed into the intestine.

6. Intestine. It is thin-walled U-shaped tube. It has two limbs: a proximal descending limb that runs toward the ventral side and a distal ascending limb that extends toward the dorsal side parallel to the other limb. The intestine opens into the rectum.

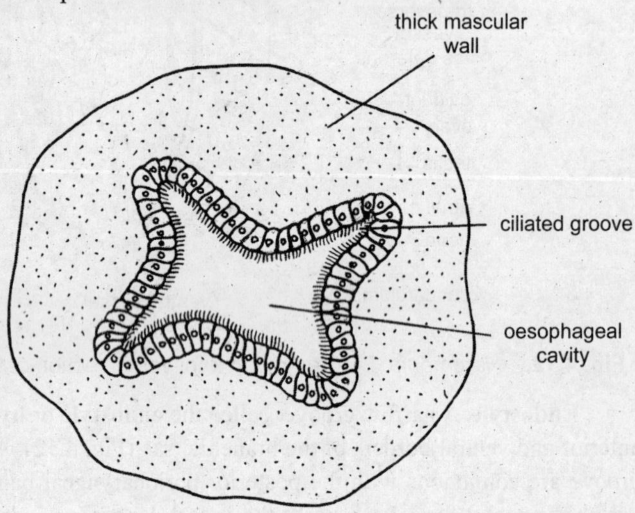

Fig. 4.15. *Herdmania*. T.S. of oesophagus.

7. Rectum. The rectum is short, relatively narrow tube that slightly curves dorsally and opens into the atrium or cloaca by the anus.

8. Anus. Anus is guarded by four lips.

Oesophagus, stomach, intestine and rectum lie on the left side of the branchial sac.

II. Histology of Alimentary Canal

All parts of the alimentary canal have a wall composed of three types of tissues: inner endodermal epithelium of columnar cells supported on a basement membrane, middle loose connective tissue with large blood sinuses and outer ectodermal (atrial) epithelium of thin cells (Fig. 4.16). The inner epithelium is ciliated in the pharynx and oesophagus, flagellated in the rectum and glandular in the oesophagus and stomach. Scattered in the wall of the stomach are numerous granules with concentric striations. These look like starch granules and probably represent stored food of *Herdmania*.

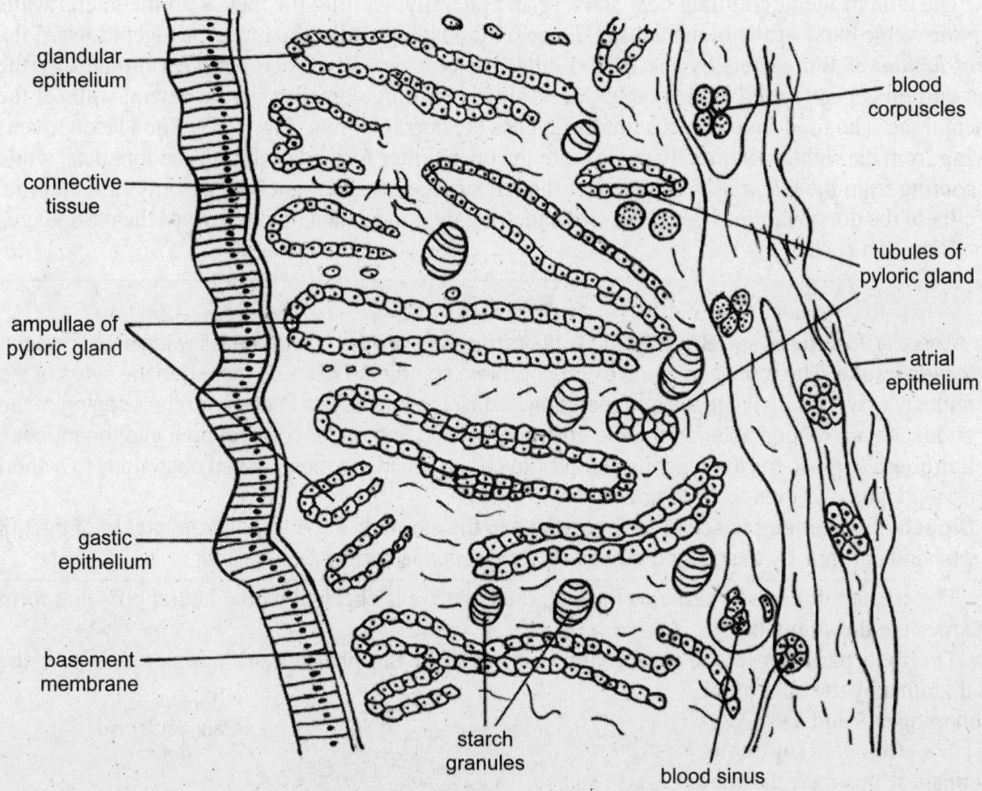

Fig. 4.16. *Herdmania*. T.S. of stomach.

III. Digestive Glands

The digestive glands associated with the alimentary canal in *Herdmania* include the liver and pyloric gland.

(*i*) Liver. The liver is a large, chocolate-coloured gland consisting of two lobes: (*a*) the larger left lobe and (*b*) smaller right lobe; both of which lie on the sides of the stomach. Liver secretes a dark, yellowish-brown fluid, which is drained into the stomach by 10 to 11 fine ciliated ducts. The liver's secretion contains enzymes such as powerful **amylase**, a **protease** and a weak **lipase**.

(*ii*) Pyloric gland. It consists of much-branched tubules ramifying in the walls of the stomach and intestine. The tubules of the gland lead into a number of ducts that unite into a single duct. The latter opens into the middle of the proximal limb of the intestine on the branchial side. The secretion of the pyloric gland resembles that of the vertebrate pancreas.

Mode of ingestion and digestion. *Herdmania* is a suspension, ciliary or filter feeder, *i.e.*, it filters out its food from a current of water passed through the pharynx. The food of *Herdmania* consists of microscopic planktonic organisms such as diatoms, algae and infusorians.

1. Action in pharynx. The water current is caused by the outward beating of the powerful *lateral cilia* bordering the stigmata. The water current enters the pharynx via mouth and buccal cavity and leaves through the stigmata into the atrium, whence it passes out by way of atrial aperture. The *branchial* tentacles form a sieve in the passage of the water current, allowing only the microscopic particles to filter through its fine meshes for entery into pharynx. The tentacles probably also taste the water current and its food particles. Occasionally, the tentacles are cleared of the large particles entangle in them and obstructing the water current by forcing out a strong **reverse current**. This is done by closing atrial siphon and contracting the body muscles whereby water of the pharynx is expelled via mouth.

In the *branchial sac*, the glandular tracts of the *endostyle* secrete a large amount of sticky mucus (Box 4.2), which is deflected sideways by the movements of the long cilia of the central tract. The cilia of the lateral tract beat actively and laterally, wafting the mucus on the lateral walls (*i.e.*, transverse bars) of the branchial sac. Once on the lateral walls, the mucus is swept toward the *dorsal lamina* as thin sheets by the frontal cilia that beat dorsally. Food particles brought by the water current get entrapped in the mucus sheets slowly shifting dorsally over the lateral walls of the branchial sac. The food-laden mucus finally reaches the dorsal lamina (Fig. 4.17). Food laden mucus arriving from the right wall directly passes into the open gutter formed of the curved longuets, while that coming from the left wall is driven over the left surface of the languets by activity of their cilia. The cilia of the dorsal lamina beat backward and drive the food-laden mucus toward the oesophagus in the form of a cord.

Box 4.2

According to **Brusca** and **Brusca** (2003), during digestion in *Herdmania* the following event occurs:

1. **Iodine intake.** The mucus, a complex mucoprotein containing iodine is moved to the sides of the endostyle by the basal flagella and then outward by the lateral cilia. Cells near the opening of the endostyle are responsible for binding environmental iodine and incorporating it into the mucus.
2. **Languets** serve to roll the mucous sheets into cords, which are then passed posteriorly to a short oesophagus and then to a stomach.
3. **Digestive enzymes** of ascidia are secreted into the stomach lumen by secretory cells of the gut wall and perhaps by associated glands, and *digestion* is largely extracellular.

The escape of the food particles through the stigmata is checked by the lateral cilia that form a fine mesh in the stigmata.

The food particles left in the prebranchial zone of the pharynx are perhaps sent into the dorsal lamina by the cilia of the peripharyngeal bands.

2. Action is oesophagus. Oesophageal lips pass the food cord into the oesophagus. In this organ the food cord is twisted and more mucus is added to it. Finally, it is sent into the stomach.

3. Action of stomach. In the stomach, digestion of food takes place with the help of enzymes contained in the secretion of the liver. The enzymes include amylase that digets starch, protease which breaks down proteins and lipase that acts on the fats.

4. Action in intestine. In the intestine, the process of digestion is completed by the secretion of pyloric gland, and the digestion products are absorbed. Indigested remains are changed into faeces which passes into the recturm.

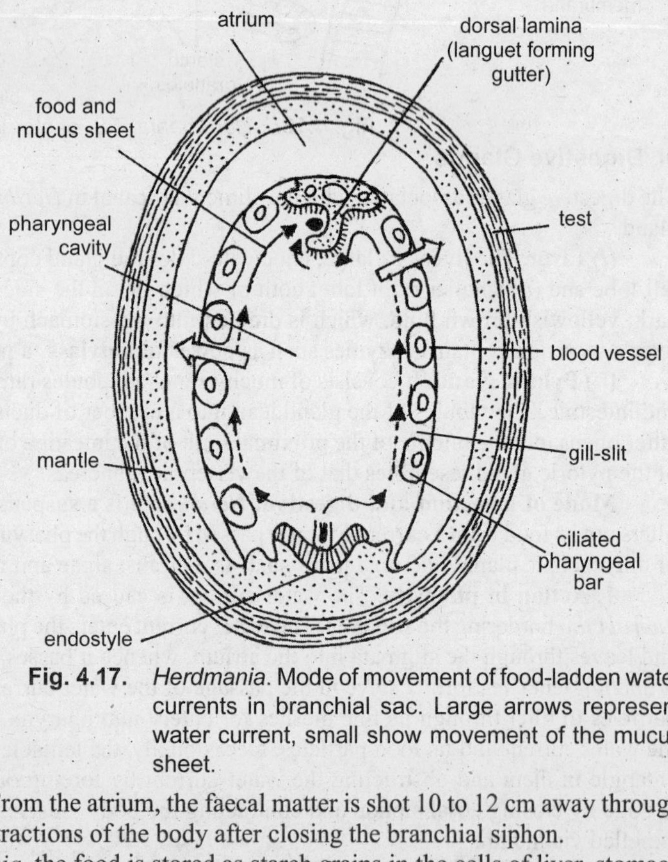

Fig. 4.17. *Herdmania*. Mode of movement of food-ladden water currents in branchial sac. Large arrows represent water current, small show movement of the mucus sheet.

Defaecation or egestion. From the atrium, the faecal matter is shot 10 to 12 cm away through atrial aperture by occasional contractions of the body after closing the branchial siphon.

Food storage. In *Herdmania*, the food is stored as starch grains in the cells of liver, stomach and intestine.

4.7 RESPIRATORY SYSTEM

The respiratory system of *Herdmania* include branchial sac, trabeculae and ampullae of test.

1. Branchial sac. The main respiratory organ in *Herdmania* is the branchial sac. The wall of this sac has a very rich blood supply and constantly comes in contact with fresh sea water that flows through the sac as a regular current. The water current is maintained by the outward lashing of the cilia bordering the stigmata. The sea water contains dissolved oxygen which is exchanged with the carbon dioxide of the blood flowing in the wall of the branchial sac. The wall of the blood vessels facing the branchial cavity is very thin, and this facilitates the gaseous exchange. Efficiency of respiration is further increased by the longitudinal folds of the inner surface of the branchial sac by increasing the respiratory surface enormously.

2. Trabeculae. Exchange of gases also takes place in the vascular trabeculae, which stretch between the pharyngeal wall and the body wall.

3. Vascular ampullae in test. The vascular ampullae present in very large numbers in the superficial part of the test bring the blood very close to the surrounding sea water for gaseous exchange.

Large surface area of pharynx, large volume of water circulated, and low activity of the *Herdmania* exclude the need for a respiratory pigment.

4.8 BLOOD VASCULAR SYSTEM

The blood vascular system of *Herdmania* is well developed and highly specialized due to the presence of a well developed branchial sac and test which act as the respiratory organs. It consists of following three organs: 1. Pericardium and heart; 2. Blood vessels and 3. blood.

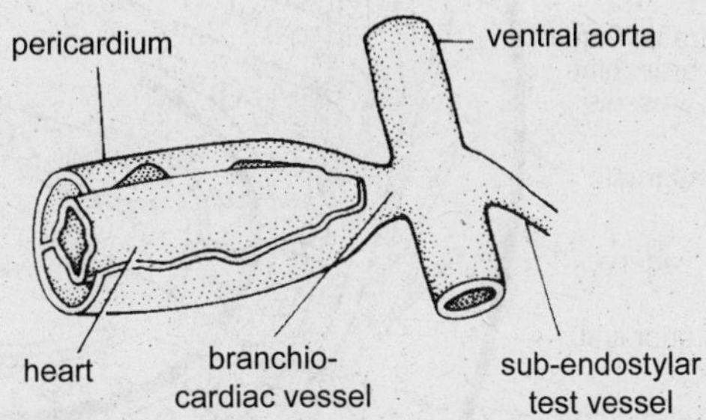

Fig. 4.18. *Herdmania.* Wall of pericardium cut open to show the attachment of the heart.

A. Pericardium and Heart

1. Pericardium. It is 7 cm long and 3 mm wide transparent and closed tube. Pericardium lies obliquely in a dorso-ventral direction in mantle below the right gonad on the right side of the body. Its walls are thick and non-contractile (Fig. 4.18). The wall contains an outer layer of connective tissue and blood sinuses, and an inner layer of epithelial cells. The pericardial cavity contains colourless pericardial fluid which has cells and corpuscles similar to those of blood.

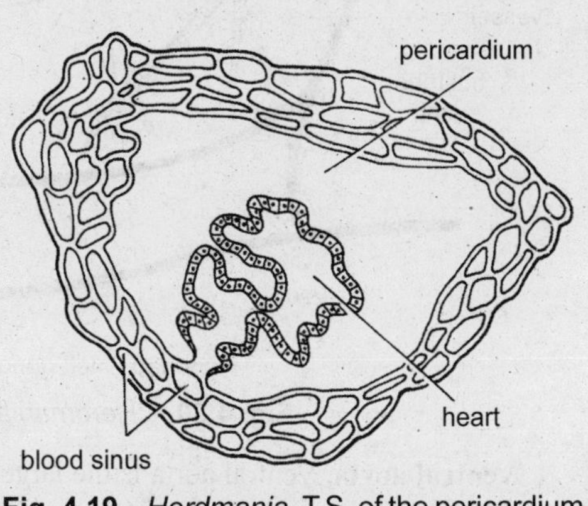

Fig. 4.19. *Herdmania.* T.S. of the pericardium and the heart.

2. Heart. The heart of *Herdmania* is contractile and tube-like. It remains enclosed within the pericardium. Heart is attached to the wall of pericardium along its entire length by a thin flap of connective tissue. In fact, the heart is formed by the infolding of pericardium. Its wall is thin and convoluted. It is composed of a single layer of epithelial cells surrounded by a thin layer of striated muscle fibers (heart is the only organ to have striated muscles in *Herdmania*). The two ends of the heart lead into main blood vessels. There are no valves. A freely movable pear shaped body lies outside the heart in pericardial cavity. This body probably regulates the direction of flow of blood.

B. Blood Vessels

Herdmania has well developed blood vessels. The larger blood vessels have a lining of endothelium, whereas the smaller ones are mere spaces in the connective tissues. There are four main blood vessels: ventral aorta, dorsal aorta, branchiovisceral vessel and cardiovisceral vessel (Fig. 4.20).

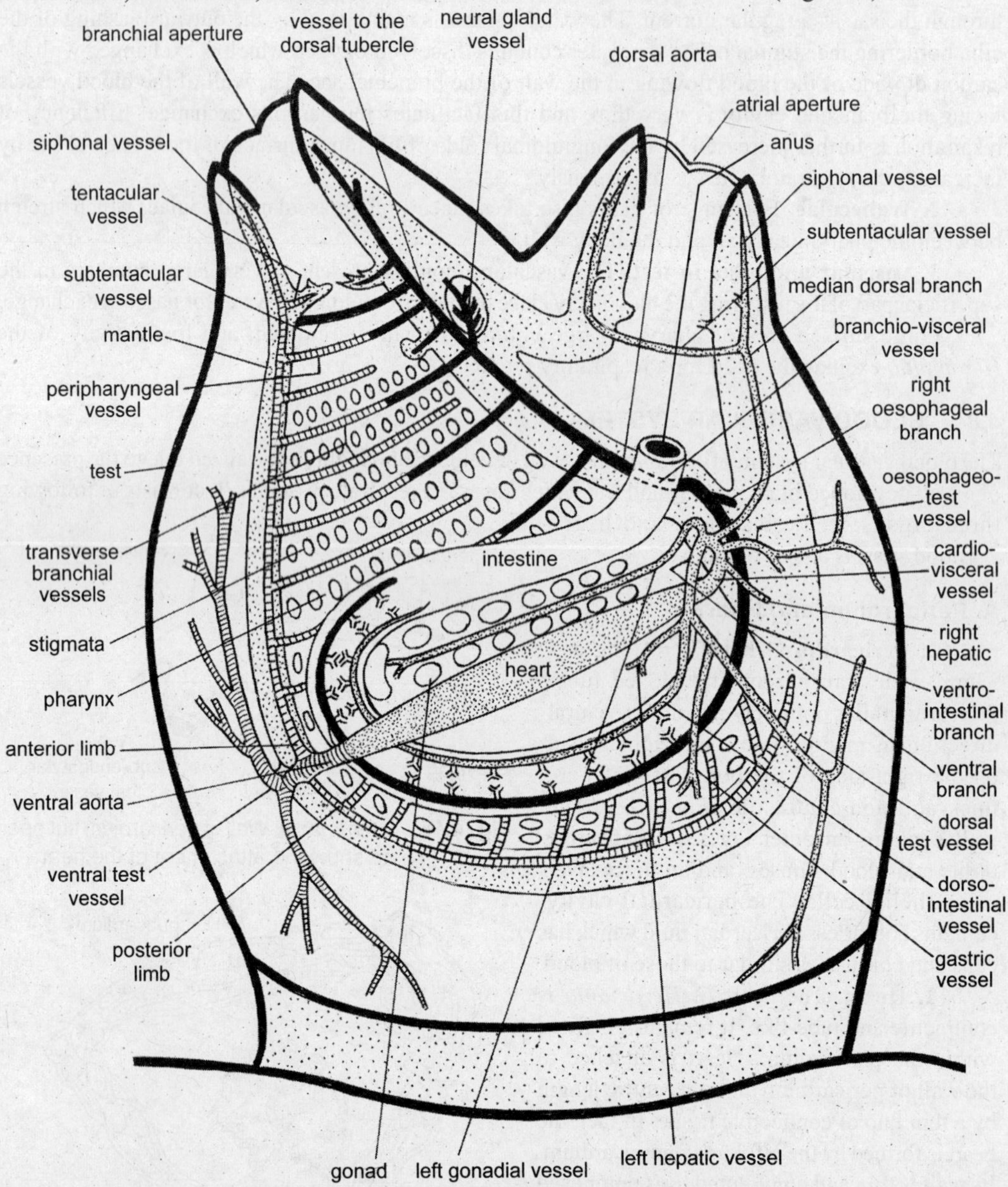

Fig. 4.20. *Herdmania*. Blood vascular system.

1. Ventral aorta. Ventral aorta is the largest vessel. It arises from the ventral end of the heart and immediatly divides into three branches: (*a*) **anterior limb**, (*b*) **posterior limb** and (*c*) **ventral test vessel**. The anterior and posterior limbs lie in the pharyngeal wall just below the endostyle (Fig. 4.21). Both of these branches give off several paired **transverse branchial branches** (40 to 56 pairs) which runs through the transverse bars of branchial sac between rows of stigmata. These limbs send many fine branches to the endostyle and mantle also.

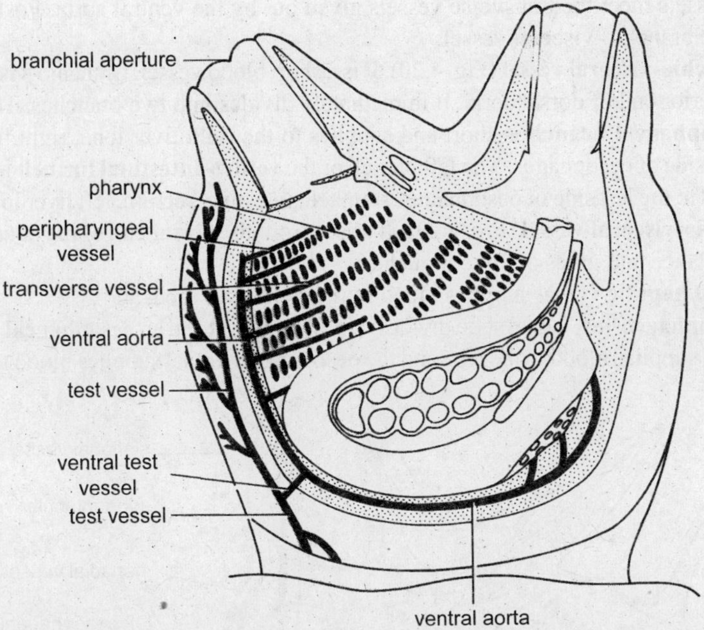

Fig. 4.21. *Herdmania.* Ventral aorta and its branches.

The **anterior limb** of ventral aorta extends forward and joins two circular vessels at the base of the branchial siphon. 1. It joins a **peripharyngeal vessel** which lies below the peripharyngeal groove of the pharynx, and 2. a **subtentacular vessel** which runs below the ring of branchial tentacles. The peripharyngeal vessel sends branches to the mantle and paired longitudinal branchial vessels cross each other and communicate together at each crossing. The **subtentacular vessel** sends a **tentacular vessel** into each branchial tentacles and 6 to 8 **siphonal vessels** into the branchial siphon.

The **posterior limb** of the ventral aorta extends upto the oesophageal area as a thin vessel. The **ventral test vessel** supplies the blood to the test of the ventral side. It ramifies profoundly and each branch terminates below the epidermis as an ampulla.

2. Dorsal aorta (Fig. 4.22). Dorsal aorta is a large and stout vessel. It lies just above the dorsal lamina in the dorsal wall of the branchial sac. It is not connected to the heart. In the anterior region it supplies blood to the neural gland, dorsal tubercle and inter-siphonal area and finally joins the peripharyngeal and subtentacular vessels. In the middle it sends many paired transverse vessels in the transverse bars of the branchial

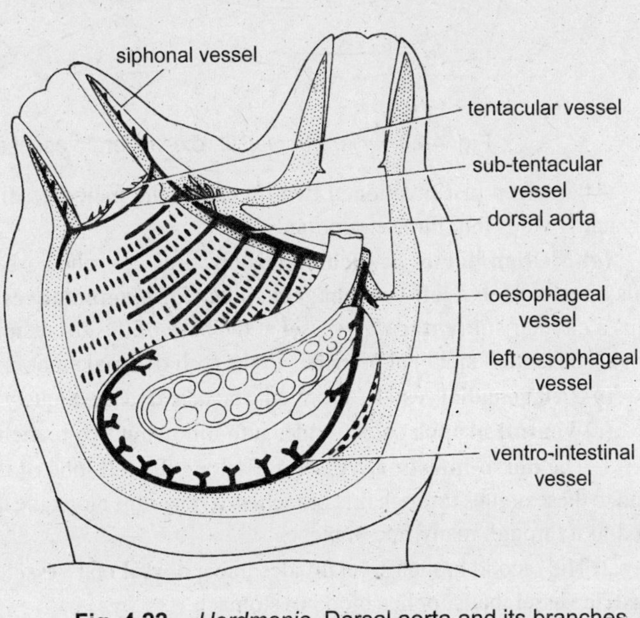

Fig. 4.22. *Herdmania.* Dorsal aorta and its branches.

sac. These vessels meet the transverse vessels given out by the ventral aorta. Posteriorly the dorsal aorta joins the branchio-visceral vessel.

3. Branchio-visceral vessel (Fig. 4.20). It is a short blood vessel. Branchio-visceral vessel arises from the posterior end of dorsal aorta. It immediately divides into two branches. The right branch or the **right oesophageal branch** is short and supplies to the right liver lobe, right lip of oesophageal area and right side of oesophagus. The left branch or the **ventro-intestinal branch** is stout and long. It supplies blood to the left side of oesophagus, stomach, intestine, rectum, left liver lobe and left gonad.

4. Cardio-visceral vessel (Fig. 4.23). It arises from the dorsal end of the heart. It runs dorsally for a short distance and gives out two branches:

(*i*) Right hepatic vessel supplies blood to the right liver lobe, and

(*ii*) Oesophageo-test vessel soon divides into a **test vessel** and an **oesophageal vessel** (Fig. 4.20). The test vessel supplies blood to the test, and the oesophageal vessel supplies blood to the oesophagus.

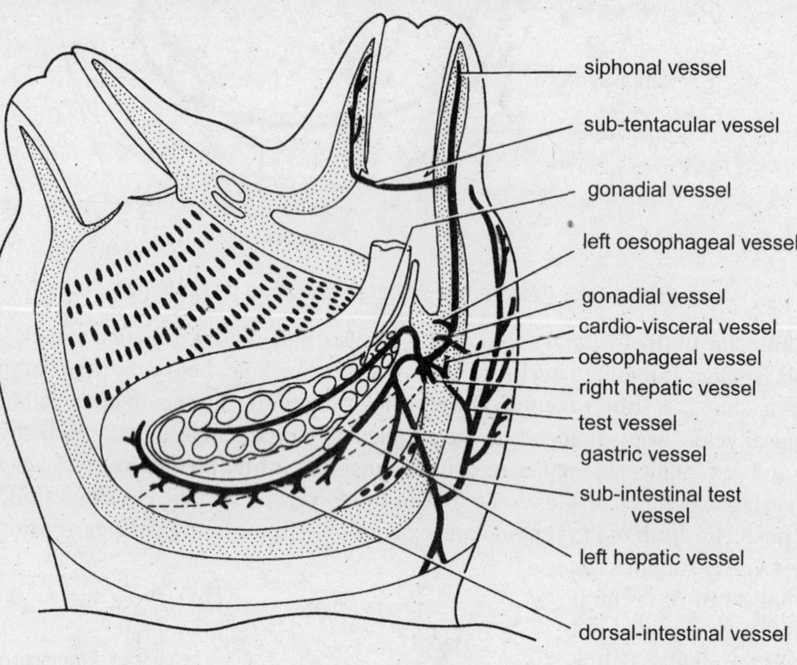

Fig. 4.23. *Herdmania*. Cardio-visceral vessel and its branches.

After given just mentioned two branches, the main vessel turns towards the left side of the body and divides into three stout branches:

(*a*) Median dorsal branch climbs upwards upto the base of the atrial siphon. In the way it sends a **test vessel**, a **left oesophageal** and a **right gonadial vessel** to respective organs. Finally it forms a circular **subtentacular vessel** at the base of the atrial siphon (Fig. 4.20). The subtentacular vessel sends 6 to 8 siphonal branches to the wall of atrial siphon.

(*b*) Left gonadial vessel is the middle branch. It runs obliquely and supplies blood to the left gonad.

(*c*) Ventral branch soon divides into following two branches:

1. The **dorso-intestinal vessel** runs along the left lobe of the liver and stomach and supplies blood to these organs through fine branches. It now runs along the dorsal side of intestine and supplies blood to it through many fine branches.

2. The second branch soon divides into a **dorsal test vessel** that supplies blood to the test, and a **gastric vessel** that supplies blood to stomach.

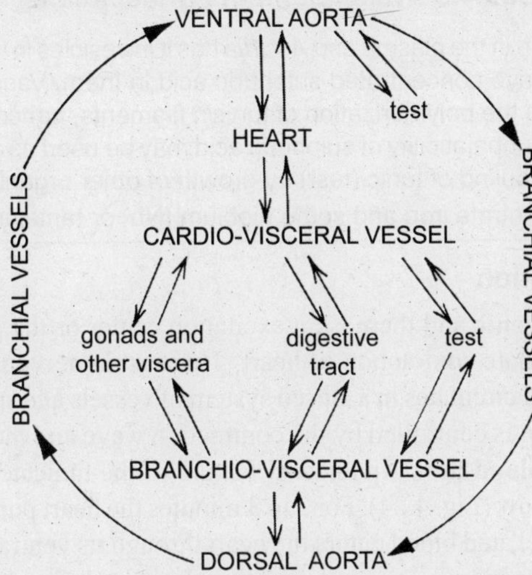

Fig. 4.24. *Herdmania*. Course of blood circulation.

C. Blood

Blood of *Herdmania* is slightly reddish, transparent and hypertonic to sea water. It has colourless plasma having eight types of cells or corpuscles: many pigmented and non-pigmented **corpuscles** and a few amoeboid colourles **cells** (leucocytes) (Fig. 4.25).

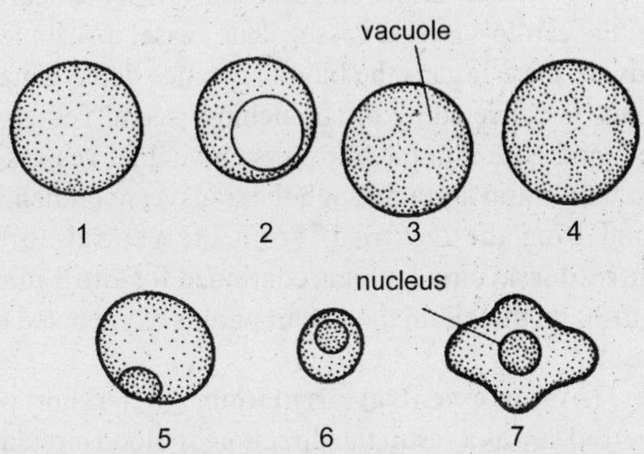

Fig. 4.25. *Herdmania*. Blood corpuscles and blood cells.

1. Orange corpuscles. These are large almost spherical and non-nucleated corpuscles with orage coloured pigment granules.

2. Signet corpuscles. These are spherical, non-nucleated and colourless corpuscles. These contain a single large vacuole.

3. Green corpuscles. These are yellowish-green, spherical, non-nucleated corpuscles and contain 4 to 6 large and small vacuoles.

4. Compartment cells. These are very large, non-nucleated, light yellow coloured corpuscles and contain yellow pigments and a large number of small vacuoles.

5. Eosinophilous corpuscles. These cells are with an excentric nucleus and brown pigments.

6. Lymphocytes. These are very small blood cells with a large central nucleus and brown pigments. These are slightly deformed, ovoid or oblong. These are produced in the mesenchyme of the body wall and give rise to all other types of cells.

7. Macrophages or Amoebocytes. These are nucleated amoeboid leucocytes (colourless). These contain blunt pseudopodia and can move. These are nutritive and phagocytic in function.

8. Storage cells or the nephrocytes. These cells are vacuolated. The vacuoles are filled with crystals of various colours, xanthine and urate crystals and other waste materials.

Most tunicates can extract the rare element, **vanadium**, from the sea water and store it in their **green blood corpuscles** or **vanadocytes**. Real function of vanadium in tunicates is not known. However, *Herdmania* does not extract vanadium (Box 4.3).

Box 4.3 Trace Elements in Ascidians

Ciona contains vanadium in the plasma and *Ascidia* has it in vesicles in vanadocytes. The vesicles containing vanadium also have concentrated sulphuric acid in them. Vanadium has two functions. Its reducing power is used in the polymerization of tunicin filaments, hence in **tunic synthesis**. The toxicity of vanadium and the unpalatibility of sulphuric acid may be used as **antibiotics** to discourage predation on tunicates and fouling of tunic (test) by growth of other organisms on it.

Certain ascidians concentrate iron and some niobium (Nb) or tantalum (Ta) in their bodies.

Course of Blood Circulation

Contraction of heart is **myogenic** and there is an excitation center or the pacemaker at each end of the heart. These centers regulate contraction of heart. There is no nervous innervation of the heart (**Goodbody** 1974). The blood circulates in a closed system of vessels and spaces. There are no valves in the heart. Blood circulation is controlled by the contraction wave and the small pear-shaped body.

Periodic reversal of blood flow. An unusual feature of the tunicate circulatory system is the **periodic reversal of blood flow** (Fig. 4.24). For 2 to 3 minutes the heart pumps blood out of its dorsal end into cardio-visseral vessel, and blood enters the heart through its ventral end. After a short period of rest, the direction of contraction is reversed, and blood is pumped out in the opposite direction. Need and importance of such periodic reversal of blood circulation is unknown.

(*i*) Ventro-dorsal circulation. When the heart contracts **ventro-dorsally** the blood is pumped into the cardio-visceral vessel. This vessel distributes the blood to the test and visceral organs of the body. In these organs the blood becomes deoxygenated. The deoxygenated blood from the visceral organs is collected by the branchio-visceral vessel. It carries blood to the dorsal aorta from where the blood goes into the transverse branchial vessels and is oxygenated. The oxygenated blood goes into the ventral aorta through the transverse branchial vessels. Ventral aorta collects the oxygenated blood from the test also. From the ventral aorta oxygenated blood reaches the heart. The **ventro-dorsal** circulation is continued for 2 to 3 minutes and the number of heart beating vary from 20 to 30. In this circulation heart pumps oxygenated blood to all body parts through the cardiovisceral vessel.

(*ii*) Dorso-ventral circulation. After a short period of rest the direction of heart contraction is reversed and as a result the direction of blood circulation is also reversed. It becomes **dorso-ventral**. The blood is pumped out into the ventral aorta which distributes it to the test and branchial sac for oxygenation. The oxygenated blood is collected by the dorsal aorta and distributed to all visceral organs through the branchio-visceral vessel. The deoxygenated blood from the viscera is collected by the cardio-visceral vessel which carries it to the heart. Thus in this circulation cycle the heart collects the deoxygenated blood through the cardio-visceral vessel and distributes it through the ventral aorta. The **dorso-ventral circulation** is also continued for 2 to 3 minutes and the number of heart beatings vary from 35 to 54. In this way the heart of tunicates pumps out oxygenated and deoxygenated blood at regular intervals. In other words, the heart functions as a "**systemic heart**" at one time and as a "**branchial heart**" at another a phenomenon not found in any group of animals.

Peculiarities of the circulatory system.

1. The heart beat is **myogenic**, *i.e.*, the heart muscle fibers pulsate without nervous excitation.
2. Each end of the heart has a **pacemaker** excitatory center from which the heart beat is initiated.
3. The heart expels blood from one end for 2 to 3 minutes, then its beat **reverses** and blood is pumped through the opposite end. The two pacemakers function in alternation.
4. Due to reversal of heart beat direction of blood circulation is reversed at regular intervals.
5. The heart pumps oxygenated and deoxygenated blood at regular intervals. It works as a **systemic heart** at one time and as a **branchial heart** at another—a phenomenon not found in any other group of animals.

4.9 NERVOUS SYSTEM

Nervous system of *Herdmania* is very simple. It includes central nervous system and peripheral nervous system (Fig. 4.26).

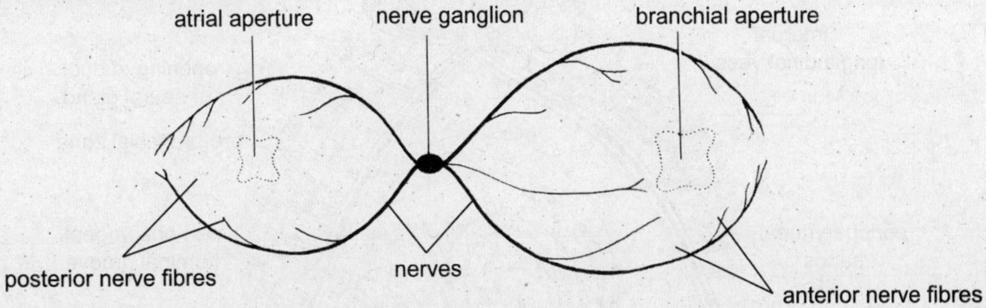

Fig. 4.26. *Herdmania*. The nerve ganglion and its nerves.

1. Central nervous system. It consists of a single pinkish, solid, elongated, brain or **nerve ganglion**. It lies in the middorsal line of the intersiphonal region of the body wall. The nerve ganglion has bipolar and multipolar **neurons** or **nerve cells** with interlacing **nerve fibers** surrounded by large **ganglion cells**. The nerve ganglion mainly controls the body reflexes.

2. Peripheral nervous system. The peripheral nervous system comprises three **anterior nerves** and two **posterior nerves** arising from the nerve ganglion. All the nerves are mixed (sensory as well as motor). Of the three anterior nerves, two are stouter than the third, and almost encircle the base of branchial siphon (Fig. 4.26). The third nerve runs obliquely toward the mouth. The anterior nerves send fine branches to the tentacles, muscles and inner epithelium of the branchial siphon. They also send some fibers into the test. The two posterior nerves surround the base of the atrial siphon (Fig. 4.26). They supply fine branches to the tentacles, muscles and epithelium of the atrial siphon. Both of them also send some branches into the test and very fine fibers to the gut and other viscera. The nerves entering the test branch repeatedly to form interlacing system of fibrils, which become specially concentrated around the receptor cells.

4.10 SENSE ORGANS

Herdmania possesses receptors for contact (the **tangoreceptors**), chemicals (**chemoreceptors**), water current (**dorsal tubercle**) and temperature (**thermoreceptors**). Tentacles act as **olfactoreceptors** (testing quality of incoming water currents). **Rheoreceptors** which are sensitive to water currents in the sea, occur in rim of branchial and atrial siphon. Here detailed account of only dorsal tubercle has been given.

Dorsal Tubercle

Dorsal tubercle is a **olfactory-cum-gustatory receptor** of *Herdmania*. It is a large, hemispherical or horse-shoe-shaped body situated in the pre-branchial zone of the pharynx in front of the union of peripharyngeal groove and the dorsal lamina (Fig. 4.27). It is distinguished into a broad concave base and two spirally coiled conical projections one on either side. Each projection consists of three loops. Lying along the base of dorsal tubercle is a shallow channel which receives the ciliated funnel-like opening of the duct of neural gland (Fig. 4.27). It divides into two and continues as a spiral channel in both conical projects.

The interior of dorsal tubercle is filled with loose connective tissue strewn with nerve fibers and blood sinuses. The entire surface of the tubercle is covered by columnar ciliated epithelium richly supplied with nerve fibrils from the nervous system. This epithelium is considered sensitive to chemicals serving to taste and smell the water current entering the branchial sac.

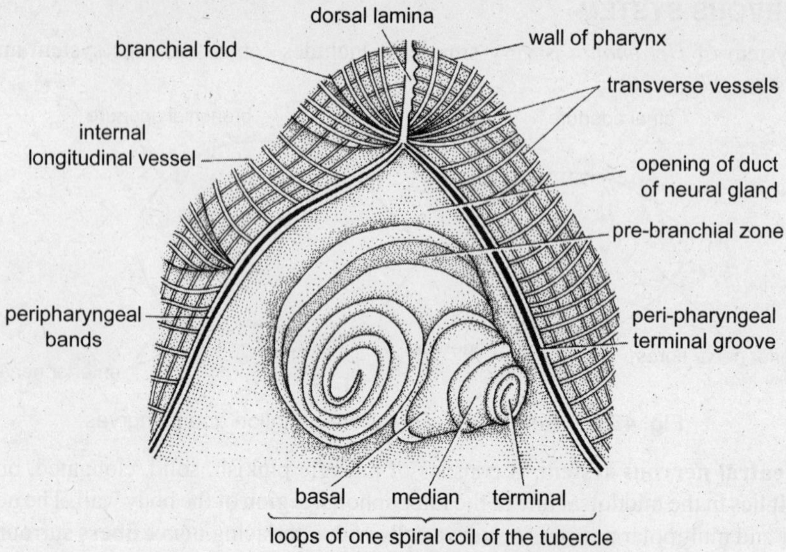

Fig. 4.27. *Herdmania*. Peripharyngeal region showing dorsal tubercle.

4.11 EXCRETORY ORGAN

Herdmania, like the other tunicates, is primarily **ammonotelic**, that is, it excretes protein nitrogen as *ammonia*. Ammonia diffuses across the pharynx into the outgoing water (**Brusca** and **Brusca** 2003). Other metabolic wastes, such as *uric acid* and *calcium oxalate* (urates) are not excreted but stored in the **nephrocytes** (of blood). The latter finally accumulate in certain tissues, such as mantle, pericardium and are lost on the death of the animal. This phenomenon is called **storage excretion**. Nephrocytes are also manipulated by the *neural gland* which is regarded as excretory organ of *Herdmania* (**Dass** 1936).

Neural Gland

Neural gland (Fig. 4.28) acts as a excretory, osmoregulatory and endocrine gland. It lies embedded in the mantle of the intersiphonal region just above the nerve ganglion. It like the ganglion, arises from the embryonic neural tube. The neural gland has a few **central tubes**, which gives off a large number, of branching **peripheral tubules** and open into a large canal (Fig. 4.29). The latter runs along the entire length of the gland and leads into the **neural gland duct** at the anterior end of the gland. The duct passes between the anterior nerves and opens into the peribranchial zone of the pharyx by a ciliated, funnel-shaped aperture situated at the middle of the base of the dorsal tubercle. The cilia lining the funnel are about the length of the cells that bear them (Fig. 4.30). The tubules, tubes and the canal are lined by small, rounded, non-ciliated cells (Fig. 4.30B) and lie embedded in a loose connective tissue. Containing large blood sinuses.

The cilia of funnel create an incurrent of water from the pharynx into the neural gland lumina and then across their walls into the branchial blood vessels. The cilia on the dorsal tubercle prevent large particles, such as sand grains, from entering the gland duct, whereas the **phagocytic** lining the gland removes bacteria and other small particulates from the water. The neural gland and dorsal tubercle serve to restore and maintain fluid volume of the blood. Blood's nephrocytes loaded with excretory crystals are excreted by the neural gland.

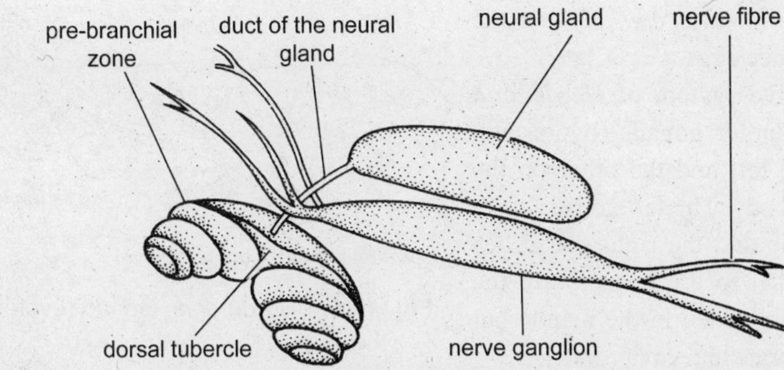

Fig. 4.28. *Herdmania*. Neural gland complex.

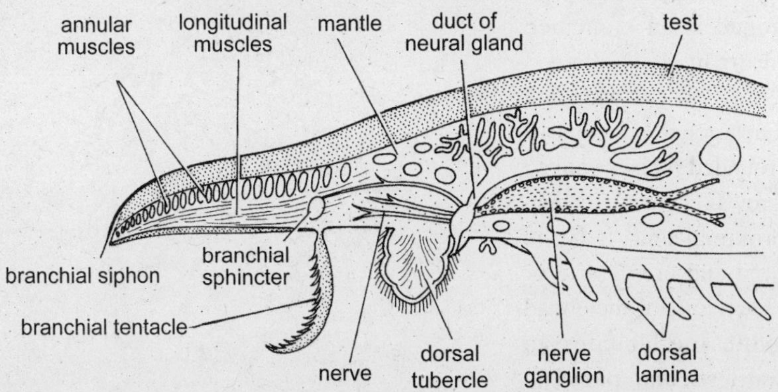

Fig. 4.29. *Herdmania*. L.S. neural complex.

According to **Julin** (1881), **Florkin** and **Dass** (1956) neural gland also secretes a hormone that controls the discharge of gametes, development and metamorphosis. However, the role of the gland in reproduction remains to be confirmed (**Goodbody** 1974). Neural gland is considered homologous to the pituitary gland of the vertebrates but according to **Dodd** and **Dodd** (1966) it does not produce any antidiuretic or melonocyte stimulatory hormone.

The neural gland, nerve ganglion and dorsal tubercle together from the **neural gland complex** of *Herdmania* (Fig. 4.28).

4.12 REPRODUCTIVE SYSTEM

Herdmania is **bisexual** or **hermaphrodite**. Self-fertilization is prevented by maturation of ovaries and testes at different times. The ovaries ripen earlier while the animal is still young. This condition is called **protogyny**. It ensures cross-fertilization. *Hermaphroditism* increases the rate of reproduction, as all the individuals are capable of laying eggs, whereas in unisexual

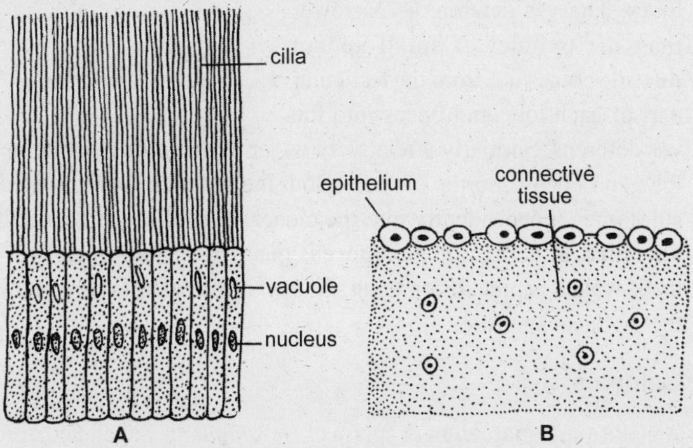

Fig. 4.30. *Herdmania*. Epithelium of funnel (A) and duct (B) of neural gland.

animals only the females, *i.e.*, 50% of the individuals, produce eggs.

Reproductive system of *Herdmania* consists of two larger gonads (ovotestes), one lying on the left and the other on the right side of the body. The left gonad lies in the intestinal loop and the right gonad lies parallel and dorsal to the heart. Both the gonads remain embedded in the mantle but bulge in the peribranchial cavity.

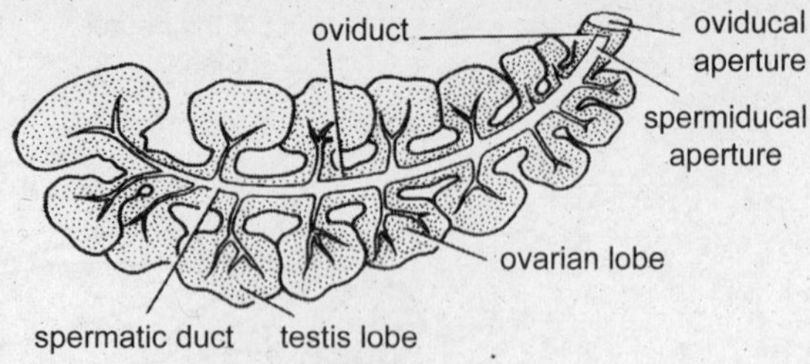

Fig. 4.31. *Herdmania.* Gonad (ovotestes).

Structure of Gonads

Each adult gonad is an elongated and lobulated organ. It is about 3 to 4 cm long, 1 to 1.5 cm wide and 2 mm thick. Each gonad contains 10 to 25 oval or rounded lobes arranged in two parallel rows. There is a median lobe at the proximal end. It is the largest, unpaired and crescentric or bean-shaped. Each lobe of the gonad is hermaphrodite and contains an outer peripheral **testicular part** and an inner **ovarian part**. The testicular part is large and brick red in colour, while the ovarian part is light red or pink in colour.

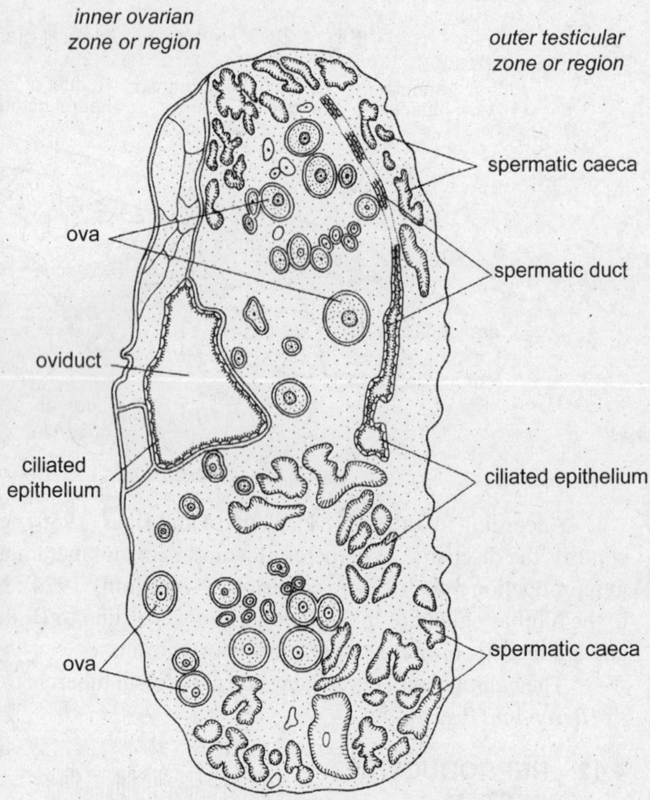

Fig. 4.32. *Herdmania.* T.S. gonad.

Gonoducts. Each gonad consists of an **oviduct** and a **vas deferens** or **sperm duct**. The two ducts run parallel to each other between the two rows of the gonadal lobes. The vas deferens is narrower than the oviduct. A small **sperm ductule** comes out from the testicular part of each lobe and opens into the vas deferens. Similarly a relatively wider **ovarian ductule** comes out from the ovarian part of each lobe and opens into the oviduct. Both the gonoducts are ciliated internally. The gonoducts of both sides open independently into the cloacal part of the atrium a little behind the anus. The **oviducal aperture** or the **female gonopore** is guarded by four thick lips. The vas deferens opens on a small thick-walled papilla. Its opening, the **spermiducal aperture** or the **male gonopore**, lies a little behind the oviducal aperture.

Gametes

A mature **spermatozoon** or sperm of *Herdmania* is flagellated, microscopic and about 4 mm in length. Each has an anterior broad and nucleated **head** capped by a beak-like **acrosome** in front, a middle piece with **neck** and a very long straight **tail**. The sperms of *Herdmania* are **polymorphic** with at least three types of sperms having acrosome either shorter, equal or longer than head.

The **ova** or **unfertilized eggs** of simple ascidians are **microlecithal** and **bipolar** having distinct animal and vegetal poles. They measure between 0.1 to 0.3 mm in diameter. The ripe ovum of *Styela partita* is laid in the germinal vesicle stage and is full of ooplasm that is well differentiated into variously coloured distinct regions. This is called **egg cytoplasmic localization**. For example, ooplasm rich in slaty gray yolk granules is located in the centre and is called **endoplasm** since it contains materials for the development of prospective **endoderm** and **notochord**. Surrounding the yolky endoplasm is a peripheral or cortical layer of protoplasm in which are embedded yellow pigment granules and many minute mitrochondria. This yellow cortical layer of ooplasm is called **myoplasm** since it is found to contain materials for prospective **mesenchyme** and **muscle cells**. A region of clear cytoplasm surrounds the germinal vesicle (or nucleus of the ovum) at the animal pole and is called **ectoplasm**, since it contains materials for the prospective **epidermal** and **neural cells**. The position of egg nucleus close to the animal pole stresses the polar organization of the egg, but there is no trace of a bilateral organization : all meridians of the ovum are exactly alike.

Mature ovum of *Herdmania* (Fig. 4.33) is about 0.3 mm in diameter and contains an eccentric nucleus with a prominent nucleolus. It is surrounded by the following three types of egg membranes : 1. **Vitelline membrane**; 2. **Inner chorion** and 3. **Outer chorion**. The thin vitelline membrane is secreted by the ovum itself and closely applied to it. The follicular cells of the ovary secrete the two chorion membranes which are **primary egg membranes**. The ovum lies eccentrically in the **perivitelline fluid** enclosed by the space between the vitelline membrane and the inner chorion. The perivitelline space also contains a large number of small-sized, red-granules-containing cells, called **inner follicle cells** or **test cells**, some of which float in the fluid but most of them attached to the vitelline membrane. These cells nourish the ovum and secrete an enzyme which digests the chorions to facilitate hatching. Attached to external surface of the outer chorion are many large-sized vacuolated **outer follicle cells**. These cells serve to keep the ovum floating in water.

Fertilization

In ascidians, **cross-fertilization** and **external-fertilization** take place, when mature sperms of older individuals and ripe ova of younger individuals are liberated in their cloaca and pass out through their atrial apertures into the surrounding sea water. During fertilization, the spermatozoon penetrates the ovum at any meridian nearer to the vegetal pole and becomes the **male** or **sperm pronucleus**. Ascidian ova (or eggs) rely on the tough chorion to prevent multiple sperm entry or **polyspermy**.

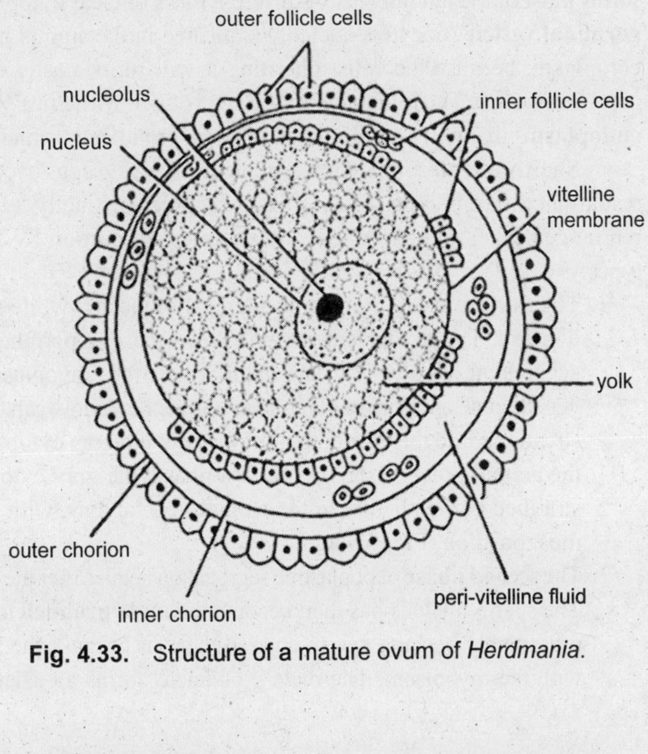

Fig. 4.33. Structure of a mature ovum of *Herdmania*.

Five minutes after sperm entry (*i.e.*, fertilization), various cytoplasmic regions are rearranged by a process known as **ooplasmic segregation** (Box 4.4). Meanwhile, the ovum completes its maturation division and releases the first or primary and second or secondary **polar bodies** at the animal pole. The resultant haploid nucleus of the ovum is called **female** or **egg pronucleus** which descends to meet the sperm pronucleus just below the equator.

Thus, due to ooplasmic segregation, fertilized egg of *Styela* becomes **bilaterally symmetrical** and **mosaic** or **determinate type** and it shows the following four presumptive areas with predictable future : 1. an upper hemisphere of clear cytoplasm containing little yolk granules, few mitochondria and materials released by germinal vesicle (GV); 2. a mitochondria-rich yellow crescent postero-ventrally; 3. a gray crescent antero-dorsally and 4. a vegetative area of slaty gray yolk cytoplasm.

Box 4.4

Ascidian eggs acquire their animal-vegetal polarity during oogenesis and their antero-posterior polarity upon fertilization. The entry of sperm triggers a dramatic reorganization of cytoplasmic components which are segregated along the two egg axes. The segregation process depends on microfilaments in the egg cortex and on microtubules in the aster surrounding the male pronucleus. The segregated cytoplasmic components are enclosed by different blastomeres, whose fate they seem to determine.

Ooplasmic Segregation

Eggs of *Herdmania*, *Styela* and other ascidians are described as **mosaic** or determinate type. Adult ascidian shed their eggs as primary oocytes are still arrested in meiotic prophase I. These oocytes are radially symmetrical relative to the animal vegetal axis and several organelles in the egg cortex are distributed with an animal-vegetal polarity. Maturation of oocyte is triggered by sperm entry, which may occur anywhere on the egg surface. In newly fertilized eggs of *Styela partita*, a layer of cytoplasm containing yellow pigment granules lie just beneath the plasma membrane (Fig. 4.34A). This yellow cytoplasm is called **myoplasm**, because it is enclosed by cells (blastomeres) that later forms most of the tail nucleus. A different mass of clear cytoplasm, derived from the contents of the **germinal vesicle** (oocyte's nucleus) is located in the animal half of the egg. This material is called **ectoplasm**, because the cells inheriting it will form mostly ectodermal structures (epidermis and nervous system). The remainder of the egg is filled with gray, yolk-rich cytoplasm (sometimes called **endoplasm**) that will be segregated in endodermal cells which form the gut.

Shortly after fertilization, the contents of *Styela* eggs become dramatically rearranged, with the result that different types of cytoplasm becomes distinctly localized (Fig. 4.34). Such cytoplasmic rearrangements are referred to as **ooplasmic segregation** by **Jeffery** (1984). It is of following two types:

1. The **first phase** of ooplasmic segregation occurs while the egg is still completing its meiotic division. The myoplasm streams down the egg periphery toward the vegetal pole, where it accumulates as a yellow cap. In the wake of the myoplasm, islets of ectoplasm also flow to the vegetal pole and form a clear layer above the yellow cap. As both myoplasm and ectoplasm are streaming into the vegetal half of the egg, the gray cytoplasm (endoplasm) is displaced toward the animal pole. At the end of first phase, the gray cytoplasm, ectoplasm and myoplasm are stratified perpendicular to the animal-vegetal axis, with the myoplasm occupying the vegetal most position (Fig. 4.34C).
2. The **second phase** of ooplasmic segregation begins after the meiotic divisions have been completed. The sperm nucleus has then become the **male pronucleus** and is located near the vegetal pole, presumably because sperm entered there or because the sperm nucleus was swept there along with the myoplasm. The male pronucleus forms an aster of microtubules and migrates along

the periphery of the egg in the direction of the animal pole. Both myoplasm and ectoplasm move along with the male pronucleus. At a position below the egg equator, the myoplasm is left behind and forms a **yellow crescent** which marks the future posterior end of embryo. After the yellow crescent has formed, the ectoplasm returns to the animal pole along with the male pronucleus. Here the ectoplasm will form a specialized region around the fused pronuclei. At the same time, major portion of the gray cytoplasm returns to the vegetal hemisphere. A fourth cytoplasmic region, the **chordoplasm**, forms opposite the yellow crescent and is later included into the blastomeres that give rise to an embryonic skeletal rod, the **notochord**.

At the close of ooplasmic segregation, the ascidian egg has two axes of polarity. One is the **animal-vegetal axis**, with the vegetal pole predicting the site of **blastopore** formation. In addition to the animal-vegetal axis, the ascidian egg has acquired an **anteroposterior axis**, with the posterior pole marked by the yellow crescent.

Box 4.5 Role of Cytoskeleton in Ooplasmic Segregation

To analyze the mechanism of yellow crescent formation, **Jeffery** and **Meier** (1983) used biochemical methods and scanning electron microscopy. They concluded that the yellow pigment granules are attached, presumably by intermediate filaments, to a **plasma membrane lamina (PML)**. The PML forms a thin meshwork right underneath the plasma membrane and includes microfilaments as major components. In unfertilized eggs, the PML is present in the entire cortex. When the yellow granules retreat to the vegetal pole after fertilization, so does the PML. The authors propose that contraction of PML provides the driving force for the movement of the myoplasm to the vegetal pole. They also suggested that PML contraction drives the gray cytoplasm into its temporary position near the animal pole.

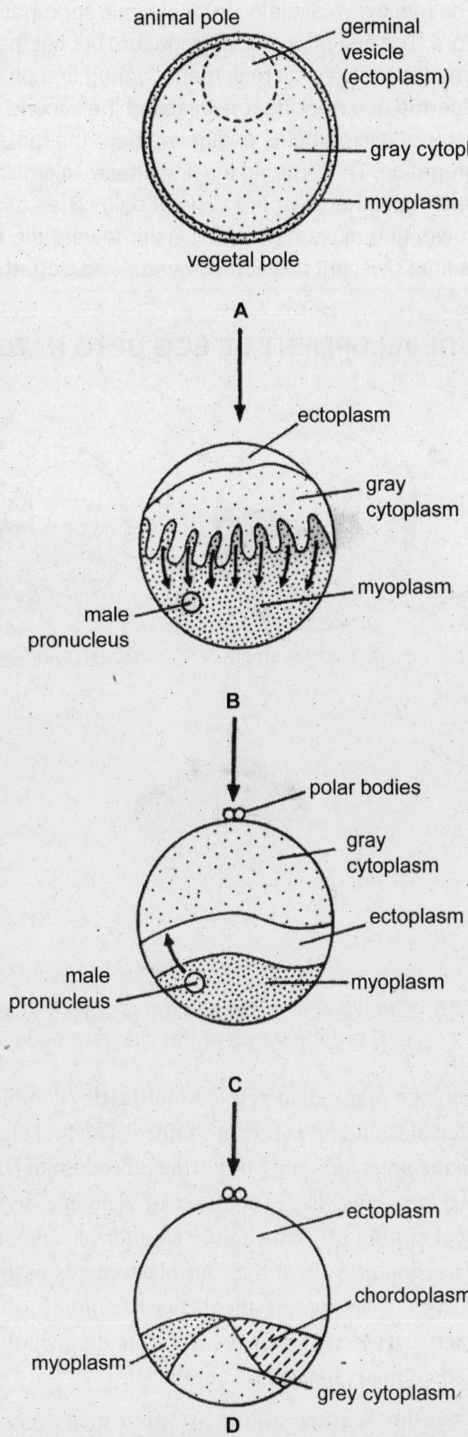

Fig. 4.34. Ooplasmic segregation in *Styela partita*. A—Unfertilized egg; B to C—Fertilized egg during the first phase of ooplasmic segregation. D—Second phase of ooplasmic segregation.

The role of cytoskeleton in ooplasmic segregation was also tested by application of appropriate inhibitors. The first phase of segregation, but not the second, was inhibited by **cytochalasin B**. This confirmed the role of microfilaments during first phase. Conversely, the microtubule inhibitors such as **colcemid** and **nocodazole** inhibited the second phase of ooplasmic segregation but not the first. The aster of microtubules originating near the male pronucleus is necessary for the second phase of segregation. Thus, during the first phase, microfilament action sweeps the sperm nucleus and the myoplasm together near the vegetal pole. Then as the microtubules of the sperm aster assemble, the growth and movement of the aster toward the animal pole might serve to bring the myoplasm into its final crescent position (**Sawada** and **Schatten** 1988).

4.13 DEVELOPMENT OF EGG UPTO HATCHING

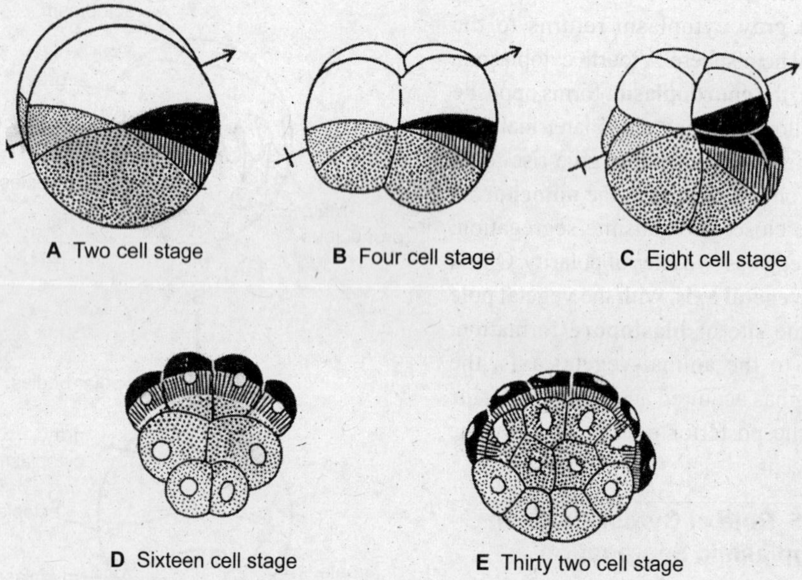

Fig. 4.35. Cleavage and blastulation of *Styela*, A to C are the side views of cleaving eggs, while, D to E are the views of the cleaving eggs from vegetal pole.

The **cleavage** of ascidian egg is **holoblastic** (complete), somewhat **unequal** and **determinate**, *i.e.*, fate of the blastomeres is predetermined. The first cleavage plane of *Styela* is meridional and it passes through the polar axis and bisects the mesodermal (*i.e.*, yellow crescent) and chordaneural (*i.e.*, gray crescent) crescents. The second set of cleavage divisions occurs at the right angles to the first, but is vertical cutting off two slightly smaller blastomeres at the posterior side and two larger ones in front. Division of each of the four blastomeres in the horizontal plane, produces an eight cell stage that consists of an upper (apical) tier of slightly smaller blastomeres and four larger lower (basal) blastomeres. The eight-celled embryo undergoes three more sets of cleavages and consequently a 64-celled compact **blastula** is produced.

Peculiar feature about ascidian cleavage. In fact, that an egg of *Styela* has very small amount of yolk, but its holoblastic cleavage, instead of producing equal-sized blastomeres, produces unequal-sized blastomeres, reflects a significant aspect about the cleavage of ascidians. The cleavage in *Styela* and other ascidians merely serves to parcel into individual blastomere areas of cytoplasm of distinctive biochemical composition and, thus, of precise developmental potentiality. That this

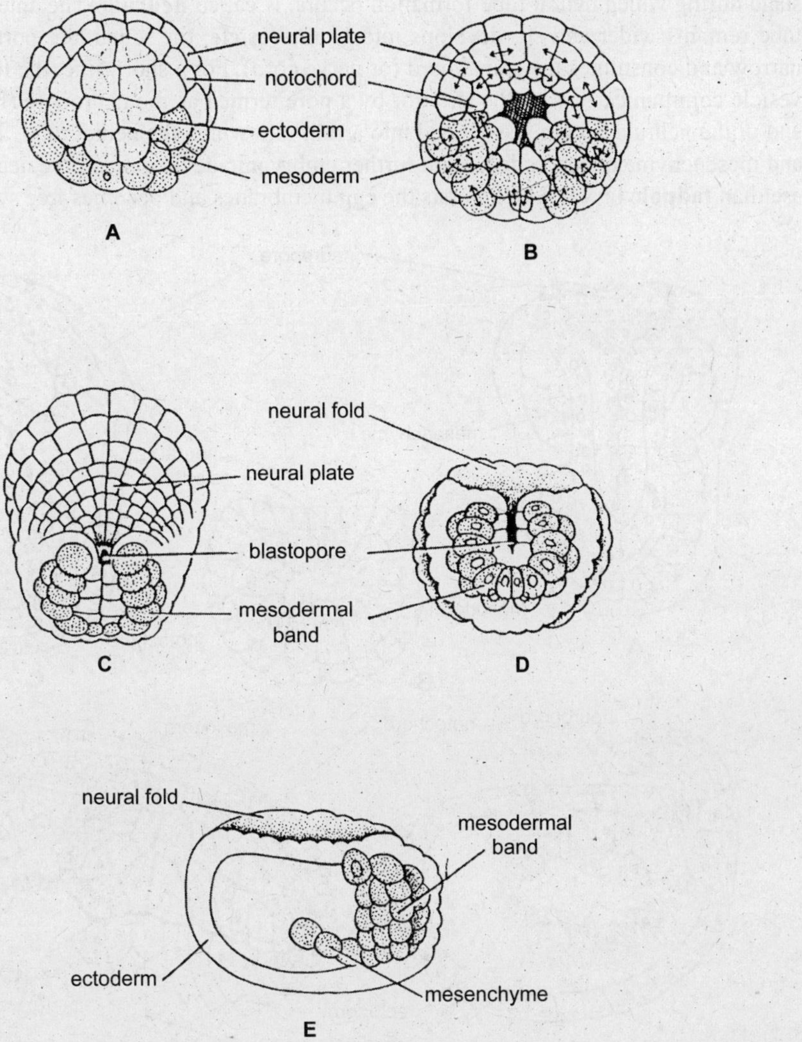

Fig. 4.36. Gastrulation in *Styela*. A—Stage in which the semicircular dorsal lip of the blastopore is just beginning to form; B—Stage in which the invagination of the endoderm cells is well advanced; C—Dorsal view of advanced gastrula; D—Posterior view of advanced gastrula, showing the neural folds and E—Lateral view of advanced gastrula, showing closure of blastopore by upgrowth of ventral lip.

parcelling is real indeed has been demonstrated by experiments in which certain blastomeres are injured so as to prohibit them from further developmental participation. If, for instance, the four right-hand blastomeres of an 8-cell stage are injured, the left-hand blastomeres produce only a left half embryo, or if the four rest blastomeres are immobilized, the four in the front produce only those parts they are normally destined to produce, namely, epidermis, neural tube, notochord, and endoderm.

Gastrulation by **invagination** occurs following the sixth cleavage and is completed by the time of the seventh cleavage, *i.e.*, between the 64-cell and 128-cell stage (at the mid-gastrula stage, there are about 76 cells). On the completion of gastrulation, the large ventral opening, the **blastopore** of the gastrula closes, and the embryo begins to lengthen along the embryonic axis indicated by the plane of the first cleavage. The upper surface or **neural plate** of the elongating gastrula then invaginates along the mid-line to form a groove which subsequently becomes a tube, the **neural tube**, and the stage during which neural tube formation occurs, is called **neurula**. The anterior portion of neural tube remains wider and so, develops into **brain-vesicle**; the remainder portion of neural tube is narrow and constitutes the spinal cord (or nerve cord). For a short time, the front part of the brain-vesicle communicates with the exterior by a pore termed as the neuropore (Fig. 4.37). The hinder end of the neurula becomes extended into a **tail**, into which the neural tube, notochord, mesoderm and mesenchyme are extended. With further embryonic development, the neurula develops into an ascidian **tadpole larva**; which bursts the egg membranes and becomes free swimming.

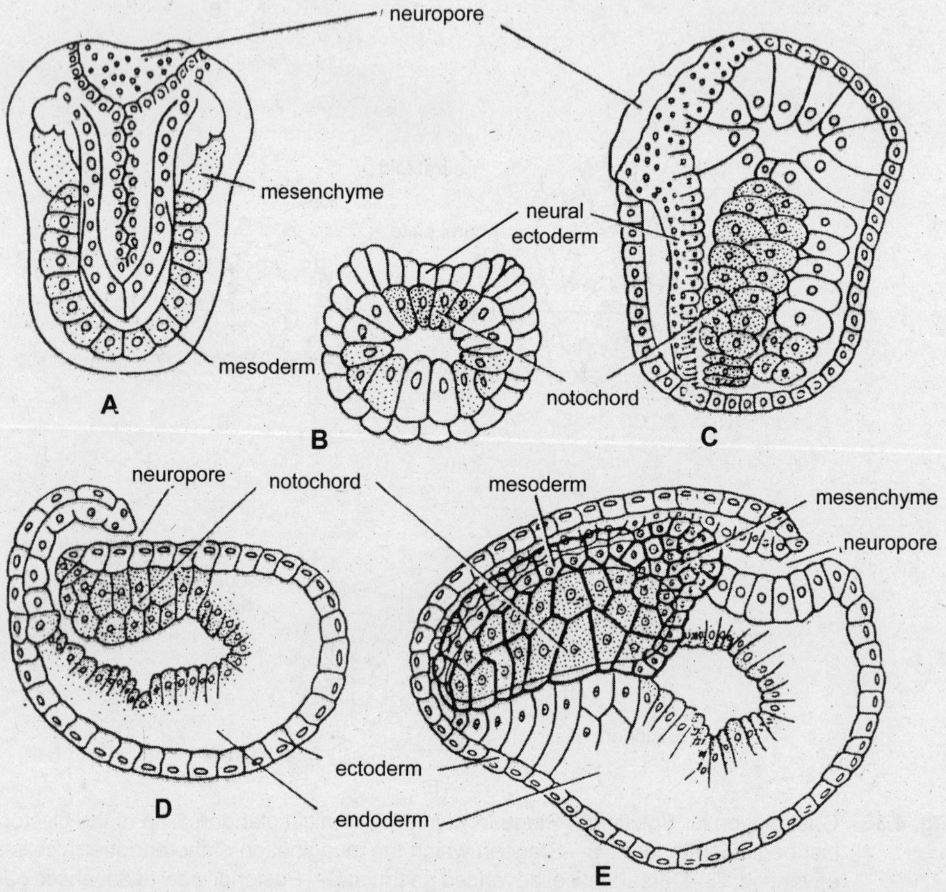

Fig. 4.37. Neurula stage of development of ascidian. A—Dorsal view and B—Side view of neurula of Styela, and C—Cross section of open neural-plate stage or younger embryo; D—Early and E—Late neurula of *Clavelina* from side, showing neuropore, neural tube, notochord, and superimposed tail mesoderm and mesenchyme.

4.14 TADPOLE LARVA AND RETROGRESSIVE METAMORPHOSIS

1. Structure of tadpole. The fully developed tadpole larva of ascidians is transparent, minute, about 1.5 mm long and highly motile (Fig. 4.38). Its entire body is covered by a thin tunic or **test** secreted by the ectoderm. The larval body shows two distinct divisions : an anterior short oval **head** or **trunk** and a posterior long **tail**.

(*i*) Head or trunk. The cylindrical trunk of larva has at its anterior end three stalked **adhesive papillae**, two being dorso-lateral and one ventro - lateral. These are made of secretory ectodermal cells and serve for attachment during metamorphosis. Situated in the dorsal part of trunk is an oval, hollow **sensory vesicle**, **brain vesicle** or **brain** which is an anterior extension of nerve cord or spinal cord of the tail. Its wall is made of a single layer of cells. The sensory vesicle encloses two types of larval sensory organs: 1. One **ocellus** or simple eye which is a photoreceptor having a cornea, a lens, and pigmented retina to distinguish light from dark surfaces. Tadpole larva of *Herdmania* contains two ocelli : one postero-dorsal and other antero-ventral. 2. One gravity-receptor, the **statolith**, **otocyst** or **otolith** which serves to maintain equilibrium. These organs make up the sensory ganglion that controls larval behaviour. In front of sensory vesicle is a conical solid mass of nerve cells, called **cerebral cone**. Behind the vesicle is a thickened solid mass of nerve cells called **visceral ganglion**, which extends posteriorly up to the tip of the tail in the form of nerve cord. **Mouth** is situated antero-dorsally leading into a rudimentary **alimentary canal** comprising a narrow branchial siphon, a large sac-like pharynx, a short narrow oesophagus, a swollen stomach, a slender intestine and a small rectum. The **anus** opens into the left side of atrium. The **pharynx** has a well developed ventral endostyle and initially one pair of large **gill-slits** on its lateral sides. Each gill-slit ultimately splits to form six stigmata on either side. The **atrium** develops as an ectodermal pouch to surround the pharynx laterally and dorsally and leads outside by a posterior mid-dorsal **atrial aperture** or **atriopore**. The **pericardium** containing **heart** is located postero-ventrally in between pharynx and stomach. **Mesenchyme cells** occur scattered all over the body beneath the ectoderm and also in a mass in the posterior region of trunk.

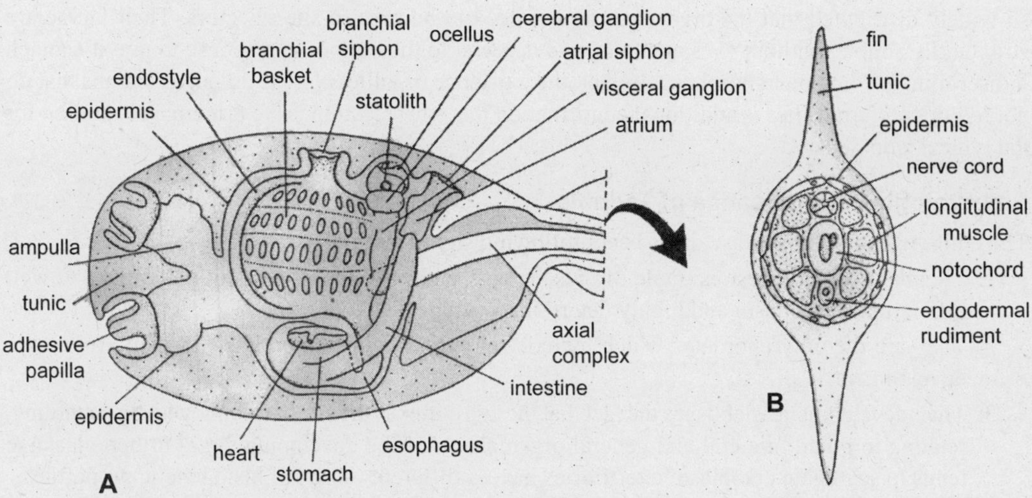

Fig. 4.38. Structure of a generalized ascidian tadpole. A—Lateral view (left side) of the body and B—Cross section of tail.

(*ii*) **Tail.** The locomotory organ of tadpole larva is its powerful transparent tail which is laterally compressed and pointed terminally. Tail is fringed with a continous vertical **tail fin** formed by the test and marked with oblique striations looking like **fin-rays**. Tail of ascidian larva is typical chordate-like containing a central **notochord**, a dorsal tubular **nerve cord** and strong lateral myotome-like **muscle bands**. The **notochord** forms the supporting axial skeletal rod restricted to tail. It is formed by a single row of turgid vacuolated cells. The anterior end of notochord slightly projects into the trunk.

Habit and Habitat of Larva

Just after hatching the tadpole larva of ascidians is **photopositive** and **geonegative**. It cannot feed because its mouth is closed by test and it gets its nourishment from the meagre yolk reserves of endodermal cells of the alimentary canal. After a short active free-swimming (in fact, **pelagic**) existence lasting 3 to 4 hours, the larva becomes **geopositive**, **photonegative** and descends. The tail permits sudden horizontal forward movements as the larva descends from surface to bottom. It is these movements and function of sensory organs that bring the larva in contact with hard vertical surfaces or under the shaded sides of rocks to which it attaches with its adhesive suckers and immediately proceeds for a peculiar type of degenerative metamorphosis, called **retrogressive metamorphosis**.

Larva as a site selector. The pelagic larvae of ascidians are very selective when it comes to invading a new site. The site must be of the right consistency (*i.e.*, mud or sand for molguloid ascidians and rock for other ascidians); otherwise they cannot attach and undergo metamorphosis. For many ascidian larvae (*e.g.*, *Herdmania*, *Ascidia*; etc.) vertical surfaces such as pilings are favoured sites. These larvae are adapted to select a preferred site, a role often superseding all other roles. Such larvae are locomotory but do not feed. They are dispersed by currents, but their pelagic lives are too short for extensive dissemination. Because it cannot feed, a larva must find a site and metamorphose within a few hours or days; otherwise it dies.

Thus, the ascidian larva has evolved as a most efficient site selectors. This tadpole-like larva with neural tube, myotomes, pharyngeal gill-slits, eye and otolith resembles a chordate body plan as it probably is the ancestral type for all vertebrates.

Further, the success of the ascidian species which are ecologically adapted for hard vertical surfaces, depends on the facility with which these larvae find and colonize a new site. However, some ascidians such as the molguloids, colonize sand or mud bottoms. These exposed surfaces are so widely distributed that the organisms do not need to be precise site selectors. Their larvae are structurally simple, without eyes and tails, and descend to the bottom in response to gravity, attach indiscriminately and metamorphose. In fact, the existence of tailless larvae among ascidians that do not require preferred sites is additional confirmation that site selection is the principal adaptation for the typical tadpole larva.

Embryological Significance of Tadpole

The embryogeny of simple ascidians bears following significant points:

1. Ascidians provide best example of mosaic eggs with well organized, pre-patterned and well differentiated ooplasm and highly determinative type development.
2. They are the only chordates which include retrogressive metamorphosis in their embryonic development.
3. Their development establishes the fact that the egg cortex is the site of morphogenetic patterning, relating to polar, bilateral and general organization of the developing egg. Further, cleavage tends to segregate cytoplasmic territories having different (cortical) histogenetic properties.
4. Ascidian larvae have provided the clue for their inclusion in phylum Chordata.
5. On the basis of recapitulation theory, the ascidian larva possessing the chordate features is considered as the relic of the free-swimming ancestral vertebrates.
6. The adult ascidians being sedentary, the pelagic habit of the larva provides the only means of dispersal of the species. It also provides chances of selecting better sites regarding food and protection, thus, ensuring survival of the race.

Importance of Larva in its Systematic Position

The tadpole larva of ascidians has a more complex organization than its adult. It show the following chordate characters:

1. Possession of **notochord**.
2. Presence of hollow **nerve cord**.
3. Presence of tail and **tail fins**.
4. Possession of a **pharynx** perforated with **stigmata** which are comparable to the gill-slits of higher chordates.

On the basis of above characters, **Kowalevsky** had recognized the chordate nature of ascidians and placed them in phylum Chordata. Previously ascidians were grouped in phylum nonchordata. This discovery influenced **Haeckel** to formulate his **recapitulation theory**. According to this theory the ascidian larva is regarded as a reminiscent of the past free-swimming ancestor which did not fulfil the promise of youth but slided back to give rise to the degenerated adult. The adult, thus, became secondarily sessile.

Thus, the ascidian larva helps in tracing the relation of ascidians to Chordata and establishes their systematic position.

4.15 SYSTEMATIC POSITION OF *Herdmania*

The true systematic position of *Herdmania* or Urochordata remained controversial for a long time. As a group the urochordates (*e.g.*, *Herdmania*) may be considered as degenerative or specialized members of the chordates because they lack many of the common characteristics of chordates. Tunicates are sedentary like the sponges. Due to this habit these animals were believed to be related to sponges for a long time. **Lamarck** (1816) called them Tunicates, but classified them between Radiata and Wormes. **Cuvier** (1817) classified them with Mollusca. **Milne Edwards** (1843) created a new class **Molluscoidea** to include molluscs with tunicates. **Kowalevsky** (1866), a Russian embryologist studied the development of a simple ascidian and observed important similarities between the larva of ascidian and the vertebrate embryo. On the basis of these similarities the tunicates were included in the phylum Chordata.

4.16 AFFINITIES OF *Herdmania*

A. Relationship with the Nonchordates

I. Affinities or Similarities

1. Urochordates, sponges and coelenterates are sedentary.
2. Filter feeding and respiration through water current is similar in urochordates, sponges, molluscs and lophophorates.
3. Eyes and otocysts are found in the larvae of many nonchordates.
4. Formation of zooids through budding is common in urochordates, coelenterates and annelids.
5. Typhlosole is found in the intestine of many invertebrates.

II. Dissimilarities

1. Nonchordates do not contain a notochord.
2. Nonchordates do not contain pharyngeal gill-slits.
3. Nonchordates do not contain dorsal nerve cord.

The differences between the urochordates and nonchordates are more striking, so these cannot be classified along with the nonchordates.

B. Relationship with the Hemichordates

I. Affinities

1. Both contain pharyngeal gill-slits.
2. Development of nervous system is similar.

II. Dissimilarities

1. Hemichordates do not contain notochord.
2. Body of hemichordates is divisible into proboscis, collar and trunk. Body of urochordates is not similarly divisible.

Hemichordates are confirmed nonchordates. The similarities of both groups are probably because of their distant relationship with the common ancestral stock.

C. Relationship with the Chordates

I. Affinities

1. Mode of cleavage and gastrulation is similar in both groups.
2. The tadpole larva of urochordates is in fact a typical chordate. It shows the following typical chordate characters:
   (*i*) It contains a notochord in its tail.
   (*ii*) It contains pharyngeal gill-slits.
   (*iii*) It contains a dorsal tubular nerve cord.

II. Dissimilarities

1. Coelom and true metameric segmentation is not found in the urochordates.
2. Notochord is restricted in the tail region and that too only in larva.
3. Nerve cells are found in the peripheral region of the nerve ganglion in the urochordates.
4. Urochordates contain a test. It is not found in other chordates.

Similarities and differences between the tadpole larva of urochordates and chordates prove that urochordates are definite chordates and both groups have originated from a common ancestor.

D. Relationship with the Cephalochordates (*Amphioxus*)

I. Affinities

(*a*) Affinities in adults. The adults of urochordates and cephalochordates show the following similarities:

1. Food gathering mechanism is similar.
2. The branchial tentacles of urochordates are comparable with the velar tentacles of *Amphioxus*.
3. In both, the branchial gill-slits open into the atrial cavity.
4. Endostyle and associated structure are similar.
5. Both contain atrium.

(*b*) Affinities of tadpole larva and adult *Amphioxus*. Notochord, dorsal tubular nerve cord and pharyngeal gill-slits are found both in the tadpole larva of urochordates and *Amphioxus* (*Branchiostoma*). Some other similarities between the two are the following:

1. Early stages of development are similar.
2. Tail contains dorsal and ventral fins.
3. Median eye, otocyst and statocyst are similar.
4. Pharynx contains similar endostyle.
5. Atrium is similar.
6. Nervous system consists of similar brain vesicle, neuropore and neurentric canal.

II. Dissimilarities

Urochordates and cephalochordates show the following differences (Table 4.1).

Table 4.1. Differences between urochordates and cephalochordates.

Character	Urochordates	Cephalochordates
1. Habit of the adult	Sedentary	Free-swimming
2. Segmentation	Absent	Present
3. Coelom	Absert	Present
4. Test	Present	Absent
5. Liver	Present	Midgut diverticulum
6. Heart	Present	Present
7. Nephridia	Absent	Present
8. Sex	Bisexual	Unisexual
9. Metamorphosis	Retrogressive	Progressive

Affinities between the urochordates and the cephalochordates show that they evolved from a common ancestor. Their well marked differences require their inclusion in two separate subphyla of the phylum Chordata.

E. Relationship with Vertebrates

I. Affinities. Tadpole larva of urochordates can be compared with the larva of a fish (a vertebrate). It resembles the vertebrates in having following features:

1. A notochord.
2. Dorsal tubular nerve cord.
3. Pharyngeal gill-slits.
4. Tail with the tail-fin.
5. Cleavage and gastrulation are similar in both groups.

The adult urochordates show the following **similarities** with the vertebrates.

1. Neural gland of urochordates is homologous with the pituitary gland of vertebrates.
2. Endostyle is homologous with the vertebrate thyroid gland.

Conclusion

It was believed in the past that the motile cephalochordates and the vertebrates arose from a larva similar to that of urochordates by neoteny. However, there is little evidence to support such an evolution. The ancestral chordates may well have motile and ciliary feeders (**Gans** and **Northcult** 1982; **Nielson** 1985). If such a case, the urochordates undoubtedly departed early from the chordate line of evolution. Most of their adult peculiarities are associated with a sedentary mode of existence. Complete absence of segmentation, even in embryonic development, indicates that the ancestral chordate was not a metameric animal and that metamerism evolved in the line leading to the cephalochordates and vertebrates after the departure of the tunicates.

QUESTIONS

Long Answer Questions

1. Describe habits, habitat, distribution and external features of *Herdmania*.(*Bundelkhand 2002*)
2. Describe internal anatomy of *Herdmania*.
3. Describe in detail the structure and functions of the test and spicules of *Herdmania*. (*Garhwal 2000; Purvanchal 2003*)
4. Give an account of digestive system and the mechanism of feeding in *Herdmania*. (*Gorakhpur 2000; Purvanchal 2000; CCSU Meerut 2001*)
5. Describe the structure of alimentary canal of *Herdmania* and explain the muco-ciliary mode of feeding and digestion. (*Garhwal 2001*)
6. Give the account of alimentary canal of *Herdmania*. (*CCSU Meerut 2008, 11, 13*)
7. Describe the blood vascular system and course of circulation in *Herdmania*. (*Rohilkhand 2003; Purvanchal 2003; Gorakhpur 2003; CCSU Meerut 2006, 13*)
8. Give an account of the blood vascular system of *Herdmania* and describe the peculiarities in the course of blood circulation. (*CCSU Meerut 2012*)

Short Answer Questions

1. Describe structure and function of excretory organ or neural gland of *Herdmania*. (*Garhwal 2002*)
2. Describe structure of pharynx or branchial basket of *Herdmania*. (*Kanpur 2003; Parvanchal 2003*)
3. Discuss the affinities and systematic position of *Herdmania*. (*Avadh 2001*)

4. Give structure of Ascidian tadpole. *(Purvanchal 2003)*
5. Explain feeding mechanism of *Herdmania*. *(Garhwal 2003)*
6. Write short notes on the following:
(*i*) Neural complex in *Herdmania*. *(CCSU Meerut 2007, 13)*
(*ii*) Dorsal tubercle of *Herdmania*. *(Garhwal 2001; Gorakhpur 2003; CCSU Meerut 2012)*
(*iii*) Endostyle of *Herdmania*. *(CCSU Meerut 2010)*
(*iv*) Blood circulation in *Herdmania*.
(*v*) Blood of *Herdmania* or tunicates.
(*vi*) Branchial sac of *Herdmania*. *(Kanpur 2001)*
(*vii*) Test of *Herdmania*. *(Gorakhpur 2003)*

Very Short Answer Questions

1. Neural complex. *(CCSU Meerut 2006, 07)*
2. What is common name given to Urochordates?
3. Why urochordates are called tunicates?
4. Name the structure used by an adult ascidian to capture food.
5. What is regulated as fore runners of the vertebrate thyroid gland in *Herdmania*.
6. Name organ of *Herdmania* which is involved in excretion.
7. Name a ciliary feeder chordate.
8. What are the vanadocytes?

Fill in the Blanks

1. Larva of *Herdmania* is called ________.
2. Todpole of *Herdmania* undergoes ________ metamorphosis.
3. *Herdmania* is commonly called ________.
4. The test of *Herdmania* is composed of ________.
5. The spicules of *Herdmania* are formed of ________ and are of two types: ________ and ________.
6. Atrium is lined by ________.
7. Pharynx of *Herdmania* is divided into two very unequal regions by ________ bands.
8. The endostyle of *Herdmania* is also called ________.
9. Anus of *Herdmania* opens into the ________.

Multiple Choice Questions

1. *Herdmania* is
(*a*) hemichordate (*b*) urochordate
(*c*) cephalochordate (*d*) nonchordate

2. Larva of *Herdmania* is called
(*a*) tornaria (*b*) auricularia
(*c*) appendicularia (*d*) trochophore

3. Excretory organ of *Herdmania* is
(*a*) dorsal tubercle (*b*) pericardium
(*c*) neural gland (*d*) none of them

4. Tadpole or appendicularia of *Herdmania* has
(*a*) notochord (*b*) nerve cord
(*c*) gill slits (*d*) all the above

5. Retrogressive metamorphosis is characteristic of
(*a*) toad (*b*) *Balanoglossus*
(*c*) *Amphioxus* (*d*) *Herdmania*

6. The only organ of *Herdmania* having the striated muscle is
   (*a*) pharynx (*b*) heart
   (*c*) body wall (*d*) endostyle
7. Green blood corpuscles of Ascidians store which metal from the following:
   (*a*) gold (*b*) silver
   (*c*) copper (*d*) vanadium

ANSWERS

Very Short Answer Questions

1. The neural complex of *Herdmania* comprises of the neural gland, nerve ganglion and dorsal tubercle.
2. Urochordates are commonly called **sea squirts** because some of them discharge water jet through the excurrent siphon whenever irritated.
3. Urochordates are called **tunicates** due to the tunic (test) that surround them. Tunic is secreted by epidermal cells and composed of proteins and a complex polysaccharide tunicin.
4. The adult ascidian capture food by pharyngeal basket. They are suspension feeders.
5. The endostyle is *Herdmania* is considered as the fore runner of the vertebrate thyriod gland. It secretes mucus, a complex protein that contains iodine bound to amino acid tyrosine.
6. Neural gland and nephrocytes.
7. *Herdmania.*
8. Cells that extract vanadium from sea water and store it.

Fill in the Blanks

1. Tadpole (appendicularia). **2.** Retrogressive. **3.** Sea squirt (sea potato). **4.** Tunicin. **5.** Calcareous matter, microscleres, megascleres. **6.** Ectoderm. **7.** Peripharyngeal. **8.** Hypopharyngeal groove. **9.** Atrial cavity (cloaca).

Multiple Choice Questions

1. (*b*) **2.** (*c*) **3.** (*c*) **4.** (*d*) **5.** (*d*) **6.** (*b*)
7. (*d*)

5

Urochordata: General Characters, Classification and Types

Urochordates or tunicates have little resemblance with the other chordates. Suphylum Urochordata includes four classes: the sessile filter feeding sea squirts, or ascidians (Class Ascidacea); the pelagic tunicates larva-like tunicates called appendicularian or larvaceans (Class Appendicularea); and the abyssal (benthic) ascidia-like sorberaceans (Class Sorberacea). Urochordata are composed of about 3000 species (**Brusca** and **Brusca** 2003).

Most significant point about the urochordates is that they are more chordate-like in the larval stage than in the adult stage.

5.1 GENERAL CHARACTERS

1. Marine suspension feeders (deposit feeders, carnivores and herbivores). Solitary or social (colonial). Sedentary (sessile) or planktonic.
2. Body varies considerably in size, form and colour. It is unsegmented (without obvious trimeric organisation) and lacks a head and a tail.
3. Many species are brightly coloured. Some maintain symbiotic relationship with prokaryotic or eukaryotic algae.
4. Entire body is covered by a characteristic exoskeleton, called **test** or **tunic** which is composed of a cellulose-like carbohydrate **tunicin**.
   Tunic varies from thick to thin and from smooth and slick (*i.e.*, smooth, wet and slippery) to wrinkled and leathery.
5. Adult urochordate does not contain any appendage. However, larva of urochordate has medial vertical tail fin which helps in swimming.
6. Body wall is called **mantle**. It consists of (*i*) simple, one layered epidermis of flat, hexagonal cells, (*ii*) dermis of connective tissue and un-striped (smooth) muscles, and (*iii*) atrial epithelium of flat ectodermal cells.
7. Adults lack notochord but some of them may contain calcareous spicules. Notochord which is a mid-dorsal elastic supporting rod is present in the tail of larva (or in adult larvaceans). It is lost (in many tunicates) during retrogressive metamorphosis.
8. Coelom is absent. Instead of it, a wide ectoderm-lined atrial cavity surrounds the pharynx. The gill-slits (stigmata), anus and genital ducts open into the atrial cavity. Atrium leads to exterior by a dorsal or posterior atrial aperture existing on the atrial siphon.
9. Alimentary canal is complete and U-shaped. It contains a large **pharynx** or the **branchial sac** which remains perforated by two to many pairs of **gill slits** or **stigmata**. Filter feeding (Box 5.1) and respiration occurs through the gill slits.

Box 5.1

Most tunicates are suspension feeders and use various kinds of mucous nets to filter plankton and organic detritus from sea water.

Food laden and oxygen rich water enters the pharynx via the oral siphon flows through the stigmata, into the cloacal chamber and out the atrial siphon. Water passing through the stigmata also passes through the mucous sheets (secreted by endostyle and pharynx), on which food particles are retained. Dorsal lamina or languets or both of pharynx serve to roll the mucous sheets into cords, which are then passed posteriorly to oesophagus and stomach for digestion (**Brusca** and **Brusca** 2003).

10. The mouth and atriopore are surrounded by muscular **incurrent** and **excurrent siphons** respectively.
11. Digestive system consists of an **endostyle** which secretes mucous (= mucus) and concentrates iodine.
12. Respiration occurs by gill slits in pharyngeal wall and vascular area of test.
13. Circulatory system is open. There is a simple, tubular, ventral heart, which alternately reverses the blood flow, a feature unique in the animal kingdom. Heart beat is myogenic.
14. Blood corpuscles are of many kinds. Certain blood corpuscles contain vanadium, iron, niobium or tentalium extracted from sea water.
15. Nervous system is represented in the adult by a single dorsal ganglion. Larva has a dorsal hollow nerve cord.
16. There are no paired sense organs such as eyes, ears, olfactory sacs. Only simple receptors are present.
17. Tunicates are primarily ammonotelic. Ammonia diffuses across the pharynx into the outgoing water. Other nitrogenous wastes such as uric acid and urates are stored in the **nephrocytes** that accumulate in certain tissues and are released when the tunicates die (called **storage excretion**). The **neural gland** restores and maintain the fluid volume of the blood.
18. Urochordates are hermaphrodite but usually self-sterile. Only cross fertilization occurs. These are only chordates that regularly reproduce asexually by budding leading to colonies.
19. Fertilization is external and development is indirect. They produce lectitropic free-living tadpole larva.
20. The free-swimming larva contains all chordate characters, but it undergoes retrogressive metamorphosis to produce a dgenerate adult.

5.2 CLASSIFICATION

The subphylum Urochordata is divided into four classes: Ascidiacea, Thaliacea, Appendicularia (Larvacea) and Sorberacea. More than 90 per cent of the known species of tunicates are ascidians.

Class I. Ascidiacea

1. The ascidians are commonly called **sea squirts**.
2. They are bottom dwelling, sessile tunicates found in all the seas.
3. They are solitary, colonial or compound (Box 5.2).

Box 5.2

Most of the large (upto 60 cm long) species are called **solitary ascidians** because they live singly and unattached to one another (*e.g.*, *Ciona*, *Molgula*), although many are highly gregarious. **Social ascidians** tend to live in clumps of individuals that are vascularly attached to one another at their bases (*e.g.*, *Clavelina*). Finally, a great number of species are **compound ascidians** and are characterized by many small individuals (zooids) living together in a common gelatinous matrix (*e.g.*, *Botryllus*).

4. Form and size of the body vary a good deal. Solitary species may be spherical, purse-like, cylindrical, potato-like or irregular.
5. Test or tunic is well developed and permanent composed of protein and a complex polysaccharide tunicin.
6. Branchial aperture is anterior and atrial aperture is dorsal in position.
7. Numeerous gill slits or stigmata are present.
8. Adults are without notochord, hollow nerve cord and tail. They however retain larval pharyngeal gill slits.
9. Hermaphrodite. Life-history includes a typical tadpole larva. Retrogressive metamorphosis occur.
10. Asexual reproduction occurs by budding (Box 5.3). Colonies and compound forms are produced by budding.
11. Tadpole larva contains chordate features such as a tail, notochord and a dorsal nerve cord.

Box 5.3

According to **M. Nakauchi** (1982) ascidians have following two types of buddings:

1. **Propagative budding.** It generally occurs during favourable environmental conditions and serves to increase colony size and exploit available resources.
2. **Survival budding.** It tends to take place during the onset of adverse conditions and may be viewed as an overwintering or other survival device. In this case colony size is generally reduced by the resorption of zooids, leaving potential buds of presumptive germinative tissues.

In general, initial buds are formed by a sexually produced individual (called **oozoid**), then the asexually produced individuals (called **blastozooids**), produce additional buds (**Brusca** and **Brusca** 2003).

12. Adult tunicates squirt water jets from both siphons.

The class Ascidiacea is divided into two orders: Enterogona and Pleurogona.

Order 1. Enterogona

1. Gonad is unpaired and lodged in the intestinal loop or projecting behind it.
2. Neural gland is ventral to nerve ganglion.
3. Larva has two sense organs: cerebral eye (ocelli) and statocyst (otolith).

This order includes following two suborders:

Suborder 1. Phlebobranchia

1. Solitary or colonical.
2. Accessory longitudinal blood vessels present in pharynx.
3. Gonads lie within the gut loop.
4. Gut loop lies either behind or to one side of the pharyngeal basket.

Examples: *Ascidia*, *Ciona*, *Phallusia* and *Corella*.

Suborder 2. Aplousobranchia

1. Colonial.
2. Pharynx without longitudinal blood vessels. Pharyngeal basket simple.
3. Gonads lie within the gut loop.

Examples: *Clavelia* and *Didemnum*.

Order 2. Pleurogona or Stolidobranchia

1. Solitary or colonial.
2. Pharyngeal wall folded. Longitudinal blood vessels present in the pharyngeal basket.
3. Neural gland dorsal or lateral to nerve ganglion.
4. Gonads two or more and embedded in mantle.
5. Gut loop lies to one side of the pharyngeal basket.

Examples: *Herdmania*, *Molgula*, *Botryllus* and *Styela*.

Class II. Thaliacea

1. The salps are free-swimming, pelagic (inhabiting open sea away from shore) and hypopleustonic (floating below the surface). Some of them swim by forcing a jet of water through the posterior end. Others are wafted about by water currents.
2. Filter feeders. Solitary or colonial.
3. Adults barrel-shaped and planktonic.
4. Incurrent and excurrent siphons are located at the opposite ends of the body.
5. Test or tunic permanent, gelatinous and transparent.
6. Mantle contains circular muscle bands.
7. Pharynx or branchial sac is small and contain two large or many small gill slits.
8. Adults without notochord, nerve cord and tail.
9. Hermaphrodite.
10. There may or may not be a tailed larval stage.
11. Asexual reproduction occurs by budding.
12. They are dimorphic and show alternation of generations.

The class Thaliacea is divided into three orders: Pyrosomida (Meromyaria), Doliolida (Cyclomyaria) and Salpida (Hemimyaria).

Order 1. Pyrosomida (Meromyaria)

1. Free swimming, pelagic, colonial forms. Colony closed at one end.
2. Compact cylindrical colony. Zooids of the colony are embedded in a common test. Branchial apertures of the zooids exists on outer side and atrial apertures occur on the inner side.
3. Muscle bands are found only at the two ends of body.
4. Gill slits are long and numerous.
5. No larval stage. Development is internal.
6. They are luminescent.
7. Most primitive thaliaceans.

Example: *Pyrosoma*.

Order 2. Doliolida (Cyclomyaria)

1. Free swimming pelagic forms.
2. Body is barrel-shaped. Branchial and atrial apertures exist at opposite ends of body.
3. Test moderately developed.
4. There are 8 to 9 complete circular bands of muscles around the body hence the name of order cyclomyaria.
5. Gill slits few to many.
6. Tailed larval stage with notochord is present.
7. There are two phases in life history: the solitary sexual phase called the **gonozooid** or **blastozooid** and the colonial or gregarious asexual phase, called **oozooid**. These two regularly alternate.

Example: *Doliolum*.

Order 3. Salpida (Hemimyaria)

1. Free swimming pelagic forms.
2. Body is fusiform.
3. Muscle bands are incomplete ventrally (hence the name of the order Hemimyaria) and may converge dorsally.
4. There is a single pair of very large lateral gill slits.
5. Tailed larval stage is absent.
6. Life history has two alternating phases—a solitary **asexual** or **oozoid phase** and colonial or gregarious **sexual** or **gonozoid phase**.

Example: *Salpa*.

Class III. Larvacea (Appendicularia)

1. Larvacea are small, transparent, free-swimming, pelagic, planktonic and solitary tunicates. They are luminescent creatures.
2. Body is small (upto 5 mm) and comma-like. It resembles an ascidian larva, hence the name Larvacea.
3. True tunic absent. A very loose gelatinous casing or **house** is secreted by a special part of skin. The house is frequently thrown off and renewed. The animal can freely move in the house.
4. Mouth is terminal, permanently open and on the anterior side. There is no atrium and atrial aperture. The anus opens to the exterior directly on the ventral side.
5. The pharynx has only two gill slits, one on each side, and these open to the exterior directly. There is no dorsal lamina.
6. Notochord, nerve cord and muscular tail are retained throughout their life.
7. Larva does metamorphose but develop gonads and starts reproduction like an adult. This phenomenon is called **neoteny** or **paedomorphosis**.
8. There is no budding. Only sexual reproduction occurs. Hermaphrodite.
9. They are primarily herbivorous, feeding on minute phytoplankton and bacteria down to a size of 0.1 μm.
10. Their feces and abandoned houses constitute important sources of food for the marine life.

The Larvacea is divided into two orders: Endostylophora and Polystylophora.

Order 1. Endostylophora

1. House bilaterally symmetrical, with separate inhalent and exhalent apertures.
2. Pharynx with endostyle.

Examples: *Oikopleura* and *Appendicularia*.

Order 2. Polystylophora

1. Bilaterally symmetrical house with single aperature.
2. Pharynx with endostyle and four longitudinal ciliated rows.

Example: *Kowlevskia*.

Class IV. Sorberacea

1. Benthic, abyssal, ascidian-like urochordates.
2. Adults possess dorsal nerve cords.
3. Lack perforated branchial sac (*i.e.*, no gill slits).

Example: *Octacnemus*.

5.3 TYPE STUDIES

1. *Ciona*

Classification

Phylum	—	**Chordata**
Subphylum	—	**Urochordata**
Class	—	**Ascidiacea**
Order	—	**Enterogona**
Genus	—	***Ciona***

Habit and Habital

Ciona is commonly called **tube sea squirt**. It is a solitary and sedentary animal. *Ciona* is found attached to rocks and ships in all temperate seas. It occurs from the tidal zone to a depth of 500 meters.

Structure

The body of *Ciona* is cylindrical sac which is 10 to 12 cm long. It remains attached to the substratum by a base. The body is covered with a thin, flexible and transparent **test** or **tunic**. Test protects and supports the body. It is composed largely of fibers of cellulose-like polysaccharide, the tunicin, mucopolysaccharide and some protein. The test has two openings—a terminal **mouth** and a somewhat dorsal **atriopore** (Fig. 5.1). Both the apertures are surrounded with muscular siphons. The **branchial siphon** or **inhalant siphon** exist around the mouth and is longer than the **atrial siphon**. Both the siphons are prominent and lie close to each other. Mouth is eight lobed and atriopore is six lobed. Entire test is lined with a transparent **mantle**. It contains longitudinal and circular muscle fibers and secrete the test. When these muscles contract the animal ejects jets of water from both the siphons. Due to this habit the animal is commonly called the **sea squirt**.

The mouth opens into a large pharynx which extends almost upto the base. It is attached to the mantle along its midventral line. Pharynx is surrounded dorsally and laterally by a cavity, the **atrium**, which opens to outside through the **atriopore**. The wall of the pharynx are perforated with numerous **gill slits** or **stigmata** through which water passes out the pharyngeal cavity into the atrium. The margins of gill-slits bear cilia. Beating of the cilia draws water current into the pharynx that brings oxygen and food organisms. *Ciona* is a **filter-feeder**. On the ventral side of the pharynx there is a deep groove, the **endostyle**. It is the forerunner of the vertebrate thyroid gland. It secretes mucus and binds iodine. The alimentary canal is U-shaped and consists of mouth, pharynx, oesophagus, stomach and intestine. The intestine opens into the atrial cavity. Digestion is extracellular.

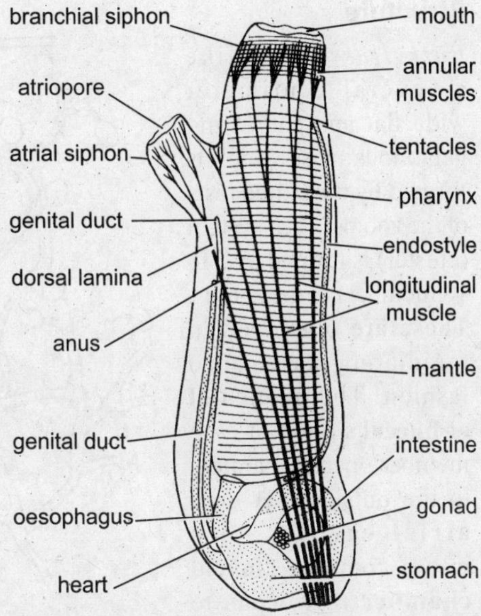

Fig. 5.1. *Ciona.*

The stomach, intestine, gonad and heart lie below the pharynx. Heart is V-shaped. Intestine has a typhlosole. Dorsal tubercle is horse-shoe shaped. Gonads lie in the intestinal loop. *Ciona* is hermaphrodite. Its free swimming highly organized larva contains a notochord and dorsal tubular nerve cord in the tail. Retrogressive meta-morphosis occurs in *Ciona*.

2. *Botryllus*

Classification

Phylum	—	**Chordata**
Subphylum	—	**Urochordata**
Class	—	**Ascidiacea**
Order	—	**Pleurogona**
Genus	—	*Botryllus*

Habit and Habital

Botryllus is a composite or compound ascidian (Fig. 5.2). It inhabits shallow water of intertidal zone in the Atlantic and Mediterranean seas. Its groups (colonies) are found attached to rocks, wharfs (dockyards), piers (posts) and ship bottoms, etc. fertilization and development occur in special cup-like cavities in the atrial wall.

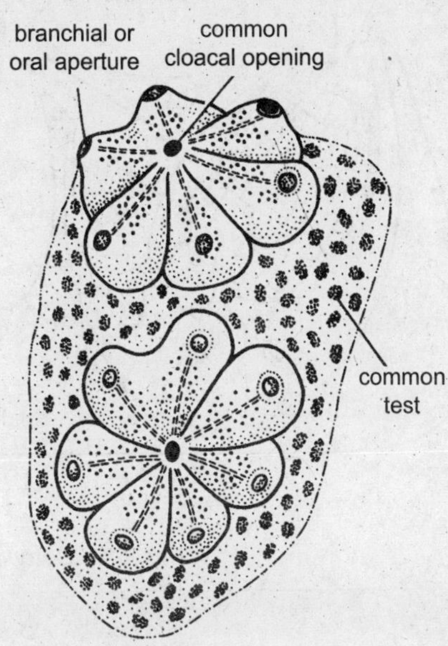

Fig. 5.2. *Botryllus*. Two composite systems from the top.

Structure

Botryllus looks like a several centimeter wide flat and encrusting gelatinous mass which is formed by the united tests of the zooids. Each system (cluster) is formed by 5 to 18 members or the zooids. These are organized in a radiating star shaped fashion. The **branchial** or **buccal siphon** of each member opens separately to the outside, but their **atrial cavities** open into a common **cloacal chamber** that oepns to the outside by a single **atrial aperature** in the middle of the colony or cluster. Each zooid is like a simple ascidian (*e.g., Herdmania*). The pharynx has numerous gill-slits or stigmata. The neural gland lies above the neural ganglia. Each zooid is hermaphrodite.

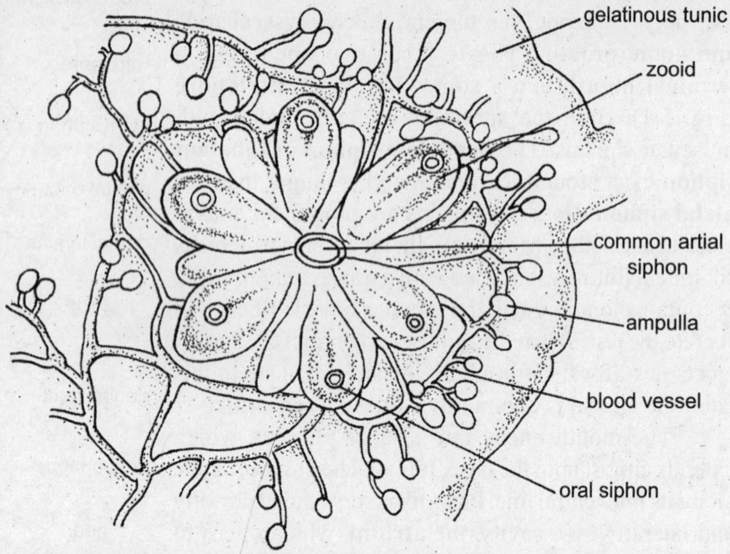

Fig. 5.3. *Botryllus*. Zooid clusters of the compound ascidian.

Fertilization and development occur in cup-like cavities present in the wall of atrial cavity. Fully developed tadpole larvae are released in the common cloacal chamber, from where they escape in the sea. Asexual reproduction and growth of colony occur by budding. The buds develop as hollow outgrowths from the wall of atrium.

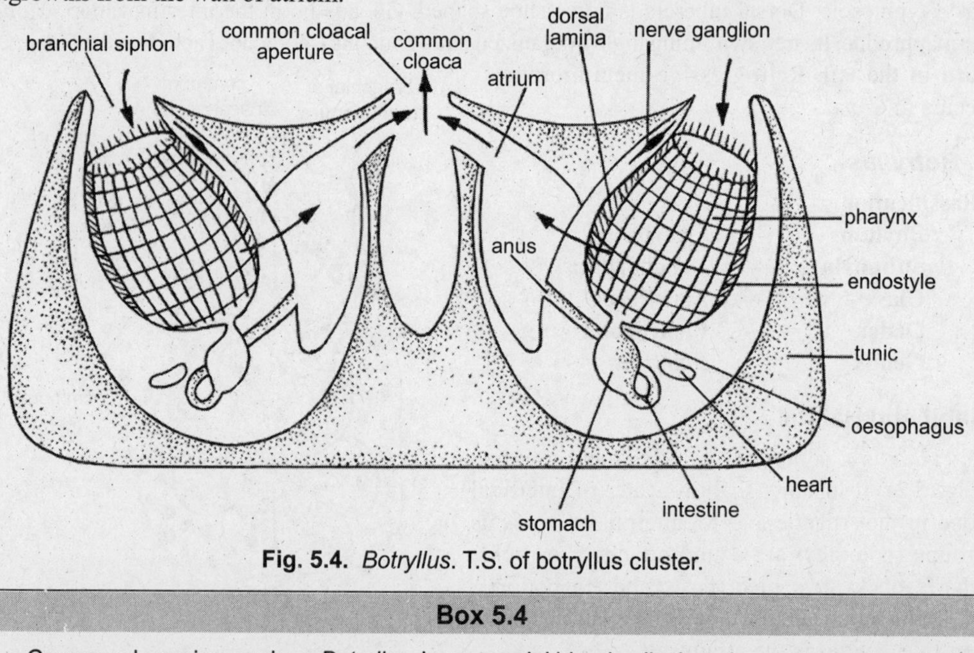

Fig. 5.4. *Botryllus*. T.S. of botryllus cluster.

Box 5.4

Compound species, such as *Botryllus*, have special blood cells that play a vital role in rejecting adjacent conspecific, but nonclone colonies (allogenic colonies) (**Brusca** and **Brusca** 2003).

3. *Pyrosoma*

Classification

Phylum	—	**Chordata**
Subphylum	—	**Urochordata**
Class	—	**Thaliacea**
Order	—	**Pyrosomida**
Genus	—	*Pyrosoma*

Habit and Habitat

Pyrosoma (Gr., *pyro* = fire) is found in warmer parts of Atlantic, Indian and Pacific oceans at the depth of 503 meters. It is a colonial, free-swimming, pelagic, phosphorescent tunicate. It swims forward horizontally by forcing out a jet of water backward through its posterior terminal opening. Velocity is increased by narrowing the terminal aperture by means of a diaphragm. Reproduction is sexual as well as asexual. Development is internal and direct. No larval form is produced by *Pyrosoma*.

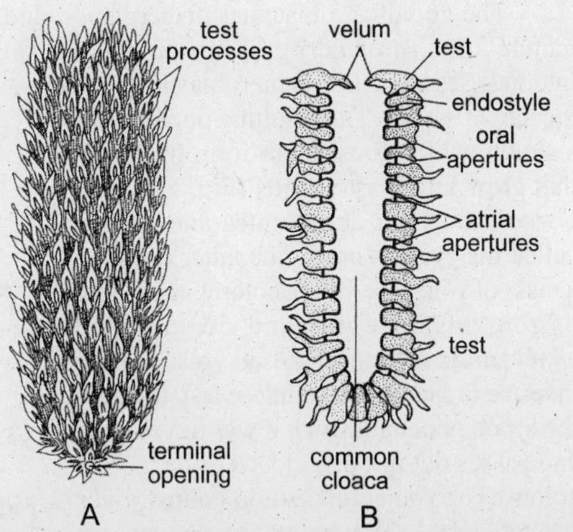

Fig. 5.5. *Pyrosoma*. A—Colony. B—Sagittal section of colony.

Structure

The colony of *Pyrosoma* has the form of hollow cylinder or barrel 1.2 cm to 2.5 long, closed at the anterior end and open at the posterior end. The individuals, called **blastozooids**, are embedded in the wall of the cylinder, which actually consists of their united tests (Fig. 5.5). The central cavity of they cylinder serves as a **common cloaca** for all the zooids. The branchial and atrial aperatures of the zooids are at the opposite ends and are provided with sphincters. The zooids are so arranged that their branchial apertures open on the surface of the cylinder and atrial apertures lead into the central common cloaca. A tongue-like process of the test, called **buccal appendage**, overhangs the branchial aperture to each zooid.

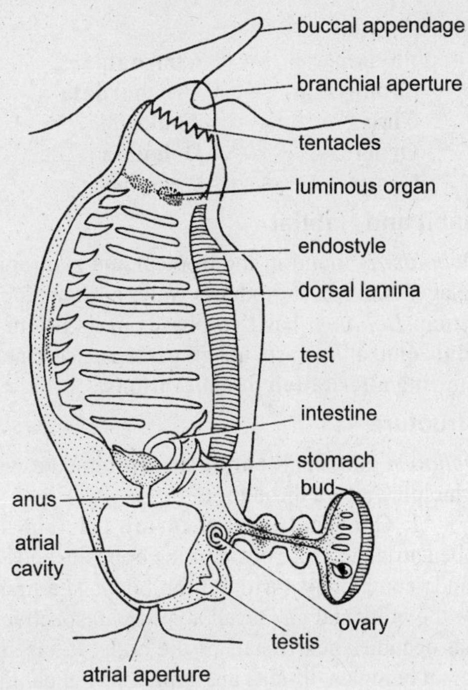

Fig. 5.6. *Pyrosoma*. L.S. of single zooid.

The zooids of *Pyrosoma* are independent of one another (Fig. 5.6). Each has a large pharynx with numerous gill-slits (stigmata), endostyle, peripharyngeal bands and dorsal lamina with languets. Stomach, intestine and gonads lie internal to the pharynx. Heart and stolon are near the end of the endostyle. Subneural gland and ocellus are also present in *Pyrosoma*. Musculature is weak. On each side of pharynx, just above the peripharyngeal bands, is situated a flat mass of specialized mesodermal cells enclosed within a blood sinus. These are the **luminescent organs**. Light is produced by the bioluminescent symbiotic bacteria that live in the cells of the organs. These bacteria invade eggs thus transferring power of luminescence to next generation. Luminescence of *Pyrosoma* is perhaps a protective mechanism, giving warning of danger to the enemies.

Each blastozooid of *Pyrosoma* sieves water from the sea into the common cloacal cavity of the colony, then a stream of water is shot out from the open posterior end and slowly moving the colony forward by propulsion. Thus, colony locomotes by jet propulsion. As in the ascidians, water movement in pyrosomes is generated entirely by ciliary action of the individual zooids.

The zooids are bisexual or hermapprodite, each having an ovary and a lobed testis. Testis matures first (protandry). Ovary produces a single large, yolky ovum at a time. Fertilization is internal. Segmentation is meroblastic. Development is direct without a larval stage and occurs within the parent's body. The resulting oozooid has a stolon, which constrict to form four buds that grow into blastozooids (Fig. 5.7). The oozooid later on degenerates and is then called the **cyathozooid**. The latter encloses a mass of yolk. The young colony, consisting of four blastozooids and degenerated cyathozooid with its enclosed yolk mass, is invested in a common cellulose test. It leaves the brood pouch in which it was developing and passes out through cloaca of the parent colony. The young **tetrazooid colony** gives rise to a new colony by budding on stolons of its zooids.

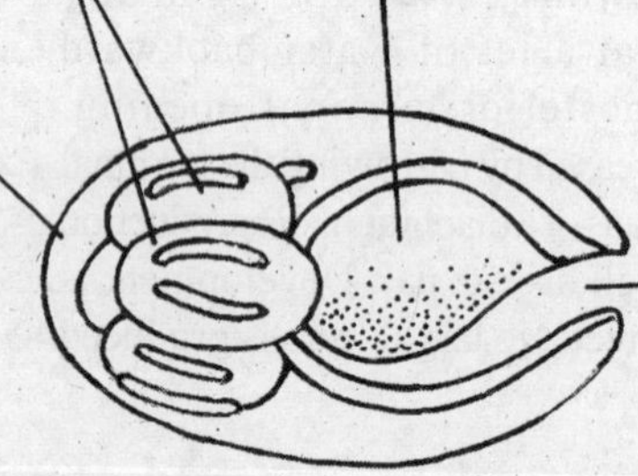

Fig. 5.7. *Pyrosoma*. Oozooid with blastozooids and yolk mass (tetrazooid colony).

4. *Doliolum*

Classification

Phylum	—	**Chordata**
Subphylum	—	**Urochordata**
Class	—	**Thaliacea**
Order	—	**Doliolida**
Genus	—	*Doliolum*

Habit and Habitat

Doliolum is found in the tropical and subtropical seas down to 200 meters. It is a free-swimming pelagic tunicate. It produces water currents for feeding and gas exchange (respiration) by the ciliary action. *Doliolum* has two phases in its life history: a sexual solitaria phase called **blastozooid** or **gonozooid** and an asexual gregaria phase termed as **oozoid**. These two phases alternate in life history, showing **alternation of generations**.

Structure

Doliolum is a small barrel-shaped tunicate with thin, delicate, transparent test that shows all the principle viscera through it.

1. Gonozoid (Sexual form). It is 1 to 1.5 cm long. The 12 lobed branchial aperture and 8 lobed atrial aperture occur at the opposite ends of the barrel-shaped animal. Eight prominent muscle bands completely surround the body. The first and last of these muscle bands serve as sphincters for the branchial and atrial aperture, respectively. The branchial aperture leads into a large pharynx that occupies nearly half of the body. It has an endostyle and peripharyngeal bands, but lacks the dorsal lamina. Gill-slits are numerous but confined to the posterior part of the pharynx. Oesophagus, stomach, intestine, digestive gland and heart lie close together at posteroventral side of pharynx. The gonozooid is protogynous (*i.e.*, the female reproductive structures mature before the male structure of the hermaphrodite animal). Fertilization is internal. Zygote is released into the sea water. It develops into a tailed larva which metamorphoses into the asexual phase, the oozooid.

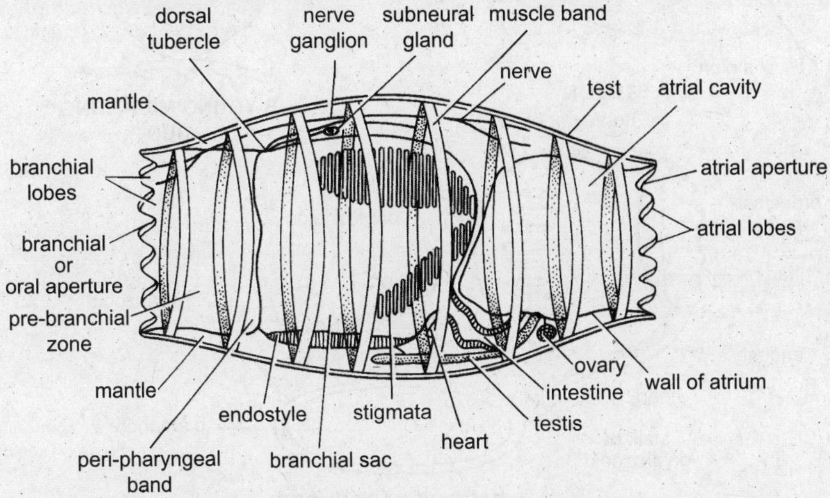

Fig. 5.8. *Doliolum*. Gonozooid.

2. Oozooid or Asexual form. Oozooid (Fig. 5.9) represent the asexual phase of *Doliolum* which is **polymorphic** having three types of zooids: natritive gastrozooids, foster phorozooids and sexual gonozooids. The oozooid is similar to gonozooid in form and general antomy, except the following differences: (*i*) the branchial and atrial apertures are 10 lobed and 12 lobed respectively; (*ii*) there are nine muscle bands; (*iii*) gill-slits are fewer but larger; (*iv*) gonads are lacking; (*v*) there are two processes: a small ventral **stolon** near the heart and a large dorsal appendix, called **cadophore** growing backward from the posterodorsal part of the body surface (*i.e.*, in between seventh and eight muscle bands). The ventral stolon cuts off from its end, small buds (Fig. 5.10), called **probuds**. The probuds tend to migrate by pseudopodial action of their ectodermal cells onto the **cadophore**. The probuds that arrive first fix to the cadophore in two lateral rows. Those probuds which are arriving later join the cadophore in two median rows. As this happens, the oozooid loses its gill-slits, endostyle and gut and it develops its muscle bands and nervous system. Oozooid has statocyst on the lateral side. At this stage, oozooid of *Doliolum* serves to propel the colony through the water. Ultimately, the lateral buds grow into **trophozooids** or **gastrozooids** which have gill-slits and carry on nutrition and respiration of the colony. The median buds give rise to **phorozooids** which develop 8 muscle bands and help the oozooid in locomotion of colony. They also serve as foster mothers of the buds and called **gonozooids**; they arrive last and fix to stalks of phorozooids. Phorozooids with buds (gonozooids) on their stalks break off and swim about freely. The buds on their stalks grow into gonozooids (solitaria phase) which finally become free.

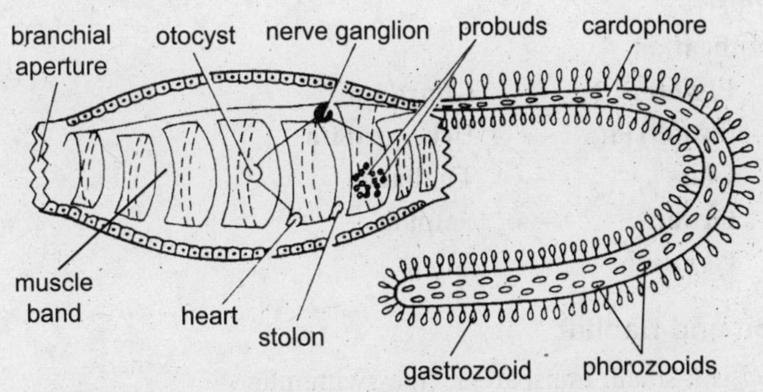

Fig. 5.9. *Doliolum*. Oozooid (asexual gregarian phase).

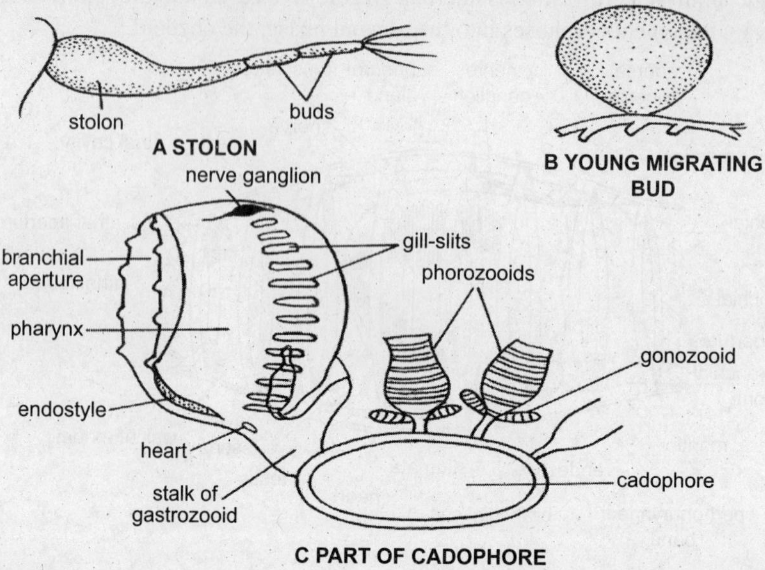

Fig. 5.10. *Doliolum*. A—Stolon; B—Migrating bud and C—A part of cadophore.

5. *Salpa*

Classification

Phylum	—	**Chordata**
Subphylum	—	**Urochordata**
Class	—	**Thaliacea**
Order	—	**Salpida**
Genus	—	*Salpa*

Habit and Habitat

Salpa is a small, transparent, free-swimming and pelagic tunicate. It is found in the upper levels of all seas but is more common in warmer seas. *Salpa* produces water current for feeding, gas exchange and locomotion by muscular contractions. It shows alternations of sexual (blastozooid) and asexual (oozooid) generations.

Salpa is a small tunicate with thin, delicate and transparent test.

1. Oozooid, solitaria or asexual phase. The oozooid is prism-shaped or cylindrical and 1.5 to 15 cm long. Branchail aperture (mouth) and atrial aperture exist at opposite ends of body. The test is closely adherent the mantle internally. The branchial aperture has dorsal and ventral lips and the atrial aperture has a pair of lateral test processes. There are nine muscle bands. The first and last

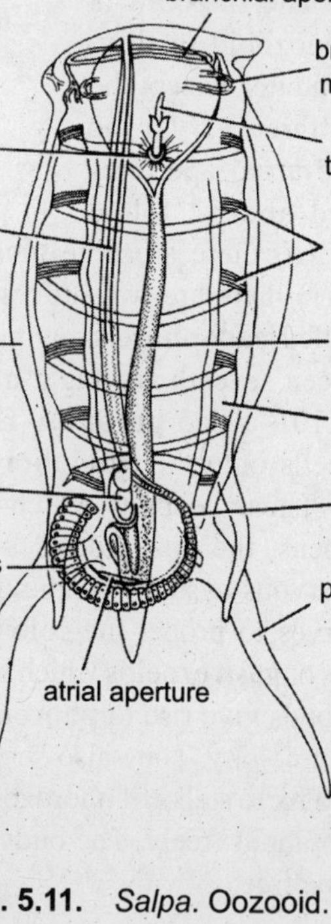

Fig. 5.11. *Salpa*. Oozooid.

muscle bands are complete and form sphincters for the branchial and atrial aperture respectively. Other muscle bands are incomplete ventrally and may converge dorsally. The oozooid swims forward by closing the branchial aperture and contracting muscle bands, thus forcing a jet of water backward through the atrial aperture. The pharynx has a single very large gill-slit on either side (Fig. 5.12). Pharynx also has an endostyle, a pair of peripharyngeal bands and a dorsal lamina at their usual positions. A single tentacle is present on the dorsal side of pharynx, stomach, intestine form a small oblique mass, called the **nucleus** or **visceral mass**, lying posteroventral to the pharynx. Heart is located ventrally behind the endostyle. Nucleus is correlated with the free swimming mode of life. A nerve ganglion and an eye (a red pigmented band) lie antero-dorsally. Infront of the pharyngeal band is dorsal tubercle as a short diverticulum of pharynx. Ventral to the ganglion is the subneural gland. A complex stolon arises from the region between the heart and endostyle and segments into **buds** that become sexual blastozooids. A chain of these new zooids projects near the atriophore. The stolon ultimately breaks off from the asexual parent and starts the gregaria sexual phase.

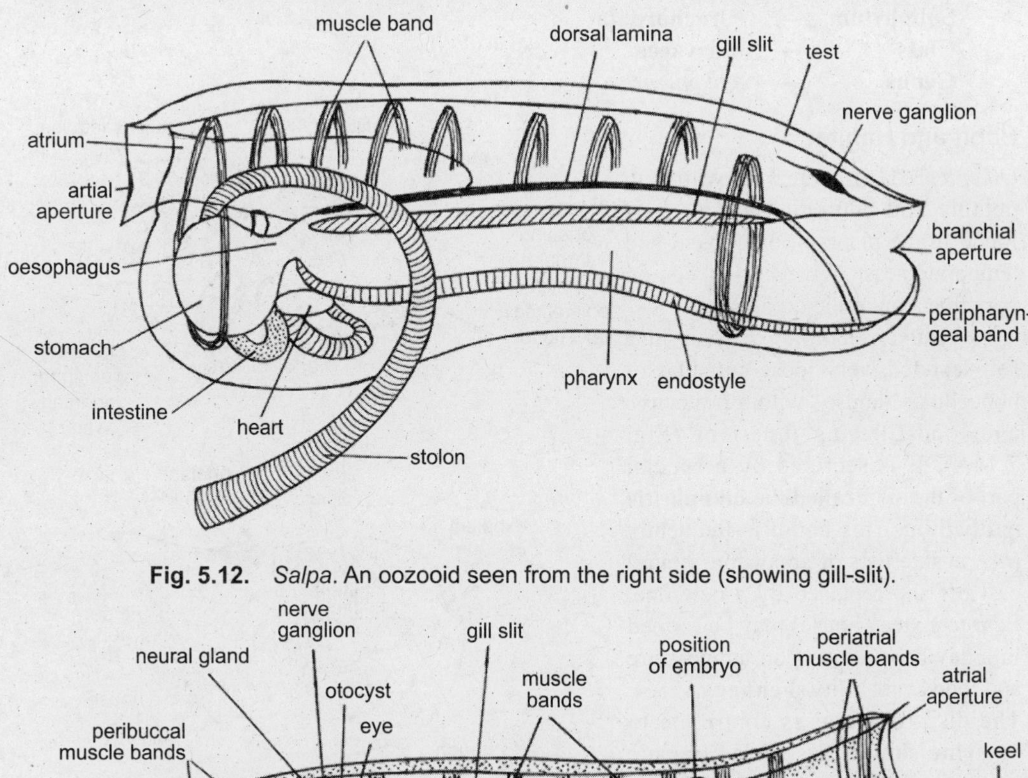

Fig. 5.12. *Salpa*. An oozooid seen from the right side (showing gill-slit).

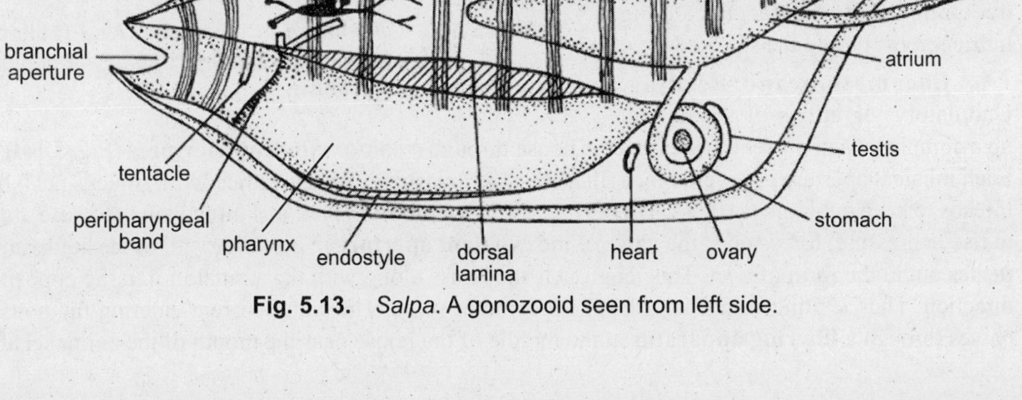

Fig. 5.13. *Salpa*. A gonozooid seen from left side.

2. Gonozooid, blastozooid, gregaria or sexual phase. The gonozooid is smaller than the oozooid. It has fewer muscle bands, lacks stolon, and possesses gonads: an ovary which forms a single ovum and a testis. The gonads lying in the nucleus; the gonads open separately into the atrium. The gonozooid is protogynous and viviparous. The egg develops inside the atrial cavity of parent. The embryo is nourished by the formation of the **placenta**, in which the parental and embryonic vessels lie close together. There is no tailed larval stage, embryo develops directly into an oozooid with a stolon. The oozooid escapes to the exterior and becomes free-swimming. Later on, this oozooid develops blastozooids on its stolon. Thus, the asexual and sexual phases alternate in life history of *Salpa* as in *Doliolum* and *Pyrosoma*.

6. *Oikopleura*

Classification

Phylum	—	**Chordata**
Subphylum	—	**Urochordata**
Class	—	**Larvacea**
Genus	—	*Oikopleura*

Habit and Habitat

Oikopleura is a minute free-swimming pelagic and neotenic animal. It is found in the plankton of tropical and temperate seas. *Oikopleura* feeds on very fine planktonic organisms.

House. *Oikopleura* lives in a self-secreted, very loose cuticular or noncellular "house" with a protective and food-filtering functions (Fig. 5.14A). It is secreted by a special part of the skin called the **oikoplastic epithelium**. This house is frequently (*i.e.*, at intervals of about four hours) cast off and replaced by a new one. Four to sixteen houses may be formed in a day depending upon temperature and food availability (**Fennaux** 1985). The discarded houses contribute to "**marine snow**" (suspended organic matter looking like snow) which is decomposed by bacteria, causing nutrient recycling in the sea.

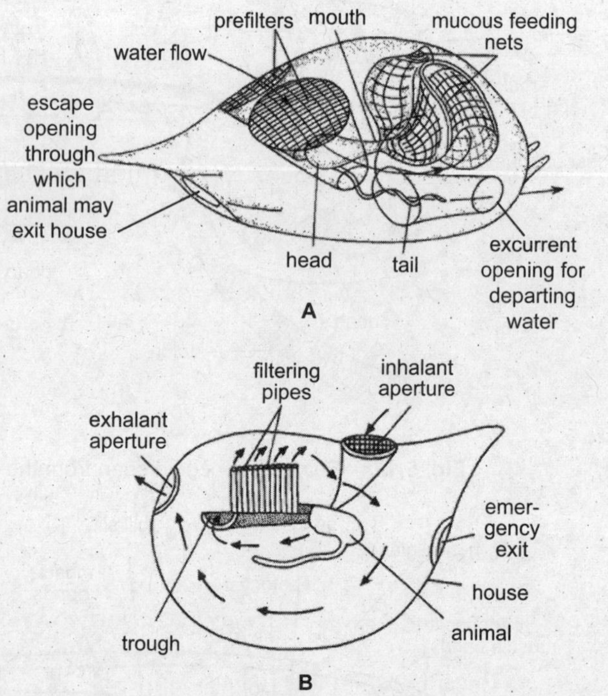

Fig. 5.14. *Oikopleura* in its house. A—Profilters and mucous feeding nets. B—Feeding mechanism.

Locomotion and Feeding. Undulatory vibrations of the tail set up a complex water current that enters the house through a pair of **inhalent apertures** (Fig. 5.14B). Each inhalent aperture is covered by a filter (mesh) consisting of a fine lattice work (mucus nets) of threads to keep out large particles (**Diebel** *et al.*, 1985; **Flood** 1991). Water already in the house, due to rise in pressure, forces open the valve of the **exhalant aperture** in posterior wall of the house and rushes out in the form of a jet. This jet propels the house along with the animal in it in the opposite direction. Thus, exiting water provides the locomotor force. The water current entering the house passes through a **filtering apparatus** at the middle of the house near the mouth of the animal. This apparatus consists of a series of pipes arranged in a trough-like oral tube. As water rises up the pipes, the food particles are left in the oral tube, whence they pass into the mouth, which sucks them in by ciliary action of gill slits. Food particles are entangled in mucus produced by the endostyle, the food-laden mucus cords are drawn into the stomach. The house has an **emergency exit** for the escape of the animal at the time of danger or when apertures are blocked.

Reproduction. *Oikopleura* is protandrous (*i.e.*, animal is first a male and then becomes a female). Fertilization is external. The newly hatched larva is like an ascidian tadpole. Its tail soon bends under the trunk and a house is secreted. It undergoes no other metamorphic change.

Structure. Animal is tranparent and about 5 mm long. Its body is differentiated into a 1 mm long bulbous **trunk** and a slender and laterally compressed muscular **tail**. The tail is narrow at the base and contains a ganglionated **nerve cord, notochord** (It is also called **urochord** since it is limited to the caudal region.) of 20 cells and two rows of 7–10 **striped muscle cells**. Tail has a broad tail-fin that helps in locomotion and feeding. The tail is twisted through 90° (Fig. 5.15).

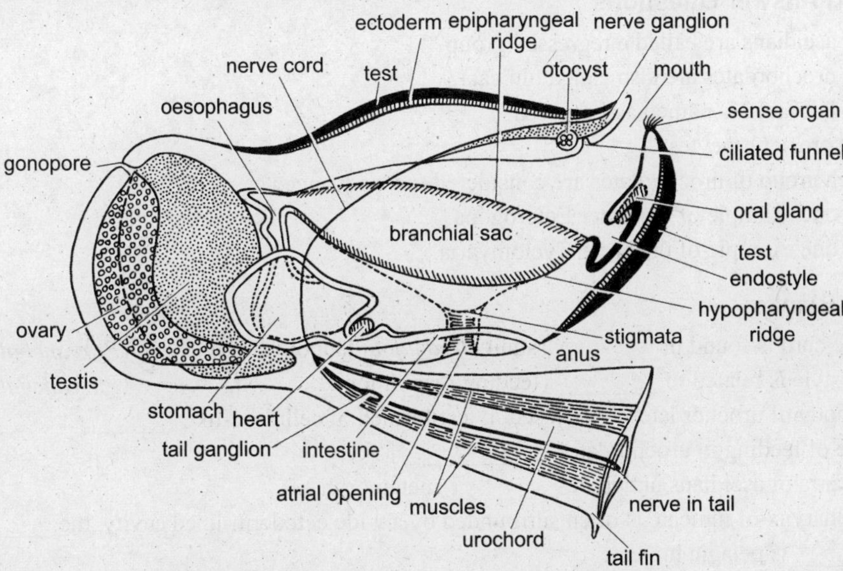

Fig. 5.15. *Oikopleura*. Animal in sagittal section.

The **mouth** (oral aperture) is located at the anterior end of body and the intestine opens directly to the outside on the ventral side. Pharynx is reduced. There are only two pharyngeal **gill-slits** (stigmata) one on each side and each opens directly to the exterior. In addition to gill-slits, the pharynx contains a short **endostyle**, two strongly ciliated **peripharyngeal bands**. A fusiform **cerebral ganglion** lies dorsal to the anterior part of the pharynx. It bears an **otocyst** (pigment spot) and remains connected with the first ganglion of the nerve cord of the tail. **Heart** lies beneath the stomach. There are no blood vessels but blood lacunae are present.

At the posterior side of the body is a large hermaphrodite gonad which contains a single **ovary** and a single or paired **testes**. *Oikopleura dioica*, however, is not hermaphrodite. It has separate males and females. In hermaphrodite animals testis matures earlier than the ovary (protandrous condition).

QUESTIONS

Long Answer Questions

1. Give the distinctive characters and classification of the Urochordata with examples. (*Garhwal 2000*)

2. Give the characters of the subphylum Urochordata, and classify it giving the diagnostic characters and examples of various groups. (*Purvanchal 2000; Bundelkhand 2000*)
3. Write a brief ecological note on the following animals: *Ciona*, *Botryllus*, *Doliolum*, *Salpa*, *Pyrosoma* and *Oikopleura*.

Short Answer Questions

1. Name the orders of class Ascideiacea. Cite one example of each.
2. Give the distinguishing characters of the class Ascidiacea.
3. Classify the following animals giving their characteristic features:
   (*i*) *Oikopleura* (*ii*) *Botryllus*
   (*iii*) *Ciona* (*iv*) *Doliolum*
   (*v*) *Salpa* (*vi*) *Pyrosoma*

Very Short Answer Questions

1. Why ascidians are called a regressed group?
2. Why urochordates are also called tunicata?
3. Why Larvacea is named as Larvacea?
4. What are nephrocytes?
5. Which group of urochordates are considered most advanced?
6. Give one example of the order Enterogona.
7. Give one example of the order Cyclomyaria.

Fill in the Blanks

1. Nerve cord is found in ________ (adult *Herdmania*/tadpole larva). (*Purvanchal 2000*)
2. Endostyle is related to ________ (feeding/respiration). (*Purvanchal 2001*)
3. The body of urochordates is enclosed in test formed of cellulose-like ________.
4. Mode of feeding in urochordates is ________.
5. The larva of ascidians undergoes ________ metamorphosis.
6. The pharynx of tunicate is often surrounded by a wide ectoderm-lined cavity, the ________.
7. ________ is pelagic tunicate.
8. ________ lives in a temporary gelatinous house.

Multiple Choice Questions

1. Which of the following types of zooids is not produced by the asexual form of '*Salpa*'?
   (*a*) gonozooid (*b*) phorozooid
   (*c*) gastrozooid (*d*) dactylozooid
2. In urochordates respiration occur through
   (*a*) test (*b*) test and gill-slits
   (*c*) gill-slits (*d*) branchial basket
3. The tube sea squirt is the popular name of
   (*a*) *Herdmania* (*b*) *Ascidia*
   (*c*) *Ciona* (*d*) *Molgula*
4. *Botryllus* is a
   (*a*) simple ascidian (*b*) composite ascidian
   (*c*) colonial ascidian (*d*) none of these

ANSWERS

Very Short Answer Questions

1. The ascidians are cosidered as regressed group of animals because most of them as adults are attached (sedentary), but have evolved from free-moving ancestors.
2. The name tunicate is suggested by the non-living tunic (test) which surrounds them and is made up of proteins and complex cellulose-like polysaccharide, the tunicin.
3. The members of class Larvacea are small, tadpole-like, under 5 mm in length and appear to be ascidian tadpoles. They represent persistent larval forms which have become **neotenous**.
4. Cells in tunicate blood with excretory role.
5. Larvacea.
6. *Ciona*.
7. *Doliolum*.

Fill in the Blanks

1. Tadpole larva. **2.** Feeding. **3.** Tunicin. **4.** Ciliary/filter. **5.** Retrogressive. **6.** Atrium. **7.** *Doliolum*. **8.** *Oikopleura*.

Multiple Choice Questions

1. (*d*) **2.** (*b*) **3.** (*c*) **4.** (*b*)

6

Branchiostoma (Amphioxus or Lancelet)

Systematic Position

Phylum	—	**Chordata**
Subphylum	—	**Cephalochordata**
Class	—	**Laptocardii**
Family	—	**Branchiostomida**
Genus	—	***Branchiostoma***

Cephalochordates demonstrated several qualities intermediate between those of invertebrates and vertebrates. They represent living descendants of the ancestors of the vertebrates.

Cephalochordates are the animals which retain the diagonostic chordate characters throughout the life. Their notochord extends right from the anterior end to the hind end (Gr. *kephale* = head; *chorde* = string or cord). There are about 25 species of cephalochordates.

Why Amphioxus is called *Branchiostoma*?

To get answer to this vital question, we have to look at the tale of discovery of this remarkable chordate. *Branchiostoma* was picked up in 1774 on the coast of Cornwall and sent to the Russian naturalist **Pallas**. He mistook it for a mollusc (slug) and accordingly named it *Limex lanceolatus* (1778). It was recognised as a lower chordate in 1834 by the Italian naturalist **Gabriel Costa**. He named it *Branchiostoma lanceolatum* (*Branchia* = gill + *stoma* = opening). The name *Amphioxus* (*amphi* = both + *axus* = axia), often used for it was given in 1836 by **Yarrel**. He mistook it for a fish of very primtive organisation. The name *Amphioxus* which was coined by Yarrel has been given up since it is being invalid according to the law of priority in nomenclature. Anglicised form of this name, amphioxus, is now used as an alternative common name. There are nine species of *Branchiostoma*. Of these *Branchiostoma lanceolatum* is best known. Detailed anatomy of *Branchiostoma* was worked out by **Muller** and supplemented by **Huxley** and others.

6.1 HABIT AND HABITAT

Branchiostoma is a widely distributed marine animal found in the coastal waters of Atlantic Ocean, Indian Ocean, Mediteranean sea and along the southern coast of China.

Branchiostoma is a burrowing animal, greater part of its body is buried and only head or anterior end projecting into water. It burrows in sand with its head first by means of vibratory action of the entire body. It is also able to swim feebly in water by means of lateral strokes of the posterior portion of the body. It comes out of the burrow and swims particularly during night and the breeding season.

Branchiostoma is a typical ciliary feeder. It feeds on planktonic microorganisms brought along with water which constantly enters the mouth placed on the projecting anterior end and leave through the atriopore. Sexes are separate. Both sexes release their gametes in water where fertilisation takes place. Development is indirect with a free-swimming larval stage which gradually develops into adult.

6.2 EXTERNAL FEATURES

Shape, Size and Colour. *Branchiostoma* has a pinkish translucent body about 4.5 to 5 cm long. It is slender, elongated, laterally compressed and spindle-like, being streamlined for burrowing and swimming. Both the ends of body are pointed and lance-like, hence the common name of *Branchiostoma* is **lancelet** (a little lance). (**Note.** *Lance* means cut open with a lancet which is a pointed, two-edged knife used by surgeons). The name amphioxus also indicates the same "sharp at both ends". The posterior end is more tapering and pointed than the anterior end. Symmetry is bilateral. The body surface is slippery.

Division of Body

A true head is absent in amphioxus. The body is divisible into two regions: the **trunk** and the **tail**. The greater anterior region constitutes the trunk, and a much smaller posterior region is tail. The anterior end of trunk extends in front as a pointed **snout** or **rostrum**.

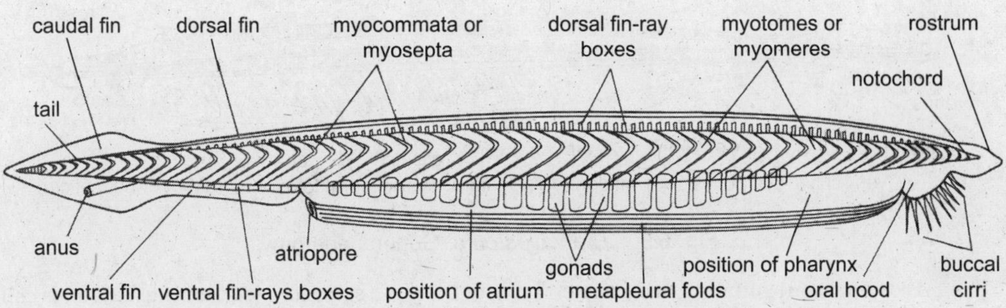

Fig. 6.1. *Branchiostoma*. Lateral side

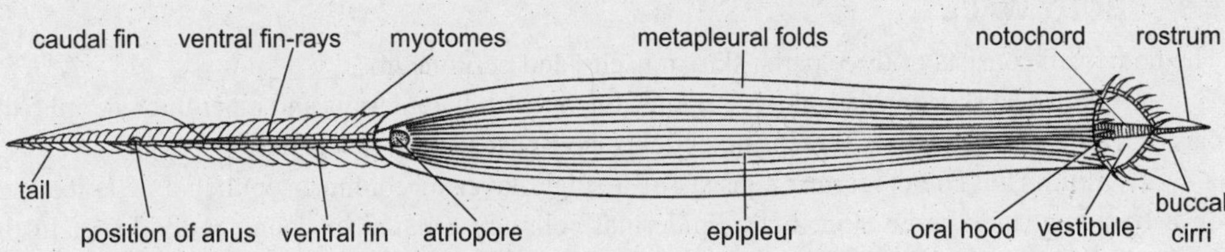

Fig. 6.2. *Branchiostoma*. Ventral side.

Apertures

The trunk bears three apertures: mouth, atriopore and anus.

1. Mouth. Below the rostrum there is a large median opening, the **mouth**. It remains surrounded by a frill-like membrane or the **oral hood**. It is formed by the dorsal and lateral projections of the anterior end of the trunk. The edge of the oral hood bears about 22 tentacles or **cirri**, called **oral cirri** or **buccal cirri**. The oral hoood encloses a cup-shaped cavity, the **vestibule** or the **stomodaeum**.

2. Atriopore. It is a small mid-ventral aperture. It is situated just in front of the ventral fin. A spacious atrial cavity or the atrium, present around the pharynx, opens to outside through this aperture.

3. Anus. It lies at the base of the caudal fin on the ventral surface, a little to the left side of the mid-ventral line. The small posterior part of the body behind the anus is called **tail**.

Fins. *Branchiostoma* bears three unpaired longitudinal median fins, *i.e.*, **dorsal**, **caudal** and **ventral**. The dorsal fin runs along the mid-dorsal line along the entire length of trunk. It is continuous with somewhat broader caudal fin posteriorly around the tail. The ventral fin runs mid-ventrally from the caudal fin upto the atriopore. It is slightly wider than the dorsal fin.

The dorsal and ventral fins are supported respectively by one and two series of **fin-ray boxes**. These are formed of a gelatinous substance and connective tissue.

Paired fins are absent in *Branchiostoma*. From the oral hood upto the atriopore each ventral-lateral edge of the body hangs down to form a pair of hollow *metapleural folds* (Fig. 6.3). The metapleural folds are continuous with the lateral edges of the oral hood, and unite together a little posterior to the atriopore. Metapleural folds help in burrowing in sand rapidly due to turgescence caused by flow of body lymph in their cavities.

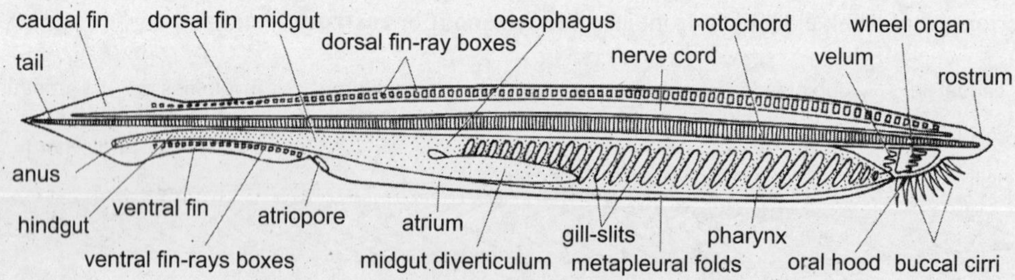

Fig. 6.3. *Branchiostoma*. General anatomy.

Myotomes and gonads. On each lateral side of body is a series of <-shaped muscle bands called **myotomes** or **myomeres**. These can be seen through the transparent body wall. Below the myotomes on both sides are present a series of gonads between mouth and atriopore (Fig. 6.1).

6.3 BODY WALL

The body wall consists of three parts: skin, muscles and peritoneum.

1. Skin. The skin consists of two regions: the outer thin epidermis and inner thick dermis (the classical cutis and subcutis). There is no cuticle (see **Brusca** and **Brusca** 2003).

(*i*) Epiderm. The epidermis consists of a single layer of columnar epithelial cells that rest on a basement membrane (Fig. 6.4). Epidermal cells are ciliated in young individuals. In the adults, they lose cilia. The epidermis contains gland cells and sensory cells scattered among its supporting cells. It lacks pigment cells. Thus, the epidermis of amphioxus is simple and a single layered similar to that of nonchordates (= invertebrates). But in higher chordates epidermis is stratified (many layered).

(*ii*) Dermis. The dermis is composed of connective tissue. It contains tissue fibers (collagen), connective tissue cells, blood vessels and nerve fibers. The collagen and other fibers are arranged in successive sheets. In each sheet, the fibers are oriented in a different direction from than in outer one. This arrangement of fibers give strength and iridescence (rainbow-like colours).

2. Muscles. Muscles lie just beneath the skin. All the muscles of body of amphioxus are striated (*i.e.*, voluntary). These form the greater part of the body wall. Muscles are very thick in the dorsal and dorso-lateral regions. These muscles enclose the nerve cord and notochord. Muscles show distinct metameric segmentation and are arranged in a series of <-shaped muscle blocks or segments on each

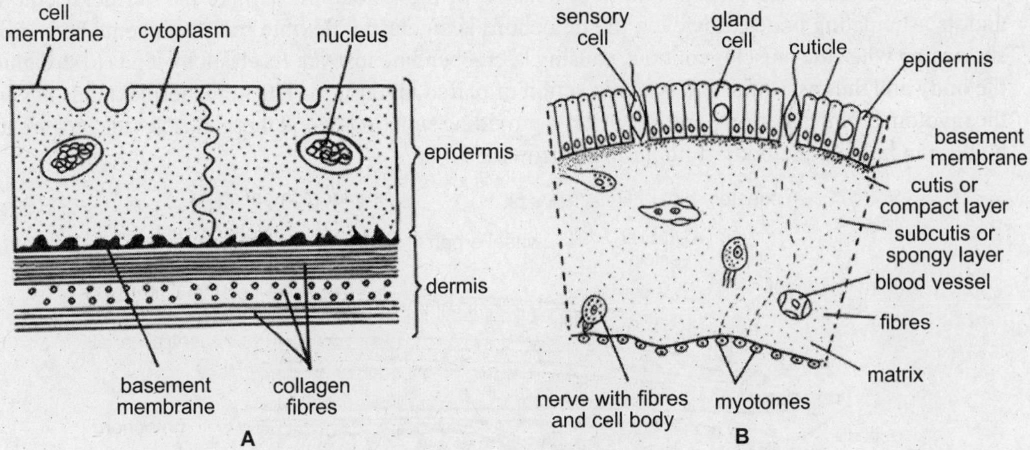

Fig. 6.4. *Branchiostoma*. V.S. of body wall. A—Skin showing epidermis and dermis; B—Classical view depicting cutis and subcutis of dermis.

side (Box 6.1). The muscle blocks are called **myotomes** or **myomeres**. The myomeres have short upper and long ventral limbs. The apices of the myotomes point forward. There are about 50 to 75 myotomes on each side and the myotomes of left side alternate with those of right side. Each myotome is enclosed in a complete envelope of connective tissue. The anterior and posterior walls of the envelope are called the **myosepta** or **myocommata**. The muscle fibers run longitudinally parallel to the long axis of the body and each end is inserted in the myocommata. Circular muscles are absent.

Box 6.1

The body wall muscles of *Branchiostoma* are distinctly vertebrate-like and occur as cheveron-shaped blocks called myotomes which are arranged longitudinally along the dorsolateral aspects of the body (Fig. 6.1). Chevron is the bent-line or stripe (V or Λ) worn by soldiers, policemen, etc., to show rank.

The myotomes help in lateral undulating movements of body as they can be twisted sideways rapidly. The **transverse muscles** running across the ventral surface of trunk between two metapleural folds are thin, unsegmented and help in drawing water of atrial cavity out of atriopore.

The amphioxus for the first time in the chordates shows a clearcut body segmentation and the consequent ability to swim by lateral body flexion (bending).

3. Peritoneum. The peritoneum covers the muscles internally. It consists of a layer of thin mesodermal cells resting on a basement membrane. In the pharyngeal region, the peritoneum is restricted to certain small tracts.

Functions. The body wall is protective, secretory, sensory, locomotory and supportive in function. Protection is provided by mucus, sensitivity is afforded by the sensory cells of the epidermis, locomotion is brought about by the myotomes and support is provided by the various connective tissue sheets. The peritoneum secretes the coelomic fluid.

6.4 SKELETON

There is no exoskeleton in *Branchiostoma*. The bony skeleton as found in vertebrates is also absent in amphioxus. The only supporting skeletal elements found are notochord, dense fibrous connective tissue (of oral hood, etc.), rods and poles of pharynx and finray boxes.

1. Notochord. It is the main skeletal structure of *Branchiostoma*. Notochord has the form of a cylindrical rod that extends the whole length of the body along the middorsal line between the nerve cord above and the alimentary canal below (Fig. 6.5). The name cephalochordata, meaning "head chordates" refers to the unusal anterior extension of the notochord. Unlike the vertebral column and its articulating bones, however, the notochord is an elastic, flexible rod. It prevents body from shortening when the muscles contract, causing lateral bending instead. Its elasticity tends to straighten the body and thus assist the antagonistic action of paired myotomes. The notochord extends beyond the myotomes both anteriorly and posteriorly, providing support beyond those muscles and apparently helping in holding the body rigid during burrowing.

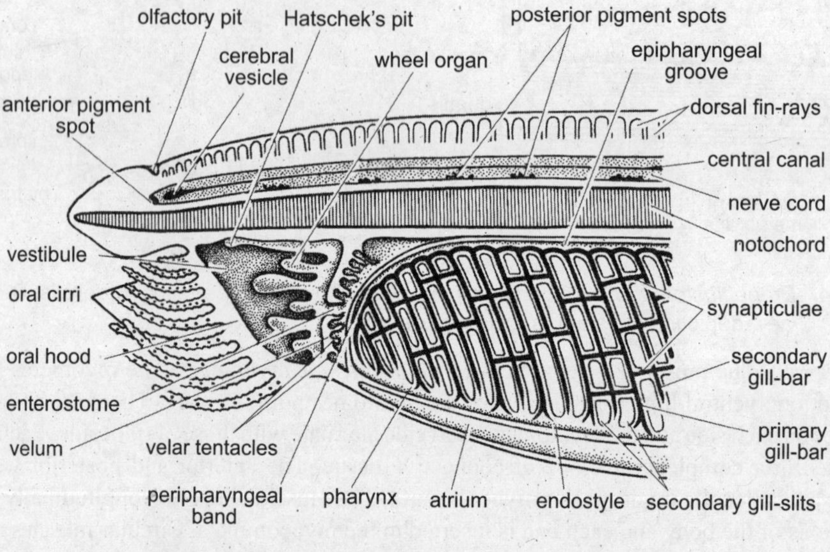

Fig. 6.5. *Branchiostoma*. L.S. of anterior part.

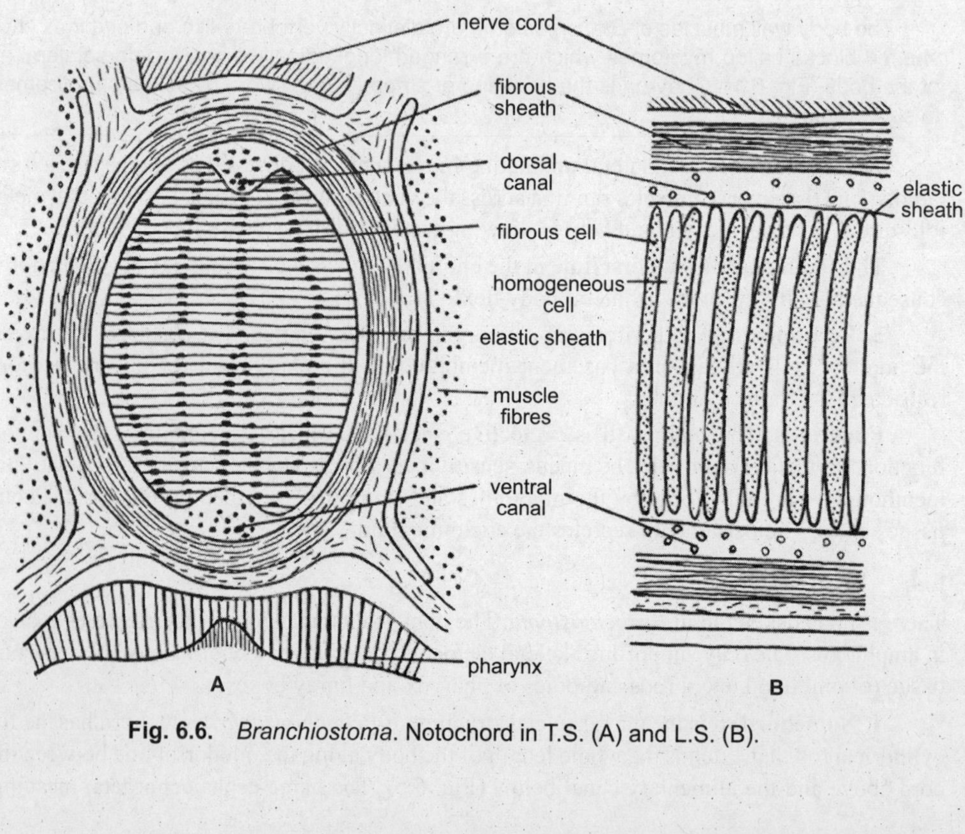

Fig. 6.6. *Branchiostoma*. Notochord in T.S. (A) and L.S. (B).

Although the notochord of cephalochordates is homologous with the same structure in other chordates, including vertebrates, it displays some unique and rather remarkable structural and functional characteristics associated with its persistence in the adult. It is not a homogeneous structure of predominantly cartilage-like matrix material. Rather, the notochord of lancelets is built of discoidal **lamellae** that are stacked like so many pocker chips along its length and surrounded by a sheath of collagenous connective tissue (**Brusca** and **Brusca** 2003). The lamellae are composed of muscle cells whose fibers are oriented transversely. Furthermore, a significant amount of extracellular fluid exists in spaces and channels around and between the lamellae within the collagenous sheath. These muscle cells are innervated by motor neurons from the dorsal nerve cord. Upon contraction, the hydrostatic pressure in the extracellular spaces increases, thereby resulting in increased stiffness of the whole notochord complex.

Muscles are not attached to the notochord. Also the viscera of amphioxus does not hang from the notochord. In this way, the notochord does not serve as a support for the myotomes and viscera.

Locomotion. The notochord also plays a major role in locomotion in lancelets. As a result of the action of segmental myotomes, the swimming action of cephalochordates is much like that of fishes, consisting basically of lateral undulations that drive water posteriorly and provide a forward thrust. The propulsive action of these body movements is enhanced by a vertical caudal fin.

2. Dense fibrous connective tissue. This tissue provides firmness to the body by filling up spaces between its organs. It is continuous throughout the body. It forms a layer inside the epidermis and outside the parietal peritoneum, surrounds the notochord and covers the nerve cord. This tissue also encloses the myotomes, whose fibers are inserted into it. It also surrounds the fin-ray boxes.

3. Rods and plates. Rods support the gill bars of the pharynx, oral cirri and free anterior edge of the oral hood (Fig. 6.5). The **plates** support the floor of the endostyle. The rods and plates are made up of agglutinated elastic fibers and are firm though flexible.

4. Fin ray boxes. The fin ray boxes support the dorsal and ventral fins. There is a single row of such boxes in the dorsal fin, but two rows (right and left) in the ventral fin. These boxes are blocks of gelatinous matter wrapped by connective tissue. The connective tissue septa betwen fin ray boxes form the **fin rays**. These fin ray boxes are not metamerically arranged, there being 4 to 5 per segment. Dorsal fin contains more than 200 fin ray boxes. No muscles are attached to them.

In fact, the dorsal and ventral fins are the dorsal and ventral **storage organs**. They are not homologous to the fins of fishes and their function appears to be housing a build-up of nutritional reserves for gamete formation (**Holland** and **Holland** 1990, 1991).

6.5 COELOM

Branchiostoma possesses a true coelom lined by somatic or parietal peritoneum and splanchnic or visceral peritoneum of mesodermal origin, and filled with coelomic fluid. Coelom exists in the adult as many components.

(*i*) In the intestinal region, coelom occurs as spacious **perivisceral cavity** or **splanchnocoel** (Fig. 6.7). However, it occurs in its typical state only in a small portion in front of the anus. At this place, coelom uniformly surrounds the intestine, except on the dorsal side, where the dorsal mesentery suspends the intestine into the coelom. In front of this portion upto the atriopore, the coelom is partly obliterated by atrium.

(*ii*) In the pharyngeal region, the coelom is greatly reduced in the adult stage. It is represented by a pair of longitudinal **dorsal pharyngeal coelomic canals** above the pharynx; a longitudinal **midventral coelomic canals** or **subendostylar canal**, beneath the pharynx and a double series of

vertical coelomic canals in the primary gill bars. Such a peculiar coelomic canal system in pharynx is not found in higher chordates.

(*iii*) Small closed coelomic spaces occur around the midgut diverticulum, in the gonads as **gonocoels**, in myomeres as **myocoels**.

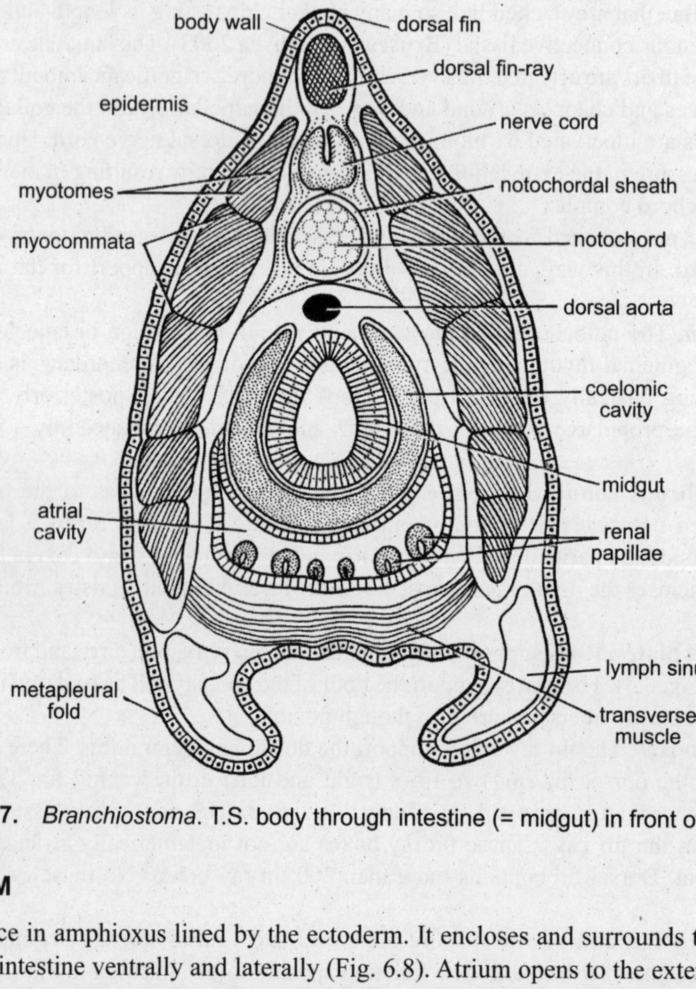

Fig. 6.7. *Branchiostoma*. T.S. body through intestine (= midgut) in front of anus.

6.6 ATRIUM

Atrium is a space in amphioxus lined by the ectoderm. It encloses and surrounds the pharynx and anterior part of intestine ventrally and laterally (Fig. 6.8). Atrium opens to the exterior at about the level of 36th myotome by atriopore. The atriopore is a mid-ventral aperture situated immediately in front of caudal fin. The atrium extends far behind the 36th myotome in the form of backwardly directed blind diverticulum as far as anus is situated on the right side of the intestine. The dorsal wall of the atrium is folded and reflected on each of the primary bars.

Development of atrium. In the larva of *Branchiostoma* atrium develops at the cost of coelom. It begins in the form of a pair of longitudinal ridges the **metapleural folds**, on the ventral side of the body behind the gill slits. These gradually extend forward dorsal to gill slits. On the inner side each fold gives out a subatrial ridge. The ridges of two sides unite forming a canal, which marks the beginning of atrium. It grows on each side to attain the full size.

Function. The atrium receives the mature gametes and expels them out through the atriopore. Similarly, the water current which enters the pharynx passes out through gill slits into the atrium and finally leaves through the atriopore.

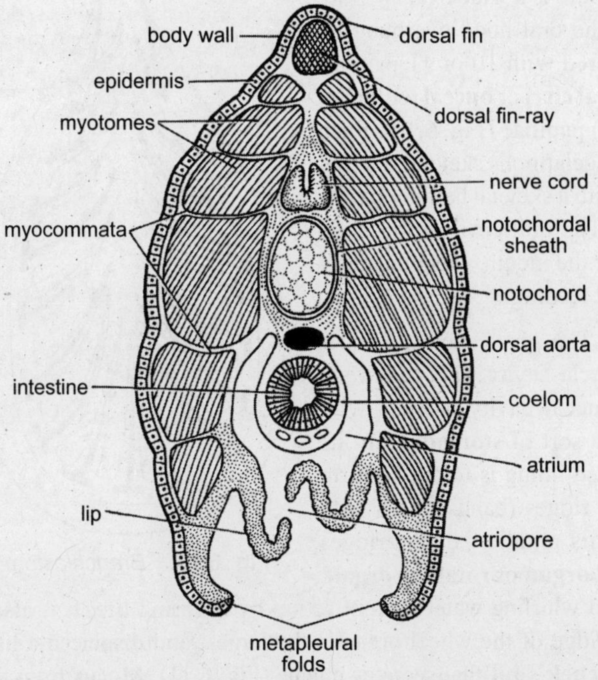

Fig. 6.8. *Branchiostoma*. T.S. body through atriopore.

6.7 DIGESTIVE SYSTEM

The digestive system of *Branchiostoma* consists of alimentary canal and digestive glands.

A. Alimentary canal. The alimentary canal of *Branchiostoma* is complete. It is a straight tube having following seven regions: mouth, buccal cavity, enterostome, pharynx, oesophagus, intestine and anus (Fig. 6.9).

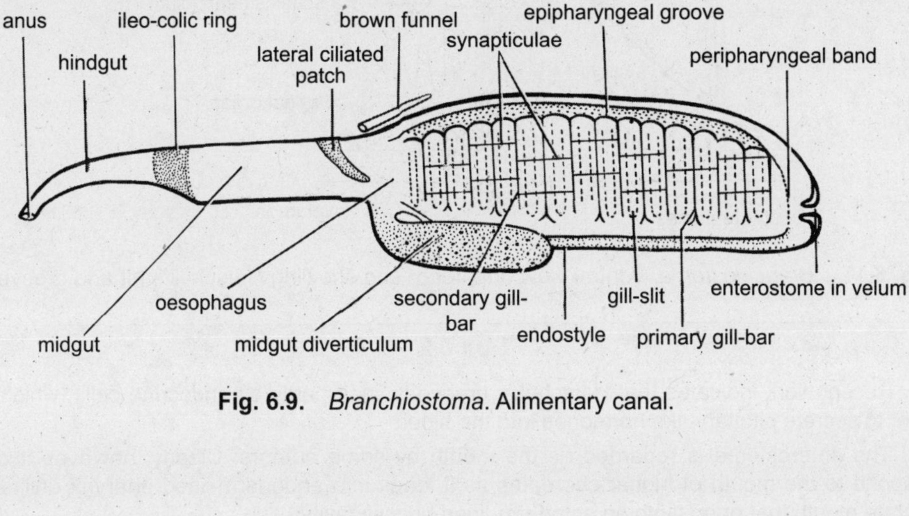

Fig. 6.9. *Branchiostoma*. Alimentary canal.

1. Mouth. The mouth is a wide oval aperture at the anterior end of the oral hood overhung by the rostrum. It is bordered with 10 or 11 pairs of slender processes, the **oral cirri** or **buccal tentacles**, bearing tactile (sensory) papillae (Fig. 6.10). Each cirrus is supported by a gelatinous skeletal rod that articulates at the base with a skeletal half hoop (*i.e.*, a rigid circular band). The oral cirri of oral hood form a coarse sieve over the mouth, checking entry of large particles during feeding.

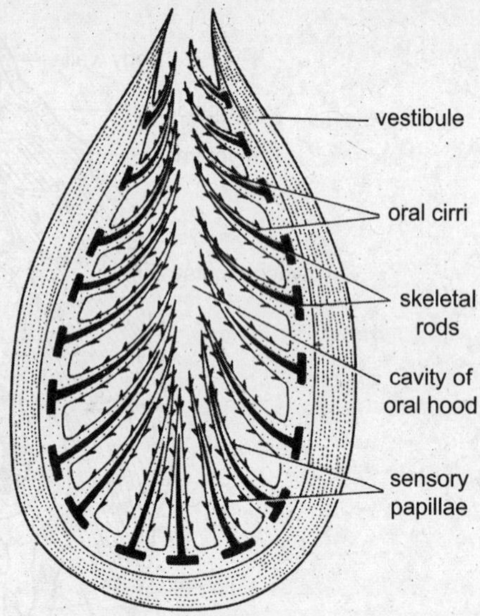

Fig. 6.10. *Branchiostoma*. Oral hood.

2. Buccal cavity. The space enclosed under the oral hood is the buccal cavity, also called the vestibule (Fig. 6.5). Buccal cavity is lined with ectoderm, constituting a sort of **stomodaeum**. Its brown, thickened, ciliated lining is folded to form a number of finger-like ridges (bands), each with a groove (tract) along its middle. All the ridges together form the **wheel organ** or **rotatory organ**, because of its shape and whirling water current set up by it during life. It is also called **Muller's organ**. The middorsal ridge of the wheel organ is the largest and displaced a little to the left. Its groove, called the **Hatschek's pit** that secretes mucus (Fig. 6.11). Mucus from this pit flows over the wheel organ. A thin, circular vertical partition, the **velum**, bounds a circular aperture, called the **enterostome**, that leads into the pharynx (Fig. 6.12 and Box 6.2).

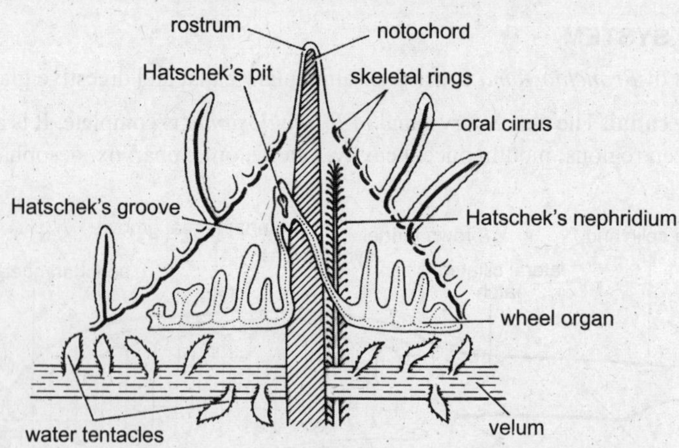

Fig. 6.11. *Branchiostoma*. Ventral view of anterior end showing Hatschek's pit and groove.

Box 6.2

1. Recent work indicates that Hatschek's groove is lined partly by endocrine cells, which are thought to secrete pituitary-like hormones into the blood.

2. The enterostome is regarded as the mouth by some authors. Clearly, however, it does correspond to the mouth of higher chordates as it leads into endoderm-lined pharynx unlike the vertebrate mouth that opens into an ectoderm-lined buccal cavity.

3. Enterostome. The enterostome is regulated by a sphincter in the velum. It is fringed with 12 slender, ciliated processes, the **velar tentacles**. The latter arise from the posterior wall of the velum and project backward into the pharynx. The velar tentacles, like the oral cirri, bear sensory papillae and act as a sieve during feeding.

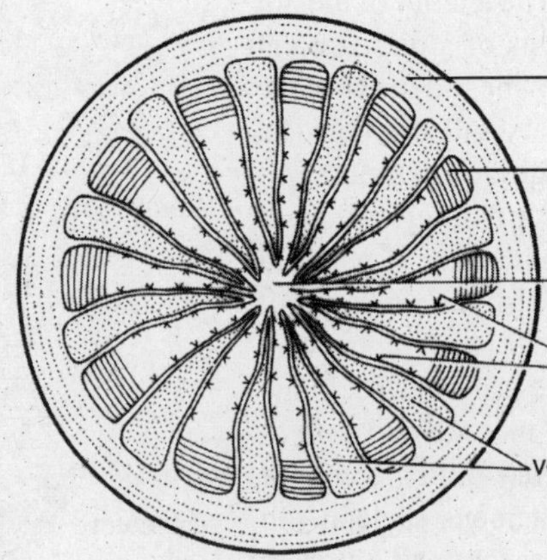

Fig. 6.12. *Branchiostoma*. Velum.

4. Pharynx. Pharynx is the largest part of the alimentary canal and occupies nearly half of the body. This laterally compressed tube has a complicated structure. The pharynx has two very unequal regions: anterior very small prebranchial region and a posterior large branchial region.

***(i)* Branchial region.** Lateral walls of pharynx are perforated by 150 to 200 pairs of narrow and obliquely set (*i.e.*, sloping antero-posteriorly) apertures called **gill clefts** or **gill slits** or **branchial apertures** (Fig. 6.5). Their number increases with the age of the animal. The pharyngeal cavity opens into the surrounding atrial cavity through the gill slits. Parts of pharyngeal wall between gill slits are called **branchial lamellae** or **gill bars**. Gill bars are supported by cartilaginous rods. The gill bars are of two types: **primary** and **secondary**. Both of these alternate with each other and differ in structure and mode of development. The primary gill bars lie in between the two gill slits while the secondary gill bar develops as a tongue bar from the dorsal wall of the gill slit. It grows downwards and divides the gill slit (*i.e.*, the primary gill slit) into two gill slits, now called **secondary gill slits** (Fig. 6.13).

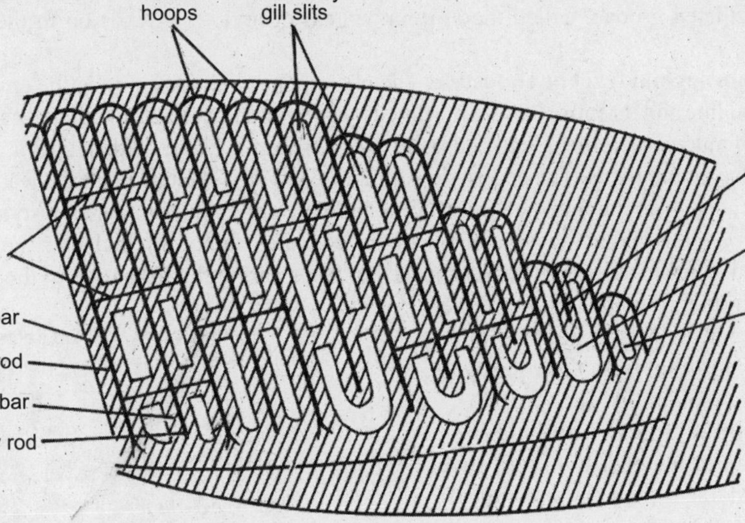

Fig. 6.13. *Branchiostoma*. A part of pharyngeal wall showing formation of gill slits.

The outer or the atrial surface of each gill bar is covered with atrial epithelium (*i.e.*, the ectoderm) and the inner or pharyngeal side and anterior and posterior surfaces of gill bars are covered with endoderm. Entire gill bar bears cilia but the ciliation is not uniform. The inner or the pharyngeal surface bears specially large cilia called **frontal** or **pharyngeal cilia**. The lateral or the anterior and the anterior and posterior surfaces bear long and dense **lateral cilia**. The outer or the atrial surface bears few cilia. The interior of the gill bars consists of a mesodermal core. The mesodermal core of a primary gill bar consists of cartilaginous skeletal rod called the **primary gill rod**, fibrous connective tissue and blood vessels. The primary gill rod is forked at its ventral end and encloses a narrow coelomic canal. The mesodermal core of the secondary gill bars consist of a cartilaginous skeletal rod called the **secondary gill rods** and its ventral end is not forked. All the gill bars are united dorsally by half loops (or half hoops), but are free ventrally (Fig. 6.13).

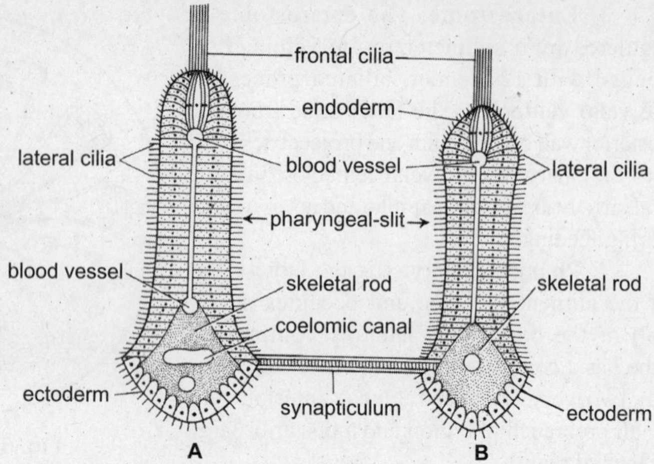

Fig. 6.14. *Branchiostoma*. T.S. of gill bars. A—Primary gill bar and B—Secondary gill bar.

The primary and secondary gill bars are interconnected by many short, horizontal **transverse bars** or the **synapticulae** and the pharynx looks like a basket. Each synapticulum contains a skeletal rod and a blood vessel.

Gill slits correspond in number with the myotomes in the larva, but outnumber tnem in the adult. Thus, they are fundamentally metameric.

Branchial region of pharynx also consists of following three types of ciliary grooves:

1. Epipharyngeal groove. The mid-dorsal region of the entire roof of the pharynx bears a prominent ciliated groove, called the **epipharyngeal groove**. It extends up to the oesophageal opening.

2. Hypobranchial tract or endostyle. The pharynx has a shallow gutter, the **endostyle**, along the mid-ventral line and its entire length. It is lined by 5 longitudinal tracts of ciliated cells alternating with 4 tracts of mucus-secreting gland cells (Fig. 6.15). The cilia of the median tract are much longer than those of the lateral tracts. The endostyle is supported by two skeletal plates beneath which is the subendostylar coelom containing the ventral aorta. It is noteworthy that similar endostyle occurs in the tunicates (*Herdmania*) and in the larva of agnathous craniates (lamprey). In the larva of the lamprey, the endostyle disappears during metamorphosis, but takes part in the formation of the thyroid gland of the adult. Further, like the thyroid of the craniates, the endostyle concentrates iodine in itself and the extract from the endostyle stimulates the action of thyroid hormore.

3. Peripharyngeal bands. From the anterior end of the endostyle start two narrow, ciliated tracts, the **peripharyngeal bands**, that pass upward and backward along the lateral pharyngeal walls to meet the **epipharyngeal groove**.

***(ii)* Prebranchial region.** The small region of the pharynx in front of the peripharyngeal band lacks gill slits and is known as the **prebranchial region**.

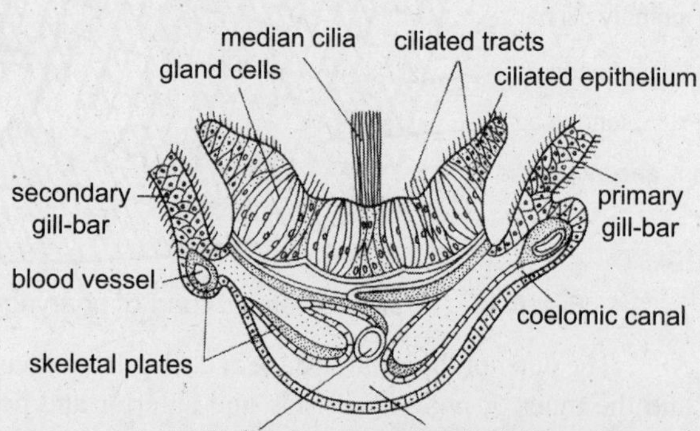

Fig. 6.15. *Branchiostoma*. T.S. of endostyle.

5. Oesophagus. It follows the pharynx. Oesophagus is a short, narrow, ciliated tube and leads into the intestine.

6. Intestine. The intestine is a tubular organ about as long as the pharynx. It shows three regions: anterior wide **midgut**, middle short **ileocolonic ring** and a posterior tapering **hind gut**. The midgut has a lateral ciliated tract on its right wall. The cilia of this tract beat downward toward a groove that starts just within the midgut diverticulum or hepatic caecum. The groove is lined with a tract of cilia that beat forward. The ileocolonic ring is heavily ciliated. The cilia beat obliquely backward and downward on the left side and forward and upward on the right side. This ciliary action serves to rotate the food cord on its longitudinal axis. A small terminal part of the hindgut is heavily ciliated and is called **rectum**. The latter opens out by anus.

7. Anus. The anus is a small, circular aperture at the base of the caudal fin on the ventral side, but a little to the left of the median line. The anal opening is controlled by a sphincter muscle.

B. Digestive glands. *Branchiostoma* has following two types of digestive glands.

1. Epithelial gland cells. These gland cells are scattered in the epithelium of whole intestine and secrete enzymes.

2. Midgut diverticulum or digestive caecum or hepatic caecum. It is a large, hollow, laterally compressed organ. It arises from the ventral side between the pharynx and oesophagus and extends forward through the atrial cavity along the right side of pharynx, ending blindly infront. The caecum has along its roof and floor ciliated grooves. The cilia of dorsal groove beat forward and those of ventral groove backward. The lateral walls of caecum are also ciliated but ciliary beat is weak. The secretion of the hepatic caecum contains digestive enzymes such as lipase, amylase and protease and it is driven into midgut along the floor by ciliary action.

Recent studies indicate that the hepatic caecum also functions in lipid and glycogen storage and protein synthesis. It is regarded as the evolutionary precursor of the vertebrate liver and perhaps the pancreas (**Brusca** and **Brusca** 2003).

Physiology. 1. Food and feeding. *Branchiostoma* is ciliary-mucus suspension feeder. It feeds on microscopic organisms such as diatoms, desmids, small protozoans, found in the plankton. These organisms are filtered from a current of sea water passed through the pharynx. This mode of capturing food is also called **filter-feeding**.

2. Creation of feeding current. At the time of feeding, amphioxus stays quietly in the burrow, keeping the anterior end (oral hood) projecting out. Water current (*i.e.*, feeding current) is set up mainly by the outward beating of the powerful lateral cilia of the gill bars with a metachronal rhythm (Box 6.3). Frontal cilia of gill bars and cilia of epithelium covering gonads too help in creating the feeding current. The water current enters the pharynx via mouth, buccal cavity and enterostome and leaves via gill slits to reach atrium. From atrium this water current is passed out through the atriopore.

Box 6.3

1. Cephalochordates employ a food-gathering mechanism similar to that of tunicates.
2. Unlike the gill ventilation current of aquatic vertebrates that are generated by muscular action, the feeding currents of lancelets are driven by pharyngeal cilia, a condition similar to tunicates.
3. The gill slits of cephalochordates are committed largely to feeding and have little to do with gas exchange (*i.e.*, respiration).

3. Filtering. Food of the feeding current is passed through two filters. The oral cirri bend over the mouth and cross one another, forming a sort of sieve that allows only extremely fine particles to filter through the meshes. The water current is subjected to further filtering while passing through the enterostome. Here, the velar tentacles, like the oral cirri, form a sieve over the aperture.

4. Entrapping and formation of mucus cord. In the pharynx, the glandular tracts of the endostyle secrete abundant mucus. The mucus is swept on to the ventrolateral walls of the pharynx by the lateral tracts of cilia, which beats outward from the midline. The long cilia of the median tract supply mucus to the lateral tracts of cilia. From the ventrolateral walls of the pharynx, the mucus is sent up the gill bars by their frontal cilia (which beat upward and forward). Epithelium of the pharynx also secretes mucus, which is also wafted upward. Food particles brought by the water current get entrapped in the mucus streams slowly moving upward. The food-laden mucus streams finally enter the epipharyngeal groove (Fig. 6.16). The cilia of the epipharyngeal groove beat backward and push the food-laden mucus into the oesophagus in the form of a cord, called **food cord**.

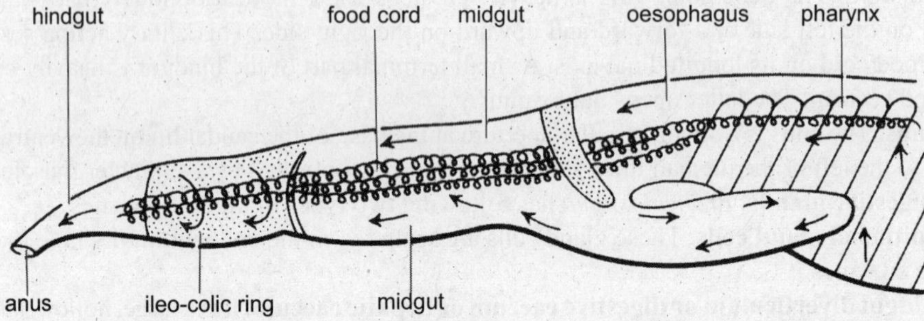

Fig. 6.16. *Branchiostoma*. Path of food current in the post-pharyngeal region.

5. Digestion. The food cord is carried back through the oesophagus and midgut by ciliary action. In the ileocolonic ring, the food cord is rotated by the strong ciliation of this region. Rotation of food cord serves three important functions: (*i*) it delays its passage to allow time for digestion in the short intestine; (*ii*) it brings about better contact between the food and the digestive secretions and (*iii*) it causes fragmentation so that small masses of food pass back.

Digestive secretion of hepatic caecum and epithelial gland cells contain amylase, lipase and protease enzymes which act on starch, fat (lipid) and protein contents of food respectively. The digestion occurring in the intestinal lumen is called **intercellular** or **extracellular digestion**. In *Branchiostoma*, intracellular digestion also occurs. Small food particles resulting from the partial extracellular digestion in the lumen of gut and caecum are engulfed by the epithelial cells of intestine or caecum for intracellular digestion by lysosomal enzymes.

6. Egestion. Absorption of the digested food takes place chiefly in the hind gut and only partly in the midgut. Movement of food masses through the hindgut also occurs by ciliary action. Elimination of faeces takes place through the anus. It is aided by an area of strong ciliation at the hind end.

7. Food storage. In addition to storage in th hepatic caecum, food reserves accumulate in longitudinally arranged dorsal and ventral storage chambers along the dorsal midline and along the ventral body margin posterior to the atriopore.

6.8 BLOOD VASCULAR SYSTEM

The blood vascular system of *Branchiostoma* is of **closed** type, well developed and peculiar. Although *Branchiostoma* does not contain a heart, its circulatory system is almost similar to that found in the vertebrates. All blood vessels have muscular walls and are contractile in nature. Only the dorsal aorta has an endothelial lining. Nearly, all the blood vessels are similar in structure but, due to their certain homologies with the blood vessels of vertebrates, some of these are called **arteries** and some **veins**. There are no true capillaries. Main vessels and their branches are as follows:

1. Sinus-venosus. The sinus-venosus is a thin-walled sac situated below the posterior end of the pharynx. It receives blood returning from various parts of the body through veins and pours it into **ventral** or **branchial aorta**.

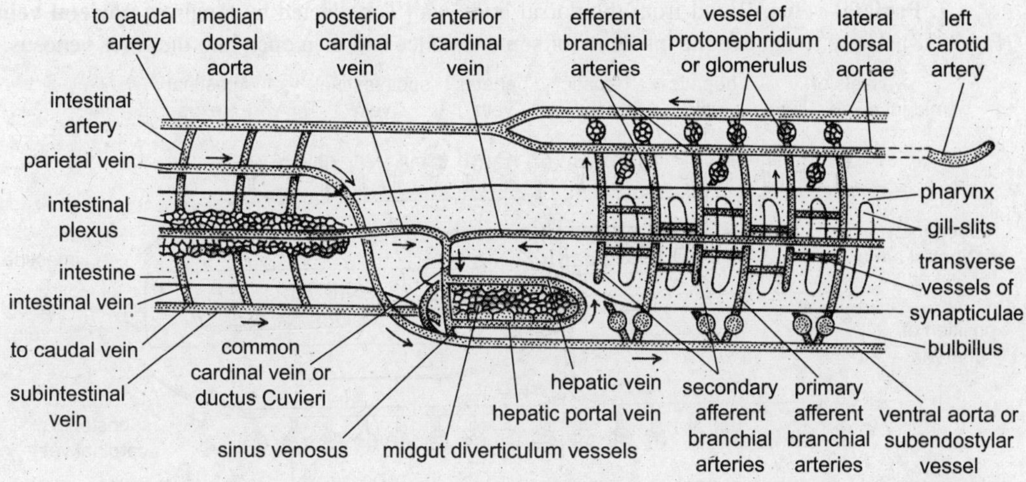

Fig. 6.17. *Branchiostoma*. Blood vascular system in lateral view.

2. Ventral aorta. From the sinus venosus runs forward a large highly contractile longitudinal blood vessel called the **ventral aorta**. It runs mid-ventrally in the wall of **pharynx**, below the endostyle and within the subendostylar coelomic canal. It pumps the blood forwards. It gives off a series of paired lateral **afferent branchial arteries**. These run through each primary gill bar of the pharynx. At the base of each primary gill bar each branchial artery has a little contractile **bulb** or **bulbillus** which functions as a branchial heart. Arteries of the primary gill bars remain connected with the vessels running through the secondary gill bars by **transverse vessels** present in the synapticulae. The bulbs pump into the gill bars.

Dorsally, a series of **efferent branchial arteries** collect blood from the primary and secondary gill bars and empty it into a pair of dorsal aortae. Before opening into the dorsal aortae, each efferent branchial artery splits to form a plexus or network of capillary-like vessels. These are called the **glomus**.

3. Dorsal aorta. The right and left dorsal aortae run on either side of the epipharyngeal groove of the pharynx. Anteriorly both dorsal aortae continue as **internal carotid arteries** and reach the snout or rostrum. These arteries supply blood to the oral hood region. The right aorta is much dilated. Posteriorly both aortae unite together just behind the pharynx and form a single **median dorsal aorta**. The aorta runs backwards upto the tail in the middle line between the notochord and intestine. In the tail region, this aorta is called the **caudal artery**. The paired aortae and median dorsal aorta supply blood to the body wall through numerous small vessels. The median dorsal aorta sends several **intestinal arteries** to the intestine. In the intestinal wall, these arteries break up to form a plexus (network) of microscopic capillary-like vessels. The blood flows backwards in the dorsal aortae.

4. Sub-intestinal vein. From the tail region the blood is collected by the **caudal vein**. It continues ahead below the intestine, and in the region of hind gut and mid gut it is called the **sub-intestinal vein**. For the greater part of its length the sub-intestinal vein forms a plexus of capillary-like vessels which receive blood from the gut wall. The blood flows forward (toward the anterior end) in it.

5. Hepatic portal system. In the region of midgut diverticulum the sub-intestinal vein is called the **hepatic portal vein**. It runs along the mid-ventral border of the midgut diverticulum and splits up into capillary-like vessels which ramify in the wall of diverticulum. In this way a forerunner of vertebrate hepatic portal system is formed in *Branchiostoma*. From the midgut diverticulum, the blood is collected by a **hepatic vein**. Hepatic vein runs along the dorsal side of the diverticulum, turns downward and joins the sinus venosus.

6. Cardinal veins. Paired **anterior** and **posterior cardinal veins** collect the blood from the muscles and body wall of the ventro-lateral regions. These run along the lateral body wall. The anterior and posterior cardinals of each side unite behind the pharynx and form a **common cardinal** or **ductus cuvieri**. The two ductus Cuvieri open into the sinus venosus.

7. Parietal veins. Blood from the dorsal body wall is collected by a pair of **parietal veins** (Fig. 6.18). These run above the intestine for some distance and then open into the sinus venosus.

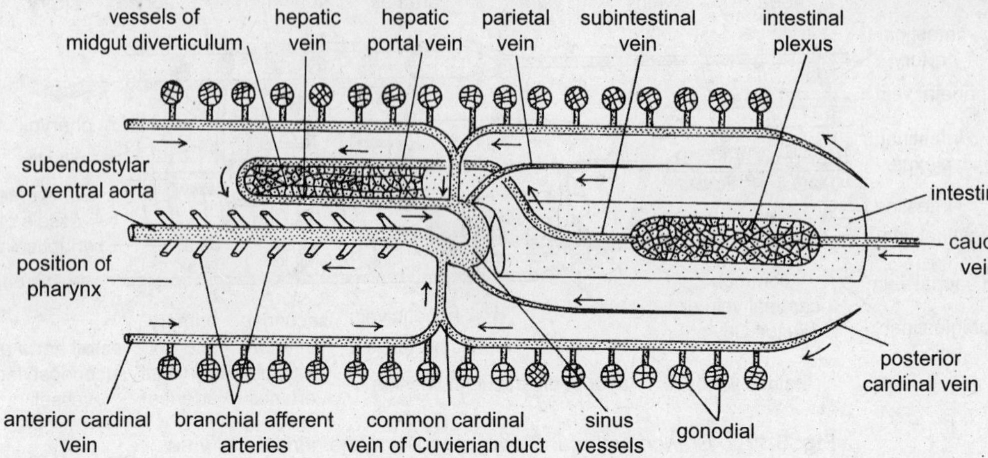

Fig. 6.18. *Branchiostoma*. Blood vascular system in dorsal view.

Blood. The blood of *Branchiostoma* is colourless and without corpuscles or cells. It does not contain any respiratory pigment. Probably the limited amount of oxygen which can be absorbed by blood is sufficient for the small energy needs of the animal which spends most of its life at rest. The main function of the blood is transportation of food and excretory products.

Course of Blood Circulation

Slow waves of contraction arise independently in the sinus venosus, branchial bulbs, sub-intestinal vein and other vessels to drive the blood forwards in the ventral vessels and backwards in the dorsal vessels. The contraction rhythms are slow (once in two minutes), irregular and coordinated by any control system. Blood pressure is low. The main function of the blood is the transport of food. It plays very little role in transport of oxygen. The course of blood circulation has been shown in Fig. 6.19.

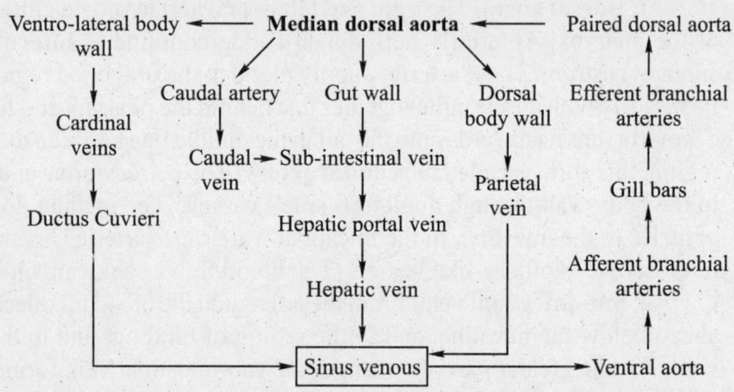

Fig. 6.19. *Branchiostoma*. Course of blood circulation.

Lymphatic system. It is represented by lymphatic spaces or sinuses present inside fins and metapleural folds. These are filled with colourless blood lacking in leucocytes. Most of the oxygen is obtained by the animal by gaseous exchange between outside water and blood in the sinuses.

Peculiarities of circulartory system. 1. Absence of heart. 2. Major blood vessels have contractile walls and the ventral vessel acts as a pumping organ. 3. The dorsal aorta is represented by paired vessels in the anterior part. 4. Beginning of hepatic portal system. 5. Replacement of ventral aorta in the posterior part of the body by a subintestinal vein. 6. Flow of blood is forward in the ventral aorta and backward in the dorsal aorta. 7. Presence of afferent and efferent branchial vessels and plexus in the intestine and liver.

Comparison of blood vascular system in *Branchiostoma* and lower vertebrates. 1. Forward flow of deoxygenated blood in the ventral aorta. 2. Backward flow of oxygenated blood in the dorsal aorta. 3. Forward flow of a part of oxygenated blood to the head region from the dorsal aortae by way of carotid arteries. 4. Passage of blood from the ventral aorta into the dorsal aortae by way of afferent and efferent branchial vessels in the gill bars. 5. Return of blood from the intestine by a hepatic portal vein to the midgut diverticulum (hepatic diverticulum). 6. Return of blood from the body wall as well as the main digestive gland to the sinus venosus. 7. Position of sinus venosus ventral to the pharynx and its correspondence to the posterior part of the vertebrate heart. 8. Continuation of dorsal aorta as a caudal artery into the tail.

Reason of absence of heart in amphioxus. Lack of heart in *Branchiostoma* is correlated with absence of kidneys. The kidneys require for their functioning a high arterial blood pressure which is developed and maintained by continuous pumping of the muscular heart. In the absence of kidneys in *Branchiostoma*, the low blood pressure developed by peristaltic contraction of the ventral aorta is sufficient for the circulation of blood round the body.

6.9 RESPIRATORY SYSTEM

Branchiostoma lacks special respiratory organs. Exchange of gases takes place by diffusion through the body surfaces in contact with the sea water, namely, epidermis, branchial epithelium and atrial epithelium. Large surface bathed in water and constantly moving current provide ample oxygen to the small body of the animal.

Pharynx plays only a minor role in respiration. This is because the blood lacks a respiratory pigment to increase its oxygen-absorbing capacity, and it passes through the gill bars in direct vessels without spreading out in capillary network. The oxygen absorbed by the epithelium of the gill bars is utilised in working of their cilia so that the blood leaving the gill bars is hardly more oxygenated than that entering them.

6.10 EXCRETORY SYSTEM

Excretory system of *Branchiostoma* shows no similarity with that of vertebrates and echinoderms. It consists of nephridia (protonephridia-without nephrostomes) which are similar to those found in several invertebrates such as platyhelminthes, molluscs and annelids (polychaetes). The nephridia of *Branchiostoma* develop from the ectoderm while the pronephros (of kidneys) of vertebrates develop from the mesoderm.

A variety of organs in *Branchiostoma* have been attributed to excretory function. These are protonephridia, Hatschek's nephridium, brown funnels and renal papillae.

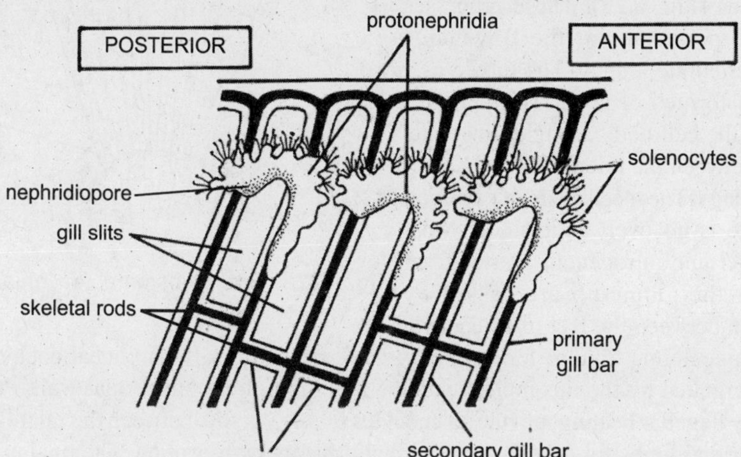

Fig. 6.20. *Branchiostoma*. Location of protonephridia.

1. Protonephridia. Amphioxus contains about 90 pairs of protonephridia. These are situated above the pharynx between the walls of atrium and dorsal coelomic canal (Fig. 6.20). These correspond in number and position with the primary gill slits and do not extend behind the region of the pharynx. Each protonephridium is a simple, sac-like, thin-walled, ciliated and bent tube (Fig. 6.21). The tube consists of a **vertical limb** and a posterior **horizontal limb**. The vertical limb lies mainly in the coelomic canal of the primary gill bar. Its ventral end terminates blindly. The horizontal limb extends upto the adjacent secondary gill bar. It opens into the atrium by a small ciliated aperture, the **nephridiopore**. The nephridiopore lies opposite the secondary gill bar. Vertical limb of protonephridium bears bunches of **flame cells** or **solenocytes** (Fig. 6.22), on projections of its anterior surface. Horizontal limb bears bunches of solenocytes on projections arising from its dorsal surface and its solenocytes extend into the dorsal pharyngeal coelomic canal. Near the nephridium, in the dorsal pharyngeal coelomic canal, lies the **glomus** or **glomerulus**, formed in the course of efferent branchial artery.

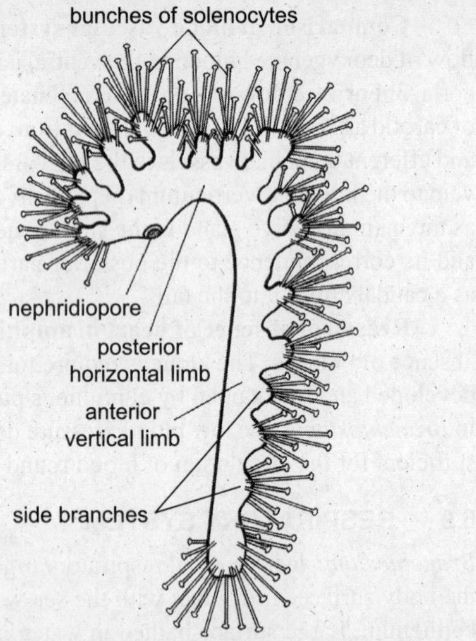

Fig. 6.21. *Branchiostoma*. An entire protonephridium.

Solenocyte. The entire protonephridium carries about 500 solenocytes. A solenocyte (Fig. 6.22) consists of subrounded **cell body** and a long fine **tubule**, about 50 μm long. From a tiny **basal granule** in cell body arises a vibratile **flagellum** that extends through the tubule and proejcts freely into the lumen of the nephridium. The cell body of a solenocyte has many foot-like processes which are applied to the glomerulus (Fig. 6.23). These processes resemble the podocytes of the Bowman's capsule of vertebrate kidney. The tubule of a solenocyte is formed of about 10 microvilli arising from the cell body. Longitudinal slits are present between the microvilli.

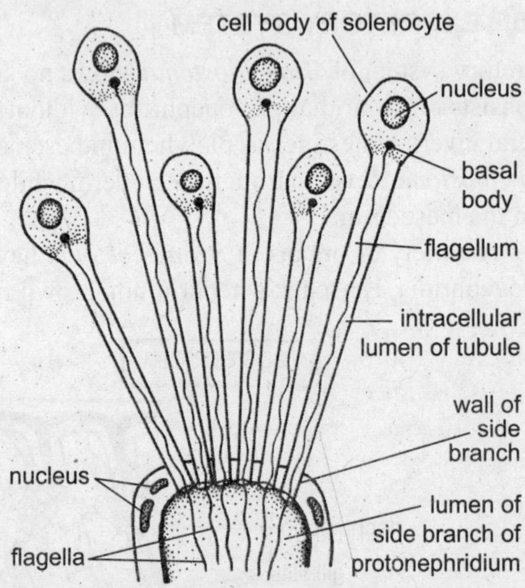

Fig. 6.22. *Branchiostoma*. An enlarged view of solenocytes.

Physiology. The mechanism of excretion by solenocyte is not well understood. Most probably water and nitrogenous wastes from the blood of the glomeruli are passed by ultrafilteration between the feet of solenocyte body into the coelomic fluid to form primary urine in the dorsal pharyngeal coelomic canal. The process is facilitated by the incomplete endothelial lining of the glomerular wall. Primary urine is then drawn by flagellar beating into the solenocytes tubule via slits between the microvilli and thence into the nephridial limbs for modification to final urine by reabsorption. The final urine is sent into the atrium. From the atrium, the wastes escape with the water current leaving the atriopore.

Branchiostoma is probably an ammonotelic animal. Uric acid, an end-product of purine metabolism, is reported to be stored in the tissues associated with the gonads. The nephridia also serve as tissues associated with the gonads. The nephridia also serve as the organ of osmoregulation.

Box 6.4

Despite the structural similarities, the homology of lancelet protonephridia with those of other invertebrates is uncertain and some specialists regard this as a case of **convergent evolution** (**Brusca** and **Brusca** 2003).

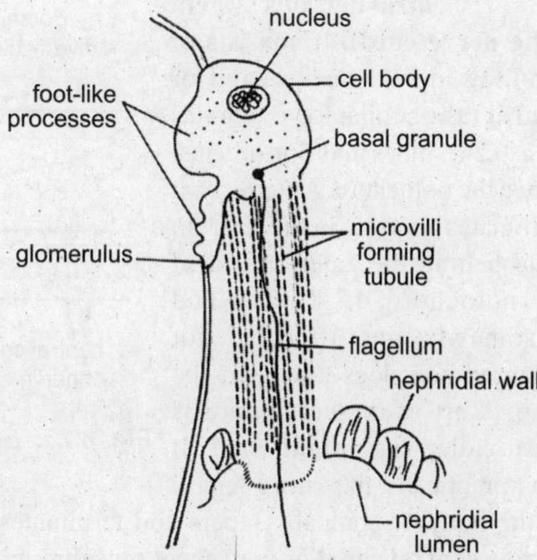

Fig. 6.23. *Branchiostoma*. Electron microscopic view of a solenocyte.

2. Hatschek's nephridium. It is a single large nephridium which is situated above the oral hood on the left side of the median line. This nephridium resembles the paired nephridia in all essential respects. Hatschek's nephridium is associated with the velum, which itself is believed to be a modified gill slit. This nephridium is a narrow tube and opens at its hind end into the pharynx just behind the velum and ends blindly just in front of the Hatschek's pit.

3. Brown funnels. These are a pair of outgrowths from the atrium lined with pigmented epithelium. They lie dorsal to the posterior end of the pharynx. They communicate with the dorsal coelomic canals by their narrow anterior ends and open into the atrium by their wide posterior ends.

4. Renal papillae. The floor of the atrium is beset with numerous small renal papillae which are presumed to help in excretion.

Primitiveness of Excretory Organs of *Branchiostoma*

Excretory system of *Branchiostoma* differs radically from that of higher chordates and echinoderms, but resembles with that of flat worms and polychaete annelids in having a series of paired ectodermal protonephridia. They are in no way comparable to the mesodermal pronephros of higher vertebrate kidney. There is no trace of any organ in *Branchiostoma* which could be compared with the vertebrate kidney. Therefore, the excretory organs of *Branchiostoma* are regarded to be of primitive nature.

6.11 NERVOUS SYSTEM

Nervous system of *Branchiostoma* resembles that of the vertebrates in its general plan. It is divisible into following three parts: central nervous system, peripheral nervous system and autonomic nervous system.

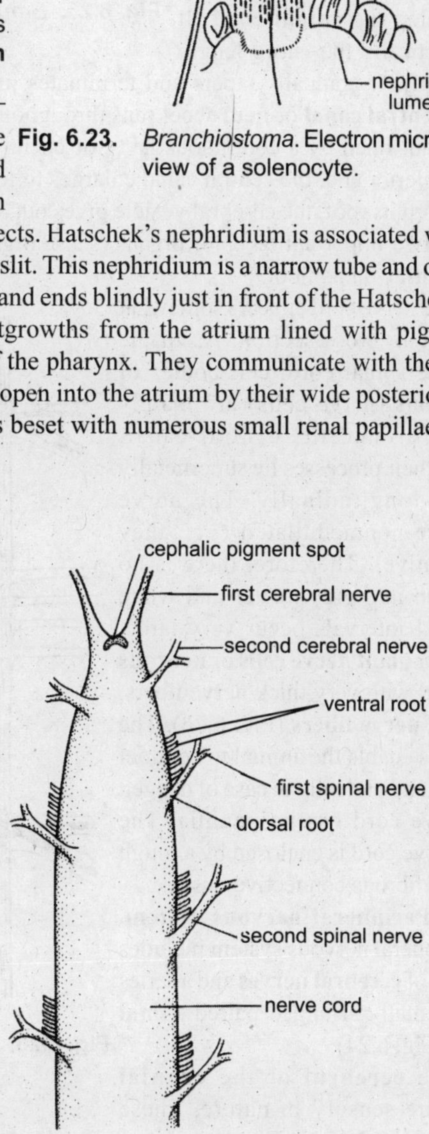

Fig. 6.24. *Branchiostoma*. Anterior part of nerve cord in dorsal view.

1. Central nervous system (The nerve cord). It consists of a hollow dorsal **nerve cord** or **neural tube** similar to vertebrates (Fig. 6.24). It lies mid-dorsally just above the notochord. Anteriorly, it terminates abruptly in the rostrum little behind the anterior end of the notochord. Its frontal end is somewhat modified but not enlarged and does not form the brain. Rest of the nerve cord is often called the **spinal cord**. It runs practically the entire length of the body. It gradually tapers and terminates just before the posterior end of the notochord. A narrow **central canal** or **neurocoel** runs throughout the length of the nerve cord (Fig. 6.25). Central canal is not lined by a continuous layer of ependymal cells. It is filled with **cerebrospinal fluid**. On the anterior end, the central canal enlarges to form the **cerebral vesicle**. It may be regarded as a brain. From its roof, the cerebral vesicle gives out a short **dorsal diverticulum**. The cerebral vesicle contains two important receptor organs (sense organs), a **pigment spot** in the anterior wall and an **infundibular** on its floor.

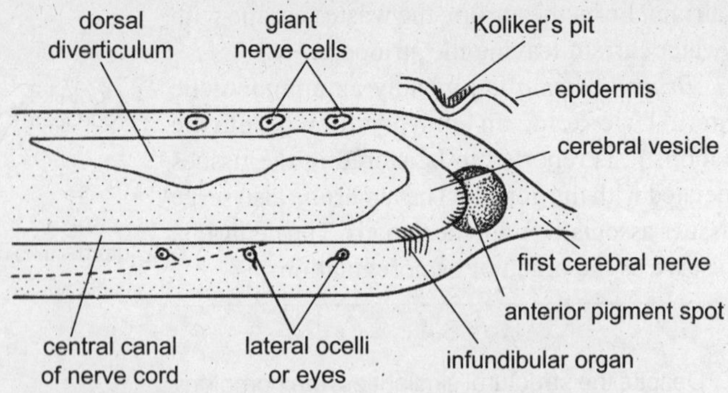

Fig. 6.25. *Branchiostoma*. V.S. anterior part of nerve cord.

The nerve cord appears somewhat triangular in cross section. As in the vertebrate spinal cord, cell bodies of the neurons (nerve cells) are mainly grouped around the central canal, whereas their processes lie superficially and run longitudinally. The nerve fibers are nonmedullated (*i.e.*, they are primitive). Therefore, there is no distinction into grey matter and white matter. At intervals occur very large, multipolar **giant nerve cells** or **neurons** connected with very thick nerve fibers, the **giant nerve fibers** (Fig. 6.26). The giant cells enable the animal to contract suddenly and violently in case of danger. The nerve cord has no ganglia. The entire nerve cord is enclosed by a tough sheath of fibrous connective tissue.

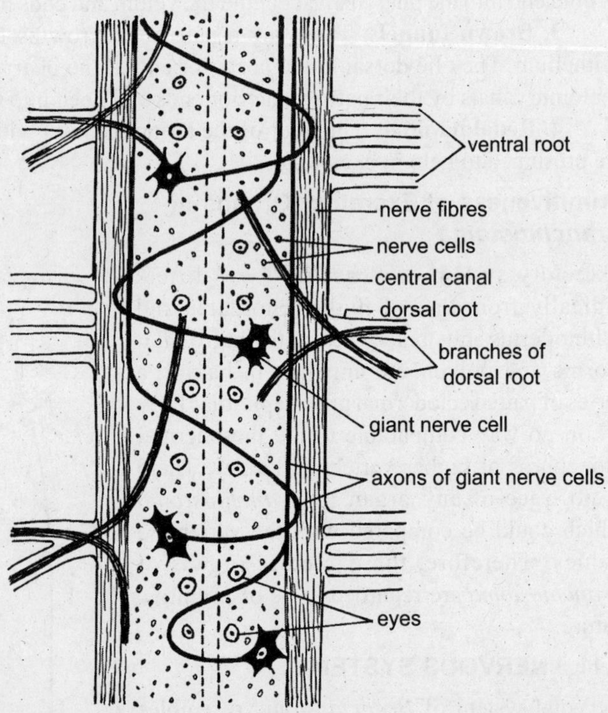

Fig. 6.26. *Branchiostoma*. L.S. of nerve cord (showing giant neurons).

2. Peripheral nervous system. The peripheral nervous system includes two pairs of **cerebral nerves** and a series of segmentally arranged paired **spinal nerves** (Fig. 6.24).

The **cerebral** or the **cranial nerves** are sensory in nature. These originate from the sense organs of rostrum, oral hood, oral cirri, Kolliker's pit and pigmented eye spot and carry sensory impulses to the cerebral vesicle. The first pair of cerebral nerves enters the anterior end and the second pair joins the cerebral vesicle on its dorsal side. These nerves lacks ventral roots.

The nerve cord behind the cerebral vesicle is called the **spinal cord**. Nerves arising from the spinal cord are called **spinal nerves** (Fig. 6.27). Each segment consists of a pair of **dorsal nerve roots** and a pair of **ventral nerve roots**. The dorsal and ventral nerve roots of a side do not unite to form a single mixed spinal nerves as happens in the vertebrates. Nerves of a pair do not originate at the same level. The dorsal nerve root of a side lies opposite to the ventral nerve root of the other side. The dorsal nerve root consists of a single nerve. It joins the spinal nerve cord a little posterior to the ventral nerve root. It is both sensory and motor (mixed type) nerve. It runs between the myotomes and divides into two branches. A dorsal **sensory branch** carries impulses from the skin and a **ventral mixed branch** innervates the muscles of the walls of the atrium and gut. The **ventral nerve root** consists of a group of nerves that are purely motor or efferent. They innervate the adjacent myotome.

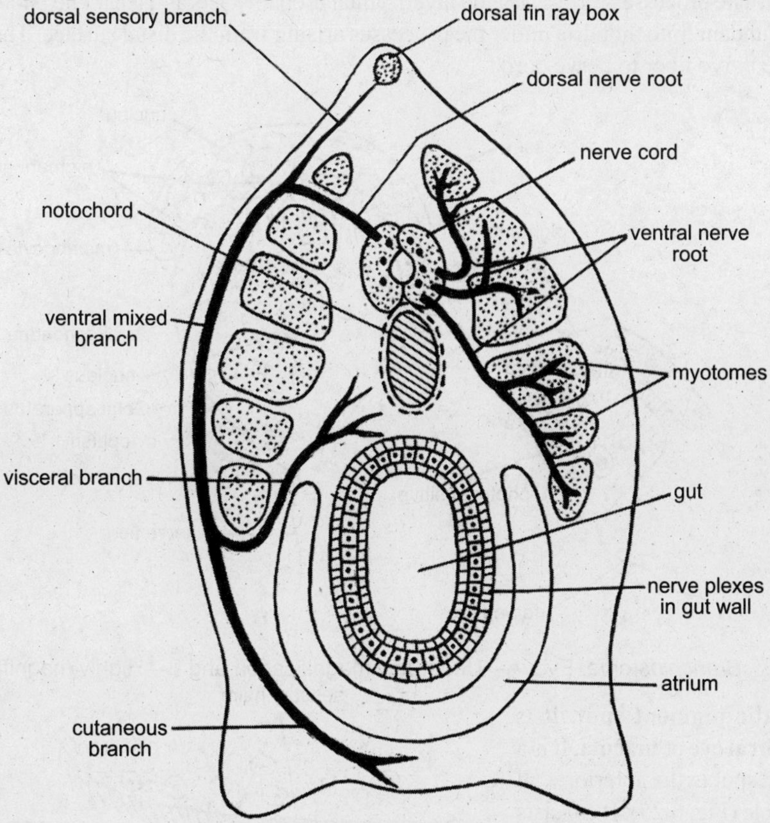

Fig. 6.27. *Branchiostoma*. Distribution of dorsal and ventral roots of spinal cord.

3. Autonomic nervous system. The unstriped muscles in the wall of gut are controlled by a simple autonomic nervous system. It consists of two **nerve plexuses** in the gut wall. These are connected to the central nervous system by **visceral nerves** which pass through the dorsal root nerves.

Peculiarities of Nervous System of *Branchiostoma*

The nervous system of *Branchiostoma* is peculiar and differ from that of the vertebrates in the following respects: 1. The brain is not differentiated. 2. The central canal is not lined by a continuous ependyma. 3. There is no distinction into grey and white matter as nerve fibers lack myelin sheath. 4. Giant nerve cells with giant fibers occur in the nerve cord. 5. Dorsal and ventral nerve roots do not join. 6. Dorsal roots are mixed in nature. 7. There are no ganglia. 8. Ventral roots arise by many nerves. 9. Asymmetrical arrangement of dorsal and ventral roots. 10. Nerve cord contains sense organs, such as cephalic pigment spot, eyes and infundibular organ.

6.12 SENSE ORGANS

Branchiostoma consists of various types of simple sense organs. These organs include eyes, cephalic pigment spot, infundibular organ, Kolliker's pit, sensory papillae, sensory cells and free nerve endings.

1. Eyes. Eyes are light sensitive receptors. They are also called **eye spots**, **ocelli** or **photoreceptors**. There are about 1500 eyes in the nerve cord behind the cerebral vesicle. An eye consists of two cells: A blackish-brown, cup-shaped **pigment cell** or **melanocyte** overlying a lens-like **photosensitive** or **receptor cell** (Fig. 6.28). The cytoplasm of the pigment cell is packed with large, membrane-bound pigment granules, each consisting of many micrograules (Fig. 6.28B). Proximal surface of the pigment cell gives off finger-like process that extend between similar processes from the receptor cell. The processes form a striated layer, which probably acts as a lens. The photosensitive cell contains abundant mitochondria under the processes arising from the distal surface. The receptor cell sends a fine nerve fiber to nerve cord.

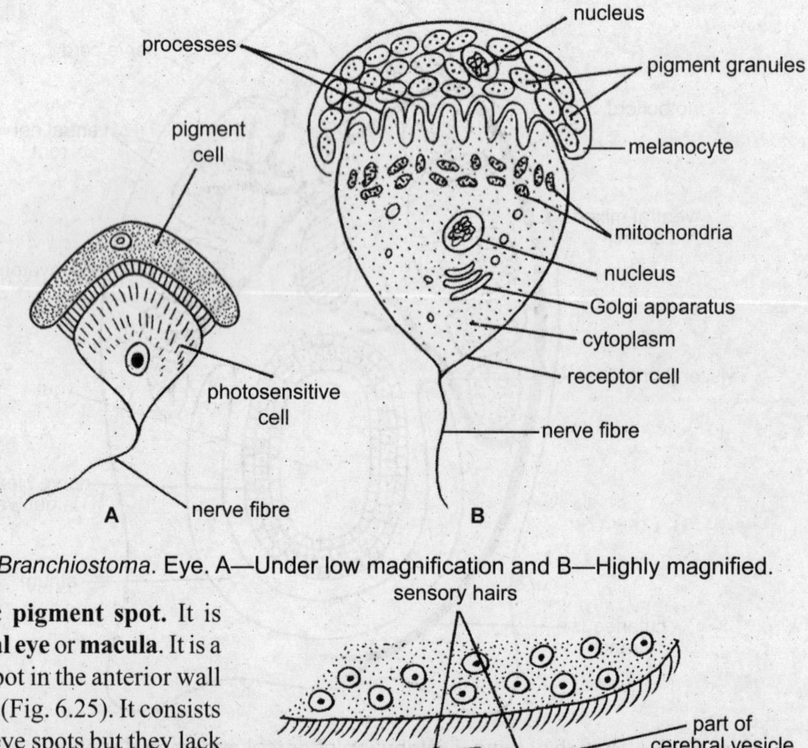

Fig. 6.28. *Branchiostoma*. Eye. A—Under low magnification and B—Highly magnified.

2. Cephalic pigment spot. It is also called **cerebral eye** or **macula**. It is a large pigmented spot in the anterior wall of cerebral vesicle (Fig. 6.25). It consists of a collection of eye spots but they lack lens and are not light sensitive. Cephalic pigment spot actually prevents light from penetrating vertically and thus, protects eye spots from direct frontal light. It probably acts as a thermoreceptor.

3. Infundibular organ. This receptor is located in the floor of cerebral vesicle. It is named infundibular organ due to its homology with the infundibulum of vertebrate brain. It is composed of tall columnar cells with long cilia (Fig. 6.29), which beat in the opposite direction to those of the rest of the vesicle. The cells

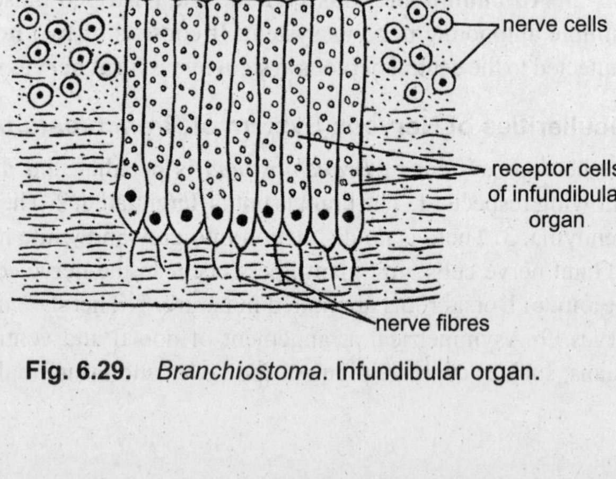

Fig. 6.29. *Branchiostoma*. Infundibular organ.

of this organ were earlier thought to detect changes in the pressure of cerebrospinal fluid in the neural tube. Later, the infundibular organ was regarded as eye stimulated by shadows cast on it by the cephalic pigment spot. Recently, these cells of this organ have been found to be secretory.

4. Kolliker's pit. It is a narrow, ciliated depression above the anterior end of the cerebral vesicle (Fig. 6.26). It opens externally on the left side of the snout. It lacks sensory cells. It was once thought to be an organ of smell. Its position, in fact, corresponds with the single nostril of cyclostomes and due to this reason it is sometimes called an **olfactory pit**. It marks the position of neuropore through which the cerebral vesicle opens to outside in the larval condition. Some workers consider it a chemoreceptor.

5. Sensory papillae. These are small groups of sensory cells on the velar tentacles and oral cirri. Those on the velar tentacles are tactile as well as chemosensory, whereas those on the oral cirri are simply tactile.

6. Sensory cells. These are found scattered in the entire epidermis of the body. These are particularly abundant on dorsal surface of body. A sensory cell is somewhat spindle-like in form and bears at its free end a short, hair-like sensory process (Fig. 6.30). A fine nerve fiber arises from the other end of the cell and passes through one of the dorsal roots to the nerve cord. Some of these cells test the nature of the sand in which the animal burrows. *Branchiostoma* is known to avoid sands that are too fine.

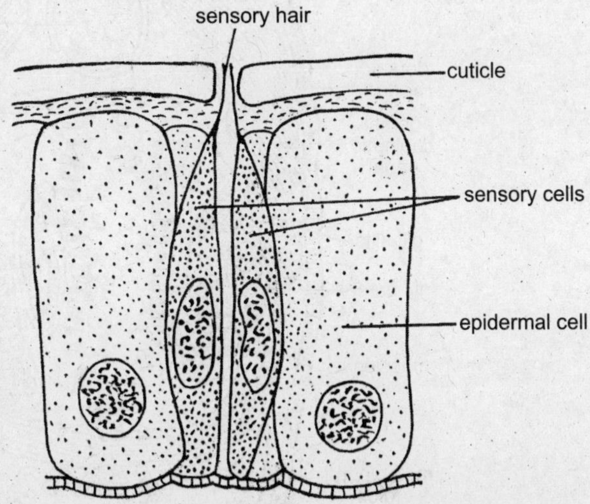

Fig. 6.30. *Branchiostoma*. Sensory cells.

7. Free nerve endings. In addition to the above mentioned exteroreceptor sense organs (which are sensitive to external stimuli) there are some **enteroreceptors** also. These are in the form of sensory nerve endings in the muscles. These are sensitive to internal stimuli such as contraction of muscles (proprioceptors).

6.13 REPRODUCTIVE SYSTEM

Sexes are separate (dioecious) in *Branchiostoma* as in all higher chordates. There is no sexual dimorphism. Male and female individuals look alike. Reproductive system of *Branchiostoma* is primitive and segmental.

Gonads. The gonads, **testes** or **ovaries**, occur in two rows, one on either side of the body. They show a metameric arrangement, there being a pair of them in each of the segments 25 to 51. Their total number is, thus, 27 pairs. They lie in the ventrolateral body wall in the pharyngeal region and prominently bulge into atrium (Fig. 6.31). There are no gonoducts. The mature gametes, spermatozoa and ova, are shed into the atrium by bursting of the inner walls of the gonads at certain points, called **cicatrices**. From the atrium, the gametes escape through the atriopore into the sea water, where fertilisation and development takes place.

Structure of gonads. Gonads are simple hollow sacs that develop from the mesoderm or the coelomic epithelium. A mature gonad encloses a cavity, the **primary gonadial cavity**, which is almost completely surrounded by a **secondary gonadial cavity** or **gonocoel** (Fig. 6.32). The germ cells develop from the cells forming the inner wall of the primary gonadial cavity. These are separated from the atrium by the ectodermal atrial epithelium and mesodermal wall of the secondary gonadial cavity. Entire gonad is bounded externally by the body wall and internally by the atrial wall. Each gonad has a large blood space that communicates with cardinal vein.

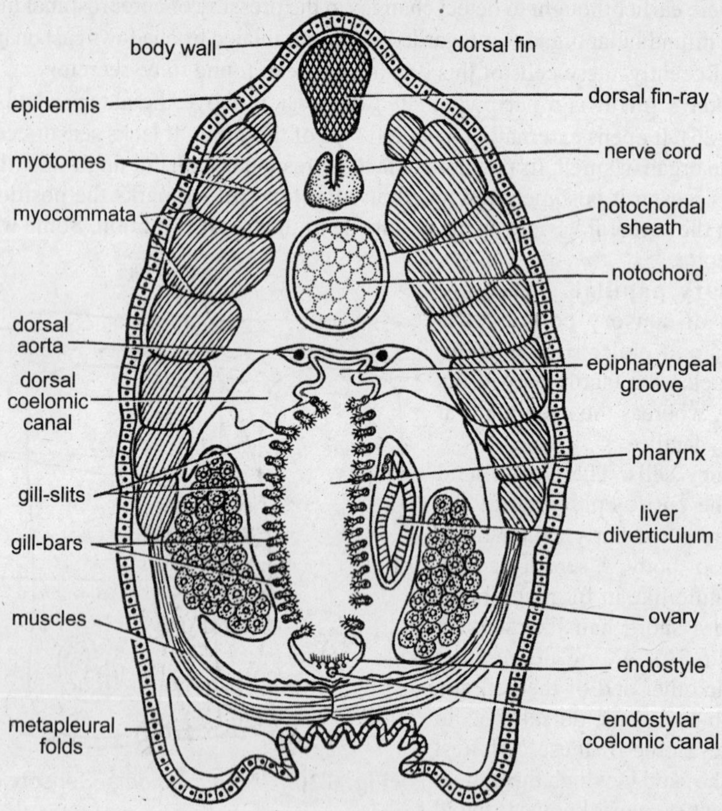

Fig. 6.31. *Branchiostoma*. T.S. body passing through posterior region of pharynx showing gonads (*i.e.*, ovaries).

Breeding. In *Branchiostoma* breeding takes place in the late spring and early summer. In the late after noons of this period, the males and females emerge from their burrows in the sand and swim about freely in the surface waters. They spawn in sunset and shed spermatozoa and ova in close proximity to ensure fertilisation.

Gametes. 1. Spermatozoa. A spermatozon of *Branchiostoma* is one of the smallest in the chordates. It is only about 18 μm long. Its head is spherical and bears a minute, bead-like acrosome in front. The middle piece is subrounded. The tail consists of a long, slender main piece of flagellum.

2. Ovum or egg. The egg (ovum) of *Branchiostoma* is spherical and about 0.12 mm in diameter. It is enclosed by a thin **vitelline envelope** which is secreted by the egg itself outside the egg's plasma membrane. The egg is **microlecithal**, *i.e.*, contains a small quantity

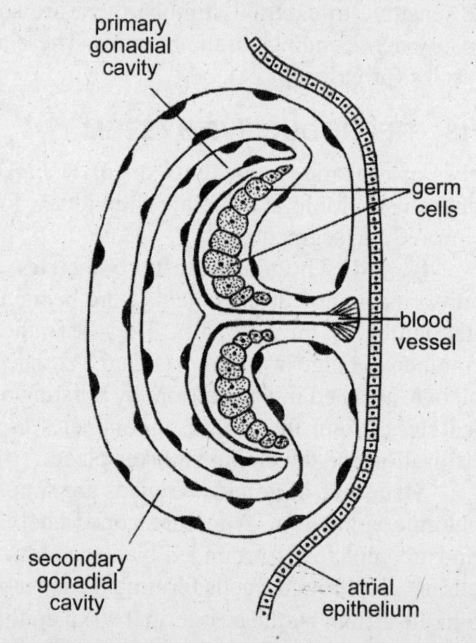

Fig. 6.32. *Branchiostoma*. V.S. mature gonad.

of yolk. The yolk is scattered unifromly throughout the ooplasm, except for a thin clear peripheral region called **egg cortex**. Thus, the egg is **homolecithal** or **isolecithal**. The germinal vesicle or nucleus lies on one side, forming **animal pole** of the egg. The opposite side of the egg is called **vegetal pole**. The egg is shed at a secondary oocyte stage and the first polar body (or polocyte) lies outside the vitelline envelope.

Development. The fertilised egg or zygote undergoes equal, holoblastic, radial cleavage to form a coeloblastula. Gastrulation takes place by invagination. Anterior coeloms, myomeres and notochord are all formed by enterocoely. Blastopore becomes the anus. A tiny, fish-like, asymmetrical, planktotrophic larva is formed. It leads a free-swimming life for about 3 months in the surface water of the sea. Then it undergoes gradual (= progressive) metamorphosis to become an adult.

6.14 PRIMITIVE, PECULIAR AND DEGENERATE CHARACTERS OF *BRANCHIOSTOMA*

Branchiostoma contains all chordate characters. It differs from the vertebrates in having relatively simple body organisation. It is regarded to be very close to the hypothetical ancestral chordate condition in body organisation. *Branchiostoma*, however, shows an interesting mixture of primitive, specialised and degenerate characters:

A. Primitive characters of *Branchiostoma*. The primitive characters are characters which appeared long ago in the ancestors but have disappeared in the existing higher members of the phylum. *Branchiostoma* shows many primitive characters as follows: 1. Absence of head or cephalisation. 2. Absence of paired fins or limbs. 3. Epidermis one cell thick, dermis absent. 4. Notochord persists throughout life. 5. Metamerically arranged myotomes from the anterior to the posterior end of the body. Vertebrates lose segmentation in the adult stage. 6. Formation of anterior coelomic pouches. 7. Lack of biting jaws. 8. Straight intestine, loops absent. 9. Liver represented by much simpler midgut diverticulum. 10. Ciliary mode of feeding. 11. Presence of endostyle. 12. Absence of heart. 13. Presence of segmentally arranged protonephridia which are not coelomoducts. 14. Segmentally arranged gonads. 15. Separate dorsal and ventral roots of the spinal nerves. 16. Absence of brain. 17. Eggs small and almost yolkless. 18. Hollow and spherical blastula (*i.e.*, coeloblastula). 19. Gastrulation by emboly.

B. Specialised or peculiar characters of *Branchiostoma*. 1. Presence of greatly enlarged and elaborate pharynx. 2. Very large number of gill slits. Their number is more than the number of metameres in which they occur. The system allows sufficient food collection. 3. Presence of well developed oral hood and velum. 4. Presence of atrium and atriopore. The coelom is displaced and reduced by the atrium. 5. Adult *Branchiostoma* has peculiar asymmetry.

C. Degenerate characters of *Branchiostoma*. These characters are present in *Branchiostoma* due to its habit and habitat. 1. The reduced brain and sense organs. This reduction or degeneration is associated with the sedentary living habit of *Branchiostoma*. 2. The forward projecting notochord may be the result of the retreat of the brain from the anterior end.

Due to these features, it is perhaps logical to consider *Branchiostoma* primitive. The specialisations indicate that *Branchiostoma* departed on a side line of vertebrate evolution.

6.15 AFFINITIES OF *Branchiostoma*

Branchiostoma is an animal of considerable phylogenetic importance. In addition to the typical chordate characters it shows many primitive and specialised characters and also some degenerate characters. It shows affinities with nonchordates, hemichordates and chordates.

A. Affinities with nonchordates. 1. The development of coelom from the archenteron (enterocoelic coelom) and asymmetries in the adult *Branchiostoma* near the echioderms. 2. Presence of metamerically arranged nephridia having flame cells and gastrulation by invagination show some similarities between the *Branchiostoma* and annelids and flatworms. However, the presence of notochord, dorsal tubular nerve cord and pharyngeal gill slits in *Branchiostoma* suggest that it is no way related to the nonchordates.

B. Affinities with hemichordates. 1. Both contain pharyngeal gill slits. 2. Both are ciliary feeders. 3. Coelom is enterocoelic. 4. Numerous gonads, no gonoducts.

Differences. 1. Muscles in the hemichordates are unsegmented. 2. Nervous system is of nonchordate type in hemichordates. 3. Notochord and tail are not found in hemichordates.

C. Affinities with Chordates.

(*a*) Affinities with urocordates. *Branchiostoma* resembles ascidians in the following features: 1. Wall of a large pharynx is performed by numerous slit-like stigmata (*i.e.*, gill slits) in the atrial cavity. 2. Presence of endostyle. 3. Ciliary mode of feeding. 4. An ectoderm lined large atrium is present. It opens to outside through the atriopore. 5. The ascidian larva contains a notochord, a dorsal hollow nerve cord and a tail with a fin, like the adult *Branchiostoma*.

Differences. 1. Urochordates are sedentary while *Branchiostoma* is capable of free swimming. 2. *Branchiostoma* lacks a test. 3. Body of urochordates do not segmented. 4. There is no coelom in urochordates. 5. Adult urochordates do not possess a notochord and nerve cord. 6. Urochordates contain a heart, this is absent in *Branchiostoma*. 7. Gonoducts are found in urochordates. These are not found in *Branchiostoma*. 8. Nephridia are present in *Branchiostoma* but absent in ascidians. 9. Alimentary canal is a straight tube in *Branchiostoma* but U-shaped in ascidians. 10. Ascidians possess power of budding and are hermaphrodite but *Branchiostoma* is unisexual with no power of budding. 11. Larva of ascidians undergo retrogressive metamorphosis, while metamorphosis in *Branchiostoma* is progressive type.

(*b*) Affinities with ammocoetes. (Ammocoetes is the larva of lamprey, an jawless cyclostome, chordates). 1. Both have a fish-shaped body with expanded caudal fin. 2. Both lack paired fins. 3. Both have a persistent notochord that extends from tail tip upto the head. 4. Both contain a longitudinal series of segmental myotomes separated by myocommata. In both, the dorsal musculature is much thicker than the ventral. 5. Both lack jaws and feed by filtering small food particles. 6. Distribution of major blood vessels and direction of blood flow is similar in both. 7. Both have an enlarged pharynx whose walls are supported by gill bars. 8. The endostyle of *Branchiostoma* and subpharyngeal gland of ammocoetes seem to be homologous. 9. Both contain a straight and uncoiled alimentary canal which opens to outside through anus at the base of caudal fin. 10. Both have similar midgut diverticulum. 11. Kolliker's pit of *Branchiostoma* seems to be homologous to the single median nostrial of ammocoetes. 12. Ventral position of mouth.

Differences. 1. *Branchiostoma* is marine; ammocoetes live in fresh water. 2. Notochord of *Branchiostoma* runs nearly the entire length of the body, whereas that of ammocoetes extend upto the level of mid-brain only. 3. Ammocoetus contains a pair of prominent eyes that develop from the brain wall. 4. *Branchiostoma* contains scattered photoreceptive cells in the nerve cord. 5. In *Branchiostoma* the pharynx is very large and has nearly 200 pairs of gill slits which open into the atrium. The pharynx of ammocetes is much shorter and contains only 7 pairs of rounded gill slits which open to the outside. 6. Ammocoetes contains gill lamellae. It is not found in *Branchiostoma*. 7. The blood of ammocoetes contains corpuscles and respiratory pigment, the haemoglobin. These are not found in *Branchiostoma*. 8. Ammocoetes contains a heart. It is absent in *Branchiostoma*. 9. Ammocoetes contains pronephric kidneys with nephrostomes. *Branchiostoma* contains metamerically arranged protonephridia which lack nephrostomes.

Branchiostoma and ammocoetes are enough alike to suggest a common ancestry.

(*c*) Affinities with other vertebrates. *Branchiostoma* resembles other vertebrates (besides cyclostomes) in having following characters: 1. Metamerically arranged myotomes. 2. True coelom. 3. Possess a tail. 4. Midgut diverticulum comparable with liver. 5. Hepatic portal system. 6. Similar arrangement of blood vessels and similar direction of blood flow.

Differences. *Branchiostoma* differs from vertebrates in not having a head, skull, vertebral column, paired limbs, heart, blood corpuscles, kidneys, brain, eyes, ears and gonoducts, etc. It also differs from vertebrates in having ciliary mode of feeding, protonephridia, atrium, numerous gonads, single cell thick epidermis, and asymmetry.

Considering its resemblances and differences with the vertebrates, *Branchiostoma* has been placed close to the vertebrates, but in a separate subphylum, the cephalochordata.

Systematic Position of *Branchiostoma*

Branchiostoma contains all the distinguishing chordate features. Its inclusion in the phylum Chordata is fully justified. However, its true systematic position in the phylum remains controversial.

Accoding to **Wiley** (1894) *Branchiostoma* is a prototype chordate and can be placed along the main line of chordate evolution. According to **Walter Garstand** (1928) and **Berrill** (1958) both cephalochordates and vertebrates evolved from a neotenic ascidian larva which failed to metamorphose. However the specialised characters of *Branchiostoma* indicate that it is not in the main line of evolution of chordates.

Gregory (1958) suggested that *Branchiostoma* may be a degenerated and specialised descendant of Agnatha. According to him *Branchiostoma* became specialised from Agnatha by multiplication of gill slits and myomeres and degeneration of head, brain, eye, auditory capsules, exoskeleton and gonoducts. But the presence of nephridia with solenocytes and the absence of many important craniate characters indicate that *Branchiostoma* is more archaic chordate.

Conclusion. *Branchiostoma* contains a mixture of primitive, degenerate and specialised characters. It can not be placed in main line of evolution of chordates. It is considered to be a primitive chordate that is very close to the ancestral chordates. Its specialised features indicate that it departed on a side line of chordate evolution and therefore, it is placed in an independent subphylum cephalochordata.

QUESTIONS

Long Answer Questions

1. Describe the habits, habitat and external features of *Branchiostoma* (amphioxus). (*Rohilkhand 2000; Avadh 2001; CCSU Meerut 2002*)
2. Describe the alimentary canal of *Branchiostoma* and explain its mode of feeding and digestion. (*Bundelkhand 2000; Purvanchal 2001; CCSU Meerut 2002; Gorakhpur 2003; Rohilkhand 2003*)
3. What is ciliary feeding and describe the digestive system of *Branchiostoma*. Explain what happens to food in the region of alimentary canal.
4. Describe structural adaptations of oral regions and pharynx to explain mode of feeding of *Branchiostoma*. (*CCSU Meerut 2003*)
5. Describe digestive system of *Amphioxus*. (*CCSU Meerut 2013*)
6. Give an account of excretory organs of *Amphioxus*. (*CCSU Meerut 2009*)
7. Give an account of blood vascular system of *Branchistoma*. (*Avadh 2001; Kanpur 2001; Gorakhpur 2002; Garhwal 2002; Purvanchal 2003; CCSU Meerut 2007*)
8. Describe the blood vascular system of *Branchiostoma*. (*CCSU Meerut 2007*)
9. What are the main characteristics of chordates? Justify the inclusion of *Herdmania* and *Amphioxus* in this phylum. (*CCSU Meerut 2006*)
10. Arrange the characters of *Branchiostoma* (amphioxus) as primitive and peculiar. (*Purvanchal 2000*)
11. Discuss the affinities of *Branchiostoma*. (*Avadh 2000; Purvanchal 2002; Garhwal 2003; Kanpur 2003*)

Short Answer Questions

1. Draw a neat and fully labelled diagram to show the internal anatomy to *Branchiostoma* or amphioxus.
2. Write a note on wheel organ of amphioxus. (*Allahabad 2002; Rohilkhand 2003; Gorakhpur 2003*)

3. Give an account of body wall of *Branchiostoma*.
4. Give an account of coelom in *Branchiostoma*. (*Garwal 2003*)
5. Give an account of skeletal system *Amphioxus*.
6. Describe the respiratory organs of *Branchiostoma* and the process of respiration in this animal.
7. Write a brief account of the excretory system of *Branchiostoma* and explain the mechanism of excretion. (*Rohilkhand 2001*)
8. What is solenocyte?
9. Give an account of nervous system in *Branchiostoma*.
10. Describe the nervous system and sense organs of amphioxus. (*Allahabad 2000*)
11. Write short notes on the following:
    (*i*) Endostyle,
    (*ii*) Velum,
    (*iii*) Atrium,
    (*iv*) Hatchek's nephridium,
    (*v*) Oral hood, (*Avadh 2002; Kanpur 2003*)
    (*vi*) Protonephridium. (*Purvanchal 2003; CCSU Meerut 2008*)
    (*vii*) **Diagrammatic questions:** Draw neat and labelled diagram of following:
    (*a*) T.S. of endostyle of *Amphioxus*. (*CCSU Meerut 2012*)
    (*b*) T.S. *Amphioxus* through gonads. (*CCSU Meerut 2007, 2008*)
    (*c*) T.S. female Amphioxus through branchial region. (*CCSU Meerut 2006*)

Very Short Answer Questions

1. What is lancelet? (*CCSU Meerut 2011*)
2. Define notochord. (*CCSU Meerut 2009*)
3. Mention the main skeletal structure of the amphioxus.
4. Name the passage between oral hood and pharynx.
5. Which gill bars have bifurcated skeletal rods in them?
6. Where is the ventral aorta located?
7. Name the cilia which are mainly responsible for setting up a water current in lancelet.
8. What types of digestion occur in *Branchiostoma*?
9. What is wheel organ? (*CCSU Meerut 2008, 2009*)
10. Write a note on protonephridia in *Amphioxus*. (*CCSU Meerut 2007*)

Fill in the Blanks

1. The popular name of *Branchiostoma* is ________.
2. *Branchiostoma*, like ________, is a ciliary feeder.
3. Mouth of *Branchiostoma* is fringed with oral ________.
4. The muscle blocks of the body of amphioxus are also called ________.
5. *Branchiostoma* has two types of body cavities: ________ and ________.
6. Wheel organ is located in the ________ of *Branchiostoma*.
7. Primary and secondary gill bars are interconnected by ________.
8. Renal papillae are involved in ________.
9. In *Branchiostoma*, blood returned from the intestine is supplied to the ________.
10. Main excretory organs of *Branchiostoma* are ________.
11. Eyes of *Amphioxus* lie within the ________.
12. Respiratory pigment in the blood of amphioxus is ________.
13. Endostyle lies along the midventral region of ________.
14. Gonoducts are present in ________ but not in ________.

15. *Branchiostoma* shows metamerism but ________ does not.
16. Hepatic portal system appears in ________ in the animal kingdom.
17. In amphioxus ________ and ________ pigment are absent. (*Purvanchal 2001*)
18. *Amphioxus* is a ________. (*Purvanchal 2001*)
19. Brown funnels are ________ organs of *Amphioxus*. (*Garhwal 2003*)

Multiple Choice Questions

1. *Branchiostoma* is a
(*a*) Hemichordata (*b*) Urochordata
(*c*) Cephalochordata (*d*) Vertebrata

2. Both coelom and atrium are present in
(*a*) lancelet (*b*) sea squirt
(*c*) both (*a*) and (*b*) (*d*) neither (*a*) nor (*b*)

3. Enterocoelic coelom is found in (*Rohilkhand 2003*)
(*a*) frog (*b*) birds
(*c*) *Amphioxus* (*d*) none of these

4. Muller's organ is found in (*Kanpur 2002*)
(*a*) fish (*b*) *Herdmania*
(*c*) *Amphioxus* (*d*) *Balanoglossus*

5. Notochord in cephalochordata (*Gorakhpur 2003*)
(*a*) extends all along the length (*b*) is restricted to the tail only
(*c*) is found in the anterior part of the body (*d*) is found only in the larva

6. Endostyle has ________ longitudinal tracts of cilia
(*a*) four (*b*) five
(*c*) two (*d*) three

7. Digestion of food in *Branchiostoma* is
(*a*) extracellular (*b*) intracellular
(*c*) both extracellular and intracellular (*d*) not clear

8. Main digestive gland of the lancelet is
(*a*) liver (*b*) pyloric gland
(*c*) hepatic caecum (*d*) gland cells in intestinal epithelium

9. *Branchiostoma* has
(*a*) renal portal system (*b*) hepatic portal system
(*c*) both (*a*) and (*b*) (*d*) neither (*a*) nor (*b*)

10. Giant nerve cells and giant nerve fibers are met within
(*a*) *Branchiostoma* (*b*) *Herdmania*
(*c*) all Protochordata (*d*) Hemichordata

11. Kolliker's pit is found in (*Gorakhpur 2003*)
(*a*) *Balanoglossus* (*b*) *Herdmania*
(*c*) *Branchiostoma* (*d*) Ammocoeta larva

12. Hatschek's pit of *Branchiostoma* is found in
(*a*) wheel organ (*b*) velum
(*c*) endostyle (*d*) olfactory pit

13. Hatschek's pit is found in (*Gorakhpur 2002*)
(*a*) *Peteromyzon* (*b*) *Balanoglossus*
(*c*) *Amphioxus* (*d*) *Herdmania*

14. Excretion in *Branchiostoma* occurs by
(*a*) nephridia (*b*) brown funnels
(*c*) renal papillae (*d*) all of these

15. Gonads have cicatrices in

(*a*) sea squirt (*b*) lancelet
(*c*) frog (*d*) all animals

16. Eggs of *Branchiostoma* are

(*a*) alecithal (*b*) telolecithal
(*c*) microlecithal (*d*) isolecithal

ANSWERS

Very Short Answer Questions

3. Notochord. **4.** Enterostome. **5.** Primary. **6.** In subendostylar coelom. **7.** Lateral cilia of gill bars. **8.** Extracellular and intracellular.

Fill in the Blanks

1. Lancelet. **2.** Herdmania. **3.** Cirri. **4.** Myotomes (myomeres). **5.** Coelom, atrium. **6.** Oral hood. **7.** Synapticule. **8.** Excretion. **9.** Midgut diverticulum. **10.** Nephridia. **11.** Nerve cord. **12.** Absent. **13.** Pharynx. **14.** *Herdmania, Branchiostoma*. **15.** *Herdmania*. **16.** Cephalochordata (*Branchiostoma*). **17.** Heart, respiratory pigment. **18.** Chordate. **19.** Excretory.

Multiple Choice Questions

1. (*c*)	**2.** (*a*)	**3.** (*c*)	**4.** (*c*)	**5.** (*a*)	**6.** (*b*)
7. (*c*)	**8.** (*c*)	**9.** (*b*)	**10.** (*a*)	**11.** (*c*)	**12.** (*b*)
13. (*c*)	**14.** (*d*)	**15.** (*b*)	**16.** (*c*)		

7

Cephalochordata: General Characters, Classification and Types

The **Cephalochordata** (Gr., *kephale* × head + *chorde* × chord) is a small group of about 25 species of lancelets, fish-like acraniate chordates. Cephalochordates demonstrate several characters which are intermediate between those of invertebrates and the vertebrates. *Amphioxus* may represent living descendant of an ancestor that gave rise to both the cephalochordates and the vertebrates (**Hickman** *et al.*, 1996). A notable feature of the group is the persistence of all the four diagnostic chordate characters throughout life without modification.

7.1 GENERAL CHARACTERS

1. Cephalochordates are the marine lancelets that inhabit the sandy bottoms of coastal waters around the world.
2. Body is slender, laterally compressed, eel-like, being adapted for burrowing and swimming.
3. Body lacks head but has a tail. It is segmented as in annelids and vertebrates.
4. There are no paired appendages (fins or limbs). Median fins (dorsal, caudal and ventral) are present (Box 7.1).

Box 7.1

The dorsal and ventral fin-like structures act as **dorsal** and **ventral storage organs**. They are not homologous to the fins of fishes and their function appears to be housing a build up of nutritional reserves for gamete formation (**Holland** and **Holland** 1991, 1991; **Brusca** and **Brusca** 2003).

5. Dorsolateral muscles are very thick and segmented into cheveron ($\angle$) shaped myomeres (myotomes) which produces the characteristics lateral swimming undulations of the body.
6. Skeleton of amphioxus includes notochord, cartilaginous rods and plates. Notochord of lancelets is built of discoidal lamellae that are stacked like so many poker chips, along its length and surrounded by a sheath of collagenous connective tissue. Notochord is persistent and extends forward to the anterior end of snout in front of nerve cord.
7. Pharynx large and with numerous oblique gill slits opening into the atrium. Gill slits provided with tongue bars. It serves both as respiratory and digestive structure.
8. Coelom is enterocoelous. It is however reduced in the pharyngeal region by an ectoderm-lined cavity, the atrium. In atrial space open the gill-slits, nephridia and gonads.
9. Alimentary canal complete and straight. Pharynx has endostyle and gill-slits. Midgut has a diverticulum or hepatic caecum, where smallest food particles are phagocytized and digested

intracellularly. Midgut diverticulum resembles the vertebrate pancreas in secreting enzymes (**Hickman** *et al.*, 1996). Filter feeder. As in tunicates, filtered water passes first in atrium then leaves the body by atriopore (equivalent to the excurrent siphon of tunicates).

10. Respiration occurs through general body surface.
11. Blood vascular system is well developed and closed. Blood flow pattern is remarkably similar to that of primitive fishes. Heart, erythrocytes and respiratory pigment absent. Hepatic portal system is present.
12. Dorsal tubular nerve cord present. Brain and ganglia absent. Nerve fibers are nonmyelinated. Dorsal and ventral roots of the spinal nerves do not unite.
13. Paired sense organs such as eyes, ears and nares are absent. Simple receptors are present.
14. Excretory system includes paired nephridia with solenocytes.
15. Sexes are separate. Sexual dimorphism does not occur. Gonads numerous, metamerically repeated and without gonoducts. No asexual reproduction.
16. Fertilization external.
17. Cleavage is total (holoblastic) and gastrulation is formed by invagination.
18. Development indirect. A ciliated free-swimming larva is formed during development.

7.2 CLASSIFICATION

The subphylum Cephalochordata includes a single class **Leptocardii**, a single family *Branchiostomidae* and only two genera:

1. *Branchiostoma* (amphioxus): More than 20 species of *Branchiostoma* have been identified. Gonads are paired in these species.
2. *Asymmetron*. It includes six species. These contain a single (unpaired) gonad.

7.3 ORIGIN OF CHORDATES (CEPHALOCHORDATA AND VERTEBRATES)

Zoologists have recognized the following trend of origin of cephalochordates:

1. Gill Slits (in pharynx) probably first evolved as a mechanism to facilitate water flow through the mouth and then out via the slits and pores, thereby enhancing gas exchange. The gill-slits function in this gas exchange mode in the hemichordates. The most primitive chordates (tunicates and cephalochordates) may have arisen with a shift to **internal feeding** through the use of pharyngeal gill-slits for extracting suspended food from the water.

2. In addition, **chordate origin** was marked by the appearance of a pharyngeal *endostyle* (associated with the feeding mechanism), a *notochord*, a muscular locomotor postanal *tail* and other features. These events were probably accompained by the appearance of a tadpole-like larva wherein other chordate features are developmentally manifested. The vertebrate thyroid gland is probably homologous with the endostyle of tunicates and cephalochordates (and larval lampreys-ammocetes).

3. **Walter Garstang** (1928) was of the opinion that the notochord originally evolved to give support to pelagic larvae as they developed longer and longer planktonic lives, a trend eventually leading to the **neotenic** appendicularians. As these forms grew larger in size, Gerstang reasoned, the notochord provided a basis for muscular support and locomotion supplementing previous reliance on ciliary locomotion.

4. Most workers feel that the **paedomorphic tendencies** (Box 7.2) of the tadpole larvae of the ancestral chordate played a major role in the origin of the cephalochordates and vertebrates. The evolution of segmental muscle bundles (myotomes) mark the beginning of the cephalochordate-vertebrate clade and allowed greater locomotor facility than the tail thrashing movements of the tadpole larval ancestor. The cephalochordates have retained the notochord and the use of pharyngeal gill slits for feeding. The vertebrates are, ofcourse, characterized by the development of an endoskeleton with cranium, vertebral column and ultimately, limbs. All of these features eventually provided increased body support and much more effective skeletomuscular mechanics and allowed great increase in locomotor abilities and body size.

Box 7.2

Paedomorphosis (Gr., *pais* × child + *morphe* × form), a term coined by **Garstang** (1928), describes the presence of evolutionary juvenile or larval traits in the adult body. Paedomorphosis can be produced by three different processes: neoteny, progenesis and post-displacement. In **neoteny**, the growth rate of body form is slowed so that the animal does not attain the ancestral form at the time it reaches reproductive maturity. **Progenesis** is the precocious maturation of gonads in a larval or juvenile body that then stops growing and never attains the adult body form. The **post displacement**, the onset of a developmental process is not attained at the time of reproductive maturation. Neoteny, progenesis and **post-displacement**, thus, describe different ways in which paedomorphosis can happen. Biologists use the inclusive term paedomorphosis to describe the results of these evolutionary-developmental processes (**Hickman** *et al.*, 1996).

5. Two key developmental inventions probably drove evolution of vertebrates: multiple **Hox gene clusters** and a new kind of embryonic cell, the **neural crest cell**, which led to the evolution of vertebrate complex head by way of embryonic **neural crest**. The earliest vertebrate fossils are armored jawless fishes and condonts (*i.e.*, teeth) from the Cambrian (510–530 mya).

Box 7.3

Hox gene. Abbreviation of "homeobox" genes. Large family of related genes that dictate regional identity in the embryo, particularly along the anterior-posterior axis (see **Gilbert**, 2010).

6. In the nineteenth century, **E. Geoffroy Saint-Hilarie** proposed an unorthodox view of chordate evolution. He suggested that the bodies of chordates, at least the higher chordates (cephalochordates and vertebrates) are **dorso-ventrally inverted** with respect to other bilaterians. This hypothesis lay fallow until the recent advent of molecular developmental research techniques, which have produced support for the idea from work on the expression of **dorso-ventral patterning** genes in insects and vertebrates (**Brusca** and **Brusca**, 2003).

7.4 ECONOMIC IMPORTANCE

Branchiostoma has very little economic importance. Since it is a clean, muscular, boneless creature, it is used as an article of food, fresh or dried, in some parts of China and Japan. Near Amoy in southern China, its annual catch is 35 tons, or about 1 billion lancelets (*Branchiostoma belcheri*). In parts of Brazil, chickens are herded onto beeches to feed on lancelets. Lastly, *Branchiostoma* is an important laboratory animal and is used for teaching.

QUESTIONS

Long Answer Questions

1. Discuss the general characters of cephalochordates.
2. Write about origin of Cephalochordata.

Short Answer Questions

1. What is role of paedomorphosis in the origin of chordates?
2. What are molecular biological and developmental evidences regarding origin of chordates?

Very Short Answer Questions

1. What are myocommata?
2. What is advantage of cheveron-shaped myomeres of *Branchiostoma*?
3. What is characteristics of notochord of *Branchiostoma*?
4. What type of coelom occurs in *Branchiostoma*?

Fill in the Blanks

1. Notochord persists throughout life in ________.
2. Cephalochordates lack ________ appendages.
3. Dorsolateral muscles are segmented into ∠-shaped ________.
4. Gonads of *Branchiostoma* lacks ________.
5. In cephalochordates, nerve fibers are ________.
6. Hepatic portal system appears first of all among the chordates in ________.
7. Main excretory organs of cephalochordates are ________.
8. There is no heart in ________.
9. Alimentary canal is ________ in *Branchiostoma*.
10. Coelom and atrium coexist in ________.

Multiple Choice Questions

1. *Branchiostoma* shows affinities with
   (*a*) nonchordates (b) urochordates
   (*c*) vertebrates (d) all of these
2. *Asymmetron* belongs to the subphylum
   (*a*) Urochordata (*b*) Cephalochordata
   (*c*) Vertebrata (*d*) none of these
3. In cephalochordates, the main function of gill slits is in
   (*a*) feeding (*b*) respiration
   (*c*) excretion (*d*) none of these
4. During evolution, locomotion of *Branchiostoma* is shifted from to
   (*a*) muscular (*b*) ciliary
   (*c*) both muscular and ciliary (*d*) none of these
5. Fin ray boxes of *Branchiostoma* are meant for
   (*a*) storage of reserve food (*b*) skeleton
   (*c*) both (*a*) and (*b*) (*d*) none of these

ANSWERS

Very Short Answer Questions

1. Connective tissue septa between myomeres.
2. They produce characteristic lateral swimming undulations of the body.
3. It contains discoidal lamellal which are stacked like poker chips.
4. Enterocoelus.

Fill in the Blanks

1. Cephalochordates. **2.** Paired. **3.** Myotomes (myomeres). **4.** Gonoducts. **5.** Nonmyelinated. **6.** *Branchiostoma*. **7.** Nephridia. **8.** Cephalochordata. **9.** Straight. **10.** Cephalochordata.

Multiple Choice Questions

1. (*d*) **2.** (*b*) **3.** (*a*) **4.** (*a*) **5.** (*a*)

8

Subphylum Vertebrata and Class Cyclostomata

Phylum chordata is broadly divided into two main groups: **Protochordata** or **Acrania** and **Vertebrata** or **Craniata** (Table 8.1). Protochordata includes two subphyla: Urochordata and Cephalochordata. Group Vertebrata is itself form third subphylum of Chordata.

Table 8.1 Differences between Protochordata and Vertebrata.

Group A: Acrania (Protochordata)	Group B: Craniata (Euchordata or Vertebrata)
1. Completely marine, small-sized chordates.	1. Aquatic or terrestrial, mostly large-sized vertebrates.
2. Head, appendages and exoskeleton absent.	2. Well-developed (cephalisation), usually two pairs of appendages and exoskeleton present.
3. Coelom enterocoelic, arising from embryonic archenteron.	3. Coelom schizocoelic, arising by splitting of mesoderm.
4. Notochord persistent.	4. Notochord covered or replaced by vertebral column.
5. Skull, cranium and vertebrae column absent.	5. Skull and cranium well developed.
6. Pharynx having permanent gill-slits or gill-clefts. Endostyle present.	6. Pharyngeal gill-slits or gill-chefts persist or disappear. Endostyle absent.
7. Heart without chambers when present. In blood, red blood cells absent.	7. Heart contains 2, 3 or 4 chambers. Blood contains erythrocytes or red blood corpuscles (RBC).
8. Excretory system has protonephridia.	8. Excretory system has kidneys which are mesonephros or metanephros.
9. Sexes separate or united. Reproduction asexual as well as sexual. Gonoducts usually absent.	9. Sexes separate. Only sexual reproduction. Gonoducts always present.
10. Development indirect with a free-swimming larval stage.	10. Development indirect or direct with or without a larval stage.

Vertebrates derived their name from the joined dorsal vertebrae column that has replaced the notochord as their primary support. The ancestral vertebrates lacked jaws. They probably swam over the bottom of sea, sucking up mud and straining it through the pharyngeal gill-slits to extract microscopic food particles. The vascularised tissues of the pharyngeal gill-slits also served as a gas-exchange function. These animals gave rise to jaw-less fishes called **ostracoderms** ("shell skinned"). They evolved a bony external armour that protected them from predators. These were aquatic small fish which could swim slowly above the substratum rather than having to burrow through it, as all earlier sediment feeders had done.

Vertebrates show a great diversity of form, size, colouration and behaviour. There are over 50,000 extant (currently living) species of vertebrates. They range in size from certain fishes weighing as little as 0.1 gram when fully mature to whales weighing over 100,000 kilograms. For each species there may be more than one hundred extinct species. The number of vertebrate species probably reached its maximum in the middle Miocene, 12 to 14 million years ago, and has been declining since then. The vertebrates originated during the Ordovician period, about 500 million years ago.

The vertebrates excel all other animals in the structure of the body as well as in efficiency of life functions. They occur and dominate in all habitats. Thus, fish reign supreme in the sea and inland waters (rivers, lakes, ponds, etc.). Birds are masters of air. Mammals surpass all other animals on land. The key of biological supermacy of the vertebrates is their highly developed nervous system.

Diagnostic Chordate Characters of Vertebrates

All vertebrates have the four diagnostic chordate characters—the **notochord**, paired **pharyngeal gill-slits**, dorsal hollow tubular **nerve cord** and muscular **tail** behind the cloacal or anal aperture. Presence of **vertebrae** is one of the most important characteristics that gives the subphylum its name. Another important feature is the presence of a **skull** or **cranium** from which the other name **craniata** is derived for the subphylum.

8.1 GENERAL CHARACTERS OF SUBPHYLUM VERTEBRATA

1. Lower vertebrates are aquatic. Higher vertebrates are mainly terrestrial.
2. Bilaterally symmetrical body typically consists of four regions: **head**, **neck**, **trunk** and **tail**. Some notable exceptions do exists. Fishes lack the neck, some amphibians lack both neck and tail, apes and human beings (mammals) have vestigial tail.
3. Metameric segmentation occur. The embryos of vertebrates show metameric segmentation in the form of paired mesodermal somites, and adult shows it in the arrangement of muscles, vertebrae, ribs, nerves and paired blood vessels.
4. Trunk bears two pairs of appendages which are supported by jointed skeletal elements. The appendages are in the form of **fins** (pectoral fins and pelvic fins) in the fishes and **limbs** (forelimbs, wings, or arms and hindlimbs or legs) in other vertebrates. Certain forms have secondarily lost one pair and some both pairs of appendages. The appendages primarily help in locomotion, but may support the body on land.
5. Skin or integument consists of outer ectodermal multilayered epidermis of a stratified epithelium and an inner mesodermal dermis of connective tissue. Epidermis contains glands.
6. Skin is covered by a protective exoskeleton of either scale, feathers, hairs, claws, nails or horns. Exoskeleton is secreted by epidermis, consists of nonliving material, and usually does not grow with the body. It may be shed or replaced periodically by a process called **moulting**.
7. Coelom large and schizocoelic.
8. Notochord ends beneath the forebrain. It is invested by cartilage or bone or replaced by a vertebral column.
9. They contain an endoskeleton of cartilage or bone or both. The endoskeleton grows with the body and is not shed. The endoskeleton consists of a skull, vertebrae column, girdles and limb bones.
10. Many muscles, which are striated, are attached to endoskeleton. These help in body movements and locomotion.

11. Digestive tract is complete and bears two large digestive glands—pancreas and liver. Buccal cavity contains tongue and usually teeth also.
12. Pharynx is of moderate size and is perforated by not more than 7 pairs of gill slits, except in some cyclostomes.
13. Respiration occurs by paired gills in lower aquatic vertebrates and by lungs in terrestrial higher vertebrates.
14. Blood vascular system is of closed type. Heart 2, 3 or 4 chambers. There are 2 or 3 portal systems: hepatic, pituitary and renal. Lymphatic system is also present.
Blood contains both white and red blood cells. Vertebrate blood contains a respiratory pigment, **haemoglobin** which is held in special cells, the **erythrocytes** or **red blood corpuscles**.
15. Excretory system consists of paired mesonephric or metanephric kidneys. Kidneys discharge waste products through ducts to exterior. A bladder for storage of urine occurs in many forms.
16. Anterior end of the dorsal tubular nerve cord forms a complex brain. Remaining nerve cord forms spinal cord. They have 8, 10 or 12 pairs of cranial nerves. Spinal nerves are formed by the union of dorsal and ventral roots in vertebrates except some cyclostomes. Thus, they are mixed in nature.
17. Vertebrates have in the head a pair of very complex eyes, derived in part from the brain; a pair of complex ears for hearing and equilibrium, and a pair of organs of smell.
18. An endocrine system of ductless glands regulates metabolism, growth and reproduction.
19. Sexes are separate. Gonads are paired. These discharge gametes to the exterior through ducts.
20. Excretory (urinary) and reproductive (genital) systems of vertebrates are connected to form urinogenital system.
21. In the majority of vertebrates, the excretory and genital ducts and the rectum open into a common chamber, the cloaca, which leads to the exterior by a cloacal aperture.

8.2 OUTLINE CLASSIFICATION OF CHORDATA

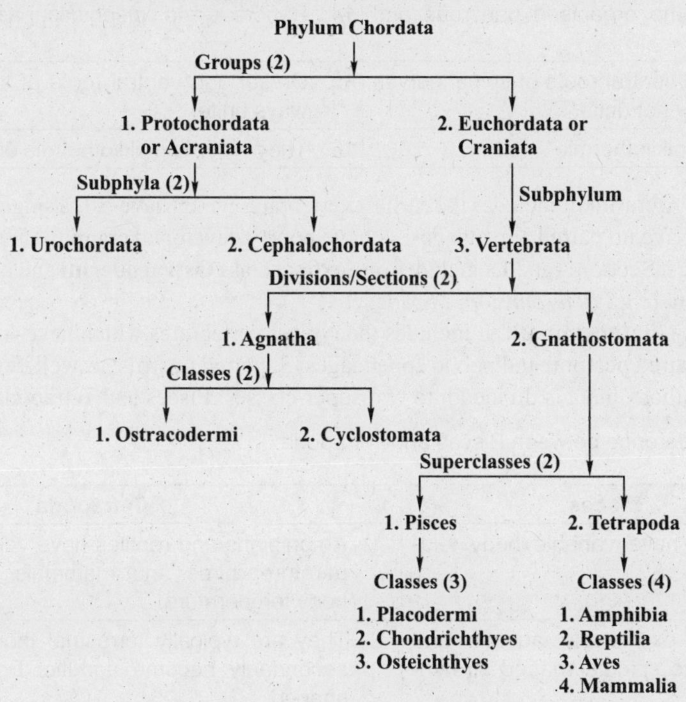

8.3 CLASSIFICATION OF VERTEBRATA

The subphylum Vertebrata is divided into two infraphyla or sections: Agnatha and Gnathostomata (Table 8.2).

Table 8.2 Differences between Agnatha and Gnathostomata.

Agnatha	Gnathostomata
1. Mouth is without jaws.	1. Mouth is provided with jaws.
2. Paired appendages are lacking	2. Paired appendages are present, may be fins or limbs.
3. Notochord is persistent.	3. Notochord is usually replaced by a vertebral column.
4. Vertebral column under-developed.	4. Vertebral column fully developed.
5. Stomach is lacking.	5. True stomach is present.
6. There is no renal portal vein.	6. Renal portal vein is mostly present.
7. There are 8 or 10 pairs of cranical nerves.	7. There are 10 or 12 pairs of cranial nerves.
8. Only a single median nostril and olfactory sac are present.	8. Paired nostrils and olfactory sacs are present.
9. Eyes lacks intrinisic muscles for the accomodation of lens.	9. Eyes has intrinstic muscles for the accomodation of lens.
10. Internal ear has 1 or 2 semicircular ducts.	10. Internal ear has 3 semicircular ducts.
11. Nerve fibers lack myelin sheath.	11. Nerve fibers have myelin sheath.
12. Lateral line receptors occur in unconnected pits.	12. Lateral line receptors occur in canals.
13. Gonad is single and without a gonoduct.	13. Gonads are paired and have gonoducts.
14. There are no organised pancreas and spleen.	14. Pancreas and spleen occur as distinct organs.
15. Dorsal and ventral roots of spinal nerves may or may not unite.	15. Dorsal and ventral roots of spinal nerves always unite.
16. They are poikilothermic.	16. They may be poikilothermic or homeothermic.

Section 1. Agnatha. It includes the earliest vertebrates which have—1. a single median nostrail; 2. lack jaws; 3. have no paired appendages; 4. have poorly developed cranium; 5. mouth is rounded in living members. Section Agnatha includes one extinct class **Ostracodermi** and one extant (living) class **Cyclostoma** (*e.g.*, *Peteromyzon*, *Myxine*, etc.)

Section 2. Gnathostomata. It includes the typical vertebrates which have—1. true upper and lower jaws; 2. paired pectoral and pelvic appendages; 3. paired nostrils; 4. well developed cranium. The section Gnathostomata is divided into two super-classes: Pisces and Tetrapoda (Table 8.3).

Table 8.3 Difference between Pisces and Tetrapoda.

Pisces	Tetrapoda
1. All Pisces have variable body temperature.	1. Amphibians and reptiles have variable body temperature; birds and mammals have regulated body temperature.
2. They are exclusively aquatic and mostly have spindle-shaped streamlined body.	2. They are typically terrestrial though some have secondarily become aquatic. Body is variously shaped.

Pisces	Tetrapoda
3. They have an exoskeleton of dermal scales of various kinds.	3. They have an exoskeleton of horny epidermal scales.
4. They breathe by gills throughout life.	4. They, as a rule, breathe by lungs. Some breathe by gills in larval stage or even throughout life.
5. Paired appendages are in the form of fins. Median fins are also present. All fins are supported by fin rays.	5. Paired appendages are in the form of pentadactyl limbs. Some have median fins without fin rays.
6. Tongue is small and bony.	6. Tongue is large and muscular, helps in manipulating in the oral cavity, capturing the prey and swallowing.
7. Salivary glands are lacking.	7. Many have salivary glands that secrete saliva (most mammals) or venom (some snakes and lizards).
8. There are no vomeronasal (Jacobson's) organ.	8. Vomeronasal organs occur in the anterior part of buccal roof for additional olfaction.
9. Heart mostly has only two chambers, an auricle and a ventricle.	9. Heart has three (two auricles and a ventricles) or four (two auricles and two ventricles) chambers.
10. There is single circulation.	10. There is double circulation.
11. There are usually 10 pairs of cranial nerves.	11. There are 10 or 12 pairs of cranial nerves.
12. Internal nares are usually absent so that olfactory chambers act only as organ of smell.	12. Internal nares are present so that olfactory chambers,besides acting as organ of smell, also aid in respiration.
13. Lateral-line sense organs are well developed.	13. Lateral-line sense organs are absent, except in a few larval amphibians.
14. External and middle ear are undeveloped.	14. External and middle ear are usually present. Middle ear open into the pharynx by eustachian canal.
15. Eyes lack eyelids and glands.	15. Eyes have eyelids and glands (tear, Harderian) to moisten, clean and nourishing the cornea.
16. Endoskeleton may be cartilaginous or bony.	16. Endoskeleton bony.
17. Locomotion is brought about by the powerful tail by using axial skeleton and muscles.	17. Locomotion is brought about by using axial skeleton, muscles and limbs.
18. Kidneys are opisthonephric.	18. Kidneys are opisthonephric or metanephric.
19. There is no urinary bladder.	19. A urinary bladder is present to store urine temporarily.
20. Fat is mostly stored as lipid droplets in the liver and muscles.	20. Fat is mainly stored in specialized cells, the adipocytes, of adipose tissue. Amphibians and reptiles store fat in distinct fat bodies in the abdomen and around the tail base. Birds and mammals store fat in subcutaneous tissue and in association with muscles,heart, gut, and other organs.
21. There is no larynx.	21. Larynx is present, formed from pharyngeal arch elements, for sound production.

Superclass 1. Pisces. The pisces are jawed fishes. They are exclusively aquatic vertebrates. They have an exoskeleton of dermal scales, have paired locomotory appendages in the form of pectoral and pelvic fins, respire throughout life with gills which are developed in visceral clefts, possess lateral line sense organs for detecting waves and currents in water.

The superclass Pisces includes three classes. 1. Placodermi (extinct species); 2. Chondrichthyes (living or extant species) and 3. Osteichthyes (living or extant species).

Class 1. Placodermi. The extinct placoderms evolved from and replaced the ostracoderms in the Devonian period. They developed primitive jaws and paired fins but retained the ostracoderm bony armour. This armour consisted of large bony plates, from which the various types of dermal scales of living fishes evolved, *e.g.*, *Climatius*.

Class 2. Chondrichthyes. These are cartilaginous fishes. They include sharks, rays, skates and chimaeras. They are mostly marine cartilaginous fishes have ventral mouth, exposed gill slits, heterocercal tail, placoid scales and cartilaginous endoskeleton. They lack swim bladder, *e.g.*, *Scoloidon* (Indian Shark).

Class 3. Osteichthyes. These are bony fishes. They include garpikes, sturgeons, codes, salmons, flat fishes, flying fishes, seahorses, catfishes, carps, and lung fishes. They occur in fresh water as well as sea water. They have anterior (terminal) mouth, gill slits covered by opercula, homocercal or other types of tails, various types of dermal scales and bony endoskeleton. Bony fishes possess swim bladder.

The class Osteichthyes is divided into two subclasses: 1. **Actinoptergii**, a large group including about 25000 species such as *Labeo*. 2. **Sarcopterygii**, a small group of only 8 species including African lung fish (*Protopterus*).

Superclass 2. Tetrapoda. They are primarily terrestrial vertebrates. They have cornified outer layer of skin, paired appendages in the form of 5-toed, or **pentadactyl limbs**, internal nares to convey air from the nasal chambers to the mouth cavity, lungs for respiration and bony endoskeleton. They lack lateral line sense organs.

The superclass Tetrapoda is divided into four classes: Amphibia, Reptilia, Aves and Mammalia.

Class 1. Amphibia. They are the first land vertebrates. They include salamanders, newts, frogs, toads and caecilians. The adult amphibians, though terrestrial, return to water for breeding. Horny layer of the epidermis is poorly developed. Digits are without horny claws. Respiration occurs by gills, skin, bucco-pharyngeal lining and lungs, singly or in combination. Embryo lacks embryonic membranes.

The class Amphibia is divided into three subclasses.

Subclass 1. Apoda (165 species), *e.g.*, *Ichthyophis*.

Subclass 2. Urodela (415 species), *e.g.*, *Salamendra*.

Subclass 3. Anura (4300 species), *e.g.*, *Rana tigerina*.

Class 2. Reptilia. The reptiles are fully adapted for land life. They include tortoises, turtles, lizards, snakes, crocodiles and aligators. Horny epidermal scales covers the gland-free skin. Digits bear horny claws. Bony endoskeleton. Gill pouches occur only in the embryo. Respiration takes place by lungs alone. Eggs are laid on land. They are large, yolky and covered by a resistant shell. Embryo develops embryonic membranes (such as amnion, chorion, allantois and yolk sac). The young resembles the parents and there is no metamorphosis.

The class Reptilia is divided into four orders:

Order 1. Chelonia (260 species), *e.g.*, *Testudo*.

Order 2. Rhynchocephalia (single species), *e.g.*, *Sphenodon*.

Order 3. Squamata (6800 species), *e.g.*, *Uromastix*, *Hemidactylus*, *Python*.

Order 4. Crocodila (22 species), *e.g.*, *Crocodylus*.

Class 3. Aves. The birds are land-dwelling animals. They are adapted to aerial life. Birds have an exoskeleton of feathers which are developed as modification of the reptilian epidermal scales. Feathers are found only in birds. The forelimbs are modified into wings to help in flight. The hindlimbs alone support the birds body on perch or land. Feet bears horny scales and usually have four toes, each ending in a horny claw. Teeth are absent. Jaws form beak. A system of air sacs is associated with the lungs to increase the efficiency of pulmonary respiration. Eggs are large, polylecithal and with shell. They are incubated by one or both parents. Embryo develop embryonic membranes like the reptilian embryo. Birds are warm blooded animals (Box 8.1).

Box 8.1

The birds have a constant body temperature much higher than that of environment and are said to be **homoiothermic**. This keeps them active throughout the year. In contrast, the body temperature of lower vertebrates (fishes, amphibians and reptiles) varies with that of environment. They are said to be **poikilothermic**.The newly hatched birds are also poikilothermic. They, however, become homoithermic before leaving the nest.

The class Aves includes two subclasses:

Subclass 1. Archaeornithes of exclusively extinct and primitive birds, *e.g.*, *Archaeopteryx*.

Subclass 2. Neornithes of extinct and extant birds, *e.g.*, *Struthio* (Ostrich), emperor penguin (*Aptenodytes*) and pigeon (*Columba*).

Class 4. Mammalia. They are terrestrial tetrapods with an exoskeleton of hair. Hair occur only in mammals. The skin contains a variety of glands. One type of gland, the mammary glands, secrete milk for the nourishment of young ones. Mammals have pinnae for collecting sound waves. They have the most highly developed nervous system among the animals. The eggs are small, micro'ecithal and without a shell. Embryo develops in the uterus and forms embryonic membranes. An organs of metabolic exchange between foetus and uterus, called **placenta**, develops for the nourishment, respiration and excretion of foetus. The young ones are born alive, making the mammals **viviparous** (L., *vivus* = alive + *pario* = to bear). Other vertebrates lay eggs and are said to be **oviparous** (L. *ovum* = egg + *pario* = to bear). The mammals are homoiothermic, active the year round and widely distributed like the birds.

The class Mammalia has two subclasses:

Subclass 1. Prototheria of primitive oviparous forms, *e.g.*, *Ornithorhynchus* (The duck bill).

Subclass 2. Theria of viviparous forms. This group is divided into 2 infraclasses: 1. **Metatheria** (or Marsupialia) of pouched mammals, *e.g.*, *Macropus* (Kangaroo). 2. **Eutheria** (or Placentalia) of typical placental mammals, *e.g.*, *Oryctolagus* (rabbit), *Rattus* (rat).

Amniota and Anamniota

According to another system vertebrates are of two types:

1. Anamniota (Lower vertebrates). It includes Cyclostomata, Chondrichthyes, Osteichthyes and Amphibia. In them extraembryonic membrane, **amnion**, does not appear during embryonic development.

2. Amniota (Higher vertebrates). It includes Reptilia, Aves and Mammalia. The extraembryonic membrane, amnion, develops during their embryonic development.

8.4 CLASS CYCLOSTOMATA

The cyclostomes may be defined as the jawless, scaleless, monorhinous (Gr., *mono* = one + *rhinos* = nose), fish-like vertebrates devoid of paired appendages.

The cyclostomata (*cyclo* = round + *stoma* = mouth) are most primitive of all existing vertebrates. They include lampreys and hags (or hag-fishes).

1. General Characters of Class Cyclostomata

1. The body is elongated, slender, eel-like rounded. It includes head, trunk and tail. Neck is absent. Tail laterally compressed.

2. Paired fins are absent. Median fins are supported by cartilaginous fin rays.
3. The skin is soft, smooth, glandular (slimy) and scaleless. Exoskeleton absent.
4. Muscles of trunk and tail are arranged in Z-shaped myomeres.
5. The endoskeleton is wholly cartilaginous. Skull is simple and primitive. Cranium does not surround brain completely. In hags (hagfishes) cranium is fibrous. A branchial basket surrounds the gill pouches.
6. Notochord persists throughout life. Vertebrae are represented by imperfect neural arches over the notochord.
7. Cyclostomes contain a true coelom.
8. Alimentary canal is straight. The mouth is jawless, and may be circular or suctorial. Tongue may bear horny teeth for rasping. Stomach absent. Intestine with **typhlosole**. Cloaca is absent.
9. Pharynx is respiratory in function. 5 to 16 pairs of gills are present in lateral sac-like pouches. Gill pouches opens externally by gill slits.
10. The heart with one atrium and one ventricle. It contains sinus venosus but lacks truncus arteriosus. Many aortic arches occur in gill region. Hags have many accessory hearts.
11. Blood with leucocytes and erythrocytes. Erythrocytes nucleated.
12. Cold blooded (poikilothermic) animals.
13. Excretory system includes a pair of opisthonephric kidneys.
14. Brain has very small cerebral hemispheres and rudimentary cerebellum. 8-10 pairs of cranial nerves. Nerves have unmyelinated fibers as in lancelets. Dorsal and ventral roots of spinal nerves may or may not unite.
15. Paired sense organs, eyes and ears, have been developed. Eyes may be functional or vestigial and sunk beneath the skin. Membranous labyrinth (internal ear) with one or two semicircular canals. Nasal sac is, however, single and median (monorhinous) condition. Olfactory duct leads into a persistent nasohypophysial sac that may end blindly or open into the pharynx. Taste buds are present. Lateral sense organs are developed, however, true lateral canals are absent.
16. Thyroid gland and pituitary body present.
17. Gonad single and without gonoduct.
18. Fertilization external. Development direct or with a long larval stage.

2. Classification of Class Cyclostomata

The class cyclostomata is divided into two subclasses: 1. Petromyzontia and 2. Myxinoidea.

Subclass 1. Petromyzontia

1. Mouth is ventral and surrounded by a large suctorial buccal funnel which bears many horny teeth. Tongue protrusible and bears horny teeth. Dorsal fin present.
2. An oral gland secretes an anticoagulent that prevents the victim's blood from clotting. They are ectoparasites of fishes and turtles, whales and porpoises.

Box 8.2

The group name Petromyzontia (Gr., *petros* = stone; *myzon* = sucking) refers to the lamprey's habit of grasping a stone with its mouth to hold position in a current.

3. There is a single nostril on the top of the head. It opens into a dorsal blind nasal sac.
4. Eyes are large, well developed and functional.
5. There are seven pairs of gill pouches that open to outside by seven pairs of gill slits. Gills have chloride cells to regulate ions and have **tidal ventilation**, in which water is both drawn and expelled via gill slits.
6. Branchial basket is complete.
7. Brain is differentiated. Large olfactory lobes, cerebral hemispheres and cerebellum are small.

Pineal body is well developed. It contains a lens and pigmented retina. It is photosensitive. Dorsal and ventral roots of spinal nerves remain separate.

8. Internal ear has two semicircular canals. Lateral line system well developed.
9. Eggs small. There is a long larval stage in the life history. Larva is called **ammocete**.
10. Lampreys have about 41 species and they occur along sea coasts as well as in rivers, lakes.

Examples: *Petromyzon marinus* (sea lamprey), *Petromyzon planeri* (brook lamprey), *Lamperta fluviatilis* (lampern or fresh water lamprey), *Ichthyomyzon*.

Subclass 2. Myxinoidea

1. The mouth is terminal and lacks buccal funnel. It has four pairs of tentacles (sensory barbels) around it. Biting mouth contains two horny plates bearing sharp, horny teeth. Tongue lacks teeth. Dorsal fin is absent.
2. There is single large nostril. It opens into a nasal canal that opens into the pharynx.
3. Eyes are vestigial and sunk beneath the skin.
4. Five to 16 pairs of gill pouches that open to outside through a pair of external gill opennings. Gills have **flow-through ventilation** in which water is drawn in via mouth and passed over the gills to move out via two gill slits.
5. Cranium is incomplete and formed of fibrous tissue rather than of cartilage. There are no vertebrae. Branchial basket is poorly developed.
6. In addition to a true heart, accessory hearts are present.
7. Brain is primitive, cerebrum and cerebellum being unrecognizable.
8. Dorsal and ventral roots of spinal nerves are united. Pineal eye is lacking.
9. Internal ear has only one semicircular canal. Lateral line system poorly developed.
10. Eggs are large and have horny shell. Development is direct (without larval stage).
11. They are scavangers. Although almost completely blind, the hagfish is quickly attracted to food, especially dead or dying fishes by its keenly developed sense of smell and touch.
12. They are marine living in dark muddy sea-bottom. They secrete an immense amount of slime around them for protection from a series of slime sacs on each side of the body, hence their name **slime eels** (Gr., *myxa* = slime).

Examples: *Myxine glutinosa* (common hagfish or slime eel) and *Bdellostoma cirhatum* (the hag fish).

8.5 TYPE STUDIES (CYCLOSTOMATA)

1. *Petromyzon marinus* (The sea lamprey)

Classification

Phylum	—	**Chordata**
Subphylum	—	**Vertebrata**
Class	—	**Cyclostomata**
Subclass	—	**Petromyzontia**
Genus	—	*Petromyzon*
Species	—	*marinus*

Distribution and Habitat

The sea lamprey or *Petromyzon marinus* is a marine and migratory (anadromus) animal. It inhabits coasts of North America, Europe, West Africa and Japan. It usually lives near the bottom.

Habits

1. Feeding. *Petromyzon* is a **sanguivorous** (blood-sucking) **ectoparasite**. It attaches to the under surface of a large healthy fish by suction of its buccal funnel and use of buccal teeth (Fig. 8.1). With their sharp horny teeth, they rasp away the flesh and suck out body fluids. To promote the flow of blood, the lamprey injects an anticoagulent into the wound. The host may die due to loss of its blood.

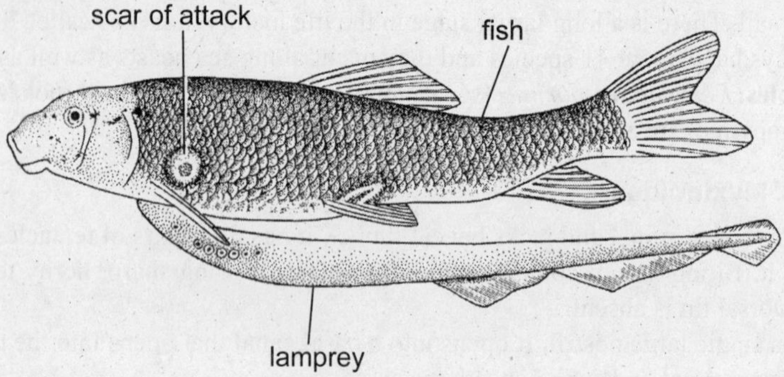

Fig. 8.1. Lake lamprey attached to a bony fish. Host also shows the scars of the attacks of other lampreys.

2. Respiration. The lamprey breathes water, but, unlike true fishes, water is drawn into the gill pouches and expelled out through the external gill slits. This method is necessary, because the lamprey, when attached to a host by its buccal funnel for feeding, cannot take water through the mouth.

3. Locomotion. The *Petromyzon* actively swims about in water by graceful lateral undulations of its highly muscular body. When moving upward in a strong current, it suddenly darts forward and clings to rocks.

4. Breeding. The lamprey breeds only once in its whole life. All lampreys do migration and ascend fresh water rivers or streams to breed. The marine forms are anadromous, that is, they leave the sea where they spend their adult lives to swim up rivers and streams to spawn (Box 8.3). The males begin **nest building** and are joined later by females. Using their oral discs to lift stones and pebbles and vigorous body vibrations to sweep away light debris, they form an oval depression. At spawning, with the female attached to a rock to maintain her position over the nest, the male attaches to the dorsal side of her head. As the eggs are shed into the nest, they are fertilized by the male. The sticky eggs adhere to the pebbles in the nest and quicky covered with sand. A large-sized sea lamprey may lay about 236,000 eggs. After spawning, the parents leave the nest and die within a few days. Being exhausted, they may be captured by large fishes.

Box 8.3

Marine fish and land animals have **rhodopsin** as visual pigment and fresh water animals have **porphyropsin** as visual pigment. As the lamprey migrates from the sea to fresh water, its visual pigment changes from rhodopsin to porphyropsin through an unknown mechanism.

Fertilization is external. Development is indirect. Life history of lamprey includes a larval stage, called **ammocoete**.

Structure

The lamprey has a long, slender, eel-like body with a soft, slimy, scaleless skin. It is a meter in length. The colour of its body is mottled greenish brown. Body of *Petromyzon* is indistinctly divided into **head, trunk** and **tail** (Fig. 8.2).

(*i*) Head. It is cylindrical and bears a mouth, a pair of large eyes, a pinean eye, a single nasal aperture and seven pairs of gill slits.

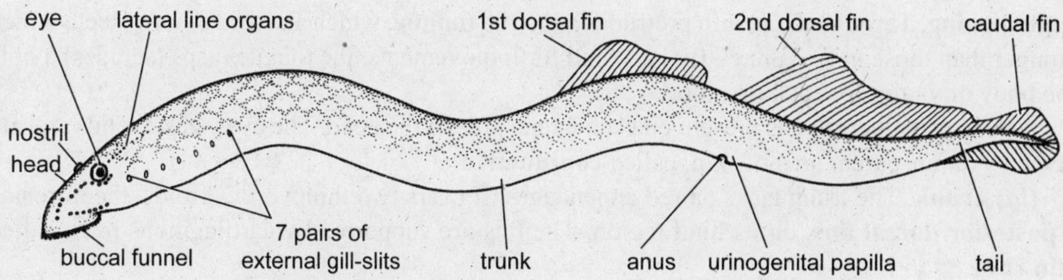

Fig. 8.2. *Petromyzon marinus*. External features.

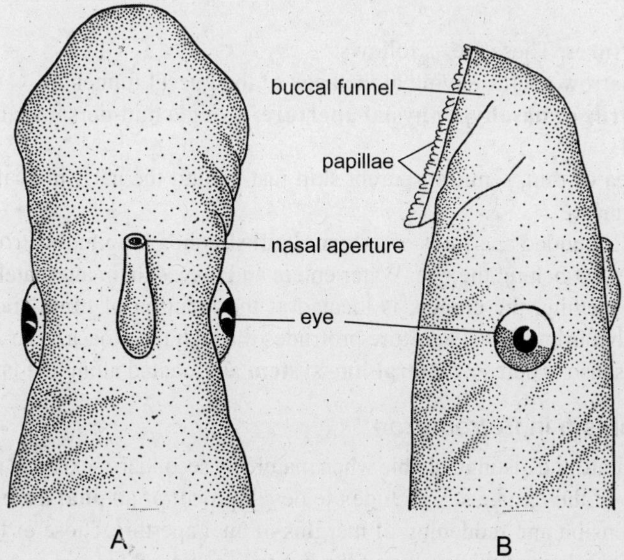

Fig. 8.3. *Petromyzon marinus*. A—Dorsal view of head showing the position of nostril; B—Lateral view of head.

(*ii*) Mouth. There is a large up-like depression on the ventral ide on the anterior end of the ead. It is called the **sucker** or **uccal funnel**. It is surrounded by fleshy marginal membrane which ears numerous small projections alled the **oral fimbriae** and **lip entacles** and some long sensory rocesses called the **cirri** (Fig. 8.4). he fimbriae help in attachment vith the host. The inner side of e buccal funnel bears radiating ows of conical, yellow, horny, pidermal teeth. Teeth have a efinite arrangement and are unlike ertebrate teeth. The teeth rest on artilaginous pads. At the center of e buccal funnel is a small **circular**

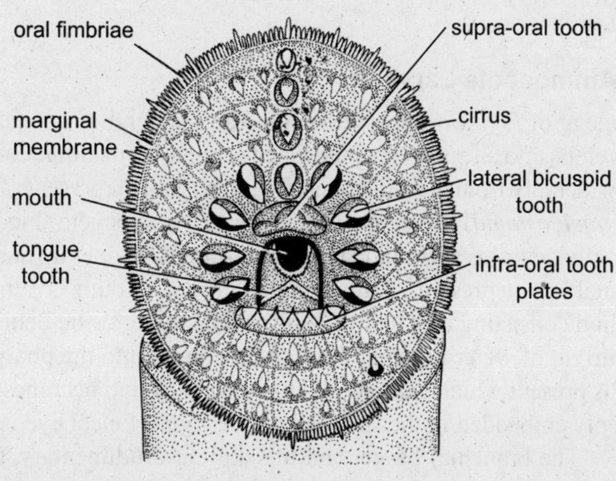

Fig. 8.4. *Petromyzon*. Ventral view of the head showing buccal funnel, mouth and tongue.

mouth opening. Through the mouth protrudes a mobile **tongue**, which also bears horny teeth which are longer than those in the funnel. By back and forth movements, the tongue rasps (scrapes) a hole in the body of victim for sucking blood.

A large prominent eye is present on each lateral side of the head. The eyes lack eyelids and are covered by a transparent area of skin, called **conjunctiva**.

***(iii)* Trunk.** The trunk lacks paired appendages. It bears two thin, vertical folds, the **anterior** and **posterior dorsal fins**, on its hind region. The fins are supported by cartilaginous rods, called the **fin rays**.

***(iv)* Tail.** The tail tapers posteriorly and bears a vertical **caudal** or **tail fin**. The dorsal lobe of the tail fin is continuous with the posterior dorsal fin. The caudal fin is also supported by **cartilaginous fin rays**.

Body Apertures. These are as follows:

(*a*) **Mouth**, a narrow aperture lying at the apex of the buccal funnel.

(*b*) **Naris, nostrils** or **nasohypophysial aperture**, a single mid-dorsal aperture between the eyes on the head.

(*c*) An oval area of thin, semitransparent skin just behind the naris indicates the position of the **pineal eye** inside.

(*d*) Seven small rounded apertures of **external gill slits** in a longitudinal row on each lateral side of the head, just behind the eye. Water enters and leaves them alternately for the respiration.

(*e*) A slit-like opening, the **cloaca**, is located at the junction of trunk and tail. A **urinogenital papilla** with a urinogenital aperture protrudes through the cloaca.

(*f*) Numerous sensory pores of **lateral line system** are located along the lateral side the body.

Sexual Dimorphism in *Petromyzon*

The sexes are separate and distinguishable when mature by secondary sexual characters. The secondary sexual characters of the female sex include the development of an **anal fin** behind the urinogenital aperture and distension and reddening of margins of anal aperture.Those of the male sex comprise formation of penial tube and thickening of the base of the dorsal fin.

Economic Importance

Lampreys kill valuable food and game fish by sucking their blood. Larval lampreys serve as a nice bait for commercial and sport fishing. Adult lampreys are used as food to some extent in some part of Europe.

2. Ammocoete Larva

The egg of *Petromyzon* is telolecithal and undergoes unequal but complete cleavage and ultimately develops into an embryo which hatches out to form ammocoete larva. The muddy brown larva differs considerably from the adult lamprey and possibly represents an intermediate stage between amphioxus (*Branchiostoma*) and a primitive craniate. Ammocoete (Fig.8.5) is about 10 mm long and the mouth (stomodaeum) is surrounded by a semi-circular upper lip and a transverse lower lip. Suctorial buccal funnel is not present. Teeth are also absent. Mouth is surrounded by several branched tentacles. Velum consisting of a pair of cup shaped structures lie behind the tentacles, and help in maintaning a current of water entering through the mouth into the pharynx. Paired fins are absent but a median fin is present which extends all along the back and becomes continuous with the caudal fin. Eyes lie deeply embedded in the skin and are vestigial. Pineal eye is well developed.

The branchial basket and the skull are rudimentary. Gall bladder and bile ducts are present Kidney is pronepheric. Seven pairs of gill sacs are present as organs of respiration; each opens to the exterior by an aperture, and internally into the pharynx. The pharynx in addition to helping in respiration, is specialized as a feeding organ, and is provided with a hypopharyngeal groove (endostyle in the midventral line (Fig. 8.6). A hyperpharyngeal groove (dorsal lamina) is also present on the mid dorsal line, and the two are joined together by a pair of peripharyngeal ciliated bands.

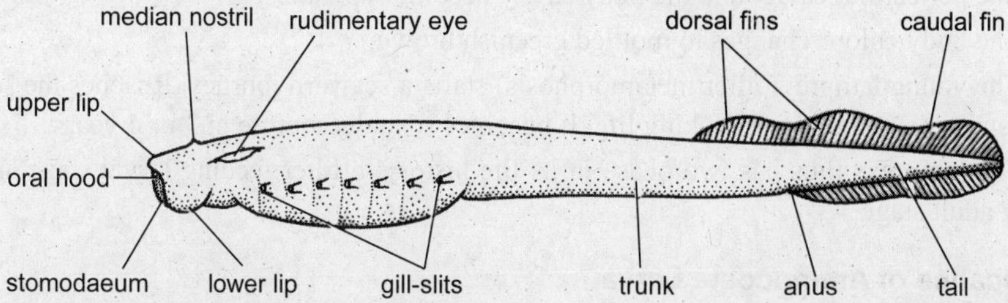

Fig. 8.5. *Petromyzon*. External features of ammocoete larva (in lateral view).

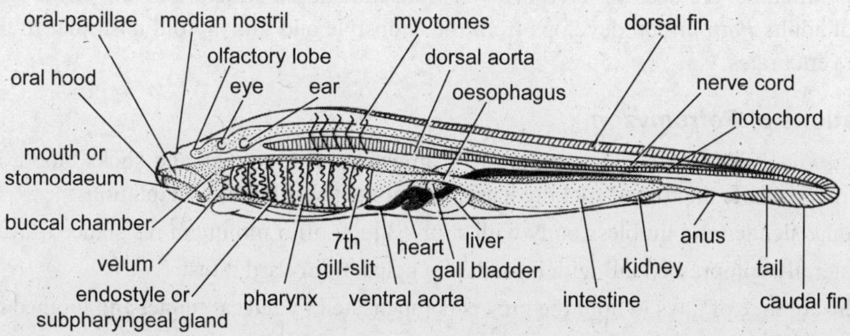

Fig. 8.6. *Petromyzon*. Internal structure of ammocoete larva.

Feeding. A current of water is maintained by the movements of the velum and the pumping action of the pharynx. This water current enters the pharynx through the mouth and brings in small food particles. Thus, the primitive ciliary mode of feeding is replaced by a muscular mode of feeding. The microscopic food particles entering the pharynx become entangled in the copious mucus secreted by the endostyle. The food particles and the mucus in the form of strands are carried to the dorsal lamina through the peripharyngeal grooves and finally enter the intestine due to the movements of cilia.

The ammocoete larva remains for most of the time in the burrow, but at night occasionally emerges to change its feeding ground. It swims with head downward and can burrow very quickly. Larval life lasts for 5 to 8 years.

Metamorphosis. Ammocoete undergoes metamorphosis losing several of its larval features and becomes the adult. During metamorphosis following important changes occur in the larva:

1. The endostyle changes into the thyroid gland.
2. Oral hood changes into buccal funnel with teeth and tongue. Thus, a suctorial funnel is formed and a rasping tongue with teeth develop.
3. Eyes become large and functional.
4. The continuous median fin breaks up into two dorsal and a caudal fin.
5. Velum is reduced. Oesophagus becomes separate from the pharynx, so that food enters the oesophagus and respiratory water goes to the ventral pharynx.
6. Gill sacs open into pharynx. Gills appear in the gill pouches. Skull, branchial basket and other cartilages develop.
7. The gall bladder and the bile duct disappear.

8. Pronephros is replaced by the mesonepheric kidney.
9. Neural arches appear in the trunk region.
10. The pericardial cavity and the body cavity become separate.
11. The body colour changes to mottled greenish brown.

The young lamprey (after metamorphosis) starts a seaward journey. Reaches the sea, it takes to free swimming and blood sucking life. It becomes sexually mature in 3 or 4 years.

The lampreys thus has two phase in its life history: a filter feeding larval stage and a blood sucking adult stage.

Significance of Ammocoete Larva

The ammocoete larva is of great phylogentic interest, because it exhibits the characters of a primitive or generalised chordate and resembles closely to the adult *Brenchiostoma*. A number of complex chordate structures are seen to have evolved from the simple structure of the larva, *e.g.*, thyroid glands of adults *Petromyzon* develops from the endostyle and this thyroid is similar to the thyroid of other vertebrates.

Adaptations of *Petromyzon*

Petromyzon is a fish-like aquatic animal. It is an ectoparasite on fish and sucks blood. Due to its aquatic mode of life and blood sucking habit, it exhibits the following adaptations:

1. Long, slender and limbless body with slimy skin to offer minimum resistance to water.
2. Laterally compressed tail with a caudal fin helps in forward thrust.
3. The absence of jaws to hold the prey is compensated by buccal funnel and toothed tongue for blood sucking.
4. The saliva contains an anticoagulant which ensures a continuous flow of blood by preventing its clotting.
5. *Petromyzon* carries on drawing in and expelling out water through the gill slits even during feeding when the buccal funnel is tightly fixed to the prey.
6. It lives in coastal water which provides easy concealment under stones and sea-weeds.
7. It escapes from its enemies by active swimming.
8. Eyes are covered by transparent skin which helps in seeing under water and protects the eyes from saltish water.
9. It lays large number of eggs. It compensates its inability to breed more than once in her life-time and counteracts mortality of larvae during migration.
10. The larvae live in burrows which protects them from enemies.
11. The filter feeding mode suits to their sedentary life in burrows.

3. *Myxine* (Hagfish)

Classification

Phylum	—	**Chordata**
Subphylum	—	**Vertebrata**
Class	—	**Cyclostomata**
Order	—	**Myxinoidea**
Genus	—	*Myxine*

Habitat. The hagfish (or hag) is found along the coasts of Northen Europe and North America.

Habits. *Myxine* is a noctural animal. It generally attacks the injured or dead fishes and the fishes caught in the nets and on lines (lines are types of nets used to catch fishes in the sea). It is predaceous or scavenger. It tends to enter the body of victim by way of its mouth or opercular outlet and consumes all the soft parts by means of its rasping tongue, leaving only a empty shell of skin and bones. The fishes usually attacked are cods and flounders. When not feeding the hagfish lives in the mud at the sea floor. Hag is capable of rapid swimming by eel like undulations. Sexes are separate. Differentiation of gonads occur late in life as in lamprey. Spawning occurs on the ocean floor. *Myxine* is not migratory. The female hagfish lays large, shelled eggs with hooked processes at each pole. By these processes, the eggs cling to one another and also to sea weeds. Development is direct. Young adult hatches out of the egg.

Structure

Hag's body is pinkish, soft, scaleless, elongated, cylinderical and even more selender and eel-like. *Myxine* reaches a length of 40 to 80 cm. Body of *Myxine* is divisible into **head**, **trunk** and **tail** (Fig. 8.7).

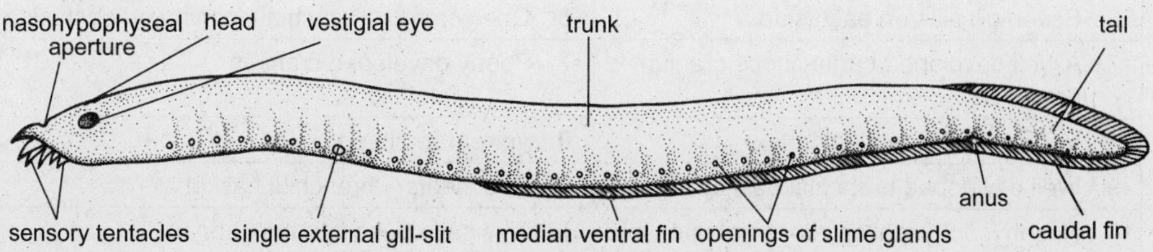

Fig. 8.7. *Myxine.* External features.

1. Head. There is no buccal funnel. The **mouth** is situated on the ventral side near the anterior end. Mouth is surrounded by soft wrinked lips like those of an old women (or hag) hence, its common name **hag-fish**. The mouth is bordered by two pairs of sensory tentacles. The tongue is protrusible. There is a single median **palatine tooth** above the mouth aperture and two rows of smaller teeth on the tongue. The tongue teeth are situated on **dental plates**. The tongue acts as powerful rasping organ. The single median **nostril** lies very close to the mouth. It is also surrounded by two pairs of tentacles. All the tentacles are supported by cartilaginous rods. The nostril opens into a **nasohypophyseal canal** which opens into the pharynx. Water reaches the gill through the nostril, nasohypophyseal canal and the pharynx. The **lateral eyes** are degenerate and covered by thick skin. The skin in the middle of head on the dorsal surface is transparent and the unpaired **pineal eye** can be seen below it.

2. Trunk and tail. Gill slits are not found in *Myxine*. Only one pair of common **branchial openings** are present on the ventral side of the body. These are situated at about one third of the body length from the anterior end. The left branchial opening is larger because a pharyngocutaneous duct also opens to outside through it. Each branchial opening communicates with six branchial pouches by six separate ducts. **Anus** or the **cloacal aperture** is slit-like. It is located just in front of the begining of tail fin on the ventral side.

Paired fins and the dorsal fin are absent. *Myxine* consists of a **caudal fin** that extends well forward on the ventral surface as the **ventral fin**. The caudal fin is of diphycercal type. **Diphycercal.** This is pertaining to a tail fin, having symmetrical upper and lower parts, and with the vertebral column extending to the tip without upturning. The fins are supported by cartilaginous fin-rays.

There is a row of small holes on both sides of the entire body. **Mucous** or **slime glands** are large sized and open on the surface of body through these holes. Enormous amount of mucus or slime is secreted by these gland.

There are some scarcely visible pits on the head that are regarded as primitive **lateral-line organs**.

Differences between lampreys (*Petromyzon*) and hag fishes (*Myxine*) have been tabulated in Table 8.4.

Table 8.4 Differences between *Petromyzon* and *Myxine*.

Lampreys (*Petromyzon*)	Hag-fishes (*Myxine*)
1. Posses one or two dorsal fins.	1. Dorsal fin is absent.
2. A suctional buccal funnel is present. Anti-coagulent is secreted.	2. No buccal funnal. Anticoagulent is not secret-ed.
3. Eyes are large.	3. Eyes rudimentary.
4. Barbels are not present.	4. Three pairs of barbels (tentacles) are present,
5. A single nasal aperture is present on the dorsal side of head.	5. Nasal aperture lies in front of head.
6. Seven gill slits on each side.	6. One common branchial opening on each side.
7. A well developed cartiaginous cranium is present.	7. Poorly developed cranium.
8. Vertebrae are cartilaginous.	8. Vertebrae absent.
9. Well developed branchial basket.	9. Rudimentary branchial basket.
10. Nasohypophyseal sac does not open into the pharynx.	10. This sac opens into the pharynx.
11. A pronephric kidney is absent in adult.	11. Pronephric kidneys are present in adult.
12. Lateral line organs are well developed.	12. These are degenerate.
13. Eggs are small and without hooks. Mete-morphosis takes place.	13. Eggs are large with hooks. There is no meta-morphosis.
14. Both marine and freshwater. Migration occur. Breeding in fresh water.	14. Exclusively marine. Breed in sea.
15. Blood sucking (sangivorous) ectopara-sites of fish and turtles.	15. Predaceous and scavengers food include fish-es, polychaetes and other invertebrates.

8.6 AFFINITIES OF CYCLOSTOMES

Cyclostomes are most ancient living craniates. They are fish like but differ from fishes in many characters. They resemble cephalochordates and ostracoderms in certain features and posses several specilized characters due to their parasitic mode of life.

A. Primitive Characters

The cyclostomes show primitiveness in having characters similar to those of cephalochordates (*Branchiostoma*).

1. Similarities with Cephalochordates

1. Absence of exoskeleton, jaws, paired fins and gonoducts. 2. Persistent and continuous (unconstricted) notochord. 3. Segmental myotomes extends from head to tail. 4. Relatively large number of gill slits

(*e.g.*, some hag fishes). 5. Alimentary canal is straight without stomach. 6. Dorsal and vental nerve roots of spinal nerves are separate in lampreys. 7. Nerves without a myelin sheath (*i.e.*, nerves are non-medullated).

In addition to the above mentioned similarities between the adults of two groups, **ammocoete larva** of lamprey resembles *Branchiostoma* in having: 1. Fish shaped body. 2. Continuous dorsal and caudal fins. 3. An mucus secreting endostyle. 4. Ciliated alimentary canal. 5. Oral-hood. 6. Both collect small food particles of food by using oral tentacles. 7. Both have an enlarged pharynx whose walls are supported by a longitudinal series of cartilaginous gill bars. 8. Distribution of major blood vessels and the direction of blood flow is similar. 9. The single median nostril of ammocoete is homologous to the Kolliker's pit of *Branchiostoma.*

2. Characters that are more Primitive than in Fishes

1. Absence of jaws, true teeth and paired fins. 2. Nearly continuous median fin folds (as in hag fishes). 3. Diphycercal caudal fin. 4. Fin rays without muscles attachments. 5. Cranium incomplete. 6. Poorly developed vertebrae. 7. Pancreas is rudimentary. 8. Poorly developed spiral or scroll valve in the intestine. 9. Single median nostril. 10. Brain relatively small. Cerebral hemispheres and cerebellum rudimentary. 11. Sympathetic nervous system and lateral line organs poorly developed. 12. Ninth and tenth nerves are not enclosed in the cranium. 13. Absence of medullated nerves. 14. Nasohypophyseal duct is larger. It is not connected with the pituitary gland. 15. Heart is a loosely twisted S-shaped tube. 16. 1 to 2 semicircular canals in the internal ear.

3. Characters which are Primitive in hag Fishes than in Lampreys

1. Median fin continuous and undifferentiated. 2. Terminal nostrils. 3. Less specialized brain. Cerebral hemispheres and cerebellum absent. 4. Cranium is less developed. 5. Only one semicircular canal in the internal ear. 6. Adult contains functional pronephros. 7. Intestine without spiral valve. 8. No vertebrae. 9. Nasohypophyseal duct opens into the pharynx. 10. Lateral-line organs less developed. 11. Relatively large numbers of gill slits.

4. Character which are more Primitive in Lampreys than in Hag-fishes

1. Roots of dorsal and ventral spinal nerves separate. 2. Small eggs with little yolk. 3. Cleavage holoblastic. 4. Larva shows many resemblances to *Branchiostoma*.

B. Specialized Characters

I. Characters of Cyclostomes that show advancement over *Branchiostoma* (orAffinities with Vertebrates)

1. A well defined head, cranium and brain is present. 2. Liver and pancreas well developed. 3. Mesonepheric kidney is present. 4. Epidermis is several layers thick. A large number of unicellular mucous glands are present. 5. Myotomes are W-shaped. 6. A second sheath around notochord. 7. Dorsal spinal nerves contain ganglia. 8. Vertebrae present. 9. Sympathetic nervous system more developed. 10. Eyes and internal ears developed. 11. Thyroid and pituitary glands developed. 12. Heart developed. Blood contains erythrocytes and leucocytes. 13. Lateral-line system present. 14. Gill slits are used for respiration, not for food collection.15. Hepatic portal and lymphatic system developed. 16. Water is drawn into the pharynx by muscular action, not by ciliary action.

II. Other Specialised Characters of Cyclostomes

Adult cyclostomes show many specializations which are probably adaptations for parasitic mode of feeding. Some of these characters are as follows: 1. Suctorial mouth with buccal funnel and horny teeth (as in lampreys). 2. Muscular protrusible and rasping tongue with teeth. 3. Production of anticogulant in saliva that keeps the wound bleeding. 4. sac-like gill pouches. 5. Complete separation of sac-like pharynx from the oesophagus. 6. A cartilaginous branchial basket to support pharynx and help in respiration. 7. Water current enters and leaves the gill pouches through the gill slits. Mouth is kept free to attach to rocks for resting or fish for feeding. 8. Dorsal position of nostril (as in lampreys). 9. Large number of mucous glands (as in hagfishes; Box 8.4) and secretion of enormous amount of slime. 10. Large, heavily-yolked eggs, meroblastic cleavage and no larval stage in hag fishes.

Box 8.4

Hag-fishes are renowned for their ability to generate enormous quantities of slime. If disturbed or roughly handled, the hagfish exudes a milky fluid from its paired mucus glands. On contact with sea water, the fluid forms a slime so slippery that the animal is impossible to grasp (**Hickman** *et al.*, 1996).

C. Degenerate Characters of Cyclostomes

1. The greatly elongated eel-like body. (Ancestral ostracoderms had fish-like shape of the body). 2. Rudimentary pair of eyes in hag-fishes. 3. Reduced liver and absence of gall bladder and bile duct in adult lamprey. 4. Absence of exoskeleton, 5. Absence of ossification of endoskeleton, 6. Absence of jaws, paired fins and girdles.

D. Relationship with Fishes

Cyclostomes are not fishes. They differ from fishes in the following ways: 1. Absence of biting jaws 2. Absence of paired fins. 3. Presence of rasping tongue. 4. Presence of a single nasal aperture. 5. Presence of a large pineal body.

Hence, cyclostomes were separated from the true fishes (gnathostomes) and placed in a separate group Agnatha. Cyclostomes are considered to be degenerate fishes by some authors due to the absence of paired jaws, paired fins and girdles. It has been suggested that the presence of vestigial eyes, median fin muscles and rudimentary branchial basket in hagfishes, show that cyclostomes are degenerate animals. Presence of small folds of skin on either side of cloaca in *Myxine* is supposed to represent vestigial pelvic fins. However, there is little evidence to support this view. The eyes of hagfishes possibly represent a primitive stage of the development of visual organs. No evidence is available to show that the ancestors of cyclostomes ever had paired limbs, functional eyes or biting jaws. Hence, cyclostomes do not appear to be degenerate descendants of fishes.

E. Relationship with Ostracoderms

Ostracoderms include the earliest known craniate agnathous chordates. They first appeared in the upper Cambrian 540 million years ago. They survived for 150 million years and became extinct in Devonian. Ostracoderms and living cyclostomes are classified together under agnatha. The two groups show following similar features: 1. Jaws absent, 2. Single nostril, 3. Pineal eye present, 4. Pouch-like gill sacs, 5. Simple structure of brain, 5. Arrangement of cranial nerves is similar.

Cyclostomes differ from the ostracoderms in the complete absence of bone, and paired fins and in the presence of suctorial mouth and a rasping tongue.

Taxonomic Position of Cyclostomata

Living cyclostomes are regarded to be highly specialized and degenerate descendants of ostracoderms. **Stensio** (1927) believed that the cyclostomes evolved from some ostracoderms by the evolution of sucking mouth and loss of bone and paired fins. Along with these features, a rasping tongue appeared in cyclostomes as an adaptation to parasitic habits. During the process bone was completely replaced by the cartilage. According to **Svertsov** (1939), the cyclostomes and ostracoderms evolved from a common ancestor. Taxonomic position of the cyclostomes remains doubtful but it is certain that the cyclostomes are closely related to extinct ostracoderms.

QUESTIONS

Long Answer Questions

1. Discuss the general characters of subphylum Vertebrata.
2. Give classification of the vertebrates upto classes only.
3. Tabulate differences between Pisces and Tetrapoda.
4. Discuss the affinities of Cyclostomata.
5. Give the distinguishing characters of cyclostomata. Classify it giving characters and examples of each group.
6. Discuss primitive, specialized and degenerate characters of Cyclostomata.
7. Describe external features of *Peteromyzon*. (*Madras 2011*)

Short Answer Questions

1. How do Agnatha and Gnathostomata differ?
2. Name the classes of tetrapoda. Give one character and one example of each.
3. Give taxonomic/systematic position of lamprey or hagfish.
4. In what respect do cephalochordates and cyclostomes resemble?
5. State briefly the adaptations for parasitic mode of life in cyclostomes.
6. What are affinities of cyclostomes with ostracoderms?
7. What are lampreys? Describe the distribution, habits and habitat of *Petromyzon*.

Very Short Answer Questions

1. What are the main respiratory organs of the vertebrates?
2. Name the portal system found in vertebrates.
3. Name the classes of the superclass Tetrapoda.
4. Name the section to which cyclostomes belong.
5. Name the vertebrates in which the notochord persists as unconstricted rod throughout life.
6. What is alternative name for Cyclostomata?
7. Mention the mode of feeding of ammocete.
8. Name a cyclostome in which development is direct.
9. The Cyclostomes are degenerate or primitive animals. Give your view.
10. Why lampreys are not true eels?
11. Why is cyclostomata also named Marsipobranchii?
12. What is branchial basket?
13. What is the role of branchial basket?
14. Why lampreys are nuisance to fish industry?
15. What is anadromus migration with reference of *Petromyzon*?
16. What are cyclostomes?
17. Name the fresh water species of lamprey.
18. Name the marine form of lamprey.
19. What causes the mouth of lamprey to open and close in the absence of jaws.
20. What are sanguivorous animals?

Fill in the Blanks

1. In cyclostomes the tail is ________.
2. In *Petromyzon*, there are ________ pairs of gill slits.
3. In the vertebrates, notochord is replaced by ________.
4. Brain is enclosed in ________ and spinal cord in ________.

5. In vertebrates, the body cavity is ________.
6. Respiratory organs of vertebrates are ________ or ________.
7. In vertebrates, respiratory pigment occurs in ________.
8. Excretory organs of vertebrates are a pair of ________.
9. In fishes, the paired appendages are ________.
10. The endoskeleton of cyclostomes is composed of ________.
11. In cyclostomes, pharynx is supported by ________.
12. The intestine of cyclostomes contains a fold called ________ or ________.
13. Lamprey's heart is ________ chambered.
14. The nerve fibers of cyclostomes are ________.
15. The zoological name of marine lamprey is ________.
16. *Myxine* is commonly called ________.
17. Larva of *Petromyzon* is known as ________.
18. *Petromyzon* breeds in ________ water.
19. Ammocoete larva is ________ feeder.
20. Lampreys feed on ________ fishes, while hagfishes attack ________ fishes.
21. Hagfishes burrow into the body of ________ fishes.

Multiple Choice Questions

1. Vertebrate body cavity is
   (*a*) haemocoel (*b*) pseudocoel
   (*c*) true coelom (*d*) coelenteron
2. Respiratory pigment in the vertebrate is
   (*a*) haemocyanin (*b*) haemoglobin
   (*c*) haemocruorin (*d*) none of these
3. Spinal neves of vertebrates are
   (*a*) mixed (*b*) sensory
   (*c*) motor (*d*) none of these
4. The excretory organs of vertebrate are
   (*a*) nephridia (*b*) neural glands
   (*c*) kidneys (*d*) proboscis glands
5. *Petromyzon* belongs to the group
   (*a*) Agnatha (*b*) Pisces
   (*c*) Gnathostomata (*d*) Tetrapoda
6. *Petromyzon* belongs to the class
   (*a*) Chondrichthyes (*b*) Osteichthyes
   (*c*) Cyclostomata (*d*) Placodermi
7. Gills are located in sac-like pouches in
   (*a*) *Scoliodon* (*b*) *Protopterus*
   (*c*) *Labeo* (*d*) *Petromyzon*
8. Organized pancreas and spleen are lacking in
   (*a*) lamprey (*b*) shark
   (*c*) carp (*d*) hagfish
9. Lampreys are
   (*a*) herbivorous (*b*) carnivorous
   (*c*) sanguivorous (*d*) detritivorous
10. *Myxine* is
   (*a*) sanguivorous (*b*) scavenger
   (*c*) herbivorous (*d*) microphagous

11. Number of semicircular canals in the internal ear of lamprey is
(*a*) one (*b*) two
(*c*) three (*d*) four

12. In Cyclostomata
(*a*) pharynx is only digestive in function
(*b*) pharynx is only respiratory in function
(*c*) pharynx is divided into respiratory and digestive parts
(*d*) pharynx is absent

13. The ammocoete larva lasts for about
(*a*) 5 to 6 days (*b*) 3 to 7 years
(*c*) 4 to 5 weeks (*d*) 4 to 5 months

14. Tail in Cyclostomata is
(*a*) diphycercal (*b*) homocercal
(*c*) heterocercal (*d*) absent

15. In cyclostomes exoskeleton is
(*a*) bony (*b*) cartilaginous
(*c*) scaly (*d*) absent

16. Kidneys in lampreys are
(*a*) mesonephric (*b*) metanephric
(*c*) pronephric (*d*) glomerular

17. Eggs of lampreys are
(*a*) isolecithal (*b*) alecithal
(*c*) polylecithal (*d*) telolecithal

ANSWERS

Very Short Answer Questions

1. Gills and lungs.
2. Renal, hepatic and pituitary.
3. Amphibia, Reptilia, Aves and Mammalia.
4. Agnatha.
5. Cyclostomata.
6. Marsipobranchii.
7. Filter (ciliary) feeding.
8. Hagfish.
9. Degenerate animals.
10. Lampreys resemble eels in appearance but are not eels, because these are without jaws, their endoskeleton is cartilaginous, respiratory organs are gill pouches and not true gills of eel. Eels are bony and highly developed and specialised fishes with bony endoskeleton.
11. In cyclostomes, the gill pouches lie in a sac-like pouch separated off from the pharynx. Therefore, these are named Marsipobranchii.
12. In *Petromyzon*, the branchial region is supported by a basket-like structure of cartilage. It is called branchial basket.
13. Branchial basket expands and contracts by the muscular activity and helps in drawing water in through mouth and pumping it out through gill slits.
14. Lampreys destroy fishes by feeding on their blood and body fluid and thus cause great harm to fish industry.
15. *Petromyzon* ascends freshwater rivers and streams for spawning (laying eggs). This type of migration from sea to freshwater is called **anadromous migration**.

16. Cyclostomes are chordate animals with a circular mouth, eel-shaped body, naked skin, single median nostril and no paired fins.
17. *Lampetra fluviatilus*.
18. *Petromyzon marinus*.
19. Mouth of lamprey is opened and closed due to back and forth (saw-like) movements of the toothed tongue.
20. Animals which survive by sucking the blood of their hosts are called **sanguivorous**, *e.g.*, *Petromyzon*.

Fill in the Blanks

1. Diphycercal. **2.** Seven. **3.** Vertebral column. **4.** Cranium, vertebral column. **5.** Coelom. **6.** Gills, lungs. **7.** Erythrocytes. **8.** Kidneys. **9.** Pectoral and pelvic fins. **10.** Cartilage. **11.** Branchial basket. **12.** Typhlosole, spiral valve. **13.** Two. **14.** Nonmyelinated. **15.** *Petromyzon marinus*. **16.** Hagfish. **17.** Ammocoete. **18.** Fresh/river. **19.** Filter. **20.** Living, dead. **21.** Dead or injured.

Multiple Choice Questions

1. (*c*)	**2.** (*b*)	**3.** (*a*)	**4.** (*c*)	**5.** (*a*)	**6.** (c)
7. (*d*)	**8.** (*a*)	**9.** (c)	**10.** (*b*)	**11.** (*b*)	**12.** (*b*)
13. (*b*)	**14.** (*a*)	**15.** (*d*)	**16.** (*a*)	**17.** (*d*)	

9

Superclass Pisces (Fishes): General Characters, Classification and Types

Fishes are the earliest known vertebrates and flourished during the Devonian period, about 400 million years ago. They form a highly successful group of animals comprising more than 40,000 species inhabiting all seas, rivers, lakes, canals, dams, muddy water, brackish water, estuaries and all places where there is water. They differ from each other in shape, size and habits, ranging from 10 millimeters to 20 meters (**Chhapgar**, 2008). A very wide distribution of fishes into a variety of habitats has resulted in numerous adaptations.

A fish may be described as an animal that live in water, has a backbone, breathes by means of gills and has paired fins to enable it to move. While some animals may have one of these characters, only a fish has all of them. Thus, a silverfish does not breathe through gills or live in water—it is as insect found making holes in books. Jellyfish, starfish, crayfish and cuttlefish do not have a backbone and therefore, are not fish. A sea snake lives in water and has a backbone, but it breathes through lungs. So also do dolphins, porpoises and whales are actually mammals that suckle their young.

Study of fishes is called **ichthyology** or **fish biology** (**Khanna** and **Singh**, 2003).

9.1 GENERAL CHARACTERS OF PISCES

Fishes exhibit the following common characters most of which are related to their permanently aquatic mode of life.

1. Fishes are aquatic cold-blooded (poikilothermic) gnathostome vertebrates.
2. They are usually with a streamlined body, but some are snake-like and a few are dorso-ventrally flattened. Body is divisible into head, trunk and tail.
3. They swim or move by seven sets of fins which are supported by fin-rays of dermal origin. Of these, four are paired, that is one on each side of the body, while the remaining three are median, being situated vertically on the midline of the body. Thus, these are pectorals, pelvics, median dorsal, ventral and caudal fins.
4. Exoskeleton of scales is usually present, but some species are naked secondarily.
5. Endoskeleton cartilaginous or bony. Jaws are hinged.
6. Notochord is partially replaced by vertebrae, which may be cartilaginous or bony.
7. Paired visceral arches present. The first pair forms the upper and lower jaws, second pair forms the suspensorium and remaining support the gills.
8. Respiration by gills.
9. A swim-or air-bladder is usually present (except in elasmobranchs).

10. Kidney is mesonephric. Excretion ureotelic.
11. Heart two chambered. Heart has one auricle and one venticle and is of venous type, pumping blood to gills for aeration.
12. Lateral line system well developed.
13. Middle ear is absent. Internal ear present.
14. There are ten pairs of cranial nerves.
15. Sexes are separate.
16. Extraembryonic membranes (*e.g.*, amnion and allantois) absent.
17. In some fishes, these are several pairs of feelers around the mouth, sensitive to touch. These are known as **barbels**.

9.2 CLASSIFICATION

J. Muller (1844) gave first classification of the lower verterates and divided *Pisces* into following 6 subclasses: 1. Dipnoi; 2. Teleostei; 3. Ganoidei; 4. Elasmobranchi; 5. Marsipobranchii (Cyclostomi) and 6. Leptocardii (Amphioxini). He included cyclostomes in Marsipobranchii and cephalochordates in Leptocardii.

Agassiz (1857) separated lampreys and hagfishes into a separate class Myzontes and divided fishes into four classes. **Boulenger** (1924) classified the teleostean fishes into thirteen orders. **C. Tate Regan** (1906–1929) gave an elaborate classification of fishes in his series of papers. **Stensio** (1931), **Goodrich** (1930), **Berg** (1940; Fig. 9.1), **Grasse** (1958) and **Romer** (1959) have given a detailed classification of fishes.

Series Pisces → **Class (7)**

1. Acanthodii
2. Coccostei (Arthrodira)
3. Pterichthyes (Antiarchi)
4. Elasmobranchii
5. Holocephali
6. Dipnoi
7. Teleostomi —Subclasses→ (*i*) Crossoptergii (lobe-finned fishes)
   (*ii*) Actinopterygii (ray-finned fishes)
   ↓
   Several orders (11)

Fig. 9.1. Scheme of classification of Berg (1940).

Berg (1940) based his classification on the work of **Regan** (1906–29) and **Stensio** (1931). He used following endings in the names of taxonomic categories:

Order	—	formes
Suborder	—	oidei
Superfamily	—	oidae
Family	—	idae
Subfamily	—	ini

Romer (1959) classified the fishes into two classes: **Chondrichthyes** (includes all cartilaginous fishes, *i.e.*, combined Elasmobranchii and Holocephali) and **Osteichthyes** (includes all *i.e.*; Dipnoi and Telostomi of Berg). Osteichthyes is divided into subclasses: Sarcopterygii (lobate fishes) and Actinopterygii (Fig. 9.2). **Parker** and **Haswell** (1962) have combined all the extinct jawed fishes under a single class: **Placodermi** or **Aphetohyoidea**. This simple division of superclass pisces into three classes: **1. Placodermi**, **2. Chondrichthyes** and **3. Osteichthyes** has been followed more or less by all the eminent authors.

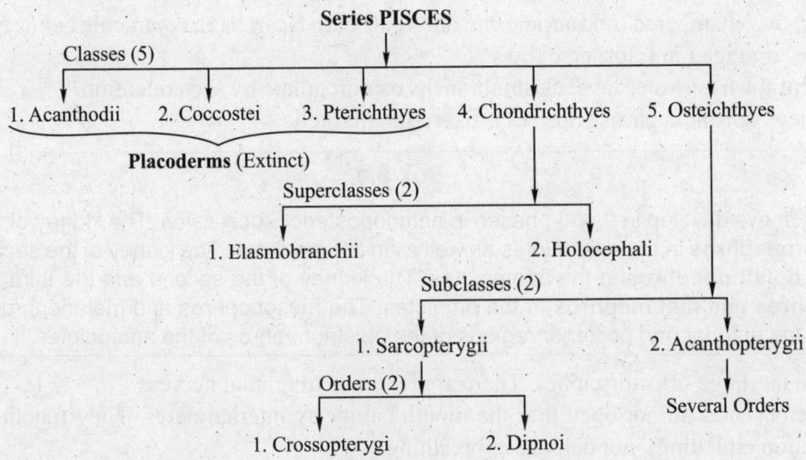

Fig. 9.2. Scheme of classification by Romer (1959).

Greenwood *et al.* (1966) have suggested some significant modifications in the scheme of classification of fishes. **Lander** and **Liem** (1983), **Nelson** (1984) and **Pough** *et al.* (1989) have suggested a new scheme, which is often referred to as the **modern classification of fishes**.

Initially, the cartilaginous fishes were supposed to be primitive fishes and the bony fishes derived from them with the development of bones; but the view is now reversed. The first vertebrates, Ostracodermi and Placodermi found in the fossil state, have a heavy armour of bone and the first bony fish is found fossil in the Silurian age. Hence the present view based on palaeontological evidence is that of absence of bone in the cartilaginous fishes is a secondary condition due to the reduction of ossification whereas the bony fishes have retained the bony skeleton and perforated it for protection (see **Ayyar** and **Ananthakrishnan**, 2008).

9.3 GENERAL CHARACTERS OF CLASS CHONDRICHTHYES

The class Chondrichthyes comprises the sharks, skates, rays and rat fishes (or chimaeras), which are predaceous. Most chondrichthyes are marine. A few are freshwater. Their body is fusiform and spindle shaped. They are characterised by the entire absence of bone, and the skeleton is composed of cartilage only; this has given the name to the class (Gr., *chondros* = cartilage + *ichthyes* = fishes). All these fishes have a covering of dermal denticles or **placoid scales**, which seem to be the remnants of the ancestral bony armour; the bones have disappeared leaving the dermal scales. Among other characteristics of the chondrichthyes are the following:

1. Skin is covered with placoid scales or naked and has mucous glands.
2. Endoskeleton cartilaginous, often calcified. Motility with the head skeleton, called **cranial kinesis**, allowing intake of large food items. Notochord persistent. Vertebrae complete and separate. Pectoral and pelvic girdles present.
3. Paired fins broad based. Fin rays of elastin ceratotrichia.
4. Mouth ventral. Jaws present. Lower jaw is composed of Meckel's cartilage. Jaw suspension is hyostylic or amphistylic. Teeth are modified placoid scales.
5. Stomach J-shaped. Intestine with spiral valve. Cloaca is present. A unique **rectal salt gland** secretes excess salts from the blood in the gut.
6. Respiration by gills. Gills lamellar. 5 to 7 gill slits are present on each side. Operculum or gill cover absent, so that gill slits open directly to the exterior.
7. A spiracle (first visceral cleft) is generally present.
8. Air bladder absent.

9. Heart two chambered (one auricle and one ventricle). Heart has a contractile conus with several valves arranged in transverse rows.
10. Except for freshwater rays, elasmobranchs osmoregulate by urea retention.
11. Kidneys opisthonephros (Box 9.1). Excretion ureotelic.

Box 9.1

The kidneys develop in 2 or 3 phases in anterioposterior succession. The kidney of first phase is called **pronephros** in the anamniotes as well as in the amniotes. The kidney of the second phase is termed **opisthonephros** in the anamniotes. The kidney of the second and the third phase are **mesonephros** and **metanephros** in the amniotes. The mesonephros and metanephros, actually represent the anterior and posterior regions of the opisthonephros of the anamniotes.

12. Brain has large olfactory lobes. There are ten pairs of cranial nerves.
13. Olfactory sacs do not open into the mouth cavity by internal nares. They function only for olfaction (smelling), not helping in breathing.
14. Vision is sharp, but cannot distinguish colours. Ear consists only of membranous labyrinth that bears three semicircular canals.
15. Lateral line sense organs are well developed. They detect waves and currents in water, and electric fields generated by other fishes.
16. Males usually with claspers, derived from pelvic fins.
17. Sexes separate, gonads paired, gonoducts open into the cloaca.
18. Fertilization internal. Oviparous or ovoviviparous. Eggs are large, yolky. Cleavage meroblastic. Development direct, without metamorphosis.

Giant Cartilaginous Fishes

There are about 840 extant species of cartilaginous fishes. They have heavy bodies due to lack of air bladder. Sharks include the largest fishes that ever lived. They are next to whales alone in size. The largest living shark is the tropical, oceanic, oviparous, plankton-feeder whale shark *Rhincodon*. It reaches about 18 meters long (weighing 13 tons). The other plankton feeder basking shark (*Cetorhinus*) is about 7.5 meters long. Following nine species of sharks are involved in attacks of humans : 1. the great white shark; 2. mako shark; 3. bull shark; 4.lemon shark; 5. tiger shark; 6. dusky shark; 7. blue shark; 8. white tip shark and 9. hammerhead shark (see **Chhapgar**, 2008). The human-eating great white shark *Carcharodon Carcharias* is 10.8 meters long. Other ferocious human-eater sharks are blue shark (*Carcharinus*) and the zebra or tiger shark (*Stegostoma tigrinum*).

9.4 CLASSIFICATION OF CHONDRICHTHYES

Class Chondrichthyes or Elasmobranchii is divided into two subclasses:
1. Selachii and 2. Holocephali.

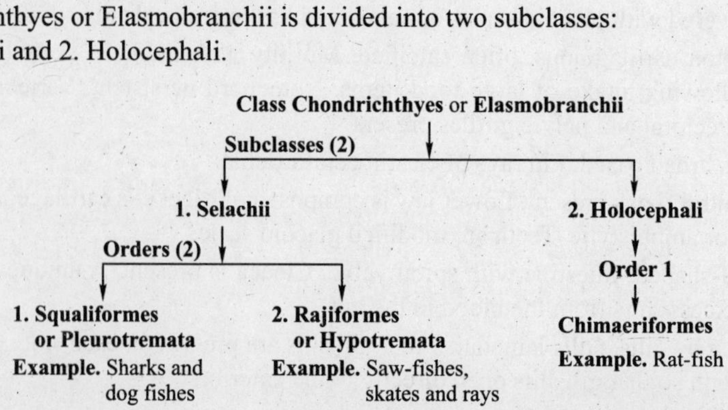

Outline classification of class chondrichthyes.

Subclass 1. Selachii

1. Paired fins without a median axis.
2. Base of paired fin constricted (narrow), pectoral radials jointed.
3. Each gill slit is protected by independent skin flaps.
4. Vertebral centra usually present.
5. Males with claspers.
6. Cloaca present.

The extinct forms are grouped under **Protoselachii** and the modern living forms under **Euselachi**. The Euselachi is divided into two orders.

Order 1. Squaliformes or Pleurotremata

(Gr. *pleuro* = side + *trema* = opening)

1. Body is typically spindle-shaped.
2. Teeth sharp and conical.
3. Gill-slits on the lateral sides of the body; 5–7 pairs gill slits.
4. Spiracles, if present, are located behind the eyes.
5. Pectoral fins are of moderate size and distinctly marked off from the head.
6. Right and left halves of pectoral girdle dorsally separated.
7. Tail heterocercal.
8. About 250 living species.
9. This order includes sharks and dogfishes.

Examples. Dogsharks or dogfishes (*Scoliodon*, etc.), spiny dogfish (*Squalus*), hammer-headed shark (*Sphyrna*), tiger shark (*Stegostoma*), whale-shark (*Rhincodon*), white-shark (*Carcharodon*), Gangetic shark (*Carcharinus*).

Order 2. Rajiformes or Hypotremata

(Gr., *hypo* = below + *trema* = opening)

1. Body depressed, flattened dorsoventrally.
2. Mouth is often highly protrusible to provide suction used to dislodge shelled invertebrates (mollusks) from the substratum. Teeth are blunt.
3. Pectoral fins enlarged, fused with sides of the head and body. Dorsal fins reduced. No anal fin.
4. Spiracles large, dorsal and functional.
5. Two halves of pectoral girdle united either to each other or to the vertebral column.
6. Tail is slender and sharply marked off from the trunk. Reduced caudal fin.
7. About 300 species. This order includes saw-fishes, rays and skates.

Examples. Sawfish (*Pristis*), Guitar fish (*Rhinobatus*), sting rays (*Trygon*, *Dasyatis*), Electric rays (*Torpedo*, *Narcine*), Eagle rays (*Myliobatus*, *Aetobatus*), Skates (*Raja*).

Subclass 2. Holocephali

(Gr., *holos* = entire + *kephale* = head)

1. Body laterally compressed.
2. Four pairs of gill slits but single gill opening on either side covered by a fleshy operculum.
3. No scales, spiracles and cloaca.
4. Upper jaw is fused with the cranium, forming holostylic suspension. Jaws with tooth plates.
5. Nasal opening single.
6. Lateral line system with open grooves.
7. Males with additional claspers. Cephalic claspers are on head. Fertilization is internal.
8. Tail long, thin and whip-like.
9. About 25 species of rat fishes. This subclass has only one order **Chimaeriformes**.

Example. Shortnose chimaeras rat fish or 'king of herrings' (*Chimaera*); *Hydrolagus* and longnose chimaeras such as *Rhinochimaera*.

9.5 TYPE STUDIES (CARTILAGINOUS FISHES)

1. *Scoliodon sorrakowah* (The Indian Shark/Dogfish)

Classification

Phylum	—	**Chordata**
Subphylum	—	**Vertebrata**
Division	—	**Gnathostomata**
Superclass	—	**Pisces**
Class	—	**Chondrichthyes (Elasmobranchi)**
Subclass	—	**Selachii**
Order	—	**Squaliformes (Pleurotremata)**
Genus	—	*Scoliodon*
Species	—	*sorrakowah*

Habit and Habitat

Scoliodon is marine. It occurs along the Indian coast. It is fast-swimmer and carnivorous. *Scoliodon* is predaceous feeding on crabs, lobsters, worms and fishes. Sexes are separate and fertilization is internal. It is viviparous.

Structure

Scoliodon (Fig. 9.3) has about 60 cm long body. Its upper surface is dark grey and the under surface is pale white. The body is long, laterally compressed spindle-shaped and tapering at both ends. **Head** is triangular, dorso-ventrally flattened and is prolonged into a wedge-shaped snout. Head bears crescentic mouth and obliquely placed crescentic nares on the ventral side and two circular eyes and five pairs of

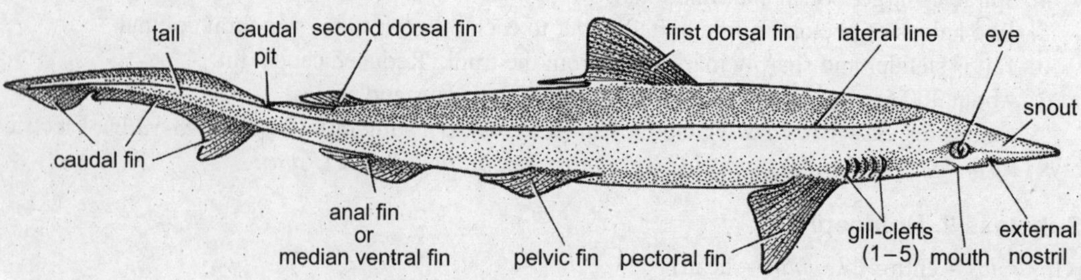

Fig. 9.3. *Scoliodon*. External features.

gill slits on the lateral sides (Fig. 9.4). Many groups of minute ampullary pores are present on the dorsal as well as the ventral surface of head. These pores lead into receptors (called ampullae of Lorenzini) for detecting the temperature variation. Laterally compressed **trunk** bears a pair of pectoral fins, a pair of pelvic fins, an anterior dorsal fin, a posterior dorsal fin and a anal fin midventrally infront of caudal fin. The fins are supported by fin rays. In the male shark, the pelvic fins bear rod-like claspers along their inner borders. Longitudinal cloacal aperture lies between the pelvic fins in both sexes. There is a small abdominal pore situated on a papilla on either side of the cloacal aperture. The coelom communicates with the exterior through the abdominal pores. Upturned heterocercal **tail** lies beyond the cloaca and bears a caudal fin. The caudal fin consists of a low dorsal epichordal lobe and a well developed ventral hypochordal lobe. The ventral lobe of caudal fin is divided by a notch into a larger anterior part and a smaller posterior part. At the root of the caudal fin, there is a caudal pit on both the dorsal and ventral surfaces. The caudal pits are characteristic of the genus *Scoliodon*. Also there is a faint receptor-containing lateral line on either side of body, extending from the head to tip of the tail. The skin of *Scoliodon* is rough in touch. The roughness is produced by the presence of placoid scales (the exoskeleton).

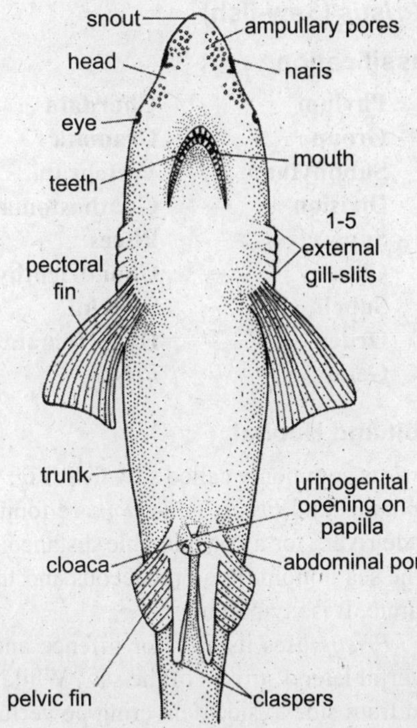

Fig. 9.4. *Scoliodon*. Male dog fish in ventral view.

2. *Sphyrna* (Hammer Headed Shark)

Classification

Phylum	—	**Chordata**
Subphylum	—	**Vertebrata**
Division	—	**Gnathostomata**
Superclass	—	**Pisces**
Class	—	**Chondrichthyes (Elasmobranchi)**
Subclass	—	**Selachii**
Order	—	**Pleurotremata**
Genus	—	*Sphyrna*
Species	—	*zygaena*

Habit and Habitat

Sphyrna (Fig. 9.5) is popularly called **hammer-headed shark**. It is found in temperate and tropical seas (*i.e.*, in Pacific ocean and Indian seas). *Sphyrna* is fast swimmer and a voracious surface feeder, feeding small fishes. It is viviparous. People have occasionally been attacked by the hammer-sharks.

Structure

Body of *Sphyrna* is elongated, measuring about 4.5–5.0 meters in length. Its dorsal side is grey and ventral surface white. Head of *Sphyrna* is broadened and set at right angles to the body, so that it forms the shape of the letter T or hammer. The eyes and nostrils are set at the two ends of the head. Swaying the head from side to side when swimming enables the fish to sample a wider path of water to see and smell its prey. Also the flattened head increases its manoeuvrability.

Crescentic mouth lies on the ventral side of the head and five pairs of vertical gill slits on the lateral sides. There is no spiracle. The large pectoral fins are somewhat ventrally inserted. The small pelvic fins enclose the cloacal aperture and bear claspers in the male. There are two spineless dorsal fins : the larger anterior dorsal fin in front of the pelvic fins and the smaller posterior dorsal fin opposite the anal fin. Tail is large and bears a heterocercal tail or caudal fin. *Sphyrna* is caught for extracting liver oil.

3. *Pristis* (Saw-fish)

Classification

Phylum	—	**Chordata**
Group	—	**Craniata**
Subphylum	—	**Vertebrata**
Division	—	**Gnathostomata**
Superclass	—	**Pisces**
Class	—	**Chondrichthyes**
Subclass	—	**Selachii**
Order	—	**Hypotremata**
Genus	—	*Pristis*

Habit and Habitat

Pristis is commonly called **saw fish** (Fig. 9.5). It is found in all tropical and subtropical seas. *Pristis cuspidatus* and *Pristis microdon* are found in the Bombay coast. The Indian species are known to invade rivers for a considerable distance. *Pristis* is a powerful swimmer but leads a sedentary life on the sea bottom. It is predaceous and feeds on small fishes, molluscs and crustaceans. Sexes are separate. It is viviparous.

Pristis uses its saw for offence and defence, slashing the body of the prey and enemy by powerful lateral strokes of the saw. While swimming through the shoals of fishes sawfish moves its head from side to side delivering powerful blows to the fishes and thus, injuring them. The injured fishes are swallowed at leisure.

Structure

Pristis grows to about 6 meters. In body form, it is intermediate between the sharks and the rays. The anterior half of the body is flattened dorsoventrally as in rays, whereas the posterior half is compressed laterally as in sharks. Rostrum is prolonged into a long flat carpenter's saw armed laterally by strong, tooth-like denticles firmly embedded in sockets. The saw may be up to 1.8 meters long and 30 cm wide and may bear upto 32 pairs of denticles (teeth). The head has on its ventral side a transverse mouth, a pair of oblique nares and five pairs of gill slits. At the dorsal side of head are located a pair of eyes followed by a pair of spiracles. The trunk bears a pair of large pectoral fins and well developed two dorsal fins : the anterior opposite to pelvics and the posterior further back. Pelvic fins have between them the cloacal aperture and bear claspers in the male. The tail is well developed and has a heterocercal caudal fin. Anal or ventral fin is absent. Oil is extracted from the liver of *Pristis*.

4. *Torpedo* (Electric-ray)

Classification

Phylum	—	**Chordata**
Group	—	**Craniata**
Subphylum	—	**Vertebrata**
Division	—	**Gnathostomata**
Superclass	—	**Pisces**
Class	—	**Chondrichthyes (Elasmobranchi)**
Subclass	—	**Selachii**
Order	—	**Hypotremata**
Genus	—	*Torpedo*
Species	—	*marmorata*

Habit and Habitat

Torpedo, the **electric ray**, is found in the Mediterranean, Red sea, Atlantic, Pacific and Indian oceans. *Torpedo marmorata* is commonly found along the Indian coasts. It lives on the bottom of the sea. Sexes are separate.

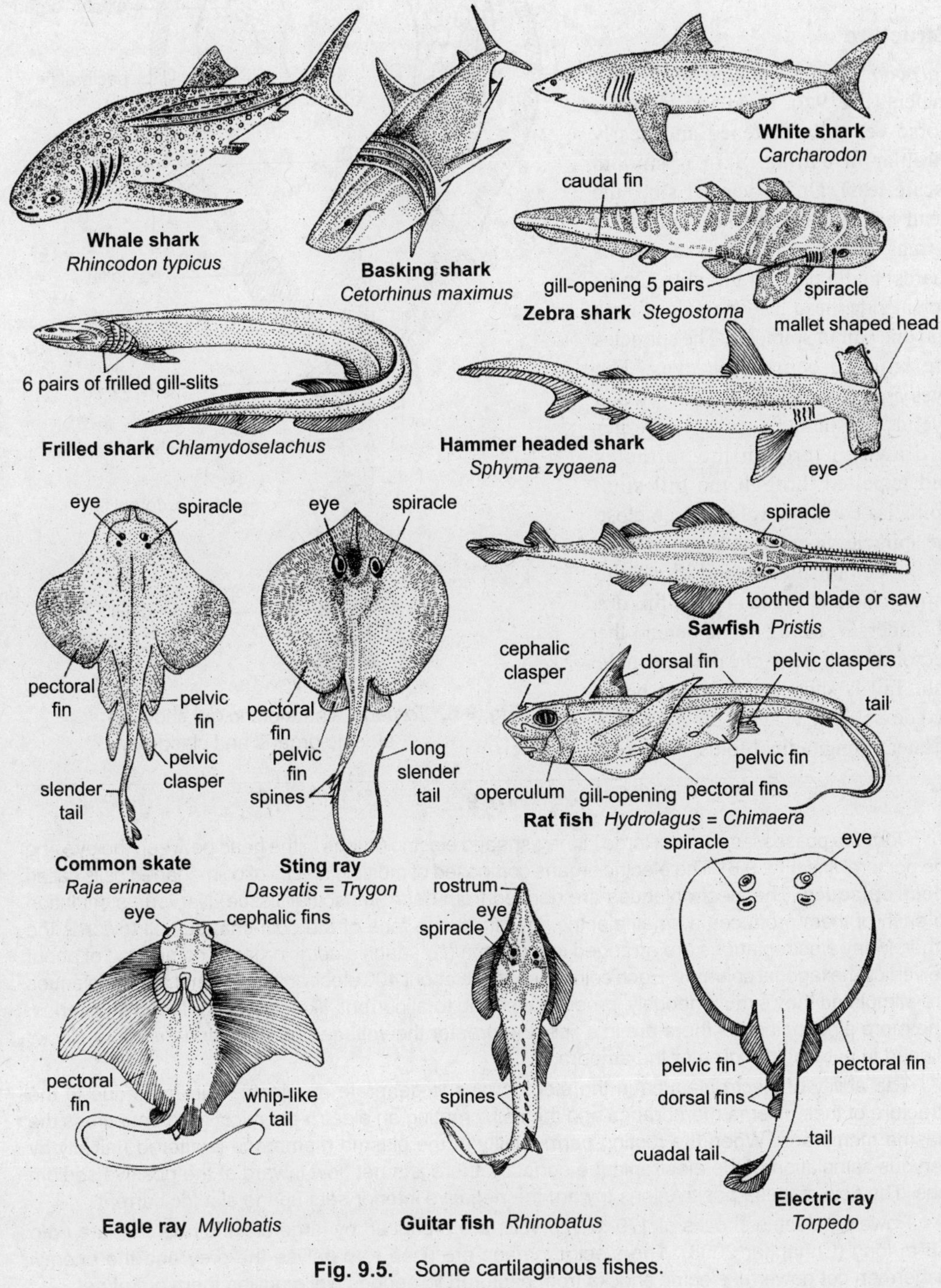

Fig. 9.5. Some cartilaginous fishes.

Torpedo swims with the help of wave-like movements of pectoral fins. It tends to numb its prey to feed them. Torpedo feeds on molluscs and crustaceans.

Structure

Torpedo reaches a length of about 1 meters (Fig. 9.6). Head and trunk are dorso-ventrally flattened and nearly circular in outline. Skin is smooth (scale less). On its ventral side, the head has a rounded snout and bears a transverse mouth, a pair of nostrils (nares) and 5 pairs of gill slits. On its dorsal side, head has one pair of eyes and one pair of spiracles. The spiracles are situated behind the eyes. The eyes are protected by well developed eyelids. Respiratory current of water is drawn in through the spiracles and expelled through the gill slits. Spiracles contain a valve which close the spiracles at expiration.

Pectoral fins extend along the entire head and trunks. Pelvic fins are of moderate size, lie close behind the pectorals and bear claspers in male fish. Tail is short and thick. It bears two dorsal fins, a caudal fin and a pair of lateral longitudinal folds.

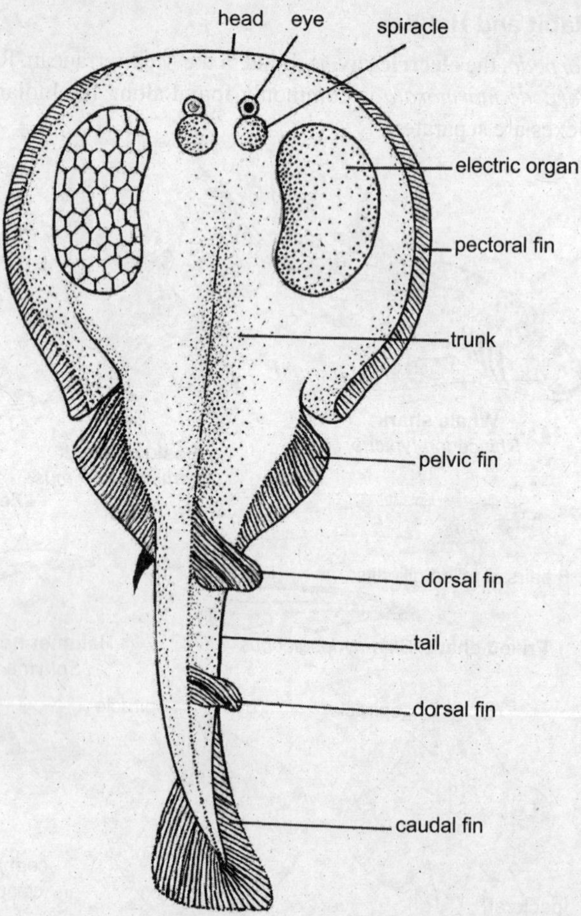

Fig. 9.6. *Torpedo*. The electric ray showing electric organs and claspers.

Box 9.2

Torpedo possesses a pair of large kidney-shaped electric organs in the head between the eye and the pectoral fins (Fig. 9.6). The electric organs composed of multiple stacks of coin-shaped cells called **electroplaques**. The electroplaques are derived from neuro-muscular tissue. Nervous stimulation to each of them produces a small electric flow, from one face of the coin-shaped cell towards the other. Many electroplaques are arranged in a column (*i.e.*, each electric organ is composed of about 45 vertical hexagonal columns. Each column contains about 400 electroplates). When electroplaques are stimulated they simultaneously, produce a strong total current, like batteries aligned in a series. The more electroplaques there are in a line, the greater the voltage produced and the more stacks aligned in parallel, the greater the amperage.

The ability of electroplaques in the electric rays to generate an electric current is due to the structure of their plasma membranes and the cell, creating an electro-chemical gradient across the plasma membrane. When the resting permeability of the plasma membrane is altered (initially by nervous stimulation of the electroplaque surface), there is a net flow inward of the positive sodium ions. The flow of these positive ions toward the negative interior sets up the electric current.

Powerful electric shocks of 170 to 220 volts are produced by large electric rays that are upto 1.8 m long (Chhapgar 2008). The electric organs are used to paralyse the prey and the enemy. Fishermen get powerful electric shocks from captured torpedoes while pulling them out of net.

5. *Trygon* (Sting-ray)

Classification

Phylum	—	**Chordata**
Group	—	**Craniata**
Subphylum	—	**Vertebrata**
Superclass	—	**Pisces**
Class	—	**Chondrichthyes (Elasmobranchi)**
Order	—	**Hypotremata**
Genus	—	*Trygon* (*Dasyatis*)

Habit and Habitat

Trygon is commonly called **sting ray** or **whip-tailed ray** (Fig. 9.5). It is found in the Mediterranean and the Atlantic, Pacific and Indian oceans. *Trygon* often lies quitely on the sea floor, becoming active only to attack a prey or in self-defense. It inflicts a painful wound with its spine. *Trygon* is viviparous. During the development of the embryo, uterine wall forms long glandular filaments, the **trophi**, which secrete a nutritive milky fluid as food for the embryo.

Characters

Trygon has a flat, kite-shaped body, about 2 meters long with smooth skin. The head bears on its ventral side, a slightly arched mouth on the ventral surface, large nares, each divided into two, in front of the mouth and five pairs of gill slits behind mouth. On its dorsal surface, head of *Trygon* have a pair of eyes and paired large spiracles behind the eyes. Pectoral fins are extensive and meet together infront of the snout. Pelvic fins are very small and bear claspers in the male. A very long, slender, whip-like tapering tail is sharply marked off from the trunk and a sharp, serrated erectile spine in place of dorsal fin. It lacks lateral longitudinal folds.

Shark, Skates and Rays

Shark, skates and rays are the members of *Neoselachi* of class *Chondrichthyes*. The sharks (pleurotremate elasmobranchs) are characterised by elongated or fusiform body, amphi- or hyostylic skull, and heterocercal tail. Their gill slits are situated on the sides of the head. The skates and rays (hypotremate elasmobranchs) have the gill openings on the ventral side of the head.

The sharks are active swimmers and predators. The heterocercal tail is a powerful organ, giving thrust over a wide range of angles, so that the fish is capable of diving and climbing in water at various speeds and attacking the prey obliquely so as to tear off flesh from its body using sharp teeth. The sharks use their various sensory organs in locating, identifying and attacking the prey. Sharks may detect their prey through mechano-receptors of their lateral line system, well developed eyes and keen olfactory organs. They have large eyes adapted for vision at low light intensity at night or in deep water. Shark's sensitive dog-like olfactory sense help in locating the prey, especially when the prey is wounded or releasing body fluids.

Species of sharks commonly found along the Indian Coast are *Stegostoma tigrenum* (the Tiger shark), *Rhincodon* (Whale shark), *Carcharius tricuspidatus*, *Sphryrna zygaena* (Hammer-headed shark) and *Scoliodon sorrakowah* (dog shark). The great white shark, *Carcharodon carcharius* kills its mammalian prey by exsanguination (bleeding it to death). The shark tends to hold a seal tightly in its jaws till it is no longer bleeding, and then bites taking a large piece of flesh. The Gangetic shark

(*Carcharinus gangeticus*) is found in the rivers of India. It migrates upstreams and attacks human beings.

Mating Behaviour of Sharks. The sharks have evolved internal fertilization and a sophisticated mechanism of breeding. Male sharks have pelvic claspers for copulation. During copulation a single clasper in bent at 90° to the long axis of shark's body, so that its dorsal groove lies directly under the cloacal papilla, from where sperms are ejected. This single clasper is inserted into the cloaca of the female and firmly locked with the help of barbs, spines and hooks at its tip. In smaller species of sharks, the male becomes secure in position by coiling round the body of the female (Fig. 9.7). However, in

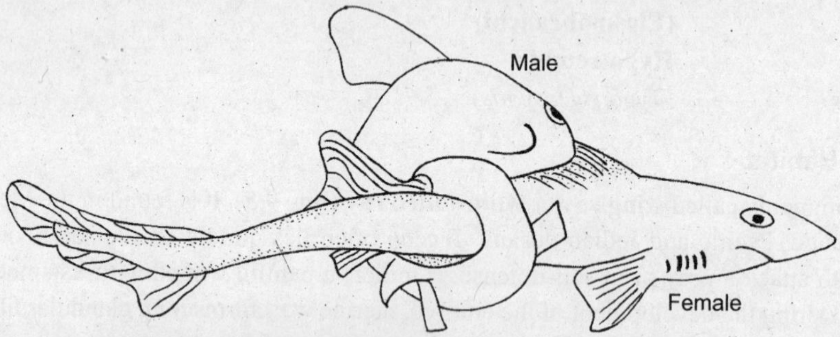

Fig. 9.7. Copulation in the dogfish (a smaller shark).

larger sharks, male and female swim side by side during copulation. During mating of some species, the male shark may bite female's flanks or holds its pectoral fin with jaws. During copulation, the sperms are ejaculated into the clasper groove. At the same time, a subcutaneous muscular sac full of sea water contracts, and washes the sperms through the clasper's groove into the female's cloace. The sperms then swim up the reproductive tract.

Sharks may be oviparous or viviparous. Some sharks and dogfish (dogsharks) lay eggs in egg cases, called **mermaid's purses** (Fig. 9.8). The corners of the egg cases have slender tendrils which get entangled amongst see weed and thus anchor them. The young may take six months to a year to hatch. Other sharks are viviparous, which means that they give birth to live young, sometimes at many as forty. In some sharks, the young are **cannibals** even before they born. A large number of eggs are found in the womb, but only two babies are born. The first baby to hatch in each compartment in the oviduct feeds on its brothers and sisters which are still in the womb (**Chhapgar**, 2008).

The hypotremata (skates, sawfishes and rays) are more diverse than the sharks, and are adapted for benthic habitat. The saw fish (*Prestis cuspidatus* and *P. microdon*) are commonly found along the Bombay coast. *Rhinobatus granulosus* is the common skate found along the Bombay coast. Another common Indian skate is *Raja ernacea* (Fig. 9.5). The saw fishes (*Pristis*) have an elongated shark like body with a heterocercal tail. But the gill apertures of these fishes are ventrally placed and the head is produced into a characteristic saw-like rostrum bearing a series of strong teeth-like denticles along its margin.

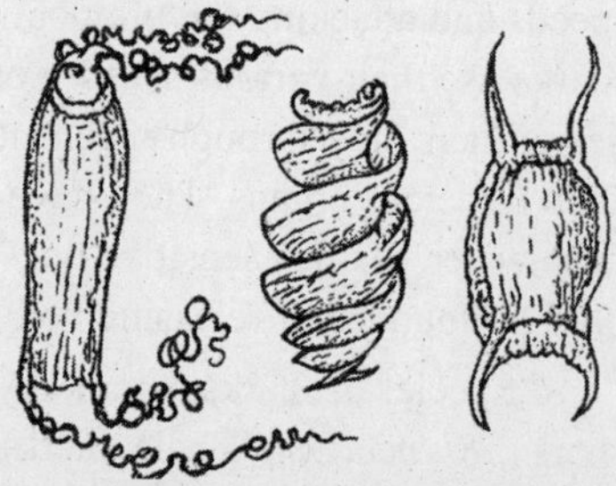

Fig. 9.8. Mermaid's purses (egg cases) of some sharks and rays.

Rays commonly found along the Bombay coast; they include sting ray (*Dasyatis/Trygon*), eagle ray (*Aetobatus*) and the electric ray (*Torpedo*). Sting ray has a kite-shaped body which ends in a long whip-like tail. This tail bears, on its upper side near the root, one to threes pines (which are modified placoid scales) with saw-toothed edges. On the lower side of the spines are two grooves containing venom glands. The ray rests on the sea bottom with its body, except its eyes, completely buried. If accidentally trodden upon or otherwise harassed, the tail is lashed from side to side and its spines can inflict a nasty cut into which venom is introduced. This causes a painful, festering wound which often requires hospitalization.

The largest of the rays, the devil ray or manta (*Manta birostris*) is found in Maldives (Sanctury Asua, Decb. 2013). It is a harmless fish feeds on minute planktonic crustaceans and fish. It gets its sinister name from the two flexible, horn-like extensions on its head which are used to funnel prey into the mouth. Eagle rays (*Aetobatus flagellum*) is found along the east and west coast of India and in chilka lake. The numb fishes or electric rays (*Torpedo* and *Narcine*) of Indian seas, though related to sting rays, have a distinct fish-like tail and flat, disc-like body. Living electric rays can deliver a powerful electric shock to paralyse their prey.

Box 9.3

Flesh of sting ray is edible. Their blood, like that of sharks, contains a lot of urea. When fishermen catch a sting ray, they usually lop off the tail to let the blood flow out. If this is not done, the urea seeps into the flesh and gives a peculiar taste. The meat of rays is relished by people from Goa and Karwar. The urea taste can be removed by soaking the flesh of sting rays in salt water or by holding it under running water for some time.

The shovel-nosed rays or guitar fish (*Rhynchobatus*, Fig. 9.5; *Rhinobatus*) are half way in evolution from sharks to rays. The front of the body is flattened into a spade-like head, the eyes and spiracles are on the upper surface and the mouth, the nostrils and gill-slits are on the lower surface. The tail is typical shark-like with two distinct fins on the back and a heterocercal tail.

Economic importance of sharks. Sharks are very useful for human beings. The dried fins can be made into delicious soup, while the liver oil is richer in vitamin A than cod liver oil. The flesh can be eaten, although it smells of urea and has to be soaked well in water before cooking. The corneas of sharks can be used for corneal transplants for humans as they do not swell when placed in varying concentrations of salt solution.

6. *Chimaera*

Classification

Phylum	—	**Chordata**
Group	—	**Craniata**
Subphylum	—	**Vertebrata**
Superclass	—	**Pisces**
Class	—	**Chondrichthyes (Elasmobranchi)**
Subclass	—	**Holocephali**
Genus	—	*Chimaera*

Habit and Habitat

Chimaera (Fig. 9.5) is popularly called shortnose chimaera, rat fish, rabbit fish or "king of herrings". It is marine found along the coasts of Europe, Japan, Australia, North America and New Zealand. *Chimaera* is found below 80 m, but move to shallow water areas to deposit eggs. It is carnivorous, feeding on small invertebrates and fishes. *Chimaera* is oviparous and lays eggs in a horny case.

Structure

Chimaera grows to 1.5 meters. It has a shark like but bizarre looking form with smooth, scale-less body (placoid scale occur in patches). Head is laterally compressed and has a blunt snout; small, transverse, fleshy-lipped ventral mouth; large, rouded subterminal nares; oronasal grooves, large laterally directed eyes and cartilaginous opercula covering the gill slits (Box 9.4). Mouth has the crushing plates, two pairs in the upper jaw and one pair in the lower jaw, which have no enamel but vitrodentine and reduced pulp cavities. Trunk bears large fan-like pectoral fins, medium-sized pelvic fins and two dorsal fins. The anterior dorsal fin is inserted over the pectorals and is bordered in front by a spine. The posterior dorsal fin is very long but low, starts some distance behind the anterior dorsal and is continued over the tail. Tail is diphycercal and in the form of a whip-like filament. Male fish bears three types of five claspers : a peculiar, club-shaped erectile **frontal** or **cephalic clasper** over the head, a pair of **anterior claspers** lying in pouches in front of pelvic fins and a pair of **posterior claspers** behind the pelvic fins (as in sharks) (Fig. 9.9). All the claspers are beset with minute denticles. The cephalic clasper is used to grip the dorsal fin of the female at the time of copulation. There is no cloaca and urinogenital aperture lies behind the anus. The lateral line system has the form of open grooves. Endoskeleton of *Chimaera* is cartilaginous. Intestine has a spiral valve.

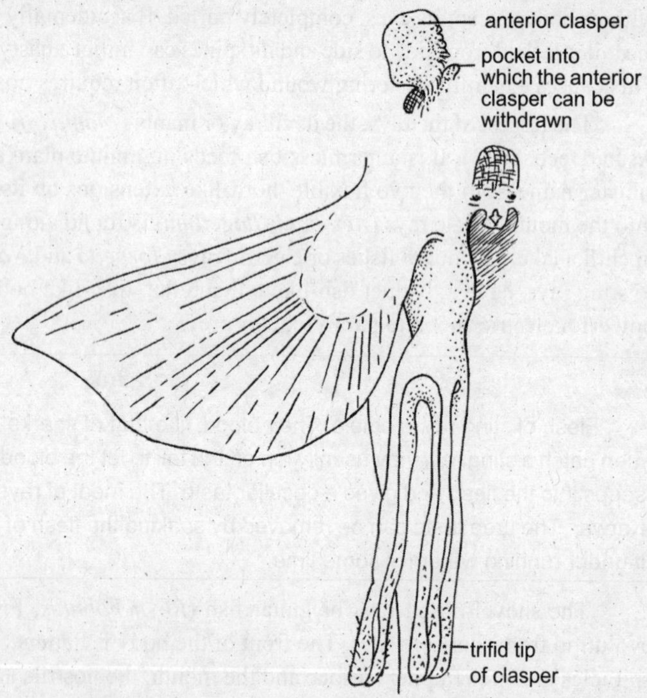

Fig. 9.9. *Chimaera*. Ventral side of pelvic fin showing anterior and posterior claspers.

Box 9.4

The first three branchial arches of *Chimaera* bear holobranchs (complete gills) while the fourth branchial arch and the hyoid arch bear hemibranchs (half-gills). The fifth branchial arch is gill less, and the interbranchial septum of the gill is reduced. All the gills are enclosed in a spacious branchial cavity which opens to the exterior by a single external branchial aperture. This is due to the development of a peculiar fold of skin over the gill slits.

Chimaera resembles the bony fishes in having small mouth with fleshy lips and only four pairs of gills covered by operculum and in lacking in spiracles and cloaca. Holocephali appear to have descended from some primitive shark-line ancestor and become specialized in several features as an adaptation to their peculiar mode of life.

9.6 CLASS OSTEICHTHYES

The class osteichthyes (Gr., *osteos* = bone + *ichtyes* = fishes) includes bony fishes. They are unique among the vertebrates in having flat, bony scales in the dermis, bony endosketeton and gill slits covered with opercula.

Osteichthyes are the most successful of all aquatic forms. This is demonstrated by the fact that they outnumber all other vertebrates combined together in the number of their species as well as in population.

General Characters

1. Marine as well as freshwater fishes.
2. Spindle-shaped streamlined body includes head, trunk and tail. Neck is absent.
3. Both median and paired fins are present. These are supported by jointed, dermal cartilaginous or bony fin-rays. There is generally single dorsal fin. Caudal fin is mostly **homocercal**, in some cases **heterocercal** or **diphycercal**.
4. Skin is soft and has many unicellular mucous glands in the epidermis. Exoskeleton in the form of dermal scales which may be **ganoid**, **cycloid** and **ctenoid**.
5. Muscles of the trunk and tail are arranged in 3-shaped myotomes.
6. Endoskeleton bony, having replacing as well as membrane bones.
7. Jaws are well developed and articulated to the skull. This is called **autostylic suspension**. Lower jaw is formed by several bones.
8. There are no salivary glands. Pharynx is large. Oesophagus is short. Intestine usually lacks spiral (scrol) valve. Alimentary canal leads to the exterior by anus, there being no cloaca.
9. Respiratory system includes 4 pairs of gills. The gills are filamentous (filiform type) and are borne on the bony gill arches. Only one pair of gill slits which are covered by **operculum** which is supported by bones. Spiracles are mostly lacking.
10. An **air** or **swim bladder** is often present. It acts as a buoyancy regulator and enables the fish to stay at a particular depth without expending energy in swimming. Swim bladders are highly vascularised or **lung-like** in lung fishes.
11. Venous heart is 2-chambered, having only one auricle and one ventricle. It contains only venous blood.
12. Brain has small olfactory lobes.
13. Jacobson's organs are absent. Lateral line sense organs are well developed.
14. Kidneys are opisthonephric. They are ammoniotelic excreting ammonia. Blood with low concentration of urea.
15. Sexes are separate. Males are without claspers. Fertilization is external. Most forms are oviparous. Ova small and numerous. Meroblastic cleavage. Development direct. Some fishes show parental care.
16. Fishes are poikilothermic.
17. There are about 18,128 species of bony fishes (**Gupta** and **Gupta**, 2006).
18. Fishes vary greatly in size. Some minnows are less than 10 mm, while sword fish may grow over 4 m. They may be herbivorous, carnivorous or omnivorous in habit.
19. Some female deep-sea fishes produce pheromones that attract the males in the dark. Some fishes can change colours (metachrosis). Many deep sea fishes are luminescent. Their light-producing organs, the **photophores** contain symbiotic bacteria (*e.g.*, *photobacterium*, *Vibrio*, etc.) that emit light.
20. The bony fishes appeared in early Devonian period and are, thus, older than the cartilaginous fishes. They arose from the placoderms, probably in fresh water. Their bony skeleton is a retention of their ancestral trait.
21. The Devonian period (about 360 to 400 million years ago) is regarded the age of fishes.

9.7 CLASSIFICATION OF OSTEICHTHYES

The class Osteichthyes include a large assemblage of the bony fishes which are both freshwater and marine. Some of the freshwater fishes are the carp, perch, bass, trout, catfish, suckers, etc.; while the marine fishes are the tarpon, meckerel, tuna, sail fish, barracula, flying fish, etc. The bony fishes were the dominant fishes of the Devonian period which is also called "age of fishes", due to their efficient locomotory and feeding mechanism. Some of these become specialised and gave rise to the land vertebrates. The class osteichthyes is divided into two subclasses by **A.S. Romer** (1959): **Sarcopterygii** and **Actinopterygii** (Fig. 9.10).

Class Osteichthyes

- **Subclasses (2)**
  - **1. Sarcopterygii**
    - **Orders (2)**
      - **1. Crossopterygii** — **Example.** 1. *Coelacanthus* (Extinct) 2. *Latimeria*
      - **2. Dipnoi** — **Example.** 1. *Neoceratodus*, 2. *Protopterus*, 3. *Lepidosiren*
  - **2. Actinopterygii**
    - **Superorders (3)**
      - **1. Chondrostei**
        - **Orders (2)**
          - **1. Polypteriformes** — **Example.** *Polypterus*
          - **2. Acipenseriformes** — **Example.** 1. *Acipenser*, 2. *Polydon*
      - **2. Holostei**
        - **Orders (2)**
          - **1. Amiiformes** — **Example.** *Amia*
          - **2. Lepidosteiformes** — **Example.** *Lepidosteus*
      - **3. Teleostei**

Superorder 3. Teleostei

It includes many orders (29 according to **Tate Regan** and 41 by **Berg**). Here, we has described only 15 orders:

1. Clupeiformes (*e.g.*, *Clupea*, *Salmo*); **2. Scopeliformes** (*e.g.*, *Harpodon*); **3. Cypriniformes** (*e.g.*, Cyprini such as *Labeo* and Siluri such as *Wallago*); **4. Anguiliformes** (*e.g.*, *Anguilla*-freshwater eel); **5. Beloniformes** (*e.g.*, *Exocoetus*); **6. Syngnathiformes** (*e.g.*, *Hippocampus*); **7. Ophiocephaliformes** (*e.g.*, *Ophiocephalus*); **8. Symbranchiformes** (*e.g.*, *Symbranchus*, *Amphipnous*-eels); **9. Mastacembeliformes** (*e.g.*, *Mastacembelus*); **10. Periformes** (*e.g.*, *Anabas*); **11. Scorpaeniformes** (*e.g.*, *Pterois*); **12. Pleuronectiformes** (*e.g.*, *Synaptura*); **13. Echeniformes** (*e.g.*, *Echeneis*); **14. Tetradontiformes** (*e.g.*, *Tetrodon*, *Diodon*); **15. Lophiiformes** (*e.g.*, *Lophius*).

Fig. 9.10. Outline Classification of class Osteichthyes.

Subclass 1. Sarcopterygii (Choanichthyes)

(Gr., *sarcos* = fleshy; *pterygium* = fin)

1. Popularly called fleshy-finned, fringe-finned, lobe-finned or air breathing fish.
2. They live in small freshwater reservoirs and had fins with lobes adapted for resting on the ground. Each of the paired fin is borne on a fleshy, lobe-like, scaly stalk, extending from the body. The lobe contains jointed skeleton and muscles, thus resembling the tetrapod limb. The lobe is fringed with dermal rays.
3. Two dorsal fins present. Heterocercal caudal fin with an epichordal lobe.
4. Olfactory sacs usually open into the mouth cavity by internal nares, hence are also known as choanichthyes (*Gr.*, *Choanae* = nares; *ichthyes* = fishes).
5. A single or double lung (air bladder) is present for air breathing.
6. Intestine has spiral valve.

The subclass sarcopterygii is divided into two orders : Crossopterygii and Dipnoi.

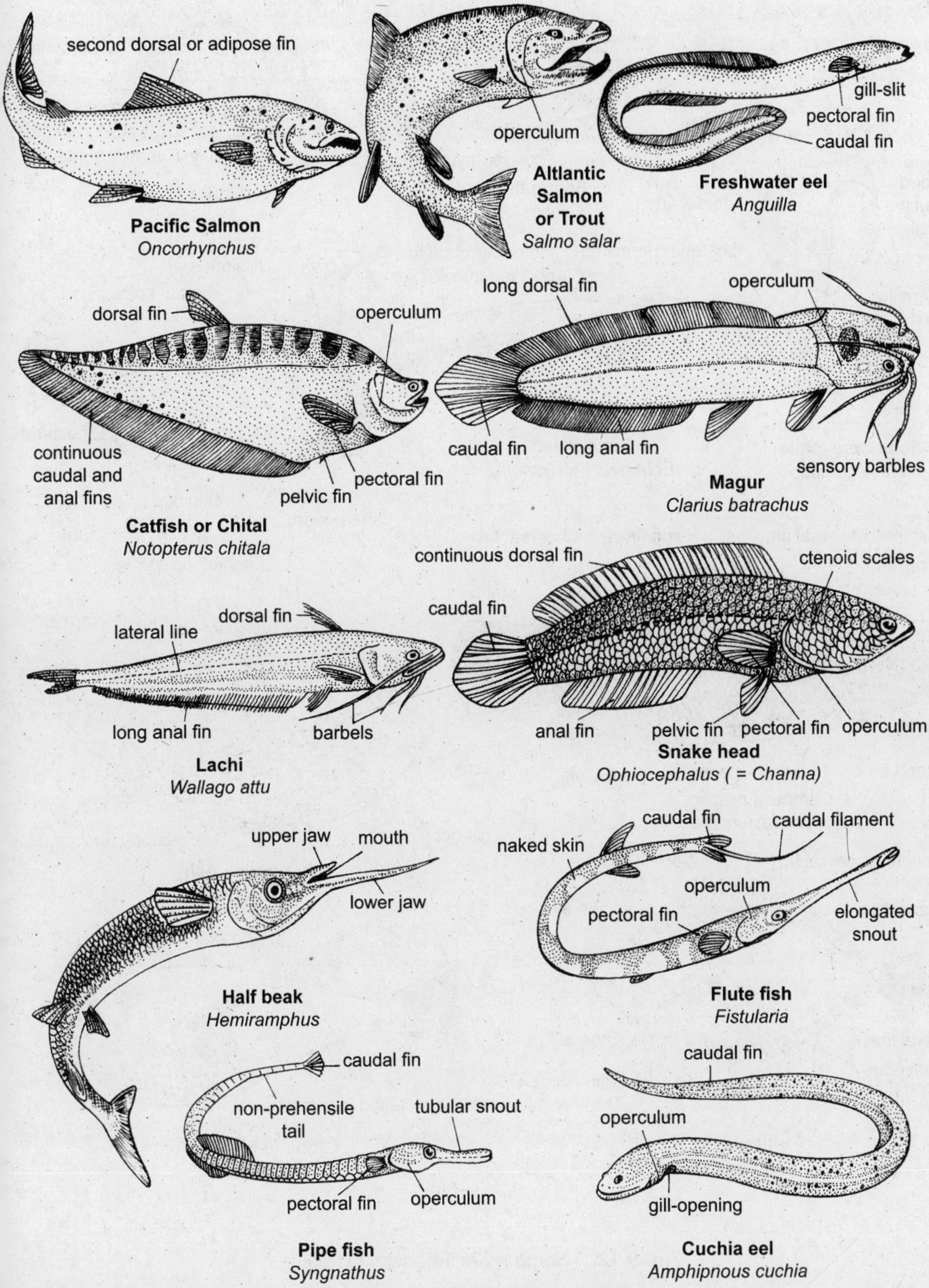

Fig. 9.11. Some Teleostei fishes.

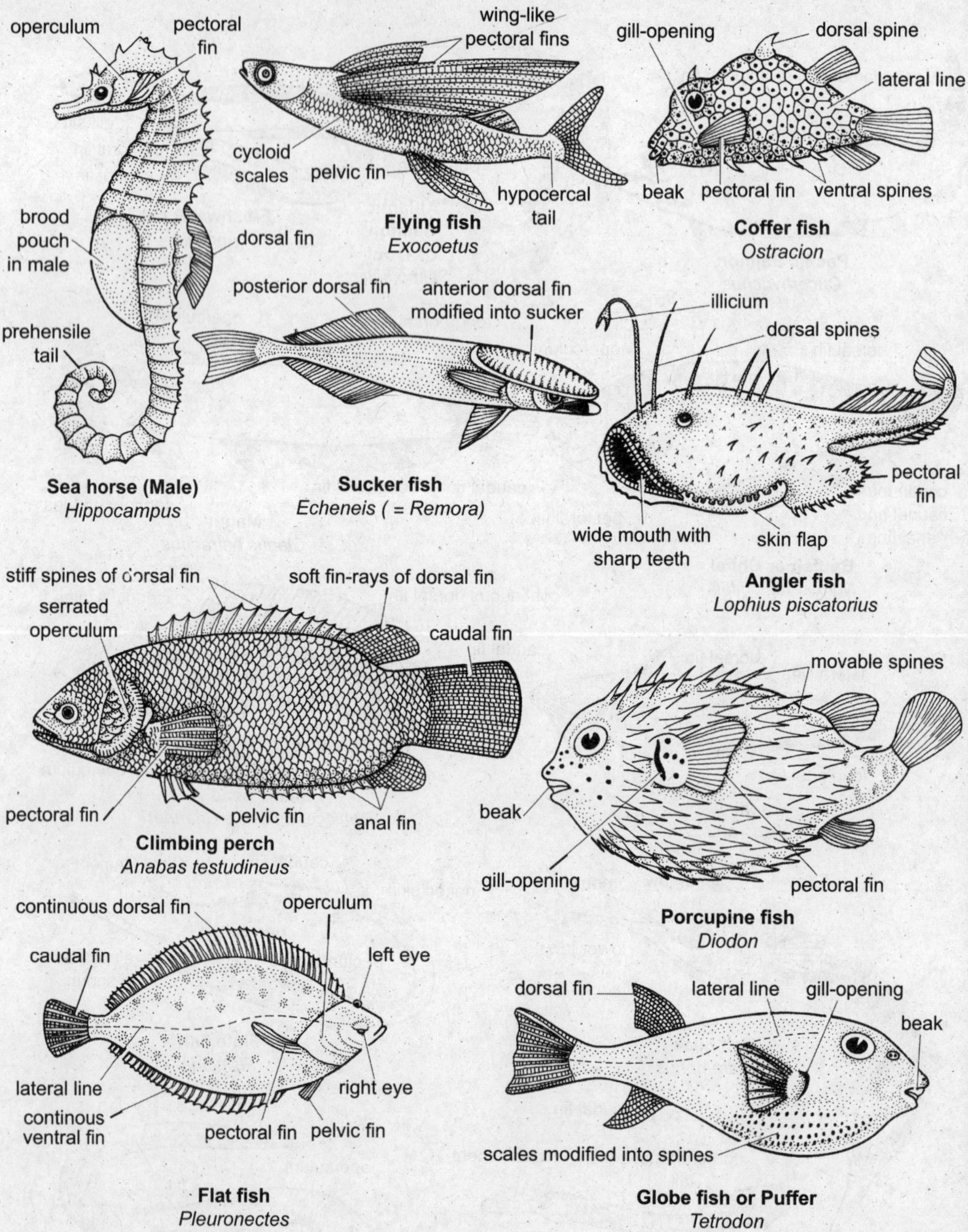

Fig. 9.12. Some more Teleostei fishes.

Order 1. Crossopterygii

(Gr., *Crossoi* = a fringe; *pteryx* = fin)

1. Paired fins are lobed covered with scales. Median fins are separate from the caudal fins. Tail fin is three-lobed.
2. Scales covered by a layer of cosmine, called **cosmoid scales**. Such scales have three layers : enamel-like *vitrodentine*, dentine-like *cosmine* and bony *isopedine*.
3. Skeleton is rich in cartilage. Premaxillae, maxillae and squamosal bones are present.
4. Air bladder is reduced and acts as a hydrostatic organ.
5. Jaw suspension is hyostylic.
6. Spiracles present.
7. Internal nares are lacking.
8. Intestine has a spiral valve.
9. They are living since the Carboniferous period.

Example. *Coelacanthus* (extinct), *Latimeria* (living fossil; Fig. 9.17).

Order 2. Dipnoi

(Gr., *di* = double; *pnoe* = breathing)

They are commonly called **lung fishes** because their air bladder acts as a lung. They have gills also and are double breathers. This lead to the name dipnoi which means double-breathing. Lung-fishes have the following general characters:

1. Freshwater fishes which are found in swamps and rivers. These fishes can thrive out of water for some time.
2. Body is long and slender.
3. Exoskeleton consisting of overlapping, cycloid, bony scales. Cycloid scales are dermal in origin, epidermis has no role in cycloid scale formation.
4. Paired fins are lobe or filament-like. Dorsal and anal fins are confluent with the caudal fin. Tail fin symmterical (diphycercal). Fin rays slender and more or less ossified.
5. Internal nares present opening into mouth. Spiracles absent.
6. Air bladder single or paired, lung-like.
7. Notochord is presistent. Vertebrae centra are lacking.
8. Premaxillae and maxillae are absent.
9. Teeth form crushing plates.
10. Jaw suspension is autostylic (*i.e.*, lower jaw plato-quadrate) is firmly fused to neurocranium.
11. Stomach absent. Intestine has spiral valve.
12. Gills are reduced and covered by operculum.
13. Heart is three chambered (almost separate right and left auricles and a partly divided ventricle).
14. Pulmonary and systemic circulation is separate. (Double type of circulation).
15. Cleavage holoblastic. Larvae may have suckers and external gills.
16. The lungfishes show discontinuous distribution. All the three living genera occur in separate continent.

Example. *Neoceratodus* (Australia; Fig. 9.19), *Protopterus* (Africa; Fig. 9.21) and *Lepidosiren* (South America; Fig. 9.23).

Subclass 2. Actinopterygii

1. Actinopterygi are popularly called the **rayfinned fishes**.
2. Paired fins have broad bases and fan-like form. They lack fleshy basal lobes and are supported by radiating dermal rays.
3. They have only one dorsal fin. It may be divided.
4. Tail is generally homocercal. Caudal fin is without epichordal lobe.

5. Scales are thin and horny. Scales may be cycloid or ctenoid. Some primitive species have ganoid scales. Some species without scales.
6. Endoskeleton bony. Jaw suspension hyostylic. Squamosal absent.
7. Only one external branchial aperture. It is covered by operculum. Spiracles absent.
8. Internal nares absent (*i.e.*, the nostrils do not open into the buccal cavity).
9. Mouth is terminal.

Subclass Actinopterygii has been divided into three superorders : Chondrostei, Holostei and Teleostei.

Superorder A. Chondrostei

Chondrostei are commonly called primitive **rayfinned fishes**. They are survivors of ancient (Palaeozoic) group of fishes, majority of which have become extinct (Box 9.5). The ray-finned fishes have the following characters:

Box 9.5. Order Palaeonisciformes

The "Palaeoniscides" are the palaeozoic ancestral fishes that lived during the Devonian and became extinct by the Triassic. They are ancestors to all the actinopterygians and are represented by the genus *Cheirolepis*, having ganoid scales and a heterocercal tail. A typical example of this order is *Palaeoniscus*.

1. They are freshwater carnivorous (predatory) fishes.
2. Mouth is large having weak and toothless jaws.
3. Scales are usually **ganoid** having rhomboidal appearance.
4. Paired fins have a broad based attachment.
5. Tail is usually heterocercal.
6. Spiracles are generally present.
7. Skeleton is largely cartilaginous. There are superficial dermal bony plates on the head.
8. Notochord is persistent and unconstricted.
9. Air bladder acts as a lung and has a pneumatic duct leading to oesophagus.
10. Intestine has a spiral valve.
11. Heart has conus arteriosus.
12. There are 5 pairs of gill slits.

The superorder Chondrostei (*i.e.*, cartilaginous ganoids) is divided into two orders: Polypteriformes and Acipenseriformes.

Order 1. Polypteriformes

1. The dorsal fin is split into eight or more flag-like finlets.
2. Body is covered with thick multilayered ganoid scales.
3. Each pectoral fin is with a small prominent scale-covered fleshy lobe. Tail is diphycercal.
4. Skeleton is well ossified.
5. Spiracles are present.
6. Air bladder is bilobed and is modified for respiration. However, internal nostrils are absent.
7. Only four branchial arches are present.
8. Spiral valve is present in intestine.
9. Larva has external gills.
10. This order includes African bichirs and reed fishes.

Examples. *Polypterus* (bichir), *Calamoichthys*.

Order 2. Acipenseriformes

1. The body is shark-like and bears five longitudinal rows of heavy (armour-like), bony, keeled, modified ganoid scales (called **scutes**).

2. Snout elongated, having barbles on the ventral surface.
3. Mouth is ventral.
4. Caudal or tail fin is heterocercal and bears ganoid scales.
5. Dorsal fin is posterior and has ganoid scales.
6. Jaws are protrusible and without teeth.
7. Notochord is unconstricted.
8. Some are freshwater of Europe, North America and Asia. Others are marine and migratory (anadromous).

   Examples. *Acipenser* (strugeon), *Polydon* (paddlefish or spoon-billed sturgeon).

Superorder B. Holostei

(Gr., *holos* = entire; *osteon* = bone)

1. Holostei are popularly called **intermediate ray finned fishes**. They represent transitional forms between Chondrostei and Teleostei.
2. Endoskeleton is exclusively bony.
3. Scale are ganoid or cycloid.
4. Tail fin hemi-heterocercal, *i.e.*, internally heterocercal but externally almost symmetrical.
5. Spiracle is absent.
6. Mouth opening small. Intestine has a vestigial spiral valve.
7. Single swim bladder with some respiratory function.
8. Most are marine and some are freshwater.

   Superorder Holostei is divided into two orders : Lepidosteiformes and Amiiformes.

Orders 1. Lepidosteiformes (Semionotiformes)

1. Elongated body is covered with an armour of thick rhomboid scales having a coat of ganonin. Scales in oblique rows.
2. Vertebrae opisthocephalus. Notochord is long during the development of vertebrae.
3. Jaws long.
4. Only one genus having seven species. They are freshwater predatory fishes of N. America.

   Examples. *Lepidosteus* (or *Lepisosteus*)–the gar pike (Fig. 9.25).

Order 2. Amiiformes

1. Trunk bears cycloid scales and head has ganoid scales. There are no scales on the fins.
2. The snout is rounded. Jaws short.
3. The dorsal fin is very long.
4. Vertebrae amphicoelous. Notochord persists in between the center of adjacent vertebrae.
5. Pyloric caeca present.
6. Presence of a single swim-bladder.
7. Only single genus is present.

   Example. *Amia calva* (Bowfin).

Superorder C. Teleostei

(Gr., *teleos* = complete; *ostgon* = bone)

Teleosti are commonly called **advanced ray finned fishes** or **higher bone fishes**. They are modern fishes and form the greater part (about 95 per cent) of world's fish fauna. Teleosts probably evolved in the sea but soon radiated into fresh water. Thus, they occur in all kinds of water and show a great variety in form and habits. Teleost fishes have the following general characters:

1. Scales cycloid, ctenoid or absent.
2. Mouth is terminal and small.

3. Paired fins are more flexible, mobile and different in shape, size and position. Paired fins have a narrow base attachment and lack scales. Pectoral fins are well up on the sides of the body. Pelvic fins are far forward.
4. Tail is generally homocercal.
5. Spiracle is absent.
6. Single external gill slit on each side of the head and it is covered by operculum.
7. Skeleton is fully ossified (bony).
8. Notochord is vestigial.
9. Air bladder is single and with hydrostatic function.
10. Intestine lacks spiral valve.
11. Heart is bulbus arteriosus in place of conus arteriosus. Bulbus arteriosus is the swollen base of the ventral aorta.
12. Gills are filamentar.

Teleostei are divided into more than 30 orders. Only 15 more important one are described here.

Order 1. Clupeiformes (or Clupeomorpha)

1. The body is strongly compressed.
2. Scales are well developed and cycloid. Head and operculum not scaled.
3. A single dorsal fin and a small ventral fin is present.
4. The anal fin is very long and continuous with the small caudal fin (gephyrocercal).
5. Dorsal fin and anal fin both do not have spine.
6. Weberian apparatus is absent.
7. Several species (*e.g.*; *Salmo*, *Ilisha*, etc.) have anadromous migration.

Examples. *Clupea* (Herringer), *Salmo* (Atlantic salmon; Fig. 9.11), *Harengula* (Herring), *Sardinella* (Sardine), *Notopterus* (Razor fish or Chital fish; Fig. 9.11), *Hilsa*, *Ilisha*, *Esox* (Pike), *Gadusia*, etc.

Order 2. Scopeliformes

1. Deep sea fishes.
2. Soft, jelly-like body with tiny eyes situated near the end of the snout.
3. On the back, well behind the rayed dorsal fin is a small adipose (fatty) fin.
4. Mouth wide and jaws are with numerous small teeth.
5. Air bladder absent.

Example. *Harpodon nehereus* (Mumbai duck), it is an important food fish of India.

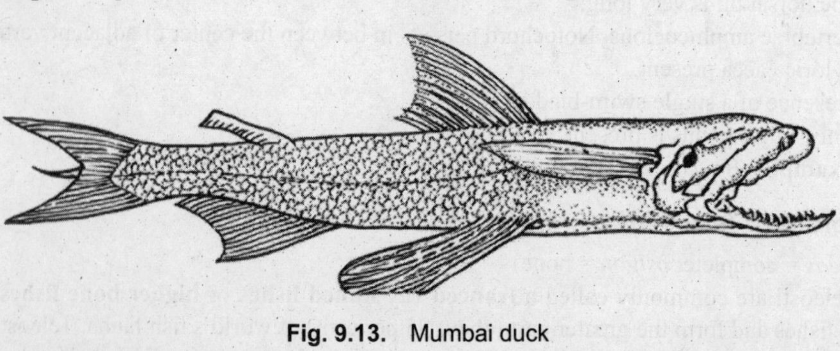

Fig. 9.13. Mumbai duck

Order 3. Cypriniformes

1. Fins either without spines or dorsal, anal and pectoral fins have a spine each.

2. Air-bladder connected with the pharynx by a duct.
3. A peculair weberian apparatus (*i.e.*, chain of small bones called weberian ossicles), connecting the internal ear with the swim-bladder present. Thus, they are more sensitive to sound.

Fishes of this order are grouped into two suborders or divisions: Cyprini and Siluri.

Division I. Cyprini

1. Body covered with cycloid scales or naked.
2. Third and fourth vertebra not fused with each other.
3. There is one dorsal fin and no adipose dorsal fin.
4. Mouth is protractile and jaws lack teeth.
5. Fins lack spines.
6. This group includes carps, barbs and loaches.

Examples. *Labeo* (Rohu; Fig. 9.27), *Catla* (Katla), *Cirrhina* (Mrigal), *Puntius* (Barbs), *Tor* (Mahseers), *Botia* (Loach), *Ctenopharyngodon idella* (Grass carp), *Hypophthalmichthys molitrix* (Silver carp), *Cyprinus carpio* (Common carp), *Carassius auratus* (Goldfish), etc.

Division II. Siluri

1. It includes catfishes. Living both in sea and fresh water.
2. Naked, scaleless skin.
3. They have four pairs of whisker-like sensory barbels (*i.e.*, nasal, mental, maxillary and mandibular) surrounding the mouth.
4. Usually have an adipose fin on the back behind the first spiny dorsal-fin.
5. Body spines (dorsal, pectorals) have venomous glands.
6. Second, third, fourth and sometimes the fifth vertebrae are generally fused.
7. Parental care occurs. Males of catfishes carry the large, yolk-filled eggs in their mouth for incubation.
8. Some catfishes such as *Clarias* and *Heteropneustes* or *Saccobranchus* have accessory respiratory organs.

Examples. *Arius* (cat fish), *Heterophenustes* (Singhi), *Mystus* (Tengra), *Clarius* (Magur; Fig. 9.11), *Wallago* (Lachi; Fig. 9.11 also called freshwater shark), *Ompok*, *Bagarius* (Goonch; freshwater catfish which grows to a length of 1.8 m), *Glyptosternum* (Hill stream catfish), etc.

Order 4. Anguiliformes

1. This order includes eels and morays (Box 9.6).
2. Body is elongated, tubular, snake-like.
3. Pectoral fins are reduced. Pelvic fins are absent.
4. Dorsal and anal fins are confluent with the caudal fin. Fins lack spines.
5. Scales are minute and embedded in the dermis so that the fish seems naked.
6. Air-bladder is with a duct.
7. Eels are catadromous in migration. They breed in sea (*i.e.*, Sargasso sea) and young then ascend rivers and streams to grow in freshwater (Box 9.6).

Box 9.6. Yawning Morays

Moray eels (such as Marbled morays, Painted moray etc. in all 120 species) lack scales as well as pectoral and pelvic fins. They swim with a sinuous motion resembling that of snakes. They breathe by gulping water into their mouth, which is then closed and the water is expelled through small gill-openings. Thus they appear to perpetually yawning. They have poor eyesight and hunt by smell, their most favoured prey being octopuses (see **B.F. Chhapgar**, 2008).

Example. *Anguilla* (Freshwater eel; Fig. 9.11), *Muraena* (Moray; Fig. 9.14), Moray eel (*Gymnothorax*).

Fig. 9.14. Moray.

Box 9.7. Eel's blood is poisonous

The flesh of eel (*Anguilla*) was considered a delicacy by the ancient Romans who used to stock them in special ponds. But their blood, if uncooked, is poisonous. Drinking eel blood results in nausea and vomiting, frothing at the mouth, blood-stained stools, distress while breathing, and sometimes, death. If a drop of eel blood gets into the eye, within minutes there will be reddening and a burning sensation. Later a feeling of some foreign body in the eye develops and lasts for days. A drop of eel's blood on our tongue makes it red and burn, and a copious saliva is released as our body tries to dilute it and wash it away (**Chhapgar**, 2008).

Order 5. Beloniformes (Atheriniformes)

1. Elongated body.
2. The scales are cycloid and extend on the head and opercula.
3. Fins are soft-rayed and spineless.
4. Pectoral fins large and located high on body.
5. Pelvic fins are abdominal.
6. Dorsal fin is opposite the anal fin.
7. Some of them are capable of jumping into the air and glide with the help of enlarged pectoral fins.

Examples. *Exocoetus* (Flying fish; Fig. 9.12), *Cypselurus* (Flying fish), *Belone* (Tylosaurm; Garfish), *Hemirhamphus* (Half beak; Fig. 9.11).

Order 6. Syngnathiformes (Gasterosteiformes)

1. Body is covered with protective scales or bony rings. Trunk is invested in an armour of bony plates.
2. The mouth is small and suctorial and lies at the end of a tubular snout.
3. Pelvic fins are absent.
4. Pectoral fins are small and a single dorsal fin is present.
5. Gills consists of peculiar tufted processes instead of filaments and are called **lophobranchs**.
6. Tail is prehensile in sea horse.
7. Males have a brood pouch on the undersurface of the abdomen or tail to carry the eggs till after hatching.
8. Physoclistic. (**Note:** *Physoclistous* means not having a channel connecting the swim bladder with the digestive tract).

Examples. *Hippocampus* (Sea horse; Fig. 9.12), *Fistularia* (Flute fish; Fig. 9.11), *Syngnathus* (Pipe fish; Fig. 9.11).

Order 7. Ophiocephaliformes (Channiformes)

1. Body covered with cycloid scales.
2. Head depressed, covered with plate-like scales.

3. Head and eyes resemble those of a snake.
4. Fins without spines.
5. Dorsal and anal fins are very long.
6. Swim-bladder very long extending into tail and without duct (*i.e.*, physoclistic).
7. There are suprabranchial chambers having accessory air-breathing organs. Due to these organs these fishes are able to survive out of water for a few hours or migrate from one pool to another.
8. These freshwater fishes have voracious appetite.

Example. *Ophiocephalus* (*Channa*)—Snake headed or Murrel (Fig. 9.11). Its flesh is highly prized for food.

Order 8. Symbranchiformes

1. Snake-like or eel-like fishes without air bladder.
2. Scales are absent.
3. Pectoral fins absent. Dorsal, caudal and anal fins continuous.
4. Fins without spines.
5. Gills reduced. Gill-slits single or join to form a transverse ventral slit.

Examples. *Amphipnous* (Cuchia eel; Fig. 9.11); *Symbranchus* (Eel).

Order 9. Mastacembiliformes

1. Eel-shaped phsoclistous fishes.
2. Elongated body with some free spines infront of dorsal fin.
3. Dorsal and anal fins are confluent. Ventral (pelvic) fins absent.
4. Buccal cavity enlarged for air-breathing.

Examples. *Mastacembelus* (Spiny eel; Fig. 9.15), *Macrognathus* (Spiny eel).

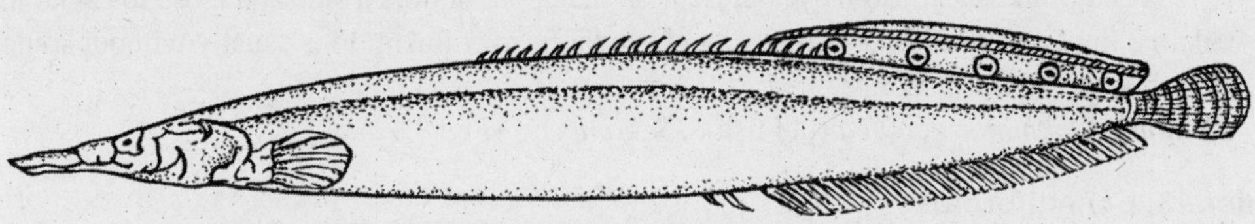

Fig. 9.15. Spiny eel.

Order 10. Perciformes (Percomorphi)

1. This group includes perches.
2. Two dorsal fins present. Ventral (pelvic) fins are thoracic and with not more than 6 rays.
3. Fin-rays hard and spiny. Fins may have spines.
4. Scales are often with serrated edges.
5. Physoclistic.

Example. *Anabas* (Climbing perch; air breathing fresh water fish; Fig. 9.12); *Perca* (Yellow perch), *Lates* (Bhetki), *Scomber*, etc.

Order 11. Scorpaeniformes

1. Marine, bottom-living and predatory fishes.
2. Enlarged heads and pectoral fins. Scales absent.
3. Projecting hollow spines from gill-covering (operculum), head and fins.

4. A highly poisonous neurotoxic subtance is produced by the venomous gland situated at the base of the hollow spines.

   Example. *Pterois* (Scorpion fish or Flying dragon; Fig. 9.16).

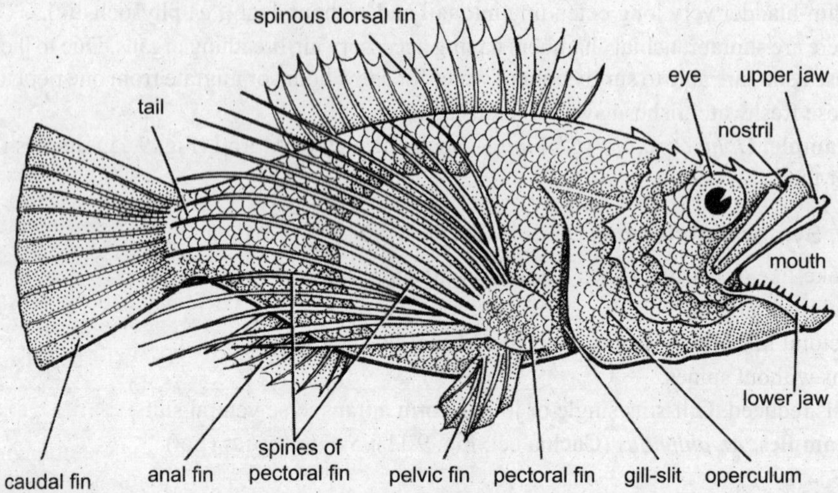

Fig. 9.16. *Pterois.*

Order 12. Pleuronectiformes

1. Bottom dwellers. Commonly known as flat-fishes.
2. Body is flat, leaf-like, lying on one side, adapted for bottom living.
3. Head is asymmetrical. Both eyes present on the upper or dorsal side and close to each other.
4. Paired fins are reduced. Dorsal, anal and caudal fins confluent. Fins usually without spines.
5. Air bladder absent in adults.

   Example. *Synaptura*, *Solea* (flat fish), *Pleuronectes* (Fig. 9.12).

Order 13. Echeniformes

1. Body is covered with a cycloid scales.
2. First dorsal fin modified into a flat oval adhesive vaccum disc or sucker on head for attachment. It possesses 12–28 transverse ridges which are modified spines.
3. No spines in second dorsal and anal fins.
4. Air bladder absent.

   Examples. *Echeneis* or *Remora* (Sucker fish, Fig. 9.12).

Order 14. Tetradontiformes

1. Body globular and spinous. Body scales modified into spines. Scutes or body plates cover the body..
2. Strong jaws with a sharp beak.
3. Gill-slits small like a hole on either side of fish infront of pectorals.
4. Dorsal fin is spiny. Pelvic fins are small.
5. Air-bladder present or absent.
6. Belly is inflatable. Some inflate by swallowing water.

Examples. *Diodon* (Porcupine fish; Fig. 9.12); *Tetrodon* (Globe fish; Fig. 9.12); *Ostracion* (Coffer fish or Trunk fish; Fig. 9.12); *Balistes* (Picasso triger fish); *Triacanthus* (File fish orTripod fish); Box fish or cowfish; *Mola mola* (sunfish), etc.

Order 15. Lophiformes

1. Marine, bottom dweller (*i.e.*, deep sea fishes) anglers.
2. Flat squat body has huge head with wide distensible mouth.
3. Scaleless leathery skin.
4. The first ray of the spiny dorsal fin is modified to form a fishing organ (angling road) on the head. It consist of a rod called the **illicium** and a bait called **esca**.
5. Pectoral fins are well developed and are used to crawl or "walk" on the bottom.
6. Pelvic fins are absent or smell in size.
7. Air bladder absent.
8. Luminescent organs present.

   Exmples. *Lophius* (Angler fish; Fig.9.12); *Antennarius* (Angler fish or fishing frog).

Some Curious and Economical Fishes

1. **Leather jacket fish** (*Aleutera*) belongs to order **Tetradontiformes** and is a marine fish. It looks fragile and harmless but has the ability to catch and eat the deadly Portuguese man-of-war (*Physalia*) whose tentacles can kill a human being with their deadly venom.
2. **Monx fish** (*Helieutea stellata*) belongs to order **Pegasiformes**.
3. **Pomfrets** (suborder **Scombriodei**) are among the most esteemed table fish of our seas. They are of three main types: **Silver pomfret** (*Pampus argenteus*), **white** or **chinese pomfret** (*Pampus chinensis*) and **black pomfret** (*Apolectus niger*).
4. All **billish** (*e.g.*, spear fish-*Makaira*; sword fish-*Xiphias*) (Suborder Scombroidei) have an upper jaw carrying a long, pointed snout. They are notorious for ramming boats and even ships with them, and then the snout often snaps off and remains buried in the wood.

9.8 TYPE STUDIES (BONY FISHES)

1. *Latimeria*

Classification

Phylum	—	**Chordata**
Group	—	**Craniata**
Subphylum	—	**Vertebrata**
Division	—	**Gnathostomata**
Superclass	—	**Pisces**
Class	—	**Osteichthyes**
Subclass	—	**Sarcopterygii**
Order	—	**Crossopterygii**
Genus	—	*Latimeria*
Species	—	*chalumnae*

Latimeria and the extinct coelacanth fishes are also called **lobe-finned fishes** or **Crossopterygian fishes**.

Discovery of *Latimeria chalumnae*—"The Living Fossil"

It was on December 22, 1938, when a trawler (Captain Goosen's boat *Nerine*) was routinely dragging its nets over the bottom in the Indian Ocean (3 miles off Chalumna River) near East London (South Africa), at a depth of about 37 fathoms. When the net was pulled aboard there was a 'weird-looking' fish ever seen by any zoologist. This fish was about 1.5 meter long, steel blue in colour and weighed 57.5kg. Unfortunately, at that time the fish could not be properly preserved and as a result much of its soft anatomy was destroyed and, therefore, to anyhow preserve it for future identification, that specimen was made into a **'Taxidermist's mount'**.

Miss **Courtney Latimer**, Curator of East London Museum (South Africa) recognised that it was a strange fish and reported the details to **Prof. J.L.B. Smith** (a Chemist by profession and an Ichthyologist by hobby) of Rhodes University, Grahmstown. Smith identified it as a coelocanth and named it *Latimeria chalumnae*, to honour Miss Latimer (The details of coelocanth, discovery are given by Smith in his book '*Old four legs*', see **Gupta** and **Gupta**, 2006). The Specific name of this fish refers to the first site of capture, offshore the mouth of Chalumna river of south Africa.

Another specimen of *Latimeria* was caught after 14 years of extensive search in 1952 from near Madagascar. Since then 150 specimens ranging in size from 75 cm to over 2 meter, weighing from 13–80 kg have been caught from the sea between Madagascar and Mozambique (**Khanna** and **Singh**, 2008). They live about 1.5 offshore at depth of 250–300 meters.

Habit. *Latimeria* is carnivorous (predator), feeding on other fishes and cephalopods. It swims by curious rotating movements of its pectoral fins, other fins aiding in steering.

Structure. *Latimeria* (Fig. 9.17) is steel blue in colour with irregular white spots and highly reflective golden eyes. Its entire fusiform body is covered with cosmoid scales. The round and heavy scales have hollow tubercles and overlap each other so as to form a strong thick protective covering. The pectoral and pelvic fins are pedunculate and lobed. The second dorsal and anal fins are also lobed. Such narrow-based paddled-fins are mechanically imperfect for aquatic progression, but well suited for locomotion on a solid or semisolid groud and thus provide information concerning the evolution of tetrapod limbs (Box 9.8). First dorsal fin is not lobed. The caudal fin is diphycercal and has a small central lobe also. The notochord is massive unconstricted and the vertebral column is cartilaginous but the neural and haemal spines are ossified. Spines present on the back of fish have coelom, thus the

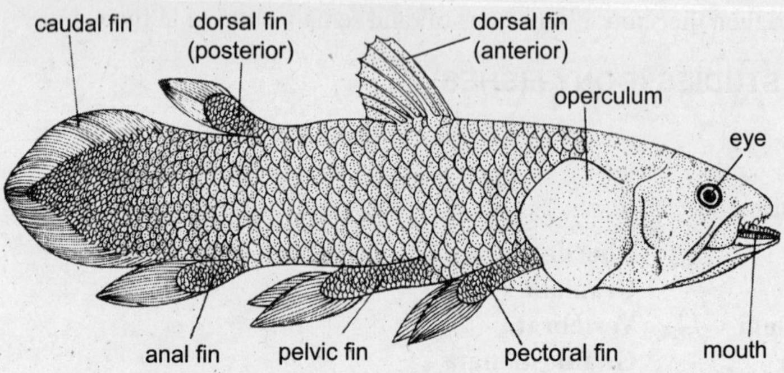

Fig. 9.17. *Latimeria chalumnae*. The living coelocanth.

Box 9.8

The fins of *Latimeria* are pedunculate each having a scale-covered fleshy lobe and fin rays arranged in a fan-like manner. The fins have an archipterygial type of endoskeleton consisting of a basal plate attached to the girdle. The basal plate is followed by a few axial rods and fin rays.

name coelocanth (hollow spine) is given. The skull has a movable hinge between the parietals and post parietals. A pair of bony plates called the 'gular plates' (Fig. 9.18) and five gill arches are present. There are two pairs of external nares communicating with the olfactory sacs. Internal nostrils are absent.

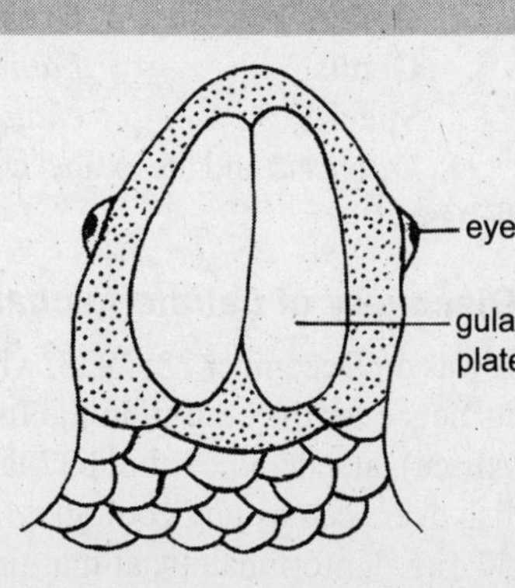

Fig. 9.18. Head of *Latimeria* in ventral view.

Alimentary canal consists of muscular oesophagus, followed by well developed stomach having gastric glands.

Intestine is a long tube having spiral valve. It opens into rectum, opening into cloaca. Rectal gland is present. Liver is a bilobed gland with a large gall bladder and the pancreas is well developed. Air bladder is small, fat covered and hydrostatic. Flesh (muscles) is oily increasing buoyancy of fish (Box 9.9). Heart is quite primitive, the auricle mainly lying behind the ventricle due to unfolding of the heart tube. Kidneys are fused to form a single median organ applied to the ventral abdominal wall. Each kidney has a ureter which is enlarged posteriorly to form a bladder. Thus, two urinary bladders are present. A pair of testes lie posterior to anus. The urinogenital aperture is situated between the pelvic fins. Copulatory organs are not present. Fertilization is internal. Eggs are of large size.

Box 9.9

Latimeria is an extremely mucilaginous species, because not only do the scales exude mucous but the body also contains a great amount of oil which oozes out continuously.

A peculiar sense organ consisting of a single median chamber full of gelatinous material in the rostral region, and opens to the exterior by means of three pairs of tubes. This rostral organ is probably an electro-receptor.

Biological significance. The discovery of a living coelocanth in 1938 was a very significant, as its group is considered to have become extinct more than 50 million years ago. Thus, *Latimeria* has survived for a much longer period than any other vertebrate, and has undergone very little change during these millions of years. Its anatomy shows the structure as existed in animals several hundred million years ago. *Latimeria* is described as the living fossil.

The discovery of *Latimeria* is of great significance, because it provides a support to the view that amphibians have enolved from some crossopterygians. The extinct crossopterygians possessed a pair of lungs. The skeletal elements of their paired fins resembled the proximal skeletal elements of limbs of tetrapods. It is now preseumed that during Devonian period certain freshwater crossopterygians started learning to clamber on to the muddy banks. By numerous slow changes fins modified into feet and the first amphibian arose some 300 million years ago. Coelocanths present some *intermediate stage*, which preferred to remain in water without undergoing any change towards terrestrial existence.

Phylogeny. The anatomy and physiology of *Latimeria* shows resemblances with the Chondrichthyes. *Latimeria* maintains high concentration of urea in the blood and is thus isotonic, as the elasmobranchs. Its pirtuitary gland also resembles with that of the cartilaginous fishes. On the other hand structure of skeleton, presence of bony scales and the swim bladder are the resemblances with bony fishes. *Latimeria* also possesses some specialized features peculiar to it. Hence there is no unanimity about the relationship of coelocanth. It appears that they have evolved independently for a very long period and have become unique and interesting issues. They live near the bottom at a depth of 250–600 m and possibly use their lobate fins to rest and "walk" on the bottom, while catching food.

2. *Neoceratodus* (Burnett Salmon or Australian Lungfish)

Classification

Phylum — Chordata
Group — Craniata
Subphylum — Vertebrata
Division — Gnathostomata
Superclass — Pisces

Class	—	**Osteichthyes**
Subclass	—	**Sarcopterygii**
Order	—	**Dipnoi**
Genus	—	*Neoceratodus*
Species	—	*foresteri*

Habit and Habitat

Neoceratodus occurs in the Burnett and Mary rivers in Queenland, Australia. It lives at the bottom of slow flowing rivers or in stagnant parts covered with vegetation. It is a sluggish fish. It swims by body undulations or crawls on the bottom with the help of pectoral and pelvic fins. When water becomes foul, this lung fish frequently rises to the surface to gulp in air for breathing with the lung. It cannot live outside water and does not aestivate unlike other lungfishes.

Neoceratodus is carnivorous feeding on molluscs, crustaceans, worms and larvae. Spawning takes place during September and October, after a complex courtship. Eggs are laid among selected vegetation and hatch in 3 to 4 weeks. No parental care has been observed. The larva is without external gills, and metamorphosis does not take place.

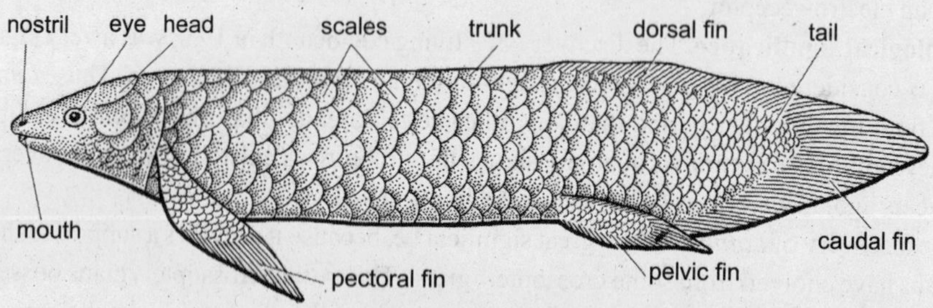

Fig. 9.19. *Neoceratodus.*

Structure

Body of fish is elongated and compressed measuring about 1.5 meters or more. It is covered by thin, overlapping cycloid scales. Paired fins are broad and paddle-like (Fig. 9.20). The dorsal and anal fins are lacking. The tail is diphycercal with symmetrical, pointed, caudal fin. The fin rays of the caudal fin are fibrous and unjointed (camptotrichia). The mouth has dental plates. The eyes are prominent. The external nares occur on the upper lip. The internal nares (choanae) open into the mouth cavity. The gill slits are covered by bony opercula. There are no spiracles. The intestine has a spiral valve. Single air bladder is vascular and acts as the lung. The lateral line is distinct..

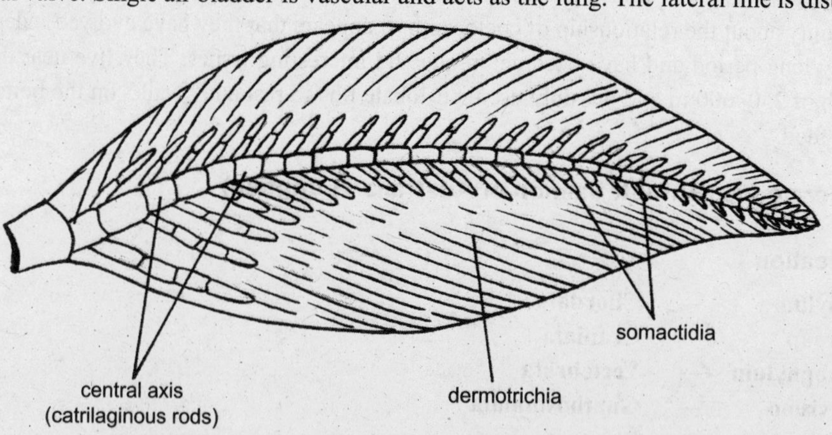

Fig. 9.20. Pectoral fin (archipterygium) of *Neoceratodus.*

3. *Protopterus* (African or Nile Lung Fish)

Classification

Phylum	—	**Chordata**
Group	—	**Craniata**
Subphylum	—	**Vertebrata**
Division	—	**Gnathostomata**
Superclass	—	**Choanichthyes**
Order	—	**Dipnoi**
Genus	—	*Protopterus*

Protopterus is widely distributed in the freshwaters of tropical and subtropical Africa. There are 4 species *viz.*, *P. aethiopicus*, *P. annectens*, *P. amphibius* and *P. dolloi*. They range from river Senegal and the White Nile on the north to Congo basin, lake Tanganyika and the Zambesi on the south.

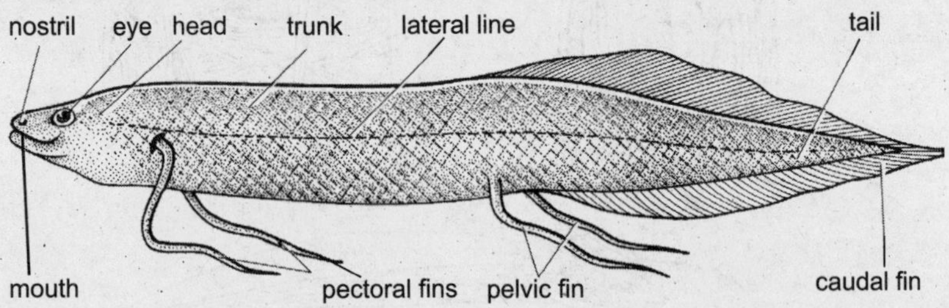

Fig. 9.21. *Protopterus.*

Habit and Habitat

Protopterus inhabits marshy land along the rivers. It aestivates during summers by embedding deep into the mud upto a depth of about 45 cm, where it coils itself in a flask-like enlargement at the bottom of the burrow lines by a capsule or "Cocoon" of hardened mucus; the respiration at this stage is through lungs only for which the air passes in from the open mouth of the burrow and then into the mouth of fish through a hole in the capsule (Fig. 9.22A). It then emerges out the capsule with the advent of the rainy season. Breeding takes place in about 30cm deep, hole-shaped, water-filled "nest" along a swamp, surrounded by tall grasses. Eggs deposited on bare mud. **Parental care** is common. Male guards the nest and furiously attacks when finds any intruder approaching the nest. Life history includes a larval form with crescentic sucker or cement gland behind and underneath the mouth, spiracles and 4 pairs of external gills. Protopterus is carnivorous and voracious feeder. It feeds mainly on molluscs, but also takes insects, and small fishes and frogs. It also exhibit cannibalism.

Structure

Protopterus has eel-like body (Fig. 9.21). It may grow upto 2 meters in length. The body of fish is covered with thin, bony, imbricate, cycloid scales. The mouth is terminal. The pectoral and pelvic fins are elongated and filamentous. They lack fin-rays. The dorsal and anal fins are continuous with the diphycercal tapering caudal fin. All these are supported by fibre-like, unjointed, partly calcified fin-rays called the **camptotrichia**. There are 6 branchial arches and 5 open gill-clefts. The gills are covered by opercula. Gill filaments developed only on 4th and 5th branchial arches. There are no spiracles.

Protopterus has a pair of vascular lungs (air bladder) with alveoli and septa on the inner surface to increase the respiratory surface. As the gills are small in size, the lungfish depends more on lungs for respiration.

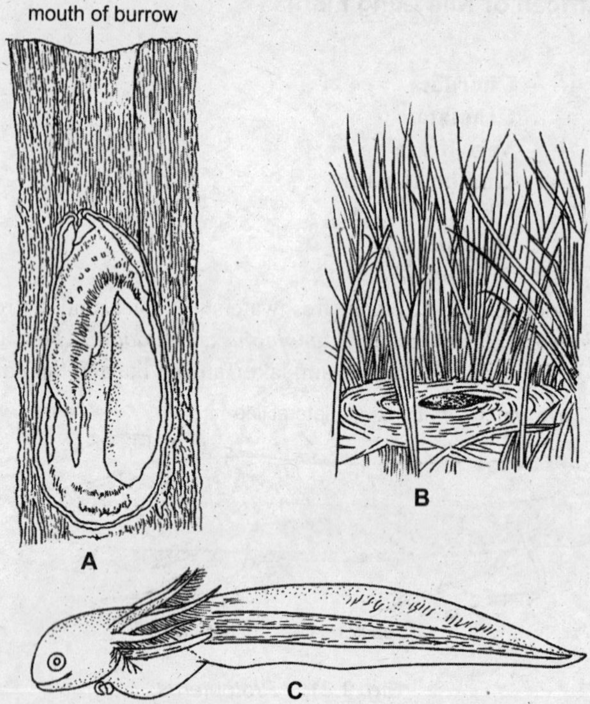

Fig. 9.22. *Protopteruis.* A—Aestivating fish; B—Nest; C—Larva.

Inhabitants of West Coast dug out the fish (*Protopterus*) of dried marshes and preserve clumps of mud for food.

4. *Lepidosiren* (American Lung Fish)

Classification

Phylum	—	**Chordata**
Group	—	**Craniata**
Subphylum	—	**Vertebrata**
Division	—	**Gnathostomata**
Superclass	—	**Pisces**
Class	—	**Osteichthyes**
Subclass	—	**Choanichthyes**
Order	—	**Dipnoi**
Genus	—	*Lepidosiren*
Species	—	*paradoxa*

Lepidosiren is commonly also called **South American Lung Fish** or **Lolach.**

Habit and Habitat

Lepidosiren lives in Amazon river of South America and its tributaries. It is found in the wide-spreading marshes and swamps along the Amazon river system. It is sluggish, wriggling slowly at the bottom, using its pelvic fins. It is omnivorous, feeding on fresh water molluscs (*Ampullaria* spp.) and algae.

Structure

Body of *Lepidosiren* is elongated, cylindrical and more or less eel-like. The skin covering the body contains the small cycloid scales. Pectoral and pelvic fins are thin and filamentous. Dorsal, anal and caudal fins are united. Tail is principal organ of locomotion. Five branchial (gill) arches and four gill-slits are covered by operculum. Air-bladder are paired (Dipneumona). It respires through gills as well as lungs. It do respiration through mouth by protruding the snout to the surface; when sunk to the bottom the bubbles of surplus air escape through the gill-clefts.

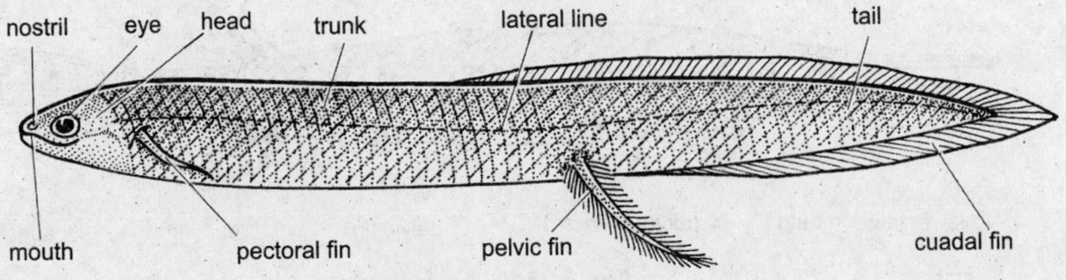

Fig. 9.23. *Lepidosiren.*

During summer, when the water dries up, the fish burrows deep into the mud and aestivates. During this period the fish remains inactive, depends on aerial respiration and used stored fat for its survival. With the arrival of the rainy season, when the river becomes flooded, the fish emerges from its burrow, feed actively and grows. A nest is prepared in the black peaty soil of swamp as a vertical tunnel, and the eggs are laid at its bottom, which is horizontal. Parental care is common. The male guards the nest till the eggs hatch. During this period, the pelvic fins of male become large and covered over by vascular filaments, which help in maintaining the surrounding water suitable by releasing oxygen. It has also been suggested that the highly vascular pevic fins may act as accessory respiratory organs, enabling the male fish to guard the nest continuously, without going up to the surface for aerial respiration. The newly hatched larva has four pairs of external gills and a **sucker** or **cement organs** (as in *Protopterus*) and undergoes metamorphosis to become the adult (Fig. 9.24). The fish is caught by spearing in the swamps for its much esteemed flesh.

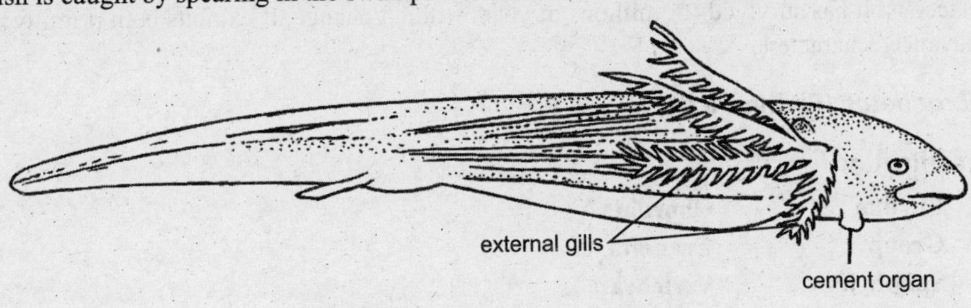

Fig. 9.24. Larva of *Lepidosiren.*

5. *Lepidosteus* (Garpike)

Classification

Phylum	—	**Chordata**
Group	—	**Craniata**
Subphylum	—	**Vertebrata**
Division	—	**Gnathostomata**
Superclass	—	**Pisces**
Class	—	**Osteichthyes**
Subclass	—	**Actinopterygii**
Superorder	—	**Holostei**

Order — **Lepidosterformes**
Genus — *Lepidosteus*

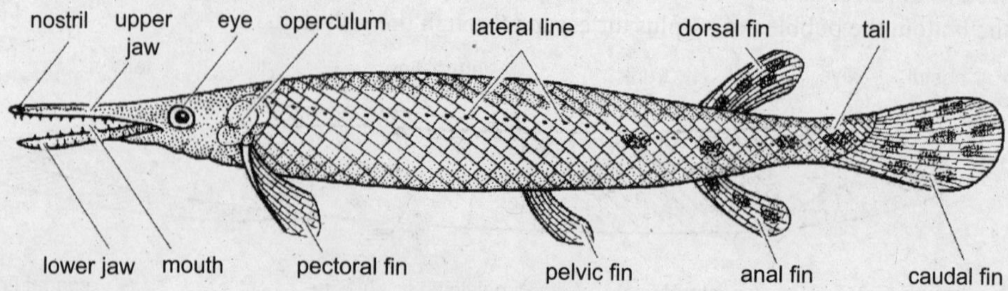

Fig. 9.25. *Lepidosteus.*

Characters

Lepidosteus osseus (long-nosed gar) and *L. platostomus* (short-nosed gar) are found in rivers and great lakes of North America. When water level falls and water becomes foul, garpike uses its air bladder as a lung. The body of garpike is elongated (upto 3 meter long) and spindle-like. It is covered with thick armour of rhomboid scales with thick covering of ganoin. Scales articulate by peg and socket joints. Jaws are produced into a long beak having sharp teeth, adapted for feeding on fishes. The maxillae are reduced and replaced by a number of tooth-bearing infraorbital plates. Cheek-plates are numerous and vertebrae are amphicoelous. Nares lie on the tip of the beak. The dorsal and anal fins are placed far behind near the caudal fin but opposite to each other. Tail is hemi-heterocercal. Spiracles are absent. The intestine has a pyloric caeca. Heart has a vascular conus. *Lepidosteus* is oviparous and the newly hatched young bears sucking disc. Scales of garpike are used in jewellery.

Lepidosteus arose in the Permian, flourished in the seas in the Jurassic, and died out in the Cretaceous. It has survived for millions of years without change. It exhibits both primitive as well as advanced characters.

6. *Exocoetus* (Flying Fish)

Classification

Phylum — **Chordata**
Group — **Craniata**
Subphylum — **Vertebrata**
Division — **Gnathostomata**
Superclass — **Pisces**
Class — **Osteichthyes**
Subclass — **Actinopterygii**
Superorder — **Teleostei**
Order — **Beloniformis**
Genus — *Exocoetus*

Characters

Exocoetus pecilopterus is an Indian species. *Exocoetus* (two winged flying fish; Fig. 9.12) is found in the tropical and subtropical seas. It is a marine pelagic fish living in shoals and feeding on prawn, young fishes and their eggs. It is oviparous.

Exocoetus often rises from the water into the air, taking a leap with sculling motion of the powerful tail, and then slowly glides through the air with the help of pectoral fins. The pectoral fins act only as parachutes as no force is required in the air. Leap in the air may be upto 6 meters. The fish takes to flight for escapting from the enemies.

Exocoetus is about 30 cm long, subcylindrical body is covered with large cycloid scales. Eyes are very large. Pelvic fins are small and lie on the ventral side nearly at the middle of the body. Dorsal and anal fins lie almost opposite to each other. The caudal fin is deeply forked, its upper lobe being smaller than the lower one. This condition is called **hypobatic**.

7. *Hippocampus* (Sea-Horse)

Classification

Phylum	—	**Chordata**
Group	—	**Craniata**
Subphylum	—	**Vertebrata**
Division	—	**Gnathostomata**
Superclass	—	**Pisces**
Class	—	**Osteichthyes**
Subclass	—	**Actinopterygii**
Superorder	—	**Teleostei**
Order	—	**Syngnathiformes (= Solenichthyes)**
Genus	—	*Hippocampus*

Hippocampus is a marine fish of small size, and is called the **sea horse** because its head resembles superficially with the head of a horse (Fig. 9.12).

Habit and Habitat

Hippocampus is found in the coastal waters of the Indian and Atlantic Oceans. It lives at the sea bottom among seaweeds. It clings to the weeds by coiling its prehensile tail (that can coil round to hold) around the twigs. Sea horse swims by means of the dorsal fin (that vibrates about 35 times a second) with the body held in vertical position (erect posture). Female lays her eggs in the brood pouch of the male. The eggs are retained in brood pouch till hatching.

Seahorse only eats tiny moving food. It feeds on crustaceans and their larvae such as shrimp-like animal and amphipods.

Structure

Indian species, *Hippocampus guttulatus* is 20 cm long. Another Indian species, *H. kuda* grows to 30.5 cm, but the largest seahorse is found in the Pacific, called *H. ingens*, it grows to 35.6 cm. The smallest is the pygmy or dwarf seahorse (*H. zosteral*) of Florida, which is only 5cm when fully grown.

The entire body of seahorse is invested by closely fitting, bony shields, which are produced into tubercles, spines or filaments of varying size at certain places. Thus, interlocking bony plates of the seahorse form a series of about 45 rings. The head is at right angles to the trunk and looks like that of a horse. It is prolonged into a tube-like snout bearing a small terminal mouth. Eyes are dorso-lateral at the base of the snout. Its both eyes can rotate independently, *e.g.*, a seahorse can look ahead with its left eye while its right eye may be focused backwards. Pectoral fins lie just behind the opercula. Pelvic fins are absent. There is a single dorsal fin at the middle of the back. Anal and caudal fins are lacking. Gills are lophobranchs, *i.e.*, gills consist of many rounded lobes attached to the gill arches. Male seahorse has a brood pouch on the ventral side of the base of the tail. The pouch is provided with an opening at its top. The female while mating, passes her eggs to this pouch where they are incubated. The young are about 12 mm long at birth and are miniature replicas of the adult.

Seahorse has a 'coronet'—a star-shaped bony crest at the back of its head. By its help, sea-horse can produce loud clicks (sounds) during its breeding season.

8. *Anabas* (Climbing perch)

Classification

Phylum	—	**Chordata**
Group	—	**Craniata**
Subphylum	—	**Vertebrata**
Division	—	**Gnathostomata**
Superclass	—	**Pisces**
Class	—	**Osteichthyes**
Superorder	—	**Teleostei**
Order	—	**Perciformes**
Genus	—	*Anabas*
Species	—	*scandens*

Characters

Anabas (Fig. 9.12) is a South Indian freshwater fish which can make migratory excursions on damp soil and paddy fields with the help of opercular spines and spiny fin-rays (of pectoral fins). The body is olive-green in colour, protected by overlapping cycloid scales. The dorsal fins are long and their anterior parts are spinous. The paired fins are large and the pelvic fins lie close to pectorals. Eyes are large and the preorbital bone lying in front of each eye bears a spine. The free margin of operculum is also produced into spines. The lateral line is interrupted and dorsally placed.

Anabas has labryinthiform accessory respiration organ, situated in the suprabranchial chamber and so it is able to breath the atmospheric air and can survive out of water. *Anabas* is a predator, feeding upon shrimps, ostracods, gastropods and young fishes. It is greatly esteemed as food in several parts of the country.

9. *Pleuronectes* (Flat-fish)

Classification

Phylum	—	**Chordata**
Group	—	**Craniata**
Subphylum	—	**Vertebrata**
Division	—	**Gnathostomata**
Superclass	—	**Pisces**
Class	—	**Osteichthyes**
Subclass	—	**Actinopterygii**
Super order	—	**Teleostei**
Order	—	**Pleuronectiformes**
Genus	—	*Pleuronectes*

Characters

There are various genera of flat fishes which have world-wide distribution.These are deep-sea fishes dwelling at the bottom with one side resting on the ground. Hence, the body has become laterally flattened and leaf-like. The side that faces bottom is unpigmented (pale or white) and lacks eye while the opposite side is pigmented and has both eyes. The flat fisher are described as **dextral** or **sinistral**, on the basis of wheather their eyes are on the right or left side of the head. The side of the body away from the eyes is known as the blind side. For example, *Pleuronectes* (Fig. 9.12) is right-handed (dextral) while *Pseudorhombus* (Flounder) is left-handed (sinistral). *Synaptura* (sole) has its eyes on the right side.

Skull of flatfishes is distorted so as to adapt the orbit to its changed position. The abdominal cavity is very small and anus is placed forward. The dorsal and anal fins extend the whole length and are some times confluent with the caudal fin. The pectoral fins are poorly developed and pelvic fins may be reduced or wanting. The scales are either cycloid or both types. The flat-fish is edible.

10. Echeneis (Remora or Sucker-fish)

Classification

Phylum	—	**Chordata**
Group	—	**Craniata**
Subphylum	—	**Vertebrata**
Division	—	**Gnathostomata**
Superclass	—	**Pisces**
Class	—	**Osteichthyes**
Subclass	—	**Actinopterygii**
Superorder	—	**Teleostei**
Order	—	**Echeneiformes**
Genus	—	*Echeneis*

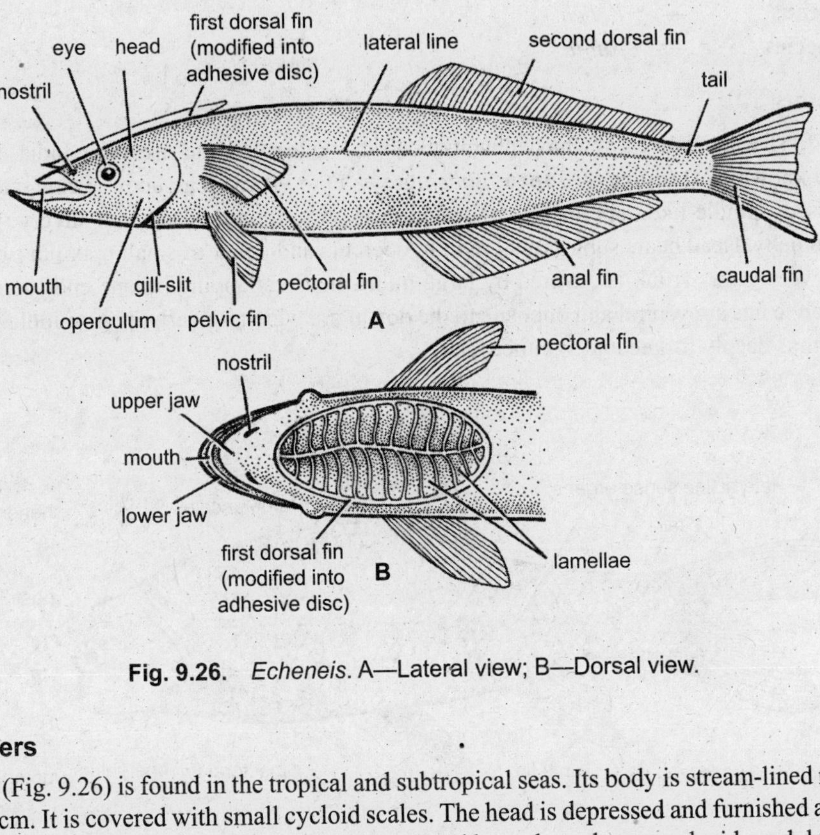

Fig. 9.26. *Echeneis.* A—Lateral view; B—Dorsal view.

Characters

Echeneis (Fig. 9.26) is found in the tropical and subtropical seas. Its body is stream-lined measuring about 70 cm. It is covered with small cycloid scales. The head is depressed and furnished above with an adhesive organ that is motified first dorsal fin. Head bears lateral eyes and wide and deep mouth. The **sucker** or **adhesive vaccum disc** is flat and oval. It consists of a double series of transverse bony plates, *i.e.*, sucker bears 21–28 transverse lamellae. The plates inside the oval disc can be erected or flattened like the slats of venetian blind. The sucker is used by the remora for the temporary attachment to the bottom of ship or the under surface of the fish (sharks, billfish, sunfish), turtles, whales, etc., and the fish can get a free ride from place to place.

In sucker fish, pectoral fins are lateral. Pelvic fins are almost below the pectorals. Second dorsal fin and anal fin are long and opposite to each other. The tail fin is entire (*i.e.*, homocercal) and slightly concave posteriorly. Sucker-fish is carnivorous feeding on small fishes. On the east coast of Africa. *Echeneis* is employed for catching the turtles.

11. *Labeo* (Rohu)

Classification

Phylum	—	**Chordata**
Group	—	**Craniata**
Subpylum	—	**Vertebrata**
Division	—	**Gnathostomata**
Superclass	—	**Pisces**
Class	—	**Osteichthyes**
Subclass	—	**Actinopterygii**
Superorder	—	**Teleostei**
Order	—	**Cypriniformes**
Genus	—	*Labeo*
species	—	*rohita*

Characters

Labeo is fresh water fish (carp) which inhabits clean and quite rivers and streams in India and Pakistan. It is also known as **Dambra** or **Tapra**. *Labeo* grows upto 90 cm. The body is elongated, laterally compressed, spindle-like, tapering at either end. It is greyish on the back and silvery white on the sides and belly. Head bears subterminal mouth, opercula and a pair of small maxillary barbels. The eyes are very large. Trunk is covered by large imbricate (overlapping by one end) dermal cycloid scales.Pelvic fins are ventral and opposite to the dorsal fin, which is nearly in the middle of the body. Caudal fin is deeply forked and symmetrical.

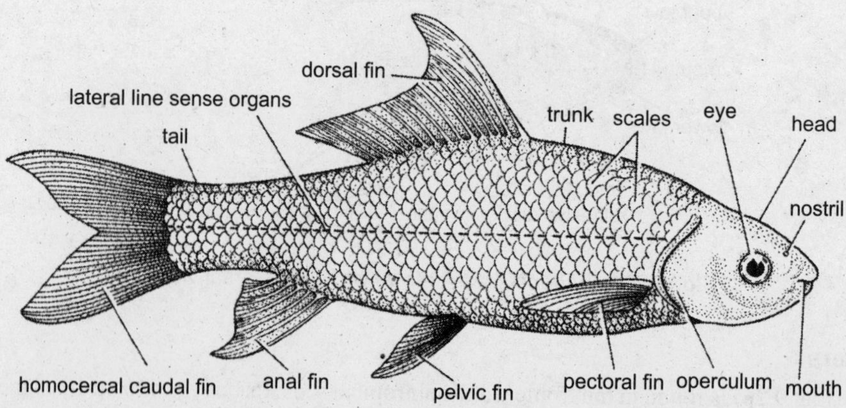

Fig. 9.27. *Labeo rohita.*

Labeo fenquently comes to the surface of water to gulp air into the air bladder through the oesophagus. It is carnivorous when young, feeding on crustaceans and insect larvae and herbivorous when adult (feeding in alage and aquatic plants). Sexes are seperate. Fertilization is external. The fish is oviparous. It breeds in July and August. *Labeo* is an excellent food-fish. It responds well to pisiculture.

9.9 ACCESSORY RESPIRATORY ORGANS OF FISHES

Majority of the teleostean fishes have well developed gills for respiration which are very efficient organs for gaseous exchange in the aquatic environment. But some tropical fishes have evolved special structures called the additional or **accessory respiratory organs**. These organs supplement or replace gill respiration and enable the fishes to live in oxygen deficient water to aestivate in dry summer and to take short excursions on land. Oxygen deficient water occurs in the swamps and muddy ponds infested with weeds. It has been shown that the absorption of oxygen in the accessory breathing organs is much greater than the excretion of carbon dioxide. Most of this carbon dioxide is ultimately excreted in the gills.

Generally, some existing respiratory and other organs of fishes have become highly vascular and modified for aerial respiratory organs. The organs so modified are skin, gills, branchial cavity, air bladder or the intestine. However, in a few species of fishes, new structures have developed for fishes, new structures have developed for accessory respiration and these are called **neo-morphic organs**. Such accessory organs are found in fish species living in shallow, stagnant water and swamps where deoxygenated water is found, but are rarely seen in marine fishes. These organs may also develop in fishes living in hill streams that are liable to dry up during summer.

1. Skin

In several species of fishes, the skin becomes highly vascular and is kept moist by mucus, so that exchange of gases takes place between blood and air.

(i) Periophthalmus, the mudskipper or goggle-eyed fish of Indian Ocean often skips mud and walk about through damp areas round the roots of mangrove trees. In mudskippers, the gill covers can completely real the gill cavities and a small amount of sea water retained inside enable the fish to remain outside water for a long time. These fish love to base in the sun at ebb tide (**Chhapgar** 2008). The old idea that the mudskippers use the vascular tail as the respiratory organ is not supported by recent ichthyologists (see **Jordan** and **Verma**, 2013).

(ii) Anguilla anquilla, the common cell, can crawl on damp grass for some distance, respiring through the skin during the venture. When on land, the cell keeps its small branchial aperture tightly closed so that water retained in the branchial chambers enables it to continue gill respiration also.

(iii) Median fin folds of many fishes are supplied with numerous blood vessels that help in cutaneous respiration.

(iv) Lepidosiren, the South American lungfish, has delicate outgrowths of the pelvic fins. These outgrowths develop in males during breeding season and serve as cutaneous respiratory organs during parented care.

(v) The highly vascular opercular folds of *sturgeons* and many cat fishes serve as accessory respiratory organs.

2. Bucco-pharyngeal epithelium

The buccal cavity of the South American fish, *symbranchus*, has a highly vascular lining capable of respiration. When water runs short of oxygen, the fish rises to the surface, fills its buccal cavity with air, closes the mouth and uses the trapped air in respiration.

In most of the fishes, the epithelial lining of buccal cavity and pharynse is usually highly vascular and permeable to gases in water. It may remain simple or may develop folds, pleats or tongues projecting into the buccal cavity and pharynx to make it an efficient respiratory organ. However, in mudskippers (*periophthalmus* and *Boleophthalmus*) the highly vascularised bucco-pharyngeal epithalium help in absorbing oxygen directly from the atmosphere.

3. Gut epithelium

In some species, a part of the alimentary canal (such as stomach or intestine) becomes modified for aerial respiration. Thus, in several fishes apithelial lining of gut becomes thin and vascular and modified to serve as a respiratory organ. Such a respiratory epithelium may be present just behind the stomach (*Misgurus fossilis*) or intestine (*Hoplosternum, Lepidocephalus guntea, Gobitus* (giant loach of Europe) or rectum (callichthyes (catfish), *Hypostomus* and Doras). Fresh air is drawn through mouth or annus and after gaseous exchange the air is voided through the anus.

4. Pharyngeal diverticula

In *Channa* (= Ophiocephalus) spp. suprabranchial cavities are formed in the roof of the pharynx (Fig. 9.28). These cavities are lined with highly vascular respiratory membrane raised into folds, pleats and tongues, and contains respiratory islets. The superbranchial cavities have alveoli also, and open antero-ventrally into the opercular cavity. All the gill arches bear gill filaments but they are reduced in size and those on the fourth arch are greatly reduced.

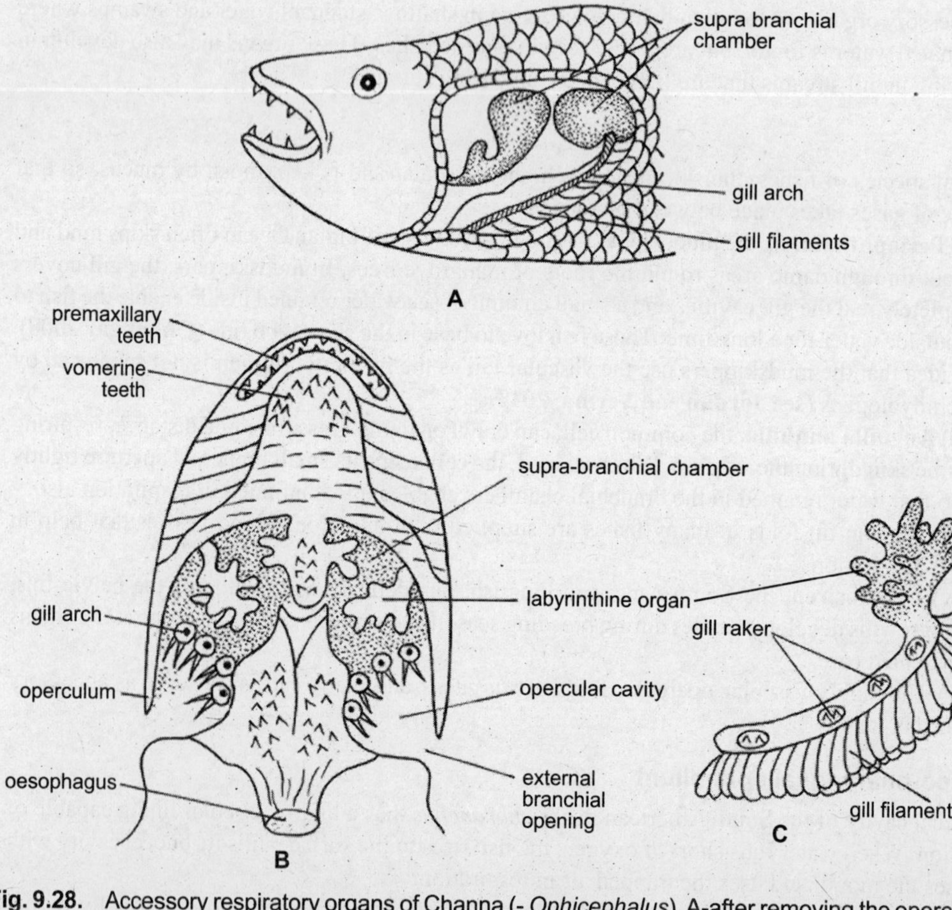

Fig. 9.28. Accessory respiratory organs of Channa (- *Ophicephalus*). A-after removing the operculum; B-Roof of the buccopharynx and C-First gill arch.

In *Channa*, processes of cartilage develop from the parasphenoid, hyomandibular and the epibranchial of the first arch and project into the supra-branchial cavity. These projections are covered by a thin, highly vascular respiratory membrane. The epibranchial of the first gill arch becomes flattened and the respiratory membrane covering it becomes folded and studded with a large number of nodules or papillar forming a dendritic labyrinthine organ. Each of these nodules is believed to be a modified gill filament. In *Channa* the roof of the buccal cavity is also covered by the vascular respiratory epithelium (or membrane), which may be over the tongue also. Thus, the entire bucco-pharyngeal cavity along with the supra-branchial chamber serves for aerial respiration and is filled with air for this purpose.

In cuchia eel, *Ampeipnous cuchia* the accessory respiratory organs consists of a pair of vascular sac-like divertical from the pharyx above the gills at both lateral sides of the head. These sacs develop

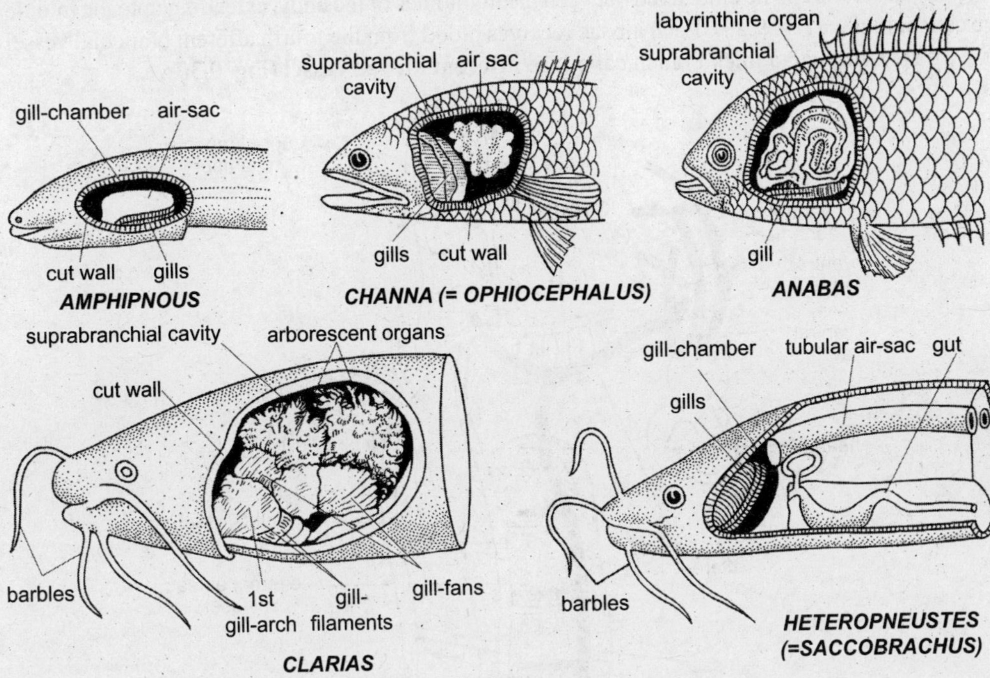

Fig. 9.29. Head of some air-breathing teleost fishes dissected to show the accessory respiratory organs.

as diverticular of the pharyx and are lined with highly vascular epithelium raised into folds. The epithelium consists of vascular areas called 'islets' bearing rosetts or papillae. Each sac has an inhalant and exhalant aperture to provide passage to air. When the muscles of the buccopharynx contract, mouth is kept closed and the air is expelled from the air sac through the exhalant aperture and goes out through the external branchial aperture. The mouth is then opened and the muscles of the buccopharynx relax, so that air enters to fill the empty space.

In cuchia eel, the gills are reduced and only the second gill arch bears filaments, while the third arch has a fleshy vascular membrane.

In *Periophthalmus* also, a small, shallow pharyngeal diverticulum lined with respiratory epithelium (vascular epithelium) is present on each side of the roof of the pharyx.

5. Opercular chamber modified for aerial respiration. In *Periophthalmus* and *Boleophthalmus* the opercular chamber serves for aerial respiration. For this purpose the opercular cavity bulges and to form two little balloon-like structures in the hinder region of the head and inhaled air is passed through the gill slits and is stored for sometime in the cavity of balloons. The membrane lining the lining the opercular cavity is thin and highly vascular. After exchange of gases, the air is passed out through the external branchial aperture.

6. Branchial diverticula. The outgrowths from gill-chambers form more complicated aerial respiratory organs than the simpler pharyngeal outgrowths in other fishes. Such air breathing organs are present in **Heteropneustes, clarias, Amabas** etc.

(i) Heteropneustes fossilis (- Sacchobranchus). The accessory air breathing organs of Indian catfish, *Heteropneusts fossilis*, are highly specialised and consists of (a) two air sacs, (b) fans and (c) respiratory membrane.

(a) Air sacs. A pair of simple sac-like structures extend posteriorly, one on each side, from the suprabranchial chamber of the opercular cavity. These sacs are thin walled, tubular structures having highly vascular walls and lie embedded between the myotomes of the body, extending upto the middle of the caudal region (Fig. 9.29). Each air sac receives blood from the fourth afferent branchial vessel of its own side, through a thick branch called the **afferent air sac vessel** (Fig. 9.30).

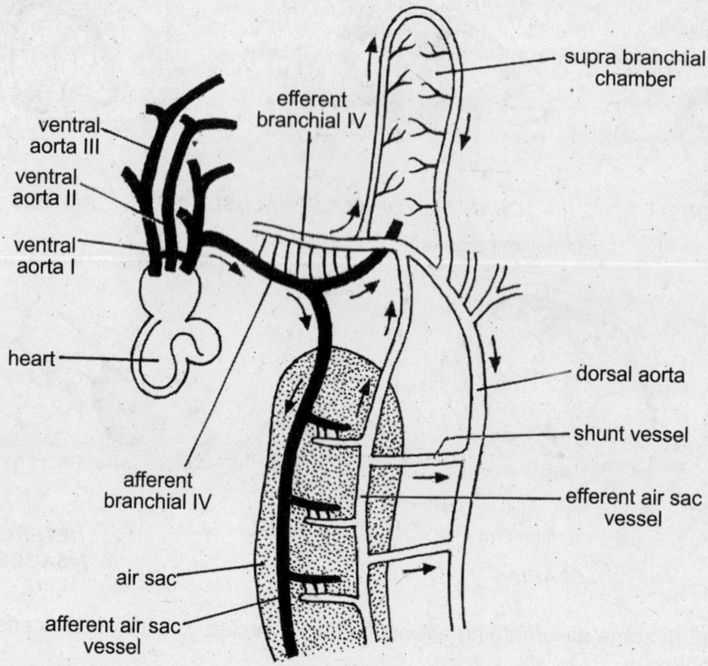

Fig. 9.30. Blood supply of the air sac and suprabranchial chamber in *Heteropneustes fossilis*.

This branch gives off several capillaries to supply blood to the respiratory islets of the sace. The blood is collected from the islets by efferent capillaries which open into an efferent air sac vessel, which in its turn carries blood to the fourth efferent branchial vessel of its own side (**Dutta Munshi** 1993).

(b) Fans. The gills of the Indian cat fish bear **fans**, Gill filaments of all the gills are reduced is size and those of the upper limb of the gill arch are fused to form fans, one of each gill arch are fused to form fans, one on each gill arch.

(c) Respiratory membrane. It lines the air sacs of Indian cat fish and is composed of vascular and non-vascular areas, and is thrown into folds and ridges to increase its surface. The vascular areas consist of a number of islets having lamellae arranged biserially are the sites of gaseous exchange. The nonvascular areas consist of lanes and possess mucous glands. The air sacs and the fans are the result of fusion and modification of gills filaments and reduction of the lamellae. During respiration, the *H.fossilis* inhales air which enter the suprabranchial chamber through the inhalant aperture and is passed into the tubular sac. After the exchange of gases, used air is expelled through the external branchial opening, and the fans help in this process.

(ii) *Anabas*. Air breathing organs of Indian climbing perch, *Anabas testudineus*, consist of a specious air chamber and a labyrinthine organ on each side (Fig. 9.29). The air chamber lies between the first gill arch and the hyomandibular and opens into the bucco-pharyngeal cavity as well as the opercular cavity. The respiratory epithelium which lines the air chamber is highly vascular and may be folded to increase the respiratory surface. A well developed **labyrinthine organ** is present on the first gill arch. This organ develops from the first epibranchial and consists of three concentric bony plates. the first plate is fused with the respiratory epithelium of the air chamber and divides it incompletely into two. The second and the third plates are greatly folded due to which the labyrinthine organ a complex structure. The labyrinthine organ is also covered by a vascular respiratory epithelium and serves to increase the respiratory area for exchange of gases.

(iii) *Trichogaster*. Air breathing organs of *Trichogester* includes a suprabranchial chamber, a labyrinthine organ and the respiratory membrane (Fig. 9.31). As in Anabas, the suprabranchial

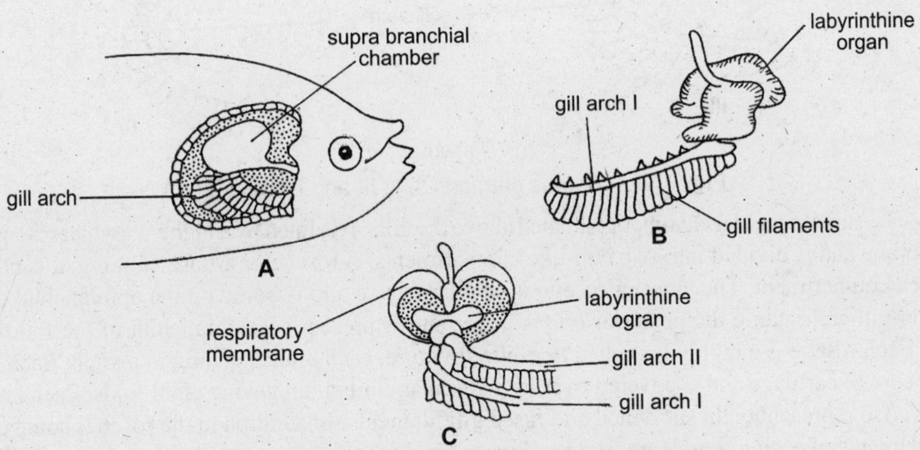

Fig. 9.31. *Trichogaster: Air breathing organs.*

chamber lies on either side above the gills, and opens into the pharynx by means of an inhalant aperture, and into the opercular chamber by means of an exhalant aperture. A labyrinthine organ has two leaf-like expansions and develops on the epibranchial of the first gill arch. It looks like a spiral organ and is simpler in organization than that of *Anabas*. The respiratory membrane covers the suprabranchial chamber and the labyrinthine organ and it consists of vascular and non-vescular areas. The vascular area contains a number of respiratory islets, which have numerous parallel blood capillaries, and are believed to be derived from the lamellae of the gill filaments.

(iv) *Clarias batrahas*. The air breathing organs of Indian cat fish *Clarias batrachus* are highly evolved and complex. These include following types of organs:

(a) a suprabranchial chamber on each side of head:

(b) two air trees, rosetter, dendritic organs or absorescent organs on each side;

(c) the fans; and

(d) the respiratory membrane.

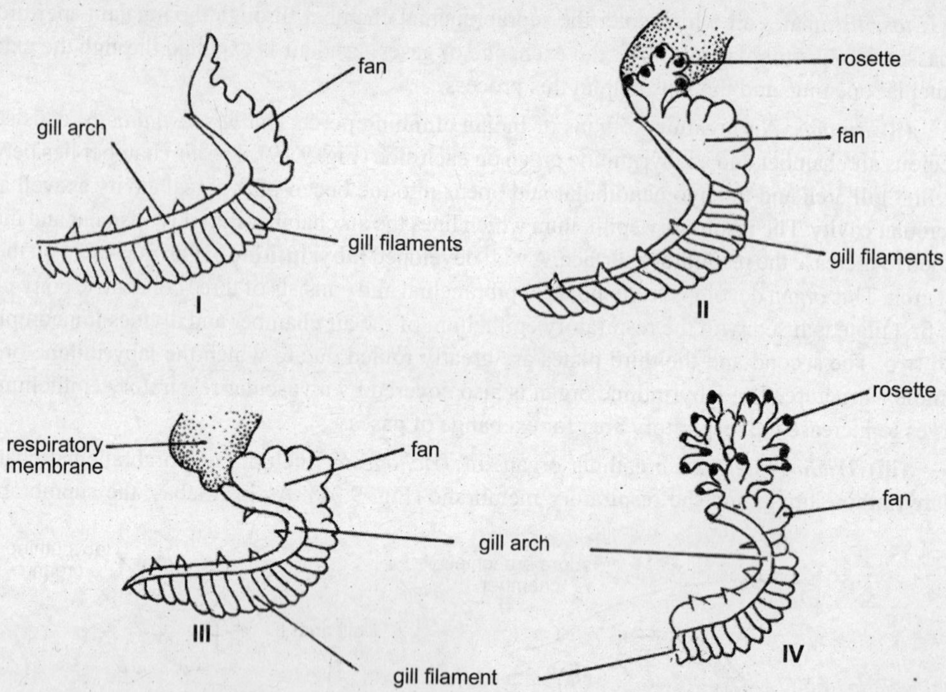

Fig. 9.32. *Clarias batracus.* I, II, III and IV gill arch.

The suprabranchial chamber is situated above the gills. It is lined by a highly vascular respiratory membrane and is divided into two cup-like compartments. A rosette or arborescent organ is present in each compartment. The anterior rosette is smaller in size and is borne on the epibranchial of the second gill arch, while the posterior one is larger and is present on epibranchial of the fourth gill arch. Each rosette is a highly branched tree-like structure, each branch ending in a knob. Each knob has a core of cartilage and is covered over by a vascular epithelium having eight folds. Each knob is believed to represent eight shortened and fused gill filaments. In addition to the rosettes some of the gill filaments of each gill arch are fused to form a "fan", so that there are four fans, one on each gill arch. The remaining filaments are reduced in size.

The suprabranchial chamber has inhalant and exhalent apertures. The respiratory membrane lining the suprabranchial chamber and covering the rosette, is composed of vascular and non-vasular areas. The vascular areas have numerous islets which serve as the sites of gaseous exchange.

The fish (*C. batrachus*) frequently rises to the surface of water and gulps in air, which enters the suprabranchial chamber through the inhalant aperture which is a narrow slit between the second and third arch. The air entering the opercular cavity is directed into the suprabranchial chamber by the action of the fans. The slit between the third and the fourth gill arches is used as the exhalant aperture, and the air, after the gaseous exchange is passed into the opercular cavity and then to the exterior.

7. Air Bladders

The air bladder is an outgrowth from the oesophagus in bony fishes. It many retain or lose communication with the oesophagus. The fishes which retain this communication are said to be **physostomous**. For example, teleost fishes of orders such as Cypriniformes, Anguilliformes and Esociformes are physostomous. An air bladder of physostomous fishes frequently rise to the water surface and gulp air into air bladder where exchange of gases takes place. Lung fishes store air in the air bladder for use in respiration during aestivation in burrows.

Diurnal fluctuation in aerial respiration in fishes. Air breathing fishes exhibit fluctuation in their breathing pattern. These fishes are nocturnal in habit and are usually inactive restive during the day, and are able to meet their oxygen requirement through aquatic respiration by gills. During night these fishes become active and need more oxygen. Which is in short supply in water during night. Hence, there fishes frequently rise to the surface to gulp in air. According to **Dutta Munshi** and **Ghosh** (1994), the two abiotic ecological factors (*i.e.*, dissolved oxygen and free carbon dioxide) act as the **steering wheel** controlling the metabolic activities of the living biota in the swamps.

9.10 PARENTAL CARE IN FISHES

Parental care behaviour is universal among fishes. Of some 250 families described in **Breder** and **Rosen's** (1966) encyclopaedic treatise on fish reproduction, about 77 per cent fish show no parental care, another 17 per cent include fish species that care for the egg only and less than 6 per cent contain species that are known to care for eggs and newly hatched young.

Fish show all grades of parental care behaviour from random spawning and from deposition of large number of uncared eggs to the protection of young. The lack of parental care behaviour is correlated with the production of great number of eggs and sperms.

Two general types of variation in parental care behaviour exist among fishes. *First* either both parents, or one alone cares for the offspring. Thus, there are **paternal, maternal** and **biparental** species. *Second,* the eggs and newly hatched young are either maintained on the substrate — that is, on plants, under stones, in excavated pits and so on (these are called **substrate-brooders** or **guarders**) or carried about in the parent's mouth (these are called **mouth-brooders** or **incubators**). Fishes have evolved many means of affording care to fertilized eggs and young ones by one or both sexes.

1. Scattering eggs over aquatic plants. In some fishes such as pikes, *Esox lucius*; carps, *Cyprinus carpio*, *Carrassius auratus,* etc., eggs are scattered usually over aquatic plants to which they are attached.

2. Depositing eggs in sticky covering. In many carps, eggs are usually laid with some special sticky covering by means of which they are attached to each other and to the stones, weeds, etc. In yellow perch, *Perca flavescens* eggs are deposited in a rope of single mass.

3. Laying of eggs at suitable places. Suitable spawning grounds are selected by anadromous fishes such as *Salmo solar, Acipenser, Oncorhyncus,* etc. They dig excavation in gravel substrate, lay their eggs in the pits, cover them with gravel and desert them. The sand gobi *Pomatoschistos minutus* lays its eggs in some protected spot where they are guarded by the male who aerates them by his movements.

4. Nest building. Only a few fishes construct nest which may be complex like that of birds or simple. Simple nests of fishes are merely hollowed out depressions in the bottom as in the lung fishes. Males of many species such as darters (*Etheostoma*), sunfishes (Fig. 9.33) and cichlids prepare a shallow basin-like nest. All stones and rock crystals are carefully removed from the bottom. The eggs are laid in the nest and the male remains on guard till the young ones are hatched.

Fig. 9.33. Shallow basin-like nest of sunfish.

The male African lung-fish, *Protopterus,* prepare a simple nest in the form of deep hole in swampy places along the river banks. After spawing he guards the nest. The South African lung fish *Lepidosiren* also prepares a nest in the form of a burrow and the male develops highly vascularized filaments on its pelvic fins for aeration. The male bowfin, *Amia calva*, of the great lakes of North America builds a crude circular nest among aquatic vegetation (Fig. 9.34). The male stands on guard till the young ones are hatched. The young ones leave the nest only under the protection of the father.

Fig. 9.34. Male *Amia calva* guarding its circular nest.

Before the onset of its courtship, the male stickleback (*e.g.,* Three-spined stickleback, *Gasterosteus aculeatus* and ten-spined stickleback, *Pygosteus pungitius*) builds a quite elaborate spherical or elongate nest. The nest is built by collecting plant fragments, rootlets and the like and then binding them together with adhesive kidney secretions. The various activities of male such as probing, boring, sucking and glueing, result in the formation of a compact nest with an internal chamber (tunnel) to receive the eggs. Male drives and induces the female into the nest for laying eggs, then chases her away, enters the nest, fertilizes the eggs and guard them from intruders (Fig. 9.35).

Fig. 9.35. Nest of ten spined stickleback;

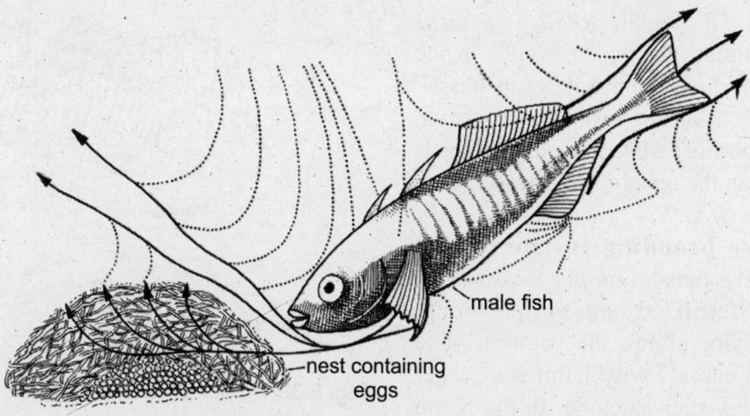

Fig. 9.36. Fanning by male stickleback at the entrance of its nest.

The most elaborate nest is made by *Apelts quadracus*; its cup-shaped nest is attached to rooted plants close to the bottom. After a clutch of eggs is laid, the male builds an extension of the nest up and over the eggs, with a concave upper surface to the extension. A second clutch of eggs is laid on the new nest floor and this procedure may be repeated several times, until the male has several clutches of eggs stacked vertically within a single multitiered nest (**Rowland,** 1974).

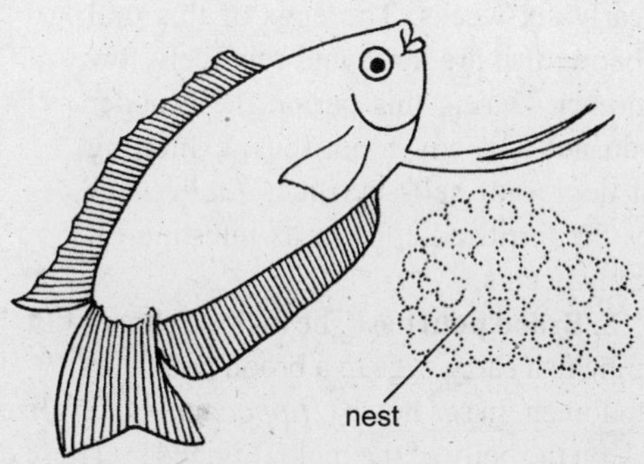

Fig. 9.37. Male Siamese fighting fish defending his floating nest.

Floating nests are made by American catfishes, in which the eggs are suspended in a mass of bubbles and mucus produced by the fish. The male Siamese fighting fish (*Betta splendens*) too builds a floating nest and sticks the fertilized eggs to the lower surface of foamy nest. He stays on guard on this nest and fights till death to defend it (Fig. 9.37). The male paradise fish, *Macropodus* also prepares a similar foamy nest.

5. Coiling round the eggs. The butter fish (*Pholis gunnellus*) rolls all eggs into a ball and curls around it (Fig. 9.38).

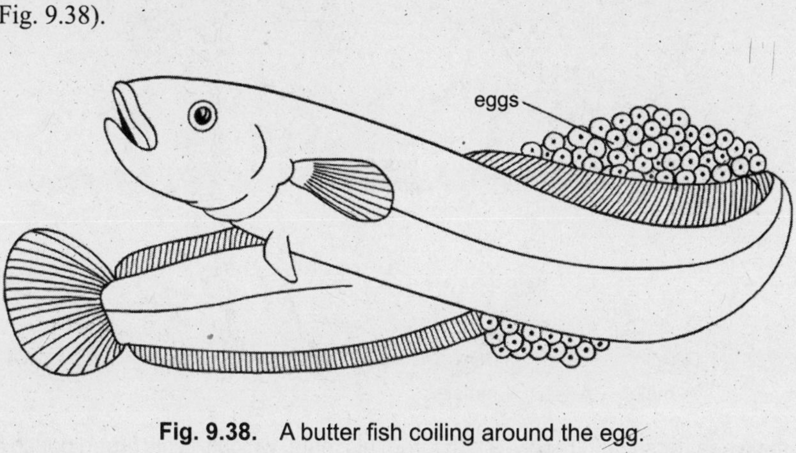

Fig. 9.38. A butter fish coiling around the egg.

6. Deposition of eggs by ovipositor in mussels. The female *Rhodeus amarus* (European bitterling) deposits eggs in the siphon of a fresh-water mussel (Swan mussel; Fig. 9.39) by means of very long urogenital papilla (ovipositor). Male immediately sheds the sperms on the opening (of mussel) over the eggs.

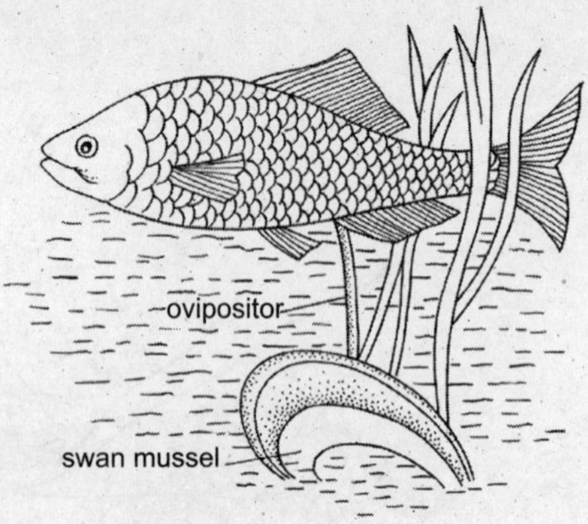

Fig. 9.39. Oviposition by European bitterling in swan mussel.

7. Egg brooding in mouth and intestine. The female *Tilapia mossambica* broods the fertilized eggs in her mouth (Fig. 9.40). She allows the young to take refuge in her buccal cavity in times of danger for some days after hatching. In the North American sea catfish (*Galeichthys felis*) the male carries eggs in the mouth for a period of nearly six weeks. The eggs of this oral incubating fish are large and relatively few in number. During this period the brooder fish do not take any food, thus, exhibiting great degree of self-sacrifice. *Tachysurus* keeps the fertilized eggs in its intestine till hatching occurs.

Fig. 9.40. Mouth brooding (oral incubation) in *Tilapia.*

8. Brood pouches. The male sea horse and pipe fish carry eggs in a brood pouch on the abdomen. In sea horse (*Hippocampus*) fertilized eggs are transferred by the female into the brood pouch on the belly of the male (Fig. 9.41). These eggs are carried by males until their hatching. Eggs become embedded in the folds of the brood pouch and for the zexchange of respiratory gases a sort of placenta is formed.

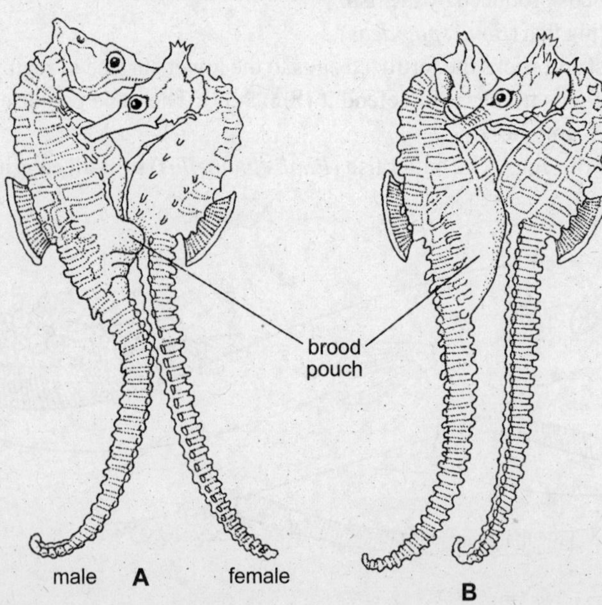

Fig. 9.41. A male sea horse (left) receives eggs from his mate, which fills his brood pouch and swell his abdomen substantially.

In the male pipe-fish (*Syngnathus acus*), a brood pouch is formed by two flaps of skin on the underside of the body on which eggs are placed by the female (Fig. 9.42).

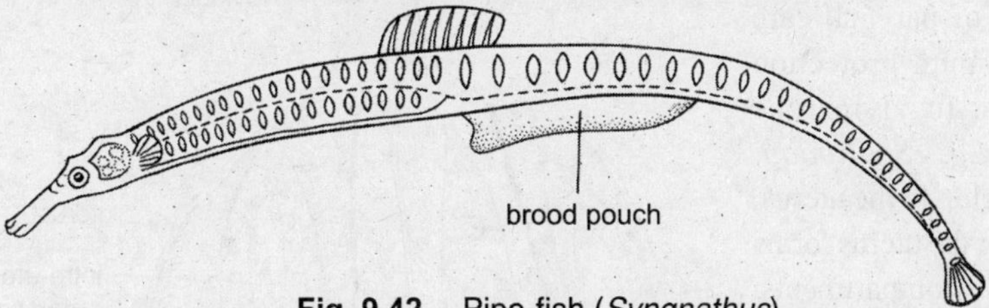

Fig. 9.42. Pipe-fish (*Syngnathus*).

The brood pouch develops an inner spongy lining which is richly supplied with blood vessels. The eggs get nourishment until hatching. Fry may return to the pouch when in danger.

9. Attachment of egg to body. The male nursery fish (*Kurtus*) of New Guinea, carries eggs held in a cephalic hook (Fig. 9.43).

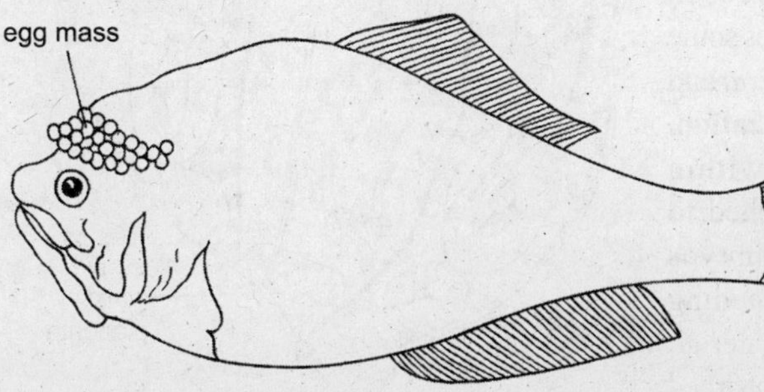

Fig. 9.43. Nursery fish (*Kurtus*) carrying egg mass entangled on a hook-like process.

10. By putting eggs in integumentary cups. In siluroids (*Aspredo* and *Platystacus*) the eggs are pressed into soft and spongy skin of female and carried about.

11. Egg capsules. In oviparous Elasmobranches such as rays and cat sharks (*Scyllium* and *Raja*) fertilized eggs are laid inside protective horny egg capsules, called **Mermaids purse** (Fig. 9.44). This capsule remains attached to the aquatic weeds by their tendrils. The development proceeds inside the capsule until the yolk has been used up. The youngs hatch out after rupturing off egg case.

Feeding the young. Like birds, some fishes help young in feeding. During its **plunge-feeding,** *Geophagus* spp., thrusts its snout vigorously into the substrate, withdraws it and expels the mouth contents. Loose particles may then be ingested. If an adult *Geophagus* is foraging in this manner with a school of its progeny, the young fish also gather and feed on the expelled material (**Loiselle** 1978).

Another form of "feeding the young" also occurs among cichlids (*Symphysodon discus, Etroplus maculatus, Chichlasoma citrinellum*). As the youngs reach free-swimming fry stage of development, they begin to graze on the mucus on their parents' sides. Mucus-producing cells in the adult's epidermis are most numerous at this stage of the reproductive cycle.

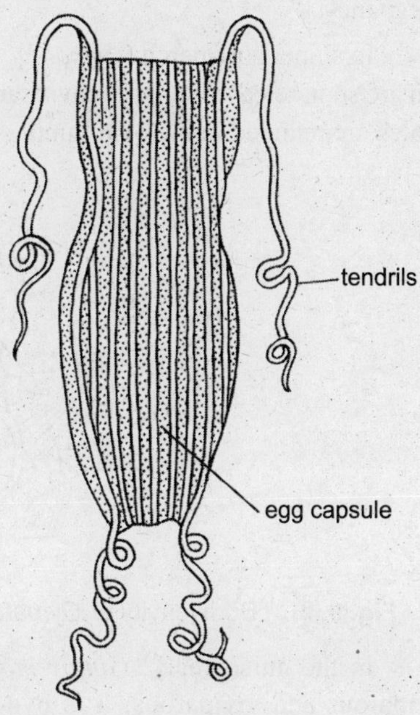

Fig. 9.44. A horny egg capsule of cat shark.

Viviparity (True internal incubation). Viviparity describes the highest degree of parental care and provides maximum protection to the young ones. In viviparous elasmobranches (*e.g., Scoliodon, Mustelus*) eggs develop in the uterus. The mucous lining of the uterus forms fluid-filled protective compartments, one for each embryo. Each embryo receives nourishment from the uterine tissues through the yolk-sac placenta (Fig. 9.45).

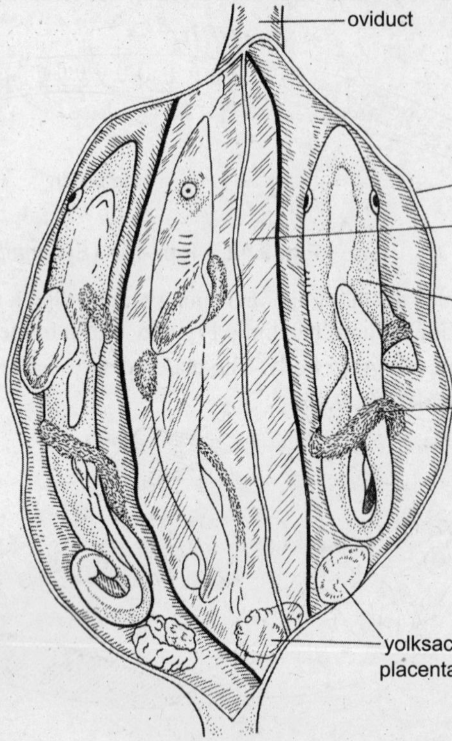

Fig. 9.45. Uterus of *Scoliodon* cut open to show the embryo with yolk-sac placentae.

In two orders of Teleosts, Cyprionodonts and Perciformes some species (*e.g., Zoarces, Gambusia* and *Poicilia*) show **internal fertilization.** In them youngs develop within ovary but they are not attached to the wall of ovary. These embryos develop freely in the sac feeding upon an "embryotrophic" material, apparently produced by the discharged ovarian follicles, which become highly vascular and remain as such throughout the several months of pregnancy.

In shiner surf-perch (*Cymatogaster aggregata*) also the eggs are fertilized in ovarian follicles but are soon released into the cavity and are nourished by a secretion from the ovary (Fig. 9.46). The males are retained in the ovary until sexually mature.

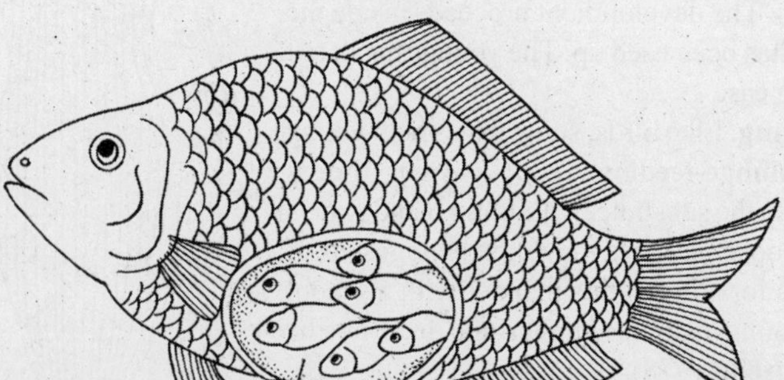

Fig. 9.46. Body cavity of *Cymatogaster aggregatus* cut open to show fully formed youngs.

In the nurse shark, *Gingly mostoma,* there occurs an intermediate condition between the oviparous and viviparous. It is **ovo-viviparous.** Its eggs are covered by a horny case and the development of the embryo occurs in the uterus.

9.11 FISH MIGRATION

Generally fishes live in a constant habitat and restrict their movement within particular territorial limits, but there are many others which migrate from one type of habitat (*e.g.*, fresh water) to another (*e.g.*, sea water or vice versa) and travel long distances. In most cases this movement of a large number of fishes may occur for the purpose of feeding, spawning or shelter and is called migration.

I. Types of Migratory Fishes

1. Diadromous fishes. These fishes migrate between fresh water and sea water. Diadromous fishes are subdivided into following three types:

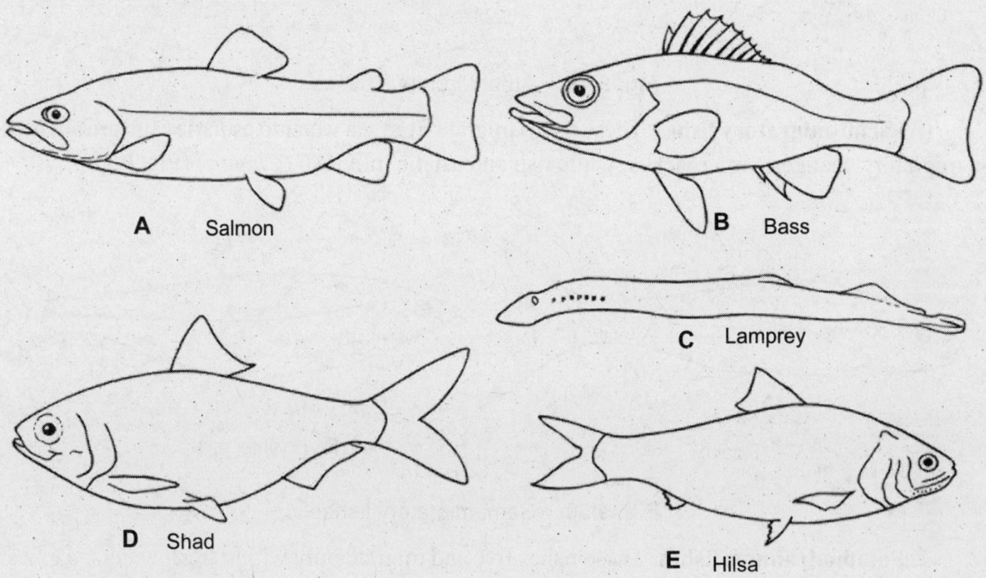

Fig. 9.47. Anadromous fishes.

***(i)* Anadromous fishes.** (Greek *Anadramein* = to run upward). These fishes migrate from sea to fresh water for spawning (breeding purpose), *e.g.*, sturgeon (*Acipencer*), Atlantic salmon (*Salmo salar*), pacific salmon (*Oncorhyncus nerka*) and *Hilsa* (Fig. 9.47).

***(ii)* Catadromous fishes.** These fishes migrate from freshwater to the sea for spawning, *e.g.,* the common or freshwater eels—*Anguilla anguilla* and *Anguilla vulgaris,* and the American eel—*Anguilla rostrata* (Fig. 9.48).

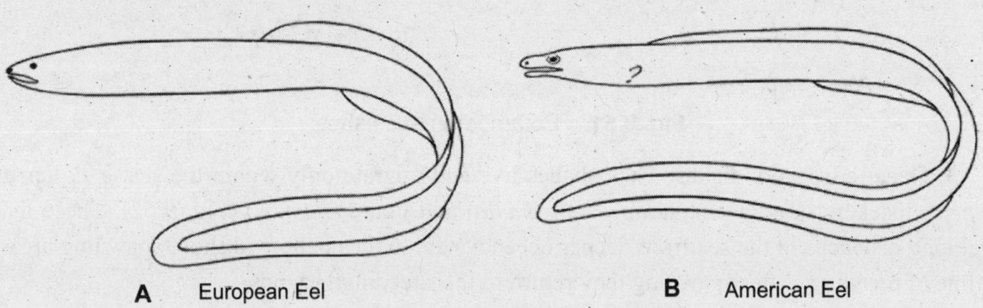

Fig. 9.48. Catadromous fishes.

(*iii*) **Amphidromous fishes.** These fishes migrate from sea water to fresh water and *vice-versa* (Fig. 9.49). This type of migration is not for the purpose of breeding but for other purposes (*e.g.*, food). This movement may occur regularly at some definite stage of life cycle, *e.g.*, gobis undertake migration for food.

Fig. 9.49. Amphidromous fishes.

(*iv*) **Semi-migratory fishes.** These fishes migrate from sea water to estuaries and are not marked by migratory changes, *e.g.*, roaches, white fish and asiatic milk fish (*Chanos*) (Fig. 9.50).

Fig. 9.50. Semi-migratory fishes.

2. Potamodromous fishes. These fishes live and migrate only within fresh waters, *e.g.*, carps, trouts (Fig. 9.51). Carps and trouts travel long distances in large rivers in search of suitable spawning grounds. These fishes return to feeding areas after spawning.

Fig. 9.51. Potamodromous fishes.

3. Oceanodromous fishes. These fishes live and migrate only within the sea, *e.g.*, herrings (*Clupea*), mackerel (*Scomber*), sardine (*Sardinella*) and tuna (*Thunnus*) (Fig. 9.52). These fishes travel long distances in the sea from deeper ocean waters to the shallow in shore spawning areas at the time of breeding. After spawning they return to their feeding grounds.

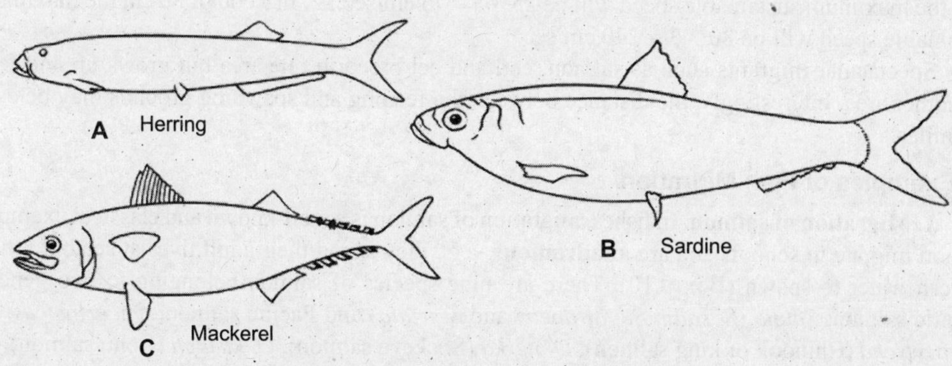

Fig. 9.52. Oceanodromous fishes.

Many marine fishes also perform seasonal migration. For example, swordfish (*Xiphias gladus*) and barracudas (*Sphyraena*), living in warm tropical seas perform latitudinal migration, moving north in spring and south in autumn. This migration occurs only for the suitable climatic condition.

Further, on the basis of the purpose, migration in fish has been classified into following types:

1. Alimental or feeding migration. This type of migration occurs for the purpose of search of food, *e.g., Chanos, Harpodon* (Bombay duck), etc.

2. Gametic or spawning/breeding migration. This type of migration is undertaken by a fish to ensure better survival and proper development of their eggs and larvae, *e.g., Hilsa ilisha.*

3. Climatic or wintering migration. This type of migration takes place to secure more suitable climatic condition, *e.g.,* sturgeons, Atlantic salmon, etc.

4. Osmoregulatory or protective migration. This type of migration takes place for osmoregulation (*e.g.,* water and minerals).

II. Methods of Migratory Movements

1. Denatant and cotranatant. Migratory movements are of two types: (*i*) along with the water currents, called **denatant** (e.g., movement of pelagic eggs and young salmons) or against the water current, called **contranatant** (*e.g.,* migration of adult salmons towards spawning ground).

2. Drifting. By this method fishes are carried passively along with the water current. The young fishes migrate to nursery grounds by drifting.

3. Swimming. This is of following two types:

(*i*) Random swimming. It leads to uniform distribution or to an aggregation.

(*ii*) Oriented swimming. The oriented swimming takes place either towards or away from the source of stimulation or at some angle to an imaginary line running between them and the source of stimulation.

III. Speed of Fish During Migration

Speed is an important factor in migrating fish. Speed at which different species of fish move during migration is influenced by a number of physiological and environmental factors:

1. Maximum speed. The maximum speed at which a fish can move is ten times a body length per second (body length × 10/sec). This speed can not be maintained by the fish for more than one minute; they have to slow down to regain stamina and may again attain the maximum speed.

2. Maximum sustainable speed. The speed can be sustained by the fish over long periods of time. It is three times a body length per second (*i.e.,* body length × 3/sec.), *e.g.,* 1. In herring of 25 cm size, the maximum sustainable speed will be 25 × 3 = 75 cm/sec.; 2. In a cod of 80 cm the maximum sustainable speed will be 80 × 3 = 240 cm/sec.

Spectacular migrants such as salmon, cod and eel breed in one area but grows up and feed in another area. Interestingly, the distance between the feeding and spawning grounds may be over 700 miles.

IV. Examples of Fish Migration

1. Migration of salmon. In fishes, migration of salmon is a well-known and classical example. Salmon migrate in schools and are **anadromous**, *e.g.,* they spend their adult lives at sea but return to fresh water to spawn (Box 9.10). There are nine species of salmon belonging to two genera, Atlantic salmon, *Salmo* (*S. salar, S. gairdneiri* and *S. trutta*) and Pacific salmon, *Oncorhyncus: O. tschawytscha* (Chinook or king salmon), *O. nerka* (Sockeye salmon), *O. kisutch* (Coho salmon), *O. keta* (Chum salmon), *O. gorbuscha* (Pink salmon), *O. masu* (Japanese masu salmon). All these nine species of salmons perform migration but there are important differences among these species. The Atlantic salmon make upstream spawning runs (breeding migrations) year after year. The Pacific salmon make a single spawning run, after which they die.

Box 9.10

A stream in which a salmon was born has proved itself suitable for reproduction. A mature salmon will therefore endure a long and difficult journey from oceanic feeding grounds to freshwater spawning grounds, risking death from exhaustation or predation, in order to have offspring in or near its birth place (**Alcock, 1998**).

The **Atlantic salmon** spawn mainly during November and December. They ascend the rivers as the breeding period approaches, travelling several thousand kilometers in sea and then inland. On entering freshwater this *contranatant* migrant stops feeding and loses weight. As a result, the energy requirement for up river movement is provided by accumulated nutrients chiefly in the form of fat deposits in body. Their bright silvery colour changes to dull reddish-brown shade. The body of the male salmon becomes spotted with red, orange and large black spots. Such male breeding salmons are

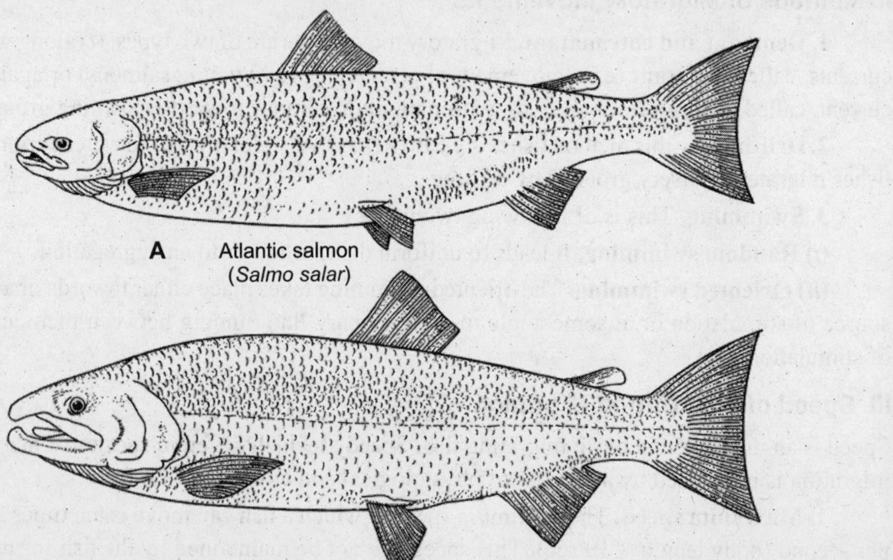

A Atlantic salmon (*Salmo salar*)

B Silver salmon (*Oncorhynchus kisutch*)

Fig. 9.53. Two common species of salmon. A—Atlantic salmon (*Salmo salar*); B—Pacific silver salmon (*Oncorhynchus kisutch*).

known as **"red fish"** and the darker ripe females as **"black fish"**. After selecting suitable spawning grounds, the fish segregate into pairs. Shallow saucer-like depressions are prepared in the river bed by the females where spawning takes place (Box 9.11).

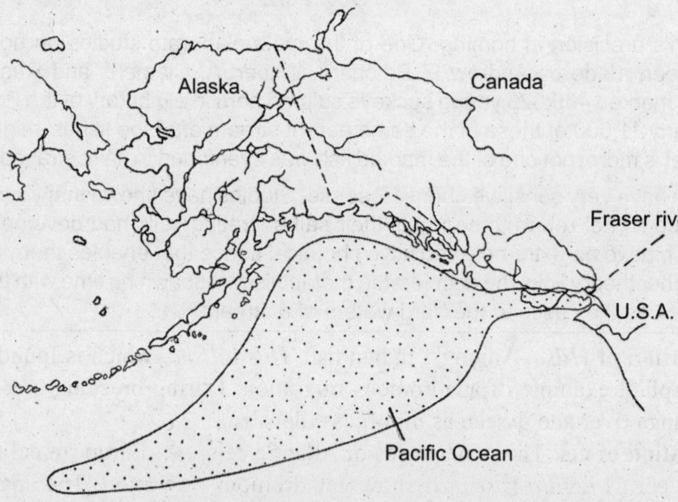

Fig. 9.54. Salmon that are born in British Columbia's Fraser river move far out in the northeastern Pacific ocean (shaded area) before returning to breed in the river where they began life.

Box 9.11

The fertilized eggs of salmon hatch into small salmons towards spring. A large amount of yolk is retained by the young salmon on emerging from the eggs; this stage is called the **'alevir stage'**. After further growth, the yolk is reabsorbed and a size of 3-5 cm is attained. This larva is now called a **"fry"** which grows further to form a **"fingerling"** and remains in the river for an year or more. During this period the young fish continues to feed and grow and transforms into the **"smolt"** (stage after attaining a certain amount of maturity). Smolts stop feeding and then starts their journey towards the sea. Some species may migrate after hatching (*e.g.,* pink salmon *O. gorbuscha*) while others take as long as 7 yrs (*Salmo salar*) before starting migration.

After spawning, the return journey to the sea begins, but few males survive to breed a second time. However, many females are able to reach the sea and start feeding. They soon recover their normal condition and silvery colour. The salmon does not usually spawn more than three times in its life span of eight or nine years.

The **king salmon** of the Pacific coast of North America travels about 3600 kilometers from the sea in order to reach the rivers to spawn. However, all of them die after spawning. No male or female succeeds in returning to the sea, as they are already too much exhausted to undertake a return journey.

Arthur Hasler (1960) has studied the migration of **sockey salmon** (*O. nerka*). This salmon hatch in streams of the United States and Canada. After the smolt stage of development, they swim down-stream to the Pacific ocean. After two or three years in sea, they become sexually mature and migrate back to the exact stream where they were spawned. The return journey to coast is probably accomplished by means of celestial cues (*i.e.,* stars or azimuth position of sun) and internal biological clocks. Once at the coast, however, they have to select the correct river and the correct tributary stream within the river system. After many years of research **Hasler** and coworkers discovered that during the smolting period the salmon **imprint** upon its olfactory system the characteristic odour of their native streams (*i.e.,* the chemicals released by the vegetation and soil in the watershed of the parent stream). So during their return journey salmon are able to discriminate between the water coming from their native stream and that from other tributaries (Box 9.12). They essentially are recognising landmarks, but because the scent of birth place flows all along the migratory route, it is not necessary for the young salmon to remember landmarks all along the outward journey.

Box 9.12

1. Salmon's precision in homing. One of the most elaborate studies on homing behaviour of salmon has been made by Andrew L. Pritchard, Wilbert A. Clemens and Russell E. Foerster in Canada. They marked 469,326 young sockeye salmon born in a tributary of the Fraser river and they recovered nearly 11,000 of these in the same parent stream after the fishes' migration to the ocean and back. What is more, not one of the marked fish was ever found to have strayed to another stream.

2. Salmon have very sensitive chemical sense. Studies have shown that young salmon hatched in freshwater could not return to sea until their salt secreting cells had developed. Some workers believe that fish have an extremely sensitive chemical sense that enables them to perform this feat (homing). Another theory links the accelerated metabolism at spawning time with the need for oxygen which increases as they ascend the hard waters of a stream.

2. Migration of *Hilsa*. Amongst Indian fish, *Hilsa ilisha,* which is found in Bay of Bengal, provides an explicit example of anadromous migration. During breeding season for spawning it ascends the Ganga river and is seen as far off as Allahabad.

3. Migration of eel. The two species of Atlantic eels, *i.e.,* European eel (*Anguilla anguilla*) and American eel (*Anguilla rostrata*) show catadromous migration. The spawning ground and spawning time of species was discovered by **Schmidt** (1967). It was found that these fish spawn near the "Sargasso sea" in the Atlantic Ocean, south-east of Bermuda. The European eel is found toward the eastern parts.

A. Migration of European eel (*Anguilla anguilla*). The European eels are found towards the eastern region of the Atlantic Ocean and in inland waters of countries near the shores of Europe some of which are Murman Coast of North Africa, Iceland, the Azores, Madeira and Canary Islands, the Baltic and Mediterranean countries, the Black sea, sea of Azor and the Red sea. Spawning of European eel occurs during spring seasons and eggs are laid at depth of 500-700 meters with the temperature ranging between 10-12°C. On hatching, the larvae ascend to a depth of about 200 metres where the temperature is warmer (*i.e.,* around 20°C). These larvae retain a yolk sac and are called **protocephaline larvae.**

Protocephaline larva transforms to **leptocephalus larva** after reabsorption of yolk content. Leptocephali are confined to a depth of about 100 metres and have a flattened leaf-like, glassy transparent body measuring about 5 cm. The skeleton remains unossified at this stage, gut has a straight tubular structure, teeth are long and pointed and eyes are large and silvery. The larvae swim by eel-like undulatory movements and is reported to show diurnal vertical migratory movements coming towards the surface only at night.

Leptocephali migrate from the Sargasso sea to the 'nursery grounds' of coastal waters by passively drifting with the warm water currents. They reach middle of the Atlantic Ocean by their second summer and finally touch the coast of Europe by the third summer. During this period the larvae grow to attain a large size. Transparency, serpentine movements and large size help the larva defence against predators on their long migratory route. During autumn and winter of the third year leptocephali metamorphose to form **"mini eels"** or **'glass eels'** called **elevers.** During metamorphosis the larvae stop feeding, their body becomes cylindrical with a decrease in length and the needle-like teeth are replaced by new ones. Elevers are initially translucent but gradually develop pigmentation on the body. Elevers may remain in coastal waters for 1-2 years, blocking the mouth of rivers till they are strong enough to swim up the rivers. They now measure about 15-20 cm long and are as thick as a pencil. The male fish prefer brackish water and stay in the estuarine region where as females prefer fresh water and are seen to ascend up river. In the stronger fish, larval teeth are lost and intestine shortens and the anal aperture moves forward. Pigmentation gradually increases, transforming the eel into yellow, cylindrical fishes, called **yellow eels.** These fish spend 8-10 years on feeding and growing they become sexually mature **silver eels** before starting the spawning migration to the sea (*i.e.,* to their breeding grounds in the Western Atlantic South of Bermuda). During maturation of eels, the following prominent nuptial changes occur: a change in colour from yellow to silvery, enlargement of eyes, development of pointed pectoral fins, shrinkage of alimentary canal and enlargement and maturation of reproductive organs. These changes trigger a movement to over 5000 to 6000 kilometers in search of more favourable conditions of salinity, depth and warmer temperature required by the gametes.

B. Migration in American eel. Interestingly, like the European eel the American eel (*A. rostrata*) is also catadromous and spawns in the 'Sargasso sea'. It is similar to the European eel in many biological aspects. The larvae called **leptocephali** drift to the American coast by water currents. In about two years they reach coastal waters (the nursery grounds) to change into **elevers** or **mini eels** before ascending the rivers. On reaching the parental habitat they feed and grow till sexual maturity is attained.

V. Factors for Fish Migration

Fish migrations are influenced by a number of physical, chemical or biological factors:

1. Physical factors. They include quality of water, bottom materials, water depths, pressure, temperature, light intensity and photoperiod, currents, tides and turbidity.

2. Chemical factors. These include salinity, pH, dissolved oxygen, types of dissolved organic substance, smell and taste of water.

3. Biological factors. These include sexual maturity, blood pressure, food, memory, effect of endocrine secretions and the related physiological changes in the body systems and behaviour (social response, biological clock). Presence or absence of predators and active competitors may also be taken as the biological factors.

VI. Theories/Hypotheses Regarding Fish Migration

Several laboratory and field studies have been conducted to solve the mystery by which a fish can select the course and subsequently maintain its direction during long migratory movements. Following theories/hypotheses have been forwarded to explain the homing/navigation during fish migration.

1. Olfactory hypothesis. According to this hypothesis, a fish may recognize its surroundings and path of migration by detecting the specific odour from a stream or water body. This odour may depend on the bottom topography, *i.e.,* soil, plants and springs at the bottom from which the water body acquires its organic quality. Scientists have also termed this as the **stream factor.**

A.D. Hasler (1983) has supported olfactory hypothesis with **imprinting** and regards it as the sensory basis for homing. He has suggested that the salmon may learn to recognize the organic quality of water in the early days of its life cycle and this may become imprinted for the rest of its life. Thus, a fish smells its way home as would a fox hound. The paleo-cortex is, therefore, a dominant part of fish brain as it receives impulses from the olfactory tissue. This region is much less significant in non-migratory fishes as well as in humans.

2. Internal biological clocks. Experimental evidences have suggested that the fish (salmon) during its migration is able to adjust its course or orientation through internal biological clocks. These adjustments to seasons (through latitudes) as well as to the time of the day (through longitudes) are according to appearance of light.

3. Sun-compass mechanism. Evidently open-water migrations in salmons may be directed by a sun-compass mechanism. These may be directed either by the altitude of the sun or by changing diurnal azimuth of the sun or by the combination of the two. To swim in a single direction, the fish probably calculates appropriate horizontal angles to the sun during different times of the day.

4. Ground water seepage hypothesis. According to this theory which was proposed by **Harden Jones** (1980), the assembly area for spawning are identified by marine fish with the help of the chemical that enter the sea by ground water seepage.

5. Pheromones. Pheromones are suggested to play an important role in identifying the spawning area. They may prevent further spawning in a particular area after a sufficient number of eggs have already been laid.

6. Celestial reference points. The use of the celestial points in maintaining a steady course in mid waters was suggested to be useful, especially in nocturnal travel. Nocturnal direction may also be assisted by gravity, magnetic field, oceanic currents, etc., especially during moonless or starless nights.

7. Tidal transport. This mode of migration is performed by Plaice of South Bight of North sea and saves energy and problems of direction are eliminated if the right tide is caught at the right time.

8. Gradient theory. This is based on the salmon's presumed ability to follow gradients of physico-chemical characteristics of water, mainly temperature and carbon dioxide. It has been shown that salmons swim in the direction of coolest waters and that high CO_2 tension repels migrating salmons.

QUESTIONS

Long Answer Questions

1. How would you identify a fish? Write classification of Pisces upto orders.
2. Classify the group pisces giving distinguishing characters and examples of two principal orders.
3. Give an account of economic importance of fishes. *(CCSU Meerut 2013)*
4. Write differences between Osteichthyes and Chondrichthyes.
5. Write an account of Dipnoi.
6. Write an account of parental care in fishes.
7. Give an detailed account of accessory respiratory organs of fishes.
8. Write an essay on Fish Migration.

Short Answer Questions

1. Summarise biological significance of Ostracoderms.
2. Differentiate between sharks and rays. *(Garhwal 2003)*
3. Give biological significance of *Chimaera* and *Latimeria*.
4. Give general characteristics and examples of Chordrichthyes.
5. Write short notes on the following:
   (*i*) *Torpedo* *(CCSU Meerut 2013)*
   (*ii*) Venous heart *(CCSU Meerut 2012)*
   (*iii*) Ovoviviparity *(CCSU Meerut 2011)*
   (*iv*) Coeloacanth or *Latimeria*
   (*v*) *Chimaera*
   (*vi*) *Sphyrna*
   (*vii*) *Pristis*
   (*viii*) *Lepidosteus*
   (*ix*) *Hippocampus* *(Kanpur 2003)*
   (*x*) *Pleuronectes*
   (*xi*) *Echeneis*
   (*xii*) *Anabas* (or climbing perch)
   (*xiii*) *Neoceratodus* *(Purvanchal 2001)*
   (*xiv*) *Protopterus*
   (*xv*) *Excoetous*
   (*xvi*) *Labeo*

Very Short Answer Questions

1. What forms exoskelton in elasmobranch fishes?
2. Name a viviparous fish.

3. Name the type of heart found in *Scoliodon*? (*Purvanchal 2001*)
4. What is scroll valve?
5. Name the vertebrate in which scroll valve is found.
6. What are neuromast organs?
7. What are rheoreceptors?
8. What is mermaid's purse?
9. What type of gill the cartilaginous fishes have?
10. Mention one character of Pleurotremata?
11. Cite two examples of Pleurotremata?
12. What is the scientific name of the hammer-headed shark?
13. Give the position of gill slits and spiracle in Hypotremata.
14. What is common name of *Pristis*?
15. To which subclass does *Chimaera* belong?
16. How does *Torpedo* defined itself?
17. Name a cartilaginous fish which resembes the bony fish.
18. Give the names of scales found in fishes. (*Purvanchal 2001*)
19. Where is *Neoceratodous* found?
20. What is the function of tail in fish?
21. What is hyostylic suspensorium?
22. What is the function of air bladder in fishes?
23. In which class of fishes there is no cloaca?
24. What is the scientific name of bowfin?
25. Which groups gave rise to amphibians, Chondrichthyes and Osteichthyes?
26. Give the name of scientist who have worked on classification of fishes (*CCSU Meerut 2013*)
27. Name the branch of Zoology dealing with the study of fishes. (*CCSU Meerut 2003*)
28. Give the scientific names of following fishes:
    (*i*) Dogfish;
    (*ii*) Electric ray;
    (*iii*) Sea horse;
    (*iv*) Sucker fish;
    (*v*) Climbing perch;
    (*vi*) Lobed fin fish;
    (*vii*) Flying fish;
    (*viii*) Hammerheaded shark;
    (*ix*) Rohu;
    (*x*) Australian lung fish;
    (*xi*) Flat fish;
    (*xii*) Saw fish;
    (*xiii*) Gar-pike;
    (*xiv*) Scorpion-fish;
    (*xv*) African lung fish.

Fill in the Blanks

1. The skeleton of Chondrichthyes is ________.
2. The cartilaginous fishes have ________ to ________ pairs of gill slits.
3. The gill in Chondrichthyes are ________.
4. The intestine of cartilaginous fishes has a ________ valve.
5. Exoskelton in Elasmobranch fishes is formed of ________ scales. (cycloid/placoid)
6. In heterocercal tail the dorsal lobe of caudal fin is ________ than ventral lobe. (small/large)
7. Placoid scales of elasmobranch fishes are regarded to be ________ derivative. (epidermal/dermal)

8. ________ cartilage supports the lower jaw. (Mekel's/ hyoid)
9. A conus arteriosus is present in the heart of ________ fishes.
10. The pelvic fin bears ________ in the males of Chondrichthyes.
11. The class Chondrichthyes is divided into two subclasses ________ and Holocephali.
12. *Sphyrna* commonly is called ________.
13. The electronic organs of *Torpedo* are modified ________.
14. Extinct fishes are called ________. (*Garwhwal 2003*)
15. Cephalic claspers are present on the head of ________.
16. Lungfish found in south America is represent by genera ________.
17. ________ are represented by a single living genus *Latimeria*.
18. Glossopharyngeal nerve is ________ cranial nerve.
19. In *Scoliodon*, development of the embryo takes place ________. (in uterus/in water)
20. The skelton of Osteichthyes is ________.
21. The bony fishes have ________ pairs of gills.
22. The gill of Osteichthyes is ________.
23. The ________ fishes generally lack a spiral valve.
24. The Osteichthyes have ________ instead of the conus arterious in their heart.
25. A buoyancy-regulating organ called ________ is present in Osteichthyes.
26. The excretory waste in bony fishes is ________.
27. There are only ________ genera of living lung fishes.
28. Lungfish found in south America is represented by genera ________.
29. Branchial chamber or opercular chamber opens outside by ________.
30. Respiration is lung fishes is carried out by ________.
31. ________ is called "king of herrings". (*Chimarea/Amia*)
32. Cat fishes are recogonised by their ________. (*Purvanchal 2001*)
33. *Anguilla* (The European eel)exhibits ________ migration. (catadromous/anadromous)
34. Light is produced in fishes by the combined action of ________ and ________. (*Garwal 2002*)
35. ________ fishes have narrow range salinity tolerance whereas ________ fishes have wide range tolerance. (*Garwal 2002*)

Multiple Choice Questions

1. Cartilaginous fishes do not have following features
   (*a*) pelvic fin (*b*) scales
   (*c*) gill slits (*d*) operculum
2. Viviparity is seen in
   (*a*) *Labeo* (*b*) *Scoliodon*
   (*c*) all fresh water (*d*) all marine teleosts
3. Venous heart is found in (*Gorakhpur 2002; Rohilkhand 2003*)
   (*a*) *Scoliodon* (*b*) frog
   (*c*) pigeon (*d*) rabbit
4. Bony fishes are classified as
   (*a*) Elasmobranchs (*b*) Osteichthyes
   (*c*) Chondrichthyes (*d*) Tetrapoda
5. Young one of a fish is called
   (*a*) fray (*b*) larva
   (*c*) nymph (*d*) juvenile fish
6. A flying fish is
   (*a*) *Acipenser* (*b*) *Amia*
   (*c*) *Exocoetus* (*d*) *Fishtularia*

7. In sucker fish, the sucker is modified form of
(*a*) dorsal fin (*b*) mouth
(*c*) gill slits (*d*) ventral fin

8. In which fish male shows parental care
(*a*) *Anabas* (*b*) *Labeo*
(*c*) *Hippocampus* (*d*) *Synaptura*

9. Lung fishes belong to order (*Kanpur 2003*)
(*a*) Holecephali (*b*) Dipnoi
(*c*) Placodermi (*d*) none of these

10. Thick rhomboid scales are present on the body of
(*a*) *Amia* (*b*) *Polypterus*
(*c*) *Polydon* (*d*) *Lepidosteus*

11. Spiracle helps in
(*a*) expulsion of excess water from mouth cavity
(*b*) testing smell of water entering mouth cavity
(*c*) regulating of intake of water inside
(*d*) serves as a nostril

12. Spiracle is absent in
(*a*) Chondrichthyes (*b*) Bradyodontii
(*c*) Teleostiformes (*d*) Osteichthyes

13. Which one of the following larvicidal fish is used in the biocontrol of mosquitoes?
(*a*) *Hilsa ilisha* (*b*) *Gambusia*
(*c*) Gold fish (*d*) Scalophagus

14. In fishes the jaws are adapted for
(*a*) defence (*b*) grasping
(*c*) manipulating objects (*d*) all of the above

15. Fishes having sensory barbles around the mouth are called
(*a*) cat fishes (*b*) dog fishes
(*c*) rat fishes (*d*) porcupine

16. Which fish is known as living fossil
(*a*) *Chimaera* (*b*) *Latimeria*
(*c*) *Pristis* (*d*) *Protopterus*

17. The tympanum (ear drum) is absent in
(*a*) skates (*b*) cartilaginous fishes
(*c*) bony fishes (*d*) all of the above

18. Scales in bony fishes are not
(*a*) cosmoid (*b*) placoid
(*c*) cycloid (*d*) ctenoid

19. Which of the following is not a bony fish?
(*a*) *Channa* (*b*) *Acipenser*
(*c*) *Chimaera* (*d*) *Pleuronectes*

20. The scientific name of the African lungfishes
(*a*) *Neoceratodus* (*b*) *Polypterus*
(*c*) *Lepidosiren* (*d*) *Protopterus*

ANSWERS

Very Short Answer Questions

1. Placoid scales.
2. *Scoliodon*.

3. Venous heart.
4. Scroll valve of a longitudinal fold attached to the inside of intestinal wall in scolidon, whose free end is rolled 2½ times aticlockwise to form a scroll. (**Note.** The scroll valve checks a rapid passage of the food through the intensine, and also serves to increas its absorptive surface).
5. *Scoliodon.*
6. Neuromast organs are groups of sensory cells in the lateral line canals, that helps fish to detect disturbances in the surrounding water.
7. Rheoreceptors are neuromast organs in fishes that detect any disturbances in the surrounding water.
8. Mermaids purse is the popular name for the egg case of oviparous dogfishes.
9. Lamelliform.
10. Gill slits are lateral.
11. *Sphyrna* and *Scoliodon.*
12. *Sphyrna* (*Zygaena*).
13. Ventral, dorsal.
14. Saw fish.
15. Holocephali.
16. By giving electric shocks.
17. *Chimaera.*
18. Placoid, cycloid and ctenoid.
19. *Neoceratodus* is found in Australia.
20. Tail and caudal fin exert propulsive force and push the fish forward.
21. In hyostylic suspensorium, the upper and lower jaws articulate with the cranium through hymandibular bones.
22. Air bladder is a hydrostatic organ.
23. Osteichthyes.
24. *Amia.*
25. Osteichthyes.
26. Tate Regan and Berg.
27. Ichthyology.
28. (*i*) *Scolidon*; (*ii*) *Astrape* or *Torpedo*; (*iii*) *Hippocampus*; (*iv*) *Echenies*; (*v*) *Anabas*; (*vi*) *Latimeria*; (*vii*) *Exocoetus* or *Cypselurnus*; (*viii*) *Sphyrna*; (*ix*) *Labeo rohita*; (*x*) *Neoceratodus*; (*xi*) *Solea*/*Pleuronectes*; (*xii*) *Pristis*; (*xiii*) *Lepidosteus*; (*xiv*) *Pterois*; (*xv*) *Protopterus*.

Fill in the blanks

1. Cartilaginous. **2.** 5 to 7. **3.** Lamelliform. **4.** Scroll (spiral). **5.** Placoid. **6.** Large. **7.** Dermal. **8.** Meckel's. **9.** Cartilaginous. **10.** Claspers. **11.** Selachi (Elasmobranchi). **12.** Hammerheaded shark. **13.** Muscle tissue. **14.** Placodermi. **15.** *Chimaera*. **16.** *Lepidosiren*. **17.** *Latimeria*. **18.** IX. **19.** In uterus. **20.** Bony. **21.** Four. **22.** Filamentar. **23.** Bony. **24.** Bulbus arteriosus. **25.** Air bladder or swim bladder. **26.** Ammonia. **27.** Three. **28.** *Lepidosiren*. **29.** Branchial or opercular. **30.** Air bladder modified into lungs. **31.** *Chimaera*. **32.** Barbles. **33.** Catadromous. **34.** Luciferin, luciferase. **35.** Stenohaline, euryhaline.

Multiple Choice Questions

1. (*d*)	**2.** (*b*)	**3.** (*a*)	**4.** (*b*)	**5.** (*a*)	**6.** (*c*)
7. (*a*)	**8.** (*c*)	**9.** (*b*)	**10.** (*b*)	**11.** (*a*)	**12.** (*d*)
13. (*b*)	**14.** (*d*)	**15.** (*a*)	**16.** (*b*)	**17.** (*d*)	**18.** (*b*)
19. (*c*)	**20.** (*d*)				

10

Class Amphibia: General Characters, Classification and Types

Class Amphibia (Gr., *amphi* = two; *bios* = life) includes amphibians, the animals which are adapted to double mode of life, *i.e.*, on land and in water. Amphibians are aquatic (fresh water) or semiaquatic, air and water breathing, carnivorous, cold-blooded, oviparous vertebrates. They are unique among the vertebrates in having soft, glandular, moist, naked (scale-less) and permeable respiratory skin. In a way, amphibians can be defined as **naked tetrapods** capable of cutaneous respiration.

Amphibians were the first vertebrates who could lead a terrestrial existence. These vertebrates are said to originate from crossopterygian ancestors in Devonian and perpetuating since Carboniferous to recent. There are about 15,000 species living today and their fossils are equally numerous.

Transition from aquatic to terrestrial life is clearly indicated in class Amphibia. Larvae of all amphibians are aquatic. In some cases, the adults have also secondarily returned to water. In others, the adults seek damp places and breed in water. Structurally the amphibians stand between fishes and reptiles.

Amphibians generally have two phases in their life history: 1. **Larval stage**, called **tadpole**, is fish-like, live in water, swim with tail and breathes with gills; and 2. **Adult stage** is terrestrial or aquatic, moves with limbs or in a serpentine manner, and breathes with lungs, buccopharyngeal lining and skin.

10.1 GENERAL CHARACTERS

The class Amphibia includes salamanders, newts, frogs, toads, caecilians, etc. and possesses the following characters:

1. They are cold-blooded or **poikilothermal** or **ectothermal** animals as the body temperature, like fishes, directly fluctuates according to the temperature of environment.
2. Body is divisible into head and elongated trunk. Neck or tail may be present or absent.
3. Head is with a pair of nostrils leading into buccal cavity. Internal nares open into anterior parts of buccal cavity.
4. Limbs usually two pairs (tetrapod), some limbless. Toes 4-5 (pentadactyle) or less. Forelimbs with four digits and hindlimbs with five digits each. Digits are without **claws**. However, African toad, *Xenopus*, has small horny claws on the digits.
5. Paired fins absent. Median fins, if present, without fin rays.
6. Skin is often smooth, moist and rich in multicellular mucous and poison glands. The mucous glands secrete mucus to keep the skin moist. The poison glands produce toxic substances to deter the predators. Skin of amphibians is highly vascular and may be respiratory.

Scales (exoskeleton) is generally absent. Some have minute dermal scales embedded in the skin. Pigment cells (chromatophores) are present.

All extant amphibians are called **Lissamphibia** because of their special skin features (Gr., ***liss*** = smooth).

7. Endoskeleton is largely bony. Cranium of skull is cartilaginous. Skull is dicondylic, *i.e.*, carries two occipital condyles. Upper jaw is fused with the skull. Jaw suspension is autostylic.
8. Mouth large. Upper or both jaws with small homodont teeth. Tongue may be broad, flat and immobile (aquatic forms) or protrusible (land forms). Alimentary canal terminates into cloaca. Fat is stored in distinct **fat bodies** (the corpora adiposa), in the abdomen and around the tail base.
9. Respiration by lungs, skin and mouth lining (= bucco-pharyngeal cavity). Larvae with external gills which may persist in some aquatic adults.
10. Heart 3-chambered (2 auricles and 1 ventricle). Sinus venosus present. Aortic arches 1–3 pairs. Renal and hepatic portal system well developed. Erythrocytes or red blood cells (RBCs) large, oval, biconvex and nucleated. Spleen present.
11. Kidneys mesonephric. Urinary bladder large. Urinary ducts (Ureters) open into cloaca. Excretion ureotelic. (**Note.** Adult amphibian kidney are termed **opisthonephros**, since, **Weichert** has reserved the term mesonephros only for the embryonic kidneys of amniotes).
12. Brain poorly developed. Cranial nerves 10 pairs.
13. External ears absent. Middle ear with a single rod-like ossicle, **columella**. Larval forms and some aquatic adults are with lateral line system.
14. Sexes separate. Males without the copulatory organ. Gonoducts open into cloaca. Fertilization external or internal. Females mostly oviparous.
15. Development indirect. Cleavage unequal holoblastic. No embryonic membranes. Development includes an aquatic larva, which undergoes progressive metamorphosis.

10.2 CLASSIFICATION

While classifying class Amphibia, **G. Kingsley Nobel** (1924) recognized 3 orders of extinct and 3 orders of living or extant amphibians. In the past, all extinct groups of Amphibia were placed under a single subclass **Stegocephalia** and all living groups of of Amphibia in another subclass **Lissamphibia** (Fig. 10.1). This simple plan of classification of Amphibia has been followed by most Indian authors.

Later on, several extinct amphibians and Lissamphibia are placed in a group termed the **temnospondyls** (**Hickman** *et al.*, 1996). This group is distinguished by having generally only four digits on the forelimbs than the five characteristics of tetrapods.

Extinct Amphibia

Subclass 1. Stegocephalia

1. Extinct Amphibia from lower Carboniferous (Permian) to Upper Triassic.
2. Earliest forms with characteristics of fish and amphibians.
3. Limbs pentadactylous.
4. Skin with scales and bony plates.
5. Skull is completely covered by dermal bones. Skull with a solid bony roof leaving openings for eyes and nostrils.

Subclass Stegocephalia is divided into 3 orders: 1. Labyrinthodontia, 2. Phyllospondyli and 3. Lepospondyli.

Order 1. Labyrinthodontia

1. Oldest known tetrapods called stem Amphibia.
2. Freshwater (aquatic), semi aquatic or terrestrial forms.
3. Salamander-like or crocodile-like fish eating amphibians.

Class Amphibia

Subclasses (2)

- **1. Stegocephalia**
  - **Orders (3)**
    - **1. Labyrinthodontia** — **Example:** *Eryops* (Extinct)
    - **2. Phyllospondyli** — **Example:** *Branchiosaurus* (*Ichthyostega*, extinct)
    - **3. Lepospondyli** — **Example:** *Diplocaulus* (Extinct)
- **2. Lissamphibia**

Subclass Lissamphibia (Modern living Amphibia)

Orders (3)

- **1. Apoda**
  Example:
  1. *Ichthyophis*
  2. *Uraeotyphlus*
  3. *Siphonops*
- **2. Urodela**
  - **Suborders (5)**
    - **1. Cryptobranchoidea** — **Examples:** 1. *Cryptobranchus* 2. *Hynobius*
    - **2. Ambystomoidea** — **Example:** 1. *Ambystoma*
    - **3. Salamandroidea** — **Example:** 1. *Salamandra* 2. *Amphiuma* 3. *Triton*
    - **4. Proteida** — **Example:** 1. *Proteus* 2. *Necturus*
    - **5. Meantes** — **Example:** 1. *Siren*
- **3. Anura or Salientia**

Order Anura

Suborders (5)

- **1. Amphicoela** — **Example:** 1. *Ascaphus* 2. *Liopelma*
- **2. Opisthocoela** — **Example:** 1. *Alytes* 2. *Pipa* 3. *Bombinator*
- **3. Anomocoela** — **Example:** 1. *Pelobates*
- **4. Procoela** — **Example:** 1. *Bufo* 2. *Hyla*
- **5. Diplasiocoela** — **Example:** 1. *Rhacophorus*

Fig. 10.1. Outline Classification of Class Amphibia. This classification is adopted from Nobel (1954) from "The Biology of Amphibia".

4. They lived from the lower Carboniferous to the Triassic period.
5. Enlarged teeth with greatly folded dentine found in both jaws, similar to their crossopterygian ancestors.

Example: *Eryops* (extinct; Fig. 10.2).

Order 2. Phyllospondyli

1. Small salamander-like extinct amphibians. Found in the carboniferous and permian deposits.
2. Heads large and flat.
3. Vertebrae tubular. Notochord and spinal cord are housed in common cavity.
4. Well marked transverse processes with stout ribs.
5. Believed to be ancestors of modern *Salientia* and *Urodela*.

Fig. 10.2. *Eryops*.

Example: *Branchiosaurus* (*Ichthyostega*; Fig. 10.3).

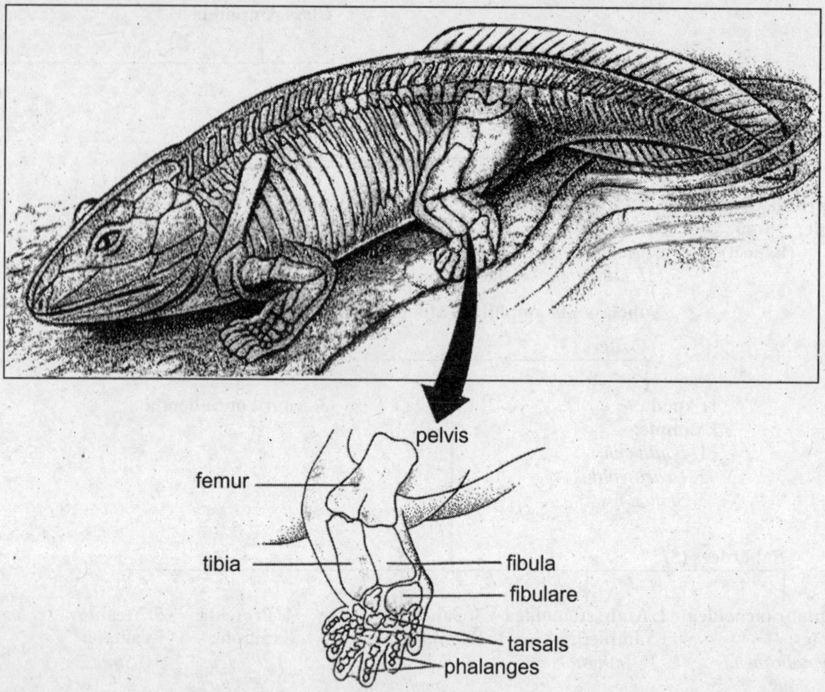

Fig. 10.3. *Ichthyostega*. An Devonian tetrapod had fully formed tetrapod limbs and must have been able to walk on land. The hindlimbs bore seven toes (the number of forelimbs is unknown).

Order 3. Lepospondyli

1. Small salamander or eel-like.
2. Carboniferous to Permian.
3. Vertebrae cylindrical, each made of a single piece. Neural arch and centrum continuous.
4. Ribs articulate with the column intervertebrally.
5. Regarded as ancestors of modern caecilians (Gymnophiona).

   Examples: *Diplocaulus* (had horned skulls and small limbs, *Lysorophus*).

Living Amphibia

Subclass 2. Lissamphibia

1. Their bodies became flatter for moving about in shallow water.
2. Porous skin of modern amphibia becomes primary or accessory breathing organ.
3. No dermal bony skeleton.
4. Teeth small and simple.

   Subclass Lissamphibia is divided into three orders: 1. Gymnophiona or Apoda; 2. Urodela or Caudata and 3. Salientia or Anura.

Order 1. Gymnophiona or Apoda (Gr., *gymnos* = naked; *ophioneos* = serpent-like) or (Gr., a = without; *podos* = foot).

1. Limbless, blind, elongate worm-like, burrowing tropical creatures commonly called **caecilians** (caecilian means blind).
2. Girdles and tail absent. Anus/cloaca terminal.
3. The skin is provided with numerous transverse grooves or wrinkles, called **annuli**.
4. Minute mesodermal scales embedded in skin of some forms. Skin contains numerous poison glands (**Khanna** and **Yadav**, 2005).
5. Skull compact, roofed with bone; 95 to 285 vertebrae.

6. Special sensory (protrusible) tentacles occur on snout.
7. Males have protrusible copulatory organs, each called **phallodeum**.
8. Some have parental care. Some are viviparous (*e.g.*, *Typhlonects* and *Scolecomorphus*).
9. Include 160 species. Occur in tropical forests of South America (their primary home), Africa and Southeast Asia.

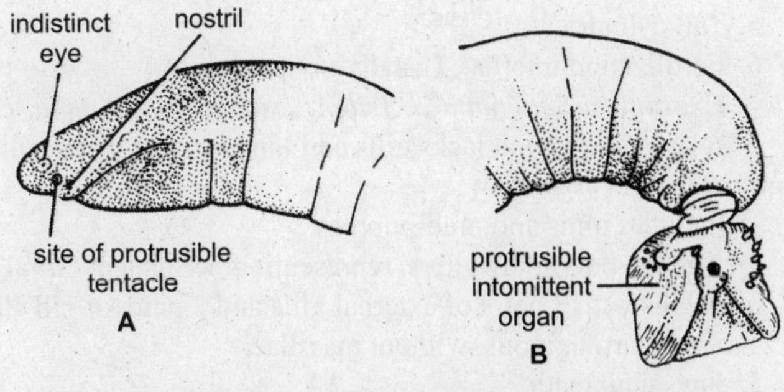

Fig. 10.4. *Caecilia tentacula*. A—Snout and B—Tail.

Examples: *Ichthyophis*, *Uraeotyphlus*, *Siphonops* and *Caecilia* (Fig. 10.4).

Order 2. Urodela or Caudata (Gr. *ura* = tail; *delos* = visible) or (L. *caudatus* = having a tail)

1. This order includes 360 species of aquatic tailed amphibians: the salamanders and newts.
2. They are carnivorous both as larvae and adults, preying on worms, small arthropods and small molluscs. Most eat only things that are moving.
3. Body with distinct head, trunk and tail.
4. Two pairs of limbs are set at right angles to the body, with forelimbs and hindlimbs of approximately equal size. In some aquatic and burrowing forms, the limbs are rudimentary or absent.
5. Skin devoid of scales and tympanum.
6. Gills are permanent or lost in adult.
7. Teeth are present in both jaws.
8. Males are without copulatory organs. Females oviparous.
9. Larvae aquatic.

The order Urodela is divided into 5 suborders: 1. Cryptobranchoidea; 2. Ambystomatoidea; 3. Salamandroidea; 4. Proteida and 5. Meantes.

Suborder 1. Cryptobranchoidea

1. Most primitive. Permanently aquatic.
2. Adults without eyelids and gills.
3. Skin is soft and flabby.
4. Premaxillary spine short.
5. Fertilization external.

Examples: *Cryptobranchus* and *Megalobatrachus*.

Suborder 2. Ambystomatoidea

1. Adults terrestrial with eyelids.
2. Premaxillary spine large.
3. Vertebrae amphicoelous.
4. Fertilization is internal.
5. Exhibits the phenomena of **neoteny** or paedogenesis, *i.e.*, axolotl larva becomes sexually mature and reproduces.

Example: *Ambystoma*.

Suborder 3. Salamandroidea

1. Includes newts, salamanders, congo eels, etc.
2. Eyelids present or absent. Maxillae absent.
3. Vertebrae opisthocoelous.
4. Gills and gill slits disappears in the adult.

5. Tail cylindrical.
6. Fertilization internal. Usually viviparous.

   Examples: *Salamandra*, *Triton*, *Triturus*, *Amphiuma*, *Desmognathus*, *Plethodon* (American salamander which lacks gills and lungs and respires with skin).

Suborder 4. Proteida

1. Includes olms and mud-puppies.
2. Aquatic bottom dwellers, representing permanent larval forms. Eyelids absent.
3. Adults with 3 pairs of external gills and 2 pairs of gill slits.
4. Skull cartilaginous, without maxillae.
5. Jaws with teeth.

   Examples: *Proteus* and *Necturus*.

Suborder 5. Meantes

1. Hindlimbs and eyelids absent.
2. Body is slender eel-like.
3. Aquatic forms; representing permanent larval forms.
4. Three pairs of external gills present.
5. Jaws with horny covering.

   Examples: *Siren* (mud eel) and *Pseudobranchus*.

Order 3. Salientia or Anura (L., *saliens* = leaping) or (Gr., *an* = without; *aura* = tail)

1. This order includes approximately 3450 species (**Hickman** *et al.*, 1997).
2. Body is short and broad and divided into head and trunk. Neck and tail absent.
3. Forelimbs are short while hindlimbs are long. Hind limbs are adapted for jumping and swimming.
4. Feet often webed.
5. Adults without gills or gill openings. Respiration by lungs.
6. Eyelids well formed. Tympanum present.
7. Skin loosely-fitting, scaleless. Mandible (lower jaw) toothless.
8. Pectoral girdle bony. Ribs reduced or absent. Vertebral column short, contains 5 to 9 vertebrae and a slender urostyle which is formed by fusion of caudal vertebrae.
9. Posterior part of tongue is free. Tongue can be thrown out as a prehensile organ.
10. Fertilization mostly external.
11. Larva is aquatic and called a tadpole.

   Order Anura is divided into 5 suborders: 1. Amphicoela; 2. Opisthocoela; 3. Anomocoela; 4. Procoela and 5. Diplasiocoela.

Suborder 1. Amphicoela

1. Vertebrae with amphicoelous centrum.
2. Ribs free. Two relict tail muscles present.
3. Fertilization internal.

   Example: *Liopelma* and *Ascaphus* (tailed-forg).

Suborder 2. Opisthocoela

1. Vertebrae with opisthocoelous centrum.
2. Ribs free.
3. Tongueless (aglossa).

   Examples: *Alytes* (midwife toad), *Bombinator* (fire-bellied toad), *Discoglossus*, *Pipa* (Surinam toad) and *Xenopus* (Box 10.1).

Box 10.1

For embryological experimentations, in the 1960s, *Xenopus laevis* replaced *Rana* frogs and the salamanders because it could be induced to mate throughout the year. Unfortunately, *Xenopus laevis* has four copies of each chromosome rather than the more usual two, and it takes 1 to 2 years to reach sexual maturity. Another *Xenopus* species, *X. tropicalis*, is often studied now, since it has all the advantages of *X. laevis*, plus it is diploid and reaches sexual maturity in mere 6 months (**Gilbert**, 2010).

African clawed frog (*Xenopus*) is found in Africa south of the Sahara Desert. Its common name is platanna (in south Africa). *Xenopus* is also medically important, since, it is used in testing for human pregnancy: urine from a pregnant women injected into a female *Xenopus* causes it to extrude eggs (**Khana** and **Yadav**, 2005).

Suborder 3. Anomocoela

1. Vertebrae procoelous (*e.g.*, sacral vertebrae) or amphicoelous.
2. Free ossified ribs absent.
3. Upper jaw with teeth.

Example: *Pelobates* and *Scaphiopus*.

Order 4. Procoela

1. Vertebrae procoelous. Ribs absent.
2. Urostyle with 2 condyles.
3. Teeth absent in *Bufo*. Teeth only in upper jaw in *Hyla*.

Example: *Bufo* (common toad), *Hyla* (tree frog/toad), *Gastrotheca* (marsupial frog), *Dendrobates* and *Rhinoderma*.

Order 5. Diplasiocoela

1. First 7 vertebrae are procoelous.
2. Eight vertebra is amphicoelous.
3. Ninth or sacral vertebra convex anteriorly (*i.e.*, acoelous) and bears two condyles posteriorly.
4. Pectoral girdle is usually fused with sternum.
5. Ribs are absent.

Example: *Rana* (Indian bullfrog) and *Rhacophorus* (flying frog).

Table 10.1 Distinction between three orders of living amphibia

	Character	Apoda	Urodela	Anura
1.	Shape	Elongated, eel-like	Long body differentiated into head, trunk and tail	Broad and flattened body differentiated into head and trunk
2.	Habitat	primitive amphibians living in burrow	Mostly aquatic	Live both on land and water
3.	Tail	Poorly developed or absent. Tail fin absent.	Long tail provided with a tail fin	No tail and neck
4.	Limbs	Absent	Two pairs, both equal in length. Both are comparatively weak and sometimes hindlimbs are absent	Two pairs and well-developed; forelimbs smaller than hindlimbs
5.	Digits	Absent	Four in fore limbs and four in hind limbs. No web found between the digits	Same (like Urolea) but web found in between the digits.
6.	Skin	Dry-rough, scaly and wrinkled skin. Small scales embedded in the dermis.	Smooth, scale-less, moist and slippery.	Scale-less but provided with warts or papillae. Moist and slippery.
7.	Segments	Pseudosegmented skin (= annuli)	Absent	Absent

	Character	Apoda	Urodela	Anura
8.	Tympanum	Absent	Tympanum and tympanic membrane is absent	Present
9.	Eyes	Poorly developed or degenerated eye which are found embedded or burried in the skin	Small, without eyelids and sometimes unfunctional.	Well developed.
10.	Vertebrae	Numerous (upto 200)	Numerous but much less than apoda.	Very less in number
11.	External gills	Absent (due to tachy-genesis*)	Many persists in adults	Gill and gill slits are absent in adult.
12.	Examples	*Ureotyphlous*, *Ichthyophis*	*Salamandra*, *Proteus*, *Ambystoma*, *Siren*, *Triton*	*Rana*, *Bufo*, *Hyla*, *Xenopus*, *Pipa*, *Rhinoderma*

*Tachygenesis. It is rapid elimination of ancestral characters by acceleration of developmental stages.

10.3 TYPE STUDIES

1. *Ichthyophis* (Blind Worm)

Classification

Class — **Amphibia**
Order — **Apoda (or Gymnophiona)**
Genus — *Ichthyophis*
Species — *glutinosa*

Habits and Habitat

Ichthyophis glutinosa is a limbless burrowing amphibian. It is found in the tropical regions of foot hills of Himalayas in India, Ceylon, Malay Archipelago and Siam. It burrows in moist ground near the stream or ponds. It is carnivorous feeding on worms and other small invertebrates. It shows parental care.

Structure

I. glutinosa is about 30 cm long. It is dark brown or bluish-black with a yellow band along the sides. The body is limbless, tailless and snake-like. The head is small and depressed and adapted for burrowing. Eyes are small, function-less and covered by skin. It is blind. Head carries a pair of nostrils and a pair of **protrusible sensory tentacles**, enclosed in pits and situated behind the nostrils. Tympanum, tympanic cavity and columella are not found.

Skin of body contains numerous transverse grooves or wrinkles or rings, called **annuli**. Minute **cycloid scales** lie buried in the skin at the bottom of grooves. The scales are arranged in transverse rows. Tail is short and vestigial.

Limbs and limb girdles are absent.Notochord is persistent. Vertebrae numerous (about 250) and amphicoelous. Right lung is much developed. The male contains a large, hooks-bearing protrusible copulatory organ, called **phallodeum**. Fertilization is internal. Female lays about 20 large eggs (size 4 × 6 mm) and coils herself around these gelatinous eggs until they hatch. During such parental care, females presumably protects her eggs, embryos against predation and also from destructive molds and desiccation. The female takes the eggs into shallow water shortly before the hatching time. The gilled larval stage is passed through the egg. Larva of blindworm bears three pairs of long and finely branched external gills which are used for the absorption of nutrients inside the egg. Gills are lost before hatching. One pair of gill slits remains open and are used for the respiration during aquatic larval life. The larva swims in water like the eel. Larval period is long.

Uraeotyphlous is like *Ichthyophis*. It is found in South East Asia.

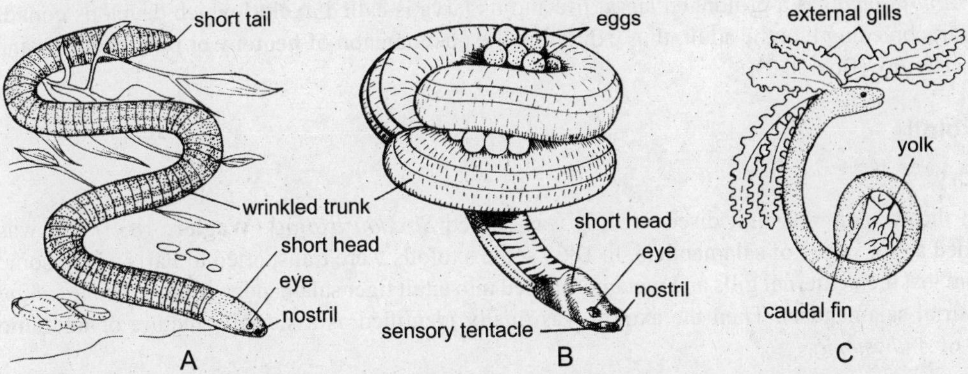

Fig. 10.5. *Ichthyophis glutinosa*. A—Male; B—Female guarding her eggs and C—Larva.

2. *Ambystoma* (Tiger Salamander)

Classification

Class	—	**Amphibia**
Order	—	**Urodela (Caudata)**
Suborder	—	**Ambystomatoidea**
Genus	—	*Ambystoma*
Species	–	*tigrinum*

Habits and Habitat

Ambystoma is terrestrial animal. It is found in North American from Southern Alaska to Mexico. Neotenic forms (*i.e.*, Axolotl larva) occur in mountain region especially in Mexico.

Structure

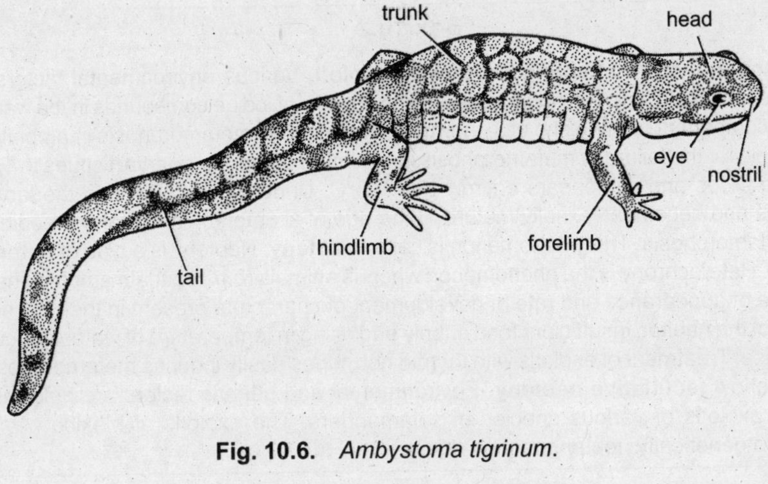

Fig. 10.6. *Ambystoma tigrinum*.

Body is elongated, lizard-like having transverse grooves (Fig. 10.6). It is about 12 cm long. Body is divided into head, trunk and tail. The colour of the body is black with yellow or organe patches. Skin is poisonous. The head is depressed with a wide mouth, well developed eyes, paired nostrils and paired parotid glands (poisonous glands). The limbs are well developed. Forelimbs bear 4 fingers and hindlimbs have 5 toes. The eyes bear eyelids, the gills and gill-slits are absent and tail is without tail fin. Sexes are separate. Fertilization is internal.

Ambystoma has a prolonged larval life and its larva is called **axolotl** which develops gonads and starts breeding like the adult, thus exhibiting the phenomenon of **neoteny** or **paedogenesis** and **heterochrony** (Box 10.2).

3. Axolotl

History

When the Axolotl was first discovered, it was named *Siredon axolotl* (**Wagler**, 1830) and was regarded a new genus of salamanders. In 1865 some axolotls were transported to Paris where some of them lost their external gills and metamorphosed into adult tiger salamander, *Ambystoma tigrinum* (terrestrial salamandar). Then the axolotl was finally identified as a sexually mature or **neotenic larva** of *Ambystoma.*

Habits and Habitat

Axolotl is the larva of salamander, *Ambystoma* (*A. tigrinum* or the tiger salamander and *A. maxicanum*). *A. tigrinum* inhabits high attitudes of Colorado of North America, and *A. maxicanum* lives in the Xochimilco lake in the highlands of Mexico. Axolotl is commonly found in the lakes of Mexico and USA. The name axolotl has been derived from Aztec Indian words—**atl** meaning water and **xolotl** meaning servant. Axolotl never leaves water and is commonly called **servant of water**. The name axolotl was given by natives of Mexico who catch them for food. Axolotls breed several times a year.

Structure

Axolotl resembles its adult in body form (Fig. 10.7). The body is divisible into head, trunk and tail. There are three pairs of external gills and four pairs of gill slits at the junction of head and trunk. The tail is long and bears a tail or caudal fin which is continuous with the dorsal fin. The fins are not supported by fin rays. The head bears a pair of eyes and a pair of nostrils and a wide mouth. The trunk bears two pairs of limbs. The forelimbs bear four digits and hindlimbs contain five toes.

Axolotl retains larval (or juvenile) characters throughout life but develops reproductive organs and reproduce (**neoteny**).

Box 10.2

Environmental factors affecting neoteny in axolotl. Various environmental factors such as abundant food supply, cold water of mountain lakes and scarcity of iodine compounds in the water (iodine is an essential component of thyroxine or T_4 hormone that induces metamorphosis in amphibians) are partly respondible for the failure of metamorphosis in axolotls. Response of larval tissues to T_4 reduces as temperature drops and disappears entirely below 5°C. Under such conditions the larva retains larval structures and becomes sexually mature at the age of six months and starts breeding without undergoing metamorphosis. This phenomenon is called **neoteny**. Neoteny is a type of **heterochrony** (**Gilbert**, 2010). Heterochrony is the phenomenon whereis animals (*e.g.*, caudate amphibians) change the relative time of appearance and rate of development of characters present in their ancestors.

Drying up of the habitat, insufficient food supply and rising in temperature of water induce axolotls to metamorphose. Treatment of axolotls with thyroid hormones easily induces metamorphosis. Thus, some axolotls show **facultative neoteny**. Environmental and genetic factors are held responsive for neoteny in axolotls of various species of salamanders. The axolotls of Mexico (*Ambystoma maxicanum*) are genetically neotenous (Box 10.3).

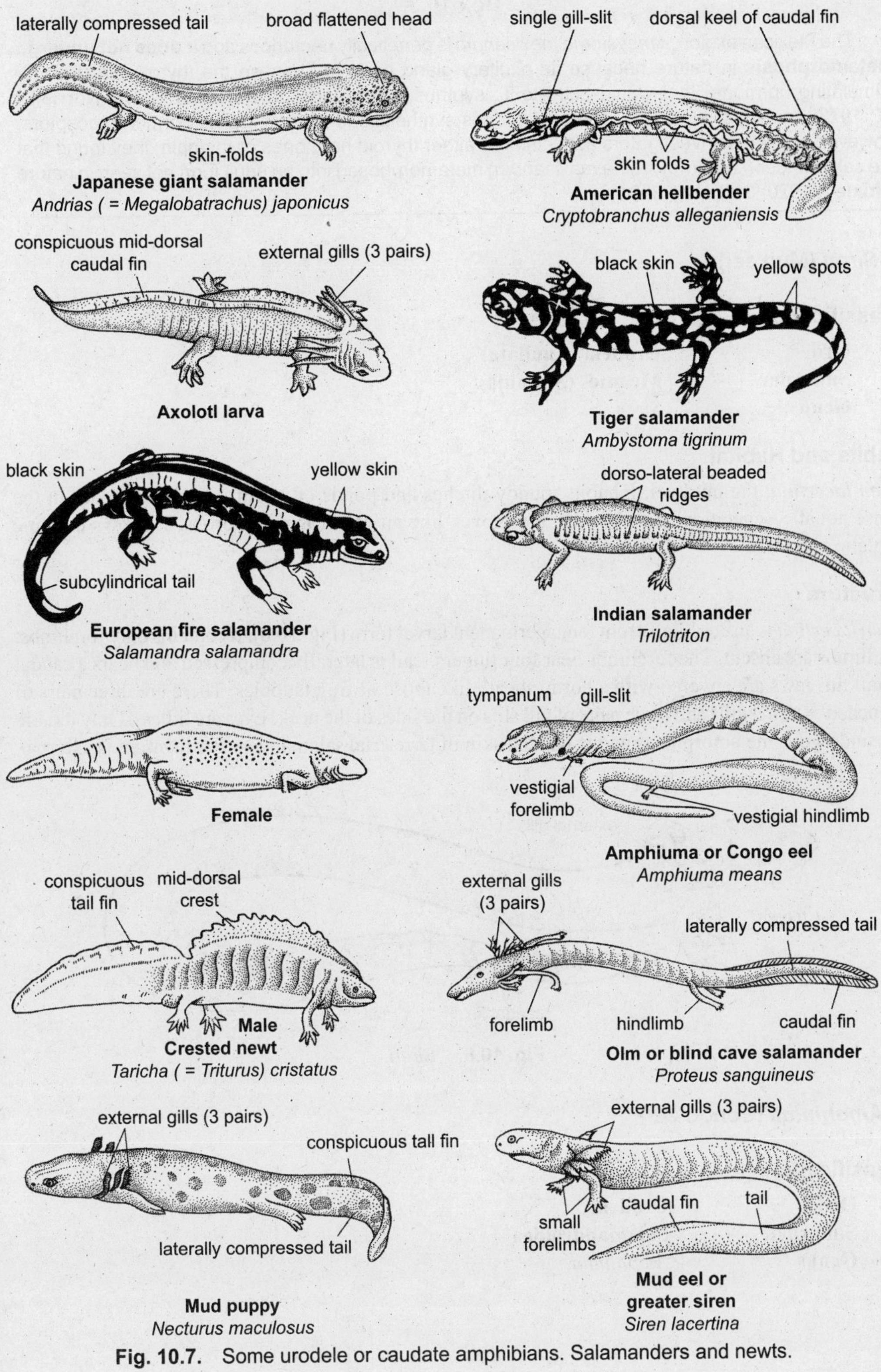

Fig. 10.7. Some urodele or caudate amphibians. Salamanders and newts.

Box 10.3

The Mexican axolotl, *Ambystoma mexicanum* is genetically neotenous and **it does not undergo metamorphosis** in nature because its pituitary gland does not release the thyrotropin (thyroid-stimulating hormone) that would activate T_4 synthesis (**Prahlad** and **DeLanny**, 1965; **Norris** *et al.*, 1973; **Taurog** *et al.*, 1974). The axolotl does synthesize functional thyroid hormone receptors, however, and when investigators administered either thyroid hormones or thyropin, they found that the salamander (*i.e.*, *Ambystoma maxicanum*) metamorphosed into an adult form not seen in nature (**Huxley**, 1920; **Safi** *et al.* 2004).

4. *Siren* (Mud eel)

Classification

Order	—	**Urodela (Caudata)**
Suborder	—	**Meantes (Sirenida)**
Genus	—	*Siren*

Habits and Habitat

Siren lacertina, the mud eel, inhabits muddy ditches and ponds of North America. It hides in the dense aquatic vegetation of the ponds and moves like an eel. It is carnivorous. It shows extreme neotelic condition.

Structure

Siren lacertina is an eel-like, 60 cm long, permanent larval form (Fig. 10.8). It contains only forelimbs, hindlimbs are absent. The forelimbs bear four fingers. Tail is laterally compressed and bears a caudal or tail fin. Jaws are covered with a horny sheath like those of frog tadpoles. There are three pairs of branched external gills and three pairs of gill slits on the sides of the neck. Eyes are lidless. Only its skin has undergone metamorphosis and is like the skin of terrestrial salamanders Fertilization is external.

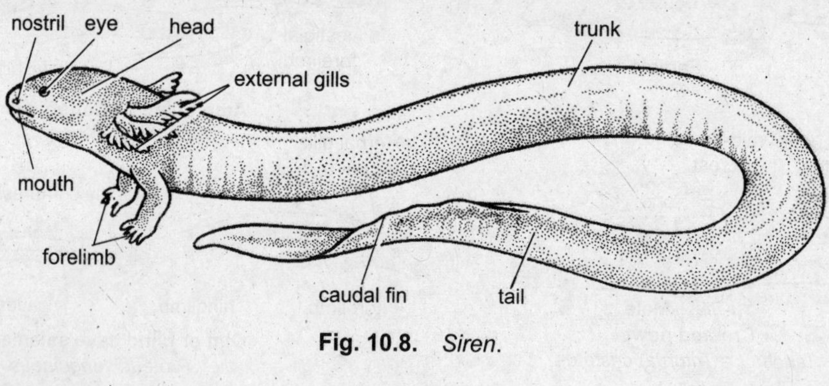

Fig. 10.8. *Siren.*

5. *Amphiuma* (Congo eel)

Classification

Order	—	**Urodela**
Suborder	—	**Salamandroidea**
Genus	—	*Amphiuma*

Habits and Habitat

Amphiuma is an eel-like salamander living in streams, swamps and rice fields in south east USA. It is commonly known as **congo eel**. It is carnivorous. It feeds on molluscs, crayfishes and small fishes.

Structure

Body is elongated and cylindrical, measuring about 90cm. Body is divided into head, trunk and tail (Fig. 10.9). *Amphiuma* is beautiful coloured with brown-black and grey belly and sides. Eyes are without eyelids and functional. Body bears two pairs of limbs. The limbs are reduced, the forelimbs with two and hindlimbs with three toes only. These vestigial limbs are of no use in locomotion. Adult is a lung breather and lacks gills. However, two pairs of gill slits are present.

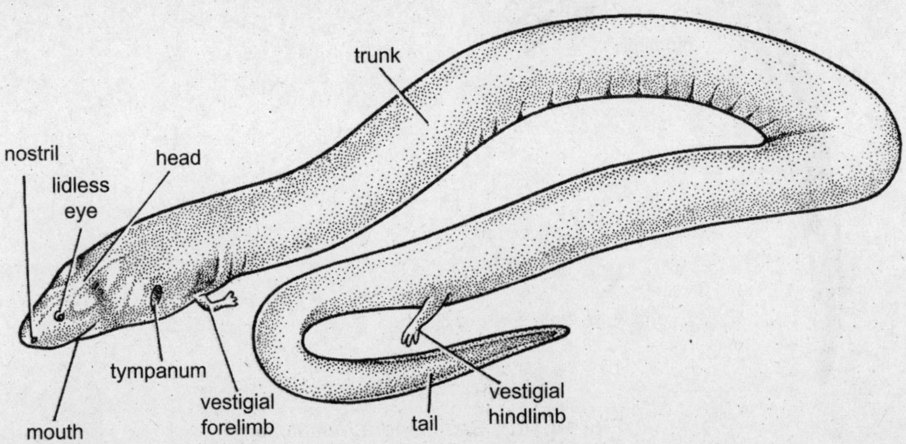

Fig. 10.9. *Amphiuma.*

The female lays eggs in rosary-like strings in a shallow nest dug in damp soil beneath a log and coils around them till hatching. *Amphiuma* has retained several larval characters, as lidless eyes, parallel arrangement of maxillary and vomerine teeth, four branchial arches, two-gill clefts and amphicoelous vertebrae. Hence it is known as **larval salamander**. *Amphiuma* has the largest erythrocytes of any animal, they are 80 μ in diameter.

6. *Salamandra* (Spotted or fire salamander)

Classification

Order	—	**Urodela**
Suborder	—	**Salamandroidea**
Genus	—	*Salamandra*

Habits and Habitat

Salamandra maculosa is commonly called **spotted salamander** or **European salamander**. It inhabits the moist shady places in Europe and Western Asia. It is a terrestrial animal and lives under the logs, stones and in the cracks and crevices. It is nocturnal. It walks slowly on land but swims actively with the help of it tail. Its main food items are worms, insects and molluscs. It has power of regeneration. Lost limbs are regenerated easily. It ejects a poisonous white fluid from its skin glands.

In *Salamandra*, the eggs are fertilized internally, usually after the female picks up a packet of sperm alled **spermatophore**) that previously has been deposited by the male on a leaf or stick. Youngs are born as larvae (viviparous) from April to June. The new eggs are fertilized in the oviducts by spermatozoa which were introduced in the previous July and stored.

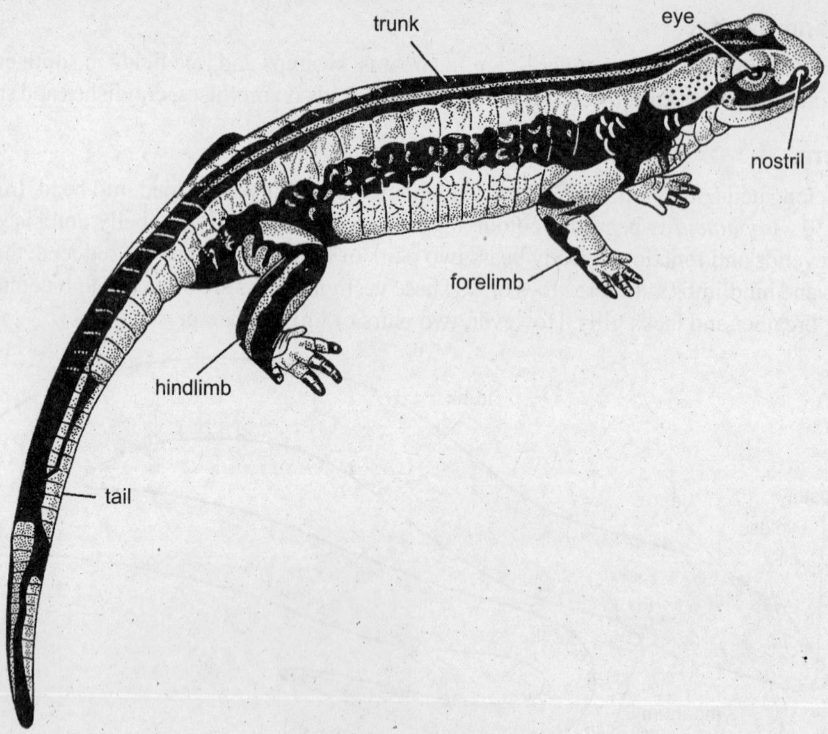

Fig. 10.10. *Salamandra maculosa.*

Structure

The 22 cm long body of *Salamandra* is strongly developed and somewhat dorso ventrally flattened with arched dorsal surface (Fig. 10.10). It is of brilliant black colour with irregular patches of yellow colour on the back. **Head** is a little depressed and contains a wide mouth, well developed eyes with movable eyelids and parotid glands behind the eyes. Both the maxillary and vomerine teeth are present. **Trunk** is also depressed and bears vertical grooves between the ribs. Gills and gill slits are absent in the adult. Limbs are well developed, forelimbs with four fingers and hindlimbs with five toes, which are neither webbed nor clawed. **Tail** is subcylindrical and tail fin is absent. The skin of *Salamandra* is smooth and slimy. The cutaneous glands secrete poisonous secretion. The young bears external gills and undergoes metamorphosis.

7. *Necturus* (Mud Puppy or Water Dog)

Classification

Order	—	**Urodela (Caudata)**
Suborder	—	**Proteida**
Genus	—	*Necturus*

Habits and Habitat

Necturus is an aquatic water newt. Its two species—*N. kunctatus* and *N. maculosus* are found in the rivers, streams and sluggish waters of North America. It feeds on invertebrates and small fishes. It is nocturnal.

Structure

Body of *Necturus* is elongated measuring 30–45 cm in length (Fig. 10.11). The dorsal surface has rusty brown colour with blackish spots. The ventral surface of body is light grey in colour. The limbs are short but stout with four toes in each limb. Adults bear a pair of lungs and three pairs of well developed, bushy external gills which are red in colour. Lateral line system is well developed. The female excavates a nest under some log or stone and lives with the egg till hatching.

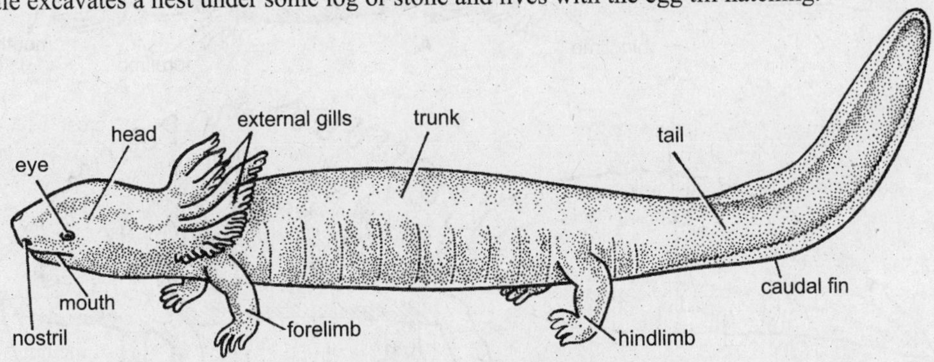

Fig. 10.11. *Necturus.*

Necturus represents permanent neotenous larva, since it has retained the larval features as lidless eyes, cartilaginous skull, external gills and flattened tail with tail fins.

8. *Triton* or *Taricha* (European Salamander or European Crested Newt)

Classification

Order	—	**Urodela (Caudata)**
Suborder	—	**Salamandroidea**
Genus	—	*Triton/Triturus/Taricha*

Habits and Habitat

It is distributed in Europe and USA. It is a terrestrial salamander.

Structure

It shows marked sexual dimorphism. The males are slightly bigger than females. During breeding season the males become brightly coloured and develop a mid-dorsal crest on body and a conspicuous median tail fin (Fig. 10.12B).

Newt's skin is soft, scaleless and covered with slime. Gills and tympanum are absent. Respiration occurs by skin and lungs. Limbs are well developed. The newt discharges a sufficiently venomous secretion from its dorsal glands to ward off enemies.

The Indian salamander, *Trilotriton*, found in eastern Himalayas, is 15 to 20 cm long. The head bears two dorso-lateral ridges which becomes beaded on trunk. There is a mid-dorsal vertebrae ridge. The tail is bilaterally compressed and carries a median vertical tail fin.

9. *Hyla* (Tree frog)

Classification

Class	—	**Amphibia**
Subclass	—	**Lissamphibia**
Order	—	**Anura**
Suborder	—	**Procoela**
Genus	—	*Hyla*

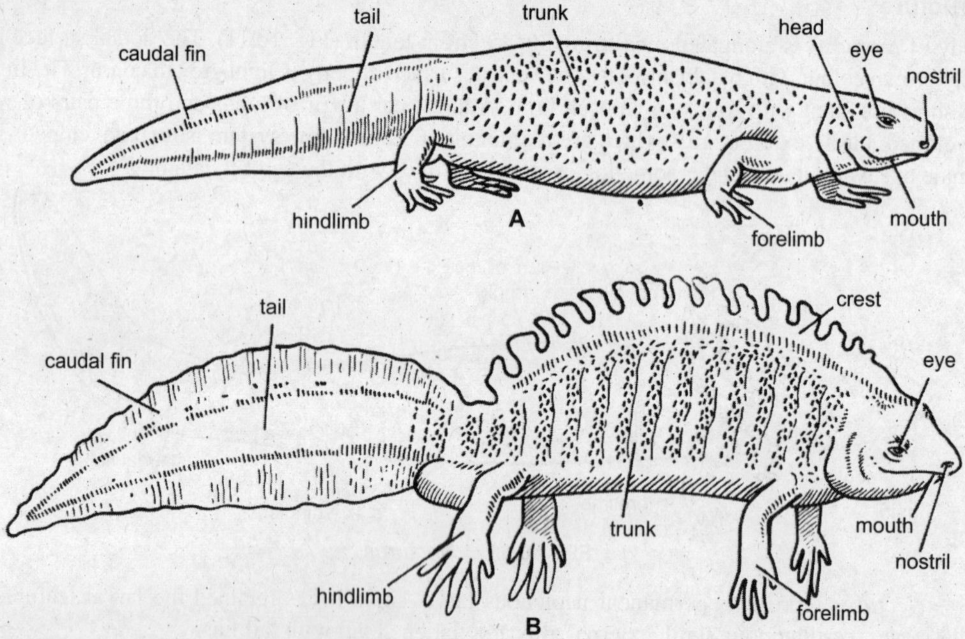

Fig. 10.12. *Triton*. A—Female and B—Male.

Habits and Habitat

Hyla arborea is a common Indian **tree frog**. Its other species are found the world over, *i.e.*, *Hyla* occur in damp forests in India, China, Africa, USA and Canada. *H. arborea* (Fig. 10.13), *H. versicolor* and *H. regita*, etc., are all tree frogs and adapted for amphibious and arboreal life. *Hyla* produce loud voice through its greatly expanded vocal sacs. In certain species of *Hyla* the eggs are carried on the back of the female (*e.g.*, *Hyla goeldii*).

Tree frogs feed on insects. Fertilization is external. Eggs are laid in a gelatinous envelope.

Structure

Hyla arborea is 5 to 8cm long. Its skin is smooth dorsally but bears papillae on the ventral surface. Its body is green in colour. The tree frogs in general are protectively coloured, often with patterns that resemble tree barks or leaves. Limbs are long, slender. Terminal bones of digits are claw-shaped. There are plate-like adhesive discs at the tip of the claw-shaped digits of all four legs. These discs are not suctorial. They possess a moist, fibrous, corrugated anti-skid external cutaneous layer. It helps the tree frogs in climbing and clinging to smooth surfaces of leaves, etc. There is an extra joint or **intercalary cartilage** between the last two phalanges of each digit facilitates adjustments of the adhesive discs for climbing trees. The forelimbs are webless, while the hindlimbs are poorly webbed. Some tree frogs can change colour (mimicry).

10. *Alytes* (Mid-wife toad)

Classification

Order	—	**Anura**
Suborder	—	**Opisthocoela**
Genus	—	*Alytes*

Habits and Habitat

Alytes is an amphibious toad. It is found in some European countries.

Structure

Alytes is 5 to 8 cm long and generally greyish or brownish in colour (Fig. 10.13). Its skin is rough, dry and papillated. Tympanum is large. This toad is characterized by having vertical pupil, disc-like non-protrusible tongue, no teeth on lower jaw, no ribs and no vocal sacs in male.

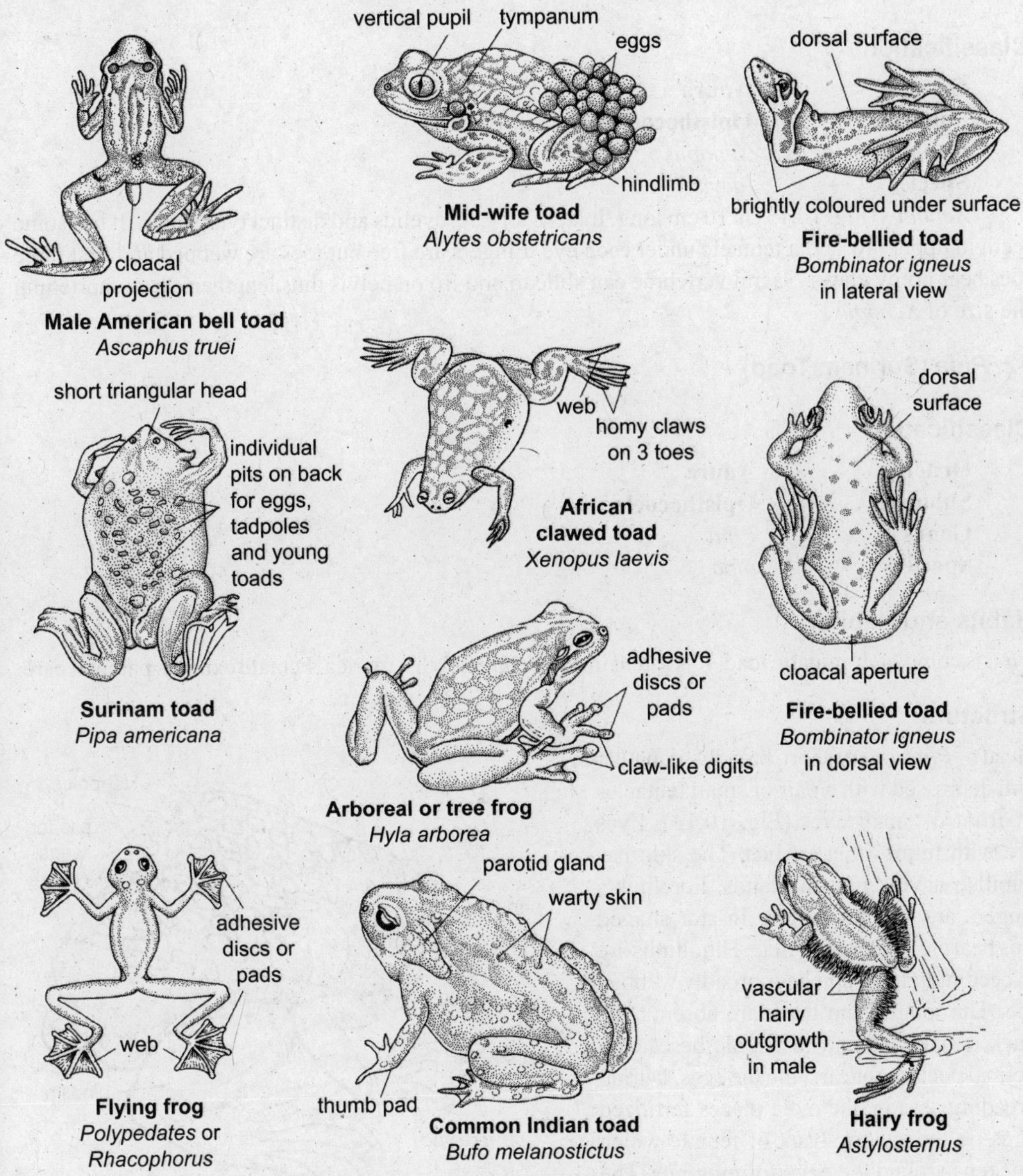

Fig. 10.13. Some anuran amphibians. Frogs and toads.

Alytes have peculiar breeding habit and parental care for the youngs. Male massages the cloaca of female for about half an hour with the toes of its hindlimb and excites the female to deposit the eggs. The eggs are laid in two strings and become attached to the waist and hindlimbs of male. The male toad with eggs retires to some moist place such as pit, etc., and burries the eggs there. When the eggs are about to hatch, the male midwife toad moves to some pond and leaves the eggs for hatching. The larvae, on coming out metamorphose into adults.

11. *Xenopus* (African Clawed Toad)

Classification

Order	—	**Anura**
Suborder	—	**Opisthocoela**
Genus	—	*Xenopus*
Species	—	*laevis*

Xenopus (Fig. 10.13) is 10 cm long. It lacks tongue, eyelids and distinct tympanum. It has some teeth in upper jaw and a tentacle under each eye. Fingers are free but toes are webbed and first three toes bear horny claws. Sacral vertebrae can slide to and fro on pelvis thus lengthening or shortening the size of *Xenopus*.

12. *Pipa* (Surinam Toad)

Classification

Order	—	**Anura**
Suborder	—	**Opisthocoela**
Genus	—	*Pipa*
Species	—	*pipa*

Habits and Habitat

Pipa is completely aquatic toad. It is found in Africa and South America. Female exhibits parental care.

Structure

Head of *Pipa* is very short, broadly triangular and depressed with a pair of small tentacles in front of small eyes (Fig. 10.14). Eyes are with immovable eyelids. The skin has papillae having poison glands. Forelimb's fingers are slender ending in star shaped tips bearing sensory papillae. Hindlimbs are exceedingly large and bear broadly webbed toe. The tongue and teeth are absent. The back is flattened and in female beset with dermal pockets for carrying the eggs. During breeding season, the male places fertilized eggs on the spongy back of female which thickens around the eggs forming pits. The eggs sink deep into these pits, which develop opercula completely covering the eggs. The skin is richly supplied with blood and the embryo establishes a metabolic exchange

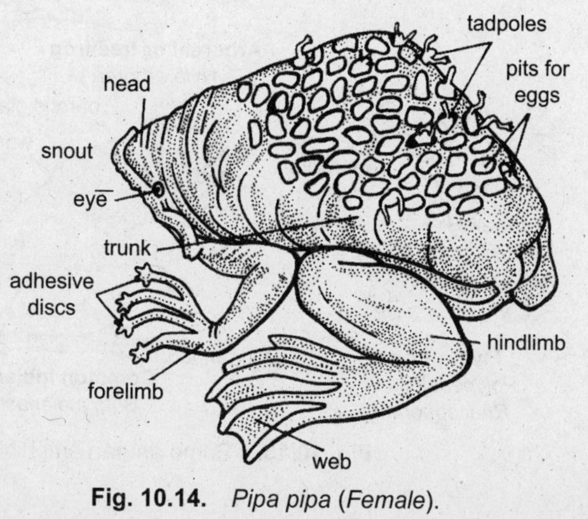

Fig. 10.14. *Pipa pipa* (*Female*).

by the development of a vascular tail. After 80 days young leap out in perfect condition. After birth of young ones the female changes its dorsal skin.

QUESTIONS

Long Answer Questions

1. Describe briefly about limbless amphibians. (*Madras, 2011*)
2. Give an outline classification of class Amphibia.
3. Classify modern amphibia upto orders giving characteristic features and suitable examples. (*CCSU, Meerut, 2010, 2013*)

Short Answer Questions

1. Classify the following animals giving their distinguishing characters:
   (*i*) *Ichythyophis* (*Avadh, 2002; Purvanchal, 2003*)
   (*ii*) *Ambystoma*
   (*iii*) Axolotl (*CCSU; Meerut, 2011*)
   (*iv*) *Amphiuma*
   (*v*) *Salamandra*
   (*vi*) *Siren*
   (*vii*) *Necturus*
   (*viii*) *Pipa*
   (*ix*) *Alytes* (*Purvanchal, 2003*)
   (*x*) *Hyla*
   (*xi*) *Rhacophorus*
   (*xii*) *Bufo*
2. Write short notes on the following:
   (*a*) Apoda or Gymnophiona (*Kanpur, 2003*)
   (*b*) Axolotl
3. Give firve main features of Urodela.
4. Describe burrowing adaptation of apoda.
5. Differentiate between Urodela and Anura.

Very Short Answer Questions

1. Name two burrowing amphibians.
2. Name two amphibians which exhibit neotenic features.
3. Name the viviparous amphibian and where it is found?
4. Define the following: (*a*) Neoteny; (*b*) Opisthocoelous; (*c*) Urodela; (*d*) Aglossa and (*e*) Anura. (*Madras, 2010*)
5. The urodeles have primitive organization and have retained a number of larval characteristics. What is this phenomenon known.
6. In which group limbless burrowing amphibians have been placed?
7. Which amphibian is called mud eel?
8. Mention the mode of respiration found in amphibians.
9. Name an amphibian with one pair of limbs only.
10. Give two examples of tongueless anurans.

Fill in the Blanks

1. Amphibians live on ___________ as well as in water.
2. Class Amphibia includes animals which have ___________ mode of life.

3. In Apoda the body is __________ and without limbs.
4. In Urodela the vertebrae are __________.
5. In Amphibia heart is __________ chambered with __________ auricles and one ventricle.
6. *Amphiuma* is commonly known as __________.
7. The common name of *Necturus* is __________.
8. Axolotl is larva of __________.
9. Surinam toad's generic name is __________.
10. The only flying amphibian is __________.
11. The delay of a larva to metamorphose while becoming sexually mature is called __________.
12. External gills are never dropped in __________.
13. Jaw suspension in amphibians is __________.
14. Middle ear develops from the __________.
15. Nephrostomes persists in the __________ kidneys.

Multiple Choice Questions

1. Amphibians are those animals that live
   (*a*) in fresh water (*b*) in estuarine water
   (*c*) on land (*d*) in fresh water on land
2. Mid-wife toad is another name of (*Gorakhpur, 2002; Kanpur, 2003*)
   (*a*) *Alytes* (*b*) *Rhacophorus*
   (*c*) *Hyla* (*d*) *Rhinoderma*
3. Animals of which order have tails throughout life
   (*a*) Apoda (*b*) Urodela
   (*c*) Anura (*d*) None of these
4. Which one of the following is tree frog
   (*a*) *Alytes* (*b*) *Hyla*
   (*c*) *Necturus* (*d*) *Rana*
5. Parotid glands are found in
   (*a*) *Bufo* (*b*) *Hyla*
   (*c*) *Alytes* (*d*) *Rana*
6. The larva of *Ambystoma* is
   (*a*) ascidian (*b*) axolotl
   (*c*) tadpole (*d*) trochophore
7. External gills are found throughout the life in
   (*a*) *Alytes* (*b*) *Ichthyophis*
   (*c*) *Pipa* (*d*) *Proteus*
8. Tailless amphibians are the members of the group
   (*a*) Anura (salentia) (*b*) Apoda (gymnophiona)
   (*c*) Urodela (candata) (*d*) Cyclostomata
9. Males are smaller and without vocal sacs in
   (*a*) *Alytes* (*b*) *Hyla*
   (*c*) *Rhacophorus* (*d*) *Amphiuma*
10. Which of the following exhibits parental care?
    (*a*) *Siren* (*b*) *Necturus*
    (*c*) *Ichthyophis* (*d*) *Amphiuma*
11. Volant adaptation is found in
    (*a*) *Hyla* (*b*) *Rhacophorus*
    (*c*) *Ambystoma* (*d*) *Ascaphus*

12. Excretory matter in amphibians is
(*a*) ammonia (*b*) urea
(*c*) uric acid (*d*) amino acids

13. In Apoda, external gills are present in the
(*a*) adult male (*b*) larva
(*c*) embryo (*d*) adult female

14. *Hyla* belongs to the subclass
(*a*) Anura (*b*) Apoda
(*c*) Urodela (*d*) Stegocephalia

ANSWERS

Short Answer Questions

1. *Ichthyophis* and *Uraeotyphlous*.
2. *Siren* and *Amphiuma*.
3. *Salamandra* found in Europe
4. (*a*) **Neoteny** is the failure or delay of a larva to metamorphose while become sexually mature.
(*b*) **Opisthocoelous** is a type of vertebrae in which anterior face of centrum is convex and posterior face is concave.
(*c*) **Urodela** is an order of class *Amphibia*, comprising of smooth-skinned, lizard-like tailed amphibians.
(*d*) **Aglossa** are anuran amphibians without tongue.
5. Neoteny.
6. Gymnophiona or Apoda.
7. *Siren*.
8. Branchial, pulmonary, cutaneous and buccopharyngeal.
9. *Siren*.

Fill in the Blanks

1. Land; **2.** Double; **3.** Worm-like; **4.** Ophisthocoelous; **5.** Three, two; **6.** Congo-eel; **7.** Mud-eel; **8.** *Ambystoma*; **9.** *Pipa*; **10.** *Rhacophorous*; **11.** Neoteny; **12.** *Necturus*; 13. Autostylic; **14.** Spiracle; **15.** Amphibian

Multiple Choice Questions

1. (*d*)	**2.** (*a*)	**3.** (*b*)	**4.** (*b*)	**5.** (*a*)	**6.** (*b*)
7. (*d*)	**8.** (*a*)	**9.** (*a*)	**10.** (*c*)	**11.** (*b*)	**12.** (*b*)
13. (*c*)	**14.** (*a*)				

11

Amphibia (Origin, Affinities, Parental Care and Heterochrony: Axolotl, Neoteny, etc.)

11.1 ORIGIN OF AMPHIBIA

Amphibians arose from some fish-like ancestor which was similar to the primitive bony fishes, namely, the **osteolepids** (family Osteolepidae). The osteolepids were members of group (subclass) **Crossopterygii**. They resembled the early amphibians in many features which showed them to be the ancestors of amphibians. They possessed internal nares and lungs and showed the fusion of the upper jaw (maxilla) with the skull. Moreover, the osteolepids had lobed paired fins with skeletal supports, which could give rise to the tetrapod limbs. These features had survival value in the Devonian period. This period experienced seasonal drought so that the forms which could live out of water for a longer period and could move on land better survived and became more and more terrestrial. In this way, the fishes gave rise to the amphibians.

Since the sea remained unaffected by drought, the amphibians might have originated from fresh water fishes. Similarities between the living amphibians and fishes confirms the origin of the amphibians from the fishes. Dermal scales still occur in the skin of early amphibians such as the Apoda (caecillians). The lateral-line sensory system and the gills, so well developed in fishes, are retained in the larvae of all and in the adults of some amphibians. Other similarities between fishes and amphibians include the possession of ventral aorta, 10 pairs of cranial nerves and functional pronephros in the larval stage, and lack of epididymes and embryonic membranes. Further, the structure of the eggs and the process of early segmentation are quite similar in the two groups.

Amphibians first appeared in the Devonian period, which was the age of fishes, and increased in number so rapidly that the next geological period, *i.e.*, the Carboniferous, became the **age of amphibians**.

The early amphibians are called **Stegocephalia** because of the solid roofing of the skull. They resembled the crossopterygial fishes in having a covering of dermal bones on the head, autostylic jaw suspension, internal nares, supracleithrum joining the pectoral girdle to the skull, and no connection between the pelvic girdle and the vertebral column. The stegocephalians inhabited fresh water and land. A group of stegocephalia are known as **Labyrinthodontia**, because of the complex folding of the enamel layers of the teeth. They were salamander-like in form and had rudimentary neck, heavy muscular tail inherited from the piscine (fish) ancestor and folded dentine in the teeth. They were mainly terrestrial animals. An important example of this group is the first recorded amphibian

Ichthyostega (Fig. 11.1) from the Devonian period of Greenland. All labyrinthodonts became extinct in the Triassic period. However, the group is very important because it included ancestors not only of other amphibia but also of reptiles, and hence of all higher tetrapod vertebrates.

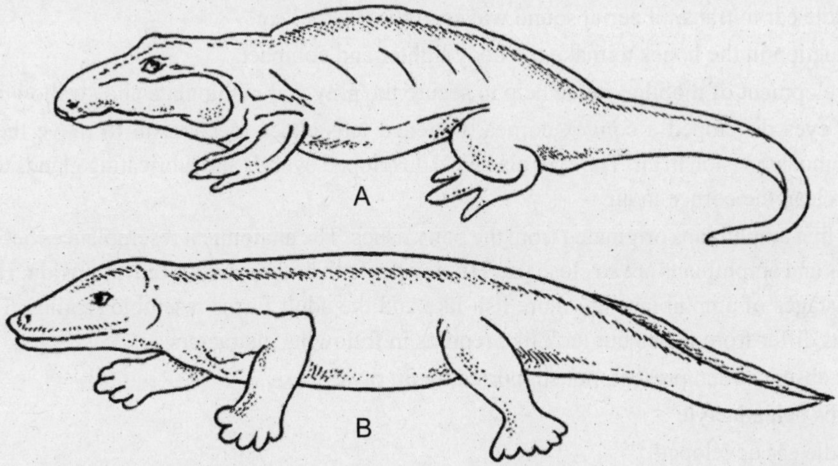

Fig. 11.1. Fossil amphibians. A—*Eryops*, about 2–7 metres long. B—*Ichthyostega*, about 1–2 metres long.

Most probably from the labyrinthodonts evolved three main lines differing in the structure of the vertebral column. One line gave rise to urodeles and apodes through small animals called **lepospondyls**. The other line produced the anurans. The third line evolved into the primitive reptiles which gave rise to the higher vertebrates.

It is now held by some evolution biologists that the amphibians are not descendants of crossopterygians, but both crossopterygians and amphibians were contempories and had a common ancestor.

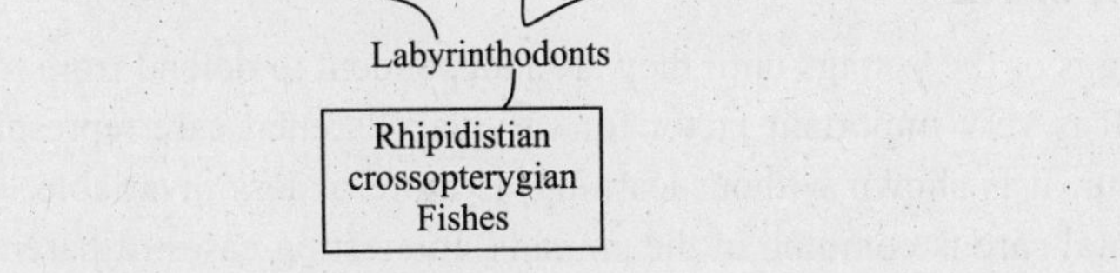

Fig. 11.2. Diagram depicting origin of tetrapod groups.

Some modern amphibians resemble the living dipnoi, but this resemblance is not taken as phylogenetic affinity between the two. It is considered to be a case of convergent or parallel evolution.

11.2 AFFINITIES OF AMPHIBIA

Amphibians exemplify the transition from aquatic life in vertebrate evolution. Their marked adaptation to land life include.

1. Limbs (pentadactyl) for movement of solid ground.
2. Lungs for aerial respiration.

3. Internal nostrils to make breathing possible with closed mouth.
4. Slimy skin for protection against desiccation.
5. Changes in the blood circulatory system to accomodate aerial respiration through lungs.
6. Middle ear to transmit aerial sound waves to the internal ear.
7. Reduction in the bones to make the body lighter and compact.
8. Development of mobile tongue help to secure the prey and manipulate and swallow it.
9. The eyes developed a convex cornea, flattened lens and a mechanism to move the lens to accomodate vision in air. The eyes also have developed eyelids and lubricating glands to protect and clean the cornea in air.

The first amphibians originated from the bony fishes. The anatomical resemblances between the bony fishes and amphibians are so close that **Huxley** put both of them under **Ichthyopsida**. However, the larval stages of amphibians are more fish-like and the adult forms resemble reptiles. The adult amphibians differ from fishes but look like reptiles in following characters:

1. Fins absent, when present, not supported by fin rays.
2. Limbs pentadactyle.
3. Middle ear developed.
4. Cloaca present.
5. Cardinal veins absent. Heart three chambered.

Most fishes are exclusively aquatic. These cannot move on land. However, the crosspterygian and lung fishes (from which amphibian originated) could breathe air through their lungs and move on mud with their lobate paired fins. The amphibians advanced a little more and are able to move on land and breathe air on permanent basis. However, amphibians require a moist environment all the time and prefer to live close to the fresh water bodies. They need to return to the water to breed. The larvae of amphibians are truely aquatic and possess gills for aquatic respiration. The mesonephric kidney of the adult still excretes water. Amphibians are thus only partially adapted to live on land. Reptiles on the other hand are truely terrestrial. They breed on land. Their young ones are able to start life on land. Their kidneys conserve water and excrete salts.

Thus, the amphibians are imperfect animals as far as true terrestrial mode of life is concerned but definitely represent transition from aquatic to land life. They can be regarded as intermediate between aquatic fishes and truely terrestrial reptiles.

11.3 PARENTAL CARE

Looking after the eggs or the youngs until they are independent to defend from predators is known as **parental care**. It is very important factor for survival. Parental care represents the instinctive (inherited) behaviour. It is shown without learning, is more or less invariable, is species-specific and adaptive. Parental care is common in the animals. Interesting cases of parental care are found in amphibians.

Perhaps no group shows more diversity in parental care than do the amphibians. Anurans show much greater diversity than urodeles and apodans. Parental care is associated only with those species that place their eggs in single clusters. The methods of parental care by amphibians generally fall under two broad categories: 1. protection by nests, nurseries or shelters; and 2. direct caring by parents. In apodans and salamanders, parental care consists only of attendance of the eggs, mostly by the female. In anurans the eggs are usually guarded by the male and the parental care involves transportation of eggs and larvae also.

A. Protection by nests, nurseries or shelters. Amphibians have evolved various curious methods of giving protection to their defenceless eggs and larvae trom the predators:

1. Selection of site. Many amphibians lay eggs in protected sites most of which are (microhabitats) on land. Many tropical trogs and toads lay eggs on land near water. Many tree frogs lay their eggs not on land, but on leaves and branches overhanging water. Species of *Phyllomedusa*, *Rhacophorus*, *Hylodes*, etc., glue their eggs to foliage hanging over water. *Rhacophorus malabaricus* in India and *Chiromantis* of Africa also deposits their spawn on trees.

2. Defending eggs or territories. In Urodela, males of *Cryptobranchus* and *Hynobius* drive away all enemies except ripe females of own species. *Andrias japonicus* shakes the eggs for proper aeration. Either sex of *Proteus anguineus* may attend the eggs. It waves its tail to direct water currents over the eggs. In terrestrial plethodontid salamanders egg clutches are usually attended by females. The female periodically rotates the eggs in the clutch in order to increase aeration and to prevent adhesive malformation.

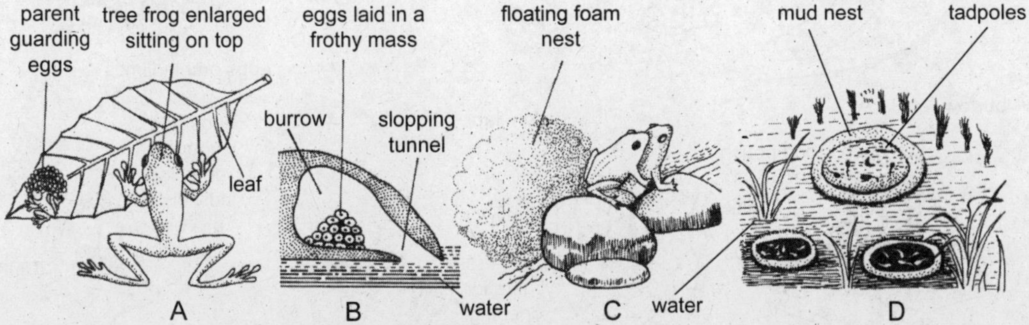

Fig. 11.3. Parental care in Amphibia. Protection by nests, nurseries and shelters. A—A tree frog guarding eggs glued to a leaf overhanging water. B—Foam nest of *Rhacophorus schlegelii* in a slopping burrow near water. C—Foam nest floating on water. D—Mud nest of *Hyla faber*.

In Anura, males of green frog *Rana clamitans* and other species maintain territories and attack small intruders to defend eggs. In *Manlophryne robusta*, the male actually sits over and holds with hands the elastic gelatinous envelope containing eggs numbering 17. Some tree frogs laying eggs above water may sit besides the eggs or rest on top of them. Males of *nest building* gladiator frog, *Hyla rosenbergi* are highly **territorial** and *aggressive*. They dig a shallow pit (a nest) near the water. Eggs are laid as a surface film on water in the nest. If the surface tension is broken, the eggs sink to the bottom of the nest and die. Male moves around the nest area and attack intruders whose entry into the nest may disrupt the surface tension. The terrestrial eggs of *Dendrobates auratus* are periodically moistened by males by periodically urinating on the egg clutch. Female of *Hemisus marmoratum* provides moisture to developing eggs. She sits on the clutch in an underground chamber and digs a tunnel leading to the water. The tadpoles escape to water from the nest chamber.

3. Direct development. In some terrestrial or tree frogs, such as *Eleutherodactylus*, *Arthroleptis*, *Hylodes* and *Hyla nebulosa*, the eggs hatch directly into little frogs thus avoiding larval mortality. In red backed salamander *Plethodon cinereus*, the hatchlings are miniatures of the adults.

4. Defending the larvae. Female *Leptodactylus ocellatus* provides protection to the tadpoles. The tadpoles move around in the dense masses. The female remains with the tadpoles and attacks birds attempting to feed on tadpoles. The tadpoles of *Dendrobates* develop in waterfilled leaf axils of various kinds of plants or tree holes, etc. Females of *D. pumilio*, *D. histrionicus* and *D. lehmanni* carry individual tadpoles to separate water-filled leaf axils and protect them. Females of *Hyperolius obstetricans* remain with their eggs on leaves overhanging the streams. Upon hatching the female kicks tadpoles out of leaves into the water.

5. Foam nests. Many amphibians convert copious mucous secretions into nests for their young. In Japanese tree frog *Rhacophorus schlegeli*, the mating couples dig a hole or tunnel into which eggs are left in a frothy mass to avoid desiccation. During rains, hatching tadpoles are washed down the slopping tunnel into pond or river water for further development.

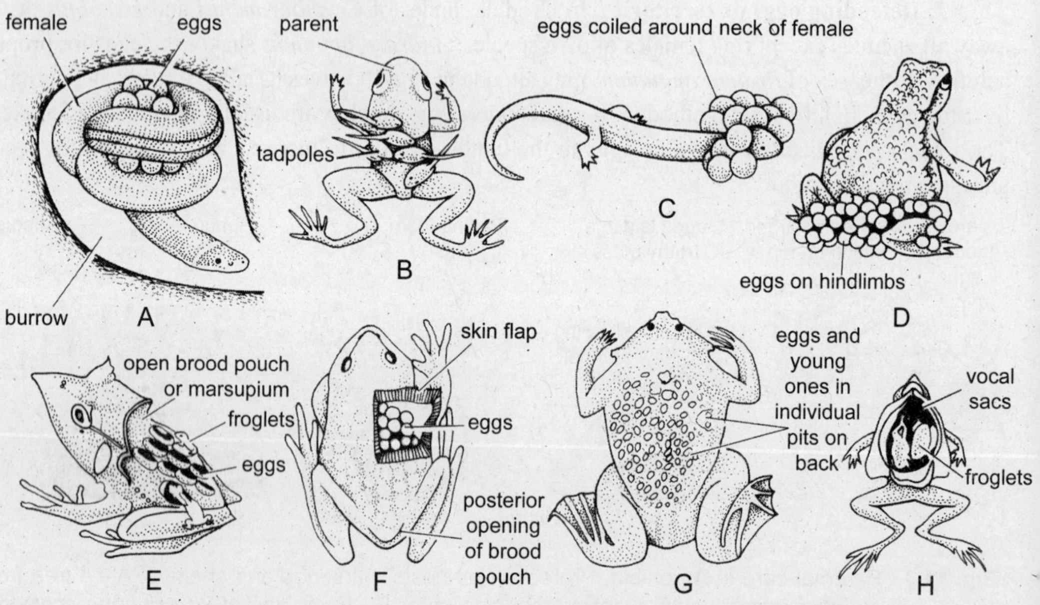

Fig. 11.4. Direct parental care in Amphibia. A—Female *Ichthyophis* coiling around eggs. B—Transportation of tadpoles attached to back of a parent C—*Desmognathus fuscus* with eggs. D—Male *Alytes obstetricans* carrying eggs around his thighs. E—A marsupial frog with eggs exposed in open brood pouch on back. F—*Nototrema* (*Gastrotheca*) with flap of dorsal brood sac cut open to show eggs. G—In *Pipa*, eggs develop completely into individual capsule on back off emale; H—Froglets inside vocal sacs cut open of *Rhinoderma darwinii.*

The female of South American tree frog, *Leptodactylus mystacinus*, stirs up a frothy mass of mucus, fills it in holes near water and lays eggs in them. The tadpoles developing in these foam nests can readily enter water. Some anurans lay eggs in nests of foam floating on water. Male frog *Adelotus brevis* actively defends the eggs laid in foam nest. The males contain long tusks on the lower jaw which are used in fighting.

6. Mud nests (nurseries). In the Brazilian tree frog, *Hyla fabre*, the male digs a little crater-like hole or **nursery** in mud in shallow water, in which the female lays her eggs. Tadpoles hatch within these mud nests and develop upto a stage when they are large enough to fend themselves.

7. Tree nests. The South American tree frog, *Phyllomedusa hypochondrales*, lays eggs in a folded leaf nest with margins glued together by a cloacal secretion. The tadpoles when formed fall straight into water below. Another tree frog, *Hyla resinfictrix*, lines a shallow tree cavity with bees wax obtained from the hives of certain stingless bees. Females lay eggs when this cavity is filled with rain water. Here, the young develop relatively free from predators.

8. Gelatinous bags. In *Phrynixalus biroi* large eggs are enclosed in a sausage-shaped transparent gelatinous membranous bag, secreted by female and left in mountain stream. *Salamandrella keyserlingi* also deposits 50 to 60 small eggs in a gelatinous bag which is fastened to aquatic plants.

9. Communal nests. The toad *Nectophyrynoides malcolmi* prepare **communal nests** in which eggs are deposited by several females. This nest is guarded by a single male.

B. Direct Caring by Parents

1. Coiling around eggs. The apodans such as *Ichthyophis glutinosa* lays eggs in a shallow hole near the water and coils herself around the gelatinous egg mass (Fig. 11.4). The attendant female eats the infected eggs and thus saves other eggs from infection. Another apodan, *Idiocranium russeli* coils around the eggs in a dense mat of grasses.

In the salamander congo eel *Amphiuma*, the female lays large eggs in burrows in damp soil and carefully guard them by coiling her body round them until they hatch (Fig. 11.5). The female salamander *Plethodon* also coils round the eggs which are laid in small packages in the hollow of a rotten log or beneath a rock. In *Megalobatrachus maximus* (Japanese giant salamander), it is the male who coils round the eggs.

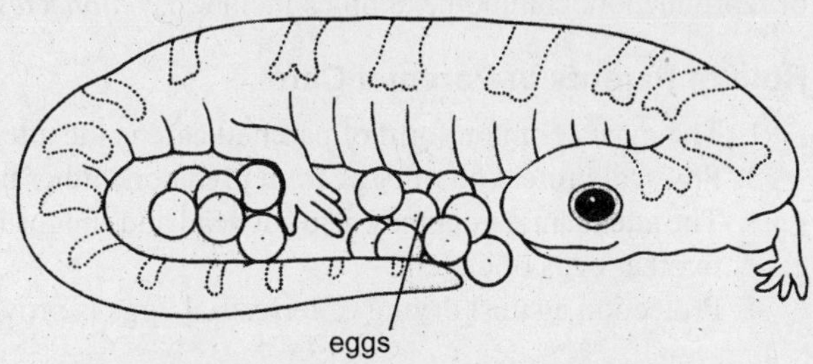

Fig. 11.5. Parental care in salamander.

2. Transferring tadpoles to water. Some species, of small frogs such as *Phyllobates*, *Arthroleptis*, *Pelobates*, *Dendrobates*, *Sooglossus*, etc., in both tropical Africa and South America deposit their eggs on ground. The tadpole hatching out fasten themselves to the back of one of the parents (*e.g.*, female in *Sooglossus seychellensis*) with their sucker-like mouth on their sticky ventral side and are transported to water.

3. Eggs glued to body. Many amphibians carry the eggs glued to their body. In the dusky salamander, *Desmognathus fuscus*. female carries the string of eggs coiled around her neck, until they are hatched. The female places her chin on the egg clutch thereby subjecting the egg to slight vibrations caused by pulsations of the throat. *Desmognathus* aggressively defends her egg. In Sri Lanka tree frog, *Rhacophorus reticulatus*, the eggs are glued to the belly of female. When female of European midwife toad, *Alytes obstetricans*, lays eggs, the male entangles them around his hindlegs. He carries them with him until they are ready to hatch. At that time he releases the tadpoles into nearest water.

4. Eggs in back pouches. In one group of tree frogs, called **marsupial frogs** or **toads**, the female carries the eggs on her back, either in an open oval depression, a closed pouch or in individual pockets. The eggs develop into miniature frogs before they leave their mother's back. In the Brazilian tree frog *Hyla goeldii* (= *Cryptobatrachus evansi*), the posterior part of the back of female forms a sort incipient brood pouch in which eggs remain exposed. In *Nototrema* (= *Gastrotheca*) *marsupium* the eggs are covered by skin forming a single large brood pouch on the back which opens posteriorly in front of cloacal aperture. The Surinam toad, *Pipa pipa* (which is completely aquatic) during breeding season, skin of female's back becomes thick, vascular, soft and gelatinous. The male presses fertilized eggs against female's back, where they sink into individual pits. A hinged cover forms over each egg enclosing it in a small capsule. Complete metamorphosis occurs within capsule and tiny tailless toads leave mother but do not enter water.

5. Organs of brooding pouches. Males of terrestrial South American Darwin's frog, *Rhinoderma darwinii*, pushes at least two fertilized eggs into his relatively large **vocal sacs**. Tadpoles complete their development and are metamorphosed in the vocal sacs and come out as mini frogs. In the West African tree frog, *Hylambates breviceps*, the female carries eggs in her buccal cavity. In *Arthroleptis*, it is male who keeps the larvae in his mouth.

The only known case of **gastric incubation/brooding** in vertebrates is found in the Australian frog *Rheobatrachus silus*. The female keeps the eggs in her stomach. The tadpoles are expelled through mouth after metamorphosis.

Feeding of larvae. *Dendrobates* is the only amphibian known to feed their youngs.

Viviparity. Some anurans are **ovoviviparous**. They retain eggs in the oviducts and the females give birth to living young. African toads *Nectophrynoides* and *Pseudophyryne* give birth to little toads. The European salamander *Salamandra salamandra* produces 20 or more small young while alpine salamander *S. atra* gives birth to one or two fully developed young.

Viviparity is wide spread in order **Gymnophiona** and three out of four families have this mode of reproduction, common examples include *Typhlonectes*, *Geotrypetes*, *Dermophis*, etc.

Role of Parents in Parental Care

1. The most common form of parental care is the attendance of the egg clutch by one parent.
2. Physical protection of eggs from predators either by driving them away or by eating them.
3. The attendant parent may also eat dead and fungus infested eggs to reduce spreading of fungus to other eggs of clutch.
4. Protection against drying of terrestrial eggs is provided by actively moistening them.

11.4 HETEROCHRONY

When relative timing of two different developmental processes is shifted independently of one another the process is called **heterochrony** (Box 11.1). Heterochrony is of three types: 1. Neoteny; 2. Progenesis and 3. Direct development.

Box 11.1 Heterochrony and Paedomorphosis

Evolutionary change in the timing of development is called **heterochrony**, a term initially used by **Haeckle** to denote exceptions to recapitulation. If the descendant's ontogeny extends beyond the ancestral one, new characteristics can be added late in development, beyond the point at which development would have terminated in the evolutionary ancestor. Features observed in the ancestor often are moved to earlier stages of development in this process, and ontogeny therefore does not recapitulate phylogeny to some degree. Ontogeny also can be shortened during evolution, however. Terminal stages of the ancestor's ontogeny are deleted, and the adults of descendants come to the resemble pre-adult stages of their ancestors. For example, aquatic axolotl retains the juvenile, aquatic morphology throughout its life unless forced to metamorphose by hormone treatment. The axolotl evolved from metamorphosing ancestors, an example of paedomorphosis. This reverse recapitulation is called **paedomorphosis**. Paedomorphosis is retention of ancestral juvenile characters by later ontogenetic stages of descendants (**Hickman** *et al.*, 1996).

1. Neoteny

Neoteny is the retention of certain larval (or juvenile) characteristics beyond the normal period or during adult life. It refers to the retention of the larval or juvenile form as a result of retarded body development relative to the development of the germ cells and gonads (which achieve maturity at the normal time; **Gilbert**, 2010).

Examples of Neoteny

1. The best example of neoteny is found in certain salamanders such as *Ambystoma tigrinum*, the tiger salamander is called **axolotl**. It has 3 pairs of branched, crimson external gills, four pairs of wide gill slits, and compressed tail bearing tail fin devoid of fin rays. Axolotl metamorphoses normally in the greater part of its geographical range. In high altitude rocky-mountain lakes, if food and water are plenty, the axolotl does not metamorphose, attains the adult size, develops gonads and starts breeding like the metamorphosed mature terrestrial adult. If exposed to air by drying up of water, the axolotl

fails to respire with gills and metamorphoses into an adult salamander by losing gills, gill slits and tail fin and by developing lungs.

2. The urodeles with persistent external gills, such as the mud puppy *Necturus maculosus* (Fig. 10.11) are in fact permanent neotenic larvae that have failed to metamorphose, have developed gonads and breed. They do not metamorphose under any condition.

3. Neoteny is also occur in the larvacean urochordates (protochordates) such as *Oikopleura*. They develop gonads and breed without undergoing metamorphosis and while retaining larval features, namely tail, notochord and nerve cord.

Causes of Neoteny

The iodine-containing thyroid hormone **thyroxine** is necessary for the amphibian metamorphosis. Certain waters inhabited by axolotls lack iodine, and cold weather at high altitudes inhibits the secretion of thyroxine. This is indicated by the fact that the administration of iodine, thyroxine or beef thyroid to a young axolotl promptly induces metamorphosis. Thus, the failure of the axolotl to metamorphose in its normal habitat is due to lack of iodine or thyroxine.

In some species, such as *Ambystoma maxicanum*, neoteny seems to be hereditary. Maxican axolotl does not undergo metamorphosis in nature, because its pituitary gland does not release the thyrotropin (thyroid-stimulating hormone) that would activate T_4 (= thyroxine or tetraiodothyronine) synthesis. The axolotl does synthesize functional thyroid hormone receptors, however, and when investigators administered either thyroid hormones or thyrotropin, they found that the salamander metamorphosed into an adult form not seen in nature (**Gilbert**, 2010).

Necturus fails to metamorphose even on the administration of thyroxine. The reason for this seems to be the inability of its tissues to respond to the thyroxine rather than to the lack of this hormone. Thyroxine is produced in *Necturus*, as its thyroid gland, when transplanted to a frog's tadpole, hastens its metamorphosis. *Necturus* was recently found to have functional thyroid hormone receptors. It appears that these receptors do not bind to those genes that initiate and promote metamorphosis (**Safi** *et al.*, 2006).

Types of Neoteny

Kollmann has recognized following two types of neoteny:

1. **Partial or temporary neoteny.** It refers to delay in metamorphosis beyond the normal period. Example. Wintering of tadpoles of *Hyla arborea*, *Rana esculenta* and *Bufo vulgaris*.
2. **Total or permanent neoteny.** It refers to lack of metamorphosis so that the larva retains gills, grow large and becomes sexually mature. It occurs in *Ambystoma maxicanum*, *Proteus*, *Siren* and *Necturus*.

Evolutionary status of neoteny. Earlier the aquatic, gill-breathing tailed amphibians were considered primitive forms. Therefore, neoteny was regarded a case of **atavism**, that is reappearance of the ancertral traits. Neoteny has also been thought as a case of retarded evolution. Now, however, the retention of gills is considered a secondary adaptive specification.

De Beer (1940) and **Gould** (1977) have speculated that neoteny is a major factor in the evolution of more complex taxa. By retarding the development of somatic tissues, neoteny may give natural selection a flexible substrate. According to **Gould** (1977), neoteny may provide an escape from specialization. Animals can discard their highly specialized adult forms, return to the lability of youth and prepare themselves for new evolutionary direction.

2. Progenesis

In **progenesis**, gonad maturation is accelerated with the rest of body develops normally to a certain stage. Progenesis has enabled some salamander species to find new ecological niches. *Bolitoglossa occidentalis* is a tropical salamander that, unlike other members of its genus, live in trees. This salamander's webbed feet and small body size suite it for arboreal existence, the webbed feet producing suction for climbing and the small body making such traction efficient. **Alberch** and **Alberch** (1981) showed that *B. subpalmata* and *B. rostrata* (whose young are small, with digits that have not yet grown past their webbing). *B. occidentalis* reaches sexual maturity at a much smaller size than its relatives, and this appears to have given it a phenotype that made tree-dwelling possible (**Gilbert**, 2010).

3. Direct Development

While some animals have extended their larval life stages, others have "accelerated" their development by abandoning their larval form for direct development. Thus there are frog species that lack tadpoles and sea urchins that have no pluteus larvae.

Elinson (1983, 87) have studied *Eleutherodactylus coqui*, a small frog that is one of the most abundant vertebrates on the island of Puerto Rico. Unlike the eggs of *Rana* and *Xenopus*, the eggs of *E. coqui* are fertilized while they are still in the female's body. Each egg is large (about 3.5 mm in diameter), roughly 20 times the volume of a *Xenopus* egg. After the eggs are laid, the male gently sits on the developing embryos, protecting them from predators and desiccation (**Taigen** *et al.*, 1984).

Early development of *E. coqui* is similar to that of most frogs. Cleavage is holoblastic, gastrulation is initiated at a sub-quatorial position and the neural folds become elevated from the surface. However, shortly after the neural tube closes, **limb buds** appear on the surface. This early emergence of limb buds is the first indication that this animal will not pass through the usual limbless tadpole stage. Moreover the development of *E. coqui* is modified such that the modeling of most of its features—including its limbs—does not depend on thyroid hormones appear to be critical for the eventual resorption of the tail (which is used as a respiratory rather than a locomoter organ) and of the primitive kidney. It appears that the thyroid-dependent phase has been pushed back into embryonic growth (**Callery** *et al.*, 2001) What emerges from the egg jelly 3 weeks after fertilization is not a tadpole but a tiny frog.

Direct developing frogs do not need ponds for their larval stages and can therefore colonize habitats that are inaccessible to other frogs. Direct development also occurs in other phyla, in which it is also correlated with a large egg. It seems that if nutrition can be provided in the eggs, the life cycle need not have a food-gathering larval stage (**Gilbert**, 2010).

QUESTIONS

Long Answer Questions

1. Discuss the origin of Amphibia.
2. Why have amphibians been considered imperfect and transitional animals.
3. Describe parental care in amphibians. (*Rohilkhand, 2000; Purvanchal, 2001; Garhwal, 2002*)
4. Write an essay on "Parental care" in Amphibia.
5. What do you mean by parental care? Write an essay on parental care. (*CCSU Meerut, 2005, 2007*)
6. Give an account of neoteny in amphibians. (*Purvanchal, 2003; CCSU Meerut, 2003*)

Short Answer Questions

1. What is heterochrony? Explain.
2. Write short note on the following:
   (*a*) Neoteny (*CCSU Meerut, 2013*)
   (*b*) Progenesis

Very Short Answer Questions

1. Name any fossil amphibian.
2. Which frog carries eggs and tadpoles in its vocal cord?
3. What is the relationship in thyroxine and neoteny?
4. What is responsible for neoteny?
5. Which frog has direct development?

Multiple Choice Questions

1. Neoteny is exhibited by
   (*a*) Axolotl (*b*) *Necturus*
   (*c*) *Proteus* (*d*) All of these
2. The most striking case of parental care is seen in
   (*a*) *Alytes* (*b*) *Rhinoderma*
   (*c*) *Pipa* (*d*) *Icthyophis*

ANSWERS

Very Short Answer Questions

1. *Eryops*.
2. *Rhinoderma darwini*.
3. The neoteny which occurs naturally in salamanders can be terminated by the administration of thyroxine hormone as it contains iodine.
4. Environmental factors such as shortage of iodine in food and water or low temperatures, may be responsible for some kind of neoteny.
5. *Eleutherodactylus coqui*.

Multiple Choice Questions

1. (*d*) 2. (*c*)

12

Class Reptilia: General Characters, Classification and Types

The class Reptilia (G., *rept* = to creep) includes reptiles which are scaly creeping creatures. Reptiles are unique among verterbrates in having a coat of horny epidermal scales and in creeping on their belly on land. The study of reptiles is called **herpetology** (G., *herpetos* = reptiles + *logos* = study). A zoologist who is expert on reptiles is known as **herpetologist**.

Reptiles can be defined as scaly tetrapodous pentadactyle animals that lay large yolky, shelled amniotic eggs (the cleidoic eggs). The embryo contains amnion and allantois, the embryonic membranes.

Reptiles are the first vertebrates which are well adapted for life in dry places on land. Some live near or in water but they too return to land for breeding. Most reptiles do not care for their eggs after laying them. Reptiles are the **amniotic** vertebrates, *i.e.*, their embryos form amnion. Amnion is a membrane enclosing and protecting the embryo and its surrounding fluid, the amnionic fluid. This fluid serves as a shock absorber for the developing embryo while preventing it from drying out. Reptiles have shelled eggs, in which horny or calcareous egg shell is relatively impermeable to water. Their skin is tough and impermeable to water and kidneys can excrete concentrated urine.

The class Reptilia includes the lizards, shakes, tortoises, turtles, crocodiles, alligators and tuatara. It has many fossil forms such as most famous dinosaurs.

12.1 GENERAL CHARACTERS

1. Reptiles are abundant in the tropical parts of the world such as India. Very few are found in the colder parts of the temperate region and none in the arctic and antarctic region.
2. Most reptiles are carnivorous or insectivorous. Tortoise are herbivorous.
3. Body form may be short and broad, long and narrow, depressed or cylindrical. Body is divisible into 4 regions: head, neck, trunk and tail. Tail fin is absent.
4. There are two pairs of pentadactyl limbs, each typically with five digits armed with horny, sharp, curved claws. They are meant for running, climbing, burrowing, swimming or soaring. Limbs are reduced in certain lizards and absent in few lizards and all snakes.
5. Skin is rough, cornified and dry (without glands). An exoskeleton of horny epidermal **scales**, bony **scutes** or bony **plates** is always present and it is periodically sloughed off in pieces or as a whole. This process is called **moulting**.
6. Endoskeleton is fully ossified (bony). Skull is with a single occipital condyle (monocondylic) and may have temporal vacuities or **fossae**. Vertebrae are generally procoelous. Sternum is greatly developed and carry external ribs.

7. Coelom consists of two unequal parts: a small anterior pericardial cavity that contains the heart and a spacious pleuroperitoneal cavity which have all the other viscera except the kidneys.
8. Mouth is large having teeth in both jaws. Teeth are homodont. Buccal cavity is roofed by an incomplete or complete secondary **palate**. Internal nares open into the anterior part of the buccal cavity. Tongue may be protrusible or nonprotrusible. Alimentary canal leads into cloaca. Distinct liver and pancreas are present.
9. Respiration takes places only by lungs (pulmonary). Gills are not developed at any stage. Turtles have cloacal respiration.
10. Heart is incompletely four-chambered, having two auricles and partly divided ventricle. There is a double circulation through the heart: pulmonary and systemic. Spleen is a distinct organ. Red blood cells oval, biconvex and nucleated.
11. There are only two meninges: 1. pia-arachnoid and 2. durameter. Cranial nerves are 12.
12. Eyes, ears and nose well developed. Jacobson's organs are well developed. Eyes generally have movable lids, Harderian and lacrimal glands. Lateral line system is absent.
13. Kidneys are metanephric. Nephrostomes absent. Glomeruli are small. Excretory waste is removed chiefly as *uric acid* to conserve water (uricotelism). The aquatic forms excrete urea (ureotelism). Urinary bladder may be present.
14. Sexes are separate. Gonads are paired. Gonoducts lead into a cloaca. Fertilization is internal. Males generally have single or double copulatory organs. Most forms are oviparous.
15. Eggs are few, large, yolk-rich and enclosed in leathery or limy **shell** to prevent mechanical injury and desiccation. They are laid and developed on dry land. Such an egg that can provide all substances necessary for the complete development of the embryo to a miniture adult within the shell is called **cleidoic egg**.
16. Cleavage is meroblastic. Development is direct. No larval stage. No metamorphosis. Embryo contains embryonic membranes such as amnion, chorion, allantois and yolk sac (Fig. 12.1). These serve for the protection, nourishment, respiration and excretion of the embryo. Amnion encloses the embryo and give it a watery environment, the amniotic fluid (Fig. 12.1). This enables the embryo to develop on land. Yolk sac contains food for the embryo. Allantois stores embryo's nitrogenous waste till hatching. The chorion surrounds the embryo, amnion, yolk sac and allantois, and controls the overall permeability of the egg. The egg is permeable to gases but not to water.

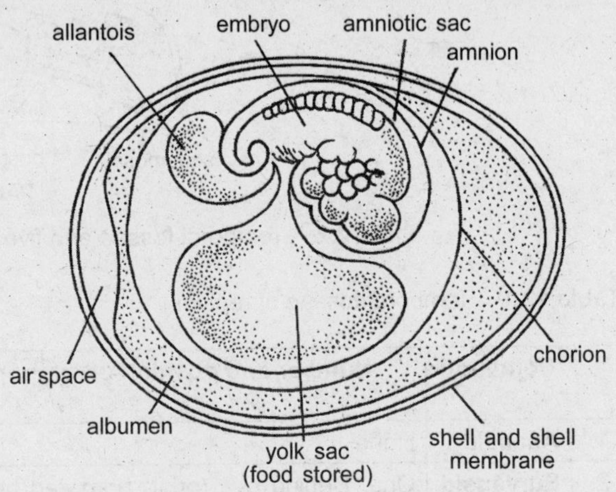

Fig. 12.1. An amniotic egg of reptiles, birds and some mammals showing four embryonic membranes.

17. Cold blooded animals. Their body temperature is variable according to environment (poikilothermy). Some reptiles raise their body temperature by exposure to sun (basking). Getting warmth from the environment is called **ectothermy**.
18. The reptiles originated from the salamander-like terrestrial amphibians sometimes in the Permian period. They reigned the world in the Mesozoic era. There are over 7000 extanct species and many extinct species.

12.2 BASIS OF CLASSIFICATION OF REPTILES

Classification of reptiles is based upon the presence or absence of **fossae** (openings or cavities) and **apses** (bony bars) between them in the temporal region of the skull (Fig. 12.2). The postero-lateral region of the skull is called **temporal** or cheek region. It is formed by post-orbital, squamosal, jugal, quadrato-jugal, parietal, post-frontal and supratemporal bones (Table 12.1). The temporal fossae enable the jaw muscles to bulge (*i.e.*, during muscle contraction) during chewing.

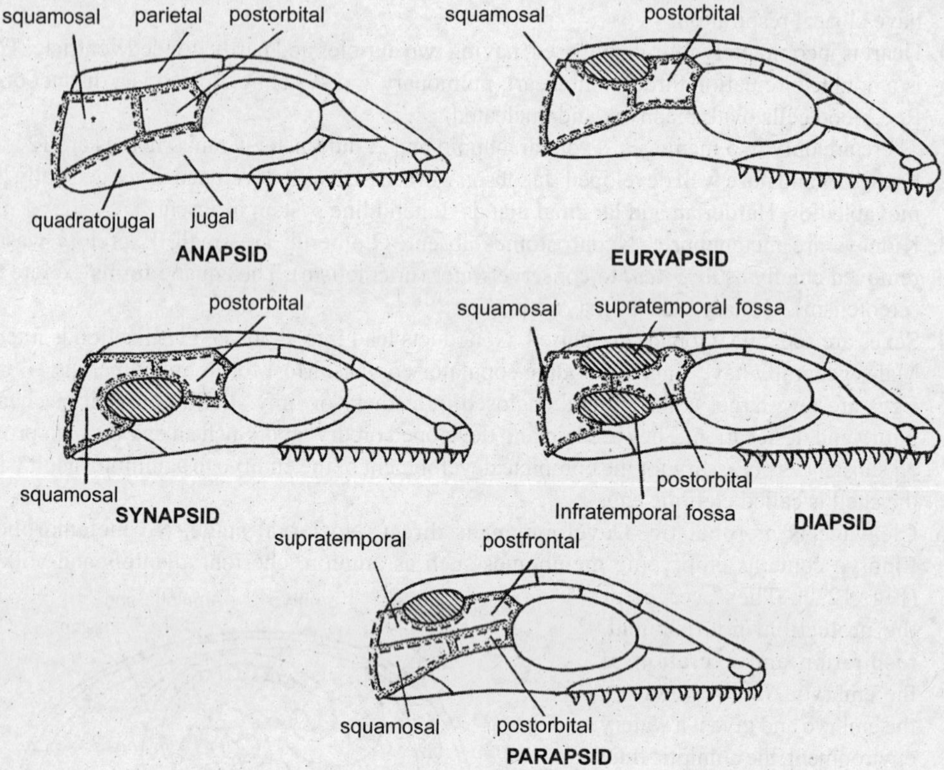

Fig. 12.2. Temporal fossae and five types of skulls in reptiles.

Table 12.1 Temporal fossae of reptiles

	Condition	Number and position of temporal fossae	Main groups showing the condition
1.	**Anapsid**	None	Cotylosauria and chelonia
2.	**Euryapsid**	One. Behind eye (orbit) bordered below the post orbital squamosal	Protorosauria and Sauropterygia
3.	**Parapsid**	One. High, behind the eye (orbit), postorbital and squamosal meet below the fossa. Fossa bordered below by post-frontal and supratemporal	Mesosauria and Ichthyopterygia
4.	**Diapsid**	Two. Postorbital and squamosal meet between the two fossae	Eosuchia, Rhynchocephalia, Squamata and Archosauria
5.	**Synapsid**	One. low behind the eye (orbit), squamosal and post orbital meet above the fossa	Pelycosauria and Therapsida

12.3 CLASSIFICATION OF REPTILIA

Five subclasses of class Reptilia are generally recognized regarding the number and location of fossae in the skull: 1. Anapsida; 2. Euryapsida; 3. Parapsida; 4. Diapsida and 5. Synapsida.

Subclass 1. Anapsida

1. Earliest primitive reptiles.
2. Skull has complete bony roof in the temporal region (behind the orbits).
3. Temporal fossae and apses are absent.

Subclass Anapsida is divided into two orders: 1. Cotylosauria (extinct); 2. Chelonia (living).

Order 1. Cotylosauria

1. Most primitive (oldest) reptiles resembling the early (labyrinthodont) amphibians.
2. They are called **stem reptiles** since from them have evolved all other reptiles, birds and mammals.

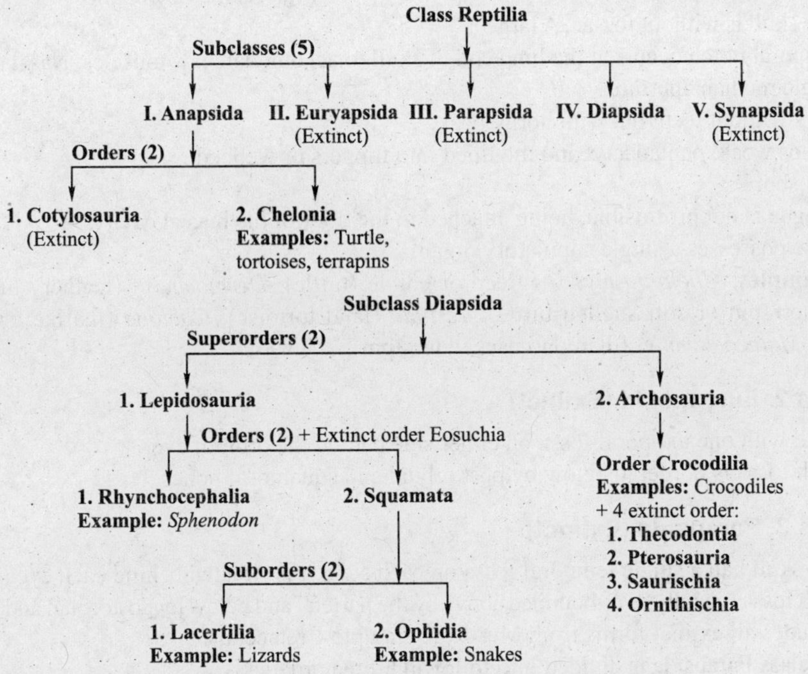

Fig. 12.3. Outline Classification of Class Reptilia.

3. The skull was heavily ossified and lacked fossae and apses.
4. The jaws and teeth were poorly developed.
5. The pelvic girdle was flat and plate-like.
6. The limbs were feeble and directed outward.
7. They appeared in the Carboniferous period and became extinct in the Triassic period.
   Example: *Seymouria* (Fig. 12.4).

Order 2. Chelonia (Gr., Chelone = turtle)

1. This order is also called **Testudinata** (Gr., *testudo* = turtle). It includes extinct as well as living forms. The living forms represent a very old group which has persisted with little change for some 175 million years (*i.e.*, form the Permian to the recent).
2. Body is divisible into head, neck, trunk and tail.
3. Trunk is short and broad and enclosed by a firm shell consisting of two parts: a dorsal arched **carapace** and a flat ventral **plastron**, joined on the lateral sides by ligament or bones.

4. The carapace and plastron are composed of bony plates which are covered by epidermal plates or scutes of horny material or skin.
5. The naked extremities, *i.e.*, neck, limbs and tail are covered with epidermal **scales** and retractile into the carapace.
6. They contain only 10 trunk vertebrae. Eight of these vertebrae and their ribs are fused with the carapace. Sternum absent. Pectoral and pelvic girdles lie within the ribs.

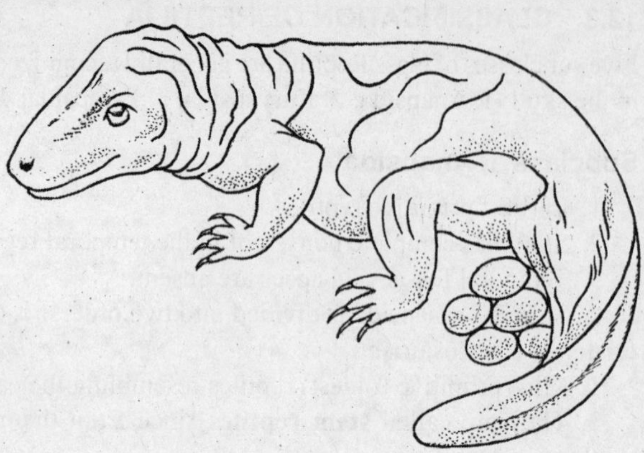

Fig. 12.4. *Seymouria.*

7. The skull is without fossae. A large temporal notch occurs at the hind end of skull to accomodate jaw muscles. Nasal opening is a single median aperture.
8. Jaws without teeth but with horny beak.
9. Limbs weak, pantadactyl and modified into flippers or webbed.
10. Tail short.
11. Tongue is not protrusible, being attached to the floor of the buccal cavity.
12. Male possesses a single copulatory organ.

Examples: *Chelonia mydas* (green or edible (turtle); *Dermochelys* (leathery turtle); *Emys Trionyx* (terrapin or soft shelled turtle); *Testudo* (land tortoise); *Chelone* (marine green turtle), *Kachuga*, *Malacoclemmys* (diamond backed terrapin).

Subclass 2. Euryapsida (Extinct)

1. Skull with one temporal fossa on either side behind the orbit of eye.
2. Each fossa is bordered below by post orbital and squamosal bones.

Subclass 3. Parapsida (Extinct)

1. The skull had a single temporal fossa on each side. It is situated behind each eye (orbit).
2. Each fossa (vacuity) was bounded above by the parietal and below by postorbital and squamosal.
3. Includes all extinct forms from Mesozoic era upto Cretaceous.

Subclass Parapsida is divided into following three orders:

Order 1. Protosauria. Terrestrial forms.

Example: *Protosaurus*.

Order 2. Ichthyosauria. They were marine forms.

Example: *Ichthyosaurus*.

Order 3. Plesiosauria

Example: *Plesiosaurus*.

Subclass 4. Diapsida

1. The skull has two lateral temporal fossae on each side. These are separated from each other by the postorbital and the squamosal.
2. The upper or superior temporal fossa is bordered above by the parietal.
3. The lower or inferior temporal fossa is bordered below by the jugal and quadratojugal.

The subclass Diapsida is divided into two superorders: 1. Lepidosauria (the scaly lizard) and 2. Archosauria (the ruling reptiles).

Superorder 1. Lepidosauria

1. Primitive reptiles with diapsid skull.
2. The pineal foramen and interparietal bone often present.
3. Palate movable articulated with the brain case.
4. Teeth occur on the palate as well as jaw margins and are usually not set in sockets.
5. Lepidosauria arose in the Triassic and form the largest group of living reptiles.
   The superorder Lepidosauria is divided into three orders: 1. Eosuchia (Thecodontia); 2. Rhynchocephalia and 3. Squamata.

Order 1. Eosuchia

1. These most primitive diapsids existed in upper Permian of South Africa.
2. First reptiles to have two temporal fossae on each side.
3. Teeth were thecodont.
4. Birds evolved from some small Eosuchia such as *Euperkaria* which had a bipedal gait.
   Example: *Youngia* and *Euperkaria*.

Order 2. Rhynchocephalia

1. Includes forms from middle Triassic to recent, mostly extinct, only single species now survives.
2. Body is elongated, depressed and lizard-like. Tail is well developed and laterally compressed. Body is covered with horny epidermal scales.
3. Quadrate immovable. Mandibles joined by ligaments.
4. Vertebrae amphicoelous and ribs single-headed.
5. Teeth are acrodont (*i.e.*, they are fused by their bases to the summit of the jaw bones).
6. Cloacal opening transverse.
7. Urinary bladder is present.
8. A median **parietal** or **third eye** occurs in the middle of the forehead. This eye has lens, retina and a nerve.
9. Limbs are relatively weak.
10. Hemipenis absent.
    Example: *Sphenodon* (tuatara; Fig. 12.14) which is restricted to coastal islands of New Zealand.

Order 3. Squamata

1. It includes almost all living forms commonly known as lizards and snakes (Table 12.2).

Table 12.2 Difference between lizards and snakes.

	Lizards		Snakes
1.	Body is mostly depressed (*i.e.*, dorso-ventrally flattened).	1.	Body is nearly cylindrical.
2.	Body is well differentiated into head, neck, trunk and tail.	2.	Neck is generally not distinct and the trunk insensibly passes into the tail.
3.	The entire body including head is covered with uniform scales. Scales are small and overlapping.	3.	Scales on the body exhibit regional differentiation. These occur as **shields** on the head, and as **transverse plates** on the belly. There may be **dorsal scales** in the mid-dorsal line on the trunk (vertebrals) and **ventrals** on the ventral side.
4.	Most lizards have two pairs of pentadactyl limbs	4.	All snakes lack limbs.
5.	Pectorals and pelvic girdles are present even if the limbs are absent.	5.	Pectorals and pelvic girdles are lacking. Pythons, however, have rudiments of pelvic girdle.
6.	Mouth is a wide transverse terminal aperture.	6.	Mouth is a still wider opening and can be greatly widened.

	Lizards		Snakes
7.	Eyes have movable eyelids, lacrimal glands and sclerotic ossicles.	7.	Eyes are without eyelids, lacrimal glands and sclerotic ossicles. Eyes are covered with a covering of transparent scales like spectacles.
8.	Tympanum, tympanic cavity and in some cases ear openings also are present.	8.	Tympanum, tympanic cavity and ear openings missing.
9.	Internal ear ossicles receives sound vibrations through the air.	9.	Internal ear receives ground vibrations.
10.	Eustachian tubes are present.	10.	Eustachian tubes are absent.
11.	Palate bones are fixed.	11.	Palate bones are movable.
12.	Only one temporal fossa (upper or superior) is present.	12.	Both temporal fossae are secondrily lost.
13.	Tongue is non-protrusible and nonsensory.	13.	Tongue is protrusible, sensory and bifid.
14.	Teeth are homodont, conical and are borne on premaxillae, dentary and palatines.	14.	Teeth are recurved and present on maxillae, palatine, ptreygoid and dentary. In poisonous snakes, maxillary teeth are modified into fangs which have open or closed venom canal.
15.	Sternum and episternum are present. Sternum is in the form of rhomboidal plate of calcified cartilage.	15.	Sternum and episternum are absent.
16.	Skull has jugal and epipterygoid bones.	16.	Skull lacks jugal and epipterygoid bones.
17.	Vertebrae are procoelous and lacks zygosphenes and zygaetra.	17.	Vertebrae are strongly procoelous and have zygosphenes and zygantra.
18.	Both lungs are equally developed and functional.	18.	Only right lung is functional, left one is reduced or absent.
19.	Urinary bladder is present.	19.	Urinary bladder is absent.
20.	Alimentary canal is coiled.	20.	Alimentary canal is nearly straight.
21.	In brain, cerebral hemispheres are short.	21.	In brain, cerebral hemispheres are elongated and extend even beyond orbits.
22.	Tail may be detached and regrown.	22.	Tail is not detachable.

2. Body is elongated, depressed or cylindrical.
3. Skin has horny epidermal scales. Some lizards have dermal bony plates under the scales.
4. Skull has secondarily lost one (lizard) or both the temporal fossae (snakes).
5. Quadrate movably articulates with squamosal.
6. Vertebrae are procoelous, ribs single headed.
7. Teeth are acrodont or pleurodont (*i.e.*, they are fixed by their sides to the lateral surface of the jaw).
8. Cloacal aperture is transverse.
9. Copulatory organs are paired and reversible. They are known as **hemipenes**.
10. Tail is fairly long.
11. Tongue is often protrusible.
12. They are most recent and successful of all reptiles.

The order squamata is divided into two suborders: 1. Lacertilia (Sauria) which includes the lizards; 2. Ophidia which includes snakes.

Suborder 1. Lacertilia

1. Elongated depressed body.
2. Tail is often readily detachable. New tail regenerates under the influence of ependymal cells from the adjacent spinal cord.
3. Body with two pairs of typical **pentadactyl limbs**, the digits ending into sharp incurved claws. Limbs are used for running, climbing or burrowing. Some lack one or both the pairs of limbs. Girdles are always present, though reduced in limbless forms.
4. Eyes generally have movable eyelids and nictitating membrane.
5. Tympanum and tympanic cavity well developed.
6. Sternum and episternum generally present.
7. Skull with a pair of superior temporal fossae.
8. Maxillae, palatine and pterygoids are fixed and hence mouth is capable of opening to only a moderate extent.
9. Tongue is usually well developed, protrusible and entire.
10. Urinary bladder is present.
11. The lizards love warmth and sunshine. **Metachrosis** (ability to change body colour) is common.
12. All are carnivorous and oviparous.

Example: *Hemidactylus* (wall lizard or gecko; Fig. 12.17), *Calotes* (garden lizard), *Draco* (flying lizard; Fig. 12.16), *Varanus* (monitor or Goya), *Phyrnosoma* (horned-toed), *Chameleon* (girgit; Fig. 12.15), *Uromastix* (spiny tailed lizard), *Heloderma* (Gila monster; Fig. 12.18), *Mabuia* (skink), *Rhineura* (limbless lizard), *Ophisaurus gracilis* (Indian glass snake).

Suborder 2. Ophidia

1. Snakes abound in tropical and subtropial countries, except New Zealand. They occur in all habitats: land, water (sea, fresh water, brackish), trees, etc. They are nocturnal and carnivorous. Prey is swallowed entire.
2. The snakes have a long, slender, cylindrical body.
3. Skin bears rows of scales that overlap like shingles on a roof. On ventral surface of all snakes, there is a single row of large transverse scales or scutes from chin to cloaca.
   They cast off the horny layer of the skin as a continuous slough many times a year.
4. Snakes lack limbs as well as girdles. Pythons and boas, however, possess vestigial hindlimbs and pelvic girdles.
5. These are no eyelids. Eye is covered by a transparent scales forming **spectacles**. Such eyes enable the snake to give a steady, unwinking stare.
6. Ear opening and tympanum are lacking.
7. Sternum and episternum are absent. Vertebrae are numerous (may be over 400) and procoelous. They have additional articular processes. These include paired wedge-shaped processes, the **zygosphenes**, at the anterior ends of the neural arches and corresponding depression, the **zygantra**, at the posterior ends of the neural arches. Articulation of these processes allows a greater lateral flexion of the body during locomotion.
8. Skull without supratemporal ossification and hence temporal fossae and temporal arcades are absent.
9. Maxillae, palatines and pterygoid are freely movable and the rami of mandibles are connected together by elastic fibers and hence the mouth is capable of forming a wide gape.
10. Teeth are acrodont and occur at the jaw margins as well as on the palate. Two fangs occur on maxillae of poisonous snakes. These are grooved or tubular to inject poison into the victim.
11. Tongue is slender and forked. It can be protruded through a notch in the lower jaw without opening the mouth. Tongue serves as an additional organ of smell, bringing odours from food or mate to the Jacobson organs (the olfactory organs).
12. Sexes are separate. Fertilization is internal. Snakes are oviparous. Some snakes such as vipers and sea snakes, are viviparous.

Example: *Naja* (cobra), *Viper russeli* (Russell's viper), *Bungarus* (krait), *Eryx johnii* (sand boa or double-mouthed snake); *Python* (Ajgar), *Typhlops* (blind snake), *Crotalus horidus* (rattle snake), *Ptyas* (rat snake or dhaman), *Echis carinata* (saw-scaled viper), *Enhydrina* (sea snake) and *Dendrophis*, etc.

Superorder 2. Archosauria

1. They are most advanced reptiles.
2. They have diapsid skull, but without interparietal bone and parietal foramen.
3. Mandibles with a vacuity on the outer side.
4. Antorbital fossa present in front of orbit.
5. Teeth either **thecodont** or absent.
6. Hindlimbs long and more powerful and in some cases adapted for **bipedal locomotion**.
7. Vertebrae amphicoelous or procoelous.
8. The Archosauria were the dominant land animals in the later Mesozoic era.
9. They include the living crocodilians and extinct reptiles. Mesozoic (extinct) reptiles include **dinosaurs** (which reigned the land), **plesiosaurs** and **ichthyosaurs** (which dominated the sea) and **pterosaurs** (which ruled the air).

   The superorder Archosauria is divided into five orders: 1. Pseudosuchia (Thecodontia); 2. Crocodilia (Loricata); 3. Pterosauria (Pterodactyla); 4. Saurischia and 5. Ornithischia. All, except crocodilia, are extinct.

Order 1. Pseudosuchia (Extinct)

1. These were small carnivorous reptiles of the Triassic period.
2. Most were terrestrial but some were aquatic.
3. Long, narrow skull with sharp thecodont teeth along the jaw edges.
4. Hindlimbs were larger than the forelimbs.

   Example: *Aetosaurus* and *Phytosaurus*.

Order 2. Crocodilia

1. Includes amphibious crocodiles, alligators and gavials. They **inhabit** rivers, lakes and land. They are found in tropical parts of new and old worlds.
2. Lizard-like. Body heavily built with large and long head.
3. Skin is thick and leathery. It is covered by horny epidermal **scales**, below which are sculptured bony plates or **scutes** on the back and sometimes on the throat also.
4. Mouth gape is wide with sharp (bluntly conical) incurved teeth.
5. Eyes, external nares and ears lie in a straight line on the top of the head. This enables the animal to use its major sense organs when only a small part of the body is exposed above water.
6. Hindlimbs are larger than the forelimbs. Hand has five fingers, while foot has four toes.
7. Cloacal aperture is a longitudinal slit.
8. Tail is large, heavy and compressed. It is used for offence and defence, besides in locomotion.
9. Tongue is not protrusible.
10. Skull is diapsid. Upper temporal fenestra typically small or exceptionally closed.
11. An extensive, bony, secondary palate is formed due to union of maxillae, palatines and pterygoids in the median line. The palate separates the nasal passages above from the buccopharygeal cavity below and displaces the **internal nares** far back in the pharynx.
12. Heart is completely four chambered, but the dorsal aorta receives the mixed blood. Lungs spongy and invested with pleural membranes.
13. Quadrate is fixed.
14. Vertebrae either amphicoelous or procoelous. Ribs with two heads: capitular and tubercular.
15. Sternum and abdominal ribs present.
16. Brain with cerebral cortex and well developed cerebellum.

17. Eye has a pecten, Lacrimal and Harderian gland.
18. Tympanic membranes protected by two scaly flaps.
19. Males with a single copulatory organ.
20. They are carnivorous, feeding chiefly on fish. Fierce predators of freshwater.
21. They are oviparous and lay 30–60 leathery eggs in nests of decaying vegetation.

Example: *Crocodylus*, *Alligator*, *Gavialis*, etc.

Order 3. Pterosauria

1. The Pterosauria are the extinct flying reptiles that lived from the Jurassic to the Cretaceous period.
2. The flight membranes, the patagia, were supported by forelimbs and extended to hindlimbs and tail.
3. The bones were pneumatic and light as in birds.
4. Jaws have teeth. Neck was long and supported the skull placed on it at a right angle.

Example: *Rhamphorhynchus* (Fig. 12.5) and *Pteranodon* (Fig. 12.6).

Fig. 12.5. *Rhamphorhynchus* (the tailed flying reptile).

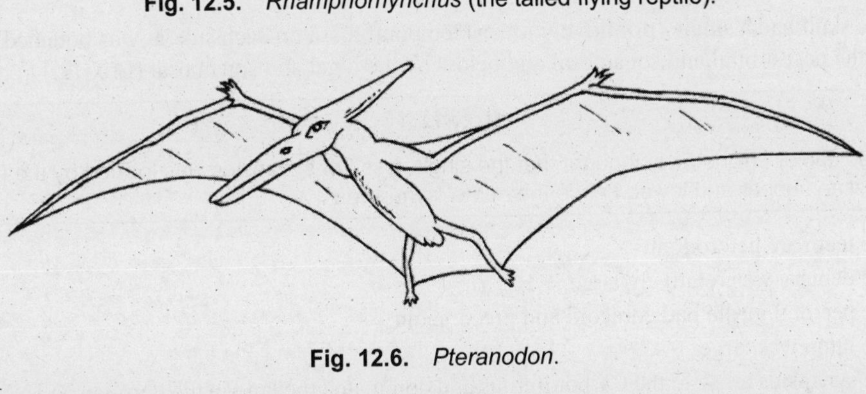

Fig. 12.6. *Pteranodon.*

Order 4. Saurischia

1. They are extinct large-sized reptiles called **dinosaurs** (Gr., *deinos* = terrible + *sauros* = lizard) a term coined by **Sir Richard Owen**.

2. Saurischia appeared in the Triassic, became numerous and disappeared in the Cretaceous.
3. The pectoral girdle was reduced to a scapula with a small coracoid. Clavicle was lacking.
4. Pelvic girdle was triradiate as in lizards (Fig. 12.7).

The order saurischia is divided into two suborders : 1. Theropopoda and 2. Sauropoda.

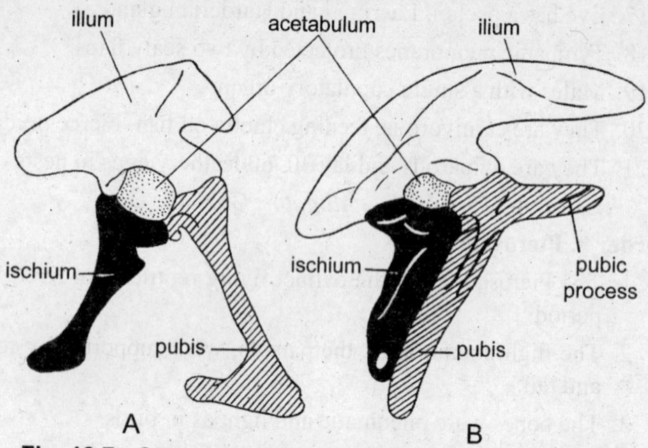

Fig. 12.7. Structure of pelvis in two orders of dinosaurs. A—Saurischia; B—Ornithischia.

Suborder (i). Theropoda

These reptiles had large skull, reduced forelimbs, large hindlimbs, bipedal gait and carnivorous habit.

Example: *Tyrannosaurus rex* (Fig. 12.8).

Suborder (ii). Sauropoda

They had small skull, long neck, long tail, quadropedal gait and herbivorous habit.

Example: *Apatosaurus* (= *Brontosaurus*, Fig. 12.8), *Diplodocus* (Fig. 12.8).

Order 5. Ornithischia

1. They are large extinct reptiles which also are called dinosaurs.
2. They arose in the Triassic, became abundant and vanished in the Cretaceous.
3. Pelvic girdle was tetraradiate (Fig. 12.7). The pubis pointed backward and the prepubic forward.
4. Forelimbs were shorter than hindlimbs.
5. Teeth were confined to the posterior region of the jaws, anterior part forming the beak.
6. The predentary bone in the lower jaw possibly supported the horny beak.
7. Ornithischia were herbivores.
8. Some of them gave rise to birds.

Example: *Triceratops* (Fig. 12.9A), *Stegosaurus* (Fig. 12.9B) and *Iguanodon* (Fig. 12.9C).

Subclass 5. Synapsida (Extinct)

1. The skull had a single (primitively lower) temporal fossa on each side. It was bounded above by the post orbital and squamosal and below by the jugal and squamosal (Box 12.1).

Box 12.1

It was earlier erroneously thought that the single fossa of synapsids was formed by the fusion of two fossae : upper and lower, hence the name "synapsida".

2. The teeth are heterodont.
3. The dentary was relatively large.
4. The pectoral girdle had coracoid and precoracoid.
5. The brain was large.
6. The synapsids arose in the Carboniferous and dominated the land in the Permian and a greater part of the Triassic.
7. They show stages through which mammals evolved. They are all extinct.

The subclass Synapsida is divided into two orders: 1. Polycosauria and 2. Therapsida.

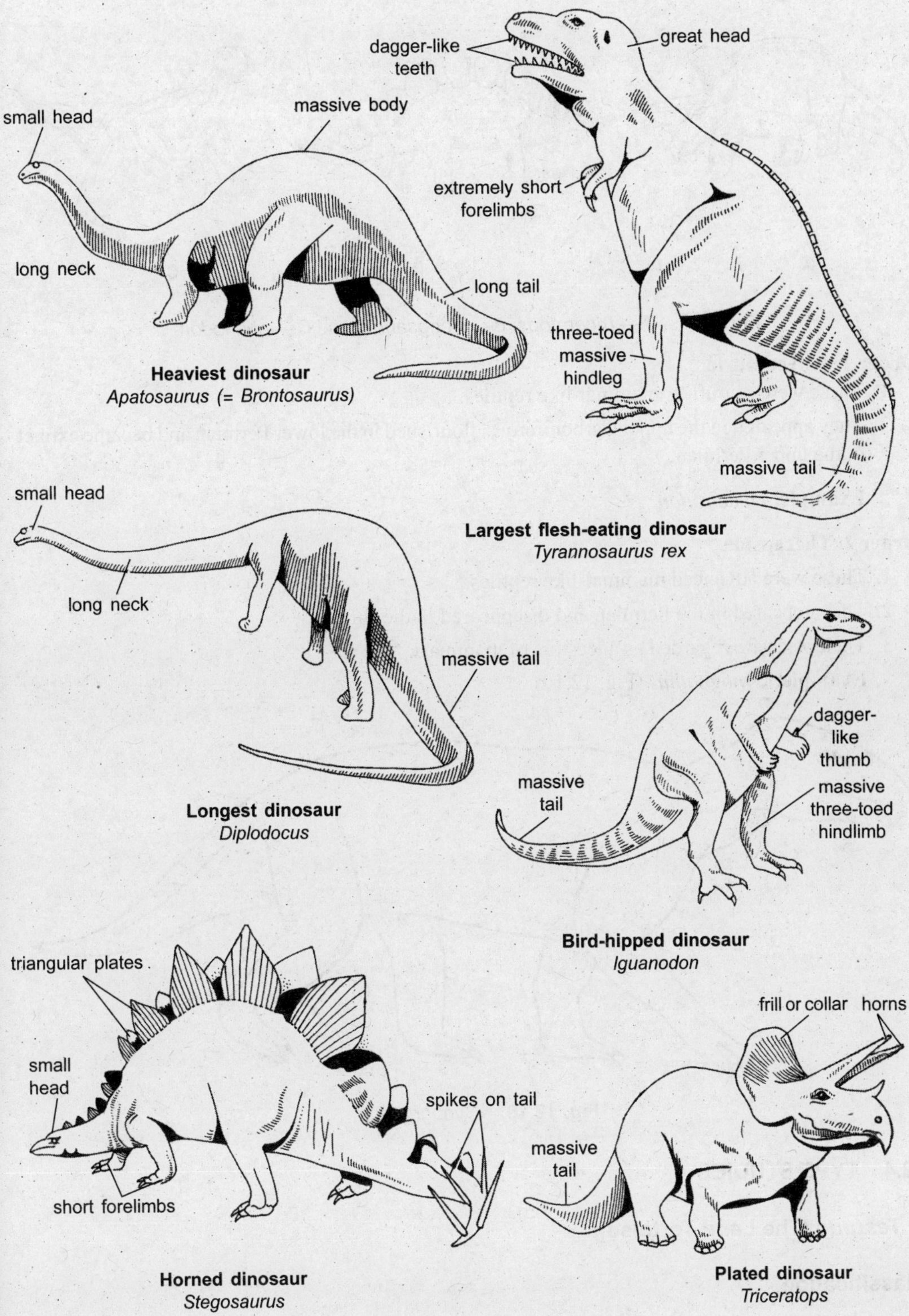

Fig. 12.8. Some types of dinosaurs.

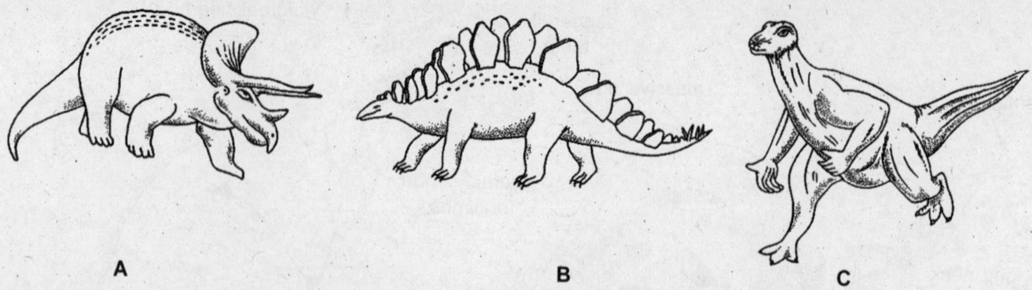

Fig. 12.9. A—*Triceratops*; B—*Stegosaurus* and C—*Iguanodon*.

Order 1. Polycosauria

1. These were primitive mammal-like reptiles.
2. They appeared in the upper Carboniferous, flourished in the lower Permian and became extinct in the upper Permian.

   Example: *Dimetrodon*.

Order 2. Therapsida

1. These were advanced mammal-like reptiles.
2. They appeared in the Permian and disappeared in the Jurassic.
3. Therapsids are regarded as ancestors of mammals.

   Example: *Cynognathus* (Fig. 12.10).

Fig. 12.10. *Cynognathus*.

12.4 TYPE STUDIES

1. *Testudo* (The Land Tortoise)

Classification

Class	—	**Reptilia**
Subclass	—	**Anapsida**

Superorder — **Lepidosauria**
Order — **Chelonia**
Genus — *Testudo*

Habits and Habitat

Testudo is widely distributed in the tropical and temperate regions of Africa, Europe and Asia.

Testudo is a terrestrial and diurnal animal. It is a common freshwater tortoise, some species of which are equally at home on dry land and enter water only to drink and bathe. It is herbivorous, though at times it also eats worms and insects. It hibernates in winter in burrows. It moves slowly. It is oviparous.

Structure

Testudo has an ovoid shell covered by well developed horny **scutes**. The carapace is highly domed. Eyes are small and have rounded pupil. Neck is long and completely retracted into the shell. Limbs are normal, each with five short, free digits terminating in horny strong claws. Tail is short. The breathing movements are produced by the protrusion and retraction of head and limbs and the pumping action of hyoid apparatus.

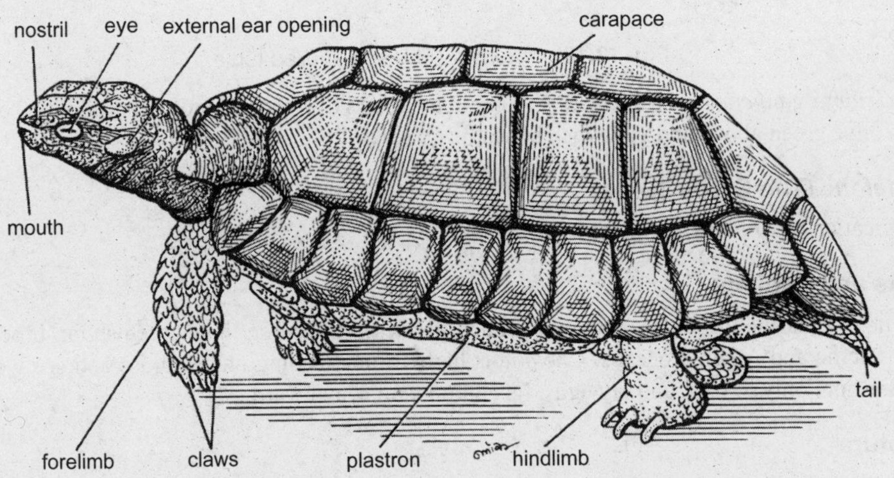

Fig. 12.11. *Testudo*—The land tortoise.

Testudo elegans is found in Southern India and Shri Lanka. It is about 30 cm long. Body of *T. elegans* is encased in a bony sheet (shell) in which head, neck, tail and limbs can be retracted at the approach of danger. The shell consists of a dorsal **carapace** and ventral **plastron** connected together by bony bridges or ligaments and is protected externally by horny epidermal scutes. The dorsal scutes are humped and bear beautiful yellowish star-like markings on a blackground. It is, therefore, known as the **starred tortoise**.

2. *Trionyx* (The Soft-shelled Turtle or Soft-shelled Terrapin)

Classification of *Trionyx* is similar to *Testudo*.

Habits and Habitat

Trionyx is found in the ponds and rivers in Asia, Africa and North America. It is carnivorous, feeds mainly on fish, frogs and molluscs.

Structure

The shell of *Trionyx* is almost circular in form and is covered with soft skin instead of horny scutes. Carapace is very low being nearly flat (Fig. 12.12). Head is large, triangular and produced into a soft proboscis. Jaws are devoid of teeth but provided with horny sheaths. Nostrils are located at the tip of proboscis. Eyes have rounded pupil. Neck forms a vertical S curve when retracted inside the shell. Flipper-like limbs are short with webed digits of which only the inner three digits bear claws.

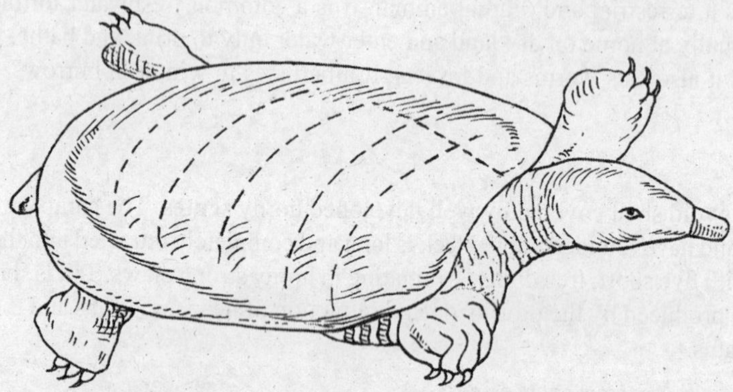

Fig. 12.12. *Trionyx*—The soft shelled turtle.

Trionyx gangeticus (Fig. 12.12) is found in the rivers of Northern India. It is about 60 cm long and is olive green above and yellowish beneath.

3. *Chelone* (The Green Turtle)

Classification of *Chelone* is similar to *Testudo*.

Habits and Habitats

Chelone is a marine turtle. It is popularly called **green turtle** owing to its greenish fat. It occurs in India, Pacific and Atlantic oceans. *Chelone* is herbivorous feeding on sea grass and see weeds. It comes ashore only to bask and lay eggs. Its gait on land is awkward.

Structure

The shell of *Chelone* is covered by large horny scutes. Capapace is low and heart shaped. It is joined to the plastron by ligaments only. Shell and skull are porous to reduce body weight. Head is short and bears eyes having eyelids and nictitating membrane. Head and neck are not fully retractile. Jaws have denticulate edges. Limbs are modified into flippers. Only first digit of forelimbs is clawed, while hindlimbs are clawed. Tail is very short.

Chelone mydas (Fig. 12.13) is found in the Bay of Bengal. It grows to 140 cm and sadly, is in great demand for its delicious meat used in turtle soup.

Box 12.2 Some Indian Turtles

1. The **leatherback turtle** (*Dermochelys coriacea*) is the largest growing to 190 cm. It is easily distinguished from other turtles by its hooked upper jaw and overlapping scutes on the back.
2. **Green turtle** (*Chelonia/Chelone mydas*). (For it account see text).
3. The hawkbill turtle (*Eratmochelys imbricata*) grows to 90 cm. It is easily distinguished from other turtles by its hooked upper jaw and overlapping scutes on the back.
4. The olive Ridley turtle (*Lepidochelys olivacea*) grows upto 70 cm and nests in enormous numbers in winter at night. Someplaces in Orissa such as Gahirmatha, are well known nesting sites where up to 200,000 turtles visit the beach in one season.

All turtles are highly endangered species (**Chhapgar**, 2008).

Fig. 12.13. *Chelone*—The green turtle.

4. *Sphenodon* (The Tuatara)

Classification

Class	—	**Reptilia**
Subclass	—	**Diapsida**
Superorder	—	**Lepidosauria**
Order	—	**Rhynchocephalia**
Genus	—	*Sphenodon*
Species	—	*punctatum*

Habits and Habitat

Sphenodon (= *Hatteria*), commonly known as "tuatara" is the sole living representative of order Rhynchocephalia. It is confined to New Zealand and some surrounding Islands. *Sphenodon* burrows into the soft soil of coastal island during the day and emerges out during night (*i.e.*, it is nocturnal). It is known to share its burrow with a sea bird called petrel, *Hydrobates pelagicus*. It is carnivorous, feeding on insects, molluscs and small vertebrates. *Sphenodon* is said to emit a soft, frog-like croak. It lays about ten eggs with hard, white shell in a hole in the ground. Eggs take 13 months to hatch. *Sphenodon* is known to live for 28 years.

Structure

The 90 cm long body of *Sphenodon* is thick, massive and lizard-like. It carries a mid-dorsal row of low spines and is protected by scales. It retains such primitive characters as two pairs of temporal fossae, primitive palate, median pineal eye, widely roofed mouth, persistent abdominal ribs and absence of copulatory organs. Vertebrae are amphicoelous with a persistent notochord. Quadrate is fixed. Dentition is acrodont in the young and is lost in the adult. There is a horny beak on the upper jaw.

Sphenodon has been described as the **living fossil** because it exhibits structural similarities with the ancient reptiles and in spite of a long evolution of 170 million years has retained many primitive features. Now it is at the verge of extinction and is protected by law.

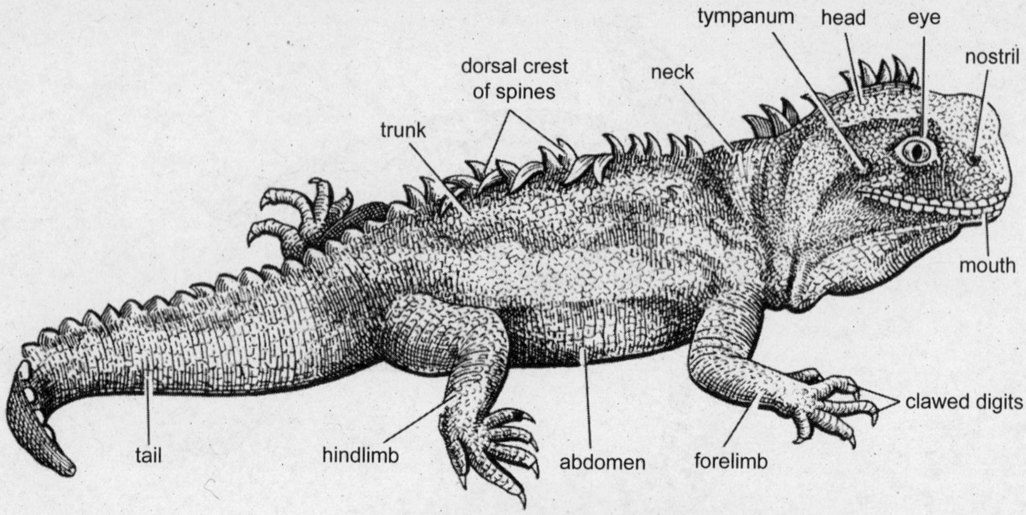

Fig. 12.14. *Sphenodon punctatus.*

5. *Chameleon*

Classification

Class	—	**Reptilia**
Subclass	—	**Diapsida**
Superclass	—	**Lepidosauria**
Order	—	**Squamata**
Suborder	—	**Lacertilia**
Genus	—	*Chameleon*

Habits and Habitat

Chameleon is found in Africa, South India, Sri Lanka and South Arabia. The Indian species *Chameleon calcaratus* is about 35 cm long and has much granulated skin.

Chameleon (Fig. 12.15) is a arboreal lizard. Its body is well adapted to live on the branches of trees and shrubs. It grips the twigs with the hands and feet as well as the tail. *Chamaeleon* moves very slowly to take up an appropriate position for offence. It is insectivorous and can catch an insect about 20 cm from it by protruding a long tongue very quickly. On seeing its enemy, *Chamaeleon* puffs up its body by filling air in the air sacs of the lungs and hisses loudly. *Chamaeleon* is known for changing colour rapidly according to its surroundings (metachrosis). It is viviparous.

Structure

Chamaeleon is about 40 cm long being greenish or blackish above and pale below. Its laterally compressed body has granular scales. The head is angular and bears a casque, hood or helmet, which projects backward over the neck. It is formed of processes of squamosal and occipital bones. Eyes are large, but the eyelids are fused together, leaving only a small hole in the center. Both eyes move independently and give the animal a very wide range of **monocular vision**. Once a prey is located, both eyes are focussed upon it and the prey is seen by **binocular vision** to judge the distance (**Harkness**, 1977). Tympanum is absent. Limbs are long and slender. Digits show **syndactyly** for grasping (*Syndactyly* is the condition characterised by union of two or more digits). In the hand three inner fingers are joined together throughout their length by a web of skin forming one group and the two outer fingers are likewise joined to form another group. In the foot, the two inner toes are united into one group and the three outer ones are united into another group. The two sets of digits in hands and feet are permanently opposed to one another and can be moved to grasp the twigs. The tail is long, cylindrical, tapering and prehensile. For getting extra grip tail coils around the branches and holds it firmly. When not in use, it remains coiled like a watch spring. There is a middorsal and a midventral row of pointed spine-like scales. Tongue is club-shaped, extremely protrusible and coated with sticky mucus.

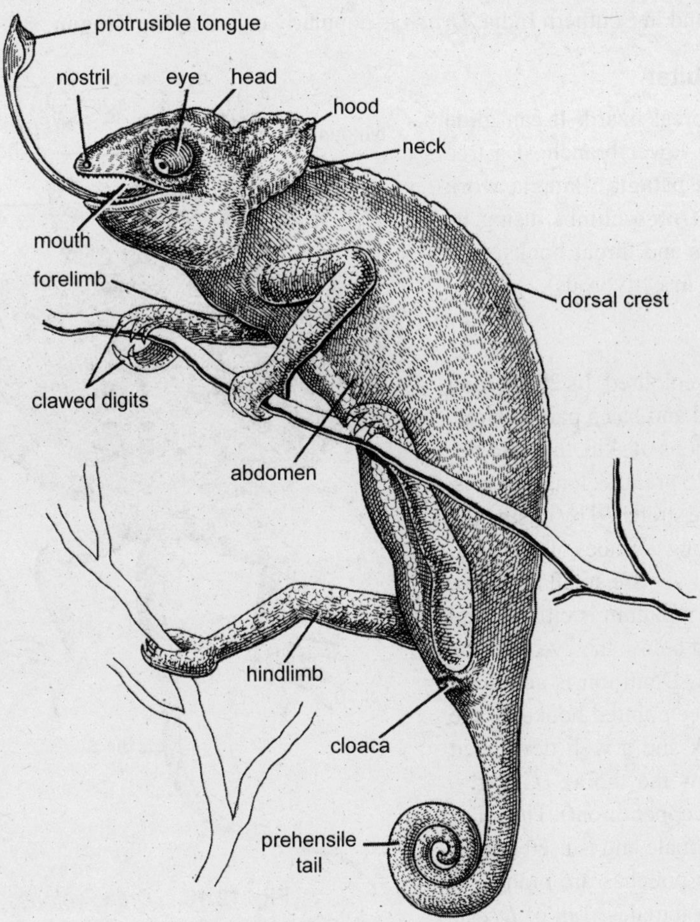

Fig. 12.15. *Chameleon.*

Teeth are **acrodont**. Vertebrae are procoelous. There is a proatlas vertebra between the skull and atlas. Several **air-sac like diverticulae** arise from the lungs and extend far back into the body cavity.

For changing its body colour *Chameleon* has **melanophores** in dermis. The colour change is under the control of autonomic nervous system.

6. *Draco* (Flying dragon)

Classification

Class	—	**Reptilia**
Subclass	—	**Diapsida**
Superorder	—	**Lepidosauria**
Order	—	**Squamata**
Suborder	—	**Lacertilia**
Genus	—	*Draco*
Species	—	*volans*

Draco volans is a flying lizard found in Assam (India), Myanmar and Malaysia. *Draco dussumieri* is found in Southern India. *Draco* is popularly called **flying dragon**.

Habits and Habitat

Draco is an arboreal lizard. It can glide from a higher to lower branch of a tree by stretching its patagia. Patagia work like parachutes. *Draco* climbs, using its pentadactyl limbs and throat hooks. It is carnivorous (*i.e.*, insectivorous).

Structure

Draco is brightly coloured. Its 25 cm long body is depressed and has a pair of large, wing-like expansions of skin, the **patagia**, supported by 5 to 6 greatly elongated ribs. The patagia can be stretched by raising the ribs and folded along the sides of the trunk, when not in use. Eyes are small and have rounded pupil. Tympanum is circular and a little behind and below the eye. Tongue is short and thick. Dentition is acrodont. *Draco* carries three pointed hooks on the underside of neck and a well developed gular pouch below the thorax (**Gular** = situated in gula or upper throat). The gular pouch is larger in male and is used during copulation. Gular pouches are orange in males and blue in females. Tail of *Draco* is slender and tapering. It is not brittle.

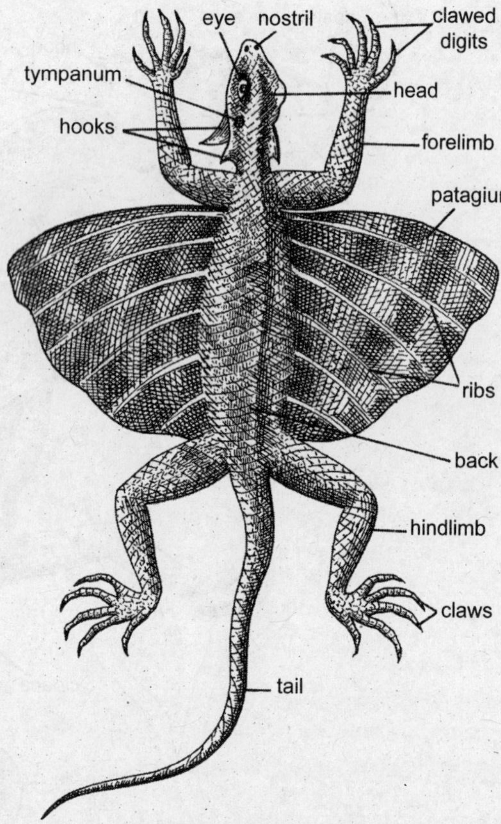

Fig. 12.16. *Draco volans*.

7. *Hemidactylus* (Common House Lizard)

Classification

Class	—	**Reptilia**
Subclass	—	**Diapsida**
Superorder	—	**Lepidosauria**

Order — **Squamata**
Suborder — **Lacertilia**
Genus — ***Hemidactylus***

Habits and Habitat

Hemidactylus flaviviridis is the common House lizard in the Northern India. It is nocturnal and insectivorous. It is often seen capturing insects with its sticky tongue at night near lights. It runs very fast. House lizard hibernates in winter in crevices and holes. It sheds its skin periodically in flakes that are eaten up. It can change the shade of its colour to match with the background. It is oviparous. It lays hard-shelled round eggs.

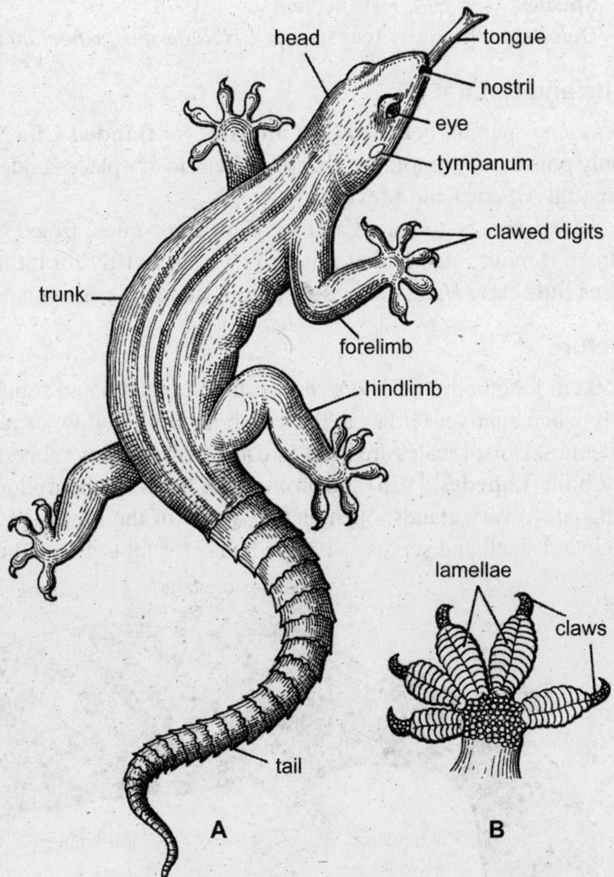

Fig. 12.17. *Hemidactylus.*
A—Dorsal view; B—Ventral view of foot.

Structure

Hemidactylus has a depressed pale-green coloured body, about 25 cm long. Scales of dorsal surface are reduced to minute granules. Head is broad and flat. Ear opening is vertical. Its tongue is short, sticky and slightly notched. Eyes have vertical pupil, lack movable lids and are permanently covered by a transparent membrane. The digits, fingers as well as toes, are dilated and bear on the ventral side adhesive pads for creeping over vertical surfaces and ceilings. The pads consist of two rows of transverse lamellae and provide adhesion by vacuum principle. They are first pressed over the surface to drive out air and then partially raised to create vacuum underneath. The lamellae are further beset with tiny hair-like outgrowths, which secure adhesion to even the slighest irregularity of surface. The claws of the digits arise from within tips of their dilated pads. Tail is thick at the base, but tapers posteriorly. Almost every cundal vertebra has a narrow transverse unossified zone, at which it breaks readily. The mechanism is used for defence. When persuaded by an enemy or caught by tail, the house lizard breaks off tail which engages the enemy by performing wriggling movements and enables the lizard to escape. The power of breaking a part of the body is called **autotomy**. A new tail regenerates, but lacks vertebrae. Male has femoral pores on the thighs.

8. *Heloderma* (The Gila Monster)

Classification

Class — **Reptilia**
Subclass — **Diapsida**

Superorder — **Lepidosauria**
Order — **Squamata**
Suborder — **Lacertilia**
Genus — *Heloderma*
Species — *suspectum*

Heloderma has only two species : *Heloderma suspectum* and *Heloderma horridum*.

Habits and Habitat

Heloderma is popularly called **Gila monster** (or **Banded Gila Monster**) or the **beaded lizard**. It is the only poisonous (venomous) lizard. It lives in dry places under rocks or in burrows in South-West and central America and Mexico.

Heloderma is nocturnal and feeds on eggs, mice, frogs, worms and centipedes. It hibernates in winter. It moves slowly, but can turn its head swiftly for biting. It adapts readily to captivity and requires little care. *Heloderma* is oviparous and lays eggs in a nest in sandy soil.

Structure

The 60 cm long body of *Heloderma* is thick, rounded and robust (Fig. 12.18) but still it can move swiftly when annoyed. It has a blackish body with yellow or pink spots. Bright colouration warns off enemies. Dorsal scales are bead-like and over lie bony tubercles. Ventral scales are flat. Tongue is fleshy, bifid (**Lapedes**, 1976) and protrusible. Teeth are pleurodont. Poison glands which are modifed sublingual salivary glands, open on the "gum" of the lower jaw. Limbs are short and rounded. Tail is thick and small and serves as a store house for fat to be used in hibernation.

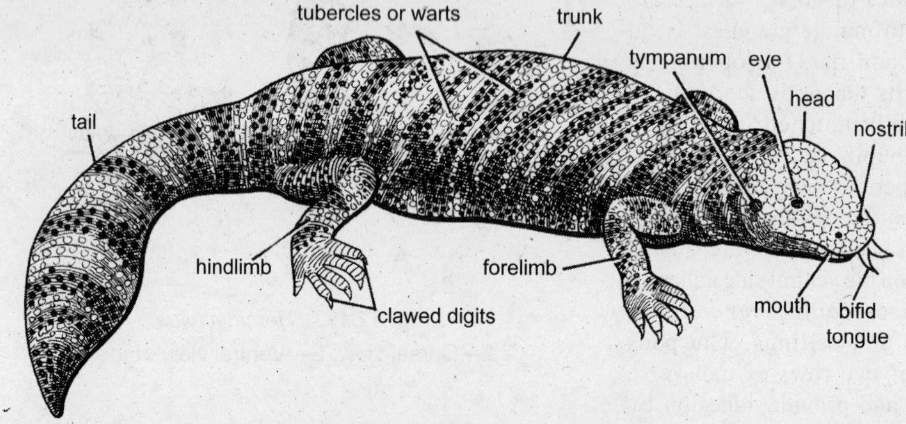

Fig. 12.18. *Heloderma suspectum.*

Some of the mandibular teeth are recurved and act as **fangs** (long sharp poison teeth). These have a groove in front for the conduction of the venom. It injects its poison into prey by chewing movement. Venom is quite potent and bite is fatal to small animals. In human beings, its venom causes severe local injury and affects the respiratory system. Its bite is occasionally fatal in human beings.

9. *Uromastyx* (The Spiny Tailed Lizard)

Classification

Class — **Reptilia**
Subclass — **Diapsida**

Superorder — Lepidosauria
Order — Squamata
Suborder — Lacertilia
Genus — *Uromastix*
Species — *hardwickii*

Habits and Habitat

Uromastix inhabits open barren lands in North India. It lives in self-dug burrow. It is chiefly herbivorous and frugivorous, taking grass and flowers and fruits of wild shrubs and trees. It hibernates in winter. It is oviparous.

Structure

Uromastix has a large, depressed body pale-grey above and whitish below. It is about 30 cm long. Head is relatively small and triangular. Nares have valves. Eyes are overhung by prominent swellings. Ear openings are oval. Neck is short. Trunk has loose, folded skin on the sides. Tail is heavy and bears transverse rows of spinous scales on the upper side.

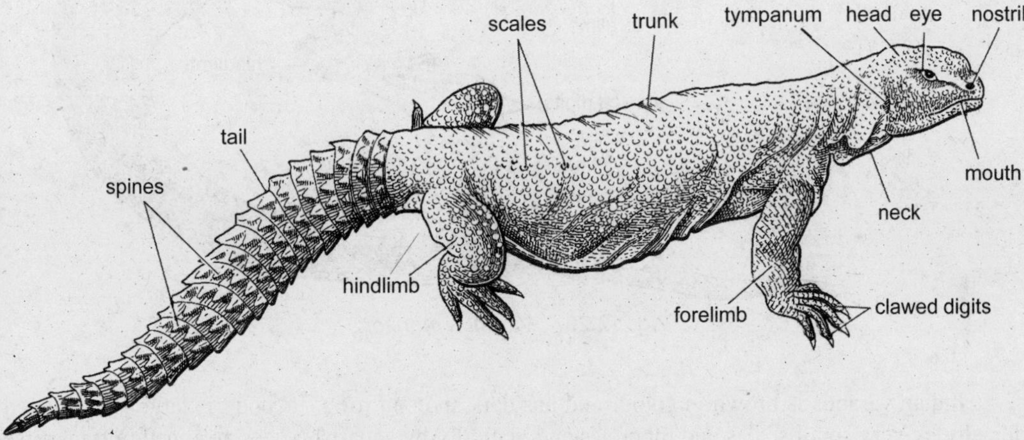

Fig. 12.19. *Uromastix.*

Uromastix is eaten by poor people. Its fat bodies yield medicinal oil.

10. *Varanus* (The Monitor Lizard)

Classification

Class — Reptilia
Subclass — Diapsida
Superorder — Lepidosauria
Order — Squamata
Suborder — Lacertilia
Genus — *Varanus*

Habits and Habitat

Varanus, commonly called the **monitor lizard**, is a largest living lizard. It is widely distributed throughout the hotter and rocky parts of Africa, South Asia, Indonesia and Australia. It lives in the burrows and on the trees. It is generally active at night. It can swim in water with its compressed tail. It is carnivorous and highly predaceous. It feeds on invertebrates, small vertebrates and carrion. It attacks poultry farms, killts the chickens and steals eggs. Some species eat the eggs and young of crocodiles. Monitor lizards are ferocious creatures and defend themselves well and by blows of the tail. It is oviparous and lays eggs in hollow tree trunks or in the ground.

Structure

Varanus flavescens and *V. monitor* are found in India. These measure 60 to 90 cm in length. The **Komodo dragon**, *Varanus komodoensis* of Komodo Island of Java is the largest. It attains a length of 3 meters and weigh over 100 kg.

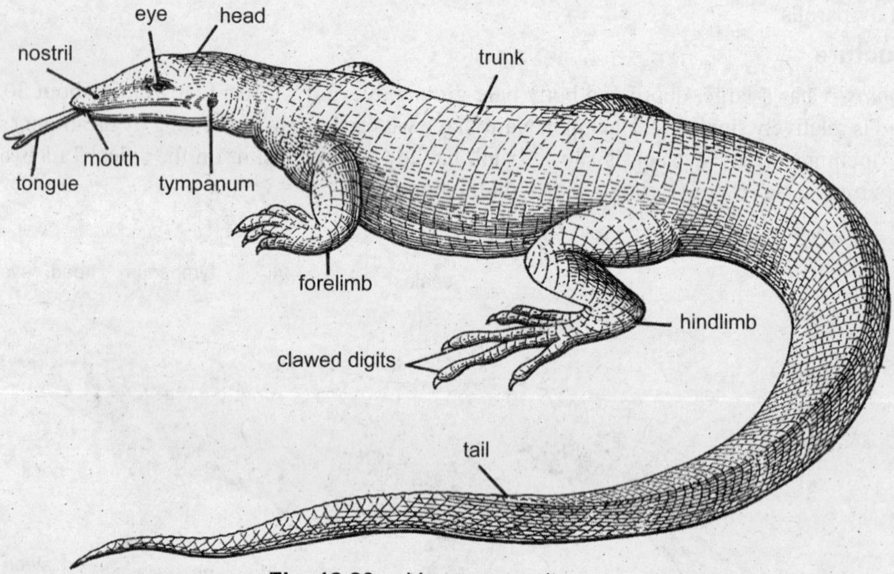

Fig. 12.20. *Varanus monitor.*

Indian varanus is brown in colour and has dark spots in rows. Its body is large and covered dorsally by very small scales and tubercles and ventrally by squarish scales arranged in transverse rows. Its neck is long and very mobile. It raises its head and neck, on the fore limbs and looks around. The eyes are conspicuous and give the lizard an alert expression. The tongue is long, forked and retractile. It darts the tongue out and in. Tongue of varanus is a sensory organ as in snakes. Teeth are large, pointed and pleurodont. The tail is laterally compressed and very powerful. The digits are armed with heavy claws.

In southern China and Malaysian region *Varanus* is hunted by dogs and used for food.

11. *Ophisaurus* (The Indian Glass Snake)

Classification

Class	—	**Reptilia**
Subclass	—	**Diapsida**
Superorder	—	**Lepidosauria**
Order	—	**Squamata**
Suborder	—	**Lacertilia**
Genus	—	*Ophisaurus*
Species	—	*gracilis*

Habits and Habitat

Ophisaurus is commonly known as **Indian glass snake**. It is a limbless lizard. It inhabits bushy areas in Eastern Himalayas, living under leaves or in burrows.

Structure

This limbless lizard moves about by lateral undulations of body very similar to a snake. It can be distinguished from snake by the presence of narrow gape of mouth, movable eyelids and tympanic cavity and tympanic opening. It differs from the limbless amphibians in the presence of epidermal scales, movable eyelids, ear openings and a long tail. *Ophisaurus* carries a pair of minute spikes at the sides of cloacal aperture. The body is greenish-brown with longitudinal stripes. Scales are small and are present above the bony plates. Tail is brittle and usually break off when the animal is roughly handled. The teeth are pleurodont. The vertebrae are procoelous. Girdles are present.

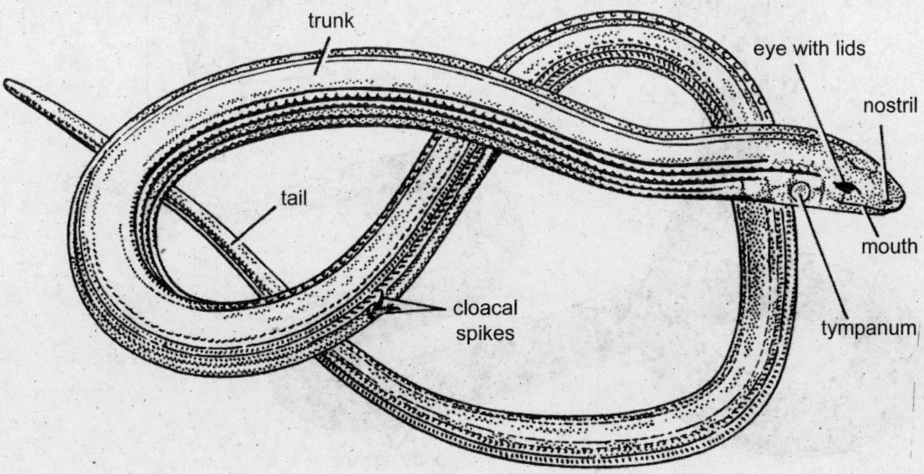

Fig. 12.21. *Ophisaurus.*

Ophisaurus is carnivorous and feeds on worms, insects, snails, small lizards and mice. It is oviparous.

12. *Naja* (The Cobra)

Classification

Class	—	**Reptilia**
Subclass	—	**Diapsida**
Superorder	—	**Lepidosauria**
Order	—	**Squamata**
Suborder	—	**Ophidia**
Genus	—	*Naja*

Habits and Habitat

The genus *Naja* has two Indian species: *Naja naja* (cobra or Nag) and *Naja hannah* (king cobra). *Naja naja* (or *Naja tripudians*) is found is Southern Asia and Africa. It prefers places where it can easily retire (*e.g.*, deserted ant hills, heaps of stones, piles of logs, rat-infested houses, thick vegetation, etc. *Naja* is a highly **poisonous snake**. It feeds on frogs, lizards, birds, rats and other snakes. It hunts in the late afternoon and in the evening and is, thus, a **diurnal** snake. It is not dangerous by nature and attacks only in self defence. Animals such as peacock and mongoose are its enemies. It hibernates in winters and mates during rainy season. The cobra is oviparous.

The cobra is known for its aggressive posture when annoyed, it lifts the front part of the body vertically and expands the neck into a thin, broad hood by raising its large cervical ribs. It hisses through the nares and sways its head back and forth to staike. It bites with a chewing motion, causing a number of wounds. Its venom is neurotoxic and more deadly that that of the viper.

Structure

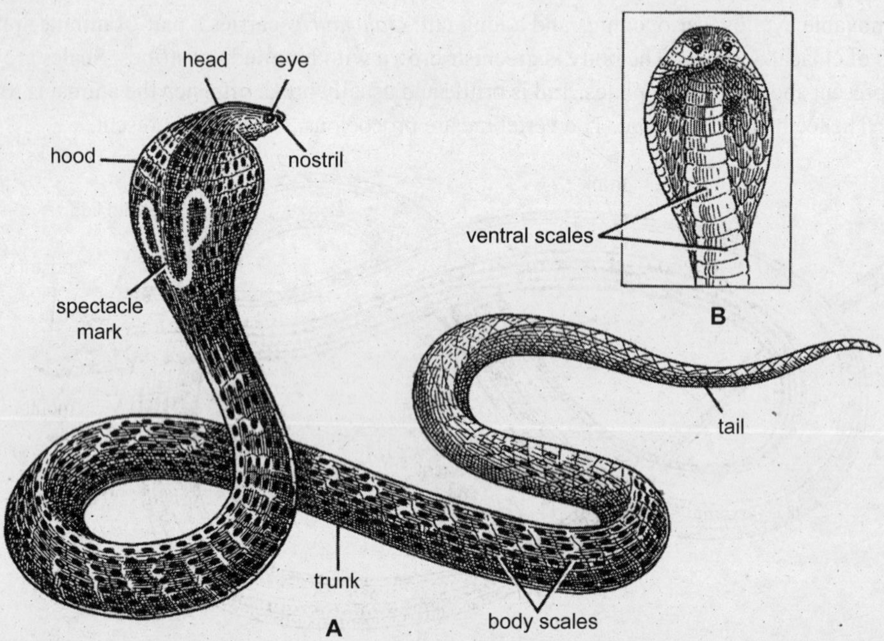

Fig. 12.22. *Naja tripudians* (*N. naja*). A—The common Indian Cobra (with expanded hood); B—Ventral view of hood.

The body of Nag (*Naja naja*) is elongated measuring 180 cm in length and divided into head, neck, trunk and tail. It has a black or brown body colour. Head is hardly distinct from the neck. Eyes are of moderate size and have rounded pupil. Nares are large and vertically elliptical. Body is covered with smooth oblique scales without pits and are arranged in 15-25 rows. The head is covered with small scales and large shields. Ventral scales run across the belly. In cobra, shields on the head and ventral scales have a definite arrangement. The third **supralabial shield** of the upper lip is big touches the eye and the **nasal shield** (*i.e.*, the shield containing the nostril).

There is a tiny tringular shield between the fourth and fifth infralabial shields (*i.e.*, shields of lower jaw). There are three small postocular scales behind the eye. The subcaudals (ventral scales of the tail) are in two rows. Scales on the dorsal side of trunk and tail are smooth and uniform.

Three varieties of Indian cobras are recognised by markings on the dorsal surface of the hood (Fig. 12.23): **binocellate** having a spectle-like mark of black and white colours (these cobras occur in Maharashtra); **monocellate** bearing a black ring (these cobras occur in Bengal) and **nonocellate** without a mark (these cobras are found in Rajasthan, Madhya Pradesh and Gujrat). (*Ococellate mark* means a black spot surrounded by white ring like the cornea and conjunctiva of eye). The ventral surface of hood of cobra bears two dark spots surrounded by white lateral borders.

Naja hannah is the **king cobra**. Its native names are **Nag Raj** or **Shesh Nag**. It grows to 500 cm length. It is black when young and yellowish with crossbars when adult. It is found in Bengal, Assam and South India. It is more dangerous than the common Indian cobra.

13. *Bungarus* (The Krait)

The classification of *Bungarus* is similar to *Naja*.

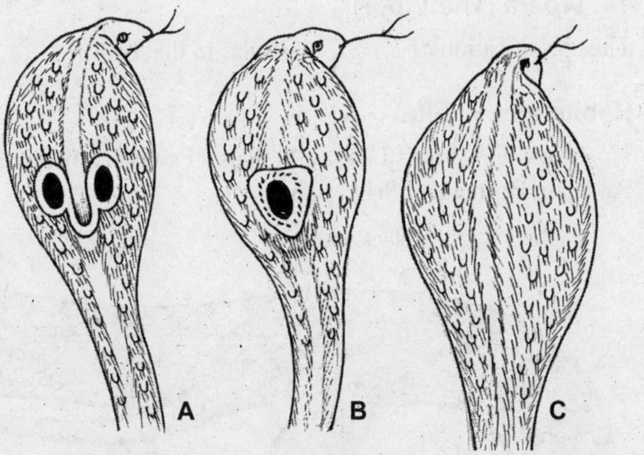

Fig. 12.23. Three types of hoods of Indian Cobra. A—Binocellate; B—Monocellate and C—Nonocellate.

Habits and Habitat

Bungarus caeruleus is the Indian species of krait. It is found in Northern India and Pakistan.

Bungarus is a highly poisonous snake. However, it is a timid and usually inoffensive snake. It does not bite unless trodden on, annoyed or injured. Its venom is *neurotoxic*. Krait often creeps in houses and feeds on toads, frogs, mice, lizards and other snakes. It is nocturnal in habit and lives in cracks in the walls and under heaps of stones or bricks.

Structure

Body of *Bungarus* is elongated and slender (Fig. 12.24). It grows to 120 cm. It is black or steelgray on the back with thin, white crossbars and uniformly white on the belly. Head is not distinct from the neck and bears large shields. Eyes are moderate in size and have a rounded pupil. Fangs are small. Back of krait is covered by smooth scales, the vertebrals (the scales in middorsal line) being a little larger in size than others and hexagonal in form. Belly carries broad plates. Tail is moderately long, cylindrical and tapering, and has a single row of scales on the undersurface.

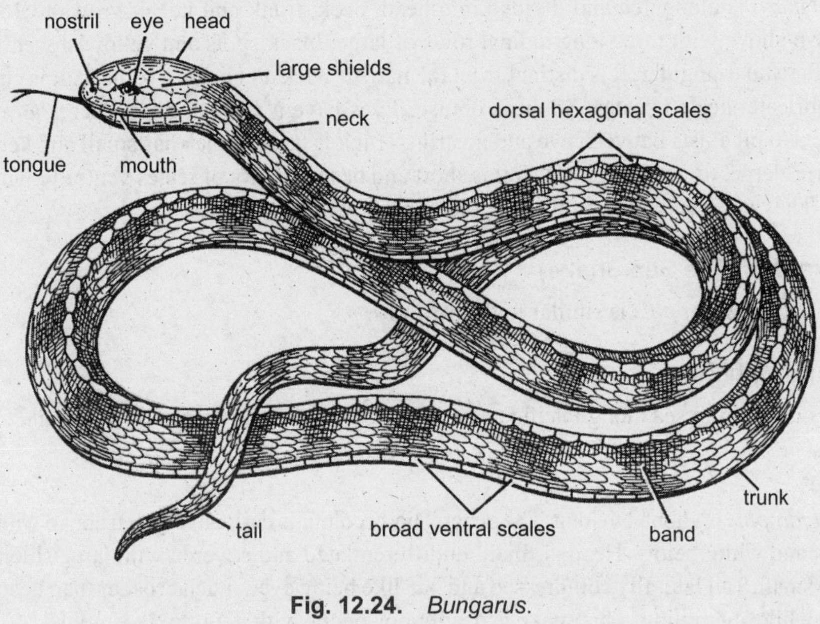

Fig. 12.24. *Bungarus.*

14. *Vipera* (The Viper)

The classification of *Vipera* is similar to that of *Naja*.

Habits and Habitat

Vipera russelli (Fig. 12.25) is the **Russel's viper**, **pitless viper** or **daboia**. It is found all over India, Pakistan, Burma and Sri Lanka.

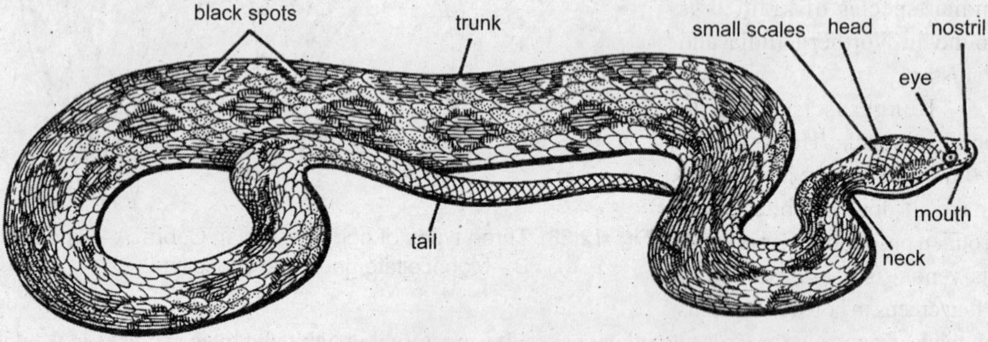

Fig. 12.25: *Vipera russelli.*

Vipera is found is rocky and bushy regions where its colour matches with the surroundings. It feeds on small mammals such as mice and rats and lizards and birds. It is Viviparous.

Vipera is nocturnal. It lies in wait for the prey, strikes it very quickly, wait for it to die, tracks it down and then swallows. It is very **poisonous**. Its venom is **haemotoxic**. It hisses loudly before striking and also swells its body.

Structure

Body of *Vipera* is elongated and divided into head, neck, trunk and tail. It grows to 150 cm. It is light-brown above with three longitudinal rows of large, black spots and yellowish ventrally. Head is broad, flat and triangular. It is distinct from the narrow neck and bears a V mark. It is covered with small, imbricate, keeled scales. Snout is obtuse. Eyes have a vertical pupil. Nares are lateral and very large. No pit exists between eye and nostrils. Trunk is heavy. Back has small and keeled scales. Belly carries large, transverse plates. Tail is short and has two rows of scales ventrally. The fangs are movable. They are erected for biting and folded back, when not in use.

15. *Hydrophis* (The Sea Snake)

Classification of *Hydrophis* is similar to that of *Naja*.

Habits and Habitat

Hydrophis is found in sea along Pacific coast, Bay of Bengal and Malaya-Archipelago.

Structure

Body of *Hydrophis* is about 2 m long. The general body colour is dark olive-green above with yellowish cross bars and white below. Head is small, undifferentiated and covered with large shields. Ventral scales are small. Tail laterally compressed and oar-like being even much broader than head. It is used as paddle while swimming. This snake is deadly poisonous with neurotoxic venom.

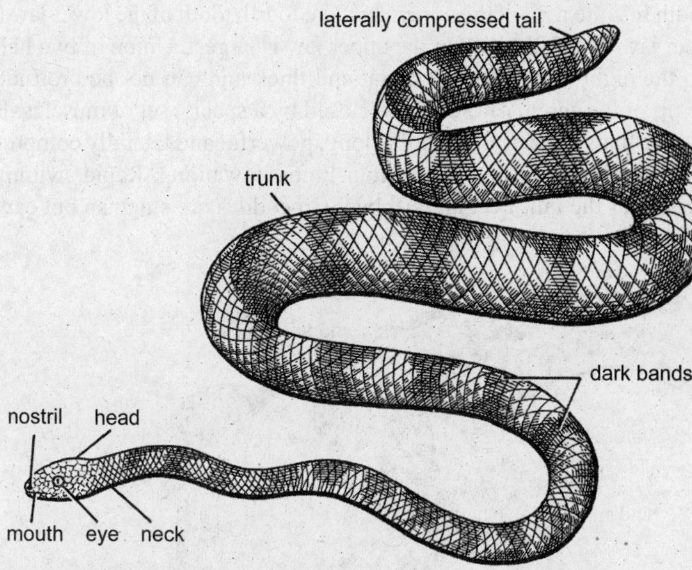

Fig. 12.26. *Hydrophis.*

16. *Crocodylus* (The Marsh Crocodile or Muggar)

Classification

Class	—	**Reptilia**
Subclass	—	**Diapsida**
Superorder	—	**Archosauria**
Order	—	**Crocodilia**
Genus	—	*Crocodylus*
Species	—	*porosus*

Habits and Habitat

Crocodylus is found in Southern Asia, Africa, Australia and Central America. *Crocodylus porosus* lives in salt water estuaries in North-West pacific, India and Southern China. It grows to a length of 7.5 m. It attacks and kills human beings.

Crocodylus palustris, the Indian crocodile, is 4–5 m long and has wide distribution and found in India, Sri Lanka, Myanmar, Malacca, Malaya, Islands and Beluchistan.

Crocodylus, commonly called **mugger**, is amphibious. In summer, when the ponds and marshes dry up, the crocodiles move on land in search of water. If unable to find water, they aestivate in the mud and come out of it in the rainy season. They are carnivorous and predaceous. They mainly feed on fish and crabs but also capture mammals that visit the rivers. Valve in the pharynx enables the crocodiles to feed in water without filling the water in the lungs. They are social animals. They defend their territories by fighting and roaring. They make a 9 to 12 m long tunnel like nest in the bank. The tunnel opens below the water level. The nest of mugger is flask-shaped. The female lays eggs in the large chamber of the nest. Mother looks after the eggs and the newly hatched youngs.

Structure

Body is stout and covered with an exoskeleton of rows of sculptured bony scutes. Epidermal scales are also present. The head is Triangular. Snout is relatively narrow. Teeth are thecodont (lodged in sockets), large, unequal in size and irregularly arranged. Each jaw contains 19 teeth on either side.

The first lower tooth fits into a pit of the upper jaw. The fourth tooth of the lower jaw fits into a groove in the side of upper jaw. The fifth tooth of the upper jaw is largest. Union of two halves of the lower jaw extends upto the eight tooth. Tongue is flat and thick and can not be protruded. The **nostrils** are located at the tip of the snout. These can be closed by a special set of muscles. Ear openings are protected by movable flaps of scales. The tail is long, powerful and laterally compressed. Fore limbs are shorter than the hind limbs. They use all four limbs in walking. Rapid swimming is produced by lateral movements of the tail. Feet are webbed. Crocodiles are sluggish but can run quickly for short distances on land.

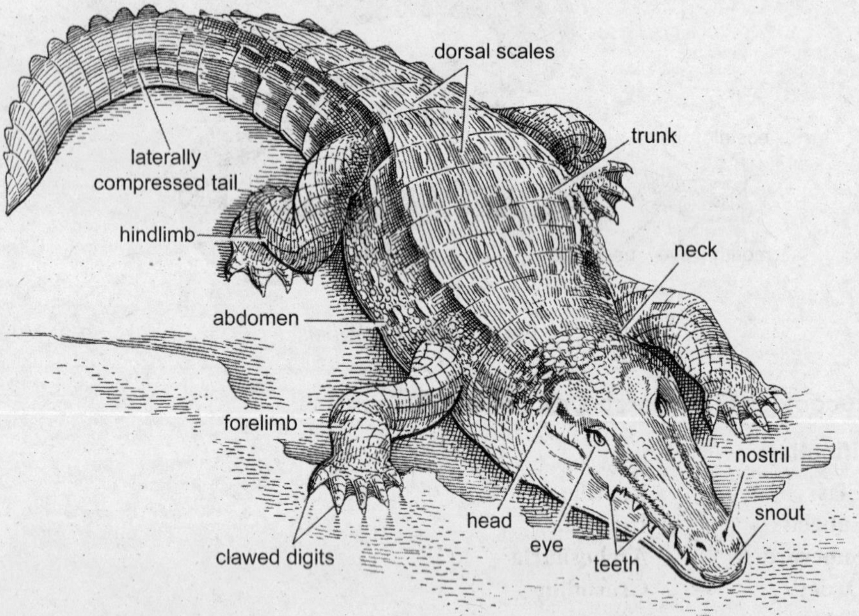

Fig. 12.27. *Crocodylus porosus* (Indian freshwater or marsh crocodile).

There is secondary palate in crocodiles and their heart is four chambered. They have acute hearing and keen eyesight. Males contain a simple grooved copulatory organ.

12.5 TORTOISE SHELL

The whole body of chelonians is encased in a strong bony case. Its dorsal part is called **carapace** and ventral **plastron**. The bony shell is formed in the dermis, outside this lies a horny exoskeleton of **epidermal plate**.

1. Carapace. The **bony dermal plates** of the carapace (Fig. 12.28C) include a **median** series of eight **neural plates**. These are surrounded by the right and left series of **coastal plates** and usually a series of **marginal plates** surrounding the coastal plates. Thus, there are in all five rows of bony plates in the carapace. The median plates are differentiated into an anterior most **nuchal plate**, usually eight **vertebral plates** and one to three **supracaudal plates**. The nuchal plate is not articulated with the dorsal vertebrae above which it lies. The eight neural plates are fused with the flattened neural spines of 2nd to 9th trunk vertebrae. The marginal plates are vary in number but usually include 11 paired plates and one median **pygal plate** at the hind end.

The horny epidermal scales or **corneoscutes** of the carapace (Fig. 12.28A) includes five **neural shields** and an anterior **nuchal shield** in the middle line, four pairs of **costal shields** on the sides of the neural shields and twelve pairs of **marginal shields** at the periphery, the last pair often forming a median **pygal shield**.

2. Plastron. The bony dermal plates of the plastron (Fig. 12.28D) include nine bony plates : one median **entoplastron** which corresponds to the interclavicle and lies a little behind the anterior end, and four paired plates known as the **epiplastra** which correspond to the clavicles and lie infront of the entoplastron and the **hypoplastra**, **hyoplastra** and **xiphiplastra** situated behind the entoplastron in the order named. The hypoplastra, hyoplastra and xiphiplastra are regarded as modified abdominal ribs.

The corneoscutes of the plastron (Fig. 12.28B) occur in six pairs of large shields, which are known from anterior to posterior direction as **gulars**, **humerals**, **pectorals**, **abdominals**, **femorals** and **anals**. Boundaries of bony plates are visible through the corneoscutes.

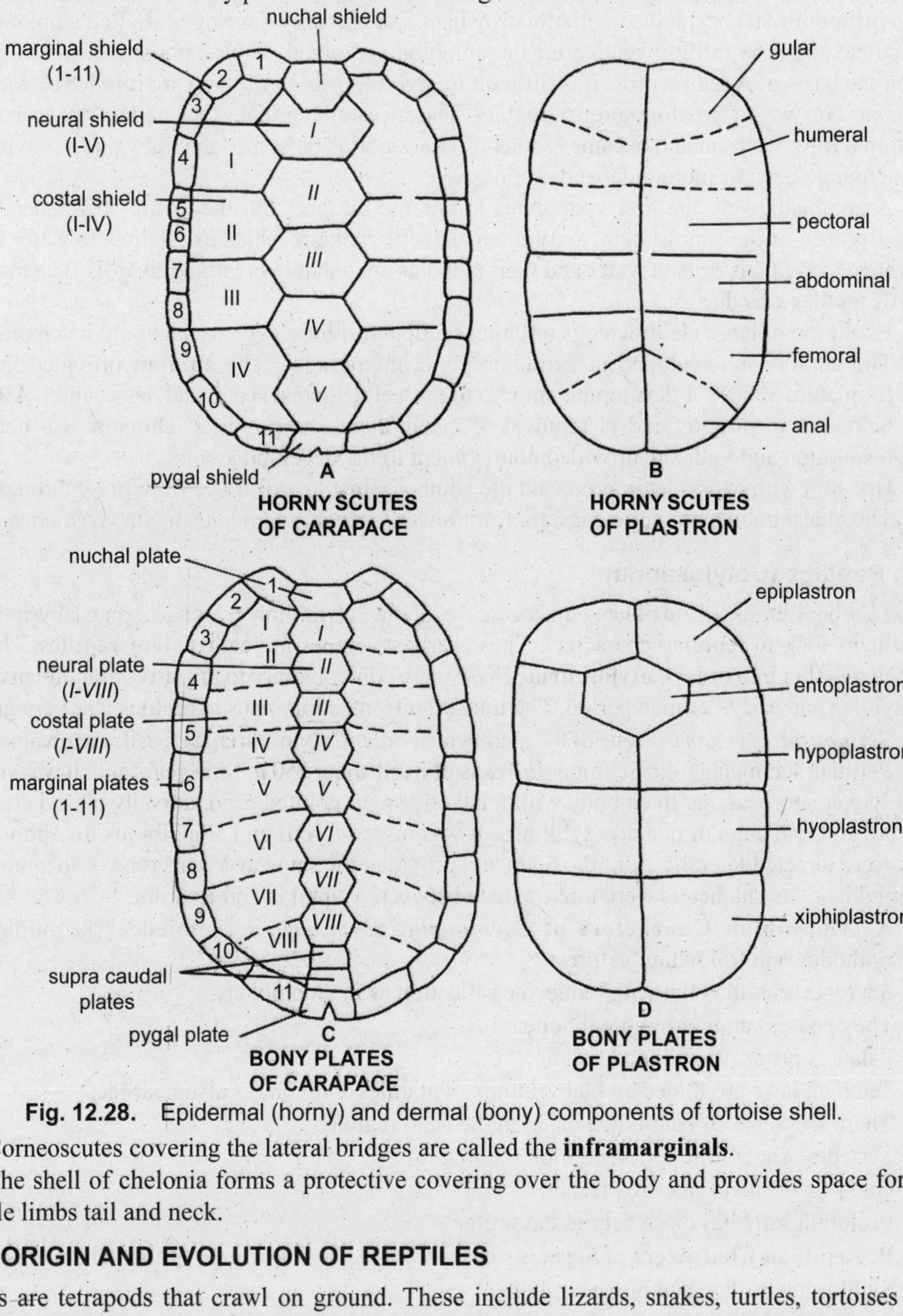

Fig. 12.28. Epidermal (horny) and dermal (bony) components of tortoise shell.

Corneoscutes covering the lateral bridges are called the **inframarginals**.

The shell of chelonia forms a protective covering over the body and provides space for the retractile limbs tail and neck.

12.6 ORIGIN AND EVOLUTION OF REPTILES

Reptiles are tetrapods that crawl on ground. These include lizards, snakes, turtles, tortoises and crocodiles. Evolution of reptiles is of special importance because it was they who developed the first land eggs and finally made the vertebrates independent of water where they originated. Thus, reptiles became the first vertebrates to wander at will on land without fear of dehydration and without needing water for laying eggs.

Reptiles first appeared in the beginning of Palaeozoic, enjoyed a dominant status in Mesozoic and then faded out in Coenozoic era. There are several presumptions about their origin.

Origin from Amphibia

It is generally accepted that reptiles originated during late Carboniferous from primitive **Labyrinthodont** stegocephalian amphibians which had probably developed shelled amniotic eggs. They arose about 50 million years after the amphibian evolved. Their emergence was so gradual that on the basis of fossil records, it is difficult to ascertain which of them are reptiles or which are amphibians on way to transformation to reptiles. The ancestral reptilian stock underwent an extensive radiation during Permian and became extinct in Triassic. It is presumed that they had a polyphyletic origin having dozen or more independent lines.

Amphibians were the first vertebrates to emerge on land, but they were still dependent on water. Because of their moist skin, amphibians have to make repeated excursions to water bodies. They have to lay their eggs in water and their tadpoles are aquatic, breathing by gills. The transition towards reptiles was due to:

1. Evolution of large **cleidoic eggs** or large, shelled **amniotic eggs** which could hatch on land.
2. The embryo is enveloped in extraembryonic membranes. The **amnion** provided the pool (miniature pond) for developing embryo to protect it against shock and desiccation. **Allantois** helped in respiration and in removal of metabolic waste products, **chorion** also helped in respiration and **yolk sac** provided nourishment to the developing embryo.
3. Dry skin with exoskeleton protected the adults against loss of water from body through skin. The oldest known amniotic eggs are from lower Permian sediments in North America.

Stem Reptiles (Cotylosauria)

During Carboniferous period of late Palaeozoic era, about 250 million years ago, some labyrinthodont amphibians took to reptilian characters. These earliest reptiles are called **stem reptiles**. They are grouped together into order **Cotylosauria**. These were contemporary to primitive amphibians of late Pennsylvian and early Permian period. The transition from amphibians to reptiles was very gradual.

Seymouria. *Seymouria* is one of the members of order Cotylosauria. Its fossils are obtained from lower Permian sediments from Seymour in Texas. It lived about 250 million years ago. It was sluggish lizard-like creature having thick body with relatively small pointed head, dorsally placed nostrils, a short tail and two pairs of pentadactyl limbs. It was insectivorous and amphibious. Its short stubby limbs were directed laterally from the body and provided the same awkward type of locomotion as in amphibians. Its characters were intermediate between Amphibia and Reptilia.

A. Amphibian Characters of *Seymouria*. *Seymouria* resembled early amphibians (Labyrinthodont) in following features:

1. Seymouria skull is flat with reduced ossification as in amphibians.
2. They possess an intertemporal bone.
3. Palate is primitive and incomplete.
4. Teeth on jaws are folded and labyrinthine. Palatines and vomers also bear teeth.
5. There are traces of lateral line canal in the head region.
6. Vertebrae show little differentiation into regions.
7. One pair of sacral ribs are present.
8. Pectoral girdle lies close behind the skull.

B. Reptilian Characters of *Seymouria*. *Seymouria* had following reptilian characters:

1. Skull is anapsidian and monocondylic.
2. Limbs are pentadactyl and muscular and arise mid-ventrally.

3. First two cervical vertebrae modified into atlas and axis and two sacral vertebrae present.
4. Number of phalanges is usually in all digits as in reptilians.
5. Neural arches or vertebrae were convex dorsally as in reptiles.
6. Pectoral girdle had a long interclavicle.
7. Blade of ilium was expanded for attachment of highly developed muscles used for walking.
8. It laid cleidoic eggs.

Later, forms gradually pulled their legs under their bodies and evolved into birds and mammals.

Conclusion. It is concluded that *Seymouria* existed too late in time to become ancestor of reptiles. At the time when it existed, the reptiles were already present for about 50 million years. **Romer** treated *Seymouria* as a reptile and placed it under *Cotylosauria*, whereas other placed it under order **Seymouriamorpha** in Amphibia. It is now regarded as a connecting link between Labyrinthodontia (Amphibia) and Cotylosauria (Reptilia).

Limnoscelis

Romer (1964) suggested that *Limnoscelis*, a captorhinomorph cotylosaur, is the primitive reptile. Its fossils were obtained from late Carboniferous or early Permian in New Mexico. It has a number of reptilian features in primitive condition:

1. Its skull was of anapsid type and laterally compressed.
2. The skull was without otic notch but its disappearance is indicated.
3. It is believed that during Permian, the early reptiles splitted into following two divergent evolutionary groups:
   (*i*) **Therapsida.** It led to mammal-like reptiles which evolved into mammals.
   (*ii*) **Sauropsida.** These evolved into different type of reptiles.

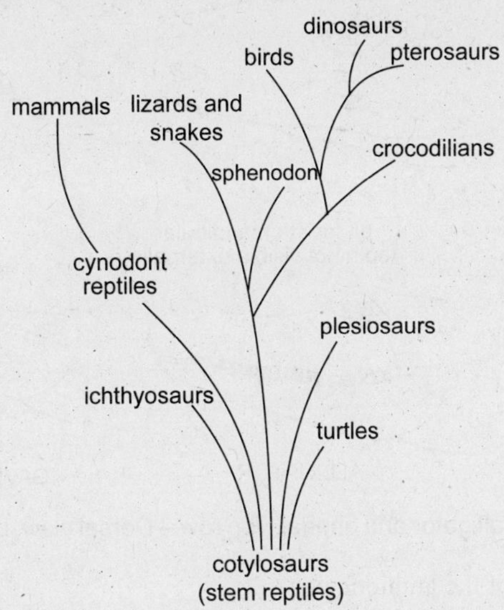

Fig. 12.29. Diagram showing possible direction of evolution of reptiles, birds and mammals from stem reptiles or Cotylosaur.

Recent investigations suggested that both Therapsida and Sauropsida evolved from Captorhinomorph cotylosaurs.

Evolution of Reptiles During Mesozoic Era

Reptiles evolved in early Mesozoic Era and become master of the land but became extinct in the late Mesozoic Era. Their numerosity can be estimated by studying their fossil forms. Based on the morphology of skull, they have been classified into following five lines: 1. Anapsid; 2. Parapsid; 3. Synapsid; 4. Eurypsid and 5. Diapsid.

QUESTIONS

Long Answer Questions

1. Give an outline classification of the class Reptilia. Mention one example of each order.
2. Enumerate the distinguishing characters of class Reptilia and classify it upto orders giving characters and suitable examples. (*Purvanchal 2000, 2001; Gorakhpur 2003*)
3. Write general characters of class Reptilia and classify it on the basis of temporal region of skull into its subclasses.

Short Answer Questions

1. Name three different genera of class crodilia and give one distinguishing feature of each.

Ans. The three genera of class crocodilia are:

(*i*) Crocodile or Magger : *Crocodylus*
(*ii*) Alligator : *Alligator*
(*iii*) Gharial : *Gavialis gangeticus*

These three genera can be identified by the shape of the snout (Fig. 12.30):

(*i*) In *Crocodylus*, the snout is moderately long and pointed.
(*ii*) In *Alligator*, the snout is short and rounded.
(*iii*) In *Gavialis*, the snout is very long and slender.

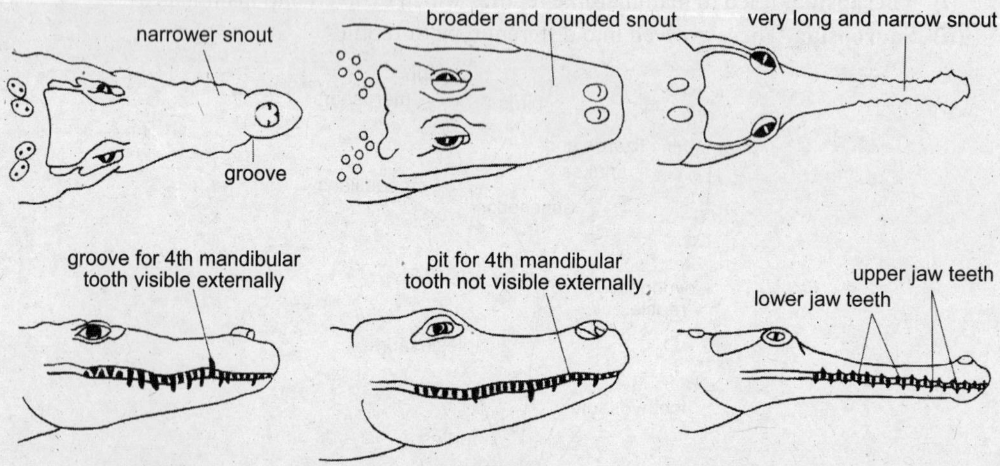

Fig. 12.30. Heads in crocodile, alligator and gharial. Top row—Dorsal view; Lower row—Ventral view.

2. What is the significance of the amniotic egg?
3. What are the characteristics that made reptiles less dependent on a source of external water?
4. What do you know about turtles and tortoises?
5. What do you know about crocodiles and alligators?
6. How do the lizards differ from the snakes? (*Rohilkhand, 2000*)
7. Give distinguishing characters of order chelonia. Give two examples.
8. Give distinguishing characters of following orders:

(*a*) Rhynchocephalia (*b*) Squamata
(*c*) Crocodila

9. Write short notes on the following:
   (*i*) *Testudo*
   (*ii*) *Sphenodon*
   (*iii*) *Chameleon* (*CCSU Meerut, 2011*)
   (*iv*) *Draco*
   (*v*) *Varanus*
   (*vi*) *Heloderma*
   (*vii*) *Ophisaurus*
   (*viii*) Cobra (*CCSU Meerut, 2012*)
   (*ix*) Rattle snake
   (*x*) *Hydrophis* (sea snake)
   (*xi*) *Crocodilus*
   (*xii*) Dinosaurs
   (*xiii*) *Brontosaurus*

Very Short Answer Questions

1. What is basis of classifying reptiles?
2. What is the role of keratin in the reptiles life?
3. Reptiles are ectothermic. Explain.
4. What is tuatara?
5. What are ruling reptiles?
6. Name the flying lizard.
7. How many sacral vertebrae the reptiles have?
8. Name a reptile that has a proatlas bone.
9. Give the name of a poisonous lizard.
10. Which reptile has thecodont teeth?
11. Give the scientific name of the glass-snake lizard.
12. Is the sea snake poisonous?
13. How does *Phrynosoma* defend itself?
14. Which is largest snake?
15. Name a reptile that carries on cloacal respiration.
16. Give scientific name of cobra. (*CCSU Meerut, 2010*)
17. What is the difference between turtles and tortoises?
18. In which order of class Reptilia body surface is covered with sculptured bony scutes?
19. Why *Sphenodon* is called living fossil?
20. What are the advanced features of Crocodilia?
21. Which major adaptation enables reptiles to live on land?

Fill in the Blanks

1. The epidermal horny scales appear for the first time in ________.
2. ________ is the only example of order Rhynchocephalia. (Lizard/*Sphenodon*). (*Garhwal, 2003*)
3. Skull of ________ is monocondylic.
4. Ophidia are commonly known as ________ (lizard/snake).
5. Among reptiles, the heart is completely four chambered in ________.
6. Excretory substance is ________ in the land reptiles.
7. ________ the first vertebrates to have cleidoic eggs.
8. The best known stem reptile is ________.
9. The shell in turtles differentiated into a dorsal carapace and a ventral ________ (plastron/carapace).
10. The largest extinct reptile was ________.
11. The limbs of *Chelone* are modified into ________.
12. The reptile confined in New Zealand is ________.

13. In ________ the cloacal aperture is longitudinal.

14. The Indian glass snake is ________ (*Ophisaurus*/*Gecko*).

15. Flying lizard is scientifically known as ________ (*Draco*/*Sphenodon*).

16. ________ is a chemoreceptor found in the roof of mouth in reptiles (Jacobson's organ/neuromast organ).

17. Anapsid skull is devoid of ________ fossae.

18. Lower temporal arcade is incomplete in lizard and ________ in snakes (present/absent).

Multiple Choice Questions

1. *Sphenodon* belongs to the order
(*a*) Squamata (*b*) Rhynchocephalia
(*c*) Chelonia (*d*) Crocodilia

2. Two temporal fossae are present in
(*a*) lizards (*b*) snakes
(*c*) *Testudo* (*d*) *Sphenodon*

3. Hands and feet are grasping in
(*a*) *Chamaeleon* (*b*) *Calotes*
(*c*) *Phrynosoma* (*d*) *Draco*

4. Which of the following is poisonous
(*a*) *Varanus* (*b*) *Phrynosoma*
(*c*) *Heloderma* (*d*) *Calotes*

5. Which of the following is a nonpoisonous snake?
(*a*) *Crotalus* (*b*) *Eryx*
(*c*) *Bungarus* (*d*) *Vipera*

6. Quadrate bone is not movable in
(*a*) Chelonia (*b*) Rhynchocephalia
(*c*) Crocodilia (*d*) All of these

7. Which of the following is a poisonous snake?
(*a*) *Python* (*b*) *Eryx*
(*c*) *Enhydrina* (*d*) *Ptyas*

8. Which of the following is an extinct reptile?
(*a*) *Ancistrodon* (*b*) *Iguanodon*
(*c*) *Typhlops* (*d*) *Trionyx*

9. Heart in a crocodile is
(*a*) two-chambered (*b*) three-chambered
(*c*) incompletely four chambered (*d*) four-chambered

10. Skin is dry and scaly in
(*a*) fishes (*b*) amphibians
(*c*) reptiles (*d*) horn lizard

11. Jacobson's organ is related to
(*a*) taste (*b*) sight
(*c*) touch (*d*) smell

12. *Sphenodon* is found in
(*a*) Australia (*b*) South Africa
(*c*) New Zealand (*d*) All of these

13. The oldest (*i.e.*, most ancient) living reptiles belong to order
(*a*) Rhyncocephalia (*b*) Chelonia
(*c*) Crocodilia (*d*) Squamata

14. Urinary bladder is absent in
(*a*) *Chameleon* (*b*) snakes
(*c*) snakes and crocodiles (*d*) wall lizard

15. In reptiles the vertebrae are
(*a*) amphicoelous (*b*) amphiplatyan
(*c*) platycoelous (*d*) procoelous

16. Animals that belong to Chelonia are
(*a*) lizards (*b*) bats
(*c*) turtles (*d*) tuatara

ANSWERS

Very Short Answer Questions

1. The reptiles are classified on the basis of the presence or absence of **fossae** or cavities in the temporal region of skull.

2. The reptiles have thick **scaly** skin which is keratinised. Keratin is a protein found in hair, finger nails and feathers. It is impermeable to water. Thus, the protective skin of reptiles prevent water loss.

3. Reptile's body temperature is variable according to environment. This is called **poikilothermy**. They are cold-blooded animals, *i.e.*, reptiles do not generate required body heat by metabolism like the birds and mammals. Some reptiles (*e.g.*, lizards, tortoise, crocodiles) raise their body temperature by exposure to sun. Getting warmth from the environment is called **ectothermy**. The ectothermic feature of reptiles allows them to survive on a fraction of food required by birds and mammals.

4. Tuatara is the popular name of *Sphenodon*. It is the only living member of the order *Rhynchocephalia* of the subclass Diapsida, Class Reptilia. *Sphenodon* or tuatara is popularly called the **living fossil lizard**.

5. The dianosaurs are called ruling reptiles. These extinct reptiles dominated the land for about 170 million years during Mesosoic era and then mostly died.

6. *Draco volans*.

7. Two (they provide support to pelvic girdles).

8. *Sphenodon*.

9. *Heloderma*.

10. Crocodile.

11. *Opisaurus gracilis*.

12. Yes.

13. By throwing jets of blood from the eyes.

14. *Python*.

15. Aquatic chelonians.

16. *Naja naja*.

17. Turtles are marine and their limbs are modified into flippers or paddles. The tortoises are found on land and their limbs are elephant-like and pentadactyl with clawed digits.

18. In crocodiles of order Crocodilia, the body surface is covered with rows of sculptured bony scutes.

19. *Sphenodon* is called living fossil because: (*i*) It has stegocephalian feature : pterygoid joining vomer. (*ii*) It has amphibian features : retention of conus arteriosus, ductus caroticus and ductus arteriosus.

20. Crocodiles show following advanced features : they have thecodont teeth, relatively complete bony palate, four-chambered heart and a muscular diaphragm to separate pleural cavities from the body cavity.

21. Evolution of amniotic or large, yolky and shelled cleidoic eggs enabled them to thrive on land.

Fill in the Blanks

1. *Sphenodon*. **2.** Reptiles. **3.** Reptiles. **4.** Snake. **5.** Crocodilia. **6.** Uric acid. **7.** Reptiles. **8.** *Seymouria*. **9.** Plastron. **10.** *Brontosaurus*. **11.** Paddles. **12.** *Sphenodon*. **13.** Chelonians. **14.** *Ophiosaurus*. **15.** *Draco*. **16.** Jacobson's organ.

Multiple Choice Questions

1. (*b*)	**2.** (*d*)	**3.** (*a*)	**4.** (*c*)	**5.** (*b*)	**6.** (*d*)
7. (*c*)	**8.** (*b*)	**9.** (*d*)	**10.** (*c*)	**11.** (*d*)	**12.** (*c*)
13. (*b*)	**14.** (*c*)	**15.** (*d*)	**16.** (*c*)		

13

Poisonous and Non-poisonous Snakes; Biting Apparatus and Biting Mechanism

A snake is said to be **poisonous** if its bite prove fatal for human victim. A snake whose bite does not normally kill a human being is **non-poisonous**. Of 2500 species of snakes, about 300 are poisonous. These are found on all continents and on many large islands, except Madagascar and New Zealand. **Poison** is used by the snakes for capturing prey and for defence against predators and human beings.

13.1 IDENTIFICATION OF POISONOUS SNAKES

All snakes are considered poisonous by people of our country. This belief is not true. Majority of snakes are non-poisonous and harmless. Of about 330 species of snakes found in India, only 69 are poisonous. These poisonous snakes include 29 species of sea snakes and 40 species of land snakes, namely vipers, cobras, coral snakes and kraits (Table 13.1).

Table 13.1 Poisonous and non-poisonous snakes.

	Common name	Zoological/Scientific name	Status
1.	Cobra or Nag	*Naja naja* or *Naja tripudians*	Poisonous
2.	King cobra or Nagraj	*Naja hanna*	Poisonous
3.	Russell's viper or Kander	*Vipera russelli*	Poisonous
4.	Himalayan pit-viper	*Ancistrodon himalayansis*	Poisonous
5.	Saw scaler viper (Phoorsa)	*Echis carinatus*	Poisonous
6.	Common krait	*Bungarus caeruleus*	Poisonous
7.	Banded krait (Ahiraj or rajsanp)	*Bungarus fasciatus*	Poisonous
8.	Kala karyat	*Bungarus niger*	Poisonous
9.	Bamboo snake	*Lachesis graminus*	Poisonous
10.	Coral snake	*Callophis melanurus*	Poisonous
11.	Sea snake (Samundri sarp)	*Hydrophis caerulescens*	Poisonous
12.	Blind snake (Andha sanp)	*Typhlops braminus*	Non-poisonous
13.	Python or Ajgar	*Python molurus*	Non-poisonous

	Common name	Zoological/Scientific name	Status
14.	Rat snake or Dhaman	*Ptyas muscasus*	Non-poisonous
15.	Double mouth snake (Dumuha or sand boa)	*Erynx johnii*	Non-poisonous
16.	Water snake	*Natrix ptolata*	Non-poisonous

The identification of poisonous and non-poisonous snakes found in U.P., Delhi, M.P., Rajasthan and Punjab is very simple. The key shown in Fig. 13.1 will serve to distinguish poisonous snakes from non-poisonous snakes.

1. If the tail is laterally compressed and widens posteriorly, the snake is poisonous and is a sea snake.

2. If the tail is cylindrical and tapers posteriorly, the snake may be poisonous or non-poisonous. Then examine the belly.

I. If the belly has small scales similar to those on the back, or the ventrals do not extend fully across the belly and have small scales on their sides, the snake is non-poisonous.

II. If the ventrals extend fully across the belly and the small back scales are not visible on the undersurface, the snake be poisonous or non-poisonous. Then head of the snake should be examined.

(*i*) If the head has small scales, the snake is poisonous and is a viper.

(*ii*) If the head has large scales, called the **shields**, the snake may be poisonous or not.

(*a*) If there is a pit or hole between the eye and the naris, the snake is poisonous and a **pit viper**.

(*b*) If the 3rd **supralabial scale** (the anterior median scale is not counted) touches the eye as well as the naris-bearing scale, the snake is poisonous and is **cobra**, **kingcobra** or **coral** snake. Cobra have a hood with or without marking, and coral snakes have special spots on the belly.

(*c*) If the **vertebrals** (scales of the middle row on the back) are hexagonal and larger than other backscales and the 4th **infralabial scale** (the anterior median scale is not counted) is the largest, the snake is poisonous and a **krait**.

(*iii*) If the head has large shields, but lacks the features mentioned in points (*a*), (*b*) and (*c*) above, the snake is non-poisonous.

13.2 BITING APPARATUS OF POISONOUS SNAKES

All poisonous snakes possess a pair of poison apparatus in their heads. This apparatus is not found in non-poisonous snakes. The poison apparatus includes a pair of **poison glands, their ducts, fangs, associated muscles** and certain **bones of the skull** and **jaws**.

1. Poison glands. Two sac-like poison glands are situated one on either side of each upper jaw, behind and below the eyes. These are modified parotid salivary glands. The poison glands are small and oval in sea snakes and large and tubular in vipers. Each gland is covered with a thick capsule of fibrous connective tissue. This tissue is connected with a fan-shaped **constrictor muscle**, often called **temporal** or **masseter**. The muscle stretches during biting and press the poison gland which delivers the poison to the fangs.

2. Poison ducts. A narrow poison duct leads from each poison gland to the base of a fang. The duct opens into the groove or canal of the fang.

3. Fangs. Fangs are the enlarged and modified teeth in the upper jaw of poisonous snakes. These are born on the maxillary bones of the upper jaw. Fangs are long, curved, sharp and pointed and used to inject the poison into the body of the prey. They act like hypodermic needles. All poisonous snakes contain two functional fangs. The fangs may have an open gutter-like channel or a hollow canal for the conduction of the poison. Both, the groove and the canal open at the base and the tip of the fangs. The poison duct opens into the groove or the canal at its base. The fangs are situated in the front region of the jaw in cobras and vipers (Fig. 13.2), but lie slightly behind in most sea snakes.

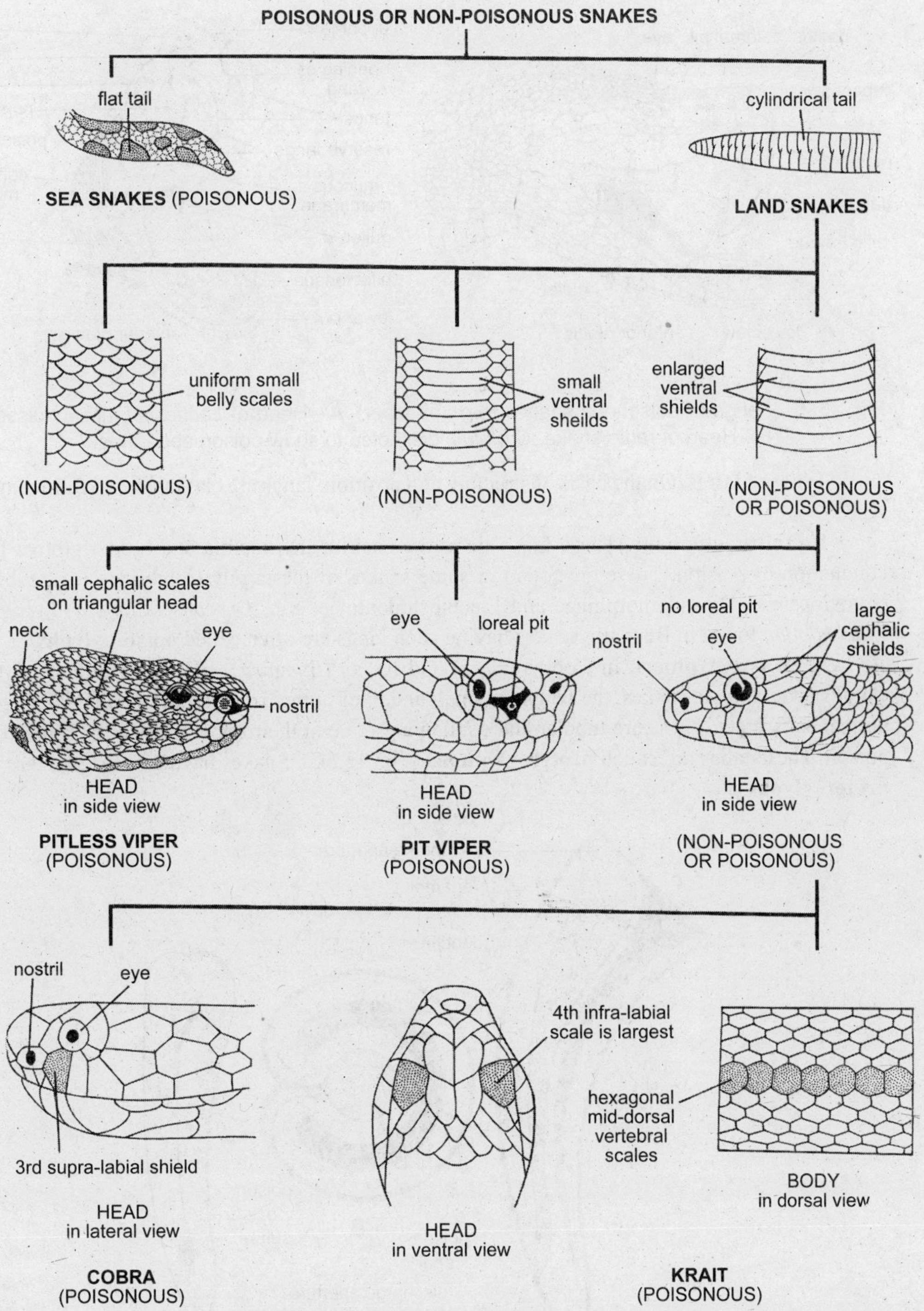

Fig. 13.1. Identification of poisonous and non-poisonous snakes.

The old and broken fangs are replaced like the other teeth. There are usually two sockets of fangs side by side at the front part of each maxilla. As the old fang is shed, its function is taken over by its neighbour. Behind these lie several buds of fangs in various stages of development. These move forward into position as the front ones are shed.

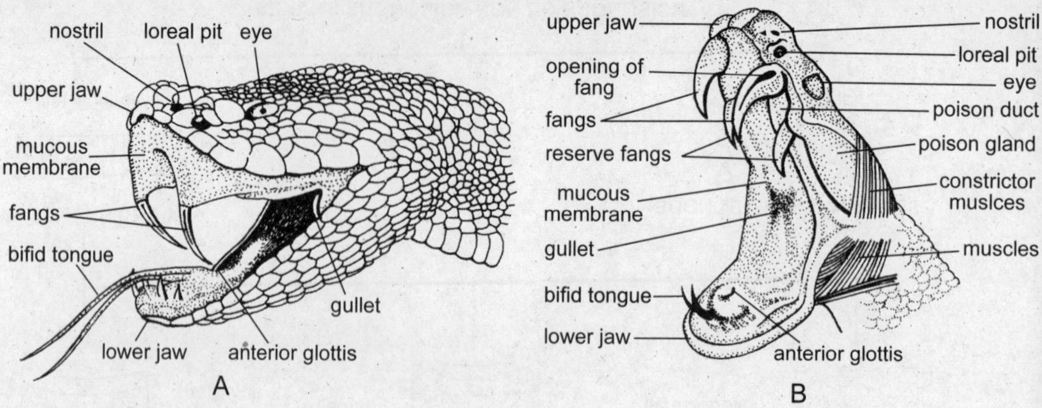

Fig. 13.2. Fangs of poisonous snakes (*e.g.*, pit viper). A—Head of *Lachesis muta* (undissected); B—Head of rattle snake, *Crotalus*, dissected to show poison apparatus.

Types of fangs. On the basis of structure and position, fangs are classified into following three types:

1. Opisthoglyphous. These fangs lie at the back of the maxilla and bear a groove for the conduction of poision. These are found in some snakes of the family Colubridae (*i.e.*, subfamily Homalopsinae, Dipsadomorphinae and Elachiostodontinae, *e.g.*, *Dryophis*, *Elapops*, *Lycognatus*, *Elachistodon* found in Bengal). Snakes having such fangs are often called **opisthoglyphs**.

2. Proteroglyphous. In Cobras (*Naja*) and Kraits (*Bungarus*) of Asia; mambas of Africa; Coral snakes and sea snakes; the fangs are small and fixed at the front part of maxilla. Fangs remain permanently erect. These are tubular and contian a groove in their upper surface for conduction of poison. These fangs are called **proteroglyphous** (Fig. 13.3C). Snakes having such fangs are called **proteroglyphs**.

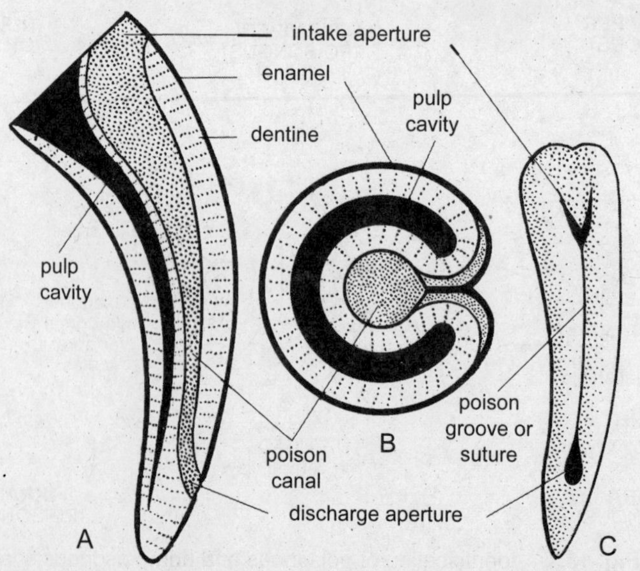

Fig. 13.3. Snake fangs. A—Solenoglyphous fangs of viper in L.S.; B—Solenoglyphous fang in T.S.; C—Entire grooved (= proteroglyphous) fang of cobra.

3. Solenoglyphous. In vipers and rattle snakes fangs are large and situated in the front part of the upper jaw. These contain a canal like that of an injection needle for the conduction of poison (Fig. 13.3A, B). The fangs are movable and lie along the roof of the mouth when not in use. These are erected by the rotation of the maxillae at the time of biting and yawning. The bases of these fangs remain covered with a sheath. The snakes having such fangs are called **solenoglyphs**.

Muscles of biting apparatus (Fig. 13.4). Following muscles are associated with the biting apparatus.

(*a*) Digastric muscle. This muscle is attached to the squamosal bone of the skull at one end and articular bone of the lower jaw at the other end. Its contraction helps in opening of the mouth.

(*b*) Sphenopterygoid and protractor-pterygoid muscle. This muscle is attached to the sphenoidal region anteriorly and dorsal surface of the pterygoid bone posteriorly. Its contraction pulls the pterygoid forward.

(*c*) Anterior, middle and posterior temporal muscles. These muscles are attached to the side walls of the cranium and the lower jaw. Their contraction closes the mouth. The anterior temporal muscle is so folded around the poison gland that when the snake raises up the lower jaw to close the mouth, it presses two-third of the body of poison gland and the poison is ejected from the gland.

4. Skull and jaw bones. The skull and jaw bones of poisonous snakes are unique and highly specialized. Several bones such as palatines, pterygoids, ectopterygoids, squamosals, quadrates and maxiallae are movably articulated and help in biting. Quadrates and the pterygoids act as levers and plays an important role in manipulating the other skull bones to open the mouth and help the fang during biting.

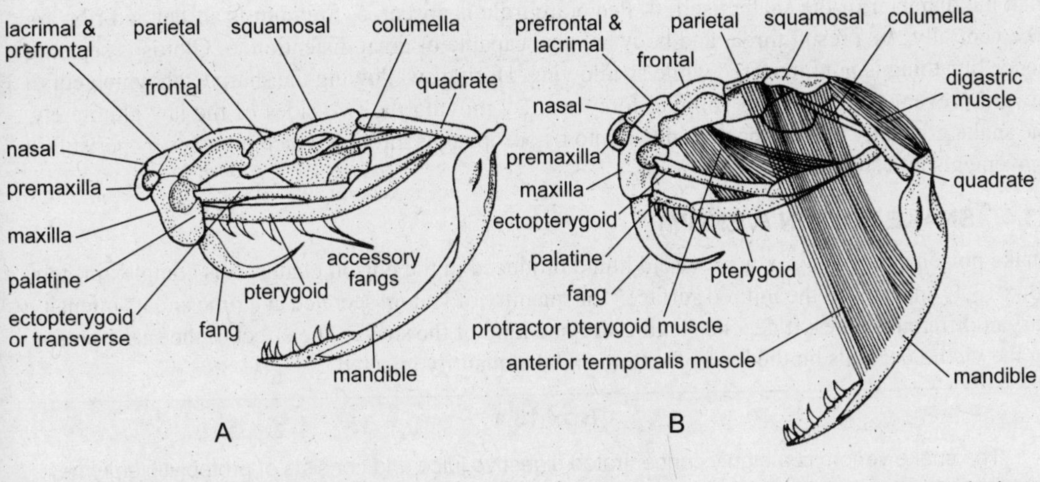

Fig. 13.4. Skull of a viper showing mechanism. A—Mouth closed at rest. B—Mouth opened when striking the prey.

13.3 BITING MECHANISM

Biting mechanism in vipers involves the following steps:

1. Body posture. The viper prepares to bite by raising its head a little above the rest of the body. The neck and the upper end of the trunk lie in two or three curves around the coiled body. In the next phase, the trunk muscles of the upper part of the snake contract suddely. It strightens the body abruptly and gives it a forward thrust to bite the prey.

2. Opening of the mouth. Contraction of digastric muscles lower the lower jaw and the mouth is opened.

3. Rotation of maxillae. Lowering of the mandibles (lower jaw) pushes the **quadrates** forward which pushes the **pterygoid** forward. Opening of the mouth is accompanied by the contraction of the **sphenopterygoid muscles** (or **protractor pterygoid muscles**) which also pulls the pterygoid forward. The pterygoids in its turn push the **ectopterygoids**. It causes the fangs bearing maxillae to rotate through 90° on its own axis around the lacrymal bone and as a result the fangs are erected.

4. Closing of the mouth. The erected fangs penetrate the flesh of the prey and immediately the sphenopterygoid muscles relax and the ectopterygoids and palatines are pushed backwards. This draws the fangs violently backwards and pushes the fangs deeper in the wound. At the same time the lower jaw is closed upon the wound. It is caused by the contraction of the three temporal muscles (*i.e.*, anterior temporalis muscles). The anterior temporal muscles squeeze the poison glands and the poison is injected into the body of the prey through the fangs.

The whole process consisting of penetration and deepening of the wound and ejection of poison is done in an instant. It takes longer time to open the mouth than to close it.

5. Disentangling. Head remains passive during disentangling process. Relaxation of neck muscles and continuous pull force exerted by the trunk muscles withdraws the fangs.

Swallowing Mechanism of Snakes

Snakes do not chew their food but swallow it whole, even if it is larger than the normal size of the mouth. Several structural adaptations make it possible: 1. The two rami of lower jaw are loosely connected anteriorly by an elastic ligament which permits lateral expansion. 2. The lower jaw is also loosely attached posteriorly to the quadrate bones which in turn are loosely attached to the skull. 3. Bones of the palate are also movable and this allows the mouth to expand several times than the diameter of the snake itself. 4. Pectoral girdle is absent. 5. Sternum is absent and ribs are free ventrally. As a result throat and body are also capable of great distention. 6. Glottis is separate hence breathing is not interfered while swallowing. During swallowing the sharp teeth which curve inward prevent the prey from slipping foreward. By moving the two sides of the jaw alternately, the snake gradually pushes the prey down into its oesophagus through which it passes by peristaltic movements into stomach.

13.4 SNAKE POISON (VENOM)

Snake poison or venom is a clear, viscid liquid produced in the poison glands. It is complex protein (Box 13.1) and acts on the injured surface. The quantity of venom secreted at a time and its intensity vary in different snakes. It depends on the size and kind of the snake. The effect of the snake poison on the victim depends on the kind of venom and the quantity injected.

Box 13.1

The snake venom is a highly concentrated digestive juice and consists of proteolytic enzymes, phosphatidases, neurotoxins and many other enzymes. Proteolytic enzymes, phosphatidases and neurotoxins produce toxic effect.

Phosphatidases cause haemolysis and effect heart and circulation. They attack capillary endothelium and cause haemorrhage.

Neurotoxins affect nervous system and cause paralysis.

Proteolytic enzymes domage muscles and endothelium. Other enzymes in the venom include proteases, erepsin, cholinesterase, hyaluronidase, ribonuclease, deoxyribonuclease and ophio-oxidase, etc. Hyaluronidase makes tissues more permeable to other venom components. Cobra and sea snake venom also contain basic polypeptides that block neuromuscular junctions, causing death by respiratory paralysis.

On drying, it forms fine needle-like crystals which easily dissolve in water. Venom can be preserved for indefinite period in dried form. The venom can be handled safely but it should not come in contact with blood or the mucous membrane. Since, it is a digestive juice, it is capable of digesting every thing of animal except hairs, feathers and claws.

Types of venom. The snake venom is of two types: Neurotoxic and Haemolytic.

1. Neurotoxic venom. It acts on the nervous system. It affects the optic nerves, causing blindness, or the phrenic nerves of the diaphragm, producing respiratory paralysis. *The venom of krait, cobra and sea snake is neurotoxic.*

2. Haemolytic venom. It breaks down the red blood corpuscles and destroys the endothelium of the smaller blood vessels, causing bleeding into the tissues and coagulation of blood there. *The venom of viper is haemolytic.*

13.5 SNAKE BITES AND ITS TREATMENT

Before treatment of a snake bite, it is necessary to know whether the bite is by a poisonous or non-poisonous snake, and if by a poisonous snake, whether poison has been injected or not.

The poisonous snakes have in the upper jaw a pair of large teeth, the fangs, in addition to small teeth. When they bite, they have two large, circular punctures made by the fangs besides rows of pricks caused by the small teeth (Fig. 13.5B, C). The non-poisonous snakes have only small teeth. Hence, they leave rows of small pricks only (Fig. 13.5A).

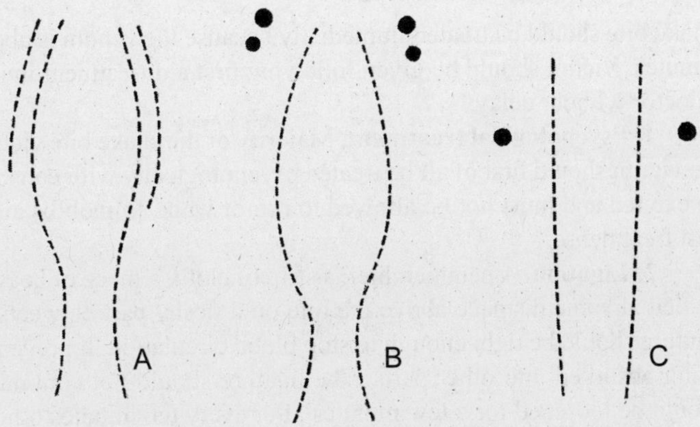

Fig. 13.5. Impression of teeth of snake. A—Non-poisonous; B—Poisonous; C—Poisonous.

Effects of venom. The effects of venom of three common poisonous snakes of India are the following:

1. Symptoms of cobra venom. Venom of cobra is most toxic (or virulent). Its venom is neurotoxic attacking nerve centers. After cobra bite local symptoms start to appear within six to eight minutes.

(*i*) The spot develops a reddish wheel and becomes tender. There is piercing pain and burning sensation at the site of bite. This is followed by numbness of bitten part which turn bluish or blackish. Soon after inflammation or swelling occurs and blood-colored fluid oozes out of the puncture.

(*ii*) After about 25 minutes the victim feels sleepy and intoxicated. He feels weak in legs and does not wish to stand or move.

(*iii*) After 35 to 50 minutes there is profound salivation, dropping of eyelids and vomiting (nausea), contraction of pupil. This is followed by paralysis and swelling of the tongue and larynx. Victim cannot swallow or speak properly.

(*iv*) After a few hours, the paralysis is complete. Respiration becomes slower, noisy and difficult. Heart beats rapidly. The victim cannot speak.

(*v*) Finally respiration stops and death occurs.

2. Symptoms of krait's venom. The krait venom is also neurotoxic similar to kobra. But it is deadly poisonous because krait injects a very large quantity of poison (about 3 times more than cobra's). Symptoms of krait's bite are very similar to those of cobra's bite, except that the victim complaints of unbearable abdominal pain due to internal haemorrhage. However, there is no swelling or burning pain at the site of bite and convulsions are milder. The feeling of drowsiness and intoxication is more intense. Destruction of red blood corpuscles (RBCs) and paralysis of trunk and limbs occurs. Death occurs within 6 to 24 hours.

3. Symptoms of viper's venom. The venom of vipers is **haemotoxic** and damages the circulatory and nervous systems more severly. Symptoms of viper's bite are the following:

(*i*) Viper's bite leads to great swelling (within 15 minutes) and discolouration of bitten parts, which spreads gradually all around.

(*ii*) There is an acute burning pain at the bitten part and and a red fluid oozes out from the wound due to massive tissue destruction (necrosis).

(*iii*) Pupils dilate, pulse rate increases, profuse vomiting occurs and victim suffers from cold sweats and finally loses consciousness.

(*iv*) Death may result due to paralysis of vasomotor centers and exhaustion from profuse bleeding.

Majority of the snakes are non-poisonous and it is the fear and ignorance that gives psychological shock and even death. Also a poisonous snake may not have injected a lethal dose of venom and in such cases, the victim may not show any symptoms. A non-poisonous snake does not inject any venom but can certainly transfer tetanus or gangerene in the wound.

13.6 FIRST-AID MEASURES

Snake bite should be treated immediatly because the venom is abosorbed into the body within a few minutes. Victim should be given following first aid treatment immediately and should be carried to a doctor without delay.

1. Psychological treatment. Majority of the snake bite victims die because of fear. Therefore, the victim should first of all be treated psychologically with consoling words. The victim should not be excited and must not be allowed to run or walk. Immobilisation of bitten part and rest in bed is best treatment.

2. Ligation. A handkerchief, a strip of cloth, a piece of heavy cord or a rubber ligature should be tied at some distnace above the bite on a fleshy part between the bitten part and the heart. The ligature should be tight enough to stop blood circulation. It prevents the venom from circulating and being absorbed into other parts. The ligature should not continue for more than half an hour and should be loosened for a few minutes after every ten minutes otherwise the parts below the ligation become gangrenous due to stoppage of blood circulation.

3. Incision. The wound should be washed with some antiseptic lotion or clean water or a weak solution of potassium permaganate ($KMnO_4$) (or of bleaching powder or calcium hypochlorite or of solution of Condy's fluid) which neutralises the venom by oxidation and is an antiseptic also. The fang wounds should be opened at the earliest by making 1 cm deep cut near the wound by a sterilized stainless steel knife or blade and the blood should be allowed to flow freely. In this way, some poison will flow out along the blood. The cut should be made lengthwise (not X-shaped) to avoid cutting of many or large blood vessels. Now the wound should be washed with antiseptic lotion or $KMnO_4$ solution.

4. Suction. Poison may also be sucked out from the site of snake bite by a suction cup but **never by mouth**.

5. Cryotherapy. The bitten parts are tried to be frozen immediately to prevent further absorption of venom in the tissue.

Anyhow, victim should be taken to a hospital at the earliest.

Antivenin Treatment

Antivenin is the only remedy for the snake bite and it should be administered within 5 hours of the bite. Various herbs, stones, ete., possessed by sadhus and others have been tested and found useless in curing snake poisoning.

Production of antivenin. A horse is made immune to a snake venom by injecting into its blood gradually increasing doses of the venom. The immune horse does not show any ill effect even if a large amount of venom is injected into it. The blood is drawn from a vein of such an immunised horse and allowed to clot. The clot is removed. The serum is left behind contains antibodies of the snake venom and is the antivenin. It is packed in vials and marketed.

The serums of the horses made immune to the venoms of cobra, krait and viper are mixed to make the antivenine effective against the bite of all poisonous snakes.

Antivenins are prepared in India at **Hopkin's Research Institute, Mumbai** and **Central Research Institute, Kasuali** (Himachal Pradesh).

Medicinal Uses of Snake Venom

1. Snake venom is used to prepare antivenin.
2. It is used in relieving pain and muscular disabilities.
3. The injection of cobra venom solution gives relief from pain in patients of neural leprosy, arthritis, epilepsy, etc.
4. Cobra venom solution provides relief from the unbearble pain caused by cancer, neurites, neuralgia, migraine and nerve leprosy,etc.
5. Venom of Russel's viper is used as a blood coagulating agent in cases of haemophilia and haemorrhages, etc.
6. In the Homeopathic and Aurvedic systems several medicines are prepared from snake venom.

13.7 PREVENTIVE MEASURES OF SNAKE BITE

To minimize the chancess of snake bite following precautions should be taken.

1. Dogs and cats dislike snakes and attack them. They rush at a snake and make a noise which attract our attention and enable us to notice the presence of snake. The dogs and cats may be tamed in houses, at least in snake infested area.

Many mammals are said to be immune for snake bites, *e.g.*, mongoose, hedgehog and pig.

2. Certain snakes live in or near houses. So, it is necessary to keep the compounds of houses free of tall grasses, jungle, nests, broken masonary, rubbish heaps, etc.

3. Thatched roofs should always be considered unsafe, as snakes are fond of living in them.

4. Snakes usually inhabit old deserted buildings, holes, crevices and overgrown grasses. While going in such places one should wear long boots and keep his hands away from crevices, holes, etc.

5. Sleeping on ground in snake infested areas should be avoided.

6. One should carry a torch when going out at night.

7. Villagers going out at night usually carry a stout stick (lathi/danda) with which they strike heavily on the ground. It is a useful precaution. A snake hears sound conveyed through the ground and moves away from one's path.

8. When going in forests, etc., one must wear long boots.

9. One should not move without shoes in snake infested areas.

10. In snake infested areas antivenin should be kept by every doctor.

QUESTIONS

Long Answer Questions

1. Give zoological names of the snakes of your locality, stating which are poisonous. (*Garhwal, 2002*)

2. Write an essay on poisonous snakes of India.

3. How will you distinguish between poisonous and non-poisonous snakes. Give the zoological names of five poisonous snakes of your locality.

4. Describe the poisonous apparatus of snake. Explain the biting mechanism of a poisonous snake. (*CCSU Meerut, 2007*)

5. What is snake venom? Distinguish between poisonous and non-poisonous snakes with examples. (*CCSU Meerut, 2013*)

6. Describe the structure of biting apparatus of a poisonous snake and explain its mode of working. (*Purvanchal, 2000*)
7. What is antivenin? How it is prepared? Give preventive measures of snake bite.
8. What is snake venom? Enumerate the symptoms of snake bite. What are the preventive measures and how is antivenin is prepared.

Short Answer Questions

1. Write short note on: Biting apparatus of poisonous snake. (*CCSU Meerut, 2013*)
2. Distinguish between poisonous and non-poisonous snakes.
3. Mention symptoms of bite of cobra and viper
4. Describe first aid meausres of snake bite.

Very Short Answer Questions

1. Name four poisonous snakes. (*CCSU Meerut, 2006*)
2. Give zoological name of krait.
3. Name two non-poisonous snake found in India.
4. By seeing the belly plates, how will you identify whether a snake is poisonous or non-poisonous?
5. Give scientic name of king cobra.
6. Which is the largest and deadliest venomous snake?
7. What are fangs?
8. What are solenoglyphous fangs?
9. Which snake has proteroglyphous fangs?
10. Which muscles help snake in biting and injecting poison in the body of victim?
11. What is the effect of neurotoxic venom?
12. Venom of which snake is neurotoxic?
13. What is haemolysis? Venom of which snake causes haemolysis?
14. What is haemotoxic venom?
15. What is antivenin?
16. How it can be identified whether a person is bitten by a poisonous or non-poisonous snake?
17. How can a sea snake be identified?
18. What is the use of tongue for a snake.
19. Identify the snake which has complete belly plates and small scales on the head and diamond-shaped patches on the body?
20. How snakes are able to engulf prey several times larger than their own diameter?

Fill in the Blanks

1. In poisonous snakes, the first pair of maxillary teeth are modified into ________. (Fangs/ Salivary gland)
2. Fangs are proteroglyphous in kraits but solenoglyphous in ________. (Viper/Cobra)
3. Rattle snake has ________ type of fangs. (Solenoglyphous/Opisthoglyphous)
4. In snakes, poison glands are modified ________ glands. (Parotid/Thyroid)
5. All marine snakes are ________. (Poisonous/Non-poisonous) (*Garhwal, 2003*)
6. In snakes, tongue serves as an organ of ________ and taste. (Olfaction/Hearing)
7. Venom of snakes is a clear, yellowish liquid formed of enzymes and ________. (Toxins/Proteins)
8. Venom of Cobra is ________ and most virulenet. (Neurotoxic/Haemotoxic)
9. Krait's bite injects ________ times more than Cobra's bite. (Two/Three)
10. Non-poisonos snakes neither have ________ not the poison apparatus. (Fangs/Teeth)
11. In India, antivenin is prepared at ________ Institute at Mumbai. (Hopkin's/Pusa)

Multiple Choice Questions

1. The most poisonous snake is
(*a*) krait (*b*) rat snake
(*c*) tree snake (*d*) python

2. In snakes, poison glands are modified (*Gorakhpur, 2002*)
(*a*) lingual glands (*b*) sublingual glands
(*c*) superior labial or parotid glands (*d*) mandibular glands

3. Poisonous fangs of a snake are modifed
(*a*) premaxillary teeth (*b*) maxillary teeth
(*c*) palatine teeth (*d*) canines

4. Teeth conducting poison in a snake are called
(*a*) incisors (*b*) canines
(*c*) fangs (*d*) molars

5. In a snake tail is cylindrical and ventral scales do not extend the entire width of belly. It is
(*a*) non-poisonous (*b*) definitely poisonous
(*c*) either poisonous or non-poisonous (*d*) deadly poisonous

6. Large size plate-like scales that fully extend from side to side on the belly will be seen in
(*a*) python and krait (*b*) krait and cobra
(*c*) dhamin and python (*d*) krait and sea snake

7. Loreal pit of pit viper is
(*a*) thermoreceptor (*b*) chemoreceptor
(*c*) olfactoreceptor (*d*) rheoreceptor

8. Third supralabial scale is largest and touches the scales bearing eye and nostril in
(*a*) krait (*b*) cobra
(*c*) viper (*d*) all coral snakes

9. Kraits are identified by
(*a*) a row of hexagonal ventebral scales (*b*) body colour
(*c*) presence of hood (*d*) large belly plates

10. In viper the scales on the head are
(*a*) small and keeled (*b*) large and shield like
(*c*) small and overlapping (*d*) absent

11. Pear-shaped head, sharply separated from the trunk and covered with small scales is a feature of
(*a*) krait (*b*) vipers
(*c*) cobras (*d*) sea snakes

12. Scales on the dorsal surface of cobra are
(*a*) small (*b*) hexagonal
(*c*) keeled (*d*) shield-like

13. Freshwater snakes are generally
(*a*) viviparous (*b*) poisonous
(*c*) non-poisonous (*d*) deadly poisonous

14. Venom of snake is
(*a*) haemolytic (*b*) neurotoxic
(*c*) neurotoxic as well as haemotoxic (*d*) either neurotoxic or haemotoxic

15. Venom of cobra is

(*a*) neurotoxic (*b*) haemotoxic
(*c*) vasotoxic (*d*) proteolytic

16. Venom of viper affects

(*a*) circulatory system (*b*) digestive system
(*c*) exceretory system (*d*) nervous system

17. Antivenin injections for snake bite are prepared at

(*a*) IARI, New Delhi (*b*) NDRI, Lucknow
(*c*) Hopkin's Research Institute, Mumbai (*d*) IVRI, Bareilly

18. Proteolysin is present in the venom of

(*a*) cobra (*b*) krait
(*c*) viper (*d*) sea snake

19. The best medicine to be injected in case of snake bite is

(*a*) penicillin (*b*) antivenin
(*c*) antiprotein (*d*) streptomycin

20. Poisonous fangs are modified

(*a*) Mandibles (*b*) maxillary
(*c*) nasals (*d*) canines

21. Fangs in viper are (*Rohilkhand, 2003*)

(*a*) solenoglyphus (*b*) proteroglyphus
(*c*) opisthoglyphus (*d*) none of these

ANSWERS

Very Short Answer Questions

1. (*a*) *Naja naja* (Cobra); (*b*) *Vipera russelli* (Russell's viper); (*c*) *Callophis melanurus* (Coral snake) and (*d*) *Hydrophis caerulescens* (Sea snake).

2. *Bungarus caeruleus*.

3. (*a*) Dhaman (*Ptyas*) and (*b*) Ajagar (*Python molurus*).

4. Belly plates are incomplete in non-poisonous snakes but if belly plates are complete or extend along the entire width of belly, the snakes may be poisonous or non-poisonous.

5. *Naja hannah* or *Ophiophagus hannah* is king cobra.

6. King cobra is the deadliest poisonous and largest Indian snake.

7. Fangs are modified maxillary teeth found in poisonous snakes which act as hypodermic needle for injecting poison in the body of victim.

8. Solenoglyphous fang is a hollow pipe-like structure with poison canal lined by enamel which opens at the tip of fang.

9. *Proteroglyphous* fangs are found in cobra, kraits, coral snakes and sea snakes.

10. The muscles that help in biting and injecting poison are **digastric muscles** that cause erection of fangs, **constrictor muscles** force the poison from poison gland into fang and **temporal muscles** that close the mouth.

11. The neurotoxic venom of cobra and krait affects nervous system and causes paralysis of respiratory muscles. Death is caused by asphyxiation.

12. Venom of cobra, krait and sea snakes is neurotoxic.

13. Haemolysis is destruction of RBCs. Krait venom causes haemolysis.

14. Haemotoxic venom causes widespread haemorrhage.

15. Antivenin is antivenom serum wich has sufficient antibodies to neutralize tha effect of proteins of venom of a particular snake.

16. The bite of a poisonous snake leaves only two large punctures made by the fangs. The bite of a non-poisonous snake leaves many small pricks and no punctures.

17. Sea snake has a broad and flattened or oar-like tail.

18. snake's tongue is able to collect odour and vibrations from the surrounding air and is sensitive to touch.

19. Pitless viper (*Vipera russelli*).

20. Snakes are able to engulf prey several times their own diameter because the two rami of lower jaw are loosely connected by elastic ligament; the lower jaw is loosely attached to movable quadrate and bones of palate are also movable.

Fill in the Blanks

1. Fangs. **2.** Viper. **3.** Solenoglyphous. **4.** Parotid. **5.** Poisonous. **6.** Olfaction. **7.** Toxins. **8.** Neurotoxic. **9.** Three. **10.** Fangs. **11.** Hopkin's.

Multiple Choice Questions

1. (*a*)	**2.** (*c*)	**3.** (*b*)	**4.** (*c*)	**5.** (*a*)	**6.** (*b*)
7. (*a*)	**8.** (*b*)	**9.** (*a*)	**10.** (*c*)	**11.** (*b*)	**12.** (*c*)
13. (*c*)	**14.** (*d*)	**15.** (*a*)	**16.** (*a*)	**17.** (*c*)	**18.** (*c*)
19. (*b*)	**20.** (*b*)	**21.** (*a*)			

14

The Pigeon: *Columba livia*

14.1 CLASSIFICATION (SYSTEMATIC POSITION)

Phylum	—	**Chordata**
Subphylum	—	**Vertebrata (Craniata)**
Section	—	**Gnathostomata**
Superclass	—	**Tetrapoda**
Class	—	**Aves**
Subclass	—	**Neornithes**
Superorder	—	**Neognathae (Carinatae)**
Order	—	**Columbiformes**
Family	—	**Columbidae**
Genus	—	***Columba***
Species	—	***livia***

The bird, called pigeon, is a nice representative of the class Aves. It is easily available, harmless and fairly large for laboratory study. There are many domestic varieties of pigeon (*Columba livia* var. *domestica*) such as **carrier**, **tumbler**, **fantail**, **pouter**, etc. These varieties of pigeon differ from one another in size, proportions, colouration and details of feather arrangement. All of them have been produced from the *Columba livia*, the common blue or wild **rock pigeon**, by animal breeders through artificial selection.

The rock pigeon originated in Western Asia or Egypt. It has two races in India: smaller and darker *intermedia* found in plains and larger and lighter *neglecta* found in North West India upto 4000 meters in the Himalayas.

14.2 HABITAT

Columba livia is common over most of Europe and Asia. Its fully wild stock inhabits rocky and inaccessible places such as caves, coastal clifs, old ruins, etc. Its semiwild variety is almost commensal of human beings. It is commonly seen in grain markets, warehouses, railway and old buildings. It often nests in houses occupied by human beings.

14.3 HABITS

The pigeon is diurnal, gregarious bird, forming small to large flocks. Often it feeds in flocks on the ground. It has the following habits:

(*i*) Feeding. Pigeon commonly feeds on grains, seeds and young green shoots. Occasionally, this bird takes insects and snails. It dips its entire beak under water and drinks like a horse (*i.e.*, pigeon does not sip water like other birds).

***(ii)* Nest.** Pigeon does not nest on trees, but in old buildings (ledges). The pigeon's nest is very crude and artless. It is made of dry grass, small twigs and roots.

***(iii)* Locomotion.** The pigeon is not capable of rapid locomotion on land because of short legs and rather weak feet. Its flight is however, swift.

***(iv)* Breeding.** The pigeon is monogamous, *i.e.*, one male lives in company of a single female. It displays a brief courtship before copulation. Pigeons have characteristic sexual and advertising displays, such as bowing and special display flights.

Only two, rarely one, elliptical plain white eggs are laid by the female in spring or summer. They are incubated by both parents. The period of incubation extends over fourteen to eighteen days. The youngs (called **squabs**) are fed by the both parents on "pigeon's milk", which is whitish fluid secreted in their crop during the breeding season.

***(v)* Sound.** Pigeons do not sing but they can produce sound. Pigeon's sound is a deep *gootr-go, gootr-goo*.

***(vi)* Homing.** The carrier pigeon has been bred for speed and stamina. It has a marked homing instinct and is used for carrying messages.

***(vii)* Enemies.** Pigeons are attacked at night by cats and owls. They are also attacked by hawks and snakes. Human beings are also its enemies. They kill many pigeons for food, fun and study.

Economic Importance

Pigeon is useful as well as harmful. It forms an important laboratory animal, a nice pet, a good food, and a reliable messenger; but it consumes large quantities of grains and sometimes becomes a source of nuisance in the houses.

14.4 EXTERNAL FEATURES

1. Shape, Size and Colour

The body of pigeon is boat-shaped and streamlined and offers little resistance to air (*i.e.*, wind) during flight (an adaptation for flight). Like the other birds, the length of pigeon's body is reduced so that there is equal distribution of weight in flying as well as in walking.

The body of pigeon is about 32 cm long. Its entire body, except the lower part of hind legs (feet) and beak, is covered with backwardly directed and overlapping feathers. The body colour is slaty grey with glistening metallic green, purple and magenta sheen around the neck and upper breast. There are two black bars on the wings and a band across the end of tail. The eyes and feet of pigeon are pink.

2. Division of the Body

The body of the pigeon is divisible into *four* regions: head, neck, trunk and tail (Fig. 14.1).

***(i)* Head.** Head is small in proportion to the body. It is round and mobile. Head consists of the following parts:

***(a)* Beak.** Head bears in front a pointed **beak**. The beak is short, slender and weak. It is adapted for picking up seeds and grains and not for cracking them. The beak is formed by the elongation of upper and lower jaws. The beak is toothless and is covered with a horny sheath called **rhamphotheca**. The **mouth** is terminal and has a wide gape. It extends along the entire border of the beak. On each side of the base of the upper beak is a sensitive patch of naked, soft, white swollen patch of skin, called **cere**.

***(b)* Nostrils.** There are two slit-like external **nares** or **nostrils** at the base of the upper beak.

***(c)* Eyes.** On either side of the head is a large, circular **eye**. Each eye is provided with **upper** and **lower eyelids** and a well-developed, semitransparent **third eyelid** or **nictitating membrane**. The third eyelid can be pulled across the eyeball from the anterior corner of the eye. Pupil is rounded.

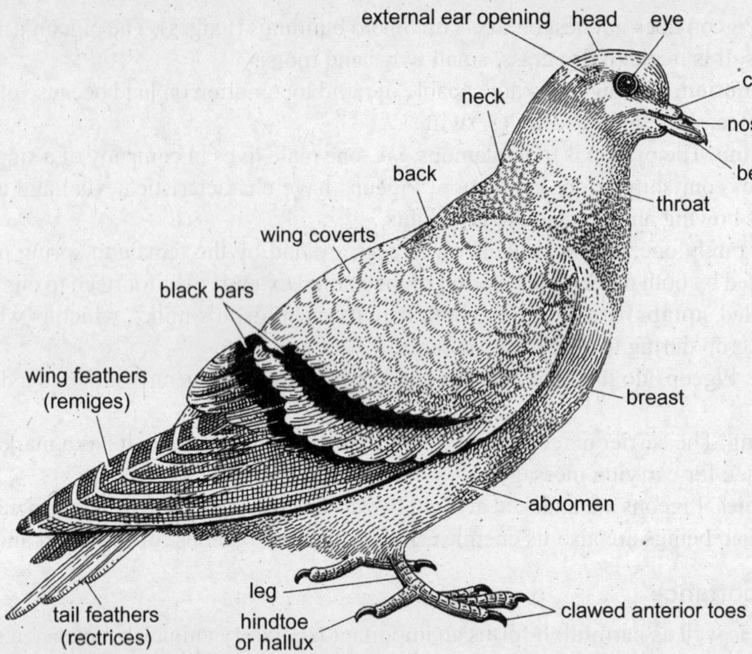

Fig. 14.1. Pigeon (*Columba livia*). External features.

(*d*) **Ear openings.** Just behind and little below each eye lies a small, rounded external ear opening or the **external auditory aperture**. It remains hidden under special **auricular feathers**. The aperture leads into a short tube, called **external auditory meatus**. The **tympanic membrane** lies at the other end of this tube.

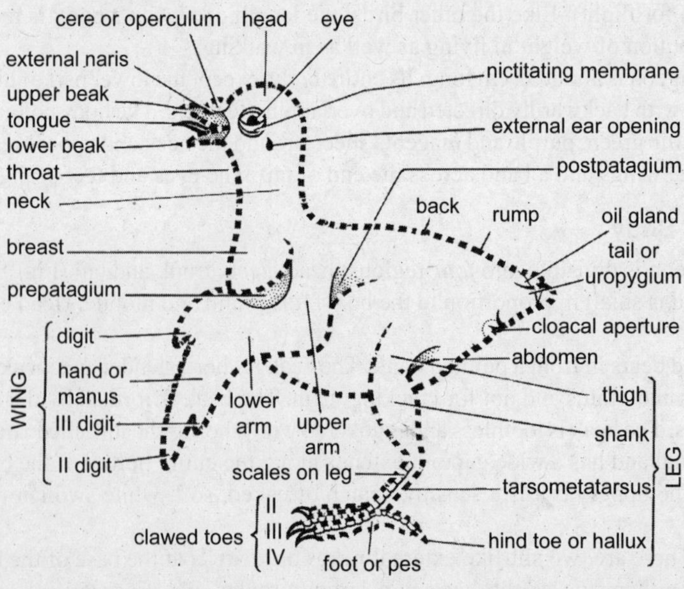

Fig. 14.2. Pigeon. Bird after removing the feathers (in left lateral view).

***(ii)* Neck.** The neck connects the head with the trunk. It is long, cyndrical and mobile (*i.e.*, highly flexible). Neck of pigeon is almost vertical and slightly widens toward the trunk.

***(iii)* Trunk.** The boat-shaped trunk is large, heavy and somewhat spindle like. It bears two pairs of limbs. It has a prominant midventral ridge, the **keel** or **carina**. Trunk itself is differentiated into an anterior **thorax** to which the wings (forelimbs) are attached, and a posterior **abdomen** to which the legs (hindlimbs) are attached. A large, transverse, slit-like **cloacal aperture** is situated ventrally at the hind end of the abdomen. Cloaca is a common outlet for the faeces, urine and genital products. It is guarded by a sphincter.

***(iv)* Tail.** The true tail is very short and stumpy, and can be seen only in a plucked bird. It is known as the **uropygium**. It bears 12 long tail feathers (called **rectrices**) to which the word "tail" is usually applied. On its upper surface is a small papilla, which bears the opening of the **preen** or **oil** or **uropygial gland**. This is the only integumentary gland in birds. The oily secretion of this gland renders the feathers waterproof.

14.5 EXOSKELETON

The birds are clothed in an exoskeleton consisting of: 1. *Claws*; 2. *Scales*, both are present on the feet. 3. *Horny beaks*. These all are of epidermal origin and formed from the stratum corneum. They are obviously indicative of reptilian descent. 4. *Feathers* which are distributed on the great part of the body and are believed to be homologous to reptilian scales but resemble them only in very earliest stages of formation.

1. Claws. The digits of the foot are provided with pointed, sharp and horny claws at their tips. Claws serve as an organ of offence and defence and help during perching and walking.

2. Scales. The scales or corneoscutes cover the shanks and digits of pigeon (*i.e.*, exposed parts of the hind limbs). These scales represent cornified epidermis and remain arranged in the tile-like fashion. They are protective in function.

Scales of pigeon are homologous with those of reptiles, thus indicating relationship of the two groups.

3. Horny beaks. The upper and lower jaws are provided with horny sheaths or beak which are horny exoskeleton derivative. These are structures of procurement of food and also act as an organ of offence and defence. At the base of beak is a naked swollen patch of skin known as **cere**. Beak is used during ingestion, preening of feathers and fighting.

4. Feathers. Feathers are light, flexible, elastic and water-proof derivatives of the corneal layer of epidermis of the skin, derived from the reptilian scales and forming the outer protective covering. The possession of feathers is an unique feature of birds. They cover the most part of the body and form the wings and tail. They are arrangement in definite tracts called **pterylae**, which are separated from the featherless tracts called **opteria** (Fig. 14.3). The distribution of feathers on the body is called **pterylosis**.

The feathers may be white due to the presence of air spaces or coloured due to presence of pigments (Box 14.1). The irridescent metallic tinge is found in feather, specially in the neck region. These are due to prismatic effects as the barbul's air spaces on which sun's rays act.

Box 14.1 Pigmentation of Feathers

The colourations of birds are for concealment, recognition and sexual dimorphism and form the basis for their social life. Feather's colours are produced partly by pigments and partly by reflection and diffraction effects (called **structural colouration**). The most common pigments of feathers are **melanins**, ranging from black through brown to yellow. Various colours are due to three factors: 1. Pigment is deposited in the feathers during development by special cells in the papilla. The processes of these amoeboid **chromatophores** convey granules of pigment into the barbs and barbules, where they are laid down in layers between those of keratin. Carotenoid pigments are also found such as the yellow **zooxanthin** and red **astaxanthin**. The colour that is seen, is due to absorption of some wavelengths of light by the pigment, thus, black, red, brown, yellow and orange colours are seen. White colour is not due to white pigment but is caused due to reflection of light from the feather without absorption of any wavelengths of light rays. The blue colour incident light is reflected from a layer of spongy keratin, pierced by holes 1–2 μm in diameter which absorbs the red and reflect blue light. 2. Structural arrangement or **striations** of feather's surface are prismatic, these cause iridescence due to reflection of light, thus, producing iridescent hues, metallic colours, gray and some shades of blue. In iridescent feathers interference of light is thin surface film. 3. A combination of pigments and prismatic striations of the feathers produces green in which yellow pigment combines with the structural blue.

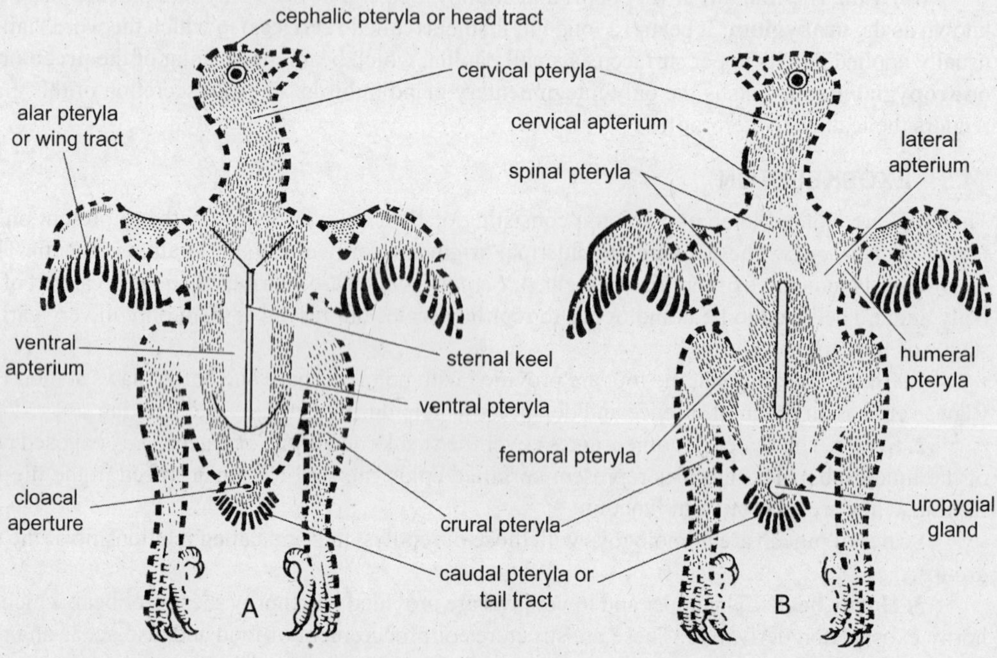

Fig. 14.3. Pigeon. Pterylosis (feather tracts) A—Ventral view and B—Dorsal view.

Structure of a Typical Feather

A typical feather of pigeon such as contour type quill or flight feather has the following parts (Fig. 14.4).

1. Scapus or stem. It is **central axis** of the feather which is a stiff axial rod running the whole length of the feather. It consists of two unequal and dissimilar portions.

***(i)* Stem or calamus.** It is the hollow, tubular semi-transparent proximal portion, the base of which is inserted into the skin. At the proximal end of the calamus there is a small aperture, the **inferior umbilicus** and at its distal end, between the quill and the vane, there is a second aperture, the **superior umbilicus**.

***(ii)* Rachis or shaft.** It is the distal solid portion of the stem and is continuous with the calamus. It is a tapering, flexible elastic rod, square in transverse section and grooved along its ventral surface. The rachis or shaft forms the longitudinal axis of the vane. It differs from the calamus in being stiff, solid, opaque, roughly quadrangular in transverse section and is filled with a closely packed mass of "pith" cells. A longitudinal furrow, known as **umbilical groove**, runs along the inner or ventral surface of the rachis throughout its length.

2. Vexillum or vane. It is the distal expanded fan-shaped part of the feather, attached along the sides of the rachis. It is made up of barbs, barbules and barbicels or hooklets.

(*a*) Barbs. There are a series of narrow, parallel and closely spaced delicate thread, like laminae attached by their bases along two sides of the rachis, perpendicular to the plane of the vane. The size of barbs gradually decreases towards the two ends of the rachis.

(*b*) Barbules. These are much smaller double row of many extremely delicate, oblique, filamentous processes which form fringes along the sides of the barbs. The barbules are of two types which are **distal barbules** directed towards the tip of the feathers and **proximal barbules** directed towards the base of the feather.

(*c*) Barbicels or hooklets. The distal barbules of each barb bear minute **hooklets** or **barbichamuli** which fit into grooves or **flanges** of the proximal barbules on the next adjacent barb. Thus, all the barbs are firmly fastened together and form a single continuous surface to offer resistance to air (Fig. 14.5). Thus, vane of either side of the axis forms a firm flexible and unbroken surface which can offer great resistance to the air.

The feathers of ostriches and kiwis lack this interlocking mechanism of feathers of the pigeon.

3. Aftershaft. A small tuft of down feathers possessing hairy processes, is present in the neighbourhood of superior umbilicus and is termed as **after shaft** or **hyporachis**. In the pigeon, it is minute but in cassowary and emu and after shaft is as long as the main feather. Hyporachis have not hooklets.

Each feather has its own basal muscles in the skin to erect it in flight, cold weather and emotional excitement.

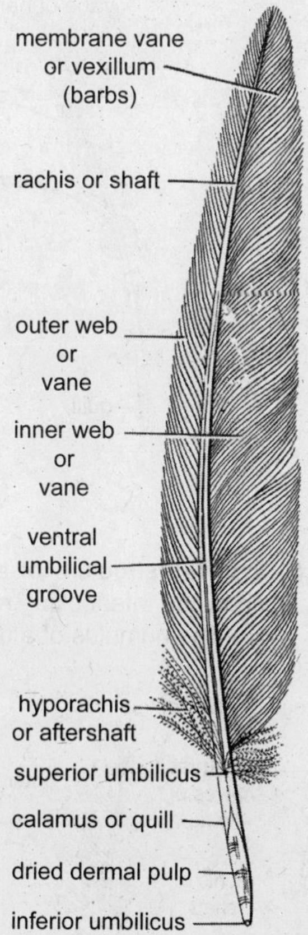

Fig. 14.4. Pigeon. A typical feather (*i.e.*, quill or flight feather) in ventral view.

Kinds of feathers. Depending upon the position, appearance and function, the feathers of a bird are distinguished into following categories.

1. Down feathers or plumules. The down feathers cover the newly hatched bird. These are called **nestling downs**. Each consists of a very short calamus and a thin reduced rachis bearing long, flexible barbs with short barbules. These are termed as the **permanent** or **powder downs**. Each consists of a short calamus bearing at its top long, flexible barbs with barbules devoid of interlocking hooklets.

In a young one of pigeon, the nestling down feathers cover the body. In adult pigeon, the nestling down feathers are replaced by contour feathers. But they persist as an undercoat beneath the contour feathers in many aquatic birds, such as ducks and swans, serving to increase the thickness of insulating layer.

2. Filoplumes. They are also called **hair feathers** or **pin feathers**. These are minute delicate hair-like rudimentary feathers left in the skin after the bird has been plucked. Each filoplume consists of a short calamus and a long delicate shaft bearing a tuft of fine barbs at the top.

3. Rictal bristles. These are modified filoplumes found near the mouth in fly-catchers. Each has a bristly hair-like appearance consists of a short calamus and a slender (long) rachis with few minute barbs present at the base of the rachis.

4. Quill or flight feathers. These are largest and most important flight feathers on the wings and tail of pigeon (Fig. 14.7). They may be subdivided into **wing feathers** or **remiges** (singular **remex**) and **tail feathers** or **rectrices**.

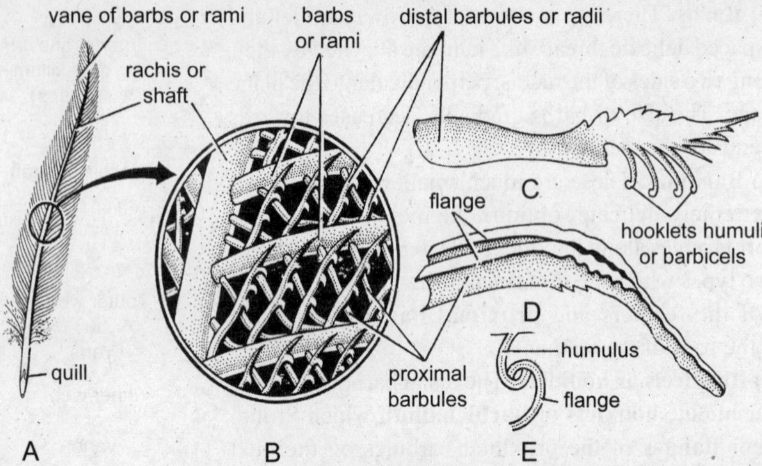

Fig. 14.5. Detailed structure of a feather. A—Complete feather; B—A small part (magnified to show the interlocking mechanism); C—A distal barbule; D—Two proximal barbules and E—The humulus of a distal barbule interlocking with the flange of a proximal barbule.

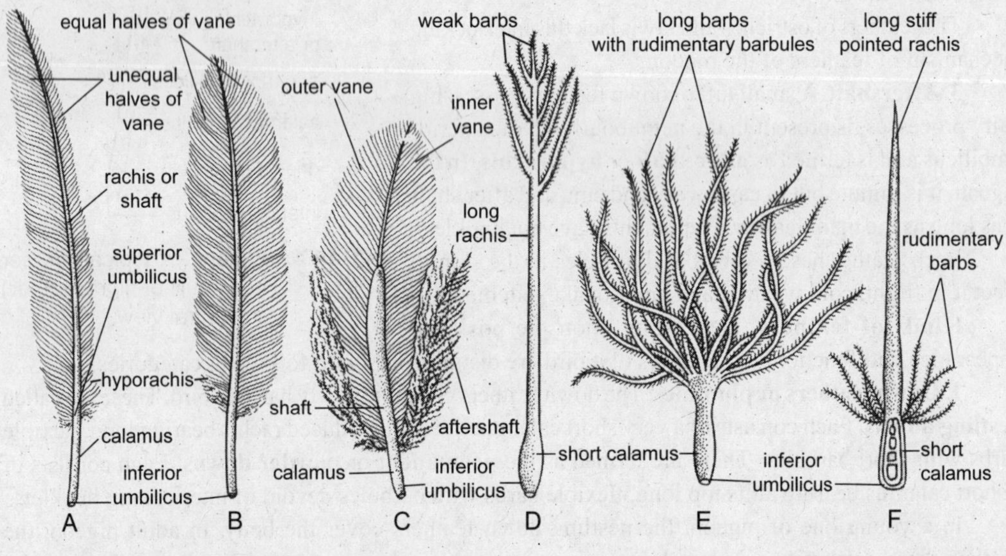

Fig. 14.6. Feathers of pigeon. A—Remex; B—Rectrix; C—Contour; D—Filoplume; E—Down and F—Rictal bristle (not found in pigeon).

(*i*) Remiges or wing feathers. The remiges are present in the wings and each wing has 23 feathers. They are placed on the hinder border of the wings and help in flight. Remiges are divided into groups in accordance with the bones to which they are attached. Twelve remiges are attached to the ulna bone of forearm and are called the **secondaries** or **cubitals**. Eleven are fixed on metacarpels and second and third digits of hand and are called **primaries** or **manual** of these, six are borne on the postaxial border of palm and are called **metacarpals** and five arise from the second and third digits and are termed the **digitals**.

The digitals are further distinguished into a single **ad-digital** attached to the single phalanx of third digit, two **mid-digitals** borne on the proximal phalanx of the second digit and two pre-digitals arising from its distal phalanx of second digit (Fig. 14.7). The first digit of the hand, *i.e.*, thumb or pollex, bears a tuft of small feathers. They form the so-called **bastard wing**, **ala spuria** or **false wing**.

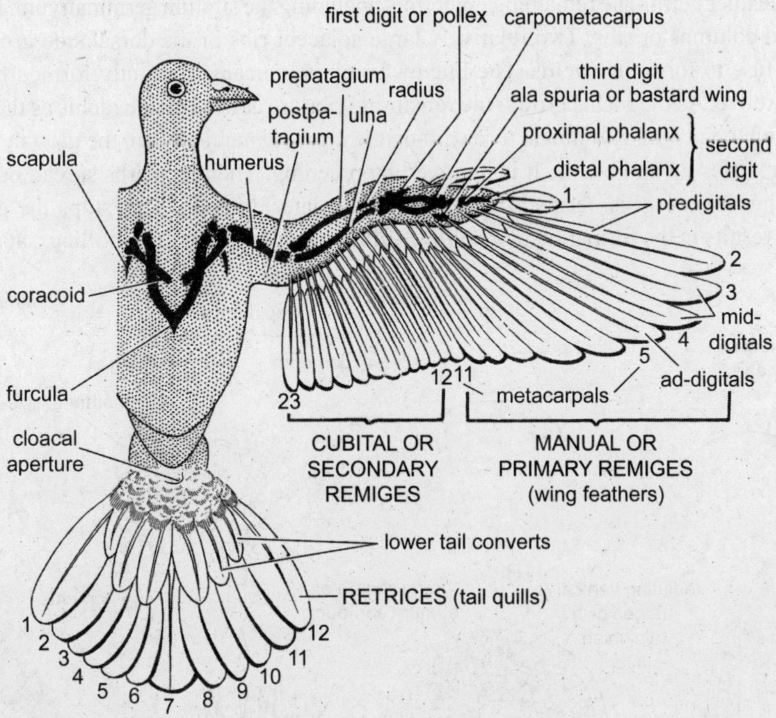

Fig. 14.7. Pigeon. Arrangement of flight feathers or remiges of a wing.

The upper arm (humerus) lacks remiges. When present, they are called **tertiaries**. In each of the remiges, the proximal portion of the vane is broader than the distal.

***(ii)* Rectrices or tail feathers.** The **quills** or **rectrices** (singular **rectrix**). There are 12 rectrices which are arranged in a semi-circle around the tail (*i.e.*, in a fan-shaped manner). In each rectrix, the parts of the vane on either side of the rachis are equally broad. Tail feathers of pigeon are used as a brake and in steering during flight.

***(iii)* Contours.** These are short, quill-like, soft feathers which form a general covering of feathers over the body. The barbules are less perfectly developed so that the barbs are easily separated from each other at the base of the vane. Contour feathers provide a smooth surfacing to the body reducing the drag or frictional resistance to motion through air, give shapes by overlapping of the vanes and makes a good heat-conserving surface.

***(iv)* Coverts.** These are quill-like smaller oval feathers which form a general covering at the basis of remiges and rectrices filling the space between them. It covers the bases of the wings and tail. Coverts close the interstices between the quills of the flight feathers thus presenting a continuous area to oppose the buoyancy of the air.

Development of Feathers

A feather, like a reptilian epidermal scale, originates as a small projection on the skin in the embryo (Fig. 14.8). The projection is called the **feather papilla**. It consists of a core of vascular dermis covered by epidermis. The epidermis gives rise to the feather and the dermis supplies material for the development. The feather papilla soon grows into a cylindrical **feather germ**. At the same time, the feather germ sinks into a tubular pit, the **feather follicle**. The latter will hold the feather in the skin. The outermost layer of the epidermis gets cornified to form a smooth sheath, called the **periderm**, around the feather germ. The remaining epidermis, including the stratum germinativum, forms parallel longitudinal columns or ribs. Two relatively large adjacent ribs of the dorsal side grow faster than others and fuse to form the **rachis**. The line of fusion is indicated in a fully formed feather by the umbilical groove. As the rachis grows, the remaining ribs are carried up the rachis as the **barbs**. New barbs arise on the ventral side and in turn shift up the growing rachis. When the growth is completed, the periderm dries up and bursts. It is removed by preening. Then the barbs spread out to form the **vane**. The nutritive dermis, termed the **feather pulp**, later dries up. This stops the growth of the feather and results in the formation of a hollow **quill** with an opening, the umbilicus, at its either end.

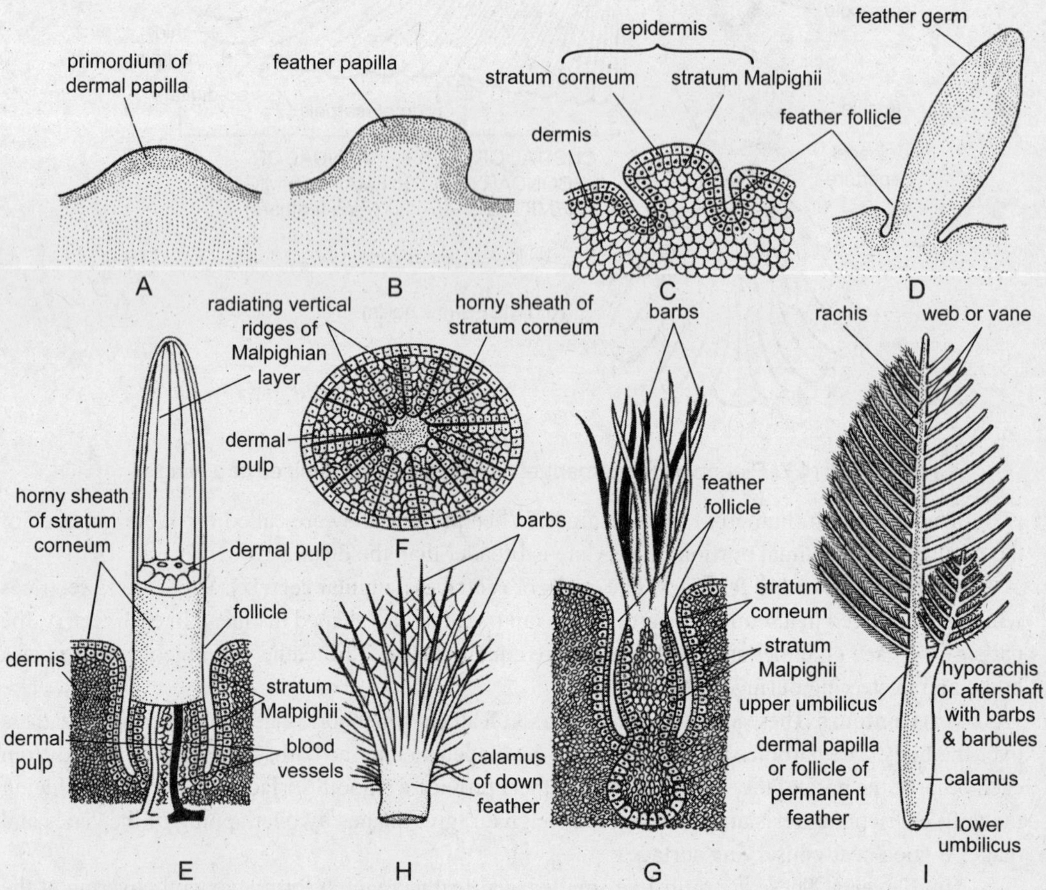

Fig. 14.8. Stages in the development of a contour feather. A—Beginning of feather papilla; B—Growth of feather papilla; C—Stage in V.S.; D—Feather germ; E—Feather germ in V.S.; F—Feather germ in T.S.; G—Down feather in the follicle; H—Mature down feather of a newly hatched bird and I—Young contour feather.

The entire feather is formed of dead, horny cells. Pigments (lipochromes and melanins) are added to the epidermal cells during growth in the follicle, but not afterwards.

Development of a down feather is relatively simple. All the ribs of the epidermis are equal and form barbs, which unfold on rupturing of the periderm.

Moulting. The feathers are moulted (*i.e.*, shed and replaced) annually. The replacement of feathers is seasonal in some birds such as peacock, while in other birds such as pigeon, it is gradual throughout the year. Moulting or ecdysis generally happens in late summer. All the feathers are not shed at the same time. Falling out of old feathers and formation of new ones in their place go on side by side. New feathers develop from the old papillae.

Succession. The first set of feathers that grows in a young bird (chick) consists of the down feathers. These feathers are called the **neossoptiles**. A growing bird develops one or more sets of juvenile feathers termed as the **messoptiles**. The feathers of the adult bird are known as the **teleoptiles**.

Functions of feathers. Feathers serve a variety of important functions:

1. Protection. Feathers provide a light weight impervious, flexible but resistant and durable body covering. They protect the underlying tender skin from wear or any mechanical and physiological injury.

2. Insulation. Feathers reduce the loss of water from the body surface and prevent water from reaching the skin.

3. Heat retention. Feathers keep bird's body warm and well protected from weather. The birds maintain extremely high body temperature, commonly 104° to 112°F (*i.e.*, 40–42°C). The feathers are bad conductors of heat. They form a very effective insulating covering and enclose numerous air spaces between themselves and the body. They prevent loss of heat from the body. In cold weather the heat loss is minimized by fluffing out the feathers. This action increases the depth of air spaces within the feathery layers. In hot weather, the feathers are held close to the body or erected to allow loss of body heat.

4. Organs of flight. Feathers make the bird's body well adapted for aerial mode of life. They are light, elastic, horny structures which make the body of birds of quite light weights. Further, the thin, flat and overlapping wing and tail feathers, with their close almost air tight linkage due to interlocking arrangement of barbules, form surfaces to support the bird in flight. Tail feathers act as a rudder in flight, a brake during landing, and a counter balance during flight and perching.

5. Camouflaging (Protective colouration). Feathers are variously coloured in birds. In certain species, the feather colour make the bird inconspicuous by their close resemblance to the surroundings. In this way, the feather colours provide protection to the birds from their enemies.

6. Sex determination and courtship. Males and females of most species of birds differ in their plumage and recognise each other by this means also. Usually the males have more showy plumage and they display it before the female in the act of courtship. On many birds the crown feathers are modified into crests (peacocks), topknots (tufted titmouse), ornamental plumes (California quail), etc., are utilized in courtship displays. Some birds have special ruffs (ruffed grouse) or pinnae (prairie chickens).

7. Natal covering. The body of newly hatched youngs of many species of birds remain covered with down feathers. These form a highly efficient insulating covering of the youngs and keep them warm.

8. Nest formation. Many species of birds use feathers in making their nests. Feathers help in conserving heat around the nestlings, and make the parent more comfortable. In ducks, a special type of down feather is grown by the female in spring and plucked for lining the nest. The **eider duck** (***Somateria mollissima***) draws a quilt of down feathers over the eggs to conserve heat, when leaves them for a meal.

9. Other uses. The powdery suubstance, produced by the distintegration of the tips of the powder down feathers in herons, bitterns (wading birds of heron family), and other aquatic birds is useful in the cleaning of plumage. The powder helps in checking the multiplication of ectoparasites.

In horn-bills the bristles guard the nostrils from the dust and serve as eye lashes. The spine-tipped tail feathers of chimney swifts and woodpeckers help in clinging to the vertical surfaces. Markings on feathers are recognition marks to their fellow birds.

Economic Importance of Feathers

Bird's feathers are used by human beings in variety of ways. Some countries even export feathers of different birds. Quill of the quill or flight feathers was used as writing pen in the past. Quill feathers are used in chemistry laboratories for collecting fine particles of materials. The feathers are used extensively for stuffing pillows, quilts, blankets, clothings and sleeping bags. The down feathers of aquatic birds such as ducks and geese, provide warmest insulation and are used for making Arctic clothing, skiattire and sub-zero sleeping bags. Feathers are used for ornamentation and decoration (*e.g.*, peacock's feathers). Many natives apply feathers in headress (war bonnet) and to their arrows so that they may go farther. Long tail feathers of rheas are used as dusters. Badminton shuttlecocks are manufactured from features of ducks, cocks and other birds. Certain feathers are still used as currency by natives. Tail feathers of peacocks has been used in classical paintings. They are also used to scare away the house lizard.

14.6 BODY WALL

The body wall of pigeon consists of the skin, muscles and peritoneum.

1. Skin. The skin (Fig. 14.9) of pigeon is thin and dry. It is firmly attached to the underlying muscles. Histologically, it consists of outer **epidermis** of ectodermal origin and an inner **dermis** of mesodermal origin.

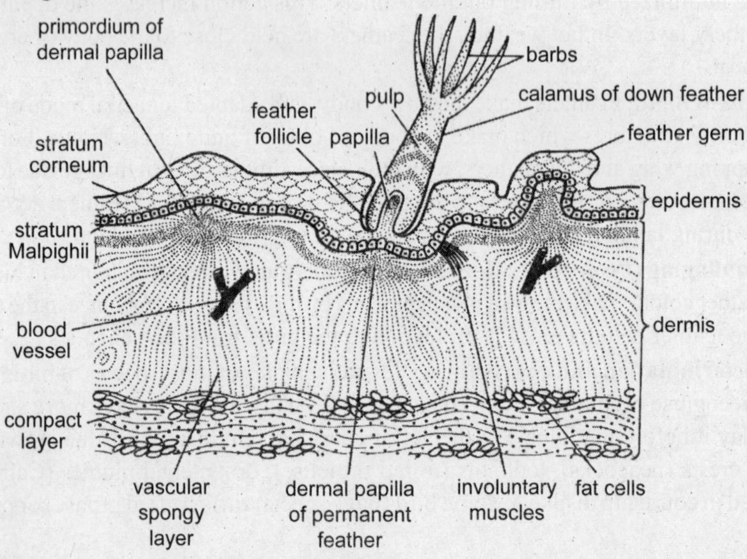

Fig. 14.9. Pigeon. V.S. of skin.

(*i*) Epidermis. It is several layer thick and composed of an outer **epitrichium**, a middle **stratum corneum** and an inner **stratum germinativum** or **stratum Malpighii**. The epitrichium is composed of a single row of flattened delicate cells. The stratum corneum is horny and protective and made of flat cornified cells. This layer is specially thickened to form the **corneoscutes** or **scutes** on the feet and then horny covering (rhamphotheca) of the beak and the claws. The corneoscutes resemble the reptilian scales in structure as well as in origin. The feathers, which cover the most of the body, are derived from the epidermis and represent the climax of modification of the stratum corneum.

The stratum germinativum is made up of large, cylindrical and continuously dividing cells. It gives origin to the cells of epitrichium and stratum corneum.

(*ii*) Dermis. It is composed of musculofibrous tissue. The muscle fibres are patricularly abundant, being used to raise and lower the feathers. Dermis also contains blood vessels and nerve fibres. Pigment is cofined to the feathers and scales.

The skin of pigeon lacks in sweat and any other type of cutaneous gland, except the **uropygial** or **preen gland** present dorsally at the base of tail. Uropygium is a simple, branched saccular gland with a septum dividing it into two halves (Fig. 14.10). The bird squeezes out a quantity of oil on its beak and applies it to the feathers to make them shining and waterproof.

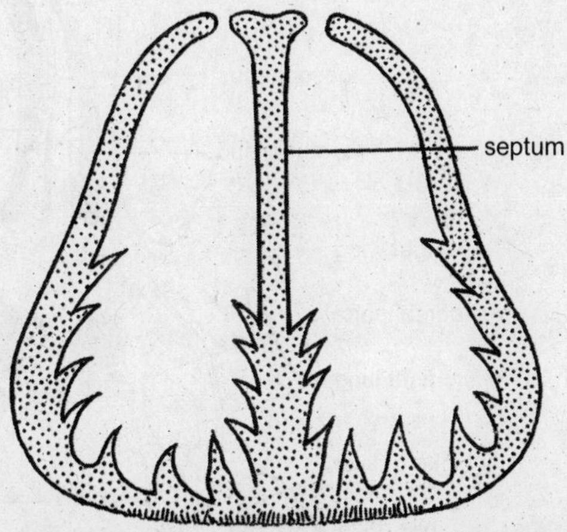

Fig. 14.10. Uropygial gland of a bird in section.

Due to lack of sweat glands, the birds cool themselves in hot weather by **panting** like the dogs.

(*iii*) Muscles. In birds, nonsegmental limb muscles are enlarged for rapid locomotion. The muscles are poorly devloped on the dorsal side of the trunk owing to the immobile trunk vertebrae. They are remarkably developed on the ventral side of the trunk for working the wings during flight. The main muscles here are the *pectoralis major* and *pectoralis minor* (described later on). Thighs are well muscled. There are few muscles on the shank and feet which, therefore, look slender and delicate. The abdominal musles form the lateral and ventral wall of abdomen. They meet in the midventral line, forming the **linea alba**. The endoskeleton is embeded in the body wall muscles.

(*iv*) Peritoneum. An epithelium is thin, close-fitting cells cover the inner surface of the muscles. It is known as **peritoneum**.

14.7 COELOM AND VISCERA

The body wall of pigeon encloses a spacious cavity in the trunk. It is known as the **coelom**. It is divided into four compartments: 1. The small anteroventral **pericardial cavity** enclosing the heart; 2. a pair of dorsolateral **pleural cavities** containing the lungs and the interclavicular and thoracic air-sacs and 3. a large **peritoneal** or **perivisceral cavity** housing the remaining visceral organs such as digestive, excretory, reproductive organs suspended by mesenteries (Fig. 14.11). The pleural cavities are separated from the perivisceral cavity by **oblique septa**. All parts of the coelom contain a watery coelomic fluid. Coelom is bounded by peritoneum.

14.8 MUSCULAR SYSTEM

In pigeon, the muscular system of neck, wings, tail, legs and ventral side of the body is well developed. Here, only appendicular muscles will be described; they provide movement to forelimbs (wings) and hindlimbs and are of following two types:

A. Muscles of Flight

The muscles of forelimbs (wings) and breast are highly developed and greatly modified in birds in accordance with the flying habit. They are called **flight muscles**.

1. Pectoral muscles. The most significant flight muscles are pectoral muscles. These muscles remain attached to the keel of sternum and provide up and down movements to the wings. They are of following two types:

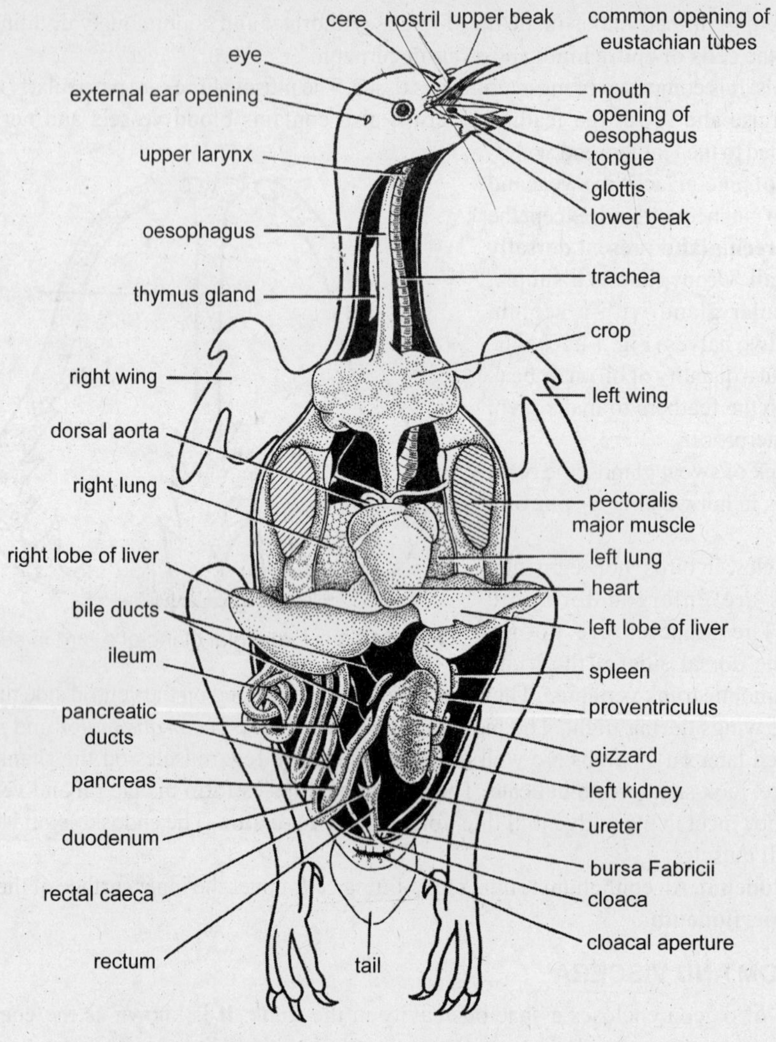

Fig. 14.11. Pigeon. Various organs present in the coelom.

(*a*) Pectoralis major. It is a very large triangular muscle having about one-fifth of the total body weight. The flight muscle is red in colour, being highly vascularised. Its fibres arise from the whole of the keel of the sternum and from the clavicle, and converge for insertion over a small area on the ventral side of the humerus near its head. The pectoralis major muscle has a broad flat tendon. Contraction of this muscle pulls the humerus downwards, causing a powerful **downstroke of the wing**. Hence, the muscle is also called the **depressor muscle**. Downstroke of the wings lifts the bird's body in flight.

2. Pectoralis minor (Supracoracoideus). It is much smaller muscle. Its fibres arise from the anterior part of the body of the sternum, dorsal to the pectoralis major and converge to a strong tendon. The latter passes through the **foramen triosseum** (an aperture between the clavicle, scapula and coracoid) and is inserted on the dorsal side of the humerus, again near its head. The foramen acts as a pulley, changing the direction of the action of the muscle. With the result, the contraction of this muscle pulls the tendon, which lift the wing, causing an **upstroke of the wing**. The pectoralis minor muscle is, therefore, also called the **elevator muscle**. In pigeon, the pectoralis minor is specially developed and causes quick take of the bird during the flight.

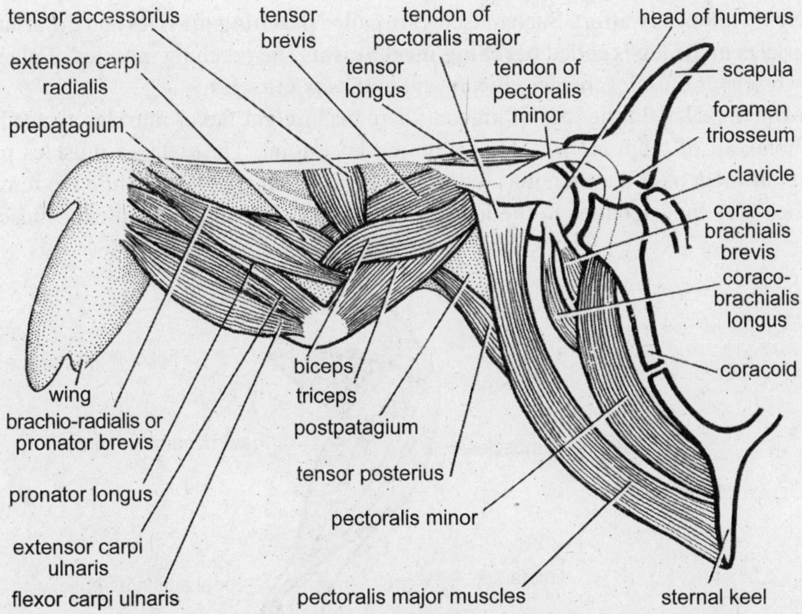

Fig. 14.12. Pigeon. Flight muscles.

3. Accessory muscles. Besides pectoral muscles there are certain other muscles which also elevate or depress the wing during flight. These muscles are as follows:

(*i*) Coraco-brachialis (Accessory depressors). These are two small triangular muscles consist of **coraco-brachialis longus** and **coraco-brachialis brevis**. They arise chiefly from the coracoid, in part from the sternum and its tendon is inserted into the ventral surface of the humerus. These assist the action of pectoralis major. The coraco-brachialis longus muscle draws the humerus medially and also depresses the wings. The coraco-brachialis brevis muscle assists the action of coraco-brachialis longus.

(*ii*) Scapulo-humerals and Coraco-humerals. They arise from the outer surface of the scapula and coracoid of the pectoral girdle to the humerus. They produce adduction (*i.e.*, movement of one part of body toward another) and lateral rotation of the humerus resulting in slight elevation of the wing.

(*iii*) Tensor muscles. These are complex group of muscles and include **tensor petagii**. There are three tensores patagii muscles known as **tensor longus**, **tensor brevis** and **tensor accessorius** which help the prepatagium tensely stretched when the wing is extended in flight. They elevate humerus. A similar **tensor posterioris alae** keeps the post patagium tensed.

(*iv*) Intrinsic muscles. The intrinsic muscles of the arm also plays an important role in flight. The **extensor carpi-radialis** and the **extensor carpiulnaris** of the forearm aid in stretching and folding the wing in cooperation with the upper-arm muscles called **biceps** and **triceps**. The extensor muscles extend from the radius and ulna to the carpels. The biceps and triceps arise from the scapula and are inserted on to the radius and ulna respectively.

The muscles of the digits, though reduced, alter the deposition of individual parts of the wing and even of individual feathers during flight.

B. Muscles of Hindlimbs and Perching Mechanism

The musculature of hindlimbs (legs) of pigeon and other birds resembles in every respect with that of mammals. But certain muscles in the legs of a bird and their tendons have a special arrangement so that when a bird sits on a perch (branch of a tree, wire or rod), its toes are mechanically flexed and grasp the perch without any effort. Such muscles are called **perching muscles** and the arrangement of perching muscles in the leg is called **perching mechanism**. The perching muscles of pigeon include following two sets of muscles, namely **flexor** and **extensor muscles**.

1. Flexor muscles. In the legs of pigeon, there occur eight flexor muscles on the back of the tibiotarsus bone, all of them are inserted up on the knee-joint. These flexor muscles perform the perching. Six of them pass their tendons to the phalanges of anterior toes, while the remaining two flexor muscles pass their tendons to the hind toe or **hallux**. The significant flexor muscles are the following:

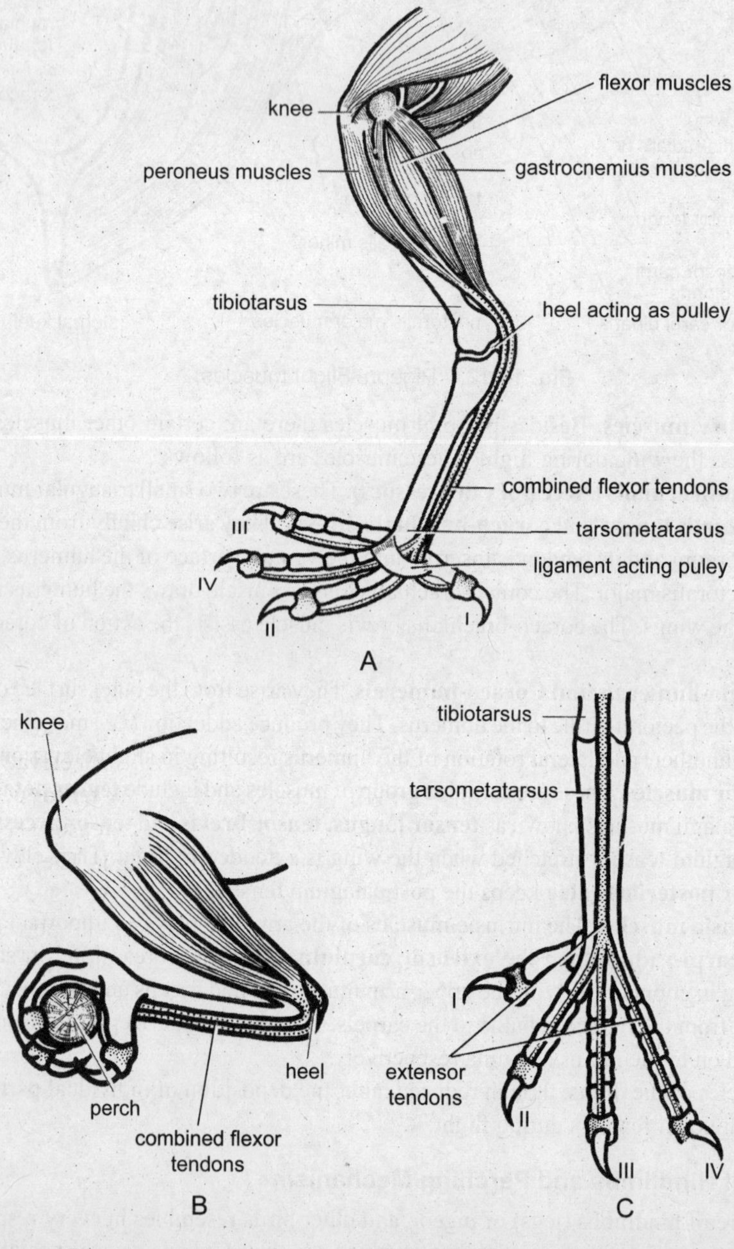

Fig. 14.13. Pigeon. Perching mechanism. A—Location of flexor muscles and their tendons in the leg; B—Leg bent with toes flexed around a perch and C—Location of extensor tendons.

(*a*) **Ambiens.** In some birds, an ambiens flexor muscle arises from the ilium and passes along the inner surface of thigh. Its long tendon runs beneath the patella bone round to the outer side of the knee enclosed in a special sheath and passes to the outer side of tibiotarsus to join the upper end of flexor muscle of the second and third toes. This muscle has little role in perching.

(*b*) **Peroneus medius.** This flexor muscle exists singularly on the anterior aspect of the shank, attached to the upper part of the tibiotarsus bone. The tendon of this muscle divides into three tendons, going to the three front digits.

(*c*) **Gastroenemius.** It is a calf muscle mainly concerned with producing flexion of the toes in the act of perching. It exists on the back of tibiotarsus. Tendon of this muscle passes behind the ankle and trifurcate to supply the three anterior toes along with peroneus medius muscles. These tendons often act as a single unit.

(*d*) **Flexor perforans.** This muscle is attached to the upper part of the tibiotarsus (above the knee). Tendon of this muscle goes to hallux and is joined by a slip with the peroneus medius muscle. Thus, a pull of one tendon flexes all the toes.

Mechanism of perching. When pigeon or any bird sits on its perch the legs are bent at the knee and the ankle, thus, the tendons are flexed (*i.e.*, contracted) and the digits are automatically closed around the perch. A pull on any tendon will close all four digits. Flexing (*i.e.*, bending) of digits is due to a **grip reflex** on the lower surface of the foot, the moment the foot touches the perch. When a bird sleep with body relaxed, the weight of the body bends the ankles more and the tendons become tighter ensuring a firmer grip of the perch, so that a sleeping bird cannot fall. This is automatic perching mechanism.

There is another perching mechanism which is a **locking device** that holds the toes flexed. The lower surface of flexor tendon is ridged (banded) at the metatarso-phalangeal joint where the weight of the body presses it against a branch. The upper surface of the tendon sheath is also ridged and as the bird settles on its perch, the two sets of ridges (bands) are interlocked.

2. Extensor muscles. Extension or opening of toes is caused by the extensor muscles situated in front of the tibiotarsus limb bone. The tendon of **tibialis anterior** muscle runs downwards along the front side of the ankle and then trifurcates to supply one branch to each anterior toe. Tendons are attached to the upper surfaces of the phalanges. When the bird raises its shank in order to leave the perch, there is a pull on the extensor muscle causing contraction of tendons and opening of toes.

14.9 DIGESTIVE SYSTEM

The digestive system of birds is modified for rapid and efficient digestion. It includes the *alimentary canal* and the *digestive glands*.

A. Alimentary Canal

The alimentary canal (Fig. 14.14) comprises mouth, buccopharyngeal cavity, oesophagus, stomach, small intestine, rectum (large intestine) and cloaca. All the segments of alimentary canal fall into three divisions: **foregut** or **stomodeum**, **midgut** or **mesenteron** and **hindgut** or **proctodeum**. Foregut includes portion of alimentary canal from mouth to stomach region. Midgut includes small intestine. Hindgut includes rectum and cloaca.

1. Mouth. The mouth has a wide gape. It is bounded by the upper and the lower horny toothless beaks. It leads into the buccopharyngeal cavity.

2. Buccopharyngeal cavity. The buccopharyngeal cavity (Fig. 14.15) is a narrow space in the head bounded dorsally by the skull and ventrally by the throat. A pair of **palatal folds** grow medially from the sides of the upper jaw, but do not meet in the middle line, leaving a gap, the **palatal cleft**. The palatal folds represent the incomplete hard palate. The anterior part of the buccopharyngeal cavity is known as the buccal cavity and the posterior part as the pharynx.

(*a*) **Buccal cavity.** The buccal cavity is narrow in front and widens posteriorly. It contains tongue, internal nares and salivary glands, but lack teeth.

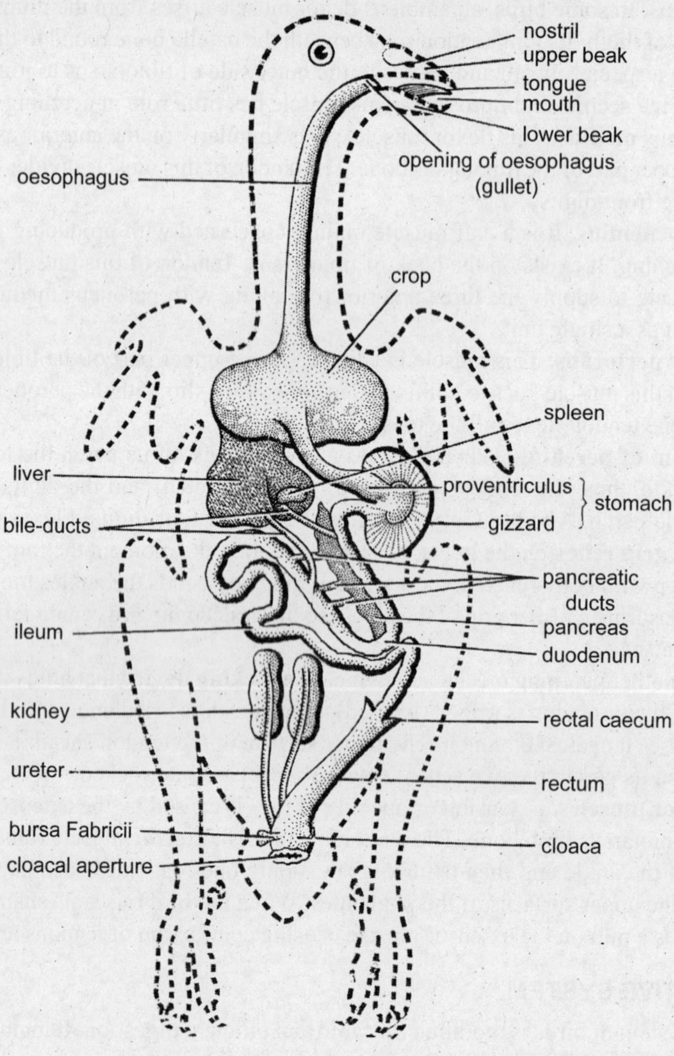

Fig. 14.14. Pigeon. Alimentary canal.

***(i)* Tongue.** The tongue is a flat, fleshy organ on the floor of the buccal cavity. It is pointed in front and bifid behind. It practically laks intrinsic muscles. It may, however, be moved by altering the position of the hyoid apparatus lying under its base. Togue has taste buds and mucous glands. It helps in swallowing of food.

***(ii)* Internal nares.** The internal nares are slit-like apertures lying side by side on the hind part of the roof of the buccal cavity. They are largely covered by the palatal folds. They play a role in respiration.

***(iii)* Salivary glands.** The salivary glands include median **sublingual glands** discharging on the floor and paired **angle glands** opening at the angles of the mouth. These glands secrete saliva which is enzyme free and only helps in moistening the food. (*Note*: Some authors have claimed about the presence of amylase enzyme in mucous secreted by some minute buccal glands; see **Jordan** and **Verma** 2013).

***(b)* Pharynx.** It is situated immediately behind the buccal cavity. The roof of pharyngeal cavity carries the median aperture of the paired eustachian tubes. Posterior to this opening the roof bears a pair of membraneous folds with fimbriated borders and hanging like a curtain constituting

soft palate. The **glottis** is situated at the base of the tongue on a hardened elevation, the laryngeal prominence. The walls of the glottis are supported by laryngeal cartilages. A large wide opening of **gullet** is situated behind the glottis which is last part of the pharynx. Posteriorly, the pharynx opens into the oesophagus.

(*i*) Oesophagus. The oesophagus is a long, narrow, distensible tube, which extends downward through the neck dorsal to the trachea, and then passes backward in the thorax to join the stomach. It is thin-walled in the thorax. It dilates in front of the sternum to form a large, thin-walled, bilobed sac, the **crop** that lies between the skin and the muscles.

The crop enables the pigeon to pick up and store quickly a large amount of food which can be later digested at leisure. It may contain over 500 cereal grains (Box 14.2). The crop plays dual role in the life of pigeon and other birds. It acts as an organ of food storage and organ of maceration (*i.e.*, grains are moistened and softened in it). During breeding season, the crop of pigeon secretes a nutritive (*i.e.*, white, slimy, proteinous, fatty) fluid, the **pigeon's milk** which is regurgitated and fed to the young. It is produced by the degeneration of epithelial cell lining by two large patches of peculiar tissue called **crop glands** (Fig. 14.16). Pigeon's milk is secreted by both the sexes in the breeding season. Its secretion is controlled by **prolactin** hormone, which is secreted by anterior pituitary. Crop's milk is more nutritions than mammalian milk. It can double the hatchling weight of the young squabs in just two days. A neonatal pigeon doubles its weight in 6 days. Pigeon's crop milk contain about 65–80% water, 13–18% protein (casein), 7–13% fat and 1.5% lactose.

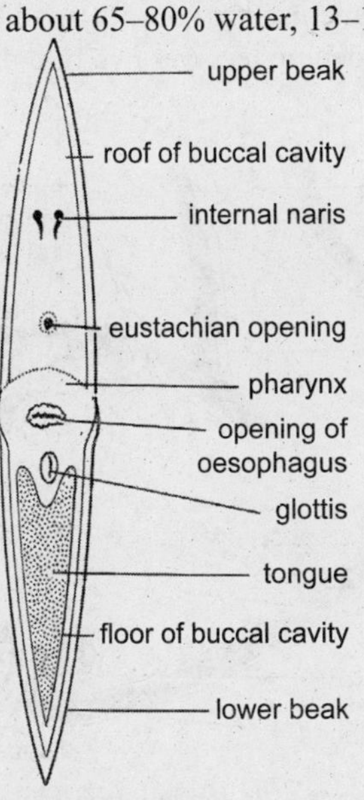

Fig. 14.15. Pigeon. Bucco-pharyngeal cavity.

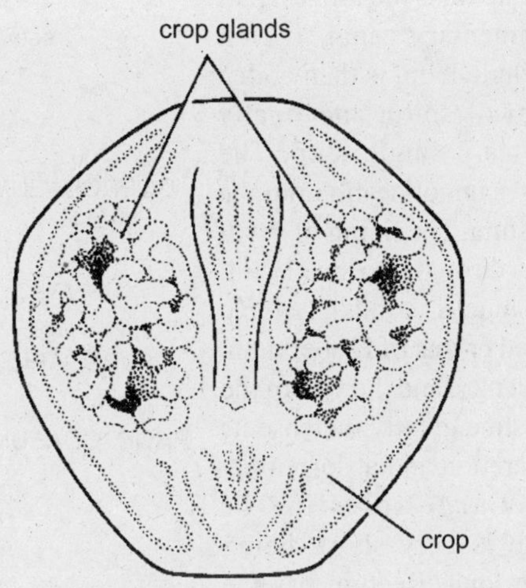

Fig. 14.16. Pigeon. Crop cut open.

Box 14.2

Crop is especially large in graminivorous (grain-eating) birds such as pigeons, finches, buntings, parrots, etc. It is small or lacking in insectivorous and carnivorous birds. Some piscirvorous birds have a very large crop.

Stomach. The stomach has two distinct chambers: anterior *proventriculus* and posterior *gizzard* or *ventriculus*.

(*a*) Proventriculus. It is slightly wider than the oesophagus. Proventriculus is glandular, having numerous gastric glands in its mucous membrane. Some digestion takes place here. A red coloured small spleen remains morphologically attached to the outer side of proventriculus.

(*b*) **Gizzard or ventriculus.** It is large, biconvex and hard. Gizzard has a very thick, muscular wall and a narrow lumen. Its lining epithelium has numerous minute tubular glands, the fluid secretion of which becomes thick, horny, yellowish or greenish lining of gizzard. The gizzard has small pebbles or stony pieces in it. Pebbles are swallowed by pigeon from outside (Fig. 14.17). These help in mechanical grinding of the grains with muscular contractions of the wall of the gizzard. The gizzard thus becomes a powerful **grinding mill** and compensates the role of teeth (molars) in pigeon. The gizzard leads into the small intestine by a passage guarded by a **pyloric valve** or **pyrolus**. In carnivorous birds the gizzard is not so muscular.

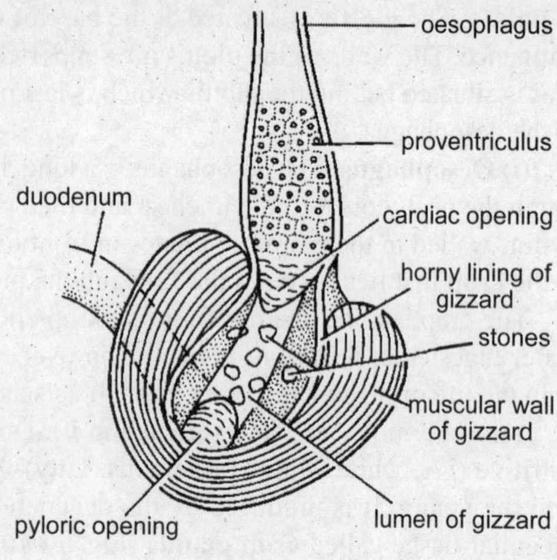

Fig. 14.17. Pigeon. Sagittal section of stomach.

(*i*) **Small intestine.** The small intestine is a long (about 75 cm), narrow tube differentiated into two regions: The anterior **duodenum** and posterior **ileum**. The duodenum is U-shaped. It encloses pancreas between its two limbs. It leaves the gizzard close to the entry of the proventriculus. The duodenum receives two bile-ducts and three pancreatic ducts. It passes unnoticeably into the ileum.

The ileum is the longest part of the alimentary canal. It first forms a single loop, is then coiled into a sort of spiral and finally again forms a single loop. The latter passes unnoticeably into the large intestine. The lining of small intestine is thrown into small villi. Duodenal mucosa contains simple or branched crypts of Lieberkuhn. The great length and the villi of the small intestine greatly increase its area of secretion and absorption.

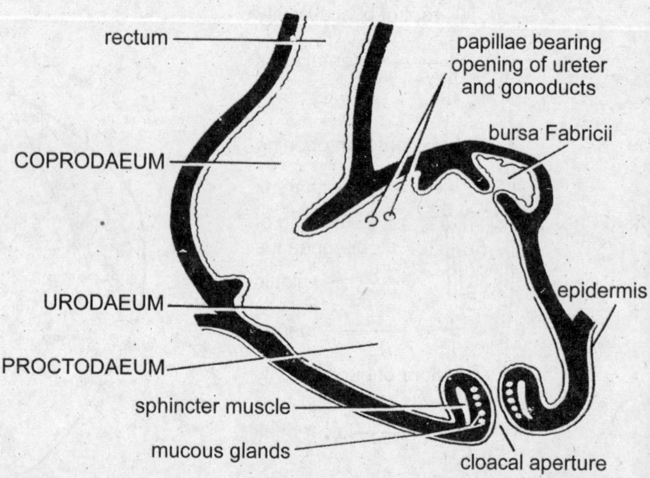

Fig. 14.18. Pigeon. Cloaca in longitudinal section.

(*ii*) **Large intestine or rectum.** It is very short, being only 4 cm long. Rectum has the diameter as the small intestine. It bears at the beginning a pair of small, lateral, blind pouches, called **colic caeca**. These caeca perhaps absorb water from the food. The rectum opens behind into the cloaca by way of the **anus** guarded by an anal **sphincter**.

(*iii*) **Cloaca.** The cloaca is a prominent chamber with three divisions, the anterior large **coprodaeum** receiving the rectum in front, the middle very short **urodaeum** receiving the renal and genital ducts, and the posterior large **proctodaeum** opening out by cloacal aperture. Elaborate cloaca serves to absorb water from the urine and faeces. The proctodaeum has on its dorsal side a thick-walled, blind sac, the **bursa fabricii** which has much lymphoid tissue, producing **lymphocytes** and serves to fight infection in the cloaca. It tends to atropy in the adult bird, so is described as **cloacal thymus**.

Digestive Glands

The digestive glands of pigeon include gastric glands, liver, pancreas and intestinal glands.

1. Gastric glands. The gastric glands are numerous microscopic, simple or branched pits in the mucous membrane of the proventriculus. They pour **gastric juice** into the lumen of the proventriculus.

2. Liver. It is a large, compact, dark red and bilobed gland. The right lobe is larger than the left. Each liver lobe gives off a bile duct. The left bile duct opens into the proximal limb, near the pylorus, and the right bile duct into the distal limb of the duodenum. There is no gall bladder. The liver secretes bile.

3. Pancreas. The pancreas is a compact, pinkish gland situated in the loop of the duodenum. It sends three pancreatic ducts, which open into the distal limb of the duodenum. Pancreas secretes **pancreatic juice**, which contains many enzymes. The hormone **secretin**, which in mammals influences the secretion of pancreatic juice, has been obtained from the wall of the pigeon's duodenum.

4. Intestinal glands. The intestinal glands lie in the mucosa of the small intestine. They are innumerable simple or branched pits, the **Crypts of Lieberkuhn**, between the villi. They secrete **intestinal juice** into the lumen of the duodenum and ileum. This juice contains a number of enzymes.

Physiology of Digestion

Food and feeding. The pigeon is grain-eating or **graminivorous** bird and it feeds chiefly on cereals, pulses, seeds, etc. These seeds are picked up with the beak, manipulated by tongue and lubricated and moistened inside the buccal cavity with the secretion of buccal glands. The food is swallowed into the crop without mastication.

Digestion. Swallowed food is passes through the oesophagus into crop, where it is stored. Inside the crop, the food is softened by joint action of body warmth, water, mucus, amylase enzyme of buccal secretions and also due to bacterial autolytic action. From crop the food passes into proventriculus where it is chemically acted by peptic enzyme of gastric juice. From proventriculus the food enters the gizzard where mechanical breakdown of food takes place by muscular contractions of gizzard wall and grinding action of small stones inside the lumen of gizzard. In gizzard little digestion of food occurs and it is made acidic by secretions of its walls. The partly digested food, the **chyme**, passes through the pylorus into the duodenum. In the duodenum, the chyme is mixed with bile, pancreatic juices and intestinal secretions. The chemical conversion of fats, proteins and carbohydrates into glycerol and fatty acids, and monosaccharides, respectively, takes place in duodenum similar to other vertebrates. Bile salts help in the digestion and absorption of fat. Thus, digestion, absorption and assimilation of food occurs in the small intestine.

The undigested portion of food passes to the rectum, where absorption of remaining water takes place. The faecal matter which is almost dry due to absorption of water is finally ejected through the cloacal aperture.

The large surface area, high temperature and great activity of birds necessitates a high food intake. This is due to rapid passage of food through the gut. In birds, the food is very efficiently utilised inspite of the rapid passage. Birds are able to convert 1/3 of their food into energy. The mammals convert only 1/10 of their food into energy.

14.10 RESPIRATORY SYSTEM

Birds respire through lungs, however, mammalian muscular diaphragm is lacking in them. Respiratory system of birds is unique and quite advanced and modified for aerial mode of life. It occupies almost 20% of the volume of the body; it is only 5% in human beings. Besides its respiratory functions, respiratory system of birds ensure cooling and regulation of water loss by panting.

Respiratory system of pigeon consists of *respiratory tract* and *respiratory organs* such as *lungs* and *air sacs*.

1. Respiratory tract. It includes nares, nasal sacs, glottis, larynx, trachea and syrinx.

(*a*) Nares, nasal sacs and glottis. The respiratory tract begins from the **external nares** which are the oblique slits situated at the base of upper beak just infront of the cere. They lead into the **olfactory sacs** enclosed in the olfacory capsules of skull. The olfactory sacs open behind by a pair of slit-like **internal nares** into the middle part of the buccopharyngeal cavity. The buccopharyngeal cavity has on its floor just behind the root of the tongue a median oval aperture with swollen lips. This is called the **glottis** and opens into the trachea. At the anterior end of trachea is present small chamber, the larynx.

(*i*) Larynx. The larynx is an expanded, voiceless chamber (*i.e.*, it lacks vocal cords). It occurs at the anteriormost margin of trachea and is greatly reduced in birds. However, larynx is supported by the mammal-like cartilages, namely a pair of **arytenoids** and a **cricoid**. The cricoid is triangular having four pieces, the **procricoids**. Larynx opens into a long trachea.

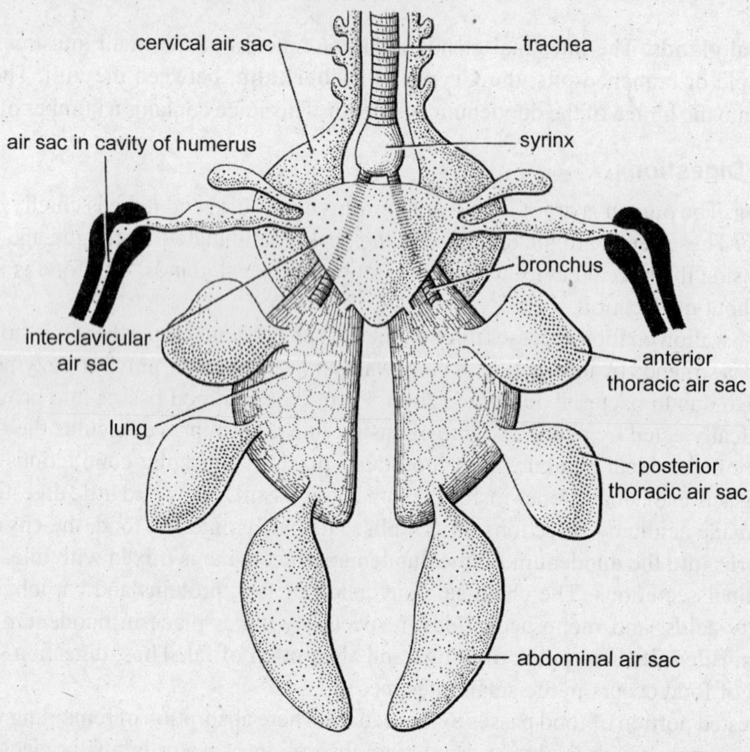

Fig. 14.19. Pigeon. Respiratory tract and respiratory organs *i.e.*, lungs and air sacs.

(*ii*) Trachea. The trachea is a long, cylindrical and flexible tube. Trachea extends down the neck ventral to the oesophagus, except for a short distance, where it is displaced to the left by the crop. The wall of trachea is supported throughout its length by complete **bony rings** that prevent its collapsing. At the lower end, the trachea bends backward to enter the thorax, where it bifurcates into two short, narrow tubes, the **primary bronchi**. The walls of bronchii are supported by *cartilaginous rings* incomplete mesially. The first bronchial ring is, however, bony. Each primary bronches enters the lungs of its side.

(*iii*) Syrinx. Where the trachea bifurcates into bronchi is a slight expansion known as **syrinx** (Fig. 14.20). The syrinx of birds functions as vocal organ or sound box. It is unique for the birds and is not found in any other class of Chordata. However, syrinx is absent in ostriches, stroks and some vultures. In pigeon, the syrinx is of **broncho-tracheal** type as it is formed by the trachea and bronchi both.

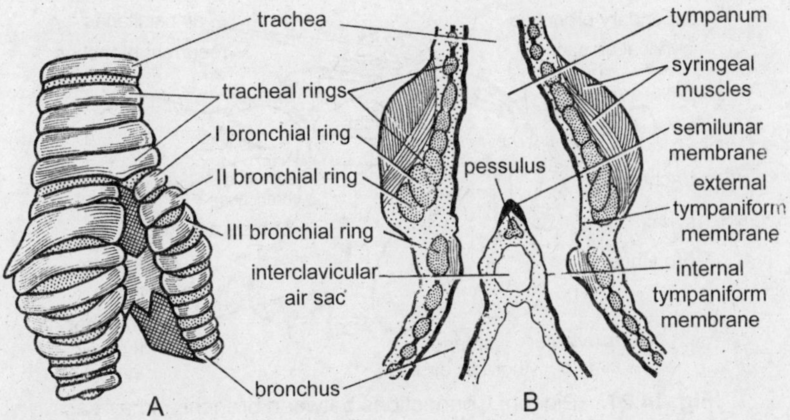

Fig. 14.20. Pigeon. Syrinx. A—External view and B—In longitudinal section.

Thus, last 3 or 4 rings of the trachea and the first or bony half ring of each bronchus take part in the formation of the syrinx. Cavity of syrinx is called **tympanum** and it is lined by a mucous membrane, which forms a cushion-like thickening on each side. The mucous membrane of the median wall of each bronchus forms the **internal tympaniform membrane** or **membrane tympaniform interna**. From the angle of bifurcation of the trachea, a thin fold of mucous membrane extends forward into the syrinx. This fold is supported by a bar of cartilage, the **pessules** and is known as the **membrana semilunaris**. Sound is produced by vibration of this membrane, when air is forcibly expelled by varying the tension of this membrane through the action of two sets of paired muscles: 1. The **intrinsic syringeal**, which arise from the sides of the trachea and are inserted on the **syrinx** and 2. The **sternotracheal**, which arise from the sternum and are inserted on the trachea infront of syrinx.

2. Respiratory organs. The respiratory organs of pigeon are a pair of **lungs**. In birds, there is only one means of respiration, *i.e.*, the **pulmonary respiration**. Lungs are bright red, small, compact, solid, slightly distensible and highly vascular. They differ from those of other vertebrates such as frog and lizard in that they do not store air (*i.e.*, they are not elastic sacs with sacculated walls). Lungs lie in the **pleural cavities** which are separated from the abdominal coelom by an oblique septum. The dorsal surface of the lungs is closely applied to the ribs and thoracic vertebrae and has no peritoneal covering. Their ventral surface is free and covered by a special fibrous peritoneal membrane called **pleura** or **pulmonary aponeurosis**. Small fan-like muscles called **costopulmonary muscles** arise from the junction of vertebral and sternal ribs and are attached to the pulmonary aponeurosis.

Lungs of pigeon have a very complex system of respiratory tubes or bronchi inside them. This is called the **bronchial intercom**. Thus, the bronchus entering a lung from the antero-ventral side does not branch immediately as in other tetrapoda and is called **primary bronchus**. Within the lung; it loses its supporting cartilaginous rings and continues upto its distal or posterior end and is called **mesobronchus**. The mesobronchus gives off branches to air sacs and also gives off several **secondary bronchi**, which may be dorsal and ventral then branch into **tertiary bronchi** or **parabronchi**. The parabronchi form an intricate system of branching and anastomosing thin-walled **air capillaries** forming loops and also join **recurrent bronchi** (Fig. 14.21). The lung parenchyma has hexagonal areas, each has a central parabronchus, surrounded by a system of air capillaries and interlacing blood capillaries. Thus parabronchi and air capillaries do not end blindly. There is no dead space in the lungs. Each air capillaries are lined with a vascular respiratory membrane and they come into contact with blood capillaries for an exchange of gases. Thus in bird's lungs the branches do not have blind endings in alveoli as in lungs of mammals. They form an inter-communicating system of bronchi, parabronchi and air capillaries forming complete air circuits in which air circulates.

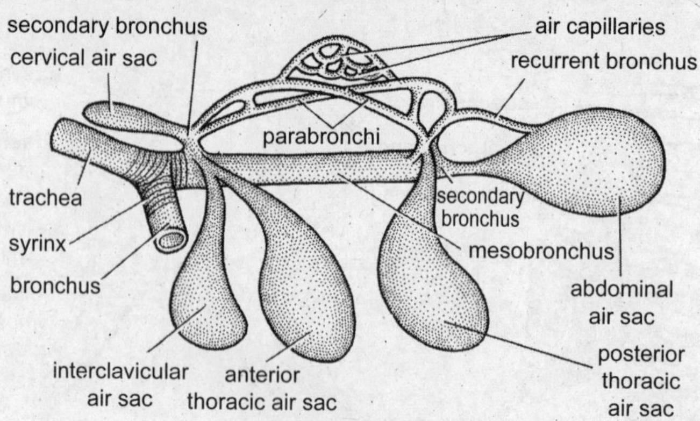

Fig. 14.21. Pigeon. Connections between bronchi and air sacs.

3. Air sacs. Associated with the lungs are characteristics large bladder-like, distensible **air sacs** formed as expansions of the mucous membrane of the secondary bronchi (Fig. 14.21). They have thin, transparent, non-muscular, walls. The wall of air sacs consists of double epithelium: the inner mucous membrane and the outer peritoneum, with supporting connective tissue in between. There are 9 such air sacs, four paired and one median. Air sacs are named according to their position in the body.

(*a*) Interclavicular. It is a median, unpaired, somewhat triangular air sac which is connected to secondary bronchi of both lungs. It lies in between the angle of two limbs of the interclavicular air sac arises two tubular **axillary air sacs**, one of which enters the humerus bone through the pneumatic foramen as **axillary** or **extra-clavicular air sac** and a **clavicular air sac**.

(*b*) Cervical. A pair of small cervical air sacs arises anteriorly, one from each lung. They lie at the base of the neck, dorsal to the interclavicular and along the side of the vertebral column. They give out minor saccular branches in the neck.

(*c*) Anterior thoracic. From the side of each lung arises an anterior thoracic air sac which lies at ventral side of lung in the anterior portion of the thorax, in close contact with the ribs and the pericardium. Ventral wall of each anterior sac is covered by the oblique septum.

(*d*) Posterior thoracic. A pair of small, posterior thoracic air sacs is found in the posterior part of the thoracic cavity just in front of abdominal sacs. Each sac overlaps the posterior end of its lung and communicates with its outer posterior angle.

(*e*) Abdominal. From the distal end of each lung arises a large abdominal air sac. This sac lies along the dorsal wall of the abdomen, ventral to the kidneys, amongst the coils or the small intestine.

Functions of air sacs. The air sacs are less vascular, so do not play any role in gaseous exchange. They perform the following important functions of pigeon:

1. Accessory respiratory organs. Air sacs act as reservoirs for the air, drawing air at the time of inspiration and expelling it out at the time of expiration, so that the lungs get fresh air during expiration also (Box 14.3). Thus, the exchange of gases occurs in lungs not only during inspiration but at expiration also. It this way air sacs increase the efficiency of lungs.

Box 14.3

It is believed that the anterior air sacs are **expiratory** (*i.e.*, interclavicular, cervical and anterior thoracic) and they are more active during flight. The posterior air sacs are **inspiratory** (*i.e.*, posterior thoracic and abdominal), they are more active when the bird is not flying.

2. Temperature regulation. Air sacs cool th body by evaporation of fluid from their surface and dissipation of heat into the air. This role of the air-sacs is important in the absence of sweat glands.

3. Lightness. The air sacs act as **balloons** giving buoyancy in flight and reducing the specific gravity of the bird due to the contained warm air. It has been observed that best fliers among birds are equipped with highly developed air sacs.

Respiratory mechanism. Pigeons like other birds use different mechanisms for respiration at the rest and during flight.

***(i)* Breathing or ventilation at rest.** A resting bird lowers and raises the sternum alternately by the action of the intercostal and abdominal muscles respectively. Lowering of the sternum enlarges the body cavity, creating a negative pressure in it. This causes the fresh air to be sucked through the respiratory tract into the lungs. Some of the incoming fresh air passes directly into the air-sacs, which expand and some goes into the air-capillaries, where exchange of gases takes place. Raising of the sternum reduces the body cavity and causes pressure over the air sacs. This pressure forces the air from the air sacs partly through the recurrent bronchi into the air-capillaries for exchange of gases and partly through the secondary bronchi and respiratory tract to the exterior.

Most remarkably the bronchioles of birds receive fresh air during both inspiration and expiration, and whole of their air changed, leaving non-residual air in them. As a result of this, the aeration of blood is very complete and this is responsible for the high metabolic activity and high temperature of the birds. Birds have a high respiratory rate. For the pigeon, it is 25–29 times per minute at rest in contrast to 14-20 times per minute in human beings. Such a quick respiration of pigeon (birds) explain how the bird fulfil with their small lungs their high oxygen requirement due to high metabolic rate.

***(ii)* Breathing or ventilation during flight.** During flight when sternum is rendered immovable, the breathing movements are produced by the elevation and depression of back and the movement of pectoral muscles.

In birds the air is constantly renewed and passes several times over the respiratory epithelium of air capillaries, because of intercommunicating system of tubules. The aeration of blood is absolute.

Peculiarities and efficiency of avian respiratory system. The entire respiratory system of birds (*i.e.*, pigeons) is highly efficient and peculiar in various ways:

1. There is unidirectional flow of air through the parabronchi and air capillaries of lungs. The direction of blood flow in the blood capillaries of lungs is crosswise to the flow of air (Fig. 14.22). It maintains a steep concentration gradient and allows the blood to become fully oxygenated. The system of gaseous exchange is called the **cross-current exchange**.

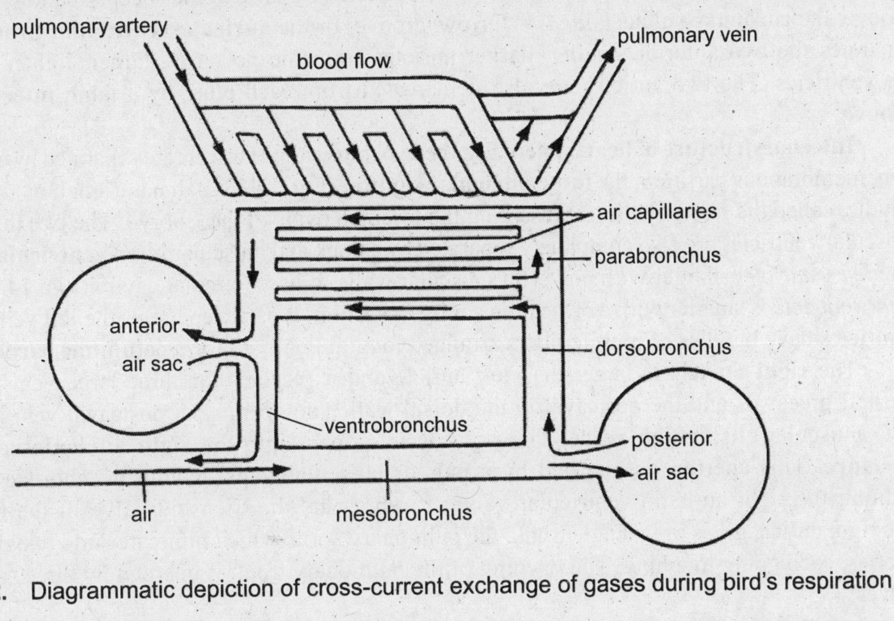

Fig. 14.22. Diagrammatic depiction of cross-current exchange of gases during bird's respiration.

2. The nearly continuous flow and cross-current exchange allows the extraction of more oxygen, especially at high altitudes.

3. There is no dead space in the lungs.

4. Exchange of gases occurs both during expiration and inspiration.

5. It takes two full breathing cycles to move a single inhaled volume of air through its complete path in the respiratory system.

6. The respiratory rate in birds is quite high in comparison to other animals and varies under different conditions. In pigeon, it is about 29 per minute at rest, 180 while walking and 450 during flight.

14.11 CIRCULATORY SYSTEM

The circulatory system of pigeon includes *blood vascular system* and *lymphatic system*. The birds have an efficient circulation, a high rate of metabolism and a high and constant temperature. In birds circulatory system maintains a perfect **double circulation** of blood. In fact, the birds and mammals are the only vertebrates that have a complete separation of the respiratory and systemic circulation, making possible a high arteriolar pressure which allows materials to reach the tissues rapidly.

A. Blood Vascular System

It includes **heart**, **arteries**, **veins** and **blood**.

1. Heart. The avian heart is relatively large and perfectly four-chambered. It consists of two **auricles** and two **ventricles**. The sinus venosus is absent being absorbed into the right auricle. Truncus arteriosus too is lacking in the avian heart.

External structure of heart. Pigeon's heart is a reddish, conical, muscular organ, unusually large for the size of the body (Fig. 14.23A). It occupies a midventral position in the anterior part of the thoracic cavity. Its broad **base** is directed forward and upward and the narrow **apex** backward, downward and slightly to the left. Heart is enclosed in a two-layered, transparent sac, the **pericardium**. The inner and outer layers of the pericardium are respectively called the **visceral pericardium** and the **parietal pericardium**. Between the two layers is a narrow space the **pericardial cavity**, which contains a watery (serous) fluid, the **pericardial fluid**. This serum-like fluid is secreted by the pericardium itself and protects the heart from shocks and allows its free movement during its beating.

The auricles remain well marked from ventricles by an external deep, shallow, transverse groove, the **coronary sulcus** (*sulcus* = furrow, groove) or the **auriculo-ventricular groove**, which separates the two anterior smaller, darker auricles from the posterior, larger, lighter in colour, the ventricles. The two auricles are also demarcated from each other by a faint, **inter-auricular groove**.

Internal structure of heart. Internally, the two thin-walled auricles are separated by a complete, thin, membranous partition, the **inter-auricular septum** (Fig. 14.23C). It has a small oval area in the middle, called the **fossa ovalis**, representing the foramen ovale of the embryo. The two thick-walled muscular ventricles are also completely separated by a thick and muscular **inter-ventricular septum**. The *left ventricle* is larger with powerful muscular walls. It has a circular cavity (Fig. 14.23D). The *right ventricle* is smaller with a crescentic cavity in section. It partly encircles the left ventricle. The ventricles have bundles of muscle fibers arranged longitudinally to form **columnae carneae**.

The **right auricle** is larger than left auricle and it receives the three large veins, the right and left precavals and the postcaval in the dorsal wall. The opening of postcaval vein is guarded by a muscular **eustachian valve**. The right auricle opens into the **right auriculo-ventricular aperture**. This aperture is guarded by a pair of muscular flaps, having no chordae tendinae, forming the right auriculo-ventricular valve. It represents the tricuspid valve of the mammals. The right auricle gives off a single trunk, the pulmonary aorta, which bifurcates into two pulmonary arteries, each going to a lung. The opening of the pulmonary trunk is guarded by three **semi-lunar valves**.

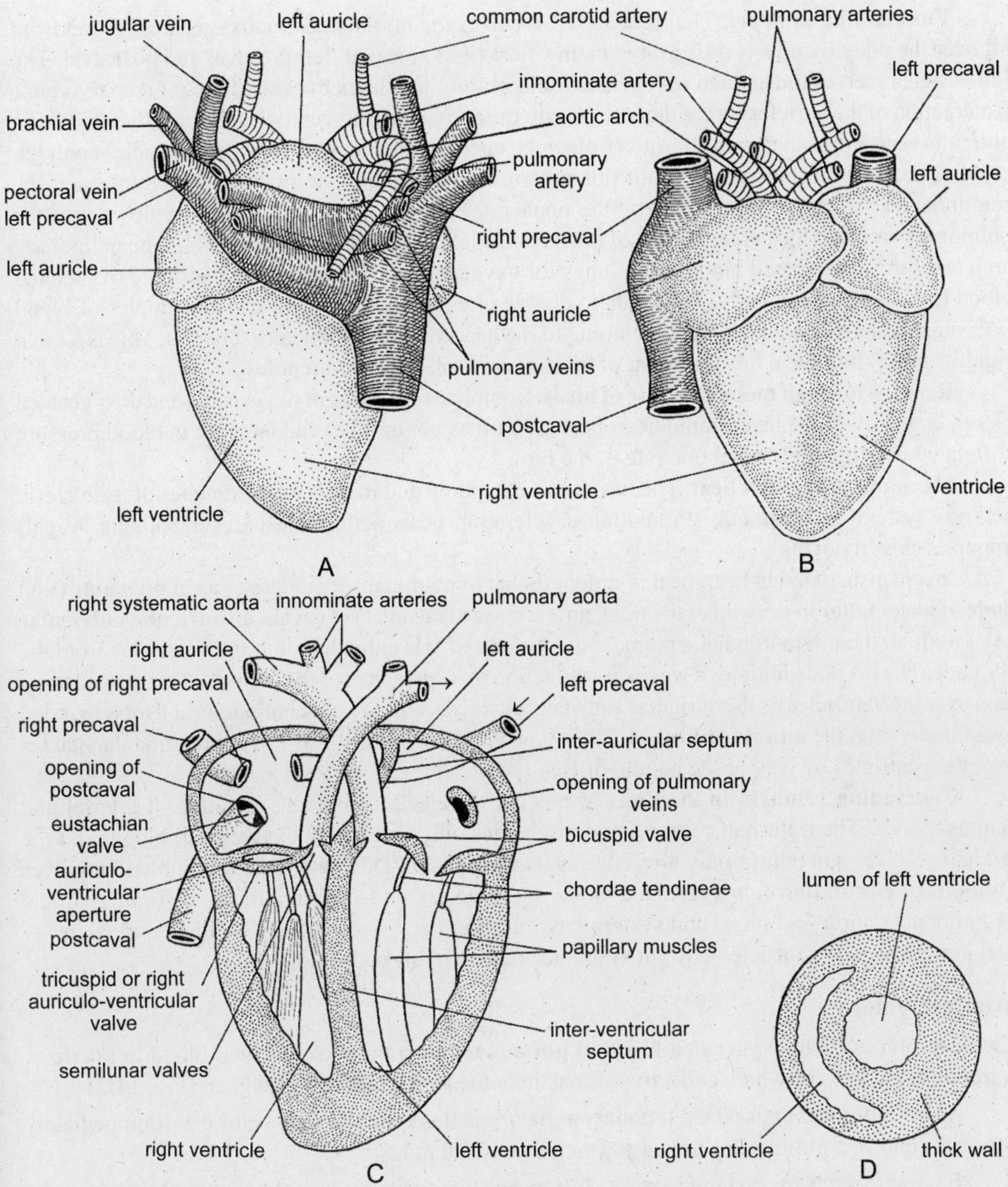

Fig. 14.23. Pigeon. Structure of heart. A—Dorsal view; B—Ventral view; C—Internal structure in ventral section (V.S.) and D—T.S. of ventricle.

The **left auricle** is smaller and receives two pulmonary veins from the lungs. It opens into the left ventricles by a circular **left auriculo-ventricular aperture**. This aperture is guarded by two membranous flaps forming the **left auriculo-ventricular valve**, which represents the bicuspid or mitral valve of mammals. The flaps of bicuspid valve are connected to the thick **papillary muscles**, which arise from the wall of the left ventricle by the **chordae tendinae**. The left ventricle gives rise to the single **right systemic** or **aortic arch**, which crosses over to the right side of the heart and continues as dorsal aorta. The left aortic arch which is present in the frog and lizard, is absent in birds. The opening of right aortic arch is guarded by three semilunar valves. The cavities of both ventricles are traversed by bars of muscles, called **trabeculae** or **columnal carneae**.

Functioning of heart. The right auricle receives the non-aerated (unoxygenated) blood from all over the body by means of the three main veins: right precaval, left precaval and postcaval. The left auricle receives the aerated (oxygenated) blood from the lungs by way of the pulmonary veins. Contraction of the auricles forces their blood into the corresponding ventricles through the respective auriculo-ventricular apertures, the valves of which open for the purpose. Now the ventricles contract, causing pressure in their blood. With this pressure, the auriculo-ventricular valves close and the semilunar valves open. Consequently, the nonaerated blood of the right ventricle flows into the pulmonary arch and the aerated blood of the left ventricle passes into the aortic arch. The pulmonary arch takes the non-aerated blood to the lungs for oxygenation. The aortic arch distribute the aerated blood to all parts of the body. Thus, due to double circulation of avian heart two samples of blood (*i.e.*, unoxygenated and oxygenated) coming to the heart remain completely separate. This is a great significant physiological advancement of birds over amphibians and reptiles.

Reasons for high metabolic rate of birds. Complete separation of oxygenated and doxygenated blood, large size of the heart, rapid heart beat (190 times per minute) and increase in blood pressure correlated with the high metabolic rate of the bird.

Mechanism of heart beat. The heart is a force pump and it beats on the dictates of its intrinsic nervous system. Sympathetic stimulation acceleration heart action and parasympathetic (vagal) impulses slow it down.

Mechanism of avian heart beat is similar to that of mammals. It comprises a **sinauricular (SA) node of pacemaker** in the wall of the right auricle close to the inlets of cavals, an **auriculoventricular (AV) node** in the interauricular septum, and a **bundle of His** extending into the ventricles from the AV node. The SA node initiates a wave of contraction that spreads over the auricles. This wave cannot pass over the ventricles as the auricular and ventricular muscles are not continuous and are separated by a fibrous ring, the **annulus fibrosus**. The AV node sets up a fresh wave of contraction that passes over the ventricles by way of the bundle of His.

Concluding remark. In the heart of pigeon, there is a complete separation of arterial and venous bloods. The right half receives and discharges only venous (unoxygenated) blood, and the left half receives and pumps only arterial (oxygenated) blood. Double circulation of blood involves **pulmonary circulation** (unoxygenated blood → Three cavals → Right auricle → Right ventricle → Pulmonary aorta → Lungs) and **systemic circulation** (oxygenated blood → Pumonary veins → Left auricle → Left ventricle → Right systemic arch → To all body parts).

Arterial System

The heart pumps its unoxygenated blood to **pulmonary aorta** and oxygenated blood to **aortic** or **systemic arch**, both of which collectively constitute the arterial system of pigeon (Fig. 14.24).

A. Pulmonary aorta. The pulmonary aorta originates from the right ventricle. It immediately bifurcates into two **pulmonary arteries** which supply the lungs.

B. Aortic arch or systemic trunk. In pigeon, like other birds, only right arch persists, the left one is absent. It originates from the left ventricle and curves over the right and gives rise to the following arteries:

1. Innominate arteries. A pair of innominate arteries separate from the aortic arch immediately after its emergence. These supply on the left and right sides of the anterior part of the body. Each innominate artery issues the following branches during its course.

(*a*) Carotid artery. It runs into the neck anteriorly bending inwards towards the middle and gives rise to the following vessels:

(*i*) **Vertebral artery** to the vertebral column.

(*ii*) **Oesophageal artery** to the oesophagus and crop.

(*iii*) **Cervico-cutaneous** to the skin and muscles of the neck region.

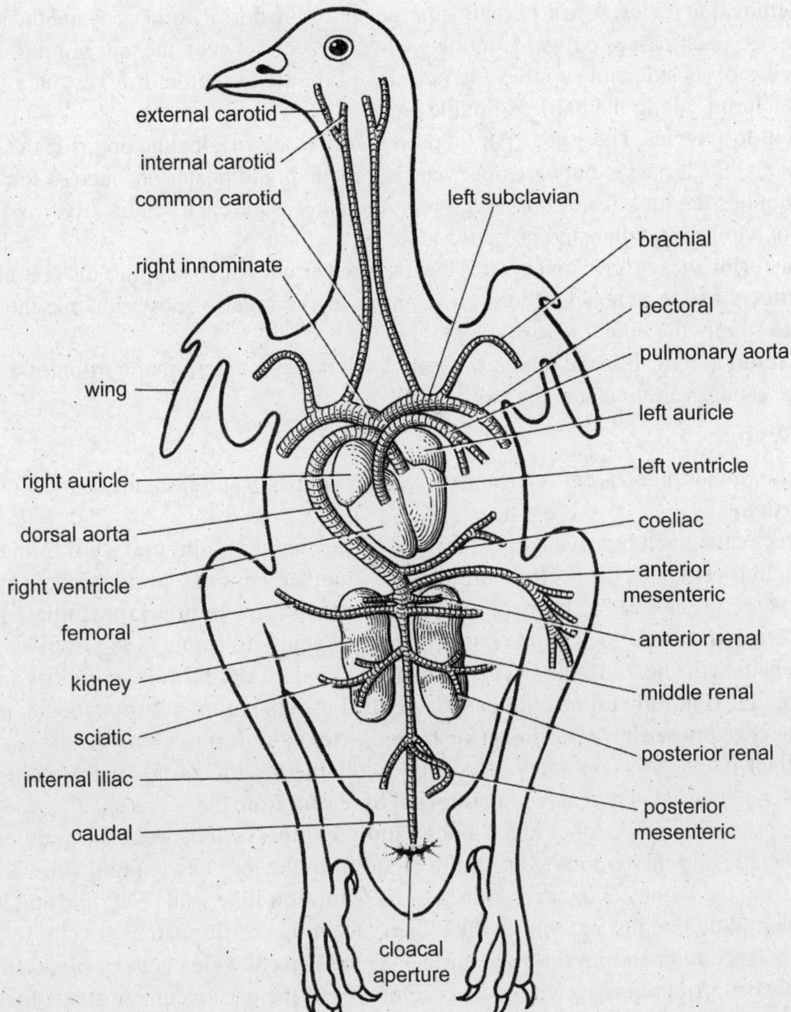

Fig. 14.24. Pigeon. Arterial system.

On reaching the angle of jaws, the carotid artery divides into an (*i*) outer **external carotid** supplying blood to the tongue, muscles of the jaws and other parts of head and (*ii*) an inner **internal carotid**, which supplies the brain.

(*b*) Subclavian artery. It originates slightly posterior to the carotid and runs forward and outward. It divides into two branches:

(*i*) **Brachial artery** to the wing and pectoral girdle.

(*ii*) **Pectoral artery** to the pectoralis major muscles (*i.e.*, flight muscle).

2. Dorsal aorta. The aortic arch after issuing innominate curves over the right bronchus to reach the dorsal body wall and then passes backward middorsally along the vertebral column as **dorsal aorta**. From the dorsal aorta originate following arteries:

(*a*) Coeliac artery. This short and thick abdominal artery emerges from the dorsal aorta and supplies branches to proventriculus, gizzard, duodenum, liver and spleen.

(*b*) Anterior mesenteric artery. It originates slightly behind the coelic artery, runs through the mesentary and supplies the intestine.

3. Anterior renal artery. A pair of them are issued from the aorta slighly in advance of the kidney and supply the kidneys. Each renal artery gives a small **gonadial branch** to the gonads of its side. In female only the left gonadial artery is present.

4. Femoral arteries. A pair of them emerge out of the dorsal aorta opposite the anterior lobes of the kidneys. Each passes outwards in the groove present between the anterior and median lobes of the kidneys of its side and supplies the muscles of the thigh region. It gives out a feeble **public artery**, which runs along the border of pubic bone.

5. Sciatic arteries. These are a pair of prominent vessels originating opposite the median lobes of the kidneys. Each passes outwards between the median and posterior lobes of the kidney of its side and supplies the muscles of thigh region on the inner side. Each sciatic artery issues two renal arteries supplying the median and posterior lobes of the kidney.

6. Posterior mesenteric artery. It is the unpaired artery which supplies the rectum and cloaca.

7. Internal iliac arteries. These artries arise at the level of posterior mesenterics and run backward to supply the hinder part of the pelvis.

8. Caudal artery. It is the median vessel which represents terminal portion of dorsal aorta. It supplies the caudal vertebrae and the pygostyle.

Venous System

In pigeon, the venous blood is collected and conveyed to the right auricle of the heart by two **precavals** and one **postcaval**.

1. Precavals. Each precaval vein is formed by the union of a **jugular vein** from the head and neck, a **brachial vein** from the pectoral muscles. The jugular veins of two sides unite with each other beneath the base of the skull to form a loop called the transverse **jugular anastomosis**.

2. Postcaval. It is formed by the union of several veins collecting blood from various regions of the body behind the heart. It starts in the form of an unpaired **caudal vein**. It divides anteriorly into the right and left **renal portal** or **hypogastric veins**. From the point of bifurcation arises a median vein, the **coccygeomesenteric** (or **posterior mesenteric**) **vein**. It runs along the cloaca and rectum, collecting their blood. It is continued anteriorly to join the hepatic portal vein. Before entering the kidney, each renal portal vein receives an **internal iliac vein** from the pelvis and delicate **sciatic vein** between the posterior middle lobes of the kidney and a **femoral vein** between the middle and anterior lobes. The sciatic and the femoral veins bring blood from the leg. The femoral vein also receives a **renal vein** from the kidney and leaves kidney as the **common iliac vein**. The right and left common iliac join the middle line just anterior to the kidneys to form a short **postcaval vein**. It runs through the liver substance to open into the right auricle. Three **hepatic veins** convey blood from the liver into the postcaval. An **epigastric vein** collects blood from the peritoneum or great omentum.

3. Pulmonary veins. Two pulmonary veins collect oxygenated blood from each lung and join with their fellows of the other side to open into the left auricle by a common opening.

Hepatic Portal System

As in other vertebrates, in pigeon also the blood from the alimentary canal reaches the liver by a **hepatic portal vein** formed by the union of **gastro duodenal, anterior mesenteric** and the **posterior mesenteric veins**. The blood after filtering through the liver is conveyed to the postcaval vein through hepatic veins.

Renal Portal System

In birds, the renal portal system is greatly reduced. The renal portal veins instead of breaking into capillaries in the kidneys, send off only a few small branches. Thus blood reaches directly to the heart without passing through the kidneys.

Blood

The **erythrocytes** are oval, biconvex and nucleated. **Leucocytes** (*i.e.*, lymphocytes, monocytes, basophills, eosinophils and neutrophils) are present and are unusually active in destroying the microbes and repairing the wounds. Thrombocytes also occur. These bring about quick clotting of blood. There are no platelets. Blood temperature is maintained at a very high level of 42°C.

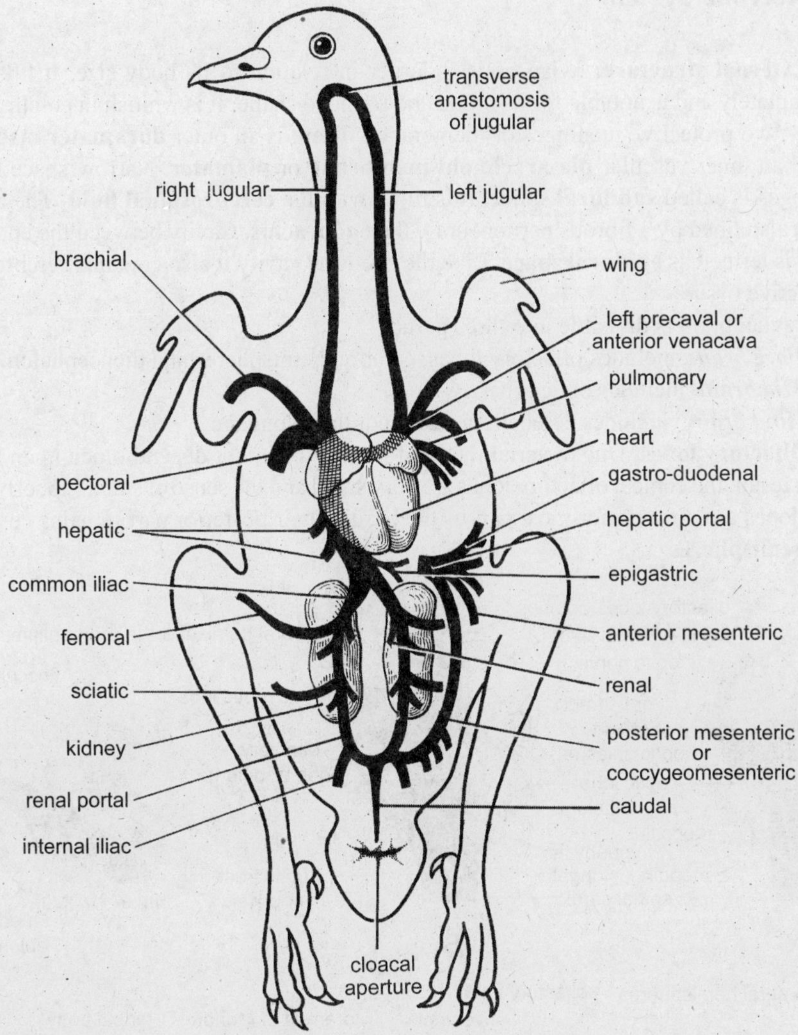

Fig. 14.25. Pigeon. Venous system.

Lymphatic System

An extensive lymphatic system is present in association of blood vascular system of pigeon. It includes the lacteals coming from the small intestine. The lymphatic vessels ultimately open into a pair of **thoracic ducts**, which discharge into the precavals. Lymphatic nodes occur in the course of the larger lymphatic vessels. Lymphatic hearts are absent.

Spleen. Which is an lymphatic organ is small, oval, red organ. It is attached to the right side of the proventriculus by a fold of peritoneum. Spleen plays a role in final destruction of red blood cells (erythrocytes), filtration and storage of blood and production of lymphocytes.

14.12 NERVOUS SYSTEM

The nervous system of birds is similar to that of the reptiles, but it shows distinct advancements over the reptilian brain. There is a reduction in the size of olfactory bulbs, but cerebral hemispheres, optic lobes and cerebellum are comparatively better developed. These changes in the brain put birds on a higher evolutionary level.

Like all other vertebrates, nervous system of birds comprises of **central nervous system** which includes brain and spinal cord; **peripheral nervous system** which includes cranial and spinal nerves, and **autonomic nervous system** including sympathetic and parasympathetic systems.

Central Nervous System

A. Brain

1. External structure. Avian brain is larger in relation to its body size. It fills the cranial cavity completely and is notable for its broad and rounded shape. It is whitish in colour and remains covered by two protective meninges or membranes. There is an outer **duramater** made of fibrous tissue, and an inner vascular **pia-arachnoid membrane** or **piamater**. Narrow space between the two meninges is called **subdural space**. It is filled with the **cerebrospinal fluid**. The cavity of the cranium is also lined by a fibrous peritoneum—the **endorachis**. Cavity between the endorachis and duramater is termed as **epidural space**. Like the subdural cavity it also contains cerebrospinal fluid and connective tissue.

The avian brain is divisible into three parts:

(*a*) ***Fore brain.*** includes olfactory lobes, cerebral hemispheres and diencephalon.

(*b*) ***Mid brain.*** includes optic lobes.

(*c*) ***Hind brain.*** includes cerebellum and medulla oblongata.

(*i*) **Olfactory lobes.** Due to aerial mode of life, birds do not depend much upon the sense of smell. As a result, the conical olfactory lobes are very small and the sense of smell is poorly developed. Olfactory lobes are the anterior most part of the brain. Their posterior part remains covered by the **cerebral hemispheres**.

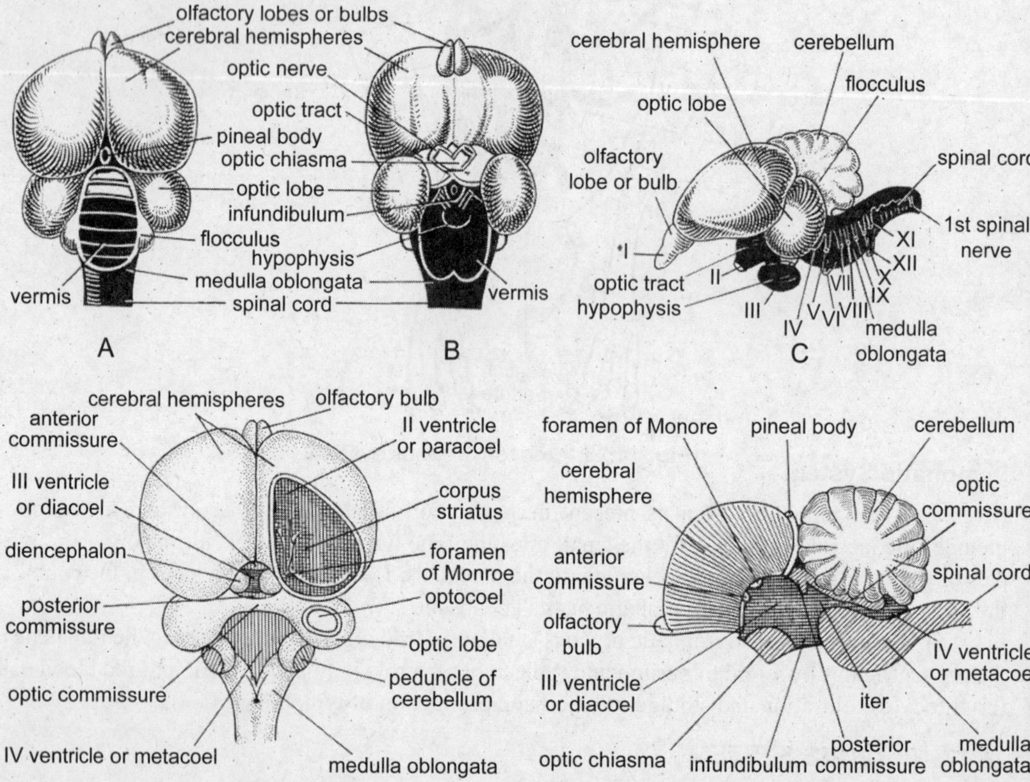

Fig. 14.26. Pigeon. Structure of brain. A—Dorsal view; B—Ventral view; C–Left lateral view; D–Brain cut open to show the ventricles; E–In sagittal section.

(*ii*) **Cerebral hemispheres.** Olfactory lobes are followed by a pair of very large and pyriform cerebral hemispheres. They are somewhat convex, narrower in front and broad behind. The two hemispheres are demarcated by a deep **sagittal fissure**. Unlike mammals, the surface of cerebral hemispheres is smooth, devoid of convolutions. The hemispheres overlap the diencephalon dorsally. Though the hemispheres are very big, their cortex, the seat of mental activities, is greatly reduced. Hemispheres consist of large masses of tissues and send projections back to the thalamus, mid brain, medulla, cerebellum and spinal cord.

***(iii)* Diencephalon.** It lies behind the hemispheres. Anteriorly its dorsal surface is overlapped by the hemispheres and posteriorly by the cerebellum, due to which in the dorsal view it is hardly visible. There is no parietal organ. A single, delicate, median and glandular **pineal body** is present on the dorsal side just posterior to the sagittal fissure. On the ventral surface, the two optic nerves cross each other to form **optic chiasma** as in other vertebrates. There is a median projection– the **infundibulum**, behind the optic chiasma. Infundibulum distally bears a large **pituitary body**.

***(iv)* Optic lobes.** Birds depend much upon the sense of sight. As a result, optic lobes are very prominent. The avian brain is often called as the eye brain. Paired rounded optic lobes occupy lateral position instead of the usual dorsal position due to the great development of cerebellum and cerebral hemispheres. A transverse **optic commissure** connects the two optic lobes.

***(v)* Cerebellum.** Cerebellum is very large and comparatively better developed than in reptiles and even mammals. Anteriorly, it displaces optic lobes from their usual dorsal position and posteriorly overlaps medulla oblongata. It is differentiated into a large, oval **central lobe** or **vermis** and a pair of small conical **lateral lobes** or **flocculi**. The vermis is further differentiated by two indistinct fissures into anterior, mesial and posterior lobes. The surface of the cerebellum is folded, and contains much grey matter. Much folded cerebellum is correlated with the great power of muscular coordination and equilibrium.

***(vi)* Medulla oblongata.** It is the last part of the brain. It is broad in front and tapers gradually towards the distal end. Distally its narrow end remains continuous with the **spinal cord**. Junction of medulla and spinal cord is marked by a **ventral flexure**.

2. Flexures. The brain is curved and shows usual flexures as found in other amniotes. **Primary flexure** is observed in the mid brain. It bends the posterior part of the brain nearly at right angles to the anterior. **Nuchal flexure** bends medulla oblongata at an angle to the spinal cord. **Pontal flexure** is present ventral to the cerebellum and bends brain in opposite direction from the previous two flexures.

3. Histology. Brain is a compact mass of two types of nervous tissues: the **grey matter** comprising nerve cells, and the **white matter** consisting of nerve fibres. Grey matter is external and the white matter is internal in the forebrain and cerebellum. The arrangement is reversed in optic lobes and medulla oblongata.

4. Internal structure and the ventricles or cavities. Brain is a hollow structure. Its cavities are called the **ventricles**. In birds the ventricles are greatly reduced. There is no cavity in olfactory lobes. The two cerebral hemispheres are hollow and their cavities are termed as **lateral ventricles** or **paracoels**. Paracoels have a very thin roof with very poorly developed **cerebral cortex**, while the floor and lateral walls are very thick and called **corpus striatum**. Corpora striata of the two hemispheres are connected together by an **anterior commissure**.

Cavity of diencephalon is called the **third ventricle** or **diacoel**. Roof of diacoel, the **epithalamus**, is very thin. Part of the epithalamus just in front of the pineal body consists of an **anterior choroid plexus** (tissue having rich supply of blood). Like hemispheres, its floor, the **hypothalamus** is thick. Lateral walls of the diacoel are also very thick and are termed **thalami** or **optic thalami**. Diacoel is connected with the two paracoels through a small common aperture, called the **foramen of Monro** (interventricular foramen).

Cavities of the optic lobes are called **optocoels**. Besides optocoels, mid brain also encloses a narrow, median, longitudinal canal–the **iter**. Both the optocoels are connected with the iter dorsally. Iter remains separated from the cavities of diencephalon (diacoel) by a **posterior commissure**. Thick longitudinal bands—**crura cerebri**, composed of nerve fibres, run beneath the optic lobes and connect medulla oblongata with the diencephalon.

Medulla oblongata encloses the **fourth ventricle** or the **metacoel** or the **myelocoel**. It is connected with the iter. Lateral walls and floor of metacoel are thick, while the roof is thin and richly supplied with blood capillaries which form the **posterior choroid plexus**. Metacoel is broader anteriorly and tapers behind to converge in the cavity of spinal cord.

Cavity of the cerebellum is obliterated.

All the cavities of the brain remain filled with cerebrospinal fluid which is secreted by the choroid plexuses.

5. Functions of the brain

(*i*) **Olfactory lobes**—smell. Sense of smell is poorly developed.

(*ii*) **Cerebral hemispheres**—control voluntary movements, conditioned reflexes, intelligence, will, memory and emotions etc.

(*iii*) **Diencephalon**—conveys impulses to the cerebral hemispheres and acts as integrating centre of autonomic nervous system. It is also concerned with the perception of sensations of extreme pain, cold, and heat etc.

(*iv*) **Optic lobes**—sight.

(*v*) **Crura cerebri**—carry sensory impulses to optic thalami from the medulla oblongata and spinal cord.

(*vi*) **Cerebellum**—coordinates voluntary movements and maintains equilibrium.

(*vii*) **Medulla oblongata**—involuntary actions.

6. Aerial adaptations in the brain. The birds depend upon sense of sight in the air and require well developed centres of muscular coordination, equilibrium and instinct. The brain of the birds possess large and well developed optic lobes. The cerebral hemispheres are much larger than any other part of the brain. The cerebellum is large, folded and well developed. It controls the timing and movement in all planes of space during flight.

B. Spinal cord

Spinal cord in birds is comparatively shorter due to their short vertebral column. It is the posterior prolongation of the medulla oblongata. The cavity of the spinal cord or the **central canal** remains continuous with the cavity of the medulla oblongata. Spinal cord passes through the neural canal of the vertebral column throughout its length. The external (duramater) and internal (pia-arachnoid membrane) meninges and epidural and subdural spaces of the brain continue as such in the spinal cord. Internal lining of the cranial cavity, the endorachis, also extends into the neural canal.

Spinal cord is a long, whitish and somewhat flat tube. Dorsally it bears a faint median groove, called the **dorsal fissure**. The median ventral groove or the **ventral fissure**, is very prominent. These fissures extend throughout the length of the spinal cord. Just below the dorsal fissure, there is a thin longitudinal vertical septum. Spinal cord bears two enlargements, in the region of fore limbs and hind limbs. These are termed as the **brachial** and **lumbar enlargements** respectively.

Central cavity of spinal cord remains filled with cerebrospinal fluid like the cavities of the brain. Central canal is closed posteriorly.

Histology. In the spinal cord, **grey matter** is internal and 'H' shaped. **White matter** is outer and surrounds the grey matter. Grey matter is produced into very prominent **dorsal** and **ventral horns** due to which white matter is divided into paired **dorsal**, **ventral** and **lateral funiculi** (longitudinal columns). In the region of lumbar enlargement, dorsal funiculi of white matter is greatly reduced and a wide elliptical space is left between the dorsal columns of grey matter. This space is called the **sinus rhomboidalis**. It remains continuous with the central cavity and is roofed over only by the meninges.

Functions. Spinal cord controls reflex actions and conducts sensory and motor impulses to and from the brain.

Peripheral Nervous System

A. Cranial nerves. Pigeon consists of twelve pairs of **cranial nerves** which arise from different parts of the brain and leave cranium through definite holes and the spinal nerves. All the cranial nerves are described as under in the tabular from:

Table 14.1 Cranial Nerves of Pigeon

Name and number	Origin	Distribution	Nature	Functions
Olfactory, I	Olfactory lobe	Olfactory sac	Sensory	Smell
Optic, II	Optic thalami	Retina of eye	Sensory	Sight
Oculomotor, III	Ventral surface of mid brain	Inferior oblique, anterior, posterior and internal rectus eye muscles	Motor	Movement of eyeball, iris, lens, eyelid
Trochlear or pathetic, IV	Dorsal surface of mid brain	Superior oblique eye muscles	Motor	Rotation of eyeball
Trigeminal, V	Lateral sides of medulla oblongata	Eye ball, skin and muscles of upper and lower jaws	Sensory and motor (mixed)	Movement of tongue and jaw muscles
Abducens, VI	Medulla oblongata	External rectus eye muscles	Motor	Rotation of eyeball
Facial, VII	Lateral sides of medulla oblongata	Muscles of face, jaws and taste buds	Mixed	Taste, facial expression, movement of neck
Auditory, VIII	Just behind VIIth nerve	Internal ear	Sensory	Hearing, equilibrium
Glossopharyngeal, IX	Ventrolateral border of medulla oblongata	Throat and tongue	Mixed	Taste and touch, swallowing, movement of pharynx
Vagus or pneumogastric, X	Behind IXth nerve	Heart, lungs, stomach and sense organs, etc.	Mixed	Lungs, respiratory reflexes, peristaltic movements, voice, swallowing, gastric secretion, inhibition of heart beat
Spinal accessory, XI	Floor of medulla oblongata	Neck region, palate, larynx	Motor	Muscles of pharynx, larynx, neck, shoulder movements
Hypoglossal, XII	Floor of medulla oblongata	Muscles of tongue	Motor	Movements of tongue

B. Spinal nerves. Spinal nerves of pigeon follow the typical vertebrate pattern. The **brachial enlargement** gives off a **brachial plexus** of nerves which innervate wings. The **lumbar enlargement** gives off a **lumbar plexus** of nerves which divides into **lumbar**, **sacral** and **pudendal plexuses**. Their nerves innervate the thigh, rest of the leg, and cloacal region and tail respectively.

14.13 SENSE ORGANS

A sense organ is a group of specially modified receptor cells. Pigeon possesses all the important sense organs such as touch, smell, vision, hearing and taste. Tactile, gustatory (taste) and olfactory senses are poorly developed while auditory and visual organs are well developed in the pigeon.

1. Tactile organs. Tactile organs are poorly developed in the birds. The **beak** acts as an important tactile organ. The **operculum** or **cere** (a soft fold of skin) situated at the base of the upper beak is a sensory organ and plays a stimulatory role during love making.

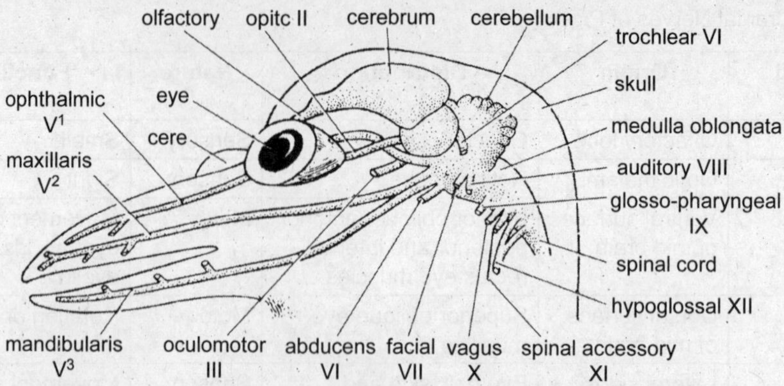

Fig. 14.27. Pigeon. Cranial nerves.

2. Gustatory organs. Sense of taste is very poorly developed in pigeon and other birds. Only a few **taste buds** are found scattered on the tongue (24 in hen, compared to 10,000 in human being).

3. Olfactory organs. Sense of smell is also poorly developed. Due to aerial mode of life, this sense is not important for birds. Nostrils lead into the small, paired **olfactory** or **nasal sacs** at the base of the upper beak. These open into the posterior part of the buccal cavity through the internal nares. The two sacs are separated from each other by the mesethmoid, and bounded laterally by ectoethmoids. These form three scroll-like processes, called the **turbinal processes** which increase the olfactory surface. The anterior part of each sac acts as a non-sensory respiratory part or **vestibule.** The posterior part is sensory. It contains the middle and posterior turbinal processes, which remain covered by one-layered epithelium or the **Schneiderian membrane**. The epithelium is innervated by the fibres of olfactory nerve.

4. Eyes. Being aerial, a bird depends solely upon the sense of sight for its existence. As a result, eyes are well developed and large. These are antero-posteriorly flattened, concave in front and convex behind. Structure of eye of birds is like that of a human being. The eye contains the following structures:

(a) Eyelids. There are three eyelids. All are without eye lashes. The **upper eyelid** is small and only slightly movable. The **lower eyelid** is better developed and freely movable. It rises upwards to close the eye. The eyelids do not blink. A **third eyelid** or the **nictitating membrane** lies as a fold at the anterior angle of eye between the upper eyelid and the eyeball. It can be stretched over the eye and pulled back with great rapidity. Nictitating membrane is semitransparent. It cleans and protects the eye. During flight, nictitating membrane protects the eye against dust etc. In nocturnal birds, it cuts sunlight during day; and in aquatic birds, it protects eyes from water.

(b) Glands. The nictitating membrane is cleaned and lubricated by the oily secretion of a **Harderian gland** situated in the inner angle of the eye. **Tear** or **lacrymal** glands are also well developed. These lie below the outer angle of the lower eyelid. Its secretion (tears) nourishes the cornea, sweeps away the dust etc., keeps the cornea always wet and thus, maintains its transparency.

(c) Eye muscles. The eyeball is supported and moved in the orbit by the **intrinsic** and **extrinsic muscles**. A set of radiating intrinsic muscles, controlled by sympathetic system, enlarge the pupil, while another ring of intrinsic muscles, governed by parasympathetic system, contract the pupil.

Extrinsic eye musculature consists of six muscles in the orbit to move the eyeball. These muscles receive their nervous supply from third, fourth and sixth cranial, nerves. Rapid movement of nictitating membrane is controlled by **quadratus** or **bursalis** and **pyramidalis** muscles. Both these muscles are situated close to the inner orbital surface.

(d) Sclerotic or the fibrous tunic. It is the outermost whitish covering of the eye ball and is divisible into **cornea** and **sclera**. Cornea is the small, exposed, convex, transparent outer part of the fibrous tunic (Fig. 14.28). It is non-vascular, and is nourished by the secretion of lacrymal glands. Cornea remains covered by a thin transparent epithelial membrane, the **conjunctiva** which is a continuation of the mucous lining of the eyelids.

The inner posterior, large, opaque, hard and cartilaginous **sclera** remains hidden in the orbit. The small part of sclera visible in the eye is called **white** of the eye. Ten to twelve small, over lapping, bony **ossicles** or **sclerotic plates** are situated in a ring at the junction of the cornea and sclera (Fig. 14.29). These strengthen the sclera.

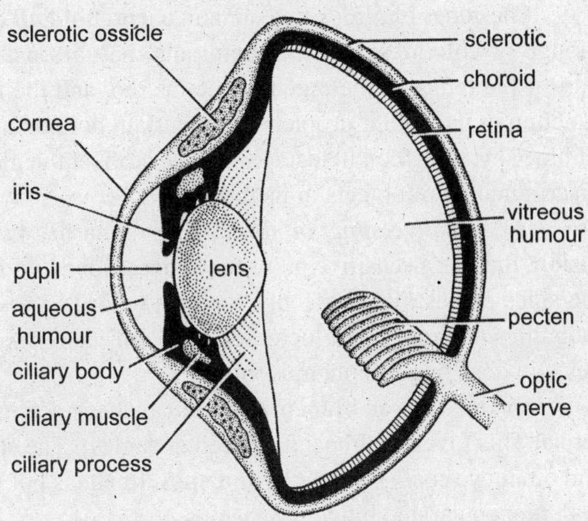

Fig. 14.28. Pigeon. Eye in sagittal section.

Fibrous tunic is protective, and provides surface for the attachment of eye muscles. Sclerotic plates keep the eye in proper position. Forward bulging of eye is also caused by the sclerotic plates.

(e) Choroid. The sclerotic is lined by a thin, pigmented and vascular layer, called the **choroid**. Anteriorly, it forms a circular, pigmented diaphragm or the **iris** under the cornea. It is pierced by a rounded aperture, the **pupil**, below the cornea, and above the lens. The iris consists of radiating and circular intrinsic muscle fibres which regulate the size of pupil to control the amount of light entering the eye. The pupil contracts to cut the light and dilates in the dark to allow more light rays to enter the eye. At the base of the iris, the choroid becomes much thicker, muscular, glandular, vascular and folded to form the **ciliary body**. **Ciliary processes** or the **suspensory ligaments** extend from the ciliary body to the lens. The ciliary body contains circular and meridional **ciliary muscles**. Meridional muscles connect the choroid with the sclerotic at the junction of the cornea and sclera.

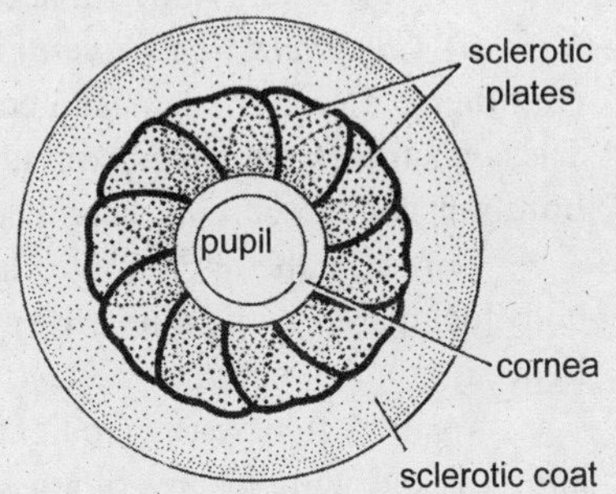

Fig. 14.29. Pigeon. Eye in outer view showing sclerotic plates.

(f) Retina. It is the innermost, light-sensitive and avascular layer of the eye. It is much thicker in birds (twice than that of human beings). It extends upto the margin of the pupil. Its part in contact with the choroid upto the ciliary body is thick and light-sensitive. This part is called **pars optica**. Its part lining the ciliary body and the iris is thin and nonsensory. It is called the **pars ciliaris**. The junction of these two parts is irregular and called the **ora serrata**.

Retina is composed of two layers. Next to choroid lies the **pigment layer** of retina which consists of heavily pigmented cuboidal cells arranged in a single row. It is only this layer that extends over the ciliary body. The inner layer or the **receptor layer** of retina lies next to the pigment layer. It is composed of light receptor visual cells—the **rods** and **cones**, and nerve cells. Rods and cones produce nerve fibres which remain connected with nerve cells, which in turn are connected with large **ganglion cells** in the retina. Nerve fibres from large ganglion cells pass over the inner surface of retina and collect together to form the **optic nerve** at the back of the eye. Being diurnal, pigeon has more cone cells in the retina. Birds recognize colours. **Bowmaker** (1979) has identified three groups of cone pigments that absorb maximally in the yellow, green and blue range. No other animal, except primate mammals, show such responsiveness to colours.

The cones of birds often contain **carotenoid oil droplets**. In diurnal birds they are red, yellow, orange or colourless, but in nocturnals they are pale yellow or colourless. The dorsal-posterior part of the retina of the pigeon contains red, and the remaining part contains yellow droplets. The function of the colour droplets is not certain but perhaps they act as filters and increase the contrast of blue sky and green fields. Red droplets offset the glare of water and perhaps provide finer colour discrimination of objects in feeding.

(g) Sensitive spots or fovea. The retina has two sensitive spots. For lateral or **monocular vision**, there is a sensitive pit or the **central fovea** within the central area of the eye. In many birds, the sides of this pit are steeply curved. According to **Walls** (1937, 1942), the curvature of the pit magnifies the image and increases acuity of vision. According to **Pumphrey** (1948), the curvature helps in detecting minute movements.

For forward or **binocular vision** there is a **temporal fovea** on the temporal surface of the retina. This fovea is almost flat (deep in eagles). The spot at which the optic nerve pierces the retina and other eye coats is termed **blind spot**. Image is not formed at this spot. Most birds are capable of both monocular and binocular visions.

(h) Lens. A colourless, transparent, biconvex and crystalline **lens** remains suspended by suspensory ligaments behind the iris (Fig. 14.28). The posterior side of the lens is more convex than its anterior side. A thin fibrous capsule surrounds the lens. The shape of the lens can be changed quickly by the ciliary body for accommodation.

(i) Chambers and humours. The lens divides the cavity of the eye into a small **aqueous chamber** between the lens and cornea, and a big **vitreous chamber** between the lens and retina. These chambers remain filled with a fluid. The aqueous chamber is filled with a watery **aqueous humour**, and the vitreous chamber contains a gelatinous **vitreous humour**.

Cornea, aqueous humour, lens and vitreous body together form the **dioptric apparatus** to focus light rays on the retina.

To allow free passage to light rays, the cornea and lens are without any blood supply. Aqueous humour compensates the loss of blood supply, and nourishes the lens and cornea. Cornea is also nourished by the secretion of lacrymal glands. Aqueous humour is constantly renewed. It is secreted by the ciliary body and is carried away, after bathing lens and cornea, by a vein.

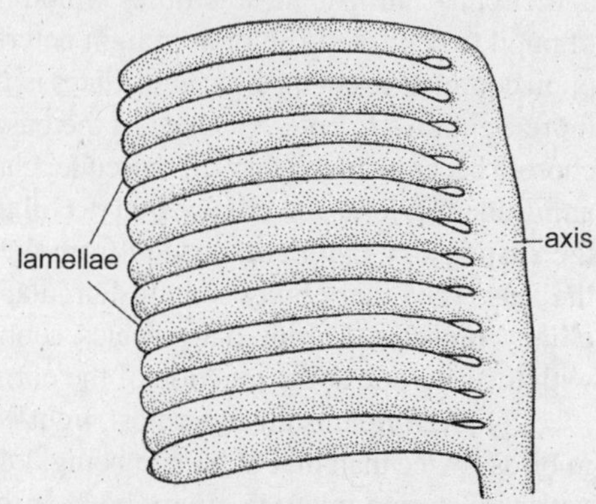

Fig. 14.30. Pigeon. Pecten.

(j) Pecten. It is the most puzzling organ of the eye of a few reptiles and all birds (except kiwi). It is a highly vascular, pigmented, folded, and comb-like structure (Fig. 14.30). It projects into the vitreous humour from the retinal surface ventral to the blind spot (Fig. 14.30). It is soft and almost rectangular.

Functions. Several functions have been assigned to the pecten:

(*i*) Probably its main function is to bring oxygen and nourishment to the retina, which is avascular in birds. It allows exchange of materials between the blood stream and the vitreous humour.

(*ii*) It is possible that the irregular shadow cast by this organ on the retina helps in detection of movements of an object. The function is supported by the observation that it is larger and much folded in predatory birds, and is small and smooth in nocturnal birds.

(*iii*) It may in some way be connected with accommodation. It probably regulates liquid pressure inside the eye which is changed by the change in the position of lens during accommodation. Last two functions are yet to be confirmed.

(*k*) *Vision and accommodation.* Being lateral in position, eyes of pigeon have a wide angle of monocular vision and only a narrow zone of binocular vision. Their fields of vision overlap a little. Extrinsic muscles, which move the eyeball, are smaller in birds and move the eye a little or not at all. Birds compensate this deficiency by turning their heads.

Birds have great power of accommodation (*i.e.*, adjustments in the eyes to see objects at various distances). It is effected by changing the position of the cornea and shape of the lens (especially the curvature of its anterior surface). For these purposes the ciliary body contains two sets of muscles. These are (*i*) **Crampton's muscles**, which pull the cornea to reduce its curvature in order to see the objects at a longer distance, and (*ii*) **Bruck's muscle**, which draws the lens forward to increase the curvature of its anterior surface in order to see near objects. Contraction of iris helps in the process.

This double method of active accommodation allows accommodation of 40–50 diopters (unit for the calibration of lenses) in the diving cormorant birds and 20 diopters in the birds in general (about 10 in human beings).

Birds recognize colours and judge distances accurately.

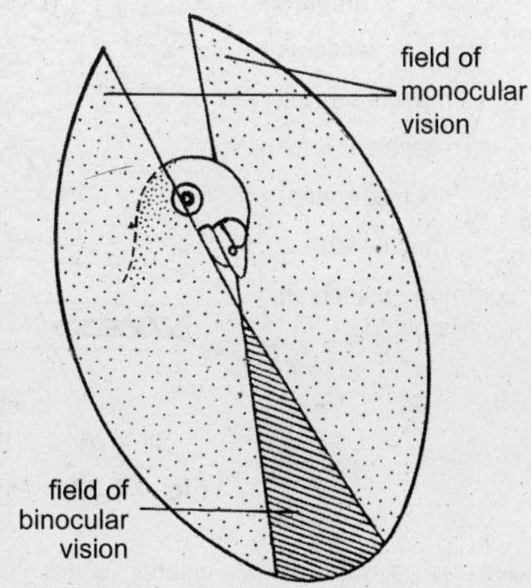

Fig. 14.31. Pigeon. Field of vision.

Ears. Birds have a well developed sense of hearing. Their ears are structurally similar to those of mammals, but there is no pinna. Bird's ear is divisible into **external, middle** and **internal parts**.

(*a*) *External ear.* It includes a circular **external ear opening** on each side of the head behind the eye. It remains covered by a special group of feathers, called the **auriculars**. The feathers do not obstruct hearing. The external ear opening leads into a short canal, called the **external auditory meatus**. At the distal end of the canal lies a thin membrane, called the **tympanum** or **ear-drum**.

(*b*) *Middle ear.* It includes an air filled **tympanic cavity**. It is bounded proximally by the **tympanum** and distally by the **auditory capsule**. A slender rod-like bony **columella auris** extends from the tympanum upto the **fenestra ovalis** of the internal ear. The inner disc-like bony portion, the **stapes**, of the columella plugs the fenestra ovalis, and its outer three-rayed cartilaginous portion, the **extracolumella**, attaches to the inner surface of the tympanum. The columella transmits sound waves from the tympanum to the internal ear.

From the tympanic cavity a narrow air passage, the **eustachian tube**, extends upto the pharynx and unites with its fellow of the other side and then opens at the back of the palate. Eustachian tube equalizes air pressure across the tympanum.

(*c*) *Internal ear or the membranous labyrinth.* It has the same typical structure as found in all other vertebrates. It remains enclosed in a **bony labyrinth**. The space between the bony labyrinth and membranous labyrinth remains filled with a fluid, called the **perilymph**, while the space within the membranous labyrinth is filled with dense fluid, called the **endolymph**. The membranous labyrinth consists of the usual **utriculus**, **sacculus**, three **semicircular canals** and **cochlear duct** or **lagena**. Three semicircular canals (two vertical and one horizontal) open into the utriculus, while lagena is formed from the prolongation of sacculus. Utriculus and sacculus remain continuous with each other. One end of each semicircular canal is swollen to form the **ampulla**.

Part of the bony labyrinth that encloses the **cochlear duct** or **lagena** is called the **cochlear canal**. The cochlear duct and cochlear canal together are termed **cochlea**. Cochlea consists of three chambers—an upper **scala vestibuli**, middle **scala media** and lower **scala tympani**. Only the scala media contains endolymph, the other two remain filled with perilymph. Floor of scala media is formed by the **basilar membrane**, and the roof is formed by the **Reissner's membrane**. The **basilar membrane** consists of rows of upto 40 tall, hairy **auditory cells** which form the **organ of Corti**. A **tectorial membrane** lies above the organ of Corti. Nerve fibres arise from the basal ends of the cells. These unite to form the cochlear branch of the auditory nerve. Different segments of the basilar membrane vibrate according to the frequency of the sound. Organ of Corti is sensitive to the sounds of higher frequencies.

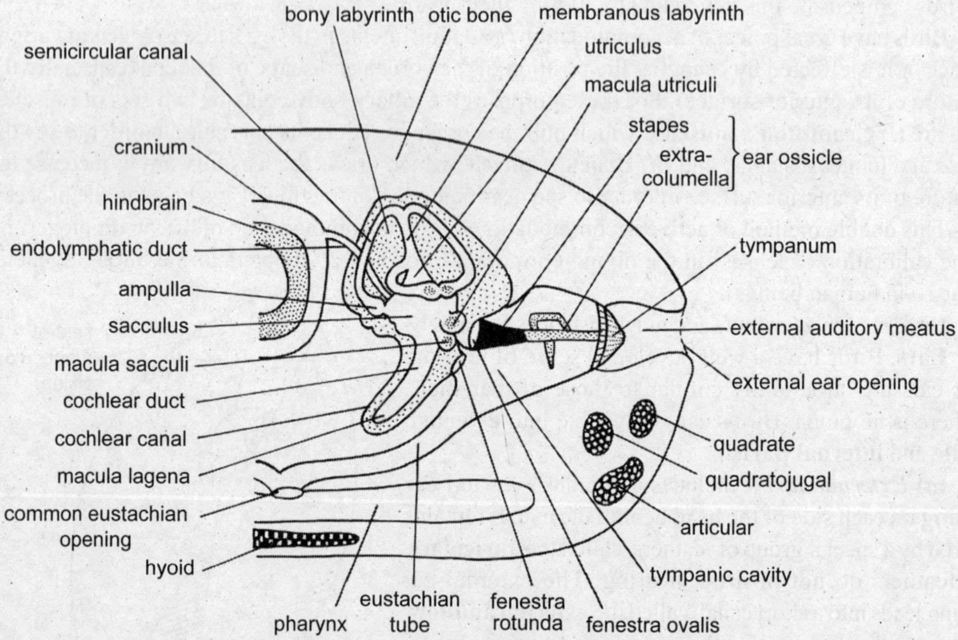

Fig. 14.32. Pigeon. T.S. through internal ear.

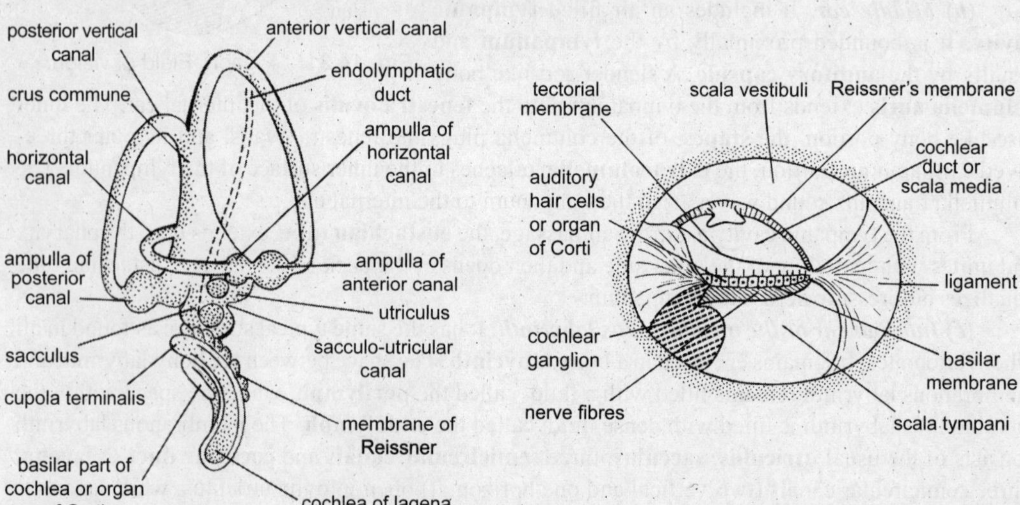

Fig. 14.33. Pigeon. A—Internal ear (membraneous labyrinth) and B—T.S. Cochlea.

Groups of auditory cells are called **cristae** and **maculae**. Those present in the ampullae are called **cristae ampullares**. The **macula utriculi**, lying in the utriculus, is horizontal and **macula sacculi** lies vertically in the sacculus. **Macula lagena** is curved, and lies at the tip of the lagena. The gravity receptors of the maculae maintain natural position of the body and head. Macula lagena is sensitive to the sounds of low frequencies.

The range of sound frequencies heard by the birds is rather narrower than that of most mammals (0.3–7.5 KHz in pigeons; upto 21 KHz in song birds). Their capacity for frequency discrimination is not much greater than ours but their capacity for time resolution is ten times better.

We cannot follow some of the rapid changes that make up a bird's song. Ability to localize sound is high. Owls and other night birds find their prey largely by ear. The long cochleas help in finding the direction of sound. A few cavebirds such as, *Steatornis*, the oil bird; *Collocalia*, the swiftlet, have the power of echolocation.

14.14 URINOGENITAL SYSTEM

The excretory and reproductive organs of pigeon are very closely connected with each other structurally, so are studied together under the heading urinogenital system. Urinary or excretory system is meant for excretion and genital or reproductive system has the function of multiplication.

1. Excretory system. The excretory system is similar in both the sexes and includes a pair of kidneys and a pair of ureters.

There is no urinary bladder. Ostrich has a urinary bladder. Lack of urinary bladder reduces the body weight and helps in flight.

(*i*) Kidneys. The kidneys are **metanephric**. The mesonephros, which serves as the embryonic kidney, disappears fully. The adult kidneys have a very characteristic form. They are large, flattened dark-red, bodies fitted in the hollows of the pelvis. Kidneys are divided into 3 lobes each by deep fissures. They are covered by peritoneum only on the ventral side. Each lobe of kidney is further differentiated into smaller lobules. Each lobule is formed of numerous convoluted **uriniferous tubules** compactly packed together in connective tissue. Each uriniferous tubule has long **loops of Henle** for the reabsorption and conservation of water from the glomerular filterate. With the result, concentrated urine, consisting largely of **uric acid** precipitate and water just sufficient to carry it down uterus, leaves the kidneys.

(*ii*) Ureters. The ureters are short, narrow tubes. They arise from the ventral surface of the kidneys between the anterior and middle lobes, run straight backward over the middle and posterior lobes and open into the urodaeum (middle compartment of the cloaca), piercing through its roof (dorsal wall). In the urodaeum of cloaca more water is reabsorbed from the urine, which changes into a white paste. With the faeces this forms a soft half-black, half-white mass which is expelled from the body.

2. Reproductive system. The reproductive system is different in the two sexes of pigeon.

A. Male reproductive system. The reproductive system of male pigeon (Fig. 14.34) consists of a pair of testes and a pair of vasa differentia (Wolffian or mesonephric ducts). There is no copulatory organ. However, ostrich and duck have erectile penis.

(*a*) Testes. The testes are large, ovoid bodies varying considerably in size according to the season. The left testis is often slightly larger than the right. Each testis is attached to the ventral surface of the anterior part of the kidney of its side by a double fold of peritoneum the **mesorchium**. Each testis is composed of convoluted, sperm-producing **seminiferous tubules** having groups of hormone (testosterone) producing **Leydig's** or **interstitial** cells between them (in the connective tissue). Testes stay permanently in the abdomen. Abdominal air-sacs are believed to extend in the testicular region to keep the testes cool in the absence of temperature-regulating scrotum of mammals.

***(b)* Vasa deferentia.** These are narrow, opaque, convoluted ducts arising from the inner sides of the posterior parts of the testes. They run backwards parallel to and outside the ureters. The anterior part of each vas deferens is more closely convoluted to form the **epididymis** which receives fine tubules, the **vasa efferentia**, from the testis. The posterior end of each vas deferens is slightly dilated to form a small **vesicula seminalis**, or **seminal vesicle** which opens into the urodaeum on the tip of a small, erectile papilla that projects into the urodaeum (of cloaca).

B. Female reproductive system. The reproductive system of the female pigeon (Fig. 14.34) consists of a single ovary and oviduct (Mullerian duct) situated on the left side. Both ovaries are present in the embryo, but later on the right ovary and right oviduct disappear. Loss of one ovary and oviduct reduces body weight to help in flight (Box 14.3).

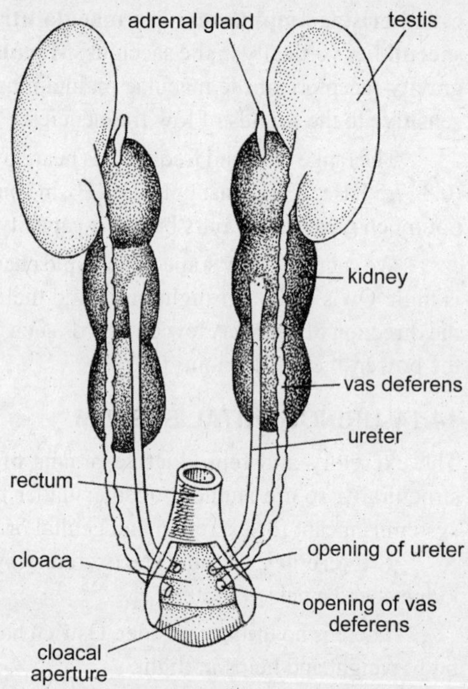

Fig. 14.34. Pigeon. Male urinogenital system.

Box 14.3

There is another advantage of one ovary and oviduct of the birds. As much as one-third of the calcium present in the two eggs it lays from one ovary. There may not be enough calcium for the shell of four eggs if both ovaries are present.

A reserve calcium is collected as the ovarian follicles mature. The oestrogen they produce increases the uptake of calcium from the food and stimulates its deposition in the bones. After ovulation the oestrogen level falls, the calcium is then mobilised from the bones, probably by parathyroid hormone in the blood becomes very high, until used by the egg.

***(i)* Ovary.** The left ovary is a large, irregular body varying in size according to the season (Fig. 14.35). It is attached to the ventral surface of the anterior part of the left kidney by a double fold of peritoneum, the **mesovarium**. The surface of ovary is studded with numerous rounded **follicles** of different sizes, each containing a single ovum. The ovary of pigeon also produces female sex hormone **oestrogen**. A vestige of right ovary is frequently present.

***(ii)* Oviduct.** The left oviduct is a long, wide, muscular, convoluted tube suspended from the dorsal body wall by a double fold of peritoneum called the **mesotubarium**. It is differentiated into several regions. Its anterior end is enlarged to form a wide membranous **oviducal funnel** with a large medially facing opening, the **ostium**, having fimbriated margin. It receives the ripe ova extruded from the follicle. As the ovum passes down the ostium, its walls secrete the thin **chalaziferous layer** of dense albumen around the egg. The succeeding part of the oviduct, called **glandular part** or **magnum**, has tubular glands which secrete the **albumen** or **egg white**. The magnum is followed by **isthmus** which secrete a parchment-like double **shell membranes** around the albumen. The next slightly swollen portion of oviduct is **uterus**. Uterus is thin walled and is lined by **nidamental glands** which form an outer white hard **calcareous shell** around the shell membrane. In many birds the egg shells are coloured, pigments derived from bile by breaking down haemoglobin, are deposited in the shells by nidamental glands of oviduct.

A small terminal part of the oviduct is called the **vagina**. It secretes mucus to help laying of egg. Vagina opens into the urodaeum of cloaca.

A vestige of the right oviduct is found attached to the right side of cloaca.

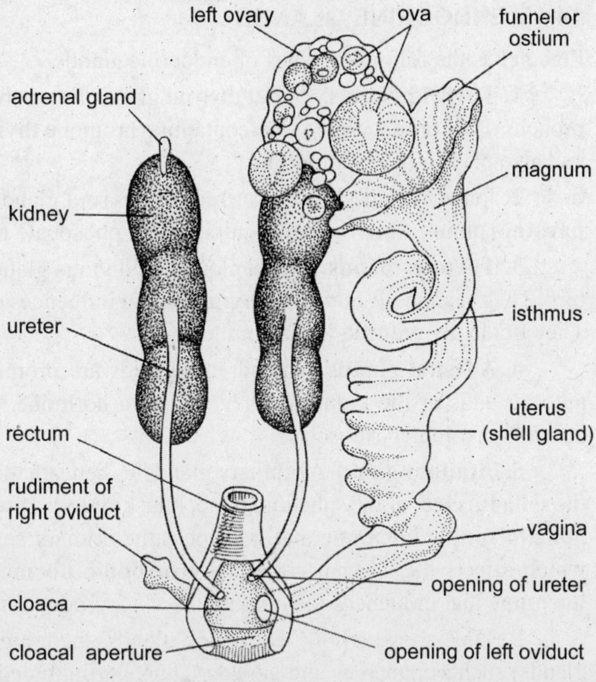

Fig. 14.35. Pigeon. Female urinogenital system.

Reproduction

(*i*) Fertilization. In pigeon, fertilization is internal and occurs in the upper thin-walled part of the oviduct. This follows copulation, which is preceded by courtship. The male treads on the back of the female, holding the skin and feathers or her neck with his beak. The female lifts up her tail and both press their everted proctodaea closely together. Now the male sheds the sperms into the urodaeum of the female and retires. From this place, sperms are squeezed into the oviduct by contractions of the urodaeum. A band of cilia in the oviduct guides the sperms forward to the site of fertilization.

(*ii*) Ovulation and formation of cleidoic eggs. Two ova are ovulated in close succession by bursting of mature follicles at a preformed non-vascular band, the **stigma** or **cicatrix**, on their surface. Ovulation depends on the cyclic release of luteinizing hormone (LH). After ovulation, the follicles quickly shrink. No corpus luteum is formed unlike the mammals. The eggs directly pass into the enlarged ciliated and muscular oviducal funnel which is closely applied to the ovary at this time. Each ovum is, at this stage, only a mass of **yolk**, having one end a small **germinal disc**, which will give rise to an embryo. Ovum is enclosed by a **vitelline envelope**. After fertilization, the ovum (the "yolk") rolls down the oviduct in a spiral manner with peristaltic contractions of the wall of the oviduct. During its passage, first a thick layer of **albumen**, the two thin, firm **shell membranes** and finally the calcareous shell having numerous fine pores are added around the ovum.

(*iii*) Hatching and parental care. Two large white, polylecithal eggs are laid in a clutch. The second egg is laid two days after the first one. Eggs are laid in a rough nest and incubated at body temperature (42°C) by constant sitting one or other of the parents. Incubation extends over fourteen days. Embryonic membranes such as amnion, chorion, allantois and yolk sac are formed during embryonic development. At the end of incubation, the chick (young bird, called squab) breaks the egg shell with a special **egg tooth** or **caruncle** developed at the end of the beak. The chicks are **altricial**. They are helpless with eyes closed and bodies clothed in fine, yellow "down". This is called **hatching**. They are fed by both the parents on "pigeon's milk". In about three weeks, the contours and flight feathers are developed and the nestlings are ready to leave the nest. They are taught to fly by the parents before they venture out to look after themselves.

Breeding continues throughout the year, but slows considerably in the rainy season.

14.15 ENDOCRINE GLANDS

Pigeon has the following types of endocrine glands.

1. Thyroid glands. Paired thyroid glands lie on the sides of the trachea near its division into bronchi. They produce an iodine-containing hromone **thyroxine**, which regulates general metabolism and also plays a role in moulting.

2. Parathyroid glands. Parired parathyroid glands occur close to the thyroids. They secrete **parathormone**, which controls calcium and phosphate metabolism.

3. Thymus glands. Paired elongated thymus glands lie on the sides of the throat in a young bird. They secrete the hormone **thymosin** that influences growth in early life and produces immunity. They gradually reduced in size in the adult.

4. Adrenal glands. The adrenal glands are prominent, yellowish, irregular bodies near the anterior ends of the kidneys. They secrete a hormone, called **adrenaline** or **epinephrine**, which controls involuntary muscles.

5. Pituitary gland. A pituitary gland lies beneath the diencephalon protected in a bony pockets, the **sella turcica**, of the sphenoid bone. It lacks intermediate lobe. It produces several hormones, *e.g.*, **gonadotrophic hormone** that influences the primary sex organs, **adrenocorticotrophic hormone** which affects the adrenal cortex, **thyrotrophic hormone** which controls thyroid and **prolactin hormone** that influences the crop glands (*i.e.*, production of pigeon milk).

Besides above wholly endocrine glands, there are some partly endocrine glands (or double glands) such as pancreas and gonads. Thus, during breeding season, the size of the ovary increases considerably due to influence of FSH and LH of anterior pituitary. The ovary secretes **oestrogen** which modifies the accessory sexual organs and behaviour of female birds. In males, the **testosterone** is produced by Leydigs cells of testes. This hormone controls the accessory sexual organs and behaviour of the male birds.

14.16 ADAPTATIONS OF PIGEON

Pigeon, like any other flying bird, shows adaptations mainly for flight, homoiothermy, feeding, water conservation and cleidoic development.

A. Flight adaptations. Adaptation for flight are found practically in all he systems of a bird.

1. **Streamlined body.** Birds have a boat-shaped body, which offers minimum resistance to air during flight. All contour feathers lie flat and are directed backward for the same purpose.
2. **Flight organs.** The flight organs of a bird are a pair of large **wings**, which are modified forelimbs beset with special firm feathers, the long **quills** and the small **coverts**. The wings have thick, strong, leading edges; thin, flexible trailing edges; convex upper surfaces and concave lower surfaces. This form of the wings efficiently parts the air in front causes a minimum turbulence behind.
3. **Flight muscles.** Enormous muscles are present to drive the wings during flight. These include the pectoralis major, which brings about a powerful downstroke, and the pectoralis minor that effects an upstroke.
4. **Skeletal surface for flight muscles.** Large sternum produced into a large midventral keel provides sufficient hard surface for the attachment of the enormous flight muscles.
5. **Energy for flight.** Respiratory and circulatory systems efficiently meet the energy requirement of the flight muscles. Blood is oxygenated during both inspiration and expiration. Unusually large, fast-beating, four-chambered heart supplies oxygenated blood to the wing muscles as well as to the rest of the body by way of a single aortic arch. Another aortic arch, if present,

would have mixed the two samples of blood and this would have reduced the metabolic rate.

6. **Lightness.** Lightness of body of pigeon and other flight birds is essential for the flight. It has been achieved by birds in various ways. Paper-thin skull bones; pneumaticity of all long bones and many skull bones; complete loss of some bones: two fingers, toe, claws of fingers and all teeth; reduces tail; short rectum; lack of urinary bladder and loss of one ovary are some of the measures to reduce weight of the body. Pigeon is hardly 500 gm. in weight, and a greater part of it is contributed by the flight muscles.
7. **Rigidity.** Rigid body can successfully withstand the pressure of the air currents during flight. Rigidity has been acquired in birds by fusion and flattening of bones without affecting lightness in any way. All vertebrae, except the cervicals, are fused (see Chapter 19), Cranium is almost without sutures. Thoracic vertebrae are compressed to form a firm fulcrum for wing strokes. Double-headed ribs, uncinate processes (of ribs) and sternum form a strong thoracic basket. Fusion of the clavicles adds to the strength of the pectoral girdle.
8. **Equilibrium.** Equilibrium is another essential requirement for flight in the air. Attachment of wings high up on the trunk; high position of light organs, such as lungs; low position of heavy flight muscles and digestive organs lower the center of gravity and prevent turning over of the body in the air.
9. **Sense organs.** Olfactory sense organs have become reduced in view of the fact that the birds flying high in the air cannot depend on smell. The organs of sight and hearing are, on the contrary, of high order, being capable to guide the bird form a distance.
10. **Well developed nervous system.** Birds have a well developed nervous system to coordinate the complex movements of wings and their feathers during flight, particularly of the flapping type.
11. **Steering device.** Tail feathers assist in steering, so essential during flight.
12. **High body temperature.** Birds are homoiothermic vertebrates and maintain a very high body temperature with the aid of brisk metabolism and an insulating coat of feathers. Only at a high temperature the strenuous activity sch as prolonged flight is possible.
13. **Take off.** The legs strengthened by the fusion of thier bones are not only effective in bipedal locomotion but also provide a forceful take off for flight.

Storage of food in the crop and water conservation by reabsorption in long Henle's loops and urodaeum enable the birds to undertake long flights, prolonging the need for food and water.

B. Adaptation for homoiothermy. Homoiothermy, *i.e.*, maintenance of constant body temperature of birds, depends on factors such as sufficient supply of food to the tissues, efficient respiratory and circulatory systems and means to check heat loss.

1. There are many devices in pigeon to supply sufficient food to the tissues—(*a*) It takes all kinds of grains and seeds it comes across and stores them in a large quantity in the crop to ensure uninterrupted supply to the tissues and (*b*) Long ileum with villi brings about maximum absorption of nutrients from the food passing through it.
2. Efficiency of respiratory system has been enhanced by the fact that—(*a*) aeration of blood occurs during inspiration as well as expiration and (*b*) by the enormous surface provided by the millions of air capillaries of the lungs for the exchange of gases.
3. Efficiency of circulatory system lies in the fact that the large fast-beating heart quickly supplies fully aerated blood to all parts of the body. This facilitates oxidation of food for release of energy, part of which warms up the body.
4. The loss of heat from the body is prevented by—(*a*) An insulating body coat of feathers on the skin, (*b*) Lack of sweat glands checks loss of heat by evaporation and (*c*) Absence of muscles in the unfeathered feet also reduces loss of heat.

C. Adaptation for feeding.

1. Pointed beak enables the pigeon to pick up grains and seeds conveniently.
2. Large crop provides storage for rapidly-swallowed unmasticated for the lack of teeth by grinding the food between pebbles through muscular contractions.

D. Adaptation for water conservation.

1. Absence of sweat glands checks loss of water.
2. Thick, feathery coat prevents evapoartion from the body surface.
3. Kidney tubules have a long loop of Henle for reabsorption of water from the glomerular filterate.
4. Water is also reabsorbed from the urine in the urodaeum. With the result, almost solid urine is passed out with the faeces.

E. Adaptation for cleidoic developement.

1. Large eggs with abundant food and calcareous shell are laid. Lack of pubic and ischial symphysis enables the bird to lay large-sized eggs conveniently.
2. Shell protect the embryo from mechanical injury and desiccation, at the same time permitting gaseous exchange through its pores.
3. Abundance of food enables the embryo to complete its development into a young bird, excluding a larval stage, which cannot survive on land.
4. Embryonic membranes provide aquatic medium within the egg by acquiring fluid in them also help in respiration and storing waste matter.
5. Insoluble nitrogenous end product requires little storage space in the egg.
6. Egg-tooth enables the young bird to break the shell for escape from it.
7. Crop secretion provides nourishing diet for the nestling.

QUESTIONS

Long Answer Questions

1. Describe the structure of a feather. Point out different kinds of feathers and their adaptations for different purposes. *(CCSU Meerut 2007)*
2. Enumerate the different types of feathers in birds. Describe the structure and development of contour feather. *(Purvanchal 2000)*
3. Write an essay on exoskeleton in birds. *(Bundelkhand 1999)*
4. Give an illustrated account of the digesive system of a bird and comment upon its special adaptive features. *(Purvanchal 2003)*
5. Describe the alimentary canal of *Columba* and mention the significance of its different structures. *(Gorakhpur 2002)*
6. Give an account of the organs of respiration and their mode of working in a bird. *(Gorakhpur 2003)*
7. Give an account of air sacs in pigeon. Comment on their function. Explain how they are useful in flight. *(Kanpur 2001; CCSU Meerut 2006)*
8. Describe the structure and mechanism of pigeon's heart. *(Bundelkhand 2000)*
9. Describe in detail the structure of brain and spinal cord of *Columba* and show how it is adapted to its aerial mode of life.
10. Describe the peripheral nervous system of pigeon.
11. Name the sense organs found in a pigeon. Describe the structure and functioning of eye. Add a note on pecten.
12. Describe the ear of pigeon.
13. Give an account of the urinogenital system of the pigeon and discuss the significance of the presence of only one ovay in most of the birds. *(CCSU Meerut 2003)*
14. Given an account of the flight muscles of pigeon.

15. Discuss adaptations of flight.
16. Discuss adaptations for pigeon.

Short Answer Questions

1. Give an account of the habitat and habits of pigeon.
2. Give systematic position of pigeon.
3. Describe the external feature of pigeon.
4. Describe the structure and function of crop and gizzard of pigeon. (*Purvanchal 2003*)
5. Give an account of perching mechanism of a bird. (*CCSU Meerut 2006*)
6. Describe body wall of pigeon.
7. Describe functions of feathers.
8. How are remiges arranged in a wing of pigeon?
9. Give an account of a quill feathers.
10. What is 'pecten'? (*CCSU Meerut 2003*)
11. What do you know about sound production in a bird? (*CCSU Meerut 2007*)
12. Write a short notes on the following:
    (*a*) Quill feather (*b*) Air sacs
13. Draw well labelled diagrams of the following:
    (*i*) Flight muscles of *Columba* (*ii*) A typical quill feather
    (*iii*) Arterial system of pigeon (*iv*) V.S. of eye of pigeon
    (*v*) Female urinogenital system of pigeon

Very Short Answer Questions

1. Write scientific/zoological name of pigeon. (*CCSU Meerut 2006*)
2. What is pecten? (*CCSU Meerut 2005*)
3. In which branch of zoology birds are studied?
4. Define a bird.
5. What is pterylosis?
6. Where does pigeon makes its nest?
7. Where is pigeon's milk is manufactured?
8. How many pancreatic ducts do you find in pigeon?
9. Which ovary is functional in birds?
10. What is the role of air sacs?
11. What is function of uropygial gland?
12. What is the function of pectoralis major muscle?
13. What is the body temperature in birds?
14. Which types of muscles are associated with perching?
15. What is bursa fabricii?
16. Give the name of vein that is present only in birds.
17. Where are organs of corti located?
18. What is Schneidernian membrane?

Fill in the Blanks

1. A wing of pigeon has ___________ quill feathers.
2. Uropygial gland lies on the ___________ of pigeon.
3. Pigeon keeps ___________ in its gizzard for grinding the grains.
4. The ___________ feathers cover the ear opening of pigeon.
5. The horny sheath covering the pigeon's jaws is called ___________.

6. Fold of skin stretching between the upper arm and the forearm in the pigeon is called __________.
7. The sound box of pigeon is __________.
8. Feathers in birds are __________ derivatives.
9. In pigeon, there are __________ major air sacs.
10. In pigeons, the eggs hatch into naked, blind and helpless young ones. This is described as __________.
11. The tracts or distinct patches on the body of a bird on which feathers are arranged, are called __________.
12. The tendon of pectoralis minor passes through __________ to be inserted on the trochanter of humerus.

Multiple Choice Questions

1. Flight muscles are attached to
   (*a*) clavicles (*b*) keel of sternum
   (*c*) sternal ribs (*d*) coracoid of pectoral girdle
2. In birds, sound is produced by the (*Gorakhpur 2002, Kanpur 2003*)
   (*a*) lungs (*b*) syrinx
   (*c*) laryn (*d*) air sacs
3. Feathers that help in flight are called
   (*a*) down feathers (*b*) contour feathers
   (*c*) coverts (*d*) remiges
4. The thick, rough and horny linings of yellow or green colour is found in
   (*a*) crop (*b*) gizzard
   (*c*) trachea (*d*) proventriculus
5. Complete double circulation is found in
   (*a*) birds (*b*) reptiles
   (*c*) amphibians (*d*) fishes
6. Coprodaeum is the part of cloaca of bird. It receives
   (*a*) rectum (*b*) ureter
   (*c*) urinary bladder (*d*) genital duct
7. In birds, gall bladder is
   (*a*) large and bilobed (*b*) small and oval
   (*c*) absent (*d*) highly specialised
8. Birds are
   (*a*) uricotelic (*b*) ureotelic
   (*c*) ammonotelic (*d*) ureco-ammonotelic
9. Crop glands secrete
   (*a*) mucus (*b*) enzymes
   (*c*) milk (*d*) hormones
10. Aftershaft is located at the
    (*a*) tip of vexillum (*b*) base of vexillum
    (*c*) base of calamus (*d*) middle of vexillum
11. Pigeon's trachea is supported by
    (*a*) incomplete cartilaginous ring (*b*) complete cartilaginous ring
    (*c*) incomplete bony ring (*d*) complete bony rings
12. Pigeon's kidneys are
    (*a*) unilobed (*b*) bilobed
    (*c*) trilobed (*d*) multilobed

13. The backwardly directed toe in the pigeon's foot is number
(*a*) one (*b*) two
(*c*) three (*d*) four

ANSWERS

Very Short Answer Questions

1. *Columba livia.*
2. Pecten is soft and almost rectangular pleated structure which projects into vitreous humour from the retinal surface ventral to the blind spot in the eye of all birds except kiwi. It supplies oxygen and nourishment to the retina.
3. Ornithology.
4. The feathered bipeds.
5. Pterylosis is arrangement of feathers on the body of the bird.
6. Pigeon makes its nests in the old buildings but never on trees.
7. In crop by the crop glands of both sexes during breeding season.
8. Three.
9. Left.
10. Increase the efficiency of respiration and cool the body.
11. Uropygial gland produces oily secretion which is used for lubricating or dressing the feathers and keeping the horny beak nonbrittle.
12. Pectoralis major pulls the wing downward and forward so that bird's body is lifted up and propelled forward in the air.
13. 104–112°F.
14. Flexor muscles and extensor muscles.
15. Bursa fabrici or cloacal thymus is a small thick walled glandular blind pouch that lies against dorsal wall of proctodaeum only in young ones and contain lymphatic tissue. Its endothelial lining produces lymphocytes.
16. Jugular anastomosis.
17. In basilar membrane of cochlear duct of internal ear.
18. Schneidernian membrane is specialized one-layered epithelium that invests the middle and posterior turbinals present in the olfactory part of nasal chambers.

Fill in the Blanks

1. Twenty three. **2.** Uropygium (tail). **3.** Pebbles. **4.** Auricular. **5.** Rhamphotheca. **6.** Prepatagium. **7.** Syrinx. **8.** Epidermal. **9.** Nine. **10.** Altricial species. **11.** Pterylae. **12.** Foramen triosseum.

Multiple Choice Questions

1. (*b*)	**2.** (*b*)	**3.** (*d*)	**4.** (*b*)	**5.** (*a*)	**6.** (*c*)
7. (*c*)	**8.** (*a*)	**9.** (*c*)	**10.** (*b*)	**11.** (*d*)	**12.** (c)
13. (*b*)					

15

Skeleton of Bird (The Fowl)

The skeleton (*i.e.*, endoskeleton; Fig. 15.1) of *Gallus domesticus*, the domestic fowl, is larger than that of pigeon and hence more convenient for study in the laboratory. The endoskeleton of birds is comparatively delicate but fully ossified and built on "hollow girder principle." It is light but strong and modified to suit aerial mode of life, bipedal locomotion and laying of large-sized eggs. The larger bones are without marrow and often pneumatic (have air spaces). Most of the bones of fowl are fused to provide rigidity.

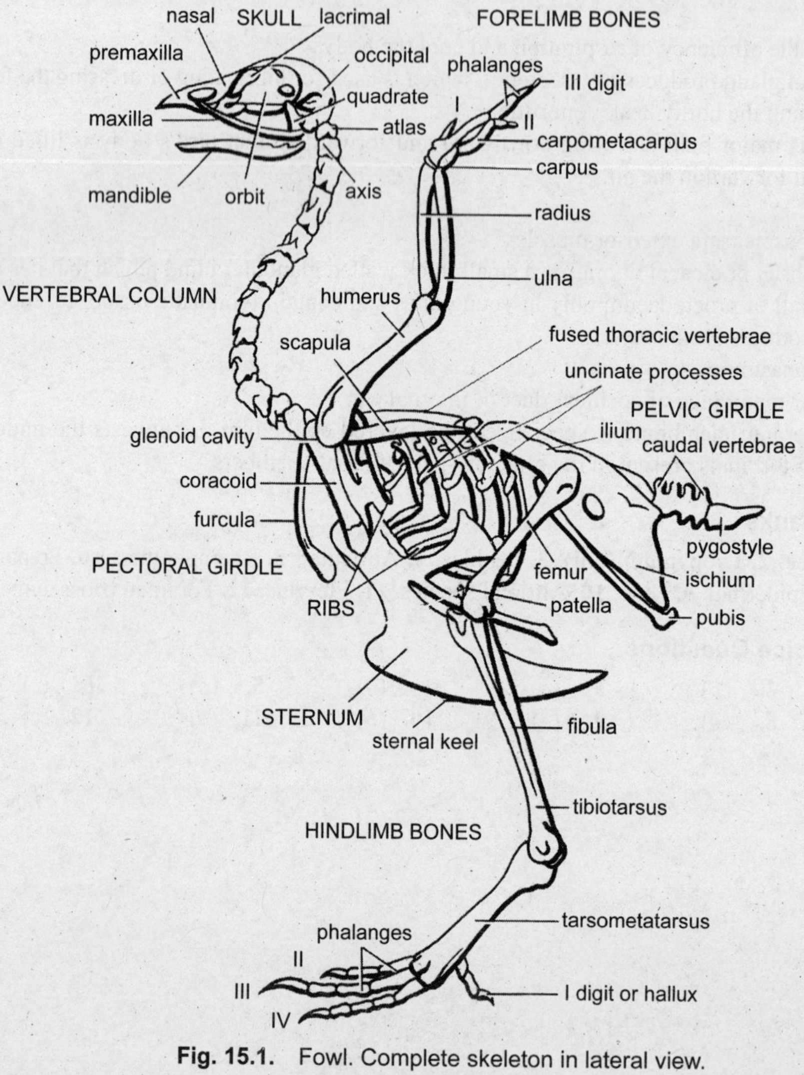

Fig. 15.1. Fowl. Complete skeleton in lateral view.

The endoskeleton of fowl, like that of other vertebrates, is divisible into two parts: *axial skeleton* and *appendicular skeleton.*

15.1 AXIAL SKELETON

It forms the longitudinal axis of the body and includes **skull**, **vertebral column**, **sternum** and **ribs**.

1. Skull

The skull of fowl (bird) shows following characteristic features:

1. The skull is light in weight.
2. Due to the fusion of most of the bones, the sutures disappear in the adult skull.
3. The jaw bones are modified to form a toothless **beak**.
4. To accommodate the larger brain and eyes the brain-case and orbits are greatly developed.
5. The skull is **monocondylic** and **tropibasic**.
6. Foramen magnum is directed downwards.
7. The jaw suspension is **autostylic**.
8. Palate is **schizoganthous**. It is formed by vomers, palatines, pterygoids and palatal processes of maxilla.

Regions of Skull

The skull of fowl consists of following three parts:

1. **Cranium.** It houses the brain.
2. **Sense capsules.** These include three pairs of sense capsules enclosing the organs of smell, hearing and sight.
3. **Visceral skeleton.** It includes the upper jaw, the lower jaw and the hyoid apparatus.

1. Cranium

It can be divided into the following three regions:

1. Occipital region. It forms the posterior part of the skull. It consists of a **basioccipital** at the base, two **exoccipitals** on the lateral sides, the **supra-occipital** at the top. All these bones enclose a large rounded opening, the **foramen Magnum**. Just below the foramen is a single **rounded occipital** condyle formed by the basioccipital. **The presence of a single occipital condyle distinguishes the avian skull from the amphibian or mammalian skull in which case the skull articulates with the atlas vertebra by two occipital condyles formed on the exoccipitals.** A prominent ridge of bone, the lamboidal crest, is present where the supra-occipital joins the parietal.

2. Parietal region. It lies in front of the occipital region. Its roof is formed by the **parietals** and the **squamosals**, the sides by **alisphenoids** and the floor by a large **basisphenoid**. Ventrally, the basisphenoid is covered over by a broad membrane bone, the **basitemporal** or **parasphenoid**. The basisphenoid is continued forward in the form of a slender **parasphenoid rostrum**. The squamosal of each side is produced outwards into a **zygomatic process**.

3. Frontal region. The roof of frontal region is formed by a pair of large **frontal bones**. The lateral posterior end of each frontal is drawn out in a strong **zygomatic** or **post-orbital process** which joins the zygomatic process of the squamosal. The alisphenoids are continued forward into the **orbito-sphenoids** to form the sides of the frontal region. The base of the frontal region is formed by a poorly developed **presphenoid** lying above the parasphenoid rostrum.

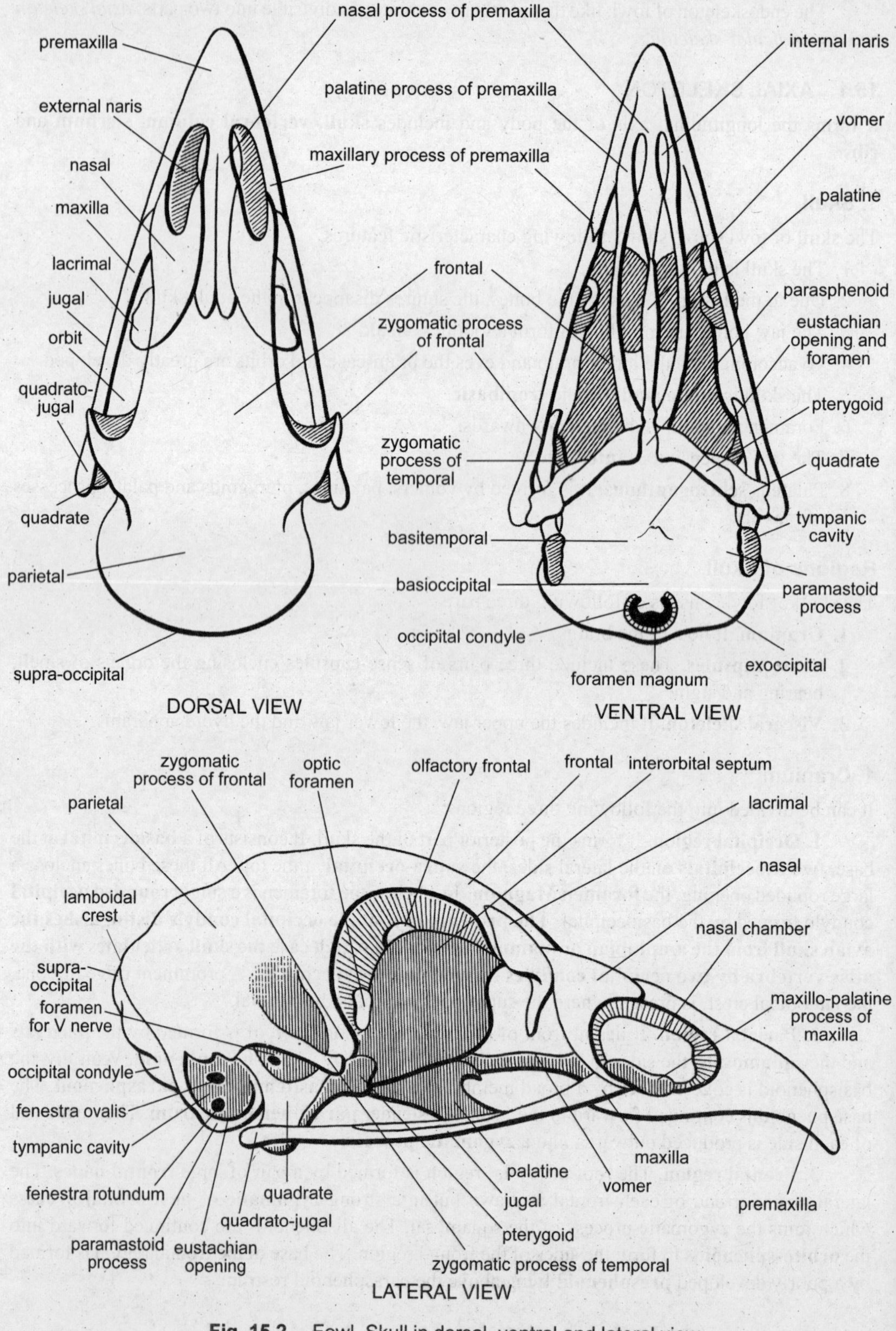

Fig. 15.2. Fowl. Skull in dorsal, ventral and lateral view.

2. Sense Capsules

1. Auditory capsules. These are a pair of auditory capsules enclosing the auditory organs (organs of hearing), attached behind one on either lateral side of the occipital region of the skull. Each capsule is formed by the **prootic bone**. A large cup-shaped **tympanic cavity** of the middle ear is present on each posterolateral side of the skull bounded by the squamosal below. The tympanic cavity contains a tympanic membrane attached within its prominent outer edge and a slender rod-like **columella** made of bone and cartilage. Tympanic cavity has several openings of which two lie in the middle of the cavity. The upper one is **fenestra ovalis** and the lower one, the **fenestra rotunda**. The tympanic cavity communicates with the mouth through a **eustachian canal**. It runs forward and inwards between the basitemporal and basisphenoid, joins with its fellow of the opposite side to open by a common median aperture in the roof of the posterior part of the mouth cavity.

2. Optic capsules or orbits. The two orbits are compartively large to accommodate massive eyes. They are separated by the **inter-orbital septum** formed by a **mesethmoid** together with **parasphenoid**, **parasphenoid rostrum** and **orbitosphenoids**. Each orbit is bounded dorsally by the **frontal**, anteriorly by the **frontal** and large **lacrimal** and posteriorly by the **post-orbital process** of frontal and the alisphenoid. The inter-orbital septum is perforated by many openings. The **olfactory foramen** lies in front of the cranium between the septum and the frontal bone. A large median optic **foramen** lies in the posterior part of the orbit. The small openings for the 3rd and 4th cranial nerves lie immediately lateral to the optic foramen.

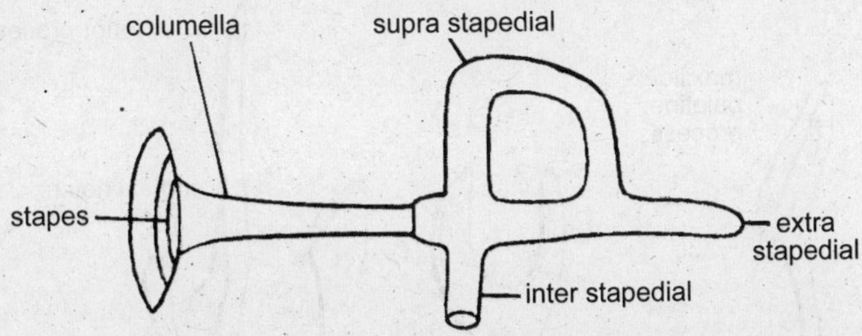

Fig. 15.3. Fowl. Columella auris.

3. Olfactory capsules. These are greatly reduced and cartilaginous. The **turbinals** are poorly developed. The nasals are a pair of thin, triangular plate-like bones forming the roof and sides of the nasal chambers. Each nasal sends three processes, a posterior one uniting with the frontal and two anterior ones furnishing the boundaries of the external nare and uniting with the process of premaxilla. The two vomers fuse to form a small median, slender bone lying in continuation with the parasphenoid rostrum and separating the two nasal chambers at the base.

3. Visceral Skeleton

It includes the following parts:

1. Upper jaw or maxillo-palatine apparatus.
2. Lower jaw or mandible.
3. Hyoid apparatus.

1. Upper jaw or maxillo-palatine apparatus. The upper jaw on each side forms two bony arcades the inner arcade consisting of **pterygoid** and **palatine**, and outer arcade formed by **premaxilla**, **maxilla**, **jugal**, **quadratojugal** and **quadrate**.

***(a)* Pterygoids.** Each pterygoid is a short, stout and rod-shaped bone lying somewhat obliquely. It articulates posteriorly with the inner surface of the quadrate and anteriorly with the palatine and the parasphenoid rostrum along which it can slide.

***(b)* Palatines.** The palatines are elongated, slender and rod-shaped bones lying immediately in front of the pterygoids. Each palatine forms the major part of the inner arcade. Its posterior end is broadened into triangular plate and articulates with the vomer and parasphenoid rostrum. Its anterior end is narrow and attached with the ventral surface of the processes of maxilla and premaxilla.

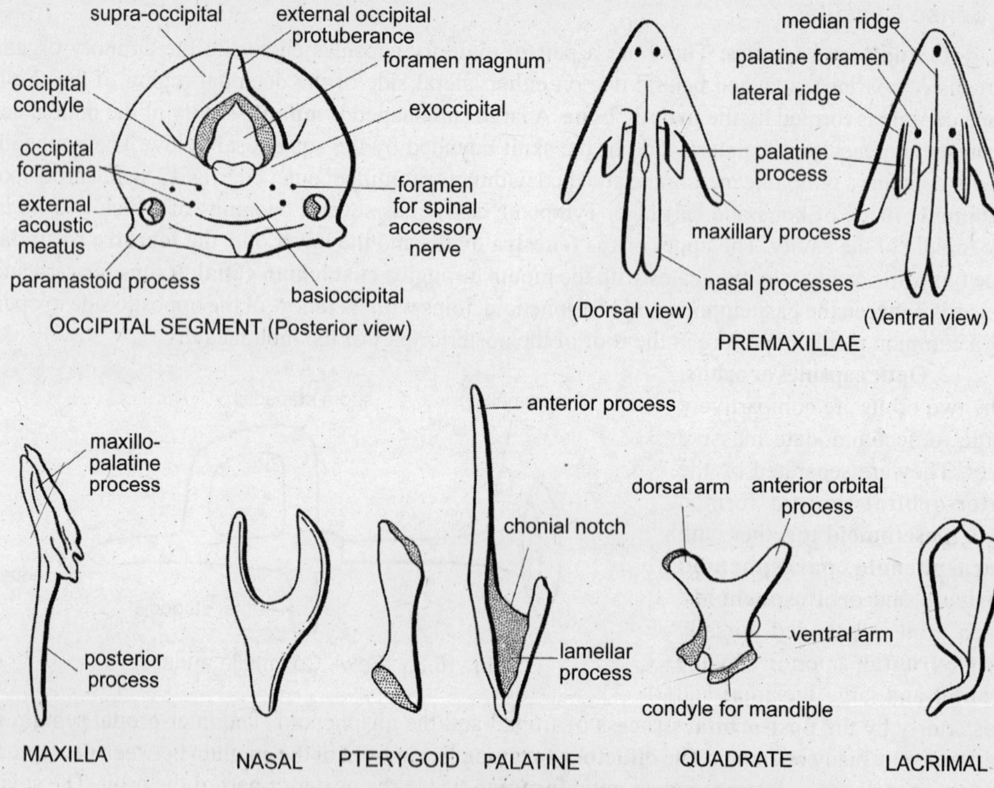

Fig. 15.4. Fowl. Disarticulated (loose) bones of skull.

(c) Premaxillae. It is the anteriormost part of the upper jaw which forms the upper beak. In fact it is formed by the complete fusion of two premaxillae in the embryonic stage. It is a triradiate bony mass extending posteriorly into three pairs of processes. The **nasal processes** run behind in close proximity to join the frontals and complete the upper boundaries of the external nares. The maxillary processes run backward and outward to form the anterior part of the upper jaw. The palatine processes unite with the palatines to form the anterior part of the palate.

(d) Maxillae. Each maxilla is a thin rod-shaped bone forming the anterior part of outer arcade of the upper jaw. Its posterior part is expanded into a spongy plate, the **maxillo-palatine process**. The maxillo-palatine process of two maxillae do not unite mesially so that the bony palate is incomplete. The posterior zygomatic process of the maxilla forms the anterior part of the **outer arcade** or **infra-orbital arcade** or the **zygomatic arch**.

(e) Jugal. It is a slender and rod-like bone forming the middle part of the suborbital bar lying dorsally above the maxilla and quadrato-jugal.

(f) Quadrato-jugal. It is also a slender and rod-like bone. It forms the posterior part of the suborbital bar of outer arcade. Its hind end is thickened and articulates with the quadrate.

(g) Quadrate. It is a stout, triradiate bone lying in front of the tympanic cavity and forming the suspensorium with the squamosal bone of the skull. Its dorsal arm articulates with the squamosal slightly above the anterior margin of the tympanic cavity. Its ventral arm is flattened and runs ventrally along the outer side of the pterygoid and articulates with the quadratojugal while its expanded inner end provides the articular surface for the lower jaw. Its anterior arm runs inwards parallel to the pterygoid.

2. Lower jaw or mandible. The lower jaw or mandible appears to be a single V-shaped bone, but it is really composed of two long and laterally compressed rami which are firmly united in front. Each ramus is composed of five bones *i.e.*, one replacing bone, the **articular**, and four investing bones, the **angular**, **supraangular**, **dentary** and **splenial**. The articular forms the posterior expanded part of the ramus. On dorsal surface it bears an elongated mandibular condyle for the articulation with the quadrate. Posteriorly, it is produced into medial and angular processes. The slender **angular** lies beneath the articular and along the lower border of the jaw. The **supra-angular** forms the dorsal part of about one-third of the posterior part of the ramus. It bears a coronoid process. **Dentary** is the longest bone and forms the anterior half of the rami or mandible. In front it fuses firmly with its fellow of the opposite side forming the **mandibular symphysis**. The **splenial** is a thin splint-like bone lying along the inner surface of the middle part of the ramus (mandible). A **mandibular foramen** is present between the dentary and other posterior bones.

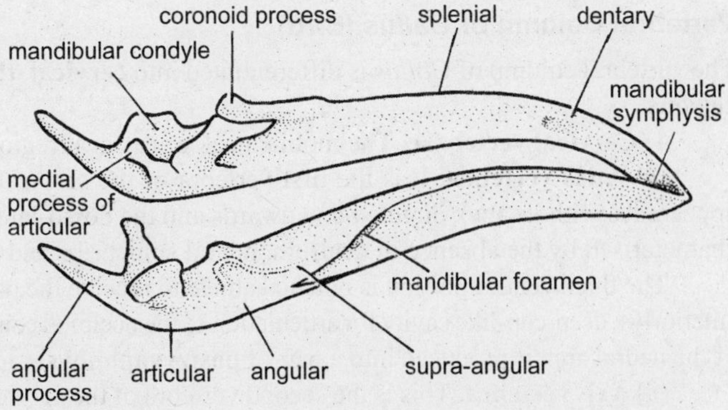

Fig. 15.5. Fowl. Lower jaw or mandible (lateral view).

3. Hyoid apparatus. The hyoid apparatus lies in the floor of the mouth and forms the skeleton of tongue. The median arrow-shaped body consists of three bones: an anterior cartilaginous **entoglossum**, a median body, **basihyal** and a posterior **urohyal**, placed end to end. The basihyal bears a pair of jointed and backwardly directed cornua, the **thyrohyals**. Each thyrohyal consists of two pieces, the proximal **basibranchial** and the distal **ceratobranchial**.

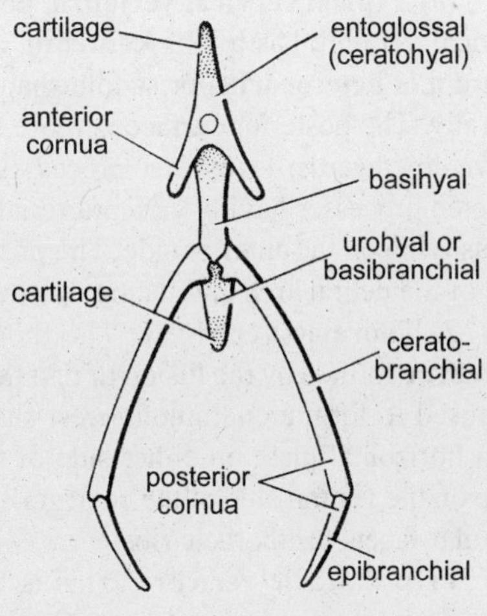

Fig. 15.6. Fowl. Hyoid apparatus.

Palate. The palate region of the skull of pigeon is of **schizognathous** type. It is formed by the **vomers**, **palatines**, **pterygoids**, and the **palatal process of maxillae**. The vomers are well developed and occupy the space between the two palatines, which thus remain widely separated.

Vertebral Column

Due to aerial mode of life, the vertebral column of a bird shows following characteristics:

1. The long and mobile neck has resulted in an increase in the number of cervical vertebrae.
2. All the thoracic vertebrae except the last one are fused into a single mass, providing a firm fulcrum for the action of wings in striking air. It is remarkable to note that in flightless birds the thoracic vertebrae are free.
3. The last thoracic, all lumbers, sacrals and few caudal vertebrae are fused to form the synsacrum which provides a compound plate-like girder to support the entire weight of the bird's body.
4. The posterior caudal vertebrae are fused to form a pygostyle which provides surface for the attachment of tail feathers.

Vertebral Column of *Gallus* (Bird)

The vertebral column of *Gallus* is differentiated into **cervical**, **thoracic**, **lumbar**, **sacral** and **caudal regions**.

1. Cervical vertebrae. The mobile neck or cervical region consists of 14 cervical vertebrae.

(*a*) Atlas vertebra. It is the first vertebra of the cervical region. It is small and ring-like in appearance with its apex directed downwards and the dorso-lateral surface convex and smooth. It is characterised by the absence of centrum, neural spine, ribs and transverse processes.

The thick ventral portion is notched above to receive the odontoid process of the axis and bears anteriorly a deep-cup-like cavity for articulation of the occipital condyle of the skull. The posterior border of the neural arch is produced into a pair of **postzygapophyses** for articulation with the axis vertebra.

(*b*) Axis vertebra. This is the second vertebra of the cervical region. The transverse processes, and cervical ribs are absent. The centrum is produced in front into an **odontoid process** for articulation with the atlas vertebra. **Pre** and **postzygapophyses** are present but hypophyses are rudimentary.

(*c*) Typical cervical vertebra. 6th to 10th vertebrae are typical cervical vertebrae. A typical cervical vertebra has a long **centrum**, a short **neural arch** and a rudimentary **neural spine**. The centrum is **heterocoelus** or saddle shaped. Its anterior end is convex dorso-ventrally but concave laterally. The posterior surface is more prominent and convex from side to side but concave from above downwards. Transverse processes are fused with the cervical ribs forming sharp backwardly directed processes having vertebrarterial canal at the bases. The **prezygapophyses** are flat articular facets lying on the anterior side. The **postzygapophyses** are projected backwards from the posterior side of the neural arch and facing downwards.

2. Thoracic vertebrae. The thoracic region is represented by a mass of **fused thoracic vertebrae** formed by the fusion of first three thoracic vertebrae with last cervical. Their neural spines are fused to form a continuous crest-shaped vertical plate. Their transverse processes also fuse to from horizontal plate on either side of the centra. The hypapophyses are fused to form a vertical plate on the ventral side of the centrum. The centrum and transverse processes of each vertebra bear articular facets for thoracic ribs.

Free thoracic vertebra. This is the fourth vertebra of the thoracic region. The centrum is heterocoelous and neural arch is well-developed. The neural spine is well developed and pointed. Transverse processes are flattened and outwardly directed. **Hypapophysis** is prominent and ventral in position. Pre and postzygapophyses are present on the anterior and posterior surfaces respectively. Centrum and transverse processes bear articular facets for the capitular and tubercular processes of thoracic ribs.

3. Synsacrum. The thoracic region of vertebral column is followed by a composite bone, **synsacrum**. It gives support to the iliac bones of the pelvic girdle. It is formed by the fusion of last thoracic vertebra, 5–6 lumbar vertebrae, 2 sacral vertebrae and 7–8 caudal vertebrae.

(*a*) Thoracic vertebra. The last thoracic vertebra forms the first vertebra of synsacrum. It bears articular surfaces for a pair of thoracic ribs.

(*b*) Lumbar vertebrae. The six vertebrae of the lumbar region and the last thoracic vertebrae are firmly fused together. They have well-developed **transverse processes** which articulate with iliac bones of pelvic girdle. Their **neural spines** form a continuous crest or **longitudinal ridge** which supports the dorsal margin of iliacs.

(*c*) Sacral vertebrae. The lumbar vertebrae are followed by two sacral vertebrae. The transverse processes of sacral vertebrae are fused to form bony plates which support the ilia along their outer margins.

(*d*) Anterior caudal vertebrae. The posteriormost part of the synsacrum is formed by the 7 or 8 **anterior caudal vertebrae**. Their transverse processes (except the last one) are bifurcated into dorsal and ventral processes. All the dorsal processes are fused to form a horizontal plate on either side of the centrum. These plates are continuous with the transverse plates of sacral vertebrae. The ventral portion of each vertebra is rod-shaped and is regarded as a rib.

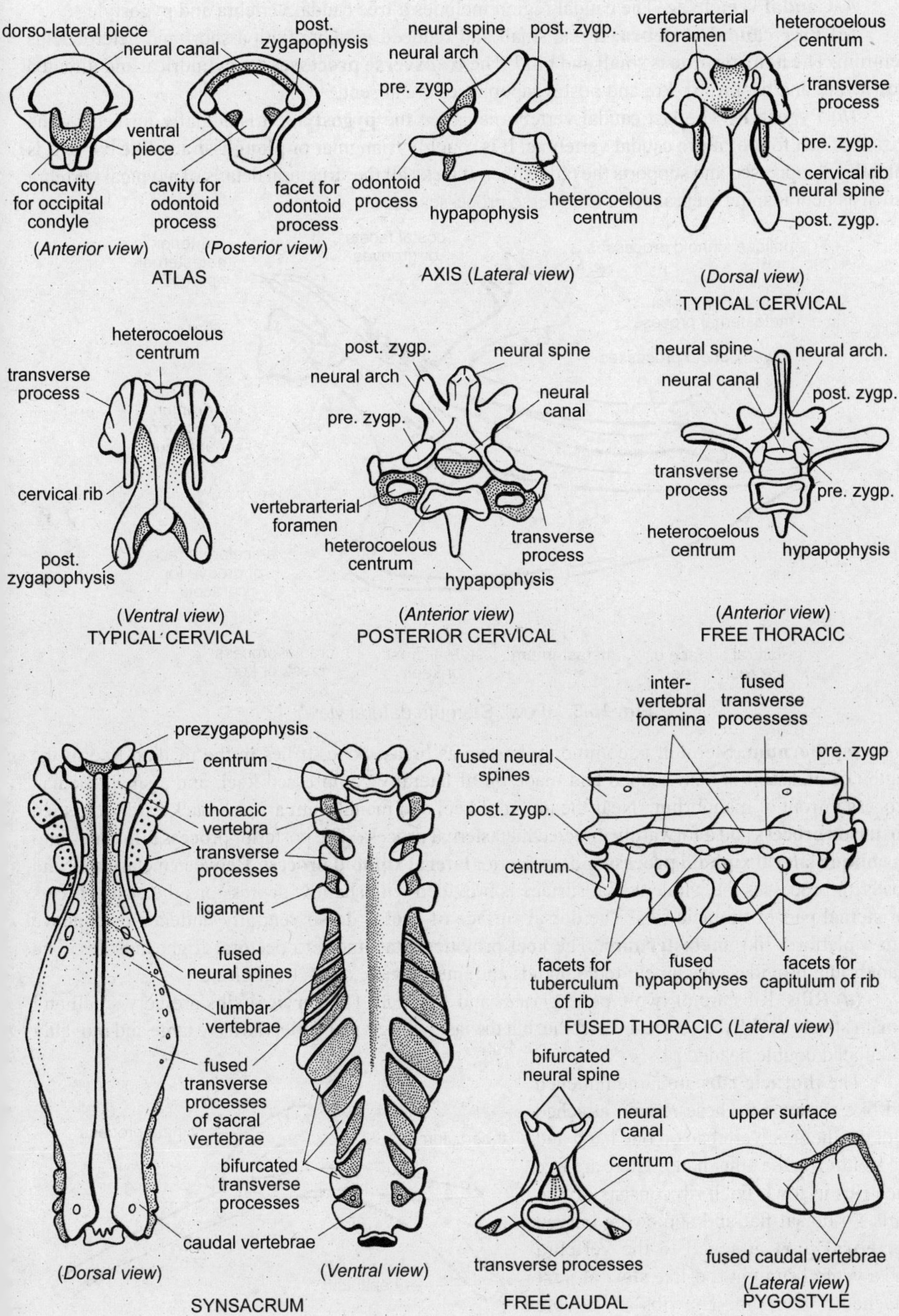

Fig. 15.7. Fowl. Vertebrae.

4. Caudal vertebrae. The caudal region includes a free caudal vertebra and pygostyle.

(*a*) Free caudal vertebra. It is a small and reduced vertebra with a short and heterocoelus centrum. The **neural spine** is small and bifid. The **transverse processes** are cylindrical and directed downward and backward. Pre and postzygapophyses are absent.

(*b*) Pygostyle. The last caudal vertebra is called the **pygostyle**. It is actually formed by the fusion of last four or more caudal vertebrae. It is roughly triangular or **plough-share-like** bone. It is laterally compressed and supports the tail feather. It lacks all the structural details of a typical vertebra (such as neural spine, neural arch, transverse processes, etc).

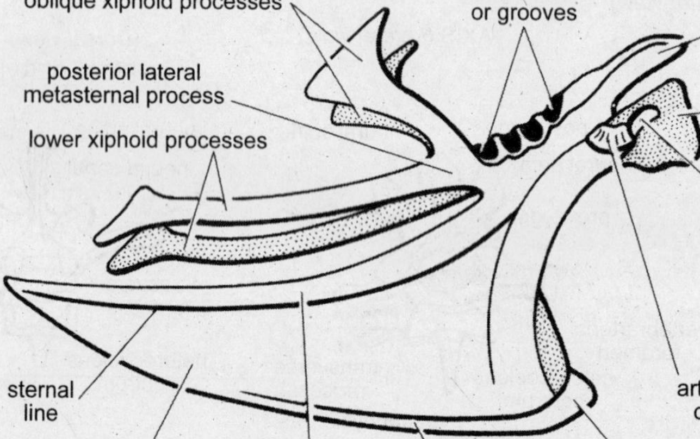

Fig. 15.8. Fowl. Sternum (lateral view).

(*c*) Sternum. Sternum is commonly known as breast bone. It lies in the median line in the region of breast. It is boat-shaped and made up of laterally compressed **keel**, and a small median process infront of manubrium. Near the union of keel and manubrium arises a backwardly directed **posterior process** and a forwardly directed metasternal process. The posterior process bifurcates into an **oblique lateral xiphoid process** and a posterior **lateral xiphoid process**. Between metasternal and posterior processes the whole dorsal surface is provided with shallow depressions the **costal facets** for sternal part of thoracic ribs. The dorsal surface of keel is dorso ventrally flattened or concave into a platform-like **metasternum**. The keel provides attachment to pectoral flight muscles. The manubrium provides attachment to coracoids and interclavicle of pectoral girdle.

(*d*) Ribs. Ribs are of two types: *cervical* and *thoracic*. The **cervical ribs** are very small and remain attached with the cervical vertebrae, but the last two cervical vertebrae have large and movably articulated double headed ribs.

The **thoracic ribs** are large flattened rod-like structures. These remain attached with the thoracic vertebae on one hand and the stratum on the other hand. There are five such ribs in fowl. Each rib consists of two parts : a dorsal flat and somewhat curved **vertebral part** attached to the vertebra and a ventral straight rod-like **sternal part** attached to sternum. The ribs are double headed or **bicephalons**. The proximal end of

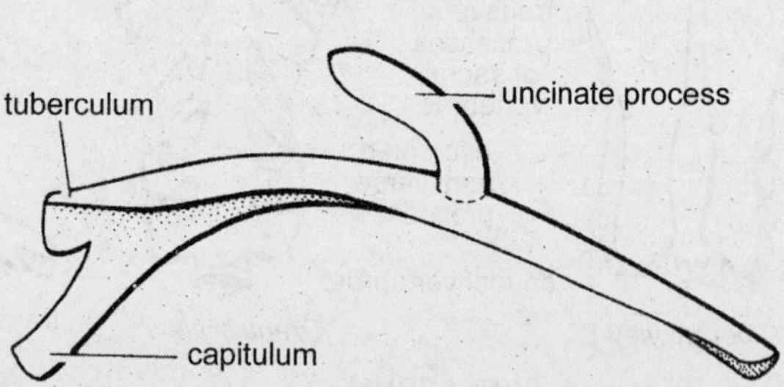

Fig. 15.9. Fowl. A thoracic rib.

the vertebral part consist of two processes, the **capitulum** attached to centrum, and the **tuberculum** attached to the transverse process. The vertebral portion of each thoracic rib, except the last and also the rib of the last cervical vertebra has a backwardly directed, **bony uncinate process** which overlaps the next rib behind. The uncinate processes increase the rigidity of the skeleton.

15.2 APPENDICULAR SKELETON

Appendicular skeleton lies on either side of the longitudinal axis and includes **girdles** and **limb bones.**

Girdles. They include pectoral girdle and pelvic girdle.

Pectoral Girdle

The pectoral girdle is divisible into two halves, each half consisting of a **coracoid**, a **scapula** and a **clavicle**. The two halves unite firmly with the sternum and form a strong bony structure supporting the wings. It differs from that of rabbit in having the large coracoid and a fully developed clavicle.

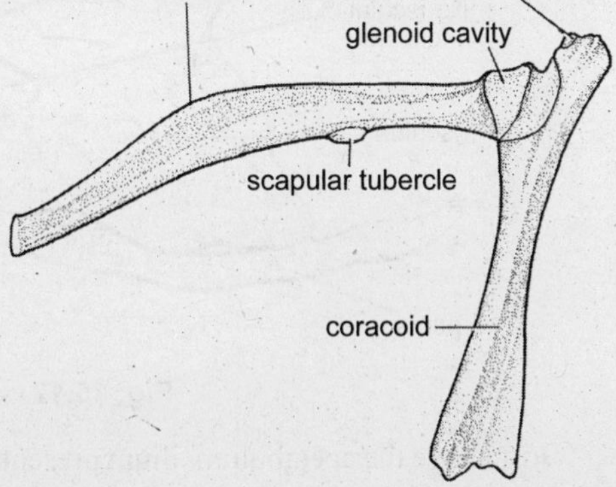

Fig. 15.10. Fowl. Pectoral girdle (right half).

1. Coracoid. It is a large, stout and straightened bone directed downwards running upto the base of manubrium. Its lower end is broad and flattened with a saddle-like articular surface for articulation with coracoid groove on the antero-lateral border of the sternum. Its lower end bears a **foramen** communicating with the air cavity found inside the bone. The upper end articulates at right angle to the scapula. The outer surface bears a cup-shaped depression which forms greater part of the **glenoid cavity** to receive the head of humerus. Its dorsal end extends beyond the glenoid cavity forming a hook-like **acrocoracoid process** directed upward and inward. This joins with the acromian process of the scapula and the upper end of the clavicle to enclose the **foramen triosseum** through which the tendon of the second pectoral muscle passes dorsally to its insertion on the back of humerus.

2. Scapula. It is a long flattened and slightly curved bone lying over the thoracic ribs and running parallel to the vertebral column. Its broad anterior end bears on its outer-surface a **cup-like depression** which takes part in the formation of glenoid cavity. It is firmly united with the coracoid by ligaments. Its inner surface is produced into an **acromian process**. Ventrally it bears a **scapular tubercle** in the middle region.

3. Clavicles. Each clavicle is a slender, rod-like and slightly curved bone. It articulates with the acromian process of the scapula and acrocoracoid process of coracoid by its upepr expanded end and encloses the **foramen triosseum**. Ventrally the two clavicles of the two sides fuse with a small **interclavicle** and form a median laterally compressed disc, called the **hypocleidium**. This disc is connected by a ligament with the rostrum of sternum. The two clavicles fuse with the interclavicle and form a composite V-shaped structure known as the **furcula**.

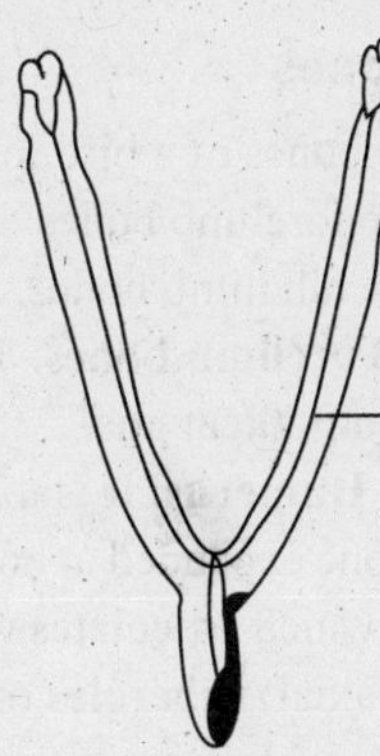

Fig. 15.11. Furcula of fowl formed by the clavicles and interclavicle.

Pelvic Girdle

The pelvic girdle of bird is made up of identical halves or os-innominates, which remain widely separated midventrally and are firmly united with the synsacrum. Thus a symphysis is absent.

Each innominate is deeply concave and is formed of usual three bones, the **ilium**, **ischium** and **pubis** which are firmly united and sutures between them are indistinct. The **acetabulum** is formed by all the three bones and has an acetabular foramen.

1. Ilium. It is greatly expanded, elongated and lamellar bone, extending both in front and behind the acetabulum. The part anterior to the **acetabulum** is known as **preacetabular ilium** and the posterior as **postacetabular ilium**. Its inner border is almost fused with the synsacrum and forms a sort of shield for the internal organs while the inner surface carries two deep concave depressions to accommodate the kidnyes of its side. The outer surface of the preacetabular ilium is concave, while that of postacetabular ilium convex.

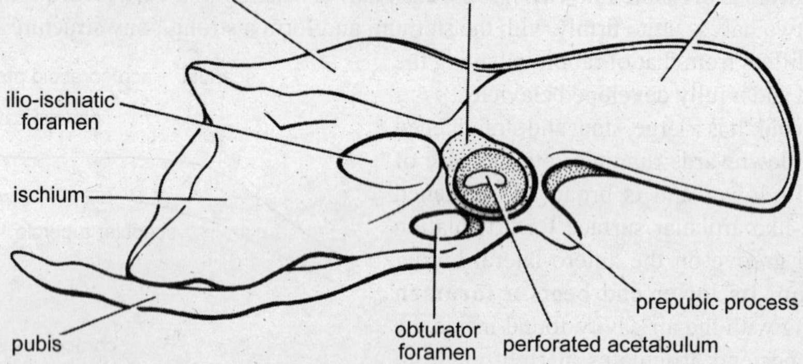

Fig. 15.12. Fowl. Pelvic girdle.

Just above the acetabulum, ilium presents an **antitrochanter projection**, which bears articular surface for the greater trochanter of femur. The postacetabular ilium is fused with ischium.

2. Ischium. Ischium lies parallel to and fused with the postacetebular ilium. An oval ilio-ischiatic foramen separates the two in the front part close to the acetabulum.

3. Pubis. It is an elongated, slender, strip-shaped bone directed backwards along the outer margin of ischium with which it fuses at certain points. Just behind the acetabulum, the pubis and ischium are separated by a small oval **obturator foramen**. In front of the acetabulum it is produced into a blunt **prepubic process**.

Limb Bones

The limb bones of a bird include:

1. The forelimb bones or bones of a wing.
2. The hindlimb bones.

1. Forelimb bones. The forelimb bones of the bird include **humerus, radius-ulna, carpals** and **carpometacarpus**.

***(a)* Humerus.** It is the bone of the upper arm of forelimb. It is a stout, elongated and slightly curved bone expanded at both the ends. The proximal end is expanded into a smooth convex surface, the **head** which articulates with the glenoid cavity. The head is lodged in the glenoid by two (preaxial and post-axial) **tubercles** or **tuberosities**. The preaxial tuberosity is continued into a **deltoid ridge**. The postaxial tuberosity is larger and bears a pneumatic foramen which leads into the air cavity of the shaft. The distal end bears a pulley-like trochlea having a pair of convex **condyles** for the radius and ulna.

***(b)* Radius and ulna.** These are the bones of forearm. The radius is short, slender and straight, and is preaxial in position. Its proximal end bears a shallow cup for articulation with the trochlea of humerus. Its distal end articulates with the carpus.

Ulna is slightly curved and stouter and larger than the radius. It is postaxial in position. Its proximal end bears an articular facet for the iner condyle of humerus and is further projected.

(*c*) Carpals. The wrist or carpus of adult bird contains two irregular bones. Of these, **radiale** is smaller and articulates with the radius while the **ulnare** is larger and articulates with the ulna. These represent the proximal row of carpals. The distal row of carpals fuses with the metacarpals in the adult.

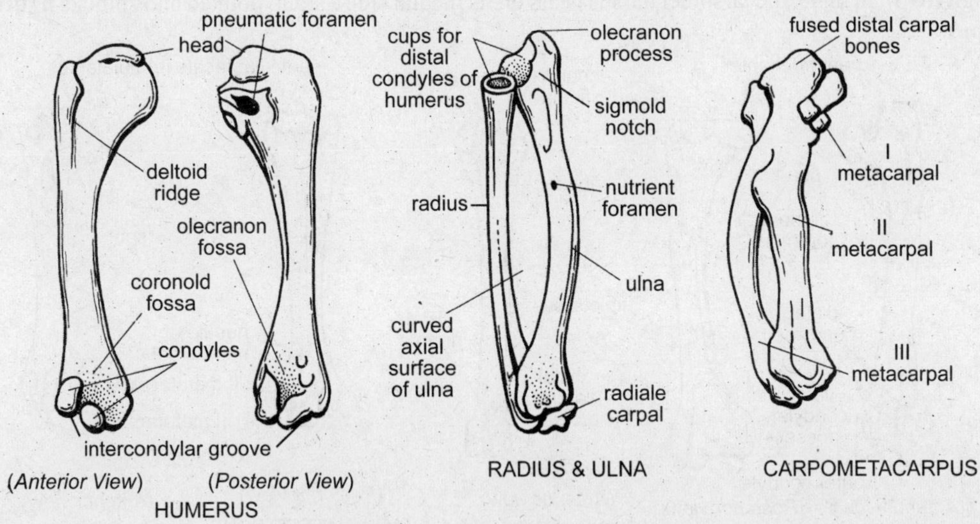

Fig. 15.13. Fowl. Forelimb bones.

(*d*) Carpometacarpus. It is a compound bone of the wrist region of forelimb. It is formed by the fusion of distal row of carpals (3) with the metacarpals which are also three in number. The carpals form the proximal end of the bone. The first metacarpal is very short and stumpy and fuses with the second metacarpal.

The 2nd metacarpal is stout and straight and is produced into a spine on the inner side. The 3rd metacarpal is slender and curved and is fused at both the ends with second metacarpal.

The digits are reduced in number. There are only three digits, the first and third digits are formed of one phalanx each, whereas, the 2nd digit consists of two phalanges.

2. Hindlimb bones. The hindlimb bones of fowl (bird) are femur, tibiotarsus and fibula, tarsals and tarsometatarsus.

(*a*) Femur. It is the bone of thigh region. It is a short but stout and slightly curved bone enlarged at both the ends. Proximally it bears on the inner side a rounded head which fits into acetabulum and on the outer side, a prominent irregular process, the **great trochanter**. Between the head and the trochanter is an articulating surface for the trochanter of ilium. Its dorsal ends bears in front a deeply grooved and pulley-like surface for the **patella** which is a sesamoid bone forming the knee joint. Two proximal condyles are also present laterally for the articulation with tibia and fibula.

(*b*) Tibiotarsus and fibula. These are the bones of shank region of hindlimb. The tibiotarsus is stout, long and nearly straight bone. It is formed by the fusion of **tibia** with the proximal row of **tarsals** (astragalus and calcaneum). Its broad proximal end bears on its anterior face a **cnemial crest** and two articular surfces for the condyles of femur. Its distal end articulates with the tarsometatarsus by a pulley-like articular surface.

The **fibula** is a small slender bone closely applied to the outer surface of the tibiotarsus. Its proximal end articulates with the femur.

(*c*) Tarsals. In adult there are no free tarsals, the proximal row of tarsals being fused with the tibia and the distal row with the metatarsals.

(*d*) **Tarso-metatarsus.** The pes or foot contains a single compound bone, the **tarsometatarsus**. It is a compound bone formed by the fusion of distal row of tarsals with the second, third and fourth metatarsals. A reduced nodule-like first **metatarsal** bone is attached by ligament to the inner surface of the tarso-metatarsus near its distal end. Proximatly, tarso-metatarsus bears two cup-like facets for tibiotarsus. At its distal end it bears three metatarsals each articulating with the corresponding digit (two). In male, the tarsometatarsus bears on its medial side a stout, pointed and spinous **fighting spur**.

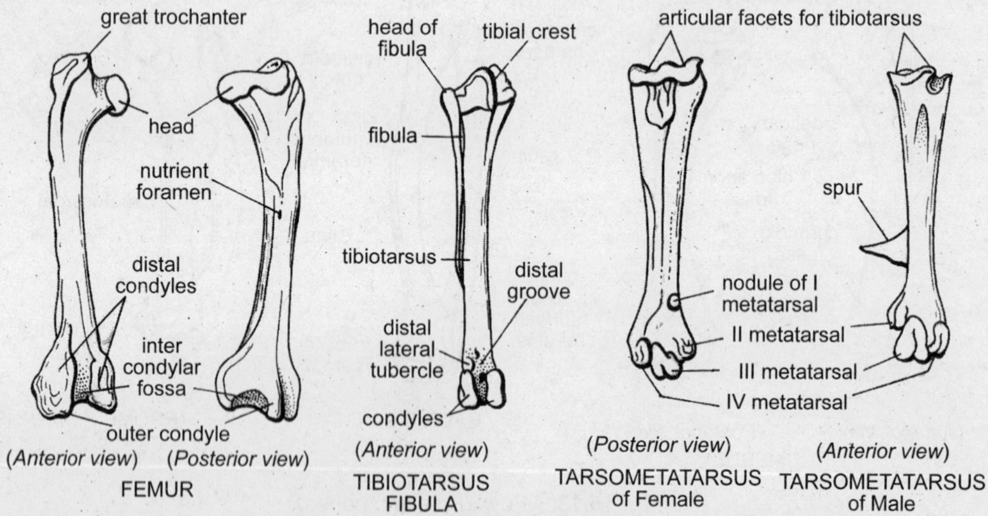

Fig. 15.14. Fowl. Hindlimb bones.

Phalanges are never more than four. The first one or hallux is usually directed backwards and the other three forward.

QUESTIONS

Long Answer Questions

1. Describe the skull of *Gallus* (fowl) or Pigeon and compare it with that of any lizard you have studied.
2. Mentioning the chief peculiarities describe the vertebral column of a bird you have studied.
3. Describe the sternum of pigeon.
4. Give an account of the pectoral and pelvic girdles of a bird.
5. Describe pectoral girdles of birds. (*Bundelkhand 2001*)
6. Give an account of the limb bones of a bird.
7. Give an account of the hindlimb bones of a bird.

Short Answer Questions

1. Describe the palatal bones of pigeon or fowl.
2. List the characteristic features of the avian skull.
3. Name the components of fowl's hyoid apparatus.
4. Describe the columella auris of fowl.
5. Write short notes on the following:
   (*a*) Carpometacarpus
   (*b*) Pygostyle
   (*c*) Sternum
   (*d*) Synsacrum
   (*e*) Furcula
6. Draw a labelled figure of the skull of *Gallus* as seen from the ventral side.

Very Short Answer Questions

1. What is palate? Which bones form the palate?
2. Why pelvic girdle is so large and posteriorly and downwardly placed?
3. Why a keel is absent in running or flight-less birds?
4. What type of sternum is found in birds?
5. Which bone forms the occipital condyle in fowl?
6. What type of centrum the cervical vertebrae of fowl have?
7. What is a pygostyle?
8. The sternal parts of the ribs are bony in reptiles/birds. Which is correct?
9. In which vertebrates the acetabulum is perforated?
10. Give the phalanges formula for avian *Galus*.
11. Name the bones which form the synsacrum.
12. What are wish bones?
13. What is fighting spur?
14. On what principle the endoskeleton of birds is built and what is its significance?
15. What are pneumatic bones?
16. Why metacarpals are reduced or fused with carpals in birds?

Fill in the Blanks

1. There is extensive fusion in the ________ skeleton.
2. Avian skeleton is light due to ________ in many bones.
3. The skull is monocondylic in ________ and ________.
4. The ________ thoracic vertebra is free in the fowl.
5. Sternum bears a midventral ________ in flying birds.
6. Just behind the ________ of pelvic girdle of fowl, the pubis and ischium are separated by a small oval ________ foramen.
7. Each half of ________ girdle, has a coracoid, a ________ and a clavicle.

Multiple Choice Questions

1. Uncinate process is associated with
   (*a*) vertebrae of birds (*b*) sternum of birds
   (*c*) ribs of mammals (*d*) ribs of birds
2. Skull in birds is
   (*a*) monocondylic (*b*) dicondylic
   (*c*) noncondylic (*d*) streptostylic
3. How many caudal vertebrae are present in synsacrum? (*Gorakhpur 2003*)
   (*a*) 2 (*b*) 5
   (*c*) 7 (*d*) 11
4. Tarsometatarsus is found in
   (*a*) forelimbs of birds (*b*) hindlimb of rabbit
   (*c*) forelimbs of rabbit (*d*) hindlimbs of birds
5. Tail vertebrae in birds are fused to form
   (*a*) urostyle (*b*) pygostyle
   (*c*) coccyx (*d*) none of these
6. Which of the following bones constitute furcula in birds
   (*a*) clavicles (*b*) interclavicles
   (*c*) both of these (*d*) none of these

7. Centrum of avian vertebrae is
   (*a*) procoelous
   (*b*) amphicoelous
   (*c*) heterocoelous
   (*d*) amphiplatyon

ANSWERS

Very Short Answer Questions

1. Palate is the bony roof of buccal cavity. It is formed of vomers, palatines, pterygoids and maxillopalatine processes of premaxilla.
2. The large pelvic girdle in birds provides more space to accommodate viscera which is shifted downward and backward.
3. The flightless birds such as ostrich, the flight muscles are poorly developed and so the keel of sternum is absent because keel provides attachment to the flight muscles.
4. In birds the sternum is large and boat-shaped and has a greatly expanded and prominent midventral keel.
5. Basioccipital.
6. Heterocoelous.
7. Pygostyle is a triangular bone at the end of the caudal vertebrae in birds.
8. Birds.
9. Aves.
10. 1, 2, 1, 0 and 0.
11. Synsacrum of birds is formed by the fusion of last thoracic, five or six lumber, two sacral and seven caudal vertebrae.
12. The V-shaped composite bone, formed by the fusion of clavicles and interclavicles, is commonly called **furcula**, **merry-thought** or **wish-bone**, in birds.
13. Male birds have a stout, conical, slightly curved bony projection called **fighting spur**, located medially on tarso-metatarsus.
14. In bird's endoskeleton is built after the "hollow girder principle". It combines strength with lightness, an essential adaptation for flight.
15. Pneumatic bones are filled with air sacs instead of bone marrow. In birds long bones are pneumatic.
16. To provide surface for attachment of flight feathers, the metacarpals are fused with large carpals.

Fill in the Blanks

1. Avian. **2.** Air-spaces. **3.** Reptiles, birds. **4.** Fourth. **5.** Keel. **6.** Acetabulum, obturator. **7.** Pectoral, scapula.

Multiple Choice Questions

1. (*d*) **2.** (*a*) **3.** (*c*) **4.** (*d*) **5.** (*b*) **6.** (*c*)
7. (*c*)

16

Class Aves (The Birds): General Characters, Classification and Types

Birds are feather-clad, warm-blooded, air-breathing, oviparous, bipedal flying vertebrates. They are unique among animals in having a coat of feathers and in walking on the hindlimbs alone clear of the ground. Alomost every part of their organisation is modified to aid in aerial life. The organisation of existing birds is remarkably uniform in its essential features. Most of them can fly. A few have lost the power of flight.

Birds show great diversity in size, the largest of the living birds is ostrich (*Struthio camelus*) which is nearly 213 cm tall; whereas the smallest is bee humming bird (*Mellisuga helenae*) of Cuba which is only 5.70 cm long (Box 16.1). Birds are generally brilliantly coloured and show variety of shades.

Box 16.1

In art, mythology and folklore, birds have been used for centuries to symbolise wisdom and truth, strength and power, birth and death. However, nothing about birds have tantalised human beings more than the gift of flight. There are birds that "fly" through water others that can attain incredible speeds while swooping on prey and some, such as humming birds that are extraordinary agile and can even fly backwards. (**D.R. Khanna** and **P.R. Yadav, 2005**)

Nearly all the orders of birds had appeared by the end of the Eocene. It is estimated that the total number of bird species that have existed is 1,54,000 of which 9,600 are alive today (**Brodkorb**, 1971).

Bird is an English term, in Greek it is termed *Ornis* and in Latin *Avis*. The study of birds is called **ornithology** (Gr., *ornis*, *ornithos* = a bird). A specialist on birds is known as an **ornithologist** (Gr., *ornithos* = bird + *logos* = discourse). An enclosure or a large cage for keeping live birds is termed an **aviary** (L., *avis* = bird). Aviary is a means of **ex-situ conservation** of wild life (**Agarwal** and **Gupta**, 2004). India has produced a world famous ornithologist **Salim Ali** who authored the famous book "*The bird of Indian Birds*".

16.1 GENERAL CHARACTERS

1. **Warm-blooded** (homoiothermous) vertebrate specially adapted for aerial mode of life.
2. Epidermal horny exoskeleton in the form of **feathers** all over the body and a few scales on the limbs.
3. Jaws produced into a toothless beak. Beaks show a great diversity in form, size and role.
4. Birds are bipeds, walk on their hind limbs or legs with the help of toes (digitigrade). Three toes directed froward and one (called **hallux**) is directed backward.

5. Body is streamlined and spindle-shaped, and divisible into head, neck and tail. **Neck** is long and flexible. Tail is **short**.
6. Birds have two pairs of limbs. Forelimbs are modified into **wings** for flying and contain only three digits. Hind limbs or **legs** bear body weight. Legs are large and adapted for walking, running, swimming, scratching, perching, food capturing, etc.
7. **Exoskeleton** includes **feathers** on the body including wings, **scales** on the legs and feet, **claws** on the toes and **horny sheath** on the beak. Feathers vary in size, structure and colouration. They insulate body and help in flying. Feathers are **moulted** regularly.
8. Skin is thin and dry without glands. However, oil glands occur at the base of tail.
9. **Endoskeleton** is **pneumatic** (*i.e.*, bones are light, being hollowed out by air spaces).
10. **Endoskeleton** is fully ossified, light but strong. Long bones have no marrow.
11. Skull is **diapsid** (contains, two temporal fossae) and **monocondylic** (having a single occipital condyle). Skull is without distinct **sutures** between the bones.
12. Vertebral column is short and compact. Vertebrae contian **heterocoelous** (saddle-shaped) centrum. Thoracic vertebrae fused to form fused **thoracic mass** and the sacravertebrae together with a thoracic, lumber and a few caudal vertebrae form **synsacrum**. Last few caudal vertebrae fuse to form a **pygostyle**.
13. Sternum is large and has a large keel in flying birds. Massive flight muscles are attached to the sternum and its keel.
14. Ribs are bicephalous (double-headed) and bear an **uncinate** process.
15. Clavicles and interclavicles fuse to form a V-shaped **furcula**.
16. Pelvic girdle is greatly elongated, tetra-radiate and fused with the synsacrum.
17. Carpals and metacarpals fuse to form **carpo-metacarpus** and the distal rows of tarsals and metatarsals unite into a **tarsometatarsus**. The proximal row of tarsals fuses with tibia to form tibio-tarsus bone.
18. There are no teeth in birds. A part of the stomach is modified to form a gizzard which grinds the food.
19. Cloaca is three chambered.
20. Respiration takes place only by lungs which are compact, spongy, little distensible and attached to ribs. Lungs are provided with air sacs.
21. Larynx is without vocal cords, so does not act as the sound box. Birds has a sound producing **syrinx** at or near the junction of trachea and bronchi. Birds are capable to sing and produce melodious voices.
22. Heart is four-chambered. It contains two auricles and two ventricles. Such a heart keeps the oxygenated and unoxygenated blood fully separate.
23. Only right aortic or systemic arch persists and form the dorsal aorta. Renal portal system greatly reduced.
24. Red blood cells are nucleated.
25. Brain large but smooth. Cerebral hemispheres, cerebellum and optic lobes greatly developed. Cranial nerves 12 pairs.
26. Olfactory organs poorly developed but sight is specially developed. Eyes with comblike structure, the **pecten**.
27. Kidneys are metanephric. Uriniferous tubules have long loops of Henle to reabsorb water from the glomerular filterate. Uricotelic. Urinary bladder is absent, except in American rhea (**Rhea americana**).
28. Sexes are separate. Sexual dimorphism is well marked.
29. Females usually contain only one ovary and oviduct (on the left side).
30. Fertilization is internal. All birds are oviparous (= egg laying). Eggs are large, yolky (polylecithal) and shelled. The eggs are incubated by one or both parents.

31. Cleavage is meroblastic. Embryo with extra-embryonic membranes such as amnion, chorion, allantois and yolky sac.
32. No larval stage in bird's life cycle (*i.e.*, development is direct). Yound may be feathered and able to move about and feed at hatching (called **precocial** or **nidifugous**, *e.g.*, chicken, duck, quail) or naked, helpless, dependent for a time on parents for food supply (called **altricial** or **nidiculous**, *e.g.*, sparrow, robin, pigeon).

16.2 OUTLINE CLASSIFICATION

Class-Aves

Subclasses (2)

- **1. Archaeornithes** (Extinct)
  Examples: *Archaeopteryx*
- **2. Neornithes** (New birds)
  - **1. Odontognathae** (Extinct)
    Examples: 1. *Hesperornis* 2. *Ichthyornis*
  - **2. Palaeognathae (Ratitae)**
    Orders (7)
    - **1. Struthioniformes**
      Examples: *Struthio camelus* (Ostrich)
    - **2. Rheiformes**
      Examples: *Rhea americana* (Rhea)
    - **3. Casuriiformes**
      Examples: 1. *Casuarius* (cassowary) 2. *Dromaeus* (emus)
    - **4. Apterygiformes**
      Example: *Apteryx autralis* (kiwi)
    - **5. Tinamiformes**
      Example: *Tinamus* (tinamous)
    - **6. Diornithiformes** (Extinct)
      Example: *Dinornis*
    - **7. Aepyornithiformes**
      (Extinct elephant birds)
      Examples: 1. *Aepyornis* 2. *Mulleornis*
  - **3. Impennae or Sphenisciformes**
    Example: *Apenodytes* (Penguins)
  - **4. Neognathae**
    Orders (21)
    - **1. Gaviformes**
      Example: *Gavia immer* (diver or loon)
    - **2. Podicipeformes**
      Example: *Padiceps ruficolis* (dabchick or little grebe)
    - **3. Procellariiformes**
      Examples: 1. *Diomedea exulans* (wandering albatross) 2. *Fulmarus* (Petrel)
    - **4. Pelicaniiformes**
      Examples: 1. *Pelecanus philippensis* (spotted billed or grey pelican) 2. *Phalacrocorax* (cormorant) 3. *Sula* (gannet)
    - **5. Ciconiiformes**
      Examples: 1. *Ardea cinerea* (grey heron) 2. *Bublcus ibis* (cattle egret) 3. *Ciconia* (stork) 4. *Phoenicopterus* (flamingo)
    - **6. Anseriformes**
      Examples: 1. *Anser indicus* (bar-headed goose) 2. *Cygnus* (swan) 3. *Anas* (duck) 4. *Anas crecca* (common teal)
    - **7. Falconiformes**
      Examples: 1. *Milvus migrans* (periah kite) 2. *Falco jugger* (lagger falcon) 3. *Pseudogyps bengalensis* (white backed vulture) 4. *Neophron* (vulture) 5. *Aquila* (eagle)

(Contd...)

(*Contd...*)

8. Galliformes
Examples: 1. *Gallus domesticus* (domestic fowl)
2. *Pavo cristatus* (peafowl)
3. *Phasianus* (pheasant)
4. *Perdix* (partridge)

9. Gruiformes
Examples: coots, cranes, rails, moorkens, etc.
1. *Grus grus* (common crane)
2. *Grus antigone* (sarus crane)
3. *Fulica atra* (coot)
4. *Crex* (corn crake)

10. Charadriiformes
Examples: Terms, gulls, lapwings, plovers, etc.
1. *Numenium* (curlew)
2. *Capella* (snipe)
3. *Calidris* (sandpiper)
4. *Larus* (gull)

11. Columbiformes
Examples: Pigeons and doves.
1. *Columba livia* (blue rock pigeon)
2. *Streptopelia* (dove)
3. *Rhaphus* (dodo, an extinct bird)

12. Psittaciformes
Example: *Psittacula* (parakeet or parrot)

13. Cuculiformes
Examples: 1. *Eudynamis scolopaceus* (koel)
2. *Cuculus* (cuckoo)
3. *Opisthocomus* (Hotzin)

14. Strigiformes
Examples: 1. *Tyto alba* (barn or screech owl)
2. *Bubo bubo* (great horned owl)
3. *Athene* (little owl)
4. *Strix* (tawny owl)

15. Caprimulgiformes
Exmaple: *Caprimulgus asiaticus* (common Indian night jar)

16. Apodiformes (Micropodiformes)
Examples: 1. *Micropus affinis* (house swift)
2. *Trochilus* (humming birds)
3. *Apus* (swift)

17. Colliformes
Example: *Colius* (mouse bird)

18. Trogoniformes
Example: *Trogon elegans* (copper-tailed trogon)

19. Coraciformes
Examples: 1. *Halcyon smyrnensis* (white breasted kingfisher)
2. *Alcedo atthis* (common kingfisher)
3. *Corcias benghalensis* (blue jay or neelkanth)
4. *Upupa epops* (hoopoe)
5. *Merops* (bee-eater)

20. Piciformes
Examples: 1. *Brachypternus benghalensis* (golden-backed woodpecker)
2. *Picus* (woodpecker)

21. Passeriformes (perching birds)
Examples: 1. *Passer domesticus* (house sparrow)
2. *Corvus splendens* (house crow)
3. *Corvus macrohynchos* (jungle crow)
4. *Ploceus philipinus* (baya or weaver bird)
5. *Orthotomus sutorius* (tailor bird)
6. *Pycnonotus cafer* (red-vented bulbul)
7. *Sylvia* (warbler)
8. *Fringilla coelebs* (chaffinch)

16.3 CLASSIFICATION

The class Aves is divided into two subclasses: Archaeornithes and Neornithes.

Subclass 1. Archaeornithes

(Gr., *archios* = ancient + *ornithos* =bird)

1. Primitive and extinct birds of the Upper Jurassic Period of Mesozoic age, about 155 million years old.
2. These fossil birds had a mixture of avian and reptilian characters.
3. They had feathers which were similar in structur to those of modern birds.
4. Skull was diapsid.
5. Jaws were prolonged into beaks. Both the jaws had sharp homodont teeth. Teeth embedded in sockets (thecodont).
6. Forelimbs were modified into wings which had little power of flight.
7. Tail long and tapering with more than 13 (*i.e.*, 18 to 20) caudal vertebrae bearing retrices (tail feathers) arranged in two lateral rows.
8. Pygostyle was absent.
9. Vertebrae were amphicoelous. Thoracic ribs were without uncinate process. Abdominal ribs present.
10. Sternum was weak and lacked keel.
11. Carpals and metacarpals were free. Fingers were free and had claws.
12. Foot consisted of tarsometatarsus with 4 clawed digits.
13. Bones were hollow but not pneumatic.
14. Skull was without rounded cranium having fused bones and single occipital condyle.

The subclass Archoeornithes includes a single order, **Archaeopterygiformes**. This order has only two genera, each having a single species, *viz.*, *Archaeopteryx lithographica* and *Archaeornis siemensi*.

Subclass 2. Neornithes

1. Includes living as well as few extinct birds.
2. Jaws lack teeth, except a few extinct forms.
3. Wings usually well developed used for flight with a few exceptions.
4. A few terminal caudal vertebrae fuse to form pygostyle.
5. Tail is short, ends in a short pygostyle. Rectrices (tail feathers) are arranged in a semicircle around the pygostyle.
6. Metacarpals fuse.
7. Digits of forelimbs rarely with claws.
8. Sternum is well developed and mostly keeled.
9. Ribs have uncinate processes.
10. Vertebrae are heterocoelus in living forms.

The subclass Neornithes is divided into four superorders: 1. Odontognathae; 2. Palaeognathae (Ratitae); 3. Spenisciformes (Impennae); 4. Neognathae (Carinatae).

Superorder I. Odontognathae

(Gr., *odontos* = teeth)

1. Extinct marine piscivorous birds from North American Upper Createaceous strata.
2. Includes toothed birds. These birds had teeth set in groovers or shallow sockets in the jaws.
3. Wings were poorly developed. Bones of forelimbs reduced.
4. Tail was short. Pygostyle was absent.
5. The vertebrae were mostly amphicoelous.
6. Legs were long and used for swimming.

The superorder Odontognathae includes two orders: 1. Hesperornithiformes and 2. Ichthyornithiformes.

Order 1. Hesperornithiformes

1. Large flightless marine diving and swimming birds.
2. Jaws with sharply pointed teeth which occurs in grooves.
3. Sternum without keel.
4. Shoulder girdle was much reduced and birds were unable to fly.
5. Vertebrae amphicolous.

   Examples: *Hesperornis*, *Enaliorinis*, *Baptorinis* and *Neogoeornis*.

Order 2. Ichthyornithiformes

1. Wings were developed.
2. Sternum with keel.
3. Jaws with small teeth arranged in sockets.
4. Neck vertebrae were biconvex.
5. Brain was a avian type.

   Examples: *Ichthyornis*, and *Apatornis*.

Superorder II. Palaeognathae or Ratitae

(Gr., *palaios* = old + *gnathos* =jaws; L., *ratis* = raft)

1. Large sized, flightlessless, walking or running birds. They evolved in the Cretaceous period from flying birds but have lost the power of flight.
2. Teeth are absent.
3. Wings are reduced, vestigial or absent. Feathers are without interlocking mechanism.
4. Sternum lacks a keel. It is raft-like.
5. Coracoid and scapula are small.
6. Clavicles are absent.
7. Tail vertebrae are usually free. Pygostyle is rudimentary.
8. Oil gland is usually absent.
9. Syrinx is absent.
10. Male has an erectile penis. Female has clitoris.
11. The young are precocious.

The superorder Palaeognathe is divided into 7 orders: 1. Struthioniformes; 2. Rheiformes; 3. Casuariiformes; 4. Apterygiformes; 5. Tinamiformes; 6. Diornithiformes and 7. Aepyornithiformes.

Order 1. Struthioniformes

(Gr., *structhio* = ostrich + form)

1. Includes large flightless terrestrial birds such as ostriches.
2. Feathers lack aftershaft.
3. Head, long neck and legs are sparsely feathered.
4. There are only two toes, third and fourth, on each foot.
5. Pubic symphysis is present.
6. Sternum without keel and pygostyle absent.

   Examples: *Struthio camelus* (true ostriches of Africa and Western Asia).

Order 2. Rheiformes

(Gr., *Rhea* = Mother of Zeus in Greek mythology)

1. Sternum is without keel.
2. Wings are relatively large but non-functional.

3. There are 3 toes (clawed digits) on each foot.
4. Head and neck are partly feathered.
5. Feathers are without aftershafts.

Examples: *Rhea americana* (American ostrich found in South America).

Order 3. Casuariiforms

1. There are three toes on each foot (2, 3, 4).
2. Head is sparsely feathered. Neck and body are densely feathered.
3. Feathers have aftershafts each of which is equal to the main shaft.
4. Sternum is without keel.
5. Wings small or rudimentary.

Examples: *Casuaris* (Crossowaries of Australia); *Dromaeus* (Emus of New Zealand).

Order 4. Apterygiformes

(Gr., *a* = not + *pteryx* = wing + form)

1. Beak is long and slender with terminal nares.
2. Each hindleg with four toes.
3. Sternum short and without keel.
4. Wings vestigial, hidden and non-functional.
5. Feathers are hair-like and without after shaft.

Example: *Apteryx* (kiwi; Fig. 16.24).

Order 5. Tinamiformes

1. Small, terrestrial birds which possess power of flight but are great runners (cursorial habit).
2. Sternum is keeled.
3. Wings are well developed.
4. Hindlimbs lack perching mechanism.
5. Vegetarian (eat fruits and grains).
6. Small pygostyle is present.

Examples: *Tinamus* (Tinamou of Mexico), *Rhyncotus*.

Order 6. Dinornithiformes

1. Giant flightless, terrestrial extinct birds of New Zealand.
2. Feathers with large aftershaft.
3. Wings (forelimbs) almost absent.
4. The beaks were short.
5. Hindlimbs very large and massive with 3 or 4 toes.
6. Coracoid and scapula reduced.

Examples: *Dinornis maximus* (Moas of New Zealand).

Order 7. Aepyornithiformes

1. Extinct flightless terrestrial elephant birds.
2. Sternum short and without keel.
3. Wings vestigial.
4. Legs powerful with 4 toes.

Examples: *Aepyornis* (elephant bird) of Madagascar.

Superorder III. Impennae

Order 1. Sphenisciformes

(Gr., *spheniscus* = wedge + form)

1. Includes the penguins.

2. These are modern aquatic (marine) flightless birds of Antarctic region.
3. Forelimbs (wings) are modified into paddles or flippers for swimming.
4. Feet are strongly webbed and have four toes, all directed forward.
5. Feathers are small and scale-like and densely cover the entire body.
6. Teeth are absent. They are carnivorous and feed on fish, cuttle fishes and crustaceans. Beak is of various types.
7. Sternum bears a keel.
8. There is a thick layer of fat under the skin to prevent loss of body head.

Examples: *Aptenodytes foresteri* (emperor penguin, Fig. 16.1) and *Spheniscus*, etc.

Fig. 16.1. *Aptenodytes foresteri* (Emperor penguin).

Superorder IV. Neognathae

1. It includes modern birds which are flying birds. They are widely distributed.
2. These birds evolved in the Cretaceous and have become dominant birds.
3. Teeth are absent.
4. Beak is variously modified.
5. Wings are well developed.
6. Palate may be schizognathous, desmognathous or aegithognathous.
7. Sternum has a strong keel.
8. Scapula and coracoid are large and joined by ligaments, with coracoscapular angle less than a right angle.
9. Clavicles are present and by union with an inter-clavicle, form a V-shaped furcula.
10. Tail has a few (5 to 6) free vertebrae and a pygostyle formed by the fusion of a few terminal caudal vertebrae.
11. Adult skull lacks sutures between the bones.
12. In the pelvic girdle, ilium and ischium are united.
13. Rectrices (tail feathers) are arranged in a semicircles around the pygostyle.
14. Body feathers are arranged in the distinct feather tracts or pterylae separated by featherless tracts or apteria.
15. Feathers have barbs, barbules and hooklets. Barb are firmly united by hooks on the barbules.
16. Oil gland is usually present.
17. Syrinx (vocal organ) is present at the junction of the trachea and bronchi.
18. Feets are variously modified.
19. Male birds lack a copulatory organ.
20. The youngs are usually altricial (*e.g.*, rock pigeon) rarely precocious (*e.g.*, duck).

The superorder Neognathae or Carinatae is divided into 21 ordrs as follows:

Order 1. Gaviformes

1. Marine birds with webbed toes on short legs.
2. Neck is short.
3. Tail feathers are short.

Examples: *Gavia immer* (Common loon), *Columbius*.

Order 2. Podicipeformes (or Colymbiformes)

1. Freshwater divers.
2. Legs are far back on the body.
3. Toes are lobate.
4. Tail is a tuft of downy feathers.

Examples: *Podicipes ruficolis* (the little grebe or dabchick), *Podilymbus*.

Order 3. Procellariiformes

1. Nares (nostrils) are tubular.
2. Wings are long and narrow.
3. Rhamphotheca (horny sheath of bill) consists of several plates.
4. Halluxis (hind toes) are vestigial or absent.
5. Feathers are compact and oily.
6. Sea birds, nest usually on islands.

Examples: Albatrosses (*Diomedea*), fulmars (*Fulmarus*), and petrels (*Procellaria*), Puffin (shear water), *Diomedia exclans* (wandering albatross).

Order 4. Pelecaniiformes

1. Aquatic birds with short legs. Good divers, piscivorous.
2. All the four toes are included in the web.
3. Beak is long with a wide gape.
4. They nest in colonies.
5. Includes pelicans, cormorants, gamets, darters and boobies.

Examples: *Pelecanus philippensis*—the spotted billed or grey pelican (it has a buccal pouch for food storage), *Phalacrocorax* (Cormorant), *Anhinga* (Indian darter).

Order 5. Ciconiiformes

1. Tropical and subtropical marshy (wading) birds with long legs and long necks. They are good fliers.
2. Web is vestigial or lacking, except in flamingos.
3. Beaks abruptly decurved at middle in flamingos.
4. Young born naked.
5. Food comprises fish and aquatic animals.
6. Many nest in colonies on the trees.

Examples: *Aedea* (Great blue heron), *Ubulcus* (Cattle herons), *Bubulcus* (Egret), *Phenicopterus* (Flamingo), *Nycticorax* (Night heron), Platalea (Spoonbill), *Ciconia* (Stork).

Order 6. Anseriformes

1. It includes aquatic birds, such as ducks, geese and swans. They are strong fliers.
2. These birds vary in feeding habit, may be herbivorous, fish-eaters or filter-feeders.
3. Beak is broad and covered with soft, cornified epidermis containing special tactile receptors. A hard cap however, covers the tip of the beak. Margins of the beak bears numerous transverse horny ridges or lamellae.
4. Tongue is thick and fleshy.
5. Legs are short. Toes are webbed. Tail is small.

Examples: *Anas* (Wild duck), *Nettion crecca* (Common teal; Fig. 16.34G), *Anser indica* (Bar-headed goose; Fig. 16.35F), *Cygnus* (Swan).

Order 7. Falconiformes

1. This order includes carnivorous birds sch as vultures, kites, hawks and falcons.

2. Diurnal and strong fliers.
3. Beak is short, strong and curved (hooked) at the tip for tearing the prey.
4. Mandibles are sharp edged.
5. Toes bear sharp, curved claws and are wide-spread for catching and carrying the prey.
6. Most are birds of prey. They are predaceous, feeding generally on living vertebrates.
7. Some take carrion and a few eat insects.

Examples: *Milvus migrans* (Periah kite), *Pseudogyps bengalensis* (Lagger falcon), *Halister indus* (Brahmini kite), *Astur badius* (Sparrow hawk), *Falco pelegrinus* (Falcon), *Aquila* (Eagle).

Order 8. Galliformes

1. This order includes terrestrial game birds such as fowls, pheasants, turkey, quails, partriges and peafowls.
2. Beak is short and seed-crushing.
3. Feathers have after shaft.
4. Feet are large and adapted for scratching and running.
5. Mainly gramivorous but take insects too.
6. Their flight is powerful but short.
7. Many birds are gregarious.
8. Youngs are precocious.

Examples: *Gallus domesticus* (Domestic fowl), *Pavo cristatus* (Pea fowl; Fig. 16.25), *Francolinus* (Black partridge), *Coturnix* (Gray Quail).

Order 9. Gruiformes

1. Shore birds such as coots, cranes, rails, moorkens and bustards. They live near the banks of rivers and lakes.
2. These birds inhabit marshy places. They run, swim and dive well, but are poor fliers or flightless.
3. They feed on fishes, molluscs, etc.

Examples: *Fulica atra* (Coot), *Anthropoides virgo* (Crested crane), *Grus* (Crane), *Antegone antegone* (Sarus crane), *Choriotis* (Bustard).

Order 10. Charadriiformes

1. Inhabitants of open watery and marshy places.
2. Feathers are dense and firm.
3. Legs are usually long. Toes often webbed.
4. Beaks are mud probing.

Examples: *Lobivanellus indicus* (Red-wattled lapwing), *Hydrophasianus chirugus* (Pheasant tailed jacana), *Metopidius indicus* (Bronze winged jacana; Fig. 16.2), *Tringa glariola* (Sand-piper), *Capella* (Snipe), *Larus* (Gull).

Order 11. Columbiformes

1. It includes pigeons and dove.
2. Gramivorous (grain-eating) or frugivorous (fruit-eating) birds. Slow-walkers
3. Skin thick and soft.
4. Beak is usually short and slender and bears a cere at the base.
5. Tarsus is usually shorter than the toes.
6. Crop is large and secretes a "pigeon milk".
7. Youngs are naked at hatching.

Examples: *Columba livia* (Blue rock pigeon), *Streptopelia risoria* (Ring turtle dove), *Gourea cristata* (Crowned pigeon), *Streptopelia chinensis* (spotted dove), *Rophus* (extinct bird the dodo).

Fig. 16.2. Bronze Winged Jacana (*Metopidius indicus*).

Order 12. Psittaciformes

1. It includes parrots, parakeets and budgerigers. These are frugivorous and gregarious birds.
2. Beak is adapted to husk the fruits and to tear dead bark for searching boring insects. It is stout, sharp-edged and hooked at the tip. Upper beak is movably articulated with the frontal bone of the skull. Cere is present.
3. Feet are adapted for grasping, having two toes (1st and 4th) directed backward and two toes forward. Such a foot is said to be zygodactylous.
4. Outer hind toe is not reversible.
5. Feathers are green, blue, yellow or red in colour.

Examples: *Psittacula eupatria* (Large Indian parakeet), *Psittacula krameri* (Green parrot or rose-ringed parrot; Fig. 16.26), *Melopsittacus* (Budgeriger).

Order 13. Cuculiformes

1. It includes cuckoos and koels.
2. Beak is of moderate size. Tail is long.
3. Feet are zygodactylous similar to parrots. However, hindtoe is reversible. Feet are not adapted for grasping.
4. Many birds of this order are nest parasites, the female laying eggs in nest of other birds. **Salim Ali** had counted up to 13 eggs of koels in a single crow's nest.

Examples: *Cuculus conorus* (Cuckoo), *Eudynamis scolopaceous* (Koel, Fig. 16.3), *Centropus sinensis* (Coucal or Crow-pheasant, Fig. 16.4).

Fig. 16.3. Male and female koel. A nest parasite.

Fig. 16.4. Coucal or crow-pheasant.

Order 14. Strigiformes

1. This order includes owls and owlets which are nocturnal birds of prey and feed on vertebrates, especially mammals. Prey is usually swallowed entire.

2. Beak is short and hooked. Feet with sharp claws and adapted for grasping the prey.
3. The head is large and rounded.
4. Eyes are large, immovable and directed forward, each surrounded by a ring of radiating feathers. Retina mainly contains rods.
5. Ear opening is large and often with a flap-like cover.
6. Plumage is soft and noiseless in flight.

Examples: *Bubo bubo* (Great horned owl; Fig. 16.5), *Tyto alba* (Barn owl; Fig. 16.6B), *Ketupa zeylonensis* (Brown fish owl), *Athena brama* (Spotted owlet; Fig. 16.6A).

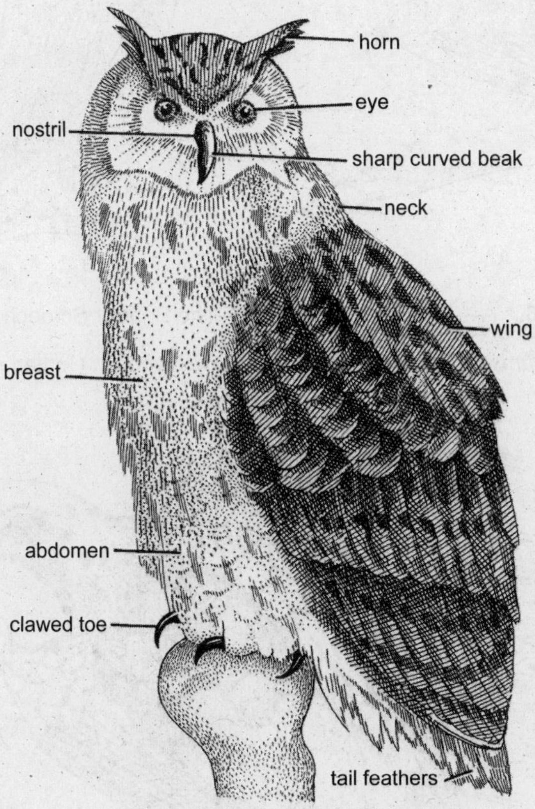

Fig. 16.5. The Great-horned owl.

Order 15. Caprimulgiformes

1. Beak is small and delicate.
2. Mouth is wide and often bordered by sensory bristle-like feathers.
3. Plumage is soft and loose.
4. Legs and feet are weak.
5. Nocturnal and insectivorous.

Examples: *Caprimulgus asiaticus* (Common Indian Night Jar), *Chordeiles* (Night hawk).

Order 16. Apodiformes (Micropodiformes)

1. Tail is short and deeply forked.
2. Beak is variable, it is small and weak in swifts and long slender in humming birds.
3. Legs are very short and feet are small.

Fig. 16.6. A—The Spotted owlet; B—The Barn or Screech owl.

Examples: *Apus affinis* (House swift; Fig. 16.7), *Micropodus* (Indian swift), Humming bird (*Trochilus*).

Fig. 16.7. House swift (*Apus affinis*).

Order 17. Coliiformes

1. First and fourth toes are reversible.

2. Tail is very long.
3. These birds feed on insects and fruits.
4. They creep on tree branches, using conical beak and peculiar feet.

Examples: *Colius* (Mouse bird which is confined to Africa).

Order 18. Trogoniformes

1. Beak is short and stout with bristles at the base.
2. Feet are small and weak.
3. Plumage is soft and loose, but brilliant in colour.

Examples: *Trogon elegan* (Copper-tailed trogon).

Trogon violaceous of Central and South America takes over the a wasp's nest, eats the adult wasps, then digs out the comb to make a nest cavity (**Khanna** and **Yadav**, 2005).

Order 19. Coraciiformes

1. This order includes kingfishers, blue jay (the roller), hoopoes, bee-eaters and hornbills. These birds have wide distribution.
2. Beak is strong.
3. Third and fourth toes are fused at the base. Such feet are called syndactylus.

Examples: *Halcyon smyrnensis* (White breasted kingfisher; Fig. 16.8), *Alcedo atthis* (Common kingfisher), *Coracias benghalensis* (Neelkanth, blue jay or roller), *Upupa epops* (Hoopoe), *Tockus birostris* (Grey hornbill, found in Corbett National Park, **Khati**, 2008), *Melophus orientalis* (Little green bee-eater; Fig. 16.9).

Fig. 16.8. The white breasted kingfisher.

Fig. 16.9. Little green bee-eater.

Order 20. Piciformes

1. It includes woodpeckers.
2. Beak is strong and chisel-shaped.
3. Tongue is long, flat, protrusible, bearing at the tip short-curved hooks.
4. Feet are zygodactylus, with first and fourth toes facing backward and the second and third forward.

Examples: *Brachypternus bengalensis* (Golden backed woodpecker), *Picus* (Woodpecker), *Dryobates* (Yellow front pied woodpecker), *Dendrocopos*, Barbets (Fig. 16.10), Fakamars, Puffbirds, Honey guide, Toucans, etc.

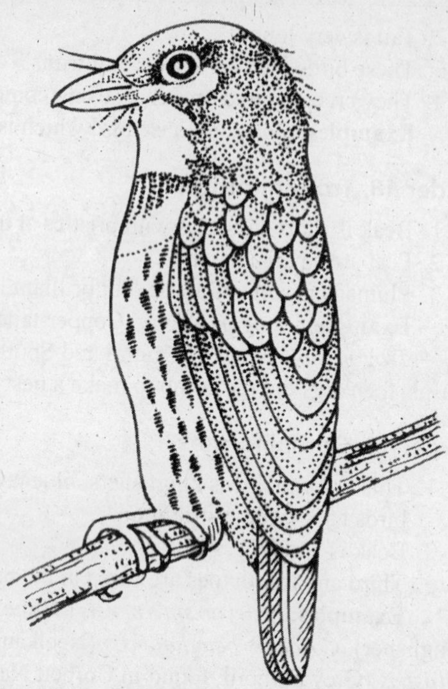

Fig. 16.10. The Crimson-breasted barbet or Coppersmith (*Megalaima haemacephala*).

Order 21. Passeriformes

1. This is largest order of birds.
2. It includes perching birds.
3. Feet are adapted for perching with hallux facing back, other toes in front. Toes are neither reversible nor fused.

Examples: *Passer domesticus* (House sparrow; Fig. 16.11), *Corvus splendens* (House crow; Fig. 16.12), *Corvus macrorhynchus* (Indian jungle crow; Fig. 16.13), *Ploceus philippinus* (Weaver bird or Baya; Fig. 16.14), *Orthothomus sutorius* (Tailor bird; Fig. 16.15), *Pycnonotus cafer* (Red-vented bulbul; Fig. 16.16), *Acridotheres tristis* (Common myna; Fig. 16.17), *Saxicoloides fulicata* (Indian robin), *Oriolus oriolus* (Golden oriole), *Dicrurus macrocercus* (Black Drongo; Fig. 16.18), *Muscicapa* (Fly catcher), Bulbul (*Molpestes*), Swallows (*Hirundo*; Fig. 16.19), *Pica* (Magpie), *Loxica* (Crossbill), *Sturnus* (Starling), *Alauda* (Larks), *Fringilla* (Finches), *Anthus* (Pipits), *Motacilla* (Wagtail), *Certhia* (Tree-creepers), *Parus* (Tits), *Lanius* (Shrikes), *Sylvia* (Warblers), Babbler (*Turdoides caudatus*; Fig. 16.20).

Fig. 16.11. Male House Sparrow

Fig. 16.12. House crow.

Fig. 16.13. Jungle crow.

Fig. 16.14. Male and Female Baya weaver bird.

Fig. 16.15. Tailor bird.

Fig. 16.16. Red-vented bulbul.

Fig. 16.17. Common myna.

Fig. 16.18. Black drongo.

Fig. 16.19. Common swallow.

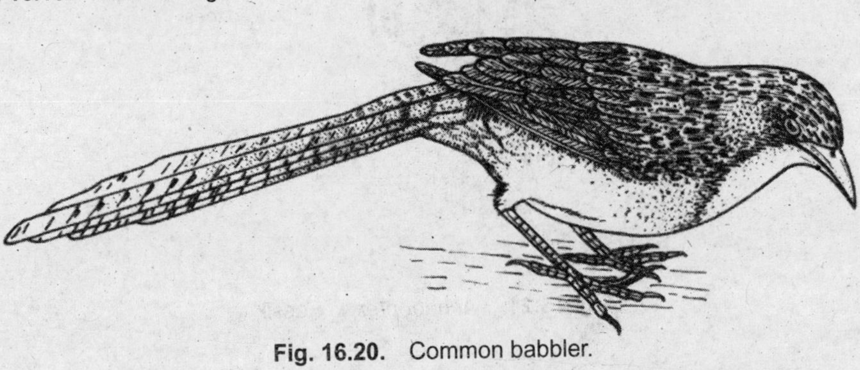

Fig. 16.20. Common babbler.

16.4 TYPE STUDIES

1. *Archaeopteryx*

Classification

Class	—	**Aves**
Subclass	—	**Archaeornithes**
Genus	—	*Archaeopteryx*
Species	—	*lithographica*

Archaeopteryx lithographica (Fig. 16.21) is the earliest known bird in the fossil records. Its fossil was discovered from a slate quarry of Solenhofenin, Bavaria (Germany) in 1861 by **Andreas Wagner** from the lime-stone rocks of Jurassic Period about 140 million years old. Six skeletons and one feather imprint of *Archaeopteryx* are now known. *Archaeopteryx* was a terrestrial bird, about the size of a crow and with a long tapering tail like that of dinosaur. This primitive bird is considered to be a **connecting link** or **transitional stage** between reptiles and birds as it possesses both reptilian as well as avian characters.

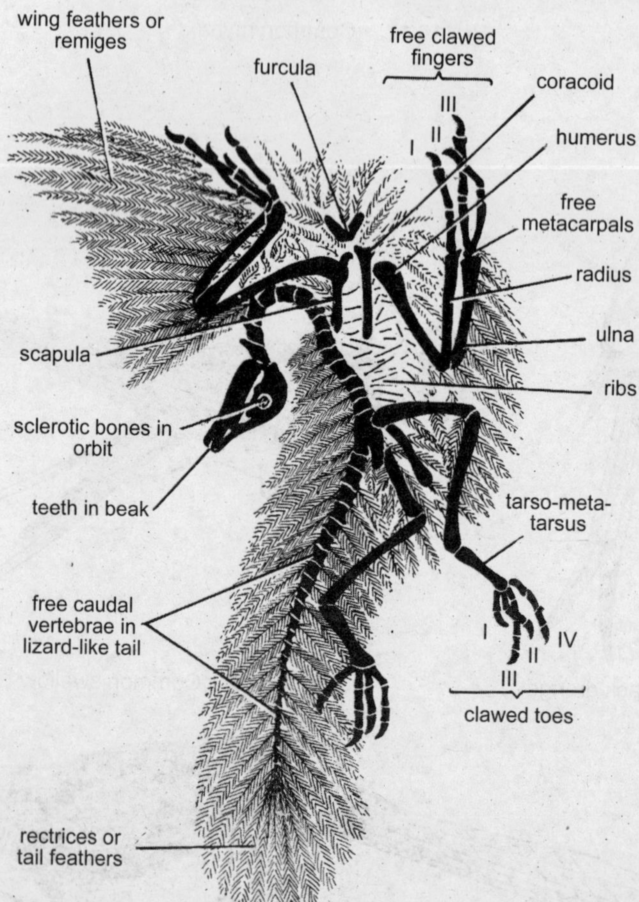

Fig. 16.21. *Achaeopteryx*. Fossil.

A. Reptilian Characters of *Archaeopteryx*

1. Elongated lizard-like body with long jointed tail.
2. The length and position of hindlimbs are in relation with forelimbs.
3. Bones solid and not pneumatized.
4. Presence of sclerotic plates around the eyes.
5. Presence of jaws with homodont dentition.
6. Vertebral centra are amphicoelous or biconcave as in *Sphenodon*.
7. Tail is supported by 13–20 free caudal vertebrae which do not fuse into a pygostyle. The tail tapers distally.
8. Ribs are single-headed and without unicinate processes.
9. Abdominal ribs are similar to *Sphenodon* and crocodiles.
10. Sternum is flat and without keel which indicates that it was not furnished with any strong flight musculature. T-shaped interclavicle is present.
11. Presence of elongated ilium and backwardly directed pubis.
12. The skeleton of hand presents reptilian pattern phalanges being 2, 3 and 4 respectively in first, second and third digits. Every digit terminates into a claw.
13. The metacarpals are separate and carpo-metacarpus is absent.

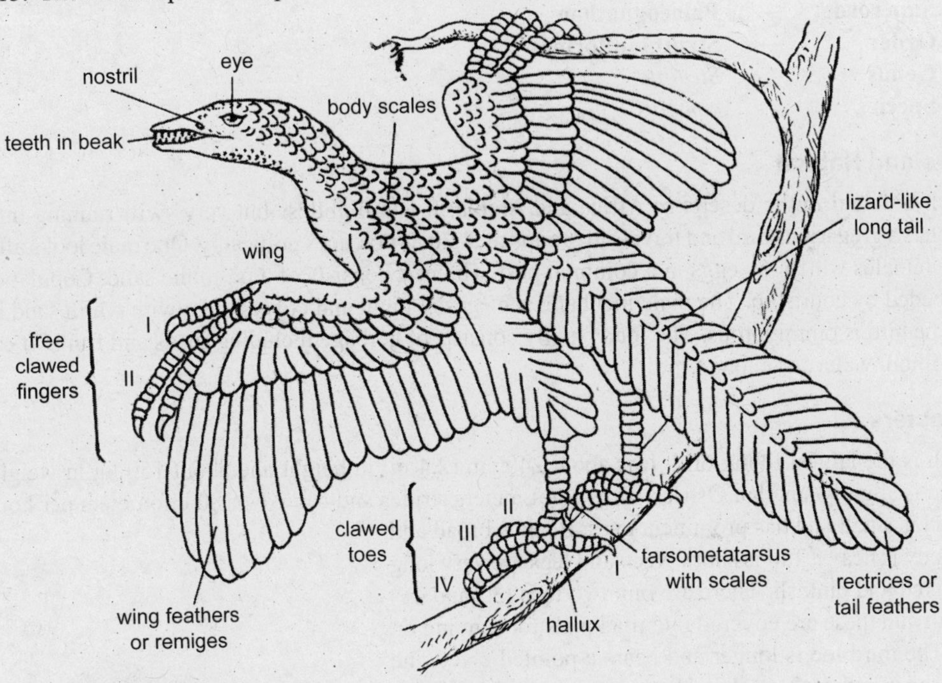

Fig. 16.22. Restoration of *Archaeopteryx lithographica*.

B. Avian Characters of *Archaeopteryx*

1. Various types of feathers were present on the body, tail, ulna and hands such as tail feathers, contour feathers, wing coverts, etc.
2. Forelimbs modified into wings. Wing consists of 17 feathers (the remiges).
3. Compact fusion of bones.
4. Proportionately large and rounded brain-case with large orbits and single occipital condyle.
5. Jaws drawn out into a beak.

6. Presence of V-shaped furcula.
7. Scapula slender and curved.
8. Pubis backwardly directed.
9. Each foot consists of tarsometatarsus and four clawed digits. These end into claws.
10. The hand had three digits.
11. Bones are hollow.

Flat keel-less sternum, small wing-span and long tail suggest that *Archaeopteryx* could not fly efficiently. It only volplaned (glided) from tree branches. Its clawed fingers indicate its scrambling progression among branches. It seems that it was warm-blooded and inhabited forests.

Archaeopteryx is called the "**lizard-bird**" because of its reptilian affinities including fewer sacral vertebrae, abdominal ribs, long tapering tail and small cerebellum lying behind the midbrain. *Archaeopteryx* is regarded as the immediate ancestor of all modern birds.

2. *Struthio* (Ostrich)

Classification

Class	—	**Aves**
Subclass	—	**Neornithes**
Superorder	—	**Palaeognathae**
Order	—	**Struthioniformes**
Genus	—	*Struthio*
Species	—	*camelus*

Habits and Habitat

Ostrich is found in the deserts of Africa and Arabia. It is a flightless but very swift running bird. Ostrich is a gregarious bird and travels in groups of 3–20. It practices polygamy. One male looks after 4 or 5 females which lay eggs in a common nest. The nest consists of a pit in the sand. Copulation is preceded by courtship. The male incubates the eggs at night and covers them with warm sand by day. The bird is ommivorous. The food chiefly consists of herbage including seeds and fruits. It can live without water for a long time.

Characters

Ostrich is the largest living bird. It is about 240 cm (2.4 m) in height and about 150 kg in weight. Female is slightly smaller. Ostrich takes eight meters strides and can cover 80 kilometers per hour. Its head is small and has prominent eyes. Beak is broad and flat. Body is heavy. Tail is short. Neck and legs are very long and naked and pinkish. There are ony two toes (3rd and 4th) per foot and these are covered by broad pads for running on sand. The third toe is longer and bears a pointed claw. The fourth toe is without claw. Hind limbs are long and powerful, they kick very hard. Feathers are without aftershaft. Skin of ostrich is so thick that it can be made into leather. Wings are rudimentary and nearly kept folded during running, but may be spreaded to work as a break or steering. The male has a single solid penis which can be retracted into a pouch of the cloaca. Feathers of adult bird are used by women as decoratives plumes. Ostriches are domesticated and there exist many ostrich farms in Africa and USA.

Fig. 16.23. African ostrich—*Struthio camelus.*

3. *Apteryx* (The Kiwi)

Classification

Class	—	**Aves**
Subclass	—	**Neornithes**
Superorder	—	**Palaeognathae**
Order	—	**Apterygiformes**
Genus	—	*Apteryx*
Species	—	*australis*

Habits and Habitat

Kiwi inhabits damp forests of New Zealand and adjoining islands. It is a swift and noiseless in movement. Kiwi is nocturnal, spending the day in a burrow, which serves as a nest. It feeds mainly on insects and worms, captured by the long, probing beak. Eggs are laid in a tunnel-like burrow. Male incubates the eggs.

Fig. 16.24. *Apteryx hoasti*—Greater spotted kiwi.

Structure

Kiwi is the smallest flightless bird (Fig. 16.24) having the size of a common fowl. Body of kiwi is covered by peculiar hair-like plumage. There are no tail feathers. Beak is long, slender and slightly curved downward. It bears nares at the tip instead of at the base. Legs are quite stout. Of the 4 toes per foot, 3 are long, forwardly directed and meet the ground, while one is small, directed backward and raised from the ground. Toes are long and bear curved claws. Kiwi is exceptional in birds in having small orbits and olfactory chambers extending backward between the eyes.

4. *Pavo cristatus* (Peafowl: Peahen and Peacock)

Classification

Class	—	**Aves**
Subclass	—	**Neornithes**
Superorder	—	**Neognathae**
Order	—	**Galliformes**
Genus	—	*Pavo*
Species	—	*cristatus*

The *Pavo cristatus* (Fig. 16.25) is the **national bird of India**. It is a protected bird. It is commonly called peafowl (*i.e.*, peacock) and **mor** or **mayur** in Hindi. It is a large game bird.

Habits and Habitat

The peafowl (= peacock) occurs all over India and adjacent countries. It inhabits forest clearings, fields and forests. It is shy and a polygamous bird. Peafowl usually lives in flock or drove of 1 cock and 4 to

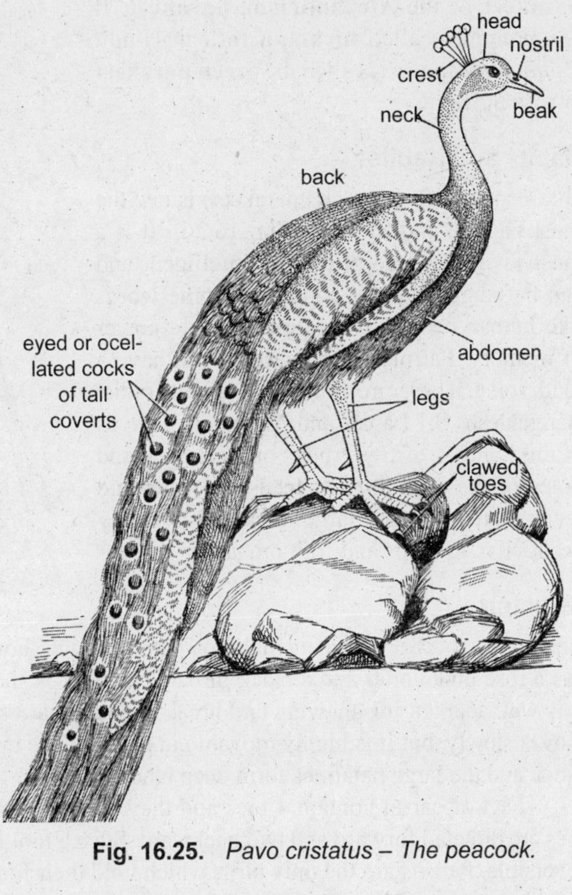

Fig. 16.25. *Pavo cristatus – The peacock.*

5 hens. It feeds on grains, vegetable shoots, insects, lizards and snakes. It feeds on the ground and rests at night on large trees. It is a good runner and capable of short and rapid flights only. Male (peacock) emits a loud, harse, screaming sound. The male dances before the females with his showy tail feathers erected and spread fanwise. Its nest is a shallow scrape in the ground in dense vegetation. It lines the nest with sticks and leaves. Three to five eggs are laid. Only the mother incubates the eggs and cares for the young.

Peafowl has a stout curved upper beak, long neck and short but strong feet with sharp claws. The male (peacock) has a brilliantly blue-coloured body with a crest of feathers on the head, fighting spurs on the legs and bright, erectile tail feathers with tips having eyelike marks (these feathers are called **coverts**).

The female (peahen) is smaller and comparatively dull coloured. It lacks spurs and train of tail feathers.

Legs of peafowl are adapted for running and scratching.

5. *Psittacula* (Parrot or Parakeet)

Classification

Class	—	**Aves**
Subclass	—	**Neornithes**
Superorder	—	**Neognathae**
Order	—	**Psittaciformes**
Genus	—	*Psittacula*

Psittacula eupatria is the **large Indian parakeet** or the **Alexandriane parakeet**. It is commonly called **hiraman tota** in Hindi. *Psittacula krameri* is called the **green parakeet** or **tota** in Hindi.

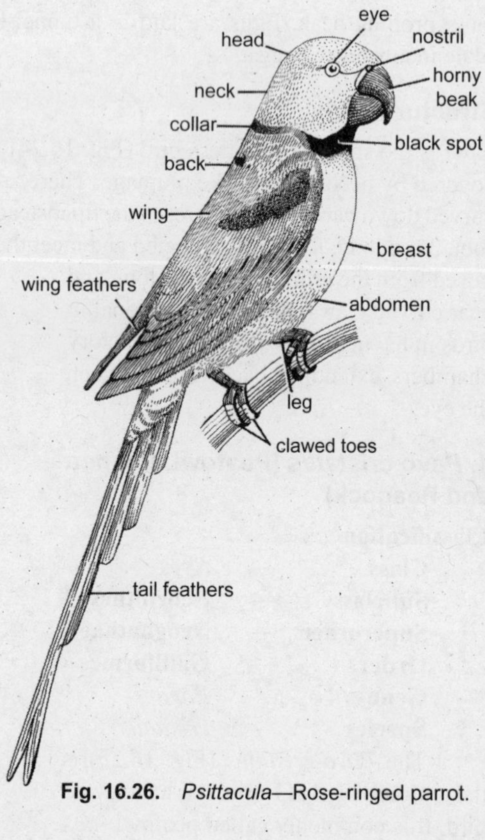

Fig. 16.26. *Psittacula*—Rose-ringed parrot.

Habits and Habitat

Hiramen tota (Rose-ringed parakeet) is one the most familiar Indian birds (Fig. 16.26). It is a most favourite cage bird, very intelligent and can be taught to copy and speak some words like human beings. It has considerable power of memory. Parrots are gregarious and have a loud voice. These are found throughout India, Bangladesh, Sri Lanka and Myanmar. It lives in the hollows of tree trunks and crevices and holes of the buildings. It feeds on fruits and seeds. Parrot forms a large flock and is highly destructive to crops and fruit orchards.

Structure

Body of parrot is a prominent mahroon patch on each shoulder. Female is green all over, but the male has a rose pink collar and a black throat. The beak is short, stout, sharp edged and deeply hooked. It is well adapted for gnawing and break open the hard seeds and nuts. In all birds, the upper beak moves slowly, but it is highly movable in parrots. The upper beak articulated with the frontals by a hinge and the large palatines form deep blades.

Feet of parrot contain 4 toes and they are adapted for grasping, holding and climbing. Two toes are directed forward and two backward. Such a foot is called **zygodactylus**. The outer toe is not reversible. Parrots are the only birds which hold their food in the foot while eating.

6. *Tyto alba* (The Barn Owl)

Classification

Class	—	**Aves**
Subclass	—	**Neornithes**
Superorder	—	**Neognathae**
Order	—	**Strigiformes**
Genus	—	*Tyto*
Species	—	*alba*

Habits and Habitat

The barn owl (Fig. 16.6B) is common all over India. It is like a jungle crow in size. It is a nocturnal bird of prey. It feeds on rats, mice, hares, birds, frogs and insects. It, like other owls, drops down from a perch to catch mammals and insects on the ground. Hard remains of food are vomited out via mouth. Once or twice a day owls regurgitate a pellet containing fur, most bones, chitinous insect remains and other indigestible parts of their prey. Barn owl is a solitary bird. The sexes are separate.

Structure

The owl has a large, rounded head with a prominent collar of stiff feathers round the flat face (Fig. 16.6B). The bird is golden buff above and silky white below. The eyes are of large size, placed side-by-side and directed forward. They mostly contian rods. During the daytime, the pupil gets reduced to a mere dot and cannot see easily. At night, the pupils expands tremendously to enable the bird to see in the darkest forest. The ears are asymmetrical (Box 16.2). The bird's entire body (even legs) are clothed in a soft, fluffy plumage, which renders it noiseless during flight. This enables it to reach the prey without being noticed. It has strong legs, stout claws and sharp curved beak for capturing and tearing the prey. Small prey is often swallowed whole.

The owl is a useful bird as it destroys rats, mice and other rodents which are enemies of agriculture.

Box 16.2

Barn owls are nocturnal hunters. They have the most exceptional hearing. Barn owls are able to catch prey in total darkness, guided by sound alone (**Payne**, 1962). They have facial masks to catch sound and have asymmetrically placed ear openings. Sounds reach one ear at a slightly different time to the other. By turning its head the owl can locate the source of a low sound, such as mouse chewing grain, very precisely (**Khanna** and **Yadav**, 2005; **Agarwal**, 2012).

16.5 BEAKS OF BIRDS

The beak or bill of a bird works as hands and mouth. It compensates the task of teeth which are universally lacking in modern birds. Birds beak is used to obtain and handle the food, to preen the feathers, to collect and arrange the nest materials, and for other purposes including feeding the young, courtship and defence. Beak consists of upper and lower jaws covered with a horny sheath, the **rhamphotheca**. The latter grows continually to replace loss through wear.

Types of Beak

The beak of bird is adapted for taking particular type of food (Box 16.3). Thus, because of the functional diversities, the beaks have undergone the range of modifications in different birds (Fig. 16.27). Some of the most important and common types of beaks of birds are described here.

Box 16.3

Bird's bills or beaks display adaptation best. The long dagger-shaped bill of the herons, the huge pouches on the pelican's bill, the hooked bill of predatory birds (owls, eagles, vultures) and the deep, heavy bills of fruit/seed eating parrots and finches are all adaptations to their food. Differences between beaks of related species reflect differences in ecology. The blue lit (*Parus caeruleus*) in Europe has a deeper bill than the coal tit (*Parus ater*) as it takes insects from broad leaves trees, while the latter feeds in conifers; a fine beak offers an advantage when probing into pine needles (**Khanna** and **Yadav** 2005).

1. Seed-eating beak (or Seed-crushing beak). It is short, stout, conical and pointed at tip. This type of beak is used to pick up and break open the seeds. Such type of beak occurs in seed-eating or graminivorous birds such as **pigeons**, **sparrows**, **finches**, **buntings** and **cardinals**, etc., The weaker breaks are employed to pierce the small seeds. While powerful beaks crush large, hard-shelled seeds and fruit stones.

2. Fruit-eating beaks. It is short or long, sharp, powerful, curved (hooked) and movable. It is used to cut fruits and is found in **parrots**, **cockatoos** and **crossbills**. In fruit-eating beaks, the upper beak is movable on the skull and is well adapted for tearing fruits and gnawing hard nuts and seeds. The beak of hornbill is large and heavy having cellular structure in its inner region that acts as resonator to produced loud sound.

3. Cutting and biting beak. It is long, sharp and strong and provided with sharp horny edges. It is used for various purposes such as tearing prey or carcass, breaking eggs, cutting fruits etc. It is found in **crows** (*Corvus*) and **rovens**.

4. Insect-catching beak. Insectivorous beaks are found in **swifts**, **swallows**, **flycatchers**, **robins**, **night jars** and **hoopoe**. In hoopoe, the beak is long, slender, weak and slightly curved adapted for turning the leaf litter and probing into soil for insect larvae, pupae and mites. (It is also called **ground probing beak**). In **swallows**, **frog-mouth** and **swifts**, the beak is small, wide and delicate to prey upon flying insects. In **flycatchers**, the beak is small, strong with notch mandibles at the tip end. **Kiwi** has a beak which is long, slender and curved down–ward. By the help of this probing beak, it feeds on insects and worms.

The shallows fly about in the evening, catching insects with their gaping beaks. Bristle-like feathers at the base of the beak help them in capturing insects.

5. Piercing and tearing beak. It is short, hooked and powerful. Such as beak is operated by well developed mandibular muscles and is used to pierce and tear apart small prey or large carcass. It occurs in **kites**, **hawks**, **eagles**, **owls** and **vultures**.

6. Fish-catching beak. It is long, powerful, sharply pointed and spear-like. It is found in **storks**, **herons**, **kingfishers**, etc. The fish is speared and deftly flips into the mouth. In **cormorant** the beak is long, narrow and slightly curved at the tip (Fig. 16.28) and have lateral tooth-like processes adapted for capturing the fish (It is also called **fish-holding beak**). In **snake-bird (Indian darter)**, the lateral tooth-like processes over the beaks are in the form of needle-like processes. In **king fisher** beak is long, powerful and sharply pointed to capture fish, frogs, toads, tadpoles, molluscs and other aquatic animals. **Storks** have long, thick bills that allow them to catch and kills snakes, frogs and insects. **Puffins** have short triangular beaks, that can grasp small fish tightly (Box 16.4).

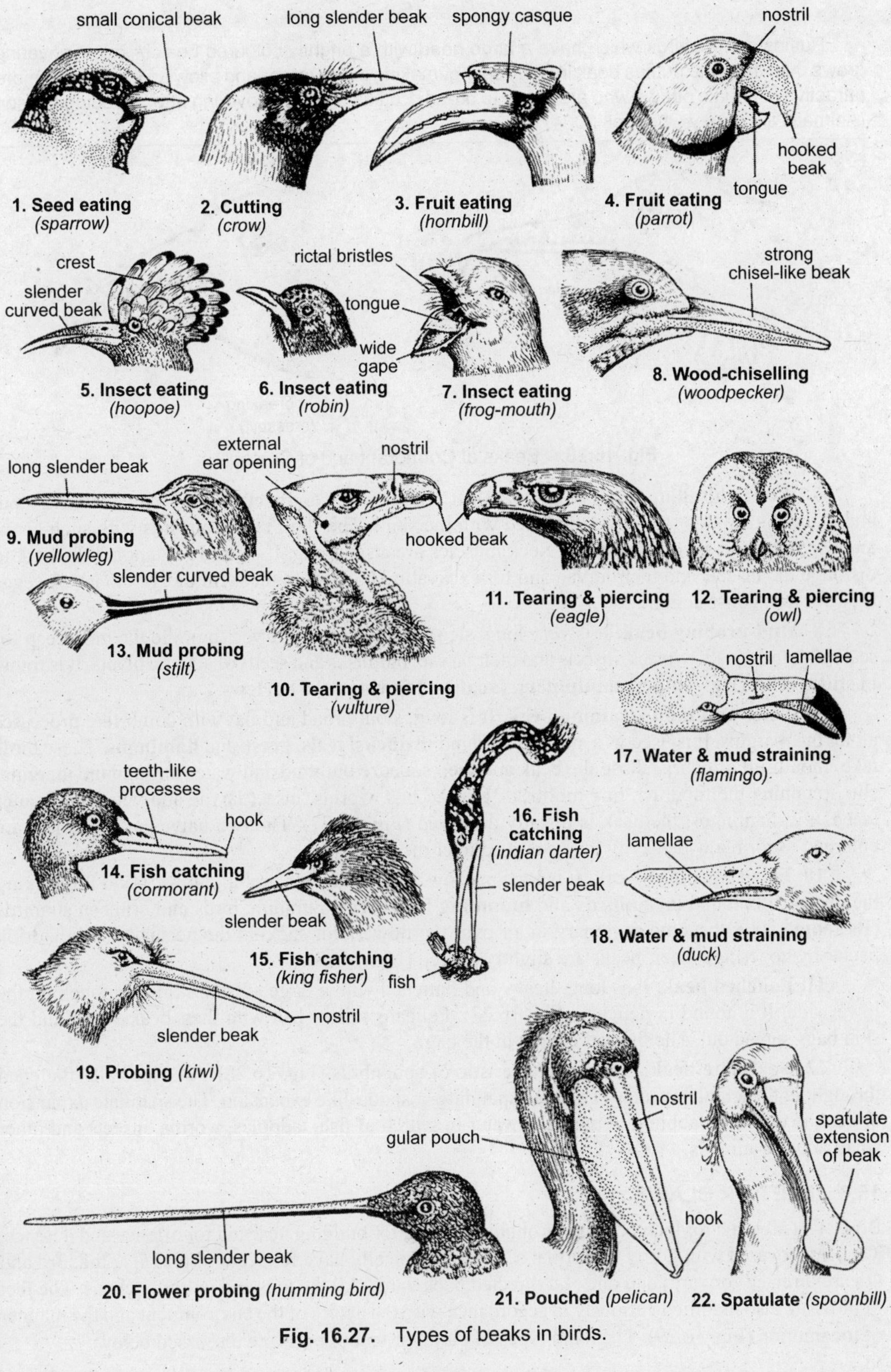

Fig. 16.27. Types of beaks in birds.

Box 16.4

Puffins are seabirds which have a large head with a brightly coloured beak. A bright covering grows over the male puffins beak in early spring, making it look larger and allowing it to appear more attractive to female puffins who are ready to breed. Puffins shed the covering after breeding season (Animals, Discovery Channel, 2004).

Fig. 16.28. Beaks of Cormorant and of Crossbill.

7. Wood-chiselling beak (Fig. 16.28). It is found in **woodpeckes**. It is long, straight, stout and chisel-like and is strongly articulated with the skull. The skull bones of birds with such beaks are also thick and shock-absorbent. Neck muscles are also strong. It is used for tapping the bark to dislodge the insects and their larvae and for excavating hollows in the stem of trees to lay the eggs (*i.e.*, for nest construction).

8. Mud-probing beak. It is very long, slender and slightly curved to probe the mud deep for searching the aquatic worms, insects and their larvae, mollusks and seeds of aquatic plants. It is found in **stilts**, **yellowlegs**, **snipes**, **sandpipers**, **jacanas**, **lapwings** and **curlews**.

9. Water and mud straining beak. It is long, stout broad and flat with comb-like processes along the margins. It is used as a sieve. It is found in **ducks, teals**, **geese** and **flamingos**. These birds take mouthfull of debris, close the beak and then squeeze out mud and water through the marginal slits, retaining the food, ie., tiny mollusks, crustaceans, worms, insect larvae and vegetable matter (seeds). In flamingos, the beak is distally decurved (Fig. 16.27). The two halves of lower jaw are enlarged to form a wide cavity that is curved over the upper jaw.

10. Flower-probing beak. It is long, narrow and pointed for probing flowers and for sucking the nectar. It is found in **sunbirds** and **humming birds**. The humming birds and crimson sunbirds (**Banenjee** 2008) become stationary in air over the flowers for sucking the nectar (*i.e.*, they drink nectar by hovering). Their beaks are slightly curved (Fig. 16.28).

11. Pouched beak. Is is long, heavy and flattened with a large, elastic skin bag pouch on the lower side. It is found in pelicans (Fig. 16.27). Pelicans follow fish with their beaks open and the skin bags spread out. The fish get caught in the bag.

12. Spatulate beak. It is a characteristic of spoonbills (Fig. 16.28). It is specially flattened throughout the length but ends in a broad spatulate or spoon-like expansion. The spatulate expansion of beak is used for dabbling in mud and water in search of fish, tadpoles, worms, insects and other small aquatic animals.

16.6 FEET OR CLAWS

Birds feet are used for locomotion, for obtaining food, for building nest and for offence and defence. They are covered with horny epidermal scales and typically have four toes–three (*i.e.*, 2nd, 3rd and 4th are directed forward) and one (1st directed backward). All the toes end in horny claws. The feet of birds are also modified variously in accordance with the nature of the environment and the manner of locomotion (Fig. 16.29). The main types of claws or feet of birds are discussed below.

1. Cursorial or running feet. The feet of running birds have long tarsometatarsi and relatively short and fewer digits. **Ostrich** has only two small and nail-less toes (*i.e.*, 3rd and 4th), where as Ratites, *e.g.*, **Rhea**, **Crossowary**, **Emu** and **Courser** and **Bustard** have three toes (2nd, 3rd and 4th), all facing forward.

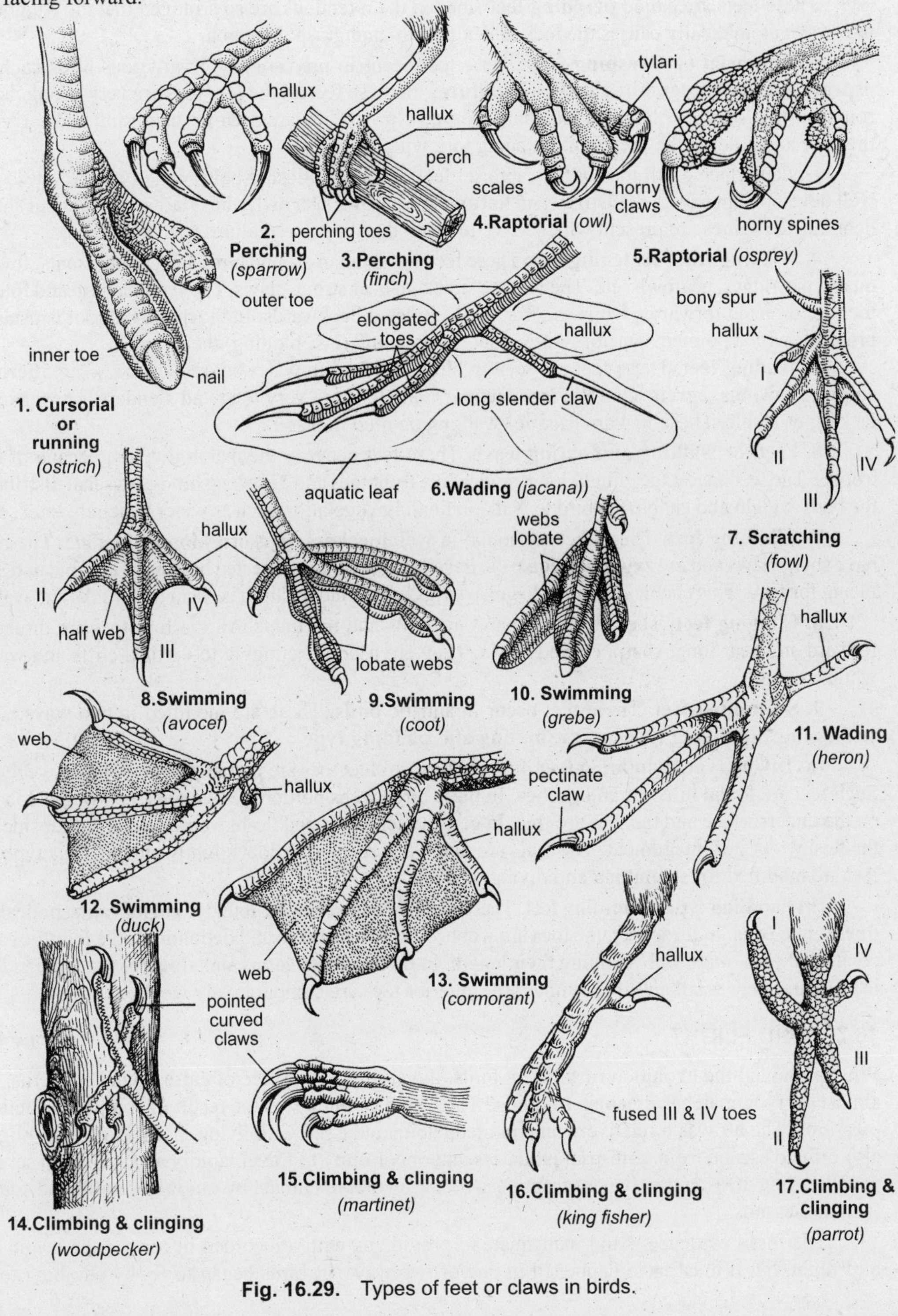

Fig. 16.29. Types of feet or claws in birds.

2. Perching feet. These are found in a majority of birds (passer birds), e.g., **Sparrows, Crows, Bulbuls, Koels, Mynas, Robins, Finches, Sunbirds**, etc. The feet are relatively small and have slender anterior toes and strong posterior toes or **hallux**. Hallux is long, strong and sharply clawed. Because of opposable toes these birds can fasten the foot to a branch of tree.

These feets are called **perching feet** since in them tendons are so arranged that the weight of the body automatically causes the toes to flex (curve) and grasp the perch.

3. Raptorial or Grasping feet. These feet occur in predatory or carnivorous birds such as **ospreys, hawks, eagles, kites, owls** and **vultures**. In these feet all the four toes are very strong, have rough surface and bear large, stout, sharp, curved claws. They are used for capturing and carrying the prey and holding the prey while it being torn with beak.

Undersurface of all toes of the raptorial feet have large fleshly bulbs called **tylari**. These are well developed in hawks. In **osprey** and **ketupa** (*i.e.*, brown fish owl), the tylari are absent but sharp denticulated spines are present which serve to catch the slippery creatures (fish and frogs).

4. Walking and Scratching feet. These feet are found in ground birds such as pheasants, fowls, quails, partridges, peafowls, etc. The feet are stout and bear strong claws. The second, third and fourth toes are directed forwards, while the first toe is directed backwards. In the male, the foot is usually provided with a pointed spur for fighting and for treading (*i.e.*, ploding) the females.

5. Wading feet. These feet are seen in birds that feed in water but rest on trees, *e.g.*, **herons snipes, lapwings, egrets**, etc. The tarsometatarsi and toes are very long and slender. The webs are lacking or feeble. These feet are used for walking on mud in water.

6. Feet for walking on floating leave. These feet occur in the marsh-dwelling jacana of the tropics. The feet have exceedingly long, spider-like front toes and very long hind claws that distribute the body weight and enable the bird to walk on floating vegetation such as water hyacinth, lotus, etc.

7. Climbing feet. These feet are found in **woodpeckers, parrots, kingfishers**, *etc*. The feet have sharp claws and are **zygodactylous**, *i.e*, have 1st and 4th toes directed backward and 2nd and 3rd facing forward. This enables the birds to grasp twigs firmly and to climb vertical tree trunks and walls.

8. Clinging feet. These feet are found in swifts and martinets. All the four toes are directed forward and bear long, sharp, curved claws. They are used for clinging to vertical cliffs and walls while feeding.

9. Swimming feet. These feet occur in aquatic birds. These are modified in two ways, *i.e.*, **swimming and diving type** and **swimming and padding type**.

(a) Diving type swimming feet. In these types of feet toes are partially or completely webbed. Such feet are found in **coots** and **grebes**. In these birds webs border the free toes, *i.e.*, web is lobate on the undersurface and the toes are free. In gerbes each toe is paddle-like bordered on either side of the fleshy lobes of integument. The nails are flat in a gerbe and in coot slightly curved. This type of feet are modified for swimming and diving in water.

(b) Paddling type swimming feet. These paddle-shaped feet are found in **cormorants, pelicans, ducks**, and **teels**. In these feet, the toes are webbed. In **cormorants** and **pelicans** all the four toes are connected with tough web all along their length. In **ducks, teals, loons** and **albatrosses**, the first toe is comparatively small and free, while three anterior toes are comparatively webbed.

16.7 BIRD FLIGHT

What prompted the evolution of flight in birds, the ability to rise free of earth bound concerns, as almost every human has dreamed of doing? The origin of flight was the result of complex selective pressures. The air was a relatively unexploited habitat amassed with flying insects for food. Flight also offered escape from terrestrial predators and opportunity to travel rapidly and widely to set up new breeding areas and to benefit from year-round favorable climate by migrating north and south with the seasons.

The fossil evidence is too inadequate to provide us with a recorded history of the origin of bird flight, but it must have happened in one of two ways: 1. birds began to fly by climbing to a high place and gliding down, or by flapping their way into the air from the ground. The **ground-up hypothesis** holds that birds were ground-dwelling runners with primitive wings used to snare insects. With continued enlargement the protowings eventually enabled the running animal to flap its way into the air. The more widely favored **trees-down hypothesis** suggests that birds passed through an arboreal apprenticeship of tree climbing, leaping through the trees, parachuting, gliding and finally fully powered flight. One thing seems certain: *feathers were an absolute requirement for flight*. The evolutionary origin of feathers preceded flight; feathers arose for their thermoregulatory role and made possible the subsequent evolution of flight. There is absolutely no support for the idea that bird ancestors were originally membrane-winged flyers, like bats, that later developed feathers.

Bird wing as a lift device. The bird wing is an airfoil that is subject to recognized laws of aerodynamics (Box 16.5). It is streamlined in cross section, with a slightly concave lower surface

Box 16.5

Air foil is a winglike surface designed to provide the maximum aerodynamic advantage for an airoplane in flight.

Aerodynamics is the branch of physics that treats of the laws of motion of gases especially atmospheric, under the influence of gravity and other mechanical forces, and of the mechanical effects produced by such motion.

Camber means to have a slight upward convex curve.

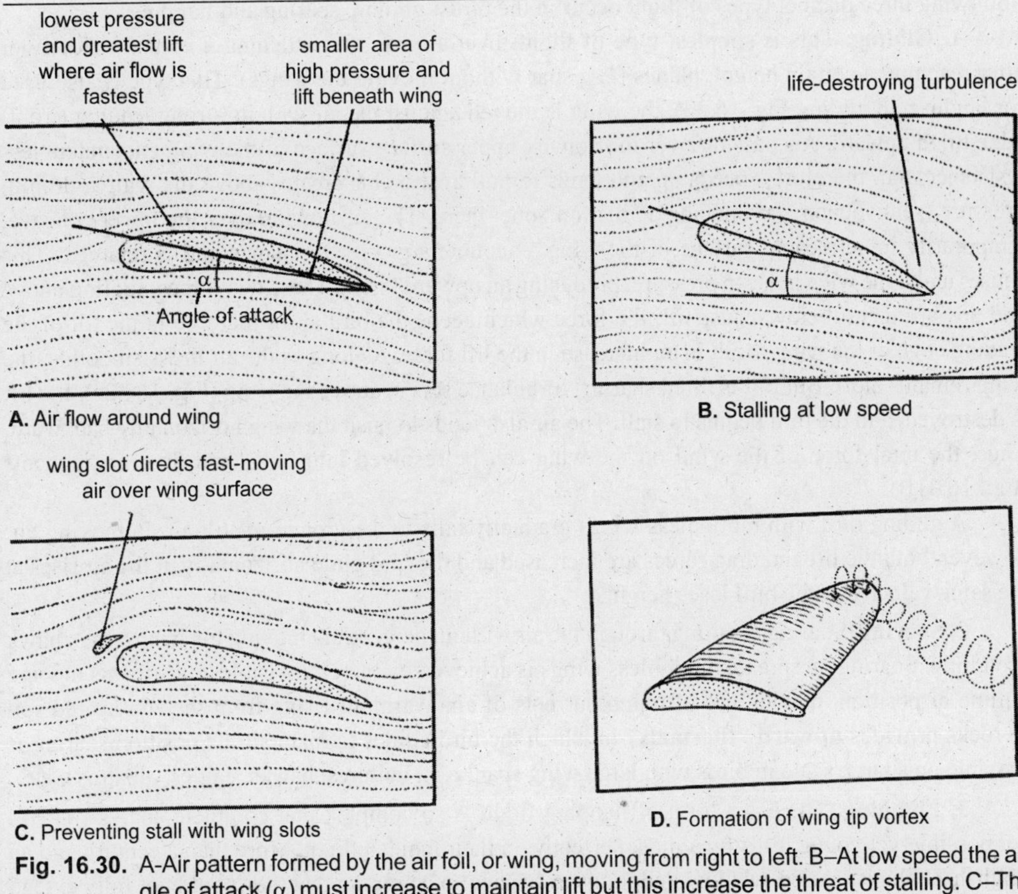

Fig. 16.30. A-Air pattern formed by the air foil, or wing, moving from right to left. B–At low speed the angle of attack (α) must increase to maintain lift but this increase the threat of stalling. C–This figure shows how low–speed stalling can be prevented with wing slots. D–Wing tip vortex, a turbulance that tends to develop at high speeds, reduces flight efficiency. The effect is reduced in wings that sweep back and taper to a tip (after Hickman, *et al.*, 1996).

(cambered) and with small, tight-fitting feathers where the leading edge meets the air (Fig. 16.30). Air slips smoothly over the wing, creating lift with minimum drag. Some lift is produced by positive pressure against the undersurface of the wing. But on the upper side, where the air stream must travel farther and faster over the convex surface, a negative pressure is created that provides more than two-thirds. The lift-to-drag ratio of an air foil is determined by the angle of tilt (angle of attack) and the airspeed (Fig. 30.16A). A wing carrying a given load can pass through the air at high speed and small angle of attack or at low speed and larger angle of attack. As speed decreases, lift can be increased by increasing the angle of attack, but drag forces also increase. Finally a point is reached at which the angle of attack becomes too steep; turbulence appears on the upper surface, lift is destroyed, and stalling occurs. Stalling can be delayed or prevented by placing a **wing slot** along the leading edge so that a layer of rapidly moving air is directed across the upper wing surface. Wing slots were and still are used in air crafts traveling at a low speed. In birds, two kinds of wing slots have developed: 1. the **alula**, or group of small feathers on the thumb which provides a midwing slot, and **2. slotting between the primary feathers**, which provides a wing-tip slot. In a number of song-birds, these together provide stall preventing slots for nearly the entire outer (and aerodynamically more important) half of the wing.

Mechanism of Flight

Following three distinct types of flight occur in the birds: gliding, soaring and flapping.

1. Gliding. This is simplest type of flight. In it a bird, after attaining a certain velocity or after reaching a certain height, planes (*i.e.*, soar without moving the wings). This type is also used for landing. In gliding Fig. 16.32), the wing is moved against the air with its strong leading (front) edge titled upward. Air passing over the convex upper surface surface of the wing encounters less resistance and, therefore, speeds up and tends to pull away. This creates above the wing a drop in pressure (semivacuum), producing a 'suction zone' there. The air flowing over the lower concave, temporarily impermeable surface of the wing, encounters greater resistance and is retarded. This causes under the wing a rise in pressure, producing an upward thrust. These two forces, **suction** above and upward thrust below, cause lift, the force which keeps the bird aloft. Increase in the tilt of the leading edge of the wing result in an increase in the lift force. As long as the air flows smoothly, the wing remains aloft. But if it is tilted sharply, turbulance sets in above the wing (Fig. 16.32c), the lift is destroyed, and the bird begins to stall. The air also tends to push the wing horizontally backward. Thus, the total force of the wind on the wing can be resolved into "lift" and drag components (Fig. 16.31B).

A gliding bird with motionless wings gradually falls to the ground in still air. In moving air, however, both the lift and drag forces are increased and the bird glide horizontally, it the air rises at the same velocity as the bird loses height.

2. Soaring. Soaring is sailing through the air without perceptibly moving the wings, as in hawk or vulture. Soaring on spread, motionless wings is achieved by minor, invisible adjustments in wing outline or position, making use of rising currents of air. Warm air rising from the sun-heated soil or rocks provides **upward "thermals"** in which the bird circles to maintain its position soaring is possible on a large scale in birds with large wing span, *e.g.*, vultures, hawks, eagles, albatross, etc.

3. Flapping. The basic form of flapping flight is so complex that complete analysis is still not possible-yet young birds fly almost perfectly on their maiden flight. More than a century ago an English Zoologist reared swallow fledglings in a space so confining that they could not fully extend their wings. Yet when released at the age when swallows normally fly, they flew immediately and without practice.

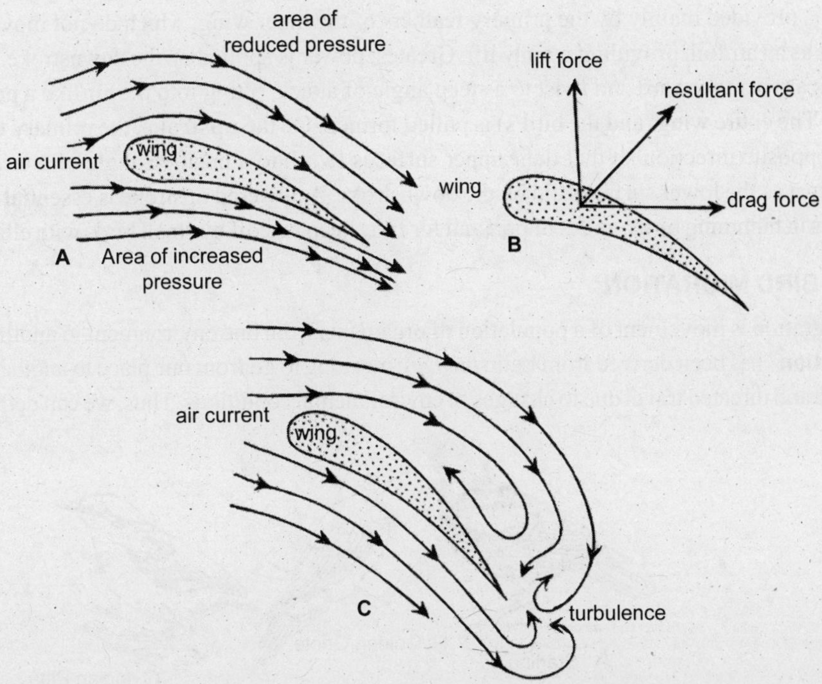

Fig. 16.31. Position of wing in gliding A-Favourable, B-Forces acting on the wing, C-Unfavourable.

Two forces are required for flapping flight: a **vertical lifting force** to support the bird's weight and a **horizontal thrusting force** to move the bird forward against the resistive forces of function.

Fig. 16.32. The normal flapping flight of strong fliers such as ducks, the wings sweep down-ward and forward fully extended. Thurst is provided by the primary feathers at the wing tips. To begin the upbeat, the wing is bent, bringing it upward and backward, bringing it upward and backward. The wing then extends, ready for next downbeat.

Thrust is provided mainly by the primary feathers of the inner wing, which do not move so far or so fast, act as an air foil, providing mainly lift. Greatest power is applied on the downstroke. The primary feathers are bent forward and twist to a steep angle of attack, biting into the air like a propeller (Fig. 16.33). The entire wing (and the bird's) is pulled formed. On the **upstroke**, the primary feathers bend in the opposite direction so that their upper surfaces twit into a positive angle of attack to produce thrust, just as the lower surfaces did on the downstroke. A powered upstroke is essential for hovering flight, as in humming birds, and is import and for fast, steep takeoff by small birds with elliptical wings.

16.8 BIRD MIGRATION

The migration is movement of a population of organisms from one environment to another. The word "**migration**" has been derived from Latin *migrare* meaning to go from one place to another. It is sort of periodic and directed travel due to changes in environmental conditions. Thus, we can define migration

Fig. 16.33. Some migratory birds.

as mass directional movement of large numbers of a species from one habitat/location to another. The term therefore applies to classic migrations (the movements of locust swarms, the inter continental journeys of birds, the transoceanic movement of eel) but also of less obvious examples such as to-and-fro movements of shore animals following tidal cycle. The study of periodic phenomena (such as migration) in animals in relation to changes, climatic and other ecological factors is called **phenology**.

The bird migration is a seasonal movement from one habitat to another and back again, to get the advantage of favourable conditions. It is traditional, hereditary (instinctive) and of regular occurrence at definite intervals every year. Usually, the birds migrate to the northern hemisphere in spring to breed and return to southern hemisphere in autumn to pass the winter (Box 16.6).

Box 16.6

Birds, with their marvellous ability to fly and with their unparalleled mobility, have explored almost the entire globe searching for food, territory, nesting sites and other necessities for survival. Because of their high metabolism and very high energy consumption, birds must have abundant and unfailing sources of food. In geographical regions with seasonal climatic changes, ecological changes occur which may require the birds to move away if they have to survive. In colder countries the intense chill of winter closes lakes and streams and snow fall covers the land and its vegetation. This compels the birds to move to warmer regions. In countries around equator, summer or dry season may reduce the plants, insects, worms and food sources; the temperature gets very high and to escape that the birds have to emigrate to some more favourable place.

Thus, birds migrate in order to secure optimum conditions for breeding and feeding, *i.e.*, to utilise the trophic niches of both the hemispheres and to utilise the old and protected nesting places year after year.

I. Migratory Status

A migrating bird is called a **migrant**. Depending on the seasons of migration of different birds, different names are given to the migrants. These are as follows:

1. Winter visitors (N $\rightleftharpoons$ S). These include those species which move from their breeding grounds to spend their wintr in a more suitable place, where food is also in plenty. For example, fieldfare, snow-bunting and redwing arrive in autumn chiefly from the north, stay throughout winter and fly northward again in spring. Greylag goose, pintail, common teal, gadwal, shoveller, common pochard and European starling are the winter visitors (migrants) visiting Keoladeo National Park, Bharatpur (Fig. 16.34). The wagtails (*Motacilla* spp.) appear in India in the beginning of winter season and at its end they leave this place, migrate to Siberia to spend summer and to lay eggs.

2. Summer visitors (S $\rightleftharpoons$ N). From spring onwards, south starts getting too hot, while luxurious summer conditions prevail in the north. The birds of south leave it in spring for north to spend the summer, breed and return to south again in autumn, when the north gets cold, *e.g.*, swifts, swallows, nightingales, cuckoos, etc. The cuckoo (*Eudynamis scolopaccus*) arrives in Uttar Pradesh in the beginning of March and breeds there. During the month of August it leaves this place and migrates to South India and Ceylon to spend winter there.

3. Transient visitors or passage migrants. These are summer and winter visitors while migrating from south to north and vice versa, stop at some places for the sake of rest only. For example, gargany teals migrate from Mangolia or Siberia to Bharatpur to go South India which is their actual breeding ground. While returning they again stop over at Bharatpur.

4. Permanent residents. These birds are found in a particular area throughout the year and do not migrate from one place to another. For example, cotton teal, spotbill duck, whistling teal, barheaded goose, mallards, comb duck (Fig. 16.34) are all permanent residents of Keoladeo National Park, Bharatpur.

II. Types of Migration

1. Daily migration. Many birds make daily movements from their nest in response to environmental forces such as light, darkness, temperature, humidity and food availability. Birds may make daily migrations from their resting sites to feeding areas. Crows, house sparrows, starlings, rockery herons, etc., show this type of migration.

Fig. 16.34. Permanent resident birds at Keoladeo National Park, Bharatpur.

2. Local migration. Because of heavy rains, floods, excessive heat and cold, birds leave an area for some time and return when the crisis is over. Flowering of certain plants and ripening of seeds also stimulate some birds to migrate locally.

3. Seasonal migration. Migration of birds in response to changes in the season is called **seasonal migration**. In tropical and subtropical areas, this occurs in the beginning or end of which rest in temperate zones, go to warmer areas. Such migrations of birds are known in northern hemisphere, Europe, North America and Asia.

4. Moult migration. In most ducks, males and juvenile birds migrate short distances northward for *moulting* leaving behind females and young birds in the breeding ground.

5. Cyclic migration. Some migrations of birds are seasonal but do not occur at regular intervals. The cyclic migrations of the snowy owl in search of leming (chief food of snowy owl) in United States in winter occurs in three to five years.

6. Partial migration. In certain cases all the birds of a group of migratory birds do not leave the native land and hence are always represented by certain individuals. But these individuals are not always the same, *e.g.*, songthrus, redbreast, titmouse, finch, etc.

7. Vagrant or irregular migration. Some of the birds (such as herons) sometimes disperse for a short or long distance for the sake of food and safety. The birds can also be swept away by powerful winds and hurricanes (oceanic storms) to very long distances.

8. Altitudinal or vertical migration. The birds living at high altitudes (hills) descend at lower altitudes in winter to save themselves from the intense cold of high altitudes. They return again to high altitudes with the advent of summer. Coots of Andes mountain in Argentina move downwards; mountain quails which nest at elevation upto 9500 feet in central California mountains, migrate downward to area below 5000 feet in the fall and in the summers they go back to the hills. Violet green swallows of U.K. and willows in Siberia are also vertical migrants. The Himalayan white caped redstarts (*Chimarrhornis*) resides at a height of 8000 to 14000 feet. During winter it descends to the height of 2000 to 8000 feet.

In certain cases, such as **blue grouse** of USA, the direction of migration is reveresed with the season. It passes winter high up in the rocky mountains probably to avoid competition and predators.

9. Latitudinal or equatorial migration. The most familiar migrations are those from north to south and vice versa. A large number of Eurasian and North American birds cross the equator to spend the winter in Africa or South America. The golden plover of the Arctic tundra passes winter 12,800 km down south in the pampas of Argentina. Cuckoo and storks pass summer in northern hemisphere and winter in south.

10. Longitudinal migration. Movement of birds from east to west or vice-versa is known as longitudinal migration, *e.g.*, evening grossbeaks (finches-like birds) that nest in northern Michigan spend the winter in New England; California gulls that breed in Utah migrate westward to winter on the Pacific Coast, starlings move from east Europe or Asia to the Atlantic Coast.

Diurnal and Nocturnal Migration

On the basis of their wing powers and method of getting food, **William Brewster** divided the birds into following three categories:

1. **Night fliers.** Small birds such as passerine birds (sparrows, titmice, jays, crows, etc.), cuckoos and wood-peckers.
2. **Day fliers.** Large birds such as **hawks**, **pigeons**, **swallows** and **robbins**, etc.
3. Few birds such as **geese** and **ducks**, migrate both by night and day.

The day fliers are usually restless and strong winged. These can feed while on wings, have least danger of enemies and can migrate to long distances. **Turkey vultures** (or Turkey buzzard—*Cathartes aura*) cover a distance of 3600 miles without a stop. Broad winged birds such as **eagle**, **storks** and **cranes** use thermal currents for soaring by which they spend less energy for flight.

Night fliers are usually small birds. Darkness provides them protection from large predatory birds. It also gives birds opportunity of using all the daylight hours for feeding thereby enabling them to build up sufficient energy resources for sustaining long distance flights.

Further, bird migration does not always take place by means of flight. **Penguins** (*Spheniscus*) breeds on the land of Antarctica continent. In the winter they move hundreds of kilometers by floating on packs of ice and return south in summer. They cover a greater path of the distance by *swimming* in the sea and *march* to their nesting sites. Migratory **guillemots** (*Uria*) mostly swim and **American coot** (*Fulica*) walks for kilometers across the country.

Fig. 16.35. Long distance migrant birds.

III. Range of Migration

The distance travelled by migratory birds depends upon the local conditions and the species concerned. The longest distance of about 17600 kms is covered by the **arctic tern** (*Sterna paradisea*; Fig. 16.30) which migrates from north to the edge of Antarctica (During summer arctic tern is found in Canada and Greenland which are its breeding grounds; during winter it migrates to Atlantic Islands in South via South America and Africa. On advent of summer it returns back to its breeding grounds). Likewise, **golden plover** migrates from the Arctic tundra to the pampas of Argentina. The **pectoral sandpiper** that breeds in Arctic tundra migrates to its winter home in South America. The most spectacular of all migrants in Europe is **white stork**. Storks spend summer in Europe but spend the winter in South Africa. Birds of north India and of the base of Himalayas visit the places in the south periodically to avoid the winter and go back in summer.

IV. Altitude of Migration

Formerly, it was thought that birds while migrating travelled very high and with a very fast speed. However, recent knowledge obtained by **telescope**, **radar** and **radio telemetry** has pointed out a great variation. Some birds fly at sea level, some fly very close to the height of Mount Everest. Most birds fly less than 7,400 feet above sea level.

V. Flying Speed during Migration

Flying speeds of migratory birds have been measured by **Doppler radar**. It was found out that the speed ranged from 32 to 64 kmph in small **song birds**, while in larger birds, such as **cranes** speed varies from 40 to 96 kmph. Smaller perching birds can fly 32–59 kmph whereas **falcons**, **ducks** and **geese** fly at the speed of 77–96 kmph, **humming bird** fly 32 kmph and **sandpipers** fly 96 kmph.

VI. Origin of Migration

Several hypotheses have been proposed to explain origin of migration but none is sufficient to explain the phenomenon of migration:

1. Northern ancestral home hypothesis. According to this view, in the prepleistocene period birds inhabited northern hemisphere and were non-migratory. In the Pleistocene, when the glacier forced its way from north to south, the birds migrated to south to save themselves from intense cold. In the following summer when the ice melted, birds returned to their nests in north. Ultimately this practice became hereditary in Lamarckian fashion.

2. Southern ancestral home hypothesis. According to this hypothesis, the bird population in the southern tropics became congested in the past, so that some of the birds migrated to the north for breeding purposes and returned to the south as soon as breeding was over. By moving northward they took advantage of longer days (16 to 18 hrs) which was necessary for gathering food, etc., during the breeding period.

3. Hypothesis of periodic response. It is understood that the tropical birds inhabited colder northern parts in the past where plenty of food was present. But these had to fly to south when the cold increased too much in winter and food became scarce due to hibernation of insects. **Woodburry** explained that by this migration, since the birds survived from the intense cold, it became the habit of the birds and ultimately in course of time, the habit was **naturally selected**.

VII. Causes of Migration

1. Shortage of food supply on the breeding ground. With the end of summer in the northern hemisphere, where the birds breed, food supply falls short due to increase in the number of birds (as a result of breeding) and secondly due to the beginning of hibernation of insects. It causes birds to migrate south.

2. Environmental factors. According to some workers fall and rise in the temperature are responsible for migration. These environmental conditions change the endocrine state and metabolic conditions of the birds in such a way as to cause migration.

3. Internal factors. Some birds migrate due to physiological changes in the gonads and sex hormones which occur with change in the season.

4. Photoperiodism. The effect of length of day or photoperiodism triggers bird migration. The day length affects pituitary and pineal body in the brain which in turn prepare birds for migration. The migratory restlessness is initiated from this stage. For example, in India, increase in day length (starting from February/March onward *induces* the Siberian birds to migrate to their breeding habitats in Siberia. On the contrary, decrease in photoperiod (starting from September-October onward) induces these Siberian birds to migrate to Indian Subcontinent.

5. Fat deposition. According to **Wolfson**, the substantial subcutaneous and visceral fat deposits play an important role in migration of the species. This fat deposition is followed by the development of migratory restlessness (zugunsuhe). Fat provides metabolic water as well as energy in route.

VIII. Navigation

Way-finding or navigation in migratory birds is done with the help of **landmarks** such as great river valleys, coastal lines, chains of oceanic islands and mountain ranges. However, a large number of bird migrants travel at night when they cannot easily make use of landmarks. There it has been suggested that nocturnal (*i.e.*, night fliers) migrants take the help of **constellations of stars**. It is proposed that birds are guided by "tradition" based on experience. [During migration birds seek guidance for one year from those birds who have followed well in the previous years.]

A German ornithologist, **Gustav Kramer**, has shown that migratory birds which travel by day orient themselves by the position of sun (**Sun compass mechanism**). However, the sun compass orientation is an **instinctive behaviour** since young birds that have never migrated before make similar navigational orientations to sun when travelling independent of their parents.

IX. Advantages of Migration

1. Securing a **better climate** for living by avoiding unfavourable climatic conditions (such as intense cold, snowfall, hot summers, stormy conditions, etc.) and food shortage by migrating.
2. By alternately exploiting two different habitats for food due to migration, more birds are able to exist.
3. Geographical wandering or change in habitat provides greater **variety in birds diets**.
4. The long summer days provide birds with **long working hours** to gather food to feed young ones.
5. In their breeding grounds the migratory birds arrive in large numbers hence the **pressure of predation** is divided among a greater number of eggs and young and this results in greater individual survival.
6. Migration provides certain evolutionary benefits. A z, is subjected to different kinds of natural selection pressures in its breeding grounds, its winter quarters and while on journey between two habitats. Indeed migratory birds have a greater range of adaptability than the resident birds.
7. Migration promotes the geographical dispersal of birds. Dispersal may isolate small populations of a species from each other. It tends to increase the rate of evolution (**Isolating mechanisms**).

X. Disadvantages of Migration

Migration is disadvantageous to the birds in the following ways:

1. A journey without rest is tiresome for the birds and most of the passive bird succumb at sea.
2. Sudden changes in weather, *i.e.*, pouring of heavy rain or snow or blowing of stormy wind sometimes perish a number of migrants.
3. The young and defenseless birds are exposed to various natural enemies (predatory birds and hunters).
4. The telegraphic wires, towers, lighthouses, etc., sometimes take a great toll of the migratory birds.

Methods of Study of Bird Migration

Following methods are used to study migration in birds:

1. Banding or ringing. It is most popular method of studying bird migration (Box 16.4). Large scale ringing has enabled ornithologists to plot migratory path or routes of many bird species. There are various methods of banding migratory birds throughout the world. Ringing methods involve the use of coloured metal or plastic leg bands, systematically numbered and having the address of the branding agency to which the recovery of the banded bird may be reported. The birds are first caught or trapped alive in large number by using mist nest or boom-net. This net is thrown into air by explosives and spread out by metal weights which soars over a flock of birds.

The bird banding laboratory at Lawrel and Maryland in USA have maintained more than thirty million banding records and add a million new ones every year. In India, at Keoladeo National Park, Bharatpur extensive research on migratory birds is in progress, especially on their ecology, behaviour and migratory routes. The work is primarily being conducted by **Bombay Natural History Society (BNHS)** in collaboration with State Forest Department. *BNHS is the only authorized institution to carry out bird banding studies in India.* More than 100,000 birds have been ringed in India between 1973–1998.

Box 16.4 Tagging of other animals

For tracking animals during homing and migration, plastic or metallic bands are placed around the legs of birds or around the wing bones of bats. These bands can be colour-coded and may have identifying number on them. The movement of fishes may be followed by capturing them and marking their fins with numbered tags. Sea turtles have been tracked using balloons attached to their shells; the balloons bob along the surface of the water as the turtle swims.

In one of the studies Spanish and Turkestan sparrows ringed at Bharatpur were recovered in Kazakhistan. By using ringing technique it has been possible to establish the migratory routes of birds, *e.g.*, the main migratory routes of birds in India run down the valleys of the Indus in the west and Brahmputra in the east. Ringing at Bharatpur has established that European swallows return not only to same locality but even to same place for nest building purposes every year, covering a distance of thousands of kilometres.

Migratory birds of Keoladeo National Park. A total of 360 species of birds have been recorded from Keoladeo National Park and a great majority of them are migratory. Most birds migrate from Himalayas, Ladakh, Tibet, Central Asia, Siberia and Eastern Europe. Some of the most commonly seen migratory bird species are **greylag goose, pintail, common teal, gadwall, shovellers, tuffed duck, crested pochard, common pochard, mallard, wigeon, common shelduck, marble teal** (Fig. 16.37). In Bharatpur, winter migration of birds starts in the August with the arrival of **gargany teals** and ends in December by the arrival of **Siberian cranes**. The return migration of most bird species starts in March.

2. By collecting dead migratory birds killed at television towers or by airport ceilometers or other human-made high towers and poles interfering with migration. (**Note:** *Ceilometer* is a photoelectric device used with a mercury-vapour lamp to determine the ceiling height of clouds under all weather conditions).

3. By tape-recording or chip-counting. The bird calls through a tape recorder, with the help of a parabolic receiver, microphone and amplifier. This is used to estimate the number of migratory birds by tabulating the call notes heard at night.

4. Radio-tracking. Modern technological methods of radar and radio-tracking have largely superceded our earlier means of observing birds with binoculars. Now individual birds can be spotted on their migratory route, and their course and behaviour can be determined over longer distances with a great deal of precision. **Miniature transmitters**, weighing 2–3 g, have been developed in the USA. Easily carried by small birds, these transmitters send a signal that may then be located by means of stationary direction-finding aerials or by vehicles and aeroplanes. The routes of many migratory birds have been tracked with this method. Researchers are now working on a largely automatic method of tracking birds on their migrations, with the aid of **satellites**.

Fig. 16.36. Migratory birds of Keoladeo National Park, Bharatpur.

5. Radar tracking. With the use of **radar tracking instruments** much knowledge has been gained about what happens to birds on their migration. In contrast with the **flight-monitoring radar instruments** used earlier, this new device detects individual birds and automatically "keeps an eye" on them. When a migratory bird passes the instrument at a greater or lesser angle, the exact range, bearing, and altitudes as well as ongoing changes in these values can be recorded or printed out, and the animal's course of flight relative to the ground can be plotted on a chart. Furthermore, the birds wing-beat frequency can be determined; this identifies or at least indicates the species and ensures that the same individual is still being tracked. Finally, the exact migratory routes of individual birds as well as other bits of information can be recorded from the natural course of events during the flight as they occur within the range of radar instrument.

6. Laboratory methods. There are some aspects of bird migration that we can study in the laboratory, where the animals live in cages or aviaries. Here we can conduct conditioning experiments or we can perform tests on birds in the state of intense activity characteristic of migratory restlessness (zugunruhe). One of the results found with conditioning experiments was that birds use the sun as a compass.

Migratory Restlessnes

In experimenting with birds that are ready to migrate, researchers take advantage of the fact that even with the best care these animals will become restless, jumping and fluttering about in their cage as the time of migration approaches. The activity of such a bird can be recorded by an ingenious method. The bird sits on an ink pad located inside a "funnel" or cone covered with blotting paper. It can look at the sky through the wire-mesh cage cover. While attempting to leave the cone in an upward flight direction, the bird stains the blotting paper with its footprints. The experimenter then assesses the amount of ink staining on the paper. Under certain conditions, activity during migratory restlessness becomes **directional**, usually corresponding to the species-typical direction of migration.

Homing

Homing is a complex instinctive behaviour of migratory fish, birds and other animals. It compels an animal to return to its original locality, after taking migratory journeys. The homing instinct is seen most prominently in the salmon. Fish inhabiting smaller water bodies such as lakes, ponds and pools also have this remarkable instinct and if displayed, they can return to their home territories. **Gerking** (1959) has aptly defined homing as "*the return of animal to a place formerly occupied, instead of going to other equally probable new places*".

If pigeons, for example, are sent by a rail in closed basket to several kilometers away and there set at liberty, they will return to their loft which they can recognize without any error. Thus, they share with the rest of the migratory birds, a certain sense of orientation and homing.

QUESTIONS

Long Answer Questions

1. Describe the distinctive features/general characters of class Aves.
2. Classify the class Aves giving important characters and examples of each group.
3. Give a detailed account of migration in birds.

Short Answer Questions

1. Write down the outline classification of birds.
2. Write short notes on the following:
   (*i*) Barn owl
   (*ii*) Kiwi
   (*iii*) Crow

(*iv*) Koel
(*v*) Parrot
(*vi*) Peacock

3. Write short note on (*a*) Homing; (*b*) Navigation during bird migration.

Very Short Answer Questions

1. Give the name of the smallest flightless bird.
2. Name the subclass of modern birds.
3. Cite one example of Impennae.
4. Name any fossil bird.
5. What is the order of woodpecker (*Picus*)?
6. Give an example of a bird having zygodactylus feet.
7. Name a bird that lays eggs in some other bird's nest.
8. Give an example of aquatic bird with short legs and webbed toes.
9. What are running birds? Name two such birds.
10. What is rhamphotheca?
11. Name the constituent bone of furcula.
12. What is tibiotarsus?
13. What is uricotelism?
14. What is guano?
15. Give the names of two flight muscles of pigeon.
16. State two important functions of air sacs in birds.

Fill in the Blanks

1. Birds are defined as the feathered ________.
2. Birds bear only ________ fingers.
3. Bird body often has feathered tracts called ________ separated by featherless areas termed ________.
4. Avian skull, like that of ________ is monocondylic.
5. Vertebrae have ________ centra in birds.
6. Avian oesophagus is frequently dilated to form ________.
7. The ________ is the sound-producing organs in birds.
8. ________ has sinus rhomboidalis in the lumber region in birds.
9. Bird kidneys are ________ lobed.
10. Excretory matter in birds is ________.
11. In bird's heart is ________.
12. The comb-like structure hanging from retina into the vitreous chamber of eye is ________.
13. ________ is the arrangement or distribution of feathers on the body of bird. (pterylosis/plumage)
14. Birds maintain a body temperature as high as ________ °F. (104/110)
15. Pecten occurs in the eyes of all the birds except in ________. (pigeon/kiwi)
16. The endoskeleton of birds is built on the "________".

Multiple Choice Questions

1. An example of the subclass Archaeornithes is

(*a*) penguin (*b*) heron
(*c*) *Archaeopteryx* (*d*) *Hesperonis*

2. Wings are modfied into swimming paddles in
   (*a*) ducks (*b*) waterfowls
   (*c*) geese (*d*) penguins
3. Which of the following is a bird of prey?
   (*a*) *Pavo* (*b*) *Bubo*
   (*c*) *Eudynamis* (*d*) *Picus*
4. The largest bird is
   (*a*) *Struthio* (ostrich) (*b*) *Aptenodytes* (penguin)
   (*c*) *Phonicopterus* (flamingo) (*d*) *Aepyornis* (giant elephant bird)
5. All flightless bird comprise a group called.
   (*a*) Carinatae (*b*) Impennae
   (*c*) Odontognathae (*d*) Ratitae
6. The smallest bird is
   (*a*) emu (*b*) swift
   (*c*) humming bird (*d*) shrew
7. In which period the birds originated?
   (*a*) coenozoic era (*b*) cretaceous era
   (*c*) jurassic period (*d*) triassic period
8. Homoiothermic animals are
   (*a*) amphibians (*b*) mammals
   (*c*) birds and mammals (*d*) none of these
9. Rectrices are the feathers of
   (*a*) wings (*b*) tail
   (*c*) body (*d*) legs
10. In birds palate is highly variable and forms one of the important feature for
    (*a*) its correlation with feeding habit (*b*) is advanced feature
    (*c*) classification (*d*) identification of different genera
11. Raptorial foot is found in
    (*a*) fast running birds (*b*) predatory birds (birds of prey)
    (*c*) clinging birds (*d*) humming birds
12. Birds are glorified reptiles, this statement was made by
    (*a*) T.H. Huxley (*b*) Charles Darwin
    (*c*) Romer (*d*) Ernst Mayr
13. Which is the national bird of India
    (*a*) peacock (*Pavo cristatus*) (*b*) koel (*Eudynamis scolopaceous*)
    (*c*) red jungle fowl (*Gallus gallus*) (*d*) blue rock pigeon (*Columba livia*)
14. Connecting link between birds and reptiles is
    (*a*) *Dodo* (*b*) *Demetrodon*
    (*c*) *Sphenodon* (*d*) *Archaeopteryx*
15. Which one of the following is a flightless bird?
    (*a*) emu (*b*) kiwi
    (*c*) ostrich (*d*) all of them
16. In birds which one of the following is absent?
    (*a*) gall bladder (*b*) urinary bladder
    (*c*) teeth (*d*) all of the above
17. Which one of the following birds has a subcutaneous fat layer?
    (*a*) penguin (*b*) ostrich
    (*c*) swan (*d*) water fowl

ANSWERS

Very Short Answer Questions

1. *Apteryx* (kiwi).
2. Neornithes.
3. *Aptenodytes* (emperor penguin).
4. *Archaeopteryx*.
5. Piciformes.
6. *Psittacula* (parakeet).
7. *Eudynamis* (cuckoo).
8. Pelicans.
9. Flightless birds are running birds. Examples are ostrich and *Rhea*.
10. Horny sheath of beak.
11. Clavicles and interclavicles.
12. Tibia plus proximal tarsals.
13. Animals that excrete uric acid as nitrogenous waste are called uricotelic animal and this mode of excretion is uricotelism.
14. Guano is a whitish substance excreted by birds. It includes urates and uric acid.
15. Pectoralis major and pectoralis minor.
16. Air sacs provide lightness and accessory respiratory organs.

Fill in the Blanks

1. Bipeds. **2.** Three. **3.** Pterylae, apteria. **4.** Reptiles. **5.** Heterocoelous. **6.** Crop. **7.** Syrinx. **8.** Spinal cord. **9.** Three. **10.** Uric acid. **11.** Four. **12.** Pecten. **13.** Pterylosis. **14.** 104. **15.** Kiwi. **16.** Hollow girder

Multiple Choice Questions

1. (*c*)	**2.** (*d*)	**3.** (*b*)	**4.** (*a*)	**5.** (*d*)	**6.** (*c*)
7. (*c*)	**8.** (*c*)	**9.** (*c*)	**10.** (*c*)	**11.** (*b*)	**12.** (*a*)
13. (*a*)	**14.** (*d*)	**15.** (*d*)	**16.** (*d*)	**17.** (*a*)	

17

Class Mammalia:

General Characters, Classification and Types

The class Mammalia includes mammals. Mammals are warm-blooded and air-breathing animals that suckle their young on milk and have a body covering of hair or fur, prominent external ears, and a mouth armed with teeth. They constitute a wide and heterogenous group of animals distributed throughout the world.

The mammals are unique among the vertebrates in fact among all the animals, in having a coat of hair and in walking on all the four limbs clear of the ground. Hence they are briefly described as the **hairy quadrupeds**. Their class name—the mammalia—refers to milk-producing mammary glands present on the ventral side of the trunk in the females for suckling the young for sometime after birth (L., mamma = breast udder).

Mammals are said to have arisen somewhere in late Permian or early Triassic Period about 230–250 years ago from some cotylosaurian reptile. As they evolved, mammals spread and adapted to many different habitats and now occupy a wide variety of niches, from small flying insect eaters to large terrestrial grazers and from tiny burrowing flesh eaters to the largest of living animals, the plankton feeding aquatic (marine) whales.

The study of mammals is called **mammalogy**. An expert on mammals is called a **mammalogist**.

17.1 GENERAL CHARACTERS

Today, nearly 5500 species of mammals are known which have the following general characters:

1. Body is variously shaped. It is divisible into head, neck, trunk and tail. The neck is flexible. Taii is often long and movable.
2. There are two pairs of pentadactyl limbs positioned underneath the body. These are variously adapted for walking, running, jumping, climbing, burrowing, swimming or flying. Each foot bears 5 toes provided with horny claws, nails or hoofs. Cetaceans and sirenians have no hindlimbs.
3. Mammals show *three* main types of foot posture (Fig. 17.1): (*i*) **Plantigrade**, *i.e.*, walking on entire hands and feet, as in bear and a human being, (*ii*) **Digitigrade**, *i.e.*, walking on the digits, keeping palms and soles raised, as in cat and dog, and (*iii*) **Unguligrade**, *i.e.*, walking on the tips of the digits, as in horse, cow and deer.
4. Skin is thick and water-proof. It is highly glandular and mostly covered by a horny epidermal exoskeleton of **hair**, which conserve body heat like avian feathers. Hair might occur all over the body or restricted to particular areas.
5. Skin has abundant multicellular glands such as sudoriferous or sweat glands, sebaceous or oil glands, ceruminous or wax glands in ears, meibomian (or tarsal) glands in eyelids in the both sexes and mammary glands mostly in females. Many mammals have scent glands also.

   The sweat produced by the sweat glands cools the body through the evaporation if it gets overheated.

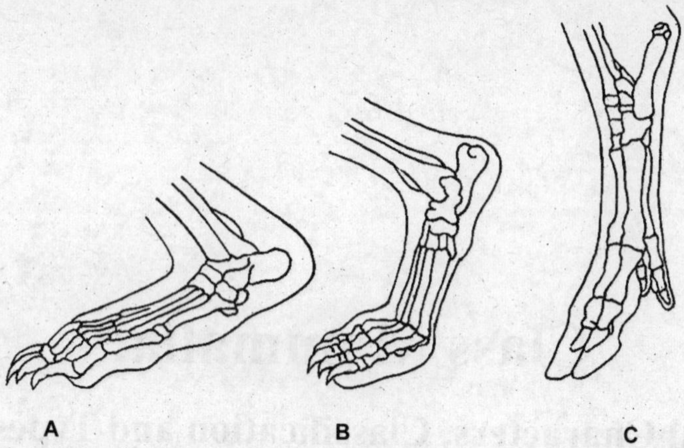

Fig. 17.1. Three foot postures in mammals: A—Plantigrade, bear; B—Digitigrade, dog; C—Unguligrade, horse.

6. The tail of opossum and many rodents (rats) is covered by epidermal scales similar to those of lizards, but lacking the hard **beta keratin** found in reptiles and birds.
7. Mammals contain special muscles for facial expression. They are the only animals that can scowl (*i.e.*, bad-tempered look), sneer (*i.e.*, ridicule) and smile.
8. The endoskeleton is fully ossified. Skull has two occipital condyles (dicondylic).
9. Each half of lower jaw consists of a single bone, the dentary and articulates with squamosal instead of quadrate, forming craniostylic jaw suspension.
10. A complete secondary palate has been formed by the fusion of processes from the premaxillae, maxillae and palatines. This fully separates the respiratory passages from the food passages.
11. Teeth are generally **heterodont**, *i.e.*, differentiated into incisors, canines, premolars and molars; **thecodont**, *i.e.*, embedded in sockets of jaw bones; and **diphyodont**, *i.e.*, grow in only two sets: milk set replaced by permanent set.
12. The long bones contain diaphyses and epiphyses, that help in the growth of bones. The ribs are bicephalous, *i.e.*, contain two heads. Cartilaginous discs (or intervertebral discs) lie between the centra of adjacent vertebrae. There are seven cervical vertebrae in nearly all mammals (6 or 9 in sloths or 6 in sirenians). The pelvic girdle is strengthened by the complete fusion of the bones.
13. Coelom is completely divided into anterior smaller thoracic cavity and posterior larger abdominal cavity by a muscular partition, the **diaphragm**. Diaphragm tends to increase efficiency of lungs in respiration.
14. Mouth has movable lips. Tongue is usually mobile. Intestine has villi which help in quicker absorption of digested food. Mammals are the only vertebrates which masticate the food and swallow a discrete bolus of food. A special adipose tissue, called brown fat in present; it has numerous mitochondria and is meant for rapid generation of heat.
15. Respiration occurs only by lungs which are lobed, spongy and elastic organs. Lungs are enclosed in special pleural cavities. Opening of trachea into the buccal cavity is protected by epiglottis.
16. Heart is 4-chambered, having two auricles and two ventricles as in birds. It keeps oxygenated and deoxygenated blood completely separate. The sinus venosus absent. Circulation is double: pulmonary and systemic. Only the left systemic arch present. Renal portal system is absent. Blood contains platelets. Red blood corpuscles are circlar, biconcave and without nucleus in most cases.
17. Brain is highly developed. Cerebrum is large, lobed and with highly developed cerebral cortex. Cerebellum is highly developed. Corpus callosum (The great band of commissural fibers uniting the cerebral hemispheres of the brain) present. Optic lobes four in number.
18. Eyes have movable lids. Humans, apes, monkeys and cats have colour vision.

External ear has fleshy pinna. Internal ear with a long and spirally coiled cochlea. Ear with three bony ossicles: malleus, incus and stapes.

19. Kidneys are metanephric and bean shaped. Excretory matter is urea (ureotelism). Urine is fluid.
20. Sexes are separate. Fertilization is internal. No larval stage exists.
21. Males with a distinct copulatory organ, the penis, and extra-abdominal testes, contained in the scrotal sacs.
22. Eggs mature in the Graafian follicles of the ovary.
23. Anus and urinogenital apetures are separate. There is no cloaca.
24. Mammals are mostly viviparous (except prototheria). Development of egg occurs in the uterus. Embryo has the extra-embryonic membranes such as amnion, chorion and allantois. The embryo is attached to the uterine wall by placenta. (No placenta in monotremes and most marsupials).
25. Exhibit postnatal care.
26. Body temperature is regulated (homoiothermy) like that of birds. This gives the mammals a high rate of metabolism and makes them active throughout the year like the birds. Mammals, like the birds, are endothermic (Box 17.1).

Box 17.1

Mammals maintain a constant and stable body temperature (except during hibernation). They generate body heat by muscular contraction including shivering and by metabolising **brown fat**. Heat is lost by sweating and dilation of dermal blood vessels. Heat is retained in the body by a covering of fur, layer of subcutaneous white fat and by counter current heat exchange systems exist in the extremities.

27. Mammals are **primary terrestrial** vertebrates. They occur in all sorts of habitats from the polar regions to the tropics including the densest forests and driest deserts. Some mammals, such as whalls, have **secondarily** become aquatic and live in the sea. Mammals show a great diversity in the habits and include fossorial (burrowing), **saltatorial** (leaping), **cursorial** (running), **scansorial** (climbing), **volant** (flying), **natatorial** (aquatic) and cave-dwelling forms. In present time, mammals are dominant animals and have the capacity to learn because of better developed brian.

17.2 CLASSIFICATION

Simpson (1945, 1975) based his classification of mammals on the mode of parental care, dentition, foot posture, type of exoskeleton, complexity of nervous system and phylogenetic relationships.

Outline Classification

Class-Mammalia

Subclasses (2)

I. Prototheria

Order 1. Monotremata

Example: 1. *Tachyglossus* or *Echidna* (Spiny ant-eater)
2. *Ornithorhynchus* (Platypus)

II. Theria

Infraclasses (3)

1. Pantotheria (Extinct)

2. Metatheria

Order 1. Marsupialia

Example: 1. *Didelphis* (Opossum)
2. *Petaurus* (Flying opossum)
3. *Phalanger* (Cuscus)
4. *Dasyurus* (Tasmanian tiger cat)
5. *Caenolestes*
6. *Notoryctes* (Marsupial mole)
7. *Peramels* (Bandicoot)
8. *Macropus* (Kangaroo)
9. *Phascolarctos* (Koala)
10. *Phascolomys* (Wombat)
11. *Thylacinus* (Marsupial wolf)
12. *Myrmecobius* (Banded anteater)
13. Wallaby

3. Eutheria

Infraclass 3. Eutheria (Placentalia)

Orders (16)

1. Insectivora
Example:
1. *Paraechinus micropus* (Hedgehog)
2. *Talpa* (Mole)
3. *Echinosorex* (Shrew)
4. *Crocidura murina* (Musk shrew or Chhachhunder)
5. Tenrec
6. Solenodon

2. Dermoptera
Example:
1. *Galeopithecus* (Flying lemur)
2. *Cynocephalus volans* (Flying lemur or colugo)

3. Chiroptera

Suborders (2)

1. Megachiroptera
Example:
1. *Pteropus kuhli* (Flying fox)
2. *Xanthrpyia*
3. *Cyanopterus*

2. Microchiroptera
Example:
1. *Desmodus* (Vampire bat of South America)
2. *Megaderma* (Indian vampire)
3. *Scotophilus medius* (Small Indian bat)
4. *Myotis* (Brown bat)
5. *Rhinolophus*
6. *Eptesicus*
7. *Vespertilio*

Order 4. Primates

Suborder 1. Lemuroidea. Example: 1. *Lemur* (Lemur)
2. Bush baby
3. Potto
4. Indris
5. Chirogale
6. *Chiromys* (Aye aye)
7. *Loris tardigradus* (Slender Loris)

Suborder 2. Tarsioidea. Example: *Tarsus spectrum* (Tarsus)

Suborder 3. Anthropoidea:

Infraorder 1. Platyrrhina.

Example: (New world monkeys)
1. *Ateles* (Spider monkey),
2. *Alouatta* (Howling monkey),
3. Marmoset

Infraorder 2. Catarrhina

Example: I. Old world monkeys.
1. *Macaca mulatta* (Rhesus monkey)
2. *Semnopithecus* (*Presbytis*) *entellus* (Langur)
3. *Papio* (Baboon)
4. *Mandrillus* (Mandril)

II. Apes. 1. *Hylobates hoolock* (Gibbon)
2. *Simia* (*Pongo*) *pygmaeus* (Orangutan)
3. *Pan satyrus* (Chimpanzee)
4. *Gorilla gorilla* (Gorilla)
5. *Homo sapiens* (Human beings)

Order 5. Edentata (Xenarthra)

Example: 1. *Dasypus novemcinctus* (Armadillo)
2. *Bradypus tridactylus* (Three-toed sloth)
3. *Myrmecophaga jubata* (South American Giant ant-eater)

Order 6. Pholidota. Example: *Manis* (Pangolin or Scaly ant-eater)

Order 7. Lagomorpha. Example: 1. *Oryctolagus cuniculus* (Rabbit)
2. *Lepus ruficaudatus* (Hare)
3. *Ochotona/Lagomys* (Pika)

Order 8. Rodentia. Example: 1. *Rattus rattus* (House rat); 2. *Mus musculus* (House mouse); 3. *Cavia* (Guinea pig); 4. *Funambulus palmarum* (Stripped squirrel); 5. *Hystrix leucura* (Porcupine); 6. *Castor* (Beaver); 7. *Petaurista petaurista* (Giant flying squirrel); 8. *Cynomys* (Prairie dog); 9. Hamster; 10. Jerboa; 11. Capybara.

Order 9. Cetacea: (2 Suborders)

Suborder 1. Mysticeti (Whalebone whales)

Example: 1. *Balaenoptera indica* (Indian whalebone whale); 2. *Balaena* (Arctic whale or Right whale); 3. *Delphinus delphis* (Common dolphin); 4. *Phocaena* (Porpoise); 5. *Orcinus* (Killer whale)

Suborder 2. Odontoceti (Toothed whale)

Example: 1. *Physeter catodon* (Sperm whale); 2. *Platanista gangeticus* (Ganges dolphin); 3. *Delphinus delphis* (Common dolphin); 4. *Phocaena* (Porpoise); 5. *Orcinus* (Killer whale)

Order 10. Carnivora: (2 Suborder)

Suborder 1. Fissipedia

Example: 1. *Felis domesticus* (House cat); *Felis chaus* (Indian jungle cat); 3. *Panthera leo* (Lion); 4. *Panthera tigris* (Tiger); 5. *Acinonyx* (Cheetah); 6. *Prionailurus bengalensis* (Leopard cat); 7. *Panthera pardus* (Leopard/Panther); 8. *Canis aureus* (Wolf); 9. *Canis familiaris* (Dog); 10. *Canis aureus* (Jackal); 11. *Ursus torquatus* (Himalayan black bear); 12. *Vulpes bengalensis* (Indian fox); 13. *Hyaena striata* (Striped hyaena); 14. *Lutra lutra* (Indian otter); 15. *Aelurus fulgens* (Panda); 16. *Ailuropoda* (Panda); 17. *Herpestes* (Mongoose); 18. *Meles* (Badger); 19. *Procyon* (Racoon); 20. *Mustela* (Mink); 21. *Mephites* (Skunk); 22. *Ailurus fulgens* (Red Panda found in Sikkim)

Suborder 2. Pinnipedia

Example: 1. *Otaria jubata* (Sea-lion); 2. *Otario stelleri* (Sea lion); 3. *Odobenus* (Walrus); 4. *Trichechus* (Walrus); 5. *Phoca vitulina* (Seal)

Order 11. Tubulidentata

Example: *Orycteropus afer* (Aardvark)

Order 12. Proboscidea

Example: 1. *Elephas indicus* (Indian elephant); 2. *Loxodonta africana* (African elephant); 3. *Elephas cyclotis* (African pygmy elephant)

Order 13. Hyracoidea

Example: 1. Hyrax, *Procavia* (Conies of South Africa)

Order 14. Sirenia

Example: 1. *Halicore* (Dugong); 2. *Trichechus* (*Manatus*) (Manatee)

Order 15. Perissodactyla

Example: 1. *Rhinoceros indicus* (or *Rhinoceros unicornis*) (Indian or one-horned rhinoceros); 2. *Equus cabalus* (Horse); 3. *Equus vulgaris* (Indian donkey); 4. *Equus zebra* (African zebra); 5. *Tapirus* (*Tapir*)

Order 16. Artiodactyla: (3 Suborder)

Suborder 1. Suiformes

Example: 1. *Sus cristatus* (Indian crested wild boar); 2. *Hippopotamus amphibius* (Hippopotamus)

Suborder 2. Tylopoda

Examples: 1. *Camelus dromedarius* (One humped or Arabian camel); 2. *Camelus bactrianus* (Two-humped or turkish or Bacterian camel); 3. *Llama vicugna* (Llama)

Suborder 3. Ruminantia

Example: 1. *Capra* (Goat); 2. *Ovis* (Sheep); 3. *Antelope* (Antelope); 4. *Bos indicus* (ox/cow); 5. *Bubalus bubalus* (buffalo); 6. *Bison bison* (Bison); 7. *Cervus axis* (Spotted deer or Chital); 8. *Giraffa camelopardalis* (Giraffe); 9. *Okapia* (Okapi); 10. *Bos grunniens* (Yak); 11. Nilgai (*Boselamphus tragocamelus*)

17.3 DETAILED CLASSIFICATION

Class Mammalia is divided into two subclasses: 1. Prototheria and 2. Theria.

Subclass 1. Prototheria

1. It includes primitive egg-laying mammals.
2. This subclass is represented by a few living representatives such as *Tachyglossus* (*Echidna*) (Spiny anteater) and *Ornithorhynchus* (duck-billed platypus). Both of them found exclusively in Australia.

3. Mammary glands present but are without teats (nipples). The glands open in a depression on the skin in which milk is collected and is sucked up by the young ones.
4. Body is covered with a dense coat of hair. Some hair on the back are modified into spines in *Tachyglossus*.
5. Jaws form a flat, duck-like beak in *Ornithorhynchus* and a hard rostrum in *Tachyglossus*.
6. Limbs are short and bear strong claws on the digits. The digits are webbed in duck-billed platypus.
7. The males have a horny **spur** on each ankle (of hindleg) bearing an opening of poison gland (Fig. 17.9B).
8. Corpus callosum either absent or poorly developed.
9. Coracoid and precoracoid bones of the pectoral girdle well developed.
10. Epiphyses either absent or poorly developed.
11. Ear pinna is absent and cochlea is less coiled.
12. Cloaca present.
13. Testes in males are abdominal and the oviducts in female remain separate throughout, *i.e.*, vagina and uterus are absent.
14. Eggs are large, yolk, and shelled. There is no placenta.
15. Imperfectly warm-blooded. They hibernate in winter.
16. They are nocturnal, fossorial and carnivorous.

The subclass Prototheria has a single order named **Monotremata** (Gr., *mono* = one + *trema* = hole). It has all characters of the class. It includes following two families:

Family 1. Ornithorhynchidae. It has a single genus *Ornithorhynchus*.

Family 2. Tachyglossidae. It has two close allied genera: *Tachyglossus* (*Echidna*) and *Zaglossus* (Fig. 17.9A).

Subclass 2. Theria

1. This subclass includes modern young-producing (viviparous) mammals.
2. Ear usually has a pinna.
3. Brain is mostly with corpus callosum.
4. Vertebrae generally bear epiphyses.
5. Coracoid is reduced to a process of scapula.
6. Testes commonly descend into scrotal sacs.
7. Cloaca is lacking. Digestive and urinogenital systems open out by separate apertures.
8. Each oviduct is differentiated into anterior Fallopian tube and posterior uterus. Uteri join to form a vagina.
9. Teeth usually occur in both the young and the adults.
10. Mammary glands open on teats.
11. Eggs are small with little yolk and lack shell. They develop in the uterus and living young are produced.
12. Penis is used for the passage of both urine and sperms.

The subclass Theria is divided into two infraclasses: 1. Metatheria (Marsupialia) and 2. Eutheria (Placentalia).

Infraclass 1. Metatheria

1. Primitive mammals in which young ones are born in an extremely premature condition and undergo further development in an ventral abdominal pouch, the **marsupium**.
2. Marsupium occurs only in females and encloses the nipples and opens to outside either forward (in Kangaroos) or backward (in opossums and Tasmanian wolf).
3. Helpless youngs are transferred by the mother to the marsupial pouch. Here each young is attached to a nipple by its mouth and remains there as "**mammary foetus**" till fully formed.

The young is incapable of sucking, and the milk is forced down its throat by the contraction of maternal muscles. When fully developed, the young ones leave the marsupium. Thereafter, they return to the marsupium for shelter.

4. Body is covered with fur. Pinnae are present. Hindlimbs are often longer thant the forelimbs. Tail is usually long. It may be prehensile or act as a balancing organ while running.
5. Herbivores often lack canines. Teeth monophyodont, except last molars which have milk predecessors.
6. In brain, olfactory bulbs are relatively large. Cerebellum is small.
7. Testes descend into scrotum, which lies in front of the penis. Glans penis is generally bifurcated.
8. They contain two uteri and two vagina.
9. Metatheria are viviparous. Majority form simple yolk sac placenta. some *e.g.*, *Permeles* form true allantoic placenta.
10. The Metatheria arose in Cretaceous period. At present, majority of them are found in the Australian region except New Zealand and a few occur in North and South America.
11. Kangaroo (*Macropus*) is terrestrial. Marsupial mole (*Notoryctes*), bandicoot and wombat are fossorial (adapted for digging). American opossum and koala are arboreal. Flying opposum is volant. Water opossum is aquatic.
12. Kangaroo and cuscus are herbivorous. Flying opossum is insectivorous. Marsupial wolf and Tasmanian tiger cat are carnivorous. Banded anteater feeds on ants. Wombats takes roots. *Tarsipes* extracts honey from the flowers with its extensile tongue.

    Examples: The infraclass Metatheria includes a single order **Marsupialia**, containing opossum, kangaroo, bandicoot, koala, etc.

Infraclass 2. Eutheria (Placentalia)

1. These are true mammals having a true **allantoic placenta**. They give birth the fully developed young ones.
2. External ear has pinna in most cases.
3. Mammary glands open on teats (nipples).
4. Testes usually descend into scrotal sacs which are at the sides of or behind the penis.
5. There is no marsupial pouch.
6. Dentition is heterodont and diphyodont.
7. Dental formula is $\frac{3}{3}, \frac{1}{1}, \frac{4}{4}, \frac{3}{3} \times 2 = 44$.
8. Pefectly warm blooded. Body temperature is regulated (**homeothermic**). It varies from 35 to 40°C.
9. Corpus callosum is well developed.
10. Coracoids vestigial.
11. Epiphyses develop at the ends of long bones and in between the vertebrae.
12. Cloaca absent. Anus and urinogenital apertures are separate.
13. Vagina is single and the two uteri are almost united.
14. Eggs are small and yolkless. Embryo develops inside the uterus and is nourished by a true allantoic placenta.

    The infraclass Eutheria includes the following 16 orders: 1. Insectivora; 2. Dermoptera; 3. Chiroptera; 4. Primates; 5. Edendata; 6. Pholidota; 7. Lagomorpha; 8. Rodentia; 9. Cetacea; 10. Carnivora; 11. Tubulidentata; 12. Proboscidea; 13. Proboscidea; 14. Sirenia; 15. Perissodactyla; 16. Artiodactyla.

Order 1. Insectivora

1. Include insect-eating nocturnal mammals such as shrews, hedgehogs, tenrecs, moles.

2. Small sharp-snouted animals with primitive characters that spent a great part of their lives underground.
3. Teeth adapted for insectivorous diet. Dental formula: $\frac{3}{3}, \frac{1}{1}, \frac{4}{4}, \frac{3}{3} \times 2 = 44$.
4. Skin bears soft fur in most cases.
5. Mammae are numerous and occur all along the ventral surface from the axillary to the inguinal region.
6. Scrotum is lacking. Testes abdominal. Uterus bicornuate, *i.e.*, uteri are over halfway fused (Fig. 17.2).

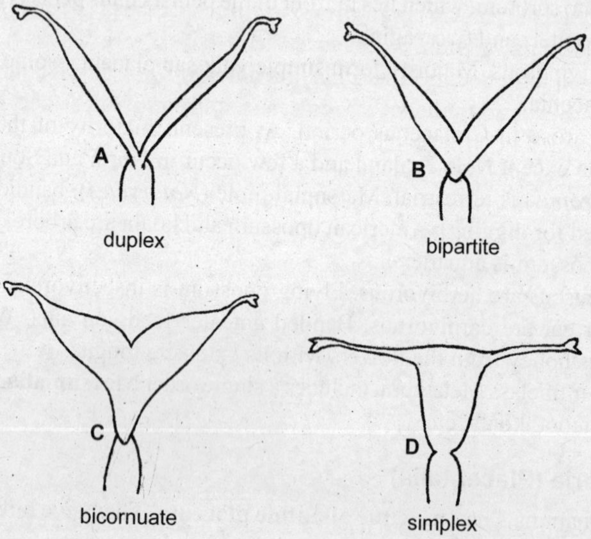

Fig. 17.2. Four types of uterus in Eutheria.

7. Skull is elongated and tubular with cranial and facial regions in a straight line, and with a small cranium. Palate often has vacuities.
8. Tympanic bone has the form of an incomplete ring. It does not form tympanic bulla so that tympanic cavity is open.
9. In the brain, olfactory bulbs are large; cerebral hemispheres are small and smooth. Corpus callosum is poorly formed.
10. The insectivora have worldwide distribution, except the Australian region and New Zealand.
11. Many are **cursorial** (fast running), some **fossorial** (burrowing, *e.g.*, moles), a few arboreal (Tree shrews) and a few are aquatic.

Example: 1. *Hemiechnus* (*Erinaceous*) *collaris* (Hedgehog found in Punjab); 2. *Paraechinum micropus* (Hedgehog found in South India); 3. *Talpa micrura* (short tailed mole); 4. *Crocidura murina* (Musk shrew); 5. *Tupaia* (Tree shrew).

Order 2. Dermoptera

1. It includes flying lemurs which are related to true bats. They are not lemurs (which are primates) and cannot fly in the strict sense of the word, but glide.
2. They are found in Malay peninsula in the East Indies.
3. Nocturnal and frugivorous.
4. A thin, wide, hairy fold of skin, the **patagium**, extends along either side of the body from the neck to the tip of the tail, enclosing both the limbs in it.
5. All the four limbs are equal in size. The digits are flattened. The soles of the paws form sucking discs for adhesion.
6. Teeth are sharp. Dental formula: $\frac{2}{3}, \frac{1}{1}, \frac{2}{2}, \frac{3}{1} \times 2 = 30$.

7. Contains only two species.
   Example: 1. *Galeopithecus* (flying lemur); 2. *Cynocephalus volans* (colugo or flying lemur).

Order 3. Chiroptera

1. It includes true flying mammals with a pair of wings. Chiroptera have true flight by flapping of wings.
2. Wings are in the form of a pair of folds of integument extending between the second digit of the forelimbs and tail. Feet are included in the patagium. Tail may be present (in small bats).
3. Forelimbs are much elongated and the metacarpals and proximal phalanges of second, third, fourth and fifth digits enormously elongated to support the patagium or wing.
4. Hindlimbs are short and weak with backwardly directed knee and five clawed digits. Walking of bats is difficult and awkward.
5. Bones are light, sternum is keeled and flight muscles are well-developed.
6. Senses of touch, hearing and smell are highly developed.
7. Snout may be short or elongated. Some bats have highly sensitive **nose leaves** around the nares.
8. Ear pinnae are well developed and sense of hearing and touch acute.
9. Eyes are small and have weak vision.
10. Bats posses echo-location apparatus (sonar and radar system) which enables them to locate the position of objects or structures in the way (Box 17.2).

Box 17.2

Bats produce a shrill ultrasonic squeak which is echoed back from the objects in the way and is picked by nose leaves and pinnae. Some large bats lack ecolocation capacity.

11. Teeth are diphyodont. Dental formula is $\frac{2133}{2133}$.
12. Testes are abdominal and uterus is bicornuate. Testes descend into the perineum during breeding season.
13. Nocturnal, gregarious, insectivorous (Microchiroptera) or frugivorous (Megachiroptera).
14. A single young is usually born at a time. It is carried by the mother on the belly.

Order Chiroptera is divided into two suborder: **1. Microchiroptera. Examples.** *Desmodus* (vampire bat), *Scotophilus medius* of India.

2. Megachiroptera. Examples. *Pteropus kuhli* (Fruit bat or flying fox of India).

Order 5. Edentata (Xenarthra)

1. The Edentata (Xenarthra) are known since the Eocine.
2. This order includes three types of South American mammals such as the armadillos, the ant-eaters and the sloths (Fig. 17.3). They show a lot of variation in habits and morphology.
3. The **armadillos** are nocturnal with a short, broad and depressed head and a long protrusible tongue for catching insects. Body covered dorsally by epidermal horny plates and ventrally by hairs. Tail and limbs with bony scutes. Molars present.
4. The **ant-eaters** (*i.e.*, arboreal silky ant-eaters and ground-based giant anteaters) have a long and narrow snout. Teeth are absent and tongue is long and sticky. Body is coverd with coarse hairs. Prehensile tail is found in silky anteaters.
5. The **sloths** are soft-skinned, slow-moving, arboreal creatures with short and rounded face. Pinnae are small. Molars present. They feed on leaves, fruits and birds.
6. These three types of animals of Edentata have many common features.
7. The brain case is low, the brain is small and the olfactory region is large.
8. Dentition is incomplete. Some lacks canines and incisors. The teeth are altogether absent in ant-eaters or represented only by molars which lack enamel and roots.

9. The lumber vertebrae have extra articulations (name of the order, the Edentata, means extra-jointed).
10. There is union of ischium and caudal vertebtrae.
11. Feet have well developed claws. Big and sharp claws of front feet of giant ant-eaters are used in breaking anthills and termite mounds and in defense. Sloths cling to tree branches upside down by means of long and curved claws. Claws of front feet of the armadillos are used in digging. Giant anteaters tend to walk on their knuckles.

Example: *Dasypus novemcinctus* (Armadillos), *Bradypus tridactylus* (Three-toed sloth), *Myrmecophaga jubata* (South American giant ant-eater), *Choloepus* (Ant-eater), *Tamandua* (Ant-eater).

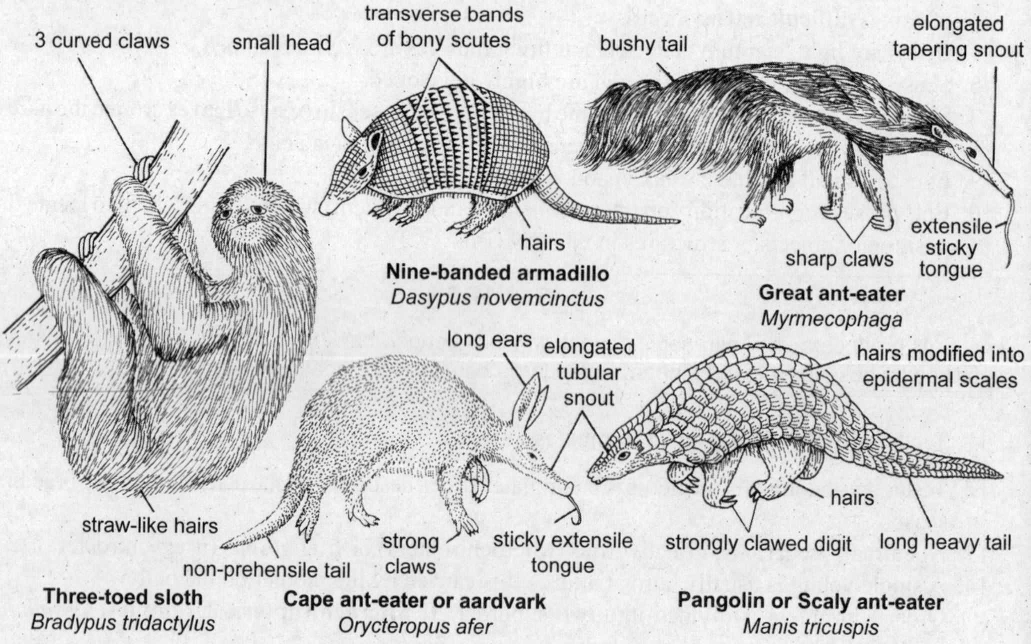

Fig. 17.3. Ant-eaters (Mammals).

Order 4. Primates

1. Pentadactyl limbs with 5 fingers and 5 toes. Long and flexible arms and legs. Most have opposable thumbs to grasp branches of trees or tools. First digits of hindlegs are opposable.
2. Plantigrade locomotion. Large feet for bipedal locomotion.
3. Hair covers the entire body except palms, soles and parts of the face.
4. There are 4 incisors in each jaw. Premolars and molars are bunodont (*i.e.*, with separate cusps).
5. The eyes are directed forward instead of sideways to give stereoscopic (binocular) vision for precise judgement of distances, a necessity for arboreal (= tree) life.
6. Neck is very mobile so that eyes may see in all directions.
7. Their hand, sense organs and brain are best developed among all mammals.
8. To accommodate the large brain, the cranium is large and rounded with foramen magnum on the lower side.
9. There is a single pair of thoracic mammae.
10. Testes descend into scrotum.
11. Penis is pendent.

12. Uterus is simplex, *i.e.*, complete fusion of uteri (Fig. 17.2D).
13. Usually a single young one is born at a time. It is brought up with a good deal of care and affection from the mother.
14. The primates are primarily arboreal mammals. Some have become secondarily terrestrial.
15. They are omnivorous and gregarious.

The order Primates (Fig. 17.4) is divided into 3 suborders: 1. Lemuroidea or Lemuriformes; 2. Tarsioidea or Tarsiformes and 3. Anthropoidea or Pithecoidea.

Box 17.3

G.G. Simpson (1945) has divided primates into two suborders: 1. **Prosimii**, *e.g.*, lemurs, bush babies, tarsiers, lorises and pottos; 2. **Anthropoidea**, *e.g.*, monkeys, gibbons, apes and human beings.

Suborder 1. Lemuroidea

1. It includes lemurs, lorises, bush baby and potto.
2. They are primitive primates and resemble the tree shrews with which they seem to have a common origin.
3. They have claws on second toes but have nails on others.
4. Some retain long snout and laterally directed eye.
5. Hindlimbs are larger than the forelimbs.
6. Fourth digit is longest.
7. Lips are joined to the gums and cannot be protruded.
8. Brain has large olfactory lobes. Cranium is small. Orbits are not completely separated from the temporal fossae behind.
9. Tail, if present, is not prehensile.
10. Cheek pouches are lacking.
11. Uterus is bicornuate. Placenta is diffuse and indeciduate. Allantois is large and filled with fluid.

Example: 1. *Lemur* (Lemur, Box 17.4); 2. *Loris tardigradus* (Slender loris); 3. *Nycticebus* (Indian lemur).

Box 17.4

Lemurs are nocturnal, arboreal, omnivorous, plantigrade primates of Madagascar. These forest dwellers have long tail, fox-like faces and scent glands on the shoulder region and wrists.

Ring-tailed lemur baby clings to its mother's back for months after it is born. These lemurs communicate with a variety of hoots. They also send messages with scent. Males often peforms "**stink fights**" to drive away another male. He first rubs his bushy tail on the smelly glands under his arms, then he waves the odorous tail in other male's face (see Time Life Mammals, 1997).

Suborder 2. Tarsioidea

1. It includes tarsiers which are intermediate between the lemuroids and the anthropoids.
2. Similar to lemuroids, some digits have claws; the hindlimbs are longer than the forelimbs, tail is long and uterus is bicornuate.
3. In brain, the cerebral hemispheres are large and cover the cerebellum and midbrain. The cranium is large.
4. The snout is short, the eyes are large and forwardly directed and protruding. Retina has rods only, no cones.
5. Pinnae are large. Neck is very mobile.
6. The placenta is discoidal and deciduate and the allantois is much reduced as in the anthropoids.
7. The orbits are largely separated from the temporal fossae.

Fig. 17.4. Some primate mammals.

8. The tarsi (Calcaneum and navicular bones) are very long.
9. Arboreal, nocturnal, insectivorous primate.
   Example: *Tarsius spectrum* (Tarsier; Fig. 17.4).

Suborder 3. Anthropoidea

1. It includes the monkeys, apes and human beings.
2. The digits have flat nails except in the marmosets which have a nail on the hallux only and claws on all other digits.
3. The head is rounded. The face is almost flat and has muscles for emotional expression.
4. There is no moist, naked rhinarium (Rhinal means nasal). The nostrils are completely bordered by naked skin.
5. The lips are free and protrusible.
6. The eyes are large and face forward, having binocular vision. The pinnae are small with their edges bent inward. Tympanic bulla and bony auditory meatus are formed.
7. The forelimbs are generally longer than the hindlimbs. The pollex and hallux are opposable, except the human hallux.
8. The tail is present in monkeys, absent in apes and human beings.
9. The cranium is spacius to accommodate the large brain.
10. Brain is relatively much larger than in lemurs or tarsiers; the olfactory lobes are small and the cerebral hemispheres are very large, have greatly convoluted surface and completely cover the cerebellum.
11. There is a pair of pectoral mammae.
12. The uterus is single, devoid of cornua.
13. The placenta is discoidal and deciduate. The allantois is much reduced.
14. The dental formula is $\frac{212 \text{ or } 33}{212 \text{ or } 33} \times 2 = 32$ or 36.
15. The anthropoids are arboreal or terrestrial, mostly diurnal, omnivorous primates.
16. Locomotion is generally quadripedal and plantigrade. There is a tendency to walk erect.
17. They lead a social life based on sight instead of smell as in other mammals.
18. The anthropoids arose in the oligocene and are flourishing since then. They mostly occur in warmer regions of both Old and New World.

The order Anthropoidea includes two **infraorders:** 1. Platyrrhina and 2. Catarrhina. The main differences between these two infraorders are given in the Table 17.1.

Table 17.1 Differences between infraorders: Platyrrhina and Catarrhina.

Infraorder-Platyrrhina (New World Monkeys)		Infraorder-Catarrhina (Old World Monkeys)	
1.	Arboreal.	1.	Arboreal or terrestrial.
2.	Comparatively smaller in size.	2.	Relatively larger in size.
3.	Nose is flat, internasal septum is wide, nostrils are far apart and face forward.	3.	Nose is raised, internasal septum is narrow, nostrils are close together and face downward.
4.	Cheek pouches are absent.	4.	Cheek pouches are present, except apes and human beings.
5.	Pollex is reduced and not opposable.	5.	Pollex is opposable.
6.	Ischial callosities (*i.e.*, thick, bare, coloured pads on buttocks) are lacking.	6.	Red ischial callosities are present, except apes and human beings.
7.	Tail is long and may be prehensile or non-prehensile.	7.	Tail, if present, is not prehensile.

Infraorder-Platyrrhina (New World Monkeys)		Infraorder-Catarrhina (Old World Monkeys)	
8.	There is no bony external auditory meatus.	8.	Bony external auditory meatus is present.
9.	Jugal meets the parietal.	9.	Jugal does not meet the parietal.
10.	There are 3 premolars in each half of the upper and lower jaws. Dentition is $\frac{2132}{2132} \times 2 = 32$ or $\frac{2133}{2133} \times 2 = 36$.	10.	There are two premolars in each half of the upper and lower jaws. Dentition is $\frac{2123}{2123} \times 2 = 32$.
11.	Found in Central and South America.	11.	Found in Asia and Africa.
12.	Includes two families: 1. Hapalidae (Marmosets); 2. Cebidae (Capuchin, Spider monkey, Howler monkey).	12.	Includes three families: 1. Cercopithecidae (Old World monkeys); 2. Simiidae (Apes); 3. Hominidae (Humans).

I. New World Monkeys (Infraorder Platyrrhina)

New World monkeys of Infraorder Platyrrhina belong to following two families: Hapalidae and Cebidae.

Family 1. Hapalidae

1. It includes small, squirrel-sized primates called **marmosets**. They are arboreal, frugivorous or insectivorous.
2. They bear claws on all digits except the hallux, which has a nail. The claws help them in climbing.
3. Pinnae are large and hairy.
4. Forelimbs are shorter thant the hindlimbs.
5. Pollex is long and not opposable.
6. Tail is long, bushy and non-prehensile.
7. Dental formula is 2132/2132 × 2 = 32.
8. Cerebrum is large but non-convoluted.
9. They produce 2 or 3 offspring.

   Example: *Callithrix* (Marmoset).

Family 2. Cebidae

1. Arboreal and frugivorous primates having long prehensile tails which are used in grasping a tree branch or a fruit.
2. All digits in both limbs have nails.
3. Pinnae are small and non-hairy.
4. Limbs are long. Both pollex and hallux are opposable.
5. Dental formula is 2133/2133 × 2 = 36.

   Example: *Cebus* (Capuchin), *Ateles* (Spider monkey), *Aloutta* (Howler monkey).

II. Old World Monkeys (Infraorder Catarrhina)

In real sense old world monkeys belong to family **Cercopithecidae** of infraorder catarrhina. Other two families are **Simiidae** having the apes (gibbon, orangutan, chimpanzee and gorilla) and **Hominidae** having human beings.

Family 1. Old World Monkeys (Family Cercopithecidae)

1. They include baboons and mandrils of Africa and macaques and langurs of Asia.
2. They are arboreal or terrestrial primates, having quadrupedal plantigrade gait and omnivorous or herbivorous diet.
3. Muzzle is long and like that of a dog.
4. Arms and legs are nearly of equal length.
5. Hands are smaller than feet.
6. All digits of both the limbs bear nails. Thumbs are opposable.
7. Tail is long and non-prehensile. It may be short or absent.
8. Dental formula is 2123/2123 × 2 = 32. Canines are large.
9. External nares close together. Many have internal cheek pouches.
10. Stomach is simple or sacculated. Caecum is small and lacks vermiform appendix.
11. There is monthly menstrual bleeding.
12. Placenta is deciduata and haemochorial.
13. Some parts of body such as nose, cheecks and rumps (buttocks) may be brightly colored.
14. There are calloused ischial tubersities on the buttoucks.

Example: *Macaca mulatta* (Rhesus monkey); *Presbytis* (= *Semnopithecus*) *entellus* (Leaf monkey or langur); *Papio* (African baboon); *Mandrillus* (Mandril); *Cynocephalus* (Savage mandril); *Nasalis* (Proboscis monkey of Borneo); Japanese macaque or Snow monkey.

Family 2. Pongidae or Simiidae (Apes)

1. Apes (Box 17.5) are found in East Asia and Africa.

Box 17.5

The apes are human-like primates without muzzle, cheek pouches, ischial callosities and tail. In contrast to monkeys, the apes have a large head, longer neck and shorter trunk with the wide chest.

2. They are mainly arboreal, diurnal, gregarous and omnivorous.
3. They have a loud voice and use it very often.
4. They walk on the outer edges of the feet and knuckles of the hands. This gives them a semierect, shuffling, bow-legged gait.
5. The apes are very intelligent and they lead a social life.
6. They can show happiness, anger, etc., by facial expression.
7. They cannot walk on the branches of the trees like the monkeys as they are larger than the latter. They swing from the branches with the powerful arms. This mode of progression is called **brachiation** (Box 17.6).

Box 17.6

Gibbons are swingers. They move through the trees with an arm-over-arm motion called **brachiation** (Fig. 17.5). Curling their long, narrow fingers, gibbons swing from branch to branch using their hands like hooks, without actually gripping the wood. They can also walk up-right on the ground, like the great apes. When they do, they hold their long arms high in the air to keep them from dragging in the dirt.

8. The forelimbs are much longer than the hindlimbs and both are adapted for grasping than for walking.
9. Brain is large and has a pattern of convolutions much like that of human brain.
10. All the digits bear nails, except in *Hylobates*. Both pollex and hallux are opposable.
11. Dentition is $\frac{2123}{2123} \times 2 = 32$ as in humans. A small diastema exists between incisors and canines. The canines are projecting and used for attack and defence.

Fig. 17.5. Brachiation by gibbon.

12. Alimentary canal has a simple stomach and a small caecum with vermiform appendix as in human beings.
13. Body is covered with hair and does not show bright colour.
14. Bony external auditory meatus is present. There is no tympanic bulla.

Example: *Hylobates hoolock* (Gibbon), *Simia* (= *Pongo*) *pygmaeus* (Orangutan), *Pan* (= *Anthropopithecus*) *satyrus* (Chimpazee); *Gorilla gorilla* (Gorilla; Box 17.7).

Box 17.7

Primatologists are scientists who study primates. Let us know the painstaking contributions of following three female primatologists:

1. The most famous **orangutan** primatologist is **Brute Galdikas**. She has studied wild orangs in Borneo for almost 30 years, patiently tramping through the forest with notebook, binoculars and a day's rations to observe these shy creatures.

2. For 20 years primatologist **Dian Fossey** studied **gorillas** in the mountains of Rwanda, Africa. When she began her observations, the gorillas ran from her. In time they accepted her being near him.

3. English primatologist **Jane Goodall** was the first person in 30 years to study **chimpanzees** in wild (Tanzania). She made two amazing discoveries about chimps: that they made tools and that they eat meat.

Family 3. Hominidae

1. Hominidae includes human beings. Humans differ from the members of family Pongidae in being more erect, in having shorter arms and larger thumbs, and in having lighter jaws and smaller front teeth.
2. Many differences between the human and the anthropoid apes are associated with greater human intelligence, human speech centers in brain and the absence of arboreal habit.

Examples: *Homo sapiens* (Human beings).

Order 6. Pholidota

1. It includes the scaly anteaters or **pangolins.** Of old World (Asia and Africa).
2. Pangolins are nocturnal, fossorial, insectivorous mammals.
3. Body has on the upper side an **exoskeleton** of large, horny epidermal scales, which overlap front to back. The scales are formed by fusion of hair. A few coarse hair project between the scales. The cheeks and under surface of the trunk lack scales and bear stiff sparse hair. Tail has scales on the lower side also.

4. Head is relatively small with a pointed snout, small eyes and reduced pinnae.
5. Limbs are short but strong, each with 5 clawed digits. Fingers have stronger claws than the toes.
6. Tail is long, flat and tapering.
7. Jaws lack teeth.
8. Tongue is very long, worm-like, protrusible and sticky to pick up ants. Tongue is retracted into a sac when not in use.
9. Submaxillary (salivary) glands are well developed. Stomach is simple.
10. Testes are inguinal.

    Example: *Manis crassicaudata* (Indian pangolin).

Order 7. Lagomorpha

1. It includes rabbits, hares and picas. They are herbivours digitigrade and prolific breeders, producing several young ones in a litter.
2. Dentition is specialized for gnawing and chewing. There are two pairs of incisors in the upper jaw, a pair of small incisors behind the pair of long (main) incisors. Lower jaw has a single pair of incisors. Incisors are sharp, chisel-shaped and rootless. They have thinner enamel coat on the hind surface than on the front surface. They wear off with gnawing and remain sharp but keep growing continuously.

   Canines are absent, leaving diastema. There are 3 premolars in the upper jaw and 2 in the lower jaw on each side.
3. Dentries are joined in front. Upper lip is cleft in the middle. Pinnae are long.
4. Hindlimbs are much larger for leaping. Digits bear claws. Palm and soles are hairy.
5. Tail is short or absent.
6. Testes permanently descend into the scrotum.
7. Teats are many and abdominal.
8. Uterus is duplex (Fig. 17.2A).
9. Placenta is discoidal and haemoendochorial.

Example: *Oryctolagus cuniculus* (Rabbit), *Lepus ruficaudatus* (North Indian hare); *Lagomys* (Pica).

Order 8. Rodentia

1. Small-sized herbivorous animals such as rats, mice, squirrels guinea-pigs, beavers, porcupines, jerboas, gerbills, lemmings and voles.
2. Forelimbs are smaller than hindlimbs. With plantigrade foot posture, digits ending in claws.
3. Gnawing animals with chisel-shaped ever growing incisors, being one pair in each jaw.
4. Canines are absent, leaving toothless space the **diastema** in the jaw. Premolars and molars are suited for chewing the food.
5. Mandibles free anteriorly and not united in a symphysis.
6. Tail may be long, short or lacking.
7. Skin bears soft fur but has spines in some cases.
8. Testes are extra-abdominal.

    Example: Order Rodentia is divided into following 6 families:

    Family 1. Sciuridae. Examples. *Funambulus palmarum* (Squirrel), woodchucks.

    Family 2. Muridae. Examples. *Rattus rattus* (rat), *Mus musculus* (House mouse).

    Family 3. Castoridae. Examples. Castor (Beaver).

    Family 4. Erethizontidae. Examples. *Hystrix leucura* (Porcupine).

    Family 5. Geomyidae. Examples. Pocket gophers.

    Family 6. Cricetidae. Examples. Hamsters, Deer mice; *Gerbillus indica* (Gerbil); voles; *Lemmus lemmus* (Lemmings).

Order 9. Cetacea

1. It includes highly modified mammals such as whales, dolphins and porpoises. They appeared in the early Miocene.
2. Body is large and fish-like, being streamlined for aquatic life.
3. Head is large with a wide mouth.
4. There is no distinct neck.
5. Skin is smooth, hairless and devoid of sweat glands and oil glands. In some species, a few sensory bristles occur on the snout. In fact, the skin of foetus is uniformly hairy.
6. Beneath the skin is a thick layer of fat, the **blubber** which insulates the body and lowers the specific gravity.
7. Eyes are very small, low in position, being near the angles of the moth. Nictitating membrane is absent. Lacrimal glands are reduced or absent.
8. Ear opening are very small and without pinnae.
9. Nares lie far posterior on the top of head. They have valves which close under water.
10. Forelimbs are modified into paddles or flippers. The digits are five in number but indistinct. In forelimbs, only joint between scapula and humerus is movable, other bones are immovable.
11. Hindlimbs are absent. Whalebone whales may have vestiges of femur and tibia.
12. An adipose dorsal fin is present in some species.
13. Tail is flattened and ends in a horizontal (not vertical as in fish) tail fin called **fluke**.
14. Both dorsal fin and caudal fins lack skeletal support.
15. Locomotion is brought about by up and down movements (not lateral as in fish) of tail, the propulsive trust being produced mainly by the fluke.
16. Skull bones are spongy and contain fat. Cranium is large and rounded. Cervical vertebrae are thin, disc-like and fused together. Sternum is reduced. Ribs are highly mobile and plays an important role in respiration. Clavicle is lacking. Pelvic girdle is reduced.
17. Teeth, if present, are numerous, conical and homodont and monophyodont.
18. Diaphragm is considerably extended horizontally backward to enable the lungs to draw maximum quantity of air in a short time.
19. Blood vessels form **retia mirabilia** (networks) in various parts of the body to store extra oxygen for use when the animal is under water.
20. Brain is large. Cerebral hemispheres are highly convoluted. Olfactory lobes are reduced or absent.
21. Testes are abdominal.
22. There are two inguinal mammae.
23. Uterus is bicornuate. Placenta is diffuse and indeciduate.
24. The cetaceans are carnivorous, gregarious aquatic mammals. They occur in all sea. They do not visit land. They mate and give birth to young ones in water. Usually a single young is born at a time.

    The order Cetacea is divided into 2 suborders: 1. Mysticeti and 2. Odontoceti.

Suborder 1. Mysticeti

1. It includes whalebone whales.
2. They have a pair of nares.
3. They lack teeth, but possess a whale bone or **baleen** consisting of rows of parallel horny plates hanging from the palate into the buccal cavity. This bone acts as a sieve during feeding. For feeding, the whale opens the mouth widely and fills it with water containing small planktonic animals such as jelly fishes and crustaceans (krill) which form the food. The mouth is now closed and water expelled through narrow chinks of the whalebone. The food is left behind and swallowed.

4. Hand has 3 digits.

Example: *Balaenoptera indica* (Indian whalebone whale); *Balaenoptera musculus* (Blue whale), *Megaptera novaeangliae* (Humpback whale); *Balaena* (Right whale or Arctic whale).

Suborder 2. Odontoceti

1. It includes toothed whales, dolphins and porpoises.
2. They lack whalebone and have numerous homodont and monophyodont teeth.
3. They feed on fishes and cuttle fishes.
4. They have a single nare (= naris).
5. Hand has 5 digits.

Example: 1. *Physeter catodon* (Sperm whale); *Orchinum* (Killer whale); *Platanista gangeticus* (River dolphin); *Delphinus delphis* (Common dolphin); *Phocaena* (Porpoise).

Order 10. Carnivora

1. The order Carnivora includes dogs, wolfs, jackals, foxes cats, lions, tigers, leopards, bears, mongooses, otters, walruses and seals. The are terrestrial as well as aquatic.
2. They are usually flesh eating animals (predators). Some are omnivorous and even herbivorous (*e.g.*, Panda).
3. Dentition is adapted for tearing and cutting the flesh. Incisors are small but occur in full number, *i.e.*, 3 on each side in each jaw. Canines are large and sharp pointed. Premolars and molars specialized for cutting the flesh into small morsels that are swallowed without mastication.
4. Muscles of the jaws and neck are very well developed.
5. Digits are armed with strong claws.
6. Lower jaw has only up and down movement and no rotary movement to prevent its slipping when the prey is seized.
7. Tympanic bulla is large.
8. Intestine is short.
9. In brain, cerebral hemispheres are well convoluted.
10. Testes descend into the scrotum.
11. Teats are abdominal.
12. Uterus is bicorunate.
13. Placenta is zonary and endotheliochorial.
14. Carnivores are agile, bold, ferocious and intelligent animals.

The order Carnivora is divided into two suborders: Fissipedia and Pinnipedia.

Suborder 1. Fissipedia

1. It includes terrestrial carnivores.
2. The digits are separate and bear claws which may be retractile into sheath.
3. Locomotion is digitigrade, but may be plantigrade or semiplantigrade.
4. Last premolar in the upper jaw and first molar in the lower jaw bite on each other like a pair of scissors for cutting the flesh. These two teeth are called **carnassial** or **sectorial** or **shearing**, teeth. There are 3 pairs of incisors in each jaw.

Suborder is divided into following five families:

Family 1. Canidae (Dog family). Examples. Dogs, wolves, foxes and coyotes.

Family 2. Felidae (Cat family). Examples. Domestic cat, tigers, lions, puma (*Felis concolor*) and lynxes.

Family 3. Ursidae (Bear family). Examples. Bears.

Family 4. Procyonidae. Examples. Raccons.

Family 5. Mustelidae (Fur-bearing family). Examples. Martens, skunks, weasels, otters, badgers, minks and wolverines.

Suborder 2. Pinnipedia

1. The Pinnipedia are aquatic (mainly marine) carnivores. Their food mainly consists of fish.
2. Body is streamlined.
3. Nostrils can be closed by special muscles.
4. Pinnae are reduced or absent.
5. Limbs are short, with their proximal parts enclosed in the skin.
6. Digits are fully webbed in both hands and feet.
7. There are no carnassial teeth. Incisors are never more than two in each half of each jaw. In walruses, upper canines are long and project downward as **tusks** in both sexes.
8. Tail is short and not used in swimming.
9. Venous plexuses in the body increase the duration of remaining under water.
10. Mating and birth of the young one occur on land.

Example: *Phoca vitulina* (Seal); *Odobenus* (Walrus); *Otaria stelleri* (Sea lion); *Callorhinus* (Fur seal).

Order 11. Tubulidentata

1. This order includes **arrdvark**. The name "aardvark" is Dutch for earth pig, a peculiar animal with a piglike body found in Africa.
2. The body is stout, sparsely hairy and of the size of pig.
3. Back is highly curved. Snout is long, mouth is tubular, pinnae are large, tongue is long, slender, sticky and protrusible.
4. Milk teeth are more than permanent teeth. Each tooth consists of many hexagonal columns of dentine separated by tubules of pulp, hence Tubulidentata. The teeth lack enamel.
5. Forelimbs and hindlimbs bear 4 and 5 digits respectively. The digits bear nails that are used for digging.
6. Brain is small and primitive.
7. Their fossils are known from the Miocene of Africa and the Pliocene of Eurasia.

Example: *Orycteropus afer* (Aardvark or Cape ant-eater).

Order 12. Proboscidea

1. The Proboscidea includes the elephants, the largest land animals.
2. Skin is very thick, loose and with scanty hairs.
3. Head is very large, with small eyes and large fan-like pinnae.
4. Nose and upper lip are prolonged into a cylindrical, highly mobile and prehensile **proboscis** or **trunk** have external nares at the tip. The trunk has one or two sensory lobes at the tip (Fig. 17.6). The trunk is used to grasp food and transfer it to the mouth. It is also used for taking water for drinking and other uses.
5. Lower jaw is much shortened.
6. Neck is short.
7. Limbs are thick and pillar-like. They do not bend at the elbow and knees. Each forelimbs and hindlimbs bears 5 short, stout digits. Each digit ends in a nail-like hoof.
8. Tail is thin and short.
9. Dentition is 1033/1033. Upper jaw has only second pair of incisors. These are rootless and grow continuously, forming large **tusks**. The tusks are formed of solid dentine (called **ivory**), the enamel being restricted only to the apex from where it soon wears off.
   There are no canines. Premolars fall off early. Molars appear in series. At a time each half of the jaw has one or two molar in actual use. To begin with there is only one molar in each half of the jaw. A second molar develops beind it and shifts forward as the first one is worn out and falls off. The this molar repeats the process. To allow forward shifting, the molars, though

thecodont, occur not in alveoli but in a gutter in the jaw bone. The molars are greatly elongated to have a very large area of mastication. They are **lophodont** (*i.e.*, their cusps fused to form ridges or lophs) and have numerous trnsverse ridges for chewing.

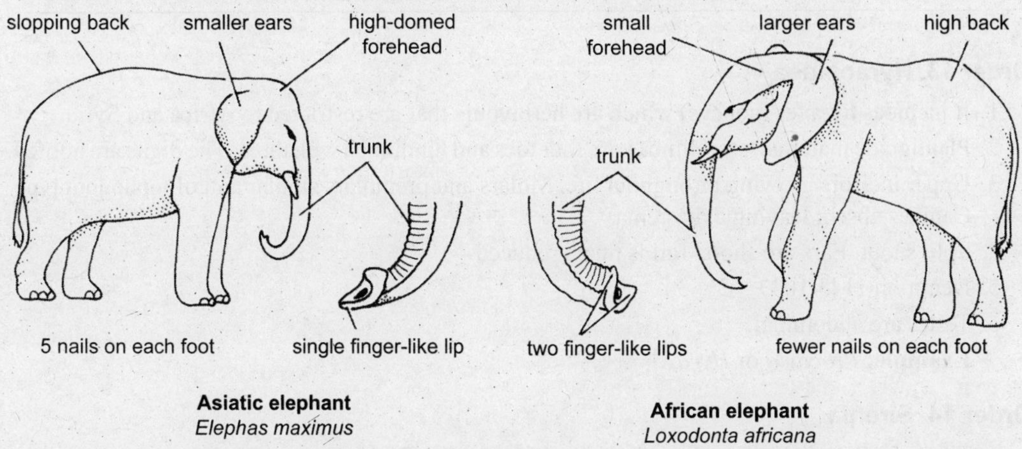

Fig. 17.6. Comparison between Indian and African elephants.

10. Skull is large with reduced facial parts. Many skull bones have air spaces which are continuous with the nasal passages. This reduces weight. Premaxillae are large to carry the huge tusks.
11. Gall bladder is absent.
12. Cerebral hemispheres are large and greatly convoluted but do not cover the cerebellum.
13. Testes are permanently abdominal.
14. Uterus is bicornuate.
15. There are two pectoral mammae.
16. Placenta is Zonary and indeciduate.
17. Only one young is born at a time. Gestation period is of about 22 months.
18. The elephants inhabit forests and tall grass in India and Africa. They are herbivorous and gregarious. They roam about in herds of 10 to 100. They have a life span of 100 to 120 years and are slowest breeders.

Example: *Elephas indicus* (Indian elephant), *Laxodonta africana* (African elephant) and *Elephas cyclotis* (African pygmy elephant).

Table 17.2 Difference between Indian and African elephants.

	Elephas indicus (Indian elephant)		*Loxodonta africana* (African elephant)
1.	Body 2.7 m high.	1.	Body 3.5 m long.
2.	Tusks 3.3 m long, occur only in male. In female, tusks do not extend beyond the lips.	2.	Tusks 2.7 m long occurs in both sexes.
3.	Pinnae of moderate size.	3.	Pinnae very large, cover shoulders.
4.	Head is rounded.	4.	Head is sloping.
5.	Back is arched.	5.	Back is depressed.
6.	Trunk ends in a single sensory lobe.	6.	Trunk ends in two opposite sensory lobes.
7.	Molars have many parallel trnasverse ridges.	7.	Molars have fewer diamond-shaped transverse ridges.

	Elephas indicus (Indian elephant)		*Loxodonta africana* (African elephant)
8.	Hands has 5 nails, foot 4.	8.	Hand has 4 nails, foot 3.
9.	Easly tamable.	9.	Not easily tamable.

Order 13. Hyracoidea

1. It includes hyraxes (coneys) which are herbivours that are restricted to Africa and Syria.
2. Plantigrade mammals, forelimbs with four toes and hindlimbs with three. The digits are hoofed.
3. Upper incisors growing throughout life. Molars and premolars similar and of lophodont type. Canines absent leaving a diastema.
4. Split snout. Ears are short. Tail is much reduced.
5. Dentition 1143/1043 × 2 = 34.
6. Testes are abdominal.

Example: *Procavia* or *Hyrax*.

Order 14. Sirenia

1. The order Sirenia includes large, herbivorous, aquatic (freshwater and marine) mammals called **sea cows**.
2. Body is streamlined for aquatic life.
3. Skin is thick and almost hairless. It lacks sweat and oil glands.
4. A thick layer of fat, the blubber, is present under the skin for insulation.
5. Muzzle is rounded and upper lip is cleft for grazing.
6. Nares are dorsal and can be closed under water. Eyes are small and have muscular lids. Pinnae are lacking.
7. Forelimbs are modified into paddles for swimming. Hand has 5 digits enclosed in a common web. Hindlimbs are absent.
8. Tail has a flattened horizontal tail fin.
9. Dentition is reduced.
10. Diaphragm is very muscular and oblique.
11. Brain is small and poorly developed.
12. Teats are thoracic.
13. Testes are abdominal.

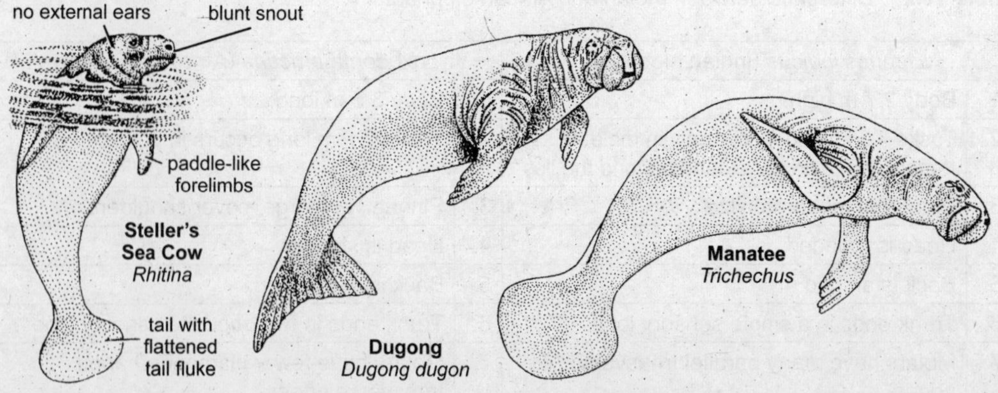

Fig. 17.7. Examples of order Sirenia.

14. Uterus is bicornuate.
15. Bones are heavy so that the animal may stay at the bottom to feed on sea weeds. Pelvic girdle is vestigial. Sternum is reduced.

    Example: *Halicore* (Dugong); *Manatus* (= *Trichechus*) (Mantee); *Rhitina* (Sea cow).

Order 15. Perissodactyla

1. It includes odd-toed (Box 17.8), hoofed, terrestrial and herbivorous mammals such as horses, asses, zebras, tapirs, rhinoceros.

Box 17.8

The odd-toed hoofed mammals have an odd numbr of toes (one or three), each with a cornified hoof. Both the Perissodactyla and the Artiodactyla are often referred to as **ungulates** or hoofed mammals with teeth adapted for chewing.

2. Limbs are long and adapted for fast running. Unguligrade foot posture.
3. Hoofed formed of **uneven number of toes**.
4. Axis of foot passes through third toe.
5. True horns with a bony core are absent.
6. Herbivorous animals. Incisors are used for chopping, premolars and molars are alike and lophodont with transverse ridges. Canines are reduced or absent.
7. Gall bladder is absent. Stomach is simple. Intestine is long.
8. Teats are inguinal.
9. Placenta is diffuse and epichorial.

Example: *Eguns cabalus* (Domestic horse); *Eguns vulgaris* (Indian donkey); *Eguns zebra* (African zebra); *Rhinoceros indicus* (or *unicornis*) (Indian or one-horned rhinoceros); *Taperus* (Tapir).

Order 16. Artiodactyla

1. It includes even-toed hoofed mammals such as swine (pigs), cattle, sheep, goats, deer, antilopes, camels, hippopotami, llama and okapi. Like Perissodactyla, they are strictly herbivorous.
2. Most of these ungulates have two toes, although the hippopotamus and some others have four.
3. Each toe is sheathed in a cornified hoof.
4. Many, such as the cow, deer and sheep have true horns with a bony core or antlers.
5. Many are ruminants, that is animals that chew the curd.
6. Incisors and canines are often absent in the upper jaw. Premolars and molars are not similar, and are bunodont or solenodont (with crescentic cusps).
7. Stomach is complicated and divided into many chambers for bacterial digestion of cellulose. Intestine is short. Gall bladder is present.
8. Cerebral hemispheres are well convoluted and cover cerebellum considerably.
9. Testes descend into the scrotum.
10. Teats are inguinal or abdominal.
11. Placenta may be diffuse or cotyledonary and epitheliochorial or syndesmochorial.

    Order Artiodactyla is divided into 3 suborders: 1. Suina or Suiformes; 2. Tylopoda and 3. Ruminantia.

Suborder 1. Suiformes

Animals of this suborder have a relatively simple stomach and do not chew the cud.

Example: *Sus cristatus* (Indian crested wild boar or pig); *Hippopotamus amphibius* (Hippopotamus).

Suborder 2. Tylopoda

1. Horns are lacking.
2. Upper lip is divided like that of rabbit.
3. The feet form broad cushions.
4. The stomach has three chambers: rumen, reticulum and abomasum. The first two chambers have a number of pouches, called **water cells** which serve to store food (not water).
5. The red blood blood corpuscles are oval.

Example: *Camelus dromedarius* (One-humped or Arabian camel); *Camelus bactrianus* (Two-humped or turkish or bacterian camel); *Llama vacugna* (Llama).

Suborder 3. Ruminantia

1. Their complicated stomach have four chambers: rumen, reticulum, omasum and abomasum.
2. While feeding, they swallow their food which is regurgitated, a mouthful at a time, chewed thoroughly and reswallowed. They are said to chew the cud or ruminate, hence ruminants.

Suborder Ruminantia is divided into following three families:

Family 1. Bovidae. Examples. *Babalus bubalus* (Buffalo); *Bos indicus* (Cow); *Bison bison* (Bison); *Capra* (Goat); *Ovis* (Sheep); *Antelope* (Antelope).

Family 2. Cervidae. Examples. *Cervus* (Deer); *Rangifer* (Reindeer or Caribou); *Alces* (Moose); *Moschus* (Musk deer).

Family 3. Giraffidae. Examples. *Giraffa* (Giraff), *Okapia* (Okapi).

17.4 TYPE STUDIES

1. *Echidna* (*Tachyglossus*) (Spiny Ant-eater)

Classification

Class	—	**Mammalia**
Subclass	—	**Prototheria**
Order	—	**Monotremata**
Genus	—	*Echidna* (*Tachyglossus*)
Species	—	*aculeatus*

Habit and Habitat

Echida aculeatus is commonly called the **short-nosed spinyant-eater**. (*Note*: There is one more species, called **long-nosed spiny ant-eater**, *Zaglossus bruijni*, found in New Guinea; Fig. 17.9A). *Echidna* is found in New Guinea, Tasmania and Australia. *Echidna* live in burrows and hollow logs. It is terrestrial, fossorial, insectivorous and nocturnal. It eats termites and ants by its long protrjusible sticky tongue.

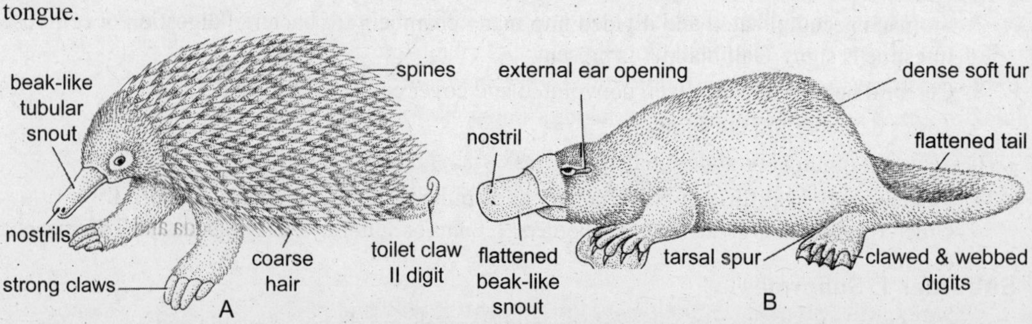

Fig. 17.8. Egg-laying mammals (Monotremes) A—*Tachyglossus aculeatus* (Australian spiny ant-eater; B—*Ornithorhynchus anatinus* (Duck-billed platypus).

The female lays a single egg in the nest (burrow). Egg is incubated outside the body by the mother and then is transferred into a small abdominal pouch, called **incubatorium** which she develops in the breeding season. The young hatches out in the pouch and feeds on milk. It is later transferred into a burrow. Male also produces milk, a condition called **gynaecomastia**.

Like reptiles, *Echidna* hibernates in winter and can fast for a long time. It can roll into a ball for protection. On seeing an enemy, it quickly digs out burrow in the ground using all four limbs and gets into it, protruding a few spines out of the burrow.

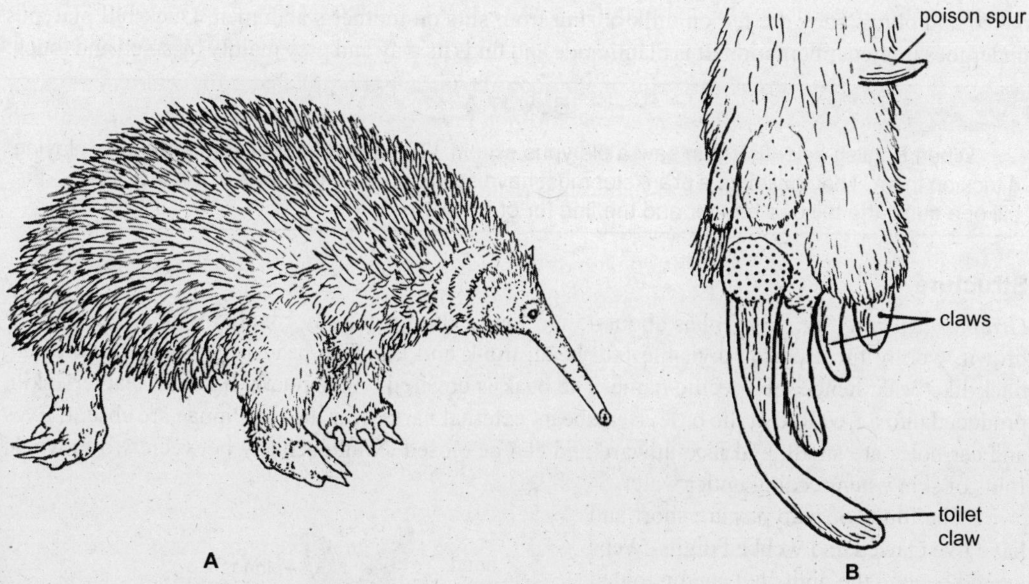

Fig. 17.9. A—*Zaglossus bruijni* (Long-nosed spiny ant-eater); B—Foot of male of short-nosed spiny ant-eater.

Structure

Echidna (Fig. 17.9A) has a dense coat of hair all over. Some hair on the back are modified into short but sharp spines. Thus, it has a spiny fur like a hedgehog. It is about 42 cm long. Body is divisible into head, neck and trunk. The small head tapers into a hard, naked, sensitive, long tube-like snout or rostrum. Snout has near the tip a small, ventral mouth and small dorsal nares. The eyes are small and without nictitating membrane. Teeth are absent. Teeth are monophyodont and occur only in the young. There are no pinnae. Tail is vestigial.

The tongue is long, cylindrical, sticky and highly protrusible. The tongue bears on its posterior part horny serrations which masticate the ants against ridges of the palate. Large salivary glands occur in the buccal cavity. The four limbs are short and held laterally. They bear strong claws to clean the hair and spines. The male has a horny grooved **spur** on each tarsus (Fig. 17.9B). Spur is connected by a duct with a poison gland located in the back of the thigh. *Echidna* lacks sweat glands, panting mechanism or heat reduction by vasodilation, so it has to hibernate. Teats are absent. Mammary glands pass out milk through pores and is licked up by the young. *Echidna* is a connecting link between reptiles and mammals having characters of both the classes.

2. *Ornithorhynchus* (Duck-billed Platypus)

Classification

Class	—	**Mammalia**
Subclass	—	**Prototheria**
Order	—	**Monotremata**
Genus	—	*Ornithorhynchus*
Species	—	*anatinus*

Habit and Habitat

Ornithorhynchus lives in burrows on the banks of rivers, lakes, ponds and lagoons. It is found in Eastern Australia and Tasmania.

It is aquatic in habit. Its burrow has two outlets, one under water and other above the water. It feeds on crayfish, shrimps, larvae of water insects, snails, tadpoles, worms and small fish. *Ornithorhynchus* is oviparous. Two or three eggs are laid in spring in a special burrow, where male is not allowed. Eggs are rounder, having a tough, leathery shell. The eggs of platypus stay stick together so they cannot roll away. Incubation is done by the mother alone (Box 17.9). The young are naked and blind. They are fed on milk oozing from slits on mother's abdomen. Duck-bill platypus undergoes limited hibernation. It is plantigrade and finds its way and prey mainly by smell and touch.

Box 17.9

When English scientists first saw a platypus skin in 1789, they thought some one was playing a trick on them. They believed a prankster must have sewn together the parts of other animals—the bill of a duck, the tail of a beaver and the fine fur of a mole.

Structure

Ornithorhynchus (Fig. 17.8B) has about 45 cm long cylindrical body covered with a dense, dark-brown, woolly fur. Body is divisible into head, trunk and tail. The head is prolonged into a flat, duck-like beak, hence the specific name. The beak is covered with smooth, hairless, leathery skin, produced into a free fold at the base. Beak bears external nares near the tip. Pinnae are absent. Eyes and ear-holes are small, and face upward and can be closed simultaneously between two movable folds of skin when feeding under water.

The limbs of platypus are short and have five clawed and webbed digits. Webs extend beyond the digits but can be folded back for walking and digging. Males have a horny spur on each ankle, bearing an opening of a poison gland. Tail is well developed and flat. It forms an efficient organ for swimming. The adult lacks teeth. The young has teeth whch are soon shed and are replaced by the underlying horny plates. These horny plates are strong enough to break the shell of mollusks. Cloaca present. Testes are abdominal and penis conduct only sperms. Uterus or vagina is absent.

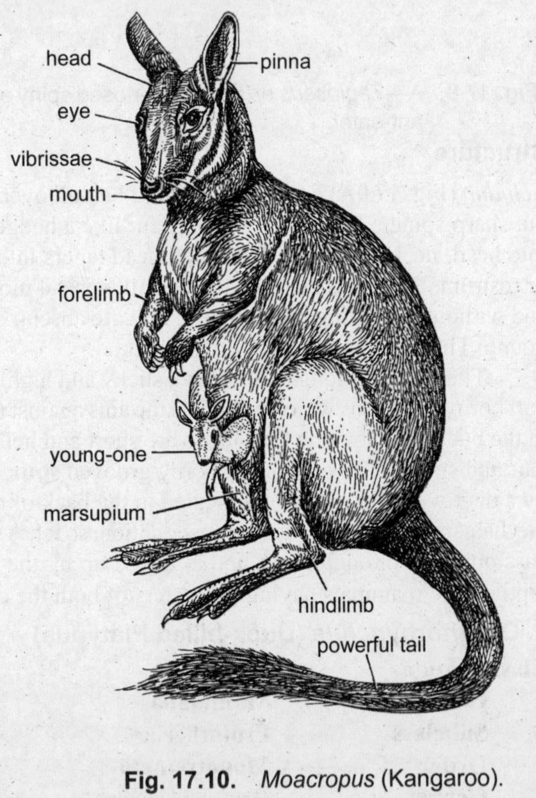

Fig. 17.10. *Moacropus* (Kangaroo).

3. *Macropus* (Kangaroo)

Classification

Class	—	**Mammalia**
Subclass	—	**Theria**
Infraclass	—	**Metatheria**
Order	—	**Marsupialia**
Genus	—	*Macropus*

Habit and Habitat

Macropus is found in East Australia and Tasmania. It is terrestrial, gregarious and herbivorous. Organised groups of 2–10 or more, called **mobs**. **Gray kangaroos** live in forests and woodlands while **red kangaroos** exists in grassy plains or savannas.

Kangaroos are the great leapers of the animal world. A kangaroo runs by taking long leaps on the hindlimbs alone. The leap may be 9 metres long. When they are just ambling along grazing and browsing on plants, they have a "five-legged" walk: they hold themselves on their small forelegs and big tail while pulling their hindlegs along.

When they are born, marsupial babies are tiny (2.5 cm long), naked, deaf, and blind. Still, in the next four minutes they must drag themselves, inch by inch, over their mother's belly to pouch of skin, the **marsupium**. Once inside the pouch, they latch onto a teat and start to drink milk. Milk is virtually forced down the throat of the young one by contraction of certain muscles of the mother's belly (Box 17.10).

Box 17.10

Just before her baby is born, the mother Kangaroo licks a path along her belly upto the pouch. Sniffing along the path, the baby climbs upwards and into the pouch. Once arrived, the baby will suckle on a teat for its food.

Male kangroos are known as **boomers**. They sometimes have fights that combine wrestling, boxing, kicking and just standing around. They will punch with their forelegs, shove each other backward and kick with their powerful hind legs. The loser male hops away.

Structure

Kangaroo is the largest marsupial. It is upto 2 meters while standing. Male is larger than the female. Body is divisible into head, neck, thorax, abdomen and tail. Head is small with long pinnae. Tail is long, muscular, non-prehensile and is used as third leg. The forelimbs are small having five clawed digits. Hindlimbs and feet are small having four clawed digits. They are powerful for leaping. Fourth toe is very long. Hallux is absent.

Dentition of kangaroo is unique. It is almost monophyodont. Number of incisors is differnt in upper and lower jaws, *i.e.*, in *Macropus* there are 5 incisors in each upper jaw and 3 in lower jaw. Also there are 3 premolars and 4 molars in each half. Cheek teeth have grinding surfaces. Digestive, respiratory, circulatory, excretory systems are typically like the eutherian mammals. Brain shows some reptilian feature, *i.e.*, cerebral hemispheres are much smaller and smooth.The female contains two vagina and two uteri. *Macropus* is viviparous. But egg is yolky and covered with albumen and a small membrane. It develops yolk sac type placenta. Birth occurs after a short uterine gestation period (33 days). The young is born in a highly immature stage but contains clawed forelimbs. The baby clutches the fur of mother through the claws and reaches the pouch of the mother and takes a nipple of the mammary gland in its mouth. The nipple swells within the mouth of the baby and baby cannot leave it now. The baby completes its development lying attached to the nipple. The baby stays in the pouch for 8–9 months. The fully developed baby leaves the pouch but may return into it for safety.

4. *Sorex* (Shrew)

Classification

Class	—	**Mammalia**
Subclass	—	**Theria**
Infraclass	—	**Eutheria**
Order	—	**Insectivora**
Genus	—	*Sorex*
Species	—	*araneus*

Habit and Habitat

Shrews are found in Asia, Europe and North America. The Indian shrew belongs to genus *Soriculus*, commonly called **Chhuchhunder**. *Sorex* lives in tunnels and comes out in the night and feeds on insects, worms, and mice. Shrews have a higher rate of metabolism than other mammals. They cannot live without food fore more than 10 hours. Because of their great expenditure of energy, shrews daily consume twice as much as food as their weight. Cat do not eat shrews because of their rancid odour (Box 17.11).

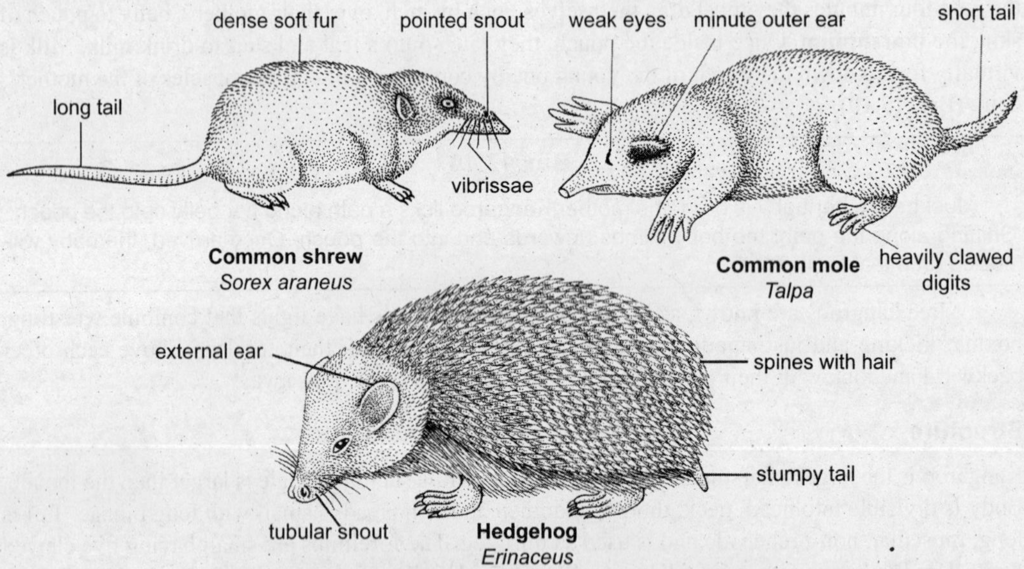

Fig. 17.11. Insect-eating mammals (Class Insectivora).

Box 17.11

Shrews are tiny sharp-nosed animals. They are so nervous that they can literally die of fright at a loud noise. When skittish little baby shrews feel scared, they form a "carvan". One baby grabs the mother's fur in its mouth, the second grabs the first one, and soon, until the mother pulls them to safety.

Structure

Sorex (Fig. 17.11) is rat-like. It is 7.50 cm long excluding the tail. It is reddish-brown and pale. Snout is elongated and body is covered with silky fur. Eyes and ears are small. First pair of incisors is very long, lower canines are absent. All teeth are red. The dentition is 3133/2013 × 2 = 32. Limbs are short and have five clawed digits.

5. *Hemiechinus* (= *Erinaceus*) (Hedgehog)

Classification

Class	—	**Mammalia**
Subclass	—	**Theria**
Infraclass	—	**Eutheria**
Order	—	**Insectivora**
Genus	—	*Hemiechnus* (*Erinaceus*)
Species	—	*collaris*

Habit and Habitat

The hedgehog occurs in almost any type of country, except marshes. *H. collaris* is common in Panjab and adjacent states. *Paraechinum micropus* is the South Indian species.

The hedgehog often lives among bushes. It is nocturnal and solitary. At the time of danger it rolls itself into a ball and spines project outward causing harm to the attacking animal. It feeds on insects, worms, slugs, frogs, lizards and snakes. It also eats fruits and roots. Hedgehog is immune to snake's bite. It hibernates in winter.

Structure

The hedgehog (Fig. 17.11) has a small (about 23 cm long), somewhat globular, dirty-brown body, pointed head with small black eyes and prominent pinnae. Neck is not distinct. Tail is small stump-like. Limbs are short and bear five clawed digits on each foot. The head, throat and belly are covered with fur, while the rest of the body is armed with short, hard, erectile spines.

6. *Crocidura* (= *Suncus*) (Musk Shrew)

Classification. Same as that of *Sorex*.

Habit and Habitat

The musk shrew (Fig. 17.12) is found almost all over India. It is nocturnal and insectivorous. It locates insects (such as cockroaches) by smell. It has a light musky smell due to the secretion of a pair of glands present behind armpits. The secretion becomes stronger in breeding season. The shrew emits a sharp, shrill cry as it runs about at night in the houses.

Structure

The musk shrew is a rat-like animal. It is 10–15 cm long and dark-grey in clour. Head has long, tapering snout and small eyes and pinnae. Tail is almost naked. The feet are fleshy. Snout, pinnae, tail and feet are pinkish.

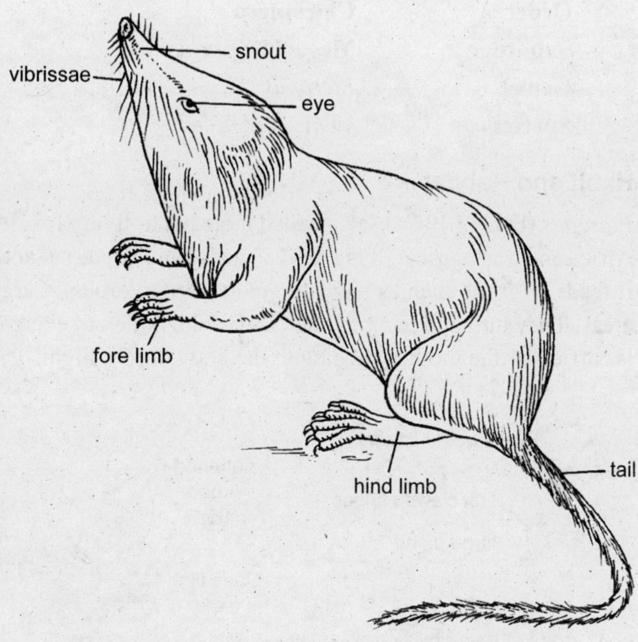

Fig. 17.12. The musk shrew (*Crocidura murina*).

7. *Talpa* (Short-tailed Mole)

Classification. Same as that of *Sorex*.

Habit and Habitat

Talpa (Fig. 17.11) is burrowing, live underground in burrows excavated by itself (Box 17.12). Food chifly consists of earthworms.

Box 17.12

Every mole digs an underground home and patrols its tunnels regularly. It carves out rooms for drinking water and a "bathroom" for waste disposal. Another room holds a supply of food. Female moles also builds a central nest chamber where they rear their babies. If a male meets another male in a tunnel, they fight.

Structure

Moles are related to shrew but are larger in size and have enormous shoulder girdle and muscles. It is greatly specialised for a subterranean burrowing life. Body is thick, cylindrical and about 15 cm long. The head is wedge-shaped with a prenasal bone in the snout, eyes are rudimentary, and pinnae and minute. Tail is short with no hair and is sensory. It has a thick velvety fur. Forelimbs are strong, manus (forefoot or hand) is large and broad, five digits have long, broad nails, there is an extra sickle-shaped bone of the carpus called **radial sesmoid** or falciform bone. The hand (manus) is used for digging. The mole lives underground digging long tunnels rapidly. It almost never comes to the surfae and has a remarkable sense of touch.

8. *Pteropus* (Large bat or Flying fox)

Classification

Class	—	**Mammalia**
Subclass	—	**Theria**
Infraclass	—	**Eutheria**
Order	—	**Chiroptera**
Suborder	—	**Megachiroptera**
Genus	—	*Pteropus*
Species	—	*kuhli*

Habit and Habitat

Pteropus (Fig. 17.13C) is also called large bat or flying fox. It occurs in India, Shri Lank, Austrialia, Africa and Madagascar. It is arboreal, aerial and nocturnal and lives in groups and roosts in colonies. It feeds on fruits such as figs and guava (Grugivorous). Large bats hang from the trees in deserted areas. They turn and hang by the claw of the pollex to excrete. Only one young is born at a time. It is carried by the mother sticking to the body during flight.

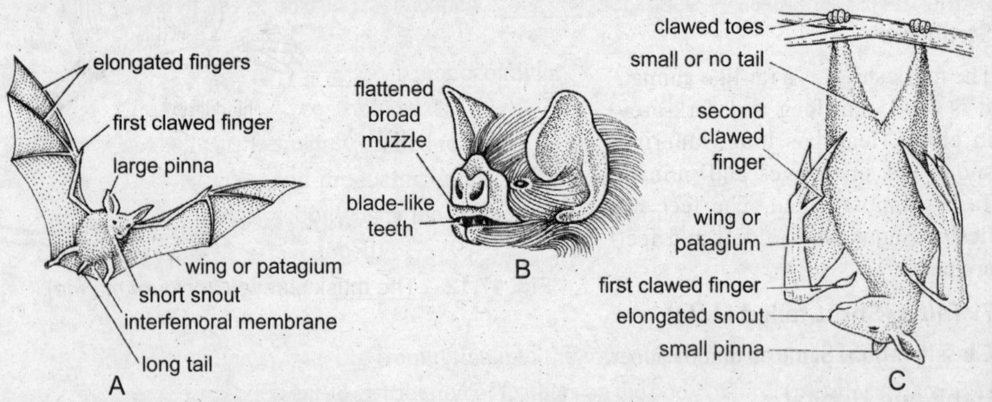

Fig. 17.13. Flying mammals or bats. A—An insect eating bat (Microchiroptera, *e.g.*, *Scotophilus medius*); B—Head of vampire bat (*Desmodus*); C—A fruit eating bat (Megachiroptera, *e.g.*, *Pteropus kuhli*).

When wakened the bats can walk, open its mouth and bite or cry, but cannot fly. It flies only after a period of few minutes of warming up by jerking the legs and shivering (**Yalden** and **Morris**, 1975).

Structure

The fruit fox or *Pteropus* is 30 cm length and its wings spread over 1.5 m. It has dog-like or fox-like snout. Head and external ears (pinnae) are usually mammalian type. It depends heavily on sight and smell and do not ecolocate. The pollex and the second finger of forelimbs bear claws.

9. *Scotophilus* (Small bat)

Classification. Same as that of *Pteropus*.

Suborder	—	**Microchiroptera**
Genus	—	*Scotophilus*

Habits and Habitat

The small bat (Fig. 17.13A) occurs in caves and old buildings all over India. It is gregarious and insectivorous. Two young ones are born at a time. The mother carries the young ones attached to her body during flight. Hibernation is common in bats of temperate zone.

Structure

The small bat is about 5 cm long. It has a small snout, small eyes, prominent pinnae, a claw only on the first digit of the forelimbs, and a tail. Patagium extends between the hindlimbs and the tail. All the toes are short and bear claws.

Some small bats such blood of homoiotherimic vertebrates including human beings and are called **vampire bats**. *Desmodus* (Fig. 17.13B) is the South American vampire bat. It cuts the skin of the victim with its sharp teeth and licks the blood. Indian vampire bat, *Megaderma*, preys upon frogs, birds and small bats.

10. *Loris* (Slender loris)

Classification

Class	—	**Mammalia**
Subclass	—	**Theria**
Infraclass	—	**Eutheria**
Order	—	**Primates**
Suborder	—	**Lemuroidea**
Genus	—	*Loris*
Species	—	*tardigradus*

Habit and Habitat

The loris (Fig. 17.4) is found in scrub jungles of South India. It is a slow moving, nocturnal primate. It is omnivorous, feeding on fruits, leaves, insects, eggs and birds.

Structure

The loris is a small (about 20 cm long), earthy brown, primate with a short, dense, soft fur. It has forwardly directed eyes, thin rounded pinnae, white rhinarium and slender limbs of nearly equal size. There are two pectoral mammae. It has a very short tail. Another species, *Nycticebus coucang* (slow loris) is found in East Indies.

11. *Manis* (Indian Pangolin)

Classification. Same at that of *Hemiechinus* (*Erinaceus*).

Order	—	**Pholidota**
Genus	—	*Manis*
Species	—	*crassicaudata*

Habit and Habitat

Manis (Pangolin) is found in the forests of South India. It can dig a burrow for itself but prefers to live in the burrows of other animals. It is nocturnal and insectivorous. It breaks open the nests of termites and ants with its strong claws and inserts its long, sticky tongue into it. The insects stick to the tongue which is then withdrawn into the mouth to swallow them. At the time of danger, the pangolin lies on its side and rolls into a tight ball. A small, naked pad under the tip of the tail hooks over the farthest scale it can reach so that the ball does not open. The pangolin, when walking, bends the claws of the hand under the latter and places the entire sole of the foot on the ground. The porcupine swats an attacker with its spiny tail, lodging barbed quills into the victim's skin.

Structure

The pangolin (Fig. 17.3) has about 90 cm long body with large, thick overlapping horny scales on the dorsal side. head is small and bears small eyes and reduced pinnae. Tongue is long, highly protrusible and sticky. Limbs are short but stout and bear strong, curved claws on all the digits. Tail is very long and depressed. *Manis* is an economically important animal since it destroys termites.

12. *Hystrix* (Porcupine)

Classification

Class	—	**Mammalia**
Subclass	—	**Theria**
Infraclass	—	**Eutheria**
Order	—	**Rodentia**
Genus	—	*Hystrix*
Species	—	*leucura*

Habit and Habitat

Hystrix or **porcupine** (Fig. 17.14) is found from East Indies to Africa, Europe and Asia. The porcupine is a terrestrial, nocturnal, gregarious and herbivorous mammals. It feeds on crops, vegetables and plant roots. The porcupine derives its name from the French word *porcupine* (meaning spiny pig), whch indicates its two features: spines on the back and grunting pig-like cries. The spines form defensive tools.

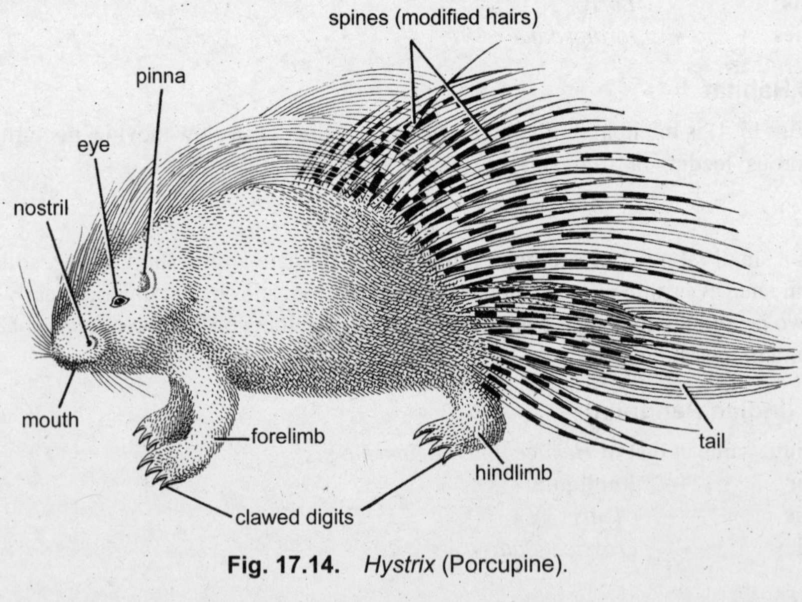

Fig. 17.14. *Hystrix* (Porcupine).

Structure

The porcupine has a stout, low body bearing on the dorsal side long, quill-like, erectile spines. Head bears a snout. A great crest of stiff long hairs is present on the head. Eyes and ear pinnae are small. Tail is short and is not prehensile. Limbs are with clawed fingers and toes and solves of feet are smooth.

The porcupine is harmful as it sometimes damages crops, vegetables and fruits.

17.5 ORIGIN AND EVOLUTION OF MAMMALS

Evidently earliest mammals made their appearance in Jurassic times as descendants of some of the true mammal-like reptiles, the Therapsida. However, zoologists have forwarded the following two ideas regarding the origin of mammals:

1. Ancestry through Amphibia. T.H. Huxley (1880) have advocated amphibian ancestry of mammals. He based his arguments on the following two points: 1. Presence of two occipital condyles in the skull of both Amphibia and Mammalia. Huxley's theory was rejected since in amphibia occipital condyles are derived from exoccipitals while in Mammalia they are derived from basioccipitals. Further, the two classes have different modes of life and show various fundamental differnces.

2. Ancestry through Reptilia. Many palaeontological evidences have established that mammals arose from reptiles. This view is also supported by the fact that the monotremes (primitive egg laying mammals) and living reptiles have close similarities in their anatomical features.

The Ancestral Mammal-like Reptile

It is universally accepted by zoologists that the mammals have originated from the stem reptiles called the **cotylosaurs**. Many related lines (*i.e.*, groups) of early reptiles developed certain mammalian features. The one line that succeeded in giving rise to mammals consisted of **cynodonts** (Fig. 17.15). From this line, during the Triassic period, arose the mammal-like reptiles which produced the reptile-like mammals that finally evolved into the true mammals. The early mammal-like reptiles (*e.g.*, *Verenope*, a fossil from Texas) were small animals (about the size of mice) but survived probably because of their speed. The later mammal-like reptiles became quite large and abundant. One of the most advanced of these later mammal-like reptile was *Cyanognathus* (dog-jaw) (Fig. 17.15) from South Africa. This animal was a 120 to 150 cm long, four-footed carnivore with a mixture of reptilian and mammalian characters. Skull had lost several bones and developed two occipital condyles. Cranial cavity was still small and brain typically reptilian. Secondary palate was present. Teeth were **heterodont**. This "**lizard-dog" animal** was, thus, intermediate between reptiles and mammals.

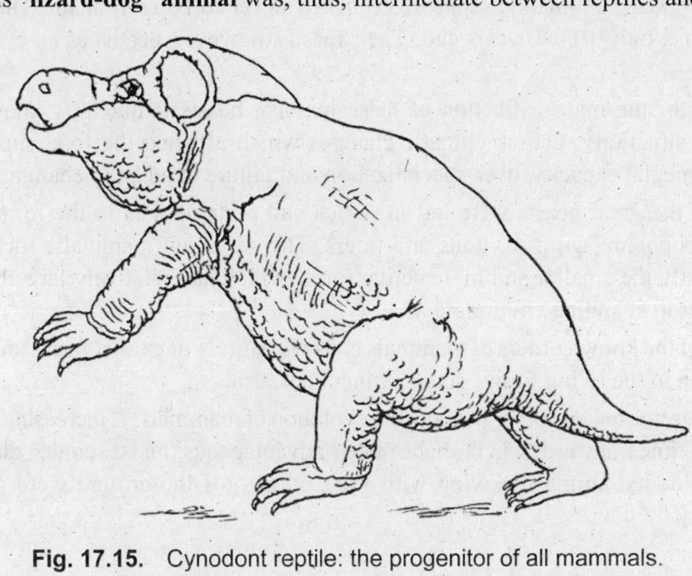

Fig. 17.15. Cynodont reptile: the progenitor of all mammals.

Evolution of Mammals

The **Prototheria** and **Theria** have separate lines of descent from the mammal-like reptiles. Which group of mammal-like reptiles gave rise to the primitive prototherian is not known. The reptile-like prototherian seem to have remained primitive due to isolation, as they occur only in Australia and New Zealand. The earliest known fossil mammals are the **Pantotheria**, which are the Jurassic members of the subclass Theria. The Pantotheria gave rise to he **Metatheria** and the **Eutheria** probably in the Cretaceous period. The marsupials began to expand first. By the close of Cretaceous, they became widely distributed, being present in Australia and elsewhere. Later, except in Australia, the marsupials became progressively less common until, at present, they occur only in the tropical and subtropical parts of New World. The decrease outside Australia is presumed to have resulted from competition with the placental mammals. The placental mammals (Eutherians) began to flourish a little later, and their descendants became widely distributed, except in Australia. Eutherians remained more successful and became the dominant animals in the Cenozoic era.

The early mammals were small shrew-like insectivores and opossum-like marsupials. They lived for millions of years in dangerous closeness to the huge, terrible dinosaurs. The probably saved themselves by leading a secretive nocturnal life in the trees or in burrows. Their original survival and ultimate dominance was due to their sensitivity to changing environments, their active life on account warm body and their remarkable ability to adapt to new and different situations by means of their precocious nervous system.

At the onset of the Cenozoic era, about 70 million years ago, the mammals flourished in both size and numbers. As the dinosaurs had perished, the mammals lost no time in exploiting the habitats vacated by them and soon evolved into a great variety of forms. First came the clumsy primitive mammals and then the modern types appeared. The Cenozoic ear (Palaeocene to Pleistocene) is called the **age of mammals** as the mammals were dominant throughout. The ancient mammals were almost as large and fearsome animals as were the dinosaurs before them. Huge woolly mammoths about 3 meter high at the shoulders, giant mastodons and sabre-toothed tigers with 15 cm long sabre-like canines came to be the undisputed rulers of the Earth. These heavy massive mammals remained confined to their own continents for a long time, but with the rise of the Isthmus of Panama and the Bering Straits in the early Pleistocene period, one million years ago, they undertook intercontinental migrations. They intermingled, interbred and flourished, repoducing the great variety of mammals that ever lived in the world. Then, as suddenly and mysteriously as the ruling reptiles prior to them, the great mammals disappeared quickly, never to be seen again. This happened in the later Pleistocene, only 10,000 years ago. Their fossils, however, persist as an evidence of their fading splendour.

The reason for the mass extinction of these massive beasts is not fully known. It may be a combination of situations such as climatic changes which affected the food supply, excessive competition, low mental capacity, over-specialization and failure to adapt to changing conditions.

Lineages of the great beasts were left in Africa and parts of Asia in the form of elephants, rhinoceroses, hippopotami, giraffes, lions and tigers. The surviving mammals, such as surviving reptiles, were mostly the smaller and more active forms which had relatively large brain and were able to adapt better to changing environments.

About half of the known orders of mammals consists entirely of extinct forms and all the others include, in addition to the living forms, some extinct ones also.

Two factors seems to have brought about the evolution of mammals: 1. increasing dry conditions from the Permian times onward, whcih made speed advantageous, and 2. colder glacial climates, in which warm blooded animals moving with their bodies off the ground were more likely to survive.

QUESTIONS

Long Answer Questions

1. Give an account of salient characters and outline classification of mammals.
2. Give a brief classification of class Mammalia stating important characters and examples of each order.
3. Write an essay on the origin and evolution of mammals.

Short Answer Questions

1. Write down the distinguishing characters of mammals.
2. What are egg laying mammals?
3. What are placental mammals?
4. Describe in brief the Chiroptera.
5. Give an account of Carnivora.
6. Examplain the differences between Old World monkeys and New World monkeys.
7. Distinguish between Indian and African elephants.
8. Write short notes on the following: (*i*) Bats; (*ii*) Duckbilled platypus; (*iii*) *Echidna*; (*iv*) Kangaroo; (*v*) Cetacea; (*vi*) *Sorex*; (*vii*) Pangolin; (*viii*) Hedgehog; (*ix*) Porcupine.

Very Short Answer Questions

1. What is toilet claw?
2. Define monotremata? (*CCSU Meerut, 2013*)
3. Write about egg-laying mammals.
4. What is a plantigrade foot? (*CCSU Meerut, 2012*)
5. What are flying foxes?
6. What is baleen?
7. Name two aquatic mammals of different orders which lack hindlimbs.
8. Cite two examples of order Sirenia.
9. Which is the order of Camels?
10. Give the name of the tallest mammal.
11. How do horse and ass differ with regard to their mane?
12. In which mammals both the pairs of limbs are paddle-like?
13. Give the scientific name of mongoose.
14. Claws are retractile in cat or dog.
15. Which teeth grow into tusks in elephants?
16. How do the Indian and African elephants differ regarding their proboscis?
17. Name a freshwater Cetacean.
18. To which order do rabbit and hare belong?
19. Which mammals have true flight?
20. Give examples with Zoological names of following?
    (*i*) Two aquatic mammals;
    (*ii*) Two flying mammals;
    (*iii*) Two egg-laying mammals;
    (*iv*) Two pouched mammals.
21. Give example with scientific name of following:
    (*i*) A mammal with scaly skin;
    (*ii*) Largest aquatic mammal;
    (*iii*) A mammal with dermal skeleton of bony scutes and horny plates.

22. Give Zoological or scientific names of the following:
(*i*) Flying squirrel;
(*ii*) Chimpazee;
(*iii*) Dog;
(*iv*) Ganges dolphin;
(*v*) Orangtun.

23. What is Chiroptera?

24. What is the role of ears in the flight of bat?

25. In which order of Mammalia sea cows or manatees are included.

26. What is the advantage of binocular vision in primates?

27. Explain the difference between old world monkeys and new world monkeys?

28. What the subcutaneous fat of whales is called?

29. What is gynaecomastism?

30. What is diastema? *(CCSU Meerut, 2005)*

Fill in the Blanks

1. Oviparous mammals are ________. *(Garhwal, 2002)*

2. ________ shows transitional link between ________ and ________. *(Garhwal, 2003)*

3. ________ is often referred as "Age of mammals".

4. Opassum is ________. (oviparous/viviparous) *(Purvanchal, 2003)*

5. The number of cervical vetebrae in all mammals is ________. (nine/seven)

6. Kangaroo belongs to the family called ________.

7. Mammals have ________ for conservation of body temperature.

8. Pinnae are found only in ________.

9. Oil glands of mammals are also called ________ glands.

10. Middle ear contains malleus, ________ and ________ in mammals.

11. Most ________ are viviparous.

12. Testes generally descend into ________ in mammals.

13. Main excretory product in mammals is ________.

14. A transverse partition, the ________, separates the thoracic and abdominal cavities in rabbit.

15. Tusks of elephants are modified incisor teeth of ________. (lower/upper)

16. In toothed whales teeth are ________ and monophyodont. (heterodont/homodont)

17. Lophodont dentition is found in ________. *(Garhwal, 2003)*

18. ________ are homoiothermic vertebrates like birds.

19. ________ is the thick subcutaneous layer of fat in aquatic mammals. (baleen/blubber)

20. Blubber provides ________ and retains ________.

21. ________ camel of cold rocky deserts has two humps. (Bacterian/Arabian)

22. The vocal cords of ________ produce ultrasonic sounds, inaudible to human beings. (Rat/Bat)

23. Marine carnivores belong to suborder ________ and their most common example is ________.

24. ________ is an oviparous mammal.

25. A common marsupial is ________.

26. There is no cloaca in ________ mammals.

27. Tail is prehensile in ________ world monkeys.

28. Flight membranes of bats are called ________.

29. The largest living mammal is ________.

30. Unguligrade locomotion is characteristic of Perissodactyla and ________.

31. Primates include humans, ________ and ________.

32. ________ are tailless primates.

Multiple Choice Questions

1. Chief features of mammals which distinguish them from other vertebrates are
(*a*) hairs and oviparity
(*b*) mammary glands,hairs on skin and diaphragm
(*c*) mammary flands and teeth
(*d*) pinna and teeth

2. The spiny ant-eater is a prototherian mammal. It is found in
(*a*) India (*b*) Africa
(*c*) Australia (*d*) Japan

3. Which of the following glands is found only in mammals?
(*a*) oil glands (*b*) sweat glands (or sudorific glands)
(*c*) secretory glands (*d*) scent glands

4. *Tachyglossus* is a representative of
(*a*) Prototheria (*b*) Metatheria
(*c*) Eutheria (*d*) Proboscidia

5. Marsupium is found in
(*a*) *Pteropus* (*b*) *Macropus*
(*c*) *Ornithorhynchus* (*d*) *Manis*

6. Marsupials are
(*a*) oviparous (*b*) ovoviviparous
(*c*) viviparous (*d*) none of the above

7. National animal of India is
(*a*) cow (*Bos indicus*) (*b*) elephant (*Elephas maximus*)
(*c*) tiger (*Panthera tigris*) (*d*) lion (*Panthera leo*)

8. A male ass is called
(*a*) hybrid (*b*) jack
(*c*) mare (*d*) mule

9. The leopard or 'Tendwa' zoologically named as
(*a*) *Panthera tigris* (*b*) *Panthera leo*
(*c*) *Panthera unica* (*d*) *Panthera pardus*

10. Carnassial teeth are found in
(*a*) herbivorous mammals (*b*) carnivorous mammals
(*c*) insectivorous mammals (*d*) ruminant mammals

11. Gnawing mammals belong to order
(*a*) Lagomorpha (*b*) Rodentia
(*c*) Both Lagomorpha and Rodentia (*d*) Rodentia and Edentata

12. Which of the following is a ruminant
(*a*) horse and ass (*b*) donkey and goat
(*c*) cow and buffalo (*d*) rhinoceros and camel

13. Body of this mammal is covered by large, thick, overlapping, horny scales in
(*a*) porcupine (*b*) hedgehog
(*c*) pangolin (*d*) hare

14. Tusks occur in
(*a*) elephant (*b*) walrus
(*c*) dugong (*d*) all of these

15. *Loris* belongs to the order
(*a*) Primates (*b*) Rodentia
(*c*) both of these (*d*) none of these

16. Hindlimbs are lacking in
(*a*) *Balanoptera* (*b*) dugong
(*c*) both of these (*d*) none of these

17. Seal belongs to the order
(*a*) Sirenia (*b*) Lagomorpha
(*c*) Carnivora (*d*) Proboscidia

18. Perissodactyla include odd-toed hoofed or ungulate mammals such as
(*a*) cows, buffalo and camel (*b*) horse, zebra and rhino
(*c*) ass, horse and camel (*d*) deer, stags and zebra

19. Locomotion in humans is
(*a*) plantigrade (*b*) digitigrade
(*c*) unguligrade (*d*) semidigitigrade

ANSWERS

Very Short Answer Questions

1. In spiny anteater (*Ecidna*), the second digit in each hindleg has a long and curved claw which is used for cleaning the spines. So, it is called toilet claw.
2. Monotremata is a order of subclass Prototheria of class Mammalia. These primitive mammals have a cloaca and lay hard shelled amniote eggs. They produce milk by mammary glands which lack teats, *e.g.*, only three genera: 1. *Ornithorhynchus* (Duck-bill platypus); 2. *Tachyglossus* or *Echidna* (Short-nosed ant-eater) and 3. *Zaglossus* (Long-nosed ant-eater).
3. Egg-laying mammals are monotreme mammals that have a cloaca and lay hard-shelled amniotic eggs, *e.g.*, Duck-bill platypus (*Echidna*) and two kinds of spiny ant-eater—*Tachyglossus* and *Zaglossus*.
4. In plantigrade foot all the five digits and metatarsals touch the ground.
5. Some fruit eating bats (Megachiroptera) are quite large in size and are sometimes called **flying foxes**, *e.g.*, *Pteropus*.
6. Baleen is a group of thin keratin plates that grow in the whale's mouth instead of teeth and act-like a stainer for catching food. Sturdy but flexible, baleen was called **whalebone** by 19th century hunters, who sold it to use in corsets, hoop skirts and umbrella.
7. Whale and dugong.
8. Dugong and manatee.
9. Artiodactyla.
10. *Giraffa cameloparadalis*.
11. Mane is hanging in horse, erect in ass.
12. Seals and walruses.
13. *Herpestes edwardsii*.
14. Cat.
15. Second pair of upper incisor.
16. Proboscis has a single sensory lobe in Indian elephant and two in African elephant.
17. Dolphin (*Platanista gangeticus*).
18. Lagomorpha.
19. Bats.
20. (*i*) *Halicore* (Dugong), *Trichechus* (Mantee). (*ii*) *Desmodus* (Vampire bat), *Pteropus* (Fruit bat). (*iii*) *Echidna* or *Tachyglossus* (Short nose spiny ant-eater), *Ornithorhynchus* (Duck-bill platypus). (*iv*) *Macropus* (Kangroo), *Dedelphis* (Opossum). (*v*) *Talpa* (Mole), *Sorex* (Shrew). (*vi*) *Loris tardigradus* (Loris), *Maccaca mulata* (Rhesus monkeys).
21. (*i*) *Manis* (Pangolin or Scaly ant-eater). (*ii*) *Balaenoptera* (Blue whale). (*iii*) *Dasypus* (Armadillo).

22. (*i*) *Galaeopithecus*. (*ii*) *Pan*. (*iii*) *Canis familaris*. (*iv*) *Platanista*. (*v*) *Pongo*.

23. Chiroptera is an order of flying mammals or bats in which forelimbs are modified into wings called patagia.

24. The ears of bats have a echo-apparatus, a kind of radar. The reflected echo is collected by the highly sensitive ears, and utilised to avoid objects during flight.

25. Sirenia.

26. In binocular vision instead of two images only one is seen due to overlapping of their fields of vision. So the object is seen in three dimensions which gives a sense of distance of the object in view.

27. Old world monkeys are found in Asia and Africa. These are relatively large with a moderate, non-prehensile tail. The new world monkeys are found in Central and South America. These are relatively small, with a long prehensile tail.

28. It is called **blubber** which insulates the body.

29. Gynaecomastism is the condition in which both male and female possess mammary gland, secrete milk and nourish the young ones.

30. The mammals included in Rhodentia, the second incisors, canines and anterior premolars are missing. They leave gaps, called **diastema**. Diastema also occurs in rabbit which is Lagomorph.

31. Vampire bats are microchiropteran bats of South America. They feed on blood of sleeping mammals such as horses, cattles and even human beings. Their upper incisors are modified into sharp cutting blades. They can transmit rabies.

Fill in the Blanks

1. Prototheria. **2.** Reptiles, mammals. **3.** Coenozoic era. **4.** Viviparous. **5.** Seven. **6.** Macropodidae. **7.** Hair. **8.** Mammals. **9.** Sebaceous. **10.** Incus, Stapes. **11.** Mammals. **12.** Scrotum. **13.** Urea. **14.** Diaphragm. **15.** Upper. **16.** Homodont. **17.** Proboscidea. **18.** Mammals. **19.** Blubber. **20.** Buoyancy, body heat **21.** Bacterian. **22.** Bat. **23.** Pinnipedia, common seal (*Phoca*). **24.** *Tachyglossus*. 25. *Macropus giganteus*. **26.** Eutherian. **27.** New. **28.** Patagia. **29.** Whale. **30.** Artiodactyla. **31.** Apes, monkey. **32.** Apes.

Multiple Choice Questions

1. (*b*)	**2.** (*c*)	**3.** (*b*)	**4.** (*a*)	**5.** (*b*)	**6.** (*c*)
7. (*c*)	**8.** (*b*)	**9.** (*d*)	**10.** (*b*)	**11.** (*c*)	**12.** (*c*)
13. (*c*)	**14.** (*d*)	**15.** (*a*)	**16.** (*c*)	**17.** (*c*)	**18.** (*b*)
19. (*a*)					

18

Dentition of Mammals

Teeth are dermal derivatives of integument developed as a result of calcification in the mucous membrane of buccal cavity in a row along the ridge of each of the two jaws. These are present in almost all mammals with few exceptions. Adult monotremes (duck-billed platypus and spiny ant-eater), pholidotes (pangolins), some edentates (giant ant-eater) and certain cetaceans (whale-bone whale) lack teeth. In tachyglossus, the teeth are absent even in embryo.

Teeth are embedded in cavities, the **sockets** or **alveoli**, of the jaw bones, namely, the premaxillae, maxillae and dentine. The upper and lower teeth are provided with projections and facets that interdigitate, so the upper set of teeth fit with the lower. This fitting is called **occlusion** and it allows even small particles of food to the caught and ground between the tooth surfaces.

18.1 STRUCTURE OF TOOTH

A typical mammalian tooth is placed in the socket or alveolus over the jaw bone and is distinguished into the three main parts (Fig. 18.1).

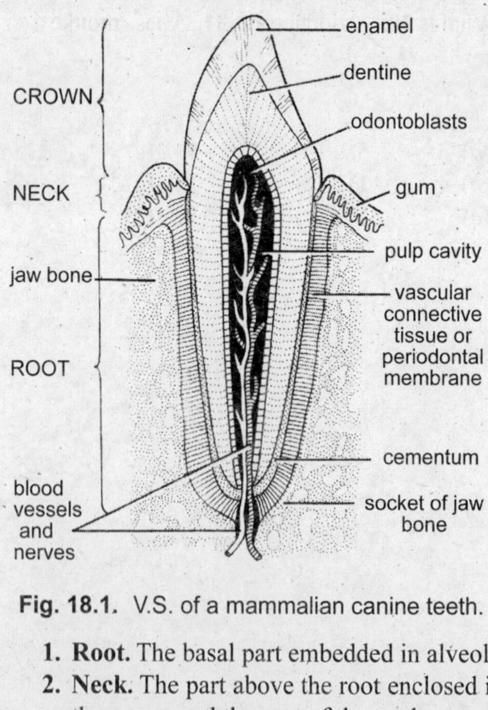

Fig. 18.1. V.S. of a mammalian canine teeth.

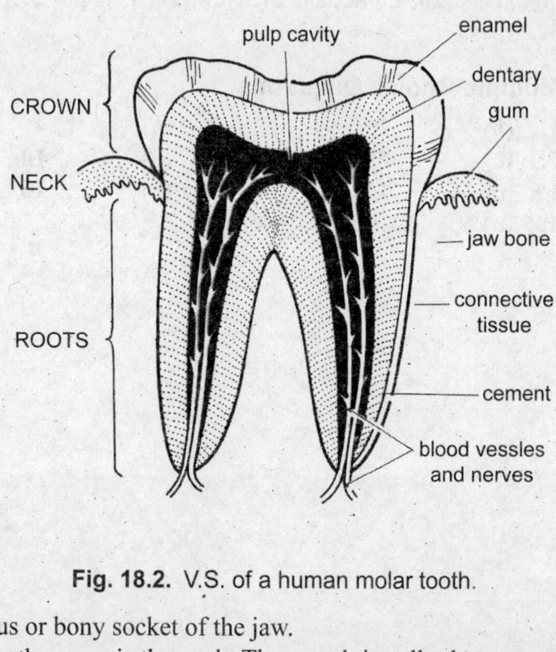

Fig. 18.2. V.S. of a human molar tooth.

1. **Root.** The basal part embedded in alveolus or bony socket of the jaw.
2. **Neck.** The part above the root enclosed in the gum, is the neck. Thus, neck is collar between the crown and the root of the teeth.

3. Crown. It is distal free part of tooth that protrudes out from the gum.

The root is separated from the alveolus by a vascular **peridontal** or **alveolar membrane**. In longitudinal section the tooth is found to consists of following structures:

(*a*) Enamel. The outer surface of the tooth is covered by a layer of enamel that is formed before eruption of the tooth by special epithelial cells called **ameloblasts**. Once the tooth has erupted, no more enamel is formed. Enamel is composed of large and dense crystals of hydroxyapatite with adsorbed carbonate, magnesium, sodium, potassium and other ions embedded in a fine meshwork of strong and almost insoluble protein fibers that are similar in physical characteristics (but not chemically identical) to the keratin of hair (see **Guyton** and **Hall**, 2011).

The crystalline structure of the salts makes the enamel extremely hard, much harder than the dentine (or dentin). Also, the special protein fiber meshwork, although constituting only about 1 percent of the enamel mass, makes enamel resistant to acids, enzymes and other corrosive agents because this protein is one of the most insoluble and resistant proteins known.

(*b*) Dentine. The main body of the tooth is composed of dentine, which has a strong, bony structure. Dentine is made up principally by hydroxyapatite crystals similar to bone but much denser (Box 18.1). These crystals are embedded in a strong meshwork of collagen fibers. Dentine is deposited and nourished by a layer of cells called **odontoblasts**, which line its inner surface along the wall of pulp cavity.

Box 18.1

The principal constituents of dentine are much the same as those of bone. The major difference is its histological organisation because dentine does not contain any osteoblasts, osteocytes, osteoclasts or spaces for blood vessels or nerves.

The calcium salts in dentine make it extremely resistant to compressional forces, and the collagen fibers make it tough and resistant to tensional forces that might result when the teeth are struck by solid objects.

(*c*) Cementum. Cementum is a bony substance secreted by cells of the peridontal membrane, which lines the tooth socket. Many collagen fibers pass directly from the bone of the jaw, through the peridontal membrane, and then into cementum. These collagen fibers and the cementum hold the tooth in place. When the teeth are exposed to excessive strain, the layer of cementum becomes thicker and stronger. Also, cementum increases in thickness and strength with age, causing the teeth to become more firmly seated in the jaws by adulthood and later.

(*d*) Pulp. The pulp cavity of each tooth is filled with **pulp**, which is composed of connective tissue with an abundant supply of nerve fibers, blood vessels, and lymphatics, The cells lining the surface of the pulp cavity are the odontoblasts, which during the formative years of the tooth, lay down the dentine but at the same time encroach more and more on the pulp cavity, making it smaller. In later life, the detine stops growing and the pulp cavity remains essentially constant in size. However, the odontoblasts are still viable and send projections into small **dentinal tubules** that penetrate all the way through the dentine; they are of importance for exchange of calcium, phosphate, and other minerals with the dentine.

18.2 CHARACTERISTICS OF MAMMALIAN DENTITION

1. Thecodont. Mammalian teeth are lodged in sockets or alveoli in the jaw bones (Fig. 18.3). (There are two more types of teeth attachments in vertebrates: (*i*) **acrodont type** which occurs in most vertebrates (shark, frog) and involves attachment of teeth with crest of jaw bones. (*ii*) **pleurodont type** which occurs in urodels and lizards and involves attachment of teeth to the shelf-like indentation at inner margins of the jaw bones).

2. Diphyodont. The mammalian teeth grow in two successive sets. The teeth of the first set are called the **milk teeth** or **deciduous teeth**. These appear early in life and are, after some time, pushed out by the second set of teeth termed the **permanent teeth** which last throughout the life of the animal. The permanent teeth are often more than the milk teeth. In tubulidentates (aardvark), however, the milk teeth outnumber the permanent teeth. In bats and guinea pigs the milk teeth are shed before birth. Insectivores, platypus, toothed whales, sloths and sirenians have **monophyodont dentition**, *i.e.*, they cut only milk teeth. Milk dentition is reduced to the replacement of third premolar in marsupials.

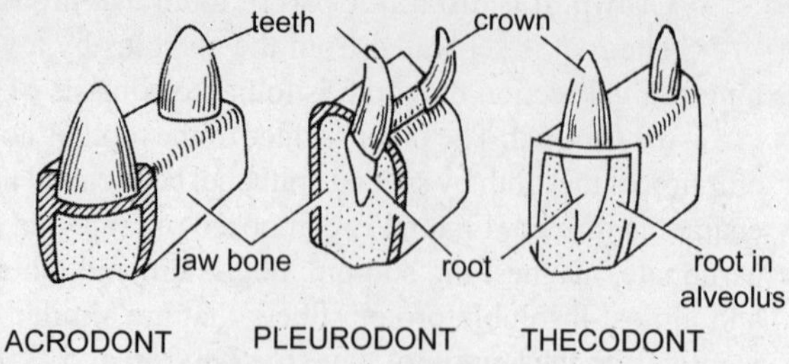

Fig. 18.3. Three ways of attachment of teeth to jaws in vertebrates.

3. Homodont. In dolphins and porpoises all teeth are similar as in fishes, amphibians and most reptiles.

4. Heterodont. Teeth of most mammals are differentiated in form and function to suite their food habits. Such teeth are said to form a **heterodont dentition**. Mammalian teeth are of following *four* types:

(*i*) Incisors. These teeth are present at the front part of each jaw, *i.e.*, on the premaxillae in upper jaw and dentary in the lower jaw. Incisors are chisel-shaped, have a single root and **monocuspid** (simple or undivided) crown with a sharp cutting edge. They are used for snipping, cropping and biting.

(*ii*) Canines. The canines occur next to the incisors, and there is a single such tooth in each half of each jaw. These are present on maxillae in the upper jaw and dentary in the lower jaw, between premolars and incisors. These used for piercing and tearing the prey, or in defence. Canines have single root and long, conical, monocuspid sharp crown.

(*iii*) Premolars. The premolars follow the canines. They are cheek teeth and borne on the maxillae. Premolars have a double root and compressed crown with one or two cusps. Premolars are used for shearing and slicing.

(*iv*) Molars. These are also cheek teeth, borne on the maxillae but occupy the posterior parts of jaws, lying next to the premolars. Molars have more than two roots and cusps. Molars always develop in the permanent set, having no milk predecessors. They are used for crushing and mastication.

Depending upon the shape, size and number of cusps, the cheek teeth (premolars and molars) are distinguished into the following types (Fig. 18.4):

1. Monocuspid teeth are with one cusp or tubercle.

2. Bicuspid teeth are with two cusps.

3. Tricuspid teeth are with three cusps. The tricuspid teeth might be **triconodont**, having all three cusps arranged in one plane otr **tritubercular** with three cusps forming a triangle.

4. Bunodont teeth with a number of distinct blunt pointed cusps, *e.g.*, human beings and monkeys.

5. Lophodont teeth are with cusps fused in the form of ridges, *e.g.*, horse and elephant.

6. Solenodont teeth are where cusps wear away, exposing the underlying dentine, which gets surrounded by crescent-shaped bands of enamel, *e.g.*, ruminats.

7. Secodont teeth are where cusps having sharp cutting surfaces, *e.g.*, dog and other carnivorous mammals.

8. Hypsodont teeth are with long crown and deeply placed neck.

9. Brachydont teeth are with short crown and neck at the surface of gum.

Diastema. The form of teeth is related to the food habits of the mammals. The insectivores (mole, bat) have narrow teeth; the carnivores have well developed canines, but poorly developed molars, except the first; the herbivores have short canines, but broad molars with enamel ridges for grinding. The cetaceans have sharp, conical, curved teeth for capturing fish.

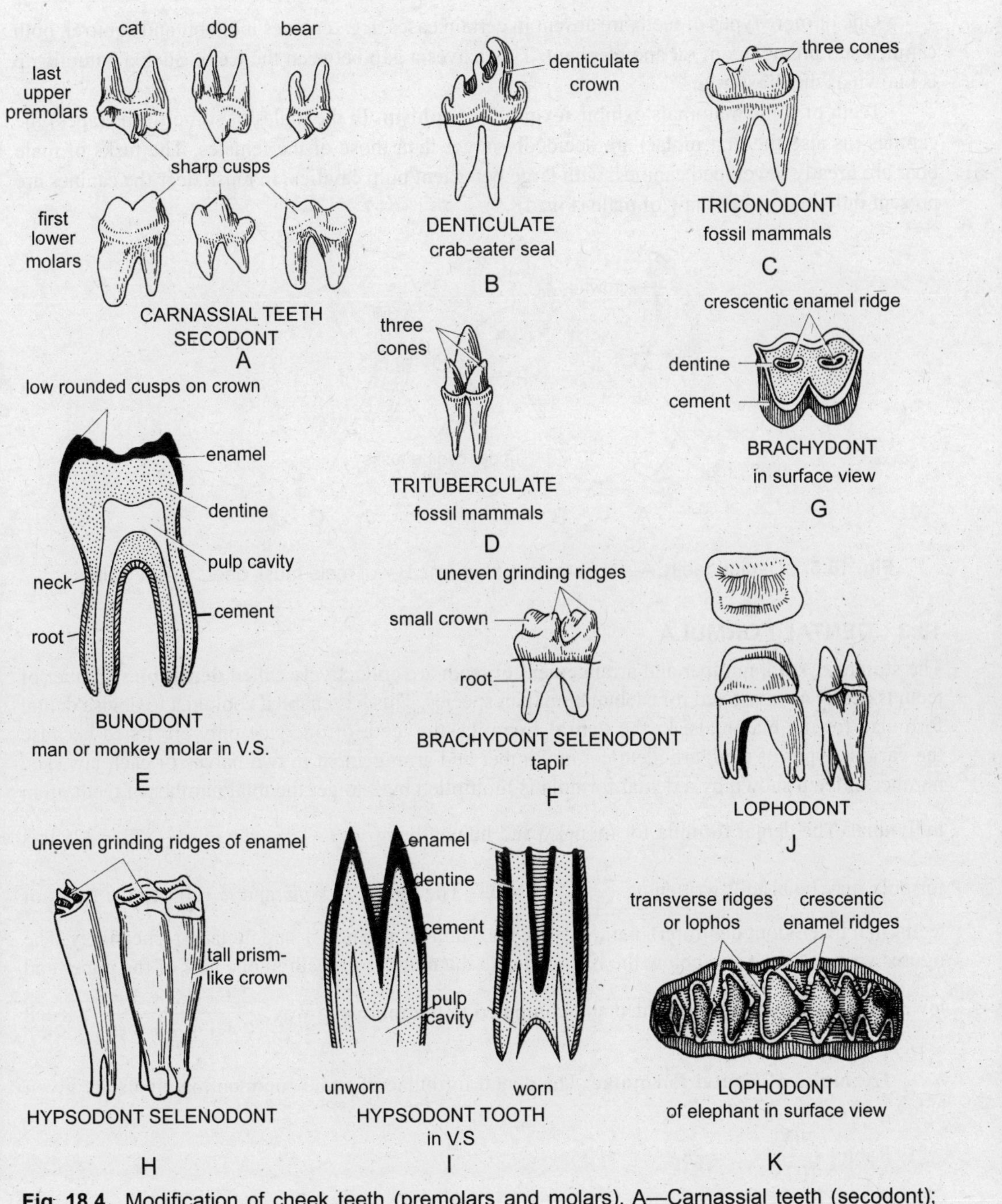

Fig. 18.4. Modification of cheek teeth (premolars and molars). A—Carnassial teeth (secodont); B—Denticulate molar; C—Triconodont tooth; D—Trituberculate tooth; E—Brachydont selenodont molar; G—Surface view of crown of brachydont molar; H—Hypsodont selenodont molars; I—Hypsodont teeth in V.S.; J—Lophodont molar; K—Lophodont in surface view.

One or more types of teeth are absent in certain cases, *e.g.*, canines in rabbit and squirrel, both canines and premolars in rat and elephant. This leaves a gap between the teeth. Such a natural gap is known as the **diastema**.

Teeth of some mammals exhibit **sexual dimorphism**. In the males of some forms (apes) the canines (as also the first molar) are decidedly larger than those of the females. The tusks of male boar are greatly developed canines, with large persistent pulp cavities. In musk deer the canines are present only in the upper jaw of male (Fig. 18.5).

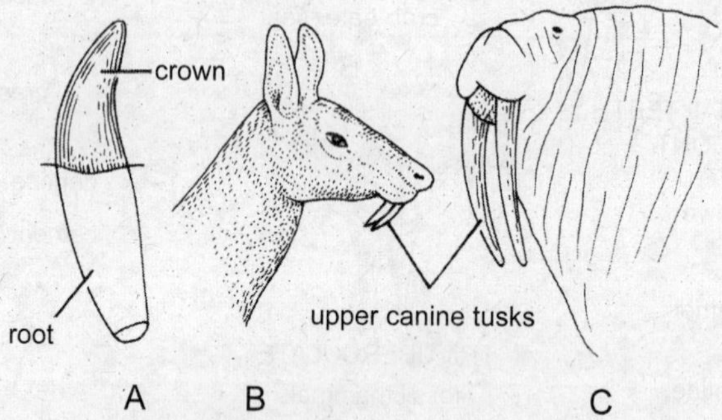

Fig. 18.5. Canine teeth. A—of carnivore (Juguar); B—of male musk deer; C—of walrus.

18.3 DENTAL FORMULA

The structure, kind, number and arrangement of teeth are collectively called **dentition.** Number of teeth is nearly always fixed for each mammalian species. This has enabled zoologist to devise dental formulae for the mammals. In the **dental formula**, the teeth of one side only are listed because the various types of teeth are identical in number and arrangement in two halves of each jaw. The number of teeth shown by a dental formula is multiplied by 2 to get the total number of teeth of an individual. The dental formula for moneky and human being is $i\frac{2}{2}, c\frac{1}{1}, pm\frac{2}{2}, m\frac{3}{3} \times 2 = 32$; this formula may be briefly written as $\frac{2123}{2123} \times 2 = 32$. The letters *i*, *c*, *pm* and *m* refer to the types of teeth (in a heterodont dentition), namely, incisors, canines, premolars and molars, respectively. The figures written above and below the lines indicate the number of teeth in one half of the upper and lower jaw. Thse numbers are added and computed in the following way: $\frac{2+1+2+3=8}{2+1+2+3=8}$; 8 + 8 = 16; 16 × 2 = 32.

Examples of Dental Formulae. The dental formulae of some common mammals are given below:

1. Rabbit $\frac{2033}{1023} \times 2 = 28$.
2. Rat $\frac{1003}{1003} \times 2 = 16$.
3. Dog $\frac{3142}{3143} \times 2 = 42$.
4. Cat $\frac{3131}{3121} \times 2 = 30$.

5. Cow, Goat and Sheep $\frac{0033}{3133} \times 2 = 32.$

6. Horse, Pig and Mole $\frac{3143}{3143} \times 2 = 44.$

7. Elephant $\frac{1003}{0003} \times 2 = 14.$

8. Kangaroo $\frac{3124}{1024} \times 2 = 34.$

The dentition of pig and horse is considered as typical in Eutheria, and that of others as modified. The toothed cetaceans (dolphins, porpoises) are peculiar among mammals in having indefinite number of teeth.

18.4 DEVELOPMENT OF TOOTH

The tooth develops partly from the epidermis and partly from the mesenchyme of dermis. Its enamel is derived from the epidermis while the remaining parts from the mesenchyme. Mammalian teeth develop in two sets. The **milk** or **deciduaous set** develops first in the buccal epithelium of the gums and is replaced later by the **permanent set**.

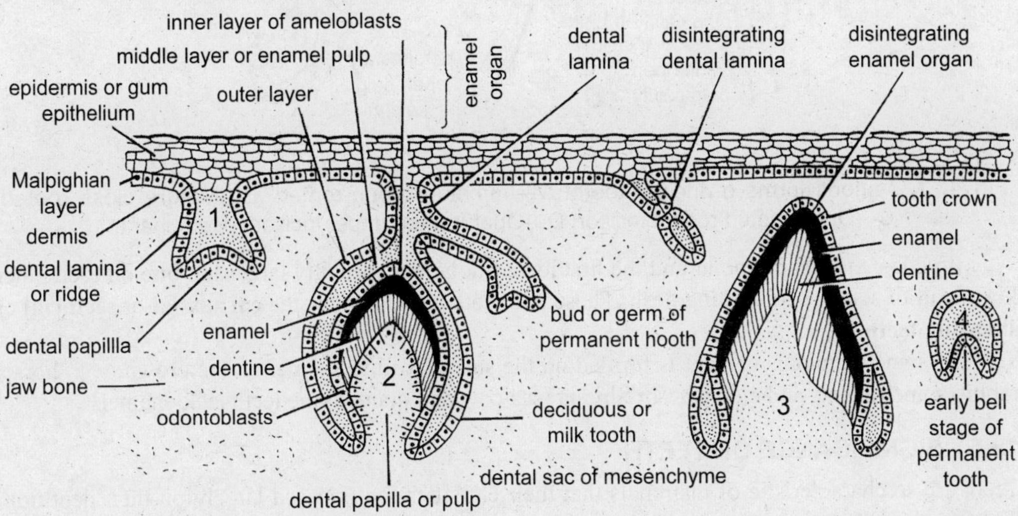

Fig. 18.6. Development of mammalian tooth. 1 to 3—Development of milk tooth; 4—Development of permanent tooth.

In the beginning the buccal epithelium of the gums sinks down forming the **dental groove** or **dental ridge**. Simultaneously, the cells of Malpighian layer of epidermis grows down into the dermis forming the **dental lamina**. Along the dental lamina, the mesenchymal cells multiply and condenses into a **dental** or **tooth bud**. This bud grows upwards and pushes the dental lamina, which assumes the form of an inverted cup. The cells of Malpighian layer become markedly columnar and are known as **ameloblasts**. The mesenchymal cells below it differentiate into a layer of **odontoblast cells**. The tooth bud is now known as **enamel organ**. The ameloblast cells secrete enamel towards their inner side, while the odontoblasts secrete dentine. The enamel surrounds the upper part of dentine like a cap. The remaining mesenchymal cells of the dental papilla form the **pulp** and the cavity of the enamel changes into the **pulp cavity**. As soon as the development of tooth is completed, the jaw bone forms a socket or **alveolus** around its base. The tooth increases in size by the secretion of more and more dentine and finally erupts out from the gum. The odontoblasts now secrete cement around the neck and root of tooth narrowing the pulp cavity.

In majority of mammals the odontoblasts stop functioning after the development of tooth is completed, and tooth ceases growth. In some cases, such as incisors of rodents and tusks of elephant, the odontoblasts are functional throughout life and teeth continue to grow. In them, the roots are not developed and the pulp cavity remains widely open being connected with the underlying mesenchyme. These are known as **rootless teeth** and lack enamel at the cutting edges. The rootless incisors of elephants and canines of wild boar and walrus are not worn away at the tip, hence they grow into large **tusks**. The latter are formidable weapons. The tusks occur in both the sexes in elephants and walrushes, but only in the males in wild boar and wart dog. On the whole, the teeth with roots have short crowns (such teeth are called brachydont) and rootless teeth have high crowns (called **hypsodont teeth**). Incisors, premolars and molars of dog are brachydont; incisors of rabbit and elephant, canines of wild boar and premolars and molars of horse are hypsodont.

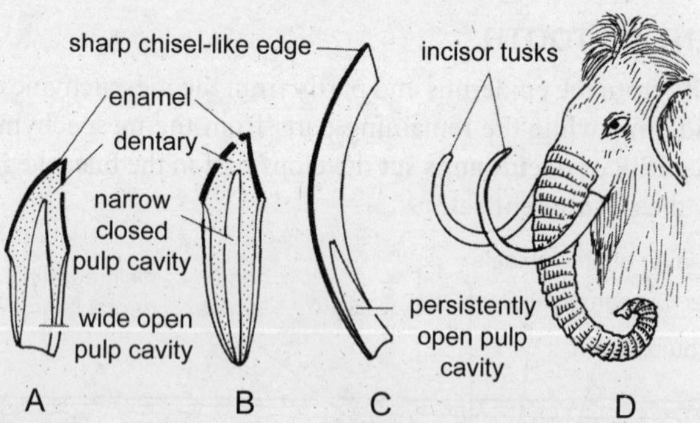

Fig. 18.7. Various forms of incisor teeth. A—Immature open-rooted; B—Mature close rooted; C—Open-rooted rodent incisor; D—Open-rooted upper incisors of elephant.

In dogs, the last premolar and the first lower molar are specially large, and bite on each other like a pair of scissors for cutting flesh. These two teeth are known as the **carnassial** or **sectorial** or **shearing teeth**.

In some mammals cement is formed on the surface of the enamel on the crown, *e.g.*, horse, rabbit, some rodents and elephant. In Sirenia (sea cow and mantee) the teeth lack enamel.

18.5 SIGNIFICANCE OF TEETH

Teeth are so characteristic of mammals that their classification is based largely on their dentition. Thus, study of dentition is important for taxonomic work on mammals. The number of teeth present gives the idea of the approximate age of mammals. Study of dentition has helped in deciding the pedgree or ancestry of certain mammals.

QUESTIONS

Long Answer Questions

1. Describe the structure of tooth. (*Garhwal, 2002*)
2. Describe the structure of mammalian tooth. How many types of teeth are there? Write your dental formula. (*Avadh, 2001*)
3. Describe the structure and development of typical tooth in mammals. (*Garhwal, 2002*)
4. What is dentition? Describe various types of teeth and their function. Write your dental formula. (*CCSU Meerut, 2012*)

5. What is dentition? Describe it in mammals. *(CCSU Meerut, 2011, 13)*
6. Write an essay on various types of dentition in mammals. *(CCSU Meerut, 2009)*

Short Answer Questions

1. Write short note on the following:
   (*i*) Heterodont dentiton *(CCSU Meerut, 2006)*
   (*ii*) Structure of tooth of a mammal *(CCSU Meerut, 2013)*
2. Draw a neat and labelled diagram of the following:
   (*i*) V.S. tooth of mammals *(CCSU Meerut, 2010)*
   (*ii*) Structure of tooth *(CCSU Meerut, 2009)*

Very Short Answer Questions

1. Why do human beings have heterodont dentition?
2. Name the molars having sharp cutting edges for tearing and cutting flesh.
3. What are tusks of elephants?
4. What is diastema? *(CCSU Meerut, 2005)*
5. What is hardest thing in human body?
6. What is study of teeth?
7. Give example of brachydont teeth.
8. Give example of hypsodont teeth.
9. What is bunodont? *(CCSU Meerut, 2008)*

Fill in the Blanks

1. The teeth absent in rodents are ________. (Molar/Incisor)
2. In carnivorous mammals last upper premolar in upper jaw and first molar in ________ jaw are called carnassial teeth. (Posterior/Lower)
3. In herbivorous grazing mammals such as horse teeth are elongated ________ shaped with high ________.
4. In mammalian teeth the crescentic highly intricate cusps of enamel connected by several transverse ridges are called lophos and are found in ________. (Secodont/Lophodont)
5. During development of teeth ________ cells secrete dentine layer and ________ form a cap of hard enamel on dentine.
6. Roots of incisors are open in ________ and lagomorpha so these continue to grow throughout life. (Rodents/Bats)

Multiple Choice Questions

1. Maximum number of teeth are found in mammals is
   (*a*) 44 as in pig and mole
   (*b*) 42 as in dog
   (*c*) 32 as in humans
   (*d*) 36 as in lemur and ant eaters
2. Molars which have sharp cutting edges of tearing and cutting flesh are called
   (*a*) selenodont
   (*b*) secodont
   (*c*) bunodont
   (*d*) hypsodont
3. Dental formula of cat is
   (*a*) $\frac{3131}{3121}$
   (*b*) $\frac{2033}{1023}$
   (*c*) $\frac{3124}{1024}$
   (*d*) $\frac{3142}{3143}$

4. Canines are modified into tusks for digging mollusks in
(*a*) tubulidentata (*b*) walrushes
(*c*) logomorpha (*d*) elephant

5. Cheek teeth in elephants are
(*a*) hypsodont (*b*) splenodont
(*c*) lophodont (*d*) brachydont

6. Horny plates or baleen are present on the jaws of
(*a*) crocodiles (*b*) whales
(*c*) turtles (*d*) sloths

7. In elephant the tusks are
(*a*) incisors (*b*) canines
(*c*) premolars (*d*) molars

8. Mammals are
(*a*) diphyodont (*b*) heterodont
(*c*) thecodont (*d*) all of these

ANSWERS

Very Short Answer Questions

1. Highly varied diet of human beings is reflected by multipurpose set of teeth as is common in omnivores.
2. Secodont.
3. These are upper incisors which are modified into tusks in elephants.
4. Diastema is a toothless space that exists between the last incisor and the first cheek tooth in rabbit.
5. Enamel of teeth's crown is the hardest thing in human body.
6. Odontology.
7. Incisors, premolars and molars of dog. They are short crowned teeth, *i.e.*, brachydont teeth.
8. Incisors of rabbit and elephant, canines of wild boar and premolars and molars of horse. They are high-crowned teeth, *i.e.*, hypsodont teeth.
9. Bunodont are molar teeth having tubercles or rounded cusps, *e.g.*, human beings.

Fill in the Blanks

1. Incisor. **2.** Lower. **3.** Prism, crown. **4.** Lophodont. **5.** Odontoblasts, amyloblasts. **6.** Rodents.

Multiple Choice Questions

1. (*a*) **2.** (*b*) **3.** (*a*) **4.** (*b*) **5.** (*c*) **6.** (*b*)
7. (*a*) **8.** (*d*)

19

Zoogeography

Animal geography or **Zoogeography** is concerned with the distribution of all the animals over the whole world (Box 19.1). There are nearly 1 million species of animals which comprise invertebrates and vertebrates, the terrestrial and aquatic creatures. It is practically impossible to cover up and study the distribution of each and every species. Most commonly the geography of land and freshwater vertebrates is taken into account, which constitutes just about 2% of the total animal strength.

Box 19.1

Biogeography is the study of the distribution patterns of organisms in various places of earth. It includes the study of the present day location of various species as well as their ancient distribution. Biogeography seeks to describe the origin of each species, its spread through time and space and to provide an explanation of the forces involved with establishing these patterns.

Biogeography of animals is called **zoogeography** while biogeography of plants is called **phytogeography**. A **biogeographical realm** is a region characterised by a distinctive biota.

Distribution of animals can be studied at three levels: **1. geographical distribution** over the whole world; **2. regional distribution** in selected segments of the world; **3. local distribution** which includes geographical distribution of species in relation to each other and to ecology and evolution. The distribution of animals is called **fauna** and that of plants is called **flora**.

The **distribution of chordates** is most important from the geographical point of view. In zoogeography the distribution of five main classes of vertebrates, *i.e.*, pisces, amphibians, reptiles, birds and mammals is studied.

In zoogeography, the area of distribution of animals is represented on the map. Such a map shows the range of the family in question, and is called a **distribution map**. Such a map can be plotted depicting the distribution of a particular group or species or it represents the distribution of all the animals. The former are called **simple maps** and latter the **compound maps**. There could be two possible methods of studying the geographical distributions of animals in the world:

1. Distribution of individual species over the world.

2. To divide the world into faunal regions (= realms) and describe their regional vertebrate fauna and their relationships.

19.1 TYPE OF DISTRIBUTION

The distribution of animals varies greatly, but when range of one particular family unit is mapped, the simple pattern that emerges falls into three main categories, *i.e.*, **continuous distribution**, **discontinuous distribution** and **bipolar distribution.**

1. Continuous Distribution

Rats and cats among mammals, and hawks and cockoos among birds have a continuous distribution and a wide rage, *i.e.*, they have world-wide distribution. Crow family is another example of world-wide distribution but it is comparatively less widely distributed being absent from New Zealand. There are families which are continuously distributed but are confined to one continent or one part of the continent only (*i.e.*, they are not found on several continents). For example, giraffe is confined to Africa (Keny-Savannas) and marmoset monkeys in South America.

2. Discontinuous Distribution

Individuals of same species or related species are found to inhibit localised areas widely separated and with or without any continuity. These are described to have **discontinuous distribution**. A number of cases of discontinuous distribution can be cited from vertebrates:

1. Lung fishes (Dipnoi). The lung fishes which belong to class Dipnoi afford the notable example of discontinuous distribution among vertebrates. The class is represented by three genera each of which occupies a different island.

(*i*) *Protopterus* inhibits swamps and marshes of Africa.

(*ii*) *Lepidosiren* is found in marshes of South America.

(*iii*) *Neoceratodus* is found in the pools and rivers of Queensland in Australia.

The fossil history of *Neoceratodus* reveals that these lung fishes had a world-wide distribution in Devonian period, being represented in India, Africa and North America.

2. Galaxies. It is a teleost fish found in South America, Africa, Australia and New Zealand.

3. Limbless amphibians (Gymnophiona). The different genera of subteranean limbless amphibians are now confined to Africa, Southern Asia, East Indies and tropical America.

4. Pleurodian turtles. These are now confined to Southern Hemisphere, whereas their fossils have been obtained even from Egypt, India, New Zealand and North America.

5. Flightless birds (Ratitae). The group is represented by five genera, each of which is confined to an isolated island having no interconnections:

(*i*) Ostriches are found in hot deserts of Africa and Arabia.

(*ii*) Rheas are confined to plains of Argentina and South Brazil in South America.

(*iii*) Cassowaries inhibit dense forests of New Guinea and North Queensland in Australia.

(*iv*) Emus are found in grassy plains and open forests of Australian region.

(*v*) Kiwis are restricted to the damp forests of New Zealand.

6. Marsupial mammals. Except opossum all other marsupials are confined to South America, North America, Australia and New Zealand.

7. Tapir. Different species of tapirs are found to inhibit Malaya, Central and South America, North America, Australia and New Zealand.

8. Camels. Camel family is represented by two stocks, the true camels and lamas. The true camels are found in Asia and lamas are found in South America.

9. Elephants and antelopes. These are found in Africa with great apes such as gorilla and chimpanzee but are absent from Brazil in South America having similar environmental conditions. Instead it possesses the tapirs, sloths and prehansile tailed monkeys.

Theories About Discontinuous Distribution of Animals

Following explanations have been forwarded for explaining causes of discontinuous distribution of animals on the earth.

1. One possibility is that the animal forms which show a discontinuous distribution at present might have been evenly distributed throughout and had a wide range distribution. But due to the changes either in the earth crust or in the climate or the intrusion of organisms of different varieties, some natural **barriers** were introduced. The barriers, therefore, prevented the migration and intermingling of the animals of the same species or same group. Therefore, a population unable to cross the barrier became **isolated** into subgroup. The animals in the regions in between either perished or became fossils due to change in the climatic conditions or were devoured by the sudden appearance or introduction of predators.

There are many instances of the continuity and even distribution of animals over a wide range, where at present they have a discontinuous distribution. The **extinction** in the region in between can be due to the changes in the condition of the environment or due to competition with highly advanced forms. The survivors of the relics of past could not enjoy that perpetuation again. There are numerous **relics,** such as glacial relics, steppe relics and marine relics, all of which are now represent areas, whereas most of their contemporary forms occur in the form of fossils.

2. Another explanation for the occurrence of discontinuous distribution can be that the distant

places which have same animal group or genera at present might have been connected together in past by some physical connection, *i.e.*, the island by some **land bridge** or sea by **sea bridge**. These bridges permitted the migration of individuals. Later on, in due course of time, these intermediates connecting bridges get submerged, separating the continuous land into two rather isolated areas. In fact, boundaries of land and seas have changed in geological time. Once North America and Eurasia were continuous across the present Berring Strait, so the fossils of tapirs are found in both the continents, where these once lived and became extinct.

3. Bipolar Distribution

A number of species are restricted to the Arctic and Antarctic waters and there is no representative in the intermediate oceans. For example, *Onchorhynchus* (Pacific salmon) and *Lamna cornubica* (shark), *Didemnus albidum* (ascidians), *Peripetus*, *Limacina helicina* and *Clione* (mollusc) and a number of colentrates such as *Sardina*, *Botrynema*, *Lampra*, etc.

19.2 HISTORICAL

The first attempt to divide the world into zoogeographical regions was made in 1957 by **P.L. Sclater**, the secretary of Zoological Society in London. His zoogeographical divisions was based upon the different kinds of birds and other terrestrial vertebrates found in different regions. Before Sclater, in 1856 **Andrew Murray** proposed only four mammalian regions. In 1869, **W.T. Blanford** proposed the name **Malayan region** for Indian region; in 1871, **E. Blyth** divided the earth's surface into seven regions, which are further divided into 26 subregions. The circumpolar distribution in zones was given by **J.A. Allen** based on climatic conditions of the world. In 1876, **A.R. Wallace** supported Sclater's zoogeographical regions. **Sclater** and **Wallace** divided the world into following six zoogeographical realms (Box 19.2).

1. Wallace's Zoogeographical Regions

Darlington (1957) described Wallace's following zoogeographical regions in a modern way:

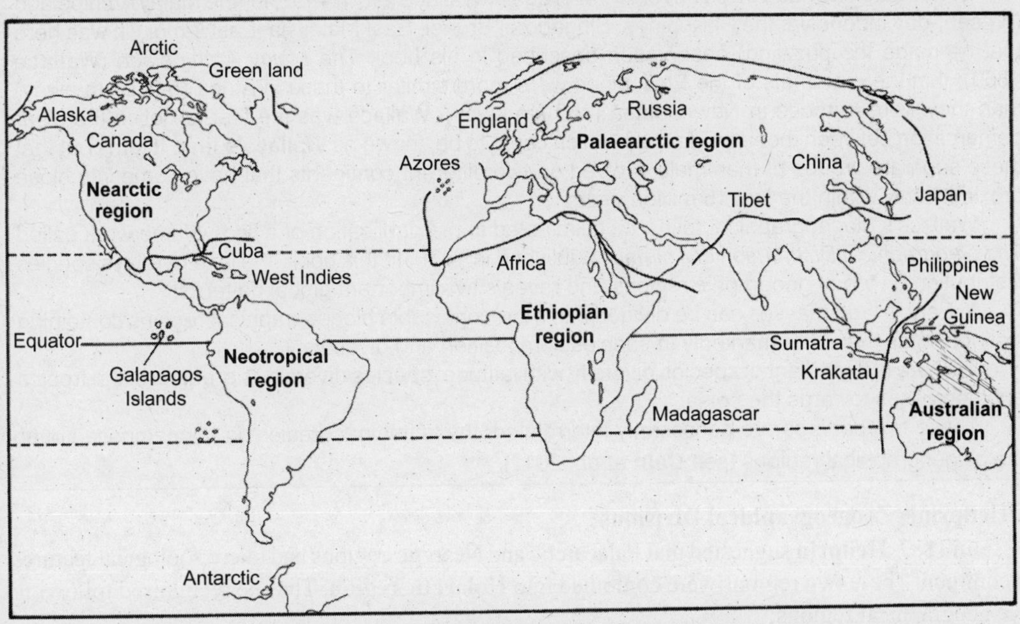

Fig. 19.1. Zoogeographical regions of the world.

1. **Palaearctic region**. It included Europe, North Africa, Asia (north and west of Himalaya) and Japan.
2. **Nearctic region**. It includes most of North America upto Mexico, Canada, Greenland and Iceland.

3. **Neotropical region**. It includes whole of South America, most of Central America, West Indies islands (Antilles islands).
4. **Ethiopian region**. It includes Africa, south of the Sahara (i.e., East Africa, West Africa, South Africa) and Madagascar and few other islands.
5. **Oriental (Sclater's) region**. It includes southern Asia from Baluchistan to Myanmar, South-East Asia and adjacent islands-Ceylon, Philippines, Borneo, Bali, etc.
6. **Australian region**. It includes Australia, Tasmania, New Guinea, Moluccans and polynesian islands such as New Zealand, Hawaii, Samoa, etc.

Since the time of Wallace several other regional classifications have been suggested but none of these has been found acceptable. This remains the basis of present-day zoogeographical division.

Box 19.2

One of the most obvious ecological patterns of Earth is the variation in species composition and diversity among geographic locations. The study of this variation is known as **biogeography** (**Cain**, **Bownan** and **Hacker** 2011).

Alfred Russel Wallace (1823–1913) is known as **father of biogeography**. Inspired by naturalists such as **Alexander Von Humboldt**, **Charles Darwin** and **Joseph Hooker**, Wallace came on the scene with considerably less wealth or education, but his intellect and motivation more than made up for what he lacked in financial resources and training. Wallace is best known, along with Charles Darwin, as the codiscoverer of the principles of Natural Selection. But his contribution was the study of species distributions across large spatial scale.

Wallace left England for Brazil in 1848 and explored the Amazon rainforest for 4 years. On his way back, the ship he was travelling on burned in the middle of the Sargasso sea, destroying all his specimens and most of his notes and illustrations. After 10 days in a lifeboat, he was rescued and made his way back to England, where he published an impressive six papers on his observation.

Even though he had vowed never to travel again, Wallace left, in 1852, for the Malay Archipelago (present-day Indonesia, the Philippines, Singapore, Brunei, East Malaysia, East Timor). It was here that he made the puzzling) observation described in his book *The Malay Archipelago* (**Wallace** 1869): that the mammals of the Philippines were more similar to those in Africa (5,500 km away) than they were to those in New Guinea (750 km away). Wallace was the first to notice the clear demarcation between these two faunas, which came to be known as **Wallace's line**. It turns out that these separate groups of mammals evolved on two different continents that have come into close proximity only within the last 15 million years.

Wallace's biogeographical research culminated in the publication of a two-volume work called *The Geographical Distribution of Animals*, published in 1876. In this book, Wallace overlays species distribution on top of geographic regions and reveals two important global patterns:

1. Earth's land masses can be divided into six recognisable biogeographical regions containing distinct biotas that differ markedly in species composition and diversity.

2. There is a gradient of species diversity with latitude: species diversity is greatest in the tropics and decreases towards the poles.

These two patterns are necessarily interrelated: the latitudinal gradient is superimposed over the biogeographical regions (see **Cain** *et at.*, 2011).

2. Heilprin's Zoogeographical Divisions

In 1887, **Heilprin** suggested that Palaearctic and Nearctic regions had many zoological features in common. These two regions were combined into **Holarctic region**. Thus he recognised following five geographical regions:

1. Holarctic Region
   (*i*) Palaearctic region
   (*ii*) Nearctic region
2. Ethiopian Region

3. Oriental Region
4. Neotropical Region
5. Australian Region

Another classification includes following zoogeographical regions:

1. Neogea = Neotropical
2. Notogea = Australian
3. Arctogea = Nearctic+ Palaearctic+ Ethiopian +Oriental

H. Robinson has presented the following zoogeographical classification:

I. **Arctogean** (Northern Land)
  (*a*) Holarctic
    (*i*) Palaearctic (East Land)
    (*ii*) Nearctic (West Land)
  (*b*) Ethiopian
  (*c*) Oriental
II. **Neogea** (New Land)
III. **Notogea** (Southern Land)
  (*a*) Australian
  (*b*) Polynesian – Hawaiian Island

19.3 SIX ZOOGEOGRAPHICAL REGIONS

I. ARCTOGEAN REALM

(The Northern Land)

The Arctogean Realm comprises North America upto Tropic of Cancer, Greenland and the northern half of old world which includes Palaearctic or Eastern Holarctic, Ethiopian and Oriental regions. It is the largest of all the realms but has least clearly defined fauna.

Characteristics of Arctogean Fauna

1. It contains highest mammals and most specialized birds. The mammalian fauna includes the most specialised ungulates and most advanced primates.
2. There are no monotremes and no marsupials (except American opossum).
3. The primitive reptiles are not represented in this realm.

The arctogean fauna represents striking homogeneity. This homogeneity is because of the land bridge existed between Asia and North America. Also there were no desert barriers as are present now which inhibit movement within Afro-Eurasian land mass. It is presumed that regionalisation of animal life within Arctogean realm is due to recent geological events, specially during pleistocene glaciation.

1. Holarctic Subregion

Geographical limits. Holarctic region comprises of Europe, Africa north of Tropic of Cancer, Asia north of Himalayan and North America.

Characteristic fauna. Holarctic region lacks monotremes, marsupials and edentates (toothless mammals). Except human beings and Barbary apes, primates are also not found in this region. Holarctic region is characterised by the most specialised forms of placental mammals.

Divisions

The Holartic region is subdivided into two regions—the eastern Palaerctic subregion and the western Nearctic subregion.

1. Palaearctic Subregion

I. Geographical Limits

Areawise this is biggest realm but without boundaries. It extends over whole of Europe, China, Japan, North of Africa (above Tropic of Cancer), north Sahara, Siberia, Mediterranian and Manchuria, Asia north of Himalayas and north of Arabia. It is bounded by sea to the west, north and east and by

Sahara desert and Himalayas to the south. The palaearctic region is, thus, a continuous land connection with two of its neighbouring regions, the Ethiopian and the Oriental.

The Palaearctic subregion is separated from Nearctic subregion by sea, but most probably had a land bridge called Bering Strait between eastern Siberia and Alaska, so that animals were able to migrate from Eurasia to North America or vice versa. Hence the animals in Palaearctic and Nearctic subregions are quite similar.

II. Climate

Climate of Palaearctic region is chiefly temperate. It includes both wet forests, dry open grasslands (steppes), Pyrenees (mountain chain between France and Spain), large areas of coniferous forests and fringe of Tundra (ice-covered regions). Eastern Asia including China and Japan contains deciduous forests. But most of the interior of Asia and North Africa is arid and open.

III. Subregions

The Palaearctic region has been subdivided into following *four* regions:

1. European subregion. It comprises of northern and central Europe, Black Sea and Caucasus. It includes the Arctic region in the north; Pyrenees grasslands; Balcans, Alps and Caucasus mountain ranges and Black Sea (or Euxine Sea) in south; Ural mountains in east and Ireland to Iceland in west. Its climate is temperate and it has coniferous forests and grasslands.

2. Mediterranean subregion. It includes areas of south of Alps, Balkans and Caucasus, portion of Arabia, whole of Mediterranean, Persia, Afganistan, Sahara desert of Africa and Baluchistan upto bank of Indus. This whole region has mountains and plateaus.

3. Siberian subregion. It extends from the Caspian Sea to Kamschatka and Bahring's straits and from arctic ocean to the Himalayas. The northern part of this subregion is occupied by the ice-covered Tundras. The remaining portion is covered by conifers, deciduous trees, grassy pastures and desert (Gobi and Turkestan deserts).

4. Manchurian subregion. It includes Korea, Japan, Manchuria upto Amur river and China upto Nanking mountains. The climate is mainly moderate.

IV. Fauna

(*i*) **Mammals**. Wolf, brown bear, sloth bear, bison, hedgehog, shrew, mole (*Talpa*), badger (*Meles),* bactrian camel (*Camelus*), roe deer (*Capreolus*), yak (*Poephagus*), chamois (*Rupicapra*), beaver, dormouse (*Myoxus*), etc., are common in European subregion. Only one genus of mammal *Myogale* (desmans) is peculiar to this subregion. In Mediterranean subregion, are found fallow deer, wild ass, elephant shrew, civet, ichneumon (*Herpestes*), stripped hyaena, hyrax and porcupine (all forming characterstic fauna of this subregion). Characteristic mammals of Siberian subregion include yalk, mole, antelopes, musk deer (*Moschus*), *Mustela*, glutton (*Gulo*), reindeer (*Rangifer*) and lemming (*Myodes*). Seal (*Phoca sibirica*) inhabits the great fresh water lake Baikal. In the Manchuria subregion are found Tibetan langur (*Rhinopithecus*), great panda (*Ailuropus*), Chinese water-deer (*Hydropotes*) and tufted deer (*Elaphodus*).

Box 19.3

Northern margins (Tundra) of Palaearctic region are inhibited by reindeer (caribou), arctic fox, arctic hare, polar bear, seal and walrus. The coniferous forest has moose, lynx and wolf and a number of other fur-bearing animals. In the deciduous forests bears, boars, deer, fox, squirrels, hedghogs and many burrowing animals are found. In steppe and semidesert areas are present herbivorous animals such as wild ass, horse, camel, sheep, goat, antelope, gazelle, jackal and corsac fox, European ground squirrel and Asiatic jerboa. Tibetan region has yak or tibetan ox. The Mediterranean area is a transitional zone and has some grazing animals and some desert creatures such as sand rats, jerboas, reptiles, eagles and hawks.

(*ii*) **Birds**. The avian fauna is characterised by the following birds: accentors, red breasts (*Erithacus*), reedlings (*Panurus*), jays (*Garrulus*), nut crackers (*Nucifraga*), finches (*Fringilla*), partidges (*Perdix*), pheasants (*Phasianus*), golden pheasants (*Chrysolophus*), grebes, loons, rains, hawks, vultures, herons, storks, ducks, wrens, cuckoos, kingfishers, swifts, wood peckers, larks, swallows, thrushes, crows, starlings, warblers, old world fly-catchers, etc.

Fig. 19.2. Fauna of Palaearctic region.

***(iii)* Reptiles**. There are no reptiles which may be restricted only to this region. However, the region is characterised by the presence of snakes (e.g., vipers, pit vipers), lizards, tortoises and alligators. Southern edge of Palaearctic region contains few other reptilian species: these are *Testudo*, *Trionyx* of eastern Asia, *Alligator* of China, *Chameleon*, *Varanus*, *Typhlops* and *Leptotyphlops* and sand boas are restricted to southern belt of Palaearctic region.

***(iv)* Amphibians**. The amphibian fauna includes the representatives of Urodela (tailed amphibians) and Anura (tail-less amphibians). Peculiar Urodela of this regon include blind white *Proteus*, *Salamandra*, *Hynobius*, *Chioglossa*, *Seirantoa*, *Onychodactylus*, *Geotrition* and *Sieboldia*. Other tailed amphibians of this region are newts, black and yellow European salamander, giant salamander from eastern Asia (5 ½ feet). Anuran fauna of this region include *Bombinator*, *Pelobates*, *Didocus*, *Alytes*, *Pelodytes*, *Discoglossus*, *Laprissa* and *Latonia*.

***(v)* Fishes**. It has about 28 specific genera of fresh water fishes belonging to families Salmonidae (*Salmo*), Cyprinoideae (Carps, *Cyprinus*), Cyprinodontidae, Cobitidinae, Percidae (Climbing perch, *Anabas*), Mestacembelidae and fishes such as catfishes, sticklebacks, sturgeons (*e.g., Acipenser*) and paddle fishes (*Polydon*).

2. NEARCTIC REGION

I. Geographical Limits

The Nearctic region include the whole of North America, Greenland, Mexican plateau, New Foundland and Aleutian islands (chain of volcanic islands, Alaska such as Fox islands). Except for a narrow strip of Central America it is completely cut off from all other regions by sea. It has extensive rocky range. In the North is the Arctic Greenland covered by thick layer of snow. This is followed by coniferous belt, deciduous or mixed forests in the eastern part of North America, extensive grasslands (prairies) in the central part and an arid zone (desert) in the south-west parts of Northern America.

II. Climate

Nearctic region has varied types of climate, *i.e.*, from arctic to tropical and has vast forests and prairies. Its climate resembles to that of the Palaearctic.

III. Subregions

The nearctic region is divided into following *four* subregions:

1. Californian subregion. It is a narrow strip of North America between Sierra, Nevada and Cascade range. In the North it includes Vancouver Island and southern part of British Columbia and in the South it extends upto head of the Gulf of California. Its climate is cool and temperate. Three families of vertebrates are peculiar to the region, namely Haplodontidae, Chamaeidae and Aniellidae. Vampire bats and free-tailed bats are characteristics of this region.

2. Rocky mountain subregion. It comprises of area east to the Californian subregion between Sanskatchewan in North and Cape St. Lucas in South. It is covered by mountain ranges and its climate is dry (desert). The characteristic animals of this region are prong buck (or prong horn, *Antilocapra*), rocky mountain goat (*Haplocerus*), American bison, praire dog (*Cynomys*) and poisonous lizard (*Heloderma*).

3. Alleghany subregion. It includes Eastern United States, *i.e.*, east of Rocky Mountain subregion and south of Great Lakes. It extends in the west across the river Mississippi, in the south near the Colorado river and to the north upto the boundary between Canada and the United States. The climate of this region is severely cold. This subregion is characterised by the following peculiar animals: opossum, star-nosed moles and vampire bats among mammals; passanger pigeons and Carolina parrots among birds. Turkeys (birds) and Sirenids (mud eels) are other two peculiar families.

4. Canadian subregion. It includes northern part of North America, Canada and Greenland. The climate is temperate and cool below freezing point. It has ice-covered arctic region and pine forests. The peculiar forms of this region are reindeer, bison, sheep, lemmings, gluttons (glutton is a voracious animal, *Gulo gulo*, of the weasel family), polor bears, elk and arctic fox.

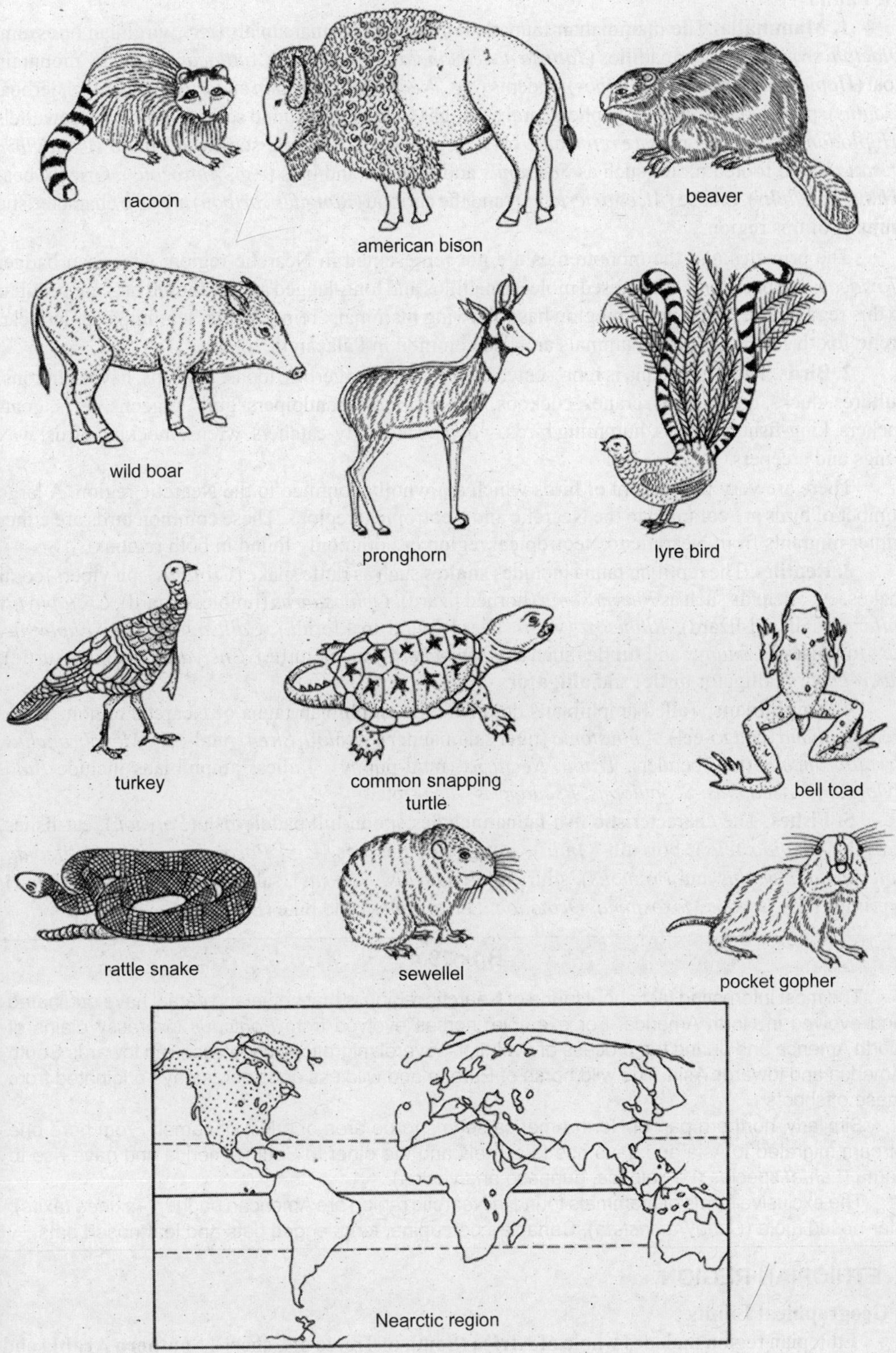

Fig. 19.3. Fauna of Nearctic region.

IV. Fauna

1. Mammalia. The mammalian fauna is represented by marsupials (*e.g.*, virginian opossum, *Didelphys virginiana*), armadillos (*Tatusia novemcincta*), prong-buck (*Antilocapra*), rocky mountain goat (*Haplocerus*), musk-ox (*Ovibos*), rodents (*e.g.*, *Neotoma*, *Sigmodon* and *Fiber*), rats (*e.g.*, jerboa, *Jaculus*), pouched rats, pocket gophers, prairie-dogs (*Cynomys*), ground squirrel (*Tamias*), sewellels (*Haplodon*), tree-porcupine (*Erethizon*), cats, dogs, bears, weasels, star-nosed mole (*Condylura cristata*), web footed moles such as *Scapanus* and *Scalops* and bats (*e.g.*, *Antrozous*). Grizzly bear (*Ursus horribilis*), moose (*Alces americana*) and the caribou (*Rangifer caribou*) are also characteristic animals of this region.

The primates and the monotremes are not represented in Nearctic region. American badger (*Taxidea*), tree porcupine, star-nosed mole, armadillos and long-lagged and leafnosed bats are peculiar to this region. The Canadian subregion has following mammals: reindeer, elk, lemmings, polar bear, arctic fox, bison, etc. These mammals are also common in Palaearctic region.

2. Birds. The avian fauna is represented by birds such as gerbes, loons, pelicans, hawks, herons, vultures, ducks, quails, rails, cranes, cuckoos, plowers, larks, sandpipers, gulls, pigeons, owls, goat-suckers, king-fishers, swifts, humming birds, wood peckers, fly-catchers, wrens, mocking birds, wax wings and creepers.

There are very few genera of birds which are wholly confined to the Nearctic region. A large number of birds are common to the Nearctic and Neotropical regions. These common birds are either winter migrants from Nearctic to Neotropical region or commonly found in both regions.

3. Reptiles. The reptilian fauna includes **snakes** such as rattle snake (*Crotalus*), pit vipers, coral snakes, etc.; **lizards** such as *Phrynosoma* (horned lizard), *Ophiosaurus* (limbless lizard), *Crotaphytus collaris* (collared-lizard), *Rhineura* (worm-lizard found in Florida), *Callisaurus*, *Uta*, *Euphryne*, etc.; **tortoises** as *Trionyx* and **turtles** such as *Aromochelys* (monk turtle), *Chelydra* (snapping turtle), *Macroclemys* (alligator turtle) and **alligators**.

4. Amphibians. Tailed amphibians dominate the amphibian fauna of Nearctic region. These are *Amphiuma* (congo-eel), *Ambystoma* (tiger salamander), *Axolotl*, *Siren* (mud-eel), *Menobranchus*, *Cryptobranchus* (hell-bender), *Triton*, *Necturus* (mud-puppy). Tailless amphibians include *Bufo*, *Hyla*, *Rana*, *Ascaphus*, *Scaphiopus*, *Pseudacris*, *Acris*, etc.

5. Fishes. The characteristic fish fauna includes spoon-bill/paddle-fish (*Polydon*), cat-fishes, gar-pike (*Lepidosteus*), bow-fin (*Amia*), Sturgeons, perches (*e.g.*, *Paralabrax*, *Huro*, *Pileoma*, *Boleosoma*, *Brytulus* and *Promotis*), silurids (*Hypodelus*, *Noturus*), salmon (*Salmo*, *Thaleichthys*), cyprinids (*e.g.*, *Meda*, *Moxostoma*, *Orthodon*, *Leucosomus*) and pike (*Esox*).

Box 19.4

The most interesting fact about fauna of Nearctic region is that horse and camel have originated and evolved in North America. For example, horses evolved from *Eohippus* in grassy plains of North America and during the process of evolution several migrations have occurred towards South America and towards Asia. The wild horse of Europe and wild ass and zebras have originated from these offshoots.

Similarly, northern part of North America seems to be area of origin of camel. From here one stream migrated to Asia and gave rise to camels and the other to South America and gave rise to llama (*Lama*) species (i.e., alpaca, guanaco and vicuna).

The exclusive forms of mammals found in Nearctic region are American badger (*Taxidea taxus*), star-nosed mole (*Condyra cristata*), Canadian porcupine, long-legged bats and leaf-nosed bats.

3. ETHIOPIAN REGION

1. Geographical Limits

Ethiopian region includes whole of **Africa** (South of Tropic of Cancer), **Southern Arabia** and **Madagascar**. **Darlington** did not include Medagascar in the Ethiopian region.

Ethiopian region has continuity with its northern neighbour (the Palaearctic Region) but Sahara desert acts as an effective barrier between the two. The remaining sides are all surrounded by sea.

2. Climate

Africa is mainly tropical country and has big rivers like those of Neotropical Region. It has large blocks of luxuriant and lofty evergreen forests in the equatorial region, mountain region and wide grassy plains (savanna) in the eastern part. Its southern part is warm temperate with mixed vegetation.

3. Subregions

Ethiopian region is divided into following subregions :

(*i*) East African subregion. It includes the Sahara, southern Arabia, north-east Africa and upto Zambesi in the south. It is moderately elevated plateau having grassy vegetation with interspersed patches of forest. It has hot (tropical) and dry climate.

It comprises of 145 families of terrestrial vertebrates of which one mammalian and one avian forms are crested rats and whale-headed bird. In addition rhinoceros, giraffes and zebras are characteristic of this region.

(*ii*) West African subregion. It includes Western Africa upto Congo south of river Gambia. It is almost wholly covered by dense forest of perennial vegetation. It has hot and moist climate.

Although it contains 134 families of terrestrial vertebrates, the peculiar families are absent. Gorillas, chimpanzees, monkeys, potamogalds (tenrecs), flying squirrels, okapi among mammals. *Pitta angolensis* (family Pittidae) are characteristic avian fauna of this region.

(*iii*) South African subregion. It includes north-west Kalahari desert, southern African Limpopo valley on the north-east and Mozambique. It has grasslands, dense forests and deserts. Its climate is warm temperate.

It includes about 13 families of terrestrial vertebrates. Family Promeropidae of mammals is the only unique family of this subregion. The other characteristic mammalian families are chrysochloridae (golden moles), Macroscelidae (elephant shrews), Pedetidae (African jumping hare), Bathyergidae (African mole rat), earth wolf and aardvark. Arian families are represented by secretary birds (Box 19.5) and ostriches.

Box 19.5

Secretary bird is a long-legged snake-eating African bird, *Sagittarius serpentarius*, with a crest likened to a quill pen stuck over a writer's ear.

(*iv*) Malagasy subregion. It includes the islands of Madagascar, Bourbon, Maurititius, Rodriguez, Seychelles and Comoro. The climate is warm temperate and region has dense forests.

It contains very interesting fauna. It contains 86 families of vertebrates. Of which 8 are confined to this region only. These are Chiromyidae (Aye-Aye), Cententidae (Trenecs) among mammals, Aerocharidae (helmet birds), Vangidae, Philepittidae, Mesoenatidae and Leptosomatidae among birds and Uropeltidae among reptiles.

There are 36 species of lemurs all belonging to peculiar genera and they are confined to Madagascar. There are 55 families of birds, five of which are peculiar.

4. Fauna

The fauna of the Ethiopian region is rich, varried and well marked.

Mammals. Mammalian fauna of Ethiopian region is most varied. There are about 38 families excluding bats. Following mammals are found to have wide distribution: shrews, rabbits, squirrels, cricetid mice (Cricetidae is a family of the order Rodentia including hamsters, voles and some mice), dogs, mustelids (such as pine marten, wolverine, skunks, polecat, weasel, common otter, stoat, badger, etc.), cats, bovids (*i.e.*, true antelopes). There are approximately twelve exclusive families and the rest are shared with either Palaearctic or Oriental region, whereas hedgehogs, porcupines, civets, hyaenas and pigs are shared with both. The Palaearctic families represented in Ethiopian region dormice (mouselike rodents of the family Gliridae, notes for their long hibernation), jerboas, coneys, rabbits or hyrax, and wild horses. The Oriental mammals of Ethiopian region are loris, Old World monkey (lemurs), apes (chimpanzee and gorilla), pangolins, bamboo rats, elephants (*Loxodonta*), rhinoceros and chevortains.

Twelve families of mammals are exclusive for the Ethiopian region. These are represented by giraffes, hippopotamus, aardvarks (Cape anteater), and warthog, three families of insectivores and six families of rodents. The insectivore families are represented by otter-shrews, golden moles, aye-aye and elephant-shrew. The rodents are Anomaluriae (*Anomalurus*), African flying squirrel, Pedetidae, African jumping hare, cane rats and rock rats. In addition different species of zebras and large number of antelopes are also unique. Rhinoceroses, horses and pangolins are distinctive of Ethiopian region.

Another characteristic of Ethiopian region is entire absence of otherwise widely distributed families such as Talpidae, Ursidae (bears), Cervidae (deer, goats and sheep) and wrens (short-winged songbirds).

Box 19.6

Essential Features of Mammalian Fauna of Ethiopian Region

1. During Pleistocene Ice Age, present Sahara Desert was a zone of steppe grassland which permitted migration of larger ungulates from Palaearctic region.
2. Creation of desert in the postglacial period served as barrier for the southward migration of recent forms such as panda, tiger, sheep and goat.
3. Endemic insectivorous mammals are golden moles, otter-shrews and elephant-shrews.
4. Grazing ungulates of Ethiopian region are antelopes, zebras, giraffes, elephants, black and white rhinoceroses and hippopotami. They occur in open grasslands and woodlands.
5. Predators such as lions and cheetah (with its spotted coat and black "tear drope" facial markings) are present in Ethiopian region.
6. The maned jackal (a small animal resembling hyaena), aardvark (Cape anteater) and the warthog (ungulate of pig family).
7. Okapi, a member of giraffe family is found only in this region.

(*i*) Birds. Ethiopian region has a rich avian fauna. Birds of this region include weaver birds, sun-birds, fruit thrushes, fly-catchers, herons, cuckoos, goat suckers, larks, thrushes, swallows, bee-eaters, ostriches, colies, hoopoes, helmet birds, secretory birds, barbets, crows, shrikes, hammer headed birds, mouse birds, plantain eaters, hawk-like birds, starlings, rollers, white eyes, hornbills, vultures and eagles. *Serpentarius* is most characteristic hawk-like bird of this region.

(*ii*) Reptiles. The reptilian fauna includes snakes, lizards, crocodiles, turtles and tortoises. Snakes belong to Colubridae (*e.g.*, *Leptorhynchus*, *Rhamnophis*, *Herpeterthiops*, *Grayia*, etc.), Dendrophidae (*e.g.*, *Hopsidrophis*, *Bucephalus*, etc.), Dryophidae (*e.g.*, *Langalia*), Dipsadidae (*e.g.*, *Pythonodipsas*), Lycodontidae (*e.g.*, *Boedon*, *Lycophidion*, etc.), Pytonidae, Elapedae and Viperidae (*e.g.*, *Atheris*).

Lizards belong to the following genera : *Monotrophis*, *Cordylus*, *Pseudoeordylus*, *Platysaurus*, *Cordylosaurus*, *Pleurostichus*, *Saurophis*, *Zonurus*, *Sphaenops*, *Scelotes*, *Sphaenocephalus*, *Sepsina*, *Pachydactylus*, *Agama*, *Chameleon*, *Mabuia* (skinks), African flying lizard (*Anomalurus*), etc.

Tortoises belong to the following genera *Cynyxis*, *Pyxis*, *Chersina* and *Cycloderma*.

(*iii*) Amphibians. The amphibians of this region are clawed-toads, frogs, tree frogs and certain caecilians. *Genus Rana* (frog) and *Bufo* (toad) are absent in Ethiopian region.

(*iv*) Fresh water fishes. Following fresh water fishes are peculiar to Ethiopian region: Nile-lung fish (*Protopterus*), *Polypterus*, *Notopterus*, *Clarias*, etc.

4. ORIENTAL REGION

I. Geographical Limits

Oriental region includes India, Ceylon (Sri Lanka), South China (Formosa or Taiwan), Indochina (Myanmar or Burma, Thailand or Siam, Laos, Cambodia, Vietnam and West Malaysia), Malaya Peninsula (or Malaysia), Indonesia or Malay Archipelago (*i.e.*, sea studded with islands) or Malayan

Islands such as Borneo, Java, Bali, Moluccans, New Guinea, Celebes (Sulawesi) and Sumatra and the Philippines.

It is bounded by the Himalayas in the north and the Indian and Pacific oceans on its other side but there is no definite physical boundary in the South-East corner where the islands of the Malay Archipelago string out until they reach Australian region.

II. Climate

Oriental region is known for its varied physical features. It presents tropical region, but northern part of India is temperate. Its eastern part including Indochina are rain forests. Western peninsular (Peninsula is a piece of land almost surrounded by water, and connected with the mainland by an isthmus) part of India is desert and southern portion of Indian subregion has luxuriant tropical forests. Similarly, Sri Lanka, Indochina and Malay contain thick forests.

III. Subregions

The Oriental region is subdivided into following *four* subregions :

1. Indian subregion. It includes whole of Indian continent including Bhutan, Sikkim, Nepal, Lakshadweep and Maldives. It has wide range of temperature variation and rainfall. It has mountain ranges (Himalayas, Vindhyas, Eastern and Western Ghats, Nilgiri hills), forests and deserts (*e.g.*, Thar desert).

It is characterised by the presence of 123 families, of which only colubrine snake (*Elachis todontidae*), antelope, black buck, India bear, shield tails, peacock and Indian bison are peculiar fauna of this subregion.

2. Ceylonese subregion. It includes the island of Ceylon (Sri Lanka). It has both mountains and forests. The climate is largely tropical.

It is characterised by the presence of shield tails, *Loris* and a species of spring rats.

3. Indo-Chinese subregion. It includes China south of Palaearctic region, major parts of Indochina Myanmar (Burma), Thailand (Siam), Laos, Cambodia, Vietnam, Formosa (Taiwan) and islands of Andamans.

It contains about 138 families. Panda, takin and Haplomyx are peculiar to this subregion. It also contains gibbons, flying lemurs, Java rhinoceros, Malayan tapir, disc-tongued frogs and *Salamander*.

4. Malayan/Indo-Malayan subregion. It includes Malaya peninsula and Islands of Malay Archipelago such as Sumatra, Java, Borneo, Bali, Philippines and Nicobar. It has tropical climate.

It contains 132 families. Orangutan (*Siama*), proboscis monkey (*Nasalis*) and Malayan badger (*Mydous*), tupaia (*tree shrew*), gibbons and flying lemurs, rhinoceros, Malayan tapir and broad bills are the forms found in Indo-Malayan subregion.

Palaeotropical Region

The fauna of Oriental region exhibits considerable resemblance with that of Ethiopian region. Due to such close resemblances between two regions, some zoogeographers have club them into one region called **Palaeotropical region**.

IV. Fauna

The fauna of Oriental region is neither rich in endemic families nor it includes wide spread families.

(*a*) Mammals. The characteristic mammalian fauna of this region includes gibbons, tarsiers, flying lemurs and tree shrews. Besides these, several well-known species and genera are quite characteristics of Oriental region. These are orangutan (*Simia satyrus*), macaque monkey (*Macacus*), tiger (*Felis tigris*), elephant (*Elephas maximus*), Malayan tapir (*Tapirus indicus*), rhinoceroses, cheverotains and pangolins. In the Indian subregion are found four-horned antelope (*Tetraceros*), black buck (*Antilope*), nylghaie/nilgai (*Boselaphus*), shield tails and sloth bear (*Melurses*). In Ceylonese subregion spiny rat and loris are characteristics. In Indochinese subregion are found takin (*Budorcas*), *Haplomys* (Muridae) and panda (*Ailurus*). In Indo-Malayan subregion are found proboscis monkey (*Nasalis*), Malayan badger (*Mydaus*), tapir, orangutan and flying lemur.

Fig. 19.4. Fauna of Ethiopian region.

Box 19.7

Oriental region is characterised by the presence of 153 families of terrestrial vertebrates. These are following:

Mammals: 1. Hylobatidae (Gibbon)
2. Tarsiidae (Tarsiers)
3. Galeopithecidae (Flying lemurs)
4. Tupaiidae (Tree shrew)

Birds: Eurylaemidae (Peacock)

Reptites: 1. Uropeltidae (Shield tails)
2. Lanthanotidae
3. Gavialidae
4. Platysternidae (Big-headed tortoise)

Four mammalian families are endemic to oriental region. These are shrews, rabbits, squirrels, cricetids, murid mice, dogs, mustellids, cats and bovides. Oriental region shares hedgehogs, porcupines, civets, hyaenas and pigs with both Palaearctic and Ethiopian regions. Elephants, rhinoceros and chevortains, etc., are shared with the Ethiopian region. But this region contains moles, bears, tapirs and deer. The exclusive families of this region are tree shrews, tarsiers, cynocephalus, spiny dormouse, flying lemurs and Indian bison.

(*b*) Birds. The avian fauna of this region include babblers (*Garrulax*, *Layardia*, *Timalia*, etc.), fork tailed *Enicurus* and rich blue *Myiophonus*, hill-tits, green bulbuls (*Phyllornis*), black and crimson minivets (*Pericrocotus*), king crown (*Dicrurus*), sunbirds (*Aethopaga*, *Nectarophila*, *Anthreptes*), flower peckers (*Piprisoma*, *Prionochilus*), weaver birds (*Ploceus*, *Munia*), glossy mynahs (*Eulabes*), swallow shrikes (*Artamus*), pittas (*Eucichla*, *Hydrornis*), broad bills (*Eurylaemus*), passers, woodpeckers, barbets, cuckoos, bee-eaters (*Nyctiornis*), kingfisher and hornbills, frog-mouthed goatsuckers (*Battrachostomus*), whiskered swifts (*Dendrochelidon*), parrots (long-tailed Palaeornis, *Loriculus*), pigeons (*Treron* and *Carpophaga*), peacocks (*Pavo*), argus pheasants (*Argusianus*), fire-backed pheasants (*Euplocamus*) and jungle fowl (*Gallus*).

(*c*) Reptiles. The reptilian fauna of this region is rich and characteristic and includes the following reptiles:

(*i*) **Snakes**. Ground snakes (Oligodontidae), fresh-water snakes (Homalosidae), tree snakes (Dendrophidae), whip snakes (Dryophidae), nocturnal tree snakes (Dipsadidae), fanged ground snake (Lycodontidae), rock snakes (Pythonidae), Venomous colubrine snakes (Elapidae), pit vipers (Crotalidae).

(*ii*) **Lizards**. *Chamaeleon*, *Calotes* (garden lizard), *Mabuia* (skink), geckoes, *e.g.*, *Hemidactylus* (wall-lizard), *Draco* (flying-lizard), *Varanus*, limbless lizard (typhlops) and eastern iguanas.

(*iii*) **Crocodiles**. Marsh crocodile (*Crocodilus porosus*), gavial or gharial (*Gavialis gangeticus*) are characteristics of the Ganges.

(*iv*) **Chelonia**. It includes turtles (*Chelone mydas*) and tortoises such as roofed-terrapin (*Kachuga tectum*), soft-river terrapin (*Trionyx gangeticus*) and land-tortoise (*Testudo*).

(*d*) Amphibia. Amphibian fauna includes toads, tree frogs and true frogs. A few salamanders and few caecilians are also present.

(*e*) Fishes. Oriental fresh water fish fauna is dominated by carps (Cyprinoids) and cat fishes (Siluroids). Most fishes belong to the following families—Nandidae, Labyrinthici, Ophiocephalidae, Siluridae and Cyprinidae.

Zoogeography of Indian Subcontinent

Geographic limits. **Blanford** in 1901 discussed the geographical distribution of animals in British India, which included not only the territory of present India, but also Pakistan and Bangladesh as these were parts of India, Nepal, Bhutan and other Himalayan states, Myanmar, and also Andamann and Nicobar Islands. At present Lacadive and Maldive islands have also been added to this.

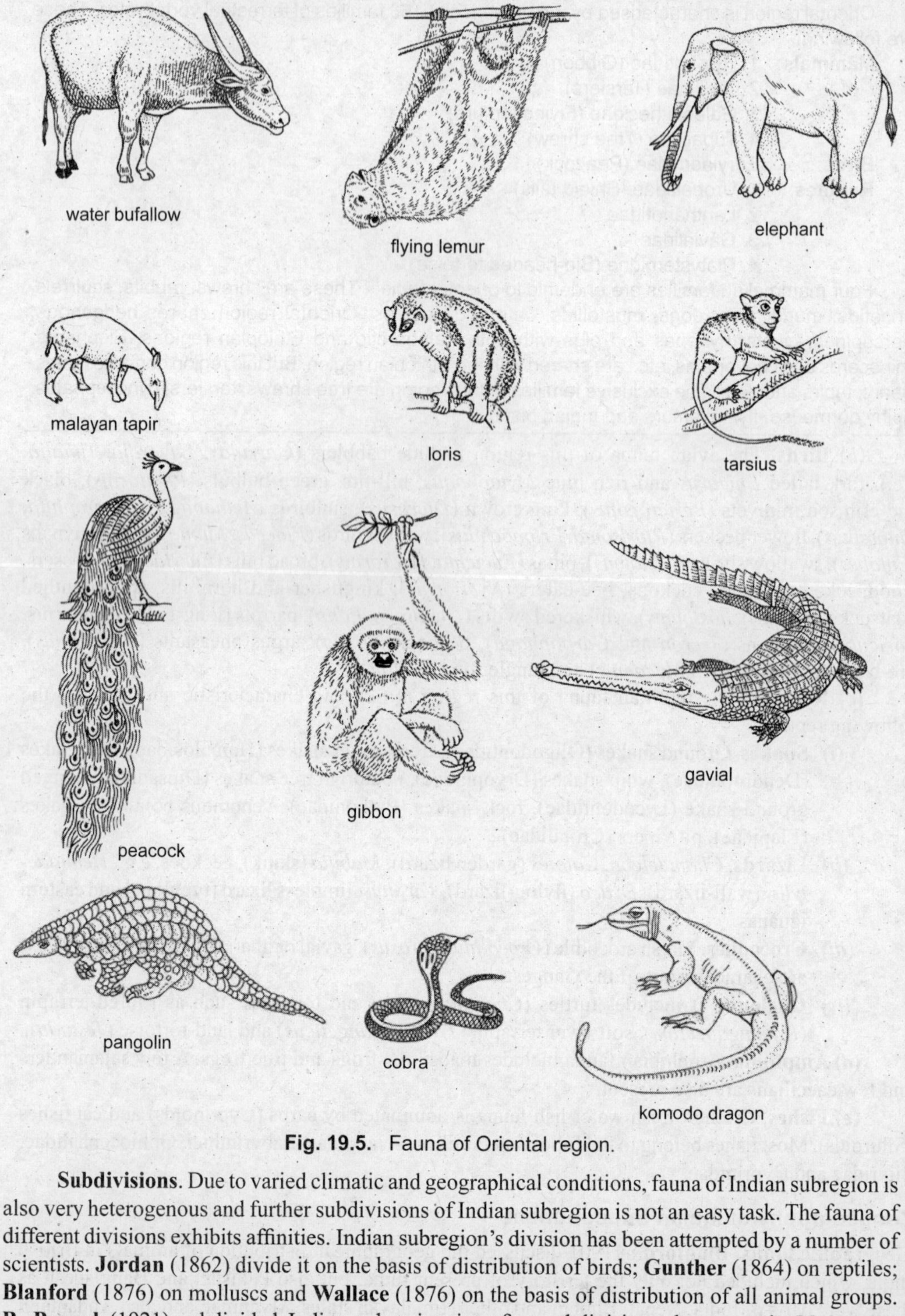

Fig. 19.5. Fauna of Oriental region.

Subdivisions. Due to varied climatic and geographical conditions, fauna of Indian subregion is also very heterogenous and further subdivisions of Indian subregion is not an easy task. The fauna of different divisions exhibits affinities. Indian subregion's division has been attempted by a number of scientists. **Jordan** (1862) divide it on the basis of distribution of birds; **Gunther** (1864) on reptiles; **Blanford** (1876) on molluscs and **Wallace** (1876) on the basis of distribution of all animal groups. **B. Prasad** (1921) subdivided Indian subregion into five subdivisions, but recently **Mahendra** (1942) has divided it into ten subdivisions:

1. The arid and semi-arid province of North-India
2. The Western Himalayas
3. Southern Myanmar province
4. Transgangetic province
5. Gangetic plain and adjacent part as far as 20° latitude
6. South India below 20° latitude excluding Travancore
7. Karnataka
8. Sri Lanka
9. Andaman Islands
10. Nicobar Islands

1. **Arid or semi-arid province of North India**. It embraces West Pakistan, Punjab, Western Rajasthan and Kutch. Fauna is not very remarkable with no endemic forms. Amphibians are few and represented with cosmopolitan genera *Rana* (frog) and *Bufo* (toad) only.
2. **Western Himalayas**. It comprises of Kashmir, Shimla, Kumaun, and Garhwal districts, adjacent mountaineous regions of western Tibet and Punjab. This region is poor in amphibians and reptiles. Fauna is markedly different and includes a number of endemic species, *e.g.*, *Gymnodactylus fasciolatus*, *G. lawderanus*, *Japalura major* and *J. Kumanonsis*, *Phrynocephalus theobaldi* and *Leiolopisma indacenses*.
3. **Transgangetic Province**. It includes northern point of Bihar, Assam and Myanmar north at 20 latitude, and Nepal. Himalayas east of Nepal, Sikkim and Bhutan. This subdivision has a characteristic reptilian fauna with a number of endemic genera and species.
4. **Southern Myanmar Province**. It includes southern part of Myanmar. It includes certain characteristics snake species and some endemic amphibians, *Gymnodactylus* sp., *Riopa anguina*, *Bungarus flaviceps*, *Dolphis intenstianlis* are characteristic snake subspecies of this region.
5. **Gangetic plain and adjacent region as far as south 20° latitude**. The boundaries of this subdivision are not very precise. This subdivision is characterised by the extreme scarcity of fauna. There is no endemic species. There is complete absence of house gecko, *Gavialis* and almost all freshwater turtles.
6. **South India below 20° latitude excluding Karnataka and Kerala**. The fauna of this subdivision has various characteristics including *Riopa lineala*, *Barkudia insularis*, *Gymnodactylus dekkanensis* and *Hemidactylus subtriedru*.
7. **Karnataka and Tamil Nadu**. It consists of hilly region of the country south of latitude 12 to 13 on the west and south of the coleroon river in the east. Its fauna is rich in endemic species.
8. **Sri Lanka**. It forms a distinct subdivision of oriental geographical realm and has clear affinities with Karnataka/Tamil Nadu subdivision.
9. **Andaman Islands**. These islands are situated in the bay of Bengal and have very characteristic fauna with clear affinities to Myanmar region on one hand and Nicobar Islands on the other hand. The endemic species are *Gymnodactylus rubidus*, *Phelsuma andamanensis*, *Calotes andamanesis*, *Typhlops andamanensis* and *Bioga andamanesis*.
10. **Nicobar Islands**. Its fauna has affinities with Andaman Islands and contains forms such as *Calotes mystaceus*, *Mabuya*, *andmanesis*, *Goniocephalus suberistatus*. The endemic species are few in number and the fauna of Nicobar appears to derived recently from Andaman Islands.

II. NEOGEAN REALM

(New Land: Central America and South America)

Neogean realm comprises only a single region, the Neotropical Region.

5. NEOTROPICAL REGION

I. Geographical Limits

The Neotropical region includes whole of South America, tropical North America (situated near Tropic of Cancer) and West Indies Islands. It is joined to Nearctic region by Central American isthmus and separated from all other regions by sea.

II. Climate

The climate is tropical in most of the region except the southern most region where it is temperate. It is noted for luxuriant forests and complete absence of desert. Extensive rain forests (called **selvas**) or evergreen forests are found in Amazon valley, tracts of drier forests and grassy plains extend in Savanna and Argentina. Subdesert areas are present in western South America. In west is seen long range of Andes which has mountain forests, plateau land and gentle slopes.

III. Subregions

The Neotropical region is divided into following *four* subregions :

1. Chilian subregion. It includes western Coast of South America and embraces the summits of Andes of Peru and Bolivia. It includes the South cold and damp forests of Tierra del Fuego, extending up to Santiago on the North-West coast, in the East barren plains of Patagonia which gradually changes into fertile but treeless pampas of La Plata in North upto 5° North latitude (*i.e.*, south of Amazon river, extending from the Atlantic to the Andes).

This subregion does not contain any peculiar family and the characteristic forms are chinchillas, llamas, oil birds and rheas.

2. Brazilian subregion. It includes area from Isthumus of Panama to La Plata in the South. In fact, it is area around Amazon river. It is covered with thick forests, pastures and open land. The climate is wet and warm.

This subregion is characterised by varied and rich fauna. All the arboreal animals of Neotropical region are confined to this subregion. American monkeys, vampire bats, American porcupines, sloths, or madillos, and opossums are found here. It also contains forms such as tapirs, cavies and spiny mice.

3. Mexican subregion. It includes Mexico and Guatemala. It is in the form of an irregular neck of land which connects North America with South America. It comprises of mostly the rocky mountains.

Some forms of the Nearctic region extend upto Guatemala, but they never reach South America. Hence it is a *transition zone* between Nearctic and Neotropical region.

Its characteristic fauna includes terrapins, tapirs, Anguidae and Plethodonidae.

4. Antillean subregion (West Indies Islands). It includes large islands of Cuba, Haiti (Island of Saint Domingue), Jamaica, Porto Rico and a chain of small islands—Saint Croix, Anguilla, Barbuda, Antigua, Barbados and Grenada with intervening islets. To the north of Cuba and Haiti are the Bahamas islands. Most of these islands are mountainous and rocky and are covered by forests. The temperature is high and uniform having enough rains.

It contains poor vertebrate fauna. Native mammals are also not found in this region.

IV. Fauna

(*a*) **Mammals**. Mammalian fauna is distinctive and varied. It includes opossum, coenolestid marsupials, anteaters, slothes, armadillos, tapirs, New World monkeys (squirrel monkeys and spider monkeys), rabbits, chinchillas, cavies, llamas (camel-like forms), deer, squirrels, New World porcupines, carnivores, rodents, bats (vampire bats), insectivores such as *Solenodon* and *Sorex* (in Guatemala and Costa Rica), peccary (*Dicotyles*).

Hedgehogs, moles, beavers, hyenas, oxen and native horses are absent from this region.

(*b*) **Birds**. Avian fauna is rich and varied. It is represented by following endemic birds: flightless or running birds such as rhea (found in Chilian subregion) and tinamous and flying birds such as todies, motmots, curassows, sun-bitterns, puff birds, oil-birds, trumpeters, homed-screamers, chaterers, tree creepers, toucans, ant-thrushes, sugar birds, plant-cutters, jacanas, hoazin, mankins, etc. Due to presence of so striking and diverse avian fauna in South America, it is regarded as "**Bird continent**". Song birds are completely absent in this region.

(*c*) **Reptiles**. Following types of reptiles are found in this region:

(*i*) **Snakes**. Snakes are represented by typhlops, leptotyphlops, many colubridae, coral snakes, boas, pit vipers. The characteristic genera are *Dromicus*, *Epicrates*, *Ungalia*, *Elaps* and anaconda (*Eunectes murinus*; a very large nonpoisonous semi-aquatic snake of tropical South America).

(*ii*) **Lizards**. Lizards are represented by *Heloderma* (gila-monster), Iguana, *Ameiva*, *Gymnopthalmus*, *Celestus*, *Diploglossus*, *Sphaerodactylus*, *Liocephalus*, *Liolaemus* and *Proctotretus*.

Fig. 19.6. Fauna of Neotropical region.

(*iii*) **Crocodiles**. They are represented by *Crocodilus* (Crocodile) and alligator (*Alligator mississipiensis*).

(*iv*) **Chelonia**. It includes tortoises such as *Dermatemys*, *Staurotypus*, *Peltocephalus*, *Hydromedusae* and *Chelys* and turtles such as *Podocnemys*.

(*d*) **Amphibians**. Amphibian fauna includes Anura such as *Pipa*, *Hyla*, *Bufo* and *Rana*; caecilians such as *Siphonopsis* and *Rhinotrema* and Urodela such as *Spelerpes*.

(*e*) **Fishes**. Fresh water fishes are very numerous. The specific fish fauna of this region includes South American lung fish (*Lepidosiren*), sting-rays, electric eels, gar-pike (*Lepidosteus*), cat fishes and characins. The carps are absent.

III. NOTOGEAN REALM
(Southern Land: Australia and Hawaii Islands)

It is most clearly defined realm. Its fauna is unique and most primitive. It occupies southern Hemisphere and is called **Southern Land**. **Newbigin** had divided the Notogean realm into three regions: Australian, Polynesian and Hawaiian subregions but **Robinson** has linked the last two divisions of Notogean realm (so that this realm has only two divisions).

6. AUSTRALIAN REGION

I. Geographical Limits

The Australian region is a very great region of the earth. It includes whole of Australia, New Zealand, New Guinea, Moluccas and all the Pacific islands. It is separated from the neighbouring Oriental region by a line called the **Wallace's line**.

II. Climate

Climatically northern part of Australia and New Guinea have very hot tropical climate. They are covered by thick rain forests. The central region of Australia is arid and hot, whereas the southern part and Tasmania are temperate and cool.

III. Subregions

The Australian region is divided into following *four* subregions :

1. Australian subregion. It includes Australian mainland and the Tasmania. It is represented by 98 families of terrestrial vertebrates of which 15 are of mammals, 67 of birds, 13 of reptiles and 3 of amphibians. Of these, six families of wombats, marsupial moles, duckbills, scrub birds, lyre birds (family Menuridae) and emus are confined to this subregion. This subregion is known as the home of marsupials. Of the 44 known genera of marsupials, 34 are confined to this region. The avian fauna too is peculiar though only three familes are confined to this subregion. The tailed amphibians are altogather absent.

The characteristic fauna of this subregion includes bandicoots, thylacines, kangaroos, honeyeaters, bower-birds, creepers, bustard, quails, swallows, shrikes, cobras, varanus, lizards and side-necked tortoises. The cobras constitute more than two third of all the snakes found in the subregion.

2. New Zealand subregion. It includes New Zealand proper and the surrounding 7 or more islands such as Norfolk Island, Kermadec Islands, Chatham Islands, Stewart Island, Auckland Island, Campbell Island and Macquarie Island.

It has 34 families of terrestrial vertebrates which include 3 familes of mammals, 27 families of birds, 3 families of reptiles and one family of amphibians. The mammalian families are represented by typicl bats and the murids. Following five families of birds are confined to this region: Xeicides, Nestorids (Nestor parrots) Stringopids (owl-parrots), Apterygids (kiwi) and Hatterids are confined to the subregion. Of lizards 13 species are peculiar in this subregion. Snakes are absent from this region and only one frog (*Liopelma*) are confined to this region.

The most interesting flightless bird of New Zealand is kiwi. There is one living fossil namely tuatara, *Sphenodon punctatus* (reptile of the family Rhyncocephalidae).

Fig. 19.7. Fauna of Australian region.

3. Polynesian subregion. It includes many islands of Pacific ocean such as Hawaiian Islands (formerly called Sandwich Islands), the Marquesas and Society Islands, Samoa Islands (formerly called Navigators Islands), Friendly Islands, Fiji Islands, New Caledonia, New Hebrides, The Ladrone and Caroline Islands.

The Polynesian subregion is characterised for its poor fauna content. Of the 53 families of terrestrial vertebrates, three are of bats, 37 of birds, 9 of reptiles and only two of amphibians. Tooth-billed pigeons are confined only to the islands of Samia, Kagu in New Caledonia and Drepanidids in Vandwish islands.

4. Austro-Malayan subregion. It includes the New Guinea (Papua) and islands of **Malaya Archipelago** such as Aru, Moluccas, Lombok, Kei, Flores, etc. All these islands are connected with new Guinea by shallow seas. This subregion has two different climates, *e.g.*, on the north-east, there is high uniform temperature, excessive rain and luxuriant forests and on the south-west arid, dry, moderately hot climate with forest rich in Australia-like vegetation such as eucalypti, acacias and thorny shrubs.

This subregion is characterised by 130 families of terrestrial vertebrates of which four are peculiar. These are crowned-pigeon and fly-river turtle, both confined to New Guinea, and two little known amphibian families, Ceratobatrachidae and Genyophrynidae from the Solomon Island and Sudest Islands, respectively. In New Guinea three marsupial genera are peculiar namely:

1. *Dendrolagus* which are remarkable for being chiefly arboreal in their habit and are representd by five species.
2. Dasyuridae or native bats which are represented by three species, and
3. True flying phalangers which are represented by one species only. Birds of paradise, honey eaters, bower birds, cassowaries, cuckoos, shrikes, fruit-eaters and the neighbouring islands the amphibians-narrow mouthed toads and tree frogs are characteristic.

IV. Fauna

(*i*) Mammalia. The mammalian fauna is represented by primitive prototherians and metatherians such as spiny anteater (*Echidna* or *Tachyglossus aculeatus*), duck billed platypus (*Ornithorhynchus anatinus*), kangaroos, phalangers, wombats, bandicoots, marsupial mole, etc. Australian region also contains some eutherian mammals such as bats (*e.g.*, *Hypoderma*, *Notopteris*, *Mystacina*) and rodents (rats and mice). All other placental mammals are absent from this region.

(*ii*) Birds. The avian fauna is represented by birds such as paradise birds, honey suckers, layer birds, scrub birds, cockatoos, broad-tailed paraquets, grass paraquets, brush-tipped-tongued parrots, mound makers, cassowaries, emu, kiwi, bower birds. Of these honey suckers and brush-tipped-tongued parrots are characteristically Australian.

(*iii*) Reptiles. The reptilian fauna is represented by pythons and poisonous cobras, scale footed lizards, monitors (*Varanus*), tuatara (*Sphenodon*) and side-necked tortoises such as *Chelodina*, *Chelemys* and *Elseya*. *Moloch horridus* (lizard) is peculiar to deserts of western and southern Australia.

(*iv*) Amphibians. The amphibian fauna is represented by genera which belong to order Anura (frogs and toads). The Urodela and Apoda are absent. The anurans are represented by such families as Bufonidae (*Notaden*, *Bufo*), Alytidae (*Helioporus*), Phryniscidae (*Pseudophryne*), etc.

(*v*) Fishes. The fresh water fishes belong to few families only such as Mugilidae, Sirenoidei, Homalopteridae, Galaxidae and Gadopsidae, etc. The only characteristic and specific fish of this region is *Neoceratodus* (Australian lung fish).

19.4 INSULAR FAUNA

The distribution of animals on the islands present complicated and interesting patterns because the islands which are even closely placed have their own distinctive fauna and their fauna is found to be different from main continents. Therefore, it is difficult to assign a particular zoogeographical region to such islands.

Types of Islands

On the basis of their origin, islands of the world have been classified into following *two* main types :

1. Continent Islands
2. Oceanic Islands

1. Continent Islands. The continent islands are those stretches of land which have somehow been separated due to invasion of the sea and which now had no connection with the mainland. Best examples of continental islands are **Great Britain** (part of British Isels which is separated from the main land by a narrow stretch of sea called strait of Dover), **Formosa (Taiwan)**, **Japan**, **New Zealand**, **Madagascar**, **Philippines**, **Java**, **Sumatra**, **Borneo**, **Celebes** and **New Guinea**. Continental islands differ from one another in the length of time being independent islands. Thus, Great Britain is a young continental island while Borneo, Formosa and Japan are older continental islands.

It is generally believed that most of the terrestrial fauna of a continental island have been reached the island across dryland when the island was connected to the continent. The degree of difference between fauna of an island and fauna of mainland depends upon the length of time, *i.e.*, when it becomes independent from the mainland. If the island is old (ancient), its fauna tends to be a little different due to evolutionary changes and the newcomers to the mainland tend to be lacking on the island when the island is young then the fauna at both places tends to be same.

2. Oceanic Islands. Oceanic islands appear independently from the oceanic floor either by the activity of some volcanoes (Box 19.8) or earth disturbances of some serious magnitude (*e.g.*, earth-quake) or by formation of coral reefs and subsequent precipitation. Due to this reason, oceanic islands never had any connection with any main continent or other land masses. Also because of their independent origin their fauna is always specific and exclusive. It has also been observed that the fauna of two such islands, even if situated very close to each other differ a great deal from each other. The examples of oceanic islands are **Bermuda** (a group of about 300 islands), **Saint Helena**, **Galapagoes Islands**, **Azores** (a group of nine islands and several islets in North Atlantic Ocean off the coast of Portugal), **Krakatau** (volcanic island in centre of Sunda Strait between Sumatra and Java, Indonesia), **Sandwich Islands** (Hawaiian Island; these are a chain of volcanic and coral island in North Central Pacific Ocean comprising 8 major and 114 minor islands), **Polynesian Islands** and **Tristan da Cunha** (volcanic islands).

Box 19.8

A **volcanic island** is formed by a submerged volcanic mountain erupting and pushing up above the surface of sea. This had happened at intervals throughout geological time and is still happening in certain parts of Pacific Ocean. Once it has broken the surface, the exposed mountain peak cools down and becomes an island capable of colonisation by animals and plants which are able to get there. Organisms may be blown to a virgin island by wind, as spores, carried by ocean currents, perhaps cling to drift wood or they may fly there if capable of doing so.

Fauna of Some Continental Islands

1. Fauna of British Isles. British Isles represent a group of Islands is Western Europe. They comprise of Great Britain, Ireland and adjacent islands. **Great Britain** comprises of **England**, **Scotland** and **Wales** having about 228,300 sq. km. area. Ireland is about 68,894 sq. km. Geological history of British Isles is complex. It is believed that during Palaeozoic and Mesozoic periods the northern and western lands were connected. There was a complete stretch of land from North America, Greenland continent to Eurasia and that this was shared with common fauna. At the end of Miocene British Isles became separated and formed a part of Europe with no land connection. After its appearance as an independent entity its fauna was exposed to assaults of nature in the form of glaciation. Its fauna is about seven thousand years old. At British Isles, there are shrews, moles, hedgehogs, rabbits, hares, squirrels, cricetids, murid rats, mice, dormice, badgers, otters, marten, stoats (ermines), weasels, cats, roe deer and red deer and few bats. Avifauna of British Isles comprises of mainly sea birds. Reptiles and amphibians are still less. There are just three species of snakes (grass snake, adder and smooth snake) and just two species of lizards (brown lizard and green sand lizard). A limbless lizard (*Anguis*) is also present. There are only six species of amphibians. Among fishes perches, pikes, carps and loaches occur predominantly.

Compared to Europe the present day fauna of Great Britain is sparse and lacks differentiation. Large sized mammals and reptiles are absent and the only endemic form is *Lagopus scoticus* (red grouse). The fauna seems to be acquired mainly across land from its parent continent.

2. Fauna of New Zealand. New Zealand consists of several islands in South West Pacific Ocean. It lies about 1000 miles South-East of Australia in the South temperate zone. The ancient island is covered by mountains (including extinct volcanoes) and open country. Island has temperate vegetation.

New Zealand contains no native mammals except two families of bats, one of which is endemic and another is Australian. A number of flightless birds inhabit New Zealand. There are several species of small kiwis, rails and the kakapo. Kakapo is a nocturnal, flightless, greenish-brown parrot (*Strigops habroptilus*). Of the flying birds some belong to world wide families, a few are Australian and there are two small endemic families—New Zealand wrens and the wattle birds. A remarkable bird of New Zealand is *Kea*, a genus of large parrot.

Among reptilian fauna New Zealand has *Sphenodon* (tuatara) which looks like a large lizard and is a living fossil. Geckos, skinks and marine snakes are found here. But there is no turtles and land snake. There is only one amphibian, the curious frog, *Liopelma*. Fresh water fishes are virtually absent in New Zealand.

The fauna of New Zealand is ancient, specialised and endemic. This suggests that entire New Zealand had a land connection with Australia in Mesozoic Era or it lies close to Australia and near Antarctic. The Mesozoic animals which might have world-wide distribution, might have migrated to New Zealand through these land connections. But by the end of Jurassic that land connection would have been lost preventing later amphibians, reptiles, birds and mammals to migrate to New Zealand. This suggests that the New Zealand represents an ancient continental island. The presence of flightless birds can be accounted by presuming that these birds might have migrated from Australia as flying birds and due to lack of competition and absence of predators secondarily lost power of flight.

3. Fauna of Madagascar. Madagascar is a large island (It is fourth largest island in the world having maximum length of 1601 km and maximum width of 579 km.) lying in the Indian Ocean about 260 miles from the South East Coast of Africa. It is separated from main continent by Mozambique Channel. It is believed that Madagascar was part of Africa until Palaeocene or even Eocene days, because Madagascar has native mammals. Thus, geologically Madagascar is an old island. It has granite plateau, swamps and sand dunes.

Flora of Madagascar includes dense fern forests and bamboo thickets. **Mammalian fauna** includes shrews, murid mice, African bush pig, bats, thirty species of Tenrecidae family of Insectivora which are only restricted in Madagascar. The insectivores of this family are spiny hedgehog tenrecs, long tailed mouse tenrecs, rice tenrecs with stout digging claws and water tenrecs with webbed feet. Primates are represented by ring tail-lemurs, sportive lemurs, gentle lemurs, woolly lemurs and silk lemurs and aye-aye. Crecetid rats and several genera of civets are also found here. Fossils of pigmy hippopotamus have been found in Pleistocene deposits.

Among the avian fauna, a large number of birds are endemic, the majority of the rest are African and a few are Oriental. Four families of birds are confined to Madagascar, two of them are flightless—the giant elephant birds which are extinct and the rail-like mesoenatids. The other two families are philepittas and the black, white and chestnut vangas. Philepittas are fruit eaters and the vangas feed mainly on chameleons and insects. African ostriches, secretary birds and mouse birds are not found in Madagascar. Hornbills and woodpeckers of Ethiopian or Oriental regions are also lacking.

Reptilian fauna of Madagascar include side neck turtles, land turtles, geckos, chameleons, skinks and spiny lizards. Agamid lizards of old world and all poisonous snakes are absent from Madagascar. No reptilian, amphibian or piscian fauna is specific to this region.

It is widely held that Madagascar had a land connection with Africa during Mesozoic which permitted African amphibians, reptiles, birds and earliest mammals to enter Madagascar. The other mammals such as ancestral insectivores, lemurs, vivertids, cricetids, side-neck turtles and chameleons travelled to Madagascar between Cretaceous and Eocene.

Fauna of Some Oceanic Islands

An oceanic island is likely to differ from the nearest continent in climate, vegetation (flora) and fauna. Due to these environmental differences their fauna tend to show divergence from the fauna of continent. But oceanic islands have certain common featurers. Both flora and fauna of oceanic islands tend to be sparse. Amphibia and fresh water fish are rare. Bats, rats, pig and small arboreal mammals are more. Carnivorous mammals tend to be lacking on oceanic islands. Due to absence of carnivorous mammals, birds gradually evolved into ground living forms such as **dodo** on Mauritius Oceanic Island, giant lizards of Galapagos Islands and giant tortoises found on several oceanic islands.

1. Fauna of Saint Helena Island. It is a small sized (122 sq. km.) volcanic island which is situated in the South Atlantic Ocean about 1950 km from West Coast of Africa. Its flora resembles to that of Africa and South America. Ebony and redwood trees are abundant on this island. Its native vertebrate fauna includes the plover (bird) which is closely related to an African species. There are no native mammals, reptiles, amphibians and fresh water fish. Some land molluscs are also found on this island.

2. Fauna of Galapagos Islands. Galapagos Islands lie in the Pacific Ocean situated on the equator and about 965 km west of Equador (South America). There are fifteen volcanic islands situated across the equator. They have diverse habitats. These islands initially were devoid of life but later became colonized by animals and plants from the mainland.

In Galapagos Islands, there are no endemic mammals and no amphibians. Reptiles, however, are abundant and gigantic land tortoises of these islands are peculiar. There are fewer species of lizards (iguanas and geckos) and two species of snakes (colubrid snakes). There are about 26 different land birds, but the avian fauna is dominated by **Darwin's finches**. In addition, a few birds resemble to Neotropical species. There are also flighless cormorants and some species of penguin. The native mammals of Galapagos are a species of bat and a genera of cricetid rat.

It is widely held that the animal forms of Galapagos have arrived from South America (Neotropical region) but only few of numerous forms are present here. This suggests that the animals have reached here by chance by crossing the sea, and there is no evidence of existence of any land bridge. For this reason only the fresh water fishes and amphibians are totally absent. Moreover, there is no shallow sea around Galapagos which can provide evidence of a land bridge. Therefore, it is safe to conclude that Galapagos islands are volcanic oceanic islands.

Theory of Island Biogeography (OR **Species Equilibrium Theory**). What determines the number of species on an island? According to **MacArthur** and **Wilson** (1967) the number of species on an island is increased by new **colonisations** but decreased by **extintions**. As long as the rate of new colonisation exceeds the rate of extinctions, the number of species grows, but when the rates become equal the number no longer changes, it is at **equilibrium**. They have suggested that **smaller islands will have great extinction rates because smaller populations are more likely to suffer extinction**.

Not all islands have the number of species that would be predicted from the overall relationship between species number and area. Mountain tops, for example, can be considered islands for species that do not inhabit lower elevations. **James Brown** (1971) found that the number of species of mammals on certain mountain peaks in Western North America is considerable lower than expected from the species equilibrium theory. He proposed that these peaks, which were uninhabitable during Pleistocene glaciations have not been colonised by certain species that should be able to survive there. That is, they have not got there yet. The fauna of these peaks (island) is not at equilibrium.

The simple model of MacArthur and Wilson to describe the number of species in a taxon on a continent was later on modified. Thus, the number of species depends on the rate of extinction and of **origination** (by speciation) rather than colonisation. Equilibrium is reached if the speciation rate and the extinction rate are equal.

Small islands may have fewer species simply because chance events are more likely to extinguish small populations than large populations. Interactions among species (Interspecific interactions) such as predation, competition, etc., can lower population sizes and thus contribute to extinction. For example, a small area offers fewer **microhabitats** and fewer "pockets" in which a prey species can escape predations. In fact, there are numerous examples in which **introduced** predators such as rats, cats and mongooses have extinguished island species of birds, snails and other creatures. A particularly dramatic and sad story is the extermination of the diverse land snails (Genus *Partula*) of the island of Moorea in the South Pacific. These snails were eradicated by a predatory snail (*Euglandina rosea*) that was introduced to control another introduced land snail that had become a pest (**Murray** *et al.*, 1988).

19.5 WALLACE'S LINE

In the middle of nineteenth century, when the world was divided into six zoogeographical regions, it was difficult to know where to draw the line between the Oriental and Australian regions. The mainland masses of two regions are separated from one another by several strings of islands, called the Malay Archipelago (Malaysia). The large islands of Sumatra, Java and Borneo belonged to the Oriental region. They are separated from Malaya and from one another by shallow sea and their animals are very similar to those of Malaya. In the East, New Guinea and Aru Islands which are separated from one another by shallow sea are included in the Australian region. They have many animals in common with that of Australia. Besides these above islands, there are Philippines, Celebes, the Moluccas, Timor and several other groups of smaller islands which are to be classified according to the zoogeographical classification scheme.

In the middle of the nineteenth century, **A.R. Wallace** studied the fauna of Malaya, Borneo and small island of Bali. In 1856 he travelled from Bali to Lombok which is 20 miles away from Bali. He noticed a marked difference between the fauna of two islands: Bali had the fauna of Oriental region and Lombok had the fauna of Australian region. Later he visited Celebes, the Kei Islands and the Aru Islands, Timor, the Moluccas and New Guinea and studied their fauna. On the basis of their fauna, he classified them as being Oriental or Australian.

Thus, in 1863, Wallace drew on the map a line which marked the boundary between two regions. This line is known as **Wallace's line**. This line runs between the Philippines and the Moluccas in the North, then South-West between Borneo and Celebes and finally South between the small islands of Bali and Lombok (Fig. 19.8).

Weber's Line

Weber's line was drawn after some years of Wallace demarcating the Oriental region and Australian region. This was actually drawn to better the demarcation made by Wallace between Oriental and Australian region. For this purpose Weber studied only mammalian and molluscan fauna of these two regions and the islands in between them. According to his studies the boundary line should pass between Celebes and Moluccas on the northern limit and Timor and Kei on the southern limit (Fig. 19.8). Weber's line exists further east than Wallace's line.

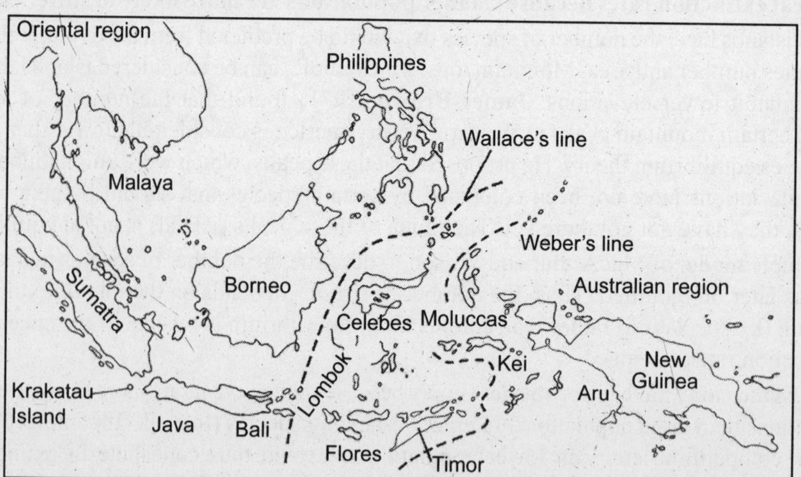

Fig. 19.8. Malay Archipelago (Malaysia) showing Wallace's line and Weber's line.

Wallacea

From the point of view of fauna, the Wallace's line forms the eastern limit of Oriental region and Weber's line forms the western limit of Australian region. The fauna of islands lying between the two lines, *i.e.*, Celebes, Flores and Lombok differ considerably from the two regions (Oriental and Australian). Since the fauna of islands, which come between the Wallace's line and Weber's line differ from both the regions and also from each other, this transitional area between the two lines has been named **Wallacea** (Fig. 19.9).

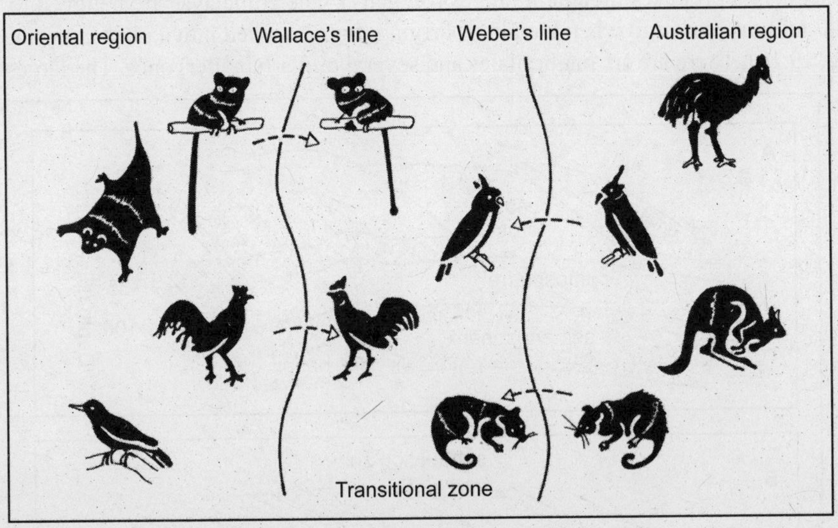

Fig. 19.9. Transitional fauna between two lines.

19.6 CONTINENTAL DRIFT AND PLATE TECTONIC

Large and small-scale events tip the balance in favour or against changes in the geographical range occupied by a species. Changes in climate and geology of the earth are the predominant forces involved. These physical characteristics of the environment determine the major conditions for survival and when they shift there may be a whole new set of rules to live by.

Sea levels rise and fall, mountain chains form and volcanic islands appear, creating and destroying barriers to distribution. But as major as these changes are, they are moderate compared with those brought about by **continental drift**. **Alfred Wegener**, in the 1915, proposed that *today's continent were once part of a single land mass which slowly split apart*. Two major lines of evidence were used to support this hypothesis. First, it was apparent that certain land masses seemed to fit together like a jigsaw puzzle, the best fit being the east coast of South America with the west coast of Africa. In addition, certain geological features were peculiar to these same regions and not others. Second, certain animal and plant species were distributed in such peculiar ways that they could best be explained by continental shift (Box 19.9).

Box 19.9

The palaeontologists with the help of fossil records compile information about the location and distribution of seas, shifting of land masses and appearance and disappearance of freshwater reservoirs in the past, *e.g.*, the occurrence of corals and echinoderms or cephalopods in a rock would indicate that in that area there was a sea or ocean in the period to which these fossils belonged. Similarly, the occurrene of fossils of bones of terrestrial vertebrates, certain gastropods and arthropods from a bed, even under the lake or sea, would indicate that in that area there was dry land in the period to which the fossils belonged.

Plate tectonics. The theory of **plate tectonics** envisions that the outer crust of the earth is divided into a number of rigid plates which can slide past each other. The plates move apart in certain regions and converge in other areas of earth. Plate separation occurs when **basalic liquid**, volcanic material deep within the earth forces itself up between the plate boundaries. These cracks exist in the ridges that run throughout the major oceanic basins.

In regions of the earth where the plates converge one plate slides down into the mantle of the earth under the leading edge of another plate. These are called **subduction zones**. When continents meet, the collision produces mountain ranges (orogeny). The Himalayas are believed to have been formed 40 million years ago when the plate carrying, India crashed into a plate carrying the rest of Asia (Fig. 19.10). There are six major plates and several minor (smaller) ones. The largest underlies

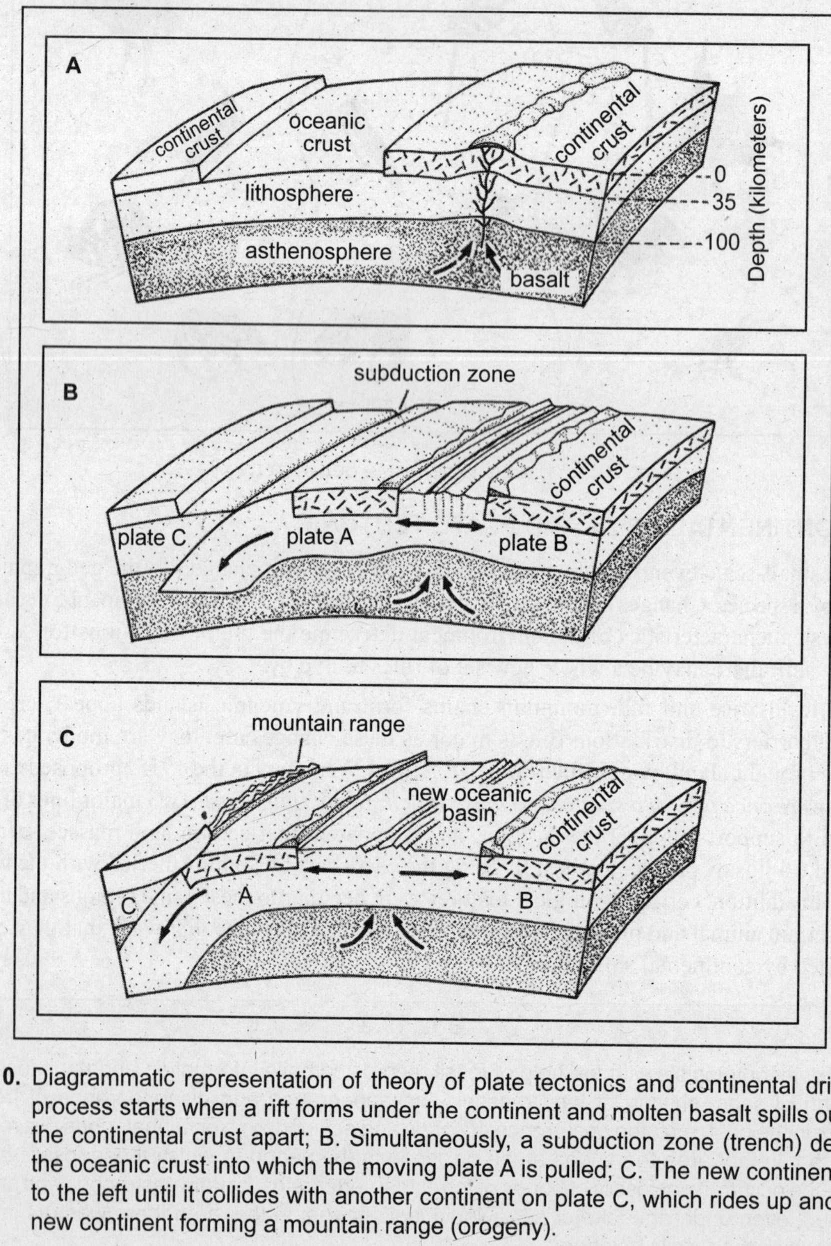

Fig. 19.10. Diagrammatic representation of theory of plate tectonics and continental drift. A. The process starts when a rift forms under the continent and molten basalt spills out, forcing the continental crust apart; B. Simultaneously, a subduction zone (trench) develops in the oceanic crust into which the moving plate A is pulled; C. The new continent is rafted to the left until it collides with another continent on plate C, which rides up and over the new continent forming a mountain range (orogeny).

almost the entire Pacific Ocean and the smallest is a plate that underlies Turkey. Plate boundaries are not necessarily coincident with continents—some plates are completely **oceanic**, others completely **continental** and still others are combination of both. Plate boundaries are characterised by *volcanic chains*, *mountain chains*, *fault lines*, *rift valleys*, *oceanic ridges* and *trenches*.

By fitting together the margins of the plates and gaining knowledge regarding the plate movements and rotation, geologists have concluded that about 200 million years ago the continents were grouped together into a super continent called **Pangaea**.

The success of plate tectonics is that it brings together such diverge geological phenomena as *continental drift*, *earth-quakes*, *mountain-chain formation*, *sea-floor spreading and volcanic activity*. Its consequences are of fundamental significance to biology and the evolution of life.

Current view of plate tectonic. The lithosphere is the solid outer layer of the earth. It bears both the continents and the crust below the oceans. Lithosphere consists of eight major and a number of minor plates that move over the denser, more plastic **asthenosphere** below. The heat of earth's core sets up convection cells within the asthenosphere. At certain regions, such as mid-oceanic ridge that runs longitudinally down the floor of the Atlantic Ocean, **Magma** from the asthenosphere rises to the surface, cools and spreads out to form new crust, pushing the existing plates to either side (Fig. 19.11). The plates move at velocities of upto 5 to 10 cm/year. Where two plates come together, the other, rejoining the asthenosphere. The pressure of these collisions is a major cause of mountain building and is responsible for the major mountain chains of the world. When a plate moves over a "hot spot" where magma is rising from the asthenosphere, **volcanoes** may be born, or a continent may be rifted apart. For example, the great lakes of eastern Africa lie in such a rift valley. Hawaiian Islands are a chain of volcanoes that have been formed by the movement of the Pacific plate over a hot spot (Fig. 19.11).

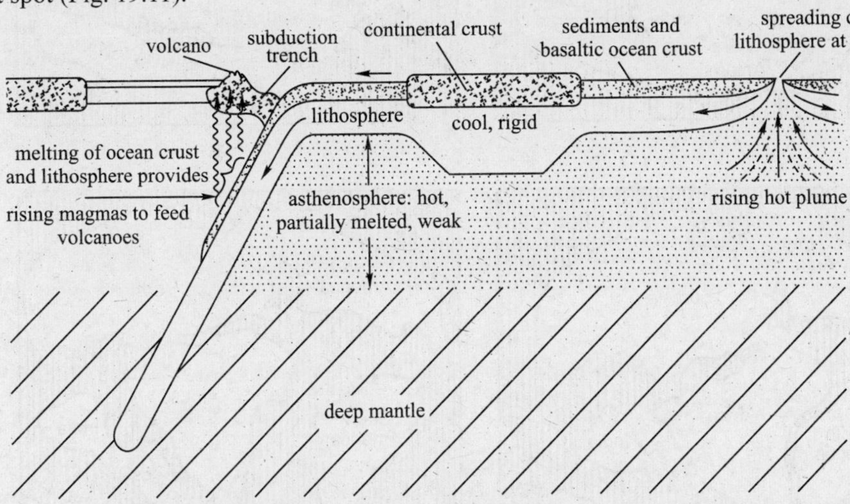

Fig. 19.11. Diagram showing the plate tectonic processes in the upper mantle of the earth. A rising hot plume of magma (at right) creates new lithosphere and pushes existing lithosphere, bearing continental land masses, to either side. When such a lithospheric plate encounters another (at left), it plunges under it (at a subduction trench), frequently causing earth quakes and orogeny (mountain formation). The heat generated melts crust and lithosphere causing volcanic activity.

Breakup of pangaea. According to the reconstruction of some geologists, Pangaea existed as an irregular land mass on one side of the earth distributed from pole to pole. It was surrounded by the universal ocean, called **Panthalassa** (Fig. 19.12A).

Pangaea began breaking up about 200 million years ago when two rifts developed. A modern rift developed from east to west near the equator, separating two subcontinents. The northern subcontinent, called **Laurasia**, included present day North America and Eurasia. The remaining land masses, called **Gondwanaland**, south of the equator were further split by another rift (Fig. 19.12B). This southern rift split South America and Africa together as a single land unit, called **West Gondwana**, leaving Antarctica, Australia and India together. Shortly thereafter, India was split off and drifted northward. These were the major land masses as of 135 million years ago, 65 million years after the breakup of Pangaea began (Fig. 19.12C).

During the next 70 million years (*i.e.*, 65 million years ago) the separation of South America away from Africa was completed. During a later stage, North America, Greenland and Eurasia drifted apart as did Australia, New Zealand and Antarctica (Fig. 19.12D). During this last period, India completed its journey to the north and collided with Eurasia. North and South America touched together at the Isthmus of Panama.

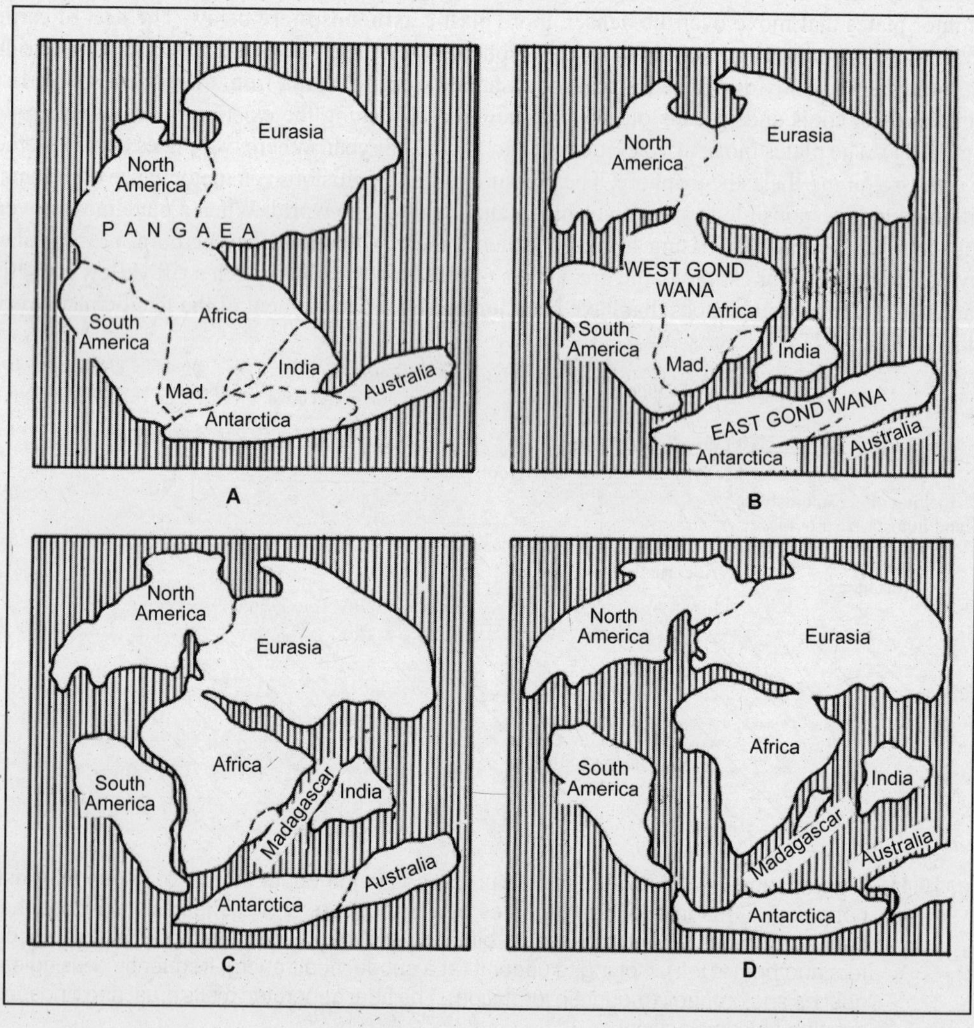

Fig. 19.12. Breakup of pangaea.

Significance of continental drift to biogeography. The newfound respectability of theory of continental drift has forced biogeographers to reanalyse certain distribution patterns. For example, primitive mammals are scattered in three widely separated areas in the Southern Hemisphere. Australia and the New Guinea region contain the monotremes (the egg-laying mammals) and marsupials. South

America has a rich record of living and fossil marsupials, plus fossils of early placental mammals. Madagascar has a collection of primitive placental mammals. Similar odd distribution (called **discontinuous** or **disjunct distribution**) occur for many animals and plants including lung fishes, hylid frogs, side-necked turtles and ratite (flightless) birds. The previous interpretation was that these organisms had evolved in one region and then by luck found their way over thousands of kilometres of water. Plate tectonic indicate that all these areas were once connected.

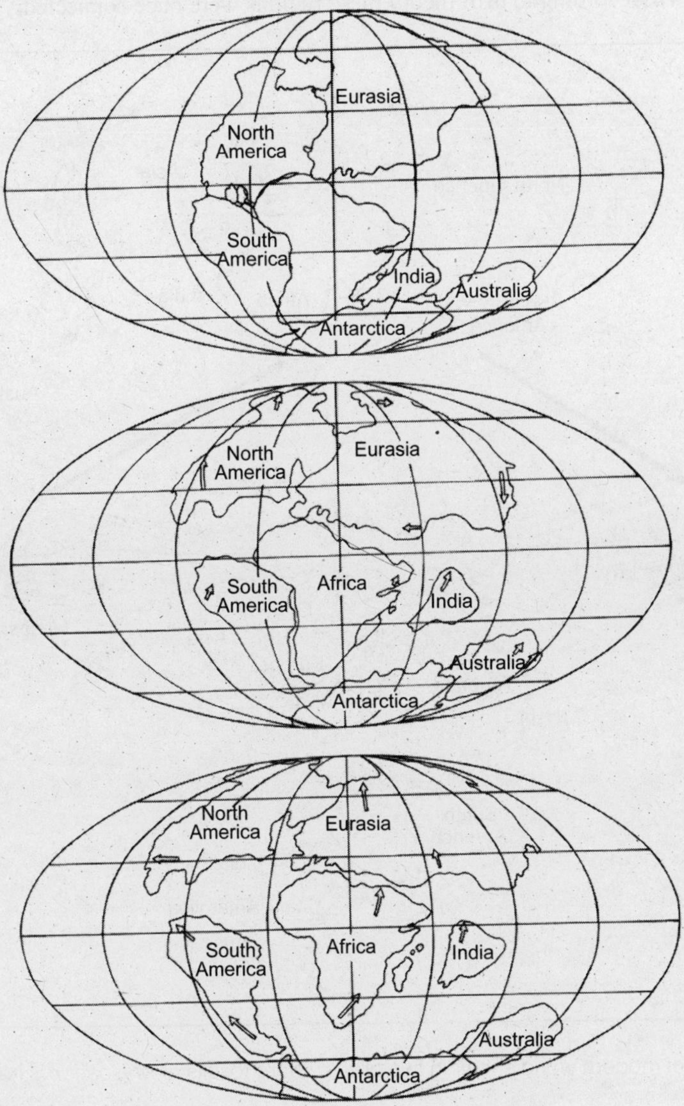

Fig. 19.13. Distribution of land masses at several points in geological time. A. Triassic; B. Late Jurassic; C. Late Cretaceous; D. Eocene. Note the aggregation into a single land mass (Pangaea) in the late Permian and Triassic (A); separation into northern (Laurasia) and southern (Gondwanaland) masses in the Jurassic, northern movement of India, which joins Eurasia by the Miocene (B); the separation between North America and South America, and the formation of the Bering land bridge between north-western North America and Eastern Eurasia (C). The different orientation and projection of the Eocene make the spatial relationship of North America and Eurasia clear (D).

Further according to continental drift theory, Antarctica used to have a very different climate and floral and faunal pattern that it does today when it exists at the South Pole. Fossil evidence indicates that this is true; specimens of ferns, amphibians and reptiles have been discovered. Another example of a strange distribution pattern that continental drift clears up is the case of extinct reptile *Mesosaurus*. Fossils of *Mesosaurus* have only been found in two regions of world, but these two areas are on opposite sides of the Atlantica along the east coast of South America and west coast of Africa (Fig. 19.14). According to drift theory these regions were once connected.

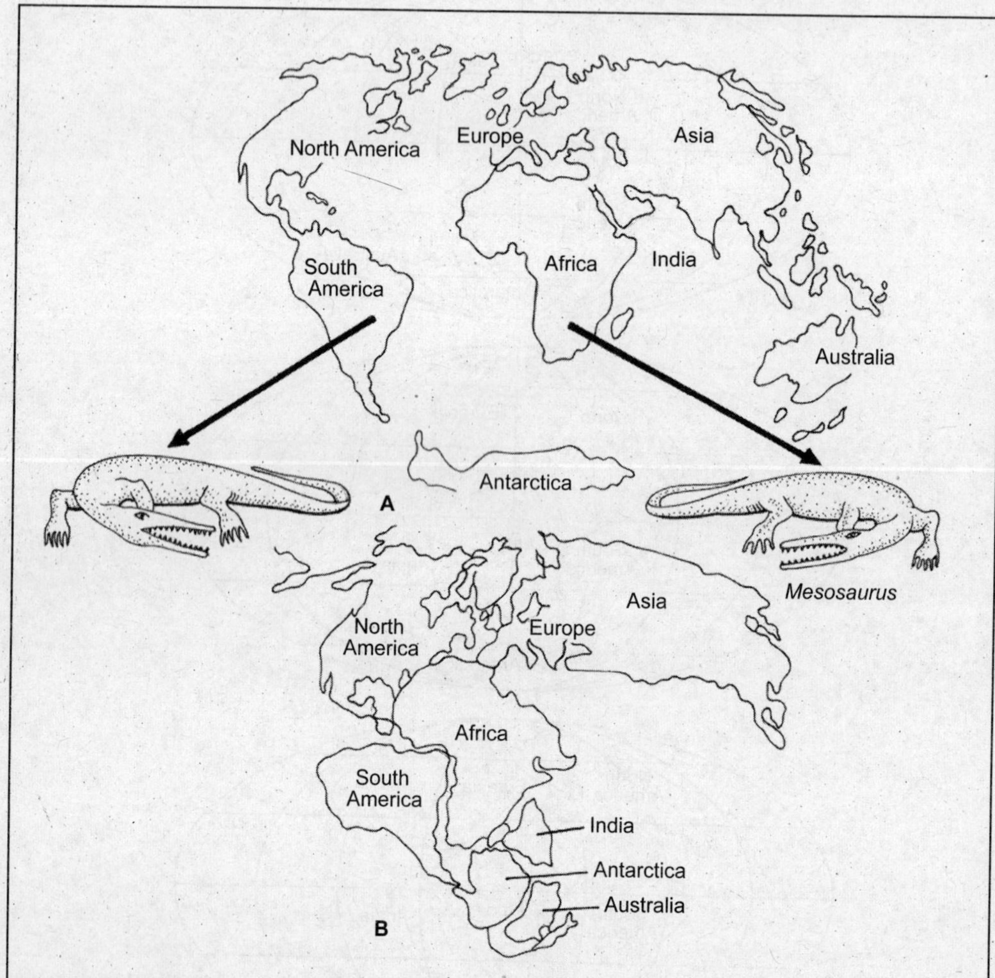

Fig. 19.14. Map of modern world showing two areas where fossils of *Mesosaurus* have been found.

QUESTIONS

Long Answer Questions

1. What is zoogeography? Describe the various zoogeographical regions with their fauna (distribution of animals).
2. What is zoogeographical realm? Who has proposed the divisions of the earth into six different realms? State the natural features and fauna of Oriental realm.

3. What are different zoogeographical regions of the world? Mention the main characteristics of the fauna of any two regions.
4. Enumerate the zoogeographical regions and mention their boundaries. Give common and zoological names of two characteristic animals of each region.
5. Describe the geographical limits, climatic conditions and biota of Palaearctic region.
6. Enumerate the various zoogeographical region and subregion of the world. Describe the limits of the Ethiopian region and its characteristic fauna.
7. Name the zoogeographical regions of the world and describe the characteristic vertebrate fauna of Ethiopian region.
8. Define zoogeography. Describe in detail the Australian region.
9. Describe various zoogeographical regions of the world, and describe the boundaries and characteristic vertebrate fauna of Australian region.
10. Name the zoogeographical regions in the world. Describe the limits and characteristic fauna of Australian region.
11. Enumerate different zoogeographical regions and mention their boundaries. Give common and zoological names of any five vertebrates of the Australian region.
12. Describe the amniote fauna and limits of any two zoogeographical region.
13. Enumerate the zoogeographical realms and mention their boundaries. Give the common and zoological names of any five characterstic vertebrates of the Australian region or Nearctic region.
14. Mention the boundaries of Oriental and Ethiopian regions. Describe the characteristic vertebrate fauna of these regions.
15. Describe the geographical limits, climatic conditions and biota of Palaearctic and Neotropical regions.
16. What is zoogeography? Give an account of the fauna of Oriental and Australian region.
17. (*a*) What is meant by zoogeographical realms?
    (*b*) Name the different zoogeographical realms stating the geographical limits of each such realm.
    (*c*) Discuss the characteristic features and faunal composition of the Oriental realm.
    (*d*) To which realm the following animals belong: 1. *Auchenia* (Llamma); 2. *Neoceratodus*?
18. (*a*) What are geographical realms? Name the primary zoogeographical realms.
    (*b*) Give an account of the realms to which the following animals belong:
    (*i*) *Tapirus indicus* (Tapir), (*ii*) *Vombatus* (Wombat).
19. Enumerate the various zoogeographical regions of the world. Describe the limits of Oriental region and its characteristic fauna.
20. In which zoogeographical region India is included? Give the boundary of this region and describe its characteristic fauna.
21. Write names of the major zoogeographical realms. Mention the physical features of the realm in which India is included.

Short Answer Questions

1. Describe zoological characteristics and geographical boundaries of Oriental region.
2. Describe fauna of Oriental region. (*Kerala 2001*)
3. Give an account of zoogeographical range, physical features and fauna of Oriental region. (*Kanpur 2001*)

4. Describe the geographical limits, climatic conditions and the typical fauna of the Nearctic region.
5. Give an account of the areas under Nearctic region, their physical features and characteristics of the fauna.
6. Write a short note on Nearctic region.
7. Discuss the continental drifts in relations to distribution of animals.
8. Write an explanatory note on continental drift.
9. Write short notes on the following:
   (*i*) Oriental region
   (*ii*) Fauna of Oriental region (*Kerala 2001*)
   (*iii*) Ethiopian region
   (*iv*) Insular fauna
   (*v*) Wallace's line (*Kanpur 2001*)
   (*vi*) Write a short note on Neotropical fauna
   (*vii*) Zoogeographical realms
   (*viii*) Fauna of Galapagos islands
10. Give an account of fauna of Oriental and Ethiopian region.
11. Describe the geographical limits, climatic conditions and typical fauna of the Neotropical region.
12. Name four mammals of Neotropical region
13. Describe the geographic range and mammalian fauna of Australian realm. (*Devi Ahilya 2001*)
14. Write short note on 'Peculiarities of Australian region'.
15. Differentiate between Nearctic region and Neotropical region.

Very Short Answer Questions

1. Which zoogeographical region includes Africa, Arabia Island and Madagascar?
2. Name the peculiar animal found in New Zealand, which is also known as "living fossil".
3. Name the running bird of Australia.
4. Mention the zoogeographical realm to which India belong.

Fill in the Blanks

Fill in the following blanks with the suitable words.

1. Kangaroo is found in ________.
2. A region characterised by a distinctive biota is called ________.
3. Six major zoogeographical regions of the world were formulated by ________.
4. Great panda is found in ________ subregion of Palaearctic region.
5. Hippopotamus belongs to ________ region.
6. *Gavialis* is characteristic animal of ________ subregion of Oriental region.
7. Giant-sized fauna on the Galapagoes islands illustrates ________.

Multiple Choice Questions

Choose the correct answer from the four alternatives given.

1. Wallace line passes between
   (*a*) Sumatra and Java (*b*) Borneo and Celebes
   (*c*) Timor and New Guinea (*d*) Moluccas and Kei

2. Evolution of unique groups of mammals in South America, Africa and Australia is an evidence for (*I.A.S. 2002*)
   (*a*) crustal movement
   (*b*) continental drift
   (*c*) glaciation
   (*d*) geographical juxtaposition
3. "Hatteria" or "Tuatara" is the oldest living lepidosaurian reptile and is known as a living fossil. The generic name of this form is (*I.A.S. 1999*)
   (*a*) *Tyranosaurus*
   (*b*) *Sphenodon*
   (*c*) *Iguanadon*
   (*d*) *Stegodon*
4. Consider the following statement
   (*i*) Palaearctic
   (*ii*) Nearctic
   (*iii*) Antarctic
   (*iv*) Ethiopian
   Which of the above statements are correct? (*I.A.S. 2000*)
   (*a*) (*i*) and (*ii*)
   (*b*) (*ii*) and (*iii*)
   (*c*) (*iii*) and (*iv*)
   (*d*) (*i*) and (*iv*)

ANSWERS

Very Short Answer Questions

1. Ethiopian region
2. *Sphenodon* (a lizard called tuatara)
3. Emu (*Dromiceius*) and cassowaries (*Casuarius*)
4. Oriental region

Fill in the Blanks

1. Australia **2.** Biogeographical realm **3.** A.R. Wallace **4.** Manchuria **5.** Ethiopian **6.** Indian **7.** Evolution

Multiple Choice Questions

1. (*b*) **2.** (*b*) **3.** (*b*) **4.** (*a*)

Physiology: Controlling and Coordinating Systems (CC-VI)

1

Tissues and Glands

1.1 TYPES OF TISSUES

A **tissue** is a group of cells of the same type or of mixed type having a common origin and performing an identical function.

Box 1.1

Historical

The word tissue was coined in 1792 by a French anatomist **Francois Xavier Bichat** (1771-1802). The term **histology** was coined by **Mayer** in 1819.

Marcello Malpighi (1628-1694) is considered as founder of histology.

Father of histology is **Bichat**.

Simple and Compound Tissues

A tissue is an assembly of cells that are arranged in a regular formation. The mode of arrangement of cells is a tissue. Tissues are of following two types:

1. Simple tissue. A tissue containing same type of cells is called **simple tissue,** *e.g.*, cartilage.

2. Compound tissue. A compound tissue contains a mixture of cells (two or more cell types), extra-cellular matrix and cell products with different functions, *e.g.*, nervous tissue which contains nerve cells (neurons), support cells (astrocytes), immune cells (microglia) and epithelial cells (ependyma).

Basic Components of Tissue

Every type of animal tissue basically contains the following components:

1. **Cells,** the living component;

2. **Ground substance** or **matrix** which is the non-living component forming **intercellular cement** or **substance.**

Classification of Tissues

Multicellular animals contain the following four basic or fundamental types of tissues:

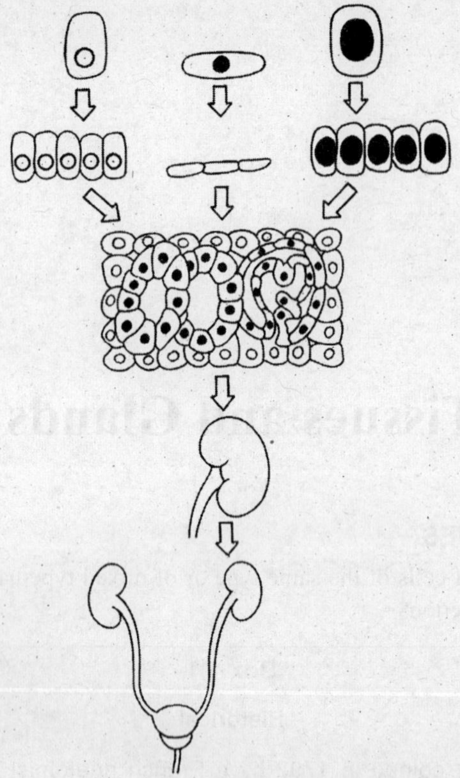

Table 1.1. Classification of tissues. Cells, tissues organs and systems

Serial Number	Type	Examples	Function
1.	**A. Epithelial tissue**	**1. Simple** (*i*) Squamous (*ii*) Cuboidal (*iii*) Columnar **2. Stratified (or Compound)** (*i*) Stratified squamous (*ii*) Stratified cuboidal (*iii*) Stratified columnar (*iv*) Pseudostratified (*v*) Transitional **3. Specialized** (*i*) Ciliated (*ii*) Neurosecretory (*iii*) Pigmented (*iv*) Germinal	Protection, absorption, sense perception, respiration, assimilation of food and removal of waste
	B. Glandular tissue	**1. Tubular glands** (*i*) Simple straight tubular (*ii*) Simple coiled tubular (*iii*) Compound	Secretion

(*Cont.*)

Serial Number	Type	Examples	Function
		2. Alveolar glands (*i*) Simple (*ii*) Compound (*iii*) Tubuloalveolar glands	
2.	**A. Connective tissue proper**	**1. Areolar tissue** **2. Adipose tissue** **3. Dense connective tissue** **4. Yellow elastic connective tissue** **5. Reticular connective tissue** **6. Mucous connective tissue**	Support, binding and storage, and protection
	B. Skeletal tissue	**1. Cartilage** (*Chondroblasts, chondrocytes and cartilage*) (*i*) Hyaline (*ii*) Elastic (*iii*) Fibrous (*iv*) Calcified **2. Bone** (*Osteoblasts* and *osteoid*) (*i*) Spongy (*ii*) Compact	
	C. Blood (Fluid connective tissue)	**1. Plasma** **2. Blood corpuscles** (Formed elements) (*i*) Erythrocytes (*ii*) Leucocytes (*iii*) Blood-platelets	Transport of food materials and gases
3.	**Muscular**	**1. Striped or Striated muscles** (Skeletal muscle) **2. Smooth or unstriated muscles** **3. Cardiac muscles**	Contraction and movements
4.	**Nervous**	**(*i*) Nerve cells (neurons)** **(*ii*) Nerve fibres** **(*iii*) Neuroglia** and **neurilemma**	Conduction and control

Box 1.2

Each of four basic tissues is defined by a set of general morphologic characteristics or by functional properties. The basis for definition of epithelium and connective tissues is primarily morphologic, whereas the basis for the definition of muscular and nervous tissues is functional. However, while muscle tissue itself is defined by its function, it is subclassified into smooth and striated categories, a purely morphologic distinction, not a functional one.

Box 1.3

Connective Tissues are Now Called Support Cells

Historically the term *connective tissue* was used to describe a wide range of living material containing cells associated with a dominant extracellular matrix component, whose function was to act as supporting stroma serving more highly specialized cell types. The original group of connective tissue included cell/matrix combinations such as bone, cartilage, tendon, fibrous tissue, bone marrow and blood and adipose tissue.

Traditionally the term **'loose areolar connective tissue'** is described as a tissue that is partly made up of support cells that produce an extracellular matrix, but also contains cells belonging to the immune system (*e.g.,* lymphocytes and macrophages), nerve cells and blood vessels.

According to **Stevens** and **Lowe** (1992) usage of the term connective tissue should be avoided since it '*underemphasizes*' the structural organization involved in such highly developed tissues. Instead of it, they use the term **support cells** which emphasizes the importance of interactions between extracellular matrix and cells.

1.2 EPITHELIAL TISSUE

Structure

Epithelial tissue is the simplest tissue which contains tightly cohesive sheets of cells. It forms continuous covering on the free surface and lining of cavities of various organs of body (Box 1.4). Epithelial tissue may be formed of one to many layers. These layers form **epithelium** which forms a protective covering.

Box 1.4

Epithelia function mainly as:

1. A covering lining for body surfaces, for example, skin, gut and ducts.
2. The functional units of secretory glands such as salivary glands and liver.

Specialities of Epithelial Tissue

1. Epithelial tissue arises from all the three germinal (or embryonic) layers; *i.e.,* ectoderm, mesoderm and endoderm.
2. Cells of epithelium are set very close to each other. They are held together with little or no intercellular matrix.
3. Epithelial cells lie on a delicate non-cellular **basement membrane** which contains a special form of matrix protein, called **collagen.** Basement membrane can be stained with PAS (Periodic Acid-Schiff) technique. It helps in the nutrition of epithelium.
4. The cells are arranged in a layer or layers forming covering around the organs or lining the body cavity and lumen of different body organs.
5. Epithelial cells are connected with other by a variety of intercellular junctions:
   (*i*) **Occluding junctions** which link cells to form an impermeable barrier, *e.g.,* tight junction.
   (*ii*) **Anchoring junctions** which link cells to provide mechanical strength, *e.g.,* adherent junctions, focal contacts, desmosomes and hemidesmosomes.

(*iii*) **Communicating junctions** which allow movement of molecules between cells, *e.g.*, gap junctions.

6. Epithelial cells may contain a variety of cell surface specializations such as **microvilli, cilia, flagella, stereocilia** and **kinocilia.**
7. Epithelium is subjected to wear, tear and injury. Its cells divide by mitosis to produce new cells to replace those that are destroyed.

Functions of Epithelial Tissue

1. It protects the underlying tissue from dessication, dehydration, infection and injuries.
2. It acts as a barrier preventing passage of certain materials across the surface and facilitating the passage of others inside.
3. Microvilli of intestinal epithelium help in food absorption.
4. Epithelium of uriniferous tubules help in filtration, selective resorption and excretion.
5. Sensory epithelium receives stimuli from the surrounding.
6. Germinal epithelium gives rise to gametocytes (spermatocytes in testes and oocytes in ovary) which divide by meiosis to produce specialized sex cells or gametes.
7. Ciliated epithelium of respiratory and genital tracts conducts mucus, fluid or gametes in desired direction.
8. Certain epithelial cells have structural specializations related to their role in the production and secretion of macromolecules such as enzymes, mucin (goblet cells), steroid hormones (*e.g.*, adrenal gland, ovary and testis).

Types of Epithelial Tissue

Mode of classification of epithelium is descriptive and based on the combination of the number of cell layers and the shape of the surface cells. The terminology is related only to structure, not to function. Thus, the epithelium are of following types:

A. Simple Epithelium

This epithelium is formed by a single layer of epithelial cells. Its every cell rests on basement membrane. Simple epithelium occurs on secretory and absorptive surface and also found in sensory organs. It can be of following types, the types are named according to the shape of cells:

1. Squamous epithelium or pavement epithelium

Structure. It is composed of a single layer of thin and plate-like flattened polygonal cells, joined to make a simple sheet. The width of the cell is greater than its height. In surface view its cells appear like mosaic tiles in the floor or pavement. The oval or spherical nucleus lies in the centre of the cell and produces a bulge in the cell (Fig 1.2):

Location and function. 1. It forms lining of vascular system, *i.e.*, heart, blood vessels and lymph vessels and is called **endothelium.** It helps in exchange and forms barrier in nervous system.

2. It forms lining of body cavity such as pleural, pericardial and peritoneal cavities and is called **mesothelium.** It helps in exchange and lubrication.

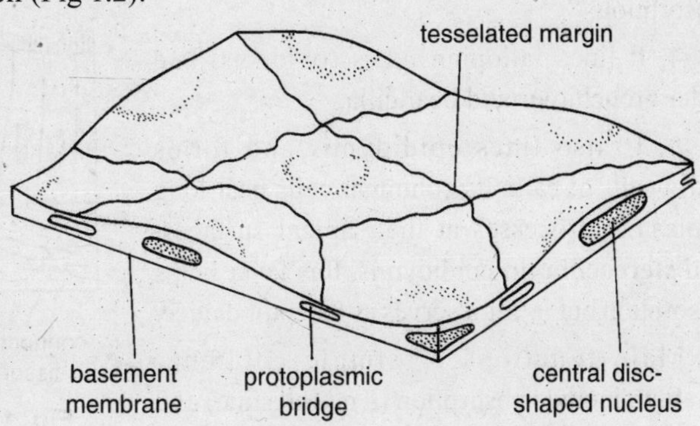

Fig. 1.2. Squamous epithelium.

3. It forms portion of uriniferous tubules of kidney such as Bowman's capsule, thin limb of Henle's loop, etc. It acts as barrier.

4. It forms lining of respiratory spaces in lung and helps in exchange.

5. It lines membranous labyrinth of the internal ear.

Thus, simple squamous epithelium is mainly protective in function.

2. Simple cuboidal epithelium

Structure. Its cells are cubical or cuboidal, *i.e.,* where the width, depth and height are approximately the same. Nucleus is located centrally in each of these cells (Fig. 1.3).

Location and function. 1. It lines small duct of exocrine glands such as sweat glands, pancreas and salivary glands and function in absorption and serve as conduit.

2. It forms germinal epithelium of gonads (testes and ovary) and acts as barrier.

3. It lines the convoluted part of uriniferous tubules of kidney and its cells bear microvilli to form **brush bordered cells.** It helps in absorption.

4. It lines thyroid vesicle and has secretory function.

5. It lines bronchioles of lungs.

6. It forms choroid plexus of eye.

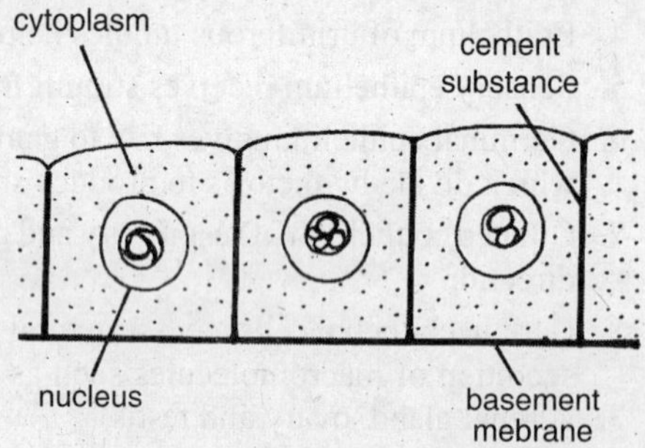

Fig. 1.3. Simple cuboidal epithelium.

Thus, simple cuboidal epithelium helps in absorption, excretion and secretion.

3. Simple columnar epithelium

Structure. It is formed of layer of tall cells, *i.e.*, height of individual cell evidently exceeds the width. The nucleus is located at the base of each cell (Fig. 1.4).

Location and function. 1. It forms lining of small intestine and colon, forming **mucous membrane.** Its cells bear microvilli forming **brush bordered columnar epithelium.** It helps in absorption and secretion.

2. It forms stomach lining (*i.e.,* mucous membrane) and gastric glands and helps in secretion.

3. It forms lining of gall bladder and helps in absorption.

4. It lines fallopian tubes (oviducts) and smaller bronchioles and bear cilia.

5. It also lines epididymis and forms sensory cells of ears. Its columnar cells bear long cytoplasmic processes at their apical surfaces, called **stereocilia.** In epididymis, this layer helps in absorption but in ear it serves as receptor device.

Main functions of simple columnar epithelium include absorption (*e.g.,* intestine) and secretion (*e.g.,* mucus by goblet cells of mucous membrane).

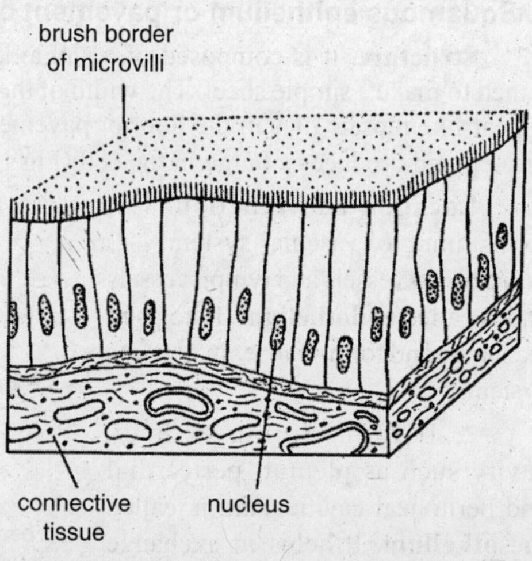

Fig. 1.4. Simple columnar epithelium.

4. Ciliated epithelium

Structure. The cells of this epithelium may be cuboidal or columnar. The free ends of these cells bear several hundred cilia, all arranged in orderly rows. Cilia arise from the basal granules and have microtubular organization. Cilia exhibit incessant movement.

Location and function. 1. It lines trachea and bronchi, where cilia play a cleansing role by sweeping mucus and trapped particulate material toward oropharynx where it is swallowed along with saliva and thus eliminated.

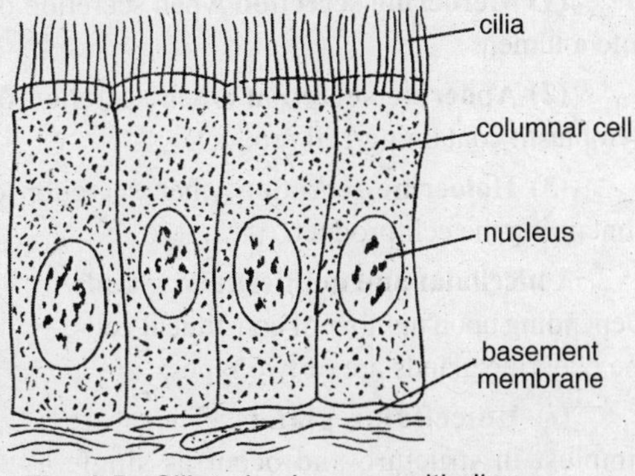

Fig. 1.5. Ciliated columnar epithelium.

2. It lines oviducts(*i.e.,* fallopian tubes) where cilia help, sweep ova and fluid through the tract toward the uterus.

5. Glandular epithelium or glands

A **gland** is an organized collection of secretory epithelial cells. Glands are composed of epithelial cells specialized to *synthesize* and *secrete* a specific product. They are classified into following two major groups reflecting how their products are distributed:

1. Exocrine glands which secrete their products from the apex of the cell onto a surface or into a lumen through ducts or tubes.

2. Endocrine glands which lack a duct system. They secrete their products from the side or base of the cell into the connective tissue from which they enter the bloodstream in order to reach their target cells (Fig. 1.6). The products of endocrine glands are called **hormones.**

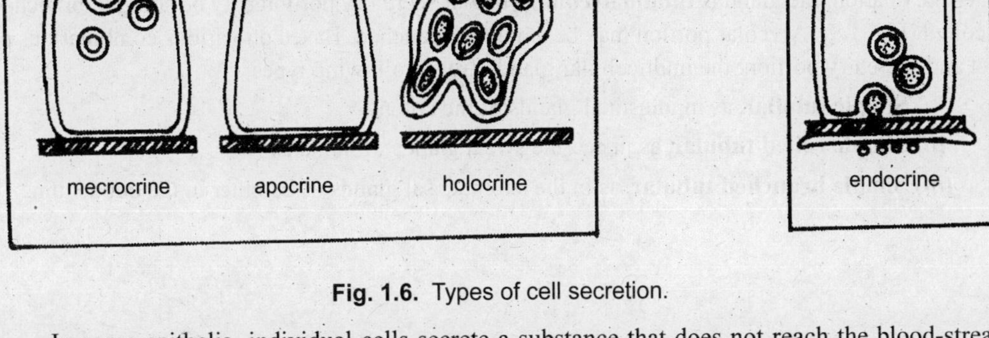

Fig. 1.6. Types of cell secretion.

In some epithelia, individual cells secrete a substance that does not reach the blood-stream but rather has an effect on other cells within the same epithelium. Such secretory activity is called **paracrine secretion.** The secretory material reaches the target cells by diffusion through the extracellular space.

Types of secretion of exocrine glands. Secretion of exocrine glands is of following three types:

(1) Merocrine secretion when secretion of cell product occurs by exocytosis from cell apex into a lumen.

(2) Apocrine secretion when secretion of cell product occurs by pinching off apical cell cytoplasm containing cell product.

(3) Holocrine secretion when secretion of cell product occurs by shedding of whole cell containing the cell product.

Unicellular and multicellular glands. Depending upon number of constituent cells, the exocrine glands are of two types:

***(i)* Unicellular glands.** These are simplest in structure and occur as single cells scattered among nonsecretory cells. A typical example of unicellular gland is **goblet cell** or **mucus secreting cell** (Fig. 1.7) which is positioned among other columnar cells. Goblet cells are located in the inner lining of large intestine and in certain passages of the respiratory tract. The mucus serves as a lubricant and protects against injurious substances.

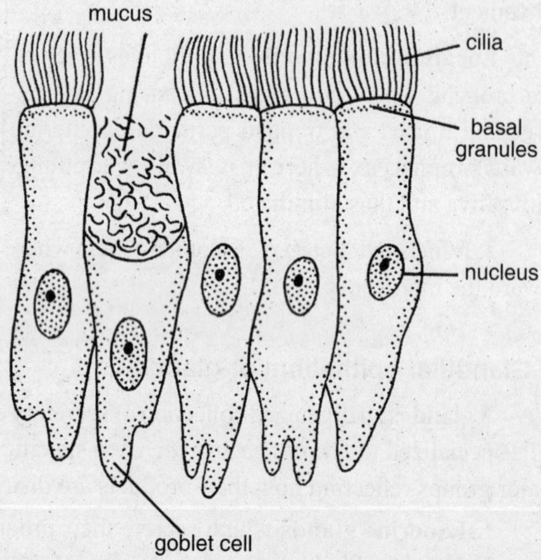

Fig. 1.7. Goblet cells.

***(ii)* Multicellular glands.** These glands are composed of more than one cell and exhibit varying degree of complexity.

The most simple arrangement of a multicellular gland is a cellular sheet in which each surface cell is a secretory cell. For example, the entire lining of the stomach with its gastric pits is a sheet of mucus-secreting cells.

Other multicellular glands typically form tubular invaginations from the surface. The end pieces of the gland contain the secretory cells (called **parenchyma**); the portion of the gland connecting the secretory cells to the surface serves as a duct. If the duct is unbranched, the gland is called **simple**; if the duct is branched, it is called **compound.** If the secretory portion of gland is shaped like a tube, the gland is **tubular;** if it is shaped like a flask, the gland is **alveolar** or **acinar;** if the tube ends in a sac-like dilation, the gland is **tubuloalveolar.** Tubular secretory portion may be straight, branched or coiled (Fig. 1.9). Alveolar portion may be single or branched. Based on various combinations of duct and secretory portion, the multicellular glands are of following types:

(i) **Simple tubular,** as in intestinal glands of the colon.

(ii) **Simple coiled tubular,** as in eccrine sweat glands of humans.

(iii) **Simple branched tubular,** as in the submucosal glands of Brunner in the duodenum.

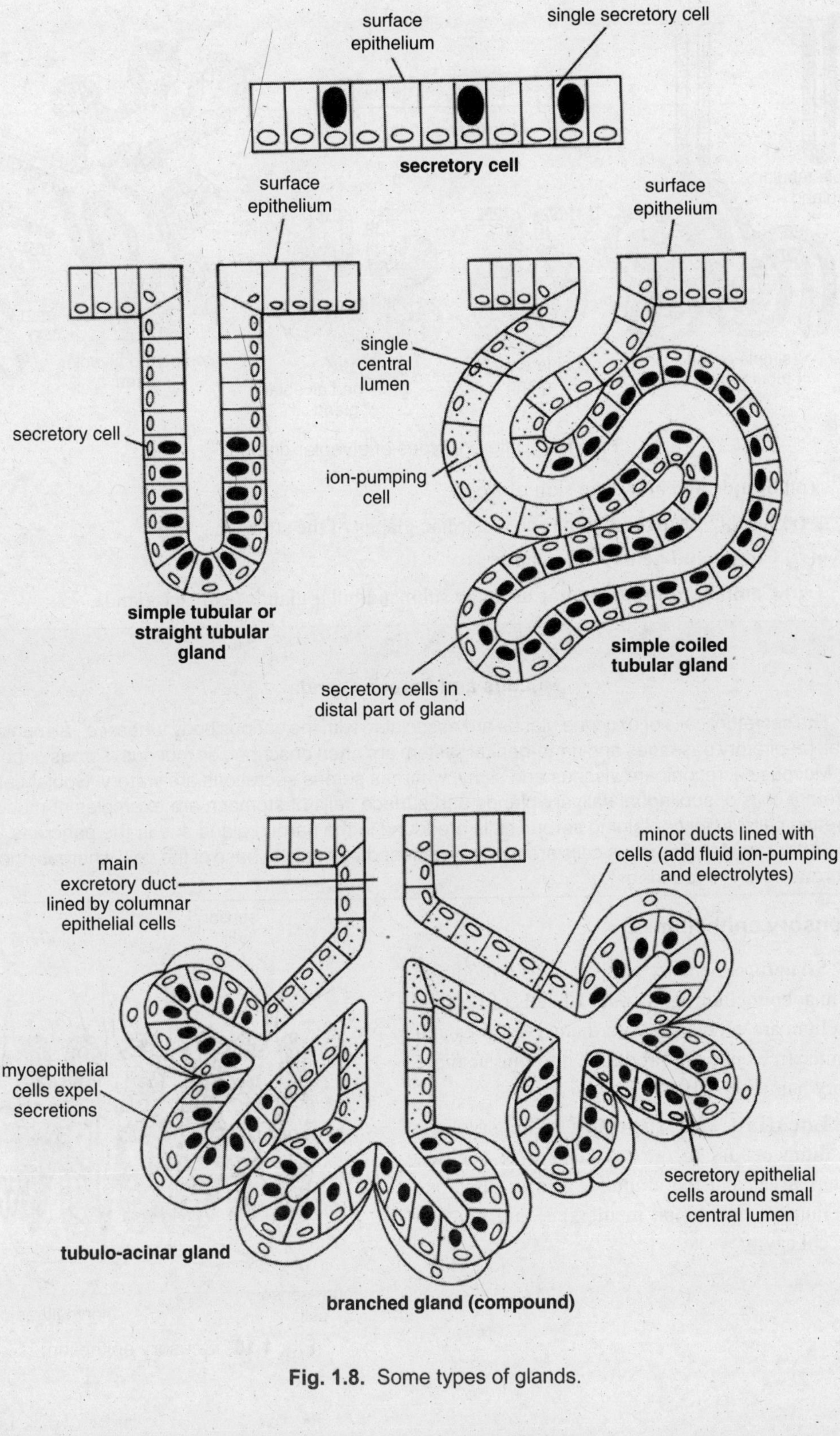

Fig. 1.8. Some types of glands.

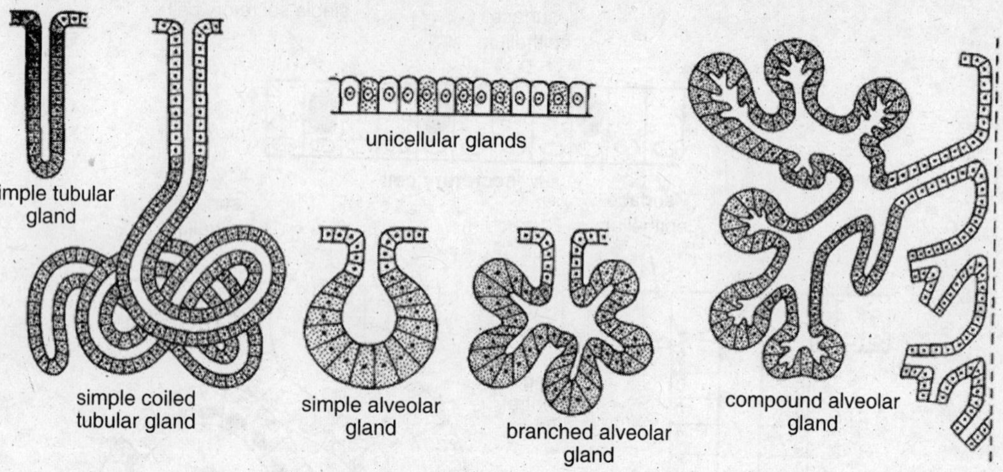

Fig. 1.9. Different types of alveolar glands.

(*iv*) **Simple alveolar** as in skin of frog.

(*v*) **Simple branched acinar** as in cardiac glands of the stomach.

(*vi*) **Compound acinar** as in pancreas.

(*vii*) **Compound tubuloacinar** as in the submandibular glands (salivary gland).

Box 1.5

Mucous and Serous Glands

The secretory cells of exocrine glands are associated with the various body tubes, *i.e.*, alimentary canal, respiratory passages and urino-genital system are often described as mucous, serous or both.

Mucous secretions are viscous and slimy, whereas serous secretions are watery. Goblet cells, secretory cells of sublingual salivary glands and surface cells of stomach are examples of mucus-secreting cells. Acini containing serous cells are found in the parotid gland and in the pancreas.

Further, nuclei of mucous cells are typically flattened against the base of the cell, whereas nuclei of serous cells are rounded.

6. Sensory epithelium

Structure. Sensory epithelium is a modified columnar epithelium. Inner ends of cells of sensory epithelium are connected with the nerve fibres and the outer free ends are produced into one or more sensory hair (Fig. 1.10).

Location and function. The sensory epithelium occurs in the retina of eye, cochlea of internal ear and the ampullae and in olfactory epithelium (Schneiderian membrane) and mucosa of buccal cavity.

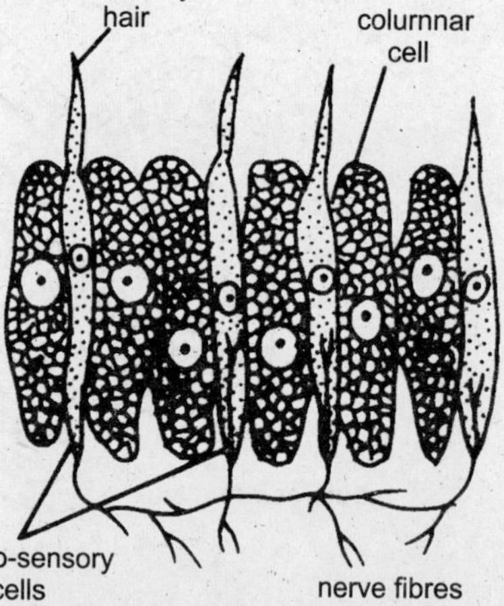

Fig. 1.10. Sensory epithelium.

7. Germinal epithelium

Structure. The lining of seminiferous tubules of testis and lining of ovary is formed of a single layer of cuboidal cells.

Location and function. The cells of germinal epithelium possess power of division and produce spermatogonia or oogonia after division.

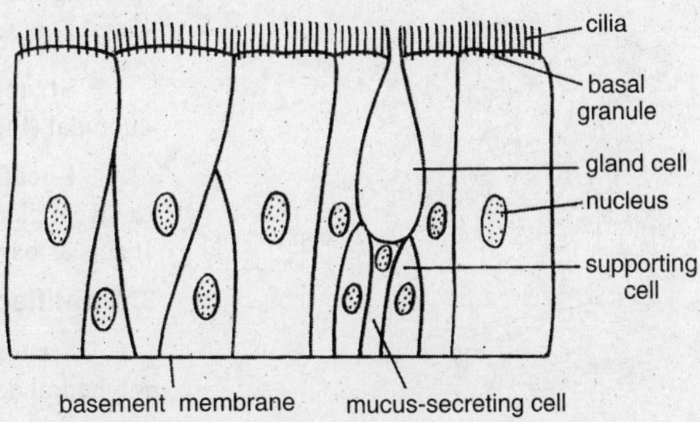

Fig. 1.11. Pseudostratified epithelium.

8. Pseudostratified epithelium

Structure. This type of epithelium is actually a simple epithelium. It is formed of a single layer of cells but all of them are not of the same length, so that it appears multilayered or stratified. All cells of this epithelium rests on basement membrane (Fig. 1.11).

Location and function. 1. This forms the mucous membrane of olfactory chamber and lines the trachea. It is either ciliated or columnar and has function in secretion and helps in conduit.

2. It forms lining of long passages, pharynx, nasal passages, trachea, large bronchi, vas deferentia, epididymis and urethera. It helps in secretion, conduit and absorption (in epididymis).

B. Compound Epithelium or Stratified Epithelium

This type of epithelium is composed of two or more cell layers. Only the cells of innermost layer rest on the basement membrane. Stratified epithelium is of the following types:

1. Stratified squamous epithelium

Structure. It is the main protective epithelium which forms barrier. It is multilayered. The number of layers varies in different places, *i.e.*, a few layers (*e.g.*, corneal epithelium) to many layers (*e.g.*, epidermis). The cells forming different layers of this epithelium are not similar, *i.e.*, deeper layers are cuboidal but cells become more flattened (squamous) towards the free surface. Stratified squamous epithelium is of two types:

(*i*) Non-keratinized epithelium which is found on wet surfaces that are subjected to considerable wear and tear and do not perform absorption.

(*ii*) Keratinized epithelium which contains **keratin** (a highly insoluble fibrous protein with water-proofing qualities) and is resistant to friction and bacterial invasion.

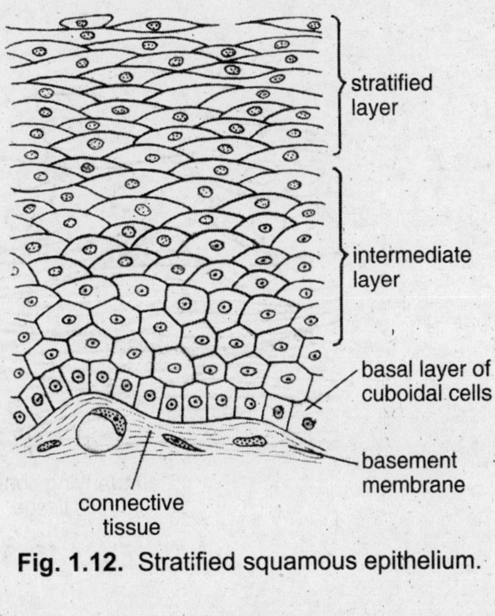

Fig. 1.12. Stratified squamous epithelium.

Location and function. (*i*) Non-keratinized epithelium. Forms lining of mouth cavity, tongue, pharynx, oesophagus, anus, cornea and conjunctiva of eye, portions of larynx, external auditory canal, vagina, vestibule and portion of urethera.

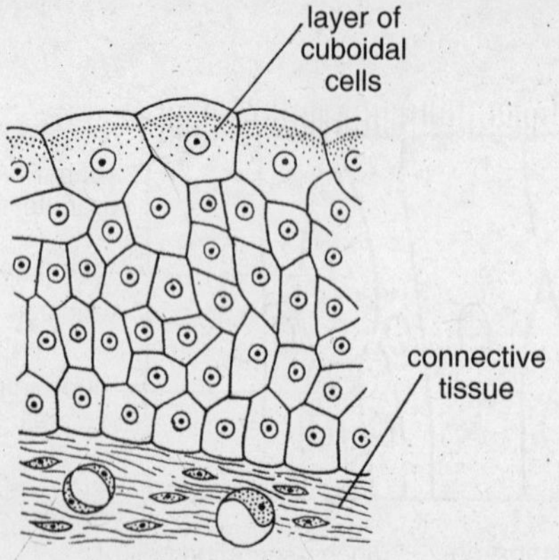

Fig. 1.13. Stratified cuboidal epithelium.

Location and function. 1. It forms lining of excretory ducts of salivary (parotid) and mammary glands.

2. It lines pharynx, larynx and anorectal junction.

It serves as barrier and conduit.

4. Transitional epithelium

Structure. It is also called **plastic epithelium.** It is thin and stretchable. This epithelium consists of 3-6 layers of partly flattened non-keratinized plastic cells. The superficial cells are in the form of large rectangular or oval, often binucleate cells and the basal cells are cuboidal. There is no basement membrane and basal cells rest on connective tissue (Fig. 1.15).

(*ii*) **Keratinized epithelium.** It forms epidermis (skin) covering the whole surface of body.

2. Stratified cuboidal epithelium

Structure. The surface cells of this epithelium is cuboidal (Fig. 1.13).

Location and function. It forms lining of ducts of sweat glands, sebaceous glands and female urethera. It acts as barrier and conduit.

3. Stratified columnar epithelium

Structure. This epithelium consists of irregular polyhedral deeper cells but columnar cells at the surface (Fig. 1.14).

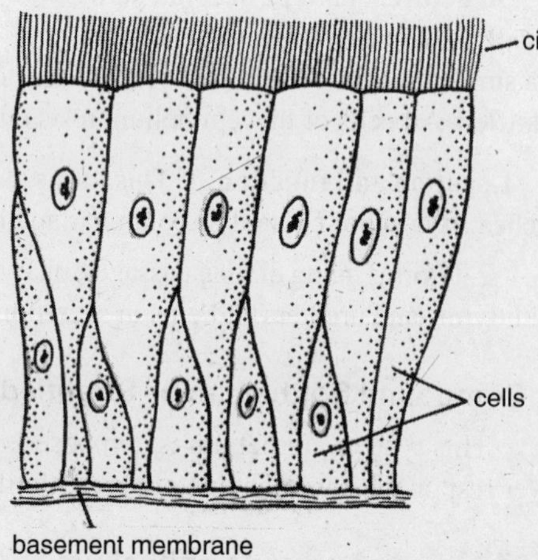

Fig. 1.14. Stratified columnar epithelium.

Location and function. It forms epithelial lining of the pelvis of kidney, the ureters, the urinary bladder and part of urethera.

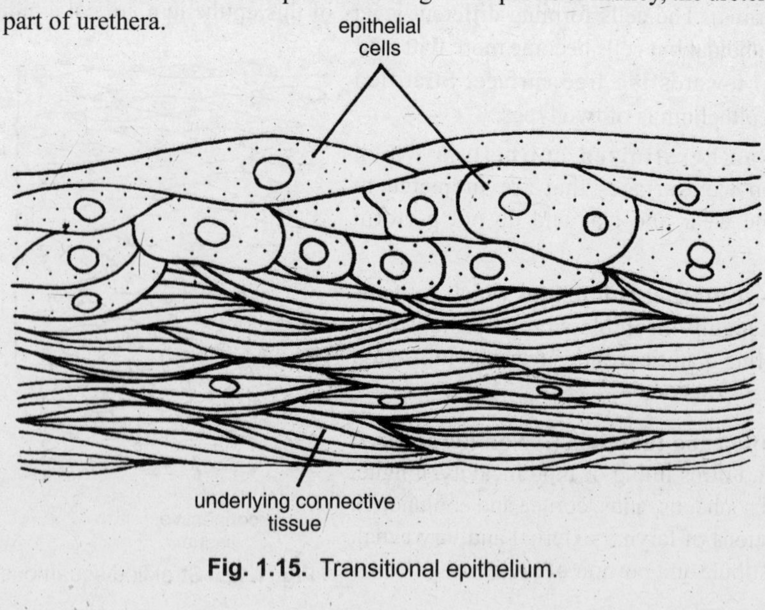

Fig. 1.15. Transitional epithelium.

1.3 CONNECTIVE TISSUES

The connective tissue is specialized to connect and anchor various body parts. It is also called **binding tissue** and it binds together different tissues in an organ to support various parts of body and forms packing around different organs. Connective tissue forms a vast and continuous compartment throughout the body, bounded by basal lamina of the various epithelia and by the basal or external lamina of muscle, nerve and vascular endothelium. It forms about 30 per cent of body mass. Connective tissue is mesodermal in origin.

Structure

The typical connective tissue has following four main components:

1. Homogenous, semifluid or gel-like **ground substance** or **intercellular matrix** which is chemically composed of **glycosaminoglycans** (GAGs) which were formerly referred to as mucopolysaccharides. Matrix is secreted by the component cells which themselves lie scattered and embedded in the former.

2. Cells are of different types such as

(*i*) **Fibroblasts** which form fibres and ground substance.

(*ii*) **Fat cells** or **adipose cells** which store fats in their vacuoles.

(*iii*) **Macrophages** which may be free-moving or fixed phagocytes (WBC). They are involved in the destruction and removal of invading bacteria, foreign bodies and damaged cells from the tissue.

(*iv*) **Mast cells** which secrete substances such as **heparin** and **histamine,** both promoting inflammation of the infected area.

(*v*) **Immunocytes** such as **lymphocytes** and **plasma cells** both producing antibodies for the immune response.

In the connective tissue, these cells lie wide apart in matrix.

3. Fibres which are insoluble protein fibres of various types such as **collagenous** or **white fibres, elastin** or **yellow fibres** and **reticular fibres,** all interspersed in the matrix.

4. Blood vessels to provide rich blood supply.

Functions

1. Connective tissue binds together different tissues and keep them in position.
2. It helps in replacing dead and damaged tissues and thus helps in regeneration and tissue repair.
3. It helps in body defence, protecting body against viruses, bacteria, and toxic substances that enter body.
4. Forms skeleton to support body.
5. Jelly-like matrix acts as shock absorber.
6. It helps in storage of fats in some parts of body.

Types of Connective Tissue

Based on the nature and functions of matrix, following three types of connective tissues have been recognized:

I. Connective tissue proper

II. Supporting or skeletal tissue (Cartilage and bone)

III. Fluid connective tissue (Blood)

In this chapter only connective tissue proper will be described.

I. Connective Tissue Proper

I. Areolar Tissue

Areolar tissue is a loose and cellular connective tissue. It is simplest and most widely distributed connective tissue which develops from the mesenchyme. This tissue binds different organs and parts of various organs together. It is a whitish viscous to jelly-like but sticky mass which fills up all spaces between the organs, muscle bundles, in the interior of organs and forms the peritoneum and mesenteries. It is also found beneath epithelia of skin.

Structure

Areolar tissue consists of following parts:

1. Matrix. It forms most abundant component of tissue. It is a white, sticky and jelly-like amorphous substance. It acts as a cementing substance.

2. Fibres. In areolar tissue fibres are thin and relatively sparse. These are of following two types:

***(i)* White collagen fibres.** These are white, thin, wavy and unbranched fibres which are arranged in thick bundles (Fig. 1.16). These are formed of a protein called **collagen** (which changes to gelatin on boiling).

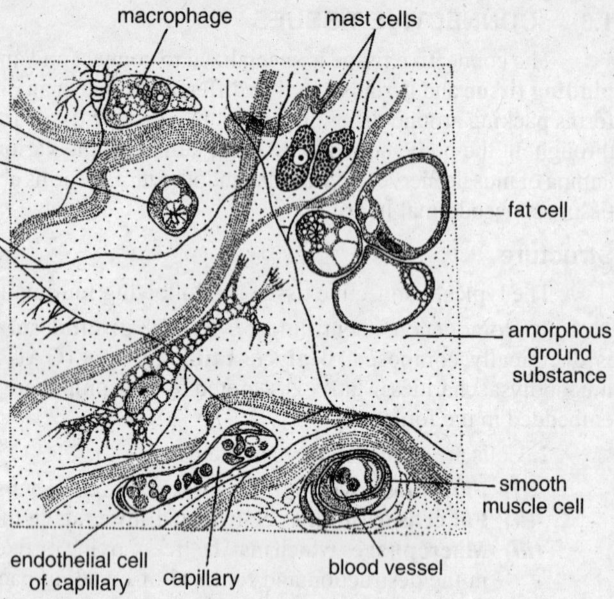

Fig. 1.16. Areolar tissue.

***(ii)* Yellow elastic fibres.** These fibres are slender, thin, branched and anastomosing forming network. These are highly elastic, formed of a protein, the **elastin.** These fibres when assembled in large number, they have yellowish colour.

3. Cellular elements. Matrix of areolar tissue contains following types of cells:

A. Fixed Cell Population

***(i)* Fibroblast cells.** These are also called **fibrocytes** or **desmocytes.** These are principal cells which are large, flattened, irregular (stellate) cells with protoplasmic processes. Each of these cells contains prominent and oval nucleus. Young fibroblasts are rounded but mature fibroblasts become elongated. These cells secrete amorphous ground substance and white and yellow protein fibres.

***(ii)* Fixed macrophages or histocytes.** These are flat, rounded or oval, phagocytic cells which are derived from monocyte (WBC) of blood and strewn singly among fibroblasts. In the region of inflammation, these cells modified into amoeboid macrophages and ingest foreign particles and bacteria.

In different parts of body, macrophages are known by different names:

***(a)* Dust cells.** These are also called pulmonary or alveolar macrophages and are found in connective tissue of septa of lung.

***(b)* Kupffer cells.** These are found in wall of liver sinusoids.

Box 1.6

Langhans Giant Cells

When macrophages encounter large foreign bodies, they may fuse to form a large cell with upto 100 nuclei that engulfs the foreign body. These multinucleated cells are called **foreign body giant cells** or **Langhans giant cells.**

***(iii)* Mast cells.** The mast cell is a large ovoid connective tissue cell with a spherical nucleus and cytoplasm filled with large basophilic granules. These granules contain following four types of substances which on secretion cause immediate hypersensitive reactions, allergy and anaphylaxis or anaphylactic shock (Box 1.7).

Box 1.7

Anaphylactic Shock

An often severe and sometimes fatal systemic reaction in a susceptible individual upon a second exposure to a specific antigen (as wasp venom or penicillin) after previous sensitization that is characterized especially by respiratory symptoms, fainting, itching and hives or urticaria.

1. Histamine
2. Slow-reacting substance of anaphylaxis (SRS-A)
3. Eosinophil chemotactic factor of anaphylaxis (ECF-A)
4. Heparin

In the immune response, histamine and SRS-A increase the permeability of small blood vessels, thereby causing edema in the surrounding tissue. Heparin is an anticoagulant. The ECF-A stimulates eosinophils (WBC) to migrate to the sites where mast cells have released their agents; the eosinophils counteract the effects of the histamine and SRS-A.

***(iv)* Adipose cells.** These cells are differentiated from fibroblasts and are found throughout the loose connective tissue as individual cells and groups of cells. They are specialized to store neutral fats in their cytoplasm.

B. Wandering Cell Population

These are cells which have migrated into the tissue from the blood in response to specific stimuli. These are of following types:

***(i)* Wandering macrophages.** These are small freely motile cells which are derived from monocytes. They act as scavengers and function in the protection of the body against infection and noxious substances.

***(ii)* Lymphocytes.** Lymphocytes are the smallest of free cells in the connective tissue, having a diameter of 6-8 μm. They have a thin rim of cytoplasm surrounding a deeply staining heterochromatic nucleus.

Lymphocytes are polymorphic containing following two functional cell types:

***(a)* T lymphocytes** which have a long life span and are involved in cell-mediated immunity.

***(b)* B lymphocytes** which have variable life span and are involved in production of **antibodies.**

***(iii)* Plasma cells.** These cells are derived from B lymphocytes, have limited migratory ability and a short life span of 10 to 30 days. A plasma cell is relatively large (20 μm) ovoid cell with considerable cytoplasm. The cytoplasm is rich in rough ER and a large-sized Golgi apparatus. The nucleus of plasma cell is small and spherical having cartwheel or clock face due to alternate peripheral arrangement of dense heterochromatin and clear euchromatin. Plasma cells are specialized in synthesizing **antibodies.**

***(iv)* Eosinophils, monocytes and neutrophils.** As a result of immune responses and tissue injury, certain cells rapidly migrate from blood to enter the connective tissue, particularly neutrophils and monocytes. Their presence generally reflects an acute inflammatory reaction. Eosinophils may be observed in normal connective tissue and are involved in allergic reactions and parasitic infections.

Functions of Areolar Tissue

1. Areolar tissue acts as a supporting and packing tissue. Matrix of this tissue is important in the **diffusion** of oxygen and nutrients from the small blood vessels that course through this connective tissue as well as in the diffusion of metabolites back.

2. It also helps in combating foreign toxins.

2. Adipose Tissue

Structure

It is a modified areolar tissue in which extra fat is stored in the form of fat globule inside cytoplasmic vacuole of the fat cell. The fat cells (or adipocytes) are large, rounded or oval in shape. The big fat droplets occupy most area of the cell. The cytoplasm and flattened nucleus are pushed to one side to give a characteristic "signet ring" appearance. Adipocytes synthesize, store and metabolise fat. Adipose tissue contains closely packed fat-filled cells with a few and very fine reticular fibres surrounding individual cells (Fig. 1.17).

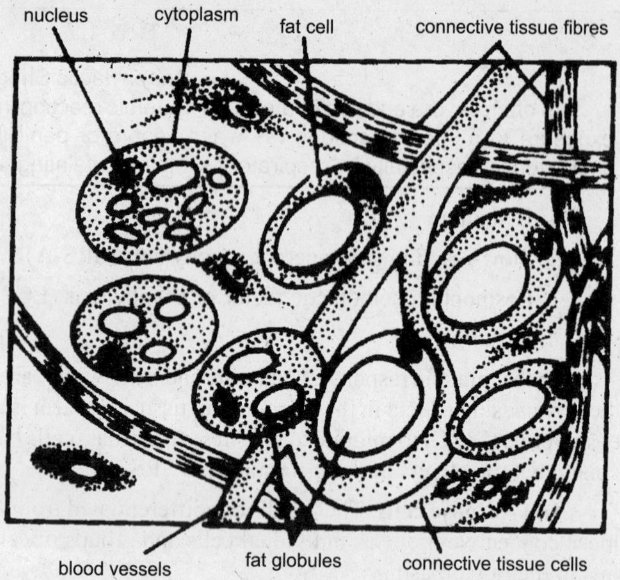

Fig. 1.17. Adipose tissue.

Location

Adipose tissue occurs beneath the skin, around heart and kidneys and also in camel's hump, blubber of whale and thick tail of merino sheep. Adipose tissue forms the so called fat deposits, *i.e.,* omentum, subcutaneous tissue, mesentery and perinephric region and the yellow bone marrow.

Functions

1. It serves as a fat reservoir.
2. It provides shape to the limbs and body.
3. Being a poor conductor of heat, it reduces heat loss, *i.e.,* it regulates body temperature.
4. It forms shock-absorbing cushions around kidneys and eye balls.
5. Keeps the visceral organs in position.

White Fat and Brown Fat

There are two types of fats in adipose tissue:

1. White fat. It contains large fat cells. Each adipose cell has a single large fat globule and is called **monolocular adipocytes.**

2. Brown fat. It contains fat cells or adipocytes with several small fat globules, which are called **multilocular adipocytes.** The fat is brown because of the presence of abundant iron-containing **cytochrome oxidase** in their mitochondria which are modified for heat-generation (thermogenesis; Box 1.8). Brown fat has exceedingly rich supply of blood capillaries and serves as a ready source of lipid, the oxidation of which produces heat. Brown fat is found in new born babies and hibernating animals.

Box 1.8

The multilocular adipocytes of brown fat contain numerous mitochondria, a small Golgi zone and only small amount of rough ER and smooth ER. These mitochondria are unique in that they do not contain elementary particles (which contain many enzymes for ATP production), *i.e.,* the energy produced by the mitochondria is dissipated or used as heat rather than stored as ATP.

II. Dense Connective Tissue or White Fibrous Tissue

It is further subclassified into two types:

1. Dense irregular connective tissue
2. Dense regular connective tissue

1. Dense irregular connective tissue

It occurs in dermis of skin and submucosa of intestine. The fibres make up the bulk of dense connective tissue. The cell population is sparse and is typically of a single type, namely the **fibroblast.** There is also relatively little ground substance. Due to presence of great amount of **collagenous fibres** in it, dense irregular connective tissue provides significant strength. Typically collagen fibres are arranged in bundles which are oriented in various directions (thus, it is irregular) to withstand stresses to which an organ may be subjected. In the wall of hollow organs (*e.g.*, intestinal tract) fibre bundles course in varying planes which permits resistance to excessive stretching and distention. In skin, this tissue is found in reticular or deep layer of dermis, and provides resistance to tearing as a result of stretching forces from different directions.

Commercial leather is prepared from dermis of mammalian skin.

2. Dense regular connective tissue

It is characterized by ordered and densely packed arrays of fibres and cells. It forms main functional component of tendons and ligaments (Fig. 1.18).

***(i)* Tendons.** Tendons are cord-like structures that join muscles to bone. They consist of parallel bundles of collagenous fibres between which are rows of fibroblasts, also referred to as **tendinocytes.** Fibroblasts are stellate having flat nuclei and they occur in far less number between fibres.

Tendons are relatively acellular and have low oxygen and nutrient demands, this is important because in order to maximize their strength they are relatively avascular, and contain only scanty capillaries.

The collagen fibres of tendons are firmly tethered to the muscle fibres in specialized areas called **myotendinous insertions**, while at the bone end, the collagen fibres of the tendon merge with the collagen of periosteum.

***(ii)* Ligaments.** Ligaments join bone to bone. They surround articular (synovial) joints, attaching one bone to the other over the outer surface of joint capsule.

Ligaments are composed of tightly packed collagen fibres all running in the same direction, with compressed fibrocytes in between, thus they resemble tendons. They differ from tendons because they contain elastic fibres and their white fibres are less regularly arranged. Ligaments strengthen the joint and they permit normal movement but prevent over-flexing or over-extension (**Sprain** is caused by excessive pulling or stretching of ligaments). The attachment of ligaments to bone is similar to that of tendons.

Ligaments associated with the vertebral column contain considerable number of elastic fibres and fewer collagen fibres. They are called **elastic liga-ments.** They provide some degree of elasticity to vertebral column.

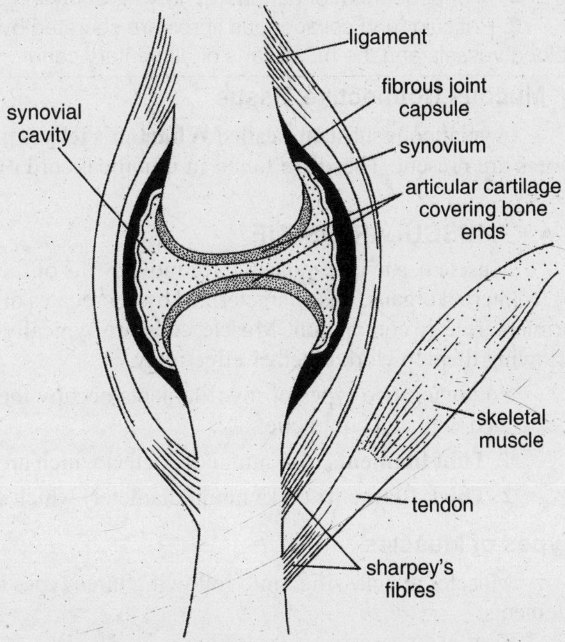

Fig. 1.18. Attachment of tendons and ligaments.

III. Yellow Elastic Connective Tissue

Structure. Yellow elastic connective tissue is a modified areolar tissue and is made up of freely branching elastic fibres so they are highly elastic. Elastic fibres chemically composed of two types of proteins: **1. Elastin** and **2. Fibrillin** forming an extracellular **microfibril.** Elastic fibres give yellow appearance to the tissues. Fibroblast is present in space between the fibres.

Location. This tissue is a component of the cartilage of larynx, the wall of the elastic arteries, the trachea, the bronchial tubes and lungs. Some special ligaments as **ligamentum flava** (joining vertebrae) and **ligamentum nuchae** which is present in the neck of grazing animals (mammals) are formed of this tissue.

IV. Reticular Connective Tissue

Structure. Reticular connective tissue is also a modified areolar tissue. It contains reticular fibres in that they both consist of collagen fibrils (it contains type III collagen). The individual fibrils that constitute the reticular fibre are always of narrow diameter (about 20 nm) and typically, the fibrils do not bundle to form thick fibres.

Location. In loose connective tissue, networks of reticular fibres are found at the boundary of connective tissue with epithelium and around adipocytes, small blood vessels, nerves and muscle cells. In all these cases reticular fibres are secreted by fibroblasts (Box 1.9).

Reticular fibres are also present as a supporting stroma in haemopoietic tissues (*e.g.,* red bone marrow) and lymphatic tissues (*e.g.,* spleen, lymph nodes and tonsils, but not in thymus). In this case reticular tissue is called **myeloid tissue.** In these reticular tissues, the collagen of the reticular fibre is produced by a special cell type, called **reticular cells.** This cell maintains a unique relationship to the fibre, surrounding it with its cytoplasm, the fibre is isolated from its environment.

Box 1.9

Besides fibroblasts and reticular cells, reticular fibres are secreted by some other type of cells:

1. In endoneurium of peripheral nerves, Schwann cells secrete reticular fibres.
2. Reticular and collagenous fibres are secreted by smooth muscle cells of the tunica media of blood vessels and the muscularis of alimentary canal.

V. Mucous Connective Tissue

A gelatinous substance called **Wharton's jelly** is present, in which few fibroblasts and collagen fibres are present. These are found in umbilical cord of embryo, cock's comb, vitreous humour of the eye.

1.4 MUSCULAR TISSUE

Muscle tissue is responsible for movements of the body and for changes in the size and shape of its internal organs. It is characterized by aggregates of specialized cells, called **muscle cells** whose primary role is contraction. Muscle cells are typically elongated and arranged in parallel arrays, allowing them to work together effectively.

Following two types of myofilaments occupy the bulk of the muscle cell cytoplasm and are associated with cell contraction:

1. Thin filaments (6-8 nm in diameter) which are composed primarily of the protein **actin.**

2. Thick filaments (~ 15 nm in diameter) which are composed of the protein **myosin.**

Types of Muscles

Muscles are classified into following three types on the basis of appearance of the contractile elements:

1. Smooth muscle or unstriated muscles in which cells do not show cross-striations.

2. Striated muscles in which the cells exhibit cross-striations at the light microscope (LM) level.

3. Cardiac muscles in which the cells show cross-striations, central location of nucleus and branching.

1. Smooth Muscles

These are also known as **plain** or **smooth, involuntary** (*i.e.*, they are never under the control of will) or **nonskeletal muscles.**

Location. Smooth muscle is the intrinsic muscle of the alimentary canal, blood vessels, genitourinary tract, respiratory tract and other hollow and tubular organs. It is also found as organized muscle bundles in the iris and ciliary body of the eye, as a thin sheet in the skin of the scrotum (the dartos muscle which is situated beneath the skin of scrotum), and as isolated fibres in association with hair follicles (arrector pili muscle).

Structure. Smooth muscle generally occurs as bundles or sheets of elongated fusiform or spindle-shaped cells. The cells, also called fibres, range in length from 20 μm in the walls of small blood vessels to about 200 μm in the wall of intestine, they may be as large as 500 μm in the wall of uterus during pregnancy. The long cigar-shaped **nuclei** of smooth muscle cells are located in the centre of the cell and often have a corkscrew appearance in longitudinal section. In TEM (=Transmission Electron Microscope), it is evident that the cytoplasmic organelles are concentrated at each end of the nucleus. In the **sarcoplasm** (*i.e.*, cytoplasm of muscle cells/fibres) adjacent to each nuclear pole, there are numerous **mitochondria,** some cisternae of **rough ER** (or rough sarcoplasmic reticulum), free **ribosomes, glycogen** granules and a small **Golgi apparatus.** The rest of the cytoplasm is seem to be filled with 6-8 nm **actin** filaments interspersed with 8-10 nm **desmin** containing intermediate filaments. The **α-actinin** protein containing **dense bodies** or **cytoplasmic densities** are found among the filaments and in relation to plasma membrane. The 16-nm **myosin** filaments can be observed in smooth muscle cells by certain special techniques.

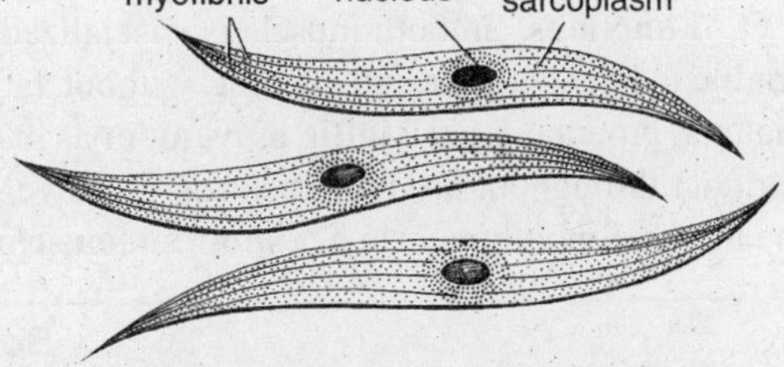

Fig. 1.19. Smooth muscle fibres.

Large number of pinocytotic vesicles are associated with the sarcolemma (= plasma membrane of muscle cells) and the smooth endoplasmic reticulum (SER). The vesicles and the SER sequester the calcium that is released to stimulate contraction. Smooth muscle cells make contact with neighboring cells by **gap junctions.** The gap junction between two smooth muscle cells is called **nexus.**

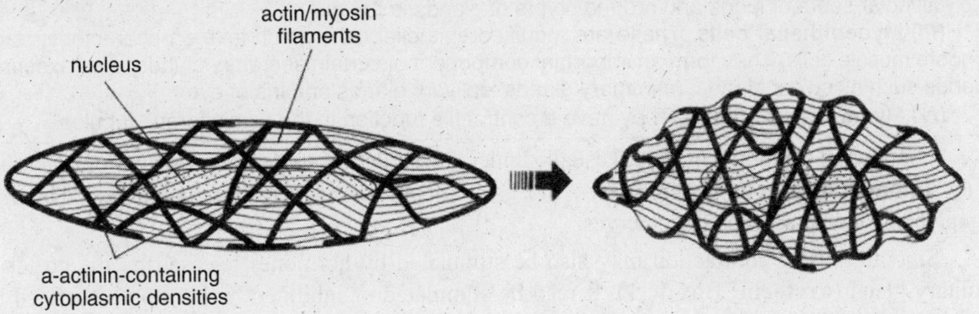

Fig. 1.20. A suggested model for smooth muscle cell contraction. Bundles of filaments containing both actin and myosin anchor on cytoplasmic densities. The densities are associated with the plasma membrane and with intermediate filaments. Because the contractile filament bundles are oriented obliquely to the long axis of the cell, their contraction shortens the cell and produces the "corkscrew" shape of the nucleus.

Box 1.10

1. Unitary smooth muscle. Smooth muscle is present in wall of hollow visceral organs where it is arranged in sheets with cells align circumferentially or longitudinally. When it contracts, it results in the reduction of lumen diameter. In the so-called **unitary smooth muscle,** cells tend to generate their own level of rhythmic contraction which is transmitted from cell to cell via the gap junctions. Such smooth cells are richly innervated by autonomic nervous system.

2. Multi-unit smooth muscle. A second arrangement of smooth muscle is typified by that in the iris of eye, vas deferens and large arteries. Here autonomic innervation precisely controls contraction, resulting in opening and closing of the pupil.

Functions. Smooth muscle is specialized for slow and prolonged contraction. It may remain contracted for long period of time without fatigue. Smooth muscle may contract in a wave-like manner, producing **peristaltic movement** as in gastrointestinal tract and male genital tract or it may contract throughout producing extrusive movements as in the urinary bladder, the gall bladder and the uterus. Smooth muscle has **spontaneous contractile** activity in the absence of nerve stimuli.

Box 1.11

Some Special Features of Smooth Muscle Cells

1. Oxytocin is a potent stimulator of smooth muscle contraction, and it is released by the posterior pituitary. It plays an essential role in uterine contraction during parturition. It is often used as an exogenous inducer or enhancer of labour.

2. Smooth muscle cells are secretory cells, they secrete tissue matrix. They synthesize both type IV (basal lamina) collagen and type III (reticular) collagen as well as laminin, elastin and proteoglycans. Except at the gap-junctions, smooth muscle cells are surrounded by a **basal lamina.** In some locations, such as the walls of blood vessels and the uterus, smooth muscle cells secrete large amounts of collagen and elastin.

3. Smooth muscle cells are not **postmitotic,** *i.e.,* they are capable of division to maintain or increase their number, *e.g.,* smooth muscle in the uterus proliferates during the normal menstrual cycle and during pregnancy.

4. Following four types of smooth muscle cells are known to develop from undifferentiated mesenchymal cells:

(*i*) Pericytes. These are stellate cells and located within the basal lamina of capillaries and venules.

(*ii*) Myofibroblasts. They have a contractile role in addition to being able to secrete collagen, *e.g.,* alveolar septa of lungs and around crypts of glands in the gut.

(*iii*) Myoepithelial cells. These are modified epithelial cells which take on characteristics of smooth muscle cells. They form an important component of certain secretory glands, *e.g.,* exocrine glands such as sweat glands, mammary glands, salivary glands and iris of eye.

(*iv*) Myoid cells of testis. They have a contractile function in the seminiferous tubules.

Contraction of smooth muscle is usually under the regulatory control of postganglionic neurons of the **autonomic nervous system** (ANS); most smooth muscle cells are directly innervated by both sympathetic and parasympathetic nerves.

Smooth muscle contraction may also be stimulated by hormones released by the posterior pituitary gland (**oxytocin;** Box 1.11). It may be stimulated or inhibited by hormones secreted by adrenal medulla (*i.e.,* epinephrine and norepinephrine).

2. Striated Muscles

Location. Striated muscle tissue forms flesh of body and its contraction is **voluntary,** under the control of will. On the basis of its location in human body, it is classified into following two types:

1. Skeletal striated muscle. It is attached to bone and responsible for movements of the axial and peripheral skeleton.

2. Visceral striated muscle. It is identical with skeletal muscle, but it is restricted to the soft tissues, namely the tongue, pharynx, diaphragm and upper part of oesophagus.

Skeletal striated muscle is responsible for gross and fine movements of limbs and digits and for the maintenance of body position and posture. The extraocular muscles of eye, too, provide precise eye movement. The muscles of tongue, pharynx, diaphragm and upper part of the oesophagus (visceral striated muscle) play essential roles in speech, breathing and swallowing.

Structure. In a skeletal muscle, each muscle cell is commonly called a **muscle fibre** and it consists of a multinucleated syncytium. It is formed by fusion of small, individual muscle cells, called **myoblasts,** during development. Each muscle cell is an elongated, straight and unbranched multinucleated cylindrical fibre which may have a diameter of 10-100 μm. Its length may be upto 4 cm in length. (**Note.** A muscle fibre should not be confused with a connective tissue fibre: muscle fibres are **cellular** elements, while connective tissue fibres are extracellular products of connective tissue cells).

Each muscle cell/fibre consists of a thin outer limiting membrane, the **sarcolemma** (Box 1-12) which encloses a coenocytic cytoplasm, known as **sarcoplasm.** The nuclei of muscle cell are peripheral in position, *i.e.,* located immediately under the sarcolemma. A large single muscle cell/ fibre may contain several hundred nuclei. In sarcoplasm lie embedded large number of **myofibrils** or **sarcostyles.** The myofibrils are relatively coarse and tightly packed in parallel bundles and are separated by thin sheets of cytoplasm. Mitochondria and glycogen deposits are located between the myofibrils in association with smooth endoplasmic reticulum (SER).

Box 1.12

Traditional histologic descriptions of skeletal muscle used the term **sarcolemma** to describe thick "membrane" as the cytoplasmic boundary of the muscle cell. It is now known that the thick sarcolemma actually represents the plasma membrane of the cell, its basal lamina and the surrounding reticular lamina.

Fibrillar connective tissue surrounds both individual muscle fibres and bundles of muscle fibres (called **fasciculi**) and is essential for force transduction. At the end of muscle, the connective tissue continues as a **tendon** that attaches the muscle to bone. A rich blood supply and a rich supply of nerves travel in the connective tissue.

The connective tissue associated with muscle is given a specific set of names such as follows (Fig. 1.21):

1. Endomysium is the delicate layer of reticular fibres that immediately surrounds *individual muscle fibres.* Only small diameter capillaries and the finest neuronal branches are present within the endomysium, running in parallel with the muscle fibres.

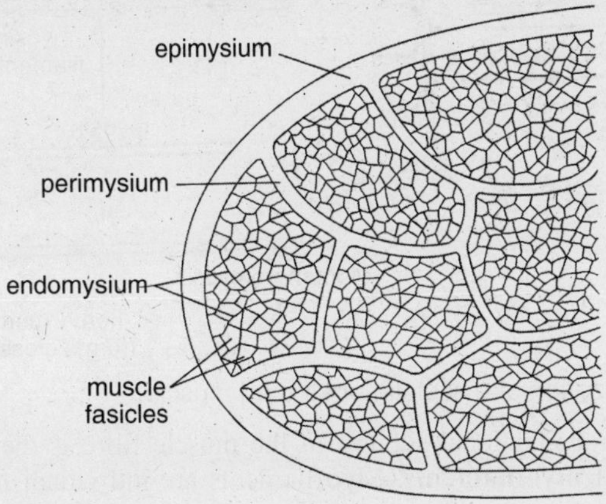

Fig. 1.21. Organization of muscle fibres into muscle.

2. Perimysium is a thicker connective tissue layer that surrounds a group of fibres to form a bundle of **fascicle.** Larger blood vessels and nerves travel in the perimysium.

3. Epimysium is the sheath of dense connective tissue that surrounds the collection of **fascicles** that constitute the muscle. The major vascular and nerve supply of the muscle penetrate through the epimysium.

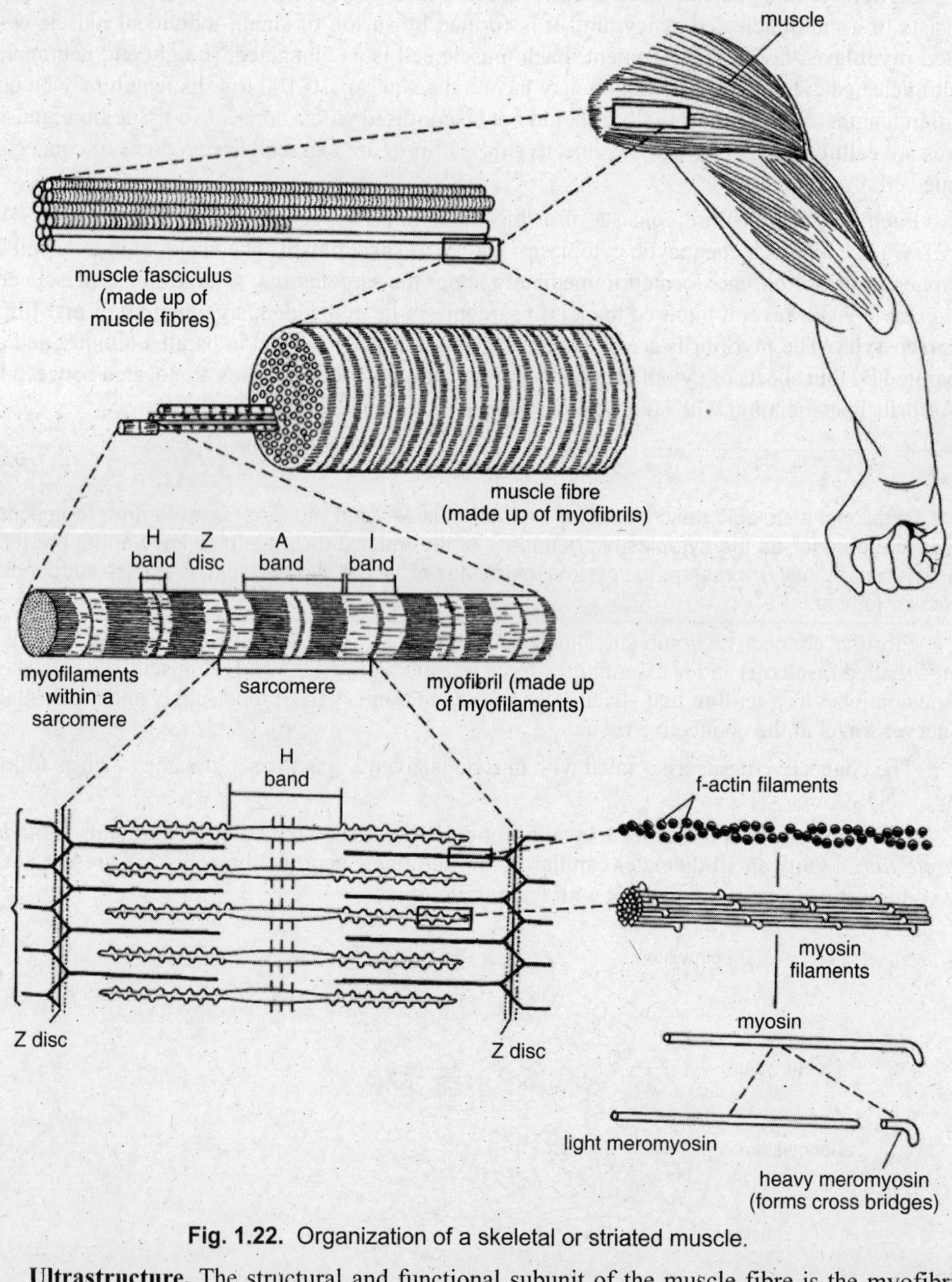

Fig. 1.22. Organization of a skeletal or striated muscle.

Ultrastructure. The structural and functional subunit of the muscle fibre is the myofibril. Myofibrils are composed of bundles of **myofilaments.** Myofilaments are individual filamentous polymers of **myosin (thick filaments)** and **actin** and its associated proteins **(thin filaments).** They are the actual contractile elements of striated muscle. **Cross striations** are the principal histological feature of striated muscle. In stained and in phase contrast microscopy, alternating light and dark bands are evident, these are termed as the **A band,** the **I band** and the **Z disc** or **Z line** (Fig. 1.22).

In polarizing microscopy, the dark bands are **birefringent,** *i.e.,* they alter the plane of polarized light. The dark bands are therefore **anisotropic** and are given the name **A band.** The light bands are **monorefringent,** *i.e.,* they do not alter the plane of polarized light. They are, therefore, **isotropic** and are given the name **I band.** Each of the major bands, the A band and the I band is bisected by a narrow zone of contrasting density.

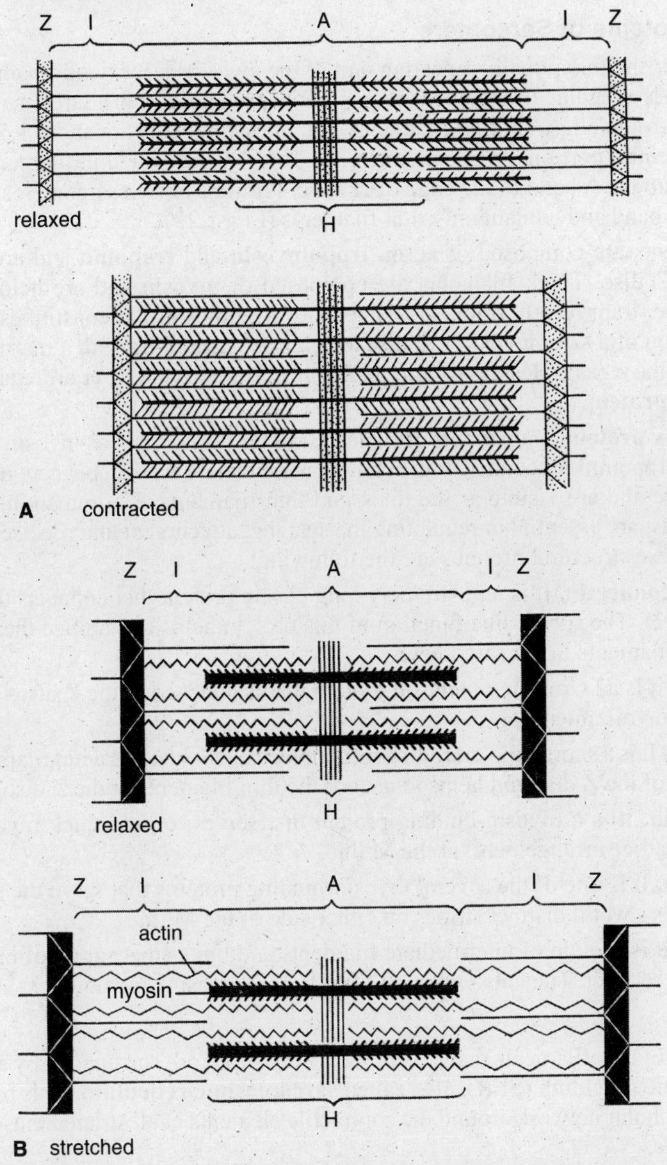

Fig. 1.23. A—Relaxed and contracted sarcomere. In the relaxed state, interdigitation of actin and myosin filament, is not complete; the H and I bands are relatively wide. During contraction, the interdigitation of the actin and myosin filaments is increased according to degree of contraction. The length of the A band remains the same; the length of the H and I bands is diminished again in proportion to the degree of contraction. B—Titin, a large elastic molecule, anchors the thick myosin filaments to the Z disc.

1. Z line or Z disc (Ger. *Zwischenscheible* = a between disc). It is a dense zone (dark refractile) that bisects the I band. It is also called **intermediate** or **Krause's membrane.**

2. H zone (Ger. *Helle* = light). It is a light zone that bisects the A band. It is also called **Hensen's disc.**

3. M line. It can be seen in the middle of the H band.

Sarcomere. The segment of the myofibril between two Z discs is called **sarcomere.** The sarcomere is the basic contractile unit of striated muscle. It measures 2-3 μm in relaxed mammalian muscle, may be stretched to more than 4 μm, and, in extreme contraction, may measure as little as 1 μm (Fig. 1.23).

Component Proteins of Sarcomere

Sarcomere is the functionally repeating unit of the myofibril; it extends from one Z disc to the next Z disc. There is a regular arrangement of contractile proteins within a sarcomere, with each thick filament being surrounded by six thin filaments. The thick filaments are about 1.5 μm long and are restricted to the central portion of the sarcomere, *i.e.,* the A band. The thin filaments attach to the Z line and extend into the A band to the edge of H zone. Portions of two sarcomeres, on either side of Z disc, form the I band and contain only thin filaments (Fig. 1.22).

Thin filaments are composed of **actin, tropomyosin** and **troponin** and are associated with **α-actinin** at the Z disc. Thick filaments are composed of **myosin** and are held in the centre of the H zone by fine, transversely oriented filaments of **myomesin** that constitutes the **M line.** In a longitudinal section of a sarcomere, the Z disc appears as zigzag line with a matrix material, the **Z matrix,** bisecting the zigzag. Besides α-actinin, Z matrix contains the other proteins such as **filamin, amorphin** and **Z protein.**

The **primary protein** components of skeletal muscle fibrils, such as myosin and actin and the tropomyosin and troponins associated with actin, constitute more than 75 per cent of the total protein of the muscle fibre and are visible as the thick and thin filaments. The remaining proteins, called **accessory proteins,** are essential in regulating the spacing, attachment and precise alignment of the myofilaments. These structural proteins are the following:

1. Titin or connectin. It is an extremely long elastic protein that connects the thick filament to Z disc (Fig. 1.22). The spring-like function of this protein helps to stabilize the centralization of the thick myosin filaments in the sarcomere.

2. Nebulin. It is an elongated inelastic protein that is attached to the Z discs and runs parallel to the thin (actin) myofilaments.

3. α-actinin. It is a short, bipolar, rod-shaped molecule that bundles actin filaments into parallel arrays at the level of the Z disc and helps to anchor the thin filaments at the Z disc.

4. Myomesin. It is a myosin-binding protein that serves to hold thick myosin filaments in register (*i.e.,* in a lattice arrangement) at the M line.

5. C protein. It is one of the several myosin-binding proteins that serve the same function as myomesin. It forms several distinct stripes on either side of the M line.

6. Desmin. It is protein of intermediate filaments and links adjacent myofibrils to each other and maintain their register. They also links myofibrils to the plasma membrane.

Triad System

The bundles of myofilaments that constitute the myofibril are surrounded by a well-developed smooth endoplasmic reticulum (SER), also called **sarcoplasmic reticulum.** This reticulum forms a highly organized tubular network around the contractile elements in all striated muscle, *i.e.,* skeletal, visceral and cardiac.

The sarcoplasmic reticulum is arranged in networks around or between a group of myofilaments (*i.e.,* a myofibril). In the most highly developed arrangements, one network of sarcoplasmic reticulum surrounds the A band, and another network surrounds the I band (Fig. 1.24). Where the two networks meet, at the junction between A and I bands, the sarcoplasmic reticulum forms a slightly more ring-like channel, called the **terminal cisterna** or **sac,** around the filaments of the myofibrils.

Invaginations of the plasma membrane of the muscle cell form the **transverse tubular system** or **T system.** The **T tubules** penetrate to all levels of the muscle fibre. They are located between the adjacent terminal cisternae.

The complex of T tubule and two terminal cisternae seen with the TEM (= transmission electron microscope) is called the **triad.** The T tubules provide for the rapid transmission of surface membrane excitation to the terminal sacs throughout the thickness of muscle fibre. In turn, the terminal sacs **release** and then reaccumulate calcium ions (Ca^{2+}). Thus, sarcoplasmic reticulum serves as a reservoir for this essential ion.

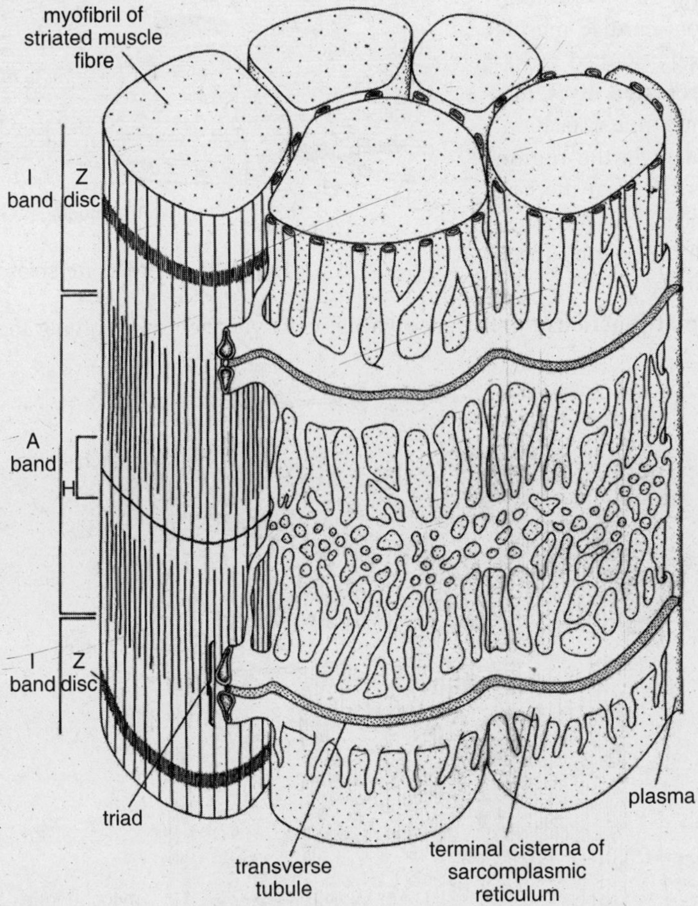

Fig. 1.24. Diagram of part of a mammalian striated muscle fibre showing the organization of a triad (there are two triads to each sarcomere).

3. Cardiac Muscle

The cardiac muscle is found in the *wall of heart* and in the *base of the large veins* that empty into heart (**Ross** *et al.*, 1995). Structurally, it shows resemblance to skeletal muscle although functionally, it is homologous to the unstriated muscles. Cardiac muscles contract rapidly, rhythmically and tirelessly, contracting endlessly from early embryonic stage until death.

Structure. Cardiac muscle consists of long fibres that appear to branch and anastomose with neighboring fibres (Fig. 1.25). Unlike skeletal (or striated) muscle fibres, the fibres of cardiac muscle are formed by individual **uninucleate** (or mononucleate) cardiac muscle cells that are joined to one another in linear array.

In the adult, a cardiac muscle cell is 15 to 20 μm in diameter and about 100 μm long with a centrally located nucleus. Cardiac muscle cells exhibit **cross-striations** like the fibres of skeletal muscle. In fact, a cardiac muscle cell has the same types and arrangement of contractile elements (**actin** and **myosin**) as in skeletal muscle. In addition, cardiac muscle fibre exhibit densely stained cross-bands, called **intercalated discs,** that cross the fibres in a linear fashion. An intercalated disc is the region of interdigitations between the cell membranes of adjacent muscle fibres. Each disc contains following three types of cell junctions:

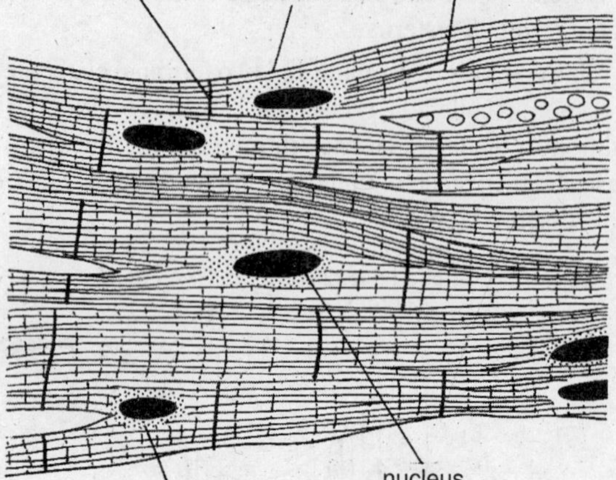

Fig. 1.25. Cardiac muscles.

1. Desmosomal junctions tightly link adjacent cells via anchors involving the intermediate filaments.

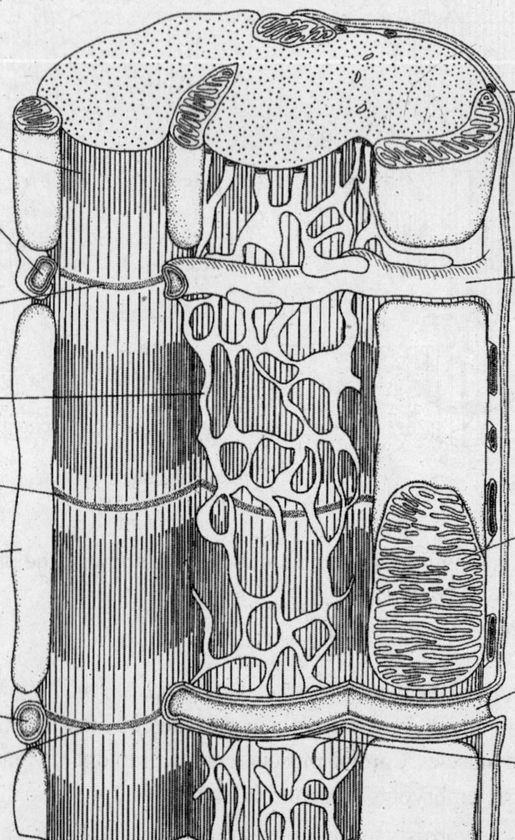

Fig. 1.26. Organization of mammalian ventricular cardiac muscle fibre.

2. Adherent type junctions anchor the actin fibres of the sarcomeres to each end of the cell.

3. Communicating **gap junctions** facilitate the passage of membrane excitation and thereby synchronize muscle contraction.

With TEM, it has been seen that the myofibrils of cardiac muscle separate to pass around the nucleus, thus outlining a biconical **juxtanuclear region** in which the cell organelles are concentrated. This region is rich in mitochondria, and contains Golgi apparatus, lipofuscin pigment granules, hormone-containing granules and glycogen.

In addition to juxtanuclear mitochondria, cardiac cells are characterized by very large mitochondria that are densely packed between the myofibrils. These large mitochondria often extend for the full length of a sarcomere and have numerous, closely packed cristae (Fig. 1.26) Transverse tubules of cardiac muscle are much larger and located at the level of Z disc. Sarcoplasmic reticulum of cardiac muscle is organized as an anastomosing network. The association of cardiac sarcoplasmic reticulum with T tubules takes the form of **diads** rather than triads (Fig. 1.24).

Mechanism of Contraction of Skeletal Muscle

The event that leads to the contraction of a skeletal muscle fibre is the arrival of a nerve impulse at the neuromuscular junction, *i.e.,* at the motor end plate (Fig. 1.27). The nerve cell conducting these impulses is called a **motor neuron.** The cell body of this neuron is within the spinal cord (or in the brain stem). It sends a long cytoplasmic process, called the **axon** or **nerve fibre,** to the skeletal muscle, where the axon branches to innervate one or more muscle fibres.

The events involved in contraction can be summarized as follows (Fig. 1.27):

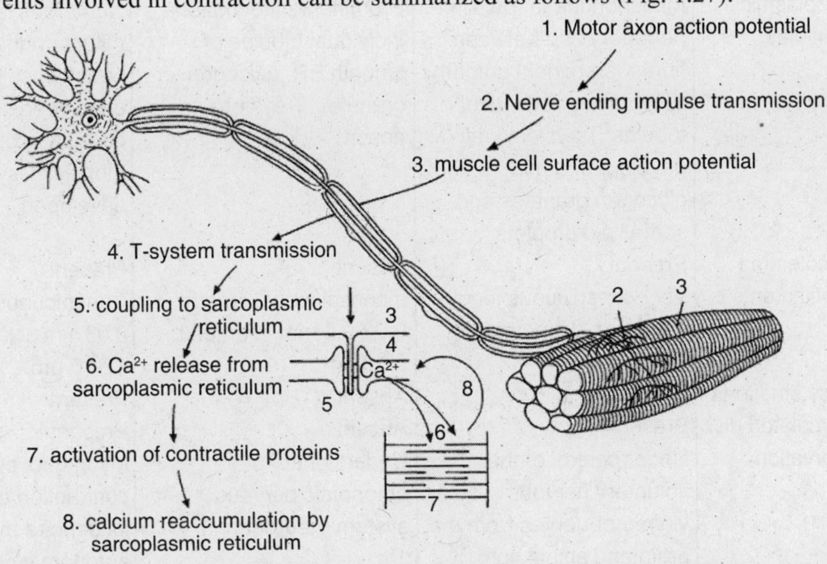

Fig. 1.27. Diagram showing summary of events leading to contraction of skeletal muscle.

1. The nerve impulses are measured as an action potential of the nerve fibre.
2-3. The nerve impulses are transmitted to the muscle cell and this results in a depolarization of the plasma membrane and in the generation of a action potential by the membrane of the muscle cell.
4. The depolarization action potential spreads over the muscle cell and continues via membranes of the T tubules into the substance of the muscle cells.
5-6. At the muscle cell triads, the T tubules are in close contact with the lateral enlargements of the sacroplasmic reticulum. At this point, the excited (depolarized) T tubule membrane causes calcium to be released from the sarcoplasmic reticulum into the cytoplasmic matrix.
7. The calcium then activates the contractile protein and contraction ensues.
8. The calcium is returned to the cisterna of sarcoplasmic reticulum at the end of these events.

Table 1.2. Comparison (*i.e.,* similarities and differences) between striated, smooth and cardiac muscles.

Feature	Striated muscles	Smooth muscles	Cardiac muscles
1. Other names	Voluntary, striped, skeletal	Involuntary, unstriped, unstriated	Heart
2. Specialization	Most highly specialized	Least specialized	More specialized than smooth
3. Structure	Very long cells, usually called fibres, subdivided into units called sarcomeres. Fibres bound together by vascular connective tissue	Consists of individual, spindle-shaped cells, associated in bundles or sheets	Cells terminally branched and connected to each other by special inter-digital surface processes (*i.e.,* plasma membrane contains desmosomes, gap junctions and tight junctions), called **intercalated discs**
4. Nucleus	Several, in variable positions near periphery of fibre	Single, elongated in shape and centrally placed	Several centrally placed
5. Cytoplasmic organelles	Mitochondria in rows in periphery and between fibres, prominent smooth ER forming network of tubules, T-system well developed forming triad, glycogen granules and some lipid droplets	Prominent mitochondria, individual tubules of smooth ER, glycogen granules, T-system absent	Numerous large mitochondria in columns between cells, poorly developed smooth ER consisting of network of tubules, T-system well developed
6. Sarcolemma	Present	Absent	Present
7. Myofilaments	Very conspicuous length 1-40 mm, diameter 10-60 μm	Inconspicuous, length 0.02-0.5 mm, diameter 5-10 μm	Conspicuous, length 0.08 mm or less, diameter 12-15 μm
8. Cross striations	Present	Absent	Present
9. Intercalated disc	Present	Absent	Present
10. Innervation	Under control of the voluntary nervous system via motor nerves from the brain and spinal cord (neurogenic)	Under control of autonomic nervous system (neurogenic)	Myogenic, but rate of contraction can be influenced by the autonomic nervous system
11. Activity	Powerful, rapid contractions, short refractory period, therefore fatigues quickly	Shows sustained rhythmic contraction and relaxation, as in peristalsis	Rapid rhythmic contraction and relaxation, long refractory period, therefore does not fatigue; contraction not sustained
12. Location	Attached to the skeleton in trunk, limbs and head	In wall of intestinal, genital, urinary and respiratory tracts, and the wall of blood vessels	Found only in wall of the heart chambers

1.5 NERVOUS TISSUE

Location

The nervous system consists of all the nervous tissue in the human body. It has been divided into the following two parts:

1. Central nervous system (CNS) which consists of the **brain** and the **spinal cord** located in the cranial cavity and spinal (= vertebral) canal, respectively.

2. Peripheral nervous system (PNS) which consists of **cranial** and **spinal nerves** (both motor and sensory), **ganglia** (*i.e.,* collection of nerve cells outside the CNS) and **motor nerve endings** and **sensory nerve endings** (receptors).

Composition

Nervous tissue consists of following two main types of cells:

1. Nerve cells or neurons. These are structural and functional units of the nervous system. Nerve cells are (*i*) specialized to receive stimuli and to conduct electrical impulses to other parts of system; (*ii*) arranged as an integrated communication network, with several neurons in a chain-like fashion typically involved in sending impulses from one part of the system to another.

The specialized *contacts* between nerve cells that provide for the transmission of information from one nerve cell to next in the chain are called **synapses.**

2. Supporting cells. These are nonconducting cells that are in close physical contact with neurons. In the CNS, these supporting cells are called **neuroglia** or simply **glia;** in the PNS, **Schwann cells** which surround nerve cell processes and **satellite cells** which surround nerve cell bodies (*i.e.,* cytons) in ganglia. Both Schwann cells and satellite cells originate from neural crest cells. Supporting cells provide (*i*) *physical support* (protection) for delicate neuronal processes; (*ii*) *electrical insulation* for nerve cell bodies and processes; and (*iii*) *metabolic exchange* pathways between the vascular system and the nervous system.

In addition to nerve cells and supporting cells, there are many blood vessels in both the CNS and the PNS. The blood vessels are separated from the nervous tissue by the interposition of **basal laminae** and variable amounts of **connective tissue,** depending on vessel size (Box 1.13).

1. Epineurium, perineurium and endoneurium. In a nerve trunk, the individual nerve fibres and their associated Schwann cells are held together by **connective tissue** that is organized into following three distinctive components:

1. Endoneurium. It is composed of longitudinally orientated **collagen** fibres, extracellular **matrix** material rich in glycosaminoglycans, sparse **fibroblasts** and **mast cells.** Endoneurium surrounds the individual axons and their associated Schwann cells as well as capillaries. It is, however, poorly vascularized.

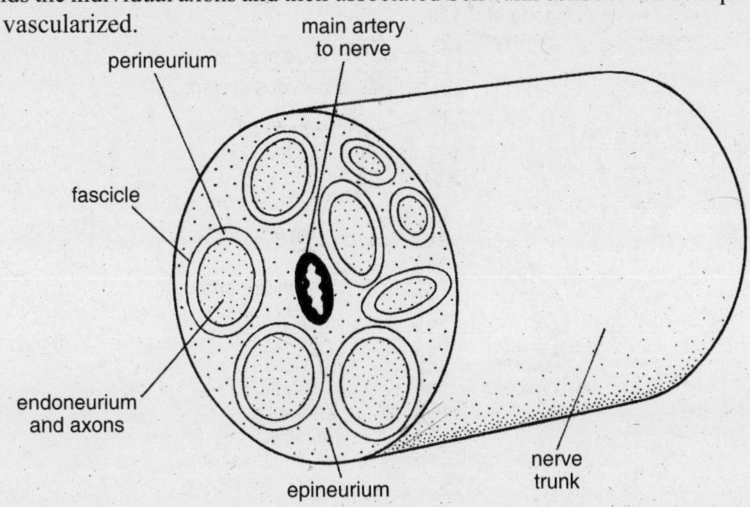

Fig. 1.28. Support (connective) tissue forming endoneurium, perineurium and epineurium in a nerve.

2. Perineurium. It surrounds groups of axons and endoneurium to form small bundles **(fascicles).** It is composed of 7 to 8 concentric layers of flattened (*i.e.,* squamous in shape) cells separated by layers of **collagen.** The cells are joined by junctional complexes and each layer of cells is surrounded by basal lamina on its both surfaces. Perineurium lacks fibroblasts and serves as a semipermeable barrier.

3. Epineurium. It is an outer sheath of dense connective tissue (*i.e.,* dense fibro-collagenous tissue) which binds individual nerve fascicles into a nerve trunk. The epineurium may also include *adipose tissue,* as well as a main muscular artery supplying the nerve trunk.

Box 1.13

Blood Brain Barrier

In the CNS, the boundary between blood vessels and nervous tissue is special in that many substances that readily leave blood vessels to enter other tissues *do not normally enter the nervous tissue.* This selective restriction of blood-borne substances in the CNS is known as the **blood-brain barrier.** The barrier has been shown to exist in the elaborate **tight junctions** between the endothelial cells of brain capillaries.

A. The Neurons (Nerve cells)

The neurone is the structural and functional unit of nervous tissue. There are more than 10 billion neurons in the human nervous system. These neurons are of following three types:

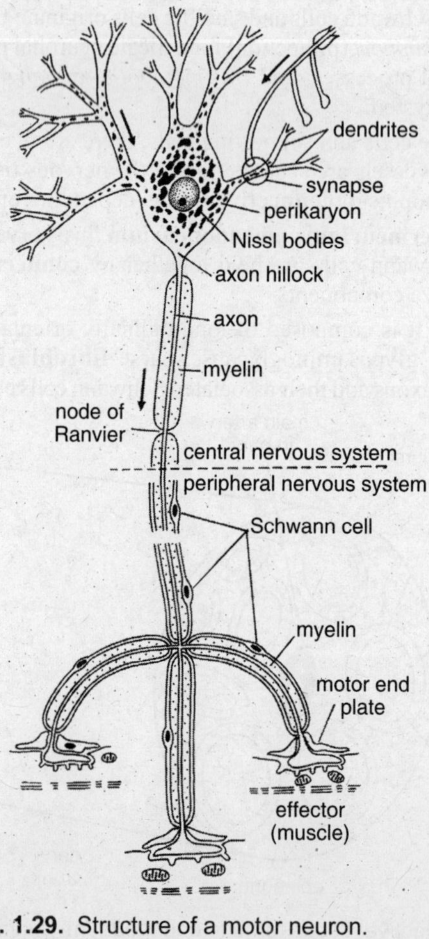

Fig. 1.29. Structure of a motor neuron.

1. Sensory neurons which convey impulses from receptors to the CNS.

2. Motor neurons which convey impulses from the CNS or from ganglia to effector cell.

3. Interneurons which are also called **internuncial neurons, intercalated neurons** or **central neurons.** They form a communicating and integrating network between the sensory and motor neurons. It is estimated that more than 99.9 per cent of all neurons belong to this integrating network.

Neurons are large, polymorphic cells consisting of **cell body** and processes such as **axon** and **dendrites** (Fig. 1.29). On the basis of the number of processes extending from the cell body, neurons are classified into following three types (Fig. 1.30):

1. Multipolar neurons. They have one axon and two or more dendrites, *e.g.,* motor neurons and interneurons.

In them dendrites and cell body form receptor portions of the cell while axon is the conducting portion. The terminal part of axon is specialized as transmitter or effector which is connected either with other nerves by synapses or with muscles or glands (effectors).

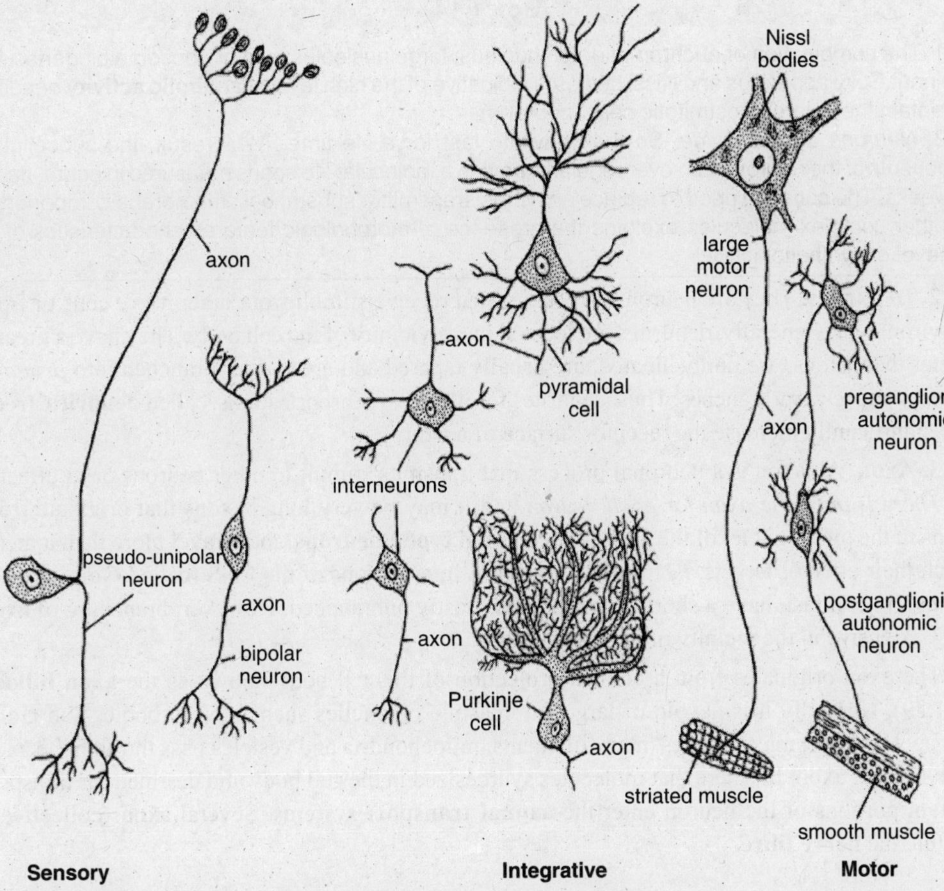

Fig. 1.30. Different types of neurons.

2. Bipolar neurons. They have one axon and one dendrite, *e.g.*, neurons of retina and VIII cranial nerve.

3. Unipolar neurons. They are, in fact, **pseudounipolar** neurons. They have one process, the axon that divides close to the cell body into two long processes, *e.g.*, sensory neurons. The cell body of a sensory neuron is situated in a dorsal root ganglion close to CNS; one axonal branch extends to periphery and one extends to CNS.

Ultrastructure

1. Cell body. It is also called **cyton** or **perikaryon.** Cell body of neuron is large, dilated region of cell and contains the nucleus and those organelles that maintain the nerve cell. Cell body contains a large oval euchromatic nucleus with a prominent nucleolus. Cytoplasm of cell body contains many darkly stained (stained by basic dyes) minute particles, called **Nissl bodies** (or **Nissl's granules**). Each Nissl body corresponds to a stack of rough endoplasmic reticulum (RER). The perikaryon also contains numerous mitochondria, a large perinuclear Golgi apparatus, lysosomes, microtubules, neurofilaments, vesicles and inclusions (Fig. 1.29). Nissl bodies (*i.e.*, RER), free ribosomes and occasionally, the Golgi apparatus extend into the dendrites but not into axon.

Box 1.14

1. The combination of euchromatin-rich nucleus, large nucleolus (site of ribosome biogenesis), prominent Golgi apparatus and Nissl bodies is indicative of the high level of **anabolic activity** needed to maintain these large postmitotic cells (neurons).

2. *Neurons do not divide.* So they have to last for a life time. As a result, the subcellular components of the neurons turn over regularly, having a molecular life spans measured in hours, days and weeks. The constant need to replace enzymes, transmitter substances, membrane components and other complex molecules explains the presence of morphologic features characteristics of a high level of synthetic activity.

2. Dendrites. They are neuronal processes that receive stimuli from other nerve cells or from the environment. Generally, dendrites are located in the vicinity of the cell body. They have a greater diameter than axons, are unmyelinated, are usually tapered and are further branched into *primary, secondary* and *tertiary* branches. Thus, dendrites form extensive arborizations, called **dendritic trees,** which significantly increase the receptor surface of a neuron.

3. Axon. An axon is a neuronal process that transmits stimuli to other neurons or to effector cells. *There is only one axon for each neuron* and it may be very long. Axons that originate from neurons in the motor nuclei of the CNS (called **Golgi type I neurons)** may travel more than a meter to reach their effector targets, *i.e.*, skeletal muscles. Interneurons of the CNS (called **Golgi type II neurons),** in contrast, have a short axon. Axon is mostly unbranched. However, branching of axon is most extensive in the vicinity of its target organs.

The axon originates from a conical projection of the cell body known as the **axon hillock** (Fig. 1.29). Normally it is devoid of large cytoplasmic organelles such as Nissl bodies and Golgi cisternae. However, microtubules, neurofilaments, mitochondria and vesicles pass through the axon hillock into the axon. It is here that molecules synthesized in the cell body and destined for transport to distant portions of the neuron enter the **axonal transport systems.** Several axons collectively constitute the **nerve fibre.**

Synapses

Neurons communicate with other neurons and with effector cells by means of **synapses.** Synapses are specialized junctions between neurons that facilitate transmission of impulses from one neuron to another. Synapses also occur between axons and effector cells such as muscle and gland cells.

Types of synapses. Synapses between neurons may be classified morphologically as follows (Fig. 1.31).

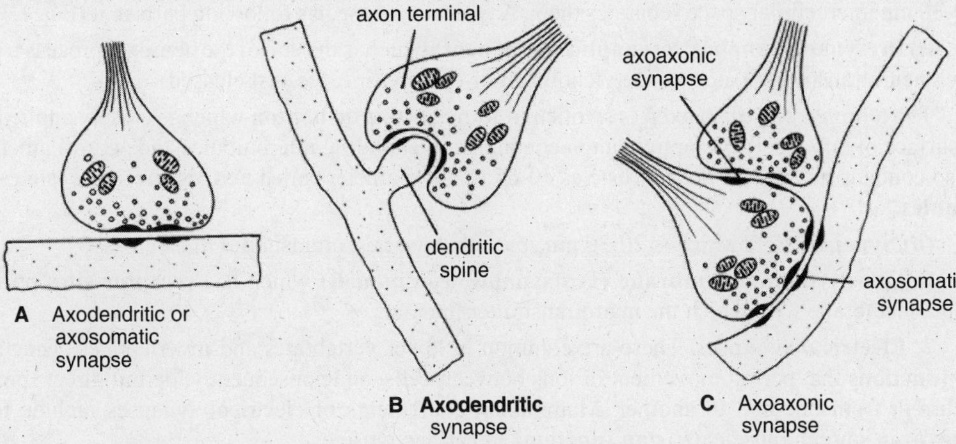

Fig. 1.31. Schematic diagram of different types of synapses. A—Axodendritic or axosomatic synapses; B—Axodendritic synapse; C—Axoaxonic synapse.

(*i*) **Axodendritic synapses** which occur between axons and dendrites.

(*ii*) **Axosomatic synapses** which occur between axons and the cell body.

(*iii*) **Axonomic synapses** which occur between axons and axons.

(*iv*) **Dendrodendritic synapses** which occur between dendrites and dendrites.

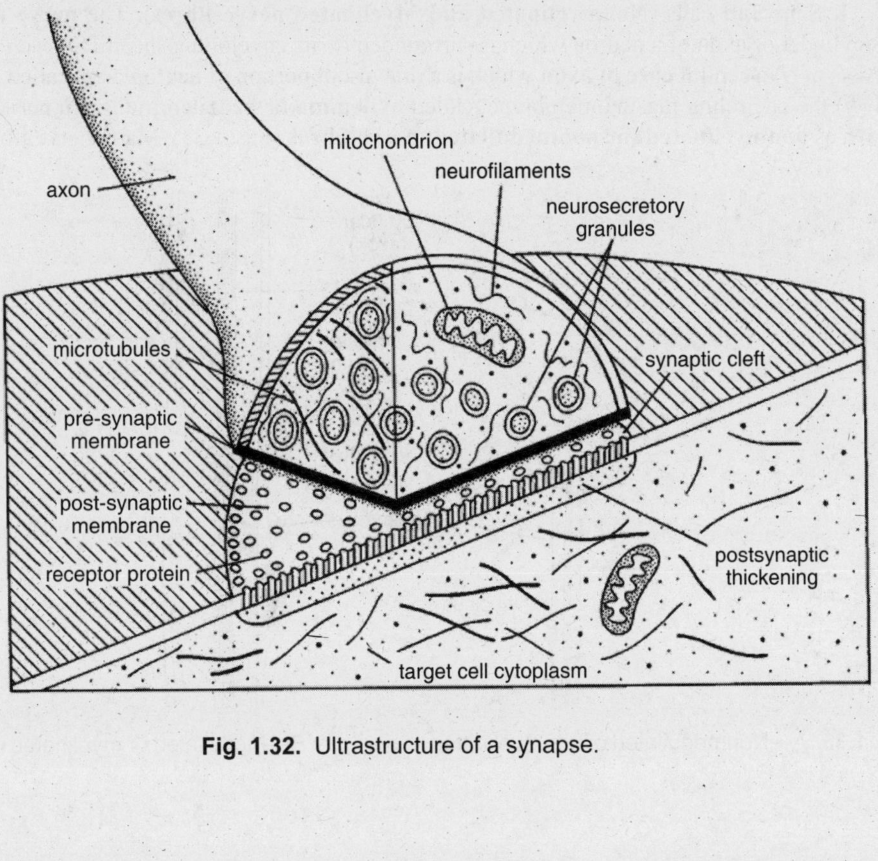

Fig. 1.32. Ultrastructure of a synapse.

Chemical and electrical synapses. Synapses may also be classified as chemical or electrical.

1. Chemical synapses. In them, neurons and their processes are closely apposed; usually only a 20-30 nm intercellular space separates them. A typical synapse has following parts in it (Fig. 1.32):

***(i)* Presynaptic knob (Presynaptic component)** which is the end of the neuronal process from which neurotransmitter (such as acetylcholine, nor-epinephrine, etc.) is released.

The terminal end of an axon is swollen to form a **synaptic bouton** which is closely applied to the surface of target cell. Synaptic bouton contains mitochondria, microtubules and neurofilaments. It also contains membrane-bound vesicles 40-65 μm in diameter called **neurosecretory vesicles** or **granules.**

***(ii)* Synaptic cleft** which is 20-30 nm space that the neurotransmitter must cross.

***(iii)* Postsynaptic membrane (Postsynaptic component)** which has **receptor sites** on the plasma membrane with which the neurotransmitter interacts.

2. Electrical synapses. These are common in lower vertebrates and invertebrates. They are **gap junctions** that permit movement of ions between cells, and consequently, permit direct spread of current from one cell to another. Mammalian equivalents of electrical synapses include the: (*i*) **nexus** in smooth muscle; (*ii*) **gap junctions** of cardiac muscle.

B. Supporting Cells

These include following three types of cells:

1. Schwann cells;
2. Satellite cells;
3. Neuroglia.

1. Schwann cells (Nonmyelinated and Myelinated nerve fibres). The **nerve fibre** is the axis cylinder or axon of a neuron which is surrounded by its enveloping sheath. So each nerve fibre consists of (*i*) a central **core of axon** which is a drawn-out portion of neuroplasm, called **axoplasm;** and (*ii*) the encircling plasma membrane, called **axilemma** or **neurilemma.** Such nerve fibres are known as **nonmyelinated (or nonmedullated) nerve fibres** (Fig. 1.33). Many nerve fibres contain

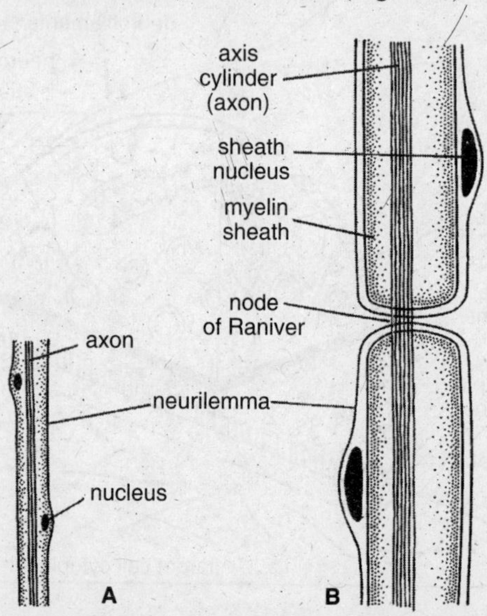

Fig. 1.33. A—Nonmedullated or nonmyelinated nerve fibre; B—Medullated or myelinated nerve fibre.

an additional covering of **medullary sheath** or **myelin sheath.** The covering is placed in between the axis cylinder and neurolemma and they are known as **myelinated (or medullated) nerve fibres.** These appear glistening white and have high contents of lipids. The myelin sheath is a sort of insulating covering and is said to enhance the speed of conduction. Immediately underneath the neurilemma is a thin sheath of cytoplasm containing nuclei. These are known as **nuclei of Schwann cells** and are concerned with the formation of neurilemma.

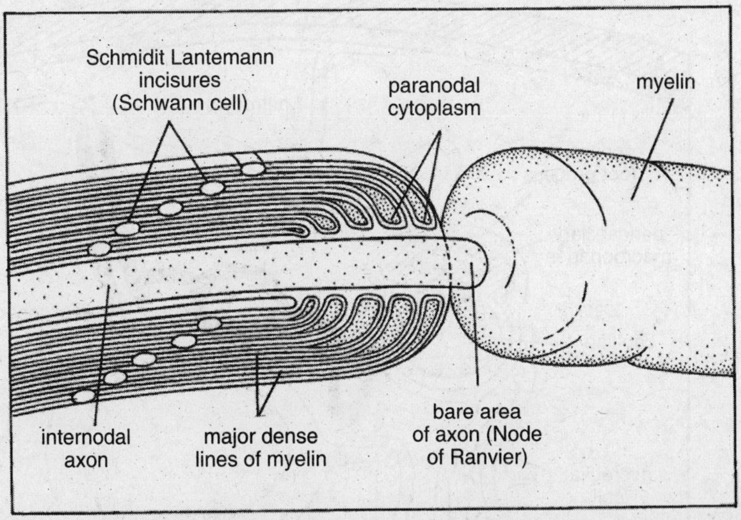

Fig. 1.34. Node of Ranvier (in L.S.).

The myelin-sheath is not continuous along the entire length of axis cylinder but is interrupted at intervals of about 1 mm. These bare areas of axon (*i.e.,* constrictions) are called **nodes of Ranvier** (Fig. 1.34). Each node of Ranvier is the region containing the **voltage-gated Na^+ channels,** which are necessary for the formation of action potential. The myelin between two sequential nodes of Ranvier is called an **internodal segment.** The medullated nerve fibres form the cranial and spinal nerves, while the non-medullated nerve fibres occur in the autonomic nervous system.

2. Satellite cells. The neuronal cell bodies of ganglia are surrounded by a layer of small cuboidal cells called **satellite cells.** In paravertebral and peripheral ganglia, neural cell processes must penetrate between the satellite cells to establish a synapse (there are no synapses in sensory ganglion). Satellite cells help to establish and maintain a controlled microenvironment around cytons of neurons in the ganglion. They provide electrical insulation and pathway for metabolic exchanges.

3. Neuroglia. Within the CNS, the supporting cells are **neuroglia** or **glial cells.** The three types of glial cells are (Fig. 1.35):

(i) **Oligodendrocytes.** These are myelin-forming cells of the CNS.

(ii) **Astrocytes.** These cells provide physical and metabolic support for neurons of CNS.

(iii) **Microglia.** These are phagocytic cells of CNS.

(i) **Oligodendrocytes.** The oligodendrocyte is the cell responsible for producing myelin within CNS. The myelin sheath in the CNS is formed by concentric layers of oligodendrocyte's plasma membrane. Each oligodendrocyte gives off several tongue-like processes that find their way to the axons. Where each process wraps itself around a portion of an axon, forming an internodal segment of myelin.

(*ii*) Astrocytes. These are largest of the neuroglial cells. Astrocytes are large and multi-processed cells having oval or slightly irregular nuclei and remarkable stellate morphology.

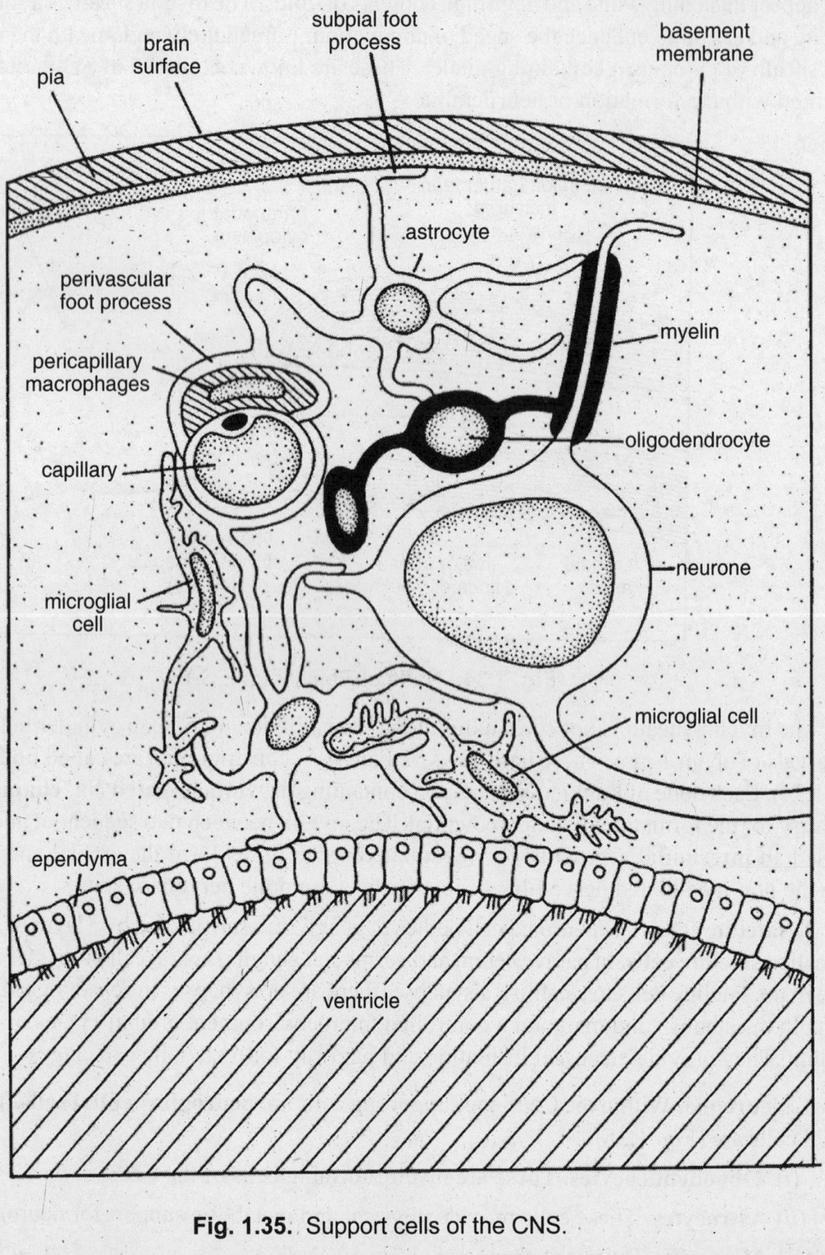

Fig. 1.35. Support cells of the CNS.

Many fine processes radiate in all directions from these cells. They contain prominent bundles of intermediate filaments which are composed of **glial fibrillary acidic protein (GFAP).** Processes of astrocytes extend between blood vessels and neurons. Their ends expand, forming **end feet (foot processes)** that may cover areas of the outer surface of a blood vessel or of the axolemma.

Astrocytes play a role in the movement of metabolites and wastes to and from neurons and regulate ionic concentrations in the inter-cellular compartment, thus maintaining micro-environment of the neurons. They also provide a covering for the "bare areas" of myelinated axons, *e.g.,* at the nodes of Ranvier and at synapses.

Box 1.15

1. Oligodendrocytes produce myelin in the CNS.
2. Schwann cells form myelin in PNS.
3. A Schwann cell myelinates only one axon, but an oligodendrocyte may myelinate several adjacent axons (Fig. 1.36).
4. Myelin is rich in **galactocerebroside,** a polymer (macromolecule) of plasma membrane of support cell.

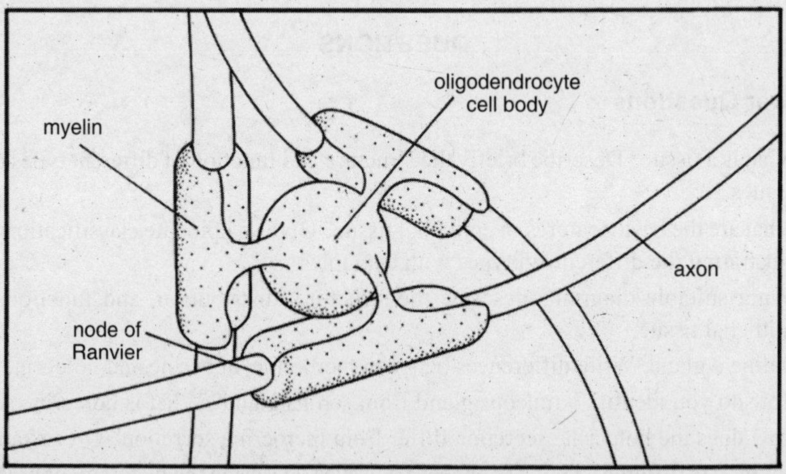

Fig. 1.36. Oligodendrocytes myelinate several adjacent axons within the CNS.

***(iii)* Microglia.** Microglia cells are phagocytic cells. They are normally present only in small numbers in the adult CNS, but proliferate and become actively phagocytic in regions of injury and disease. They are specialized macrophages and are believed to originate in the bone marrow and enter the CNS from blood.

Microglia are smallest of the neuroglial cells and have relatively small rod-shaped nuclei. They have extensive fine ramifying processes. Both processes and cell body of microglia are covered with numerous spikes. The TEM reveals in microglia, numerous lysosomes, inclusions and vesicles. However, there is little rough ER and there are few microtubules and microfilaments.

Table 1.3. Differences in medullated or myelinated and non-medullated or non-myelinated nerve fibres.

Medullated/Myelinated fibres	Non-medullated/Non-myelinated fibres
1. Glistening white fibres and constricted at regular intervals.	1. Greyish fibres, not constricted.
2. Myelin/medullary sheath lies between neurilemma and axoplasm.	2. Myelin sheath is absent.
3. Node of Ranvier is present.	3. Node of Ranvier is absent.
4. Occur in cranial and spinal nerves.	4. Occur in nerves of autonomic nervous system.
5. Transmission of impulses is from node to node (*i.e.,* saltatory or discontinuous conduction) and it is very speedy.	5. Transmission of impulses is comparatively slow.

Table 1.4. Differences in neurilemma and sarcolemma.

Neurilemma	Sarcolemma
1. It is the middle sheath around the axon of myelinated nerve fibres.	1. It represents the plasma membrane of muscle cell along with its **basal lamina.** and surrounding reticular lamina.
2. It protects and secretes myelin sheath.	2. It is also protective in function.

QUESTIONS

Long Answer Questions

1. What is a tissue? Describe briefly the structure and function of different types of epithelial tissues.
2. What are the basic features of epithelial tissue? Give an uptodate classification of epithelia illustrating the different subtypes with examples.
3. Using suitable diagrams describe the structure, distribution, and functions of simple epithelial tissue.
4. Define a gland. Write differences between merocrine, apocrine and holocrine gland.
5. How do you identify a mucous gland from serous gland? What is holocrine secretion?
6. How does the holocrine secretion differ from merocrine secretion? Give examples.
7. Enumerate different types of connective tissues. Describe the histology of areolar, fibrous and adipose tissues.
8. Describe briefly the structure and functions of various types of connective tissues met within animals.
9. Give a brief account of collagen fibres and point out differences from yellow elastic fibres.
10. Describe structure of striated muscles.
11. Give a detailed structure of striated muscle.
12. Draw and describe the ultrastructure of striated muscles.
13. Describe the histology of the nervous tissue.
14. Give a detailed account of nervous tissue.
15. Describe a neurone. Mention the different types of neurons found in animal body. State the functions and types of neuroglial cells found in animal body.
16. What are neurons and neuroglial cells. Mention different types of neurone found in animal body.
17. Describe a neurone with diagram.
18. Write the distinctive features of cardiac, smooth and skeletal muscles with diagrams. Describe changes in the skeletal muscles during contraction. (*Calcutta* (*H*) *1998*)

Short Answer Questions

1. **Write Short Note on following:**
   (*i*) Stratified epithelial cells
   (*ii*) Squamous eithelium
   (*iii*) Stratified epithelium
   (*iv*) Types of glands based on mode of secretion

(*v*) Glands
(*vi*) Collagen and elastin
(*vii*) Connective tissue
(*viii*) Areolar tissue
(*ix*) Adipose tissue
(*x*) Ligament
(*xi*) Sarcomere
(*xii*) Sarcolemma
(*xiii*) Contractile elements of skeletal muscles
(*xiv*) Sarcoplasmic reticulum
(*xv*) Cardiac muscle
(*xvi*) Nervous tissue
(*xvii*) Neuroglia
(*xviii*) Neuron or Nerve cell or Nerve fibre
(*xix*) Neurilemma
(*xx*) Dendrites
(*xxi*) Medullary sheath
(*xxii*) Node of Ranvier

2. Briefly describe the proteins of a sarcomere of a striated muscle.

3. **Draw neat and well labelled diagrams of the following :**
   (*i*) Neuron
   (*ii*) Cardiac muscle
   (*iii*) Smooth muscle
   (*iv*) Triad
   (*v*) Support cells of CNS

4. **Differentiate between the following:**
   (*a*) Simple and stratified epithelium
   (*b*) Squamous and columnar epithelium
   (*c*) Squamous epithelium and ciliated epithelium
   (*d*) Tubular and alveolar glands
   (*e*) Ligament and tendon
   (*f*) Cardiac muscle and striated muscle
   (*g*) Smooth muscle and striped muscle
   (*h*) Unstriped, Striped and Cardiac muscles
   (*k*) Epimysium, perimysium and endomysium
   (*l*) Neuroglia and neurone
   (*m*) Myelinated nerve fibres and non-myelinated nerve fibres.

   or

   Medullated and non-medullated fibres.

5. **Very Short Answer Questions**
   1. What are tissues?
   2. What are glands?

3. What is epithelium?
4. Define the connective tissue.
5. What is neuron?
6. Define the axon.
7. Define the synapse.
8. Give two examples of neurotramitters.
9. What is dendrite?
10. What is glia?
11. Define the nurotransmitters.

Fill in the Blank

Fill in the following blank with the suitable word:

In the nerve cell, ... serves to bring the impulse from outside.

(**Ans.** Dendrite)

Multiple Choice Questions

1. Epithelial tissue always has an exposed outer surface and an inner surface anchored to connective tissue by a thin, noncellular layer called the

(*a*) nonstratified layer (*b*) stratified layer
(*c*) basement membrane (*d*) fibroblast

2. Which type of tissue forms the inner lining of a blood vessel ?

(*a*) epithelial (*b*) connective
(*c*) nervous (*d*) muscle

3. Which type of tissue forms glands?

(*a*) epithelial (*b*) connective
(*c*) nervous (*d*) muscle

4. Ciliated epithelium lines the inner surface of

(*a*) urinary bladder (*b*) oesophagus
(*c*) epididymis (*d*) fallopian tube

5. Epithelium present in cornea, oesophagus, urethera and vagina is

(*a*) glandular (*b*) ciliated
(*c*) stratified columnar (*d*) stratified squamous

6. Which layer of epidermis is thickest at the sole? (*VKS (H) 1995*)

(*a*) stratum corneum (*b*) stratum granulosum
(*c*) stratum germinativum (*d*) stratum lucidum

7. Mammary glands are the modification of (*VKS (H) 1995*)

(*a*) sebaceous gland (*b*) sudoriferous gland
(*c*) scent gland (*d*) none of these

8. Cytogenous gland is (*VKS (H) 1996*)

(*a*) liver (*b*) adrenal
(*c*) pancreas (*d*) gonads

9. Areolar tissue is a

(*a*) nervous tissue (*b*) connective tissue
(*c*) muscular tissue (*d*) bony tissue

10. Tendon is a structure which connects

(*a*) a bone with another bone (*b*) a muscle with a bone

(*c*) a nerve with a muscle (*d*) a muscle with a muscle

11. Fibrous tissue which connects bones is

(*a*) tendon (*b*) adipose tissue

(*c*) connective tissue (*d*) ligament

12. Fat bodies are also known as

(*a*) corpora adiposa (*b*) haemolymph glands

(*c*) peyer's patches (*d*) none of these

13. The basic unit of muscle contraction is

(*a*) sarcomere (*b*) tropomyosin

(*c*) myosin (*d*) actin

14. The light and dark bands are visible in

(*a*) striped mucsles (*b*) unstriped muscle

(*c*) cardiac (*d*) Both (1) and (3)

15. Smooth muscles occur in

(*a*) uterus (*b*) artery (*c*) vein (*d*) all the above

16. Which of the following is contractile protein of a muscle ?

(*a*) tubulin (*b*) myosin (*c*) tropomyosin (*d*) actin

17. Fine connective tissue surrounding individual muscle fibres is

(*a*) epimycium (*b*) endomycium (*c*) perimycium (*d*) none

18. Schwann cells and nodes of Ranvier are present in

(*a*) muscles (*b*) cartilages

(*c*) bone cells (*d*) neurons

19. Node of Ranvier are found in

(*a*) cardiac muscle fibre (*b*) striated muscle fibre

(*c*) medullated nerver fibre (*d*) non-medullated nerve fibre

20. Neuroglia cells are

(*a*) astrocytes (*b*) oligodendrocytes

(*c*) microglia (*d*) all the above

21. Nissl granules found in cyton of a neuron and having affinity for basic dyes are composed of

(*a*) DNA (*b*) RNA

(*c*) proteins (*d*) amino acids

22. Axons serve to

(*a*) bring impulse to cyton (*b*) take away impulse from cytons

(*c*) bring blood to heart (*d*) none of these

ANSWERS

Very Short Answer Questions

1. Similar cells grouped together in certain areas of the body in multicellular organisms and usually specialised for a single function, *e.g*, nervous tissue, muscle tissue, epithelium, etc.

2. Glands are structures in animals that secrete special liquids. In animals the glands are divided into **endocrine glands** (*e.g.*, thyroid glands), which secrete hormones into the blood stream, and **exocrine glands**, which usually secrete materials, via ducts into internal organs or onto body surfaces. For example, in humans, skin contains two types of **exocrine glands:** sweat gland, which secrete watery fluid (prespiration), and sebaceous glands which secrete sebum. Lacrimal glands secrete tears; mammary glands secrete milk. Salivary glands secrete saliva, which partially digests food and facilitates swallowing.
3. Epithelial cells tightly linked together on a basement membrane to form a sheet or tube with little extracellular matrix.
4. Connective tissue is the tissue that joins together or gives support to the other organs of the body. It consists principally of an extracellular matrix: a gel of polysaccharides containing microscopic protein fibers, often collagen or elastin. There are relatively few cells compared with other tissues. Connective tissues include cartilage, bones, tendons, ligaments, adipose tissue (fat), the elastic supportive tissues around the lungs and the framework of organs such as liver and spleen.
5. The neuron or nerve cell is the basic unit of the nervous system (including the brain and spinal cord). Each neuron has a long axon, specialized for transmitting electrical impulses and releasing chemical neurotransmitters, at the synapse, which lies at the end of the axon. These synapses act on muscle or effector cells or other neurons (some stimulatory, some inhibitory) that have synapses acting on its cell body.
6. The axon is the fiber of a nerve cell (neuron) that conducts impulse away from the cell body. At the end of the axon is a synapse, where the nerve impulse is transmitted to another cell by means of neurotransmitters. The axons of motor nerves can be over a meter in length.
7. The synapse is the point of connection between two neurons (nerve cells) or between a nerve and a muscle. An electrical nerve impulse, arriving at the end of the axon in the pre-synaptic cell, releases a chemical transmitter (a neurotransmitter). This crosses a small gap, the synaptic cleft, and affects ion channels on the dendrites and cell body of the post-synaptic cell (neuron).
8. Acetylcholine and nor-epinephrine.
9. A dendrite is a tapering receptor branch of a neuron or, any of the branching protoplasmic processes that conduct impulses toward the body of a nerve cell (neuron).
10. Glia are supportive cells of the nervous system, derived from neuroepithelial cells.
11. Neurotransmitters are chemical messenger molecules secreted at the ends of the axons that cross the synaptic cleft and are received by the adjacent neuron, thus, relaying the neural signal.

Multiple Choice Questions

1. (*c*)	**2.** (*a*)	**3.** (*a*)	**4.** (*d*)	**5.** (*d*)	**6.** (*a*)
7. (*b*)	**8.** (*d*)	**9.** (*b*)	**10.** (*b*)	**11.** (*d*)	**12.** (*a*)
13. (*a*)	**14.** (*d*)	**15.** (*d*)	**16.** (*b*)	**17.** (*b*)	**18.** (*d*)
19. (*c*)	**20.** (*d*)	**21.** (*b*)	**22.** (*b*)		

2

Bone and Cartilage

The supporting or skeletal tissue includes **cartilage** and **bone** which form the **endoskeleton** of vertebrate body. Both of these tissues are connective tissues and are very rigid. They provide the resistant surfaces for the attachment of muscles to move limbs. They also provide protection to delicate structures (body organs) enclosed by them from mechanical injuries.

2.1 CARTILAGE

General structure. Cartilage is an **avascular tissue** (**Note.** Laryngeal and nasal cartilages, however, are vascular; **Ross** *et al.*, 1995). It consists of the following components:

1. Matrix. Cartilage has extensive matrix that is produced and maintained by the chondrocytes. It is solid, cheese-like and firm but also somewhat pliable (elastic). This accounts for its resilient nature. The large amount of glycosaminoglycans (*e.g.,* **hyaluronic acid, chondroitin sulphate** and **keratan sulphate;** Box 2.1) in the matrix permits diffusion of substances between blood vessels in the surrounding connective tissue and the chondrocytes, thereby maintaining viability of the tissue.

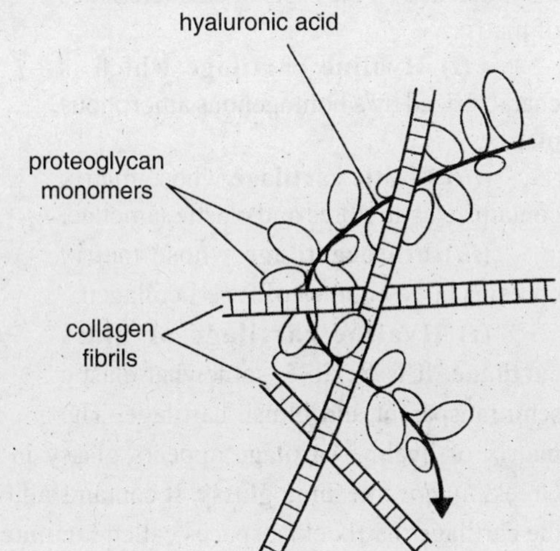

Fig. 2.1. Schematic diagram showing a large hyaluronate proteoglycan aggregate of cartilage matrix

Box 2.1

As in loose connective tissue matrix, the chondroitin sulphate and keratan sulphate of cartilage matrix are joined to a core protein to form a **proteoglycan monomer.** Within the tissue, each hyaluronic acid molecule is associated with about 80 proteoglycan units to form large aggregates whose structure is reinforced by **link proteins.** The large **hyaluronate proteoglycan aggregates** are bound to the thin collagen fibrils by electrostatic interactions and by cross-linking glycoproteins.

The presence of large amounts of hyaluronic acid in the matrix of cartilage makes it well adapted to serve in weight bearing capacity, especially at points of movements, as in synovial joints.

2. Chondrocytes or cartilage cells. These are numerous spherical cells with eccentrically located nuclei. These lie within the matrix either singly or in groups of 2 or 4. These are enclosed within lacunae. Chondrocytes multiply by simple mitosis and help in internal growth of cartilage.

These cells secrete matrix. They secrete not only the collagen of the matrix but also all of the glycosaminoglycans and proteoglycans.

Active chondrocytes contain great amount of rough endoplasmic reticulum, large-sized Golgi apparatus, numerous mitochondria, secretory granules and three types of protein fibres/tubules: actin-containing microfilaments, intermediate filaments and microtubules. Old inactive cartilage cells contain numerous inter-mediate filaments, lipid droplets and glycogen stores. They become distorted in shape due to shrinkage.

3. Perichondrium. It forms a dense fibrous sheath (connective tissue) around the mature cartilage. It contains **inner cellular layer** of closely packed fibroblast cells, **outer fibrous layer** of fibres and blood vessels (Fig. 2.2).

Types of Cartilage

Following three kinds of cartilage are distinguished on the basis of characteristics of matrix.

(*i*) **Hyaline cartilage** which is characterized by a homogenous amorphous matrix.

(*ii*) **Elastic cartilage** whose matrix contains elastic fibres and elastic lamellae.

(*iii*) **Fibrocartilage** whose matrix contains large bundles of type I collagen.

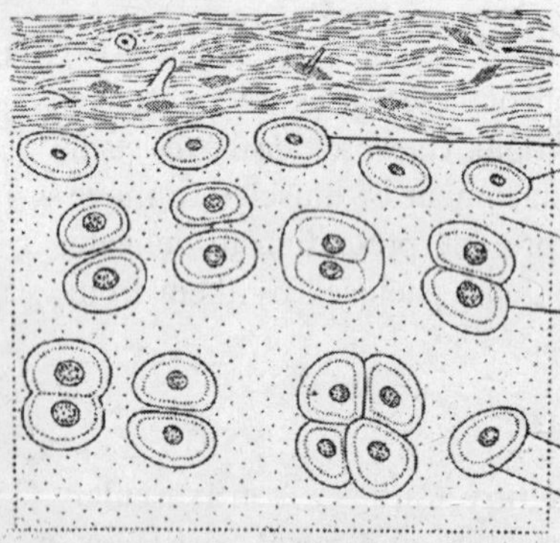

Fig. 2.2. T.S. hyaline cartilage.

(*i*) **Hyaline cartilage or glass cartilage.** It is flexible, somewhat elastic semitransparent and bluish cartilage. The matrix of hyaline cartilage appears glassy in the living state, hence the name hyaline from the Greek, *hyalos,* meaning **glassy.** It contains all basic components of cartilage (Fig. 2.2). Throughout the cartilage matrix exist spaces called **lacunae,** each having one to four **chondrocytes.** The matrix consists of two components: **collagen fibrils** (predominantly type II collagen) and ground substance. It is highly **hydrated,** *i.e.,* from 60 to 78 per cent of its net weight is water. Much of this water is bound tightly to the **proteoglycan aggregates** (see Box 2.2) which explains the resilience of cartilage. Some of the water is bound loosely enough, to allow the essential diffusion of small metabolites to and from the chondrocytes.

Hyaline cartilage is the precursor to bones that will develop in the process of **endochondral ossification.** While endochondral bone development is in progress, hyaline cartilage serves as a epiphyseal growth site. This cartilagenous **epiphyseal growth plate (epiphyseal disc)** remains functional as long as the bone grows in length. In bones of the adult, hyaline cartilage is present only at articular surfaces. Hyaline cartilage is also present in the adult as the skeletal unit in the rings of trachea, bronchi, larynx, nose and the ends of ribs (costal cartilages). It withstands forces of compression and torsion.

Box 2.2

Stains Used for Cartilages

1. Specimens of hyaline cartilage and fibro cartilage are stained with **haematoxylin** and **eosin** (H & E).

2. Elastin fibres of elastic cartilage are best demonstrated in paraffin sections with special stains such as **resorcin-fuchsin** and **orcein.**

***(ii)* Elastic cartilage.** It is opaque, yellow-coloured, more flexible and highly elastic. In addition to the normal components of the hyaline cartilage matrix, elastic cartilage matrix contains yellow elastic fibres and interconnecting sheets of elastic material (lamellae). Unlike hyaline cartilage, however, the matrix of elastic cartilage does not calcify.

Elastic cartilage is found in the external ear, in the walls of external auditory canal and the eustachian (auditory) tube, in the epiglottis and in the larynx.

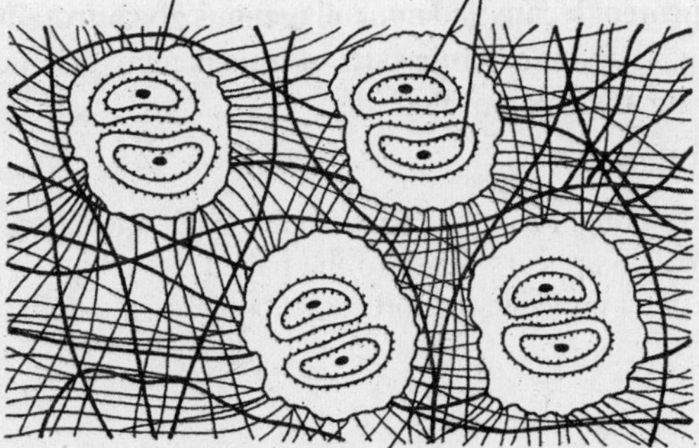

Fig. 2.3. T.S. elastic cartilage.

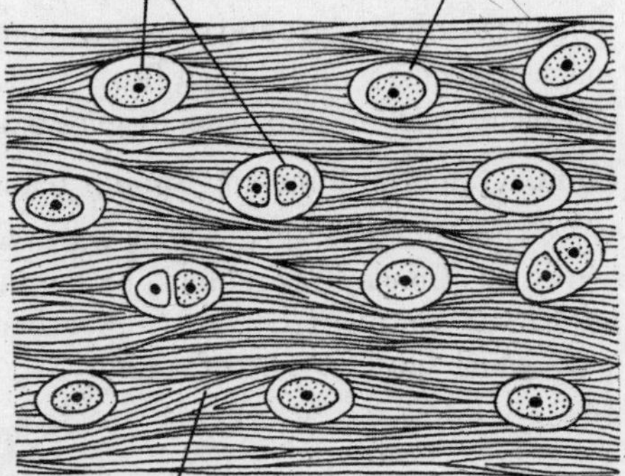

Fig. 2.4. C.S. fibrous cartilage.

***(iii)* Fibrocartilage.** It is often difficult to distinguish fibrocartilage from dense regular connective tissue. It contains numerous thick and compact bundles of white **collagen fibres** which lie parallel separated by narrow clefts. The chondrocytes are few and found in the clefts. The amorphous matrix is scanty and perichondrium is absent.

The presence of fibrocartilage indicates that resistance to both compression and shear forces is required of the tissue. Fibrocartilage is typically present in the intervertebral discs, the symphysis pubis (of pelvic girdle), the articular disc of the sternoclavicular and temporo-mandibular joints, the menisci of knee joints and certain places where tendons attach to bones (**Menisci** = Either of two crescent-shaped lamellae of fibrocartilage that border and partly cover the articular surfaces of the tibia and femur in the knee).

2.2 Bone

Bone is a solid, rigid and strong connective tissue. It is known as **osseus tissue** since its matrix consists of an organic component, called **ossein** which is a **collagen of type I.** Like other connective tissues, bone consists of cells and extracellular matrix. The feature that distinguishes bone from other connective tissues is the **mineralization** of matrix (*i.e.,* deposition of lime salts in bone's elastic matrix). This produces an extremely hard tissue capable of providing support and protection. The mineral is calcium phosphate in the form of **hydroxyapatite crystals** [$Ca_{10}(PO_4)_6(OH)_2$].

Bone forms vertebrate endoskeleton and serves following functions:

1. It provides shape to the body.
2. It provides skeletal support to the body.
3. It protects vital body organs.
4. It serves as storage site of calcium and phosphate (Box 2.3).

Bones are the organs of the skeletal system and bone tissue is the structural component of bones. Typically a bone consists of bone tissue, other connective tissues, fat tissue, blood vessels, nerves and

Box 2.3

Both calcium and phosphate can be mobilized from the bone matrix and taken up by the blood as needed to maintain appropriate levels throughout the body. Thus, in addition to support and protection, bone has an important secondary function in a homeostatic regulation of blood calcium levels.

if the bone forms a freely movable (synovial) joint, hyaline cartilage. The **bone tissue** consists of matrix, fibres and bone cells. The organic matrix forms nearly one-third of the total weight (*i.e.,* 33 per cent). It is formed of **ground substance** containing **proteoglycans** and **noncollagenous glycoproteins** and is deposited with **collagen fibres** (type I collagen) and **inorganic salts of lime** namely calcium phosphate, calcium carbonate, magnesium phosphate and calcium fluoride.

Compact (dense) Bone and Spongy (cancellous) Bone

There are several criteria by which we classify bone. From its visual appearance, one can see two types of bone; **Concellous** or **spongy bone**, which appears dense to the naked eye (Fig. 2.5). From its position, we recognize Cortical bone in the outer boundary or cortex of a bone and metuallary bone, which lines the core of bone.

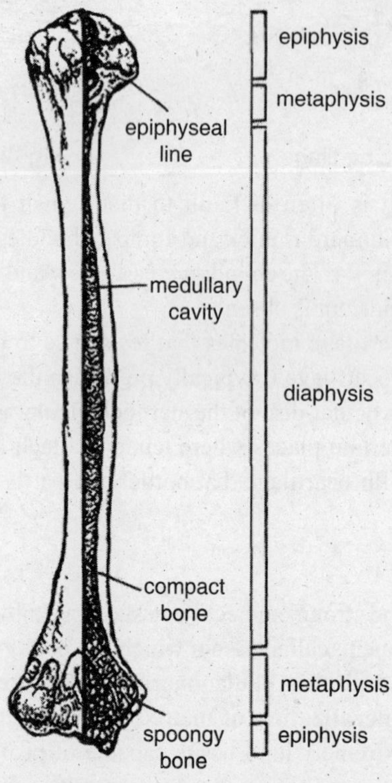

Fig. 2.5. Regions of long bone. The middle section of bone is the diaphysis (shaft) containing the medullary cavity. In mammals and a few other groups, the ends of the bone develop secondary centers of ossification, or epiphysis, although this term is sometime used loosely to mean simply the end of a bone. Between the diaphysis and the epiphysis is the metaphysis, the actively growing region of bone. Compact bone is dense; spongy (or cancellous) bone is porous. The medullary cavity and all spaces in spongy bone are filled with hemopoetic (blood-forming) tissue (after Kardong, 2006).

The presence or absence of bone cells gives us **cellular** and **acellular bone**, respectively. Bone may be described as **vascular** or **avascular** if many or few blood (= vascular) channels traverse it, respectively. The organization of the bone, especially the orientation of collagen and the ordered placement of bone cells within the matrix, is one of the most used criteria to classify bone, *e.g.*, lamellar and **Nonlamellar bone** (or **fibrolamellar bone** or **woven bone**) is characterized by disorderly, irregular arrangement of collagen with the matrix (Fig. 2.6A). It is typical of fast growing bone. **Lamellar bone** is characterized by the orderly, regular arrangement of collagenous fibers within the matrix, usually accompanied by the regular orientation of bone cells (Fig. 2.6 B). It is typical of slow-growing bone. A layer of bone matrix with its closely-packed collagen fibers aligned in parallel is termed a **lamellar** collagen fibers aligned in parallel is termed a lamellar successive layers of lamellar may have their alignment of collagen oriented at different angles to adjacent lamellae. This production a "plywood like" structure, adding strength to the bone. A special type of lamellar bone is **Haversian bone** (Fig. 2.6C). Organic salts are arranged in a regular highly ordered unit, known as osteon (Fig. 2.6C and 2.7). Each osteon is a series of concentric sings made up of bone cells.

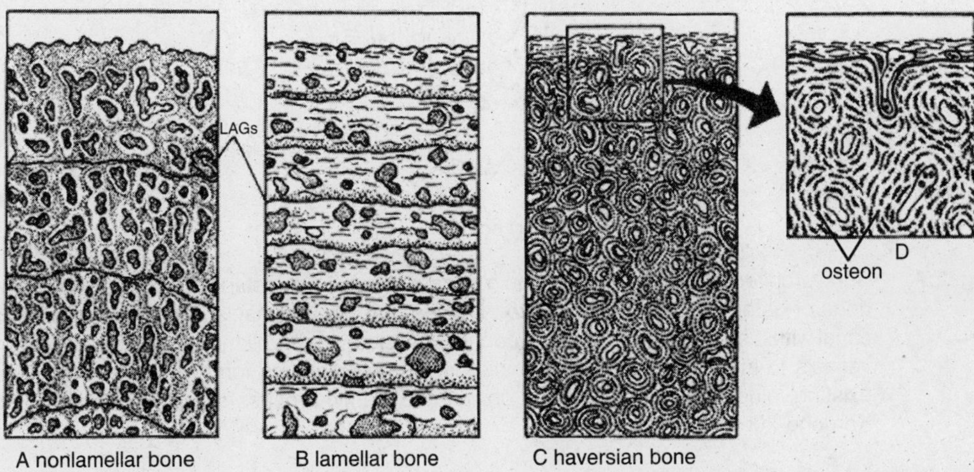

Fig. 2.6. General types of bone. A-Nonlamellar bone, based on a young American alligator. B-Lamellar bone, based upon several living and extinct turtles. C-Haversian bone, a specialized form of lamellar bone. D-Enlarged section of Haversian bone. Note lines of arrested growths (LAGs), which may appear in all types of bone. Bone interruped by LAGs is sometimes called zonal bone. (After Kardong 2006)

and layers of matrix around a central canal through which blood vessels, lymphatic vessels, and nerves travel. **Volkamann's Canals** running diagonally through this system interconnect blood vessels between osteons.

Many bones exhibit **lines of arrested growth (LAGs)** during which growth Ceases, or there may even be slight erosion or absorption of previously deposited bone (Fig. 5.22). Lags form as a result of seasonal activity, nutritional abundance or quality of food, differences in rates of growth due to environmental stress (*e.g.*, cold weather). Typically, ecotothermic animals exhibit LAGs on annual or seasonal basis, but a few endotherms such as various species of Eurasian grouse, Old world field mice, the common dolphin, various species of voles minks, Sika deer and gibbons also produce LAGs. Bone exhibiting periods of sustained deposition interrupted by LAGs or rings is termed **Zonal bone (kardong** 2006).

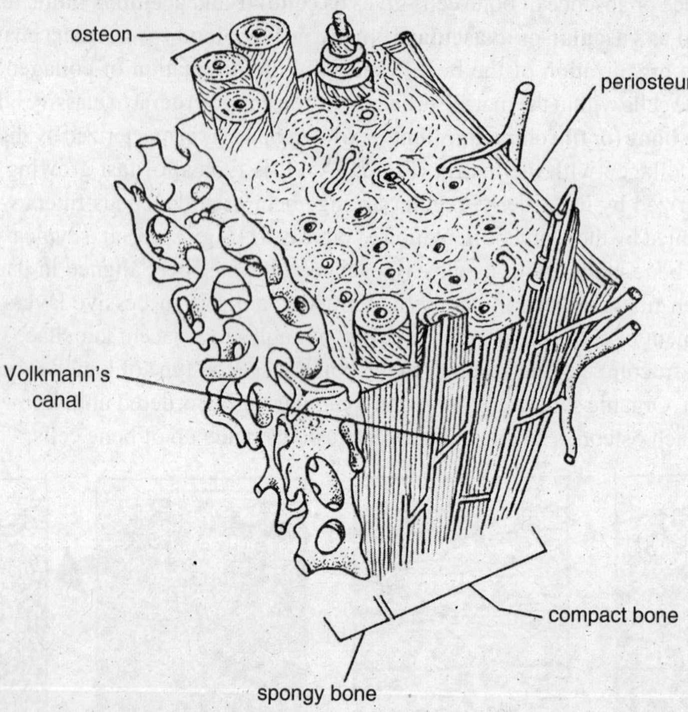

Fig. 2.7. Bone architecture, Osteons make up compact bones. Each osteon is a series of concentric rings of osteocytes and their matrix. Nerves and blood vessels pass through a central canal within each osteon. Diagonal connections, known as Volkmann's canals, allow blood vessels to interconnect between osteons. As new osteons form, they usually override existing, older osteons, as part of on going dynamic process of bone remodeling after Kardong 2006).

Classification of Bones

On the basis of shape, bones can be classified into following four groups:

1. Long bones. They are longer in one dimension than the other and consists of a **shaft** and two **ends,** *e.g.,* the tibia and metacarpals (Fig. 2.5).

2. Short bones. They are nearly equal in length and diameter, *e.g.,* the carpal bones of the hand.

3. Flat bones. They are thin and plate-like, *e.g.,* the bones of the clavarium (the skull cap) and the sternum. They consist of two layers of relatively thick compact bone with an intervening layer of spongy bone (they are called **diploid bones**).

4. Irregular bones. They have a shape that does not fit into any one of the three groups just described; the shape may be complex, *e.g.,* a vertebra.

General Structure of Bone

Long bones have a shaft called the **diaphysis** and an **epiphysis** (Fig. 2.5). The flared portion of the bone between the diaphysis and epiphysis is called **metaphysis.** It expands from the diaphysis to the epiphyseal line.

A large cavity filled with bone marrow, called thé **marrow** or **medullary cavity** forms the inner portion of the bone. In the shaft, almost the entire thickness of the bone tissue is compact, however, only a small amount of spongy bone faces the medullary cavity. At the end of bone, the reverse is true. Hence, the spongy bone is extensive and the compact bone is little more than a thin outer well.

Box 2.4

Calcination and Decalcination

1. Calcination. If the organic matrix of bone is removed by burning then the procedure is known as **calcination.** In this technique, the general shape and internal arrangement of the bone is retained.

2. Decalcification. When the inorganic materials are dissolved out by treatment with mineral acids, then the appearance of the bone is little changed but it becomes so soft and flexible that the bone can easily be cut or bent. A decalcified human fibula may even be tied into knot. Decalcified bone consists of **bone collagen,** called **ossein,** from which gelatin is made.

Except the articular surfaces that are covered by **hyaline** or **articular cartilage,** the outer surface of the bone is covered by a fibrous connective tissue capsule called the **periosteum.**

Histology of Mammalian Bone

A thin cross section of dried or calcined (Box 2.4) long bone presents the following structure:

1. Periosteum. It is the outer dense fibrous connective tissue sheath investing the bone. In it undifferentiated cells called **osteoprogenitor cells** may lie embedded. These cells are capable of undergoing division and becoming **osteoblasts.** Periosteum contains blood vessels, lymphatic vessels and nerves that enter bone.

2. Outer layer of osteoblasts. In young and developing bones the inner layer of periosteum is formed of flattened connective tissue cells, osteoblasts (Fig. 2.7). These are undifferentiated cells which give rise to bone cells.

3. Endosteum. The marrow cavities in the bone are lined by a thin layer of fibrous connective tissue containing **endosteal cells** (*i.e.,* osteoprogenitor cells).

4. Inner layer of osteoblasts. External to endosteum occur continuous layer of osteoblast cells.

5. Matrix. In the area of bone between periosteum and endosteum exists mineralized **matrix** having type I collagen and ground substance of proteoglycans. Matrix of bone is made up of concentric layers or **lamellae** of calcified interstitial substance. The arrangement of lamellae in bone matrix varies in spongy and compact bones and also in different groups of animals. Embedded within the matrix but arranged on lamellae are found numerous small cavities or spaces, called **lacunae.** Each lacuna encloses a bone cell or **osteocyte.** The wall of each lacuna is drawn into numerous fine canals or tunnels called **canaliculi** which penetrate the matrix in all directions. These canaliculi branch, anastomose and connect with canaliculi of other lacunae to form an interconnected system of spaces. Through these canaliculi, fine cytoplasmic processes of one osteocyte extend to meet (*i.e.,* to communicate by **gap junctions**) the similar processes of other osteocytes.

This extensive system of intercommunicating lacunae, canaliculi and vascular canals in the bone matrix permits the tissue fluid to permeate through all parts of the bone to supply nourishment and oxygen to osteocytes.

Haversian System

Mature bone of an adult mammal is composed of cylindrical structural units called **osteons** or **Haversian systems** (after English anatomist **Clopton Havers** who described histology of bone lamellae and canal in 1691). Each osteon consists of **concentric lamellae** of bone matrix surrounding a central canal, the **osteonal canal** or **Haversian canal,** that contains the vascular and nerve supply along with areolar connective tissue. Osteonal (= Haversian) canals run parallel to long axis of the bone. Canaliculi containing the processes of osteocytes are arranged mostly in a radial pattern with respect to canal (Fig. 2.7). The system of canaliculi that open to the osteonal canal also serves for the passage of substances between the osteocytes and the blood vessels. Between the osteons exist remnants of previous osteonal lamellae, now called **interstitial lamellae.** Bones containing this form of matrix organization is called **lamellar bone.**

The collagen fibres in the concentric lamellae in an osteon are laid down parallel to one another in any given lamella but in different directions in adjacent lamellae. This arrangement gives the cut surface of lamellar bone the appearance of a plywood and provide great strength to osteon.

Lamellar bone is also found at sites other than the osteon. **Circumferential lamellae** follow the entire inner and outer circumferences of the shaft of a long bone, appearing much like the growth rings of a tree (Fig. 2.8). **Perforating canals** or **Volkmann's canals** are channels in lamellar bone through which blood vessels and nerves travel from the periosteal and endosteal surfaces to reach the osteonal canal; Volkmann's canals also connect osteonal canals to one another. They usually run at approximately right angles to the long axis of the osteons and of the bone. The Volkmann's canal is not surrounded by concentric lamellae, a key feature in its histological identification.

Lamellar bone may be compared with **bundle/woven bone** or **nonlamellar bone,** in which the collagen fibres and bundles are at random orientations and are loosely interwined to give a woven appearance. Spongy bones may be either woven or lamellar bone; compact bone is almost entirely lamellar bone.

Bone Marrow

The medullary cavity and spaces in spongy bone contain bone marrow. Bone marrow is of following two types:

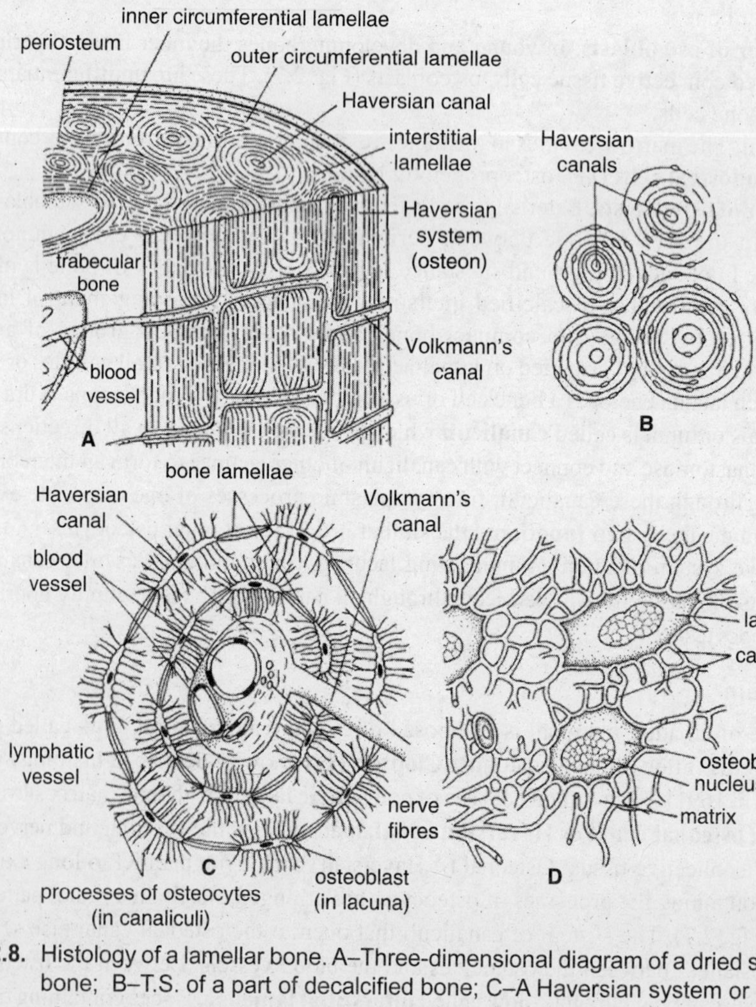

Fig. 2.8. Histology of a lamellar bone. A–Three-dimensional diagram of a dried sample of compact bone; B–T.S. of a part of decalcified bone; C–A Haversian system or osteon; D–Details of osteocyte cells in lacunae.

1. Red bone marrow. It consists of developing blood cells in different stages of development and a network of reticular cells and fibres that serve as a supporting framework for the developing blood cells and blood vessels. In the adult, red bone marrow is normally restricted to spaces in spongy bone such as vertebrae, ribs, skull, pelvis, proximal femur, etc.

2. Yellow bone marrow. As an individual grows, the amount of red marrow does not increase in proportion to bone growth. In later stages of growth and in the adult, when the rate of blood cell formation has diminished, the tissue in the medullary cavity consists mostly of fat cells (adipose tissue); it is then called **yellow bone marrow.** Under appropriate stimuli, such as extreme blood loss, the yellow bone marrow can revert to red bone marrow.

In the adult, the **red blood cells, granulocytes, monocytes** and **platelets** are formed in red bone marrow. Lymphocytes are formed both in the marrow and in the lymphatic tissues.

Types and Functions of Cells of Bone

Bone tissue is a type of connective tissue and contains the following four types of cells:

1. Osteoprogenitor cells 2. Osteoblasts 3. Osteocytes 4. Osteoclasts

With the exception of the osteoclasts, each of these cells may be regarded as a differentiated form of the same basic cell type because each undergoes transformation from less mature form to a more mature form in relation to functional activity (growth of bone).

1. Osteoprogenitor cells. Osteoprogenitor cells are found on the external and internal surfaces of bones. They are flattened cells having elongate or ovoid nuclei. They comprise following two types of cells:

(*i*) Periosteal cells which form the innermost layer of the periosteum.

(*ii*) Endosteal cells which lines the marrow cavities, the osteonal (Haversian) canals and the proliferating canals (Volkmann's canals).

Osteoprogenitor cells are derived from mesenchymal cells. They have ability to differentiate into four types of cells: 1. *osteoblasts;* 2. *adipose cells;* 3. *chondroblasts;* and 4. *fibroblasts.* Like the fibroblasts they are able to modify their morphologic and functional characteristics in response to specific stimuli. They help in healing of fractures of bone.

2. Osteoblasts. The osteoblasts are very versatile secretory cells. While retaining the ability to divide, they secrete both the collagen and ground substance that makes the initial unmineralized bone or **osteoid.** The osteoblast is also responsible for the calcification of the bone matrix (Box 2.5).

Box 2.5

Biological Mineralization

The calcification process of bone matrix appears to be initiated by the osteoblasts through the secretion into matrix of small 50-250 nm, membrane-limited **matrix vesicles** (Fig. 2.8). The vesicles are actively secreted (by exocytosis) only during the period in which osteoblasts produce the bone matrix. Matrix vesicles contain enzymes such as **alkaline phosphatase** and **pyrophosphatase** that are responsible for loading the vesicle with ions of calcium (Ca^{2+}) and phosphate (PO_4^{2-}). Subsequently, the loaded matrix vesicles rupture, producing an increase in local concentration of mineral sufficient to initiate mineralization. A glycoprotein **(osteocalcin)** in osteoid (matrix) binds extracellular Ca^{2+}–ions, leading to high local concentration.

The osteoblasts produce very thin cytoplasmic processes that penetrate through the adjacent osteoid that cell has produced and are joined by **gap junctions** to similar processes of adjacent osteocytes. It is by this method, *i.e.*, by the early establishment of junction between the adjacent osteocytes that neighboring cells within the bone tissue communicates.

3. Osteocytes. The osteocyte is the mature bone cell or differentiated osteoblast. It is enclosed by bone matrix that it previously secreted as an osteoblast. osteocytes are responsible for maintaining bone matrix. They have the capacity to synthesize matrix, as well as to resorb it (*i.e.*, **osteocytic osteolysis**). Such activities are important in contributing to the homeostasis of blood calcium. Death of osteocytes, either through trauma (*e.g.,* a fracture) or cell senescence, results in resorption of the bone matrix by osteoclast activity, followed by repair or remodelling of the bone tissue by osteoblast activity.

Each osteocyte occupies a space or **lacuna** in matrix that conforms to the lenticular shape of the cell. The osteocytes extend cytoplasmic processes through the fine tunnels or **canaliculi** in the matrix to contact, by means of gap junctions, processes of neighboring cells. The osteocyte is typically smaller than its precursor and contain stored glycogen in its cytoplasm. Following three functional states of osteocytes can be described:

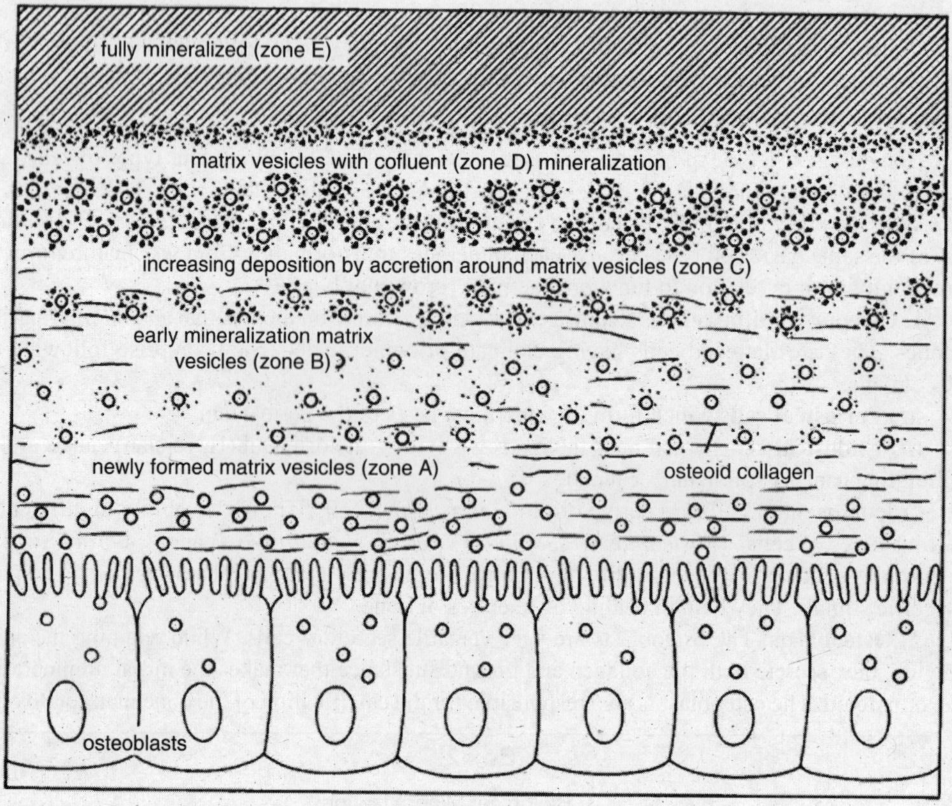

Fig. 2.9. Process of mineralization of bone.

(1) Quiescent osteocytes with less RER and small-sized Golgi apparatus.

(2) Formative osteocytes with abundant RER and Golgi apparatus.

(3) Resorptive osteocytes with well developed RER and Golgi apparatus in addition to secondary lysosomes.

4. Osteoclasts. The osteoclast is a large multi-nucleated cell having numerous lysosomes. Its function is to resorb bone.

When active, the osteoclasts rests directly on the surface of the bone where resorption is to take place. As a result of its activity, a shallow bay called **Howship's lacuna** or a **resorption bay** is formed in the bone directly under the osteoclast. The portion of the cell directly in contact with the bone is divisible into two distinct parts: 1. a central region containing numerous plasma membrane infoldings, forming the **ruffled border** and a ring-like microfilament-rich perimeter of cytoplasm, the **clear zone** which demarcates the bone area being resorbed. In the cytoplasm of osteoclast, in close proximity of ruffled border occur numerous mitochondria and lysosomes, profiles of RER, multiple stacks of Golgi saccules and many vesicles.

During osteoclast action, lysosomal enzymes such as **collagenase** are released from the ruffled border of osteoclast and digest the organic components of the bone matrix. Before this can occur, however, it is necessary to decalcify the bone matrix. Recent evidence indicates that the dessolution of the calcium salts occurs through secretion of organic acids such as **carbonic acid, lactic acid** and **citric acid** by the plasma membrane of the ruffled border.

Osteoclasts also act as phagocytic cells. They have numerous **coated pits** and **coated vesicles.** Some of the soluble breakdown products of demineralization (*e.g.*, Ca^{2+} and PO_4^{2-} ions) and protein hydrolysis (*e.g.*, amino acids) are resobed by the osteoclasts by endocytosis (*i.e.*, by receptor mediated endocytosis).

Osteoclasts arise from **monocytes** and acquire their multinucleate form either by repeated DNA replications and nuclear divisions without subsequent cell division or by fusion of many monocytes or developing osteoclasts. In both their origin and their function, they are closely related to other macrophages. Morphologically, they resemble the Langhan's giant cells formed by fusion of tissue macrophages (histiocytes) in loose connective tissue.

Hormonal control of osteoclast action. Activities of osteoclasts are regulated by two types of hormones:

1. An increase in parathyroid hormone **(parathormone)** level promotes bone resorption, increase in number and activities of osteoclasts and also of osteoblasts.

2. In contrast, **calcitonin** hormone which is secreted by parafollicular cells of the thyroid gland, has a counter-balancing effect, reducing osteoclast activity. Calcitonin inhibits bone resorption.

QUESTIONS

Long Answer Questions

1. Describe the microscopic structure of different types of cartilage tissues found in mammals.
2. With the help of suitable diagrams describe the structure of cartilage and bone.
3. Give an account of different types of skeletal tissue and state their functions.
4. Describe histology of a mammalian bone. What is the difference between a compact bone and a spongy bone.
5. Give difference between a compact bone and a spongy bone.
6. Describe histology of bone in a mammal. What is the difference between a compact bone and a spongy bone.

Short Answer Questions

1. **Write Short note on the following:**
   (*i*) Hyaline cartilage/cartilage
   (*ii*) Bone
   (*iii*) Haversian canal
   (*iv*) Haversian system
   (*v*) Microscopic structure of bone
   (*vi*) Histological structure of cartilage
2. Describe structure and function of various cells of bone.
3. **Differentiate between the following:**
   (*i*) Bone and cartilage
   (*ii*) Hyaline and fibrocartilage
   (*iii*) Compact bone and cancellous bone

4. **Draw neat and labelled diagram of the following:**
   (*i*) T.S.bone
   (*ii*) T.S. Cartilage of a mammal

Very Short Answer Questions

1. What is a bone?
2. Define the cartilage.
3. What is osteoblast.
4. Define the osteoclast.
5. What is ocetocyte?
6. What is another name of chondrocyte?
7. What is fibroblast?

Multiple Choice Questions

1. Haversian canals are found in
   (*a*) all the bones (*b*) long human bones
   (*c*) alimentary canal (*d*) none of these
2. Outer covering of a cartilage is known as
   (*a*) endosteum (*b*) perichondrium
   (*c*) periosteum (*d*) serous coat
3. Bone forming cells are called
   (*a*) chondroblasts (*b*) chondrocytes
   (*c*) osteoclasts (*d*) osteoblasts
4. Cartilage is produced by
   (*a*) osteoblasts (*b*) chondrocytes
   (*c*) fibroblasts (*d*) epithelial cells

ANSWERS

Very Short Answer Questions

1. Bone is a hard connective tissue which forms the skeleton of vertebrates other than the cartilaginous fish(sharks, rays). It is laid down by specialized) mesodermal cells called **osteocytes**. Bone comprises a matrix of collagen fibers (30% by weight, providing tensile strength) impregnated with bone salts (65% by weight, providing hardness), mostly calcium phosphate (hydroxyapatite, $Ca_{10}(PO_4)_6(OH)_2$], in which are embedded bone cells: **osteoblasts** (which secrete the matrix) and osteocytes.

   Bone is a living tissue, supplied with nerves and blood vessels. Its composition changes under hormone influence. It is constantly being reshaped by the osteoblasts and by osteoclasts, which remove calcium. Vitamin D makes calcium available from the food to the blood, and its deficiency leads to rickets.
2. Cartilage is a tough, flexible connective tissue found in all vertebrates, consisting of cartilage cells (chondrocytes) in a matrix of **collagen** fibers and a rubbery protin gel. The skeleton of the vertebrate embryo is formed wholly of cartilage, but in the majority of groups much of this is converted to **bone** during growth. Only in the cartilaginous fish (sharks and rays) does the cartilage remain unmineralized. There are 3 main types of cartilage: 1. Hyaline; 2. Elastic; 3. Fibrocartilage.
3. Osteoblast is any of the cells, found in bone that secrete collagen and other substances that form the calcified intercellular matrix of bone. Bone is constantly undergoing reshaping, and also acts as the calcium "bank" for the body so that blood calcium levels are kept constant. It is called osteocyte after calcified matrix has been set down.

4. Osteoclast is any of the multinucleate amoeboid cells, found on the bone surface, that removes (or breaks down) the calcified intercellular matrix to enable the further development and restructuring of bone during growth and repair.
5. Osteocyte is any of the cells found in bone that performs the cellular activities, such as respiration and exchange of materials with the blood, that are required for the maintenance of bone tissue. They are the mature osteoblasts.
6. Chondrocyte is a cartilage cell.
7. Fibroblast is a connective tissue cell of mesenchymal origin that secretes proteins and especially molecular collagen from which the extracellular matrix of connective tissue forms.

Multiple Choice Questions

1. (*b*) 2. (*b*) 3. (*d*) 4. (*b*)

3

Physiology of Transmission of Nerve Impulse

The nervous system is a communication and control network that allows an animal to interact in appropriate ways with its environment. The environment includes both the external environment (the world outside the body) and the internal environment (the components and cavities of the body). The general functions of the nervous system include *sensory detection*, *information processing*, and the *expression of behaviour*. Sensory detection is process whereby neurons transduce environmental energy into neural signals. Sensory detection is accomplished by special neurons called **sensory receptors**. Various forms of energy can be sensed, including mechanical, light, sound, chemical, thermal and in some animals, electrical.

The nervous system has central and peripheral parts, each with further subdivisions. The peripheral nervous system provides an interface between the environment and the central nervous system. It includes sensory (or primary afferent) neurons, somatic motor neurons and autonomic neurons.

3.1 HISTOLOGY OF NERVOUS SYSTEM

Location

The nervous system has been divided into the following two parts:

1. Central nervous system (CNS) which consists of the **brain** and the **spinal cord** located in the cranial cavity and spinal (= vertebral) canal, respectively.

2. Peripheral nervous system (PNS) which consists of **cranial** and **spinal nerves** (both motor and sensory), **ganglia** (*i.e.*, collection of nerve cells outside the CNS) and **motor nerve endings** and **sensory nerve endings** (receptors).

Cellular Components of Nervous System

Nervous tissue consists of following two types of cells:

1. Nerve cells or Neurons. These are structural and functional units of nervous system. Nerve cells are (*i*) specialized to receive stimuli and to conduct electrical impulses to other parts of system; (*ii*) arranged as an integrated communication network, with several neurons in a chain-like fashion typically involved in sending impulses from one part of the system to another.

The specialized contacts between nerve cells that provide for the transmission of information from one nerve cell to next in the chain are called **synapses**.

2. Supporting cells. These are nonconducting cells that are in close physical contact with neurons. In the CNS, these supporting cells are called **neuroglia** ("nerve glue") or simply **glia**; in the PNS, **Schwann cells** which surround nerve cell processes and **satellite cells** which surround nerve cell bodies (*i.e.*, cytons) in ganglia. Both Schwann cells and satellite cells originate from neural crest cells. Supporting cells provide (*i*) *physical support* (protection) for delicated neuronal processes;

(*ii*) *electrical insulation* for nerve cell bodies and processes; and (*iii*) *metabolic exchange* pathways between the vascular system and the nervous system.

In addition to nerve cells and supporting cells, there are many blood vessels in both the CNS and the PNS. The blood vessels are separated from the nervous tissue by the interposition of **basal laminae** and variable amount of **connective tissue**, depending on vessel size (Box 3.1).

Box 3.1 Blood Brain Barrier

In the CNS, the boundary between blood vessels and nervous tissue is special in that many substances that readily leave blood vessels to enter other tissues *do not normally enter the nervous tissue*. This selective restriction of blood-borne substances in the CNS is known as the **blood-brain barrier**. The barrier has been shown to exist in the elaborate **tight junctions** between the endothelial cells of brain capillaries.

A. The Neurons (Nerve Cells)

The neuron is the structural and functional unit of nervous tissue. There are more than 10 billion neurons in the human nervous system. These neurons are of following three types:

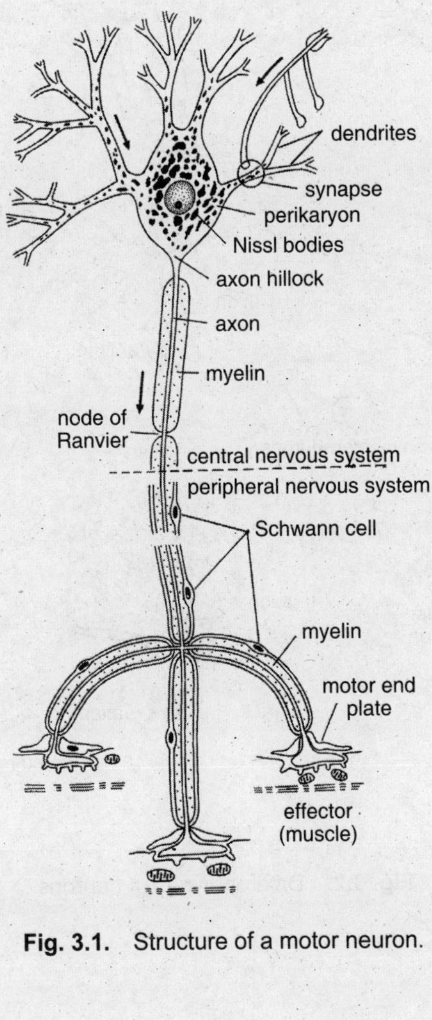

Fig. 3.1. Structure of a motor neuron.

1. **Sensory neurons** which convey impulses from receptors to the CNS.
2. **Motor neurons** which convey impulses from the CNS or from ganglia to effector cell.
3. **Interneurons** which are also called **internuncial neurons**, **intercalated neurons** or **central neurons**. They form a communicating and integrating network between the sensory and motor neurons. It is estimated that more than 99.9 per cent of all neurons belong to this integrating network.

Neurons are large, polymorphic cells consisting of **cell body** and processes such as **axon** and **dendrites** (Fig. 3.l). On the basis of the number of processes extending from the cell body, neurons are classified into following three types (Fig. 3.2):

1. Multipolar neurons. They have one axon and two or more dendrites, *e.g.*, motor neurons and interneurons. In them dendrites and cell body form receptor portions of the cell while axon is the conducting portion. The terminal part of axon is specialized as transmitter or effector which is connected either with other nerves by synapses or with muscles or glands (effectors).

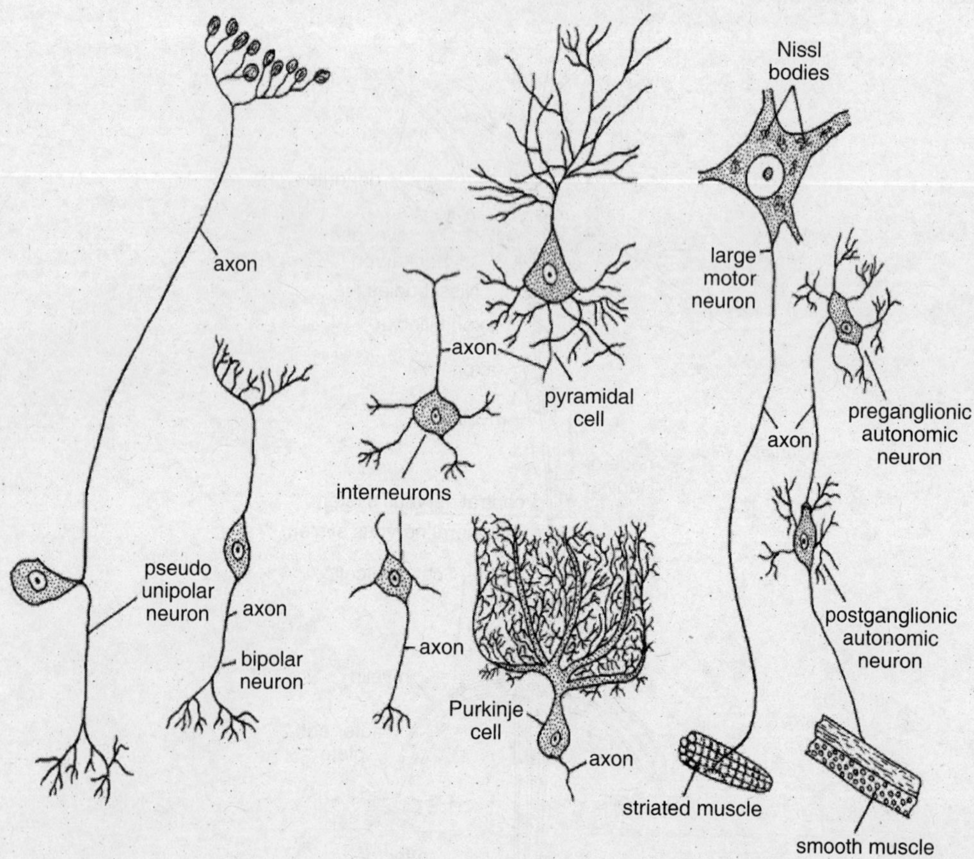

Fig. 3.2. Different types of neurons.

2. Bipolar neurons. They have one axon and one dendrite, *e.g.*, neurons of retina and VIII cranial nerve.

3. Unipolar neurons. They are, in fact, **pseudounipolar** neurons. They have one process, the axon that divides close to the cell body into two long processes, *e.g.*, sensory neurons. The cell body of a sensory neuron is situated in a dorsal root ganglion close to CNS; one axonal branch extends to periphery and one extends to CNS.

Ultrastructure

1. Cell body. It is also called **soma**, **cyton** or **perikaryon**. Cell body of neuron is large, dilated region of cell and contains the nucleus and those organelles that maintain the nerve cell. Cell body contains a large oval euchromatic nucleus with a prominent nucleolus. Cytoplasm of cell body contains many darkly stained (stained by basic dyes) minute particles, called **Nissl bodies** (or **Nissl's granules**) (Named after their discoverer **Franz Nissl**, a German neurologist, 1860–1919). Each Nissl body corresponds to a stack of rough endoplasmic reticulum (RER). The perikaryon also contains numerous mitochondria, a large perinuclear Golgi apparatus, lysosomes, microtubules, neurofilaments, vesicles and inclusions (Fig. 3.1). Nissl bodies (*i.e.*, RER), free ribosomes and occasionally, the Golgi apparatus extend into the dendrites but not into axon.

Box 3.3

1. The combination of euchromatin-rich nucleus, large nucleolus (site of ribosome biogenesis), prominent Golgi apparatus and Nissl bodies is indicative of the high level of **anabolic activity** needed to maintain these large postmitotic cells (neurons).

2. *Neurons do not divide*. So they have to last for a life time. As a result, the subcellular components of the neurons turn over regularly, having a molecular life spans measured in hours, days and weeks. The constant need to replace enzymes, transmitter substances, membrane components and other complex molecules explains the presence of morphologic features characteristics of a high level of synthetic activity.

2. Dendrites. They are neuronal processes that receive stimuli from other nerve cells or from the environment. Generally, dendrites are located in the vicinity of the cell body. They have a greater diameter than axons, are unmyelinated, are usually tapered and are further branched into *primary*, *secondary* and *tertiary* branches. Thus, dendrites form extensive arborizations, called **dendritic trees**, which significantly increase the receptor surface of a neuron.

3. Axon. An axon is a neuronal process that transmits stimuli to other neurons or to effector cells. *There is only one axon for each neuron* and it may be very long. Axons that originate from neurons in the motor nuclei of the CNS (called **Golgi type I neurons**) may travel more than a meter to reach their effector targets, *i.e.*, skeletal muscles. Interneurons of the CNS (called **Goigi type II neurons**), in contrast, have a short axon. Axon is mostly unbranched. However, branching of axon is most extensive in the vicinity of its target organs.

The axon originates from a conical projection of the cell body known as the **axon hillock** (Fig. 3.1). Normally it is devoid of large cytoplasmic organelles such as Nissl bodies and Golgi cisternae. However, microtubules, neurofilaments, mitochondria and vesicles pass through the axon hillock into the axon. It is here that molecules synthesized in the cell body and destined for transport to distant portions of the neuron enter the **axonal transport systems**. Several axons collectively constitute the **nerve fibre**.

Synapses

Neurons communicate with other neurons and with effector cells by means of **synapses**. Synapses are specialized junctions between neurons that facilitate transmission of impulses from one neuron to another. Synapses also occur between axons and effector cells such as muscle and gland cells.

Types of synapses. Synapses between neurons may be classified morphologically as follows (Fig. 3.3).

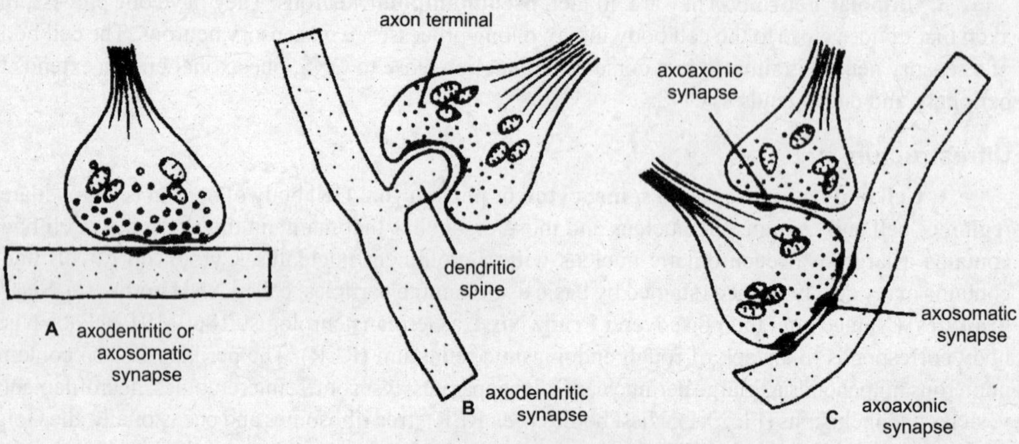

Fig. 3.3. Schematic diagram of different types of synapses. A—Axosomatic synapses; B—Axodendritic synapse; C—Axoaxonic synapse.

(*i*) **Axodendritic synapses** which occur between axons and dendrites.
(*ii*) **Axosomatic synapses** which occur between axons and the cell body.
(*iii*) **Axoaxonic synapses** which occur between axons and axons.
(*iv*) **Dendrodendritic synapses** which occur between dendrites and dendrites.

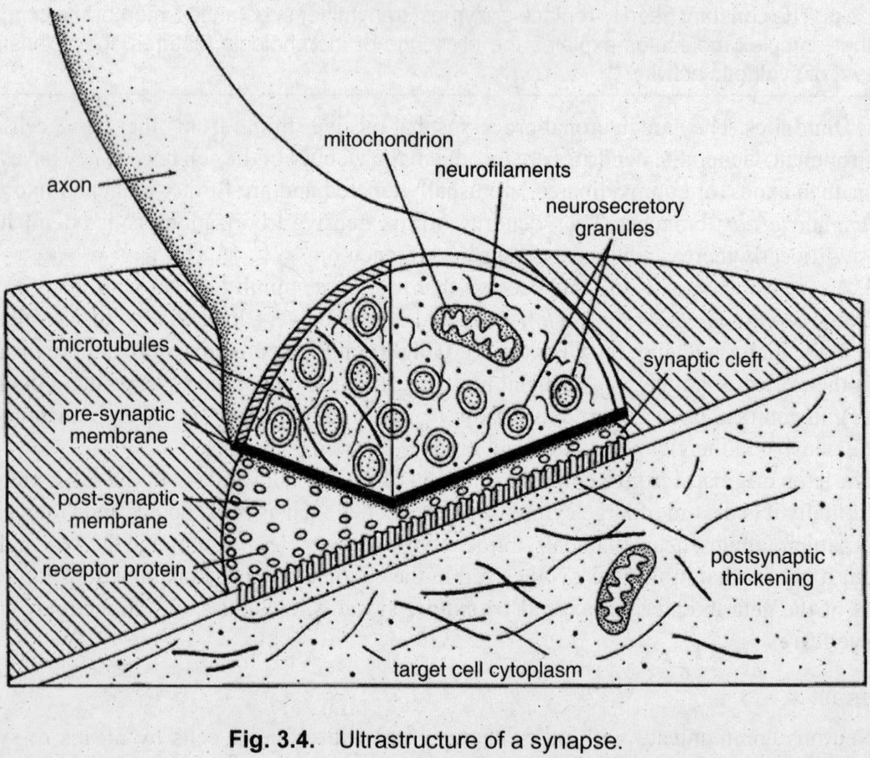

Fig. 3.4. Ultrastructure of a synapse.

Chemical and electrical synapses. Synapses may also be classified as chemical or electrical.

1. Chemical synapses. In them, neurons and their processes are closely apposed; usually only a 20–30 nm intercellular space separates them. A typical synapse has following parts in it (Fig. 3.4):

***(i)* Presynaptic knob (Presynaptic component)** which is the end of the neuronal process from which neurotransmitter (such as acetylcholine, nor-epinephrine, etc.) is released.

The terminal end of an axon is swollen to form a **synaptic bouton** which is closely applied to the surface of target cell. Synaptic bouton contains mitochondria, microtubules and neurofilaments. It also contains membrane-bound vesicles 40–65 μm in diameter called **neurosecretory vesicles** or **granules**.

***(ii)* Synaptic cleft** which is 20–30 nm space that the neurotransmitter must cross.

***(iii)* Postsynaptic membrane (Postsynaptic component)** which has **receptor sites** on the plasma membrane with which the neurotransmitter interacts.

2. Electrical synapses. These are common in lower vertebrates and invertebrates. They are **gap junctions** that permit movement of ions between cells, and consequently, permit direct spread of current from one cell to another. Mammalian equivalents of electrical synapses include the: (*i*) **nexus** in smooth muscle; (*ii*) **gap junctions** of cardiac muscle.

B. Supporting Cells

These include following three types of cells:

1. Schwann cells;
2. Satellite cells;
3. Neuroglia.

1. Schwann cells (Nonmyelinated and Myelinated nerve fibres). The **nerve fibre** is the axis cylinder or axon of a neuron which is surrounded by its enveloping sheath. So each nerve fibre consists of (*i*) a central **core of axon** which is a drawn-out portion of neuroplasm, called **axoplasm**; and (*ii*) the encircling plasma membrane, called **axilemma** or **neurilemma**. Such nerve fibres are known as **nonmyelinated (or nonmedullated) nerve fibres** (Fig. 3.5). Many nerve fibres contain an additional covering of **medullary sheath** or **myelin sheath**. The covering is placed in between the axis cylinder and neurolemma and they are known as **myelinated (or medullated) nerve fibres**. These appear glistening white and have high contents of lipids. The myelin sheath is a sort of insulating covering and is said to enhance the speed of conduction. Immediately underneath the neurilemma is a thin sheath of cytoplasm containing nuclei. These are known as **nuclei of Schwann cells** and are concerned with the formation of neurilemma.

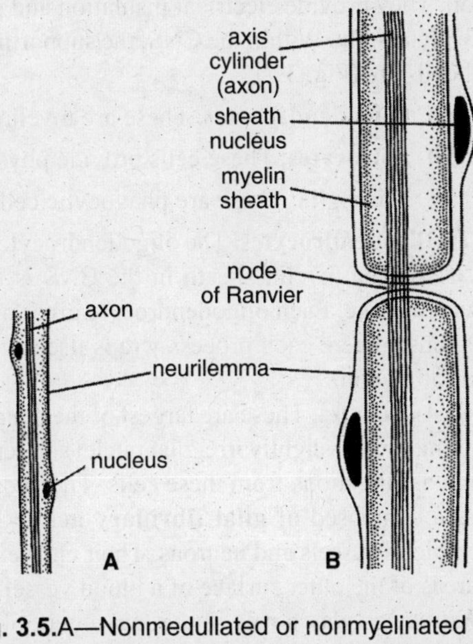

Fig. 3.5.A—Nonmedullated or nonmyelinated nerve fibre; B—Medullated or myelinated nerve fibre.

The myelin-sheath is not continuous along the entire length of axis cylinder but is interrupted at intervals of about 1 mm. These bare areas of axon (*i.e.*, constrictions) are called **nodes of Ranvier** (Fig. 3.6). Each node of Ranvier is the region containing the **voltage-gated Na^+**

channels, which are necessary for the formation of action potential. The myelin between two sequential nodes of Ranvier is called an **internodal segment**. The medullated nerve fibres form the cranial and spinal nerves, while the non-medullated nerve fibres occur in the autonomic nervous system.

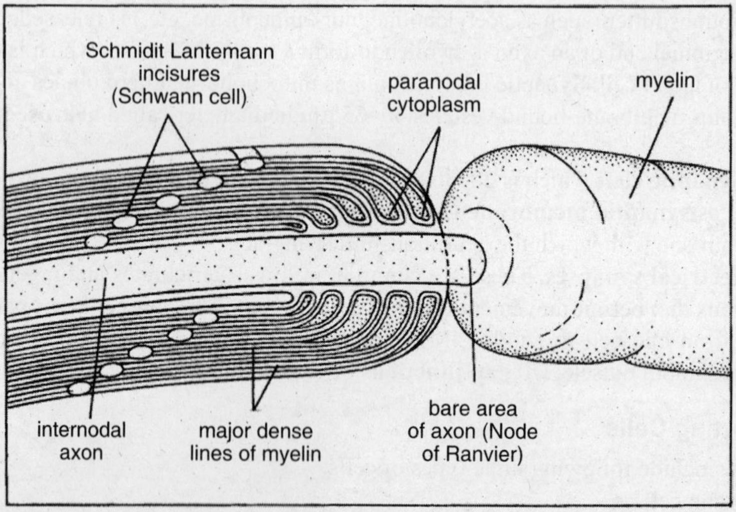

Fig. 3.6. Node of Ranvier (in L.S.).

2. Satellite cells. The neuronal cell bodies of ganglia are surrounded by a layer of small cuboidal cells called **satellite cells**. In paravertebral and peripheral ganglia, neural cell processes must penetrate between the satellite cells to establish a synapse (there are no synapses in sensory ganglion). Satellite cells help to establish and maintain a controlled microenvironment around cytons of neurons in the ganglion. They provide electrical insulation and pathway for metabolic exchanges.

3. Neuroglia. Within the CNS, the supporting cells are **neuroglia** or **glial cells**. The three types of glial cells are (Fig. 3.7):

(*i*) **Oligodendrocytes.** These are myelin-forming cells of the CNS.

(*ii*) **Astrocytes.** These cells provide physical and metabolic support for neurons of CNS.

(*iii*) **Microglia.** These are phagocytic cells of CNS.

(*i*) Oligodendrocytes. The oligodendrocyte is the cell responsible for producing myelin sheath within CNS. The myelin sheath in the CNS is formed by concentric layers of oligodendrocyte's plasma membrane. Each oligodendrocyte gives off several tongue-like processes that find their way to the axons. Where each process wraps itself around a portion of an axon, forming an internodal segment of myelin.

(*ii*) Astrocytes. These are largest of the neuroglial cells. Astrocytes are large and multi-process cells having oval or slightly irregular nuclei and remarkable stellate morphology. Many fine processes radiate in all directions from these cells. They contain prominent bundles of intermediate filaments which are composed of **glial fibrillary acidic protein (GFAP)**. Processes of astrocytes extend between blood vessels and neurons. Their ends expand, forming **end feet (foot processes)** that may cover areas of the outer surface of a blood vessel or of the axolemma.

Astrocytes play a role in the movement of metabolites and wastes to and from neurons and regulate ionic concentrations in the inter-cellular compartment, thus maintaining micro-environment of the neurons. They also provide a covering for the "bare areas" of myelinated axons, *e.g.*, at the nodes of Ranvier and at synapses.

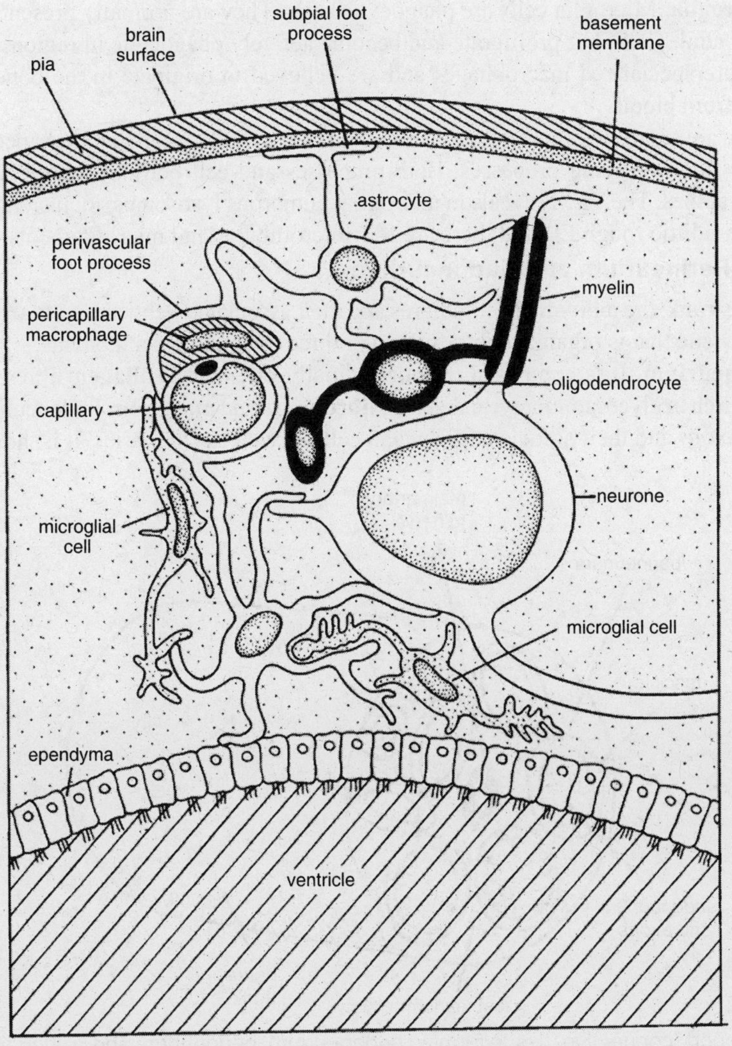

Fig. 3.7. Support cells of the CNS.

Box 3.3

1. Oligodendrocytes produce myelin in the CNS.

2. Schwann cells form myelin in PNS.

3. A Schwann cell myelinates only one axon, but an oligodendrocyte may myelinate several adjacent axons (Fig. 3.8).

4. Myelin is rich in **galactocerebroside**, a polymer (macromolecule) of plasma membrane of support cell.

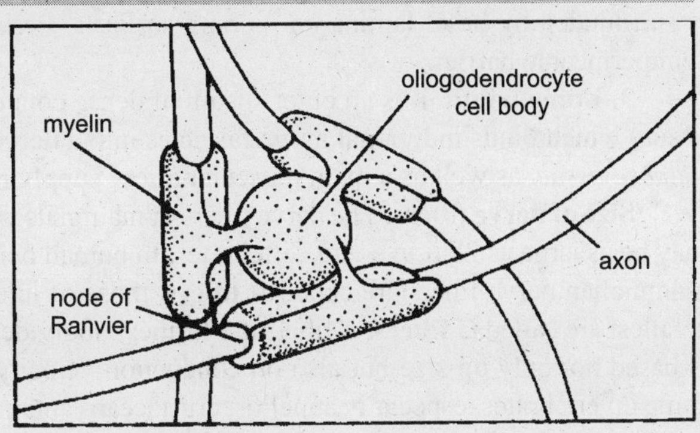

Fig. 3.8. Oligodendrocytes myelinate several adjacent axons within the CNS.

(*iii*) **Microglia.** Microglia cells are phagocytic cells. They are normally present only in small numbers in the adult CNS, but proliferate and become actively phagocytic in regions of injury and disease. They are specialized macrophages and are believed to originate in the bone marrow and enter the CNS from blood.

Microglia are smallest of the neuroglial cells and have relatively small rod-shaped nuclei. They have extensive fine ramifying processes. Both processes and cell body of microglia are covered with numerous spikes. The TEM reveals in microglia, numerous lysosomes, inclusions and vesicles. However, there is little rough ER and there are few microtubules and microfilaments.

Epineurium, Perineurium and Endoneurium

In a nerve trunk, the individual nerve fibres and their associated Schwann cells are held together by connective tissue that is organized into following three distinctive components:

1. Endoneurium. It is composed of longitudinally orientated **collagen** fibres, extracellular matrix material rich in glycosaminoglycans, sparse **fibroblasts** and **mast cells**. Endoneurium surrounds the individual axons and their associated Schwann cells as well as capillaries. It is, however, poorly vascularized.

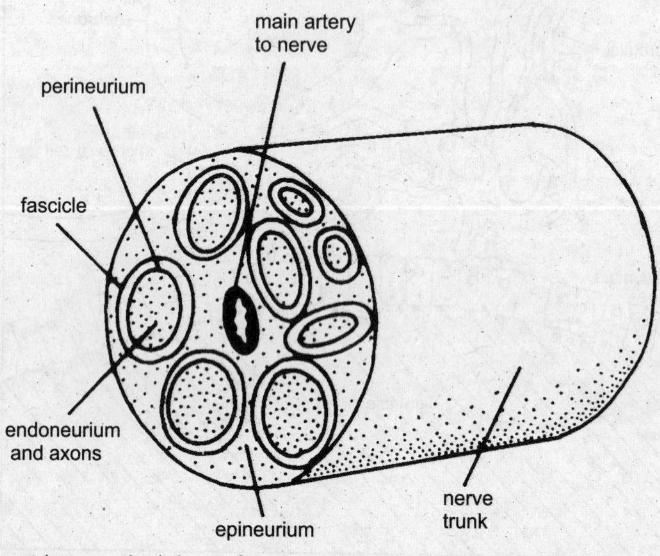

Fig. 3.9. Support (connective) tissue forming endoneurium, perineurium and epineurium in a nerve.

2. Perineurium. It surrounds groups of axons and endoneurium to form small bundles **(fascicles)**. It is composed of 7 to 8 concentric layers of flattened (*i.e.*, squamous in shape) cells separated by layers of **collagen**. The cells are joined by junctional complexes and each layer of cells is surrounded by basal lamina on its both surfaces. Perineurium lacks fibroblasts and serves as a semipermeable barrier.

3. Epineurium. It is an outer sheath of dense connective tissue (*i.e.*, dense fibro-collagenous tissue) which binds individual nerve fascicles into a nerve trunk. The epineurium may also include *adipose tissue*, as well as a main muscular artery supplying the nerve trunk.

Size of nerve fibers. The nerve fibers of mammals vary greatly in size. The diameter of an axon may be as large as 25 μ in some vertebrates. In human being it ranges from 0.1 μ to 1 μ. The largest mammalian nerve fibers are called **A fibers**, those of intermediate size are called **B fibers** and the smallest are called **C fibers**. A fibers are further subdivided into α, β and γ fibers. This classification is based not only on size but also on conduction velocity, which are closely related. The fibers of some invertebrates, especially annelids, crustaceans and molluscs are very large. The axon of *Loligo*, popularly known as **squid giant axon**, has a diameter of nearly 1500 μ. Most of the fundamental resarch on the physiology of nervous system has been carried out on this axon.

It should be understood at this point that the cell body of every nerve cell is confined to the central nervous system (brain, spinal cord and ganglia). The nerves are made up exclusively from axons. A nerve usually contains many axons. This accounts for the grey and white matter of the central nervous system. The gray colour is solely due to the cell bodies and non-myelinated processes, whereas the white matter comes from the myelinated nerve fibers as myelin has a characteristic glistening white appearance. Differences in myelinated (medulla) and non-myelinated (non-medullated) nerve fibers has been tabulated in Table 3.1.

Table 3.1. Differences in medullated or myelinated and non-medullated or non-myelinated nerve fibers.

Medullated/Myelinated nerve fibres	Non-medullated/Non-myelinated nerve fibres
1. Glistening white fibres and constricted at regular intervals.	1. Greyish fibres, not constricted.
2. Myelin/medullary sheath lies between neurilemma and axoplasm.	2. Myelin sheath is absent.
3. Node of Ranvier is present.	3. Node of Ranvier is absent.
4. Occur in cranial and spinal nerves.	4. Occur in nerves of autonomic nervous system.
5. Transmission of impulses is from node to node (*i.e.*, saltatory or discontinuous conduction) and is very speedy.	5. Transmission of impulses is comparatively slow.

Axonal Transport

Several types of axonal transport exist. Membrane-bound organelles and mitochondria are transported relatively rapidly by **fast axonal transport**. Substances that are dissolved in cytoplasm, such as proteins, are moved by **slow axonal transport**. In mammals, fast axonal transport proceeds as rapidly as 400 mm/day, whereas slow axonal transport occurs at about 1 mm/day.

Axonal transport requires metabolic energy and involves calcium ions (Ca^{2+}). Microtubules provide a system of guide wires along which membrane-bound organelles moves (Fig. 3.10). Organelles attach to microtubules through a linkage similar to that between the thick and thin filaments of skeletal muscle fibers. Calcium ion (*i.e.*, Ca^{2+}) triggers movements of organelles along the microtubules. Special microtubule-associated **motor proteins** called **kinesin** and **dynein** are required for axonal transport.

Axonal transport occurs in both directions. Transport from the soma toward the axonal terminals is called **anterograde axonal transport**. This transport involves **kinesin** and it allows the replenishment of synaptic vesicles and enzymes responsible for the synthesis of neurotransmitters in synaptic terminals (boutons). Transport in the opposite direction, which is driven by **dynein**, is called **retrograde axonal transport**. This process returns recycled synaptic vesicle membrane to the soma (or cell body) for lysomal degradation or reuse.

3.2 PHYSIOLOGY OF NEURON

Functionally, a nerve cell or neuron can be distinguished into three regions:

1. Generator region includes dendrites, cyton and colaterals. It is the part where initial polarization or **generator potential** develops.

2. Conductile region includes the axon which carries the stimulus.

3. Transmissional region comprises of the nerve ending or terminal branches of the axon which end in synaptic vesicles.

Propagation and Conduction of Nerve Impulse

The conduction of nerve impulse is an electrochemical event which involves the passage of measurable electric current along a nerve and a few metabolic activities within it at the synaptic ends.

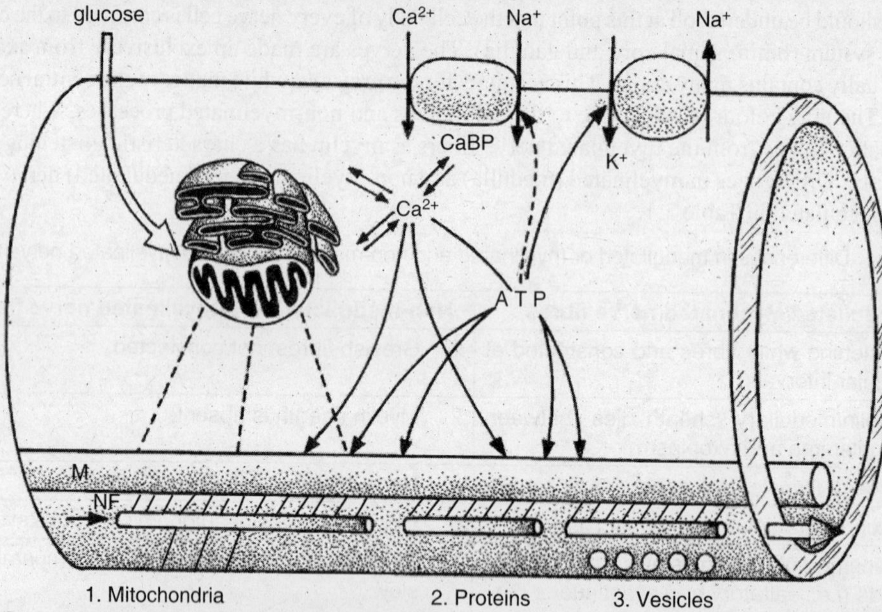

Fig. 3.10. Axonal transport depends on the movement of transport filaments. Energy is required and is supplied by glucose. Mitochondria control the level of cations in the axoplasm by supplying ATP to the ion pumps. An important cation for axonal transport is Ca^{2+}. Transport filaments move along the cytoskeleton (microtubules [M] or neurofilaments [NF]) by means of cross bridges. Transported components attach to the transport filaments. CaBP = Ca^{2+} binding protein (Koeppen and Stanton, 2010).

Like all other cells, the nerve cells exist in a fluid environment in which salts and ions are dissolved. The medium is known as **interstitial fluid**. The plasma membrane of the neuron is impermeable to most of these ions, permitting only potassium (K^+) ions to diffuse freely and keeping sodium (Na^+) and chloride (Cl^-) ions outside.

1. Resting potential. When a nerve cell is in a resting state, *i.e.*, it is not engaged in impulse transmission, its interior is negatively charged with respect to outside. How does it happen? The axon is bathed in the body fluid which contain approximately the same number of electrically charged particles or ions as contained in the axonic fluid, but the chemical composition of the two solutions is quite different. In the external solution more than 90% of the charged particles are sodium ions (Na^+) which are positively charged and chloride ions (Cl^-) that are negatively charged. Inside the nerve cell these ions together account for less than 10% of the solutes. Here the principal positive ions is potassium (K^+) and the negatively charged ions include organic molecules synthesized within the cell itself. These negative ions are large and do not diffuse easily through the axon membrane. It has been estimated that the concentration of Na^+ is about 10 times higher outside the axon, and the concentration of K^+ is about 30 times higher inside the axon. Potassium and chloride ions can move through the membrane easily as compared to sodium and large organic ions. Consequently, a large *voltage drop* is caused, about 60 to 90 mV across the membrane, with the inside of the cell being negative with respect to outside. It is suggested that to maintain these differences in ionic concentration the nerve cell's plasma membrane uses **ionic pumps** (channels and gates) that forces sodium ions uphill and outward through the membrane. The permeability of the resting cell membrane to Na^+ is quite low, hence the leakage is very small. Working of sodium pump is little understood, but it appears that for each sodium removed through the membrane, it accepts one potassium ion. The K^+ ions move freely in the axon and when the cell is resting they tend to leak downhill and outward through the membrane at a slow rate.

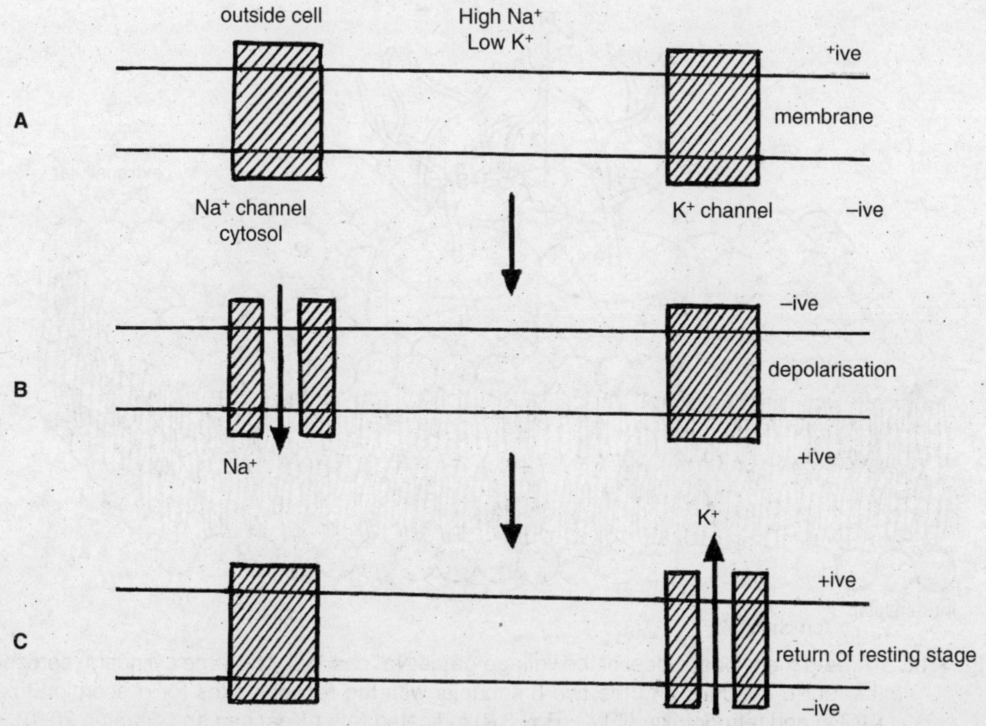

Fig. 3.11. Opening and closing of ion channels during depolarization: A—Resting state; B—During depolarization; C—Return to the resting state.

Molecular Structure of Ionic Channels and Gates

Recently it has been suggested that membrane of axon has a variety of protein channels and gates for different types of ions. For example, a **voltage-gated N^+ channel** consist of a single α subunit in association with β_1 and β_2 subunits. The α subunit has four repeated motifs of six transmembrane helices that surround a central ion channel or pore. The channel's walls are partly formed by the number 6 helices in each motif.

The functional channels of most voltage-gated K^+ channels are comprised of four separate subunits, each consisting of a polypeptide with 6 membrane-spanning segments and similar to the motifs which make up the single α subunit of the Na^+ channel.

Another important characteristic of some channels such as those that underlie the **action potential** is that they are gated by change in voltage (*i.e.*, they are **voltage-gated channel**). The gates sense the potential across the membrane and then act to either open or close the channel according to the membrane potential. The gates are formed by groups of charged amino acid residues, and the voltage dependence of the Na^+ and K^+ channels in g_{Na} and g_k that occur during an action potential.

2. Action potential. When a neuron is in a resting state, its interior is negative with respect to outside. It has been suggested that the membrane is selectively permeable to potassium ions and impermeable to sodium, chloride and other organic acids. An **electrical stimulus** depolarizes the membrane, increasing its sodium permeability due to opening up of sodium gates on either side. Consequently, influx of sodium takes place (Fig. 3.11). If the inward flow is slow, it is counter balanced by the outward flow of potassium ions which are far more mobile. If the initial depolarization is sufficiently large, sodium ions move faster than the potassium leaving the fibers. This causes the membrane potential to drop further when the outflow of potassium can no longer keep pace with the faster influx of sodium. This causes an action potential to develop.

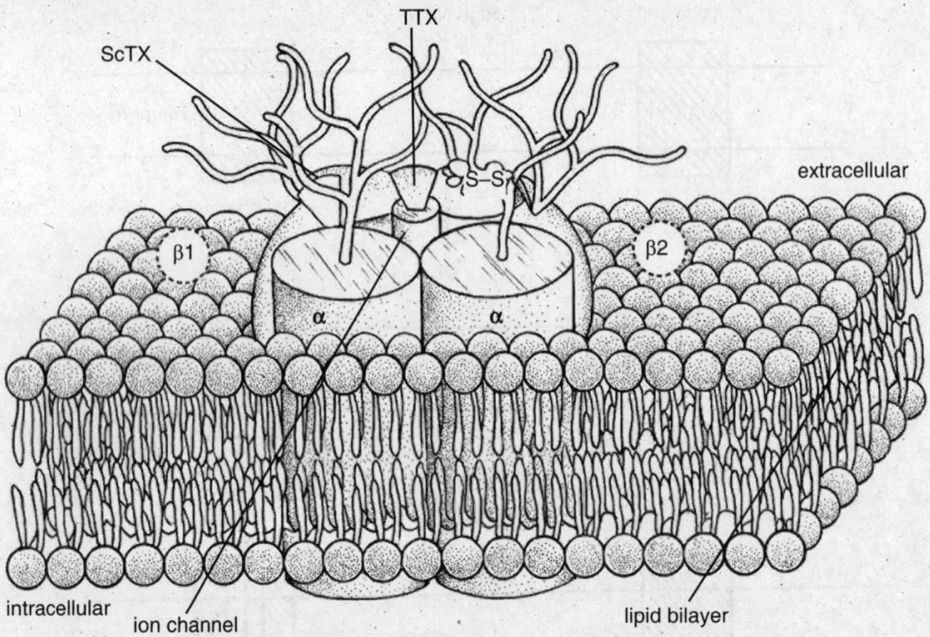

Fig. 3.12. Three-dimensional model of the voltage-gated Na^+ channel. The large cylinders represent the four α subunits and the two β subunits with the receptor sites for α scorpion toxin (ScTx) and tetrodotoxin (TTX) (Box 3.2) indicated (after Koeppen and Stanton 2010).

Box 3.2 TTX, TEA and Saxitoxin

Tetrodotoxin (TTX) is one of the most potent poisons known. It specifically blocks the *sodium (Na^+) channel*. TTX binds the extracellular side of the sodium channel (Fig. 3.12). **Tetraethylammonium (TEA^+)** is another poison which blocks potassium (K^+) channels. TEA^+ enters the K^+ channel from the cytoplasmic side and blocks the channel because TEA^+ is unable to pass through it. The ovaries of certain species of puffer fish (or blow fish) contain TTX. Raw puffer fish is a highly prized delicacy in Japan. Connoisseurs (*i.e.*, persons with good judgement on matters in which taste is needed) of puffer fish enjoy the tingling numbness of the lips caused by the iota of TTX present in the flesh. Sushi chefs who are trained to remove the ovaries safely are licensed by the government to prepare puffer fish. Despite these precautions, several people die each year as a result of eating improperly prepared puffer fish.

Saxitoxin is another blocker of Na^+ channels that is produced by the reddish dinoflagellates that are responsible for so-called **red-tide**. Shell fish eat the dinoflagellates and concentrate saxitoxin in their tissues. A person who eats these shell fish may experience life-threatening paralysis within 30 minutes after the meal.

The influx of sodium ions does not continue indefinitely and for some reason it stops near the peak of action potential (Fig. 3.13). As a result, the sodium gates which had opened during the rise of the spike are closed again and the potassium gates are opened briefly. This causes rapid outflow of the potassium ions, which restores the original negative charge of the interior of the axon. Soon after attaining the peak of the spike sodium permeability stops and potassium permeability rises above the resting state. At the end of the spike, the original level of potential is restored and the sodium permeability is zero. For a milliseconds after the membrane voltage has been driven to its initial level, it is difficult to displace the voltage and set up another impulse. This is called **refractory period**. But the ionic permeabilities quickly return to their initial condition and the cell is ready to fire another impulse.

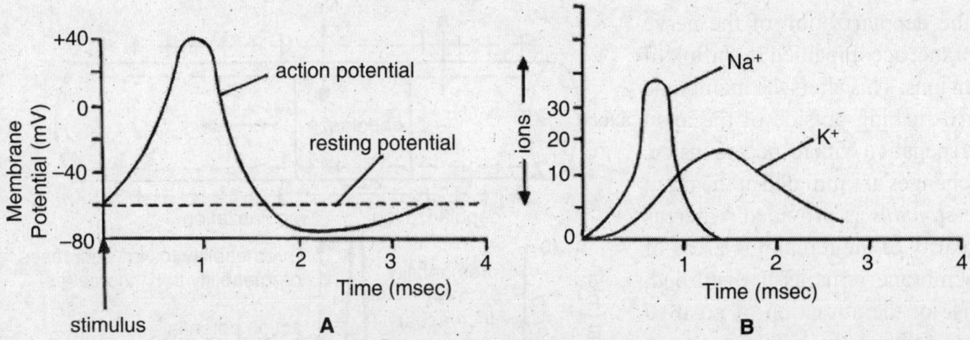

Fig. 3.13. Action potential. A—Depolarisation of the membrane potential. B—Changes in the permeability of the plasma membrane to Na^+ and K^+.

3. Impulse conduction. Conduction of impulse is a bioelectrical phenomenon and it is transmitted along the nerve undisturbed. Certain characteristics of nerve conduction should be taken into account, such as threshold of excitation, all or none response and refractory period. **Excitability threshold** is the critical level when the membrane potential is developed. If the intensity of stimulus is further increased, the height of the spike will remain the same. This is called **all or none response**. The point where an impulse has passed over the axon, a **refractory period** follows during which an action potential cannot be developed if stimulated at that point. An impulse once generated is non-decremental in its amplitude all along the nerve fiber.

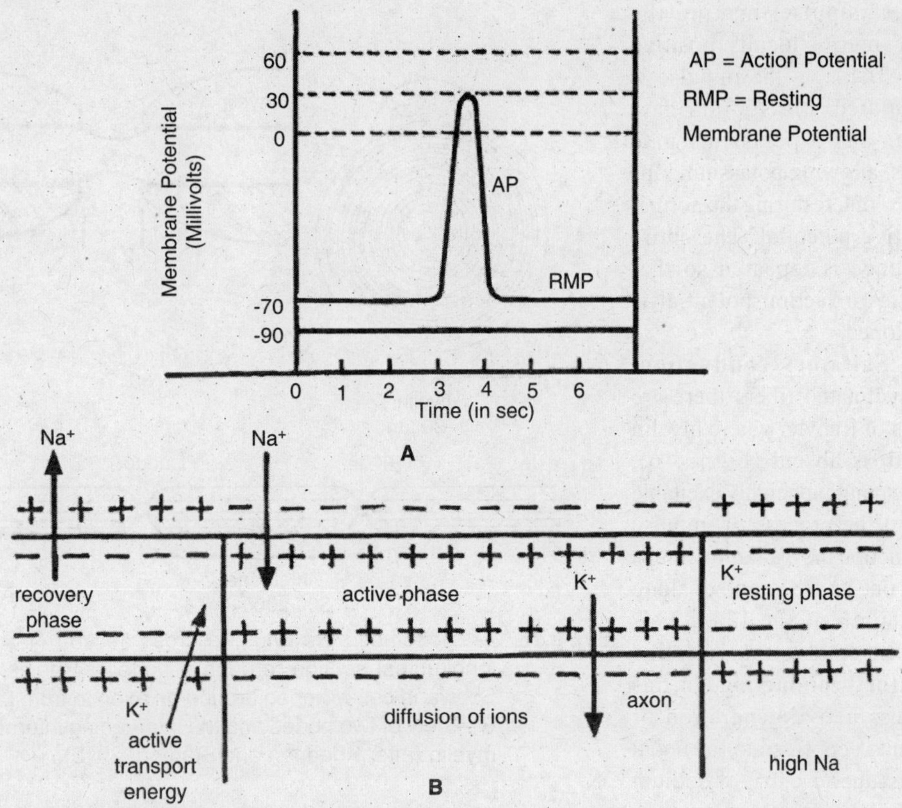

Fig. 3.14. Transmission of nerve impulse through nerve fiber. AP = action potential; RMP = resting membrane potential.

An action potential is initiated with the depolarization of the nerve membrane, accompanied by influx of sodium ions. This alters the membrane polarity, making outside of the axon (nerve) negative with respect to inside. Such changes are initiated at the point where *stimulus* is provided, whereas the polarity of the remaining areas of the membrane remains undisturbed. Because of the attraction of positive charges toward the negative ones, positive charges flow in toward the negative, setting of a flow of current (Fig. 3.14). Propagation of nerve impulse coincides with changes in the permeability of the axon membrane. Normally the axon interior is rich in potassium and poor in sodium ions, the fluid outside has a reverse composition. When a nerve impulse arises, a gate opens and sodium ions enter into the axon in advance of the impulse, making the axon interior locally positive (Fig. 3.15). As the impulse is stimulated, the sodium gate closes and a potassium gate opens, allowing potassium ions to flow out, restoring the normal negative potential. The entire sequence is repeated so that a series of action potential is developed.

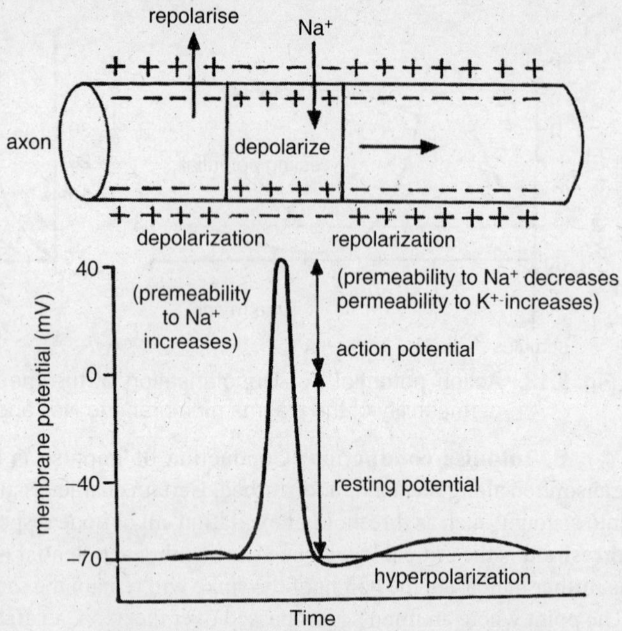

Fig. 3.15. Movement of action potential along the membrane.

Saltatory conduction. In myelinated fibers there are nodes of Ranvier where myelin sheath is absent (Fig. 3.16). These points are highly excitable and in the process of impulse conduction the excitation jumps from one node to another along the length of the fiber. This type of conduction is called **saltatory conduction**. In this process active generation of current is confined to the nodal points due to entry of sodium

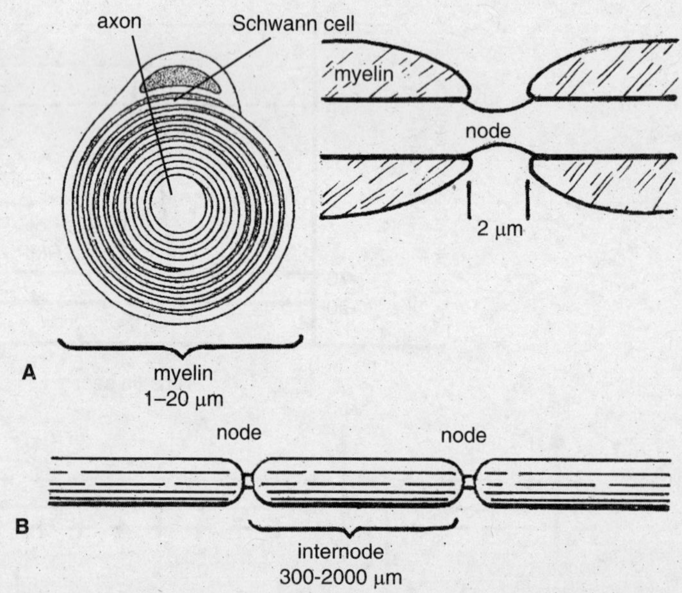

Fig. 3.16. A—Schematic drawings, in cross section (T.S.) and longitudinal section through a node of Ranvier, of a Schwann cell wrapped around an axon to from myelin. B—View of two nodes and the intervening internode of myelin (after Koeppen and Stanton, 2010).

ions, whereas the internode points are depolarized by local circuit action. The amount of sodium and potassium exchange is also reduced. The conduction velocity depends upon the diameter of the nerve and internodal distance. The impulse travels much faster in myelinated fibers. Myelin sheath increases the speed of conduction and avoids dissipation of impulse into adjacent fibers.

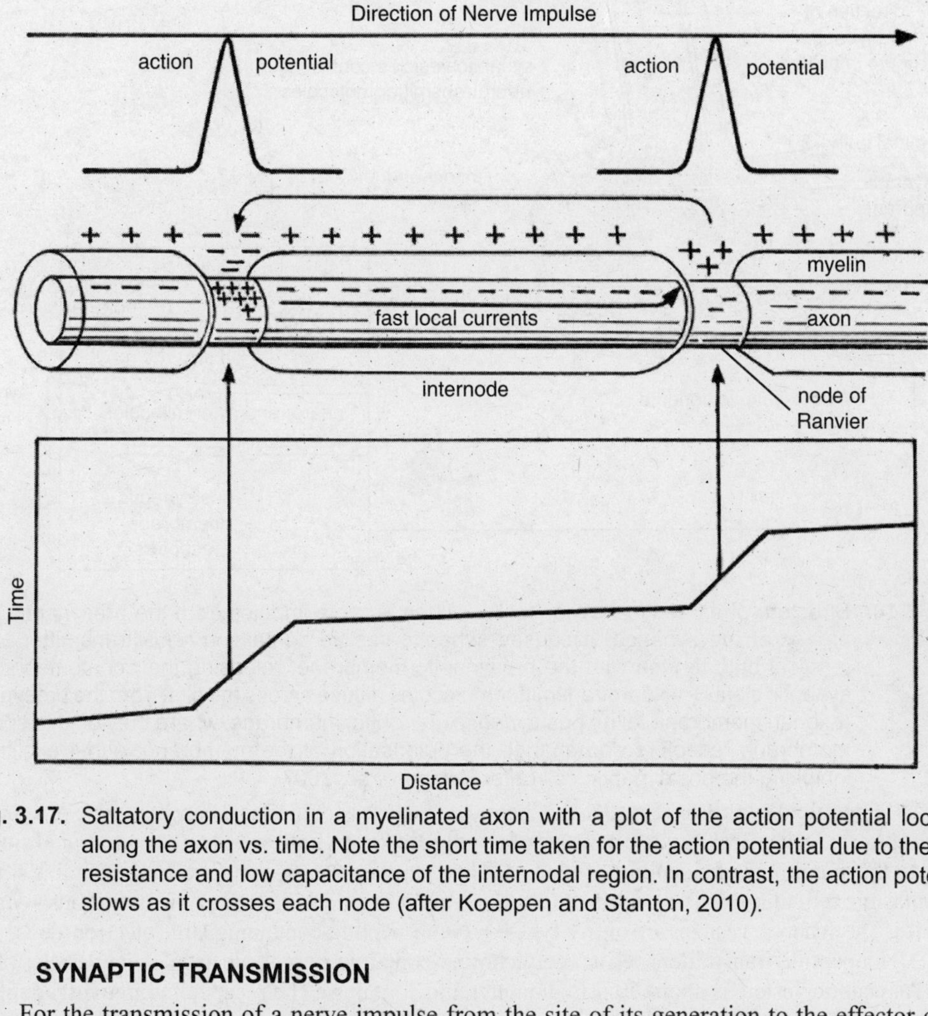

Fig. 3.17. Saltatory conduction in a myelinated axon with a plot of the action potential location along the axon vs. time. Note the short time taken for the action potential due to the high resistance and low capacitance of the internodal region. In contrast, the action potential slows as it crosses each node (after Koeppen and Stanton, 2010).

3.3 SYNAPTIC TRANSMISSION

For the transmission of a nerve impulse from the site of its generation to the effector organ needs more than one neurons that lie end-to-end in a chain, forming **synapses** in between. A synapse is the close proximity of the motor end plate present at the terminal branches of the axon of one neuron and the dendrites of the next neuron in the chain without actual contact (Fig. 3.1 and 3.4). A narrow intercellular space (about 200 A°) exists between the axon and the dendrite of the synapting neurons. This gap is called the **synaptic cleft**. A synapse between an axon and dendrites is called **axodendritic synapse** (Fig. 3.3B). A synapse may be between an axon and the soma (cell body) of two neurons. It is termed **axosomatic synapse** (Fig. 3.3A). A structurally different but functionally similar synapse occurs between the axon terminals of a motor neuron and the surface of a muscle fiber. It is known as a **neuromuscular junction** (Fig. 3.1).

There are two modes of transmission of impulses across a synapse: chemical transmission and electrical transmission. The synapses involved in such transmissions are respectively called **chemical** and **electrical synapses**.

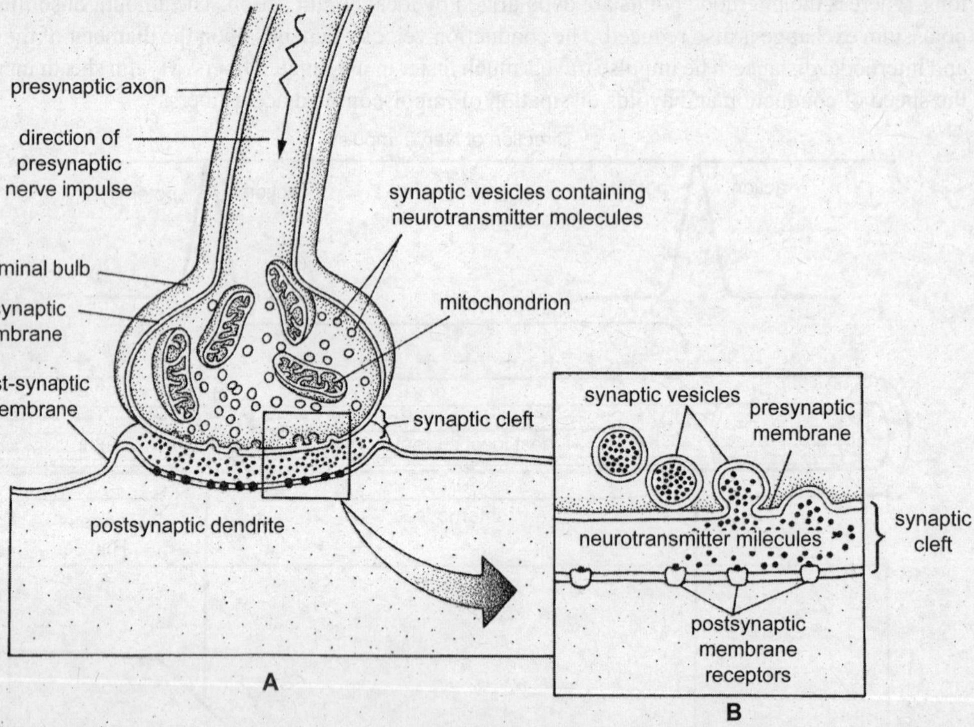

Fig. 3.18. Structure of a chemical synapse. A—When a nerve impulse from the presynaptic axon arrives at the synapse, it causes synaptic vesicle containing neurotransmitter in the terminal bulb to fuse with the presynaptic membrane, releasing their contents into the synaptic cleft. B—Neurotransmitter molecules diffuse across the cleft from the presynaptic (axonal) membrane to the postsynaptic (dendritic) membrane, where they bind to specific membrane receptors and change the polarisation of the membrane, either exciting or inhibiting the postsynaptic cell (after Becker *et al.*, 2007).

A. Chemical synapses (Fig. 3.18). Chemical synapses are the most common type of synapses in vertebrates. A chemical synapse consists of a **synaptic knob** (**bouton**) of an axon terminal situated close to the membrane of a dendrite (Fig. 3.4). The cytoplasm of the synaptic knob contains smooth endoplasmic reticulum, mitochondria, neurofilaments (*i.e.*, microfilaments) and numerous **synaptic vesicles**. The synaptic vesicles are of two types: 1. Neuropeptide-containing large electron-dense vesicles; 2. Nonpeptide-transmitter such as acetylcholine-containing small, electronlucent vesicles (Table 3.2). A synaptic vesicle is about 50 μ in diameter and contains a chemical called **neurotransmitter substance**. The latter transmits the nerve impulse across the synapse. The membrane of the synaptic knob adjacent to the synapse is thickened due to cytoplasmic condensation and is called **presynaptic membrane**. This membrane is modified for the attachment of synaptic vesicles and the release of the neurotransmitter substance into the synaptic cleft. The dendrite membrane next to synapse is also thickened and is termed the **postsynaptic membrane**. A 20 nm wide space, the **synaptic cleft**, separates the presynaptic and the postsynaptic membranes. The postsynaptic membrane

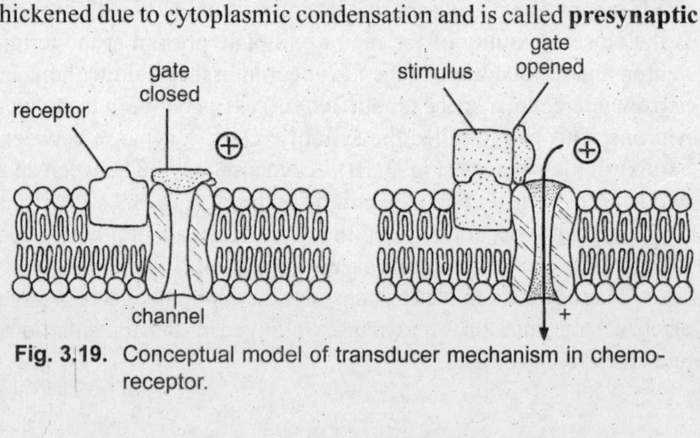

Fig. 3.19. Conceptual model of transducer mechanism in chemoreceptor.

contains large **receptor molecules** for the neurotransmitter substance, and has many **channels** and **pores**, normally closed, for the entry of ions into the dendrite (Fig. 3.19).

The neurotransmitter substance may be synthesized either in the synaptic knob or bouton (*e.g.*, acetylcholine) or in the cyton and passed down the axon to the synaptic knob by axoplasmic streaming through microtubules (*e.g.*, peptide neurotransmitters). Synthesis of peptide neurotransmitters occurs on ribosomes of rough endoplasmic reticulum (RER) and packaged by Golgi complex. In the synaptic knob, the neurotransmitter substance is also packaged into recycled synaptic vesicles and stored till needed. The vertebrates usually produce two neurotransmitters: **acetylcholine** (ACh) and **noradrenaline**. Acetylcholine was the first neurotransmitter substance to be isolated by **Otto Loewi** in 1920 from the ending of the parasympathetic neurons of the vagus nerve in frog's heart. It is synthesized from acetic acid and choline with in help of an enzyme **choline acetylase**. There are about 70,000 acetylcholine molecules in each small synaptic vesicle. The neurons releasing acetylcholine and noradrenaline are respectively called **cholinergic** and **adrenergic neurons**.

Mechanism of chemical synaptic transmission (Fig. 3.20 and Fig. 3.21). On reaching the synaptic knob, the nerve impulse depolarises the presynaptic membrane and increases the permeability to calcium ions (*i.e.*, Ca^{2+}). The ions flow into the axon terminal via calcium channels that normally are closed but open in response to the change in membrane voltage.

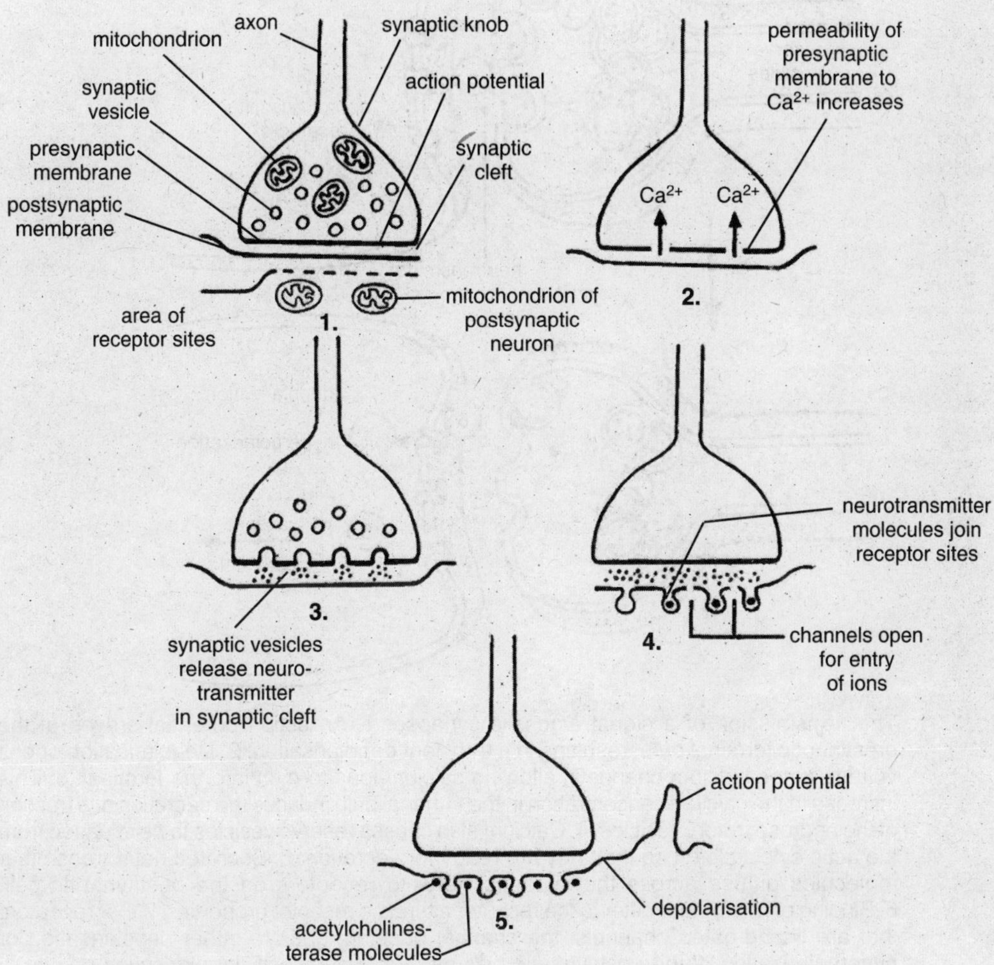

Fig. 3.20. Process of transmission of nerve impulse in the chemical synapse.

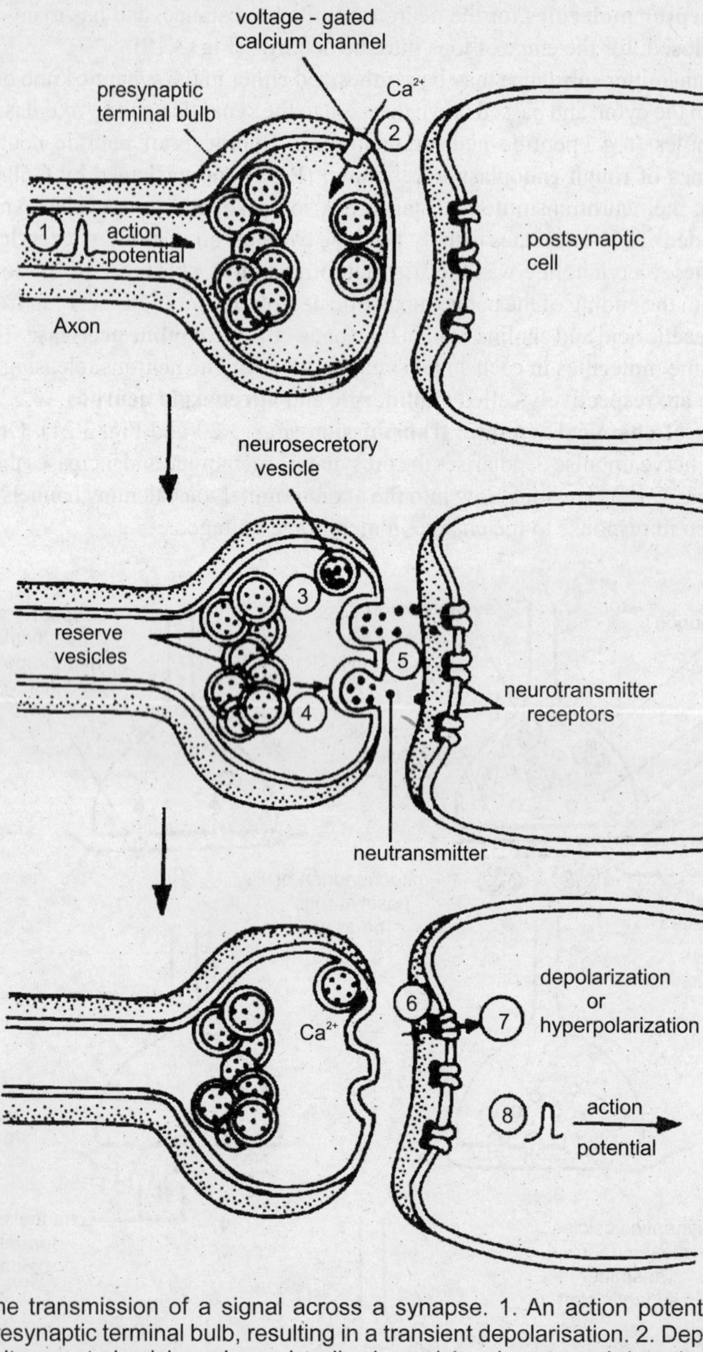

Fig. 3.21. The transmission of a signal across a synapse. 1. An action potential arrives at the presynaptic terminal bulb, resulting in a transient depolarisation. 2. Depolarisation opens voltage-gated calcium channels, allowing calcium ions to rush into the terminal. 3. This increase in the calcium concentration in the terminal bulb induces the secretion of a fraction of the neurosecretory vesicles. 4. Calcium also causes reserve vesicles to be released from the actin cytoskeleton so that they are ready for secretion. 5. Secreted neurotransmitter molecules diffuse across the synaptic clefts to receptors on the postsynaptic cell. 6. Binding of neurotransmitter to the receptor alters the receptor properties. 7. For receptors that are ligand-gated channels, the channel opening leads to either depolarisation or hyperpolarization of the postsynaptic plasma membrane. 8. If depolarisation results, a sufficient amount of excitatory neurotransmitter will result in an action potential in the postsynaptic cell (after Becker *et al.*, 2007).

Calcium channels are, thus, **voltage-gated channels**. The rise in Ca^{2+} concentration is extremely brief because the synaptic knob contains an apparatus which quickly sequesters Ca^{2+}. Entry of the Ca^{2+} into the synaptic knob causes fusion of synaptic vesicles with the presynaptic membrane and their rupture (exocytosis), releasing their contents into the synaptic cleft. The process is called **excitation-secretion coupling**. About 300 synaptic vesicles release their neurotransmitter substance for each nerve impulse.

Action of acetylcholine. Each synaptic vesicle releases about 10,000 molecules of acetylcholine. Acetylcholine diffuses across the synaptic cleft and attaches to specific **acetylcholine receptor molecules** on the postsynaptic membrane of the dendrite (Fig. 3.22). This combination changes the configuration of receptor molecules, opening up the channels in the postsynaptic membrane. These channels open in response to acetylcholine, and are, therefore, called **chemically gated** or **ligand-gated channels**, as opposed to the voltage-gated channels for calcium ions in presynaptic membrane. The channels allow the Na^+ ions (sodium ions) to enter and K^+ ions (potassium ions) to leave down their concentration gradients. This causes depolarization of the postsynaptic membrane and initiates a new action potential which passes as a wave (nerve impulse) along the next neuron. There is an interval of about 0.5 milliseconds between the presynaptic action potential and postsynaptic action.

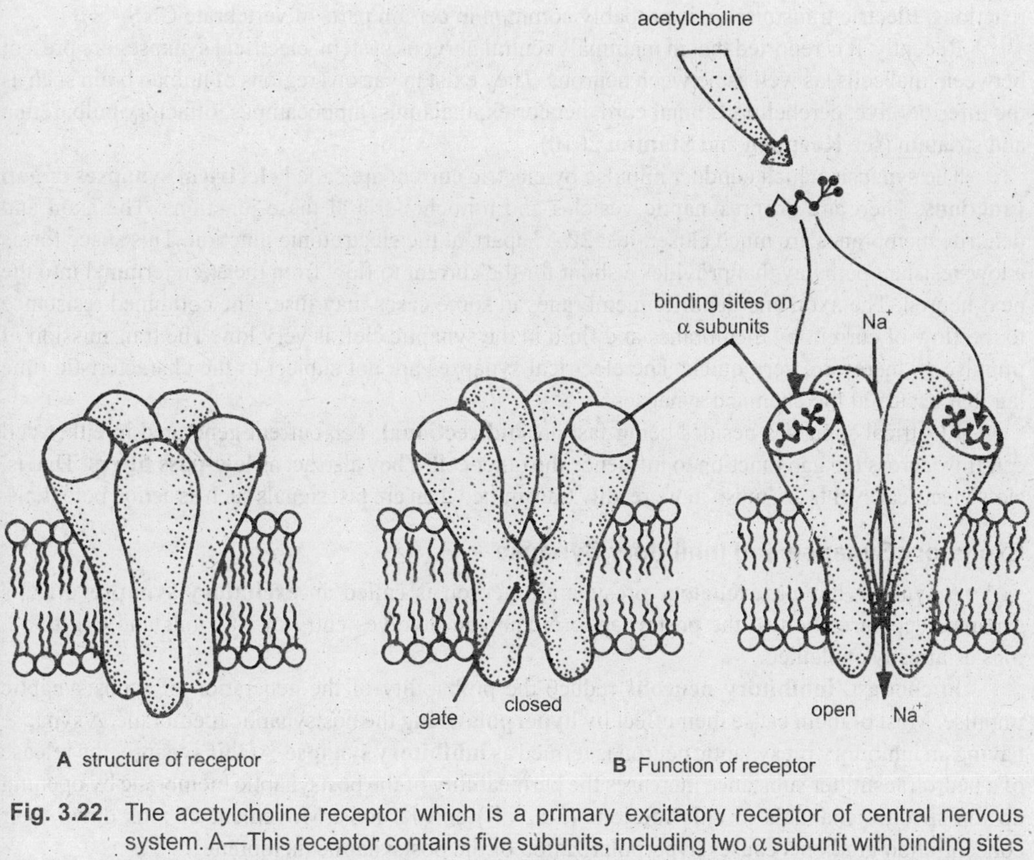

Fig. 3.22. The acetylcholine receptor which is a primary excitatory receptor of central nervous system. A—This receptor contains five subunits, including two α subunit with binding sites for acetylcholine and one each of β, γ and δ. the subunits aggregate in the lipid bilayer in such a way that the transmembrane portion form a channel. B—The channel (shown here with β subunit removed) is normally closed, but when acetylcholine binds to the two sites on the α subunits, the subunits are altered in such a way that the channel opens to allow sodium ions to cross (after Becker *et al.*, 2007).

Inactivation of acetylcholine. The acetylcholine is inactivated by an enzyme *acetylcholinesterase*, present in the postsynaptic (dendrite) membrane.

$$\text{Acetylcholine} + H_2O \xrightarrow{\text{Acetylcholinesterase}} \text{Acetic acid} + \text{Choline}$$

The enzyme splits acetylcholine by hydrolysis into its components, acetic acid and choline. Due to increased permeability of Na^+, the membrane of dendrite repolarize. The constituent of acetylcholine are **inactive** and reabsorbed by axons for the reutilization.

Thus, chemical transmission at the synapse involves two processes: **neurosecretion** by axon ending (boutons), *i.e.*, release of acetylcholine or peptides and **chemoreception** by dendrites or muscle fibers. Since synapse cannot transmit a impulse in reverse direction as the dendrites cannot secrete acetylcholine or any other neurotransmitter. Thus, the synapse acts as a **one-way valve**, permitting the conduct of impulse from axon to dendrite only. Thus, a chemical synapse is **unidirectional**.

B. Electric transmission. In case of electric transmission, the action potential on reaching the end of axon of the presynaptic neuron sets up electric currents in the external fluid of the neuron to generate an action potential. Electric transmission at the synapse was first demonstrated in the abdominal nerve cord of crayfish by **Furshpan** and **Potter** in 1959. It has since been found in many coelenterates, annelids, molluscs and arthropods. Electric transmission occurs in fish at certain large nerve cells, called **Mauthner cells** of the spinal cord, where rapid transmission is important in escape reactions. Electric transmission is probably common in certain parts of vertebrate CNS.

Recently, it is reported that in mammals central nervous system, electrical synapses are present between glial cells, as well as between neurons. They exist in various regions of human brain such as the inferior olive, cerebellum, spinal cord, neocortex, thalamus, hippocampus, olfactory bulb, retina and striatum (see **Koeppen** and **Stanton** 2010).

The synapsis which conduct impulse by electric current are called **electrical synapses** or **gap junctions**. There are no presynaptic vesicles and mitochondria at these junctions. The axon and dendrite membranes are much closer, just 20 A° apart, at the electrotonic junction. This space forms a low-resistant pathway that provides a shunt for the current to flow from the axon terminal into the next neuron. The axon and dendrite membranes in some cases may fuse. The combined resistance to the flow of current by membranes and fluid in the synaptic cleft is very low. The transmission of impulse is, therefore, very quick. The electrical synapses are not subject to the characteristic time lag that occurs at the chemical synapses.

Electrical synapses besides being fast are **bidirectional**, *i.e.*, current generated in either cell can flow across the gap junction to influence the other cell. They also act as **low-pass filters.** That is, slow electrical events are mush more readily transmitted than are fast signals such as action potentials.

Excitatory Synapse and Inhibitory Synapse

A synapse having excitatory presynaptic neuron is called an **excitatory synapse**. At this synapse, depolarization of the postsynaptic membrane involves entry of Na^+ ions and exit of K^+ ions as already explained.

In contrast, **inhibitory neurons** reduce the probability of the generation of a postsynaptic impulse. Most of them cause their effect by **hyperpolarizing** the postsynaptic membrane. A synapse having an inhibitory presynaptic neuron is termed as **inhibitory synapse**. At this synapse, the release of a neurotransmitter substance increases the permeability of the postsynaptic membrane by opening up ion-specific channels to chloride ions (*i.e.*, Cl^-) and K^+ ions. Movement of ions down their concentration gradients causes hyperpolarization of the postsynaptic membrane.

Molecular Apparatus Underlying Vesicular Release

The small vesicles that contain nonpeptide neurotransmitters can fuse with the presynaptic membrane only at specific sites, called **active zones.** To become competent to fuse with the presynaptic membrane at an active zone, a small vesicle must first **dock** at the active zone (Fig. 3.23). It must

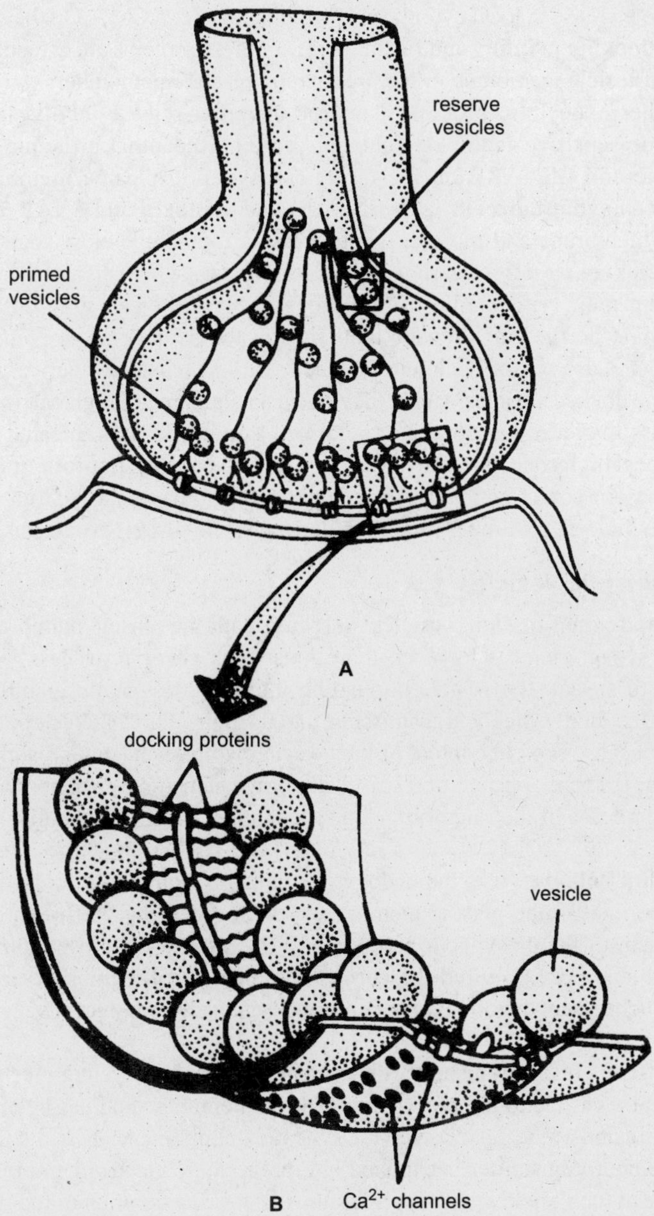

Fig. 3.23. Docking of synaptic vesicles with the plasma membrane of the presynaptic neuron. A—In response to local elevation of calcium in the presynaptic neuron, synaptic vesicles are released from the cytoskeleton, allowing them to fuse with the plasma membrane of the terminal bulb within the active zone. Vesicles bound to the cytoskeleton, via synapsin form a reserve pool of vesicles. Once released from the cytoskeleton, vesicles move near the plasma membrane and become "primed" (*i.e.*, charged or prepared). Some vesicles become tightly associated with the plasma membrane, or "docked". When nearby calcium channels open, such docked vesicles fuse with the plasma membrane, releasing their contents. B—A drawing based on an actual TEM reconstruction of the active zone of a motor neuron form a frog. Docked secretory vesicles are arranged in rows and connected by a complex of proteins associated with docking. Calcium channels lie immediately beneath the vesicles (after Becker *et al.*, 2007).

than undergo a **priming process** before the vesicle can fuse and release its transmitter into the synaptic cleft in response to increase in local cytoplasmic calcium ions (*i.e.*, Ca^{2+}). On the order of 25 proteins may play roles in **docking priming** and **fusion**. Some of these proteins are cytosolic , whereas others are proteins of the vesicle membrane or the presynaptic plasma membrane.

As with other exocytotic processes, neurotransmitter-release involves **SNARE** (**S**oluble **N**- ethylmaleimide-sensitive factor **A**ttachment protein **R**eceptor) proteins: V-SNARE in the vesicle membrane and t-SNARE on the target (presynaptic plasma membrane). Zipper-like interactions between **synaptobrevin** (a v- SNARE) and **syntaxin** and **SNAP-25** (two t-SNARE) bring the vesicle membrane and the presynaptic plasma membrane close together before fusion. The SNARE proteins are targets for various **botulinum toxins**, which disrupt synaptic transmission, thus demonstrating their critical role in this process. Nevertheless, they do not bind Ca^{2+}, so another protein must be the **Ca^{2+} sensor** that triggers the actual fusion event. **Synaptotagmin protein** is one such Ca^{2+} sensor. Calcium channels are located in the active zone membrane at sites adjacent to the docked vesicles. When they open, a small area of high calcium ions (*i.e.*, Ca^{2+}), which lasts for less than a millisecond and is termed a **microdomain**, created at the active zone (Fig. 3.23). This local high concentration of Ca^{2+} to a protein called **synaptotagmin**, and it is thought that this binding causes a conformational change in synaptotagmin protein that triggers the fusion event of a docked vesicle. Indeed, the time from Ca^{2+} influx to vesicle fusion (exocytosis) is about 0.2 msec.

Recycling of Synaptic Vesicles

During synaptic transmission, vesicles must fuse with the plasma membrane to release their contents into the synaptic cleft. However, there must be a **reverse process**; otherwise not only would it be hard to sustain the vesicle population, but the presynaptic membrane's area would also grow with each bout of synaptic transmission, and its molecular content and functionality would likewise change (because protein content of the vesicle membrane is quite distinct from that of the terminal membrane). There appear to be two distinct mechanisms by which synaptic vesicles are retrieved after and release of their neurotransmitter content: 1. fusion and collapse and 2. kiss-and-run fusion.

1. Fusion and collapse. It is the endocytotic pathway commonly found in most cell types. **Coated pits** are formed in the plasma membrane, which then pinch off to form **coated vesicles** within the cytoplasm of the presynaptic terminal. These vesicles then lose their coat and undergo further transformations (*i.e.*, acquired the correct complement of membrane and be refilled with neurotransmitter such as acetylcholine) to become once again electron-lucent synaptic vesicles ready for release.

2. Kiss-and-run- fusion. This is a well-evident most rapid recycling mechanism. It involves fusion of the synaptic vesicle to the presynaptic plasma membrane and has been called "**kiss-and-run**". In this case, fusion of the vesicle with the synaptic membrane leads to the formation of a pore through which the neurotransmitter is expelled into presynaptic plasma membrane, but there is no wholesale collapse of the vesicle into the membrane. Instead, the duration of the fusion is very brief, after which the vesicle detaches from the plasma membrane retaining its molecular identity. Its contents can then simply be replenished in soma, thereby making the vesicle ready for use again.

Neurotransmitters

Neurotransmitters are the substances that mediate chemical signaling between neurons. For a substance to be considered a neurotransmitter, it must meet several generally recognized criteria. 1. the substance must be demonstrated to be present in the presynaptic terminal; 2. the cell must be able to synthesize the substance; 3. it should be released on depolarization of the terminal; 4. there should be specific receptors for it on the postsynaptic membrane. This last criterion is certainly true for substances that act as synaptic transmitters but if we want to be inclusive and include substances that act over widespread territories rather than just at a single synapse, the last criterion needs to be relaxed to include situations in which receptors are located outside the synapse. The term *neurotransmitter* has been suggested as a general term to describe both synaptic and nonsynaptic signaling between cells.

More than 100 substances have been identified as potential neurotransmitters because they have met some (hence the "potential" qualifier) or all of these criteria. These substances can be subdivided into three major categories: **1. Small-molecule transmitters** such as acetylcholine, amino acids, biogenic amines and purines. The first three groups of small-molecule transmitters contain what are considered the **classic neurotransmitters. 2. Peptide transmitters** such as neuropeptides, opoids, peptides, substance P, etc. **3. Gaseous transmitters** such as nitric oxide (NO) and carbon monoxide (CO).

In some ways neuropeptides are like the classic neurotransmitters: they are packaged into synaptic vesicles, their release is dependent on Ca^{2+}, and they bind to specific receptors on target neurons. However, there are also significant differences, ones that have led to alternative names for the intercellular communication mediated by neuropeptides, such as nonsynaptic, parasynaptic and volume transmission. Table 3.2 summarises some of these differences between classic and peptide neurotransmitters.

Table 3.2. Distinctions between classic nonpeptide neurotransmitters and peptide neurotransmitters (source Koeppen and Stanton 2010).

Nonpeptide transmitters	Peptide transmitters
1. Synthesized and packaged in the nerve terminal.	1. Synthesized and packaged in the cell body (soma); transported to the nerve terminal by fast axonal transport.
2. Synthesized in active form.	2. Active peptide formed when it is cleaved from much larger polypeptide that contains several neuropeptides.
3. Usually present in small clear vesicles.	3. Usually present in large, electron dense vesicles.
4. Released into synaptic cleft.	4. May be released some distance from the postsynaptic cell. There may be no well-defined synaptic structure.
5. Action of many terminated because of uptake by presynaptic terminal via Na^+-powered active transport.	5. Action terminated by proteolysis or by peptide diffusing away.
6. Typically, action has short latency and short duration (msec).	6. Action may have long latency and may persist for many seconds.

Unlike classic neurotransmitters, which are synthesized at the presynaptic terminal, neuropeptides are synthesized in the cell body and then transported to the terminal (Fig. 3.24). Neuropeptides are packaged into large electron-dense vesicles that are scattered throughout the presynaptic terminal rather than in small electron-lucent vesicles that are docked at the active zone, where small-molecule transmitter are stored.

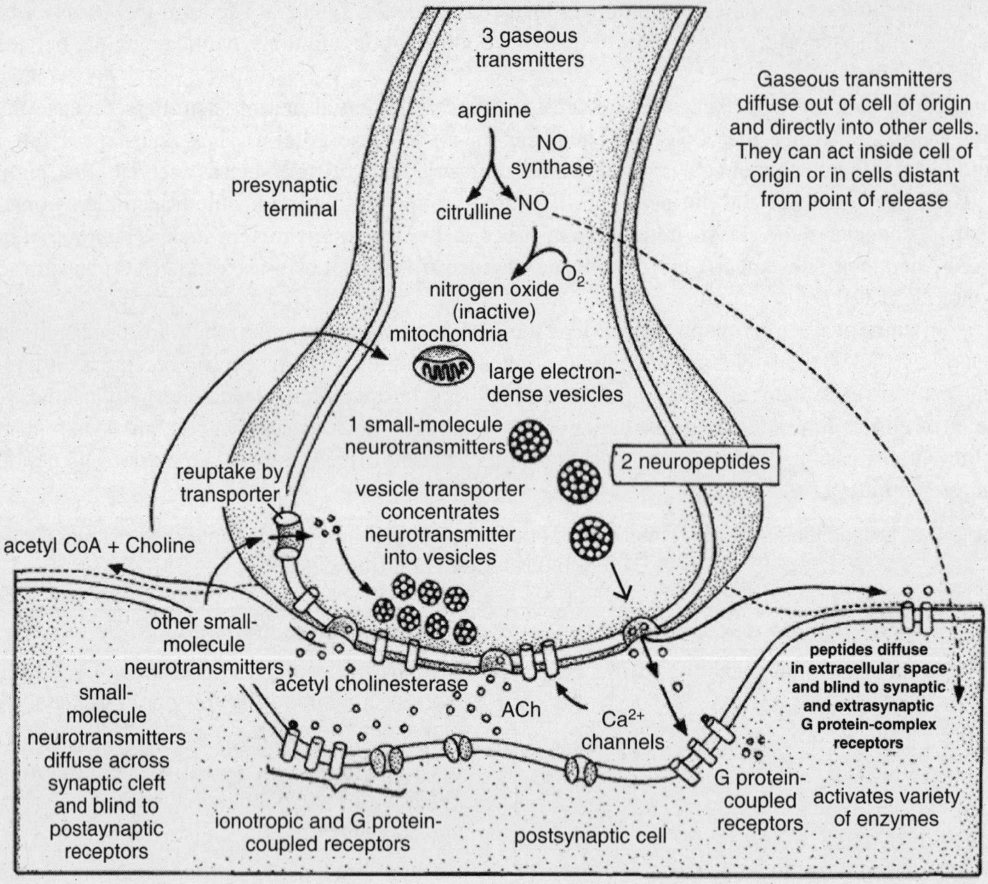

Fig. 3.24. Schematic of a chemical synaptic terminal releasing all three main classes of neurotransmitters. For each, the mechanisms of release, sites of action and mechanism for termination of activity are shown. Real synapses release transmitter from one or more classes.

QUESTIONS

Long Answer Questions

1. Describe the structure of a neuron. Explain the transmission of nerve impulse through the nerve fiber. (*CCSU Meerut 2011*)
2. Discuss the membrane theory (ionic theory) of nerve impulse generation.
3. Describe the transmission of nerve impulse across a synapse.
4. Give the structure of a synapse. Mention the various types of synapses.
5. Give an account of molecular apparatus underlying vesicular release at a chemical synapse.
6. What is nerve impulse? Describe physiology of its conduction through a neuron. (*CCSU Meerut 2014*)

Short Answer Questions

1. Explain the structure of a neuron. (*CCSU Meerut 2010*)
2. What is action potential? How does it propagate along a neuron?

3. Draw neat and labeled diagrams of the following:
   (*i*) Neuron; (*CCSU Meerut 2009*)
   (*ii*) Ultrastructure of a chemical synapse.
4. Differentiate between the following:
   (*a*) Resting potential and action potential;
   (*b*) Cholinergic and adrenergic nerve fibers;
   (*c*) Medullated and nonmedullated nerve fibers;
   (*d*) Small-molecule transmitters and peptide transmitters.
5. What is cholinesterase and what is its function?
6. Give an account of axonal transport.
7. What is nerve impulse? How is it transmitted to the muscles?
8. What is threshold stimulus?
9. Write short notes on the following
   (*i*) Neurosecretion;
   (*ii*) Nerve impulse; (*Bundelkhand 2002*; *Kanpur 2003*)
   (*iii*) Reflex action; (*CCSU Meerut 2002, 2011*; *Kanpur 2003*)
   (*iv*) Synaptic cleft; (*Kumaon 2001*; *Avadh 2003*)
   (*v*) Synaptic transmission;
   (*vi*) Synapse; (*Rohilkhand 2003*)
   (*vii*) Action potential; (*Rohilkhand 2003*)
   (*viii*) Cholinergic nerves;
   (*ix*) Sodium pump;
   (*x*) Acetylcholine;
   (*xi*) Neurotransmitters.

Very Short Answer Questions

1. What is bouton?
2. What is axosomatic synapse?
3. What causes synaptic fatigue?
4. Why chemical synapse is a one-way valve?
5. What is synaptic delay?
6. Give the role of monoamino oxidase in nerve impulse.
7. Which nerve fibers release noradrenaline into the synaptic cleft?
8. Which is narrower? A chemical synapse or an electrical synapse.
9. Name the nerve fibers which show saltatory conduction of nerve impulse.
10. What are gated channels?
11. What is all or none law in nerve conduction. (*Osmania 2005*)

Fill in the Blanks

1. Two special properties of the neuron are ________ and ________.
2. The membrane of an axon having a positive charge on the outside and a negative charge on the inside is said to be ________.
3. End-to-end orientation of the process of two neurons is called ________.
4. The membrane of an axon having a negative charge on the outside and positive charge on the inside is said to be ________.
5. A momentary depolarization of an axonal membrane is known as ________.
6. Conduction (*i.e.*, movement) of action potential along an axon is known as ________.
7. At the chemical synapse, axon releases ________ which sets up a new nerve impulse in the ________.

8. Junction between two neurons is called a ________ whereas the junction between a neuron and a muscle cell is termed a ________.
9. The action potentials are nondecremental, *i.e.*, ________.
10. Nerve impulse travels faster in ________ and ________ nerve fiber.
11. Saltatory conduction of nerve impulse is shown by the ________ nerve fibers.
12. At the synapse, the channels in postsynaptic membrane for Na^+ and K^+ ions are ________ or ________ channels.
13. ________ axonal transport involves the motor protein, called kinesin.
14. ________ axonal transport is derived by the motor protein, called dynein.

Multiple Choice Questions

1. Synapse is a close proximity of two
   (*a*) arteries (*b*) veins
   (*c*) neurons (*d*) lymphatic vessels
2. A nerve impulse leave a neuron through the
   (*a*) dendrite (*b*) cyton
   (*c*) axon (*d*) Nissl bodies
3. The dominating current during the membrane depolarization phase in an axon is
   (*a*) potassium influx (*b*) chloride efflux
   (*c*) calcium efflux (*d*) sodium influx
4. During an action potential, the rapid repolarization of the axon membrane is caused by increased permeability to ions of
   (*a*) sodium (*b*) chloride
   (*c*) calcium (*d*) potassium
5. What is common between acetylcholine, noradrenaline and serotonin?
   (*a*) all lower blood pressure (*b*) all are antidiuretic
   (*c*) all promote appetite (*d*) all are neurotransmitters
6. Myelinated nerve fibers differ from nonmyelinated fibers in
   (*a*) lacking nodes of Ranvier
   (*b*) being without Schwann cells
   (*c*) showing saltatory conduction of nerve impulse
   (*d*) slow conduction of nerve impulse
7. Electrical transmission of a nerve impulse across a synapse
   (*a*) is slower than chemical transmission
   (*b*) requires a neurotransmitter
   (*c*) travels faster than chemical transmission
   (*d*) has to cross a wider synapse than chemical transmission
8. The postsynaptic membrane has
   (*a*) nongated channels (*b*) voltage-gated channels
   (*c*) ligand-gated channels (*d*) no channels
9. Conduction of a nerve impulse is a
   (*a*) biochemical process (*b*) physical process
   (*c*) electro-chemical process (*d*) chemical process
10. When an axon is stimulated the inside of the membrane becomes
   (*a*) negatively charged (*b*) positively charged
   (*c*) filled with acetylcholine (*d*) filled with adrenaline
11. In a nonconducting axon
   (*a*) Na^+ ions predominate in intracellular fluid
   (*b*) Cl^- ions predominate in extracellular fluid

(*c*) K^+ ions predominate in extracellular fluid
(*d*) organic ions are lacking in intracellular fluid

12. Schwann cell and nodes of Ranvier are present in
(*a*) muscles (*b*) cartilages
(*c*) bone cells (*d*) neurons

13. Node of Ranvier are found in
(*a*) cardiac muscle fibre (*b*) striated muscle fibre
(*c*) medullated nerver fibre (*d*) non-medullated nerve fibre

14. Neuroglia cells are
(*a*) astrocytes (*b*) oligodendrocytes
(*c*) microglia (*d*) all the above

15. Nissl granules found in cyton of a neuron and having affinity for basic dyes are composed of
(*a*) DNA (*b*) RNA
(*c*) proteins (*d*) all the above

16. Axons serve to
(*a*) bring impulse to cyton (*b*) take away impulse from cytons
(*c*) bring blood to heart (*d*) none of these

ANSWERS

Very Short Answer Questions

1. Bouton is a terminal club-shaped enlargement of a nerve fiber at a synapse with other neuron. It is also called end foot.
2. A synapse is juction between the axon of one neuron and cyton of another neuron.
3. Depletion of neurotransmitter due to repeated stimulation of presynaptic membrane.
4. Inability of dendrite to secrete a neurotransmitter.
5. Difference in the rates of nerve impulse conduction through an axon and a synapse.
6. Inactivation of noradrenaline.
7. Sympathetic nerve fibers that make final contact with the effector organ.
8. Electrical synapse.
9. Medullated nerve fibers of vertebrates.
10. The channels that open for short periods when needed and then close.

Fill in the Blanks

1. Excitability, conductivity. **2.** Polarised. **3.** Synapse. **4.** Depolarised. **5.** Action potential. **6.** Nerve impulse. **7.** Acetylcholine, dendrite membrane. **8.** Synapse, neuromuscular junction. **9.** Non-decreasing. **10.** Giant, medullated. **11.** Medullated or myelinated. **12.** Chemically-gated, ligand-gated. **13.** Anterograde. **14.** Retrograde.

Multiple Choice Questions

1. (*c*)	**2.** (*c*)	**3.** (*d*)	**4.** (*d*)	**5.** (*d*)	**6.** (*c*)
7. (*c*)	**8.** (*c*)	**9.** (*c*)	**10.** (*a*)	**11.** (*b*)	**12.** (*d*)
13. (*c*)	**14.** (*c*)	**15.** (*b*)	**16.** (*b*)		

4

Physiology of Vision (Photoreception)

4.1 THE SPECIAL SENSES

Animals require a constant inflow of information from the environment to regulate their lives. Sense organs are specialized receptors designed for detecting environmental status and change. An animals sense organs are its first level of environmental perception; they are channels for bringing information to the brain (**Hickman** *et al.*, 1996).

A **stimulus** is some form of energy-electrical, mechanical, chemical, or radiant. The task of sense organs is to transform the energy form of the stimulus it receives into nerve pulses, the common language of the nervous system. In a very real sense, then, sense organs are *biological transducers*. A microphone, for example, is a transducer that converts mechanical (sound) energy into electrical energy. Like the microphone that is sensitive only to sound, sense organs are as a rule, specific for one kind of stimulus. Thus eyes respond only to light, ears to sound, pressure receptors to pressure, and chemoreceptors to chemical molecules. But again, all of these different forms of energy are converted into nerve impulses.

Since all nerve impulses are qualitatively alike, how do animals perceive and distinguish the different sensations of varying stimuli? The answer is that the real perception of sensation is done in localized regions of the brain, where each sensory organ has its own hook up (*i.e.,* link to electronic equipment). Impulses arriving at a particular sensory area of the brain can be interpreted in only one way.

Cephalization and Receptors

The evolution of vertebrates shows a trend called *cephalization* in which special sensory organs develop in the heads of animals, along with the corresponding development of the brain. These special sensory systems, which include the visual, auditory, vestibular, olfactory and gustatory systems, detect and analyze light, sound and chemical signals in the environment, as well as signal the position and movement of the head. The stimuli detected and transduced by these systems are most familiar to us when they provide conscious awareness of our environment, but they are equally important as sensory input for reflexive and sub-conscious behaviour.

4.2 THE VISUAL SYSTEM

Vision is one of the most important special senses in humans and along with audition, is the basis for most human communication. The visual system detects and interprets electromagnetic waves between 400 and 750 nm long, which constitutes **visible light**.

Box 4.1

The eyes of most vertebrates are placed laterally upon the head, permitting a panormic, but monocular (= uniocular) vision. In some limited groups of mammals (*e.g.*, human beings) and some birds, eyes have shifted to a somewhat frontal position so that the two uniocular fields overlap, and some part of the environment directly infront of animals is focussed by both eyes. This is **binocular vision**. Primate mammals such as monkeys, apes and human beings acquired a **stereoscopic** (*i.e.*, 3-dimensional) vision due to their most highly evolved *motor understanding* of visual sensations. In proportion to body size, deers presumably possess largest eyes amongst vertebrates.

The human eye can distinguish two aspects of light, its *brightness* (or luminance) and its *wavelength* (or colour). Light enters the eye and impinges (*i.e.,* has an effect) on *photoreceptors* in a specialized sensory epithelium, the *retina*. The photoreceptors include rods and cones. *Rods* have high sensitivity for detecting low light intensities but do not provide well-defined visual images, nor do they contribute to colour vision. Rods operate best under conditions of reduced lighting (**scotopic vision**). *Cones*, by contrast, are not as sensitive to light as rods are and thus operate best under daylight conditions (**photopic vision**). Cones are responsible for high visual acuity and colour vision.

Information processing within the retina is performed by **retinal interneurons**, and the output signals are carried to the brain by the axons of ***retinal ganglion cells***. The axons travel in the **optic nerves**; there is a partial crossing in the **optic chiasma**, the axons of retinal ganglion cells pass through the **optic tracts** and synapse in nuclei of the brain. The main visual pathway in humans is through the *lateral geniculate nucleus* (*LGN*) of the thalamus. This nucleus projects through the visual radiation to the visual cortex. Other visual pathways project to the **superior colliculus, pretectum** and **hypothalamus**, and these structures participate in orientation of the eyes, control of pupil size and circadian rhythms, respectively.

Structure of Eye

Each human eye is a fluid-filled globular sac-like structure called *eye ball*. It is safely lodged in a socket-like *eye orbit* of skull. A smaller portion of globular eye is exposed in front; the rest of it

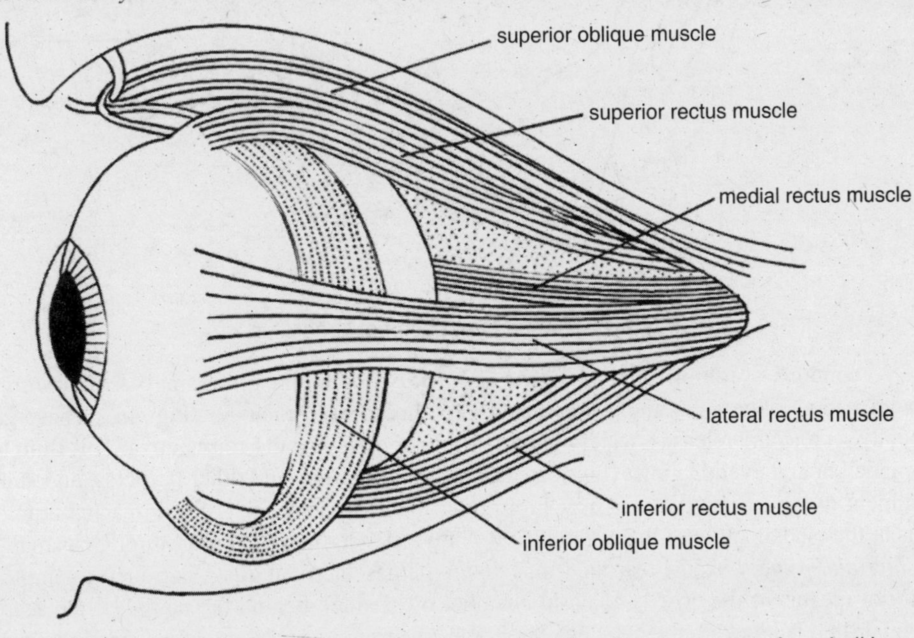

Fig. 4.1. Muscles of the eye ball. Various muscles controlling the movement of eye ball have been shown.

is hidden in the bony socket of the orbit, on a cushion of fat and connective tissue. The eye can turn freely within a certain range vertically, horizontally or obliquely through the actions of three pairs of muscles: two pairs of *retus muscles* and one pair of **oblique muscles** (Fig. 4.1). These eye muscles are innervated by the **oculomotor cranial nerve** (*CN III*), *troclear* (*CN IV*) and *abducens* (*CN VI*) nerves. Several muscles are also found within the eye (intraocular muscles)

1. Eyelids and lacrimal apparatus. The eyelids are two movable curtains in front of the eyeball. Eyelids close reflexly to protect the eye. They are provided with three or four rows of ***eye-lashes*** which serve to trap foreign particles. The eyelids are lined with delicate membrane, the ***conjunctiva***

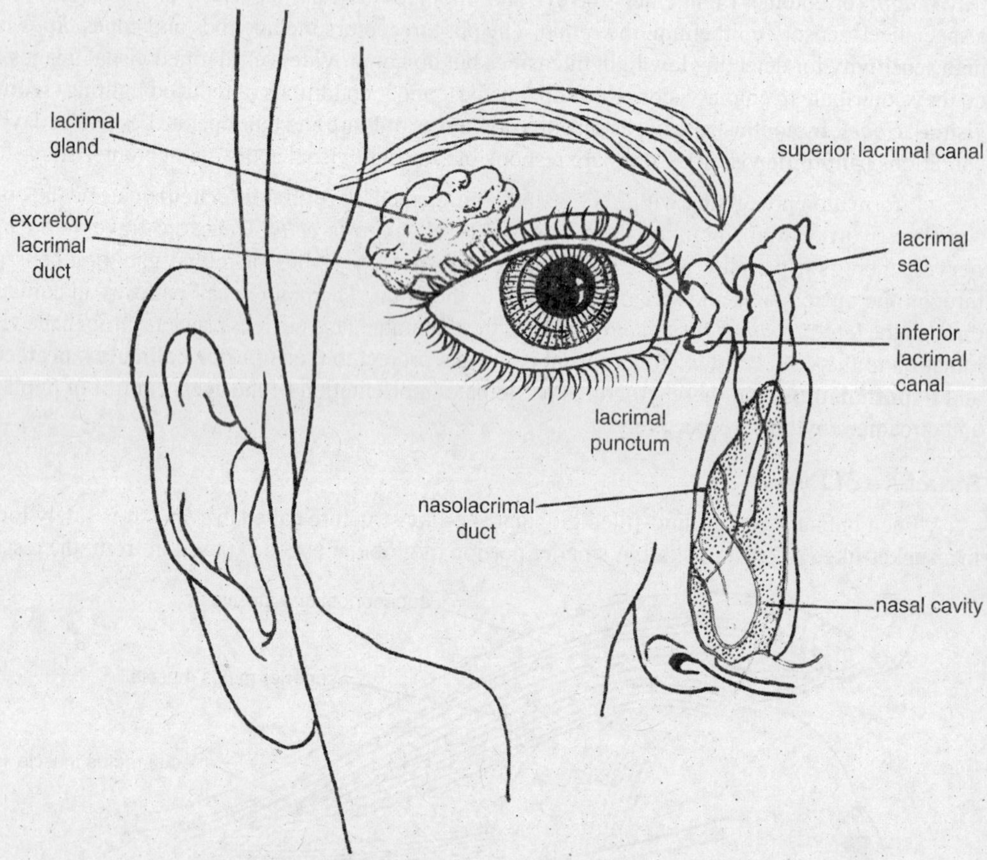

Fig. 4.2. Human eye: Lacrimal apparatus.

Eye contains a complicated *lacrimal apparatus*, including three small ***tear*** or ***lacrimal glands*** located near the outer angle (*i.e.,* upper lateral corner) of each eye (Fig. 4.2). These glands continually secrete and release a watery lacrimal fluid or tear on to the conjunctiva. This fluid keeps the conjunctiva and eyelids moist (lubricated), washes away dust and other particles, and supplies nourishment to the nonvascular cornea. From over the conjunctiva, this fluid (tear) continuously collects at the medial angle of the eye and flows into a slender **lacrimal duct** through a small pore called ***lacrimal punctum***. The tear then runs down the nasolacrimal duct (tear duct) to empty into the inferior meatus of the nose (*i.e.,* nasal chamber of its side), behind the nostril. If the secretion of lacrimal fluid is blocked, the delicate membrane covering the eye soon dries up and the eye gets inflammed and finally damaged.

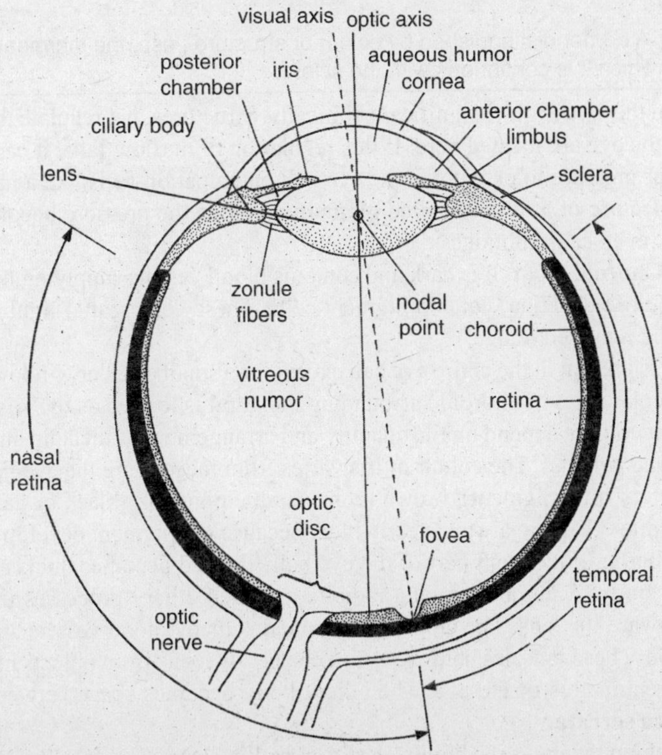

Fig. 4.3. View of a horizontal section of the right human eye (after **Koeppen** and **Stanton** 2010).

A series of small **meibomian glands** are located in eyelid perpendicularly to its free border. These glands secrete an oily substance, a thin film of which is always spread over the border of each eyelid. It prevents the tear from falling upon the cheeks instead of conjunctiva. Injury, falling of some object upon conjunctiva, smoke or irritating vapours, emotional upsets, etc., stimulate ample secretion of tear which, then overflows the cornea on to the face. In human beings, secretion of tear begins about four months after birth.

2. Coats of the eye. Each eyeball is composed of three coats or layers:

(*i*) **The sclerotic.** it is outer most layer (Fig. 4.3)

(*ii*) **The choroid.** It is a middle vascular layer.

(*iii*) **The retina.** It is the delicate and intricate layer. It is the light sensitive layer of the eye.

I. The sclerotic coat. It is outer most hard coat (= layer) which gives shape to the eye and prevents the inner parts from minor injuries. The part of the opaque sclerotic coat visible from outside is known as **white of the eye**. Sclerotic coat of the eye is formed by the tough fibrous layer of connective tissue. Tendons of the oculomotor muscles insert into the sclera. The connective tissue of the sclera is rich in collagenous fibers which run in different directions and have fibroblasts. Numerous elastic fibers are also present.

In the centre of white of eye, there is a transparent and round window known as **Cornea**. The infront of the cornea is highly curved and convex. The cornea has two properties : the high refractive index (1.38) and its small radius of curvature. These make the cornea extremely important in image formation.

The **transparency** of cornea is due to regular arrangement of its fibrous elements. The fibers in turn are probably kept in close register by the glycosaminoglycans in the cornea. The cornea is vascular and receives nutrition by diffusion from the aqueous humour (= humor) at the centre and blood vessels from the **limbus** (*i.e.*, corneal-scleral junction) at the periphery. The anatomy of limbus.

Box 4.2

Limbus. A border distinguished by colour or structure ; esp. the marginal region of the cornea of the eye by which it is continuous with the sclera.

is complex and the area is quite significant clinically. Structures that regulate the outflow of aqueous humour from the eye are located here. If this regulation of outflow fails, it can cause an increase in the intraoccular pressure to extent that the eye ball on palpation feels like a nut. This condition is a characteristic feature of a disease called **glaucoma**. When the pressure becomes abnormally high, glaucoma may even cause blindness.

II. The Choroid Coat. It is dark and contains blood vessels supplying nutrients to the eye. Its inner surface (close to retina) contain pigment cells. These cells scatter light and then increase the sharpness of the focussed image.

Choroid is present in the entire eye ball except for a small circular window-like opening, called **pupil**. The circular band of choroid surrounding the pupil is termed as the **iris**. The iris is coloured and its different colours depend on the quantity and arrangement of melanin and the thickness of the connective tissue lamellal. The colour of iris varies from race giving the eye its characteristic shade such as blue (very little pigment), brown (considerable pigment), black or hazel eyes. The eyes of babies shortly after birth are always nearly blue, because the pigment develop somewhat later.

Ciliary body. In the front part of the eye ball, choroid becomes thicker and forms the round ruffled **ciliary body.** A number of long radially arranged ciliary processes project from the thick ciliary body toward the lens. The ciliary body contains many blood vessels and **ciliary muscles** of accommodation. These muscles help the eye to change its focus from objects in the distance to those within a few centimetres of the eye. The ciliary body occupies space between the edge of visual retina called **ora serrata**.

Lens. Behind the iris is the highly elastic crystalline lens (Fig. 4.3). It is made up of intricately interlacing fibers and covered with a capsule to which fine suspensory ligaments are attached which in turn attach it to the ciliary body. Lens is located approximately 2 mm behind the cornea. It is biconvex but its rear surface is more convex than the front. The lens is perfectly transparent. However, due to disease or abnormality, it may become opaque or semi-transparent and thus cause blindness. Such a condition is called **cataract**. The lens divides the eye into two fluid filled chambers: 1. **Aqueous humour** filled **anterior** and **posterior chambers** (Box. 4.3) and **vitreous humour**. The aqueous humour is an extracellular fluid and is secreted by the epithelium of the ciliary body into the posterior chamber of the eye. It then circulates through the pupil and into the anterior chamber (*i.e.,* spaces of Fontana), where it is drained into the venous system by the **canal of schlemm** (Fig. 4.4). Aqueous humour pressure.

Box 4.3

Composition of Aqueous Humour

1. Aqueous humour is a thin watery fluid with pH 7.1 - 7.3 and S.G. 1002-1004. Being on ultrafiltrate, it is low in protein and is formed by *diffusion* from capillaries (of ciliary processes) and by *active transport* of components from the plasma.
2. It contains high content of *vitamin C*, which hels in glucose metabolism.
3. It has high concentration of NaCl because Na^+ ions are actively transported from the plasma and Cl^- ions follow them passively; accumulation of these ions results in passive transport of water due to osmosis.
4. Aqueous humour is low in *glucose* but high in *lactic acid* content because *glucose* is utilized by a vascular cornea and lens anaerobically.
5. It has high amount of *hyaluronic acid* which is kept in the fluid state by hyaluronidase present in the ciliary body.

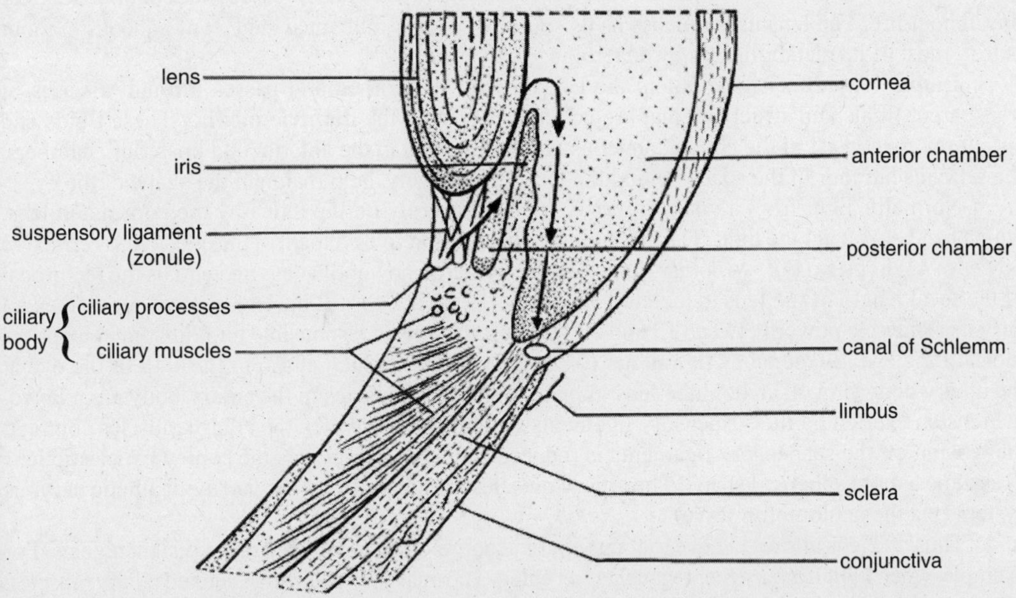

Fig. 4.4. Human eye. The Circulation of aqueous humour.

which is normally less than 22 mm Hg, determines the pressure within the eye. Canal of Schlemm is a circular canal lying in the substance of the sclerocorneal junction (at the limbus) of the eye and draining the aqueous humour from the anterior chamber into the veins (ocular veins) draining the eyeball. This keeps a continuous circulation of aqueous humour around the lens (Fig. 4.5).

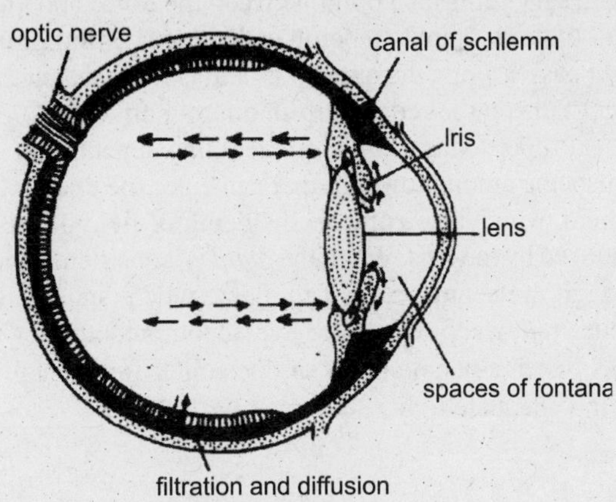

Fig. 4.5. Human eye. Diagrammatic representation of the fluid system of eye.

The vitreous humour is a gel composed of extracellular fluid that contains collagen (protein) and hyaluronic acid ; unlike aqueous humour, however, is turn over very slowly.

The diameter of lens of humans is about 7 mm. Depending on the degree of contraction or relaxation of the ciliary muscle, it is 3.7 - 4.5 mm thick. The lens proper is made up of lens fibers which are highly modified epithelial cells. Each of these fibers is a hexagonal prismatic structure that is 8-12 μm wide, about 2 μm thick and 7-10 μm long. Each lens fiber contains a nucleus, but is relatively devoid of organelles and intracellular inclusions. The plasma membranes of adjacent-lens fibres are separated by intercellular clefts 15 μm wide and these membranes are elaborately interdigitated. The substance of the lens has a very high protein context (about 35%). Human lens is made of the transparent lens - specific proteins called **Crystallins** (**Gilbert** 2010). The susmensory ligaments of lens are kept tense by the intra-ocular pressure, which tends to increase the diameter of annular ciliary body.

Although the lens has no blood supply, it is metabolically active tissue that continue to grow throughout life. The lens uses glucose as its source of energy, but since the Po_2 of aqueous humour is lost, most of its metabolism is by anaerobic glycolysis.

Dioptric media. Light enters the eye through the cornea and passes through a series of transparent fluids and structures that are collectively called the **dioptric media**. These fluids and structures are consist of the cornea, aqueous humour located in the anterior and posterior chambers, the vitreous humour in the space behind the lens, respectively, help maintain the shape of the eye.

Normally, light from a visual target is focussed sharply on the retina by the cornea and lens, which bend or **refract** the light. The cornea is the major refractive element of the eye, with a refractive power of 43 diopters (D). (A *diopter* describes the refractile power of a lens and equals the reciprocal of the focal length of the lens in meters). However, unlike the cornea the lens can change shape and vary its refractive power between 13 and 26 D. Thus, the lens is responsible for adjusting the optical focus of the eye. **Suspensory ligaments** (or **Zonule fibers**), which attach to the wall of the eye at the ciliary body (Fig. 4.3), hold the lens in place. When the muscles in the ciliary body are relaxed, the tension exerted by the suspensory ligaments flattens the lens when the ciliary muscles contract, the tension on the suspensory ligaments is reduced ; this process allows the somewhat elastic lens to assume a more spherical shape. The ciliary muscles are activated by the parasympathetic nervous system (via the oculomotor nerve).

Thus, the lens allows the eye to focus on, or accommodate to, either near or distant objects. For example, when light from a distant visual target enters a normal eye (one with a relaxed ciliary muscle), the target is focussed behind the retina (*i.e.*, the image at the retina is blurred) until **accommodation** occurs. The ciliary muscle contracts and the zonule fibers relax; the image is sharpened when the convexity of the lens increases as a result of these muscular changes.

Although the optic axis of the human eye passes through the nodal point of the lens and reaches the retina at a point between the fovea and the optic disc (Fig. 4.3), the eye is directed by the oculomotor system to a point, called the **fixation point**, on the visual target. Light from the fixation point passes along the optic axis, and is focussed on the **fovea**. Light from the remainder of the visual target falls on the retina surrounding the fovea.

Proper focus of light on the retina depends not only on the lens but also on the **iris**, which also adjusts the amount of light that can enter the eye. In this respect the iris acts like the diaphragm in a camera, which also controls the depth of field of the image and the amount of spherical aberration produced by the lens. When the *pupil* is constricted, the depth of field is increased, and light is directed through the central part of the lens, where spherical aberration is minimal. Pupillary constriction occurs *reflexively* when the eye accommodates for near vision or adapts to bright light, or both. Thus, when a person reads or does ohter fine visual work, the quality of the image is improved by having adequate light.

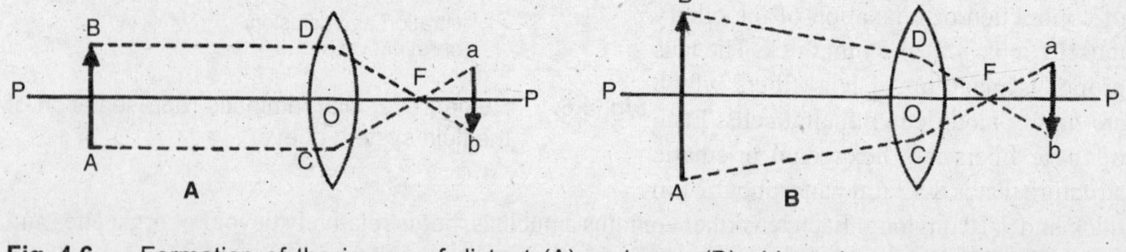

Fig. 4.6. Formation of the image of distant (A) and near (B) objects in an eye. AB = object ; AC, BD = light rays; PP = principal axis; O = optical centre of lens; F = principal focus; OF = focal distance of lens ; *ab* = image.

Inverted image. The line *AB* (Fig. 4.6) representing an object is made up of innumerable points, each reflecting a light ray independently. For each point in the object, there will be a corresponding point in the image *ab*. Because each image point appears on the opposite side of the principal axis due to *refraction* from that of its corresponding object point, the image is always **inverted**. When the sensory impulses of this image are carried to the brain by fibers of the optic nerve, the nerve centres of sight, located in thalamus and cerebral cortex, analyse and coordinate these impulses so that the human beings become aware of the normal erect pattern of the object, *i.e.*, it actually "sees" the object. Thus, it is the brain and not the eye, that is responsible for "seeing".

4.3 DEFECTS OF IMAGE FORMING MECHANISM

1. **Presbyopia** (*Means loss of accommodation*).
   The ability of the eye to accommodate decreases with advancing age due to increasing sclerosis (hardening) of the lens substance. As a result the near and far points of vision become so close to each other that reading becomes difficult. Thus, the aging can no longer accommodate for both far and near vision.

The loss of elasticity of lens is due to denaturation of its proteins by the ultra violet rays which are being absorbed by the lens. As a result of such irradiation the protein agglutinate and coagulate in the presence of Ca^{2+}. Eventually, the lens becomes swollen, hard and opaque (called **cataract**). It can be *corrected* by wearing glasses with *bifocal lens* (The upper segment focussed for distant objects and the low segment for near vision).

2. **Myopia or Short-sightedness**
   Myopia is genetic in origin. In this condition the parallel rays of light from distant object are focussed in front of the retina (Fig. 4.7A). Myopia is because :
   (*i*) either length of the eyeball is too long (called **axial myopic eye**), or
   (*ii*) refractive power of lens increases (**called refractive myopic eye**).

Characteristic features of myopia

(*i*) A person *cannot see the distant object*, thus, called **near-sightedness** or **short-sightedness**.

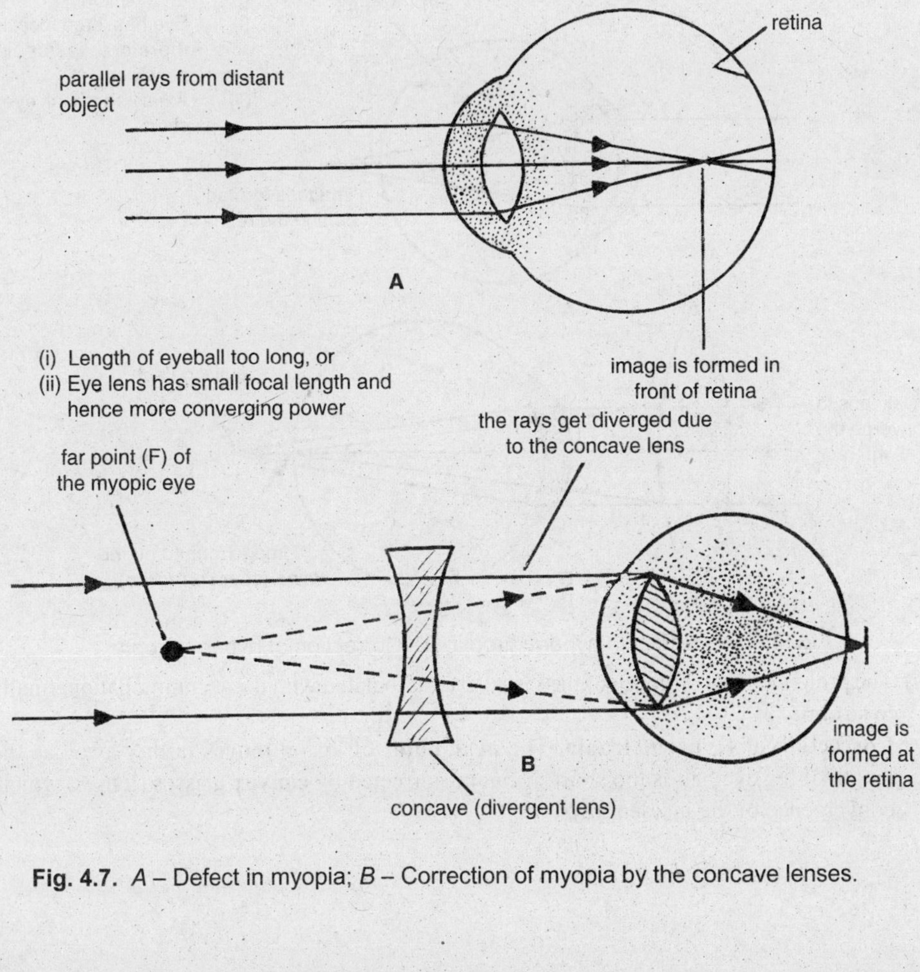

Fig. 4.7. *A* – Defect in myopia; *B* – Correction of myopia by the concave lenses.

(*ii*) **Far point** of vision is at a definite distance from the eye ; may be less than 1 metre in severe myopia. Normally it is at infinity, more than 6 metres (20 feet)

(*iii*) The **near point** of vision becomes nearer (close to the eye) as axial length of eyeball increases.

(*iv*) As a result of (*ii*) and (*iii*), 'far point' and 'near point' of the vision are closely approximated, therefore, the **range** of **accommodation** decreases.

Correction of myopia. Myopia can be corrected by concave glasses (lenses) which causes divergences of incident rays The power of the lens required gives a measure of the degree of myopia.

3. **Hypermetropia (or Hyperopia) or Long-sighted.** In this condition parallel light rays from distant object are focussed behind the retina (Fig. 4.8A). Long-sightedness is because:

(*i*) either length of the eyeball is too short (**axial hypermetropic** eye) or

(*ii*) refractive power of the lens decreases (**refractive hypermetropic** eye)

Characteristic features of hypermetropia.

(*i*) A person can see the distant objects clearly only while using some accommodation, thus, called '**far-sightedness**' or '**long sightedness**'.

(*ii*) Hypertrophy of the ciliary muscle occurs because individual will be using, the accommodation all the times for seeing the far objects. Sustained accommodation (*i.e.*, prolonged muscular effort) is tiring and may cause severe headache and blurring of vision.

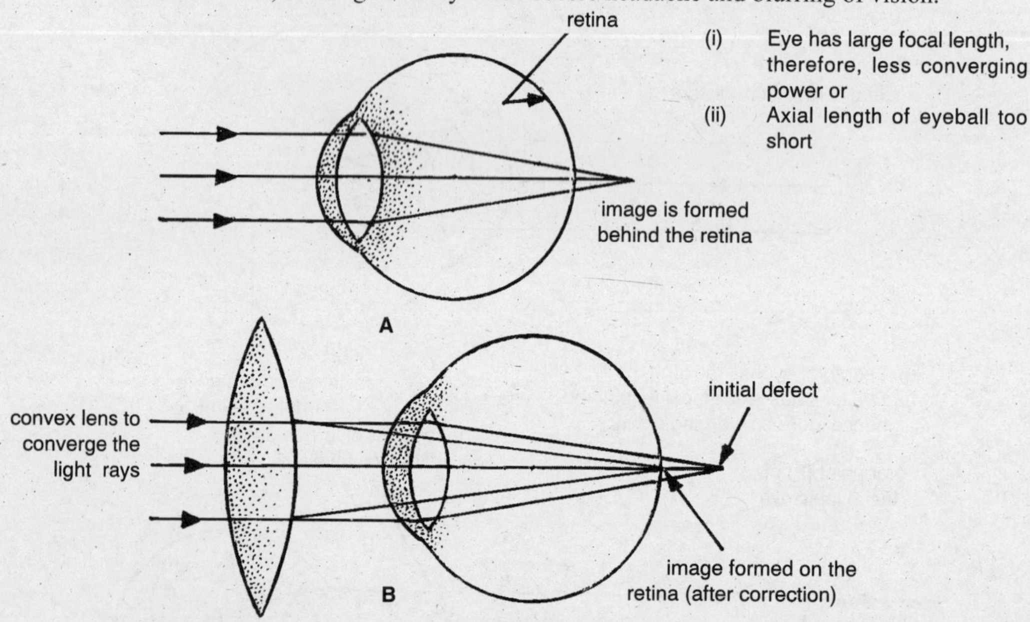

Fig. 4.8. A - Defect in hypermetropia; B - Correction of hypermetropia.

(*ii*) The prolonged convergence of the visual axes associated with the accommodation, finally leads to **squint**.

Correction of Hypermetropia. The **near point** of vision moves farther away as the axial length of the eye ball is too short. It can be corrected by **convex** glasses (lenses) which cause convergence of the incident rays.

4. **Astigmatism**

In astigmatism, a asymmetry exists in the radii of curvature of different meridians of the cornea or lens (or sometimes of the retina). Astigmatism can often be corrected with lenses that posses, appropriately matched radii of curvature.

III. Retina

A. The retina contains 10 layers except in optic disc and fovea centralis. The **photoreceptors** (rods and cones) are placed outermost towards the choroid. The layers of retina from outwards to inwards are the following (Fig. 4.9).

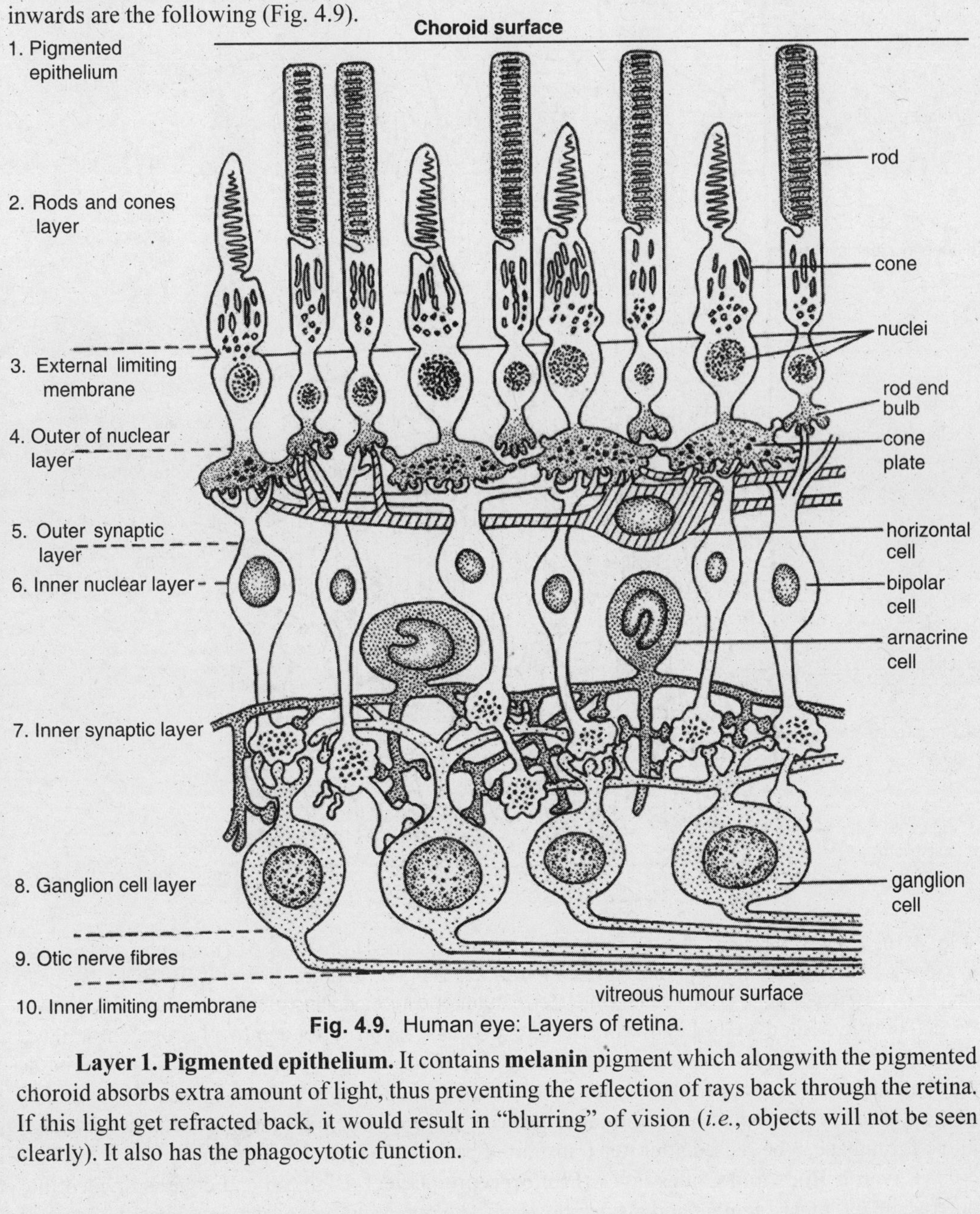

Fig. 4.9. Human eye: Layers of retina.

Layer 1. Pigmented epithelium. It contains **melanin** pigment which alongwith the pigmented choroid absorbs extra amount of light, thus preventing the reflection of rays back through the retina. If this light get refracted back, it would result in "blurring" of vision (*i.e.*, objects will not be seen clearly). It also has the phagocytotic function.

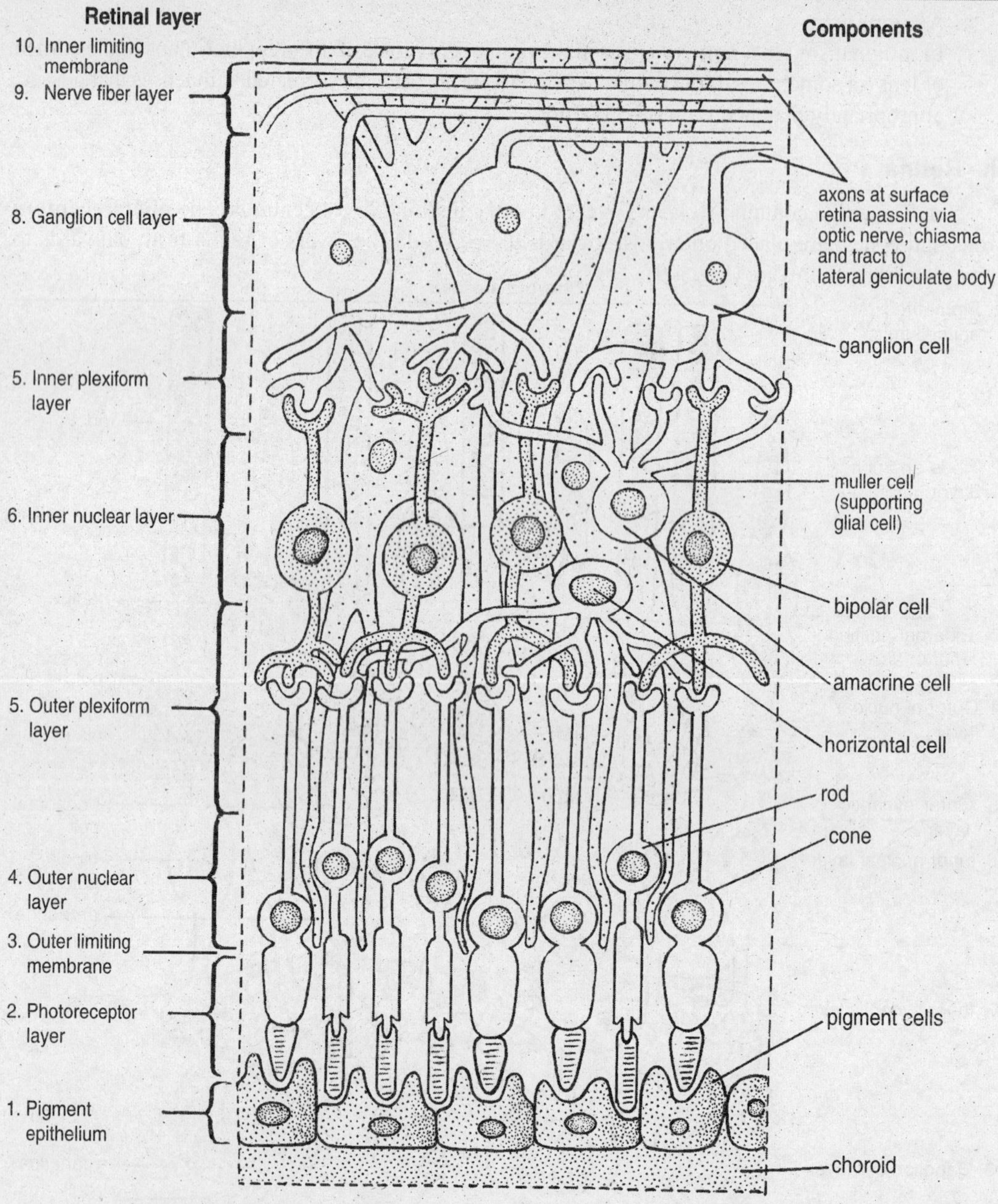

Fig. 4.10. Layers of retina. Light impinging on the retina comes from the top of the figure and passes through all the superficial layers to reach the photoreceptor rods and cones. Muller cells (*i.e.* supporting glial cells) which form outer and inner limiting membranes are also shown.

The pigmented cells have tentacle-like processes that extend into photoreceptor layer (Fig. 4.10) and surround the outer segments of the rods and cones. Thus, pigmented layer can (*i*) provide nutrients and remove waste from the photoreceptors ; (*ii*) phagocytose the ends of the outer segments of the rods and cones which are continuously shed ; and (*iii*) reconvert metabolized photopigment into a form that can be reused after it is transported back to the photoreceptors.

Layer 2. Rods and Cones layer (Photoreceptor layer). Each rod and cone is divided into outer segment, inner segment and a synaptic zone. The outer and inner segments form the layer of rods and cones.

(i) The outer segment.

(*a*) The outer segments of rods and cones are modified cilia and are made up of regularly arranged biles of fattened discs or saccules composed of membranes. In cones, the **saccules** are formed by infolding of the cell/plasma membrane but in rods, the **discs** are separated from the plasma membrane.

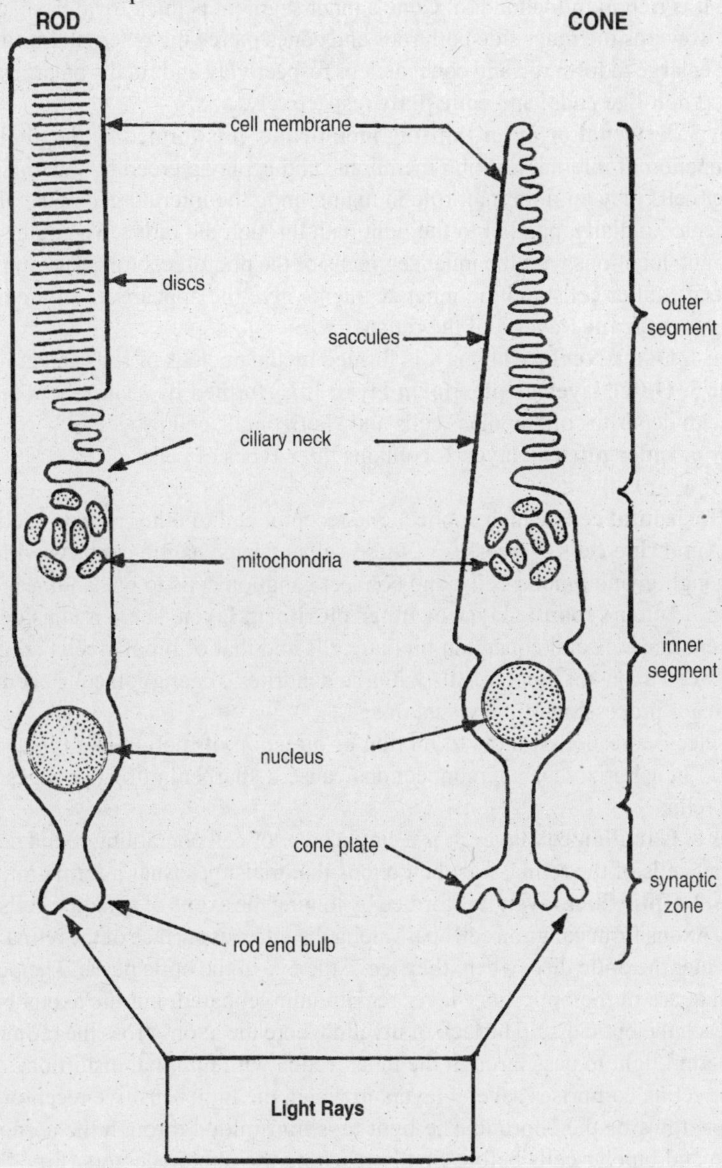

Fig. 4.11. Rods and cones of human eye. Electron microscopic view.

(*b*) The discs or saccules contain the **photosensitive pigment** (*i.e.*, **rhodopsin** in rods and iodopsin in cones). The rods are extremely sensitive to light and are receptors for **dim-light vision** (**scotopic vision**); while the cones are responsible for **bright light** (**photopic vision**), high visual activity and for colour vision.

(*c*) **Rods** are named for the thin, rod-like appearance of their outer segments (Fig. 4.11). Its outer segments are being constantly renewed by formation of new discs at the inner edge of the segments and phagocytosis of old discs from the outer tip by cells of the pigment epithelium.

(*d*) **Cones** generally have conical outer segment, its renewal is more diffused process and appears to occur at multiple sites in the outer segment.

(ii) The inner segment

(*a*) It is rich in mitochondria. Cone's inner segment is thick, oval in shape and is larger.

(*b*) Towards the inner side both rods and cones pierce the external limiting membrane, then enlarge to form rod and cone nucleus respectively and finally enlarges into **rod and bulb** (knob-like ends) and **cone plate** respectively.

Layer 3. External or outer limiting membrane. It is formed by the glial or Muller cells. It is the continuation of internal limiting membrane and is play pierced by the rods and cones.

Muller cells play an important role in maintaining the internal geometry of the retina. Muller cells are oriented radially, parallel to the light path through the retina. The outer ends of the Muller cells form tight junctions with the inner segments of the photoreceptors. The numerous connections made between Muller cells and the inner segments give the appearance of a continuous layer, the **outer lining membrane** (layer 3 of the retina).

Layer 4. Outer Nuclear layer. It is formed by the nucleus of rods and cones.

Layer 5. Outer layer or plexiform layer. It is formed by snapse between the ends of rods and cones with dendrites of "bipolar" cells and "horizontal" cell processes.

Layer 6. Inner nuclear layer. It contains three types of cells.

(i) Bipolar cells

(ii) Horizontal cells which connect one receptor cell to other receptor cell; and

(iii) Amacrine cells. Processes of these cells make synaptic contacts with dendrites of both ganglion and bipolar cells, and connect ganglion cells to one another.

Layer 7. Inner synaptic layer or inner plexiform layer. There is considerable *convergence* of photoreceptors (rods and cones) on bipolar cells and that of bipolar cells on ganglion cells. The synapse between axons of bipolar cells with the dendrites of ganglion cells occur in this layer. It is the site of major processing of the visual image.

The selective responsiveness to motion or direction of motion is a function of the amacrine cell system. Their horizontal interconnections cause a sharpening of the edges of any stimulated field on the retina.

Layer 8. Ganglion cell layer. It is a single layer of cell containing round cells. Ganglion cells are the output cells of the retina; it is their axons that transmit visual information to brain.

Layer 9. Optic fiber layer. It is formed by joining the axons of ganglion cells; here all the axons run parallel. Axons from ganglion cells pass along the vitreous surface of the retina while avoiding the fovea, and enter the optic disc, where they leave the eye in the optic nerve. The portions of ganglion cell axons that are in the optic fiber layer remain unmyelinated, but the axons become myelinated after they reach the optic disc. The lack of myelin where the axons cross the retina is a specialization that helps permit light to pass through the inner retina with minimal distortion.

B. The retina comprises several layers in which the light sensitive receptors (rods and cones) are outermost towards the choroid. The light rays transmitted through the cornea have to traverse the ganglion and bipolar cells before finally reaching the photoreceptors (Fig. 4.12). The ganglion cells and their fibers cannot themselves function as light receptors, since the point of entrance of the optic nerve (optic disc) is insensitive to light and constitute the **blind spot**.

C. The fovea contains no rods, and each foveal cone has a single bipolar cell connecting it to one ganglion cell. Thus each foveal cone is connected to a single fiber in the optic nerve. To recall, in fovea, cones are densely packed, so fovea is a point of greatest visual acuity. In peripheral portion of retina rods predominate ; rods and cones often converge on the same ganglion cell.

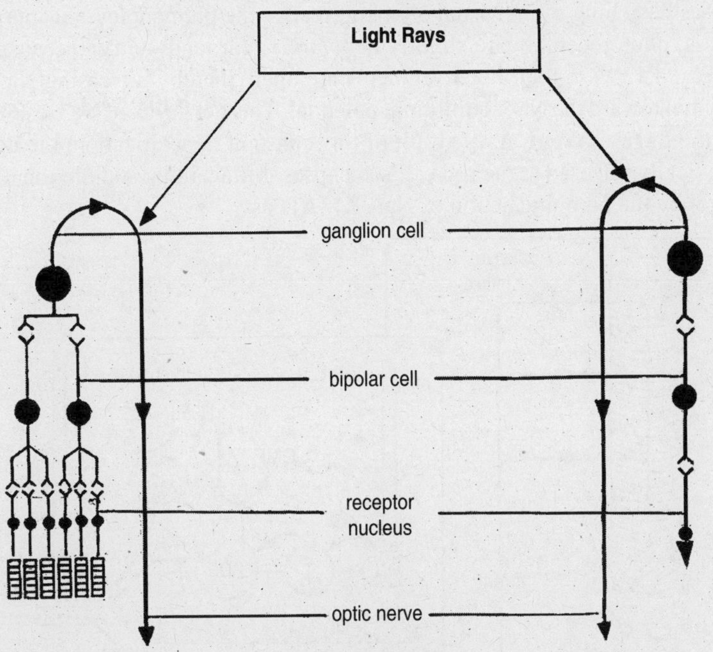

Fig. 4.12. Diagram showing how the light rays reach the photoreceptors. Note that one optic nerve fiber is serving several rods (on the left) while serving a single cone (on the right)

D. There are about 6-7 million cones and 120-125 million rods in each human eye, but only 0.8-1.2 million nerve fibers in each optic nerve; therefore, **overall convergence of receptors** through bipolar cells on ganglion cells is approximately 105-106 : 1.

The retinal blood vessels supply the bipolar and ganglion cells, but the photoreceptors (rods and cones) are nourished for the most part by the capillary plexus in the choroid.

Chemistry of Vision (Visual Transduction)

The rods and cones change the light energy into nerve impulses. Vitamin A plays an important role in this transformation and is the basic chemical which is utilized by both rods and cones for synthesizing substances which are sensitive to light. In the rods vitamin A is converted into a substance known as **retinine**. The retinine then combines with a protein **opsin** present in the rods to form a light sensitive pigment - **rhodopsin or visual purple** (so named because it has a purple appearance after green or blue light has been observed). Rhodopsin absorbs light best at a wavelength of 500 nm. (Box 4.4).

Box 4.4

Night Blindness or Nyctalopia

Rhodopsin contains a chromophore, called **retinal**, that is the aldehyde of **retinol** or **vitamin A.** Retinol is derived from carotenoids, such as β-carotene, the orange pigment found in carrots. Like other vitamins, retinol cannot be synthesized by humans ; instead, it is derived from food sources. Individuals with a severe vitamin A deficiency suffer from "night blindness", a condition in which vision is defective in low light situations.

Thus, rhodopsin is formed when a retinal isomer II-Cis *retinal*, is combined with a glyco-protein known as **opsin.** When **rhodopsin** absorbs light, it is "boosted" to a higher energy state. This boost causes a series of chemical changes that lead to isomerization of II-*cis* retinal to all *trans* retinal, release of the bond with opsin, and conversion of retinal to **retinol**. Separation of all-*trans* retinal from opsin causes bleaching of the visual pigment; that is, the pigment loses its purple colour.

In darkness, photoreceptors are slightly *depolarized* (around 40mV) because cGMP - gated Na^+ channels (Fig. 4.13A) in their outer segments are open, thereby increasing Na and driving the membrane potential toward the Na^+ equilibrium potential. This net influx of Na^+ result in a continuous current, called the **dark current**. As a result of this constant depolarization, the neutrotransmitter **glutamate** is tonically released at the rod cell's synapses. Intracellular sodium ion (Na^+) is kept at a steady-state level by the pumping action of Na^+, K^+- ATPase.

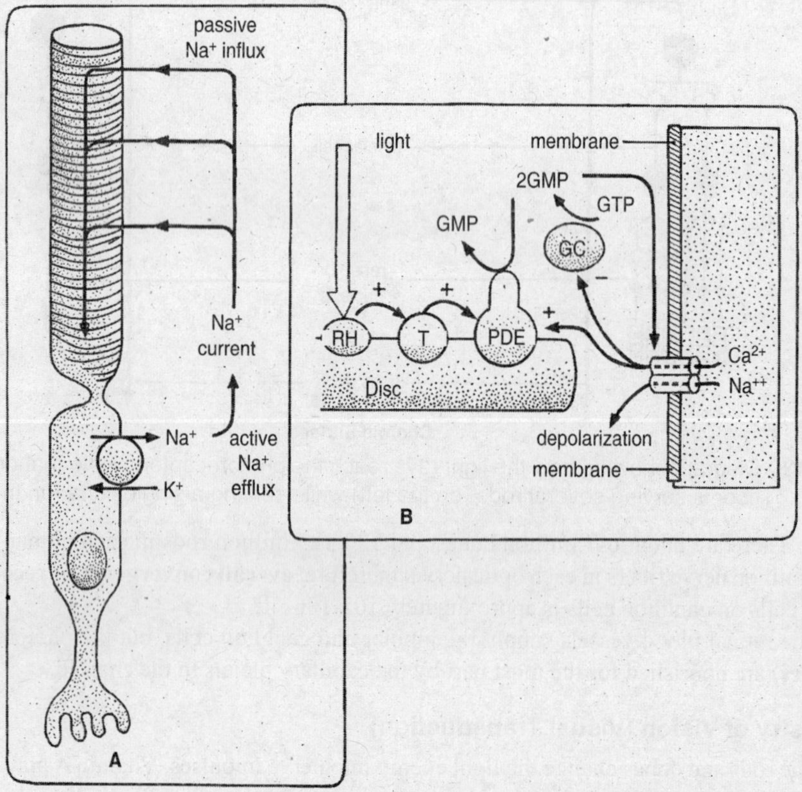

Fig. 4.13. Visual transduction is rods. *A* - The flow of current in dark is indicated, as well as the Na^+ pump. *B* - Sequence of the second messenger events that follow the absorption of light cGMP, cyclic guanosine monophosphate ; GC, guanylate cyclase ; GTP, guanosine triphosphate ; PDE, phosphodiesterase ; Rh, rhodopsin ; T, transduction (after Koeppen and Stanton, 2010).

When light is absorbed, the photo-isomerization of the rhodopsin activated a *G-protein* called **transducin** (Fig. 4.13*B*). This *G* protein, in turn, activates **cyclic guanosine monophosphate phosphodiesterase**, which is associated with the rhodopsin-containing discs, hydrolyzes cGMP to 5′-GMP and lowers the cGMP concentration in the rod cytoplasm. The reduction in cGMP leads to closing of the cGMP-gated Na^+ channels, hyperpolarization of the photoreceptor membrane, and a reduction in the release of transmitter. Thus, cGMP acts as a **second messenger** to translate reception of a photon by the photopigment into a change in membrane potential.

The extraordinary sensitivity of rods, which can signal the capture of a single photon, is enhanced by an amplification mechanism such that photoactivation of only one rhodopsin molecule can activate hundreds of transducin molecules. In addition, each phosphodiesterase molecule hydrolyzes thousand of cGMP molecules per second. Similar events occur in cones, but the membrane hyperpolarization occurs much more quickly than in rods, and requires thousands of photons. Thus, in all photoreceptors, capture of light energy leads to (i) hyperpolarization of the photoreceptor and (ii) a reduction in the release of transmitter. Note that with the very short distance between the site of transduction and the synapse, this transmitter modulation is accomplished without the generation of an action potential.

Colour Vision

Colour perception is the work of the cones of the retina. The three visual pigments in the outer segments of the cones have opsins that differ from the opsin found in rhodopsin. As a result of these differences, the three types of cone pigments absorb light best at different wavelengths. Although the cone pigments have maximum efficiency closer to violet, green and yellow wavelengths (Box 4.5), they are referred to as **blue**, **green** and **red pigments**, respectively (Fig. 4.14).

According to **trichromacy theory**, these differences in absorption efficiency are presumed to account for colour vision because a suitable mixture of three colours can produce any other colour. However, a neural system must also exist for the analysis of colour brightness because the amount of

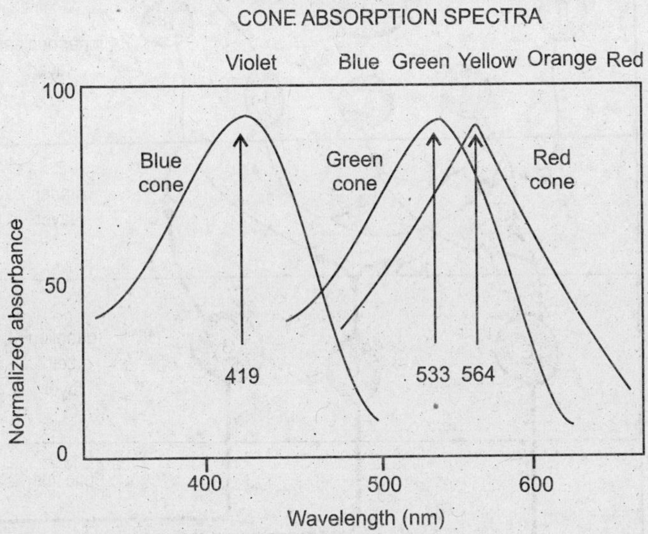

Fig. 4.14. The spectral sensitivity of the three types of cone pigments in the human retina is shown. Note that the curves overlap and that the so-called blue and red cones actually absorb maximally in the violet and yellow range, respectively.

light absorbed by a visual pigment, as well as the subsequent response of the cell, depends on both the *wavelength* and the *intensity* of the light (Fig. 4.14). Two or three of the cone pigments may absorb a particular wavelength of light, but the amount absorbed by each will differ according to their efficiencies at that wavelength. If the intensity of the light is increased (or decreased), all will absorb

Box 4.5

For the detection of different colours, different types of cones are present in the eye. In lower animals, it has been seen that there are seven types of cones which respond to the seven colours of the spectrum. In humans there are only three types of cones, showing maximum sensitivity to blue, green and red.

more (or less), but the ratio of absorption among them will remain constant. Consequently, there must be a neural mechanism to *compare* the absorption of light of different wavelengths by the different types of cones for the visual system to distinguish different colours. At least two different kinds of cones are required for colour vision. The presence of three kinds decreases the ambiguity in distinguishing colours when all three absorb light, and it ensures that at least two types of cones will absorb most wavelengths of visible light.

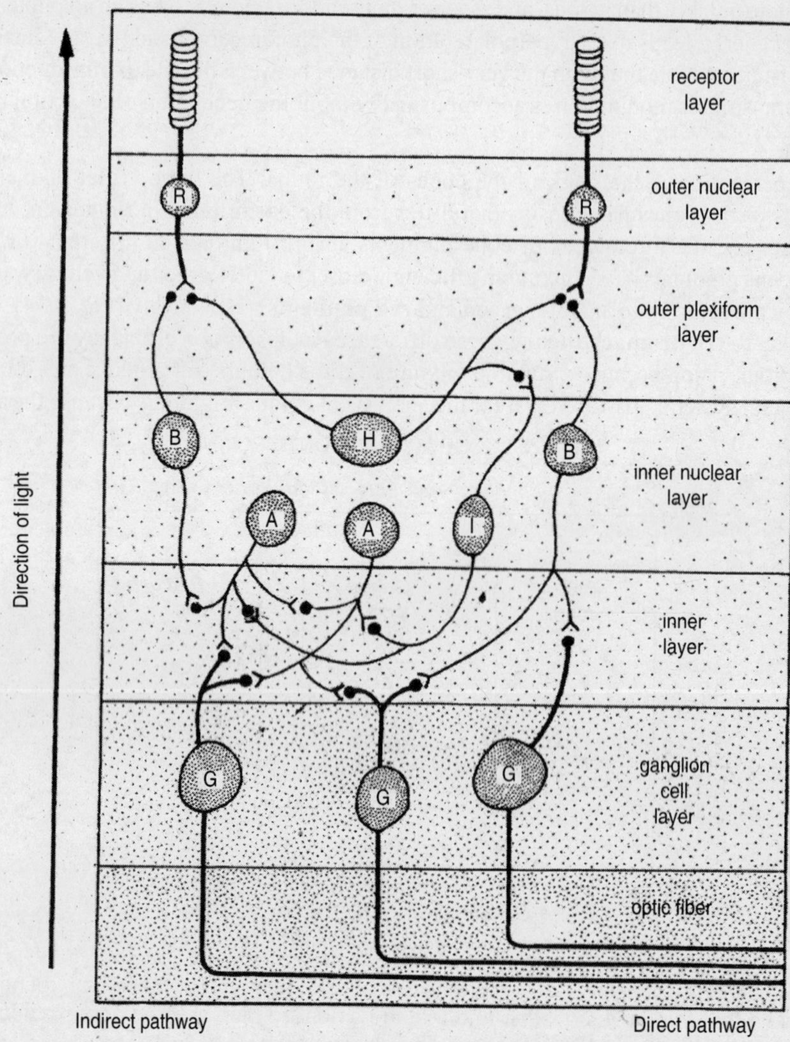

Fig. 4.15. Basic retinal circuitry. The arrow at the left indicates the direction of light through the retina. A–amacrine cells ; B–bipolar cells ; G–ganglion cells ; H–horizontal cells ; I–interplexi- form cells ; R–photoreceptors.

The **opponent process theory** is based on observation that certain pairs of colours seem to activate opposing neural processes. Green and red are opposed, as are yellow and blue, as well as black and white. For example, if a grey area is surrounded by a green ring, the grey area appears to acquire a reddish colour. Furthermore, a greenish red or a bluish yellow colour does not exist. These observations are supported by findings that *neurons activated by green are inhibited by red.* Similarly, neurons with these characteristics are found both in the retina and at higher levels of the visual pathway and seem to serve to increase our ability to see that contrasts between opposing colours.

Retinal Circuitry (The Visual Pathway)

The visual fibers arise in the layer of nerve cells (*bipolar* and *ganglion cells*) in the retina. Sensory pathway from the eye to the cerebral cortex in a 3-order neuron pathway viz., primary, secondary and tertiary neurons.

1. **First-order neurons or primary neurons.** These are the bipolar cells whose dendrites synapse with photoreceptors (rods and cones) and axons synapse with dendrites of ganglion cells.
2. **Secod-order neurons or secondary neurons.** These are ganglion cells of the retina:

(*i*) The axons of the ganglion cells pass backwards along the **optic nerve** to the **optic chiasma** where partial crossing of the fibers takes place. The fibers from the temporal side of the retinae remain "uncrossed" and those from the nasal retinae 'cross' to the opposite side (Fig. 4.16).

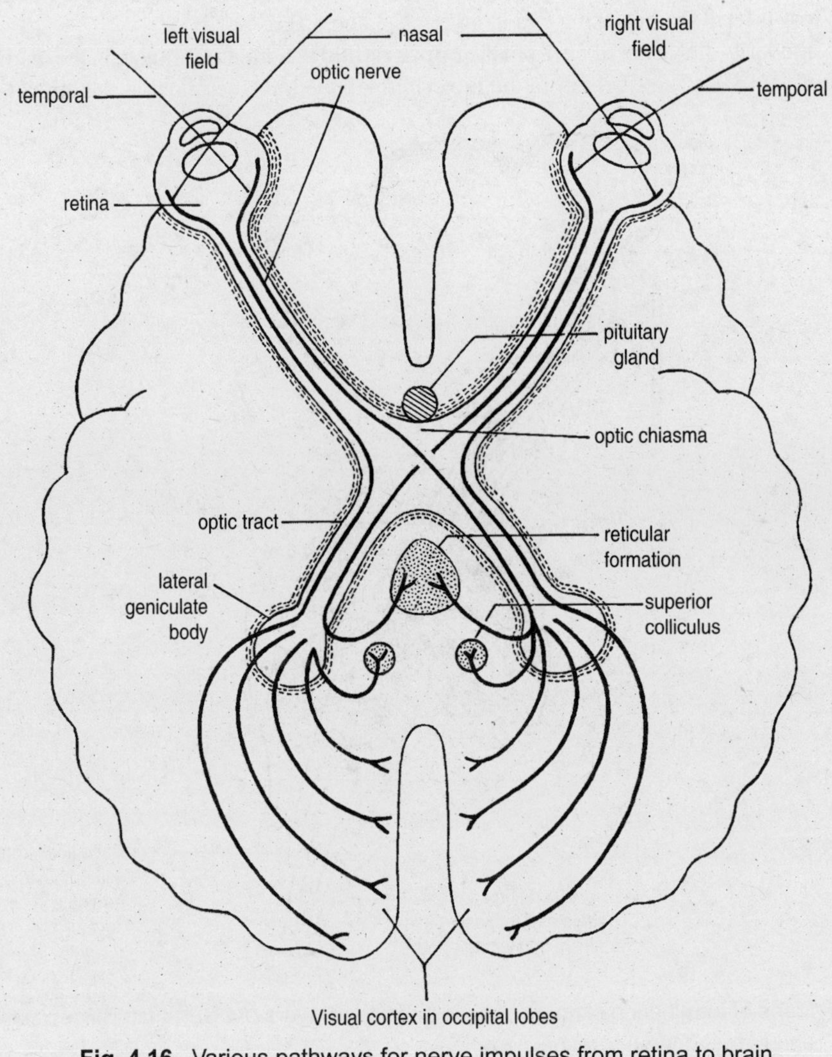

Fig. 4.16. Various pathways for nerve impulses from retina to brain

(*ii*) Thus, two **optic tracts** are formed, the left optic tract conveys fibers from the left halves of both retinae and the right optic tract from the right halves of both retinae.

(*iii*) The optic tract passes in between crura ciberium (the peduncles of the cerebrum) and in two main areas:

(*a*) majority of fibers end in the **lateral geniculate body** (LGB), a part of the thalamus;

(*b*) few fibers enter the **superior colliculus** (in mid brain); these fibers synapse with pretectal nucleus and serve as ceter for **visual reflexes**.

Box 4.6

Some axons of the ganglion cells of retina pass directly from the optic chiasma to the suprachiasmatic nuclei in the hypothalamus, where they form connections that mediate a variety of **endocrine** and other **circadian rhythm** with the light-dark cycle.

(*iv*) In the **lateral geniculate body (LGB)**, there is an orderly point to point (topographical) representation of retina.

(*a*) The grey matter in the LGB shows six clear-cut layers numbered 1 to 6. The fibers from the retina to the **contra-lateral** (opposite) side end in layers 1, 4 and 6; the fibers from the **ipsi-lateral** (same) retina end in layers 2, 3 and 5 (Fig. 4.17).

(*b*) Similarly, the visual fibers from the **upper retinal** quadrants terminate in the **medial halves** of the LGB. The fibers from the **lower** retinal quadrants terminal in the **lateral halves**.

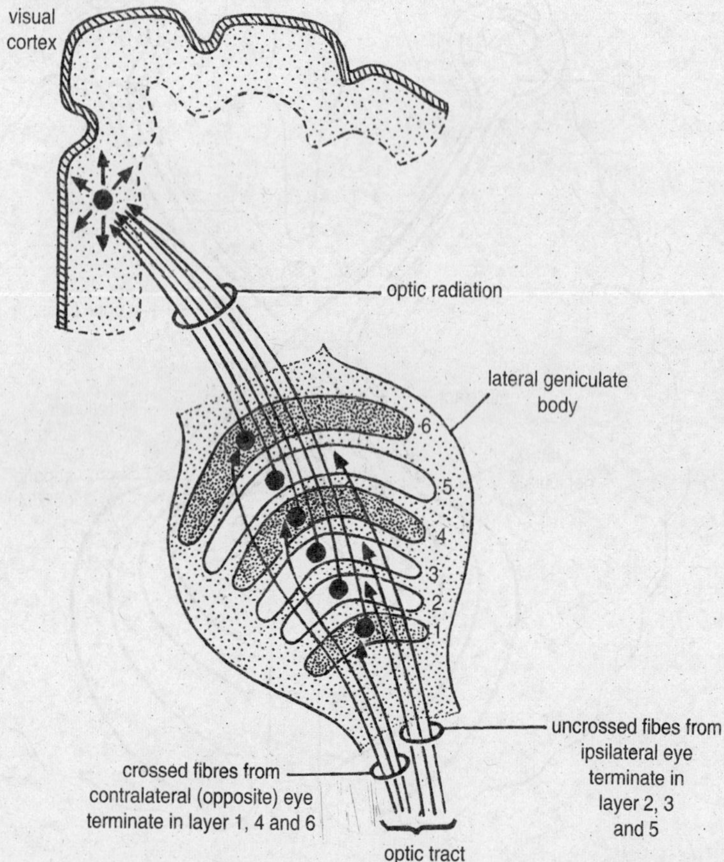

Fig. 4.17. Mode of termination of fibers of optic tract in six layers of LGB (= lateral geniculate body) and further course to visual cortex.

3. **Third order neurons or tertiary neurons**. These are neurons located in the LGB. These axons form the **geniculo-calcarine tract** which passes to the **occipital lobe** of the cerebral cortex.

4. **Visual Cortex**

(*i*) Optic radiations from the LGP pass through the internal capsule vis geniculo-calcarine tract to the **primary visual area**, also called visuo-sensory area (V-1) (area 17). It is the highest area for the perception of visual sense and is located on the medial surface of the occipital lobe, above and below calcarine fissure (Fig. 4.18). In this case also, there is an orderly **point to point** representation of the retina.

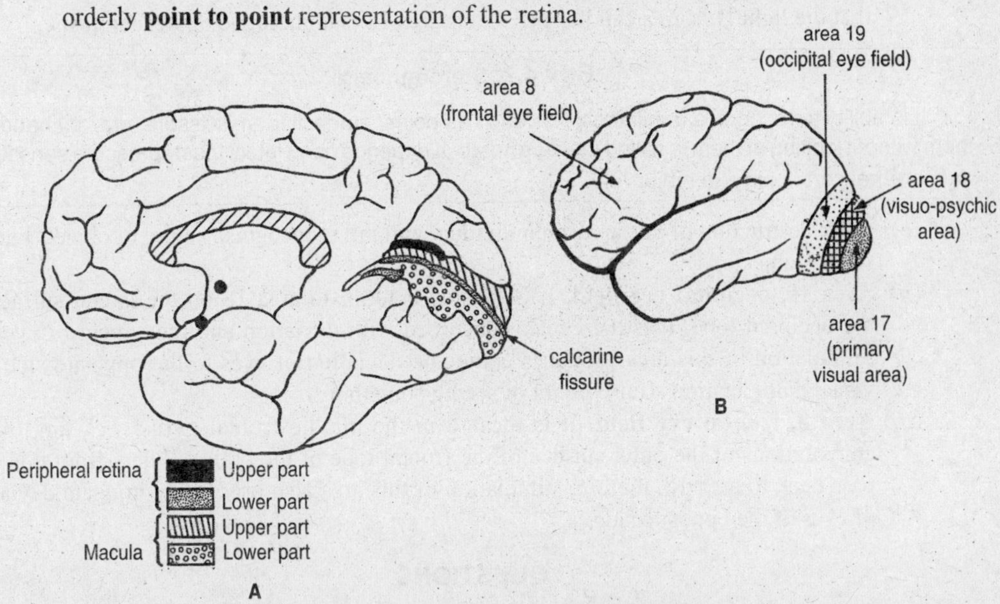

Fig. 4.18. Representation of visual fibers in occipital lobe; medial surface (A) and lateral surface (B).

(*ii*) The geniculo-calcarine fibers from **medial halves of LGB** (*i.e.,* upper retinal quadrants) terminate on the **superior lip** of the calcarine fissure, while those from the **lateral halves** (*i.e., lower retinal quadrants*) terminate on the inferior lip.

(*iii*) Furthermore, the fibers from the LGB, which subserve **macular vision** (*i.e.,* vision by macula lutea or yellow spot) separate from those that subserve peripheral vision and end more posteriorly on the lips of the calcarine fissure. Thus, the macula has a very large central representation and that of peripheral retina is small.

(*iv*) Results have been obtained by studying the *evoked potentials* in the occipital cortex resulting from stimulation of separate parts of retina. It has been seen that 1 mm of fovea retina is represented by about 16 mm in the cortex, whereas more peripheral parts of the retina receive only approximately 1/60th of this reresentation per mm.

(*v*) Each half of the retina receives light rays from the opposite half of the field of vision ; therefore, each visual cortex is called half **visual center**, as one half of the visual cortex represents half visual fields in one eye.

(*vi*) Like the rest of the neocortex (*i.e.* the dorsal region of the cerebral cortex), the visual cortex has six layers. The axons from the LGB neurons end on pyramidal cell in layer 4 which in turn project primarily to more superficial layers (layers 2 and 3). These layers contain 'clusters' of cells, called **blobs** which contain high concentration of the mitochondrial enzyme cytochrome oxidase. These are concerned with *colour vision.* The primary visual area (area 17) also plays a role in **visual discrimination**.

5. Visual Association Areas

(*i*) **Area 18, Visuo-psychic area (V-2).** It is located on lateral surface of occipital lobe above anterior to area 17.

(*a*) It is concerned with higher visual functions such as **visual orientation, depth perception** and relay of information from visual cortex to other parts of the brain.

(*b*) It is the area where visual senses are interpreted and integrated in the light of **past experience**. It also enables assessment of distance and *orientation of an object in space.*

For example, there is a light and area 17 sees the light. It is the area 18 which interprets that the light is from a candle stick. Thus area 18 is homologue of **stereognosis**.

Box 4.7 Stereognosis

It is ability to recognize familiar objects such as coins, a pencil, pen, scissors, etc., by handling them without looking at them is called **stereognosis**. It depends upon intact touch-pressure sensation, but cerebral cortex plays a major role.

(*c*) After destruction of this area, an individual will fail to recognise nature of objects but can see the objects.

(*ii*) **Area 19, occipital eye field.** It is located anterior to area 18 on the lateral surface of the occipital lobe. It is the center concerned with deviation and movement of eyeball. Stimulation of this area produces conjugate deviation of eyes to the opposite side and visual hallucinations (false belief of seeing something).

(*iii*) **Area 8, frontal eye field.** It is located in the middle frontal gyrus (*i.e.*, any of the convolutions of the outer surface of the frontal lobe of the brain). Fibers from this area pass back to area 18, thefore, stimulation of this area also produces conjugate deviation of eyes of the opposite side.

QUESTIONS

Short Answer Questions

1. **Differentiate between the following:**
   (*a*) Rods and Cones;
   (*b*) Aqueous humour and Vitreous humour;
   (*c*) Photopic vision and Scotopic vision;
   (*d*) Rhodopsin and Iodopsin;
   (*e*) Blid spot and Yellow spot
2. Compare the choroid and retina.
3. Draw labelled diagram of anatomy of eye.
4. Write short notes on the following:
   (*i*) Retina;
   (*ii*) Mechanism of vision;

Multiple Choice Questions

1. The colour of the eye is due to
   (*a*) sclera (*b*) cornea
   (*c*) pupil (*d*) iris
2. Aqueous humour differs from plasma in that it.
   (*a*) is high in protein (*b*) has high content of vitamin C and NaCl
   (*c*) has high glucose but low lactic acid (*d*) has high viscocity
3. In what layer of the retina is a large store of vitamin A found?
   (*a*) inner limiting membrane (*b*) pigment layer
   (*c*) outer nuclear layer (*d*) photoreceptor layer
4. The central point of a lens is known as
   (*a*) nodal point (*b*) principal axis
   (*c*) principal focus (*d*) visual point

5. Duplicity theory refers to
   (*a*) photopic and scotopic vision
   (*b*) double pain
   (*c*) diplopia
   (*d*) two point discrimination
6. The fovea centralis is point of greatest visual acuity because
   (*a*) rod free portion of retina
   (*b*) avascular
   (*c*) highly developed in humans
   (*d*) each foveal cone is connected to a single fiber in the optic
7. Colour perception is the function of
   (*a*) rods
   (*b*) cones
   (*c*) pigmented epithelium
   (*d*) visual association area
8. Highest refracive index is that of
   (*a*) cornea
   (*b*) aqueous humour
   (*c*) lens
   (*d*) vitreous humour
9. The region of the vertebrate eye, where the optic nerve passes out of the retina, is called the
   (*a*) fovea
   (*b*) iris
   (*c*) blind spot
   (*d*) optic chiasma

ANSWERS

Multiple Choice Questions

1. (*d*) **2.** (*b*) **3.** (*b*) **4.** (*c*) **5.** (*a*) **6.** (*d*)
7. (*b*) **8.** (*c*) **9.** (*c*)

5

Physiology of Hearing

The ear is a specialized receptor for detecting sound waves in the surrounding environment. Because sound communication and reception are the integral parts of the lives of terrestrial vertebrates, we may be surprised to discover that most invertebrates inhabit a silent world.

In its evolution, the vertebrate ear originated as a balance organ, the **labyrinth**. In all jawed vertebrates, from fishes to mammals, the labyrinth has a similar structure, consisting of two small chambers called the **saccule** and the **utricle**, and three **semicircular canals**. In fish the base of the saccule is extended into a tiny pocket (the **lagena**) that during the evolution of the vertebrates, developed into the hearing receptor of tetrapods. With continued elaboration and elongation in the birds and mammals, the finger like lagena evolved into the **cochlea** (Hickman et al., 1996).

5.1 STRUCTURE OF HUMAN EAR

The ear is the organ of **hearing** and **equilibrium**. Anatomically, it is divided into following three parts (Fig. 5.1).

1. The external ear;
2. The middle ear;
3. The internal ear.

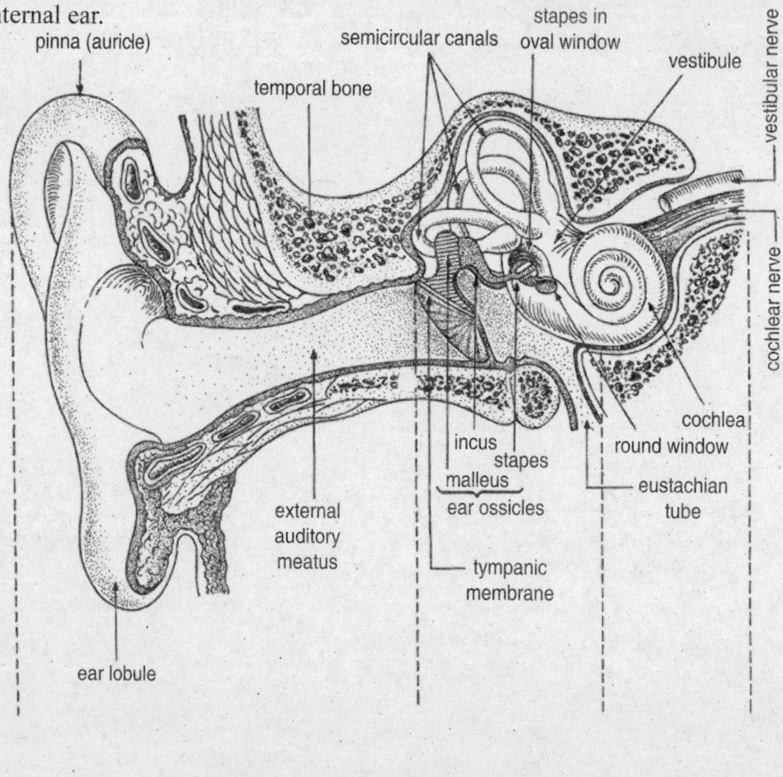

Fig. 5.1. Structure of the human ear.

A. The External Ear

The external ear includes the *pinna* and the *external auditory meatus* (= canal).

1. Pinna. It is a cartilagenous expanded portion covered with skin projecting from the side of the head. The pinna collects the vibrations in the air which produce sound. Thus, it helps to *collect sound waves* and *to localise* the source of sound.

2. External auditory meatus. It is a 2.5 cm long canal which runs forwards and inwards (medially) and helps in transporting the sound waves to the middle ear. It is lined with skin which secretes wax (from cereminous glands) and oil (from sebaceous glands) which may be trapping the foreign bodies and keep the ear canal clear. It terminates at the **tympanic membrane** (ear drum) which covers the opening of the middle ear (Fig. 5.1).

B. The Middle Ear

It is an **air filled chamber** and consists of the following parts (Fig. 5.2.).

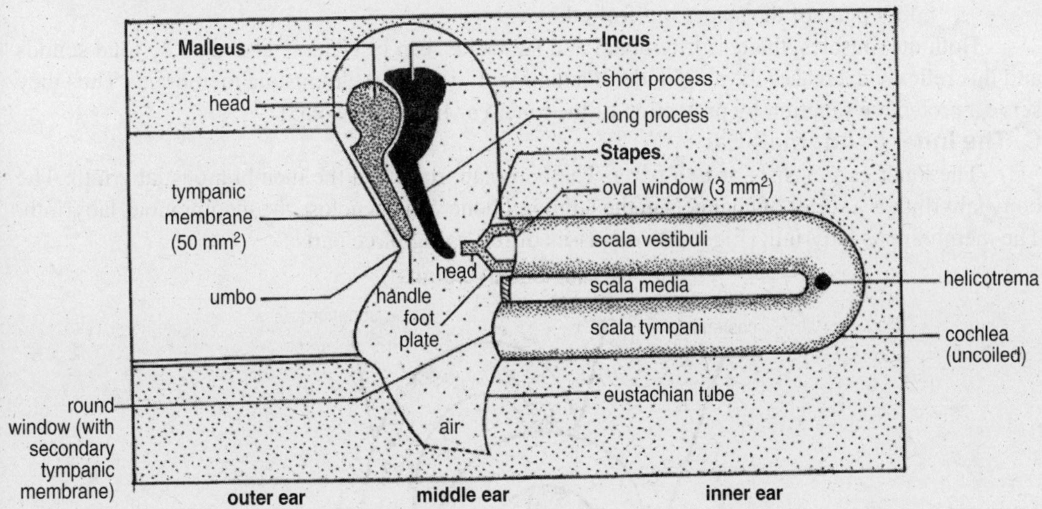

Fig. 5.2. Arrangement of ear (auditory) ossicles in the middle ear and scalae in the cochlea (inner ear).

1. Tympanic membrane. It consists of connective tissue covered with skin on the outside and mucous membrane on the inside. Tympanic membrane has the shape of shallow funnel with the centre (umbo) pointing inwards and is kept in this position by the handle of the malleus (first ear ossicle).

The presence of air at atmospheric pressure on both sides of the tympanic membrane enables it to vibrate freely when sound waves strike it, which then passed on to the ear ossicles.

2. Ear ossicles. There are following three minute bony (ear or auditory) ossicles: **malleus** (hammer), **incus** (anvil; an iron block on which a smith hammers metal into shape) and **stapes** (stirrup; a support for a horse rider's foot). These ossicles are attached to one another in a chain-like fashion and they extend across the middle cavity from the tympanic membrane and to the oval window.

(i) **Malleus.** Its **handle** is attached to the inner side of the tympanic membrane and its tip is situated at the umbo. The **head** of malleus is firmly bound by ligaments to the incus.

(ii) **Incus.** At one side incus is attached to the head of the malleus and on the other side its long process passes downwards to articulate with the head of stapes.

(iii) **Stapes.** Its **foot plate** is attached by annular ligament to the margins of the oval window of the cochlea.

The ear ossicles function to magnify the *intensity of sound* by 1.2 to 1.3 times.

The opposite wall of the middle ear contains two membrane–covered openings which face the internal ear. They are an **oval window** above and a smaller **round window** just below it.

3. Pharyngotympanic tube or eustachian tube. It is 4.5 cm in length and connects the middle ear cavity with the pharynx. Normally its pharyngeal opening is closed but it can be opened by swallowing, chewing or yawning which contracts the tensor palati muscle and air enters into the middle ear. Therefore, eustachian tube serves to equalize the pressure on the two sides of the tympanic membrane when atmospheric pressure decreases (high altitude) or increases (deep sea diving).

4. Two skeletal muscles. There occur two muscles to regulate the functions of ear ossicles: tensor tympani and stapedius.

(i) **Tensor Tympani.** This muscle is attached to the neck of malleus and innervated by **trigeminal cranial nerve** (*i.e.*, vth). Its contraction increases the tension of the tympanic membrane by pulling the handle of the malleus medially. Thus, it keeps the tympanic membrane firmly attached.

(ii) **Stapedius.** It is attached to the neck of the stapes and to the posterior wall of the middle ear. This muscle is innervated by facial (vii) cranial nerve and on contraction it pulls the foot plate of the stapes out from the oval window.

Both the muscles, the tensor tympani and stapedius, can be reflexly activated by loud sounds and this reflexly decreases the amplitude of sound vibrations of the tympanic membrane. Thus they serve a protective function by protection of the inner ear from loud sound.

C. The Inner (= Internal) Ear

The inner ear consists of **two** parts: The bony labyrinth and the membranous labyrinth. The bony labyrinth is a series of channels in the temporal bone which enclose the membranous labyrinth. The membranous labyrinth (Fig. 5.3) comprises of following three parts :

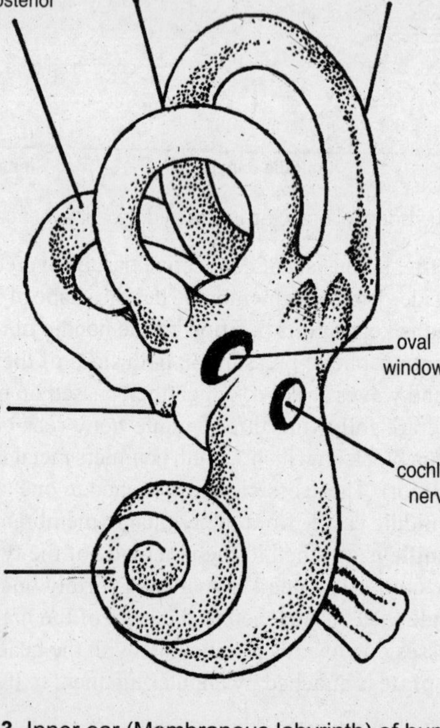

Fig. 5.3. Inner ear (Membranous labyrinth) of human.

1. One **vestibule** (**utricle** and **saccule**). It is the expanded part very close to the middle ear.

2. One **cochlea** (resembles shell of snail). It is a special tube having a broad base with a narrow apex and is concerned with hearing.

3. Three **semicircular canals** which communicate with the vestibule and are arranged at right angle to each other. Together it is called the vestibular apparatus and is concerned with the equilibrium, *i.e.*, maintenance of posture and balance of the body. It detects head motion and position and transfer these informations to a neural signal.

The Cochlea

It is a coiled tube, 35 mm long which makes 2.5 to 2.75 turns. The bony core around which the cochlea coils is called **modiolus**. The lumen of cochlea is divided throughout the length by two membranes (**Reissner's membrane** and **basilar membrane**) into three compartments, called scalae (Fig. 5.2).

(i) **Scala vestibuli.** It is a space above the Reissner's membrane and is filled with **perilymph.** Perilymph closely resembles cerebrospinal fluid.

(ii) **Scala tympani.** It is a space below the Basilar membrane and is also filled with perilymph.

(iii) **Scala media.** It is a space between Reissner's and Basilar membranes and is filled with K^+ - rich fluid called **endolymph** (Box 5.1). It is a closed space and is continuous with the membranous labyrinth.

Box 5.1

Endolymph contains high potassium ions (K^+) (about 145 mM) and low sodium ions (Na^+) (about 2mM) and has a high positive potential (about +80 mV) with respect to the perilymph. As a result, a very large potential gradient (about 140 mV) exists across the membranes of the hair cell cilia that extend into endolymph. Endolymph is secreted by the **stria vascularis** and is drained through the endolymphatic duct into the dural venous sinuses.

The *basilar membrane* is attached medially to the bony spiral lamina and laterally to the fibrous spiral ligament, whereas the *Reissner's membrane* is attached medially to the wall of the limbus and laterally to the upper margin of the stria vascularis (Fig. 5.4).

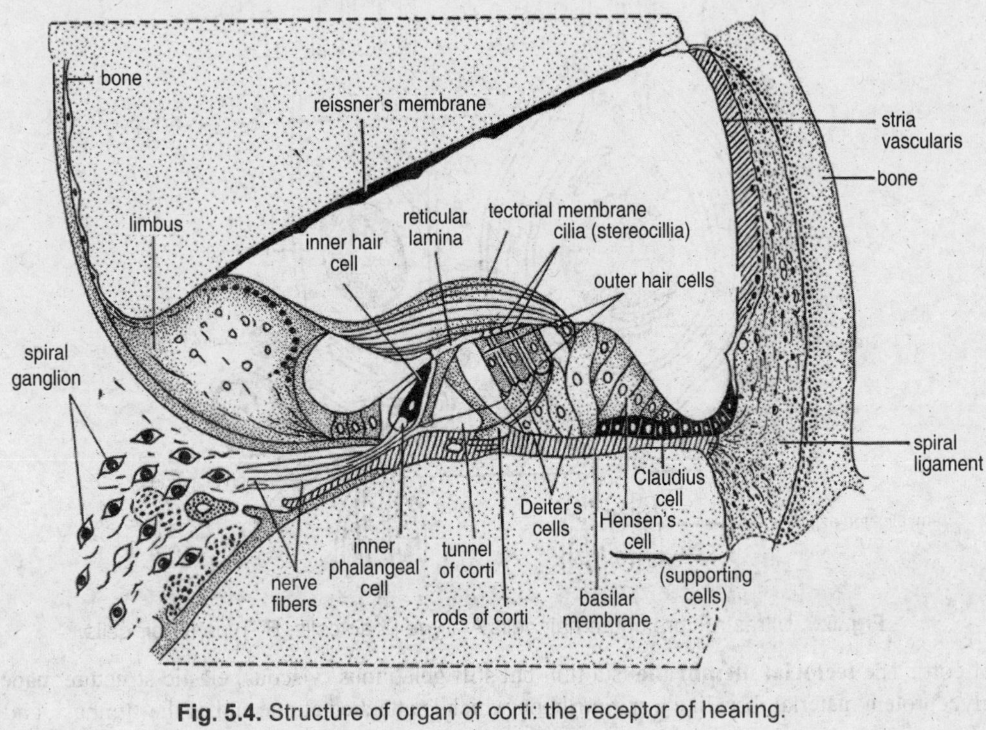

Fig. 5.4. Structure of organ of corti: the receptor of hearing.

Scala tympani and scala vestibuli communicate at the apex of cochlea by a small opening called **helicotrema** (Fig. 5.2). At the base of the cochlea, the scala vestibuli ends at the oval window which is closed by the foot plate of the stapes. The scala tympani ends at the **round window**, a foramen on the median wall of the middle ear which is closed by the flexible **secondary tympanic membrane**.

The Organ of Corti

The primary receptor of hearing is the **'organ of Corti'** which is located on the basilar membrane extending from the apex to the base of cochlea. On this membrane stand two rods (**rods of corti**) which project into the scala media. In between the two rods is a **tunnel of corti** which is filled with perilymph.

Internal to the inner rod is a single row of **inner hair cells** (= receptor cells), and external to the outer rod are three or four rows of **outer hair cells** (= receptor cells). There are in all 3500 inner hair cells and 20,000 outer hair cells in each cochlea. The inner hair cells are supported by inner **phalangeal cells** while the outer hair cells are supported by **Deiter's cells** (outer phalangeal cells).

From the upper surface of the hair cell project tiny "cilia", the **stereocilia** (8-12 μm in diameter, 4 μm long and 0.1 μm thick; Fig. 5.5) which pass through a thin dense granular *reticular lamina* and get embedded in the *tectorial membrane.* The reticular lamina is a tough membrane supported by rods

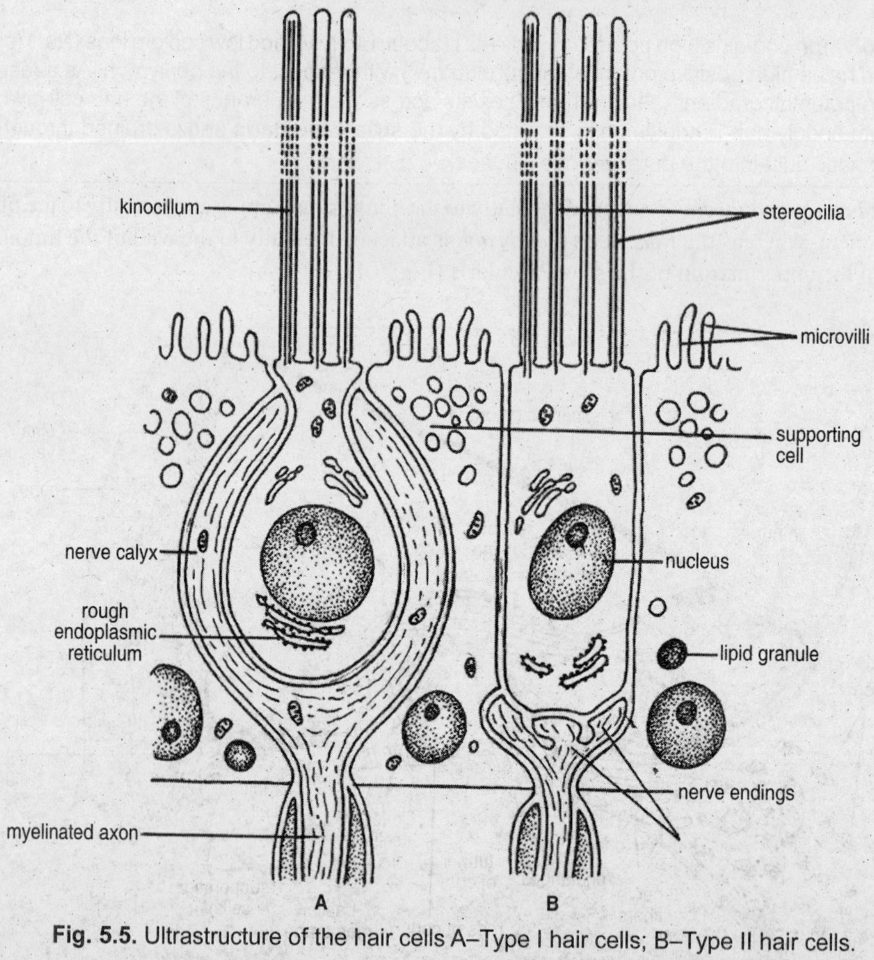

Fig. 5.5. Ultrastructure of the hair cells A–Type I hair cells; B–Type II hair cells.

of corti. The **tectorial membrane** is a thin but stiff gelatinous (viscous) elastic structure made of glycoprotein material. It is like a spiral ribbon which is attached at one end of the 'limbus', and by other end to the Hensen's cells (*i.e.*, to its outer edge). **Hensen's cells** are supporting cells of organ of corti which lie outside the outer hair cells.

Innervation of Hair Cells

(a) **Afferent innervation.** The hair cells are innervated by nerve fibers of the cochlear or auditory division of VIII cranial nerve (Auditory nerve). These fibers have their cell bodies in the **spiral ganglion**. This ganglion is located within the modiolus.

(b) **Efferent innervation.** Efferent cholinergic fibers arise from both the ipsilateral and the contralateral superior olivary nucleus via the **olivocochlear bundle.** These fibers descent to join VIII cranial nerve and end around the bases of outer hair cells.

Functions of hair cells. The inner hair cells are the *primary sensory (receptor) cells* that generate action potential in the auditory nerves and are stimulated by the fluid moving between the tectorial membrane and hair cells. These cells thus besides detecting the sound are also responsible for **fine auditory discrimination.**

The outer hair cells are responsible for detecting the presence of sound and they also *improve hearing* (amplitude and clarity of sound) by influencing the vibration patterns of the basilar membrane.

Transmission of Impulses from Cochlea to Brain

From the endings of the cochlear nerves in the organ of Corti along the basilar membrane, there is a precise orientation through all the relay centers enroute to the auditory cortex (of brain). The important relay centers are the **spiral ganglion**, the **cochlear nuclei**, the **superior nuclei** and the **inferior colliculi** (Fig. 5.6)

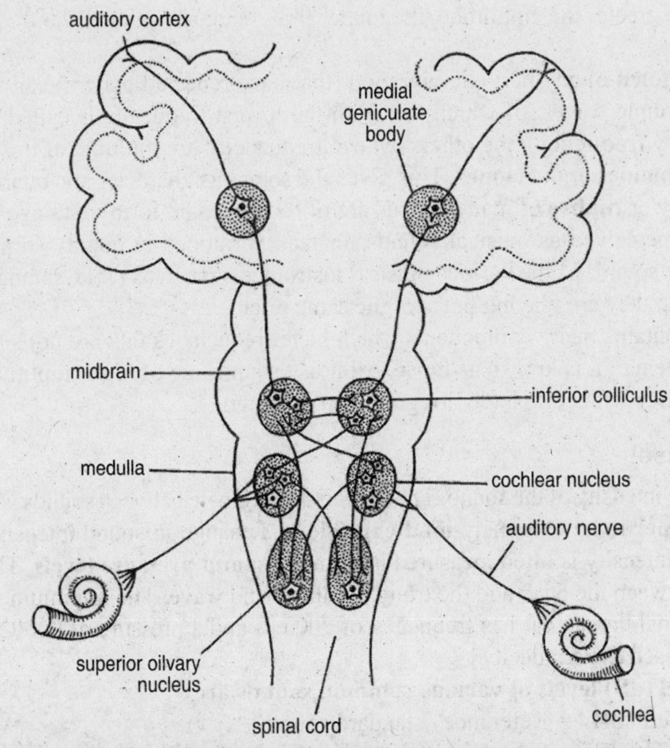

Fig. 5.6. The ascending neural pathways of the auditory system. The auditory nerves synapse in the ipsilateral cochlear nuclei. Neurons in the cochlear nuclei send axons to both olivary nuclei and to the contralateral inferior colliculus. From the inferior colliculus, fibers pass via the medial geniculate body to the primary auditory cortex.

The peripheral fibers of the cochlear nerve enter spiral ganglion in which their cell bodies are located. These bipolar neurons send fibers centrally to the cochlear nuclei in the medulla of brain. All the cochlear nerve fibers synapse in these nuclei. Most of the neurons originating in the cochlear nuclei send their axons to the superior olivary nucleus in the same side of the medulla. However, some fibers pass to the superior olivary nucleus on the other side of the medulla. This means that these olivary centers receive signals from both the ears.

Most fibers from the cells in the olivary nuclei pass to the inferior colliculi in the midbrain, where they synapse again with neurons leading to the auditory cortex, deep in the lateral fissures of the cerebral hemispheres.

Along their course, many of the nerves give off collaterals. This particularly is there in the cerebellum and the reticular formation, both of which are immediately activated by a sudden loud noise.

5.2 PHYSICAL PROPERTIES OF SOUND

1. The vibrating objects cause alternating phases of *compression* and *rarefaction* (pressure changes) in the medium surrounding the object. The compressions and rarefaction spread out as sound waves which are perceived by the auditory mechanism as sound.
2. Sound *travels in air* at a speed of approximately 330 meters / second (1100 ft/sec) at 0°C at sea level, while *in water* sound travels much faster, at a speed of approximately 1450 to 1500 m/ at 20°C (being greater in salt water as compared to fresh water). The speed of sound slightly increases with temperature and altitude.
3. In general **amplitude** (intensity) of sound waves determines the **loudness** of a sound while **frequency** of sound waves (number of waves per unit of time) determines the **pitch** of the sound. The greater the amplitude, the louder the sound; and greater the frequency, the higher the pitch.
4. The **wave form** of any periodic vibration (that have repeated pattern) can be resolved into a series of simple waves, of which that with the lowest frequency is called the **fundamental** (or **primary frequency**), the others, whose frequencies are multiple of this fundamental, are called **harmonics** or **overtones.** This gives the sound its characteristic quality.
5. The **quality** or **timbre** of many sounds are of 'complicated form' but have repeating pattern. These are perceived as musical sounds. In fact, variations in quality (timbre) permit us to identify the sounds of the various musical instruments (such as tabla, sarangi and guitar, etc.) even though they are playing notes of the same pitch.
6. If sound contains many components which have frequencies that are not simple multiples of the fundamental, it is known as **noise.** All noisy sounds are of high amplitude (intensity) and low frequency with completely irregular waves pattern.

Loudness (Noise)

Noise is the intensity of the sound. The ear can discriminate between sounds varying in intensity from a **soft whisper** to an extremely **loud explosion**. A change in sound intensity is expressed in **decibels**. Sound intensity is often measured in terms of **sound pressure levels.** The pressure being the difference between the peak and the trough of the sound wave. The minimum sound that can be heard by the normal human ear has frequency of 200 cps and a pressure of 00002 dynes/cm^2. This pressure is expressed as zero decibel.

The decibel (dB) levels of various common sounds are

'Zero' dB = Reference / standard sound
10 dB = Sound such as paper being gently folded
10 – 30 dB = Whispering (audible range 2 meters)
30 – 40 dB = Normal room noise
40 – 60 dB = Normal conversation (audible range 4 meters)

60–80 dB = Heavy traffic noise
80 – 100 dB = Loud motor horn; heavy music
100–120 dB = Noise and discomfort
120–140 dB = Painful and damaging to the organ of Corti.

5.3 SOUND TRANSDUCTION

Sound waves are **transduced** by the organ of Corti. Sound waves that reach the ear cause the tympanic membrane to oscillate and these **oscillations** are transmitted to the scala vestibuli by the ear ossicles. This creates a *pressure difference* between the scala vestibuli and the scala tympani that serves to displace the basilar membrane and with it, the organ of Corti (Fig. 5.7 A and B). Because of the shear forces set up by the relative displacement of the basilar and tectorial membranes, the stereocilia of the hair cells bend. Upward displacement bends the stereocilia toward the tallest cilium (away from the

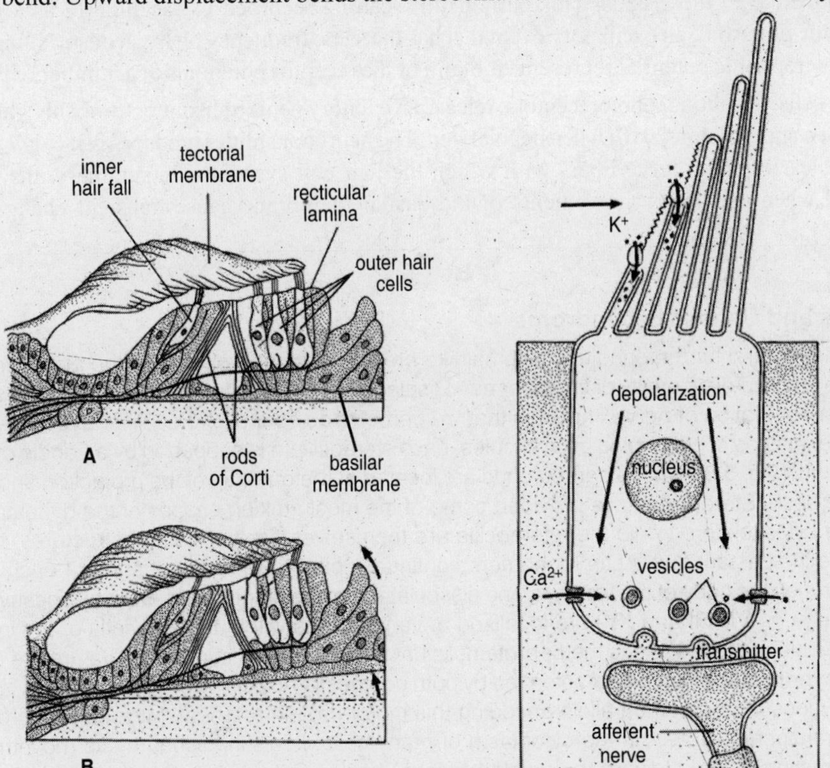

Fig. 5.7. A, B–Detailed structure of organ of corti; demonstrating how movement of the basilar membrane will cause the stereocilia to bend because of shear forces produced by relative displacement of the hair cells and the tectorial membrane; C–Diagram of a hair cell with tip-connections between the hair cell cilia to demonstrate how shear forces open mechanoreceptor channels (after Koeppen an Stanton 2010).

modiolus), which depolarizes the hair cells; downward deflection bends the stereocilia in the opposite direction, which hyperpolarizes the hair cells.

Physiology of Sound Transduction

In view of the wide range of frequencies and amplitudes of sound stimuli, hair cell transduction should provide for a fast response. The fast response to direction of the cilia is based on direct **opening of ion channels** by "tip links" that connect the tip of each stereocilium with the shaft of the next taller one (Fig. 5.7C). With deflection, the tip links are subjected to *lever action* that transiently opens the channels, permits the entry of K^+ (because of the high potassium ions and high potential of endolymph), and depolarizes the hair cell. Several mechanisms have been proposed to account for the equally important rapid adaptation necessary for a high-frequency response. A **spring response** by the tip links would allow the attachment point of the tip link to be moved along the stereocilium's shaft to reset the mechanical leverage of the tip link. (Box 5.2). In addition, it has been observed that calcium ions (Ca^{2+}) can enter and bind to the open channel, change it to require greater opening force, and thereby reduce the statistical probability of opening.

The potential gradient that induces movement of ions into hair cells includes both the resting potential of the hair cells and the positive potential of the endolymph. As noted previously, the total gradient across the apical membrane of hair cells is about 140 mV. Therefore, a change in *membrane conductance* in the apical membranes of hair cells results in a rapid current flow that produces the receptor potential in these cells. This current flow can be recorded extracellularly as a **cochlear microphonic potential**, an oscillatory event that has the same frequency on the acoustic stimulus. The cochlear microphonic potential represents the sum of the receptor potentials of a number of hair cells.

Hair cells, like retinal photoreceptors, release an excitory neurotransmitter (probably **glutamate;** see **Koeppen** and **Stanton** 2010) when depolarized. The neurotransmitter produces *generator potential* in the cochlear efferent nerve fibers with which the hair cell synapses. In summary, the sound is transduced when oscillatory movements of the basilar membrane cause transient changes in the

Box 5.2

Hair Cells and Usher 1B Syndrome

Hair cells are named for the bundle of stiff, hairlike **stereocilia** that project from the apical surface of the cell into the fluid-filled cavity of the inner ear. Displacement of the stereocilia by mechanical stimuli leads to the generation of nerve impulses that we perceive as sound. Stereocilia have no relation to the true cilia. Instead of containing microtubules, each stereocilium is supported by a bundle of parallel actin filaments (Fig. 5.8), whose barbed ends are located at the outer tip of the projection and pointed ends at the base. Stereocilia have provided some of the most striking images of the dynamic nature of the actin cytoskeleton. While the stereocilia are themselves are permanent structures, the actin bundles are in constant flux. Actin monomers continually bind to the barbed ends of each filament treadmill through the body of the filament, and dissociate from the pointed end. Several unconventional myosins (i.e., I, V, VI, VII and XV) are localized at various sites within the hair cells of the inner ear. The precise role of these various motor proteins is unclear. Mutations in myosin VIIa are the cause of **Usher 1B syndrome**, which is characterized by both deafness and blindness. As in humans, mice that are homozygous for the mutant allele encoding this motor protein are deaf. Myosin VI, a processive organelle transporter protein in the cytoplasm of many cells, is distinguished by its movement in a "reverse direction", that is toward the pointed (minus) end of actin filament. Myosin VI is thought to be involved in the formation of clathrin-coated vesicles at the plasma membrane, the movement of the uncoated vesicles to the early endosomes, and the anchoring of membranes to actin filaments. Mutations in myosin VI are the cause of several inherited diseases.

Usher's syndrome. An inherited disease that is characterized by congenital deafness or progressive hearing loss during childhood and by retinitis pigmentosa (*i.e.*, pigmentary retinopathy. which eventually leads to blindness) and that is inherited chiefly as an autosomal recessive trait.

---

transmembrane voltage of the hair cells and, consequently, the generation of action potential in cochlear afferent nerve fibers. The activity of a large number of cochlear afferent fibers can be recorded extracellularly as a compound action potential. However, not all the cochlear afferent nerve fibers discharge in response to a particular sound frequency. One factor that influences which afferent fibers discharge is their location along the organ of Corti. The location of an afferent fiber is important because for any given sound frequency, there is a site of maximum displacement as the pressure waves travels along the basilar membrane. The location varies because the width and tension along the basilar membrane vary with distance from the base.

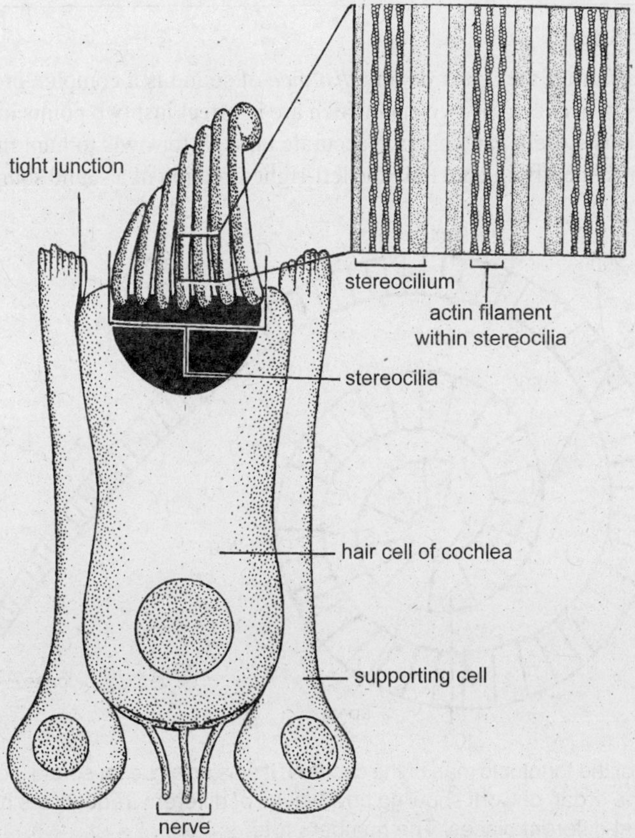

Fig. 5.8 Hair cells and actin bundles. Drawing of a hair cell of the cochlea. The inset shows a portion of several of the stereocilia, which are composed of a tightly grouped bundle of actin filaments (after Karp, 2010).

On the basis of these differences in width and tension, investigators originally concluded that different parts of the basilar membrane have different resonance frequencies. For example, the basilar membrane is about 100 μm wide at the **base** and 500 μm wide at the **apex.** It also has higher tension at the base. Thus, the base was predicted to vibrate at higher frequencies than the apex, as do the shorter strings of musical instruments. However, experiments have shown that the basilar membrane moves as a whole in travelling waves. Movement of the basilar membrane are maximal nearer the base of the cochlea during high-frequency tones and maximal nearer the apex during low frequency tones.

In effect, the basilar membrane serves as a *frequency analyzer;* it distributes the stimulus along the organ of Corti so that different hair cells will respond differentially to particular frequencies of sound. This is the basis of the **place theory of hearing.** In addition, hair cells located at different places along the organ of Corti are **tuned** to different frequencies because of differences in their stereocilia and biophysical properties. As a result of these factors, the basilar membrane and organ of Corti have a so-called **tonotopic map** (Fig. 5.9.)

Box 5.3

Tonotopic maps have been found in the cochlear nuclei, superior oliver complex, inferior colliculus, medial geniculate nucleus, and auditory cortex. A given auditory structure may, in fact, contain several tonotopic maps (**Koeppen** and **Stanton** 2010).

Localizations of Sound

The determination of the *direction* and *distance* of sound is a complex process. The audition requires a comparison between the two ears which are in effect just two points in space to locate the source of stimuli. This system can be made accurate enough for owls to hunt mice on dark nights, solely by their sounds, identifying not only the left-right direction of a sound source, but its elevation as well.

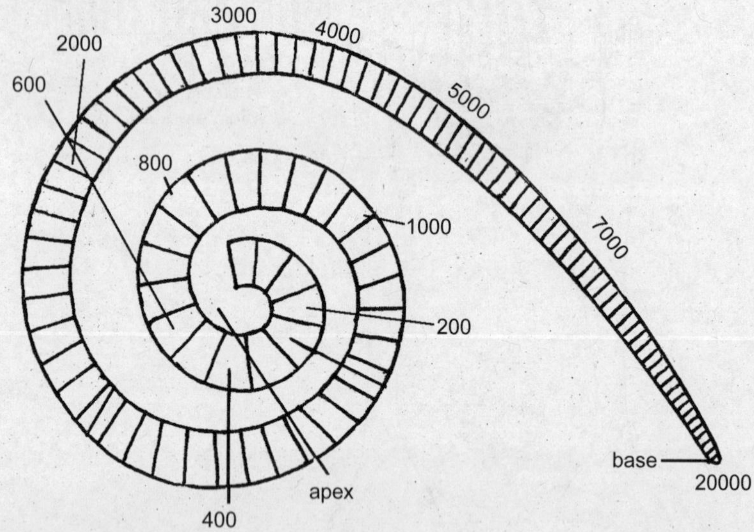

Fig. 5.9. Layout of the tonotopic map of the cochlea. It acts as a cue for sound localization. Locations along the organ of Corti showing how waves of different frequencies have their maximum effects at different places. The numbers refer to hertz.

Information from the two ears progresses through several stages on the way to the brain (Fig. 5.7). In fact, all aspects of the auditory cortex show adaptations that facilitate the localization of sounds. Following two methods are used for the localization of sound:

1. The first is the *difference in loudness* between the two ears. The head impedes (*i.e.*, blocks the process) the passage of sound waves, especially the wavelength is shorter than the width of the head *i.e.* is, for short wavelength (high frequency) sounds, the head creates a **sound shadow** consequently the sound is louder for the closer ear. In adults, this mechanism produces accurate sound localization for pitches above 3000Hz.
2. The second method of localization is the *phase difference for sound waves* arriving at the two ears. When sound waves travel farther to reach one ear than the other, they reach the two ears at slightly different times and thus are slightly out of phase. Phase difference provides accurate localization for sounds less than about 150 Hz but not for higher pitches. (**Note.** *Hertz.* It is SI unit of frequency, equal to one cycle per second).

QUESTIONS

Long Answer Questions

1. Give mechanism of vibration of basilar membrane.

2. What is cochlea? Describe its function in detail.

3. Describe the mechanism of hearing in brief.

Short Answer Questions

1. Write short notes on the following:
   (a) Organ of corti;
   (b) Role of middle ear in hearing;
   (c) Place theory.

Very Short Answer Questions

1. Name the three ear ossicles.
2. Describe sound conducting mechanism of the middle ear.
3. What are the names of fluids of inner ear?
4. What are three parts of cochlea?
5. Where is auditory receptor located in human ear?
6. How many hair calls are found in human's organ of Corti?
7. What is the basis of place theory of hearing?

Fill in the Blanks

1. The pinna and auditory canal convey airborne sound ________ to the __________ membrane.
2. A pure tone is characterized in terms of its amplitude ________ and phase.
3. The three small bones (*i.e.*, ________) of the middle ear transmit the _______ of the tympanic membrane to the ______ window of the fluid-filled _______ ear.
4. The organ of Corti lies on _______ membrane.
5. The cochlea occurs in ______________.
6. The cochlea has three compartments: the ________ the scala tympani and the _______.
7. The scala media is also called ______________.
8. The _________ connects the middle ear to the pharynx.
9. The organ of Corti consists of a few longitudinal rows of columnar sensory cells, separated by columns of supporting cells (___ ______), basal cells (_____ _________) and small fluid-filled spaces (tunnels). The fluid of spaces is endolymph, but it is called _________.

Multiple Choice Questions

1. The amplitude and frequency of sound waves determines respectively ________ and ____ of a sound.
   (a) speed and loudness (b) loudness and speed
   (c) loudness and pitch (d) pitch and speed
2. The sound becomes painful above decibel of
   (a) 70 (b) 80
   (c) 140 (d) 160
3. The arrangement of three minute bony (ear) ossicles across the cavity from tympanic membrane to oval window is
   (a) stapes, incus, malleus (b) incus, malleus, stapes
   (c) malleus, stapes, incus (d) malleus, incus, stapes
4. A structure not present in the middle ear
   (a) tympanic membrane (b) three ear ossicles : malleus, incus and stapes
   (c) eustachian tube (d) membranous labyrinth
5. Scala media is filled with
   (a) fluid having electrolyte concentration very similar to that of ECF
   (b) perilymph
   (c) endolymph
   (d) fluid rich in Na^+ and low in K^+

6. The hair cells depolarise when
   (a) stereocilia moves away from limbus
   (b) stereocilia moves towards limbus
   (c) basilar membrane moves downwards
   (d) Cl^- moves out of the hair cell membrane
7. The auditory pathway passes via all except
   (a) cochlear nuclei
   (b) superior olivary nucleus
   (c) trapezoid body*
   (d) superior colliculus

   *Trapezoid body is a bundle of transverse fibers in the dorsal part of the pons).
8. Ear is most sensitive to frequency (Hz) of
   (a) 300–500
   (b) 1000 – 3000
   (c) 1000 – 20,000
   (d) 16000 – 20,000
9. Cells of deiter occur in
   (a) retina of eyes
   (b) sebaceous glands
   (c) organ of corti
   (d) utriculus
10. Eustachian tube
    (a) equalises air pressure in middle and external ear
    (b) equalises air pressure in middle and internal ear
    (c) acts as organ of equilibrium and hearing
    (d) vibrates tympanum
11. Cochlea and organ of Corti of mammalian ear are concerned with
    (a) hearing
    (b) balance of body
    (c) both of these
    (d) perception of atmospheric pressure
12. Sounds above what level are considered hazardous noise pollution.
    (a) 30 dB
    (b) 100 dB
    (c) 80 dB
    (d) 120 dB

ANSWERS

Very Short Answer Questions

1. Malleus (hammer), incus (anvil) and stapes (stirrup).
2. There is a chain of three auditory ossicles (bones) in the middle ear. The mechanical distortion of the tympanic membrane and the vibrations of the chain of three auditory ossicles bring about the conductance of sound in the middle ear. These ossicles convey the vibrations from the tympanic membrane to the membrane that separates the middle and inner ears, the oval window.
3. Both the scala vestibuli and scala tympani are filled with perilymph. The scala media is filled with endolymph.
4. Cochlea is coiled like a snail and is a part of inner ear. It has three parts or tubes: scala vestibuli (upper), scala tympani (lower) and scala media (intermediate).
5. The auditory receptor of human is organ of Corti. It is located in the basilar membrane of scala media (or cochlear duct) or cochlea.
6. The organ of Corti in humans contain 15,000 outer hair cells and 3500 inner hair cells. Hair cells bear stereocilia.
7. The basilar membrane serves as a frequency analyzer; it distributes the stimulus along the organ of Corti, so that different hair cells will respond differentially to particular frequencies of sound. This is the basis of the place theory of hearing.

Fill in the Blanks

1. Waves, tympanic. **2.** Frequency. **3.** Ossicles, vibrations, oval, inner. **4.** Basilar. **5.** Inner ear. **6.** Scala vestibuli, scala media. **7.** Cochlear duct. **8.** Eustachian tube. **9.** Of Hensen, of Deiter, cortilymph.

Multiple Choice Questions

1. (*c*)	**2.** (*c*)	**3.** (*d*)	**4.** (*d*)	**5.** (*c*)	**6.** (*a*)
7. (*d*)	**8.** (*b*)	**9.** (*c*)	**10.** (*a*)	**11.** (*a*)	**12.** (*c*)

6

Physiology of Muscle Contraction

Muscle tissue is responsible for movements of the body and for changes in the size and shape of its internal organs. It is characterized by aggregates of specialized cells, called **muscle cells** whose primary role is **contraction**. Muscle cells are typically elongated and arranged in parallel arrays, allowing them to work together effectively.

Muscle cells are highly specialized cells for the conversion of chemical energy to mechanical energy. Specifically, muscle cells use the energy in ATP to generate force to do work. Because work can take many forms (such as locomotion, pumping blood or peristalsis), several types of muscles are evolved. The three basic types of muscles are **skeletal muscle**, **cardiac muscle** and **smooth muscle**.

Skeletal muscle acts on the skeleton. In limbs, for example, skeletal muscle spans a joint, thereby allowing a lever action. Skeletal muscle is under voluntary control (*i.e.*, controlled by the central nervous system) and plays a key role in numerous activities such as maintenance of posture, locomotion, speech and respiration. When viewed under the microscope, skeletal muscle exhibits transverse striations (at intervals of 2 to 3 μm) that results from the highly organized arrangement of actin (thin filaments) and myosin (thick filaments) molecules within the skeletal muscle cells. Thus, skeletal muscle is classified as a **striated muscle**. The heart is composed of *cardiac muscle* and although it is also a striated muscle, it is involuntary muscle (*i.e.*, controlled by an intrinsic pacemaker and modulated by the autonomic nervous system). *Smooth muscle* (which lacks the striations evident in skeletal and cardiac muscle) is an involuntary muscle typically found lining hollow organs such as the intestine and blood vessels. In all three muscle types, force is generated by the interaction of actin and myosin molecules, a process that requires transient elevation of intracellular calcium ions (*i.e.*, Ca^{2+}).

6.1 TYPES OF MUSCLES

1. Smooth Muscles

These are also known as **plain** or **smooth**, **involuntary** (*i.e.*, they are never under the control of will) or **nonskeletal muscles**.

Location. Smooth muscle is the intrinsic muscle of the alimentary canal, blood vessels, genitourinary tract, respiratory tract and other hollow and tubular organs. It is also found as

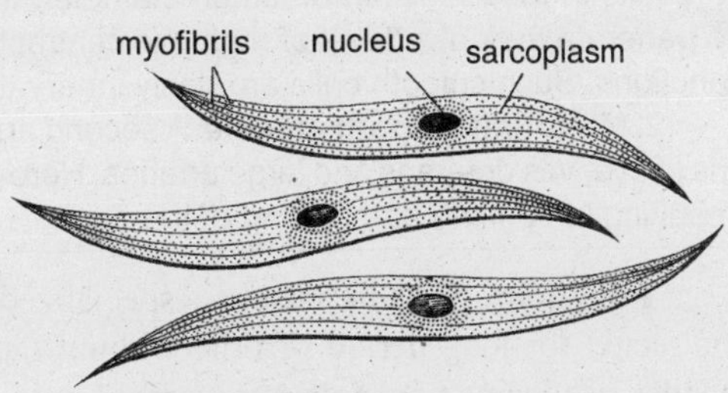

Fig. 6.1. Smooth muscle fibres.

organized muscle bundles in the iris and ciliary body of the eye, as a thin sheet in the skin of the scrotum (the dartos muscle which is situated beneath the skin of scrotum), and as isolated fibres in association with hair follicles (arrector pili muscle).

Structure. Smooth muscle generally occurs as bundles or sheets of elongated fusiform or spindle-shaped cells. The cells, also called fibres, range in length from 20 μm in the walls of small blood vessels to about 200 μm in the wall of intestine, they may be as large as 500 μm in the wall of uterus during pregnancy. The long cigar-shaped nuclei of smooth muscle cells are located in the centre of the cell and often have a corkscrew appearance in longitudinal section. In TEM (= Transmission Electron Microscope), it is evident that the cytoplasmic organelles are concentrated at each end of the nucleus. In the **sarcoplasm** (*i.e.*, cytoplasm of muscle cells/fibres) adjacent to each nuclear pole, there are numerous mitochondria, some cisternae of **rough ER** (or rough sarcoplasmic reticulum), free **ribosomes**, **glycogen** granules and a small **Golgi apparatus**. The rest of the cytoplasm is seen to be filled with 6–8 nm **actin** filaments interspersed with 8–10 nm **desmin** containing intermediate filaments. α-**actinin** protein containing **dense bodies** or **cytoplasmic densities** are found among the filaments and in relation to plasma membrane. The 16 nm **myosin** filaments can be observed in smooth muscle cells by certain special techniques.

Large number of pinocytotic vesicles are associated with the sarcolemma (= plasma membrane of muscle cells) and the smooth endoplasmic reticulum (SER). The vesicles and the SER sequester the calcium that is released to stimulate contraction. Smooth muscle cells make contact with neighboring cells by **gap junctions**. The gap junction between two smooth muscle cells is called **nexus**.

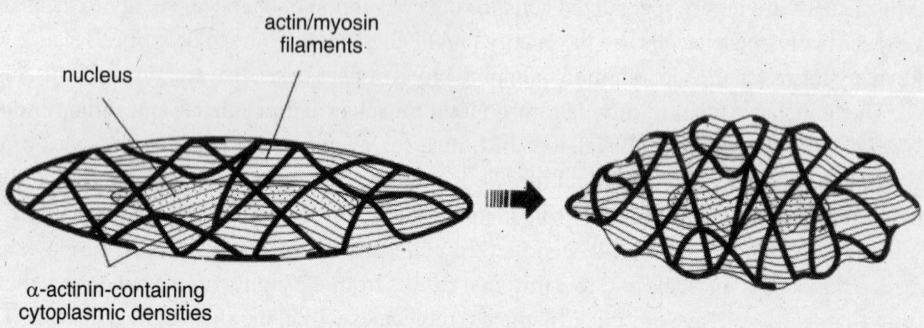

Fig. 6.2. A suggested model for smooth muscle cell contraction. Bundles of filaments containing both actin and myosin anchor on cytoplasmic densities. The densities are associated with the plasma membrane and with intermediate filaments. Because the contractile filament bundles are oriented obliquely to the long axis of the cell, their contraction shortens the cell and produces the "corkscrew" shape of the nucleus.

Box 6.1

1. Unitary smooth muscle. Most smooth muscle is present in wall of hollow visceral organs where it is arranged in sheets with cells align circumferentially or longitudinally. When it contracts, it results in the reduction of lumen diameter. In the so-called **unitary smooth muscle**, cells tend to generate their own level of rhythmic contraction which is transmitted from cell to cell via the gap junctions. Such smooth cells are richly innervated by autonomic nervous system.

2. Multi-unit smooth muscle. A second arrangement of smooth muscle is typified by that in the iris of eye, vas deferens and large arteries. Here autonomic innervation precisely controls contraction, resulting in opening and closing of the pupil.

Functions. Smooth muscle is specialized for slow and prolonged contraction. It may remain contracted for long period of time without fatigue. Smooth muscle may contract in a wave-like manner, producing **peristaltic movement** as in gastrointestinal tract and male genital tract or it may contract throughout producing extrusive movements as in the urinary bladder, the gall bladder and the uterus. Smooth muscle has spontaneous contractile activity in the absence of nerve stimuli.

Box 6.2 Some Special Features of Smooth Muscle Cells

1. **Oxytocin** is a potent stimulator of smooth muscle contraction, and it is released by the posterior pituitary. It plays an essential role in uterine contraction during parturition. It is often used as an exogenous inducer or enhancer of labour.

2. Smooth muscle cells are secretory cells, they secrete tissue matrix. They synthesize both type IV (basal lamina) collagen and type III (reticular) collagen as well as laminin, elastin and proteoglycans. Except at the gap-junctions, smooth muscle cells are surrounded by a **basal lamina**. In some locations, such as the walls of blood vessels and the uterus, smooth muscle cells secrete large amounts of collagen and elastin.

3. Smooth muscle cells are not **postmitotic**, *i.e.*, they are capable of division to maintain or increase their number, *e.g.*, smooth muscle in the uterus proliferates during the normal menstrual cycle and during pregnancy.

4. Following four types of smooth muscle cells are known to develop from undifferentiated mesenchymal cells:

(*i*) **Pericytes.** These are stellate cells and located within the basal lamina of capillaries and venules.

(*ii*) **Myofibroblasts.** They have a contractile role in addition to being able to secrete collagen, *e.g.*, alveolar septa of lungs and around crypts of glands in the gut.

(*iii*) **Myoepithelial cells.** These are modified epithelial cells which take on characteristics of smooth muscle cells. They form an important component of certain secretory glands, *e.g.*, exocrine glands such as sweat glands, mammary glands, salivary glands and iris of eye.

(*iv*) **Myoid cells of testis.** They have a contractile function in the seminiferous tubules.

Contraction of smooth muscle is usually under the regulatory control of postganglionic neurons of the **autonomic nervous system** (ANS); most smooth muscle cells are directly innervated by both sympathetic and parasympathetic nerves.

Smooth muscle contraction may also be stimulated by hormones released by the posterior pituitary gland (**oxytocin**; Box 6.2). It may be stimulated or inhibited by hormones secreted by adrenal medulla (*i.e.*, epinephrine and norepinephrine).

2. Cardiac Muscle

The cardiac muscle is found in the *wall of heart* and in the *base of the large veins* that empty into heart (**Ross** *et al.*, 1995). Structurally, it shows resemblance to skeletal muscle although functionally, it is homologous to the unstriated muscles. Cardiac muscles contract rapidly, rhythmically and tirelessly, contracting endlessly from early embryonic stage until death.

Structure. Cardiac muscle consists of long fibres that appear to branch and anastomose with neighboring fibres (Fig. 6.3). Unlike skeletal (or striated) muscle fibres, the fibres of cardiac muscle are formed by individual **uninucleate** (or mononucleate) cardiac muscle cells that are joined to one another in linear array.

In the adult, a cardiac muscle cell is 15 to 20 μm in diameter and about 100 μm long with a centrally located nucleus. Cardiac muscle cells exhibit **cross-striations** like the fibres of skeletal muscle. In fact, a cardiac muscle cell

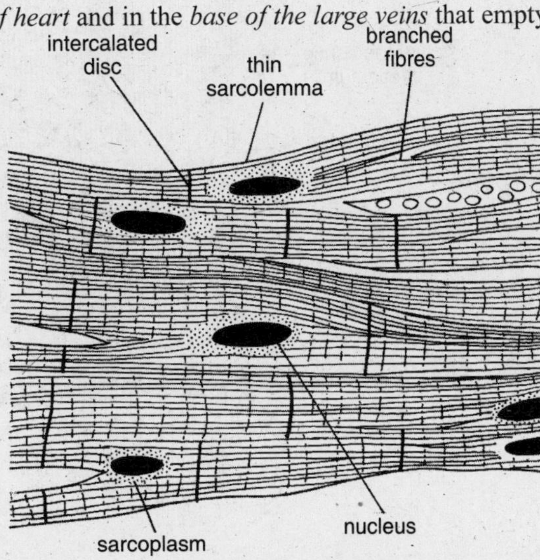

Fig. 6.3. Cardiac muscles.

has the same types and arrangement of contractile elements (**actin** and **myosin**) as in skeletal muscle. In addition, cardiac muscle fibre exhibit densely stained cross-bands, called **intercalated discs**, that cross the fibres in a linear fashion. An intercalated disc is the region of interdigitations between the cell membranes of adjacent muscle fibres. Each disc contains following three types of cell junctions:

1. Desmosomal junctions tightly link adjacent cells via anchors involving the intermediate filaments.

2. Adherent type junctions anchor the actin fibres of the sarcomeres to each end of the cell.

3. Communicating **gap junctions** facilitate the passage of membrane excitation and thereby synchronize muscle contraction.

With TEM, it has been seen that the myofibrils of cardiac muscle separate to pass around the nucleus, thus outlining a biconical **juxtanuclear region** in which the cell organelles are concentrated. This region is rich in mitochondria, and contains Golgi apparatus, lipofuscin pigment granules, hormone-containing granules and glycogen.

In addition to juxtanuclear mitochondria, cardiac cells are characterized by very large mitochondria that are densely packed between the myofibrils. These large mitochondria often extend for the full length of a sarcomere and have numerous, closely packed cristae (Fig. 6.4). Transverse tubules of cardiac muscle are much larger and located at the level of Z disc. Sarcoplasmic reticulum of cardiac muscle is organized as an anastomosing network. The association of cardiac sarcoplasmic reticulum with T tubules takes the form of **diads** rather than triads (Fig. 6.8).

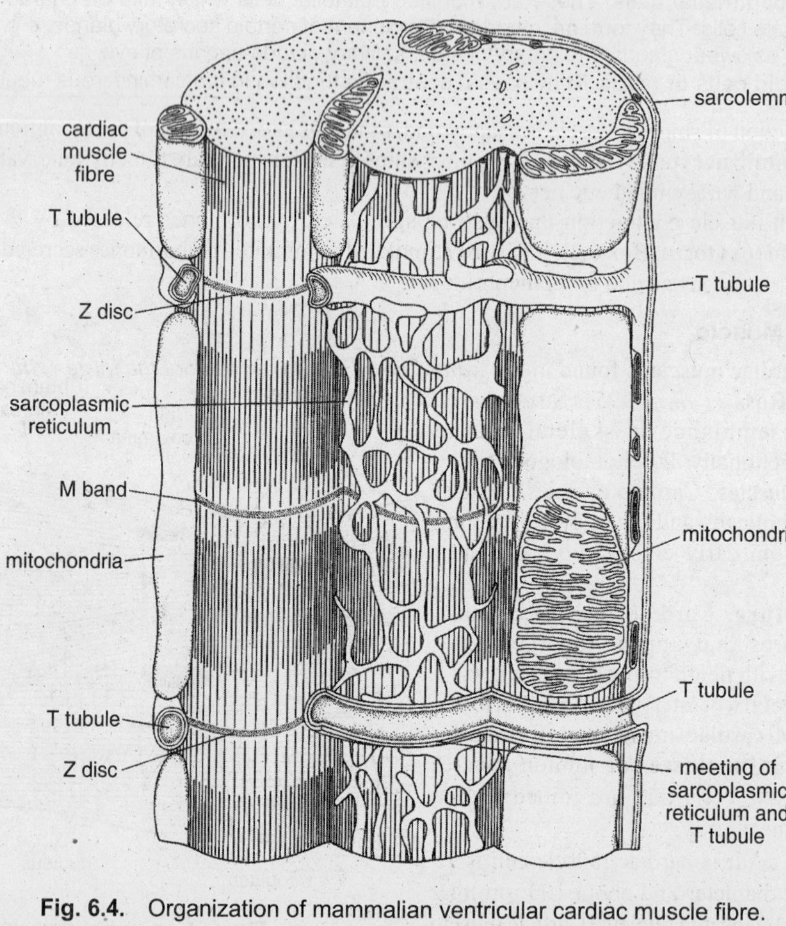

Fig. 6.4. Organization of mammalian ventricular cardiac muscle fibre.

3. Striated Muscles

Location. Striated muscle tissue forms flesh of body and its contraction is **voluntary**, under the control of will. On the basis of its location in human body, it is classified into following two types:

1. Skeletal striated muscle. It is attached to bone and responsible for movements of the axial and peripheral skeleton.

2. Visceral striated muscle. It is identical with skeletal muscle, but it is restricted to the soft tissues, namely the tongue, pharynx, diaphragm and upper part of oesophagus.

Skeletal striated muscle is responsible for gross and fine movements of limbs and digits and for the maintenance of body position and posture. The extraocular muscles of eye, too, provide precise eye movement. The muscles of tongue, pharynx, diaphragm and upper part of the oesophagus (visceral striated muscle) play essential roles in speech, breathing and swallowing.

Structure. In a skeletal muscle, each muscle cell is commonly called a **muscle fibre** and it consists of a multinucleated syncytium. It is formed by fusion of small, individual muscle cells, called **myoblasts**, during development. Each muscle cell is an elongated, straight and unbranched multinucleated cylindrical fibre which may have a diameter of 10–100 μm. Its length may be upto 4 cm in length. (**Note.** A muscle fibre should not be confused with a connective tissue fibre: muscle fibres are **cellular** elements, while connective tissue fibres are extracellular products of connective tissue cells).

Each muscle cell/fibre consists of a thin outer limiting membrane, the **sarcolemma** (Box 6.3) which encloses a coenocytic cytoplasm, known as **sarcoplasm**. The nuclei of muscle cell are peripheral in position, *i.e.*, located immediately under the sarcolemma. A large single muscle cell/ fibre may contain several hundred nuclei. In sarcoplasm lie embedded large number of **myofibrils** or **sarcostyles**. The myofibrils are relatively coarse and tightly packed in parallel bundles and are separated by thin sheets of cytoplasm. Mitochondria and glycogen deposits are located between the myofibrils in association with smooth endoplasmic reticulum (SER).

Box 6.3

Traditional histologic descriptions of skeletal muscle used the term sarcolemma to describe thick "membrane" as the cytoplasmic boundary of the muscle cell. It is now known that the thick sarcolemma actually represents the plasma membrane of the cell, its basal lamina and the surrounding reticular lamina.

Fibrillar connective tissue surrounds both individual musle fibres and bundles of muscle fibres (called **fasciculi**) and is essential for force transduction. At the end of muscle, the connective tissue continues as a **tendon** that attaches the muscle to bone. A rich blood supply and a rich supply of nerves travel in the connective tissue.

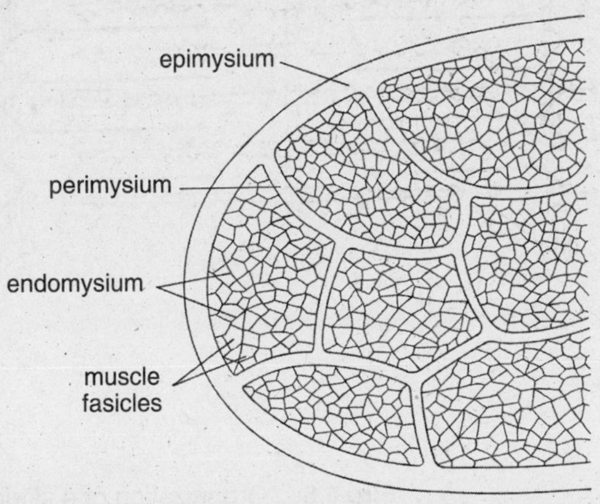

Fig. 6.5. Organization of muscle fibres into muscle.

The connective tissue associated with muscle is given a specific set of names such as follows (Fig. 6.5).

1. Endomysium is the delicate layer of reticular fibres that immediately surrounds *individual muscle fibres*. Only small diameter capillaries and the finest neuronal branches are present within the endomysium, running in parallel with the muscle fibres.

2. Perimysium is a thicker connective tissue layer that surrounds a group of fibres to form a bundle of **fascicles**. Larger blood vessels and nerves travel in the perimysium.

3. Epimysium is the sheath of dense connective tissue that surrounds the collection of **fascicles** that constitute the muscle. The major vascular and nerve supply of the muscle penetrate through the epimysium.

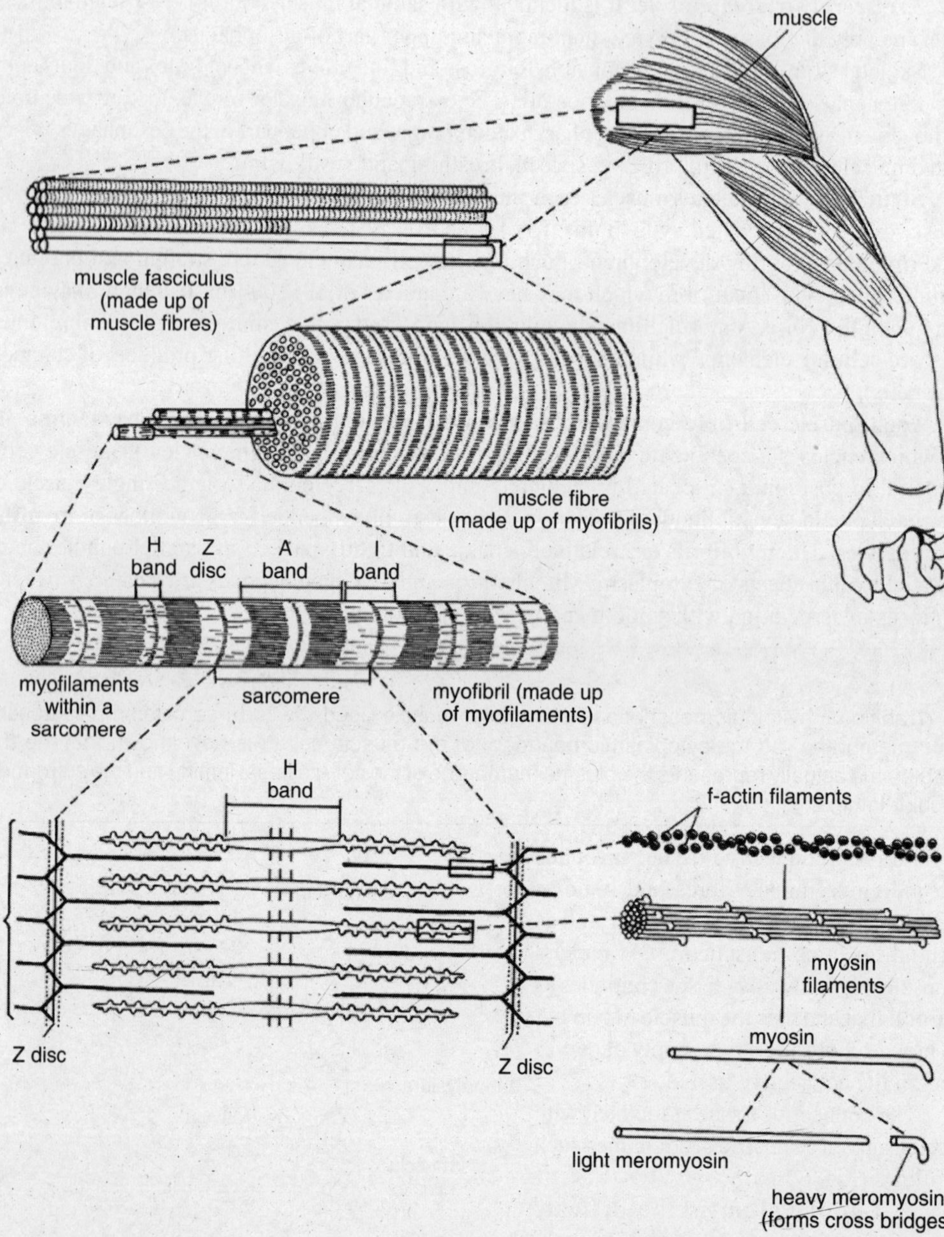

Fig. 6.6. Organization of a skeletal or striated muscle.

Ultrastructure. The structural and functional subunit of the muscle fibre is the myofibril. Myofibrils are composed of bundles of **myofilaments**. Myofilaments are individual filamentous polymers of **myosin** (**thick filaments**) and **actin** and its associated proteins (**thin filaments**). They are the actual contractile elements of striated muscle. **Cross striations** are the principal histological feature of striated muscle. In stained and in phase contrast microscopy, alternating light and dark bands are evident, these are termed as the **A band**, the **I band** and the **Z disc** or **Z line** (Fig. 6.6).

In polarizing microscopy, the dark bands are **birefringent**, *i.e.*, they alter the plane of polarized light. The dark bands are therefore **anisotropic** and are given the name **A band**. The light bands are **monorefringent**, *i.e.*, they do not alter the plane of polarized light. They are, therefore, **isotropic** and are given the name **I band**. Each of the major bands; the A band and the I band is bisected by a narrow zone of contrasting density.

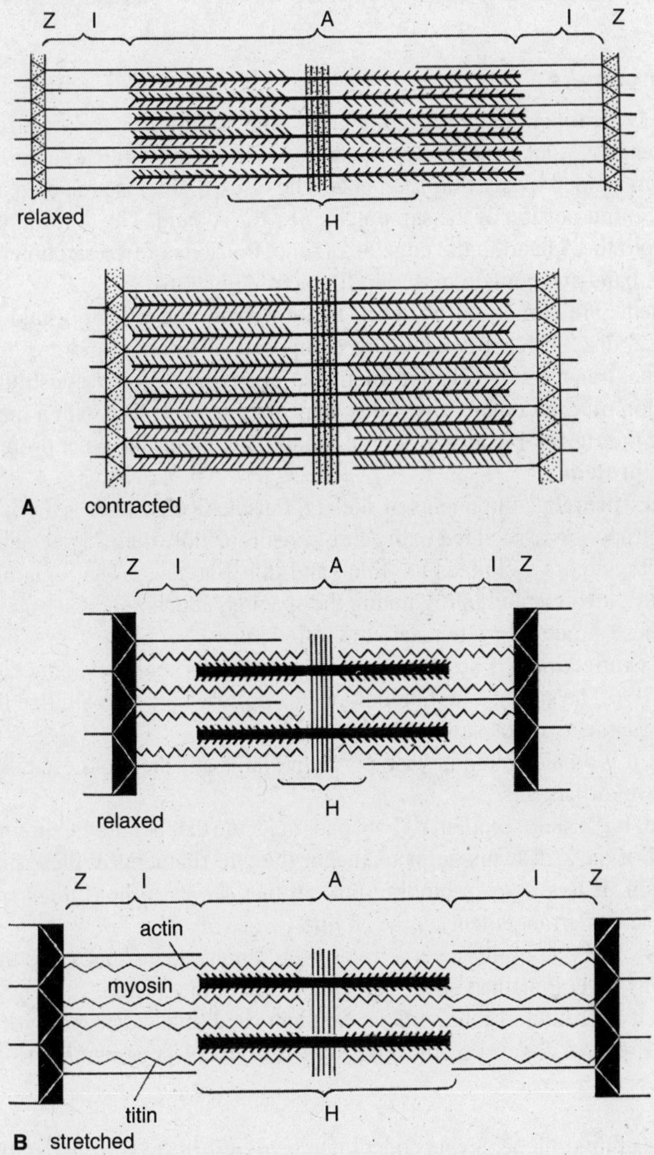

Fig. 6.7. A—Relaxed and contracted sarcomere. In the relaxed state, interdigitation of actin and myosin filament, is not complete; the H and I bands are relatively wide. During contraction, the interdigitation of the actin and myosin filaments is increased according to degree of contraction. The length of the A band remains the same; the length of the H and I bands is diminished again in proportion to the degree of contraction. B—Titin, a large elastic molecule, anchors the thick myosin filaments to the Z disc.

1. Z line or **Z disc** (Ger. *Zwischenscheible* = a between disc). It is a dense zone (dark refractile) that bisects the I band. It is also called **intermediate** or **Krause's membrane**.

2. H zone (Ger. *Helle* = light). It is a light zone that bisects the A band. It is also called **Hensen's disc**.

3. M line. It can be seen in the middle of the H band.

Sarcomere. The segment of the myofibril between two Z discs is called **sarcomere**. The sarcomere is the basic contractile unit of striated muscle. It measures 2–3 μm in relaxed mammalian muscle, may be stretched to more than 4 μm, and , in extreme contraction, may measure as little as 1 μm (Fig. 6.7).

Proteins of Sarcomere

Sarcomere is the functionally repeating unit of the myofibril; it extends from one Z disc to the next Z disc. There is a regular arrangement of contractile proteins within a sarcomere, with each thick filament being surrounded by six thin filaments. The thick filaments are about 1.5 μm long and are restricted to the central portion of the sarcomere, *i.e.*, the A band. The thin filaments attach to the Z line and extend into the A band to the edge of H zone. Portions of two sarcomeres, on either side of Z disc, form the I band and contain only thin filaments (Fig. 6.6).

Thin filaments are composed of **actin**, **tropomyosin** and **troponin** and are associated with **α-actinin** at the Z disc. Thick filaments are composed of **myosin** and are held in the centre of the H zone by fine, transversely oriented filaments of **myomesin** that constitutes the **M line**. In a longitudinal section of a sarcomere, the Z disc appears as zigzag line with a matrix material, the **Z matrix**, bisecting the zigzag. Besides α-actinin, Z matrix contains the other proteins such as **filamin**, **amorphin** and **Z protein**.

The **primary protein** components of skeletal muscle fibrils, such as myosin and actin and the tropomyosin and troponins associated with actin, constitute more than 75 per cent of the total protein of the muscle fibre and are visible as the thick and thin filaments. The remaining proteins, called **accessory proteins**, are essential in regulating the spacing, attachment and precise alignment of the myofilaments. These structural proteins are the following:

1. Titin or connectin. It is an extremely long elastic protein that connects the thick filament to Z disc (Fig. 6.7B). The spring-like function of this protein helps to stabilize the centralization of the thick myosin filaments in the sarcomere.

2. Nebulin. It is an elongated inelastic protein that is attached to the Z discs and runs parallel to the thin (actin) myofilaments.

3. α-actinin. It is a short, bipolar, rod-shaped molecule that bundles actin filaments into parallel arrays at the level of the Z disc and helps to anchor the thin filaments at the Z disc.

4. Myomesin. It is a myosin-binding protein that serves to hold thick myosin filaments in register (*i.e.*, in a lattice arrangement) at the M line.

5. C protein. It is one of the several myosin-binding proteins that serve the same function as myomesin. It forms several distinct stripes on either side of the M line.

6. Desmin. It is protein of intermediate filaments and links adjacent myofibrils to each other and maintain their register. They also links myofibrils to the plasma membrane.

Triad System

The bundles of myofilaments that constitute the myofibril are surrounded by a well-developed smooth endoplasmic reticulum (SER), also called **sarcoplasmic reticulum**. This reticulum forms a highly organized tubular network around the contractile elements in all striated muscles, *i.e.*, skeletal, visceral and cardiac.

The sarcoplasmic reticulum is arranged in networks around or between a group of myofilaments (*i.e.*, a myofibril). In the most highly developed arrangements, one network of sarcoplasmic reticulum surrounds the A band, and another network surrounds the I band (Fig. 6.8). Where the two networks meet, at the junction between A and I bands, the sarcoplasmic reticulum forms a slightly more ring-like channel, called the **terminal cisterna** or **sac**, around the filaments of the myofibrils.

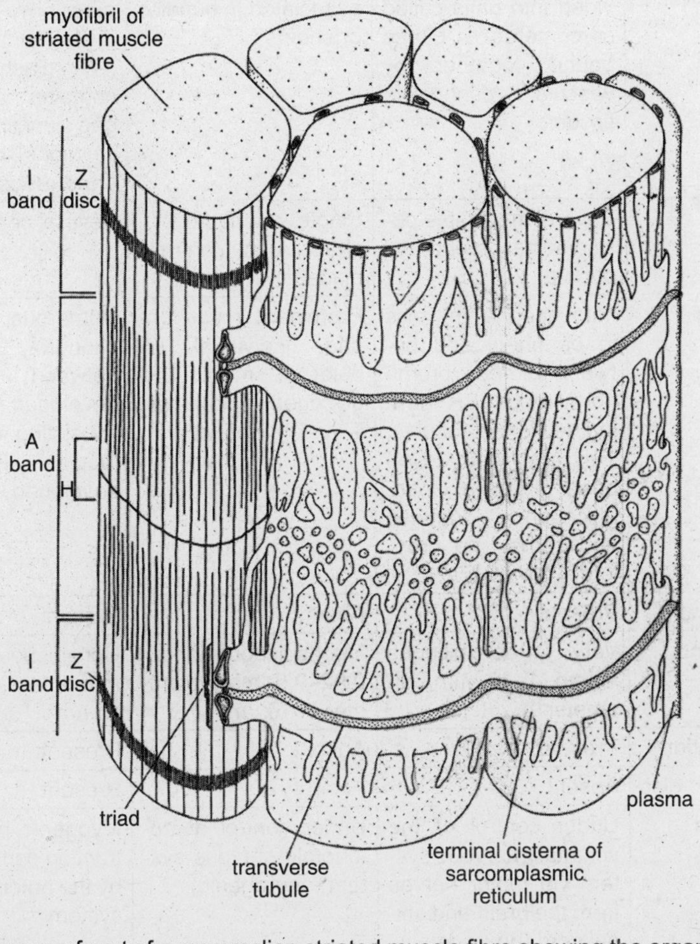

Fig. 6.8. Diagram of part of a mammalian striated muscle fibre showing the organization of a triad (there are two triads to each sarcomere).

Invaginations of the plasma membrane of the muscle cell form the **transverse tubular system** or **T system**. The **T tubules** penetrate to all levels of the muscle fibre. They are located between the adjacent terminal cisternae.

The complex of T tubule and two terminal cisternae seen with the TEM (= transmission electron microscope) is called the **triad**. The T tubules provide for the rapid transmission of surface membrane excitation to the terminal sacs throughout the thickness of muscle fibre. In turn, the terminal sacs release and then reaccumulate calcium ions (Ca^{2+}). Thus, sarcoplasmic reticulum serves as a reservoir for this essential ion (*i.e.*, Ca^{2+}).

Table 6.1 Comparision (*i.e.*, similarities and differences) between striated, smooth and cardiac muscles.

	Feature	Striated muscles	Smooth muscles	Cardiac muscles
1.	Other names	Voluntary, striped, skeletal	Involuntary, un-striped, unstriated	Heart
2.	Specialization	Most highly specialized	Least specialized	More specialized than smooth

	Feature	Striated muscles	Smooth muscles	Cardiac muscles
3.	Structure	Very long cells, usually called fibres, subdivided into units called sarcomeres. Fibres bound together by vascular connective tissue	Consists of individual, spindle-shaped cells, associated in bundles or sheets	Cells terminally branched and connected to each other by special interdigital surface processes (*i.e.*, plasma membrane contains desmosomes, gap junctions and tight junctions), called **Intercalated discs**
4.	Nucleus	Several in variable positions near periphery of fibre	Single, elongated in shape and centrally placed	Several centrally placed
5.	Cytoplasmic organelles	Mitochondria in rows in periphery and between fibres, prominent smooth ER forming network of tubules, T-system well developed forming triad, glycogen granules and some lipid droplets	Prominent mitochondria, individual tubules of smooth ER, glycogen granules, T-system absent	Numerous large mitochondria in columns between cells, poorly developed smooth ER consisting of network of tubules, T-system well developed
6.	Sarcolemma	Present	Absent	Present
7.	Myofilaments	Very conspicuous length 1–40 mm, diameter 10–60 μm	Inconspicuous, length 0.02–0.5 mm, diameter 5–10 μm	Conspicuous, length 0.08 mm or less, diameter 12–15 μm
8.	Cross striations	Present	Absent	Present
9.	Intercalated disc	Absent	Absent	Present
10.	Innervation	Under control of the voluntary nervous system via motor nerves from the brain and spinal cord (neurogenic)	Under control of autonomic nervous system (neurogenic)	Myogenic, but rate of contraction can be influenced by the autonomic nervous system
11.	Activity	Powerful, rapid contractions, short refractory period, therefore fatigues quickly	Shows sustained rhythmic contraction and relaxation, as in peristalsis	Rapid rhythmic contraction and relaxation, long refractory period, therefore does not fatigue; contraction not sustained
12.	Location	Attached to the skeleton in trunk, limbs and head	In wall of intestinal, genital, urinary and respiratory tracts, and the wall of blood vessels	Found only in wall of the heart chambers

6.2 PHYSIOLOGY OF MUSCLE CONTRACTION

At the outset let's **summarise** the main events involved in contraction of a striated muscle fiber according to the well evident sliding filament theory:

Depolarization of sarcolemma → Transfer of action potential to membranes of sarcoplasmic reticulum (SR) → Release of Ca^{2+} ions from sarcoplasmic reticulum (SR) → Binding of Ca^{2+} ions to TnC of troponin (of actin) → Formation of cross bridges between actin-myosin ATPase → Activation of myosin-ATPase → Release of energy from ATP → Breaking of actin-myosin bond → Movement of cross bridges → Sliding of thick and thin filaments past each other.

At this stage, molecular events involved in muscle contraction as follows:

1. Motor Nerve and Motor Units

Skeletal muscle is controlled by the central nervous system. Specifically, each skeletal muscle is innervated by a α **motor neuron**. The cell bodies of α motor neurons are located in the ventral horn of the spinal cord. The motor axon exit via the ventral roots and reach the muscle through mixed peripheral nerves. The motor nerves branch in the muscle, with each branch innervating a single muscle fiber. The specialized cholinergic synapse forms of **neuro-muscular junction**.

The **motor unit** consists of the motor nerve and all the muscle fibers innervated by the nerve. The motor unit is the functional contractile unit because all the muscle cells within a motor unit contract synchronously when the motor nerve fires. The size of motor units within a muscle varies depending on the function of the muscle. In the rectus muscles of the eye the motor units are small (*i.e.*, only a small number of muscle fibers are innervated by a motor neuron) and thus movement of the eye can be precisely controlled. In contrast, the motor units of the legs are large, which facilitate running. Activation of varying numbers of motor units within a muscle is one way in which the tension developed by a muscle can be controlled.

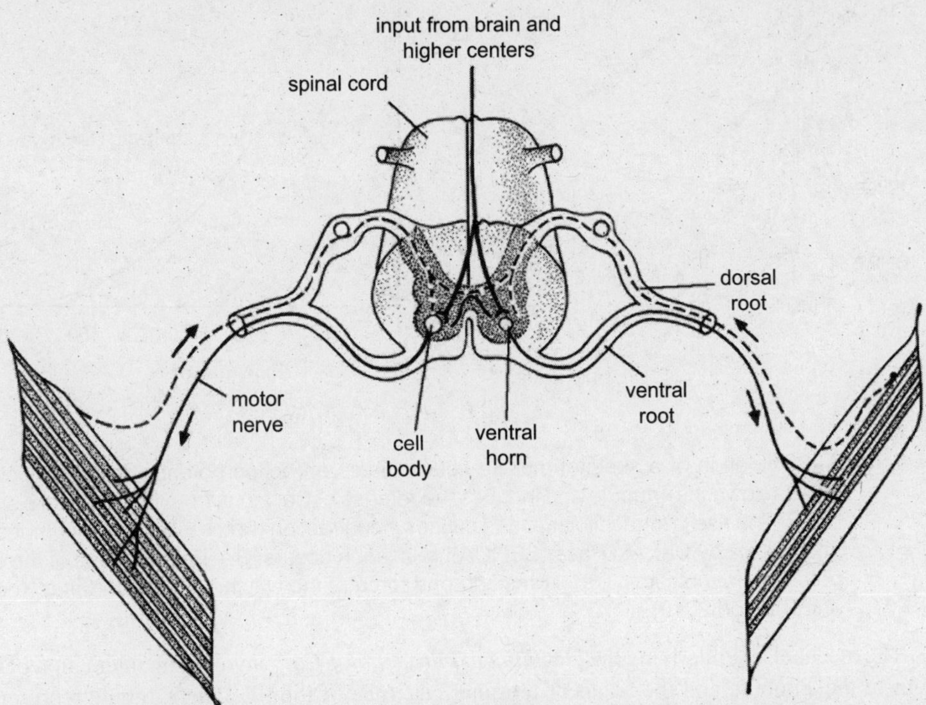

Fig. 6.9. Skeletal muscle is a voluntary muscle controlled by the central nervous system, with efferent signals (*i.e.*, action potentials) passing through an α motor neuron to muscle fibers. Each motor neuron may innervate many muscle fibers within a muscle, although each muscle fiber is innervated by only one motor neuron (after Koeppen and Stanton 2010).

The neuromuscular junction formed by the α motor neuron is called an **end plate**. Acetylcholine released from the α motor neuron at the nueromuscular junction *initiates an action potential* in the muscle fiber that rapidly spreads along its length. The duration of the action potential in skeletal muscle is less than 5 msec. This contrasts with the duration of the action potential in cardiac muscle, which is approximately 200 msec. The short duration of the skeletal muscle action potential allows very rapid contractions of the fiber and provides yet another mechanism by which the force of contraction can be increased. Increasing tension of repetitive stimulation of the muscle is called **tetany**.

2. Excitation Contration Coupling

When an action potential is transmitted along the sarcolemma of the muscle fiber and then down the T tubule, Ca^{2+} is released from the terminal cisternae of SR into the myoplasm. This release of Ca^{2+} from the SR raises intracellular Ca^{2+}, which in turn promotes actin-myosin interaction and contraction. The time course for the increase in intracellular Ca^{2+} relative to the action potential and development of force is shown in Fig. 6.10. The action potential is extremely short-lived (~ 5 msec). The elevation in intracellular Ca^{2+} begins slightly after the action potential and peaks at approximately 20 msec. This increase in intracellular Ca^{2+} initiates a contraction called a **twitch**.

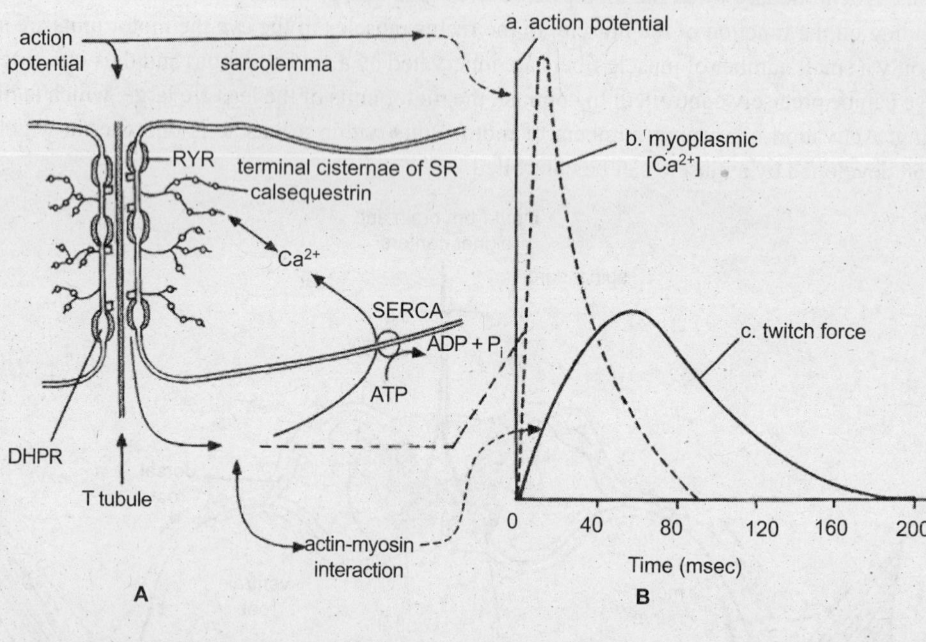

Fig. 6.10. A–Stimulation of a skeletal muscle fiber initiates an action potential in the muscle that travels down the T tubule and induces the release of Ca^{2+} ions from the terminal cisternae of SR. The rise in intracellular (Ca^{2+}) causes a contraction (twitch). As Ca^{2+} is pumped back into the SR by Ca^{2+}-ATPase (SERCA), relaxation occurs. B—Time courses of the action potential, myoplasmic Ca^{2+} transient, and force of the twitch contraction (after Koeppen and Stanton 2010).

The mechanism underlying the *elevation in intracellular Ca^{2+}* involves an interaction between protein in the T tubule and the adjacent terminal cisternae of the SR. The T tubule represents an invagination of the sorcolemma that extends into the muscle fiber and forms a close association with two terminal cisternae of the SR (Fig. 6.8; Fig. 6.10). The association of a T tubule with two opposing ιerminal cisternae is called a **triad**. Although there is a gap (≈ 15 nm in width) between the T tubule and the terminal cisternae, proteins bridge this gap. Based on their appearance on electron micrographs, these bridging proteins are called **feet** (Fig. 6.11). These feet are the Ca^{2+} release channels in the membrane of the terminal cisternae that are responsible for the elevation in intracellular Ca^{2+} in response to the action potential. Because this channel binds the drug **ryanodine**, it is commonly called the **ryanodine receptor** (**RYR**). RYR is a large protein (≈ 500 kDa) that exists as homotetramer. Only a small portion of the RYR molecule is actually embedded in SR membrane. Most of the RYR molecule appears to be in the myoplasm and spans the gap between the terminal cisternae and the T tubule (Fig. 6.12).

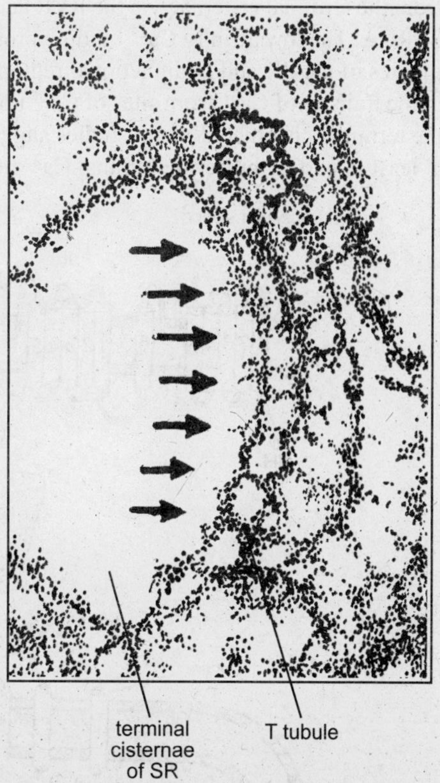

Fig. 6.11. Electron micrograph of a triad ellustrating the "feet" between the T tubule and the SR, which are thought to be the ryanodine receptors (RYRs) in the SR.

At the T tubule membrane, the RYR is thought to interact with a protein called the **dihydropyridine receptor** (**DHPR**). DHPR is an L-type voltage gated Ca^{2+} channel with five subunits. One of these subunits binds the dihydropyridine class of channel blocking drugs and appears to be critical for the ability of the action potential in the T tubule to induce release of Ca^{2+} from the SR. However, influx of Ca^{2+} into the cell through the DHPR is not needed for the initiation of Ca^{2+} release from the SR. Release of Ca^{2+} from the terminal cisternae of the SR is thought to result from a **conformational change** in the DHPR as the action potential passes down the T tubule, and this conformational change in DHPR, by means of protein-protein interaction, opens the RYR and release Ca^{2+} into the myoplasm.

Structural analysis, including the use of freeze fracture techniques, provides evidence for a close physical association of DHPR and RYR. DHPR in the T tubule membrane appears to reside directly opposite the four corners of the underlying homotetramic RYR channel in SR membrane (Fig. 6.12).

Other proteins that reside near the RYR include **calsequestrin**, **triadin** and **junctin** (Fig. 6.12). *Calsequestrin* is a low affinity Ca^{2+} binding protein that is present in the lumen of the terminal cisternae. It allows Ca^{2+} to be "stored" at high concentration and thereby establishes a favourable concentration gradient that facilitates the efflux of Ca^{2+} from the SR into the myoplasm when the RYR opens. *Triadin* and *junctin* are in the terminal cisternae membrane and bind both RYR and calsequestrin; they could anchor calsequestrin near the RYR and thereby increase Ca^{2+} buffering capacity at the site of Ca^{2+} release. **Histidine-rich calcium binding protein** (**HRC**) is another low-affinity Ca^{2+} binding protein in the SR lumen, although it is less abundant than calsequestrin. HRC appears to bind triadin in a Ca^{2+} dependent manner, which raises the possibility of a role greater than serving simply as a Ca^{2+} buffer.

Relaxation of skeletal muscle occurs as intracellular Ca^{2+} is resequestered by the SR. Uptake of Ca^{2+} into the SR is due to the action of a Ca^{2+} pump (*i.e.*, Ca^{2+} ATPase). This pump is not unique to skeletal muscle and is found in all cells in association with endoplasmic reticulum. Accordingly, it is named **SERCA**, which stands for **sarcoplasmic endoplasmic reticulum calcium ATPase**. SERCA is the most abundant protein in SR of skeletal muscle, and it is distributed thoughout the longitudinal tubules and the terminal cisternae as well. It transports two molecules of Ca^{2+} into its lumen for each molecule of ATP hydrolysed (*i.e.*, during transport of Ca^{2+}, SERCA exchanges $2Ca^{2+}$ ions for H^+ ions). Thus, the Ca^{2+} transient seen during twitch contraction reflects rlease of Ca^{2+}

from the terminal cisternae via the RYR and reuptake primarily into the logitudinal portion of SR by SERCA. The low affinity Ca^{2+} binding protein **sarcolumenin** is present throughout the longitudinal tubules of the SR and nonjunctional regions of the terminal cisternae and is thought to be involved in the transfer of Ca^{2+} from sites of Ca^{2+} uptake in the longitudinal tubules to sites of Ca^{2+} release in the terminal cisternae. Recent studies suggest that sarcalumenin increases Ca^{2+} uptake by SERCA, at least in part by buffering luminal Ca^{2+} near the pump.

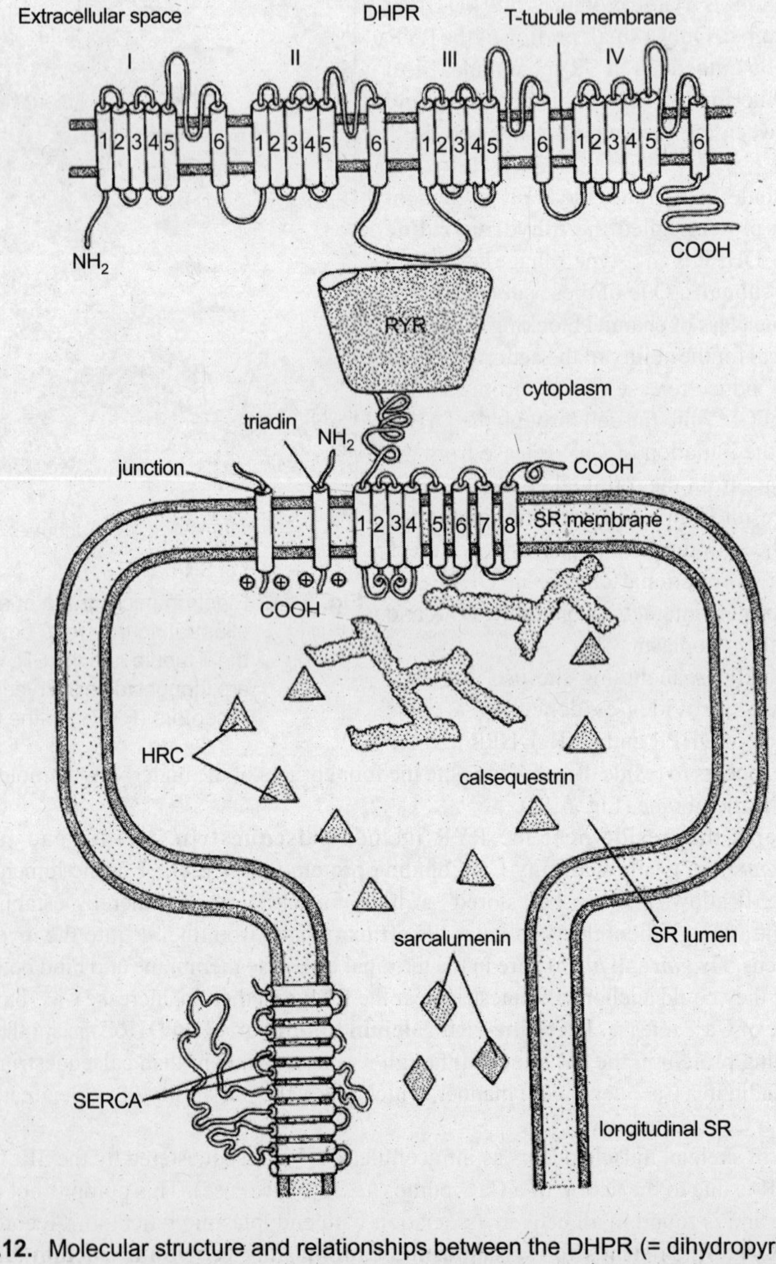

Fig. 6.12. Molecular structure and relationships between the DHPR (= dihydropyridine receptor) in the T tubule membrane and the RYR in the SR membrane. Triadin is an associated SR protein that may participate in the interaction of RYR and DHPR. Calsequestrin is a low-affinity Ca^{2+} binding protein that helps accumulate Ca^{2+} in the terminal cisternae (after Koeppen and Stanton 2010).

3. Actin-Myosin Interaction: Cross Bridge Formation

As noted, contraction of skeletal muscle requires an increase in intracellular Ca^{2+}. Moreover, the process of contraction is regulated by the thin filament. As shown in Fig. 6.13, contractile force (*i.e.*, tension) increases in sigmoidal fashion as intracellular Ca^{2+} is elevated above 0.1 μm, with half-maximal force occurring at less than 1 μM Ca^{2+}. The mechanism by which Ca^{2+} promotes this increase in tension is as follows. Ca^{2+} released from the SR binds to troponin C (TnC). Once bound with Ca^{2+}, troponin C facilitate movement of the associated tropomyosin molecule toward the cleft of the actin filament. This movement of tropomyosin exposes the myosin binding site on the actin filament and allows a cross-bridge to form and thereby generate tension. Troponin C has four Ca^{2+} binding sites. Two of these sites have high affinity for Ca^{2+} but also bind magnesium (Mg^{2+}) at rest. These sites seem to be involved in controlling and enhancing the interaction between the troponin I (TnI) and troponin T (TnT) subunits. The other two binding sites have lower affinity and bind calcium (Ca^{2+}) as its concentration rises after release from the SR. Binding of myosin to the actin filaments appears to cause a further shift in tropomyosin. Although a given tropomyosin molecule extends over seven actin molecules, it is hypothesized that the strong binding of myosin to actin results in movement of an adjacent tropomyosin molecule, perhaps exposing myosin binding sites on as many as 14 actin molecules. This ability of one tropomyosin molecule to influence the movement of another may be a consequence of the close proximity of adjacent tropomyosin molecules (Fig. 6.14).

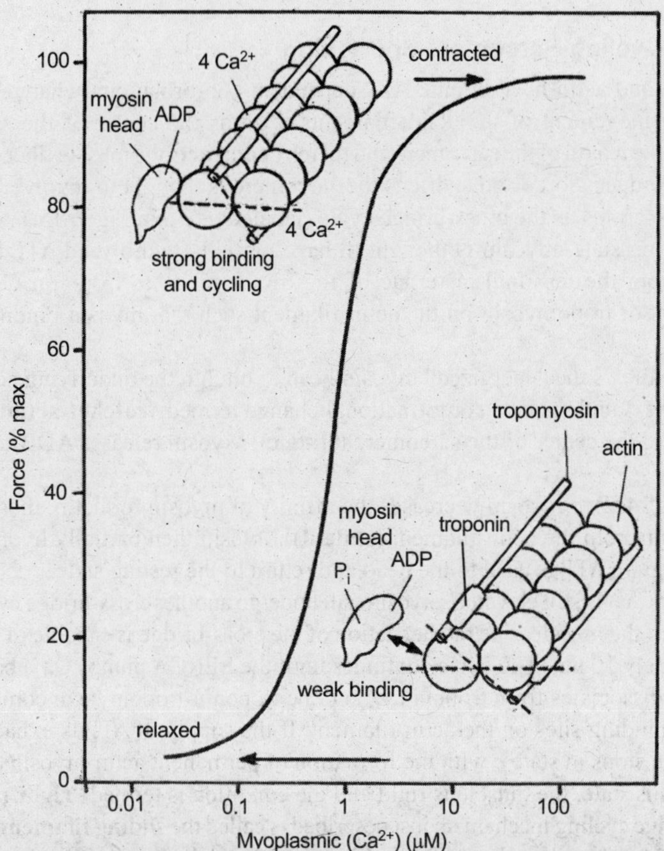

Fig. 6.13. The contractile force of skeletal muscle increase in a Ca^{2+}-dependent manner as a result of binding of Ca^{2+} to troponin C and the subsequent movement of tropomyosin away from myosin binding sites on the underlying actin molecules (after Koeppen and Stanton 2010).

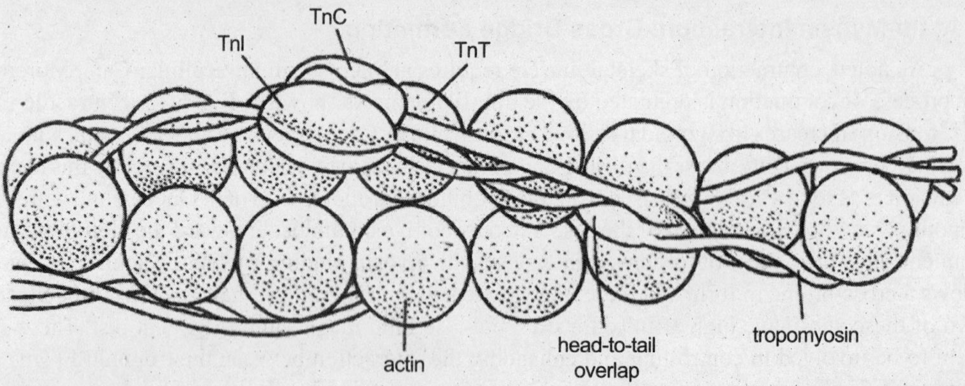

Fig. 6.14. Organization of the thin filament of the muscle cell showing a double helical array of tropomyosin on the actin filament, with sequential tropomyosin molecules arranged in a head-to-tail configuration. Such a configuration may promote the interaction of one tropomyosin unit with an adjacent tropomyosin. Also shown here is the troponin complex consisting of its three subunits: troponin C (TnC; it is a calcium binding site), troponin I (TnI; it is inhibitory subunit) and troponin T (TnT; it is tropomyosin binding subunit) (after Koeppen and Stanton, 2010).

4. Cross-Bridge Cycling-Sarcomere Shortening

Once myosin and actin have bound, ATP dependent conformational changes in the myosin molecules result in movement of the actin filaments towards the center of the sarcomere. Such movement shortens the length of the sarcomere and thereby **contracts** the muscle fiber. The mechanism by which myosin produces force and shortens the sarcomere is thought to involve **four** basic steps that are collectively termed as the cross-bridge cycle (labeled *a* to *d* in Fig. 6.15.).

(*i*) In the resting state, myosin is thought to have partially hydrolyzed ATP (**state a**). When Ca^{2+} is released from the terminal cisternae of the SR, it binds to troponin C, which in turn promotes movement of tropomyosin on the actin filament such that myosin binding sites of actin are exposed.

(*ii*) This then allows the "energized" myosin head to bind to the underlying actin (**state b**).

(*iii*) Myosin next undergoes a conformational change termed "**ratchet-action**" that pulls the actin filament toward the center of the sarcomere (**state c**). Myosin releases ADP and Pi during the transition of state c.

(*iv*) Binding of ATP to myosin decreases the affinity of myosin for actin, thereby resulting in the release of myosin from the actin filament (**state d**). Myosin then partially hydrolyzes the ATP, and part of the energy in ATP is used to the head and return to the resting state.

If intracellular Ca^{2+} is still elevated, myosin will undergo another cross-bridge cycle and produce further contraction of the muscle. The ratchet action of the cross-bridge is capable of moving the thin filament approximately 10 nm. The cycle continues until the SERCA pumps Ca^{2+} back into the SR. As Ca^{2+} falls, Ca^{2+} dissociates from troponin C, and the troponin-tropomyosin complex moves and blocks the myosin binding sites on the actin filament. If the supply of ATP is exhausted, as occurs with death, the cycle stops in state *c* with the formation of permanent actin-myosin complexes (*i.e.*, the rigor state). In this state, the muscle is rigid and the condition is termed "**rigor mortis**".

The cross-bridge cycling mechanism just described is called the **sliding filament theory** because the myosin cross-bridge is pulling the actin thin filament toward the center of the sarcomere, thereby resulting in an apparent "sliding" of the thin filament past the thick filament. The sliding filament theory was forwarded in 1965 by **A.F. Huxley** and **H.E. Huxley**.

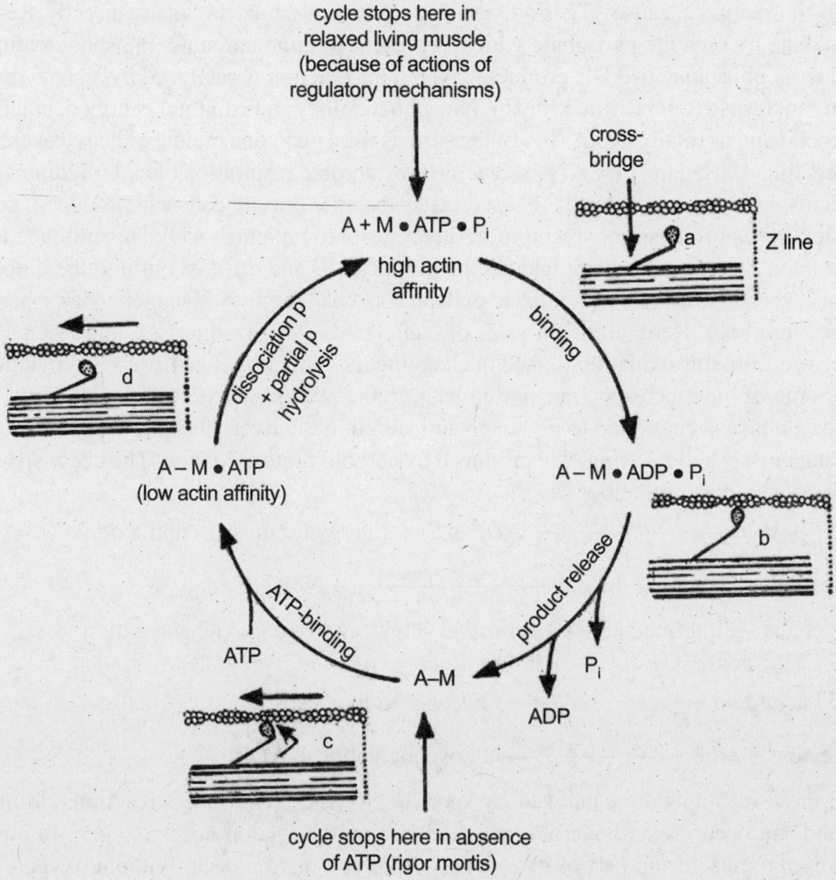

Fig. 6.15. Cross-bridge cycle. **State a**, in the relaxed state, ATP is partially hydrolyzed (M.ADP.Pi). **State b**, in the presence of elevated myoplasmic Ca^{2+}, myosin binds to actin. **State c**, hydrolysis of ATP is completed and causes a conformational change in the myosin molecule that pulls the actin filament toward the center of the sarcomere. **State d**, A new ATP binds to myosin and causes release of the cross-bridge. Partial hydrolysis of the newly bound ATP recocks the myosin head, which is now ready to bind again and again. If myoplasmic Ca^{2+} is still elevated, the cycle repeats. If myoplasmic Ca^{2+} is low, relaxation results.

6.3 ENERGY FOR MUSCLE CONTRACTION

The ultimate source of energy for muscle contraction is usually glycogen, but may be fatty acids. Metabolism of these substances during respiration generates ATP which is immediate source of energy for muscle contraction. Normally, the oxygen used in aerobic respiration is supplied by haemoglobin. But the muscles also have their own store of oxygen in the protein, **myoglobin**, which releases oxygen only when haemoglobin fails to meet its demand as in strenuous exercise.

Muscle fibers of all vertebrates and many invertebrates contain two organic phosphates, named **adenosine triphosphate (ATP)** and **phosphocreatine** or **creatine phosphate** with high energy phosphate bond. In other invertebrates, **phosphoarginine** is the energy rich compound. Such compounds with high energy for muscle contraction is provided by hydrolysis of ATP. Myosin, one of the muscle proteins, acts as an enzyme, **myosin ATPase**, to catalyze this hydrolysis which produces **adenosine diphosphate (ADP)** and inorganic phosphate (Pi). The store of ATP is quickly exhausted as it is sufficient for only ten rapid contractions. The used up ATP must be restored for additional contraction because ATP is only energy compound a muscle can use directly. Regeneration of ATP is done by creatine phosphate which is present in large amounts. Phosphocreatine donates its high energy phosphate to ADP, producing ATP. This reaction is catalyzed by an enzyme **creatine phosphotransferase** (or **creatine kinase**). Phosphocreatine can fulfil total energy demand for about 5 to 10 second and normally about 70% of the store is used up in one minute of heavy exercise. When phosphocreatine is used up, new ATP is generated by aerobic respiration of carbohydrate (glycogen) or fatty acids in muscle cells. If ATP is used faster than the muscle cells can produce it aerobically, the muscle fibers start anaerobic respiration (glycolysis) to replenish ATP. This produce **lactic acid** as a byproduct. Lactic acid accumulates in the muscle cells and diffuses out into the blood. A small portion of it reenters the relaxing muscle cells and is changed into glycogen. Major part of lactic acid passes into liver. Here, one-fifth (1/5) of lactic acid is oxidized to CO_2 and H_2O. The energy (ATP) derived from this oxidation is used in changing the remaning four-fifth of liver lactic acid into glucose. Some of the glucose is transported back to the muscle where it is changed into glycogen. The rest of glucose is converted to glycogen and stored in the liver. Phosphocreatine is regenerated in a relaxing muscle cell by using ATP produced by carbohydrate oxidation. The chemistry of muscle contraction is summarized below:

(*i*) $ATP + H_2O \xrightarrow{\text{Myosine ATPase}} ADP + Pi + \text{Energy for muscle contraction}$

(*ii*) $\text{Creatine} \sim \text{Phosphate} + ADP \xrightarrow{\text{Contracting muscle}} ATP + \text{Energy}$

(*iii*) Glycogen $\rightarrow$ Intermediate $\rightarrow$ Lactic acid + Energy (used in ATP generation)

(*iv*) 1/5 Lactic acid + $O_2 \rightarrow CO_2 + H_2O$ + Energy (ATP) in liver cells

$\quad$ $4/5 \text{ Lactic acid} + \text{Water} \xrightarrow{ATP} \text{Glycogen in liver cells}$

(*v*) $\text{Creatine} + ATP \xrightarrow{\text{Relaxing muscle}} \text{Phosphocreatine} + ADP$

All these reactions are catalyzed by specific enzymes. The first 3 reactions do not involve oxygen and can occur under anaerobic conditions. Fourth reaction needs oxygen. In other words, contraction of a muscle and part of its recovery from contraction occur without oxygen. This is of great importance as it enables us to undertake great exertions. Oxygen is needed only in the recovery from contraction.

Cori cycle. This cycle was worked by **Carl F. Cori**, U.S. Biochemist and Nobel Laureate in 1896. The lactic acid formed in reaction (*iii*) above is carried in the blood to the liver, where it is converted into glycogen. The glycogen releases glucose into the blood, which is reconverted into glycogen in the muscle. This lactic acid $\rightarrow$ Glycogen $\rightarrow$ Glucose $\rightarrow$ Glycogen cycle is called the Cori's cycle (Fig. 6.16). It shows the role of liver in the metabolism of muscle contraction.

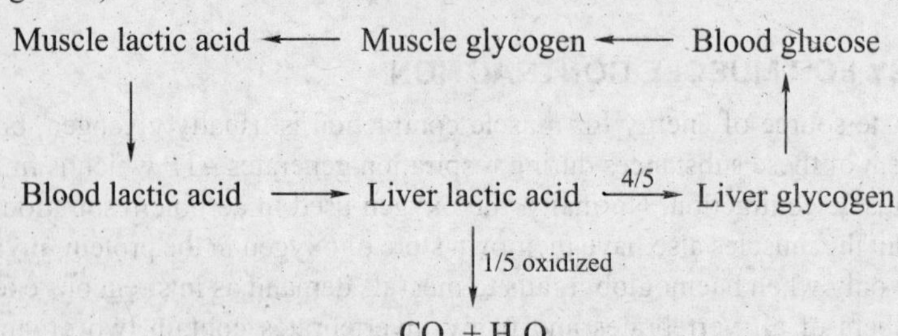

Fig. 6.16. Cori's cycle in muscle and liver cells.

6.4 TYPES OF SKELETAL MUSCLE

Skeletal muscle can be classified as either **fast-twitch** (also called **white fibers**, type II A and II B) or **slow-twitch** (also called **tonic fibers, red fibers** or type I) **muscle.** The lateral rectus muscle of the eye contracts very quickly, with peak tension attained within 7.5 msec after stimulation. The gastrocnemius muscle of the leg (*Gastrocnemius muscle* is largest muscle and most superficial muscle of calf of the human leg that arises by two heads from the condyles of the femur and has its tendon of insertion incorporated as part of the Achilles tendon.), in contrast requires 40 msec to develop peak tension. The soleus muscle of the leg requires even longer (≈ 90 msec) for peak tension to develop. Thus, the soleus muscle is classified as *slow-twitch muscle*, whereas the lateral rectus muscle would be classified as a *fast-twitch muscle*. The gastrochemius muscle contains a mixture of fast- and slow-twitch fibers and thus exhibit a weighted-average intermediate rate of tension development when the whole muscle is stimulated.

1. Tonic, slow or red muscle fibers. They are thin, dark-red and slow contracting. They have a high content of myoglobin and cytochrome pigments, hence red in appearance. They have many mitochondria, low glycogen content and poorly formed sarcoplasmic reticulum. They depend on aerobic respiration. The tonic muscle fibers are innervated by small nerve fibers have a diameter of about 5 μm and providing many end-plates along the length of the fiber. This is called **multi-terminal innervation.**

Sustained contraction provided by the tonic muscle fibers is used for the maintenance of posture by the animal. The **extensor muscle** of the back in humans remain in sustained contraction to maintain erect posture against gravity.

2. Twitch, fast or white muscle fibers. They are much thicker, lighter in colour and fast-contracting. They have a low content of myoglobin or cytochrome. They have few mitochondria, abundant glycogen granules and well formed sarcoplasmic reticulum. The twitch muscle fibers are innervated by large nerve fibers, having a diameter 10 to 20 μm and providing a single end plate to each fiber. This is called **uniterminal innervation**. These muscles depend on anaerobic respiration (glycolysis) for ATP supply. Lactic acid accumulates during sternuous work and the fibers get fatigued soon. They contract three times faster than the tonic muscle fibers. Glycogen is used extensively as a respiratory substrate. The muscles that move the eyeballs are very rich in twitch muscle fibers.

QUESTIONS

Long Answer Questions

1. Describe the ultrastructure of myofibril of striated muscle fiber and discuss mechanism of muscle contraction. (*Awadh 2004; CCSU Meerut 2013*)
2. Give an account of chemical composition of voluntary muscle. Describe various steps in contraction of voluntary muscle. (*CCSU Meerut 2009*)
3. Give an account of different types of muscles found in vertebrates. Describe physiology of muscle contraction. (*CCSU Meerut 2008*)
4. Describe the sliding filament theory of muscle contraction in a striated muscle. (*Allahabad 2003*)
5. Discuss the physico-chemical changes during muscle contraction. (*MSU Tirunelveli 2007*)
6. Write an essay on the ultra-structure of striated muscle and add a note on physiology of muscle contraction. (*Osmania 2005*)
7. Describe the ultrastructure of a myofibril of a striated muscle fiber and discuss mechanism of muscle contraction. (*CCSU Meerut 2013*)
8. Name four major proteins in muscles. (*CCSU Meerut 2013*)

Short Answer Questions

1. Define the term fatigue and rigor mortis. (*Kanpur 2003*)
2. What are the role of Ca^{2+} and ATP muscle contraction? (*Avadh 2003*)

3. Draw a neat and labelled diagram of the following:
   (*i*) Unstriated muscle fiber of vertebrate; (*CCSU Meerut 2006, 2007, 2010*)
   (*ii*) Sarcomere; (*CCSU Meerut 2014*)
   (*iii*) Triad.
4. Write short notes on the following:
   (*i*) Cori Cycle;
   (*ii*) Muscle proteins;
   (*iii*) Sarcolemma;
   (*iv*) Neuromuscular junction;
   (*v*) Oxygen debt.

Very Short Answer Questions

1. Name four major proteins in muscle. (*CCSU Meerut 2013*)
2. What are the two types of striated muscle filaments?
3. Give a term for a continued state of contraction caused by many rapidly repeating stimuli in a muscle.
4. Name the thin and thick protein of muscle fibers.
5. Name the tissue which connects muscle to a bone.
6. What property of muscle fiber enables them to cause movement.
7. Name the functional contractile unit of the muscle.

Fill in the Blanks

1. A muscle gets fatigued by accumulation of ________ in it.
2. Sarcolemma is stimulated (depolarized) by ________ released from the axon at the neuromuscular junction.
3. Depolarization of sarcolemma causes the release of ________ from the ________ of the muscle fiber.
4. A triad consists of a ________ tubule and two cisternae of ________.
5. Myoglobin found in ________ stores ________.
6. ________ muscle contracts quickly.
7. The state of sustained contraction of a muscle resulting from a rapid succession of nerve impulses is called ________.

Multiple Choice Questions

1. The basic unit of muscle contraction is
   (*a*) sarcomere (*b*) tropomyosin
   (*c*) myosin (*d*) actin
2. The light and dark bands are visible in
   (*a*) striped muscle (*b*) unstriped muscle
   (*c*) cardiac muscle (*d*) both (*a*) and (*c*)
3. Smooth muscles occur in
   (*a*) uterus (*b*) artery
   (*c*) vein (*d*) all the above
4. Which of the following is contractile protein of a muscle?
   (*a*) tubulin (*b*) myosin
   (*c*) tropomyosin (*d*) actin
5. Fine connective tissue surrounding individual muscle fibres is
   (*a*) epimycium (*b*) endomycium
   (*c*) perimycium (*d*) none

6. Which of the following yields ATP required for muscle contraction?
(*a*) myosin (*b*) myoglobin
(*c*) creatine phosphate (*d*) creatinine phosphate

7. Muscle contraction requires
(*a*) proteins (*b*) ATP, Mg^{2+} and Ca^{2+}
(*c*) stimulus (*d*) all of these

8. The contraction and relaxation phases of a muscle constitute
(*a*) twitch (*b*) beat
(*c*) condition (*d*) stimulus

9. The functional unit of the contractile system in a striated muscle is
(*a*) sarcomere (*b*) A-band
(*c*) Z-line (*d*) myofibril

10. Cori cycle is associated with
(*a*) lactic acid (*b*) pyruvic acid
(*c*) oxalic acid (*d*) citric acid

11. Substance that acumulates in a fatigued muscle are
(*a*) ADP (*b*) lactic acid
(*c*) CO_2 (*d*) pyruvic acid

12. The cross bridges of the sarcomere in skeletal muscle are made up of
(*a*) actin (*b*) myosin
(*c*) tropomyosin (*d*) troponin

13. Sliding theory was proposed by (*MSU Tirunelveli 2007*)
(*a*) William (*b*) Hill
(*c*) Huxley (*d*) Gerald

14. The muscle found in blood vessel is (*MSU Tirunelveli 2007*)
(*a*) striped muscle (*b*) unstriped muscle
(*c*) cardiac muscle (*d*) none of the above

ANSWERS

Very Short Answer Questions

1. Myosin, actin, tropomyosin and troponin.
2. Actin and myosin.
3. Tetanus.
4. Actin and myosin
5. Tendon (*i.e.*, white fibrous connective tissue).
6. Contractility.
7. Sarcomere.

Fill in the Blanks

1. Lactic acid; **2.** Acetylcholine; **3.** Calcium (Ca^{2+}) ion, sarcoplasmic reticulum; **4.** T, sarcoplasmic reticulum; **5.** Tonic/slow/red, oxygen; **6.** White/fast/twitch; **7.** Tetanus.

Multiple Choice Questions

1. (*a*) **2.** (*d*) **3.** (*d*) **4.** (*b*) **5.** (*b*) **6.** (*c*)
7. (*b*) **8.** (*a*) **9.** (*a*) **10.** (*a*) **11.** (*b*) **12.** (*b*)
13. (*c*) **14.** (*b*)

7

Sexual Reproduction (Histophysiology)

Reproduction is a biological process by which living organisms produce new individuals of their own kind. When reproduction is performed by the participation of two individuals it is called **sexual reproduction;** when it is carried out singly, it is referred to as **asexual reproduction.** Sexual reproduction is of widespread occurrence in angiospermic plants and animals; most higher animals reproduce only by sexual reproduction. The basic requirement of sexual reproduction is to achieve the union of the participating individuals at the cellular level. This is accomplished by the union of the nuclei of the sex cell or gamete, *i.e.*, **ovum** and **sperm** to form the zygote. This phenomenon is called **fertilization.** The two participants (individuals) in sexual reproduction are morphologically and functionally different —the one producing the *ovum* is called the **female** and the other producing *sperm* is known as **male.** The reproductive system of nearly all mammals is equipped to bring about the successful development of **zygote** into a new individual within the body of the female.

The reproductive system is provided with *gonads* which produce **sex cells** (*germ cells* or *gametes*). These gonads are **testes** in the male and **ovaries** in the female. The gonads are double organs : the gonads besides producing the germ cells like the **exocrine organs,** also secrete a variety of hormones like the **endocrine organs.** In addition, the reproductive system is provided with accessory glands, and ducts which transport the gametes.

7.1 GENITAL RIDGE

The human embryo starts its life sexually neutral. As it grows it develops into either a male or a female. Male embryos have an X and a Y chromosome. A gene on the Y chromosome codes for the production of **factor SRY**, which stimulates the development of the testes, so the embryo develops into a boy (SRY = **S**ex determining **r**egion of the **Y** chromosome). Female embryos have two X chromosomes and so do not have the gene that codes for SRY (Box 7.1).

Box 7.1

The story of sex determination in mammals starts in the **genital ridge**, which can become either type of gonad. Here, the genes encoding wt1, Sox9, Wnt4, Lhx9, Fgf9, GATA4, and SF1 are expressed and the loss of function of any one of them will prevent the normal development of any gonad.

In humans, the major gene for testis determination resides near the tip of the short arm of Y chromosome. In this region, **Sinclair** and colleagues (1990) found a male-specific DNA sequence that encodes a peptide of 223 amino acids, the *human testis determining factor).* The SRY gene is probably active for only a few hours. During this time, it synthesizes the Sry transcription factor whose primary role is to activate SOX9 gene, an autosomal gene involved in several development processes, most notably bone formation. *Sry* gene is found specifically in mammals, while *SOX9* gene is found throughout the vertebrate phyla. Thus, *SRY* may act merely as a "switch" operating during a very short time to activate **Sox9** and SOX9 protein to testis formation (**Gilbert** 2010).

Wnt4 and R-spondin 1. In mice, **Wnt4** is expressed in the genital ridge of both sexes, but its expression later becomes undetectable in XY gonads (which become testes), whereas it is maintained in XX gonads as they begin to form ovaries. Wnt4, although it is not the only determining factor.

R-spondin1 (Rspo1) is also critical in ovary formation, since in human case studies several XX individuals with RSPO1 gene mutations became males (**Parma** *et al.*, 2006). RSPO1 acts in synergy with Wnt4 to produce **β-Catenin** and this β-catenin appears to be critical both in activating further ovaries development and in blocking the synthesis of Sox9. **Sox9** is necessary in testes determination (Gilbert 2010).

7.2 THE MALE REPRODUCTIVE SYSTEM

The male reproductive system has evolved for **continuous, life-long gametogenesis,** coupled with occasional **internal insemination** with a *high density of sperm* ($> 60 \times 10^6$ /ml in 3 to 5 mL of semen). In adult men the basic roles of gonadal hormones are (1) support of gametogenesis (**spermatogenesis**), (2) maintenance of the male reproductive tract and production of semen and (3) maintenance of secondary sex characteristics and libido. There is no overall cyclicity of this activity in men.

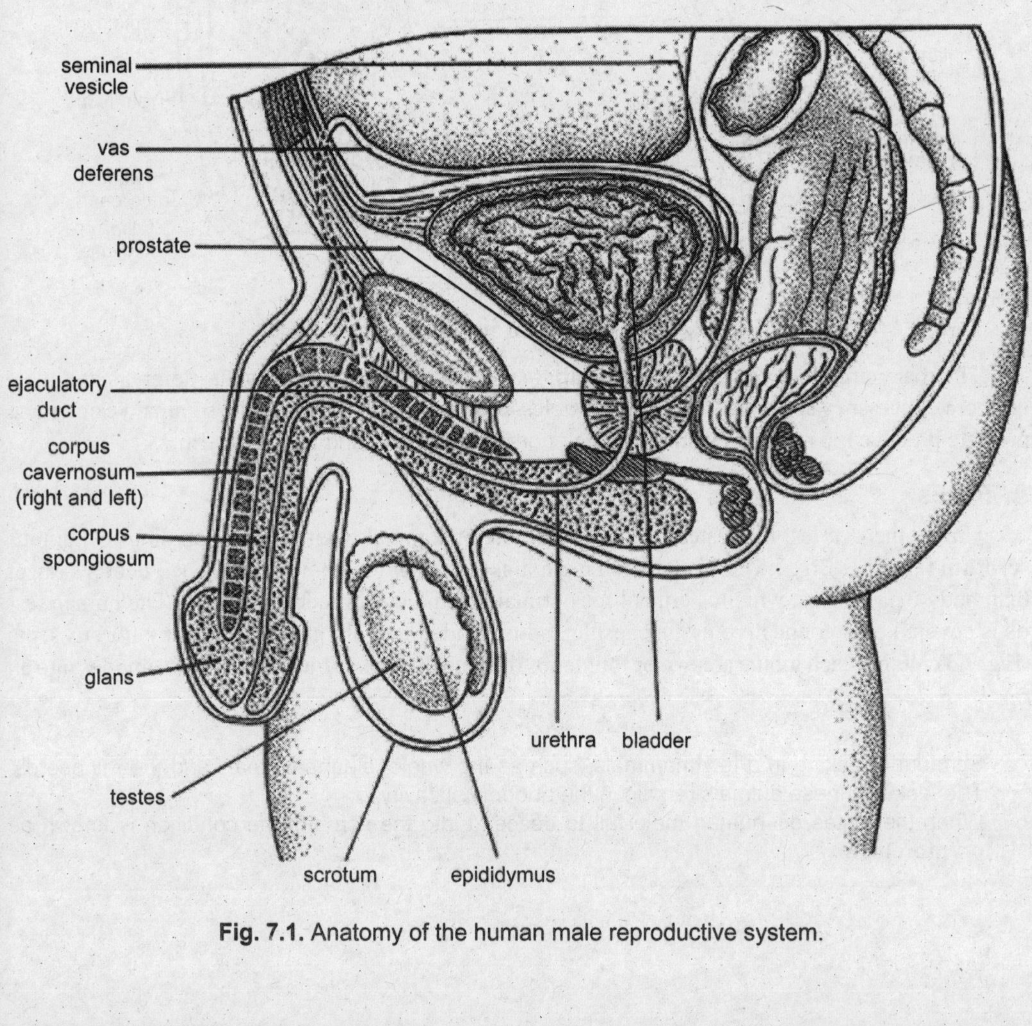

Fig. 7.1. Anatomy of the human male reproductive system.

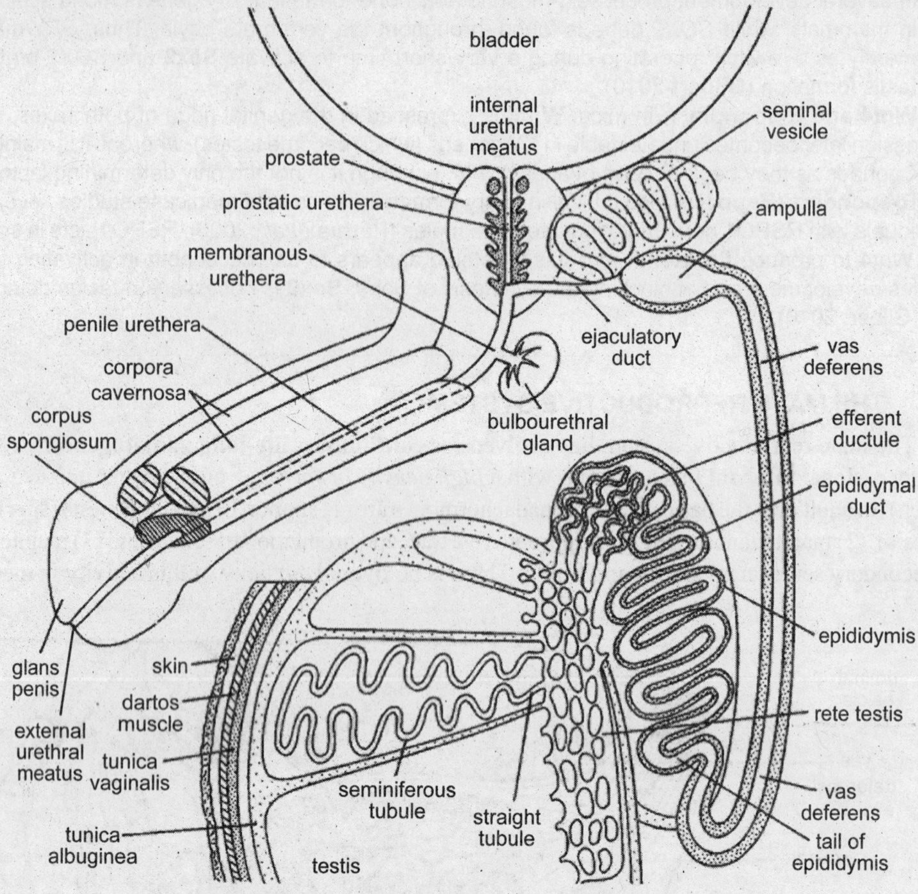

Fig. 7.2. Details of genital system of human male.

The male reproductive tract includes **tubal structures** (epididymis, ductus deferens and male urethera), **accessory sex glands** (seminal vesicles, prostate), and the **penis**. The seminal vesicles and prostate produce most of the ejaculate, which nourishes, buffers and protect sperm.

A. Testes

Unlike the ovaries, the testes reside outside the abdominal cavity in the dangling, pendulous **scrotum** (Fig. 7.1; Box 7.2). This location maintains testicular temperature at about 2 degrees lower than body temperature, which is crucial for optimal sperm development (Box 7.3). The human testis is covered by a connective tissue capsule and is divided into about 300 lobules by fibrous septa (Fig. 7.3). Within each lobule are two or four loops of **seminiferous tubules.** Each loop empties into an

Box 7.2

Scrotum is lacking in a few mammals, such as the whale, elephant, seals and the rhinoceros. The testes in these animals remain in the abdominal cavity.

When the testes of human male fail to descend into the scrotum the condition is known as **cryptorchism.**

Box 7.3

Increasing the temperature of the testes can prevent spermatogenesis by causing degeneration of most cells of the seminiferous tubules besides the spermatogonia. On cold days, scrotal reflexes cause the musculature of the scrotum (pouch of skin containing the testicles) to contract, pulling the testes close to the body to maintain this 2°C differential (**Hall** 2011).

Elevation of testes is done by **cremaster muscle,** a small circular band of skeletal muscles upon exposure to cold. Dartos (Fig. 7.2) is a thin layer of vascular contractile tissue that contains smooth muscle fibers but no fat and is situated beneath the skin of scrotum or beneath that of labia majora (in female).

Dartos causes wrinkling of the skin of scrotum.

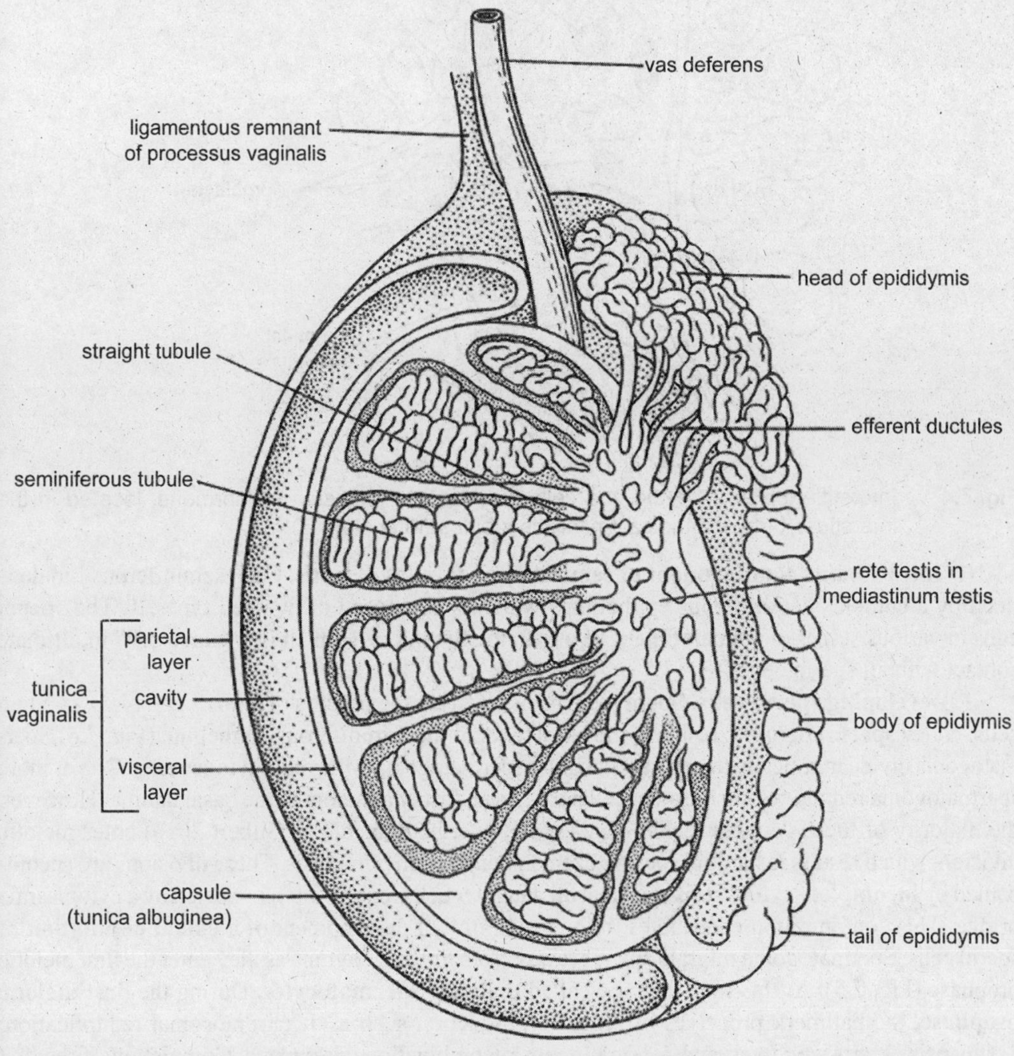

Fig. 7.3. Anatomy and organization of testis.

anastomosing network of tubules called the rete **testis**. The rete testis is continuous with small ducts, the **efferent ductules,** that lead the sperm out of the testis into the head of the **epididymis** on the superior pole of the testis (Fig. 7.3). Once in the epididymis, the sperm pass from the **head,** to the

body, to the **tail** of the epididymis and then to the **vas (ductus) deferens.** Viable sperm can be stored in the tail of the epididymis and the vas deferens for several months.

The presence of the seminiferous tubules creates two compartments: an intratubular compartment, which is composed of the **seminiferous epithelium** of the seminiferous tubule and a peritubular compartment (interstice), which is composed of neurovascular elements, connective tissue cells, immune cells, and the **interstitial cells of Leydig,** whose main function is to produce **testosterone** hormone (Fig. 7.4).

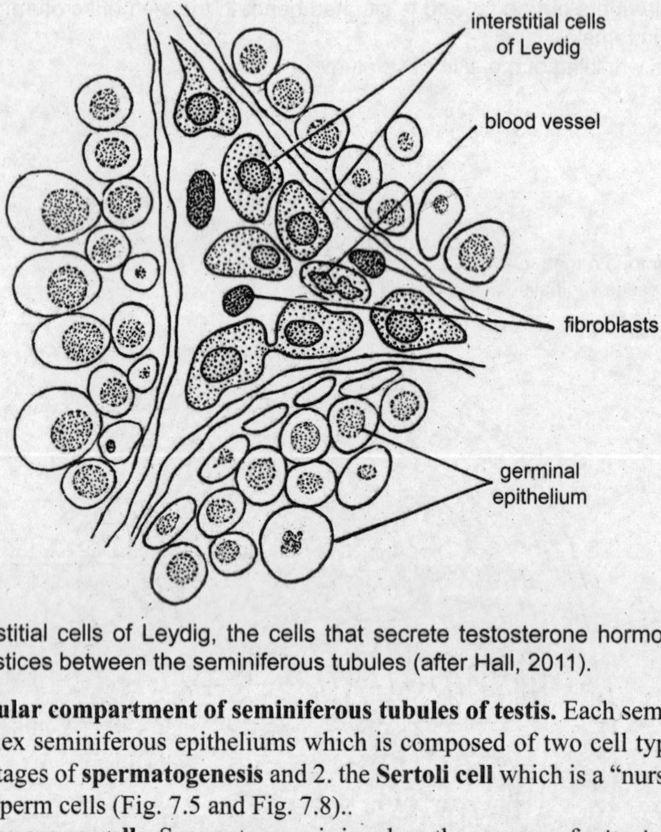

Fig. 7.4 Interstitial cells of Leydig, the cells that secrete testosterone hormone, located in the interstices between the seminiferous tubules (after Hall, 2011).

(i) Intratubular compartment of seminiferous tubules of testis. Each seminiferous tubule is lined by a complex seminiferous epitheliums which is composed of two cell types: 1. The **sperm cells** in various stages of **spermatogenesis** and 2. the **Sertoli cell** which is a "nurse cell" in ultimate contact with all sperm cells (Fig. 7.5 and Fig. 7.8)..

1. Developing sperm cells. Spermatogenesis involves the process of *mitosis* and *meiosis*. Stem cells, called **spermatogonia**, reside at the basal level of the seminiferous epithelium (Fig. 7.6) Spermatogonia divide mitotically to generate daughter spermatogonia **(spermatocytogenesis).** One or more spermatogonia remain within the stem cell population, firmly adherent to the basal lamina. However, the majority of these daughter cells (*e.g.*, Type B spermatogonia, see Gilbert 2010) enter meiotic division, which results in haploid spermatozoa on completion of meiosis. These divisions are accompanied by **incomplete cytokinesis** such that all daughter cells remain interconnected by a cytoplasmic bridge. This configuration contributes to the synchrony of development of a **clonal population** of sperm cells. Spermatogonia migrate apically away from the basal lamina as they enter the first meiotic prophase (Fig. 7.5). At this time they are called **primary spermatocytes.** During the **first meiotic prophase,** the hallmark processes of sexual reproduction such as 1. chromosomal reduplication; 2. synapsis; 3. crossing over; and 4. homologous recombination takes place. Completion of the first meiotic division gives rise to **secondary spermatocytes,** which quickly (*i.e.*, 20 minutes) completes the **second meiotic division.** The initial products of meiosis are haploid **spermatids.**

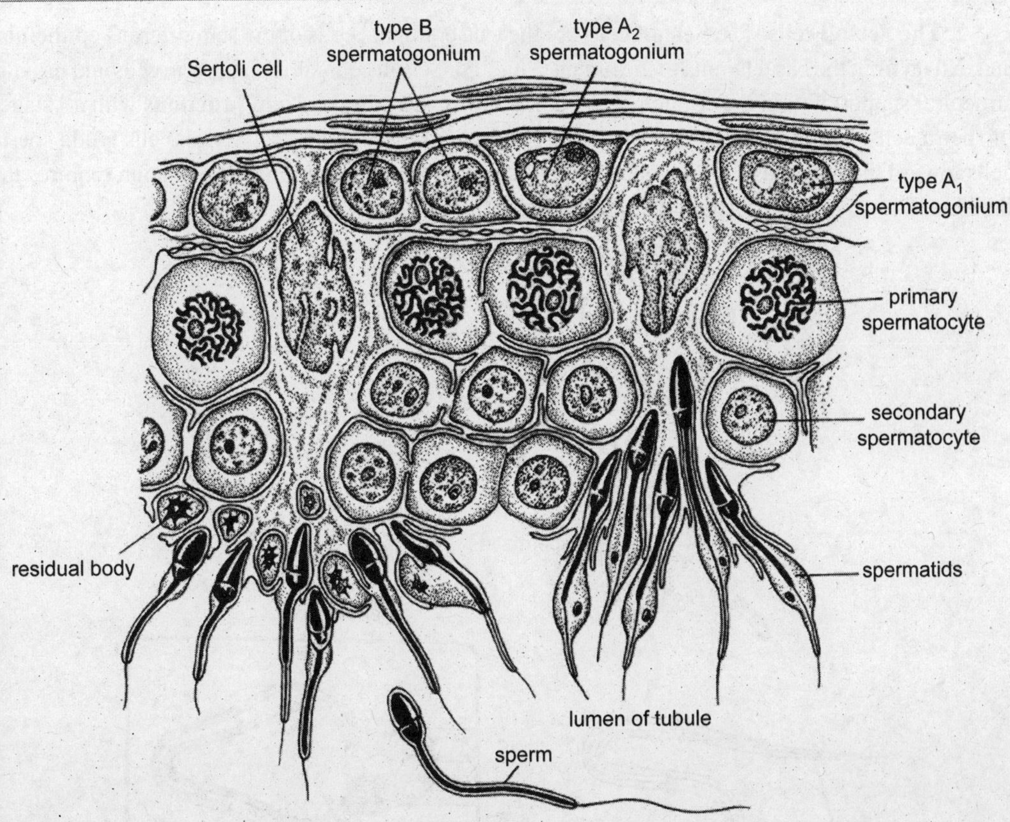

Fig. 7.5 Section of seminiferous tubule, showing the relationship between Sertoli cells and the developing sperm. As germ cells mature, they progress toward the lumen of seminiferous tubule (after Gilbert 2010).

Spermatids are small, round cells that undergo a remarkable metamorphosis called **spermiogenesis.** (Fig. 7.6). The products of spermiogenesis are the streamlined spermatozoa. As the spermatid matures into a **spermatozoon,** the size of the nucleus decreases and a prominent tail is formed. The tail of sperm contains microtubular structures that propel sperm, similar to a flagellum. The chromatin material in the sperm nucleus condenses, and most of the cytoplasm is lost. The acrosome is a membrane enclosed structure on the head of the sperm that acts as a lysosome and contains hydrolytic enzymes that are important for fertilization. These enzymes remain inactive until the acrosomal reaction occurs (during fertilization).

Spermiation. Spermatozoa are found at the luminal surface of the seminiferous tubule. Release of sperm, or **spermiation,** is controlled by Sertoli cells. The process of spermatogenesis takes about 72 days. A cohort (*i.e.*, a group of people with a shared feature) of adjacent spermatogonia enter the process ever 16 days so that the process is faltered at one point along a seminiferous tubule. In addition, the process is staggered along the length of a seminiferous tubule (*i.e.*, not all spermatogonia enter the process of spermatogenesis at the same time along the entire length of the tubule or in synchrony with every other tubule; there are about 500 seminiferous tubules per testis). Because the seminiferous tubules within one testis are about 400 m in length, spermatozoa are continually being generated at many sites within the testis at any given time (see **Koeppen** and **Stanton** 2010).

2. The Sertoli cells. The **Sertoli cells** are the true epithelial cells of the seminiferous epithelium and extend from the basal lamina to the lumen (Fig. 7.5). Sertoli cells surround sperm cells and provide structural support within the epithelium, and they form **adhering** and **gap junctions** with all stages of sperm cells. Through the *formation* and *breakdown* of these junctions, Sertoli cells guide sperm cells toward the lumen as they advance to later stages of spermatogenesis. Spermiation requires the final breakdown of Sertoli-sperm cell junctions.

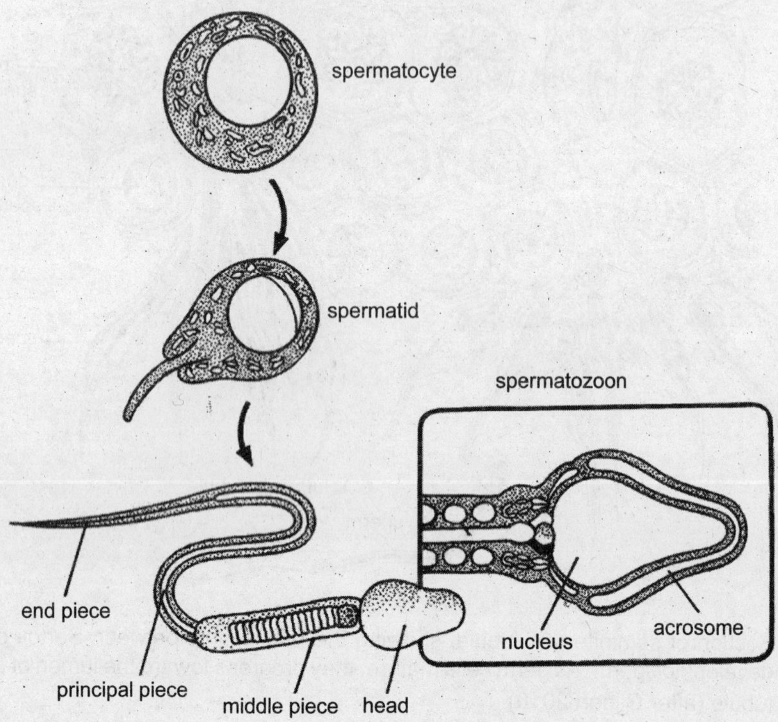

Fig. 7.6 Structure of sperm cells during the process of spermatogenesis and spermiogenesis.

Another important structural feature of Sertoli cells in the formation of **tight junctions** between adjacent Sertoli cells. These Sertoli-Sertoli cell occluding junctions divide the seminiferous epithelium into a **basal compartment** containing the spermatogonia and early stage primary spermatocytes and an **adluminal compartment** containing late-stage primary spermatocytes and all subsequent stages of sperm cells. As early primary spermatocytes move apically from the basal to adluminal compartment of Sertoli cells, the tight junctions need to be disassembled and reassembled. These tight junctions form the physical basis for the **blood-testis barrier,** which creates a specialized, immunologically safe microenvironment for developing sperm. By blocking paracellular diffusion, the tight junctions restrict movement of substances between blood and the developing sperm cells through a trans-Sertoli cell transport pathway and, in this manner, allow the Sertoli cell to regulate the availability of nutrients to germ cells.

Healthy Sertoli cell function is essential for the viability and development of sperm cells. In addition, spermatogenesis is absolutely dependent on **testosterone** hormone produced by peritubular Leydig cells, yet it is the Sertoli cells that express the **androgen receptor,** not the developing sperm cells. Similarly, the pituitary hormone FSH (= follicle-stimulating hormone) is also required for maximal sperm production, and again, it is the Sertoli cell that expresses the **FSH receptor**, not the developing sperm. Thus, these hormones (*e.g.*, testosterone and FSH) support spermatogenesis indirectly through stimulation of Sertoli cell function.

Sertoli cells have many other functions:

1. They express the **enzyme CYPI9** (also called **aromatase**), which converts Leydig cell-derived testosterone to the potent estrogen E_2 (*i.e.*, **estradiol-17β).** This local production of estrogen may enhance spermatogenesis in humans.

2. Sertoli cells also produce **androgen-binding protein (ABP),** which maintains a high androgen level within the adluminal compartment, the lumens of the seminiferous tubules, and the proximal part of the male reproductive tract.

3. Sertoli cells also produce a large amount of fluid. This fluid provides an appropriate bathing medium for the sperm and assist in moving the **immotile** spermatozoa from the seminiferous tubule into the epididymis.

4. Sertoli cells perform an important **phagocytic function** by engulfing the **residual bodies,** which represent cytoplasm shed by spermatozoa during spermiogenesis.

5. Finally, the Sertoli cell has an important *endocrine role* : (i) During development Sertoli cell produce **anti-Mullerian hormone (AMH);** also called **Mullerian inhibitory substance,** which induces regression of the embryonic Mullerian duct that is programmed to give rise to the female reproductive tract. (ii) The Sertoli cells also produce the hormone **inhibin**. Inhibin is a heterodimer protein hormone related to the transforming growth factor-β family. FSH stimulates inhibin production, which then **negatively feedback** on gonadotropes to inhibit FSH production. Thus, inhibin keeps FSH levels within a set point.

***(ii)* The peritubular compartment of testis.** The peritubular compartment of testis contains the primary endocrine cell of the testis, the Leydig cells (Fig. 7.4.) This compartment also contains the common cell types of loose connective tissue and an extremely rich peritubular capillary network that provides nutrients to the seminiferous tubules (by way of Sertoli cells) while conveying testosterone away from the testes to the peripheral circulation.

The Leydig Cells

The Leydig cells are steroidogenic stromal cells. These cells synthesize **cholesterol** de novo, as well as acquire it through low-density lipoprotein (LDL) receptors and high-density lipoprotein (HDL) receptors (also called **scavenger receptor B1 (SR-B1),** and store cholesterol as cholesterol ester hydrolase and transferred to the outer mitochondrial membrane and then to the inner mitochondrial membrane in a steroidogenic acute regulatory (St AR) protein-dependent manner. As in all steroidogenic, cholesterol is converted to **pregnenolone** by **CYP11A1.** Pregnenolone is the processed to progesterone, 17-hydroxyprogesterone, and androstenedione by **3β-hydroxysteroid dehydrogenase (3β–HSD)** and **CYP17** (Fig. 7.7). The CYP17 is a bifunctional enzyme with **17-hydroxylase activity** and **17, 20-lyase.** CYP17 enzyme displays a robust level of both activities in the Leydig cell. The Leydig cell also expresses a Leydig cell-specific isoform of **17β-HSD type 3** enzyme, which converts **androstenedione** to testosterone (Fig. 7.7).

Fates and Actions of Androgens

1. Intratesticular androgen. The testosterone produced by Leydig cells has several fates and multiple actions. Because of the proximity of Leydig cells to the seminiferous tubules, significant amounts of testosterone diffuse into the seminiferous tubules and become concentrated within the adluminal compartment by ABP (androgen-binding protein; Fig. 7.7). Testosterone levels within the seminiferous tubules that are greater than 100 times more concentrated than circulating testosterone levels are required for normal spermatogenesis. As mentioned earlier, Sertoli cells express the enzyme CYP19 (aromatase) which converts a small amount of testosterone into the high potent estrogen **estradiol-17β.** Human sperm cells express at least one isoform of the **estrogen receptor** and evidently this locally produced estrogen optimizes spermatogenesis in humans.

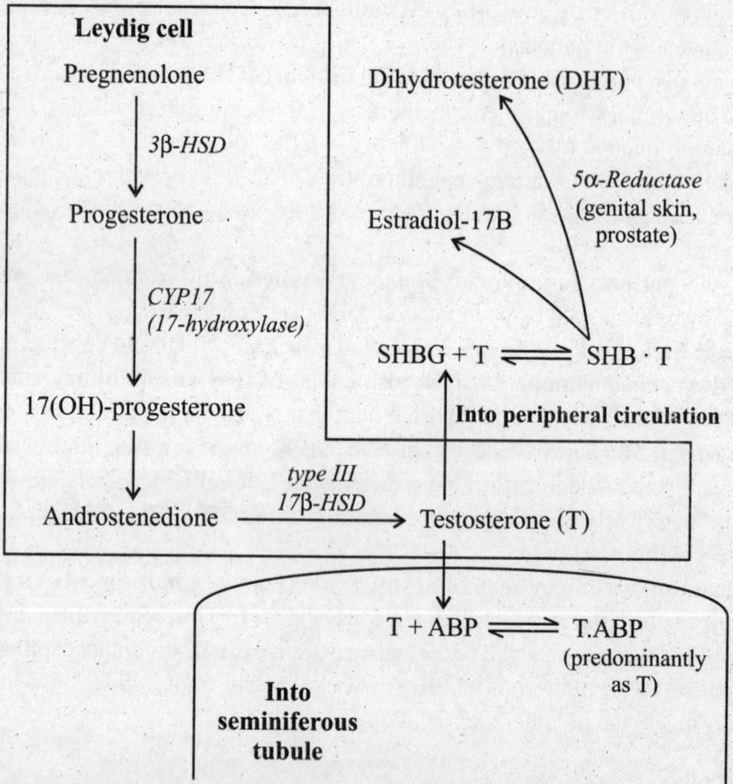

Fig. 7.7 Steroidogenic pathway in Leydig cells (the first step of converting cholesterol to pregnenolone is omitted). Testosterone is sequestered by binding to androgen-binding protein (ABP) within the peripheral circulation bound to sex hormone binding globulin (SHBG) and can be peripherally converted to dihydro-testosterone or estradiol-17β (after Koeppen and Stanton 2010).

2. Peripheral conversion of testosterone to estrogen. In several tissues (especially adipose tissue), testosterone is converted to estrogen (Fig. 7.7). Studies involving men with *aromatase deficiency* have shown that an inability to produce estrogen results in *tall stature* because of the lack of epiphyseal closure in long bones and osteoporosis. Thus, peripheral estrogen plays an important role in bone maturation and biology in men. These studies also implicated estrogen in promoting insulin sensitivity, improving lipoprotein profiles (*i.e.*, increasing HDL, decreasing triglycerides and LDL) and exerting negative feedback on pituitary gonadotropins.

3. Peripheral conversion of testosterone to dihydrosterone (DHT). Testosterone can also be converted into a potent **non-aromatizable androgen, DHT (5α-dihydro-testosterone)** by the enzyme 5α-reductase (Fig. 7.7). There are two isoforms of 5α-reductase, type 1 and type 2. Major sites of 5α-reductase 2 expression are *the male urinogenital tract, genital skin, hair follicles and liver.* 5α-reductase 2 generates DHT, which is required for masculinization of the external genitalia *in utero* and for many of changes associated with **puberty,** including growth and activity of the prostate gland, growth of the penis, darkening and folding of the scrotum, growth of pubic and axillary hair, growth of facial and body hair and increased muscle mass (Fig. 7.8). The onset of 5α-reductase1 expression occurs at puberty and contributes to sebaceous gland activity and the **acne** associated with puberty.

4. Peripheral testosterone actions. Testosterone has a direct action (*i.e.*, without conversion to DHT) in serval cell types (Fig. 7.8).

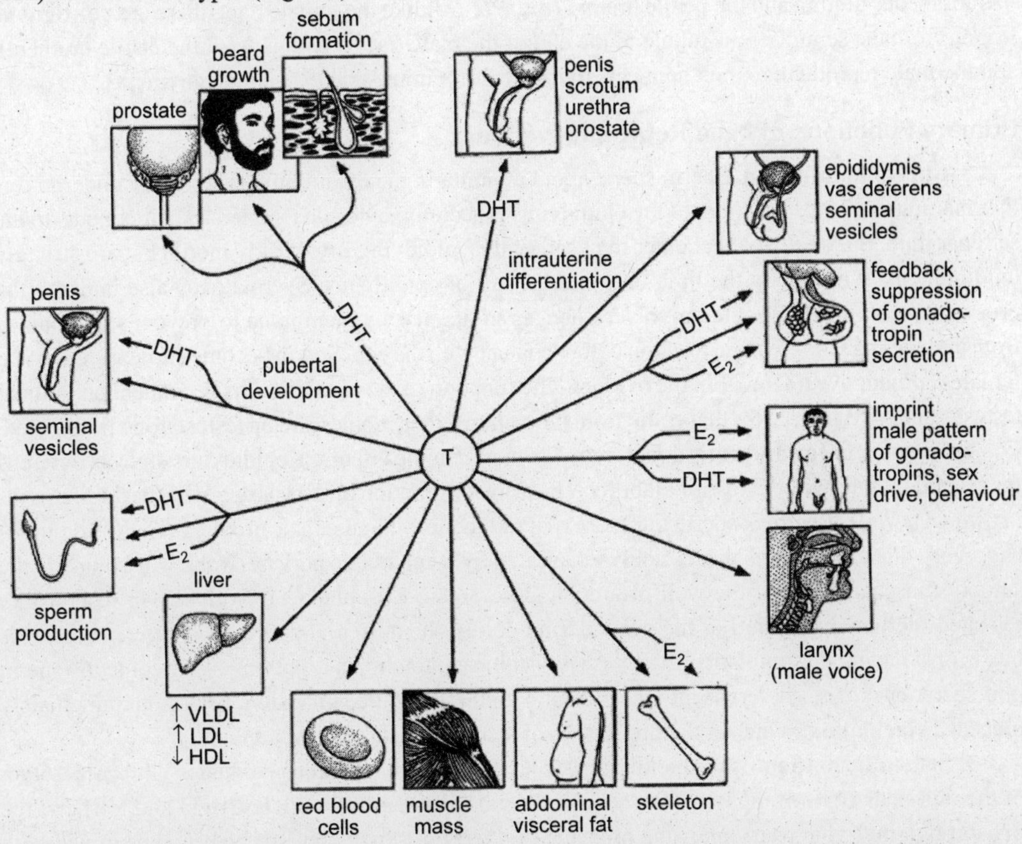

Fig. 7.8. Spectrum of effects of testosterone (T). Note that some effects result from the action of testosterone itself, whereas others are mediated by dihydrotestosterone (DHT) and estradiol (E_2) after they are produced from testosterone. VLDL, LDL, HDL, very-low density, low-density and high-density lipoproteins, respectively.

1. Testosterone regulates Sertoli cell function.

2. It induces development of the male tract from the mesonephric duct in the absence of 5α-reductase.

3. Testosterone has several **metabolic effects,** including increasing very low density lipoprotein (VLDL) and LDL, while decreasing HDL, promoting the deposition of abdominal adipose tissue, increasing red blood cell production, promoting bone growth and health, and exerting a protein anabolic effect on muscle.

4. Testosterone is sufficient to maintain erectile function and libido.

5. Mechanism of androgen action. Testosterone and DHT, both hormones act through the same androgen receptor (AR). The AR resides in the cytoplasm bound to chaperone proteins in the absence of ligand. Testosterone - AR binding or DHT-AR binding causes dissociation of the chaperone proteins, followed by nuclear translocation of androgen-AR complex, dimerization, binding to an **androgen response element** (ARE) and recruitment of coactivator proteins and general transcription factors to the vicinity of a specific gene's promoter.

B. The Male Reproductive Tract

Once spermatozoa emerge from the efferent ductules, they leave the gonad (testis) and enter the male reproductive tract (Fig. 7.1.) The segments of the reproductive tract are as follows: the **epididymis** (heads, body and tail), the **vas deferens,** the **ejaculatory duct,** the prostatic urethra, the membraneous urethra and the penile urethra (Fig. 7.2). Unlike the female tract, there is a **contiguous lumen** from the seminiferous tubule to the end of the male tract (*i.e.*, the tip of the penile urethera), and the male reproductive tract connects to the **distal urinary tract** (*i.e.*, male urethera).

Primary Functions of Male Reproductive Tract

1. Sperm maturation. Sperm spend about a month in the epididymis, where they undergo further maturation. The epithelium of the epididymis is secretory and adds numerous components to the seminal fluid. Spermatozoa that enter the head of the epididymis are weakly motile but are strongly unidirectionally motile by the time they exit the tail of epididymis. Spermatozoa also undergo the process of **decapitation** which involves changes in the plasma membrane to prevent spermatozoa from undergoing the acrosome reaction before contact with the egg. Sperm become **capacitated** by the female reproductive tract within the oviduct. The function of the epididymis is dependent on luminal testosterone ABP complexes that come from the seminiferous tubules and on testosterone from blood.

2. Sperm storage and emission. Sperm are *stored* in the tail of the epididymis and vas deferens for several months without loss of viability. The primary function of the vas deferens, besides providing *storage site*, is to *propel* sperm during sexual intercourse into the male urethera. The vas deferens has a very thick muscularis that is richly innervated by sympathetic nerves. (*Note*: Muscularis is the smooth muscular layer of the wall of a contractile organ, *e.g.*, bladder). Normally in response to repeated **tactile stimulation** of the penis during coitus, the muscularis of the vas diferens receives bursts of sympathetic stimulation that cause peristaltic contractions. Emptying of the contents (sperm plus secretions) of vas deferens into the prostatic urethera is called **emission.** Emission immediately precedes **ejaculation,** which is the propulsion of semen out of the male urethera.

3. Semen formation. During emission, contraction of vas deferens coincides with contraction of the muscular coats of the two accessory sex glands, the **seminal vesicles** (right and left) and the **prostate gland** (which surrounds the prostatic urethera). At this point, sperm becomes mixed with all the components of **semen.** The seminal vesicles secrete approximately 60 % of the volume. These glands are the primary source of **fructose,** a critical nutrient for sperm. Seminal fluid is rich in **citric acid, fibrinogen** and prostaglandins. The seminal vesicles also secrete **semenogelins,** which induce coagulation of semen immediately after ejaculation. The alkaline milky secretions of prostate, which makes up about 30 % of the volume, are high in **citrate, zinc, spermine** and **acid phosphatase. Prostate-specific antigen (PSA)** is a serine protease that liqifies coagulated semen after a few minutes PSA can be detected in blood under conditions of prostatic infection, benign prostatic hypertrophy, and prostatic carcinoma. It is currently used as one indicator of prostatic health. The predominant buffers in semen are phosphate and bicarbonate. A third accessory gland, the **bulbourethral glands** (also called Cowper's glands), empty into the penile urethera in response to sexual excitement before emission and ejaculation. This secretion is high in mucus, which lubricates, cleanses, and buffers the urethera. Average sperm counts are between 60 to 100 million/mL, less than 50 % motile sperm, or less than 60 % normally confirmed sperm are usually **infertile**?

4. Erection and ejaculation. Emission and ejaculation occur during coitus in response to a reflex arc that involves sensory stimulation from the penis (via the pudendal nerve) followed by sympathetic motor stimulation to the smooth muscle of the male tract and somatic motor stimulation to the musculature associated with the base of the penis. However, for sexual intercourse to occur in the first place, the man has to achieve and maintain an **erection** of the penis. The **penis** has evolved as an intromittent organ designed to **separate** the walls of the vagina, pass through the potential space of the vaginal lumen, and **deposit** semen at the distal end of the vaginal lumen near the cervix. This process of **internal insemination** can be performed only if the penis is stiffened (*i.e.*, long and hard) from the process of erection.

Erection is a neurovascular event. The penis is composed of three erectile bodies: two **corpora cavernosa** and one **corpus spongiosum** (Fig. 7.9). These three bodies are composed of **erectile tissues**—an anastomosing network of potential **cavernous vascular spaces** lined with continuous endothelia within a loose connective tissue support. During the **flaccid state of penis,** blood flows to the cavernous spaces. In response to sexual arousal, the parasympathetic cavernous nerves innervating the

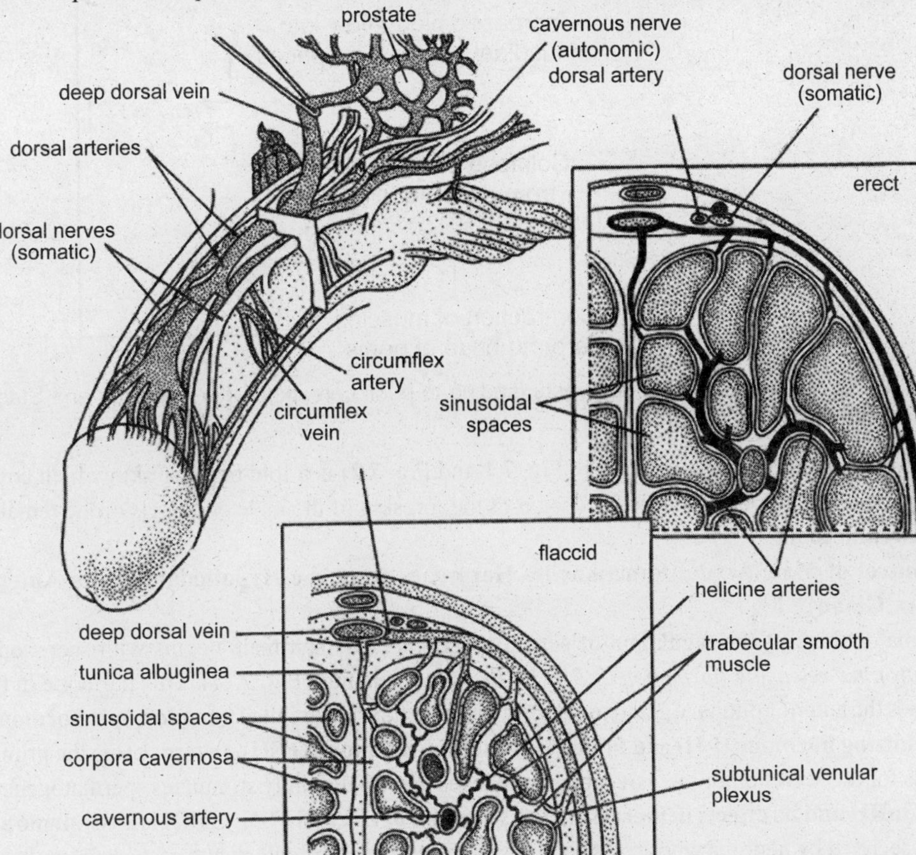

Fig. 7.9. Arrangement of the vasculature and cavernous tissue within the penis. During the flaccid state, blood flow into the cavernous spaces in limited by contraction of the helicine arteries (after Koeppen and Stanton, 2010).

vascular smooth muscle of the helicine arteries release **nitric oxide (NO).** NO activates guanylyl cyclase, thereby increasing cGMP, which decreases intracellular calcium ions (Ca^{2+}) and causes muscular relaxation (Fig. 7.10). The vasodilation allows blood to flow into the cavernous spaces to induce engorgement and erection. It also presses on veins in the penis and reduces venous drainage.

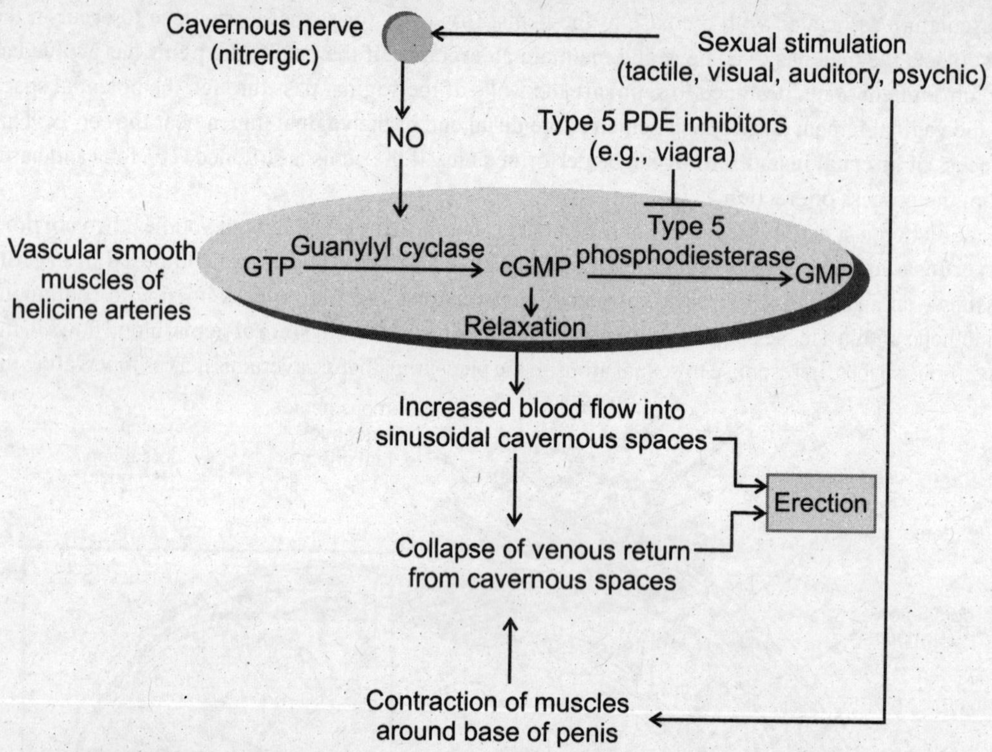

Fig. 7.10. Outline of neurovascular events leading to penile erection (after Koeppen and Stanton 2010).

Prepuce or **foreskin** (Not shown in Fig. 7.1 and Fig. 7.2) is a fold of loose skin which covers the glans penis of the penis. Surgical excision of the foreskin of the male or clitoris of the female is called **circumcision.**

Control of Male Sexual Functions by Hormones from the Hypothalamus and Anterior Pituitary Gland

A major share of the regulation of sexual functions in human male begins with secretion of *gonadotropin - releasing hormone (GnRH)* by the hypothalamus (Fig. 7.11). This hormone in turn stimulates the anterior pituitary gland to secrete two other hormones called **gonadotropic hormones: (1) luteinizing hormone (LH)** and **follicle-stimulating hormone (FSH).** In turn, LH is the primary stimulus for the secretion of **testosterone** by the testis and FSH mainly stimulates spermatogenesis.

1. GnRH and its effects in increasing the secretion of LH and FSH. GnRH is a 10–amino acid peptide secreted by neurons whose cell bodies are located in the *arcuate nuclei of the hypothalamus*. The endings of these neurons terminate mainly in the median eminence of the hypothalamus, where they release GnRH into the hypothalamic-hypophyseal portal vascular system. Then the GnRH is transported to the anterior pituitary gland in the hypophyseal portal blood and stimulates the release of the two gonadotropins, LH and FSH. GnRH is secreted intermittently a few minutes at a time once every 1 to 3 hours.

The secretion of LH by the anterior pituitary gland is also cyclical, with LH following the pulsatile release of GnRH. Conversely, FSH secretion increases and decreases only slightly with each fluctuation of GnRH secretion. Because of the much closer relation between GnRH secretion and LH secretion, GnRH is also widely known as **LH-releasing hormone.**

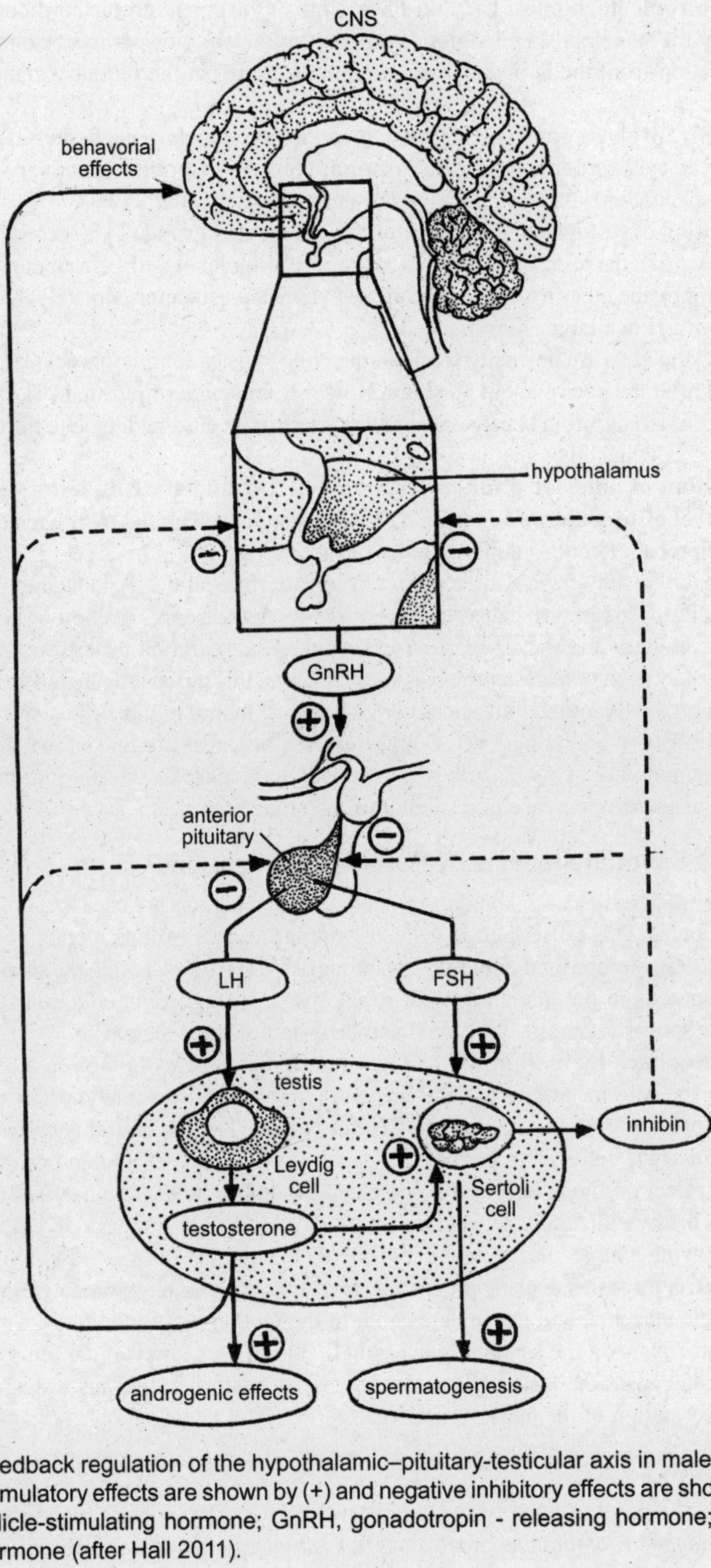

Fig. 7.11. Feedback regulation of the hypothalamic–pituitary-testicular axis in male human beings. Stimulatory effects are shown by (+) and negative inhibitory effects are shown by (–). FSH, follicle-stimulating hormone; GnRH, gonadotropin - releasing hormone; LH, luteinizing hormone (after Hall 2011).

2. Gonadotropic hormones: LH and FSH. Both of the gonadotropic hormones, LH and FSH, are secreted by the same cells, called **gonadotropes,** in the anterior pituitary gland. In the absence of GnRH secretion from the hypothalamus, the gonadotropes in the pituitary gland secrete almost no LH or FSH.

LH and FSH are glycoproteins. They exert their effects on their target tissues in the testes mainly by **activating the cyclic adenosine monophosphate (cAMP) second messenger system,** which in turn activates specific enzyme systems in the respective target cells.

***(i)* Regulation of testosterone production by LH. Testosterone** (T) is secreted by the **interstitial cells of Leydig** in the testes, but only when they are stimulated by LH from the anterior pituitary gland. Furthermore, the quantity of testosterone secreted increases approximately in direct proportion to the amount of LH available.

Mature Leydig cells are normally found in the child's testes for a few weeks after birth but then disappear until after the age of about 10 years. However, injection of purified LH into a child at any age or secretion of LH at **puberty** causes testicular interstitial cells that look like fibroblasts to evolve into functioning Leydig cells.

***(ii)* Inhibition of anterior pituitary secretion of LH and FSH by testosterone—Negative feedback control of testosterone secretion.** The testosterone secreted by the testes in response to LH has the reciprocal effect of inhibiting anterior pituitary secretion of LH (Fig. 7.11). Most of this inhibition probably results from a direct effect of testosterone on the hypothalamus to decrease the secretion of GnRH. This in turn causes a corresponding decrease in secretion of both LH and FHS by the anterior pituitary and the decrease in LH reduces the secretion of testosterone by the testes. Thus, whenever secretion of testosterone becomes too great, this **automatic negative feedback** effect, operating through the hypothalamus and anterior pituitary gland, reduces the testosterone secretion back toward the desired operating level. Conversely, too little testosterone allows the hypothalamus to secrete large amounts of GnRH, with a corresponding increase in anterior pituitary LH and FSH secretion and consequent increase in testicular testosterone secretion.

Regulation of Spermatogenesis by FSH and Testosterone

As already has described, FSH binds with specific FSH receptors attached to the Sertoli cells in the seminiferous tubules. This causes Sertoli cells to grow and secrete various spermatogenic substances. Simultaneously, testosterone (and dihydrotestosterone or DTH) diffusing into the seminiferous tubules from the Leydig cells in the interstitial spaces also has a strong tropic effect on spermatogenesis. Thus, to initiate spermatogenesis, both FSH and testosterone are necessary.

Negative feedback by inhibin. When the seminiferous tubules fail to produce sperms, secretion of FSH by the anterior pituitary gland increases markedly. Conversely, when spermatogenesis proceeds too rapidly, pituitary secretion of FSH diminishes. The cause of this *negative feedback* on the anterior pituitary is believed to be secretion by the Sertoli cells of inhibin hormone. Inhibin has a strong direct effect on the anterior pituitary gland to inhibit the secretion of FSH and possibly a slight effect on the hypothalamus to inhibit secretion of GnRH. Inhibin is a glycoprotein, like both LH and FSH, having a molecular weight between 10,000 and 30,000.

Effect of hCG on testes. During pregnancy the hormone *human chorionic gonadotropin* (hCG) is secreted by the placenta, and it circulates both in the mother and in the fetus. This hormone has almost the same effects on the sexual organs as LH. During pregnancy, if the fetus, is a male, hCG from the placenta causes the testes of the fetus to secrete testosterone. This testosterone is critical for promoting formation of the male sexual organs.

Onset of Puberty

Initiation of the onset of puberty has long been a mystery. But it has now been determined that *during childhood the hypothalamus simply does not secrete significant amounts of GnRH*. One of the reasons for this is that, during childhood, the slightest secretion of any sex steroid hormones exerts a strong inhibitory effect on hypothalamic secretion of GnRH. Yet for reasons still not understood (see **Hall** 2011), at the time of puberty, the secretion of hypothalamic GnRH breaks through the **childhood inhibition** and adult sexual life begins.

Male climacteric (or Andropause). After puberty, gonadotropic hormones are produced by the male pituitary gland for the remainder of life, and at least some spermatogenesis usually continues until death. Most men, however, begin to exhibit slowly decreasing sexual functions in their 50s or 60s, and one study showed that the average age for terminating intersexual relations was 68, although the variation was great. This decline in sexual function is related to decrease in testosterone secretion. The decrease in male sexual function is called the **andropause** (**Koepper** and **Stanton** 2010) or **male climacteric** (**Hall** 2011). Occasionally the male climacteric is associated with symptoms of hot flashes, suffocation and psychic disorders similar to the menopausal symptoms of the female. These symptoms can be abolished by administration of testosterone, synthetic androgens, or even estrogens are used for treatment of menopausal symptoms in the female.

7.3 THE FEMALE REPRODUCTIVE SYSTEM

The female reproductive system is composed of the gonads, called **ovaries,** and the female reproductive tract, which includes the **oviducts, uterus, cervix, vagina** (Fig. 7.12) and **external genitalia** (Fig. 7.14).

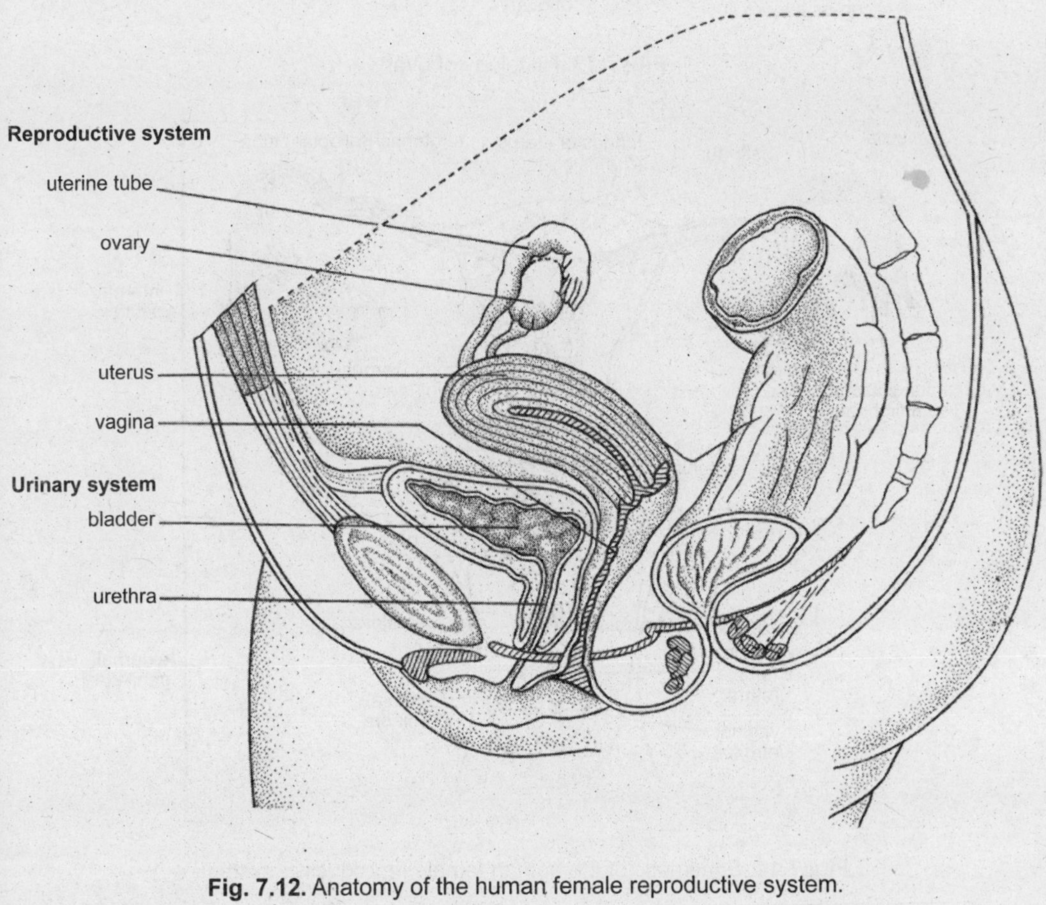

Fig. 7.12. Anatomy of the human female reproductive system.

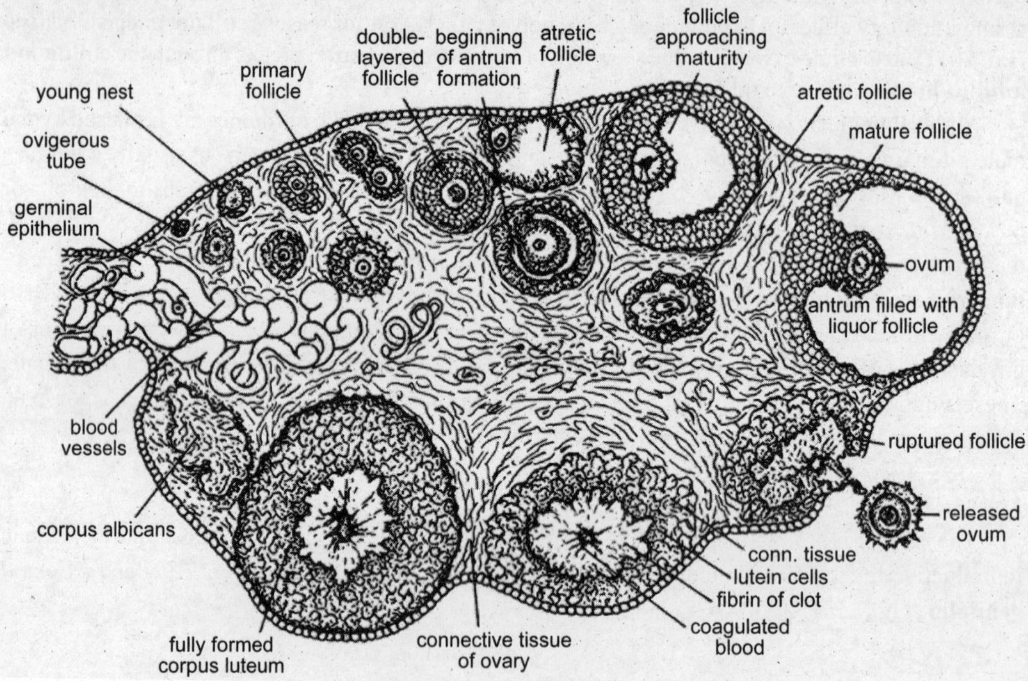

Fig. 7.13. Histology of ovary.

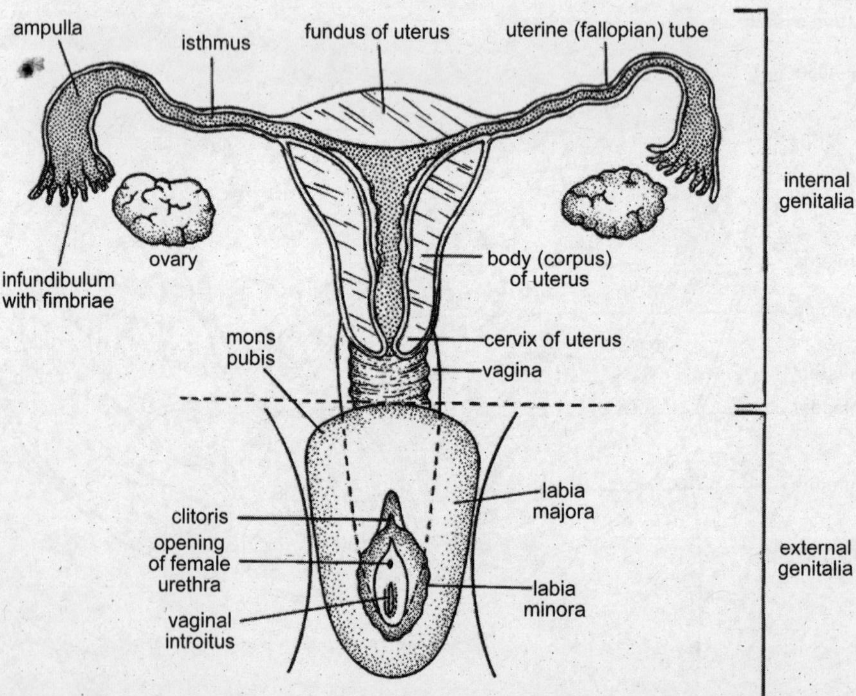

Fig. 7.14. Schematic of the human female reproductive system.

A. Ovary

The ovary is located within a fold of peritoneum called the **broad ligament,** usually close to the lateral wall of the pelvic cavity (Fig. 7.12). The ovary is divided into an outer **cortex** and inner **medulla** (Fig. 7.13). Neurovascular elements innervate the medulla of the ovary. The cortex of the ovary is composed of a densely cellular **stroma.** Within this stroma reside the **ovarian follicles,** which contain a **primary oocyte** surrounded by follicle cells. The cortex is covered by a connective tissue capsule, the **tunica albuginea,** and a layer of simple epithelium consisting of **ovarian surface epithelial cells.** There are no ducts emerging from the ovary to convey its gametes to the reproductive tract. Thus, the process of **ovulation** involves an inflammatory event that erodes the wall of ovary (Fig. 7.15). After ovulation, the ovarian surface epithelial cells rapidly divide to repair the wall.

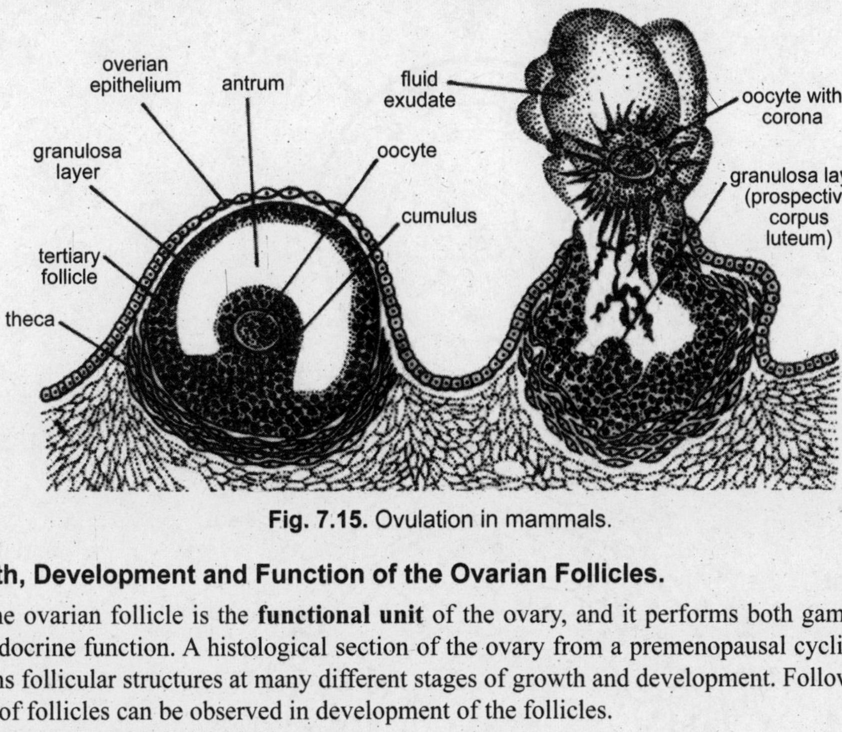

Fig. 7.15. Ovulation in mammals.

Growth, Development and Function of the Ovarian Follicles.

The ovarian follicle is the **functional unit** of the ovary, and it performs both gametogenesis and endocrine function. A histological section of the ovary from a premenopausal cycling woman contains follicular structures at many different stages of growth and development. Following seven stages of follicles can be observed in development of the follicles.

1. Resting primordial follicle.
2. Growing preantral (primary and secondary) follicle.
3. Growing antral (tertiary) follicle.
4. Dominant (preovulatory or Graafian follicle).
5. Dominant follicle within the periovulatory period.
6. Corpus luteum (of menstruation or of pregnancy).
7. Atretic follicles.

1. Resting Primordial Follicle

(*i*) Growth and structure. Resting primordial follicles (Fig. 7.16) represent the earliest and simplest follicular structures in the ovary. Primordial germ cells (PGCs) appear during midgestation period through the interaction of gametes and somatic cells. PGCs that have migrated to the gonad continue to divide mitotically as oogonia until the fifth month of gestation in humans. At this point approximately 7 million (70 lakh) oogonia enter the process of meiosis and become **primary oocytes.** During this time the primary oocytes become surrounded by a simple epithelium of somatic **follicle cells,** thereby creating the primordial follicles (Fig. 7.16A).

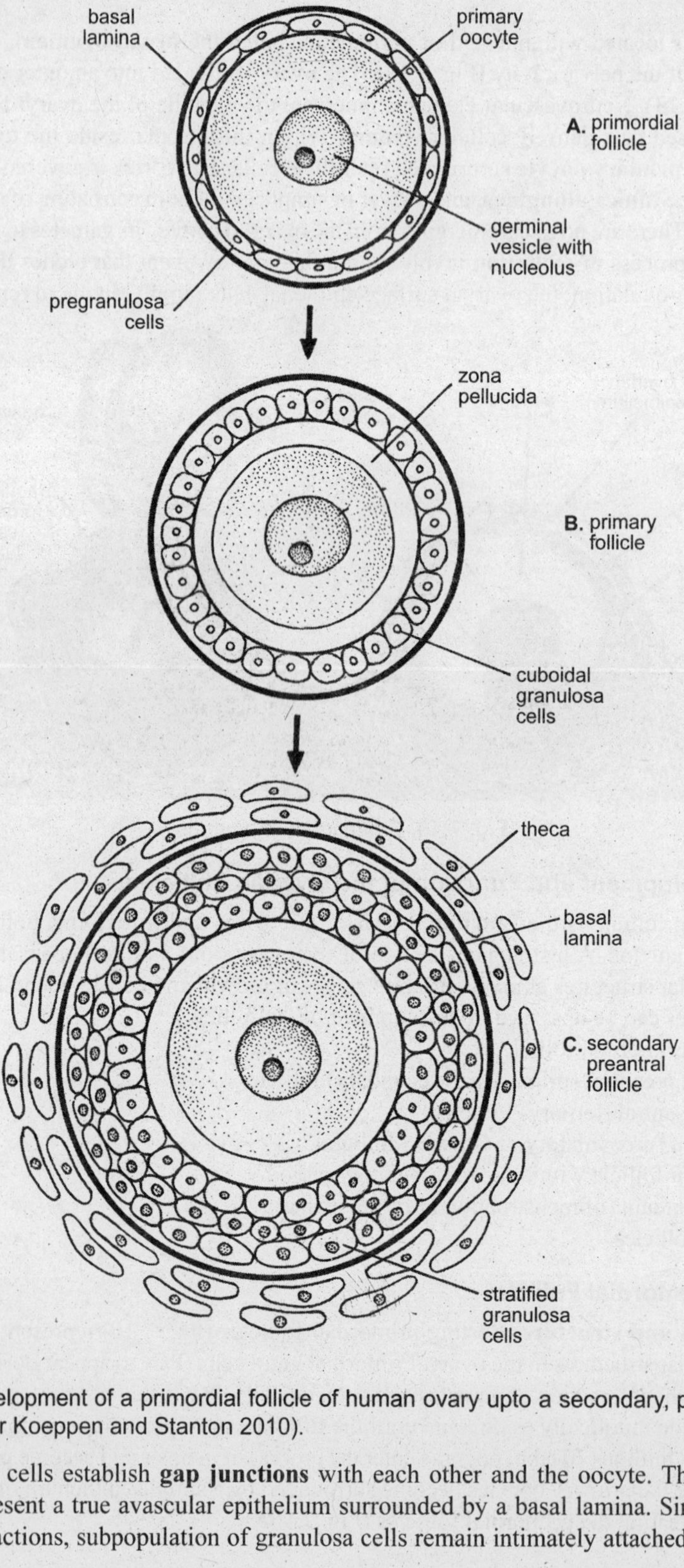

Fig. 7.16. Development of a primordial follicle of human ovary upto a secondary, preantral follicle (after Koeppen and Stanton 2010).

The follicle cells establish **gap junctions** with each other and the oocyte. The follicle cells themselves represent a true avascular epithelium surrounded by a basal lamina. Similar to Sertoli cell-sperm interactions, subpopulation of granulosa cells remain intimately attached to the oocytes throughout their development. Granulosa cells provide nutrients such as amino acids and pyruvate to support oocyte maturation.

The primordial follicles represent the ovarian reserve follicles (Fig. 7.17). This reserve is reduced from a starting number of about 7 million to less than 300,000 follicles at reproductive maturity. Of these, a woman will **ovulate** about 450 between **menarche** (first menstrual cycle) and a **menopause** (cessation of menstrual cycle). At menopause, less than 1000 primordial follicles are lost primarily from death as a result of follicular atresia. **Atresia** is disappearance of an anatomical part (as an ovarian follicle) by degeneration. However, a small subset of primordial follicles will enter follicular growth in waves. Because the ovarian follicular reserve represents a fixed, finite number, the rate at which resting primordial follicles die or begin to develop (or both) will determine the *reproductive life span* of a women. Age at the onset of menopause has a strong genetic component but is also influenced by environmental factors. For example, cigarette smoking significantly depletes the ovarian reserve. An overly rapid rate of atresia or development will depletes reserve and give rise to premature ovarian failure.

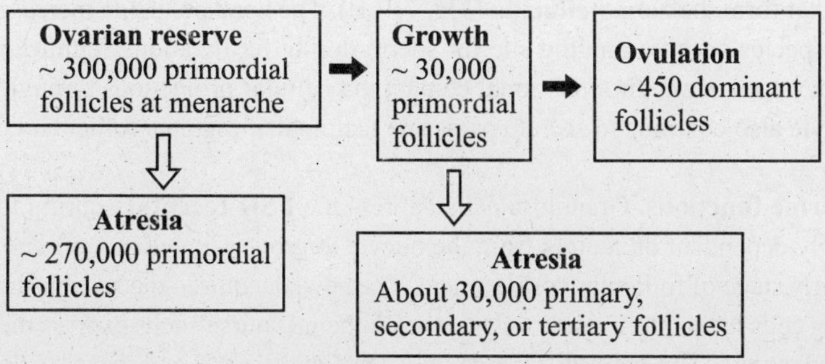

Fig. 7.17 Fate of ovarian follicles.

(*ii*) **Role of hormones.** Pituitary gonadotropins maintain a normal ovarian reserve by promoting the general health of the ovary. However, the rate at which resting primordial follicles enter the growth process appears to be independent of pituitary gonadotropin. The decision of a resting follicle to enter the early growth phase is primarily dependent on intraovarian **paracrine factors** that are produced by both the follicle cells and oocytes.

(*iii*) **Gamete.** In primordial follicles, the gamete is derived from oogonia that have entered the *first meiotic division;* such oogonia are called **primary oocytes.** Primary oocytes progress through most of the prophase of first meiosis (*i.e.*, prophase I) over a 2-week period and then **arrest** in the **diplotene** stage. This meiotic I arrest stage is called **dictyate state (Gilbert 2010).** This stage is characterized by the **transcription** needed for oocyte maturation. Meiotic arrest at this stage, which may last for upto 50 years, appears to be due to *"maturational incompetence,"* or lack of the cell cycle proteins needed to support the completion of meiosis. The nucleus of the oocyte, called the **germinal vesicle** remains intact at this stage.

2. Growing Preanteral Follicles

(*i*) **Growth and structure.** The first stage of follicular growth is **preantral,** which refers to the development that occurs before the formation of a fluid - filled **antral cavity** (antral means cavity). Antrum becomes filled with complex mixture of proteins, hormones and other molecules (**Gilbert** 2010). One of the first visible signs of follicle growth is the appearance of cuboidal granulosa cells. At this point the follicle is known as a **primary follicle** (Fig. 7.16B). Concomitant with oocyte growth is an increase in the number of granulosa cells, which form concentric layers forming **stratified epithelium)** around the oocyte. At this stage the follicle is called **secondary follicle**

(Fig. 7.16 C). The proliferation of granulosa cells is mediated by a **paracrine factor, GDF9,** a member of the TGF-β family (**Dong** *et al.*, 1996).

Once a secondary follicle acquires three to six layers of granulosa cells, it secretes paracrine factors that induce nearby stromal cells of ovary to differentiate into epithelioid **thecal cells.** Thecal cells form a flattened layer of cells around the follicle. Once a thecal layer forms, the follicle is known as **mature preantral follicle** (Fig. 7.16C). In humans it takes several months for a primary follicle to reach the mature preantral stage.

Follicular growth and development is associated with an inward movement of the follicle from the outer cortex to the inner cortex, closer to the vasculature of the ovarian medulla. Follicles release **angiogenic factors** induce the development of one or two arterioles that form a vascular wreath (wreath means curl or ring of vessels) around the follicle.

(*ii*) Gamete. During the preantral stage, the oocyte begins to grow and produce cellular and secreted proteins. The oocyte initiates secretion of extracellular matrix glycoproteins, called **ZP1, ZP2** and **ZP3,** that form the **zona pellucida** (Fig. 7.16B). The zona pellucida increases in thickness and provides a species-specific binding site for sperm during fertilization. Significantly, granulosa cells and the oocyte maintain gap junctional contact via cellular projections through the zona pellucida. The oocyte also continue to secrete paracrine factors that regulate follicular cell growth and differentiation.

(*iii*) Endocrine functions. Granulosa cells express the **FSH receptors** during this period, but they are primarily dependent on factors from the oocyte to grow. They do not produce ovarian hormones at this early stage of follicular development. The newly acquired thecal cells are analogous to testicular Leydig cells in that they reside outside the epithelial "nurse" cells express the **LH receptor** and produce **androgens.** The main difference between Leydig cells and thecal cells is that thecal cells do not express high levels of **17β-HSD** (*i.e.*, 17β-hydroxysteroid dehydrogenase) enzyme. Thus, the major product of thecal cells is **androstenedione,** as opposed to testosterone. Androstenedione production at this stage, however, is minimal.

3. Growing Antral Follicles

(*i*) Growth and structure. Mature preantral follicles develop into **early antral follicles** (Fig. 7.18). Over a period of about 25 days, during which they grow from a diameter of about 0.1 mm to a diameter 0.2 mm. Once the granulosa epithelium increases to six to seven layers, fluid-filled spaces appear between cells and coalesce in to the **antrum.** Over a period of about 45 days, this wave of small antral follicles will continue to grow to large, recruitable antral follicles that are 2 to 5 mm in diameter. This period of growth is characterized by about a 100-fold increase in granulosa cell (from about 10,000 to 1,000,000 cells). It is also characterized by swelling of the antral cavity, which increasingly divides the granulosa cells into *two* discrete populations (Fig. 7.18).

(*a*) Mural granulosa cells (also called the **stratum granulosum).** These cells form the outer wall of the follicle. The basal layer is adherent to the basal lamina and in close proximity to the outer-lying thecal layers. Mural granulosa cells become highly steroidogenic and remain in the ovary after ovulation to differentiate into the corpus luteum.

(*b*) Cumuls cells (also called the **cumulus oophorus or corona radiata).** These are the inner cells that surround the oocyte. The innermost layer of cumulus cells maintains **gap** and **adhesion junctions** with the oocyte. Cumulus cells are released with the oocyte (collectively called **cumulus - oocyte complex)** during the process of ovulation. Cumulus cells are crucial for the ability of fimbriated end of the oviduct to **"capture"** and move the oocyte by a ciliary transport mechanism along the length of the oviduct to the site of fertilization.

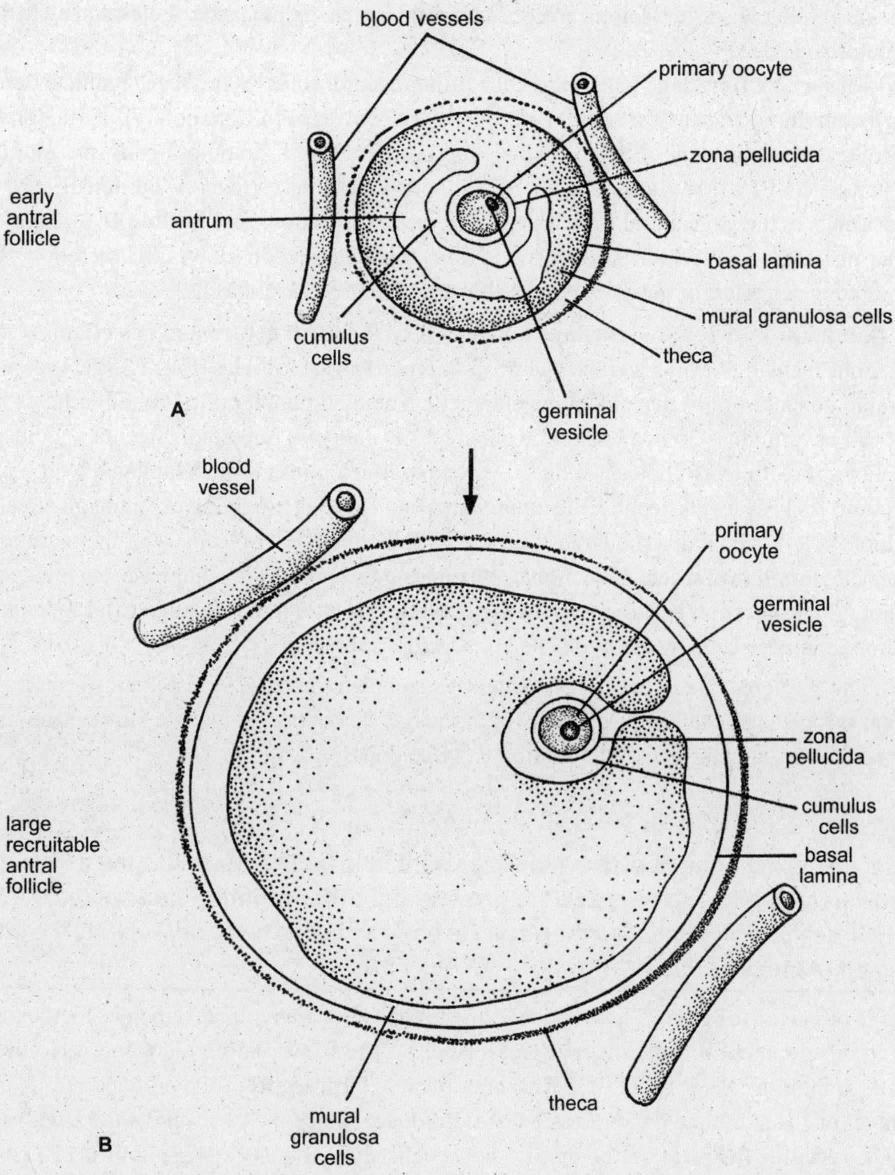

Fig. 7.18. Development of an early antral follicle to a mature preovulatory follicle.

Role of FSH in growth of antral follicles. Early antral follicles are dependent on pituitary FSH for normal growth. Large antral follicles become highly dependent on pituitary FSH for their growth and sustained viability. 2 to 5 mm follicles are recruited to enter a rapid growth phase via the short-lived increase in FSH that occurs toward the end of the previous menstrual cycle.

***(ii)* The gamete.** In the early stages of antral follicles—growth then slows in larger follicles. During the antral stages, the oocyte synthesizes sufficient amounts of cell cycle components so that it becomes competent to complete meiosis I at ovulation (note that the human egg arrests after ovulation at a second point, metaphase II, until it is fertilized by sperm). Thus, in early primary and secondary follicles, the oocyte fails to complete meiosis I because of lack of specific meiosis-associated proteins. However, larger antral follicles gain **meiotic competence** but still maintain **meiotic arrest**

until the midcycle LH surge. Meiotic arrest is achieved by the maintenance of elevated cAMP levels in the mature oocyte (Fig. 7.19).

***(iii)* Endocrine function.** The thecal cells of large antral follicles produce significant amounts of **androstenedione** and **testosterone.** Androgens are converted to estradiol-17β by the granulosa cells. However, at this stage, **FSH** stimulates the proliferation of granulosa cells and induces the expression of **CYP19 (aromatase)** required for the synthesis of estrogen. Additionally, the mural granulosa cells of the large antral follicles produce increasing amounts of **inhibin B** during the early follicular phase. Low levels of estrogen and inhibin negatively feedback on FSH secretion, thereby contributing to **selection** of the follicle with the most FSH-responsive cells.

4. Dominant follicle (or Graafian follicle). (i) Growth, structure and selection. At the end of a previous menstrual cycle, a crop of large (2 to 5 mm) antral follicles (Fig. 7.18) are recruited to begin rapid gonadotropin - dependent development. The total number of recruited follicles in both ovaries can be as high as 20 in a younger women (< 33 years old) but rapidly declines at older ages. The number of recruited follicles is reduced to the **prolifera quota** (one in humans) by the **process of selection.** As FSH levels decline, the rapidly growing follicles progressively undergo atresia until one follicle is left. Generally, the largest follicle with the most FSH receptors of the recruited crop becomes the **dominant follicle.** Selection occurs during the early follicular phase. By midcycle, the dominant follicle becomes a large **preovulatory follicle** that is 20 mm in diameter and contains about 50 million granulosa cells by the midcycle gonadotropin surge.

***(ii)* The gamete.** The oocyte is competent to complete meiosis I but remains arrested in the dominant follicle until the LH surge. Growth of the oocyte continues, but at a slower rate, until the oocyte reaches a diameter of about 140 μm by ovulation (Box 7.4).

Box 7.4

Just as the maturing oocyte synthesizes paracrine factors that allow the follicle cells to proliferate, the follicle cells secrete **growth** and **differentiation factors** (*e.g.*, TGF-β_2, VEGF, leptin, Fgf_2) that allow the oocyte to grow and bring blood vessels into the follicular region (**Antczak** *et al.*, 1997).

***(iii)* Endocrine function.** The newly selected follicle (Graafian follicle) emerges for the first time during its development as a significant steroidogenic "gland." Ovarian steroidogenesis requires both theca and granulosa cells (Fig. 7.20). Thecal cells express **LH receptors** and produce androgens. The basal level of LH stimulate the expression of steroidogenic enzymes, as well as the **LDL receptor** and **HDL receptor (SR-B1),** in the theca. Thecal cells show robust expression of **CYP11A1** (side chain cleavage enzyme), **3β-HSD** and **CYP17** (with both 17-hydroxylase activity and 17, 20-lyase activity. Androgens (primarily **androstenedione** but also some **testosterone**) released from the theca diffuse into the mural granulosa cells or enter the vasculature surrounding the follicle.

High number of **FSH receptors** occur on the mural granulosa cells of the selected follicle and are very sensitive of FSH, which upregulates **CYP19** (*i.e.*, **aromatase**) gene expression and activity (Fig. 7.20). CYP19 converts androstenedione to the weak estrogen **estrone** and converts testosterone to the potent estrogen estradiol -17β. Granulosa cells express activating isoforms of **17β-HSD,** which ultimately drives steroidogenesis toward the production of estradiol-17β. FSH also induces the expression of inhibin B during follicular phase.

During the second half of the follicular phase, FSH also induces the expression of LH receptors in mural granulosa cells (Fig. 7.20). Thus, mural granulosa cells become responsive to both gonadotropins, which allow these cells to maintain high levels of aromatase (CYP19) enzyme in the face of declining FSH levels. Addition of LH receptors also ensures that mural granulosa cells respond to the LH surge (or LH increase).

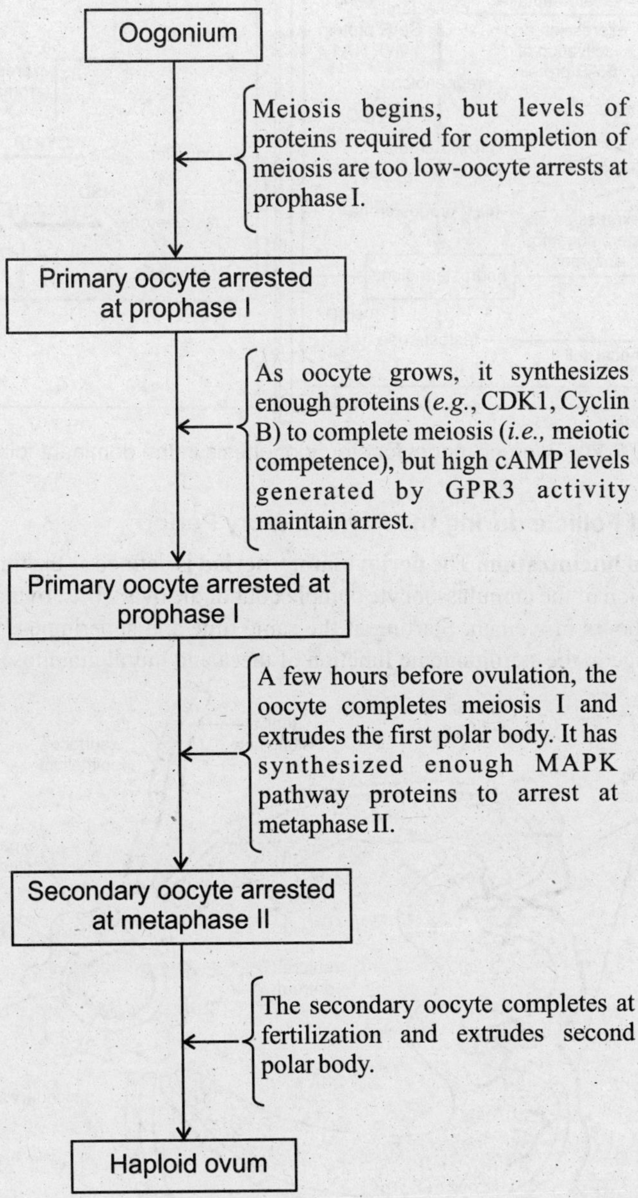

Fig. 7.19. Events involved in meiotic arrest and maturation of the oocyte. MAPK, mitogen-activated protein kinase (after Koeppen and Stanton 2010)

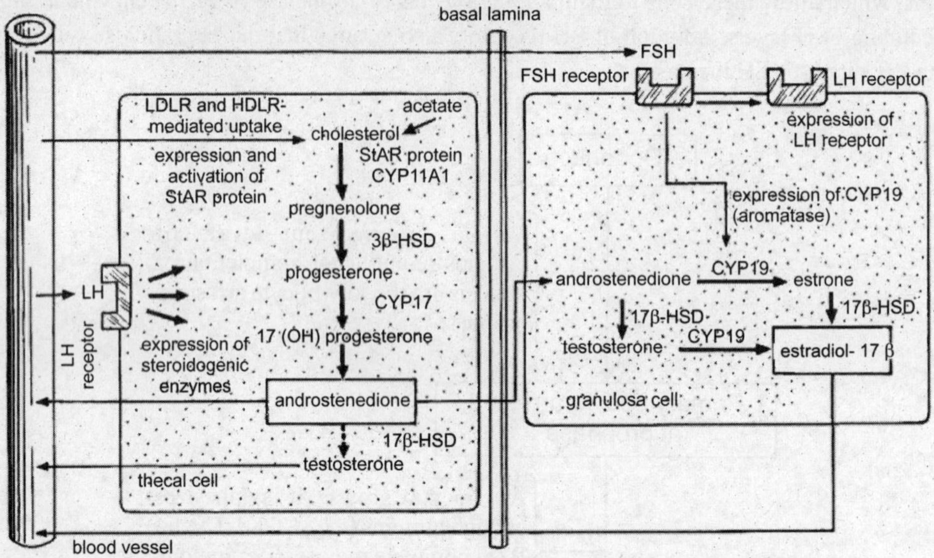

Fig. 7.20. Two-cell model for steroidogenesis in the dominant follicle.

5. The Dominant Follicle during the Periovulatory Period

Ovulation and luteinization. The **periovulatory period** is defined as the time from the onset of LH surge to expulsion of the cumulus-oocyte complex out of the ovary (*i.e.*, ovulation). This process lasts for 32 to 36 hours in women. Starting at the same time and superimposed on the process of ovulation is a change in the steroidogenic function of theca and mural granulosa cells. This process

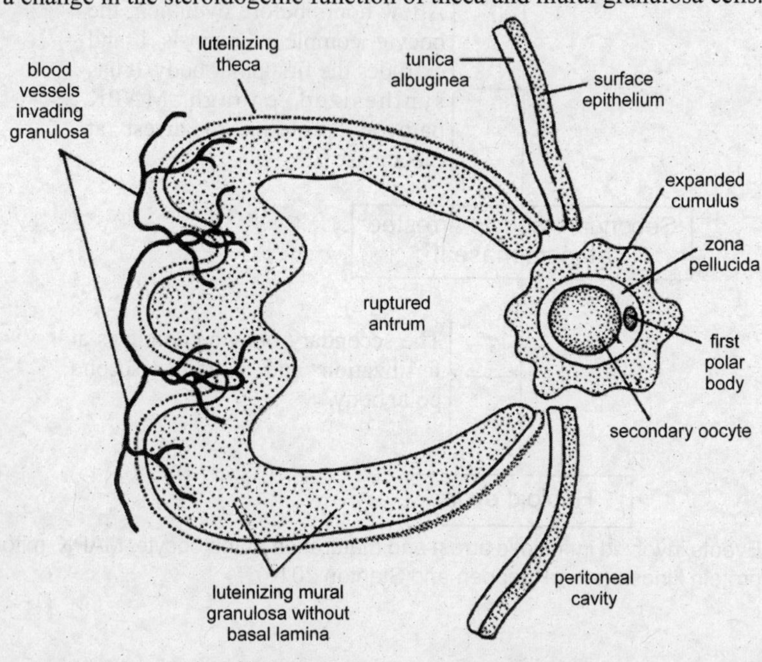

Fig. 7.21. Ovulation in human beings.

is called **luteinization** and concludes in the formation of a **corpus luteum** that is a capable of producing high amounts of **progesterone** hormone, along with **estrogen** hormone, within a few days after ovulation. Thus, the LH surge induces the onset of complex processes during the periovulatory period that complete the gametogenic function of the ovary for a given month and switches the endocrine function. The hormones of corpus luteum prepare the female reproductive tract for implantation and gestation.

***(i)* Growth and structure.** The LH surge induces dramatic structural changes in the dominant follicle that involves its **rupture, ovulation** of cumulus - oocyte complex and the **biogenesis** of a new endocrine structure called the **corpus luteum.** This gland is formed from the remaining thecal and mural granulosa cells. During this transition, following major structural changes occur:

(*a*) Before ovulation, the large preovulatory follicle presses against the ovarian surface and generates a poorly vascularized bulge of the ovarian wall called the **stigma.** The LH surge induces the release of inflammatory *cytokines* and *hydrolytic* enzymes from the theca and granulosa cells. These secreted components lead to breakdown of the follicle wall, tunica albuginea, and surface epithelium in the vicinity of the stigma (Fig. 7.21). At the end of this process, the antral cavity becomes continuous with the peritoneal cavity.

(*b*) The attachment of the cumulus cells to the mural granulosa cells degenerates, and cumulus-oocyte complex becomes free-floating within the antral cavity (Fig. 7.21). Cumulus cells also respond to the LH surge by secreting **hyaluronic acid** and other extracellular matrix components. These substances enlarge the entire cumulus-oocyte complex, a process called **cumulus expansion.** This enlarged cumulus-oocyte complex is more easily captured and transported by the oviduct. The expanded cumulus also makes the cumulus-oocyte complex easier for spermatozoa to find. Sperm express a **membrane hyaluronidase** (enzyme) that allows them to penetrate the expanded cumulus. The cumulus-oocyte complex is released through the ruptured stigma in a slow, gentle process.

(*c*) The basal lamina of mural granulosa cells is broken down so that blood vessels and outer lying theca cells secrete **angiogenic factors**, such as vascular endothelial growth factor (VEGF), angiopoietin 2, and basic fibroblast growth factors (bFGF), which significantly increase in blood supply to the new corpus luteum.

***(ii)* The gamete.** Before ovulation, the primary oocyte is competent to complete meiosis, but it is arrested in prophase I. The LH surge induces the oocyte to progress to metaphase II (Fig. 7.19). The oocyte subsequently arrest at metaphase II until fertilization. The LH receptors are present on mural granulosa cells, not on cumulus cells.

***(iii)* Endocrine function.** Both theca and mural granulosa cells express LH receptors at the time of the LH surge. The LH surge induces differentiation of the granulosa cells—a process that continues for several days after ovulation. During periovulatory period, the LH surge induces the following shifts in steroidogenic activity of the mural granulosa cells (Fig. 7.22).

1. Transient inhibition of CYP19 expression and consequent estrogen production. The rapid decline in estrogen helps turn off the positive feedback on LH secretion.

2. Breakdown of the basal lamina and vascularization of the granulosa cells. This makes LDL and HDL cholesterol available to these cells for steroidogenesis. The LH surge also increases expression of the LDL receptor and HDL receptor (SR-B) in granulosa cells.

3. Increased expression of StAR (= steroidogenic acute regulatory protein), CYP11A1 (side chain cleavage enzyme) and 3β-HSD. Because CYP17 activity, especially its 17, 20-lyase function, is largely absent in granulosa cells, these cells begin to secrete **progesterone.** The progesterone levels gradually increase over the next week.

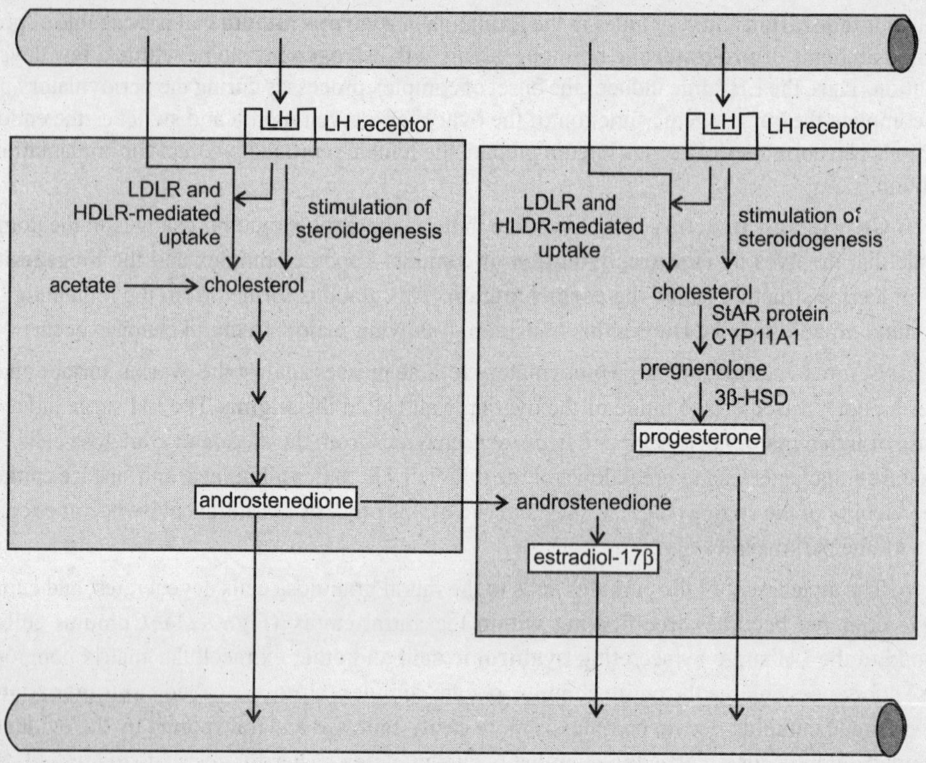

Fig. 7.22. Steroidogenic pathways in the Corpus luteum (after Koeppen and Stanton 2010).

6. The Corpus Luteum

***(i)* Growth and structure.** After ovulation, the remnant of the antral cavity fills with blood from damaged blood vessels in the vicinity of the stigma. This gives rise to a **corpus hemorrhagicum** (*corpus* means body). Within a few days, red blood cells and debris are removed by macrophages and fibroblasts fill in the antral cavity with a hyaline-like extra-cellular matrix. In the mature **corpus luteum,** the pale granulosa cells, now called **granulosa lutein cells,** enlarge and become filled with lipid (cholesterol esters). Thus, *corpus luteum* is a yellowish mass of progesterone-secreting endocrine tissue. The enlarged granulosa lutein cells collapse into and partially fill in the old antral cavity. Proliferation of these cells is very limited. The theca, along with blood vessels, mast cells, macrophages, leucocytes, and other resident connective tissue cells, infiltrate the granulosa layer at multiple sites.

The human corpus luteum is programmed to live for 14 days, plus or minus 2 days (**corpus luteum of menstruation),** unless 'rescued' by LH-like hormone **human chorionic gonadotropin (hCG),** which originates from an implanting embryo. If rescued, the **corpus luteum of pregnancy** will remain viable during the pregnancy (usually about 9 months). The mechanism by which the corpus luteum of menstruation regresses in 14 days is not fully understood. Regression appears to involve release of the **prostaglandin $PGF_2\alpha$,** from both granulosa lutein cells and the uterus in response to declining level of progesterone during the second week of the luteal phase. Several paracrine factors (such as endothelin, monocyte chemotactic protein-1) from immune and vascular cells are likely to play a role in the death and removal of granulosa lutein cells. The corpus luteum is ultimately turned into a scar-like body called the **corpus albicans** which slowly sinks into the medulla of the ovary and is slowly absorbed.

***(ii)* The gamete.** The LH surge induces two parallel events, ovulation and luteinization.

***(iii)* Endocrine function.** Progesterone production by the corpus luteum (Fig. 7.22) increases steadily from the onset of the LH surge and peaks during the midluteal phase. The main object of this timing is to transform the uterine lining into an adhesive and supportive structure for *implantation* and early pregnancy. The midluteal phase is synchronised with early embryogenesis so that the uterus is optimally primed when a blastocyst tumbles into the uterus around day 22 of the menstrual cycle. Estrogen production transiently decreases in response to the LH surge but then springs back and peaks at the midluteal phase.

Luteal hormonal output is absolutely dependent on basal LH levels (Fig. 7.22). In fact, progesterone output is closely correlated with the pulsatile pattern of LH release in woman. Both FSH and LH are reduced to basal levels during the luteal phase by negative feedback from progesterone and estrogen. In addition, granulosa lutein cells secrete **inhibin A,** which selectively repressed FSH secretion.

To support implantation and early pregnancy, the corpus luteum must generate large amounts of progesterone. Accordingly, the life of the corpus luteum is very regular, and a shortened luteal typically leads to infertility. The quality of corpus luteum is largely dependent on the size and health of the dominant follicle from which it developed, which in turn is dependent on normal hypothalamic and pituitary stimulation during the follicular phase. Numerous factors that disturb hypothalamic and pituitary output during the follicular phase, including heavy exercise, starvation, high prolactin levels and abnormal thyroid function, can lead to **luteal phase deficiency** and **infertility.**

7. Atretic Follicle

Follicular atresia refers to the death of an ovarian follicle. During atresia, the granulosa cells and oocytes undergo **apoptosis.** The thecal cells typically persist and repopulate the cellular stroma of the ovary. The thecal cells retain LH receptors and the ability to produce androgens and are collectively referred to as the **interstitial gland** of the ovary. Follicles can undergo atresia at any time during development.

B. Oviduct

***(i)* Structure and function.** The oviducts (also called the **uterine tubes** and the **fallopian tubes**) are muscular tubes with the distal ends close to the surface of each ovary and the proximal ends traversing the wall of the uterus. The oviducts are divided into four sections (going from distal to proximal end): the **infundibulum**, or open end of the oviduct, which has finger-like projections called **fimbriae** that sweep over the surface of the ovary; the **ampulla,** which has a relatively wide lumen and extensive folding of the mucosa; the **isthmus,** which has a relatively narrow lumen and less mucosal folding; and the **intramural** or **uterine segment,** which extends through the uterine wall at the superior corners of the uterus (Fig. 7.14). Oviducts have **three** main functions :

1. To capture the cumulus-oocyte complex at ovulation and transfer the complex to a midway point (the **ampullary isthmus junction),** where fertilization takes place. Oviducal secretions coat and infuse the cumulus-oocyte complex and may be required for viability and fertilizability.

2. To provide a site for sperm storage. Women who ovulate upto 5 days after sexual intercourse can get pregnant. Sperm remain viable by adhering to the epithelial cells lining the isthmus. The secretions of the oviduct also induce **capacitation** and **hyperactivity** of sperm.

3. To secrete fluids that provide nutritional support to the preimplantation embryo.

The **timing** of movement of the embryo into the uterus is critical because the uterus has an implantation window of about 3 days. The oviduct needs to hold the early embryo until it reaches the blastocyst stage (5 days after fertilization) and then allow it to pass into the internal cavity.

Histologically the wall of the oviduct is composed of a mucosa (called the **endosalpinx),** a two-layered muscularis (called the **myosalpinx**) and an outer-laying connective tissue (the **perisalpinx).** The endosalpinx is thrown into many folds, almost to the extent that the lumen is eliminated and is lined by a simple epithelium made up of two cell types: **ciliated cells** and **secretory cells.** The cilia are most numerous at the infundibular end and propel the cumulus-oocyte complex toward the uterus. The cilia on the fimbriae are the sole mechanism for transport of the ovulated cumulus - oocyte complex. Once the complex passes through the ostium of the oviduct and enters the ampulla, it is moved by both cilia and peristaltic contractions of the muscularis.

The **secretory** cells produce a protein-rich mucus that is conveyed along the oviduct to the uterus by the cilia. This ciliary–mucus escalator maintains a healthy epithelium, moves the cumulus-oocyte complex toward the uterus, and may provide directional cues for swimming sperm (*Escalator* is a moving stairs carrying people up or down). Movement of the cumulus-oocyte complex slows at the ampullary-isthmus junction, where fertilization normally takes place. This appears to be due, in part, to thick mucus produced by the isthmus and increased tone of the muscularis of the isthmus. The composition of oviducal secretions is complex and includes growth factors, enzymes and oviduct - specific glycoproteins.

C. Uterus

Uterus is the organ of gestation which receives and retains the fertilized ovum, holds the fetus during development, and becomes the principal agent of its expulsion at term.

(*i*) Structure and function. The **uterus** is a single organ that sits in the midline of the pelvic cavity between the bladder and rectum (Fig. 7.12). The mucosa of the uterus is called the **endometrium;** the three-layered, thick muscularis is called the **myometrium**; and the outer connective tissue and serosa are called the **perimetrium.** The parts of the uterus are (1) the **fundus,** which is the portion that rises superiorly from the entrance of the oviducts; (2) the **body** of the uterus, which makes up most of the uterus, (3) the **isthumus,** a short narrowed part of the body at its inferior end; and (4) the **cervix,** which extends into the **vagina** (Fig. 7.14). Cervical mucosa is distinct from the rest of the uterus and it does not undergo the process of menstruation.

The established **function** of the uterus are all related to fertilization and pregnancy. The main functions of the uterus are to

1. Assist the movement of sperm from the vagina to the oviducts.
2. Provide a suitable site for attachment and implantation of the blastocyst, including a thick, nutrient-rich stroma.
3. Limit the invasiveness of the implanting embryo so that it stays in the endometrium and does not reach the myometrium.
4. Provide a maternal side of the mature placental architecture. This includes the basal plate, to which the fetal side attaches, and large intervillous spaces that become filled with maternal blood after the first trimester.
5. Grow and expand with the growing fetus so that the fetus develops within an aqueous non-adhesive environment.
6. Provide strong muscular contractions to expel the fetus and placenta at term.

Fine structure of endometrium. The luminal surface of the endometrium is covered with a simple cuboidal/columnar epithelium (Fig. 7.23). The epithelium is continuous with mucosal glands (called **uterine glands**) that extend deep into the endometrium. The mucosa is vascularized by **spiral arteries,** which are branches of the **uterine artery** that run through the myometrium. The terminal arterioles of the spiral arteries project just beneath the surface epithelium. These arterioles give rise to a subepithelial plexus of capillaries and venules that have ballooned, thin-walled segments called **venous lakes** or **lacunae.** The lamina propria itself is densely cellular. The stromal cells of the lamina propria play important roles during both pregnancy and menstruation.

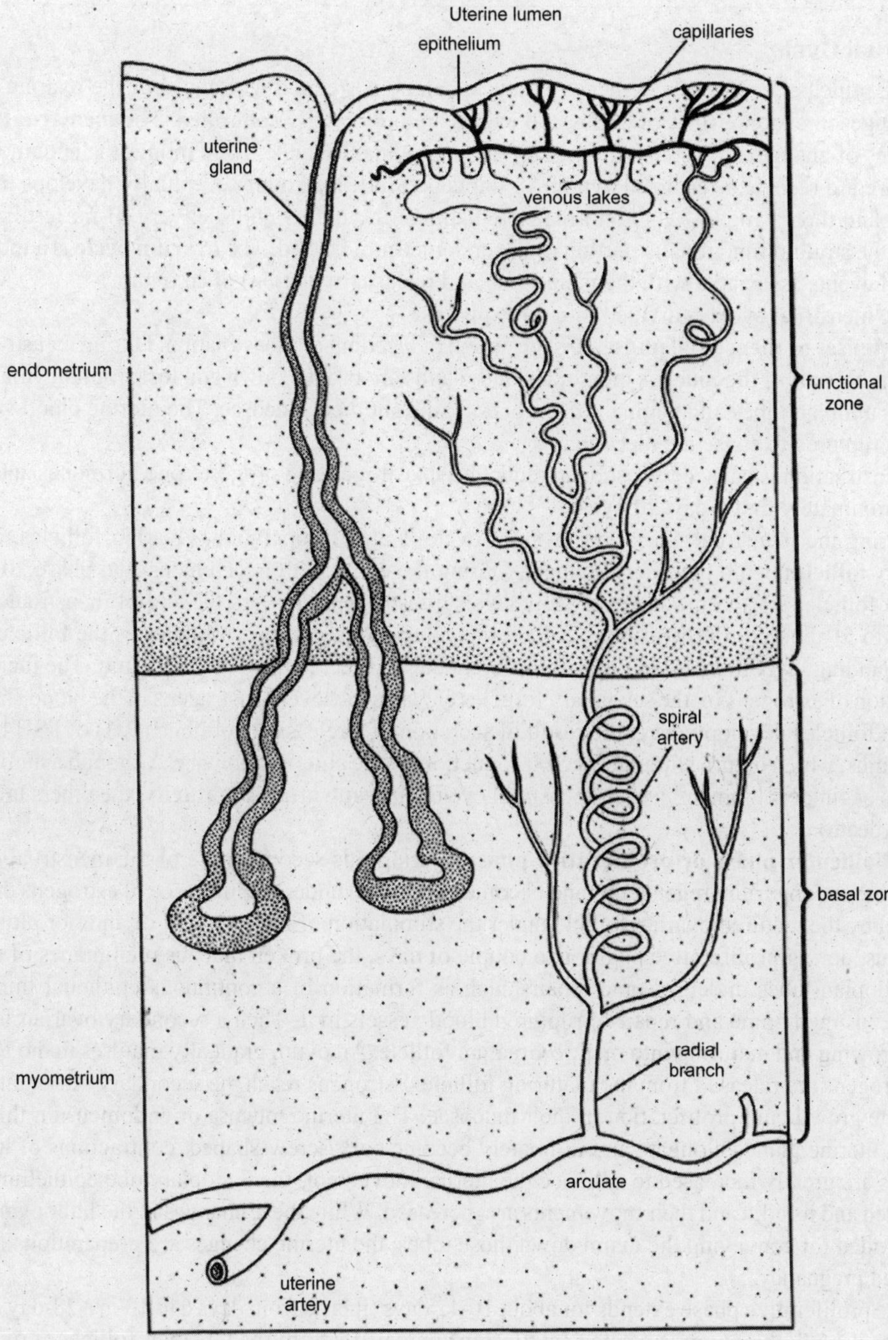

Fig. 7.23. Diagram of the organization of glands and blood flow within uterine endometrium.

About 2/3 of the luminal side of the endometrium is lost during menstruation and is called the **functional zone** (also called **stratum functionalis).** The basal third of the endometrium that remains after menstruation is called the **basal zone** (also called **stratum basale).** The basal zone is fed by straight arteries that are separate from spiral arteries, and it contains all the cell types of the endometrium (*i.e.*, epithelial cells from the remaining tips of glands, stromal cells and endothelial cells).

Menstrual Cycle

The principal events of the menstrual cycle can be correlated with those of the ovarian cycle and changes in endometrium of uterus. All events are hormonally controlled. The **menstrual cycle** is a series of changes in the endometrium of a non-pregnant female. Each month the endometrium is prepared to receive a fertilized ovum. An implanted fertilized ovum eventually develops into an embryo and then a foetus, which normally remains in the uterus until delivery. If no fertilization occurs, the stratum functionalis portion of the endometrium is shed. The **ovarian cycle** is a monthly series of events associated with the maturation and release (ovulation) of an ovum.

The menstrual cycle is divided into three phases:

1. **Menses or menstrual phase.** It is the period of bleeding or menstruation. During menstruation considerable part of the mucous membrane and glands are broken down and lost from the linings of uterus, Fallopian tubes and vagina. This process thins out their mucosae. The uterine blood vessels are also ruptured to cause the bleeding.

Menstruation is caused by a sudden reduction in estrogen and progesterone hormones and lasts for approximately first 5 days of cycle.

During the menstrual phase, the ovarian cycle is also in operation. Ovarian follicles, called **primary follicles** begin their development. During the early part of each menstrual phase 20 to 25 primary follicles start to produce very low levels of estrogens. Toward the end of menstrual phase (days 4 to 5) about 20 of the primary follicles develop into **secondary follicles** as the follicle cells of surrounding layer increase in number, differentiate and secrete the follicular fluid. The increased production of estrogens by the secondary follicles elevates the level of estrogens in the blood slightly. Ovarian follicles development is the result of secretion of a releasing hormone (RH) of FSH by hypothalamus, which in turn stimulates FSH production by the anterior pituitary. Although a number of follicles begin development in each menstrual cycle, **only one attains maturity,** the others undergo atresia (death).

2. **Follicular phase or proliferative phase.** During this second phase of the menstrual cycle, the uterine endometrium regenerates and becomes thickened under the influence of estrogens that are secreted by the matured ovarian follicle under the stimulation of FSH and LH of anterior pituitary.

Thus, during proliferative phase, in a couple of days, the broken mucous membranes of uterus and Fallopian tubes undergo some repairs such as formation of a continuous epithelial lining on the mucous membrane and repair of ruptured blood vessels in it. Then a secondary ovarian follicle starts growing and maturing into mature Graafian follicle. An ovum gradually matures in the follicle and estrogens are released from the maturing follicle. Estrogens reach the secondary sex organs and stimulate growth and proliferation in their mucosae. The uterine mucosa or endometrium thickens greatly, uterine glands elongate and ultimately become cork-screw-shaped, contractions of uterine muscles are greatly increased to enhance the uterine movement, the Fallopian tube epithelium gets thickened and its cilia and their movements are increased. While these changes in the Fallopian tubes are intended for conveying the ovum down those tubes, the uterine changes are preparations for the expected pregnancy.

The proliferative phase extends for about 10-12 days (it lasts from days 6 to 13 in a 28-day cycle) and near its end, the ovum is ejected **(ovulation)** on day 14 from the Graafian follicle of ovary in the abdominal cavity *(i.e.,* pelvic cavity). Ovulation occurs due to decreased levels of FSH and LH in the blood.

It has been found that LH is necessary for final follicular growth and ovulation. Approximately two days before ovulation, the rate of secretion of LH by anterior pituitary gland increases six to ten-fold and peaking about 18 hours before ovulation. This is called **ovulatory surge of LH**. FSH also increases about two fold at the same time, and two of these hormones act synergistically to cause rapid swelling of follicle. *LH also stimulates the granulosa to secrete less estrogen but progressively increasing amount of progesterone.*

3. Secretory or luteal phase. This phase lasts from days 15 to 28 in a 28-day cycle and represents the time between ovulation and onset of next menses. During this phase the ruptured follicle changes into a corpus luteum (yellow body) in the ovary. The corpus luteum secretes the increasing quantities of the estrogens and progesterone. Progesterone produces changes in the secondary sex organs to further the preparations for the anticipated pregnancy. So, the uterine endometrium thickens further, uterine glands become more cork-screw-shaped and begin secreting a juice in the uterus and uterine movements are reduced to keep the uterus quiet. All these uterine changes help in the implantation of the fertilized ovum on the uterine wall. Progesterone also inhibits further follicular maturation and ovulation from the ovary in this phase.

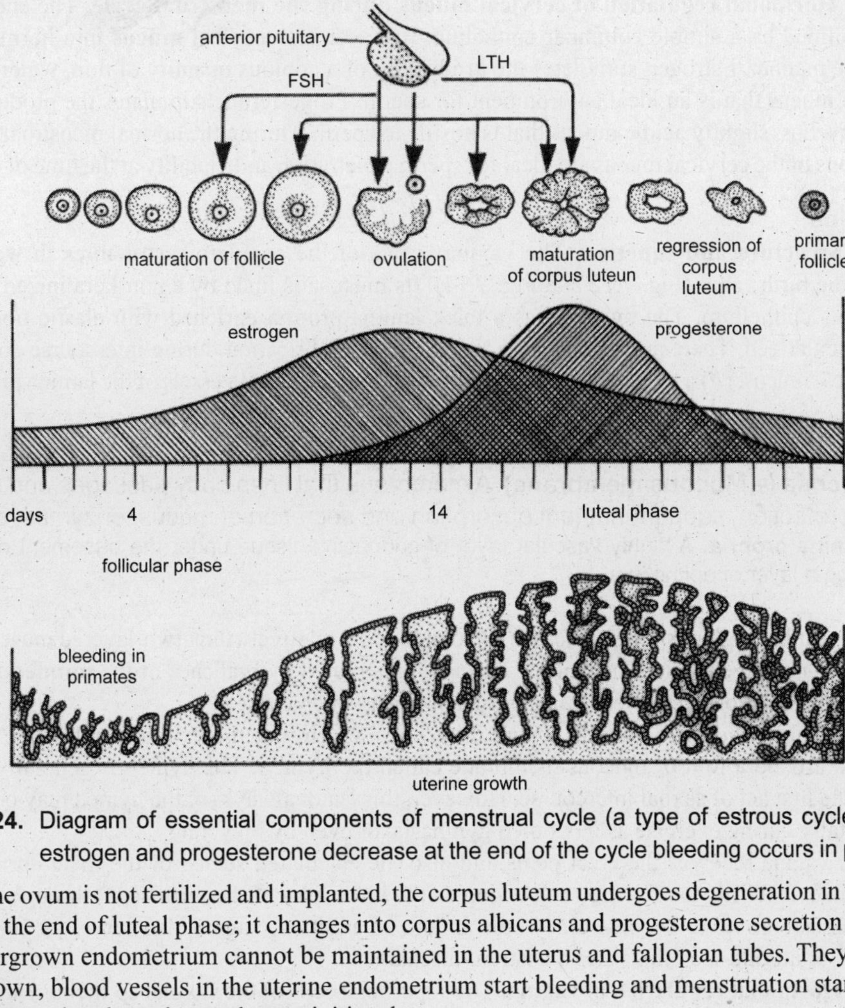

Fig. 7.24. Diagram of essential components of menstrual cycle (a type of estrous cycle). When estrogen and progesterone decrease at the end of the cycle bleeding occurs in primates.

If the ovum is not fertilized and implanted, the corpus luteum undergoes degeneration in the ovary towards the end of luteal phase; it changes into corpus albicans and progesterone secretion declines. The overgrown endometrium cannot be maintained in the uterus and fallopian tubes. They begin to break down, blood vessels in the uterine endometrium start bleeding and menstruation starts. Thus, new menstrual and ovarian cycles are initiated.

Menarche and menopause. The menstrual cycle normally occurs once each month from **menarche**, the first menses to **menopause,** the last menses. The advent of menopause is signalled by the **climacteric,** *i.e.,* menstrual cycle becomes less frequent between ages 40 to 50. In the post-menopausal woman there will be some atrophy of the ovaries, uterine tubes, uterus, vagina, external genitalia and breasts.

D. The Cervix

***(i)* Structure and function.** The cervix is the inferior extention of the uterus that projects into the vagina (Fig. 7.14) It has a mucosa that lines the **endocervical canal,** which has a highly elastic lamina propria and a muscularis with the myometrium. The part of the cervix that extends into the vaginal vault is called the **ectocervix,** whereas the part surrounding the endocervical canal is called the **endocervix.** The openings of the endocervical canal at the uterus and vagina are called the **internal cervical os** and **external cervical os**, respectively. The cervix acts as a gateway to the upper female tract–at midcycle, the endocervical canal facilitates sperm viability and entry. During the luteal phase, the endocervical canal hampers the passage of sperm and microbes, thereby inhibiting **superimplantation** of a second embryo or ascending infection into the placenta, fetal membranes and fetus. The cervix physically supports the weight of the growing fetus. At term, **cervical softening** and **dilation** allow passage of the newborn and placenta from the uterus into vagina.

(*ii*) Hormonal regulation of cervical mucus during the menstrual cycle. The endocervical canal is lined by a simple columnar epithelium that secretes **cervical mucus** in a hormonally responsive manner. Estrogen stimulates the production of a copious quantity of thin, watery, slightly alkaline mucus that is an ideal environment for sperm. Progesterone stimulates the production of a scant, viscous, slightly acidic mucus that is hostile to sperm. During the normal menstrual cycle, the conditions of the cervical mucus are ideal for sperm penetration and viability at the time of ovulation.

E. Vagina

(*i*) Structure and function. The vagina is one of the copulatory structures in women and acts as the birth canal (Fig. 7.12 and Fig. 7.14). Its mucosa is lined by a non-keratinized, stratified squamous epithelium. The mucosa has a thick lamina propria enriched with elastic fibers and is well vascularized. There are no glands in the vagina, so lubrication during intercourse comes from (*i*) cervical mucus, (*ii*) a transudate (*i.e.*, ultrafiltrate) from the blood vessels of the lamina propria, and

> **Box 7.5**
>
> **Mucosa (= Mucous membrane).** A membrane that lines body passages and functions in protection, support, nutrient absorption and secretion of mucus, enzymes, salts, etc.
> **Lamina propria.** A highly vascular layer of connective tissue under the basement membrane lining a layer of epithelium.

(*iii*) the vestibular glands. The mucosa is surrounded by a relatively thin, two-layered muscularis and an outer connective tissue. The wall of vagina is innervated by branches of the **pundendal nerve,** which contribute to sexual pleasure and orgasm during intercourse.

In females who have not had sexual intercourse (a virgin), the external end of the vagina is bordered around a fold of mucous membrane called the **hymen.** The hymen is generally ruptured during the first act of sexual intercourse. However, this is not always so; the hymen may be ruptured accidentally during exercise as in women gymnasts, or even by bicycling.

The vagina receives the erect penis and also the ejaculated semen of the male during sexual intercourse. It fulfills the function of serving as the lower part of the birth canal through which the fetus passes enroute to become a newborn infant. In addition, vagina also serves as an excretory passage for uterine secretions and menstrual discharge.

(*ii*) Hormonal regulation of vagina. The superficial cells of the vaginal epithelium are continually desquamating (*i.e.*, peeling of epithelial cells as scales) and the nature of these cells is influenced by the hormonal environment. **Estrogen** stimulates proliferation of the vaginal epithelium and increases its glycogen content. (*Note*: This process is called **cornification** in nonhuman mammals). The glycogen is metabolized to lactic acid by commensal lactobacilli, thereby maintaining **acidic environment.** This inhibits infection by noncommensal bacteria and fungi. **Progesterone** increases the desquamation of epithelial cells.

F. The Female External Genitalia

(*i*) Structure and function. The female external genitalia are surrounded by the **labia majora**

(homologues of the scrotum) laterally and **labia pubis** anteriorly (Fig. 7.14). The **vulva** collectively refers to an area that includes the labia majora and mons pubis, plus the **labia minora,** the **clitoris,** the **vestibule** of the vagina, the **vestibular bulbs** (glands) and **external urethral orifice.** The vulva is called the **pudendum** by clinicians. The structure of the vulva serve the functions of sexual arousal and climax, directing the flow of urine, partially covering the opening of the vagina, thereby inhibiting the entry of the pathogens.

Perineum is the skin covered muscular region between the vaginal orifice (vestibule) and the anus. The perineum is usually ruptured during child-birth. In order to avoid this an incision (**episiotomy**) is made in the perineal tissue just before the passage of the fetus down the vagina. This provides more room for the fetus, and minimises damage to the perineum and anal sphincters.

The clitoris is the embryological homologue of the penis and is composed of two **corpora cavernosa,** which attach the clitoris to the ischiopubic rami and a **glans.** These structures are composed of erectile tissue and undergo the process of erection in essentially the same manner as the penis. Unlike the penis, clitoral tissue is completely separate from the urethra. Thus, the clitoris is involved in sexual arousal and climax at orgasm.

***(ii)* Hormonal regulation.** The external genitalia and vagina are responsive to androgens (testosterone and dihydrotesterone) and estrogen. Androgens also act on the central nervous system (CNS) to increase libido (or sexual drive) in women.

Copulation (Or Coitus)

Internal fertilization is an essential part of reproductive cycles in terrestrial organisms and this is facilitated in many organisms, including humans, by the development of an **intromittent organ,** the **penis,** which is inserted into the vagina and releases gametes as high as possible within the female reproductive tract. Erection of the penis occurs as a result of local increase in blood pressure in erectile tissue of the penis, due to parasympathetic nervous activity producing constricting of the veins and dilation of the arterioles following some form of sexual excitement. In this state the penis can be inserted into the vagina where the friction, produced by the rhythmic movements of sexual intercourse, increases the tactile stimulation of sensory cells at the tip of the penis (i.e., glans penis). This activates sympathetic neurons which lead to closure of the internal sphincter of the bladder and contraction of the smooth muscle of the epididymis, vas deferens, and the male accessory glands, the seminal vesicle, prostate glands and bulbo-urethral glands. This action discharges sperm and seminal fluid into the proximal (internal) urethra where they mix to form **semen**. The increased pressure of these fluids in the proximal urethra leads to reflex activity in the motor neurons supplying the muscles at the base of penis. Rhythmic wave-like contractions of these muscles force semen out through the distal (external) urethra during **ejaculation** which marks the climax of copulation. The other physiological and psychological sensations associated with this climax in both male and females are called **orgasm.** Lubrication is provided during intercourse partly by a clear mucus secreted by the male bulbo-urethral glands following erection but mainly by glands in the vulva.

Capacitation of sperm and fertilization. Although the spermatozoa are said to be "mature" when they leave the epididymis, their activity is held in check by multiple inhibitory factors secreted by the genital duct epithelia. Therefore, when they are first expelled in the semen, they are unable to fertilize the ovum. However, on coming in contact with the fluids of the female genital tract, multiple changes occur that activate the sperm for the final processes of fertilization. These collective changes are called **capacitation** of the spermatozoa.

Fertilization. After the male ejaculates semen into the vagina during intercourse, a few sperm are transported within 5 to 10 minutes upward from the vagina and through the uterus and fallopian tubes to the **ampullae** of the fallopian tubes near the ovarian ends of the tube. This transport of the sperm is aided by contractions of the uterus and fallopian tubes stimulated by **prostaglandins** in the male seminal fluid and also by **oxytocin** released from the posterior pituitary gland of the female during her orgasm (**Hall** 2011). Of the almost half a billion sperm deposited in the vagina, a few thousand succeed in reaching each ampulla.

Acrosome reaction. Stored in the acrosome of the sperm are large quantities of **hyaluronidase** and **proteolytic enzymes.** Hyaluronidase depolymerizes the hyaluronic acid polymers in the intercellular cement that holds the ovarian granulosa cells together. The proteolytic enzymes digest proteins in the structural elements of tissue cells that still adhere to the ovum.

When the ovum is expelled from the ovarian follicle into the fallopian tube, it still carries with it multiple layers of granulosa cells. Before a sperm can fertilize the ovum, it must dissolute these granulosa cell layers and then it must penetrate through the thick covering of the ovum itself, the *zona pellucida*. To achieve this, the stored enzymes in the acrosome begin to be released. It is believed that the hyaluronidase among these enzymes is especially important in opening pathways between the granulosa cells so that the sperm can reach the ovum.

When sperm reaches the zona pellucida of the ovum, the anterior membrane of the sperm binds specifically with receptor proteins in the zona pellucida. Next, the entire acrosome rapidly dissolves and the all the acrosomal enzymes are released. Within minutes, these enzymes open a penetrating pathway for passage of the sperm head through the zona pellucida to the inside of the ovum. Within another 30 minutes, the plasma membrane of sperm head and of the ovum fuse with each other to form a single cell (called **zygote**). At the same time, the genetic material of the sperm and the oocyte combine to form a completely new cell genome, containing equal numbers of chromosomes and genes from mother and father. This is the process of **fertilization;** then the embryo begins to develop.

Why does only one sperm enter the oocyte? With as many sperm as there are, why does only one enter the oocyte? The reason is not entirely known, but within a few minutes after the first sperm penetrates the zona pellucida of the ovum, calcium ions (Ca^{+2}) diffuse inward through the oocyte membrane and cause multiple **cortical granules** to be released by exocytosis from the oocyte into the perivitelline space. These granules contain substances that permeate all portions of the zona pellucida (*i.e.*, elevation of **fertilization membrane** or **envelope** occurs during the slow block to polyspermy) and prevent binding of additional sperm and they even cause any sperm that have already begin to bind to fall off. Thus, almost never does than one sperm enter the oocyte during fertilization.

Transport of the fertilized ovum in the fallopian tube. After fertilization has occurred, an additional 3 to 5 days is normally required for transport of the fertilized ovum through the remainder of the fallopian tube into the cavity of the uterus. This transport is effected mainly by a feeble fluid

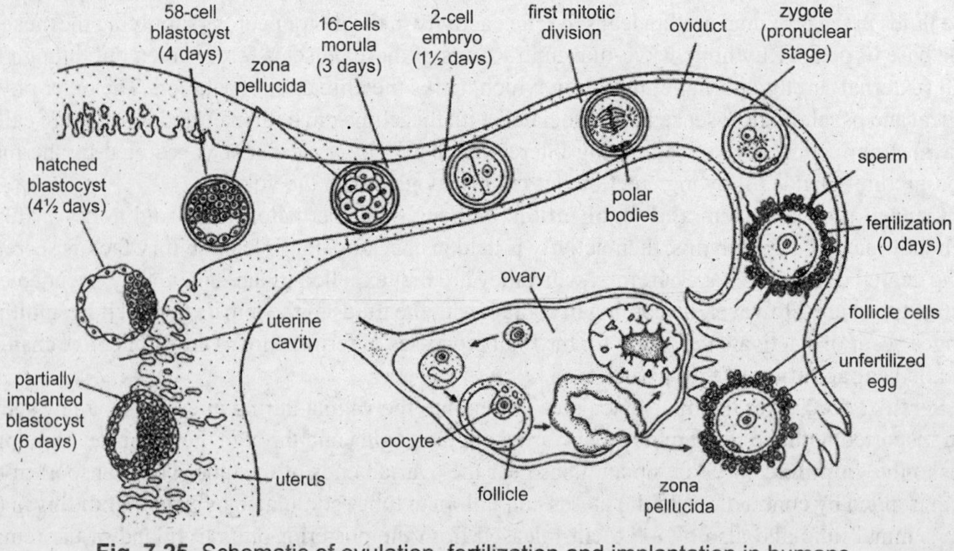

Fig. 7.25. Schematic of ovulation, fertilization and implantation in humans.

current in the tube resulting from epithelial secretion plus action of the ciliated epithelium that lines the tube; the cilia always beat toward the uterus. Weak contractions of the fallopian tube may also aid the ovum passage.

This delayed transport of the fertilized ovum through the fallopian tube allows several stages of cell division (cleavage) to occur before the dividing ovum now called a **blastocyst,** with about 100 cells–enters the uterus. During this time, the fallopian tube's secretory cells produce large quantities of secretions used for the nutrition of the developing blastocyst.

Gestation and implantation. The development of young within the female reproductive system from the time of conception to childbirth is called **pregnancy** or **gestation.** In humans, the normal period of gestation is approximately 266 days.

After reaching the uterus, the developing blastocyst usually remains in the uterine cavity an additionally 1 to 3 days before it **implants** in the endometrium. Thus, **implantation** ordinarily occurs on about the fifth to seventh day after ovulation. Before implantation, the blastocyst obtains its nutrition from the uterine endometrial secretions, called **"uterine milk".**

Implantation occurs from the action of **trophoblast cells** that develop over the surface of the blastocyst. These cells secrete proteolytic enzymes that digest and liquefy the adjacent cells of the uterine epithelium. Some of the fluid and nutrients released are actively transported by the same trophoblast cells into the blastocyst, adding more sustenance for growth. Once implantation has taken place, the trophoblast cells and other adjacent cells (from the blastocyst and the uterine epithelium proliferate rapidly forming the **placenta** and the various membranes of pregnancy.

7.4 CONTRACEPTION

Prevention of conception (pregnancy) is termed **contraception.** It is a birth control measure. An ideal contraceptive should be : (i) user friendly; (ii) easily available and cheaper; (iii) effective but also reversible; (iv) with no or least side effects; (v) should not interfere with the sexual desire and/or sexual act of the user.

Measures of contraceptions have a significant role in checking the uncontrolled growth of human population, but not regular requirements for reproductive health. These are required either to prevent pregnancy or space pregnancy.

Types of Contraceptives

1. Temporary methods. A. Mechanical barrier methods. These prevent physical meeting of ovum and sperm and are of following types :

***(i)* Condom.** These are non-porous disposable rubber or latex sheaths which are put on the erect penis during intercourse. These are popularly called **Nirodh.** These check pregnancy by preventing deposition of sperm in the vagina. Condoms are widely used contraceptive in India because they are easily available, low cost, reliable, effective and with no side effects. These can be self inserted so give desired privacy to the user. Nirodh also prevent the spread of sexually transmitted diseases (STDs) including AIDS, syphilis, etc.

***(ii)* Female condoms (Femidoms).** These are made of skin-friendly polyurethane which act as a clip on the vagina wall blocking the passage to the womb (uterus). These are used in Britain, USA, Brazil, Zimbambe, South Africa, France, Sri Lanka, Thailand and Nepal.

***(iii)* Intra-uterine devices (IUDs).** These prevent fertilization of egg or implantation of zygote in uterine mucosa by altering endometrial function. These also increase phagocytosis of sperms in the uterus. IUDs are fitted in the uterus. In India, IUDs are most widely accepted method of contraception in females for spacing the children. Several types of IUDs are available: (a) *Non-medicated IUDs, e.g.*, Lippe's loop; (b) *Copper-releasing IUDs, e.g.*, CU-T (or copper T200), Cu-7, multiload Cu-250 and multi-load 375. These release Cu-ions which decrease the sperm motility and their fertilizability. (c) *Hormone-releasing IUDs, e.g.*, Progestasert, and LNG-20, whose hormones make the uterus unsuitable for implantation and the cervix hostile to sperms.

Though IUDs form one of the most widely accepted methods of contraception in India, but these have certain drawbacks such as occasional hemorrhage, chances of infection and spontaneous expulsion (miscarriage).

Chemical barrier methods. Spermicidal tablets, jellies, pastes, creams or sponge are placed in vagina of a female about 5 to 15 minutes before coital activity. These kill sperm. Common spermicidal chemicals used are lactic acid, citric acid, potassium permaganate, zinc sulphate etc.

2. Natural contraception. (a) To abstain. It involves refraining from coital activity.

(*b*) Coitus interruptus. It involves withdrawing penis by male before ejaculation so that semen is not deposited in the vagina. It is traditional method of voluntary fertility control. This method has certain shortcomings: 1. some sperm may be deposited in the vagina even before the sexual climax. 2. It may develop psychological and physiological problems to both partners.

(*c*) Safe period or rhythm method. A week before and a week after the menstrual phase are supposed to be **safe period** for sexual intercourse. It reduces the chances of pregnancy by about 80 per cent. It is so because, 1. Ovulation generally occurs on 14th day of the menstrual cycle (range is 13th to 16th day of the menstrual cycle). 2. Viability period of ovum is of about two or three days (or 48 hours to 72 hours). 3. Viability period of sperm is of only one day (24 hours).

The period from 10th to 17th day (both days inclusive) of the menstrual cycle is called **danger** or **risk** or *fertile period* because mature Graafian follicle is present during this period in the ovary of a healthy human female. So unprotected sexual intercourse should be avoided during this period. Effectiveness of rhythm method can be increased by **basal body temperature (BBT) technology** (Box 7.6).

Box 7.6

The release of the ovum (egg) from the ovary (*i.e.*, ovulation) takes place at fairly fixed intervals. The time of ovulation can be known recording the *basal body temperature* (*BBT*), *i.e.*, recording the body temperature (oral or rectal), before getting out of bed in the morning. In preovulatory phase the oral BBT is 36.3 – 36.8 °C which increases by 0.3 to 0.5 °C one-two days after ovulation. It is probably due to the increase in progesterone hormone secretion, since progesterone is thermogenic. Progesterone increases BBT due to formation of its derivatives, etiocholanolone (a testosterone metabolite $C_{19}H_{30}O_2$ that occurs in urine) and **pregnanediol** (a metabolite produced by reduction of progesterone and is present in the urine of pregnant women in the form of its glucuronide). (**Jain** 2012).

(*d*) Lactational amenorrhoea (LAM): It has been noticed that during the period of intense lactation after the partiturition, the mother does not undergo menstruation and ovulation as high level of prolactin suppresses the secretion of gonadotropins from adenohypophysis so that the chances of conception are nearly nil. It also has no side effects but is effective only upto a maximum period of six months after parturition.

Hormonal Methods

1. Oral Methods: (a) Combined pills. Birth control pills are taken through mouth. These contain a combination of synthetic progestins (acting like progesterone hormone) and estrogen hormones; they are called **combined pills.** These progesterone-estrogen combinations inhibit secretion of FSH and LH from anterior lobe of pituitary glands, so these inhibit ovulation from the ovary and implantation. These hormones also retard the entry of sperm in the uterus by changing the quality of cervical mucus. Commonly used hormonal pills in India are: 1. **Mala-D** (daily) and 2. **Saheli** (weekly). **Mala-D** has to be taken daily at bed time for a period of 21 days starting preferably within the first five days of menstrual cycle and repeated after a gap of 7 days. **Saheli** contains a non-steroidal preparation called **centchroman.** It has to be taken "once a week." It has high contraceptive value and with very few side effects. Saheli pill was developed by scientists at **Central Drug Research Institute** located at Lucknow.

Drawbacks of pills. These birth control pills generally have some side effects such as nausea, breast-tenderness, weight gain, breakthrough bleeding (*i.e.*, bleeding between menstrual periods) and breast cancer.

(*b*) Sequential pills. For first 15 days of cycle large doses of oestrogen followed by combination of oestrogen and progesterone for a week are given. This also prevents ovulation.

(*c*) Mini pills or micropills method. It involves the administration of 'low dose of progesterone' through whole of the menstrual cycle. This prevents fertility without inhibiting ovulation. It may act on: 1. cervical mucus; or 2. endometium, or decrease the motility of fallopian tubes.

(*d*) Postcoital pills. Pills are given after coitus but befoıe 48 hours. These are called **morning after pills.** They are prepared by hormonal combinations.

The method employed is to give a double dose of "combined pill" that is 2 pills immediately followed by another 2 pills 12 hours later. It is advocated as an **emergency method,** *e.g.*, after unprotected intercourse, rape or contraceptive failure. These pills prevent implantation of fertilized ovum in uterine wall.

(*e*) Progesterone antagonist such as **Mifepristone** is helpful in producing abortion following the conception. It acts by inhibiting the progestational effects on the uterus.

(*f*) Implants. These contain either only progesterone or progesterone-estrogens combinations. These are placed under the skin and act like the oral pills. Implants have much longer effective period and are found to be very effective even within 72 hours of coitus. So alongwith oral pills are used as emergency contraceptives. **Norplant** is most commonly used implant in India. It consists of six small plastic cylinders that contain progestin. It is inserted directly under the skin of upper arm by a simple surgical procedure. Norplant prevents ovulation and thickens the cervical mucus which makes the sperm entry into the uterus more difficult.

Postconception Methods

(*i*) Endometrial aspiration. It is done by introducing a small plastic catheter into the uterus through the cervix. Aspiration of lining of uterus is done by suction. Method is safe and does not require anaesthesia. It is done also before the pregnancy is established.

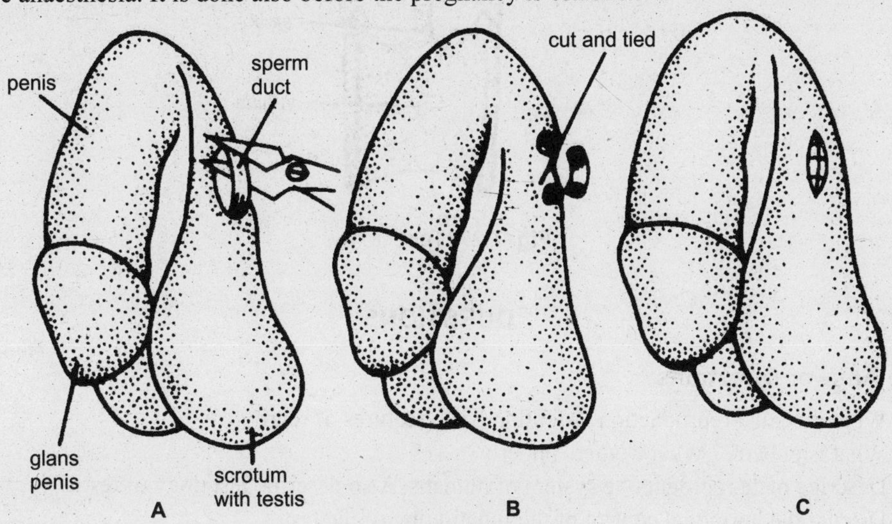

Fig. 7.26. Vasectomy.

(*ii*) **Medical termination** of pregnancy is done by using following methods: (a) Dilation and curettage (D&C); (b) Vacuum aspiration, (c) Introducing intrauterine hypertonic saline solution (20% saline or 40% urea) or prostaglandin (D4F2); (d) Pitocin drip with 5% glucose.

Permanent Methods

These are surgical methods which block the gamete transport and so prevent conception. These methods are highly effective as their failure rate is almost zero per cent but their reversibility is very low. These cannot be considered as contraceptive measures. Permanent contraceptive methods include the following:

1. Male sterilization. It is a permanent method of birth control in which either testes are surgically removed, called **castration,** or cutting or tying of the vas deferens, called **vasectomy** (Fig. 7.26.). It is done through a small cut/incision in the scrotum.

2. Female sterilization. Removal of uterus, called **hysterectomy** and removal of ovaries, called **oophorectomy** or **ovariectomy** are absolute preventive (*i.e.*, sterilization) method of female.

(*i*) **Tubectomy.** It involves cutting or tying of fallopian tubes by an small incision in the abdomen (Fig. 7.27) or through vagina.

(*ii*) **Tubal ligation.** It involves blocking of fallopian tubes by an instrument called **laproscope.**

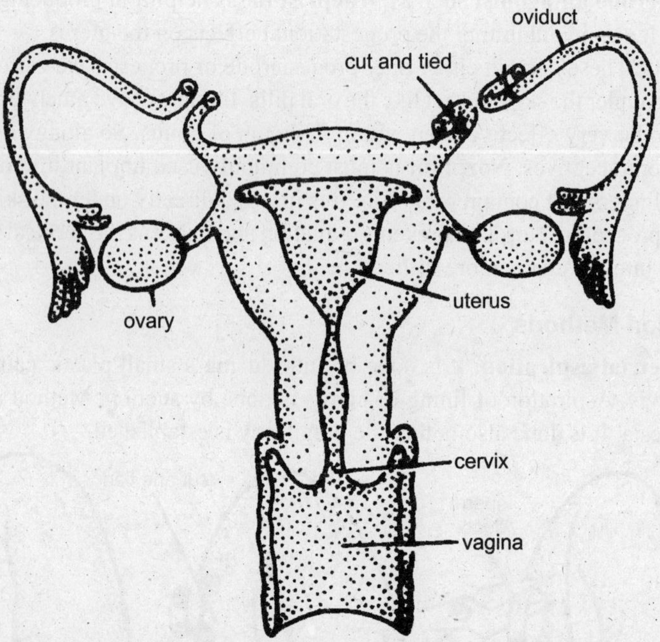

Fig. 7.27. Tubectomy.

QUESTIONS

Long Answer Questions

1. What is sexual reproduction? Mention basic features of reproduction.
2. What is puberty? What triggers puberty?
3. Describe male reproductive system of humans. Also describe histology of testis.
4. Discuss the hormonal control of the functioning of the testes.
5. Describe the structure and function of follicles of human ovary.
6. Give an account of the development of a Graafian follicle.

7. Discuss the events which takes place in menstrual cycle.
8. List the functions of oestrogen (estradiol) and progesterone.
9. Describe secretion and functions of ovarian hormones and regulation of their secretion.
10. Give a detailed account of birth control or contraception.

Short Answer Questions

1. List the functions of Sertoli cells.
2. Distinguish between
   (*a*) Spermacytogenesis and spermiogenesis
   (*b*) Corona radiata and zona pellucida
3. Write short notes on the following:
   (*i*) Ovulation;
   (*ii*) Testosterone;
   (*iii*) Contraceptives;
   (*iv*) Ovarian follicles;
   (*v*) Semen;
   (*vi*) Hormones of testis;
   (*vii*) Puberty;
   (*viii*) Ovarian hormones.
4. Differentiate between
   (*i*) Leydig and Sertoli cells;
   (*ii*) Role of FSH and LH in spermatogenesis;
   (*iii*) Testosterone and DHT
5. Describe chemical method of birth control.
6. Differentiate between tubectomy and vasectomy.

Very Short Answer Questions

1. Give the term used for the period between fertilization and birth of young one.
2. What is "after birth".
3. Give full form of IUD.

Fill in the Blanks

1. In humans, the development of young one occurs in the ______________.
2. Development of young one in the mother's womb is called ________.
3. Gestation period for the human beings is _________ days.
4. Intimate connection established between foetal membrane and uterine wall is known as _________.
5. Process of giving birth to a young one is called __________.
6. Placenta acts as the nutritive, ____________ and _______ organ of the developing placenta.
7. Secretion of milk by mammary glands after the birth of a young one is called ________.
8. Parturition involves forceful contraction called __________.

Multiple Choice Questions

1. Dartos tunic of smooth muscle fibers occurs in
   (a) uterine wall (b) urethral wall
   (c) scrotal wall (d) abdominal wall
2. Sertoli cells in the seminiferous tubules
   (a) secrete male sex hormone (b) activate the spermatozoa
   (c) nourish the developing sperm (d) clump the spermatogonia

3. Spermiation means
(a) release of mature sperms from Sertoli cell
(b) conversion of spermatids to sperms
(c) conversion of non-motile to motile sperms
(d) process by which sperms are formed

4. Not a feature of Sertoli cells
(a) nourish sperms and control their release (b) provide blood testis barrier
(c) secrete testosterone (d) secrete inhibin B

5. Primary spermatocytes to get converted into secondary spermatocytes require presence of
(a) FSH (b) LH
(c) FSH and LH both (d) testosterone

6. Optimal temperature for spermatogenesis is
(a) 37 ºC (b) 43 ºC
(c) 40 ºC (d) 32 ºC

7. Testosterone secretion from Leydig cell is under control of
(a) FSH (b) LH
(c) androgen (d) estrogen

8. pH range of semen is
(a) 6.2 – 6.5 (b) 7.0 – 7.3
(c) 7.35 – 7.5 (d) 7.5 – 7.9

9. Sperm motility is decreased in
(a) alkaline pH (b) neutral pH
(c) acidic pH (d) none of the these

10. In the testis, the cells that secrete testosterone
(a) geminal epithelium (b) Leydig cells
(c) Sertoli cells (d) sperm cells

11. Primitive gonads are identical in both sexes upto............weeks of gestation.
(a) 4 (b) 6
(c) 8 (d) 10

12. Testosterone secretion begins in .... week of intrauterine life.
(a) fifth (b) sixth
(c) seventh (d) eight

13. Puberty does not normally occur in human under the age of 8 years because
(a) the tissues are unresponsive to gonadal steroids
(b) the ovaries and testes are unresponsive to gonadotrophins
(c) the pituitary cannot manufacture adequate amount of gonadotrophins
(d) the hypothalamus fails to secrete GnRH* in a pulsatile manner.
(*GnRH = Gonadotrophin releasing hormone)

14. Hair growth in either sex of puberty is due to
(a) growth hormone
(b) insulin
(c) estrogens in females, testosterone in male
(d) androgens

15. Sperm capacitation takes place in the
(a) epididymis (b) seminal vesicle
(c) female genital tract (d) rete testis

16. Penile erection is caused primarily by
(a) reflex sympathetic constriction of the arterioles

(b) contraction of the bulbo cavernous muscle
(c) reflex-parasympathetic constriction of the venules
(d) parasympathetic induced dilation of the arterioles

17. Parasympathetic stimulation causes vasodilation in penis by release of
(a) Ach (b) nitric oxide
(c) bradykinin (d) epinephrine

18. Oogenesis begins
(a) from approximately 10th week of foetal life
(b) at term
(c) after birth
(d) at puberty

19. In large doses, oestrogen produces
(a) acts directly on hypothalamus and anterior pituitary to inhibit gonadotropin secretion
(b) positive feedback on LH secretion
(c) inhibit secretion of FSH
(d) inhibit secretion of LH

20. Oogonia reach their maximum number at
(a) five months of foetal life (b) birth
(c) puberty (12-14 years of age) (d) adolescence (16-20 years of age)
(e) adulthood (21-26 years of age)

21. Antrum is the cavity in the
(a) testis (b) ovary
(c) primary follicle (d) Graafian follicle

22. At the time of ovulation, the female gamete is at ...... stage
(a) oogonium (b) primary oocyte
(c) secondary oocyte (d) mature ovum

23. The first sign of pregnancy is
(a) absence of menstrual bleeding (b) movements of foetus
(c) enlargement of abdomen (d) lack of hCG in urine

24. Embryo reaches the uterus from the fallopian tube in about
(a) 24 hours (b) two days
(c) seven days (d) fourteen days

25. Peak secretion of estrogen hormone occurs
(a) Just before ovulation
(b) mid-luteal phase
(c) at the time of ovulation
(d) secreted at the same rate during all phases of menstrual cycle

26. Theca cells are derived from
(a) zona pellucida (b) germinal epithelium
(c) stromal cells (d) granulosa cells

27. Ejaculation of seminal fluid is under control of
(a) sympathetic nerve (b) parasympathetic nerve
(c) spinal nerves (d) all of the above

28. Ovum fertilization usually occurs in which part of fallopian tube ?
(a) fimbrial end (b) infundibulum
(c) ampulla (d) isthmus

29. The transport of spermatozoa from the cervix to fallopian tube takes
(a) 5-10 minutes (b) 10-30 minutes
(c) 30-60 minutes (d) 60-90 minutes

30. Uterine contractility during labour is increased by
(a) altered estrogen-progesterone ratio (b) oxytocin
(c) prostaglandins (d) all of these

31. The pregnancy test is based on detection of which hormone ?
(a) progesterone (b) estrogen
(c) HCG (d) LH

32. After about _____ days, blastocyst implants on the wall of the uterus.
(a) 6 (b) 7
(c) 8 (d) 9

33. Time of ovulation is indicated by
(a) recording blood levels of estrogen (b) recording of basal body temperature (BBT)
(c) recording of blood level of progesterone (d) recording of level of FSH in blood

34. Time of ovulation is usually
(a) 12 days prior to menses (b) 14 days prior to menses
(c) 16 days prior to menses (d) 18 days prior to menses

35. Estrogen acts on
(a) plasma membrane receptor (b) cytoplasmic receptor
(c) nuclear receptor (d) mitochondrial receptor

36. The viability of the spermatozoa within the female genital tract is upto .... hours.
(a) 6-12 (b) 12–24
(c) 24–48 (d) 48 – 72

37. Best contraceptive measure in males is
(a) conventional method (b) coitus interruptus
(c) vasectomy (d) drug gossypol

38. Mechanism of action of intrauterine contraceptive device (IUCD) is based on
(a) prevents its implantation of endometrium
(b) it inhibits ovulation
(c) it blocks the entry of sperms in female genital tract
(d) promotes an ovulatory cycle

39. The mechanism of action of contraceptive pills is based on
(a) by inhibiting ovulation
(b) by increasing the motility of fallopian tubes
(c) by decreasing motility of sperms
(d) it blocks the entry of sperm into fallopian tubes

ANSWERS

Very Short Answer Questions

1. Gestation period.
2. Placenta is expelled after delivery.
3. Intrauterine device.

Fill in the Blanks

1. Uterus; 2. Viviparity; 3. About 280 days/40 week; 4. Placenta; 5. Parturition; 6. Respiratory, excretory; 7. Lactation; 8. Labour.

Multiple Choice Questions

1. (c) **2.** (c) **3.** (a) **4.** (c) **5.** (a) **6.** (d)
7. (b) **8.** (c) **9.** (c) **10.** (b) **11.** (b) **12.** (d)
13. (d) **14.** (d) **15.** (c) **16.** (d) **17.** (b) **18.** (a)
19. (b) **20.** (a) **21.** (d) **22.** (c) **23.** (a) **24.** (c)
25. (a) **26.** (c) **27.** (a) **29.** (c) **30.** (d) **31.** (c)
32. (b) **33.** (b) **34.** (b) **35.** (c) **36.** (c) **37.** (c)
38. (a) **39.** (a)

8

Endocrine Glands

8.1 NEUROENDOCRINE GLANDS

Two integrating systems, viz., **nervous system** and **endocrine system** were acquired by higher metazoans, such as vertebrates during the course of evolution. These systems coordinate and integrate functions of different parts of body in accordance with the changing demands of external and internal environmental conditions by transmitting coded informations of signals from one to the other parts. The **nervous system** achieves functional coordination and integration for quick responses of body by transmitting informations in milliseconds in the form of **"nerve impulses"** like a **"high-speed"**

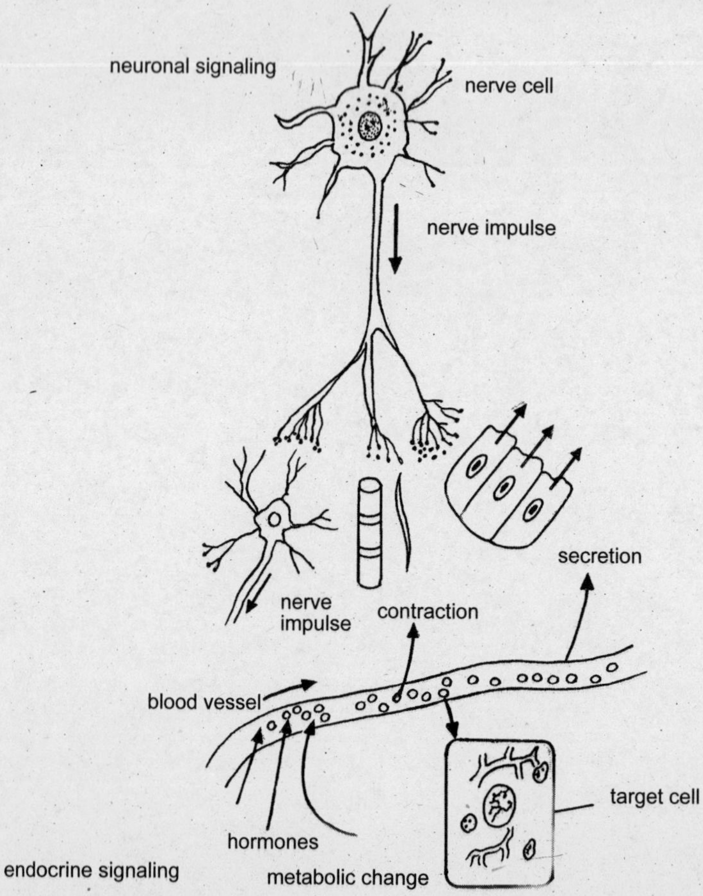

Fig. 8.1. Illustrating gross functions of nervous (A) and endocrine systems (B).

or **"no-delay"** communication service. Muscles and glands are the effectors of these responses of body (Fig. 8.1). Contrarily, the **endocrine system** achieves coordination and integration for slow and persistent responses of body by transmitting informations through **chemical messengers**, like a **'low speed communication service'**. The chemical messengers of endocrine system are called **hormones**. These are released by constituents (*i.e.,* organs and cells) of endocrine system mostly into the blood, circulate in the whole body and regulate such persistent activities as metabolism, growth, secretion, homeostasis, gastrointestinal motility and digestion, blood pressure and heartbeat, kidney functions, lactation, reproduction, etc.

8.2 HISTORY

Knowledge of hormones dates back to 1849 when **Berthold** first demonstrated that the secondary sexual characteristics are significantly affected in roosters (cockerels) by removal of testes (castration) or by grafting testes in castrated roosters. In 1854 **Schiff** reported animals dying after thyroidectomy (removal of thyroid glands). In 1855, French physiologist **Claude Bernard** first used the term **"internal secretion"** (Box 8.1). In the same year, English physician **Thomas Addison** (1793-1860) demonstrated that symptoms of a disease, now called **Addison's disease,** appear in a person if the cortical parts of his/her adrenal glands are destroyed. For his discoveries, Thomas Addison became known as **"The Father of Endocrinology"**.

Box 8.1

Claude Bernard is regarded as **Father of Experimental Medicine**. He demonstrated the role of pancreas in digestion, discussed the presence and function of glycogen in the liver (1856) and reported the existence of the vasomotor nerves.

Being the system of endocrine glands, study of endocrine system was named **endocrinology**. Certain compounds, such as epinephrine and nor-epinephrine, act both as neurotransmitter (neurohumours) and hormones. **Harris** (1955) therefore, explained that a number of physiological processes in the body are regulated jointly by nervous and endocrine systems and the hypothalamus of brain controls this coordination. That is why, endocrinology is now called **neuroendocrinology.**

Ernest Henry Starling (1866-1927), a British physiologist of London University who, in collaboration with his brother in-law **Sir W.M. Bayliss** (1860-1924), introduced the term **hormone** to describe substances that, when secreted from one part of the body, have a specific effect on another. In 1902, they discovered **secretin**, one of the first hormones to be identified. Later (in 1904), they showed that secretin could stimulate the action of a pancreatic secretion. These substances were than called **chemical messengers**.

Bayliss also introduced the **saline injection** for cases of surgical shock during World War I (1914-1918). Both Bayliss and Starling are called **Founders of Endocrinology**.

Sir Frederick Grant Banting (1891-1941), a Canadian physician who along with **C.H. Best**, is credited with curing dogs ailing with **diabetes mellitus** by giving them pancreatic extract containing **insulin**.

Charles Herbert Best (1899-1978), a US-Candaian physiologist who assisted **F.G. Banting** in the Isolation of Insulin Hormone but to Banting's annoyance, did not share in the Nobel Prize in Medicine or Physiology that Banting shared with **J.J.R. Macleod** in 1923 (**Jain** *et al.*, 2012).

Box 8.2

Banting and **Best** experimented upon a dog on 19.11.1921. The dog was depancreatized prior to administration of the pancreatic extract. Following injection, the blood sugar fell from 330 mg per cent to 170 mg per cent in one hour. This experiment demonstrated that the active principle could be extracted from the fetal pancreas with acetone and alcohol and that it was not destroyed by chloroform and ether.

Later, insulin purified from pig or cow pancreas was successfully tried on humans but it produced immunogenic reactions in them. However, in 1982, using genetic engineering (Recombinant DNA Technology) insulin was manufactured and marketed for human use. Thus **insulin became the first hormone product of genetic engineering.** The new recombinant insulin has the exact structure of human insulin and hence will not produce immunogenic reactions.

In 1977, Nobel Laureate **Roger C.L. Guillemin**, French-born American physiologist explained that neurotransmitters also function at local level. **Guillemin** has defined hormones as compounds which are released by certain cells to alter the metabolism of other cells near or far, reaching their target cells through blood stream, axoplasmic flow, or immediate intercellular space.

Kinds of Hormones

Hormones are of following two main types:

1. **Circulating (endocrine) hormones**. These diffuse into blood, circulate throughout body and usually affect the metabolism of distant target cells (Fig. 8.2A).

2. **Local hormones.** Instead of diffusing into blood, these hormones remain in ECF and affect the metabolism of neighbouring cells (Fig. 8.2B). These are of two types : paracrine and autocrine. **Paracrine hormones** affect metabolism of other, neighbouring cells, but the **autocrine hormones** affect metabolism of those very cells which secrete them.

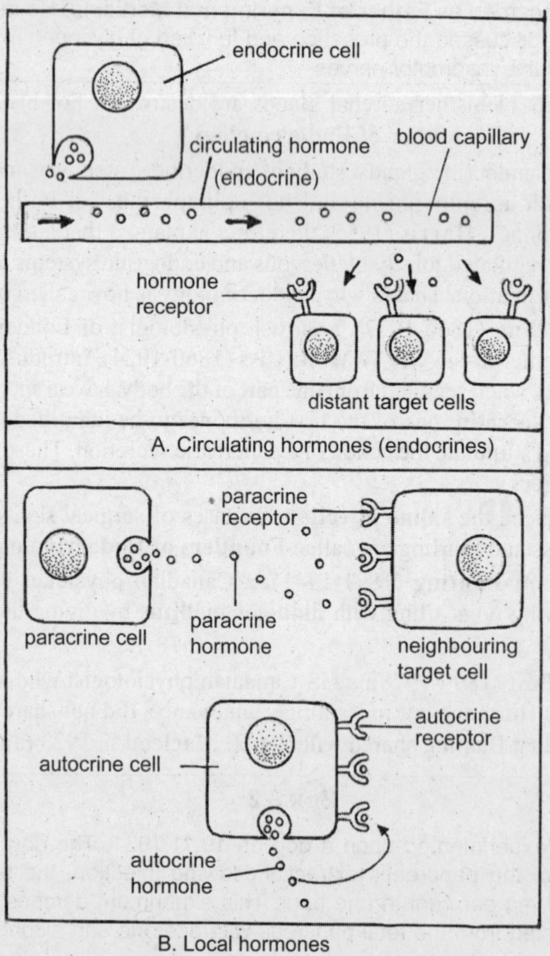

Fig. 8.2. Target cells of circulating (A) and local (B) hormones.

8.3 AUTOCRINE, PARACRINE, EXOCRINE AND ENDOCRINE CELLS

The human body is composed of billions of cells, each with a distinct function. However, the function of cell is tightly coordinated and integrated by external chemical signals, including hormones, neurotransmitters, growth factors, odorants and products of cellular metabolism that serve as chemical messengers and provide cell-to-cell communication (**Koeppen** and **Stanton**, 2010). The use of chemical signals by the body instead of electrical signals has two significances: 1. Chemical molecules can spread to all tissues in the blood. 2. Each kind of chemical molecule has a unique shape unlike any other.

The path of communication within the vertebrate body can be considered as a series of simple steps (Fig. 8.3).

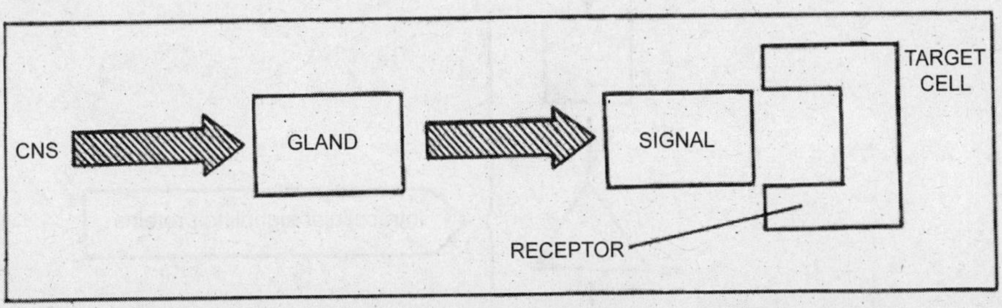

Fig. 8.3. The mode of hormonal communication in vertebrates. A command is issued by the central nervous system (CNS) that reaches the appropriate gland. A hormone is then secreted by the gland that acts as a signal to a particular receptor or target. The receptor after the hormone has bound to it, responds by changing the shape. This triggers a change in the activity of cell.

1. Issuing the command. The central nervous system is the most important regulator of body's activities although these organs can also independently control the release of chemical messengers. An area of brain called the **hypothalamus** controls the release of chemical messengers from the pituitary gland.

2. Transport of hormones in the circulation. Hormones may act on an adjacent cell, be carried throughout the body by the blood stream or even pass to different organism (via the ectohormones or pheromones).

Box 8.3

The organs or cells which are influenced by a particular hormone are called **target organs** or **target cells**. These are extraordinarily sensitive to a particular hormone.

3. Hitting the target. When a hormone encounters a cell with a matching receptor called a **target call**, (Box 8.3) it binds to the receptor.

4. Having an effect. When the hormone binds to its receptor, the protein responds by changing shapes which triggers a change in cell activity.

An overview of how cell communicate with each other is presented in Fig. 8.4. Cell communicate by releasing extracellular signaling molecules (*e.g.*, hormones and neurotransmitters) that bind to **receptor** protein located in the plasma membrane, cytoplasm or nucleus. This signal is transduced into the activation or inactivation, of one or more intracellular messengers by interacting with receptors. Receptors interact with a variety of intercellular signaling proteins, including **kinases**, **phosphatases** and GTP- binding proteins **(G-proteins)**. These signaling proteins interact with and any smaller molecule that binds to a larger molecule. regulate the activity of target proteins and thereby modulate cellular function. Target proteins include, ion channels, transport proteins, metabolic enzymes, cytoskeletal proteins, gene regulatory proteins and cell cycle proteins that regulate cell growth and division.

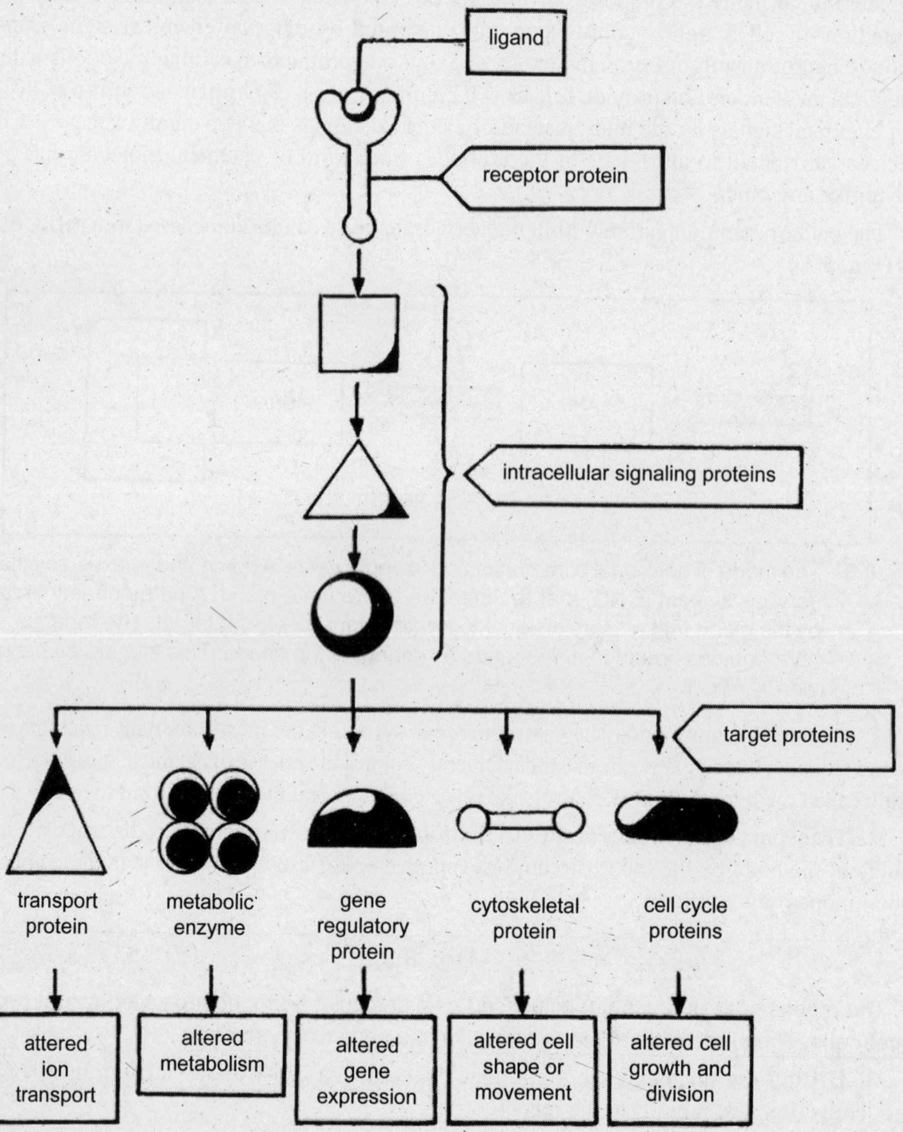

Fig. 8.4. An overview of how cell communicate. A ligand* (*i.e.*, hormone or neurotransmitter) binds to a receptor, which may be in the plasma membrane, cytosol or nucleus. Binding of ligand to a receptor activates intracellular signaling proteins which interact with and regulate the activity of one or more target proteins to change cellular function. Signaling molecules regulate cell growth, division and differentiation and influence cellular metabolism. In addition, they modulate the intracellular ionic composition by regulating the activity of ionic channels and transport proteins. Signaling molecules also control cytoskeleton-associated events, including cell shape, division and migration and cell-to-cell and cell-to-matrix adhesion (after **Koeppen** and **Stanton**, 2010). *Ligand is a generic term that refers to any smaller molecule that binds to a larger molecule.

Paracrine and Autocrine Hormones

Signaling molecules can act over long or short distances and can require cell-to-cell contact or very close cellular proximity (Fig. 8.5). Contact dependent signaling is important during development and in immune responses. Molecules that are released and act locally are called **paracrine** or **autocrine hormones**. Paracrine signals are released by one type of cell and act on another type: they are usually taken up by target cells or rapidly degraded (within minutes) by enzymes.

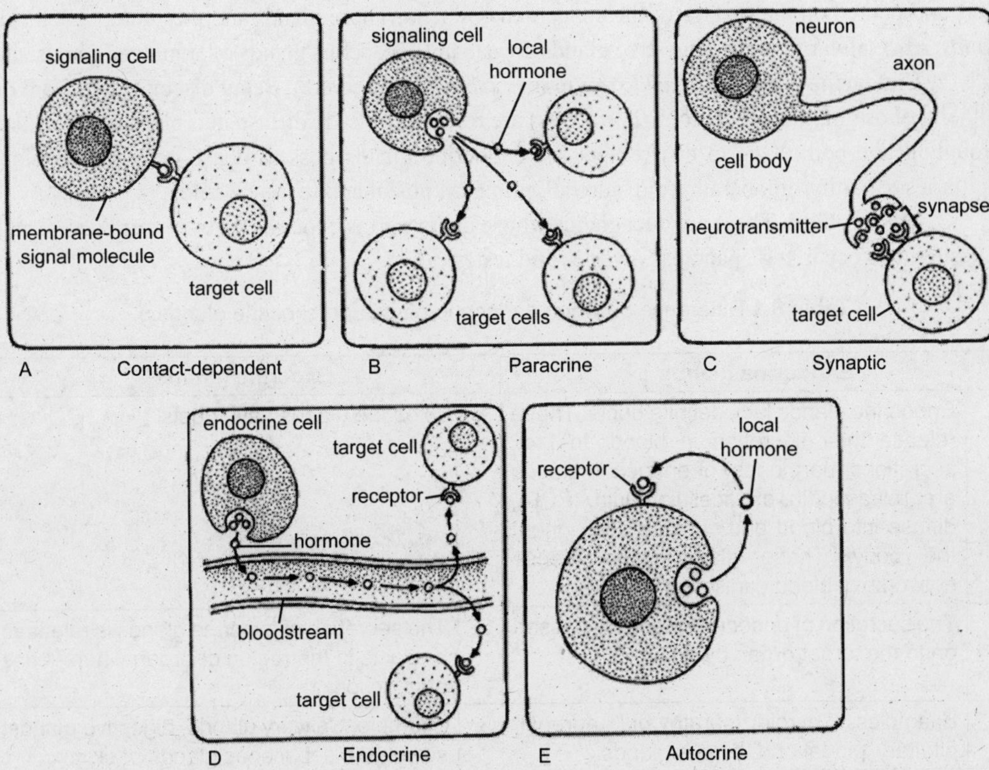

Fig. 8.5. Cell-to-cell communication is mediated by five basic mechanisms described in the text (after **Koeppen** and **Stanton**, 2010).

Autocrine signaling involves the release of a molecule that affects the same cell or other cell of the same type. **Synapting signaling** occurs when neurons transmit electrical signals along their axons and release neurotransmitters at synapses that affect the function of the other neurons or cells that distant from neuron cell body.

Neuroendocrine glands. In neuroendocrine glands the nerve cell or nerve cells releases the messenger molecule directly into the blood and is carried to a distant target organ. For example, hypothalamus of vertebrate brain is a neuroendocrine gland. Its neurosecretory cells discharge several hormones, called **releasing factors** which influence secretions of hormones from adenohypophysis of pituitary gland.

Exocrine and Endocrine Glands

A **gland** is such a cell, or such a tissue or organ whose cells synthesize and secrete some compound. Several types of glands are found in vertebrate body. **Unicellular glands** occur here and there in epithelial layers. **Multicellular glands** form by foldings of epithelia. These are of following three types:

1. Exocrine glands (Gr., *ex* = out + *krinein* = to secrete). These glands remain connected by thin ducts to the epithelia from which these form. The secretions of these glands are carried by their ducts and poured on the surface of the concerned epithelia. These glands are, therefore, called *duct glands, e.g.*, salivary glands, digestive glands, sweat and sebaceous glands of skin, tear glands, etc.

2. Endocrine glands (Gr., *endo* = within + *krinein* = to secrete). Being disconnected from the epithelia of origin, their secretions (hormones) are released in ECF, diffuse into blood and circulate throughout the body. Due to this reason, the endocrine glands possess dense networks of blood capillaries, *e.g.*, thyroid, parathyroid, adrenal, pituitary, pineal and thymus glands.

3. Mixed gland. These are duct glands whose major part is exocrine, but some endocrine cells or tissues also occur, *e.g.*, pancreas, ovaries and testes.

Table 8.1 Difference between endocrine gland and exocrine glands.

	Endocrine glands		Exocrine glands
1.	Endocrine glands lack definite ducts. These release their secretion in blood. In fact, secretions (hormones) of endocrine glands are released in extracellular fluid (ECF), diffuse into blood and circulate throughout the body. These glands have dense networks of blood capillaries.	1.	Exocrine glands have ducts.
2.	The secretion of endocrine glands is passed on to the target organ by blood.	2.	The secretion of exocrine glands is released by ducts in the region or organ where it has no function.
3.	Examples. Thyroid, parathyroid, adrenal, pituitary, pineal and thymus glands.	3.	Example. Salivary glands, digestive glands, sweat and sebaceous glands of skin.

8.4 ENDOCRINE SYSTEM

The **endorine system** (Gr., *endon* = within + *crinein* = to set apart) consists of a system of ductless glands which produce certain chemical substances, called **hormones** (Gr., *hormein* = to excite) directly in the blood stream. These physiologically active substances establish homeostasis in body by regulating various vital functions.

1. Necessity of Endocrine Glands (Role of Endocrine Organs)

The endocrine system is associated with chemical coordination of various physiological activities occurring within the body. These include general body metabolism and metabolic rate, growth, reproduction, appearance of secondary sexual characters, milk production in mammals, ion concentration in blood, blood pressure, blood sugar concentration, water balance by kidney. Therefor endocrine glands are of vital importance in our body. They maintain homeostasis of body via hormones. In insects and frogs, hormones are responsible for metamorphosis.

2. Endocrine Glands

The endocrine system comprises following three types of endocrine structures:

A. Major or Dedicated Endocrine Glands

1. Pituitary (Anterior lobe)
2. Thyroid
3. Parathyroid
4. Endocrine pancreas
5. Adrenal glands
6. Ovary (Gonads)
7. Testis (Gonads)
8. Thymus

B. Endocrine cells within organs whose primary function is not endocrine

1. Brain (Hypothalamus and Posterior lobe of pituitary)
2. Heart
3. Kidney
4. Adipose tissue
5. Stomach
6. Intestine
7. Liver

C. Peripheral conversion (*i.e.*, cell types that express intracellular enzymes, ectoenzymes or secreted substances that modify inactive precursors or less active hormones into highly active hormones)

1. Lungs (angiotensin II)

3. Feedback Loops within the Endocrine System
Negative and Positive Feedback Loops

Feedback is the return to the input of part of the output of a machine, system or process (as for producing changes in a circuit that improve performance). The predominant mode of a closed feedback loop among endocrine glands is a **negative feedback**. In a negative-feedback loop, "hormone A" acts on one or more target organs to induce a change (either a decrease or increase) in circulating levels of "component B" and the change in component B then inhibits secretion of hormone A. Negative-feedback loops confer stability by keeping a physiological parameter (*e.g.*, blood glucose) within a normal range. There are also a few examples of **positive-feedback** in endocrine regulation.

A positive closed feedback loop, in which hormone X increases levels of component Y and component Y stimulates secretion of hormone X, confers instability. Under the control of positive-feedback loops, something has "got to give". For example, positive feedback loops control processes that lead to rupture of a follicle through the ovarian wall or expulsion of a fetus from the uterus.

Table 8.2. Hormones and their sites of production in humans (Source: Koeppen and Stanton 2010).

Gland	Hormone	Chemical structure	Representative action	Mechanism of regulation
A. Hormones synthesised and secreted by dedicated endocrine glands				
1. Pituitary (Anterior lobe)	1. Growth hormone (GH)	Prolactin	Stimulates general growth, particularly skeletal, affects metabolic functions	Hypothalamic hormones
	2. Prolaction	Protein	Stimulates milk production and secretion (in humans)	Hypothalamic hormones
	3. Adrenocorticotropic hormone (ACTH)	Polypeptide	Stimulates adrenal cortex to secrete glucocorticoids (*e.g.*, cortisol)	cortisol in blood, hypothalamic hormones
	4. Thyroid-stimulating hormone (TSH)	Glycoprotein	Stimulates thyriod gland to secrete hormones	Thyroxine in blood, hypothalamic hormones
	5. Follicle-stimulating hormones (FSH)	Glycoprotein	Stimulates ovarian follicles and spermatogenesis	Estrogen in blood, hypothalamic hormones
	6. Leuteinsing hormone (LH)	Glycoprotein	Stimulates corpus luteum and ovulation in females and interstitial cells in males	Progesterone or testosterone in hypothalamic hormones
2. Thyroid gland	1. Thyroxine or tetraiodothyronine (T_4) 2. Triiodothyronine (T_3)	Iodinated amino acids	Increase oxygen consumption and heat production, stimulate and maintain metabolic processes	TSH
	3. Calcitonin	Peptide	Lowers blood calcium levels by inhibiting calcium release from bone	Ca^{2+} concentration in blood
3. Parathyroid gland	Parathyroid hormone (PTH)	Peptide	Raises plasma calcium levels by stimulating calcium release, promoting calcium uptake from bone and from gastrointestinal tract	Ca^{2+} concentration in blood
4. Islets of Langerhans (Endocrine pancreas)	1. Insulin	Polypeptide	Lowers blood sugar, increase glycogen storage in liver and stimulates protein synthesis	Glucose concentration in blood, somatostatin
	2. Glucagon	Polypeptide	Stimulates glycogen breakdown in liver	Glucose and amino acid concentration in blood

Gland	Hormone	Chemical structure	Representative action	Mechanism of regulation
	3. Somatostatin	Peptide	Suppresses release of growth hormone	Nervous control, insulin and glucagon, feedback from growth hormone
5. Adrenal glands (*i*) Medulla	1. Epinephrine	Catecholamine	Increase blood sugar, constricts blood vessels in skin, mucus membranes and kidney	Nervous system
	2. Norepinephrine	Catecholamine	Increases heart rate and force of contraction of cardiac muscles, constricts blood vessels through the body	Nervous system
(*ii*) Cortex	Glucocorticoids (*e.g.*, cortisol)	Steroids	Increase blood sugar by affecting many aspects of carbohydrate metabolism	ACTH
	Mineralocorticoids (*e.g.*, aldosterone)	Steroids	Promote reabsorption of sodium and excretion of potassium in the kidneys	K^+ in blood
6. Gonads: Testis	1. Androgens (*e.g.*, testosterone)	Steroids	Support spermatogenesis, develop and maintain secondary male sexual characteristics	FSH and LH
	2. Antimullerian hormone (AMH)			
	3. Inhibin			
Ovary (follicle)	(*i*) Estrogen	Steroids	Initiate building of uterine lining, develop and maintain female sexual characteristics	FSH and LH
Ovary (corpus luteum)	(*ii*) Progesterone	Steroids	Promote continued growth of uterine lining	FSH and LH
B. Hormones synthesized in organs with a primary function other than endocrine glands				
1. Pituitary: posterior lobe (release hormones produced by hypothalamus of brain)	1. Antidiuretic hormone (ADH; vasopressin)	Peptide	Promote reabsorption of water from collecting ducts of kidneys	Blood osmolarity

Gland	Hormone	Chemical structure	Representative action	Mechanism of regulation
	2. Oxytocin	Peptide	Stimulates contraction of uterine muscles and mammary gland cells	Nervous system
	3. Corticotropin-releasing hormone (CRH)			
	4. Thyrotropin-releasing hormone (TRH)			
	5. Gonadotropin-releasing hormone (GnRH)			
	6. Growth hormone-releasing hormone (GHRH)			
	7. Somatostatin			
	8. Dopamine			
2. Brain (pineal gland)	Melatonin	Catecholamine	Involved in circadial rhythms	Light, dark cycles
3. Heart	Atrial natriuretic peptide (ANP)			
4. Kidney	Erythropoetin			
5. Adipose tissue	1. Leptin 2. Adiponectin			
6. Stomach	1. Gastrin 2. Somatostatin 3. Ghrelin 4. Secretin 5. Cholecystokinin 6. Glucagon like peptide-1 (GLP-1)			

Gland	Hormone	Chemical structure	Representative action	Mechanism of regulation
7. Intestine	1. Glucagon like peptide-2 (GLP-2)			
	2. Glucose dependent insulinotropic peptide (GIP); gastrin inhibitory peptide			
	3. Motilin			
8. Liver	Insulin-like growth factor type-I (IGF-I)			
C. Hormone produced to a significant degree by peripheral conversion				
1. Lungs	Angiostensin-II			
2. Kidney	1,25-Dihydroxy-vitamin D (Vitamin D)			
3. Adipose (mammary glands and other glands)	Estradiol-17B			
4. Liver, sebaceous gland, other organs	Testosterone			
5. Genital skin, prostate, other organs	5-Dihydrotestosterone (DHT)			
6. Many organs	T_3			

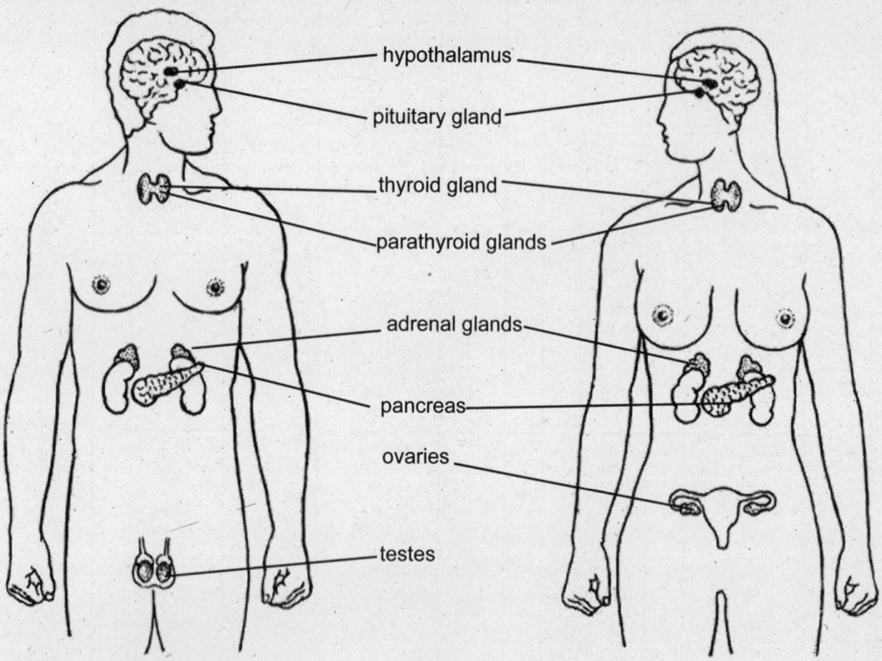

Fig. 8.6. Glands of the endocrine system of human being.

Types of Negative-Feedback Loops (Fig. 8.7)

Negative-feedback loops are of following types: 1. Physiological response-driven feedback loop and 2. Endocrine-axis driven feedback loop.

1. Physiological response-driven feedback loop. It is observed in endocrine glands that control blood glucose levels (pancreatic islets), blood Ca^{2+} and Pi levels (parathyroid glands, kidney), blood osmolarity and volume (hypothalamus/posterior pituitary) and blood Na^+, K^+ and H^+ (zona glomerulosa of the adrenal cortex and atrial cells). In the response-driven configuration, secretion of a hormone is stimulated or inhibited by a change in the level of a specific extracellular parameter (*e.g.*, an increase in blood glucose stimulate insulin secretion). Altered hormone levels lead to changes in the physiology of target organs (*e.g.*, decreased hepatic gluconeogenesis, increase uptake of glucose by muscle) that directly regulate the parameter (*i.e.*, blood glucose) in question. The change in the parameter (*i.e.*, decreased blood glucose) then inhibits further secretion of the hormone (*i.e.*, insulin secretion drops as blood glucose falls).

2. Endocrine-axis driven feedback loop. Much of the endocrine system organized into **endocrine axes**, with each axis consisting of the hypothalamus and the pituitary and peripheral endocrine glands (Fig. 8.7). Thus, the endocrine axis-driven feedback loop involves a three-tired system. The first tier is represented by **hypothalamic neuroendocrine neurons** that secrete **releasing hormones**. Releasing hormones stimulate the production and secretion of **tropic hormones** from the **pituitary gland** (second tier). Tropic hormones stimulate the production and secretion of hormones from **peripheral endocrine glands** (third tier). The peripherally produced hormones, namely, thyroid hormone, cortisol, sex steroids and IGF-I, typically have **pleiotropic actions** (*e.g.*, multiple phenotypic effects) on numerous cell types. However, in endocrine axis driven feedback, the primary feedback loop involves feedback inhibition of pituitary tropic hormones and hypothalamic releasing hormones by the peripherally produced hormone. In contrast to response-driven feedback, the physiological responses to the peripherally produced hormone play only a minor role in regulation of feedback within endocrine axis-driven feedback loops.

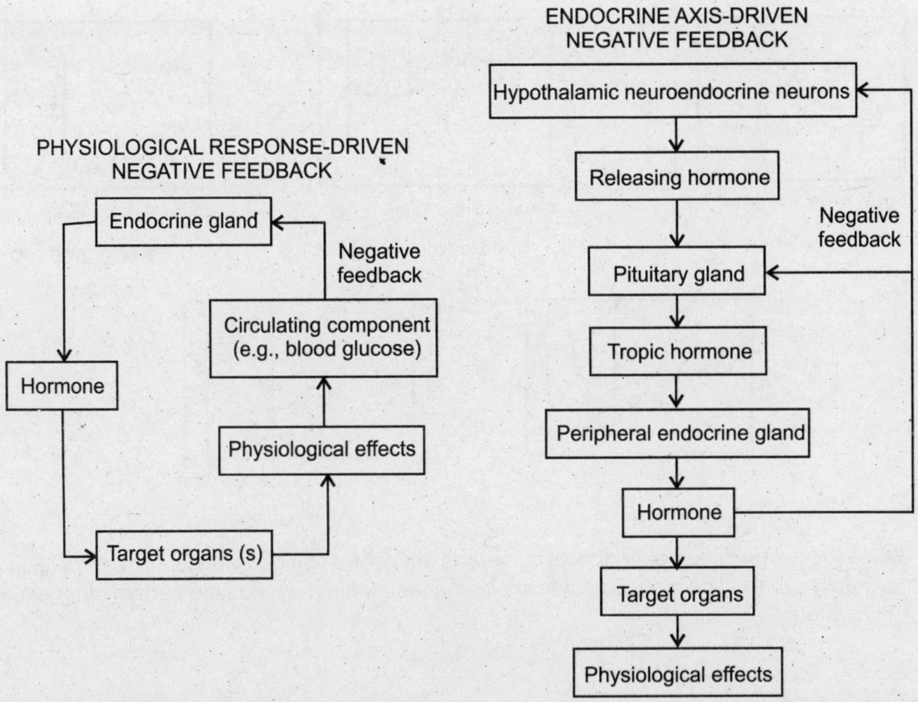

Fig. 8.7. Two types of negative-feedback loops. A—physiological response-driven; B—Endocrine axis driven.

Negative feedback and circadian rhythm. An important aspect of the endocrine axes is the ability of descending and ascending neuronal signals to modulate release of the hypothalamic releasing hormones and thereby control the activity of the axis. A major neuronal input to releasing hormone secreting neurons comes from another region of the hypothalamus called the **suprachiasmatic nucleus (SCN)**. SCN neurons impose a daily rhythm, called a **circadian rhythm**, on the secretion of hypothalamic releasing hormones and the endocrine axis that they control (Fig. 8.7). SCN neurons represent an intrinsic circadian clock, as evidenced by the fact that they demonstrate a spontaneous peak of electrical activity at the same time every 24 to 25 hours. The 24- to 25-hour cycle can be "entrained" by the normal environmental light dark cycle created by the earth's rotation such that the periodicity of the clock appears to be environmentally controlled (Fig. 8.8). Neural input is generated from specialized light- sensitive retinal cells that are distinct from rods and cones and from signals to the SCN via that retinohypothalamic tract. Under consant coditions of light and dark, however, the SCN clock becomes " free running" and slightly drifts away from a 24-hour cycle each day.

The **pineal gland** forms a neuroendocrine link between the SCN and various physiological processes that require circadian control. This tiny gland, close to the hypothalamus, synthesizes the hormone **melatonin** from the neurotransmitter **serotonin**, of which tryptophan is the precursor. The rate-limiting enzyme for melatonin synthesis is **N-acetyltransferase**. The amount and activity of this enzyme in the pineal gland vary markedly in a cyclic fashion, which accounts for the cycling of melatonin secretion and its plasma levels. Synthesis of melatonin is inhibited by light and markedly stimulated by darkness (Fig. 8.8). Thus, melatonin may transmit the information that night time has arrived, and body functions are regulated accordingly. Melatonin feedback to the SCN at dawn or dusk may also help evoke day-night entrainment of the SCN 24- to 25-hour clock. Melatonin has numerous other actions, including induction of sleep.

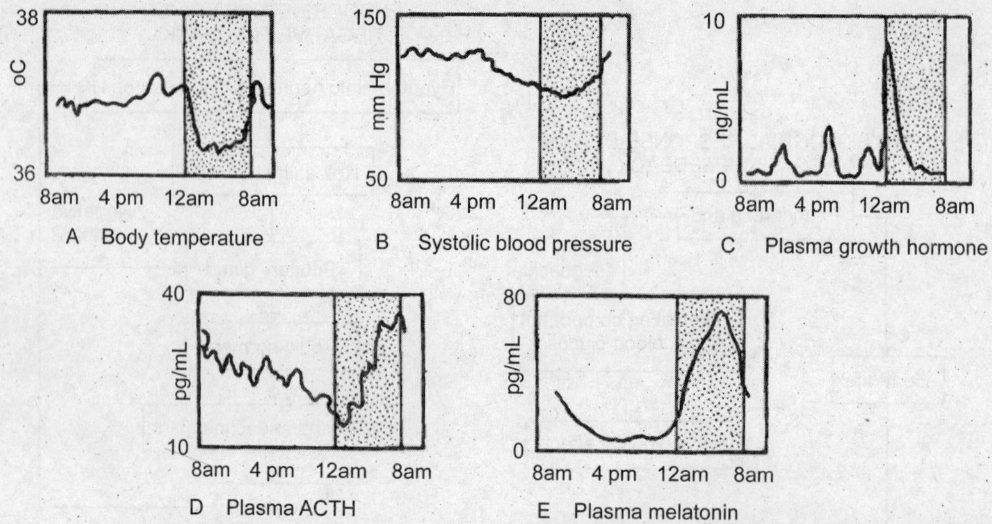

Fig. 8.8. A circadian pacemaker directs numerous endocrine and body functions, each with its own daily profile. The night time rise in plasma melatonin may mediate certain other circadian patterns.

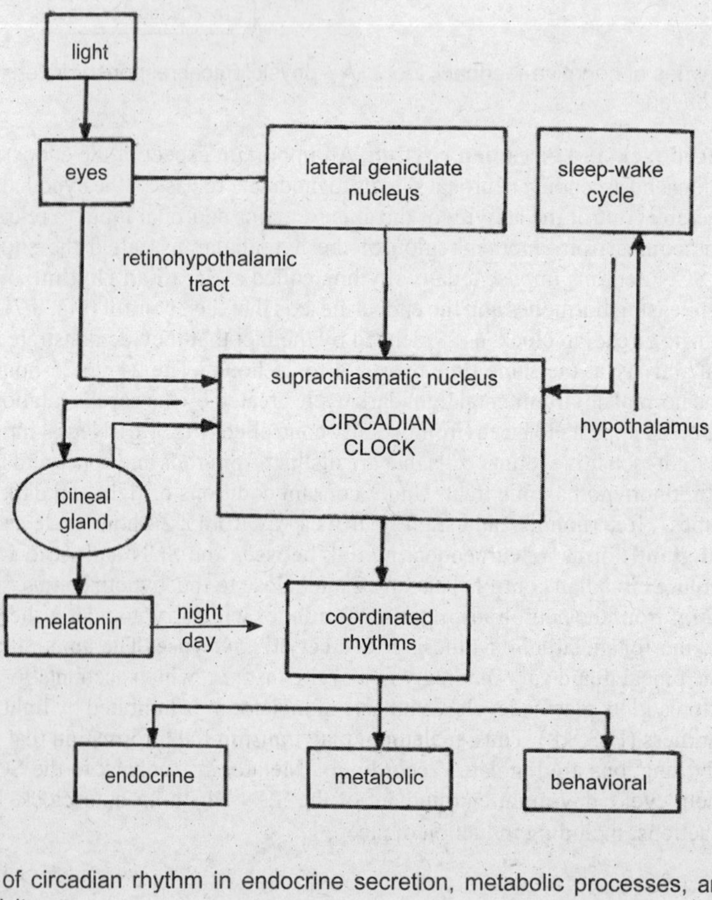

Fig. 8.9. Origin of circadian rhythm in endocrine secretion, metabolic processes, and behavioural activity.

Role of stress in negative feedback. Another input to hypothalamic neurons and the pituitary gland is **stress**, either as **systemic stress** (*e.g.*, hemorrhage, inflammation) or as **processive stress** (*e.g.*, fear, anxiety). Major medical or surgical stress overrides the circadian clock and causes a pattern of persistent and exaggerated hormone release and metabolism that mobilizes endogenous fuels, such as glucose and free fatty acids, and boosts their delivery to critical organs. By contrast, growth and reproductive processes are suppressed. Additionally, cytokines released during inflammatory or immune responses, or both, directly regulate the release of hypothalamic releasing hormones and pituitary hormones.

8.5. Hormone

Hormones are information molecules secreted by the endocrine cells in one part of the body in response to changes in external or internal environment, and carried by blood to another part where they stimulate or inhibit specific physiological processes for the good of the body as a whole. In other words, the hormones are chemical messengers that regulate the biological processes in the organisms.

1. Chemical nature of hormones. Hormones are classified biochemically as (*i*) **proteins** and **peptides**; (*ii*) **catecholamines**; (*iii*) **steroids** and (*iv*) **iodothyronines**. The chemical nature of a hormone determines: 1. how it is synthesized, stored and released; 2. how it is transported in blood; 3. its biological half life ($T_{1/2}$) and mode of clearance; and 4. its cellular mechanism of action.

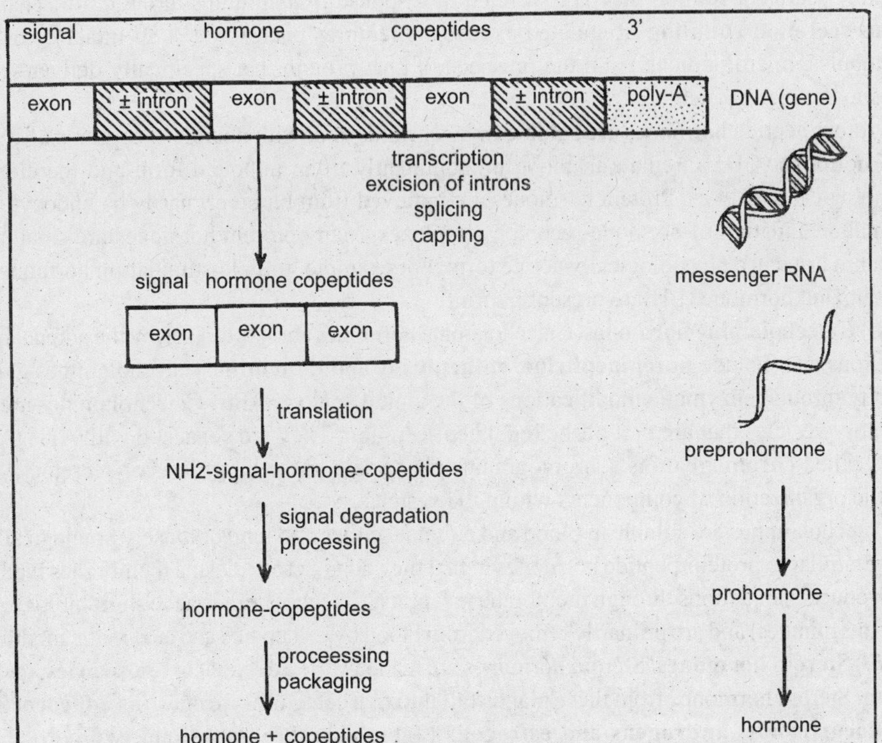

Fig. 8.10. Schematic representation of peptide hormone synthesis. In the nucleus the primary gene transcript, a premessenger RNA molecule, undergoes excision of introns, splicing of exons, capping of the 5′ end, and addition of poly (A) at the 3′ end. The resultant mature messenger RNA enters the cytoplasm, where it directs the synthesis of a preprohormone peptide sequence on ribosomes. In this process the N-terminal signal is removed, and the resultant prohormone is transferred vectorially into the endoplasmic reticulum (ER). The prohormone undergoes further processeing and packaging in the Golgi apparatus. After final cleavage of the prohormone within the granules, they contain the hormone and copeptides ready for secretion by exocytocytosis (after Koeppen and Stanton, 2010).

(i) **Protein/peptide hormones.** Protein/peptide hormones can be grouped into structurally related molecules that are encoded by **gene families**. Protein/peptide hormones gain their specificity from their primary amino acid sequence and from posttranslational modification, especially glycosylation.

Because protein/peptide hormones are destined for secretion outside the cell, they are synthesized on the polyribosome as larger preprohormones or prohormones (Fig. 8.10). The nascent peptides have at their N-terminal a group of 15 to 30 amino acids called the **signal peptide**. The signal peptide interacts with a ribonucleoprotein particle, which ultimately directs the growing peptide chain through a pore in the membrane of the endoplasmic reticulum located on the cisternal (*i.e.*, inner) surface of the endoplasmic reticular membrane. Removal of the signal peptide by a **signal peptidase** (enzyme) generates a hormone or prohormone, which is then transported from cisternae of the RER to the Golgi apparatus, where it is packaged into a membrane bound secretory vesicle that is subsequently released into the cytoplasm. The carbohydrate moiety of glycoprotein is added in the Golgi apparatus.

Most hormones are produced as **prohormones**. Prohormones harbor the peptide sequence of the active hormone within their primary sequence. However, prohormones are inactive or less active and require the action of endopeptidases to trim away the neighboring inactive sequences.

Protein/peptide hormones are stored in the gland as membrane-bound secretory vesicles and are released by **exocytosis** through the **regulated secretory pathway**. Thus, *hormones are not continually secreted.* Rather, they are secreted in response to a stimulus through a mechanism of **stimulus-secretion coupling**. Regulated exocytosis requires energy, Ca^{2+}, an intact cytoskeleton (microtubules, microfilaments) and the presence of coat protein that specifically deliver secretory vesicles to the plasma membrane.

Protein/peptide hormones are soluble in body fluids and, with notable exceptions of IGFs and growth hormone (GH), circulate in blood predominantly in an unbound form and therefore have short biological half-lives. Protein hormones are removed from blood primarily by endocytosis and lysosomal degradation of hormone-receptor complexes. Many protein hormones are small enough to appear in urine in a physiologically active form. For example, follicle-stimulating hormone (FSH) and luteinizing hormone (LH)are present in urine.

(ii) **Catecholamine hormones.** Catecholamine hormones are synthesized by the adrenal medulla and neurons and include **norepinephrine**, **epinephrine** and **dopamine**. Catecholamines gain their specificity through enzymatic modifications of the amino acid **tyrosine**. Catecholamines are stored in secretory vesicles that are part of the regulated secretion. They are copacked with ATP, Ca^{2+} and proteins called **chromogranins**. Chromogranins play a role in the biogenesis of secretory vesicles and in the organization of components within the vesicles.

Catecholamines are soluble in blood and circulate either unbound or loosely bound to albumin. They are similar to protein/peptide hormones in that they do not cross plasma membranes readily and hence produce their actions through membrane receptors. Catecholamines have short biological half-lives (1 to 2 minutes) and are primarily removed from blood by cell uptake and enzymatic modification.

(iii) **Steroid hormones.** Steroid hormones are made by the **adrenal cortex**, **ovaries**, **testes** and **placenta**. Steroid hormones from these glands fall into five categories: **progestins**, **glucocorticoids**, **mineralocorticoids**, **androgens** and **estrogens** (Table 8.3). Progestins and two corticoids are 21-carbon steroids, whereas androgens are 19-carbon steroids and estrogens are 18-carbon steroids. Steroid hormones also include the active metabolite of **vitamin D**, which is a secosteroid (*i.e.*, one of the rings has an open conformation).

Table 8.3. Some characteristics of steroid hormones.

	Family	Number of carbons	Specific hormone	Primary site of synthesis	Primary receptor
1.	Progestin	21	Progesterone	Ovary Placenta	Progesterone receptor (PR)
2.	Glucocorticoid	21	Cortisol Corticosterone	Adrenal cortex	Glucocorticoid receptor (GR)
3.	Mineralocorticoid	21	Aldosterone II- Deoxycorticosterone	Adrenal cortex	Mineralocorticoid receptor (MR)
4.	Androgen	19	Testosterone Dihydrotestosterone	Testis	Androgen receptor (AR)
5.	Estrogen	18	Estradiol-17B Estriol	Ovary Placenta	Estrogen receptor

Steroid hormones are synthesized by a series of enzymatic modifications of cholesterol and have a cyclopentanoperhydrophenanthrene ring.

Because of their nonpolar nature, steroid hormones are not readily soluble in blood. Therefore, steroid hormones circulate bound to **transport protein**, including albumin, but also the specific transport protein **sex hormone-binding globulins (SHBG)** and **corticosteroid-binding globulin (CBG)**. Excretion of steroid hormones from the body typically involves inactivating modification followed by **glucuronide** or **sulphate conjugation** in the liver. These modifications increase the water solubility of the steroid and decrease its affinity for transport proteins, thereby allowing the inactivated steroid hormone to be excreted by the kidney.

***(iv)* Iodothyronines.** Thyroid hormones are iodothyronines (Fig. 8.11) that are made by the coupling of "iodinated tyrosine residues through ether linkage." Their specificity is determined by the thyronine structure, as well as by where the thyronine is iodinated. Thyroid hormones cross cell membranes by both diffusion and transport systems. They are stored extracellularly in the thyroid as an integral part of the glycoprotein molecule thyroglobulin.

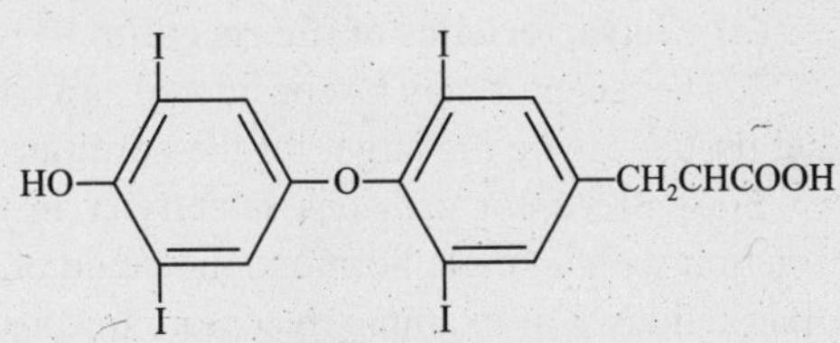

3, 5, 3′, 5′-Triiodothyonine (thyroxine, or T_4)

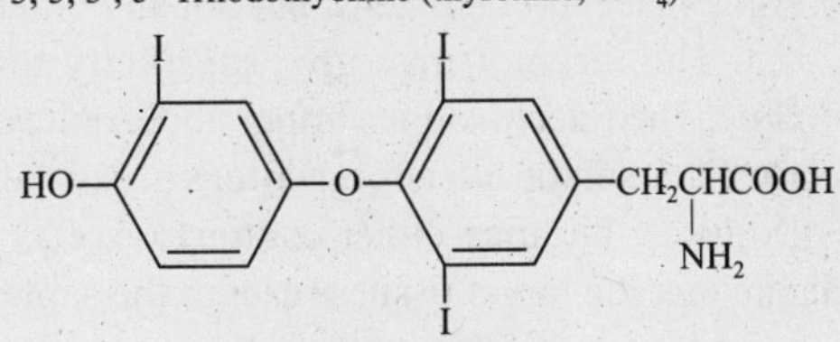

3, 5, 3′-Triiodothyonine (T_3)

Fig. 8.11. Structure of thyroid hormones, which are iodinated thyronines.

Thyroid hormones are sparingly soluble in blood and aqueous fluids and are transported in blood bound (>99%) to serum-binding proteins. A major transport protein is **thyroid hormone-binding globulin (TBG)**. Thyroid hormones have long half-lives ($t_{½}$; thyroxine (T_4) = 7 days; triiodothyronine (T_3) = 18 hours). Thyroid hormones are similar to steroid hormones in that the **thyroid hormone receptor (TR)** is intracellular and acts as a transcription factor.

2. Transport of hormones in circulation. A significant fraction of steroid and thyroid hormones is transported in blood bound to plasma proteins that are produced in a regulated manner by the liver. Protein and polypeptide hormones are generally transported free in blood. The concentrations of **bound hormone (HP), free hormone (H)** and **plasma transport protein (P)** are in equilibrium. If free hormone levels drop, hormone will be released from the transport proteins. This relationship may be express as follows:

$$[H] \times [P] = [HP] \text{ or } K = [H] \times [P]/[HP]$$

where K is the dissociation constant.

Free hormone is the biologically active form for action on the target organ, feedback control, and clearance by cellular uptake and metabolism. Protein binding serves several purposes. It prolongs the circulating half lives ($t_{1/2}$) of the hormone. Many hormones cross plasma membrane readily and would either enter cells or be excreted by the kidney were they not protein bound. The bound hormone represents a "reservoir" of hormone and as such can serve to "buffer" acute changes in hormone secretion. Some hormones, such as steroids, are sparingly soluble in blood, and protein binding facilitates their transport.

3. Hormone action. Hormones are also termed as **ligands**, in the context of ligand-receptor binding and as **agonists**, in that their binding to the receptor is transduced into a cellular response. Receptor **antagonists** typically bind to a receptor and lock it in an inactive state, unable to induce a cellular response. Loss or inactivation of a receptor results in **hormone resistance**.

It has been seen that only certain cells respond to particular hormone. It is because of the existence of specific high affinity receptors. It is now possible to explain most hormone action on the basis of hormone receptor.

A. Mechanism of hormone action. Hormones can act at very low concentrations. A given hormone can affect target cells in an animal differently, or vary in effect from one species of animal to another, *e.g.*, thyroxine. In humans and other vertebrates, this hormone is responsible for metabolic regulation, but it also plays an important role in the development of an adult frog from the tadpole stage. It triggers resorption of tadpole's tail and other morphological changes during metamorphosis.

Although the diversity of responses to hormone is seemingly endless, a hormone triggers specific change mechanisms:

(*i*) Some hormones such as steroids enter the nucleus of target cells and influence the expression of the cell's genes.

(*ii*) Most non-steroid hormones, including peptides, attach to the cell surface and influence activity within the cell through cytoplasmic intermediaries called **second messengers**.

In both of the mechanisms receptor molecules to which the hormone bind are involved. The receptors are proteins. The hormone receptor complex triggers the effect of hormone.

(*a*) Characteristics of the receptor

1. The receptor may be considered as a chemical **transducer** which must be present in a tissue if that tissue is to be the target for the hormone.

2. The receptor provides **selectivity** in the response system, because it may be capable of interaction with a single hormone molecule or, at the most, a very limited array of hormones with similar actions. For example, receptors for growth hormone can be occupied by growth hormones and some placental lactogens, but not any other type of protein hormone.

3. The receptor may give **specificity** to the nature of response to the stimulation of given hormone. The catecholamine brings about quite distinct responses in different tissues, depending on the precise nature of the **adrenoreceptors** in each of those tissues. For example, catecholamines stimulate heart muscle, but may either contract or relax vascular smooth muscle. Different adrenoreceptors indicate that the target tissues react to the same stimulus in quite different ways.

(*b*) Location of receptors. Receptors for hormones appear to be located on the membranes or within the cytoplasm of the target cell.

(*i*) Membrane receptors. The membrane receptors are presumably proteins on the external face and are freely accessible to hormones present in the extracellular fluid. Membrane receptors are of following three types: 1. Ion channel receptors; 2. Receptors linked to G proteins and 3. Catalytic receptors (Fig. 8.12).

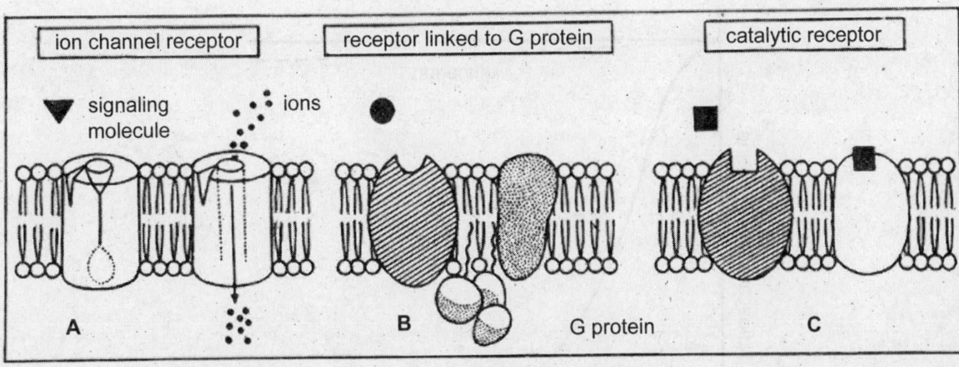

Fig. 8.12. Three types of plasma membrane receptors. A—An ion channel receptor. It is a membrane protein that function both as a receptor for a signaling ligand and as an ion channel. The permeability of the channel is controlled by binding of the signaling ligand. B—A G-protein linked receptor. It is a membrane protein that activates a G protein when a signaling ligand binds to the receptor. Activated G protein in turn interacts with other membrane proteins, which may be enzymes or ion channels. C—A catalytic receptor. It is a membrane protein that functions both as a receptor for a signaling ligand and as an enzyme. The receptor's enzymatic activity is controlled by binding of the signaling ligand.

(*ii*) Cytoplasmic receptors. The receptors for the apolar hormone molecules which are able to diffuse through membranes are located inside the target cells in the cytoplasm, *e.g.*, the steroids.

(*c*) Nature of receptors. Receptors are conformationally mobile macromolecules (*i.e.*, they change their form or shape). They are proteins of appropriate hormone, despite a vast excess of other types of molecules. They can move and change their overall shape and form.

The binding of a hormone to its receptor occurs more readily than unbinding, *i.e.*, dissociation of the receptor-hormone complex. The receptor can be the binder and the hormone can be the ligand. The interaction of receptor and hormone is passive as no energy is involved in this process. It is however, concentration dependent and reversible, with the various forms of the reactants satisfying an equilibrium

$$\underset{\text{Receptor}}{\text{R}} + \underset{\text{Hormone}}{\text{H}} \rightarrow \underset{\substack{\text{Receptor–Hormone} \\ \text{Complex}}}{\text{RH}}$$

(*d*) Regulation of number of receptors. The number of receptors changes during ontogeny. Receptors may be absent until a specific stage in development is reached. For example, despite the high concentrations of growth hormones in foetal sheep blood, there seem to be no receptor present in liver, the key target tissue for this hormone during postnatal life, until some days after birth. The reverse is also true: receptors may disappear when the animal reaches a particular developmental stage. The receptor regulation which occurs dynamically in target tissues is called **autoregulation**. In this case, this hormone itself is the cause of alteration in the number of its own receptors. They can be of two type: **(*i*) Up regulated**. In this process the presence of hormone acts to increase the number of receptors. **(*ii*) Down regulated**. In this process the presence of hormone causes a decrease in the number of receptors.

(*e*) Mode of hormone action.

(*i*) Specificity. The binding of a signaling molecule to its receptor is a highly specific interaction that resembles the recognition between an enzyme and its substrate. To assess the specificity of the binding between a signaling ligand and receptor, one measures the ability of non-radioactive substances to compete with the radioactive signaling ligand for the receptor site (Fig. 8.13).

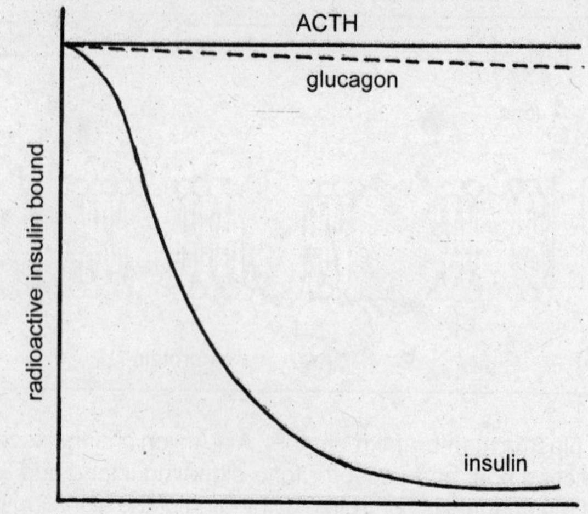

Fig. 8.13. Demonstration of specificity in the binding of a hormone to its receptor. Radioactive insulin was incubated with isolated plasma membranes in the presence of varying concentrations of either non-radioactive insulin, glucagon or ACTH. The hormones unrelated to insulin do not compete with the binding of radioactive insulin to its plasma membrane receptor.

***(ii)* High affinity binding.** By analyzing the kinetics of the binding reaction, information concerning the affinity of a ligand for its receptor is obtained. When the amount of ligand that binds to a receptor is graphed as a function of the total ligand concentration, a hyperbolic curve is obtained that is similar in shape to the curve relating enzyme activity to substrate concentration (Fig. 8.14A). The ligand concentration at which binding to receptor is half-maximal is called that **dissociation constant (kd)**. This dissociation constant (analogous to the Michaelis constant of enzymatic reactions) is inversely related to the affinity of receptor for its ligand. A high value of *kd* indicates that a large concentration of ligand is required before a significant amount of ligand will bind to the receptor. On the other hand, low value of *kd* means that ligand binds to the receptor at low ligand concentration. This indicates that the receptor has a high affinity for the ligand. Interaction between signaling ligand and its receptors typically exhibit low *kd* values (high affinity).

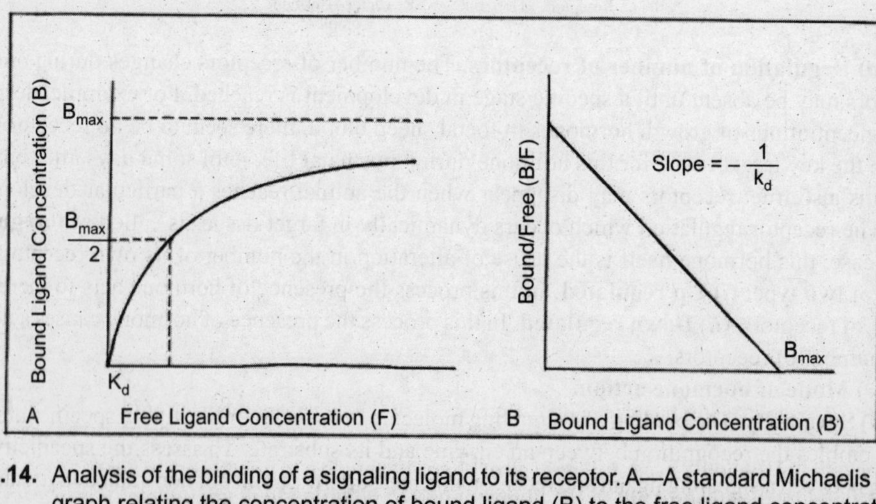

Fig. 8.14. Analysis of the binding of a signaling ligand to its receptor. A—A standard Michaelis Menten graph relating the concentration of bound ligand (B) to the free ligand concentration (F). The receptor becomes saturated with bound ligand at high concentration of free ligand. B—A scatachard plot of the same data in which the ratio of bound to free ligand concentration (B/F) is plotted as a function of the free ligand concentration (F).

3. Saturation kinetics. Because cell contains a limited number of receptor molecules, increasing the concentration of signaling ligand should eventually lead to occupation of all receptor binding sites. This is calculated by Scatchard plot (Fig. 8.14B).

This type of analysis (*i.e.*, Scatchard plot) converts the data into linear form by graphing the concentration of ligand bound to receptor against the ratio of bound to free ligand. In such a plot x intercept corresponds to the total number of binding sites, or β max, while the slope of the line corresponds to the negative reciprocal of Kd (that is –1/Kd). The actual number of receptors detected per cell for a given signaling molecule usually ranges from less than a dozen to several hundred thousand, depending on the type of receptor involved.

4. Reversibility. Usually the binding of a signaling ligand to its receptor is a rapidly reversible interaction. This characteristic helps to prevent a receptor from becoming permanently activated by a signaling ligand.

5. Physiological response. The binding of a signaling ligand to its receptor should trigger an appropriate physiological response whose extent is directly related to the number of receptor molecules containing bound ligand. For example, the hormone insulin stimulates the transport of glucose into target cells. Thus during the binding of insulin to its plasma membrane receptor in intact cells, one should observe an increase in glucose transport that is directly related to the number of receptor molecule containing bound insulin.

6. Interaction between hormones. Certain interactions occur between hormones. The example being synergism, antagonism and priming.

***(i)* Synergism.** In this phenomenon the response to the two hormones is greater than that the sum of responses to each of the hormones when they act alone. For example, hormone A alone elicits a response to 10 units and hormone B alone elicits the same kind of response, but of 6 units. Synergism exists when hormone A and hormone B together bring about a response of greater than 16 (10 + 6) units.

***(ii)* Antagonism.** When the hormones A and B evoke responses that oppose each other, they are said to be antagonistic. Hormone A might elevate plasma concentration of glucose, when hormone B might depress them: their actions are antagonistic.

***(iii)* Priming.** When a particular hormone though does not bring about a particular response, is required for another hormone to evoke its response; the process is called **priming**. For example, oxytocin (hormone) stimulates the uterine excitability and cause more frequent contractions, each of higher magnitude. It can only do so if the uterus has been exposed to oestrogen (hormone). The oestrogen primed uterus is responsive to oxytocin. It is now know that oestrogen is responsible for the binding of oxytocin receptors in uterus. Oestrogen itself cannot directly stimulate uterine contractions, but it does provide the neceptors that enable oxytocin to stimulate the muscle. Such actions as the one for oestrogen may also called permissive.

I. Mode of action of steroid hormones. In cells sensitive to a steroid hormone, receptor molecules are present. These receptors bind specifically to that hormone. The steroid receptors are specialized and attached with specific proteins, within the cells, perhaps in the nuclear membrane. Steroid pass through the target after being received by the receptors of plasma membrane and enter the nucleus, where they affect the transcription of specific genes.

A steroid hormone that crosses the plasma membrane of its target cell binds to a receptor protein in the nucleus. This hormone-protein complex then binds to another specific protein, an **acceptor protein** which is located on certain region of the chromatin. This association initiates the transcription of selected genes. The messenger RNA molecules thus produced are processed and transported to the cytoplasm, where they direct the synthesis of new proteins. For example, ostrogen induces certain cells in the reproductive system of a female bird to synthesize large quantities of **ovalbumin**, the main storage protein stockpiled in egg white. Different proteins are synthesized in the liver cells of the same animal in response to oestrogen (hormone).

In the late 1950s, it was discovered that injecting rats with the steroid hormone, **hydrocortisone**, triggers the synthesis of the liver enzymes **tyrosine aminotransferase** and **tryptophan oxygenase**. Soon it was found that other steroid hormones act in a comparable fashion, stimulating the formation of specific proteins in their respective target tissues (Table 8.4).The ability of steroid hormones to stimulate the formation of new proteins can usually be inhibited by blocking gene transcription with either actinomycin D or the RNA polymerase II inhibitor—amanitin. Steroid hormones, thus, must be inducing the transcription of their respective genes.

Table 8.4. Selected examples of proteins whose synthesis is induced by hormones.

Steroid hormone	Target tissue	Induced protein
A. Oestrogen	1. Oviduct (hen)	Ovalbumin, conalbumin, ovomucoid, lysozyme
	2. Liver (rooster)	Vitellogenin, very low density lipoportein
	3. Pituitary (rat)	Prolactin
B. Glucocorticoids	1. Liver (rat)	Tyrosine aminotransferase, tryptophan oxygenase
	2. Kidney (rat)	Phosphoenolpyruvate carboxylase
	3. Pituitary (rat)	Growth hormone
C. Progesterone	1. Oviduct (hen)	Avidin, conalbumin, ovomucoid, lysozyme
	2. Uterus (rat)	Uteroglobin
D. Testosterone	1. Liver (rat)	2-Microglobulin
	2. Prostate (rat)	Aldolase

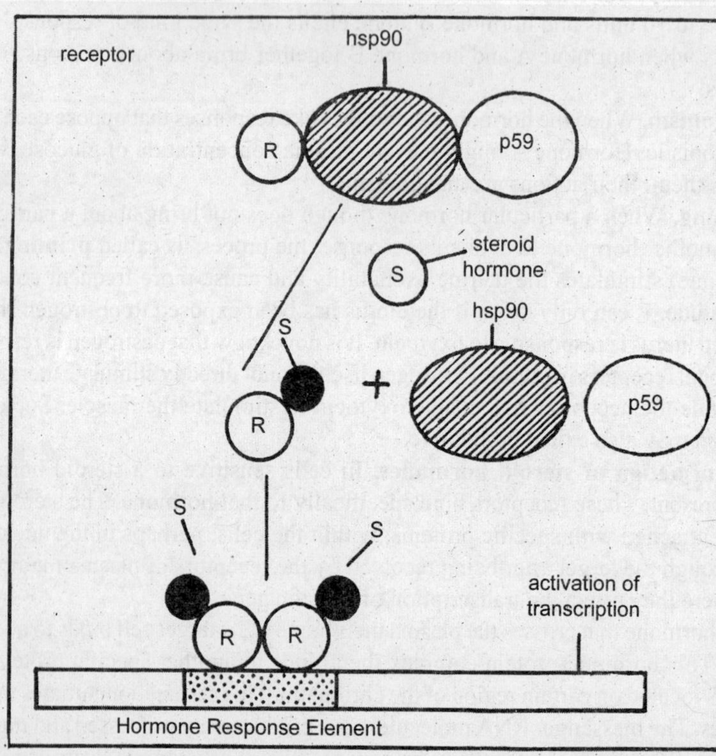

Fig. 8.15. Pathway activation of gene transcription by steroid hormones. In the absence of hormone, steroid hormone receptors are complexed with two proteins (hsp 90 and p59) that prevent the receptor (R) from interacting with DNA. The binding of a steroid hormone to its corresponding receptor displaces these two proteins, allowing the receptor to form a dimer (RS) that binds to the hormone response element and activates transcription.

Typically receptors for steroid hormones contain three distinct domains:

(*i*) a domain that binds to the appropriate steroid hormone,

(*ii*) a domain that binds to the receptor's DNA response element, and

(*iii*) a domain that is responsible for activating transcription.

The process by which hormone causes its receptor to activate transcription involves several distinct steps:

(*a*) Binding of the hormone to its receptor. Prior to this step, the receptor is unable to bind to DNA or activate transcription, because it is complexed with two other proteins called **hsp 90** (a heat-shock protein) and p59. The binding of a steroid hormone to its receptor displaces these two proteins and unmasked the receptor for domains that are required for DNA-binding and transcriptional activation (Fig. 8.15).

(*b*) Formation of a dimer. Molecules of the steroid hormone-receptor complex join together to form a dimer, which in turn binds to the receptor DNA response element. Because hormone response elements are palindromes, the two receptor molecules present in the dimer can bind to the two symmetric sequences that are present at opposite ends of the palindrome. Binding of receptor to the response elements transform the DNA at location into a **DNAI hypersensitive site**. This suggests that a nucleosome has been displaced. Nucleosome removal presumably facilitates transcription by making the promoter region more accessible to RNA polymerase and its associated transcription factors.

Steroid hormone receptors recognize their respective *DNA response elements* through a DNA-binding motif known as a **cysteine-cysteine zinc finger**. In contrast to cysteine-histidine fingers, which are present in variable numbers in protein that contain them, the cysteine-cysteine finger motif of steroid hormone receptors consists of only two adjacent zinc fingers (Fig. 8.16).

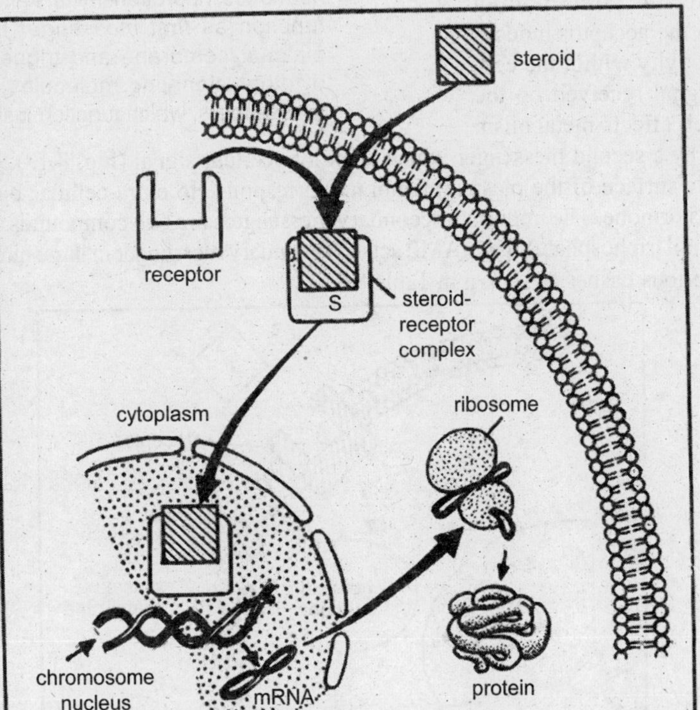

Fig. 8.16. Summary of mode of action of steroid hormones. The lipid soluble steroid hormones can readily pass through the plasma membrane of the cells into cytoplasm. These then bind to specific receptor proteins forming a complex that enters the nucleus and binds a specific regulatory site on chromosomes. The binding initiates transcription of the genes regulated by the site and thus leads to the production of specific protein.

The steroid hormone receptors usually activate transcription of genes located adjacent to hormone response elements. In some cases, however, transcription is inhibited. A prime example involves the *glucocorticoid receptor*, which can bind to two different kinds of response elements; one type of response element is adjacent to genes whose transcription is activated, and the other is adjacent to genes whose transcription is inhibited. Following binding of the glucocorticoid receptor to the inhibitor type response element, the receptor depresses the initiation of gene transcription by interfering with the binding of other transcription factors to the promoter region. The existence of separate activating and inhibiting response elements for the same hormone receptor allows a single hormone to stimulate the formation of mRNAs coding for one group of proteins, while it depresses the production of mRNAs coding for a different group of proteins.

II. Mode of action of peptide hormones. Peptide hormones and most hormones derived from amino acids are unable to pass through the plasma membrane of their target cells, but they still influence particular activity (Fig. 8.17). These hormones bind to specific receptors on the cell's outer surface. There may be as many as 10,000 receptors on the surface of target cell for a particular hormone. In extracellular surface of each receptor protein, is a binding site that specifically fits a particular hormone molecule. This binding of hormone to the acceptor induces biochemical activity within the cell. A chemical signal received on the outside of a cell affects metabolism inside the cell by a second messenger which is an intracelluar signal (Fig. 8.18). It is produced by the cytoplasmic surface of the plasma membrane in response to extra-cellular binding of the first messenger, the hormone. The important secondary messengers are two compounds. These are **cAMP** and **IP$_3$** (= inositol triphosphate). The cAMP acts as secondary messenger in large number of hormonal responses in various tissues as shown in Table 8.5.

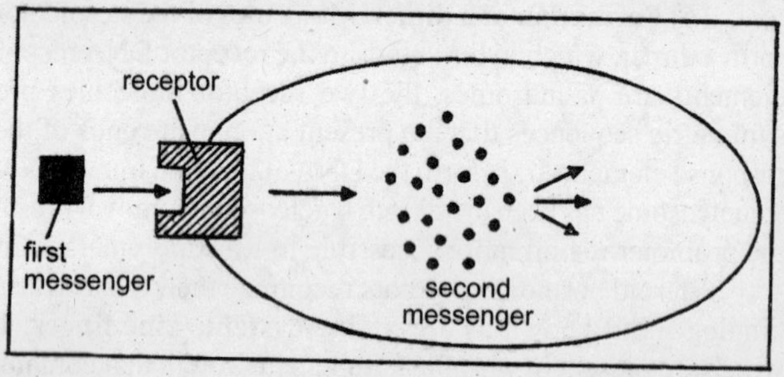

Fig. 8.17. The second messenger concept. Various hormones, neurotransmitters and local mediators function as *first messenger* that bind to the plasma membrane and trigger the formation of other signaling molecules, called *second messengers*, which function inside the cells.

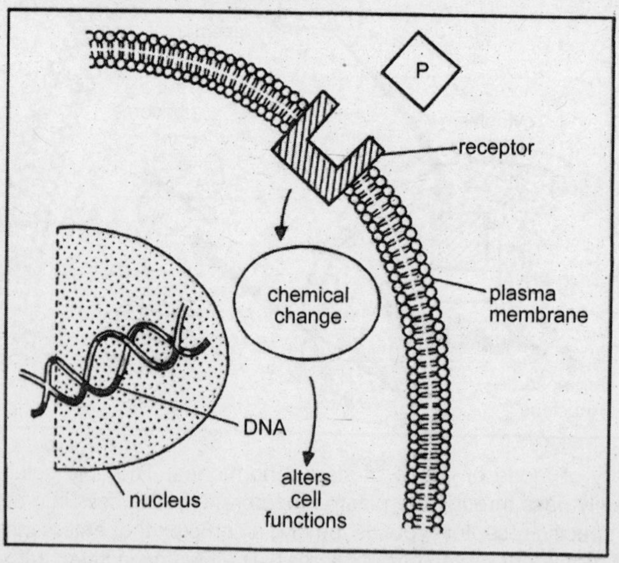

Fig. 8.18. Mode of action of peptide hormones. P = peptide hormone.

Table 8.5. Some hormonal responses associated with increased cyclic AMP levels.

	Hormone	Tissue	Response
1.	ACTH	Adrenal	Hydrocortisone formation
2.	TSH	Thyroid	Thyroxine formation
3.	LH	Ovary	Progesterone formation
4.	Vasopressin	Kidney	Water resorption
5.	Glucagon	Adipose tissues	Lipid breakdown
6.	Glucagon	Liver	Glycogen breakdown
7.	Parathormone	Kidney	Phosphate excretion
8.	Parathormone	Bone	Bone resorption
9.	Secretin	Intestine	Pancreatic enzyme release
10.	Epinephrine	Cardiac muscle	Increased contractility
11.	Epinephrine	Liver	Glycogen breakdown
12.	Epinephrine	Erythrocyte	Increased Na^+ permeability

(*i*) Role of Cyclin AMP. **E.W. Sutherland** and coworkers studied the mechanism of action of epinephrine (hormone) and found that non-steroid hormones act via second messenger (Fig. 8.19). It was observed that the binding of epinephrine to the receptors elevates the cytoplasmic concentration of cyclic adenosine monophosphate (cyclic AMP or cAMP). An enzyme, **adenylate cyclase** is a membrane protein with its active site facing the cytoplasm. The cAMP functions as a second messenger by relaying the signals to the cytoplasm. It however, itself does not persist for long in the absence of the first messenger, because it gets converted to an inactivated protein by another enzyme. Cyclic AMP is inactivated protein by phosphodiesterase, an enzyme that is inhibited by theophylline or caffeine. (*Theophylline* is a feeble basic bitter crystalline compound $C_7H_8N_4O_2$ that is present in small amounts in tea and acts as a bronchiodilator.)

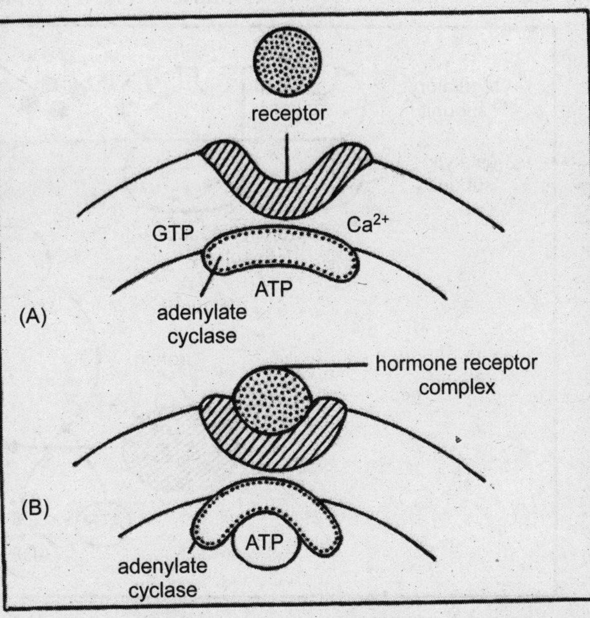

Fig. 8.19. Hormone receptor binding and mechanism of hormone action. A—Hormones that act through cyclic AMP, bind first to a receptor protein in the plasma membrane. Molecules such as Ca^{2+} and guanosine triphosphate (GTP) convey the change in receptor conformation to the membrane bound adenylate cyclase. B—The activated adenylate cyclase catalyses the formation of cAMP from AMP and ATP. This initiates the sequence of reaction considered as a generalized response of cAMP sensitive cell.

The hormone receptor and adenylate cyclase are both built into the plasma membrane. They, however, do not interact directly. The binding of hormone to the receptor, catalyses the activation of adenylate cyclase by another membrane protein which functions as an intermediary, called **G protein**, because it binds to GTP (guanosine triphosphate). GTP is an energy carrier closely related to ATP. The G protein is activated by docking of the hormone with the receptor. It in turn activates adenylate cyclase (enzyme) to produce cAMP. GTP provides the energy for converting the first message to second message, where as the G protein hydrolyses GTP to GDP (guanosine diphosphate). The exergonic reaction is basically same as the ATP $\rightarrow$ ADP + P (pyrophosphate) reaction.

The adenylate cyclase (enzyme) is inhibited by another type of G protein. When an inhibitory hormone binds to its receptor, this G protein lowers the activity of adenylate cyclase. The cytoplasmic concentration of cAMP is then reduced. It is because of these opposing effects of proteins in regulating adenylase cyclase (enzyme) that the cell is able to fine tune its metabolic response to slight changes in the proportion of antagonistic hormones.

$$\text{ATP} \xrightarrow[\text{Adenyl cyclase}]{H_2O} \text{cAMP} + \underset{\text{Pyrophosphate}}{\text{P}}$$

Fig. 8.20. The formation of cyclic AMP (cAMP). Adenylate cyclase an enzyme located in the plasma membrane catalyses the formation of cyclic AMP from ATP. Cyclic AMP is broken down to AMP by the cytoplasmic enzyme phosphodiesterase.

The cAMP affects that rest of the cell by binding to and activating an enzyme called cAMP–dependent protein kinase or **protein kinase A** (Fig. 8.21). Protein kinases are a family of enzymes that

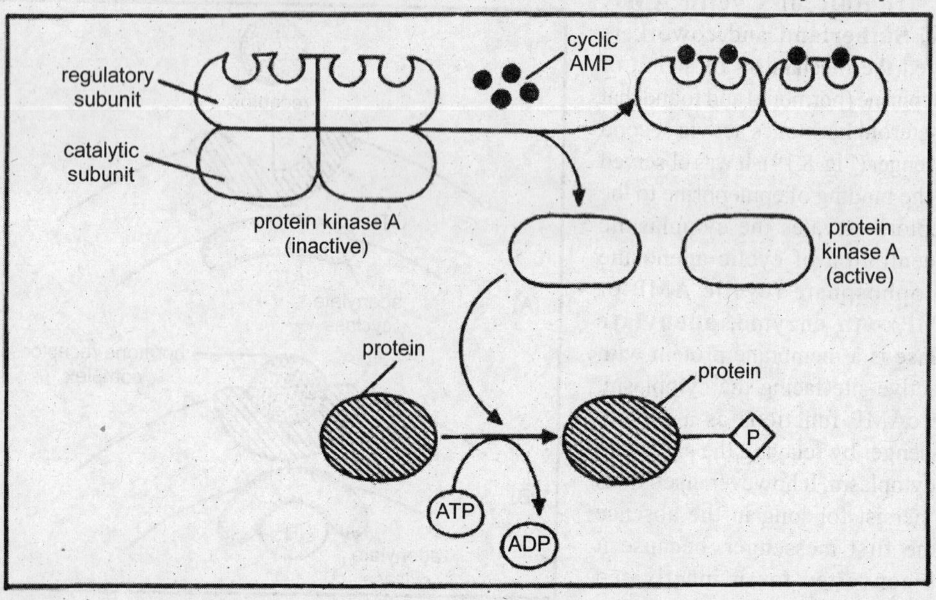

Fig. 8.21. Activation of protein kinase A by cyclic AMP. Cyclic AMP causes the regulatory subunits of protein kinase A to dissociate from the catalytic subunits; the free catalytic subunits, in turn catalyse protein phosphorylation.

catalyses the transfer of terminal phosphate group from ATP to an amino acid in a protein molecule. This process of **protein phosphorylation** regulates the activity of the protein that is phosphorylated. Cyclic AMP activates protein kinase A that phosphorylates the amino acids **serine** and **threonine** in a selected set of proteins. The protein kinase A enzyme consists of **two** regulatory and two catalytic subunits. The four subunits are bound together as a tetramer to bind to promoter. At first glance it might appear as if the cAMP mediated pathway of a signal transduction is unnecessarily complex; the binding of signaling molecule to its receptor (*i*) activates a protein (Gs) that (*ii*) Activates an enzyme (*i.e.*, adenyl cyclase) that (*iii*) catalyses the formation of a compound (cyclic AMP) that (*iv*) activates another enzyme (protein kinase A) that (*v*) catalyses the phosphorylation of other target proteins. One may question, instead of this complex cascade, why does not the signaling molecule simply interact directly with the final target protein (s)? It is possible because of signal amplification of each step in the pathway is catalytic. Amplification of the incoming signal occurs that is, each activated receptor activates many molecules of Gs, each of which can activate a molecule of adenylyl cyclase. Each adenylyl cyclase molecule in turn produce hundreds or thousand of molecule of cAMP per second, each of which can activate a molecule of protein kinase. Every activated protein kinase molecule in turn catalyzes the phosphorylation of hundreds or thousand of molecules of phosphorylase kinase, glycogen synthase hormone-sensitive lipase, or whatever enzyme is being acted upon in turn. These enzymes may convert hundreds or thousand of substrates molecule into products. In this way, a single hormone molecule acting on the cell surface is ultimately capable of triggering a response involving millions of molecules.

Multistep cascade also gives advantage of added flexibility that is achieved when multiple components are included in a regulatory pathway. Each component can serve as a separate target for control by alternate mechanisms, creating the potential for a complex interplay-between regulatory networks. For example, in addition to being activated by protein kinase A, phosphorylase kinase is also stimulated by calcium ions. Hence any stimulus altering the intercellular concentration of Ca^{2+} ions will influence glycogen break down by a mechanism that is independent of changes in levels of cAMP.

***(ii)* Role of ionositol triphosphate (IP_3) and phospholipase C.** In animals, many chemical messengers, *e.g.*, neurotransmitters, growth factors and some hormones, induce responses in their target cells by increasing the cytoplasmic concentration of calcium ions (Ca^{2+}). Binding of messenger to a specific receptor on the cell surface results in the release of Ca^{2+} from a reservoir within the cell, usually the endoplasmic reticulum. The release of Ca^{2+} to the cytoplasm alters the activities of specific enzymes, either directly or first binding to a protein called **calmodulin**. Calmodulin then binds to other proteins and changes their activities. The hormonal signal and the rise in cytoplasmic Ca^{2+} concentration is mediated by a second messenger called **inositol triphosphate (IP_3)**. Calcium therefore is the **third messenger** in the system (Fig. 8.22).

Production of IP_3 and the phospholipase C. Inositol triphosphate (IP_3) is driven from a particular type of phospholipid present in the plasma membrane. Production of IP_3 involves a reaction sequence in the membrane analogous to the mechanism that transduces a hormonal signal in the production of cAMP. A hormone binds to its specific receptor and the hormone-receptor complex activates a G protein in the membrane. This G protein is different from the one that functions in the cAMP system and it stimulates a membrane enzyme, called **phospholipase C**.

The IP_3 generated by phospholipase C serves as a second messenger that diffuses from the plasma membrane, where it is produced, to the ER, where it triggers the release of stored Ca^{2+} into the cytosol. The calcium ions bind to the calcium binding proteins, *e.g.*, calmodulin, to affect the cell metabolism.

Phospholipase C tends to cleave a phospholipid into two products, IP_3 and another compound called **diacylglycerol (DAG)**. Both these may function as second messengers. The DAG activates an enzyme, protein kinase C, which triggers some of the cytoplasmic responses to the hormone by phosphorylating specific protein similar to the action of cAMP-dependent protein kinase, but protein kinase activates a different set of cytoplasmic proteins. The IP_3 released from the plasma membrane probably interacts with the membrane of ER to cause release of Ca^{2+}. Phospholipase C which normally resides in the cytosol is recruited to the inner surface of the plasma membrane, where it catalyses the formation of DAG and IP_3. DAG functions by activating protein kinase C, while IP_3 mobilises the release of Ca^{2+} from the ER (Fig. 8.22).

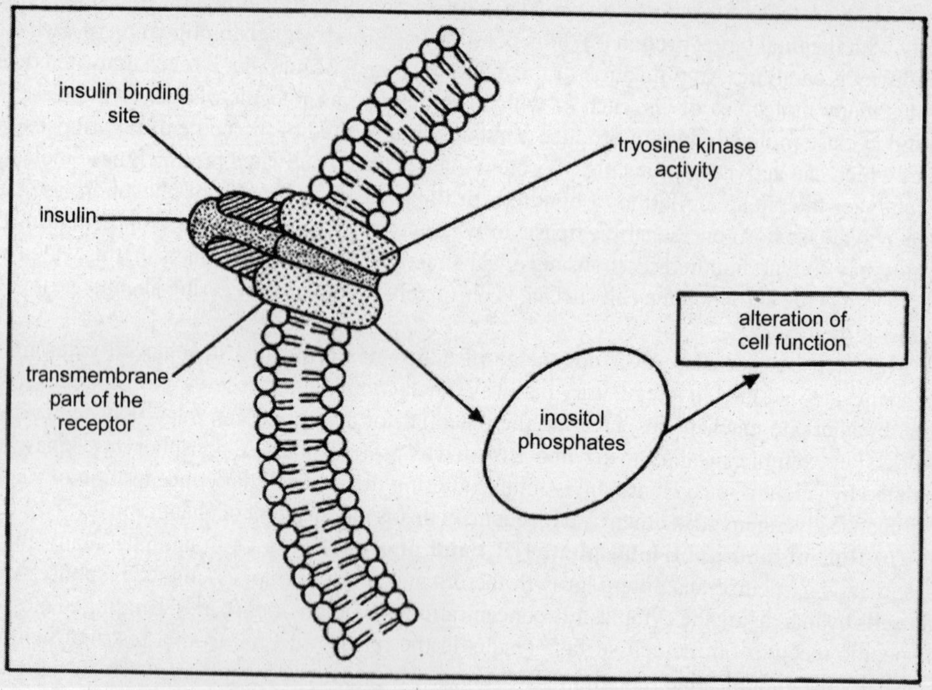

Fig. 8.22. How insulin works? Or Inositol mediated action. Insulin binds to the insulin binding site of the receptor (a transmembrane protein).

Comparison of Action of Steroid and Protein Hormones

1. Steroid hormones generally work by entering the nucleus of the cell and altering gene expression. The peptide hormones and most other hormones, bind to the cell surface (*i.e.*, to plasma membrane) and regulate cytoplasmic enzymes via second messengers such as cAMP, phospholipase C or IP_3.

2. The steroids affect the synthesis of proteins, while peptide hormones affect the activity of enzymes and other proteins already present in the cell.

3. Target cells generally respond more slowly to steroids than to peptide hormones, but the changes induced by steroids usually last longer.

The thyroxine hormone produced by the thyroid gland is a modified form of the amino acid tyrosine and contains four iodine items. Tyrosine acts at target cell in a unique way. It diffuses directly into the cell cytoplasm and then into the cell nucleus, where it interacts with a protein receptor attached to the DNA to initiate production of particular growth promoting messenger RNAs.

Hypophysis and Hypothalamus

Hypophysis is located in the sella turcica of the sphenoid bone at the base of the skull (Fig. 8.23). The pituitary gland is made up of two lobes: the anterior lobe, called **adenohypophysis** which is 75% of weight and the posterior lobe, called **neurohypophysis** which is 25% of weight.

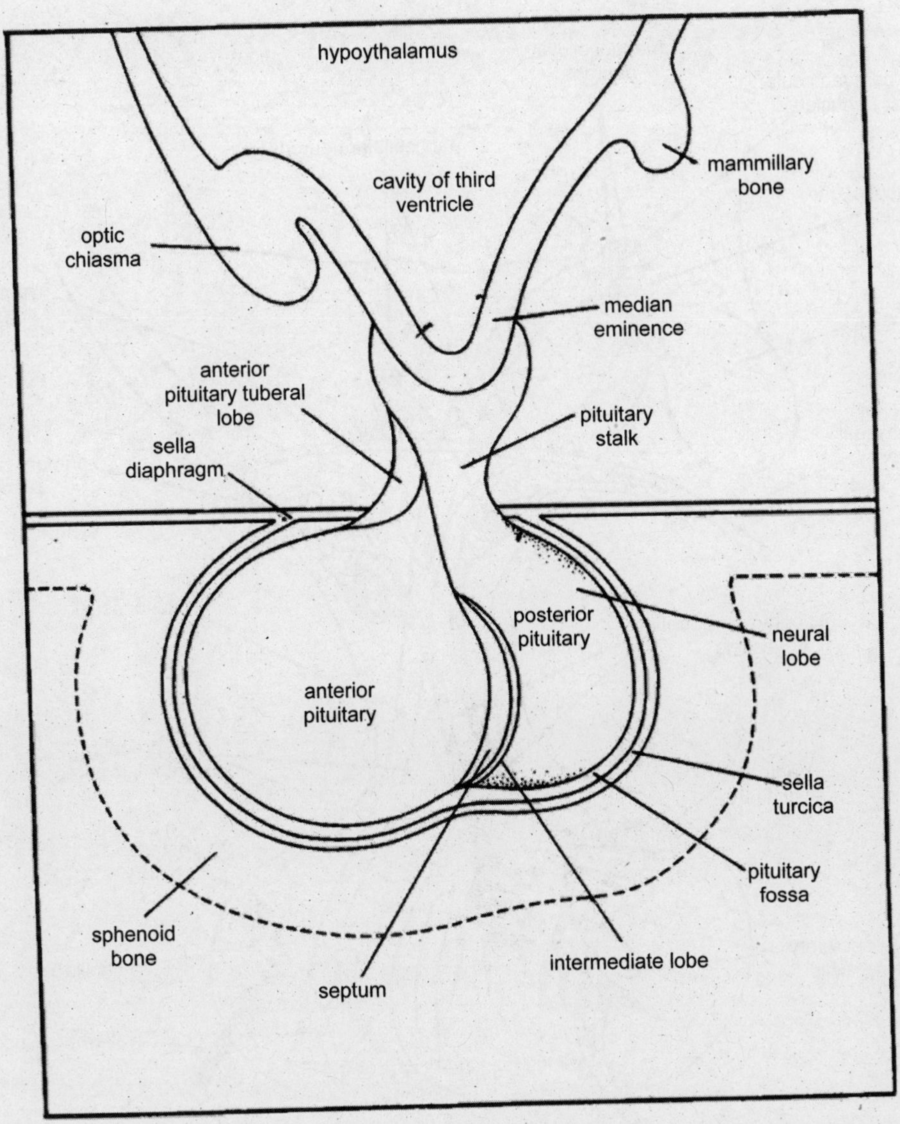

Fig. 8.23. Gross structure of the pituitary gland. The pituitary gland is below the hypothalamus and is connected to it by the pituitary stalk. The gland sits within the sella turcica, a fossa within the sphenoid bone, and is covered by a dural reflection, the diaphragma sellae. The pars distalis makes up most of the anterior pituitary.

1. Anterior Pituitary or Adenohypophysis

It is composed of two major cell types: chromophils and chromophobes (Fig. 8.25).

1. **Chromophils** are cells with affinity for dyes and they constitute 75% of cells and exist in two forms: (*a*) **Acidophils** which stain red or orange with acidic dyes, are **somatotrophs** which secrete growth hormone and **mammotrophs** secreting prolactin. (*b*) **Basophils** which stain blue or green with basic dyes include *gonadotrophs* secreting luteinizing hormone (LH) and follicle stimulating hormone (FSH), *thyrotrophs* which secrete TSH (thyroid stimulating hormone) and *corticotrophs* secreting ACTH (adrenocorticotropic hormone).

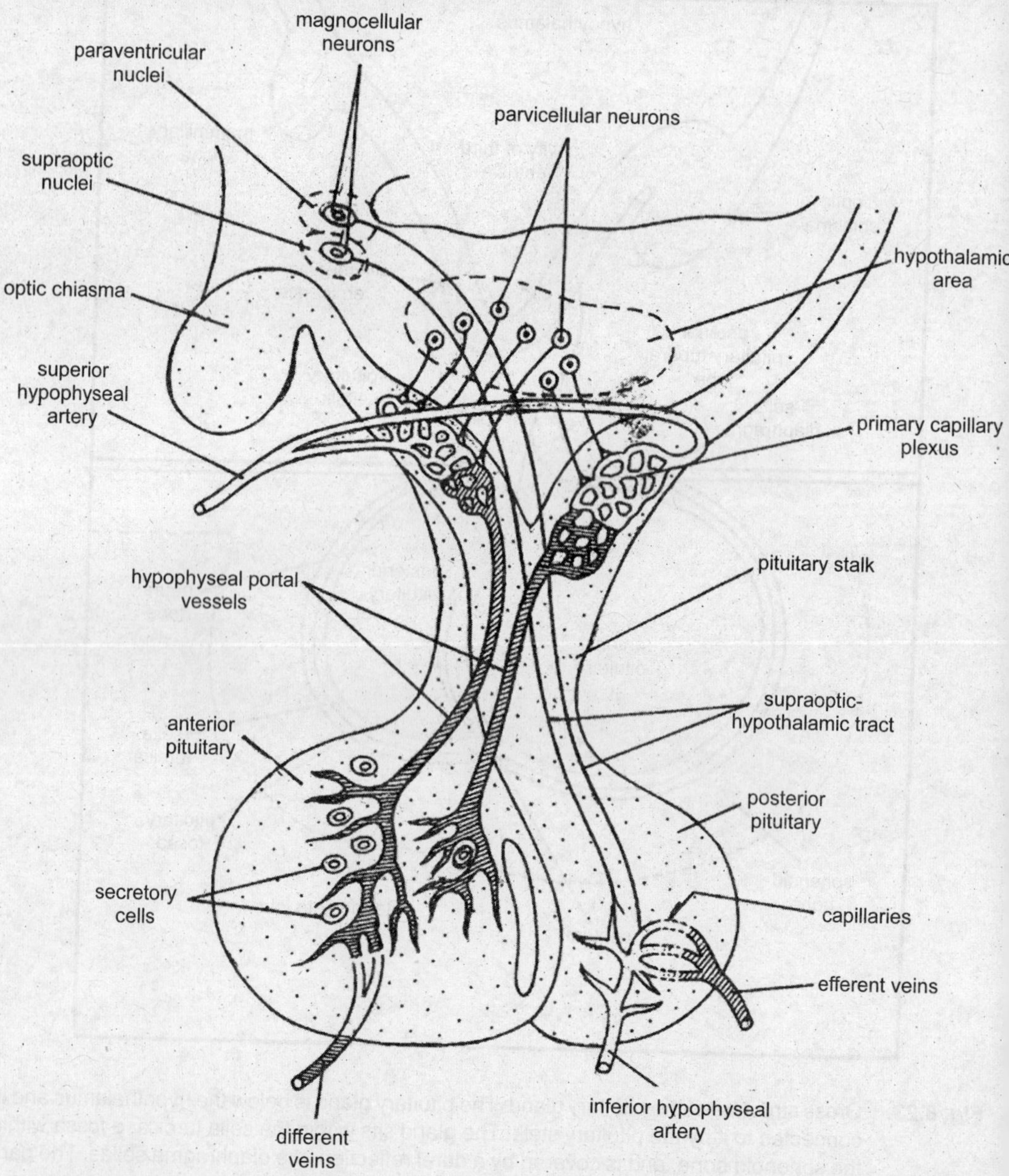

Fig. 8.24. Magnocellular neurons of the hypothalamus (paraventricular and supraoptic nuclei) project their axons; down the infundibular stalk (= process) and terminate in the pars nervosa (posterior lobe) where they release their hormones (either ADH or oxytocin) into a capillary bed.

Table 8.6. Difference in acidophils and basophils (two types of chromophils) of adenohypophysis or anterior lobe of pituitary.

	Acidophils		**Basophils**
1.	Acidophils or eosinophils are more numerous about 43% with more and large granules.	1.	Bosophils are about 7% and are large in rise. Their secretory granules are few and small.
2.	Their granular cytoplasm is stained with acid stain (eosin).	2.	Their cytoplasmic granules are stained with basic dye.
3.	Acidophils are of two types: **(a) Somatotrophs.** Medium sized, oval with large nucleus and numerous spherical secretory granules, produce STH (or GH). **(b) Mammotrophs.** Small granules large, sparse and of various shapes	3.	Basophils are of three types: **(i) Thyrotrophs.** Large sized polygonal and irregular, produce TSH. **(ii) Gonadotrophs.** Large FSH cells and small LH cells. **(iii) Corticotrophs.** Secrete ACTH.
4.	Produce two hormones: (i) Growth hormone (GH) or somatotropic hormone (STH) by somatotrophs. (ii) Prolactin (PL) or lactogenic hormone (LTH) by mammotrophs.	4.	Produce four hormones, (i) Thyroid stimulating hormone (TSH) (ii) Adrenocorticotropic hormone (ACTH) (iii) Follicle stimulating hormone (FSH) (i) Luteinizing hormone (LH)

2. **Chromophobes** are cells without affinity for dyes and constitute remaining 25% of cells of anterior pituitary. These are degranulated secretory cells.

Anterior pituitary is under control of hypothalamic hormones which reach there via hypothalamo-hypophyseal portal vessels, a circulatory system which begins and ends in capillaries without going to the heart.

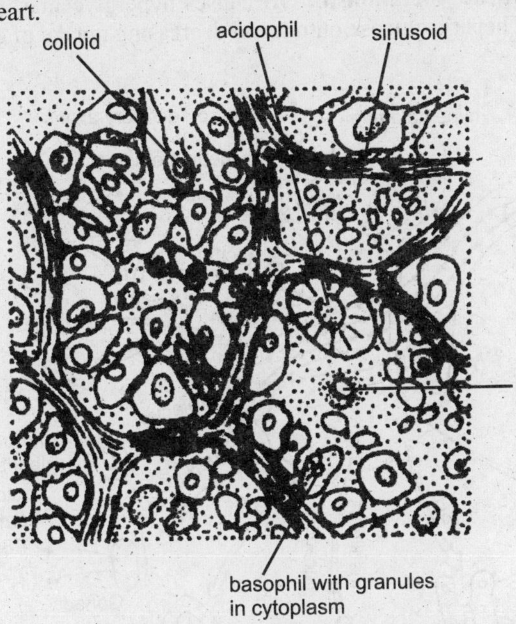

Fig. 8.25. T.S. of anterior pituitary.

Pituitary hormone secretions are regulated by negative (rarely positive) feedback from the hypothalamus get decreased when the concentrations of the target gland hormone get increased and vise versa. If the hormone secreted from peripheral glands affect both the hypothalamus and the anterior pituitary, then it makes a **long negative feedback loop** and if they affect only the anterior pituitary then it constitutes a **short negative feedback loop**.

A. Hormones secreted by anterior lobe of pituitary or adenohypophysis

1. *Growth hormone* (*GH*). It is also called **somatotropin**. Human growth hormone is a single unbranched polypeptide chain containing 191 amino acids with molecular weight of 21,500. It varies considerably in structure from species to species (of mammals).

GH is secreted episodically in periods of 20 to 30 minutes with large diurnal fluctuations. A peak of GH secretion occurs 1–2 hours after the onset of deep sleep. Secretion of GH remains high in the newborn till 4 years of age, thereafter it remains at intermediate levels till 25 years of age and subsequently its level decline with ageing.

(*a*) Control of GH secretion. GH secretion is primarily under the control of two hypothalamic hormones—**GH-releasing hormone (GHRH)** and **GH-inhibiting hormone (GHIH)** or **somatostatin**. (*i*) Stimuli that increase GH secretion include: 1. deep sleep; 2. exercise; 3. hypoglycemia; 4. estrogens; 5. androgens; 6. increase in circulating levels of amino acids; 7. glucagon; 8. emotional stress; 9. surgical procedures; 10. exposure to cold; 11. pyrogens; and 12. drugs such as L-dopa, amphetamine. (*ii*) Stimuli that decrease GH secretion include: 1. GH itself via short-loop negative feedback loop; 2. glucocorticoids; 3. hyperglycemia; 4. REM sleep; and 5. late pregnancy.

(*b*) Actions of GH.

1. Stimulation of growth of bone, cartilage and connective tissue. The effects of GH on skeletal system is mediated by a family of polypeptides called **somatomedins** which are synthesized from liver (mainly), kidneys, muscles and other tissues in response to stimulation by GH.

Thyroid hormone and insulin also play fascilitatory role in this growth. Before the closure of epiphysis, GH stimulates proliferation of chondrocytes, appearance of osteoblasts (cells which cause bone formation), stimulation of RNA and DNA synthesis and collagen formation in cartilages. This helps in linear skeletal growth. After epiphyseal closure, bone length is not increased by GH but bone thickening can occur.

2. Effect on protein metabolism. GH increases protein synthesis (anabolic effect) and promotes positive nitrogen balance.

3. Effect on carbohydrate metabolism. GH causes hyperglycemia, *i.e.*, increase in blood glucose levels by increasing hepatic glucose output and decreasing uptake of glucose in muscle and adipose tissues.

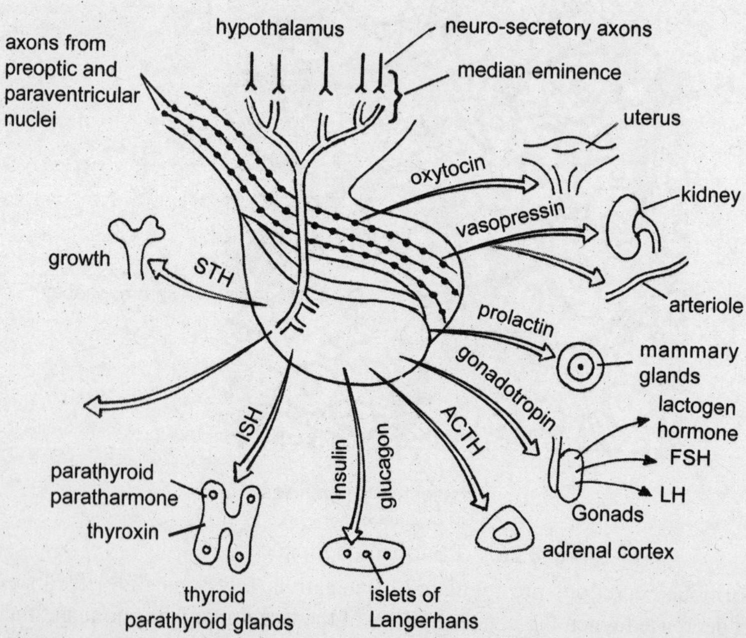

Fig. 8.26. Diagram showing hormones of pituitary gland, their target organs and relation of pituitary with hypothalamus.

4. Effect on fat metabolism. GH has catabolic effect, *i.e.*, increases mobilization of fats from adipose tissues, increases levels of free fatty acids in circulation and stimulates gluconeogenesis in liver. GH also increases hepatic oxidation of fatty acids to ketone bodies.

5. Effect on minerals. GH increases Ca^{2+} absorption from GIT (gastrointestinal tract) and decreases Na^+, K^+, Ca^{2+} and phosphorus excretion from kidneys.

(*c*) Diseases caused by hypersecretion or hyposecretion of growth hormone (GH). Excess of GH production or hypersecretion of GH before puberty leads to **gigantism** while after puberty leads to **acromegaly**. Gigantism is characterized by tall stature, breast enlargement, large hands and feet, coarse facial feature and impotence. Characteristic feature of acromegaly are elongation and widening of the mandible, prominent brows, thickening of skin and coarsening of facial features, kyphosis (bowing of spine), hypertrophy of organs such as liver, heart, kidney and spleen.

Deficiency of GH leads to **dwarfism** (short stature).

2. *Prolactin* (*PRL*). Prolactin is also called **luteotropic hormone (LTH), lactogenic hormone** or **mammotropic hormone (MH)**. It is a single peptide chain containing 198 amino acids with M.W. 25,000. PRL/LTH is secreted during pregnancy and stimulates proliferation of glandular tissue of mammary glands and helps in completing development of breast.

PRL or LTH secretion is under inhibitory control of hypothalamus by *prolactin inhibitory hormone* which is actually dopamine. The main stimulus of PRL or LTH release is suckling of breast nipple by infant during breast feeding. Blood PRL levels are higher in adult females than the males. Progressive increase occurs during pregnancy, peaking at delivery if breast-feeding is initiated soon. Thereafter, high PRL levels are maintained by suckling of breast and then declines when breast feeding is discontinued.

Prolactin plays an important role in the development of mammary gland and in milk synthesis. A *deficiency* of PRL leads to failure to lactate while *excess secretion* of PRL leads to **hyperprolactinemia** characterized by *galactorrhea* (a spontaneous flow of milk from the nipple), *amenorrhea* (cessation of menstrual periods), infertility in females. In males PRL can lead to breast enlargement (gynecomastia), loss of libido and impotence.

Factors which can cause hyperprolactinemia are chronic renal failure (due to decreased peripheral clearance of PRL), dopamine receptor antagonist drugs such as antipsychotics (tranquilisers).

PRL or LTH also initiates secretion of progesterone from corpus luteum.

3. *Adrenocorticotropic hormone* (*ACTH*) *or Corticotropin.* ACTH is a peptide hormone secreted from anterior pituitary whose primary function is to stimulate synthesis and release of **cortisol** from adrenal cortex. Hypothalamus regulates ACTH secretion by **corticotropin releasing hormone (CRH)**. Secretion of ACTH has characteristic circadian rhythm with peak at 6 am, decreasing during afternoon and reaching lowest at midnight. ACTH secreation is also stimulated by severe stress or trauma.

Excess production of ACTH leads to hypertrophy (*i.e.*, excessive development) of adrenal gland and **Cushing syndrome** characterized by moon-like face, thin limbs, protruded abdomen, acne, osteoporosis (*i.e.*, decrease in bone mass with decreased density producing porosity and fragility of bones), hypertension, peptic ulcer, poor wound healing, etc.

Deficiency of ACTH leads to adrenal atrophy (*i.e.*, decrease in size), **Simmond's disease** (a disease that is characterized by extreme and progressive emaciation (*i.e.*, losing of flesh so as to become very thin) with atrophy of internal organs, loss of body hair and evidences of premature aging).

4. *Thyroid stimulating hormone* (*TSH*) *or Thyrotropin.* TSH is a glycoprotein hormone. Its secretions are pulsatile and follow a biological rhythm with highest levels being during sleep at night. TSH stimulates thyroid gland to produce thyroid hormones—T_3 (triiodothyronine) and T_4 (thyroxine or tetraiodothyronine). TSH secretion is controlled by **thyrotropin releasing hormone (TRH)** from hypothalamus. TRH and TSH secretion are subjected to negative feedback inhibition by thyroid hormones. *Deficiency of TSH* leads to secondary hypothyroidism (Deficient activity of the thyroid gland and resultant lowered metabolic rate and general loss of vigor).

5. *Gonadotropins.* Gonadotropins such as **FSH (follicle stimulating hormone)** and **LH (luteinizing hormone)** control the activity of gonads. They are under the influence of **gonadotropin releasing hormone (GnRH)** from hypothalamus which is secreted in discrete pulses every 60–120 min.

Functions of FSH in males include stimulation and maintenance of **spermatogenesis** in seminiferous cells of testis while *in females, FSH causes follicles development in ovaries*, production of estrogen and regulation of menstrual cycle.

LH is required in females for ovulation, formation and maintenance of corpus luteum, promotion of estrogen and progesterone production and regulation of menstrual cycle. In males, LH or **ICSH (interstitial cell stimulating hormone)** helps in biosynthesis of androgens and testosterone by interstistial (Leydig) cells of testis.

Premature activation of hypothalamic-pituitary-gonadal axis leads to precocious puberty while its disorder leads to delayed puberty. *Excess production of FSH/LH* in adult women results in polycystic ovaries and ovarian enlargement conversely inadequate secretion results in amenorrhea (abnormal absence or suppression of menstruation) and infertility in women and oligozoospermia, impotence and infertility in men.

B. Posterior Pituitary (Neurohypophysis)

The pars nervosa (= posterior lobe of pituitary) that is the site of release of neurohormones adjacent to a rich bed of fenestrated capillaries. The peptide hormones that are released are **antidiuretic hormone (ADH)** or **arginine vasopressin (AVP)** and **oxytocin** (Fig. 8.24). The cell bodies of the neurons that project to the pars nervosa are located in the **supraoptic nuclei (SON)** and **paraventricular nuclei (PVN)** of the **hypothalamus**. The cell bodies of these neurons are described as **magnocellular** (*i.e.*, large cell bodies which are packaged into neurosecretory granules)

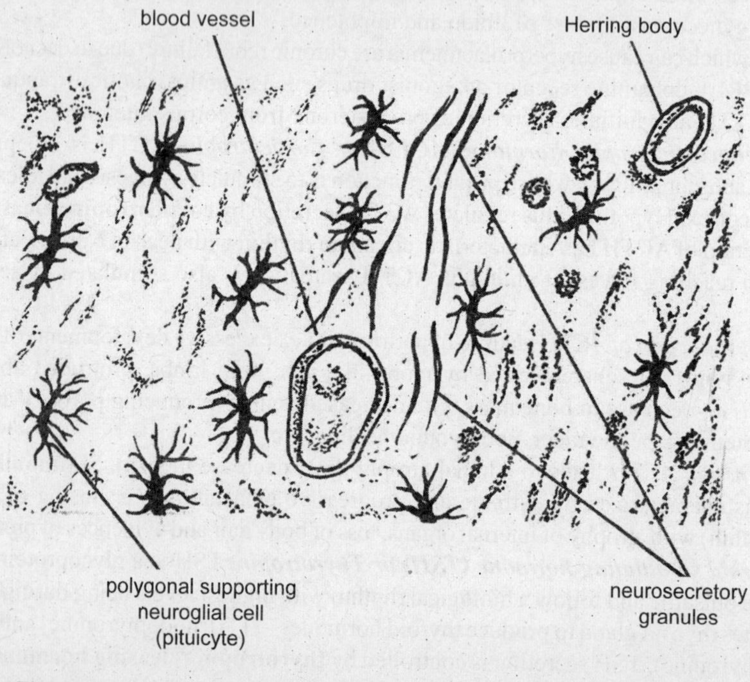

Fig. 8.27. T.S. pars nervosa or posterior pituitary (= neurohypophysis).

and project axons down the infundibular stalk as the **hypothalamohypophyseal tract.** These axons terminate in the pars nervosa (= posterior pituitary). In addition to axonal processes and termini from SON and PVN, there are glial-like supportive cell called **pituicytes** (Fig. 8.27). The posterior pituitary is extensively vascularised and the capillaries are fenestrated, thereby facilitating diffusion of hormones into the vasculature.

Synthesis of ADH and Oxytocin

Hormones—ADH and oxytocin, are nonapeptides (nine amino acids) that are similar in structure, differing in only two amino acids. They have limited overlapping activity. Both of these hormones synthesized as preprohormones (Fig. 8.27). Each prohormone harbors the structure of oxytocin or ADH and a cosecreted peptide: either **neurophysin I** (associated with ADH) or **neurophysin II** (associated with oxytocin). These preprohormones are called **preprovasophysin** and **preprooxyphysin**. The N-signal peptide is cleaved as the peptide is transported into the endoplasmic reticulum. The prohormone is packaged in the endoplasmic reticulum and Golgi apparatus in the membrane bound secretory granule in cell bodies within the SON and PVN. The secretory granules are conveyed-intraaxonally through a "fast" (*i.e.*, millimeters per hour) ATP - dependent transport mechanism down the infundibular stalk to axonal termini in the pars nervosa. During transit of the secretory granule, the prohormones are proteolytically cleaved to produce equimolar amounts of hormone and neurophysin. Secretory granules containing fully processed peptides are stored in the axonal termini. Axonal swelling because of the storage of secretory granule can be observed by light microscopy and is termed **Herring bodies**.

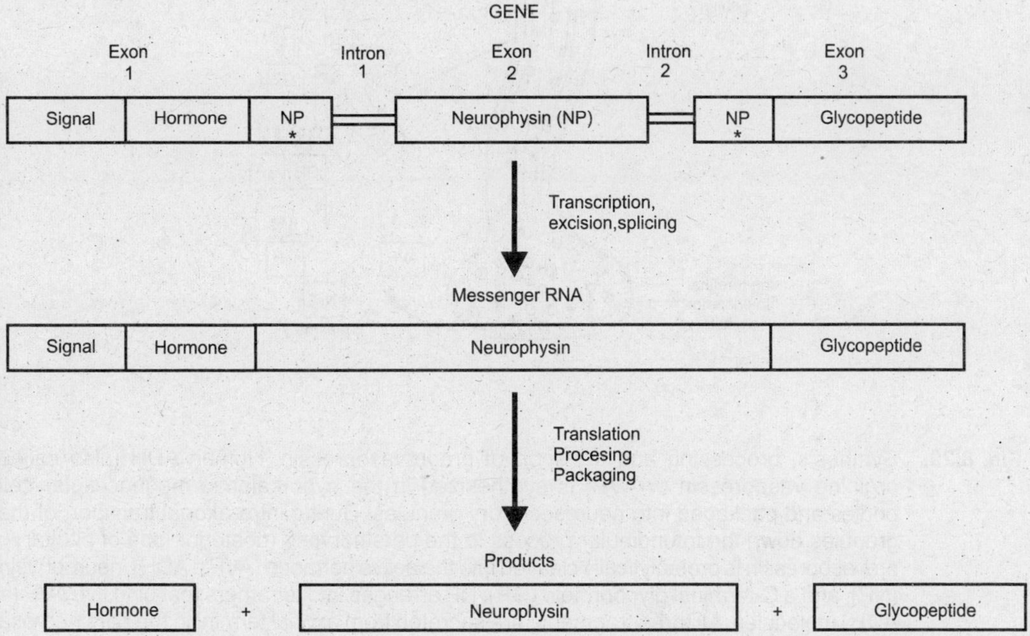

Fig. 8.28. Synthesis and processing of preprovasopressin or prepro-oxytocin.

ADH and oxytocin are released from the pars nervosa in response to stimuli that are primarily detected at the cell body and its dendrites in SON and PVN of the hypothalamus. The stimuli are mainly in the form of neurotransmitters released from hypothalmic inter-neurons. With sufficient stimulus, the neurons will depolarize and propagate an action potential down the axon. At the axonal termini, the action potential increases intracellular (Ca^{2+}) and results in a stimulus-secretion response, with the exocytosis of ADH or oxytocin, along with neurophysin, into the extracellular fluid of the pars nervosa (Fig. 8.28). Hormones and neurophysins enter the peripheral circulation and both can be measured in blood (**Koeppen** and **Stanton** 2010).

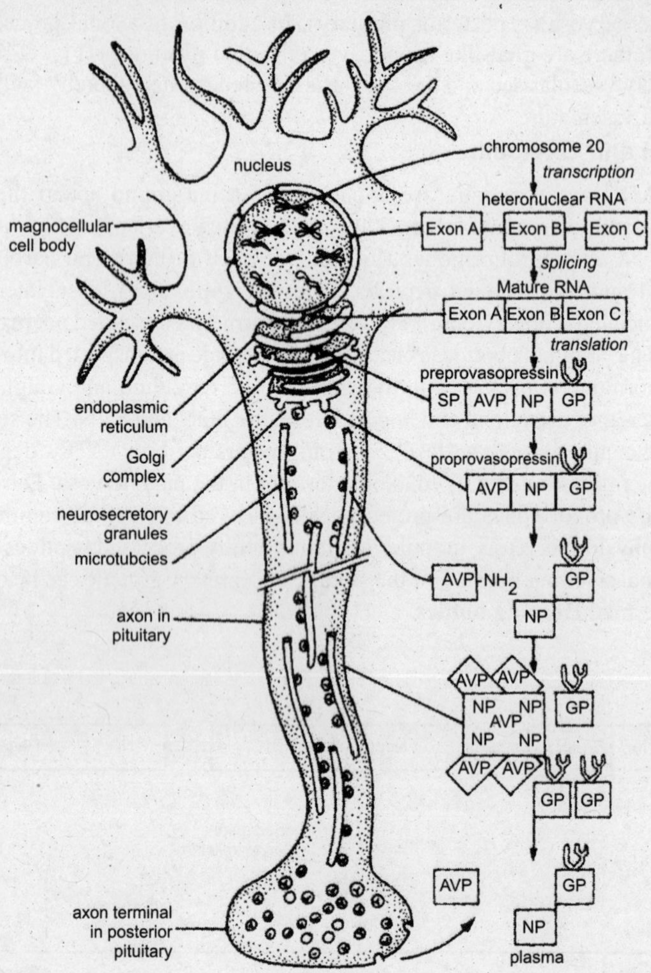

Fig. 8.29. Synthesis, processing and transport of preprovasopressin. Human ADH (also called arginine vasopressin or AVP) is synthesized in the hypothalamic magnocellular cell bodies and packaged into neurosecretory granules. During intra-axonal transport of the granules down the infundibular process to the pars nervosa (posterior lobe of pituitary), provasopressin is proteolytically cleaved into the active hormone (AVP = ADH), neurophysin (NP), and a C-terminal glycoprotein (GP). NP arranges into tetramers that bind five AVP (= ADH) molecules. All three fragments are secreted from axonal termini in the pars nervosa (posterior pituitary) and enter the systemic blood. Only AVP (= ADH) is biologically active. (after Koeppen and Stanton, 2010).

1. Anti-diuretic hormone (ADH) or vasopressin. The neurohormone ADH is a nonapeptide (containing nine amino acids) with M.W. 1000. It is **released** in response to 1. decrease in plasma volume; 2. increase in plasma osmolality; 3. pain; 4. stress; 5.anxiety; 6. nicotine; 7. drugs such as morphine, general anesthetics; and 8. chronic conditions where effective circulating blood volume is reduced as in cirrhosis of liver and cardiac failure.

ADH **acts** through three classes of receptors: V_{1a}, V_{1b}, and V_2. Acting through **V_2 receptors** present on collecting duct of kidney ADH promotes water reabsorbtion. ADH causes vasoconstriction by acting on **V_{1a} receptors** (hence the name vasopressin) and via **V_{1b} receptors** it causes release of ACTH from anterior pituitary.

Deficiency of ADH leads to disease called **diabetes insipidus** characterized by excretion of large amount of urine, dehydration, thirst and bed wetting in children.

Increased ADH secretion causes syndrome of inappropriate secretion of ADH (**SIADH**) featuring water retention, dilutional hyponatremia (low plasma sodium levels).

2. Oxytocin. The neurohormone oxytocin is also a polypeptide containing nine amino acids. Oxytocin has primary role in milk ejection in lactating mothers by causing contraction of myoepithelial cells surrounding the mammary alveoli. It also has a facilitatory role in initiation of labour and parturition. Oxytocin powerfully stimulates the gravid uterus especially towards the end of gestation.

Stimuli for release of oxytocin include: 1. stimulation of tactile receptors in the areolar region of the female breast during suckling; 2. genital tract stimulation during coitus and parturition and 3. stretch on cervix at the time of labour.

Factors inhibiting release of oxytocin: 1. emotional stress and psychic factor as fright can inhibit milk let-down; 2. ethanol inhibits endogenous oxytocin release; 3. enkephalins also inhibit oxytocin release.

Pituitary is not a master gland

Earlier pituitary endocrine gland was considered to be a 'Master Gland' because the hormones secreted by pituitary control the secretion of all the other endocrine glands of the body. After the role of hormones (*i.e.*, neurohormones) secreted by the hyhpothalamus has been known, it was discarded to be a master glands. Now it has been established that the releasing hormones (RH) secreted by hypothalamus control the quantity of pituitary secretion and hypothalamus forms a connection between nervous system and endocrine system (*e.g.*, Hypothalamic-pituitary-adrenal or HPA axis). Hypothalamus becomes active by the hyper or hyposecretion of a particular hormone (Fig. 8.30). For, example, ACTH (adrenocorticotropic hormone) or corticotropin of adenohypophysis is under stimulatory control by the hypothalamus. A subset of **parvicellular** hypothalamic neurons (*parvicellular* means neurons characterized by relatively small cell bodies) expresses the peptide **procorticotropin-releasing hormone (pro-CRH)**. Pro-CRH is processed to an amidated 41-amino acid peptide, CRH. CRH actually stimulates ACTH secretion and increases transcription of the POMC (= proopiomelanocortin) gene. The parvicellular neurons that express CRH also coexpress ADH, and ADH potentiates the action of CRH on Corticotropes (adrenal cortex).

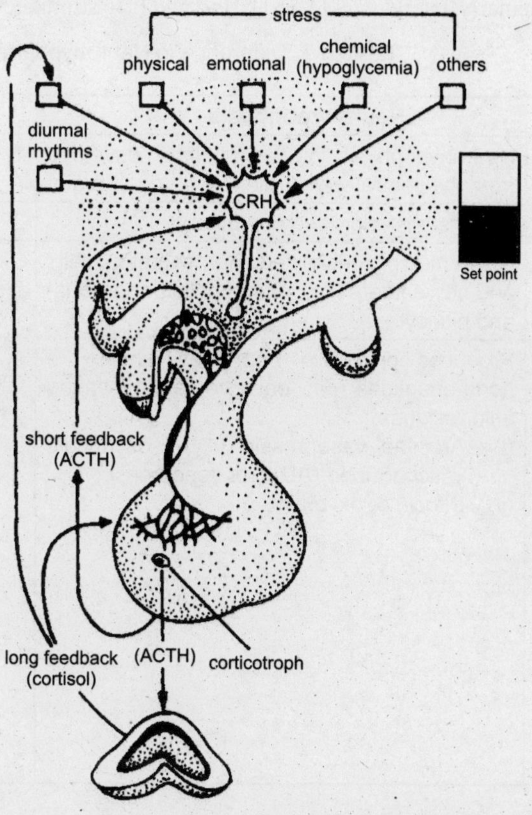

Fig. 8.30. Hypothalamic-pituitary-adrenal (HPA) axis illustrating factors regulating the secretion of corticotropin-releasing hormone (CRH). ACTH, adrenocortico-tropic hormone (after Koeppen and Stanton, 2010).

ACTH secretion has a pronounced diurnal pattern, with a peak in early morning and a valley in late afternoon. In addition, secretion of CRH-and hence secretion of ACTH-is pulsatile.

There are multiple regulators of the HPA axis, and many of them are mediated through the CNS. Many types of stress, both **neurogenic** (*e.g.*, fear) and **systemic** (*e.g.*, infection), stimulate secretion of ACTH. The stress effects are mediated through CRH and ADH and the CNS. The response to many forms of severe stress can persist despite negative feedback from high cortisol levels. This means that the hypothalamus has the ability to reset the "set point" of the HPA axis in response to stress. Severe, chronic depression can reset the HPA axis as a result of hypersecretion of CRH and is a factor in the development of tertiary **hypercortisolism**. Cortisol exerts negative feedback on the pituitary, where it suppresses POMC gene expression and ACTH secretion and on the hypothalamus, where it decreases pro-CRH gene expression and release of CRH. Because cortisol has profound effects on the immune system, the HPA axis and the immune system are closely coupled. Moreover, cytokines, particularly, interleukin-1 (IL-1), IL-2 and IL-6 stimulate the HPA axis (**Koeppen** and **Stanton**, 2010).

Table 8.7. Difference in Neurohypophysis and Adenohypophysis.

	Neurohypophysis		Adenohypophysis
1.	Neurohypophysis of the pituitary develops from the ventral wall of diencephalon of brain.	1.	Adenohypophysis of the pituitary arises from the roof of embryonic pharynx.
2.	It forms posterior lobe of pituitary.	2.	It forms anterior lobe of pituitary.
3.	It is formed of specialized neurosecretory cells and nerve fibers (*i.e.*, magnocellular neurons) and pituicytes.	3.	It is formed of acidophil, basophil and chromophob cells.
4.	Produces only two hormones which are nona-apeptides (*i.e.*, each contains 9 amino acid residues): (i) Arginine vasopressin (AVP) or antidiu -retichormone (ADH) or vasopressin (ii) Pitocin or oxytocin	4.	Adendhypophysis produces six hormones, namely (i) STH (somatotropin) or GH (growth hormone) (ii) PRL (prolactin) (iii) ACTH (adrenocorticotropic hormone) (iv) TSH (thyroid stimulating hormone) (v) ICSH or LH (interstitial cells stimulating hormone or luteinizing hormone). (vi) FSH (follicle stimulating hormone) **(Note. MSH** or **melanocyte stimulating hormone)** is produced by intermediate lobe of pituitary.

C. Intermediate Lobe of Pituitary

It is rudimentary in the pituitary of humans. In animals it secrete **melanocyte stimulating hormone** (**MSH**) which increases the synthesis of melanin in melanocytes causing dark skin. In lower chordates MSH indues spreading of melanin in pigment cells and makes integument lighter in colour.

In humans, cells in anterior and intermediate lobe of pituitary contain a large protein known as **pro-opiomelanocortin** which splits off into **ACTH**, **α-MSH**, **β-MSH** and **α-endorphin**. In humans contraction and expansion of melanocytes does not occur. MSH controls melanin production. **Albinos** are individuals with congenital deficiency of melanin pigment in skin, hair and eyes.

8.5 THYROID GLAND

The word thyroid is derived from Greek word '*thyroid*' means 'a shield'. Normal thyroid gland of the adult is the largest endocrine gland, weighing 15–25 grams and is highly vascular organ. It is situated in front of the larynx, has two lobes—right and left which are joined by thyroid isthmus (Fig. 8.31). Thyroid is made up of multiple follicles, which are spheres of the epithelial cells (cuboidal) surrounding a semifluid material called **colloid** (Fig. 8.32).

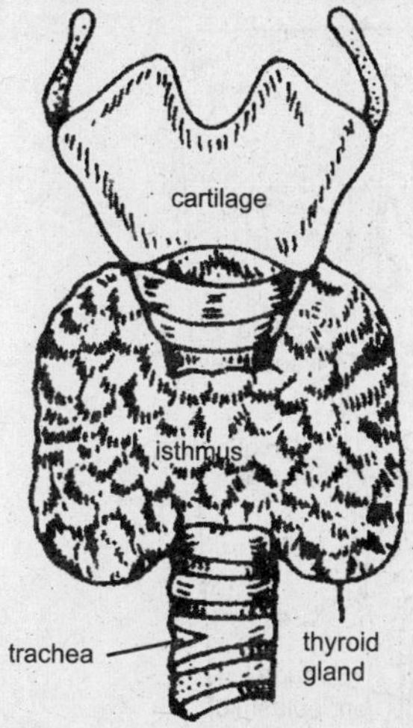

Fig. 8.31. Human thyroid gland seen from front.

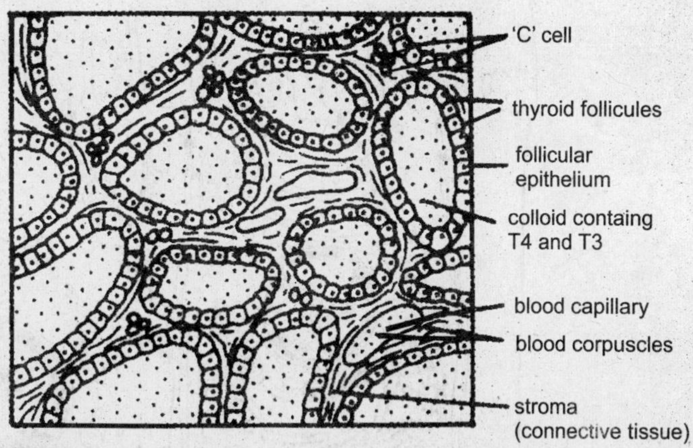

Fig. 8.32. T.S of thyroid gland.

The follicular cells under the influence of TSH secrete two thyroid hormones—**thyroxine** or **tetraiodothyronine (T_4)** and **tri-iodothyronine (T_3)** which are formed by iodination of *tyrosine* residues of thyroglobulin, a large glycoprotein present in colloid.

A few cells known as **parafollicular cells** or **C cells** lie embedded between the follicles and they produce hormone, **calcitonin** or **thyrocalcitonin**.

I. Synthesis and Secretion of Thyroid Hormones

Thyroid is only endocrine gland which synthesizes and stores large amount of thyroid hormone (Fig. 8.33). The steps of hormone synthesis are as follows:

1. Iodide trapping by follicular cells. Iodide from blood is actively transported into follicular cells by Na^+/I-symports. A minimum of 60 mg of elemental iodine is required daily for synthesis of thyroid hormones.

2. Oxidation of iodides. Negatively charged I^- cannot bind to tyrosine residues of thyroglobulin. These anions undergo oxidation to iodine ($2I^- \rightarrow I_2$) by the help of enzyme **peroxidase**. Iodine is then transported to colloid.

3. Iodination of tyrosine. Iodine attaches to tyrosine residues of thyroglobulin molecules in the colloid and **monoiodotyrosine (MIT)** and **diiodotyrosine (DIT)** are formed. This step is also called **organification of iodine**.

4. Coupling of MIT or DIT. Two DIT molecules join together to form T_4 while one DIT and one MIT form T_3. This coupling is catalysed by peroxidase enzyme.

5. Secretion and release of T_3 and T_4. T_3 and T_4 bound to thyroglobulin are taken up by follicular cells by the process of endocytosis, the endocytic vesicle fuses with lysosomes, the proteolytic enzymes act on the thyroglobulin to release free T_3, T_4 which are secreted into the plasma. The simultaneously released MIT, DIT are either recycled or metabolized.

6. Transportation of T_3, T_4 in plasma. Thyroid hormones circulate in blood bound to plasma proteins—thyroxine binding globulin (75%), thyroxine binding prealbumin (15–20%) and albumin (12–15%). More than 99% of thyroid hormones remain protein bound, 0.5% is free T_3, 0.05% is free T_4. It is free T_3 which enters the cell to produce action. The extensive protein binding of hormone acts as temporary reservoir which minimize fluctuations in levels of free T_3 and T_4.

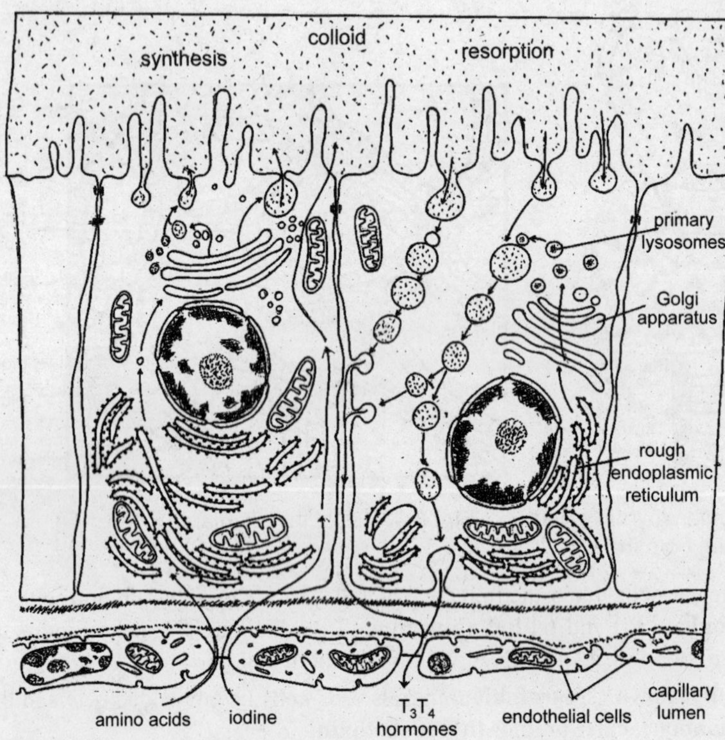

Fig. 8.33. Diagram showing structural-functional relationships in thyroid follicular cells (thyrocytes). *Arrows* indicate the general metabolic pathways. The cell under '**synthesis**' shows the pathways of *synthesis*, *glycosylation*, *secretion* and *iodination* of the thyroglobulin that is stored in colloid. Iodide is absorbed at the basal surface of the cell, oxidized in the apical portion of the cell, and secreted into the colloid, where it iodinate tyrosine residues (amino acids) in the thyroglobulin. The cell under '**resorption**' shows the pathways of colloid resorption by *endocytosis* and digestion by lysosomal enzymes that liberate iodinated tyrosine dimers [T_3 or triiodothyronine and T_4 or tetraiodothyronine or thyroxine] from the thyroglobulin. These thyroid hormones are released into the extracellular space at the lateral and basal cell surfaces and diffuse into the underlying capillary.

7. Peripheral conversion of T_4 to T_3 and metabolism of thyroid hormones. The active form of thyroid hormone is T_3 which binds to receptor to produce action. Hence T_4 is converted to T_3 by the enzyme **5′-deiodinase** enzyme type I and II (DID I, DID II). Most conversion of T_4 to T_3 by **type 1 deiodinase** occurs in tissues with high blood flow and rapid exchanges with plasma, such as the liver, kidneys and skeletal muscle. The brain maintains constant intracellular levels of T_3 by a high-affinity deiodinase called **type 2 deiodinase** (**DID-II**) that is expressed in glial cells in the central nervous system. There is also an inactivating deiodinase, called **DID-III** (**type 3 deiodinase**). DID-III converts T_4 to inactive or **reverse T_3** (**rT_3**). Ultimately T_3, r T_3, etc., are deiodinated, deaminated and decarboxylated to thyronine which is excreted from the body.

II. Regulation of Thyroid Hormone Secretion

***(a)* TSH.** Synthesis of T_3 and T_4 is regulated by TSH released from anterior pituitary which itself is controlled by TRH from hypothalamus. When circulating levels of free T3 and T4 are high, there is negative feedback **inhibition** by TRH and TSH release from hypothalamus and anterior pituitary respectively. Other factors which inhibit TSH release are somatostatin, dopamine from hypothalamus. Acute psychosis, extreme cold weather **stimulate** TRH release.

***(b)* Thyroid autoregulation.** Iodine in small doses is necessary for normal thyroid functions while in high doses, it decreases the thyroid gland functions. The latter effect is called **Wolff-Chaikoff effect** (Box 8.4) and is brought by: (*i*) prevention of oxidation of iodide to iodine; (*ii*) prevention of incorporation of iodine to tyrosine; (*iii*) decreased proteolysis of thyroglobulin; (*iv*) decreased blood flow to thyroid gland.

Box 8.4

If the intake of iodide exceeds 2 mg/day, the intraglandular concentration of iodide reaches a level that suppresses NADPH oxidase activity and the NIS and TPO genes and thereby the mechanism of hormone biosynthesis. This autoregulate phenomenon is called **Wolff-Chaikoff effect**.

III. Actions of Thyroid Hormone

1. Calorigenic effects. Thyroid hormones stimulate heat production in body, increases O_2 consumption, increases basal metabolic rate (BMR). They help in maintaining normal body temperature.

2. Effect on metabolism. Thyroid hormones increase cellular metabolism of carbohydrates, proteins and lipids. They stimulate protein synthesis and increase utilization of glucose and fatty acids for ATP production leading to enhanced cholesterol excretion.

3. Cardiovascular system. T_3 and T_4 (hormones) upregulate beta receptors in heart leading to increased sensitivity of heart to catecholamines. This leads to increase in heart rate, increase in force of myocardial contraction.

4. Growth and development. Thyroid hormones along with growth hormone and insulin accelerate body growth mainly the development and growth of central nervous system and skeletal system.

5. Mineral metabolism. Hormone thyrocalcitonin (TCT) secreted by parafollicular cells (C-cells) controls amount of calcium and phosphorus ions in the body fluid. It lowers blood calcium but increases calcification of bones. Its **hypersecretion** causes excretion of these ions in urine. It may lead to **diureses** (an increased excretion of urine) with the loss of potassium in urine.

IV. Thyroid Disorders

1. Goiter. It is enlargement (hypertrophy) of thyroid gland and substances that cause this are called **goiterogens**. In hypertrophy of thyroid, the thyroid gland increases in size due to increase in number of follicles. It causes enlargement of neck. It is described as **simple goiter**, **endemic goitre** or **colloidal goiter**. Hilly regions of India form goiter-belt because there is deficiencies of iodine in water.

Goiterogens agents include the following: (*i*) iodine deficiency (daily intake of iodine falls below 50 μg); (*ii*) excess iodine; (*iii*) monovalent ions such as perchlorate, thiocyanates; (*iv*) vegetables such as cabbages and turnip.

2. Hashimoto's disease. It is chronic thyroiditis (inflammation of the thyroid gland) characterized by goiter, thyroid fibrosis, infiltration of thyroid tissue by lymphoid tissue and the production of autoantibodies that attacks the thyroid.

Hashimoto's disease is a special condition, when thyroid secretion is too less. The medicines given to increase thyroxine secreation act as antigens and body produces antibodies to destroy them.

These antibodies destroy thyroid gland also. It is described as **suicide of thyroid** or anti-immune disease.

The hypoactivity of thyroid gland can be controlled by increasing iodine in food by using iodised salt.

3. Hypothyroidism. It is a syndrome resulting from deficiency of thyroid hormone (*i.e.*, hypoactivity of thyroid gland). It appears in following two forms.

(*i*) Cretinism. This type of hypothyroidism occurs in children. It is characterized by mental retardation, dwarfism and stunted growth, pot belly, failure of sexual development. It results due to iodine deficiency or due to presence of ectopic thyroid or presence of TSH receptor-blocking antibodies.

(*ii*) Myxodema. Hypothyroidism during the adult years produces myxodema which is characterized by following symptoms: puffiness of face with periorbital swelling, low BMR, deep hoarse voice, decreased heart rate, decreased cardiac output, constipation, low body temperature, sensitivity to cold, dry hair, deposition of myxomatous tissue under the skin (*Myxomatous tissue* consists of semifluid substance rich in protein and mucopolysaccharides), muscular weakness, lethargy, loss of memory, loss of libido, infertility and weight gain.

Hypothyroidism is treated by giving thyroid hormone preparations, commonly T_4 hormone life long.

4. Hyperthyroidism. It is the condition resulting from increased circulating levels of free T_3 and T_4 (*i.e.*, hyperactivity of thyroid gland). The commonest cause of this disease is **Grave's disease**, an autoimmune disorder in which there is production of thyroid stimulating immunoglobulins or long acting thyroid stimulator (LATS). These antibodies mimic the action of TSH on thyroid gland but are not regulated by the normal negative feedback control through T_3, T_4. The ultimate result is continuous production of thyroid hormones.

Clinical features of hyperthyroidism are diffuse goiter, protrusion of eyes (**exophthalmos** or **exophthalmia**), increased BMR, heat intolerance, weight loss, increase in heart rate, menstrual irregularities, warm moist skin, hyperglycemia, skeletal muscle weakness, fine tremors, diarrhoea, insomnia, nervousness, etc. **Treatment** of hyperthyroidism is by intake of anti-thyroid drugs.

8.6 PARATHYROID GLAND AND BONE MINERAL HOMEOSTASIS

Parathyroid hormone or parathormone (PTH), calcitonin and calcitriol (vitamin D) are principal hormone regulators of bone mineral homeostasis. Calcium (Ca^{2+}) and phosphate (PO_4^{3-}) are main minerals found in bones.

Normal serum calcium levels lie between 9–11 mg% and is regulated by parathyroid hormone and calcitonin.

A. Parathyroid Hormone (Parathormone or PTH)

It is secreted from the four parathyroid glands located posterior of the thyroid gland (Fig. 8.34). Parathyroid glands consist of densely packed masses of cords of **epithelial cells** with interspersed blood vessels and reticular fibers. These cells are of two types: 1. **Chief cells or principal cells.** These are (*i*) **light coloured polygonal cells** which have nongranular cytoplasm rich in glycogen but poor in secretory granules, and (*ii*) **dark chief cells** which are poor in glycogen but rich in secretory cells. These cells secrete parathyroid hormone called **parathormone (PTH)** or **Collip's hormone** (discovered by **Philips Collips** in 1925). 2. **Oxyphil cells** or **eosinophil cells**. These are large polyhedral nonsecretory cells with acidophilic cytoplasm. In human beings these cells appear around puberty and increase with age.

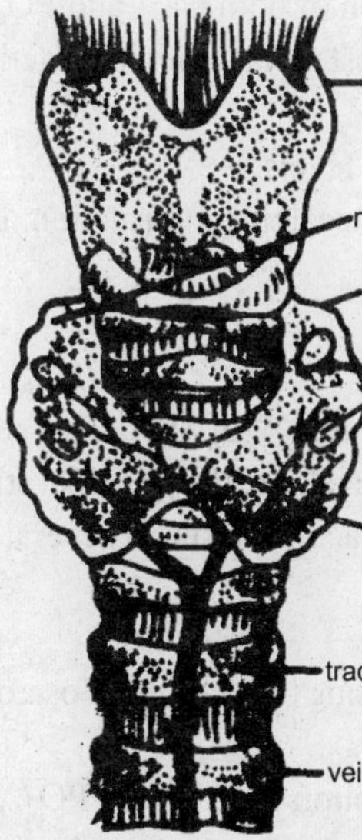

Fig. 8.34. Human parathyroids.

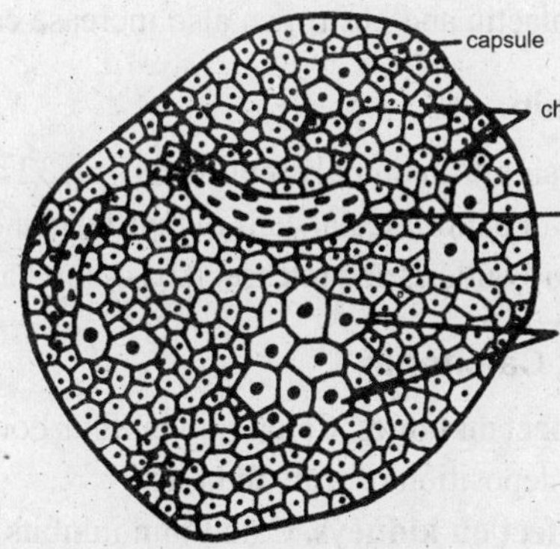

Fig. 8.35. T.S. of parathyroid gland.

PTH is a polypeptide having 84 amino acids, MW 9500, half life 2–5 min. **Hypocalcemia** (low serum Ca^{2+} levels) is the principal factor which regulates PTH synthesis and release.

Actions of PTH. PTH increases serum Ca^{2+} and decreases serum PO_4^{3+}.

1. Effect on bones. PTH has both bone resorption and formation effects but net effect of excess PTH is increase in bone resorption. In low to intermittent (discontinuous) levels, PTH increases bone formation without stimulating bone resorption. That is why recombinant PTH is used for the treatment of osteoporosis (a disease of bone characterized by defective architecture and mineralization). In case of prolonged hypocalcemia, PTH stimulates Ca^{2+} release from the bones by increasing the activity and number of bone resorbing cells, **osteoclasts**.

2. Effect of kidneys. PTH increases renal tubular absorption of Ca^{2+} and excretion of PO_4^{3-}. It also stimulates the conversion of 25 (OH) D_3 (25-hydroxyvitamin D which is also referred to as **calcifediol**) to 1,25 $(OH)_2D_3$ (1,25-dihydroxyvitamin D or calcitriol).

3. Effect on intestine. PTH indirectly increases Ca^{2+} and PO_4^{3-} absorption from small intestine by inducing formation of calcitriol.

4. Effect on lactating mammary glands. PTH decreases Ca^{2+} levels in breast milk.

B. Vitamin D (Calcitriol)

It includes two related substances-cholecalciferol (vitamin D_3) and ergocalciferol (vitamin D_2). **Cholecalciferol** is produced in skin from 7-dehydrocholesterol by the action of ultraviolet light from the sun rays. Ergocalciferol is obtained from vegetable source.

In the body, vitamin D is converted in liver and kidney to active form-1, 25 dihydroxy-cholecalciferol, *i.e.*, 1, 25 $(OH)_2D_3$ or calcitriol.

Actions of Vitamin D

(*i*) Effect on intestine. Calcitriol act on intestine to stimulate absorption of Ca^{2+} and PO_4^{3-}.

***(ii)* Effect on kidneys.** It enhances proximal tubular reabsorption of both Ca^{2+} and PO_4^{3-}.

***(iii)* Effect on bones.** Calcitriol acts on osteoblasts and increases osteoblast-mediated activation of osteoblasts.

Thus in contrast to PTH, vitamin D raises serum Ca^{2+} and PO_4^{3-} levels.

Regulation of vitamin D levels. Low serum Ca^{2+}, PO_4^{3-} stimulate calcitriol formation. During lactation prolactin and oestrogen also increase calcitriol synthesis.

C. Calcitonin

It is a single chain polypeptide having 32 amino acid residues and MW of 3600. It is secreted from C-cells or parafollicular cells of thyroid gland. Rising serum Ca^{2+} levels are the principal stimulus for calcitonin synthesis and release. It has short half life of 10 min but its actions lasts for several hours.

Actions of Calcitonin

***(a)* Effect on bones.** Calcitonin inhibits bone resorption by inhibiting the activity of osteoclasts. It promotes deposition of Ca^{2+} in bones.

***(b)* Effect on kidneys.** Calcitonin inhibits renal tubular reabsorption of Ca^{2+} and PO_4^{3-}.

***(c)* Effect on intestines.** Calcitonin inhibits absorption of Ca^{2+} and PO_4^{3-} through intestine.

D. Other Hormones Regulating Calcium Homeostasis

***(i)* Glucocorticoids.** Decrease absorption of Ca^{2+}, PO_4^{3-} from intestines and increase their renal excretion by inhibiting the action of vitamin D.

High levels of glucocorticoids can lead to osteoporosis in adults and growth retardation in children.

***(ii)* Estrogen.** Reduce the bone resorbing action of PTH, increases levels of calcitriol in the blood and prevent bone resorption by causing apoptosis of osteoclasts. Thus estrogens prevent osteoporosis.

***(iii)* Androgens.** Reduce Ca^{2+} excretion.

***(iv)* Insulin.** increases bone formation.

Disorders of Hormones Regulating Calcium Levels

1. Hypoparathyroidism. The commonest cause is accidental removal or damage to the parathyroid gland during thyroidectomy. Result of this operation is decrease in serum calcium levels to 4–8 mg/dl and ionised Ca^{2+} to 3 mg/dl leading to **tetany**. Tetany is characterized by tingling and numbness in the extremities, muscle cramps, facial nerve irritability, intestinal colic and ECG changes.

2. Hyperparathyroidism. It occurs due to diffuse hyperplasia (An abnormal increase in the elements/cells composing a part) of parathyroid glands or tumor of the gland. There is rise in serum calcium levels which results in weakness, malaise, thirst, polyuria, anorexia, nausea, vomiting, kidney stones, pain in bones due to demineralization.

8.7 ADRENAL GLANDS

The two adrenal glands are situated on top of kidneys and each consists of two distinct parts—the outer **adrenal cortex** and the inner **adrenal medulla**. Adrenal medulla is part of sympathetic system and secretes two hormone: **epinephrine** (adrenaline) and **nor-epinephrine** (nor-adrenaline).

1. Adrenal Cortex

Adrenal cortex is divided into three layers (Fig. 8.36):

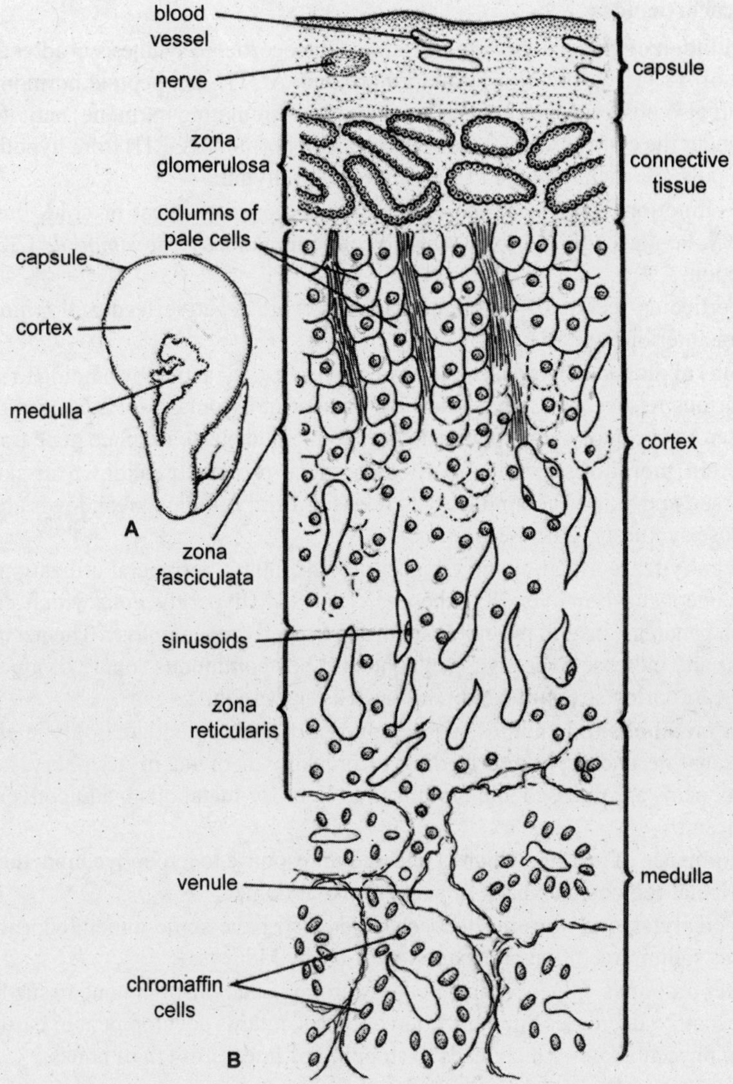

Fig. 8.36. T.S. adrenal gland. A—T.S. adrenal gland under low magnification; B—T.S. adrenal gland under high magnification.

1. Outermost is zone glomerulosa which secretes **mineralocorticoid**, *i.e.*, **aldosterone**.
2. Middle layer is zona fasciculata which secretes **glucocorticoid**, *i.e.*, **cortisol** or **hydrocortisone**.
3. Innermost layer is zona reticularis which secretes **sex steroids**.

Cholesterol is an obligatory building block in the biosynthesis of hormones of adrenal cortex. Cholesterol is transported into mitochondria where it is converted to **pregnenolone** by the enzyme **CYPIIAI**, located in inner mitochondria membrane. This step is the rate limiting step in the steroid biosynthesis.

Steroid transport in blood. Glucocorticoids are transported in systemic circulation bound to globulin, transcortin or corticosteroid binding globulin (80%) and to albumin (10%); remaining 10% circulates in free form.

Aldosterone is present in plasma in very small concentration, 60% bound to albumin and some to corticosteroid binding globulin (CBG).

1. Glucocorticoids

(*i*) Regulation of glucocorticoids secretion. Glucocorticoid synthesis in adrenal cortex is under the influence of ACTH, an anterior pituitary hormone. ACTH is a peptide hormone containing 39 amino acids. It also possesses some MSH (melanocyte stimulating hormone) activity. The secretion of ACTH is under the control of corticotropin releasing hormone (CRH) from hypothalamus. ACTH is secreted in pulsatile manner which follows circadian rhythm.

Stress—Emotional, physical (pain, exercise, exposure to heat or cold, burns), biological (surgery, injury, hemorrhage, hypoxia, hypoglycemia injections) also *stimulate* CRH → ACTH → cortisol secretion.

Glucocorticoids secreted from adrenal glands exert negative feedback *inhibition* on ACTH secretion from anterior pituitary.

2. Actions of glucocorticoids. Glucocorticoids exert their action by binding to steroid receptors present on various tissues. These receptors are present in cytoplasm of cells and on binding with steroid, the steroid-receptor complex translocates to the nucleus to regulate gene transcription.

(*i*) Protein metabolism. Glucocorticoids cause protein breakdown in skeletal muscles, connective tissue and skin. The amino acids released from proteins breakdown are transported to liver for glucose synthesis (gluconeogenesis).

(*ii*) Carbohydrate metabolism. Glucocorticoids inhibit peripheral utilization of glucose and stimulate gluconeogenesis in liver. Both these effects lead to hyperglycemia which stimulates insulin secretion from pancreas. Insulin promotes synthesis from glycogen in liver. The net result is increase in blood sugar and increase liver glycogen content. Thus continuous supply of glucose is provided to brain and heart during stressful conditions such as starvation.

(*iii*) Fat metabolism. In adipose tissue, glucocorticoids potentiate lipolytic effects of growth hormone, adrenaline and thyroxine leading to breakdown of fat in periphery, *i.e.*, extremities. Thus free fatty acids are released and mobilized to liver for metabolism and conversion to glucose (gluconeogenesis).

Under influence of insulin, which is released in response to hyperglycemia, there is deposition of fat in centripetal regions, *i.e.*, back of neck, trunk, abdomen.

(*iv*) Electrolytes and water. Glucocorticoids also have some mineralocorticoid activity by which it causes sodium retention and excretion of K^+ by kidneys.

(*v*) Effect on bones. Glucocorticoid causes protein catabolism in bone tissue leading to break down of bone matrix, decreases mineralization of bones, retards development of cartilage and causes thinning of epiphyseal plates which leads to stopping of linear growth of bones.

These hormones also inhibit Ca^{2+} absorption from intestines by antagonizing actions of vitamin D. Calcium ions (Ca^{2+}) excretion from kidneys is also increased.

(*vi*) Anti-inflammatory, Anti-allergic actions. The endogenous glucocorticoids protect the body from excessive inflammatory response by body's defence mechanisms during stress and infections. Hence glucocorticoids are used as medicines/drugs in treating unwanted inflammation and immune response.

Glucocorticoids inhibit inflammation occurring due to any cause affect both *early components*, *i.e.*, swelling, pain, fibrin deposition, hyperemia (*i.e.*, excess of blood) as well as *late components* such as wound healing, collagen deposition. They act by inhibiting the enzyme **phospholipase A_2** which synthesizes arachidonic acid from phospholipids. **Ararchidonic acid** is precursor of various mediators of inflammation such as prostaglandins, leukotrines and platlet activating factor. Glucocorticoids also block production and release of cytokines –IL-1, IL-6, IL-4, TNF-α which results in reduction of activation of T cells.

They also inhibit synthesis of adhesion molecules in endothelial cells thus impairing chemotaxis of neutrophils and other leukocytes to the site of inflammation.

Glucocorticoids decrease release of histamine and LTC_4 from mast cells and basophils during IgE (immunoglobulin E) -mediated type I hypersensitivity reactions.

(*vii*) Effect on immune response. Glucocorticoids suppress both cell-mediated and humoral immunity. Thus they are important drugs to treat autoimmune disorders such as rheumatoid arthritis, myasthenia gravis and to prevent rejection of organ transplant.

(*viii*) Effect on blood cells and lymphatic organs. Glucocorticoids increase the number of RBCs, platelets and neutrophils in circulation while they decrease the levels of eosinophils, lymphocytes and basophils.

(*ix*) Effect on GIT. Glucocorticoids increase gastric acid and pepsin secretion in stomach along with decreased gastric mucosal cells proliferation. They also suppress immune response against *Helicobacter pylori*, a bacteria implicated in causing peptic ulceration. This predisposes to peptic-ulcer formation.

(*x*) Effect on cardiovascular system. Glucocorticoids directly stimulate cardiac output and also potentiate the response of vascular smooth muscles to pressor (tending to raise blood pressure) effects of catecholamines (epinephrine, nor-epinephrine) and angiotensin II.

(*xi*) Miscellaneous effect of glucocorticoids. (*a*) Glucocorticoids are essential for maintaining normal glomerulus filtration in kidneys. (*b*) They are required for normal maturation of fetal lungs. (*c*) They decrease secretion of TSH and growth hormone.

Disorders due to Glucocorticoids

(*a*) Cushing's syndrome

It is a clinical disorder resulting from high circulating levels of glucocorticoids.

Causes of this syndrome are following:

1. Excess glucocorticoids with low ACTH levels:
   (*a*) Prolonged treatment with high doses of glucocorticoids.
   (*b*) Adrenal cortex tumours.
2. Excess glucocorticoids with increased ACTH levels:
   (*a*) Ectopic ACTH production by non-pituitary tumours, *e.g.*, cancer of the lungs.
   (*b*) ACTH-secreting pituitary tumour.

Clinical features of Cushing's syndrome

1. Protein metabolism. Retardation of growth, thinning of skin and connective tissue, muscle weakness and wasting, poor wound healing, hair become thin and rough.

2. Carbohydrate metabolism. Hyperglycemia, excretion of glucose in urine (glucosuria), precipitation of diabetes mellitus.

3. Fat metabolism. Thin extremities, deposition of fat in abdomen, back of neck (buffalo hump), on face (*i.e.*, moon like face).

4. Water and electrolytes. Oedema, hypertension.

5. Bone. Osteoporosis, increased risk of fractures.

6. Blood. *Eosinopenia* (*i.e.*, abnormal increase in number of eosinophils in the blood that is characteristic of allergic state and various parasitic infections.), *lymphopenia* (*i.e.*, reduction in the number of lymphocytes in the blood), *neutrophilia* (*i.e.*, increase in white blood cell is chiefly in neutrophils.), *polycythemia* (*i.e.*, an abnormal increase in the number of circulating red blood cells.), increased platelet count and decreased clotting time.

7. GIT. Peptic ulceration.

8. CNS. Euphoria, insomnia, anxiety, frank (clinically evident) psychosis.

9. Increased susceptibility to infections.

(*b*) Addison's Disease

It is chronic form of adrenal cortex insufficiency resulting in reduced secretion of glucocorticoids and mineralocorticoids.

A. Causes of Addison's Disease

1. Auto-immune disorders; 2. Tuberculosis; 3. Carcinoma; 4. Hemorrhage in adrenal cortex; 5. Secondary to severe infection (Water-house Friderichsen syndrome, *i.e.*, acute and severe meaningococcemia with hemorrhage into the adrenal glands).

B. Clinical Features of Addison's Disease

Hypotension, anorexia, nausea, vomiting, diarrhoea, muscular weakness, mental confusion, increased ACTH secretion which leads to generalized pigmentation of skin, mucus membrane, gums, pressure points (**Pressure point.** It is a region of the body in which the distribution of soft and skeletal parts is such that a static position (as of a part in cast or of a bed ridden person) tends to cause circulatory deficiency and necrosis due to local compression of blood vessels), decreased ability to withstand stressful conditions.

C. Acute Addisonian Crisis

This is acute form of adrenal cortex insufficiency which occurs after surgical removal of adrenal gland or sudden withdrawl of glucocorticoids after being given in high doses for long period (due to suppression of endogenous hypothalamo-pituitary-adrenal axis). It is characterized by severe hypotension, low Na^+ levels (hyponatremia), increased serum K^+ levels (hyperkalemia), extreme weakness, dehydration. This disorder can also be precipitated by exposure of patient of Addison's disease to sudden stress, infection, trauma or surgery.

Secondary and tertiary adrenocortical insufficiency occurs due to deficient secretion of ACTH (secondary insufficiency) or CRH (tertiary) with resultant decrease in glucocorticoid secretion. In this case too features of Addison's disease are present.

2. Mineralocorticoids

Aldosterone is the main mineralocorticoid secreted in body. The secretion is under the influence of renin-angiotensin system.

Renin-angiotensin system. Renin is a protein synthesized and released from juxtaglomerular cells of the kidney. Its release is controlled by following three mechanisms: (*i*) decrease in renal perfusion; (*ii*) decrease Na^+ concentration in distal tubules; (*iii*) stimulation of B_1 adrenergic receptors on the juxtaglomerular cells by norepinephrine (due to sympathetic system stimulation and PGE_2 (*i.e.*, prostaglandin E_2).

Renin acts on angiotensinogen, a circulating α_2 globulin synthesized by liver. Levels of angiotensinogen is increased during pregnancy and by glucocorticoids, estrogen and thyroid hormones.

Degradation of angiotensinogen by renin produces 10-amino acid peptide, **angiotensin I (Ang-I)** which is converted to 8-amino acid peptide, **angiotensin II (Ang-II)** by an angiotensin converting enzyme (ACE or kininase II) (Fig. 8.37). Angiotensin II acts on the receptor to cause: (*a*) generalized vasoconstriction; (*b*) stimulation of sympathetic nervous system; (*c*) secretion of aldosterone from the adrenal cortex.

Actions of mineralocorticoids. Aldosterone binds to mineralocorticoid receptor which is same as glucocorticoid receptor. It causes retention of sodium and excretion of potassium by acting on distal convoluted and cortical collecting tubules in kidneys. It also increases absorption of Na^+ from GIT, salivary and sweat glands.

(*a*) Primary hyperaldosteronism or Conn's syndrome. It is a condition in which there is prolonged, excessive secretion of aldosterone from the adrenal cortex. It is characterized by: (1) Rise in plasma sodium levels and marked fall in potassium levels. (2) Hypertension due to retention of sodium and water. (3) Prolonged low potassium levels lead to marked muscle weakness, kidney damage, metabolic alkalosis which may precipitate tetany.

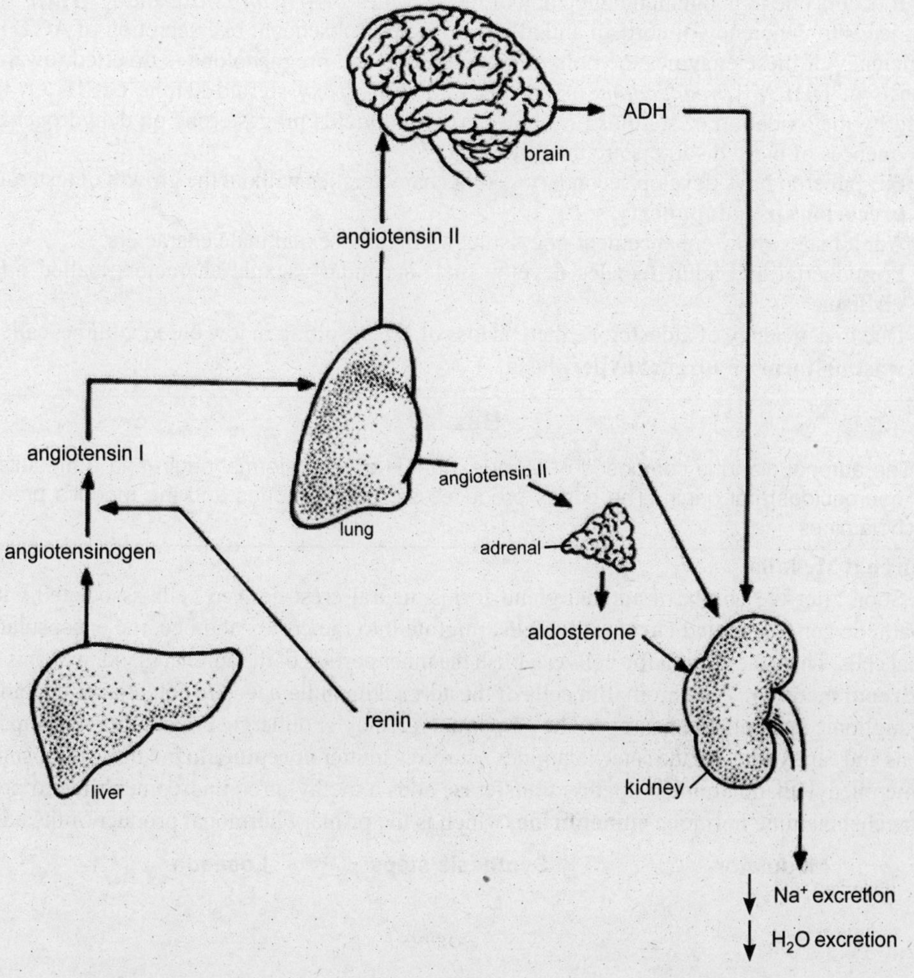

Fig. 8.37. Schematic representation of the essential components of the renin-angiotensin aldosterone system. Activation of this system results in a decrease in the excretion of Na^+ and water by the kidneys. *Note*: Angiotensin I is converted to angiotensin II by ACE (an enzyme which present on vascular endothelial cells of lung and kidney) (after Koeppen and Stanton, 2010).

(*b*) **Secondary hyperaldosteronism.** In this there is oversecretion of aldosterone due to extra-adrenal factors such as congestive heart failure, nephrosis, cirrhosis of liver and toxemia of pregnancy. It is characterized by hypertension and oedema due to Na+ and water retention. There is no hypokalemia as urinary excretion of K^+ is not increased due to reduced flow of fluid to distal tubules of kidney which reduces K^+ secretion.

3. Sex Steroids

During fetal life, adrenals are hyperplastic (an abnormal increase in cells in the glands) and produce **dehydroepiandrosterone (DHEA)** which serves as main precursor of oestrogen formation by the placenta. After birth secretion of sex hormone by adrenal cortex decreases and increases at the time of puberty. It contributes to increase in muscle mass, sexual hair and **seborrhea** (*i.e.*, abnormal increased secretion and discharge of sebum producing an oily appearance of the skin and the formation of greasy scales). Its secretion is controlled by ACTH.

Adrenogenital Syndrome

It occurs due to congenital deficiency of the enzymes *21-β hydroxylase* and *11-β hydroxylase* which leads to deficiency of cortisol and aldosterone and subsequent oversecretion of ACTH. Due to deficiency of these enzymes the bulk of progesterone and pregnenolone is diverted towards the synthesis of DHEA. [*Pregnenolone* is an unsaturated hydroxy steroid ketone C_2, $H_{32}O_2$ that is formed by the oxidation of steroids (as cholesterol) and yields progesterone on dehydrogenation]. Consequences of these disorders are the following:

1. Pre-pubertal boys develop secondary sexual characterstics without the growth of testes, called **precocious pseudopuberty.**
2. Adult males show enhancement of existing secondary sexual male characters.
3. Pre-pubertal and adult females develop male secondary sexual characters, called **adrenal virilism**.
4. Due to deficiency of aldosterone there is loss of Na^+ resulting in low blood volume, called **salt wasting form of adrenal hyperplasia**.

Box 8.5

The adrenal gland is composed of a cortex that is of mesodermal origin and a medulla that is of neuroectodermal origin. The cortex produces steroid hormones and the medulla produces catecholamines.

2. Adrenal Medulla

Soon after the cortex of adrenal gland forms, neural crest-derived cells associated with the sympathetic ganglia, called **chromaffin cells,** migrate into the cortex and become encapsulated by cortical cells. Thus, the chromaffin cells establish the inner portion of the adrenal gland, which is called the **adrenal medulla.** The chromaffin cells of the adrenal medulla have the potential to develop into postganglionic sympathetic neurons. They are innervated by cholinergic preganglionic sympathetic neurons and can synthesize the catecholamines neurotransmitter **norepinephrine** from thyrosine. The enzyme phenylethanolamine N-methyl transferase adds a methyl group norepinephrine to produce the catechcholamine hormone **epinephrine,** which is the primary hormonal product of the adrenal.

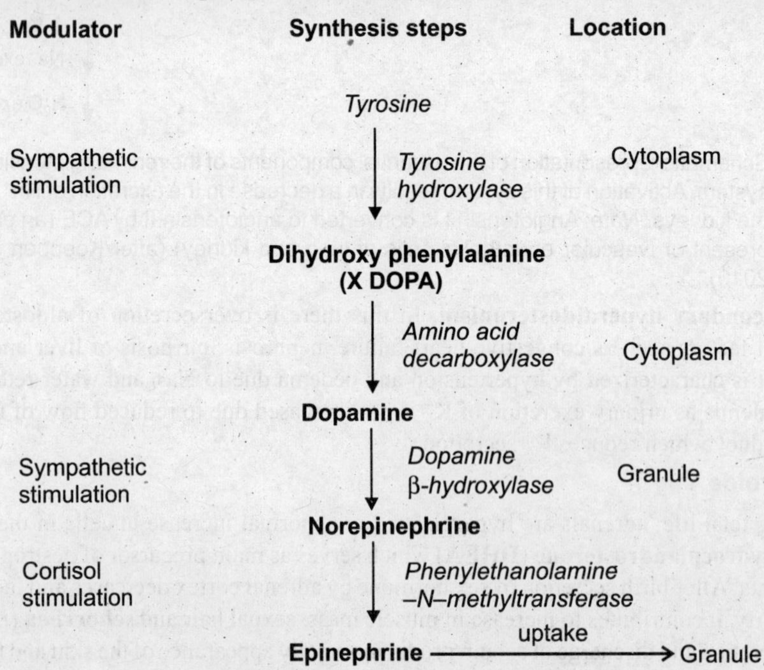

Fig. 8.38. Steps in the synthesis of catecholamines.

About 80% of the cells of the adrenal medulla secrete epinephrine, and remaining 20% secrete norepinephrine. Because the adrenal medulla is not the sole source of catecholamine production, this tissue is not essential for life.

Synthesis of epinephrine. Synthesis of epinephrine begins with transport of the amino acid **tyrosine** into the chromaffin cells cytoplasm and the subsequent hydroxylation of tyrosine by the rate-limiting enzyme **tyrosine hydroxylase** to produce **dihydroxyphenylalanine (DOPA)** (Fig. 8.38). DOPA is converted to **dopamine** by a cytoplasmic enzyme aromatic amino acid decarboxylase and is transported into the secretory vesicle (also called the chromaffin granule). Within the granule, dopamine is completely converted to **norepinephrine** by the enzyme dopamine β–hydroxylase. In most adrenomedullary cells, essentially all of the norepinephrine diffuses out of the chromaffin granule by facilitated transport and is methylated by the cytoplasmic enzyme **phenylethanolamine – N – methyl transferase (PNMT)** to form epinephrine. Epinephrine is then transported back into the granule.

Secretion of epinephrine and norepinephrine from the adrenal medulla is regulated primarily by descending sympathetic signals in response to various forms of stress, including exercise, hypoglycemia, and hemorhagic hypovolemia (Fig. 8.39). The primary autonomic centers that initiate sympathetic responses reside in the hypothalamus and brainstem, and they receive inputs from the cerebral cortex, the limbic system and other regions of the hypothalamus and brainstem.

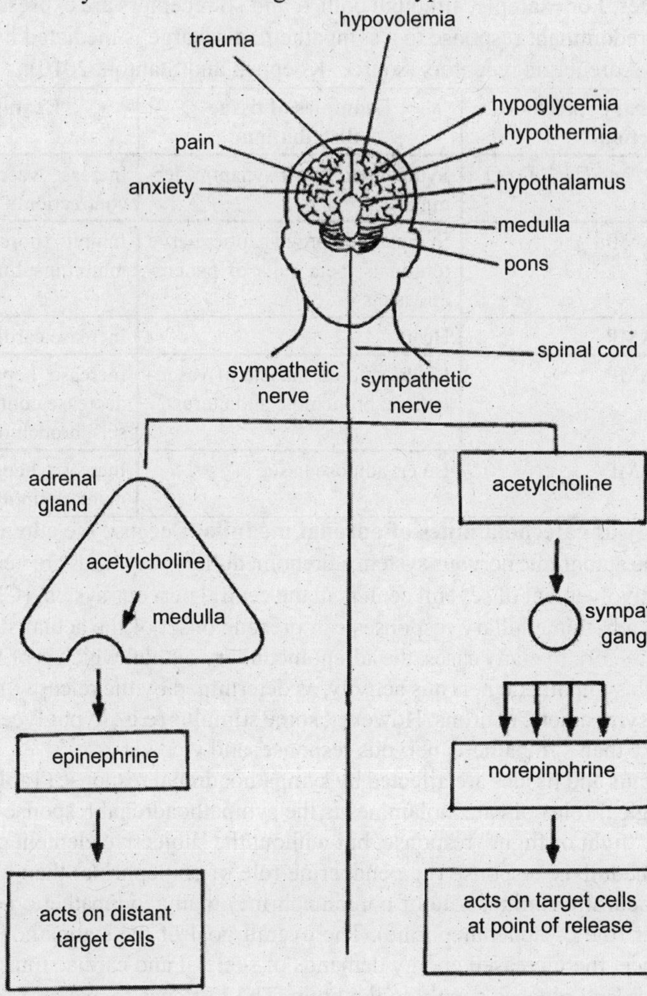

Fig. 8.39. Stimuli that enhance the secretion of catecholamines.

The chemical signal for secretion of catecholamine from the adrenal medulla is **acetylcholine** (Ach), which is secreted from **preganglionic sympathetic neurons** and binds to **nicotonic receptors** on chromaffin cells. Ach increases the activity of the rate-limiting enzyme tyrosine hydroxylase in chromaffin cells (Fig. 8.38). It also increases the activity of the dopamine β -hydroxylase and stimulates exocytosis of the chromaffin granules. Synthesis of epinephrine and norepinephrine is closely coupled to secretion so that levels of intra-cellular catecholamines do not change significantly, even in the face of changing sympathetic activity.

Mechanism of action of catecholamines. Adrenergic receptors are generally classified as α - and β - **adrenergic receptors,** with the α - adrenergic receptors further divided into α_1 and α_2 receptors and β - adrenergic receptors divided into β_1, β_2 and β_3 receptors (Table 8.8). These receptors can be characterized according to

1. *Relative potency of endogenous and pharmacological agonists andantagonists*. Epinephrine and norepinephrine are potent agonists for α receptors and β_1 and β_3 receptors, whereas epinephrine is more potent than norepinephrine for β_2 receptors.

2. *Downstream signaling pathways*. Table 8.8 shows the primary pathways that are coupled to the different adrenergic receptors.

3. *Location and relative density of receptors*. Importantly, different receptor types predominate in different tissues. For example, although both α and β receptors are expressed by pancreatic islet beta cells, the predominant response to a sympathetic discharge is mediated by α_2 receptors.

Table 8.8 Adrenergic receptors (source: Koeppen and Stanton, 2010).,

Receptor type	Primary mechanism of action	Examples of tissue distribution	Examples of action
1. α_1	↑ IP_3 and Ca^{2+}, DAG	Sympathetic post-synaptic terminals	Increase vascular smooth muscle contraction
2. α_2	↓ cAMP	Sympathetic pre-synaptic, nerve terminals; beta cells of pancreatic islets	Inhibit norepinephrine release; inhibit insulin release
3. β_1	↑ cAMP	Heart	Increase cardiac output
4. β_2	↑ cAMP	Liver; smooth muscle of vasculatuare, bronchiole and uterus	Increase hepatic glucose output; decrease contraction of blood vessels, bronchioles, and uterus
5. β_3	↑ cAMP	Liver; adipose tissue	Increase hepatic glucose output; increase lipolysis

Physiology of catecholamines of adrenal medulla. Because the adrenal medulla is directly innervated by the autonomic nervous system, adrenomedullary responses are very rapid. Furthermore, because of the involvement of several centers in the central nervous system (CNS), most notably the cerebral cortex, adrenomedullary responses can precede onset of the actual stress (i.e., they can be anticipated (Fig. 8.39). In many cases, the adrenomedullary output, which is primarily epinephrine, is coordinated with sympathetic nervous activity, as determined by the release of norepinephrine from postganglionic sympathetic neurons. However, some stimulus (e.g., hypoglycemia) evoke a stronger adrenomedullary than sympathetic nervous response, and vice versa.

Many organs and tissues are affected by sympathoadrenal response (Table 8.9). An example of major physiological roles of catecholamines is the sympathoadrenal response to exercise. **Exercise** is similar to the "fight or flight" response, but without the subjective element of fear, and involves a greater adrenomedullary response (i.e., endocrine role of epinephrine) than a sympathetic nervous response (i.e., neurotransmitter role of norepinephrine) than a sympathetic nervous response (i.e., neurotransmitter role of nonepinephrine). The overall goal of the sympathoadrenal system during exercise is to meet the increased energy demands of skeletal and cardiac muscle while maintaining sufficient oxygen and glucose supply to the brain. The response to exercise includes the following major physiological actions of epinephrine:

Table 8.9. Some actions of catecholamine hormones (Source: Koeppen and Stanton 2010).

	β: Epinephrine > norepinephrine	**α: Nor epinephrine > Epinephrine**
1.	↑ Glycogenolysis	Gluconeogenesis (α_1)
2.	↑ Gluconeogenesis (β_2)	↑ Gluconeogenesis (α_1)
3.	↑ Lipolysis (β_3) (β_2)	↑ Glycogenolysis (α_1)
4.	↑ Calorigenesis (β_2)	
5.	↓ Glucose utilization	
6.	↑ Insulin secretion (β_2)	↓ Insulin secretion (α_2)
7.	↑ Glucose secretion (β_2)	
8.	↑ Muscle K^+ uptake (β_2)	
9.	↑ Cardiac contractility (β_1)	
10.	↑ Heart rate (β_1)	
11.	↑ Conduction velocity (β_1)	
12.	↑ Arteriolar dilation: ↓ BP* (β_2) muscle	↑ Arteriolar vasoconstriction; ↑ BP (α_1) (splanchnic, renal, cutaneous, genital)
13.	↑ Muscle relaxation (β_2)	↑ Sphincter contraction (α_1)
14.	Gastrointestinal	Gastrointestinal
15.	Urinary	Urinary
16.	Bronchial	↑ Platelet aggregation (α_2)
17.		↑ Sweating ("adrenergic")
18.		↑ Dilation of pupils (α_1)

* BP, blood pressure

1. Increased blood flow to muslces in achieved by the integrated action of norepinephrine and epinephrine on the heart, veins and lymphatics and nonmuscular (e.g., splanchnic) and muscular arteriolar beds.

2. Epinephrine promotes glycogenolysis in muscle. Exercising muscle can also utilize free fatty acids (FFAs), and epinephrine and norepinephrine promote lipolysis in adipose tissue. The actions just described increase circulating levels of lactate and glycerol, which can be used by the liver as gluconeogenic substrates to increase glucose. Epinephrine increases blood glucose by increasing hepatic glycogenolysis and gluconeogenesis. The promotion of lipolysis in adipose tissue is also coordinated with an epinephrine – induced increase in hepatic ketogenesis. Finally, the effects of catecholamines on metabolism are strengthened by the fact that they *stimulate glucagon secretion* (β receptors) and *inhibit insuline secretion* (α_2 receptors). Efficient production of ATP during normal exercise (*i.e.*, a one-hour workout) also requires efficient exchange of gases with an adequate supply of oxygen to exercising muscle. Catecholamines promote this by relaxation of bronchiolar smooth muscle.

3. Catecholamines decrease energy demand by visceral smooth muscle. In general, a sympatho-adrenal response decreases overall motility of the smooth muscle in the gastrointestinal (GI) and urinary tracts, thereby conserving energy where it is not needed.

Metabolism of catecholamines. The neurotransmitter norepinephrine is degraded by the enzymes *monoamine oxidase (MAO)* and *catechol-o-methyltransferase (COMT)*. The neurotransmitter norepinephrine is degraded by MAO and COMT after uptake into the presynaptic terminal. This mechanism is also involved in the catabolism of circulating adrenal catecholamines.

8.8 PANCREAS

The pancreas has both exocrine and endocrine functions. The exocrine secretions are digestive enzymes which promote digestion of carbohydrates, proteins and lipids. Scattered among the exocrine portions (acini) of the pancreas are millions of tiny clusters of endocrine tissue known as **islets of Langerhans** or **pancreatic islets** (Fig. 8.40). These islets contain four types of cells secreting different hormones:

1. **Alpha cells (α or A-cells)** constitute 15–20% of islet mass and secrete **glucagon**.
2. **Beta cells or (β-cells or B-cells)** constitute 70–80% of islet mass and secrete **insulin**.
3. **Delta cells (D-cells)** are 1–8% and secrete **somatostatin**.
4. **F cells** are 1–2% and secrete **pancreatic polypeptide**.

Fig. 8.40. A part of T.S of pancreas showing the endocrine part.

The islet cells are innervated by both cholinergic and sympathetic nerve endings which contro secretion of hormones. The A, B, D cells are in close proximity to each other and exert control o each other's secretions, *e.g.*, insulin inhibits glucagon secretion while glucagon stimulates secretio of insulin and somatostatin. On the other hand, somatostatin inhibits secretion of both insulin an glucagon.

1. Glucagon. It is a polypeptide hormone secreted from A cells of islets of Langerhans Decreased blood glucose levels, sympathetic system stimulation and protein-rich diet **stimulat** glucagon secreation; while hyperglycemia and somatostatin **inhibit** glucagon secreation.

Actions of glucagon. Glucagon is a **catabolic hormone**. Its secretion increases during period of food deprivation, and it acts to mobilize nutrient reserves. It also mobilises glycogen, fat and eve protein (**Koeppen** and **Stanton**, 2010). Glucagon raises blood glucose levels by acting on the live and adipose tissue. Glucagon has the following functions:

(*i*) Stimulates *glycogenolysis*, *i.e.*, breakdown of liver glycogen to glucose.

(*ii*) Promotes *glyconeogenesis*, *i.e.*, formation of glucose from lactate, pyruvate, glycerol and amin acids.

(*iii*) Promotes **lipolysis** in adipose tissue to release free fatty acids and glycerol in the circulatio In the liver, excess free fatty acids are converted to ketone bodies—acetoacetic acid, aceton and β-hydroxybutyric acid.

(*iv*) Causes relaxation of GIT.

(*v*) Inhibits gastric acid secretion.

(*vi*) Stimulates heart, increases heart rete (**chronotropic effect**) and force of contraction (**inotropic effect**).

2. Insulin. Insulin is a polypeptide hormone containing 51 amino acids with MW 6000. It contains two chains, A chain contains 21 amino acids. B chain contains 30 amino acids. Both chains are connected together by two disulphide bridges.

Synthesis and secretion of insulin. Insulin is synthesized on ribosomes in the rough endoplasmic reticulum as proinsulin which is composed of two polypeptide chains A and B connected with each other by C (connecting) peptide. The proteolytic degradation of proinsulin in Golgi apparatus yields insulin and C peptide (Fig. 8.39) which are then stored in the pancreatic B-cells granules. The C peptide of proinsulin helps in correct folding of A and B chains of insulin and proper alignment of disulphide bonds. When plasma glucose levels > 70 mg/dl, there is stimulation of insulin release from β-cells. Glucose enters β-cells by **GLUT-2 glucose transporter**, gets metabolized and ATP is released which inhibits the activity of ATP-sensitive **K^+ channels**. Inhibition of these channels induced B-cell depolarization, opening of voltage-gated **Ca^{2+} channels**, increase in intracellular Ca^{2+} which causes release of insulin by exocytosis. Normal basal rate of insulin release is 1–2 U/hr.

Regulation of insulin secretion.

1. **Substrate control.** It is the most important factor in the control of insulin secretion.

(*i*) Rise in plasma glucose levels > 70 mg/dl stimulates β-cells to release insulin.

(*ii*) Several amino acids or AAs (*e.g.*, arginine and lysine) in plasma stimulates release of insulin.

(*iii*) Presence of long-chain free fatty acids (FFAs) increase insulin secretion.

2. Vagal (parasympathetic) cholinergic innervation (*i.e.*, in response to a meal) also stimulate secretion of insulin.

3. The gastrointestinal tract releases **incretin** hormones that stimulate pancreatic insulin secretion. GLP-1 and GIP are particularly potent in augmenting glucose dependent stimulation of insulin secretion.

4. Other hormones such as growth hormones, cortisol, glucagon and to a lesser extent progesterone and oesterogen also stimulate insulin secretion. Activation of α_2-adrenergic receptors **inhibit** insulin secretion.

Actions of insulin. Insulin is a primary **anabolic hormone** that is responsible for maintaining the upper limit of blood glucose. Receptor sites on plasma membrane bind insulin, causing change in membrane permeability and enzyme activity in the cytoplasm. Insulin serves the following functions:

1. It enables liver and muscles to store glucose as glycogen (**glycogenesis**) and enables the tissues to take up and use glucose as a source of energy. Insulin thus lowers the blood-glucose level.
2. It promotes uptake and utilization of glucose by muscle and adipose tissue.
3. It promotes synthesis of proteins from amino acids in the tissues.
4. It reduces breakdown of proteins to maintain the tissues.
5. It promotes synthesis of fats from fatty acids and glucose in the adipose tissue.
6. It reduces respiratory breakdown of fats in the body.
7. It increases the uptake of fat by plasma membrane.
8. It increases the formation of ATP, DNA and RNA.

Deficiency disorders of insulin. Deficiency of insulin causes **diabetes mellitus**. The patient cannot use or store glucose. Thus glucose accumulates in the blood to a level above that the kidney can reabsorb, the **renal threshold**. So it is excreted by kidneys in urine. This condition is called **glycosuria**, *i.e.*, presence of glucose in urine, causing loss of water from the body in urine. This produces excessive thirst. Degradation of fats increases, producing ketone bodies such as acetoacetate and acetone. Blood cholesterol rises. Breakdown of proteins also increases, leading to wasting of body tissues. Healing power is impaired. Injuries may change into **gangrenes** (dead tissue). The diabetic person has a blurred vision and is weak, tired, irritable, nauseated and under weight. In extreme cases, patient may pass into coma (complete loss of consciousness) and die. Administration of insulin gives relief by lowering blood-glucose level (Box 8.6).

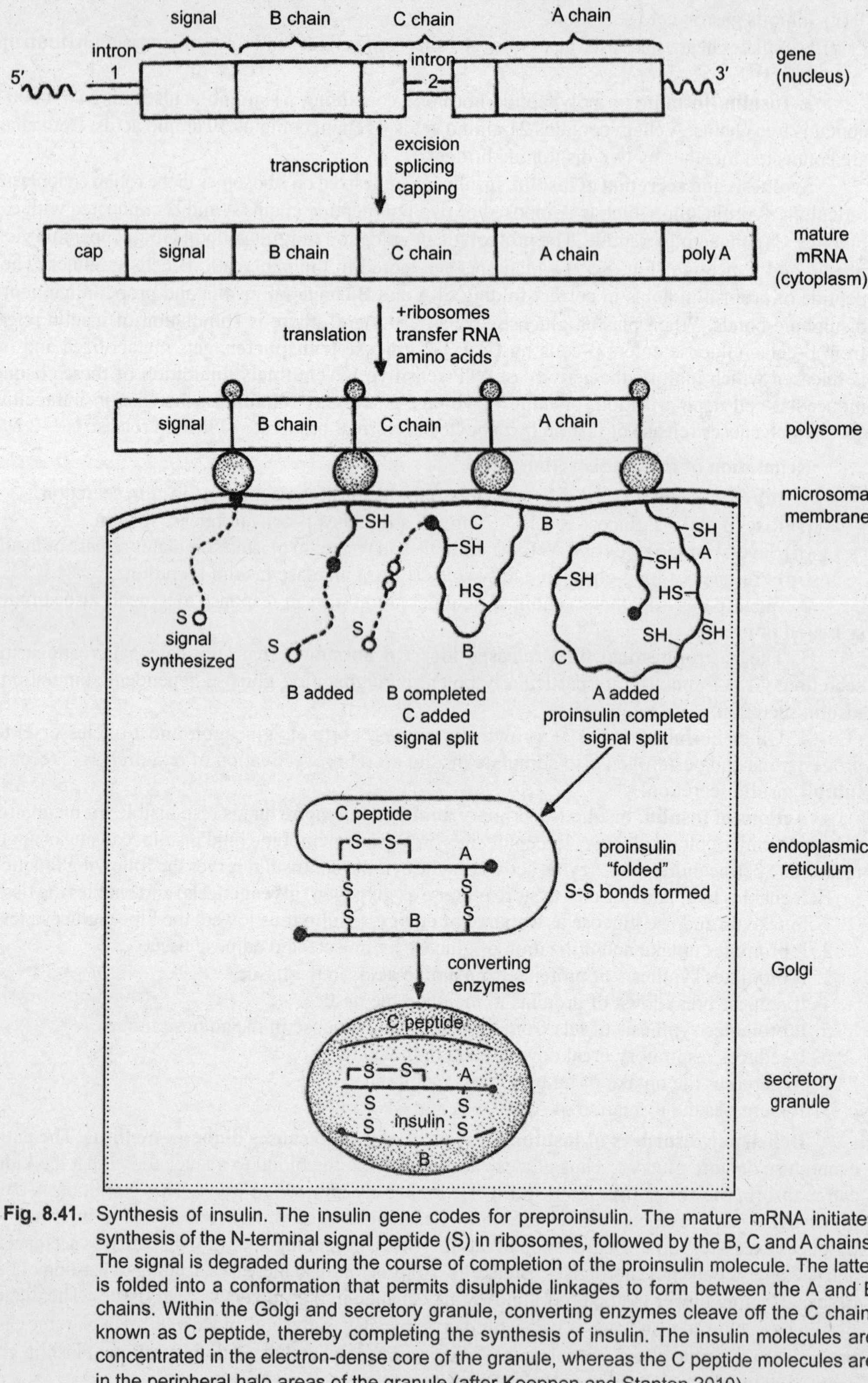

Fig. 8.41. Synthesis of insulin. The insulin gene codes for preproinsulin. The mature mRNA initiates synthesis of the N-terminal signal peptide (S) in ribosomes, followed by the B, C and A chains. The signal is degraded during the course of completion of the proinsulin molecule. The latter is folded into a conformation that permits disulphide linkages to form between the A and B chains. Within the Golgi and secretory granule, converting enzymes cleave off the C chain, known as C peptide, thereby completing the synthesis of insulin. The insulin molecules are concentrated in the electron-dense core of the granule, whereas the C peptide molecules are in the peripheral halo areas of the granule (after Koeppen and Stanton 2010).

Box 8.6 Diabetes Mellitus (DM)

Diabetes is derived from Greek word. It means siphoning. It denotes excretion of large volume of urine. *Mellitus* is a Latin word for honey and represents high sugar content of urine.

Diabetes mellitus is classified as type 1 (T_1DM) and type 2 (T_2DM). **T_1DM** or **Diabetes mellitus type 1**, is characterized by the destruction of pancreatic beta cells, and exogenous insulin is required for treatment. Major symptoms of T_1DM include hyperglycemia, polyuria, polydipsia (excessive or abnormal thirst), polyphagia (excessive appetite or eating), muscular wasting, asthenia (lack of energy), electrolytic depletion and ketoacidosis (acidosis accompanied by ketosis).

Diabetes mellitus type 2 (**T_2DM**) can be due to numerous factors but is usually characterized as **insulin resistance** coupled to some degree of beta cell deficiency. Patients with T_2DM may require exogenous insulin at some point to maintain blood glucose levels. (*Note*. With normal fasting, *i.e.*, no caloric intake for at least 8 hours) plasma glucose levels should be below 110 mg/dl. **Obesity-associated T_2DM** is currently at epidemic proportions worldwide and is characterized by insulin resistance because of lipotoxicity, hyperinsulinemia (the presence of excess insulin in the blood due to adenoma of islets) and inflammatory cytokines produced by adipose tissue. T_2DM is often associated with obesity, insulin resistance, hypertension and coronary artery disease. This constellation of risk factors is referred to as the **metabolic syndrome**.

3. Somatostatin. The delta cells of endocrine pancreas secrete the peptide hormone **somatostatin (SS)** which decreases the rate of nutrient absorption into the blood stream from the gastrointestinal tract (GIT). It also inhibits the secretion of hypophysial growth hormone (GH) and TSH. It is also an *inhibitor* of both insulin and glucagon.

4. Pancreatic polypeptide. It is secreted by F cells of endocrine pancreas. It inhibits the release of pancreatic juice.

8.9 THYMUS GLAND

Thymus is a bilobed lymphoepithelial organ. It is located in the upper part of chest covering the lower part of trachea. Thymus is conspicuous in childhood and attains maximum size during puberty. Thereafter, the thymic tissue (*i.e.*, mainly lymphocytes) is replaced by fat. By the time a person reaches maturity, the thymus gland atrophies.

Histologically, thymus is found to be enclosed in a **capsule** and is divided into **cortex** and **medulla**. Cortex has **lymphocytes** or **thymocytes** and **macrophages**. Medulla is rich in **thymic corpuscles** or **Hassell's corpuscles**.

Functions

1. Lymphopoiesis. Thymus gland is associated with formation of small lymphocytes during foetal and early life and release them in blood and lymph.

2. Immunity. Before birth of foetus, small lymphocytes from thymus move out into the blood stream and settle down in peripheral lymphoid tissue and give rise to T-cells of the body. These develop **cell-mediated immunity** or **cellular immunity**.

3. Endocrine function. Thymus secretes a hormone called **thymosin**. It accelerates cell division, thus, influencing the rate of growth during early life. It hastens attainment of sexual maturity.

4. Tissue-graft rejection. Humoral factor of thymus induces lymphatic tissue to form immunologically active antibodies which reject tissue grafts.

8.10 HORMONES OF GASTRO-INTESTINAL TRACT

The gastro-intestinal tract (GIT) hormones are secreted by the mucous membrane of stomach, duodenum and small intestine. These hormones stimulate and regulate secretion of digestive enzymes or digestive juices from the glands present in mucous membrane of GIT and also from liver and pancreas.

A. Hormones from gastric mucosa. The mucosa of the pyloric stomach near the duodenum secretes a hormone, called **gastrin** (*i.e.*, by G cells of gastric antrum). Presence of food inside stomach provides stimulus for gastric secretion. Gastrin stimulates the gastric glands to produce the gastric juice containing HCl and pepsin (*i.e.*, parietal cells which have CCK_2 receptors for gastrin secrete HCl and chief cells secrete pepsinogen). Gastrin also stimulates stomach movement.

B. Hormones from duodenal mucosa. The duodenal mucosa secretes following *three* hormones. Their secretion is stimulated by the entry of chyme (food) from stomach into duodenum.

***(i)* Secretin.** It is formed of 27 amino acids. It stimulates cells of pancreatic lobules for secreting fluid portion (water and electrolytes such as sodium bicarbonate) of pancreatic juice.

***(ii)* Pancreozymin (PZ).** It stimulates pancreatic acini to secrete enzymatic part of pancreatic juice.

***(iii)* Cholecystokinin (CCK).** It stimulates rhythmic contractions in the wall of gall bladder and thus causes release of bile juice from gall bladder into duodenum.

C. Hormones from intestinal mucosa. Mucous membrane of small intestine secretes following *three* hormones.

(*a*) Enterocrinin. It stimulates release of intestinal juice (saccus entericus) from the crypts of Lieberkuhn in intestinal mucosa.

(*b*) Durocrinin. It stimulates and controls secretion of viscous mucus from Brunner's glands into the intestinal juice.

(*c*) Enterogastrone. This hormone inhibits gastric secretion and gastric motility. It thus stops gastric secretion after the digestion in stomach is completed. By reducing gastric motility, this hormone delays emptying of stomach to ensure complete digestion in stomach. Secretion of enterogastrone is stimulated by the presence of fat in the food (chyme) that enters the intestine.

8.11 GONADS

Besides producing gametes, the gonads secrete sex hormones from the onset of puberty (sexual maturity) to control the reproductive organs and sexual behaviour. The testes of the human males and ovaries of human females secrete different hormones.

1. Testes. Testes are located in the scrotum. They secrete male sex hormones, called **androgens**, such as **testosterone**, from groups of **interstitial cells**, or **Leydig's cells**, present in the connective tissue between the sperm-producing seminiferous tubules. (Note. Leydig's cells also secrete two more sex hormones called **15-hydroxytesterone** and **6-hydroxytestosterone**). The Leydig's cells are stimulated to secrete testosterone by the luteinising hormone (LH) or **Interstitial cell stimulating hormone (ICSH)** from the anterior lobe of the piuitary gland. Rise of testosterone level in the blood above normal inhibits LH secretion by the anterior pituitary lobe. This negative feedback checks oversecretion of testosterone.

Functions of testosterone. Testosterone serves the following functions:

1. It stimulates the male reproductive system to grow fully and become functional.
2. It stimulates the formation of the sperms (spermatogenesis) in the seminiferous tubules.
3. It stimulates the development of male accessory sex characters such as hair on the face (beard and moustaches), pubic and axillary hair, growth and distribution of hair on the body, deepening of voice, broadening of shoulders, enlarged and stronger bones and muscles. Testosterone also maintain these characters.
4. It also determines the male sexual behaviour and the sex urge (libido).

In fact, all the changes a male's body undergoes at puberty are produced by the testosterone.

Disorder. Failure of testosterone secretion causes **eunchoidism**. A eunuch has underdeveloped and non-functional sex organs (*i.e.*, lack penis and testes). It lacks secondary and accessory sex characters. It does not produce sperms.

Inhibin hormone. The Sertoli cells present in the germinal epithelium of the seminiferous tubules secrete a hormone **inhibin**. Inhibin has a negative feedback control on the hypothalamus and pituitary, reducing the release of FSH and LH to decrease testosterone production.

2. Ovary. Two ovaries lie in the abdomen of human female. They secrete *three* female sex hormones: **estrogen (estradiol), progesterone** and **relaxin.**

(*i*) Estrogen. Estrogen includes 4 hormones: **estrone, α-estradiol, β-estradiol** and **estriol**. These are secreted by the cells of the Graafian (= ovarian) follicle surrounding the maturing ovum in the ovary. They stimulate the female reproductive tract to grow to full size and become functional. They also stimulate differentiation of ova (oogenesis) in the ovary. Estrogens also stimulate the development of secondary and accessory sex characters such as enlargement of breasts, broadening of pelvis, growth of pubic and axillary hair, deposition of fat in the thighs, and onset of menstrual cycle. Graafian follicle cells are stimulated to secrete estrogens by luteinising hormone (LH) from the anterior lobe of the pituitary gland. Rise of blood-estrogen level above normal inhibits the secretion of LH from the anterior pituitary. This negative feedback prevents the oversecretion of estrogens.

Estradiol binds to blood proteins for transport to the target organs. It is altered in the liver or kidney for excretion via bile or urine.

(*ii*) Progesterone. This sex hormone is secreted by a temporary endocrine gland called **corpus luteum** which is a yellowish gland formed in the empty (ruptured) Graafian follicle after the release of the ovum (ovulation). Its hormone, progesterone, suspends ovulation during pregnancy, fixes the foetus to the uterine wall, forms placenta and controls the development of the foetus in the uterus. Ovulation, formation of corpus luteum and secretion of progesterone are stimulated by the luteinising hormone (LH) from the anterior pituitary.

(*iii*) Relaxin. It is produced by the corpus luteum at the time of gestation period. It relaxes the cervix of the uterus and ligaments of the pelvic girdle for easy birth of the young one.

The ovaries also produce some androgens, perhaps by the interstitial cells present in their stroma.

8.12 PLACENTA

Placenta secretes three hormones: 1. HCG, 2. progesterone, and 3. osterogens (*i.e.*, oestriol).

HCG or Human Chorionic Gonadotropin

This hormone is secreted by chorionic villi of human foetus. HCG is a glycoprotein with molecular weight of 39,000. It is secreted by syncytiotrophoblast soon after fertilisation. Its peak is reached with 60 to 90 days. HCG appears in blood and urine during early pregnancy. It is detectable in maternal blood 8 to 9 days after conception. Its detection in urine forms the basis for immunological test. Levels of HCG in maternal blood is also the index of functional status of trophoblast. Biologic actions of HCG are almost same as those of LH.

Functions of HCG

1. Maintains the corpus luteum until placenta becomes autonomous in terms of hormone synthesis (about 6 to 7 weeks after conception).
2. Converts corpus luteum of menstruation to corpus luteum of pregnancy thereby extending its life span.
3. Stimulates corpus luteum to secrete 17-alpha hydroxyprogesterones.

QUESTIONS

Long Answer Questions

1. How the hormones differ from enzymes? Explain the feedback mechanism to control hormone level in blood. (*CCSU Meerut 2013*)

2. What are hormones? What are the structures secreting them and what are their main functions? *(CCSU Meerut 2007)*
3. What is endocrine system? Discuss role of endocrine system as a regulatory device. Also discuss its necessity and importance.
4. What are hormones? What factors control their secretion?
5. Write an essay on mechanism of biological action of hormone. How do hormones in general modify the physiological activity of various cells?
6. Discuss second messenger theory of hormone action. Discuss role of cyclic AMP in hormone action.
7. What are endocrine glands? Describe the structure and functions of pituitary gland. *(CCSU Meerut 2007, 2010)*
8. Give a brief account of hormones of adenohypophysis of pituitary in mammals.
9. Describe the position, structure and hormones secreted by the thyroid gland. Write down about the hyposecretion and hypersecretion of thyroid gland. *(CCSU Meerut 2011)*
10. Define hormones. How are they differ from enzymes? Describe the structure and functions of thyroid glands. *(CCSU Meerut 2009)*
11. Describe structure and functions of parathyroid gland. Why parathormone is essential for healthy human body? *(CCSU Meerut 2013)*
12. Describe the structure and fucntion of adrenal glands.
13. Give names of hormones secreted by pancreas and discuss their function. How these regulate sugar metabolism?
14. Give an account of sex hormones of a mammal.
15. "Pituitary is a mastergland". Justify.
16. What are hormones? Discuss different mechanisms which control their functions. *(CCSU Meerut 2014)*
17. Give an account of different cell types and functions of adenohypophysis of pituitary gland of a mammal. *(CCSU Meerut 2014)*
18. How the hormones differ from enzymes? Explain the feedback mechanism to control hormone level in blood. *(CCSU Meerut 2013)*
19. Describe structure and functions of parathyroid gland. Why parathormone is essential for healthy human body? *(CCSU Meerut 2013)*

Short Answer Questions

1. Define the following terms:
   (*i*) autocrine (*ii*) paracrine
   (*iii*) target organs
2. Write short notes on the following:
   (*a*) Neuroendocrine gland: (*b*) Fucntions of hormones;
   (*c*) Positive feedback; (*d*) Negative feedback;
   (*e*) Tropic hormones; (*f*) Somatotropic hormone;
   (*g*) Antidiuretic hormone; (*h*) Abnormalities of thyroid;
   (*i*) Thyroxine; (*j*) Myxodema;
   (*k*) Cretinism; (*l*) Hashimoto's disease;
   (*m*) Parathormone; (*n*) Emergency hormone;
   (*o*) Cortisol; (*p*) Diabetes mellitus;
   (*q*) Islets of Langerhans; (*r*) Androgens;
   (*s*) Estrogens; (*t*) HCG.

3. Write differences between the following:
   Endocrine glands and Exocrine glands.
4. What are chemical messengers? Give one example. (*CCSU Meerut 2009*)
5. Draw a neat and labelled diagram of following:
   (*i*) T.S. ovary of mammals; (*CCSU Meerut 2006, 2010*)
   (*ii*) T.S. pancreas;
   (*iii*) T.S. thyroid of mammal; (*CCSU Meerut 2007, 2008*)
   (*iv*) T.S. testis of mammal; (*CCSU Meerut 2008, 2011*)
6. Write about thyroid gland and its biological role. (*Osmania 2005*)
7. Describe islets of Langerhans and their secretions. (*MSU Tirunelveli 2007*)

Very Short Answer Questions

1. Which blood vessel, a vein or an artery, carries hormone from an endocrine gland?
2. Name the male and female sex hormones.
3. Which hormone causes release of milk after the birth of young one?
4. What causes gigantism?
5. Name any hormone secreted by corpus luteum.
6. What are chemical messengers?
7. What are secondary messengers?
8. What are releaser hormones?
9. Give the causes of diabetes mellitus and diabetes insipidus.
10. What is the study of hormones called? (*CCSU Meerut 2008*)
11. Make the list of steroid hormones.
12. Name the hormones secreted by parathyroid and thymus glands.
13. Name the three regions of adrenal cortex.
14. Hypothalamus produces two types of hormones. Give their names.
15. Name the two hormones formed in kidneys.
16. How are hyperglycemia, ketogenesis and glucosauria caused?
17. Give the expanded form of ACTH.
18. Name the hormone that regulate calcium level in blood.
19. Name the kind of cells which secrete testosterone.
20. Which hormone is known as brain hormone?
21. What is the source of aldosterone?
22. Why is oxytocin called birth hormone?
23. Name the disease caused due to deficiency of iodine. (*CCSU Meerut 2007, 2010*)
24. What will happen if thyroxine is injected inside a young tadpole of frog? (*CCSU Meerut 2007*)
25. Name the glands of dual function (*i.e.*, exocrine and endocrine). (*CCSU Meerut 2006, 2011*)
26. What are chemical messengers? Give one example. (*CCSU Meerut 2009, 2010*)
27. Write a note on adrenal gland. (*Osmania 2005*)
28. Differentiate between Diabetes mellitus and Diabetes insipidus. (*CCSU Meerut 2014*)
29. Differentiate between insulin and glucagon. (*CCSU Meerut 2014*)

Fill in the Blanks

1. Three endocrine glands are located in the head namely ________, ________ and ________.
2. Dwarfism is caused by deficiency of ________ formed in the ________.
3. The organs influenced by hormones are known as ________.
4. The glands purely endocrine in function include thyroid, ________ and pineal.

5. The glands which are partly endocrine and partly exocrine are known as ________.
6. ACTH is secreted by ________ lobe of ________.
7. ADH is secreted by ________ lobe of ________.
8. Melatonin is a secretion of ________.
9. PTH controls calcium and ________ balance.
10. Thyroid secretes three hormones: thyroxine, ________ and ________.
11. Calcitonin regulates ________.
12. Cell division is accelerated by a hormone called ________.
13. Goitre is due to deficiency of ________.
14. Hyposecretion of thyroxine causes cretinism in ________ and ________ in adults.
15. The heterocrine glands include pancreas, ________, ________, and kidneys.

Multiple Choice Questions

1. Which of the following statements is correct in relation to the endocrine system?
   (*a*) non-nutrient chemical produced by the body in trace amounts that act as intercellular messengers are known as hormones.
   (*b*) releasing and inhibitory hormones are produced by the pituitary gland.
   (*c*) adenohypophysis is under direct neural regulation of the hypothalamus.
   (*d*) organs in the body such as gastrointestinal tract (GIT), heart, kidney and liver do not produce any hormones.
2. Which of the following hormones does not have a particular target organ in the body?
   (*a*) growth hormone (*b*) TSH
   (*c*) oxytocin (*d*) FSH
3. Reabsorption of Na^+ is controlled by which one of the following hormones?
   (*a*) aldosterone (*b*) oestrogen
   (*c*) glucocorticoid (*d*) testosterone
4. Which one of the following is both exocrine and endocrine gland?
   (*a*) liver (*b*) thyroid
   (*c*) pancreas (*d*) adrenal
5. Addison's disease results from
   (*a*) hypertrophy of gonads (*b*) hyperactivity of Leydig's cells
   (*c*) hypo-secretion of adrenal cortex (*d*) none of the above
6. Grave's disease is due to
   (*a*) hyperactivity of thyroid gland
   (*b*) hyperactivity of adrenal cortex
   (*c*) hyperactivity of adrenal medulla
   (*d*) hypoactivity of islets of Langerhans
7. The production, secretion and ejection of milk by mammary gland require the synergistic effect of
   (*a*) estrogens and progesterone (*b*) prolactin
   (*c*) oxytocin (*d*) all of the above
8. Secretion of progesterone by corpus luteum is initiated by
   (*a*) testosterone (*b*) thyroxine
   (*c*) MSH (*d*) LH
9. Which of the following represents the action of insulin?
   (*a*) increases blood glucose level by stimulating glycogen production
   (*b*) decreases blood glucose level by forming glycogen

(c) increases blood glucose level by promoting cellular uptake of glucose
(d) increases blood glucose by hydrolysis of glycogen

10. An endocrine gland which atrophies in the adult human is
(a) thyroid (b) parathyroid
(c) thymus (d) pineal

11. The hormone not secreted by adenohypophysis is
(a) ACTH (b) ADH
(c) GH (d) TSH

12. The primary physiological role of insulin is increased
(a) glycogen synthesis in liver (b) glucose uptake by brain
(c) lipid synthesis in liver (d) glucose uptake by skeletal muscles

13. Maximum number of hormones are secreted by
(a) pituitary gland (b) hypothalamus
(c) adrenal gland (d) placenta

14. Cyclic adenosine phosphate (cAMP) does not help in the action of
(a) peptide hormone (b) amino acid derivative hormone
(c) steriod hormone (d) protein hormone

15. Hormone produced during emotional stress is
(a) calcitonin (b) melatonin
(c) relaxin (d) adrenaline

16. The excess secretion of a growth hormone during adulthood produces
(a) giant (b) dwarf
(c) acromegaly (d) myxoedema

17. Melatonin hormone is secreted by
(a) pineal gland (b) skin
(c) pituitary gland (d) thyroid

18. A patient of diabetes mellitus excretes glucose in urine even when he is kept in a carbohydrate free diet. It is because
(a) fats are catabolised to form glucose
(b) amino acids are catabolised in liver
(c) amino acids are discharged in blood stream from liver
(d) glycogen from muscles is released in the blood stream

19. Cashing's disease is caused by hyperactivity of
(a) GH (b) thyroxin
(c) insulin (d) glucocorticoids

20. Which hormone possesses anti-insulin effect
(a) glucocorticoids (b) calcitonin
(c) oxytocin (d) aldosterone

21. Reabsorption of water in distal parts of kidney tubule is controlled by
(a) vasopressin (b) oxytocin
(c) calcitonin (d) relaxin

22. Gluconeogenesis is controlled by
(a) thyroxin (b) cortisol
(c) corticosterone (d) all the above

23. Female sex-hormone is
(a) estrogen (b) androgen
(c) insulin (d) adrenalin

24. Neurohypophysis secretes
(*a*) vasopressin and estrogen
(*b*) oxytocin and estrogen
(*c*) vasopressin and oxytocin
(*d*) vasopressin and growth hormone

25. Urine of the would-be-mother contains
(*a*) FSH
(*b*) progesterone
(*c*) luteinzing hormone
(*d*) chorionic gonadotropin

ANSWERS

Very Short Answer Questions

1. Vein.
2. Testosterone, estrogen.
3. Oxytocin.
4. Hypersecretion of GH.
5. Progesterone.
6. Hormones.
7. Cyclic AMP.
8. Secretions of hypothalamus.
9. Insulin deficiency causes diabetes mellitus and while deficiency of antidiuretic hormone results in diabets insipidus.
10. Endocrinology.
11. (*i*) aldosterone, (*ii*) cortisol, (*iii*) testosterone, (*iv*) estrogen, and (*v*) progesterone.
12. Collip's hormone or parathormone, thymosin.
13. (*i*) Zona glomerulosa, (*ii*) Zona fasciculata, and (*iii*) Zona reticularis.
14. Releaser hormones and inhibiting hormones.
15. Renin and dihydroxycholecalciferol.
16. All by insulin deficency.
17. Adrenocorticotropic hormone.
18. Calcitonin.
19. Leydig cells.
20. Oxytocin.
21. Adrenal cortex.
22. Oxytocin stimulates contraction of uterine muscles at the time of child birth and release of milk from mammary glands.
23. Goiter.
24. Metamorphosis of tadpole is started.
25. Pancreas, gonads and GIT.
26. Chemical messengers are hormones. They are produced by endocrine glands and transported by the blood to other tissues (or organs), called **target cells** or **organs**, where these stimulate a change in some metabolic activity. Example of a chemical messenger is insulin.
27. Renal gland is either of a pair of complex endocrine gland near the anterior medial border of the kidney consisting of a mesodermal **cortex** that produces glucocorticoid, mineralocorticoid and androgenic hormones and an ectodermal **medulla** that produces epinephrine and norepinephrine.
28. Diabetes mellitus (DM) is a condition of glycosauria accompanied with hyperglycemia due to lack of insulin hormone. It is caused by degeneration or hypoactivity of β-cells of islets of Langerhans of endocrine pancreas. The patient of DM has high level of blood glucose and excretes glucose with urine.
29. **Insulin** (a hypoglcemic factor or hG factor) is a peptide horomone secreted by the β (or B) - cells of islets of Langerhans in the pancreas that act as a storage hormone. Insulin has a profound influence on carbohydrate metabolism. It reduces the level of sugar in the blood,

and is secreted in response to rise in blood sugar (e.g., after meals or in conditions of stress); the sugar is converted into glycogen in the cells of muscle and the liver under the influence of insulin. It facilitates entry of glucose and other sugars into the cells, by increasing penetration of cells membrane and augmenting phosphorylation of glucose.

The deficiency of insulin, results in the accumulation of large amounts of glucose in the blood **(hyperglycemia)** and its subsequent excretion in the urine in abnormally high concentrations **(glycosuria).** This condition is medically known as **diabetes mellitus.** Diabetes can be fatal if untreated, treatment being by injection of insulin into the blood stream. Opposed in its action to glucagon.

The **glucagon** (a hyperglycemic factor or HG factor) is a polypeptide vertebrate hormone. It is secreted by the α (or A) cells of islets of Langerhans in the pancreas in response to drop in blood glucose level.

Glucagon activates the enzyme adenyl cyclase in target cells (e.g., liver, adipose tissue) with resultant rise in cyclic AMP in those cells and appropriate enzyme activation to ensure **glycogenolysis** (with release of glucose from glycogen), **gluconeogenesis** and **lipolysis** (with release of free fatty acids), restoring levels of these metabolites in the blood plasma. It also stimulates insulin release from β cells of islets of Langerhans, the two hormones acting antagonistically in control of blood glucose (or sugar) and free fatty acid levels. Opposed in its action to insulin .

Fill in the Blanks

1. Hypothalamus, pituitary, pineal.
2. Growth hormone, anterior pituitary.
3. Target organs.
4. Parathyroid, thymus, adrenals, pituitary.
5. Heterocline glands.
6. Anterior, pituitary.
7. Posterior, pituitary.
8. Intermediate lobe of pituitary.
9. Phosphorus.
10. Triiodothyronine, calcitonin.
11. Calcium and phosphorus balance.
12. Thymosine.
13. Iodine.
14. Infants, myxoedema.
15. Gastric mucosa, intestinal mucosa, gonads.

Multiple Choice Questions

1. (*a*)	**2.** (*a*)	**3.** (*a*)	**4.** (*c*)	**5.** (*c*)	**6.** (*a*)
7. (*d*)	**8.** (*d*)	**9.** (*b*)	**10.** (*c*)	**11.** (*b*)	**12.** (*d*)
13. (*b*)	**14.** (*c*)	**15.** (*d*)	**16.** (*c*)	**17.** (*a*)	**18.** (*a*)
19. (*d*)	**20.** (*a*)	**21.** (*a*)	**22.** (*d*)	**23.** (*a*)	**24.** (*c*)
25. (*d*)					

9

Homeostasis and Thermoregulation

9.1 HOMEOSTASIS

The term *homeostasis* is used by physiologists to mean *maintenance of nearly constant conditions in the internal environment*. Essentially all organs, tissues and cells of the body perform functions that help maintain these relatively constant conditions (Box 9.1).

Box 9.1

Human body is *a social order of about 100 trillion cells* organised into different functional structures, some of which are called **organs** (**Guyton** and **Hall** 2011). Each functional structure contributes its share to the maintenance of homeostatic conditions in the extracellular fluid (interstitial fluid), which is called the **internal environment**.As long as normal conditions are maintained in this internal environment, the cells of body continue to live and function properly. Each cell benefits from homeostasis, and in turn each cell contributes its share toward the maintenance of homeostasis. This reciprocal interplay provides continuous **automaticity** of the body until one or more functional systems lose their ability to contribute their share of function. When this happens, all the cells of the body suffer. Extreme dysfunction leads to **death**, moderate dysfunction gives way to sickness.

An organism may be defined as a physico-chemical system existing in a steady state with its external environment. The success of the organisms depends upon their ability to maintain a steady state in an ever changing environment. The organisms have evolved a variety of morphological, anatomical, physiological and behavioural mechanisms for maintaining a constant internal environment. The advantage of a steady state (*i.e.*, constant) internal environment in making metabolism efficient was first recognized in 1857 by French physiologist **Claude Bernard** (1813–1878). He distinguished between the external environment in which organisms live and the internal environment (*i.e.*,tissue fluid or interstitial fluid) wherein the individual cells exist. He realised the significance of steady state condition around the cells.

In 1912, the American physiologist **Walter Cannon** (1871–1945) introduced the term **homeostasis** (G., *homoios* = same; *stasis* = standing) for Bernard's constancy of internal environment and for the mechanisms which help maintain steady state. For instance, the lungs provide oxygen to the extracellular fluid to replenish the oxygen used by the cells, the kidneys maintain constant ion concentration and the gastrointestinal system provide nutrients. The stable cellular environment makes the organisms independent of the environment. Independence of the environment is the measure of 'success' of an organism. On this basis, birds and mammals may be regarded successful animals because they remain almost equally active all the year round inspite of changes in the environment.

Cybernetics

To achieve necessary stability of cellular environment, the activities of organisms need regulation at all levels of biological organisation, from the molecular to the population. The organisms employ for keeping steady state in them a variety of mechanisms that best suit their level of complexity and mode of life. Stability is attained by control measures called **control theory**. The science **of control theory** is called **cybernetics** (G., *kybernetics* = thing pertaining to control) a term coined by **Wiener** in 1948. The biological control systems include *temperature regulation, gaseous exchange rates, heart rate, arterial blood pressure, hormone* and *metabolic levels, water and ionic balance, pH regulation,* etc. The control systems frequently employ the **feedback systems**.

9.2 TEMPERATURE REGULATION

Influence of Temperature on Animals

Heat is a form of energy necessary for the maintenance of living systems. Life occurs in regions where the air temperatures vary from –40°C, as in the Arctic, to 50°C in the deserts. Majority of living organisms, however, exist at places with a temperature range of 10 to 35°C. Sun is the **exogenous** source of heat. Organic food is the **endogenous** source of heat. Food releases heat by catabolic reactions or respiration. **Temperature** indicates the amount of heat energy in a system. It is the major factor determining the rate of chemical reactions (metabolism) in living organisms. Temperature acts by influencing the enzymes, and motility of atoms and molecules. This directly affects the rate of growth of animals. In addition, the temperture may also affect the geographical distribution and activity of animals.

Water forms a relatively stable environment with regard to temperature fluctuations. Therefore, most aquatic animals rarely need physiological and behavioural mechanisms for temperature regulation. Some very active fish, such as the tuna (*Thunnus thynnus*), can, however, maintain body temperature above that of water. Air has a relatively low specific heat and therefore, air temperature can fluctuate widely over a 24 hour period. One of the problems associated with the colonization of land the animals was adapting to temperature fluctuations. This is produced many physiological and behavioural machanisms for regulating the body temperature.

Sources of Heat Animals (Thermogenesis)

Animals get heat from two sources: **external environment** if its temperature is higher than that of the body, and **internal metabolism** including muscular activity, involuntary muscular contraction (**shivering**), and chemical reactions occurring without noticeable muscular activity. There is a small gain of heat from any hot flood or liquid taken. The ability of the animals to generate and conserve heat depends upon anatomical, physiological, and behavioural mechanisms evolved by them. The animals are often divided into two categories: **poikilothermic** and **homeothermic animals** on the basis of their ability to regulate their body temperature in the face of variations in the environmental temperature. Poikilothermic and homeothermic animals are frequently called **cold-blooded** and **warm-blooded animals**, respectively. These terms are inappropriate as on a hot summer noon the temperature of lizard may be higher than that of human being.

1. Poikilothermic animals or poikilotherms. These animals are unable to maintain their body temperature, which varies directly with that of environment (G, *poikilos* = varied; *therme* = heat). Invertebrates, fishes, amphibians and reptiles are poikilothermic animals.

2. Homeothermic animals or homeotherms. These animals maintain relatively constant body temperature (Gr., *homoios* = like; *therme* = heat). Birds and mammals are homeothermic animals.

Human Body Temperature (Heat Balance)

Human body temperature is maintained at an approximately constant level because of the balance which is maintained between the heat gained and heat loss. Human beings are homeotherms or endotherms. Their body temperature is maintained at **98.6°F** (= **Fahrenhit**) or 37°C (**Celcius** or **centigrade**; Box 9.2),. For recording human body temperature, a **thermomerter** is placed in the mouth under the tongue, in the axilla (armpits), in the groin or inserted into the rectum. The temperature recorded in the rectum gives the true **core temperature** of the body.

Box 9.2 Core Temperature

1. F is derived from the name of scientist, **Gabriel D. Fahrenhit** (1686–1736).
2. C is derived from the name of scientist, **Ander Celsius**, Swedish astronomer (1701–1744).
3. The maintenance of a constant body temperature applies only to the **core** of the body, that is, within the skull, thorax and abdomen. This temperature is normally determined by taking the rectal temperature.

Method of conversions of reading from °F to °C and vice versa. The formula of conversion is given below:

1. °F to °C: substract 32, multiply by 5, divide by 9.

$$63°F \text{ to } °C : 63–32 = 31 \times 5$$
$$= 155 \div 9 = 17.2°C$$

2. °C to °F: multiply by 9, divide by 5 and add 32

$$37° C \text{ to } °F : 37 \times 9 = 333 \div 5 = 66.6 + 32$$
$$= 98.6 °F$$

Ectothermic and endothermic animals. Recently, the above terminology has been changed. Many reptiles, which have generally been considered poikilotherms, have been found to maintain a rather stable body temperature in the wild though not in the laboratory. In the wild, they seek sunny areas when their body temperature falls, and shift to shady places when their body temperature rises. This has led to the introduction of new terms, such as ectotherms and endotherms.

***(i)* Ectotherms** (Gr., *ektos* = outside). These animals depend more on external source of heat (sun) than on internal source of heat (*i.e.*, metabolism) for raising their body temperature.

***(ii)* Endotherms** (Gr., *endon* = within). These animals rely on the internal source of heat for the maintenance of their body temperature.

Mode of Heat Loss

Mode of heat loss in both ectotherms and endotherms includes the following means: Convection, Conduction, Radiation and Evaporation.

1. Convection. When the environmental temperature is below body temperture, heat is lost to the surrounding air by convection. The amount of heat lost in convection depends on the temperature gradient between the body and the environment—the higher the difference the faster is heat loss. In convection the heat is lost by organisms to environment—air or water—in contact with their body surface. Air or water in direct contact with the organisms become warmer less dense and moves away, and is replaced by cooler air or water. The rate of heat loss by convection is related to the rate of air or water movement. It may be reduced by insulation with feathers in birds or hair in mammals or clothes in humans.

2. Conduction. Heat is lost by conduction through physical contact with the environment, such as between organism and ground. The heat loss is almost insignificant for most land animals (as air is vary bad conductor of heat), but may be considerable for aquatic and soil-dwelling organisms. Loss of heat by conduction is minimised by wearing clothes that are poor conductors of heat. Air trapped undeneath the clothes makes a good thermal insulator.

3. Radiation. Radiation is the loss of heat by electromagetic waves. The radiant heat has slightly longer wavelenght and a little less frequency than the visible light, hence called **infrared radiation**. This radition does not heat the medium through which it passes and can pass through a vacuum. The heat from the sun comes by this form of radiation. The amount of heat lost by radiation depends on the temperature of an object as well as on its colour and texture. Dark, rough surfaces radiate maximally (black body radiators). Light, shiny surfaces at the same temperature lose heat slowly. The same is true of the heat gain by radiations. A block of ice-cream melts more rapidly if it is wrapped in black paper than if wrapped in shiny aluminium foil.

The human skin acts as a black body radiator irrespective of the actual skin colour. White clothes are more suitable than black both in very hot tropics and in the very cold Antarctics.In the tropics, there is less heat gain, and in the Antarctica there is less heat loss.

The human body continuously emits heat radiations which are absorbed by the surrounding objects. The amount of heat lost by radiation depends on the surface area of the body and the skin temperature. Radiation accounts for about 50% of the total heat loss in humans.

4. Evaporation. Heat is lost from the body surface in the conversion of water to water vapour. The vapourization of 1 ml of water consumes 0.58 calories of heat. Body loses water by **evaporation**, as the skin is not an entirely water-proof covering, and by way of sweat glands. If the sweat drops fall off the skin without evaporating, there is no cooling effect. Water is also lost as vapour from lungs. The **insensible** water loss through the skin and from the lungs by evaporation is continuous and it cannot be controlled (called **unregulated heat loss**). Regulated evaporative heat loss is controlled and it involves water loss by **sweating** and **panting. Sweat glands** form the only mechanism for reducing body temperature when the environmental temperature is higher than that of body. Evaporation of water does not take place if the humidity of the air is high. The body can withstand very high enviornmental temperature well below the body temperature may be uncontrollable if the humidity is high.

Prolonged sweating leads to **dehydration**. Sweating then ceases and the body temperature may begin to rise.

Regulation of Heat Gain

Body gains heat from two sources (*i*) environment (*ii*) heat obtained from metabolism or **thermogenesis**.

1. Heat gain from environment. When the environment temperature is higher than body, the animal body gains heat from the surrounding through *convection* and *conduction*. Besides this, the body also gains heat by absorbing radiations through the skin. Another factor which decides heat gain is the pigmentation of the skin. A dark skin, which absorbs more radiations, gains more heat as compared to a light skin.

2. Heat production by the body. All the energy liberated during metabolism is not converted into ATP. Only about 40% energy get converted into ATP, the rest is released as heat. This is known as **thermogenesis.**

Table 9.1. Various mechanisms of heat gain and heat loss by the body.

	Heat gain		Heat loss
1.	From external environment	1.	To environment:
	(*a*) Convection		(*a*) Convection
	(*b*) Conduction		(*b*) Conduction
	(*c*) Radiation		(*c*) Radiation
2.	Internal metabolism—Thermogenesis	2.	Heat loss due to evaporation
	(*i*) Basal heat production		(*i*) Basal heat loss
	(*a*) ingestion of hot food and drink		(*a*) insensible prespiration

Heat gain	Heat loss
(*b*) unconscious tensing of muscles	(*b*) water evaporation through lungs
(*c*) disease	(*c*) warming of respired air
(*ii*) active heat production	(*d*) through urine and faeces
(*a*) exercise	(*ii*) active heat loss
(*b*) shivering	(*a*) sweating
	(*b*) panting

During standard resting conditions (basal conditions)—conditions under which the individual is under complete physical and physiological rest, *i.e.*, 12 to 18 hours after meal and in equable environmental temperature, the heat production is approximately, 1700 Cals in human males and about 1500 Cals in human females per day. Males and females, indulging in moderate physical activity, require about 3000 Cals and 2500 Cals respectively per day. This rises further in individual who do heavy physical work.

The heat produced by the various organs of the body depends on the degree of their activity. Although the rate of heat production by heart and brain does not vary very much, the heat produced by the skeletal muscles, on the other hand, varies a great deal, it is very low when muscles are resting and very high when muscles are active. Active muscles in fact, are the most important source of thermogenesis.

The **Due Boise** temperature balance (Fig. 9.1) shows that the following factors increases heat production.

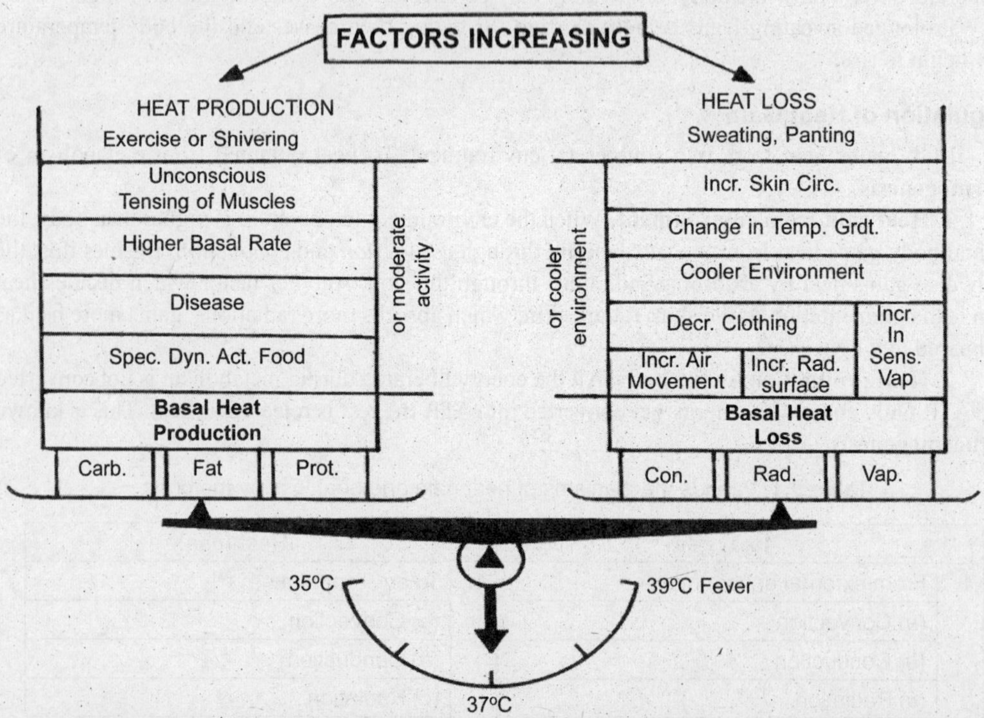

Fig. 9.1. The Due Bois temperature balance between the factors increasing heat production and heat loss.

(*a*) Active heat production (thermogenesis) is achieved through **exercise** and **shivering**.

(*b*) Basal heat production depends on—(*i*) unconscious tensing of muscles, (*ii*) disease and (*iii*) specific dynamic action of food.

Specific dynamic action of food. The rate of metabolism is influenced by the nature of the diet. In general, whenever food is taken, the rate of metabolism rises. All classes of food, however, do not stimulate metabolism to the same extent. The extent to which a particular food influences the rate of metabolism is referred to as its **specific dynamic action** (see **Agarwal** *et al.*, 2013). Proteins have a higher specific dynamic action; these raise the metabolic rate by about 30%; carbohydrates and lipids only about 4.5% and a mixed diet reaches its peak 3–5 hours after ingestion. In cold weather, therefore protein, owing to its high specific dynamic action, gives valuable assistance in warming of body. High protein diet, therefore, is not suitable for a hot climate, as the greater heat production caused by protein intake, puts an additional burden on the cooling mechanism of the body.

In homeothermic animals the thermogenesis depends on the following processes:

1. Shivering thermogenesis. A human being, when appropriately acclimatized, may show an eighteen fold increase in heat production within 12 minutes if suddenly immersed in a bath at 4°C. This record heat production is associated with the shivering or the **involuntary muscle contraction**.

Shivering can be evoked either by a slight decrease in the core temperature or by a decrease in the skin temperature. Thus, its control can be either central or peripheral or both. The conspicuousness and importance of shivering varies from animal to animal. Among mammals, the most obvious shivers are probably the fruit bats or flying foxes of the old world. Some flying foxes of the genus *Pteropus* shiver strongly and visibly at all temperatures below their thermal neutral zone.

2. Non-shivering thermogenesis. The heat produced from extramuscular source is called **nonshivering thermogenesis**. During cold stress acclimatized rats produce as much as 65% of their heat in fat and visceral organs.

Nonshivering thermogenesis plays an important role in the acclimatization (*i.e.*, to become accustomed to a new climate) of mammals to the low temperatures and in arousal from the hibernation and daily torpor. It is clear that the capacity for nonshivering thermogenesis acquired during acclimatization to cold allows rats to be active at much lower temperatures than would be the case if they had to depend on shivering alone, for exercise eliminates shivering but does not affect the nonshivering thermogenesis exercise and nonshivering thermogenesis are addtive, exercise and shivering are not. Nonshivering thermogenesis also significantly increases metabolic scope for activity. In cold acclimatized rats with a high level of nonshivering thermogenesis, the metabolic scope below 30°C is almost the twice that of the warm-acclimatized rats at the same temperature.

The increase in nonshivering thermogenesis associated with the cold adaptation in animals involves changes in the pathways and relationships of the **intermediary metabolism**. It seems probable that the cold-acclimatized rats have increased capacity for the formation and utilisation of glucose. It also seems clear that some nonshivering thermogenesis is evoked in part through the **hormonal stimulation** of lipid metabolism. For example, the thermogenic response to injection of norepinephrine in cold-acclimatized rats involves a marked increase in oxygen consumption and a rise in body temperature. Both these responses can be related to an increased capacity of the liver to from and mobilise nonesterified fatty acids. The thyroid hormone plays an important role in long term development of mammalian nonshivering thermogenesis. One of the important sites of nonshivering thermogenesis is the brown fat.

3. Special thermogenic tissue. Many mammals have a special thermogenic tissue termed as the brown adipose tissue or **brown fat**. Its brown appearance comes from abundant mitochondria and rich blood supply. Brown fat also has relatively high concentrations of myoglobin, cytochromes and flavin compounds. Brown fat was first described in hibernating marmots, *Muris alpinus*. It was called the **hibernating gland** and assumed to play an active role in hibernation. The brown fat is most conspicuously developed in many newborn mammals (including human beings), in hibernating animals at the time of arousal and in cold-acclimatized animals. Brown fat occurs in the neck and thoracic regions (Fig. 9.2) in relation to the major blood vessels so that its heat is quickly transported to those organs (heart and brain) whose continuous high temperatures are vital for a homeotherm. Brown fat differs cytologically from the white fat in that the cells are smaller, polygonal with central nulei and numerous fat droplets which are dispersed rather than concentrated in a single vacuole; mitochondria are small, numerous and closely associated with the fat droplets. Brown fat also differs biochemically from the white fat , with higher concentration of water, protein, phospholipid (UFAs), cholesterol and certain mitochondrial enzymes (*i.e.*, cytochrome oxidase).

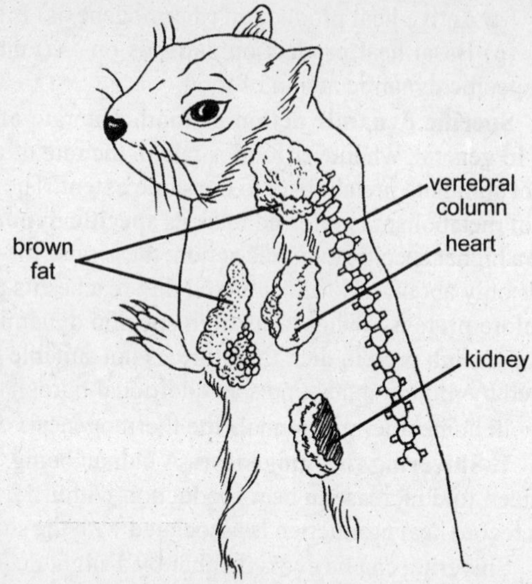

Fig. 9.2. Brown fat in a mammal. It is a thermogenic or heat-producing tissue.

In brown fat cells a unique protein called **thermogenin** uncouple oxidative phosphorylation. This means that metabolic fuels are consumed to produce heat without the production of ATP. This tissue has been regarded as a rapidly firing furnace that warms up some mammals quickly in emergency, without muscular activity or shivering.

Like a smoothly operating automatic furnace, brown fat responds to cold stress with a large burst of heat. The physiological mediator for the stimulation of brown fat appears to be the noradrenaline hormone. Sympathetic stimulation activates the **adenyl cyclase** system with rapid lipase hydrolysis and oxidation of fats.

Countercurrent heat exchange. Some vertebrates also retain heat in a cold environment by using a **countercurrent heat exchange** (Fig. 9.3). In this process, a vessel carrying warm blood from deep within the body passes next to a vessel carrying cold blood from surface of body. The warm blood going out heats the cold blood returning from the body surface, so that this blood is no longer cold when it reaches the interior (*i.e.*, to heart) of the body.

Regulation of Heat Loss

Three structures in the skin influence heat loss: hair/feathers, blood vessels and sweat glands, and one under the skin, *i.e.*, subcutaneous fat.

1. Hair/Feathers. Piloerection protects the body from the vagaries of extreme cold. All homeotherms are provided with hairs or feathers. When the weather is warm the hairs lie flat and the animals are said to wear a thin fur coat. However, when the temperature falls the hair erect and the animals appear thickly coated with fur.

In fact, hair/feathers enclose an insulating layer of still air just outside the skin. This layer of air, being poor conductor of heat, reduces the loss of body heat by radiation. The effectiveness of the insulating air coat is varied according to the environmental requirements through altering its thickness by raising or lowering the air-trapping hair/feathers with the help of their muscles. This mechanism is poor in humans due to reduced body hair. It is limited to the effect called "goose-flesh" or "goose pimples". Humans compensate for the reduction of hair by wearing clothes that provide insulation.

2. Blood vessels. There is an extensive **venous plexus** in the dermis of skin (Fig. 9.4). The rate of blood flow through the plexus can be varied upto 100 times by **dilation** and **contraction** of the arterioles supplying the plexus. Their expansion in hot weather bring more blood to the plexus for rapid loss of heat by radiation. Their constriction in cold weather reduces the blood flow to the plexus for decreasing the heat loss.

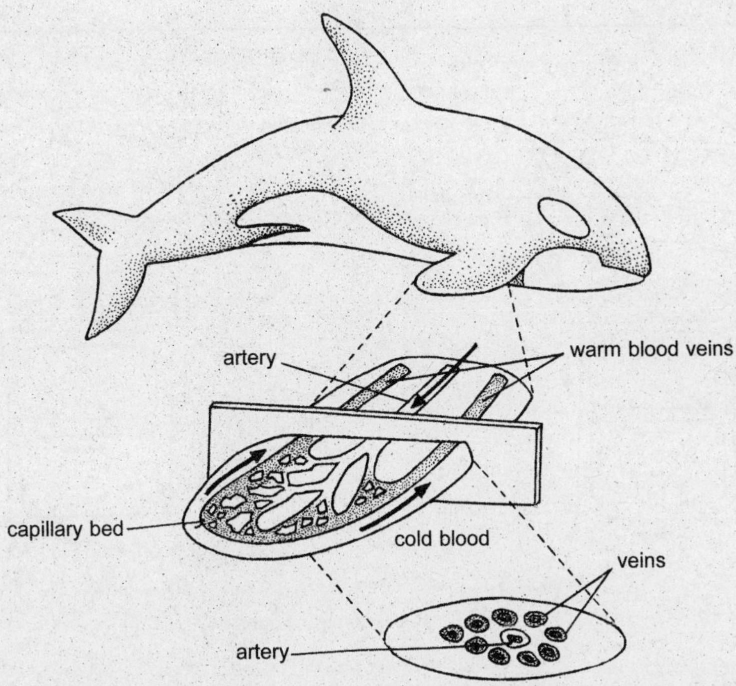

Fig. 9.3. Countercurrent heat exchange. Many marine mammals, such as this killer whale, limit heat loss in cold water by countercurrent flow. The warm blood pumped from within the body in arteries warms the cold blood returning from the skin in veins, so that the core body temperature can remain constant in cold water.

Under the capillary beds, there are **arterio-venous anastomoses** directing connecting the arterioles and the venules deep in the dermis, and 'connecting veins' directly linking arteries and veins in the subcutaneous tissue (Fig. 9.4). **Constriction** of the arteriole reduces the blood flow through the capillaries and the arterio-venous anastomoses which become narrow due to fall of blood pressure in them. Only sufficient blood passes into the skin to keep the tissues alive. Most of the blood bypasses the skin via the connecting veins. Less blood supply to the skin reduces the heat loss. **Dilation** of the arteriole increases the blood flow through the capillaries and the arteriovenous anastomoses which widen due to rise in blood pressure within them. Less blood passes through the connecting veins. Increased blood flow in the skin increases the heat loss.

3. Sweat glands. Human sweat glands may secrete about 10-15 liters of sweat per day, roughly half of the water in the body. Evaporation of this water takes heat from the skin and cools the body, About 580 kcal are used up in evaporating a liter of sweat. The rate of evaporation is reduced by low environmental temperature, high humidity and lack of wind. The sweat glands are under the control of sympathetic nervous system.

Humans, horses and pigs have sweat glands throughout the skin. Mammals having much hair on the bodies have sweat glands on the bare regions. For example, rabbits have sweat glands only in the lips, dogs and cats on the pads of the feet only, and rats on the ears only. These animals increase heat loss by licking their bodies and allowing moisture to evaporate.

4. Panting. Panting is rapid shallow breathing. It accelerates the heat loss by evaporation from the moist nasal and buccal cavity. It can be observed in dogs. Panting has two disadvantages. Increased ventilation causes excessive loss of CO_2 which can result in severe alkalosis (excess of base in body fluid). Increased ventilation requires muscular work, which in turn, increases heat production and this adds to heat load.

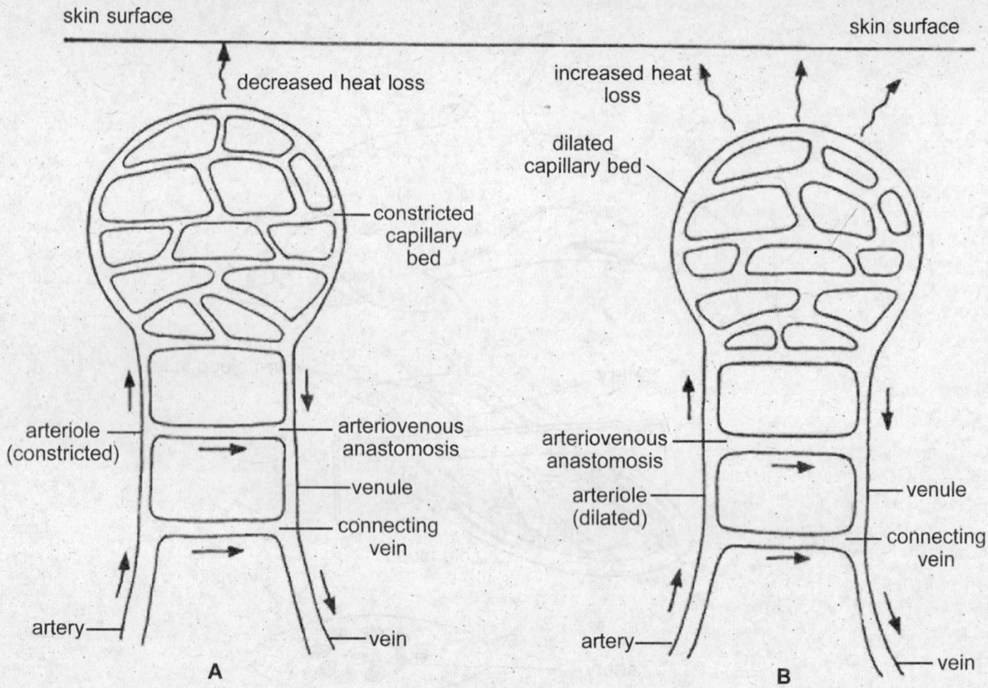

Fig. 9.4. Mechanism of regulation of blood flow and heat loss through skin.

Indeed panting is highly useful to an animal under a sudden heat stress. For example, a fast-running African gazelle pursued by a predator can acquire a high body core temperature and yet keep the brain, the most heat-sensitive organ, at a lower temperature.

5. Subcutaneous fat. The fat deposits under the skin acts as an insulating coat, preventing too rapid heat loss. Aquatic mammals, particularly those inhabiting cold waters, such as whales, sea-lions, walruses and seals, have a thick layer of **blubber**, just under the skin for thermal insulation. Polar bear also has subcutaneous blubber for insulation when in water that almost reaches the skin due to coarse open fur.

9.3 TEMPERATURE CONTROLLING SYSTEM

Hypothalamus of the brain act as the **thermostat** of the body. [Thermostat (Gr., *therme* = heat; *statos* = standing) is an apparatus for automatic regulation of heat.] It monitors the temperature of blood, which is in equilibrium with the core temperature of the body. The hypothalamus excercises its control through the sympathetic and parasympathetic nervous systems and the endocrine system. A **preoptic center** in the hypothalamus has neurons which can respond directly to temperature, giving off more impulses when the temperature of blood falls. From this temperature center, impulses pass to two other centers in the hypothalamus, the anterior **heat-loss center** that initiate cooling and the posterior **heat-production center** which initiates warming of the body.

QUESTIONS

Long Answer Questions

1. Discuss the modes of heat gain and heat loss in animals.
2. How do the homeothermic animals maintain their body temperature?

Short Answer Questions

1. Define the following terms:
   (*i*) Homeostasis;
   (*ii*) Cybernetics;
   (*iii*) Ectotherm;
   (*iv*) Endotherm.
2. Give definition of following terms:
   (*i*) Radiation;
   (*ii*) Thermogenesis;
   (*iii*) Brown fat;
   (*iv*) Countercurrent.
3. Make lists of the modes of heat gain and heat loss by animals.
4. Describe briefly about thermoregulation centers. (*MSU Tirunelveli 2007*)
5. Explain the homeostasis. (*CCSU Meerut 2013*)
6. Differentiate between poikilothermal and homeothermal animals. (*CCSU Meerut 2014*)

Very short Answer Questions

1. Who first recognised the advantage of a constant internal environment?
2. What are poikilotherms?
3. Homeothermic animals are also called endotherms. Why?
4. Name two ectothermic and two endothermic vertebrates.
5. Give examples of three poikilothermic invertebrates.
6. What is basal metabolic rate (BMR) in human beings?
7. What will happen if human beings fail to lose heat?
8. What is arteriovenous anastomosis?
9. What is core temperature?

Fill in the Blanks

1. The term homeostasis was introduced by ________.
2. The animals have anatomical, physiological and ________ mechanisms for maintaining a constant internal environment.
3. The science of control mechanism is called ________.
4. Majority of organisms can live at places with a temperature range of ________ to ________.
5. The ________ fish can maintain body temperature above that of water.
6. The main sources of heat for animals are ________ and ________.
7. Poikilothermic and homeothermic animals are now-a-days are known as ________ and ________.
8. Animals lose heat by ________, conduction, ________ and evaporation.
9. Endothermic animals are ________ and ________.

Multiple Choice Questions

1. The unique protein present in brown fat cell is
   (*a*) prolactin (*b*) thermogenin
   (*c*) melatonin (*d*) calcitonin
2. Human beings are better adapted to
   (*a*) cold climate (*b*) hot climate
   (*c*) moderate climate (*d*) no climate
3. Shivering is a
   (*a*) voluntary process (*b*) heat-losing process
   (*c*) heat-gaining process (*d*) none of these

4. Which of the following is nearest the core temperature?
   (*a*) oral temperature (*b*) axillary temperature
   (*c*) rectal temperature (*d*) groin temperature
5. Countercurrent heat transfer system comprises
   (*a*) muscles (*b*) limbs of Henle's loop
   (*c*) arteries and veins (*d*) vasa recta
6. Heat is lost by organisms as electromagnetic waves in
   (*a*) conduction (*b*) convection
   (*c*) radiation (*d*) evaporation
7. Hypothalamus has ________ center for thermoregulation.
   (*a*) three (*b*) two
   (*c*) four (*d*) many
8. During exposure to the sun in a hot desert, sweating can be kept to a minimum by
   (*a*) removing the clothes
   (*b*) removing clothes and walking slowly to increase convection current
   (*c*) sitting quitetely clothed
   (*d*) sitting quitely unclothed
9. Mechanism activated by cold weather to produce or conserve heat include all except
   (*a*) cutaneous vasodilation (*b*) shivering
   (*c*) thyroxine secretion (*d*) hair erection

ANSWERS

Very Short Answer Questions

1. Claude Bernard.
2. Animals are unable to maintain the body temperature.
3. Since homeotherms rely on internal source of heat (metabolism) for maintaining their body temperature.
4. Ectothermic vertebrates are fish and frog; endothermic vertebrates are pigeon and rabbit.
5. Earthworm, cockroach, snail.
6. About 70 calories per hour.
7. Death due to heat stroke within a few hours.
8. Direct connection between arterioles and venule.
9. Temperature in the skull, thorax and abdomen.

Fill in the Blanks

1. Walter Cannon. **2.** Behavioural. **3.** Cybernetics. **4.** 10, 35°C. **5.** Tuna. **6.** Environment, metebolism. **7.** Ectothermic, endothermic. **8.** Radiation, convection. **9.** Birds, mammals.

Multiple Choice Questions

1. (*b*)	**2.** (*b*)	**3.** (*c*)	**4.** (*c*)	**5.** (*c*)	**6.** (*c*)
7. (*b*)	**8.** (*c*)	**9.** (*a*)			

Fundamentals of Biochemistry (CC–VII)

1

Introduction to Biochemistry

1.1 DEFINITION OF BIOCHEMISTRY

Biochemistry is the study of the molecules and chemical reactions of life. It is the discipline that uses the principles and language of chemistry to explain biology at molecular level (**Horton** *et al.*, 2006).

Biochemistry is a branch of science which deals with the study of complex chemical reactions and chemical structures which give rise to life and life processes. It is application of chemistry to study the biological processes at the cellular and molecular level. In biochemistry are studied the structures and functions of biomolecules such as proteins, carbohydrates, lipids and nucleic acids which constitute cell—the basic unit of cellular organisms. A cell is nothing but an assemblage of various chemicals in a particular ratio.

Biochemists have discovered that the same chemical compounds and the same central metabolic processes are found in organisms as distantly related as bacteria, plants and humans. It is now known that the basic principles of biochemistry are common to all living organisms.

Biochemistry is an **interdisciplinary** subject which is closely related to other fields of biology such as cell biology, genetics, molecular biology, microbiology, immunology, biotechnology and biophysics (Fig. 1.1). Researches in biochemistry use techniques and ideas from different disciplines of science. With the involvement of methods of physics, chemistry and biology, biochemistry aims not only to find out the structure of complex molecules present in biological materials but the ways these molecules interact to form cell, tissues and the whole organism.

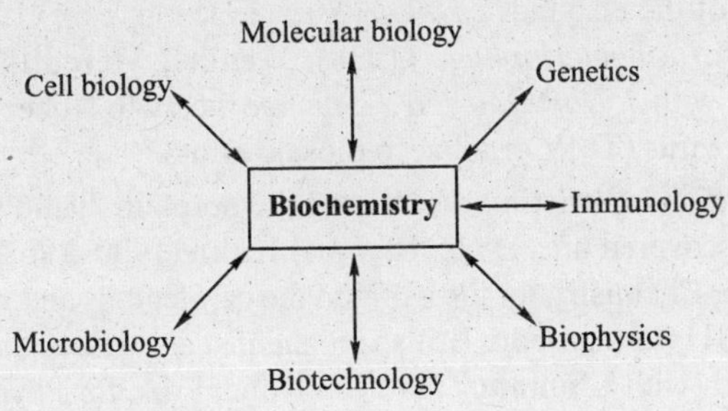

Fig. 1.1. Biochemistry: An interdisciplinary science.

1.2 HISTORY OF BIOCHEMISTRY

Biochemistry is relatively a young branch of science. It has recently emerged as a distinct discipline of biological science.

Biochemistry is started as a branch of organic chemistry. Term biochemistry was coined in 1903 by German chemist **Carl Neuberg** from two Greek words (Bios = life; chemo = chemical).

The history of biochemistry begins with the isolation of *urea* by **Rouelle** from urine in 1773. In 1780, **Scheele** isolated several organic compounds such as *citric acid, malic acid, lactic acid, glycerol* and *uric acid* from natural resources. In 1783, **Lazzaro Spallanzani** (1729–1799) Italian biologist reported that protein digestion is a chemical process. In 1789, **Antoine Lavoisier** (1743–1794) French scientist developed the concept of oxidation and respiration. He is regarded as father of modern biochemistry.

In 1828, **Friedrich Wohler** (1800–1882) German chemist who first synthesized an organic compound, urea, from ammonia and cyanate (inorganic precursors). It marked real begining of biochemistry. **Josiah Willard Gibbs** (1839–1903) US physicist best known for his pioneering work in Chemical Thermodynamics. **Payen** and **Persoz**, in 1833, purified diastase (amylase) of wheat, showed importance of enzymes in biochemistry. In 1878, **F.W. Kuhne** coined the term *enzyme*.

Emil Hermann Fischer (1852–1919) German organic chemist and pioneer of biochemistry, awarded the 1902 Nobel Prize for Chemistry for his work on the structure of *sugars* and *purines*, simple members of both of which he synthesized. He also synthesized *peptides* and studied enzyme action in the breaking down of proteins. The breakthrough in enzyme chemistry was made by **Eduard Buchner** (1860–1917) German organic chemist, awarded the 1907 Nobel Prize in Chemistry for his discovery that *fermentation* does not require the presence of complete yeast cells but only extract containing the enzyme *zymase*. In 1905, **Knoop** worked out β-oxidation of fatty acids. **Albrecht Kossel** (1853–1927) German biochemist awarded the 1910 Nobel Prize for Physiology or Medicine for his work on *proteins* and *nucleic acids*, principally for showing that "nuclein" from cellular sources was composed of a protein and a nonprotein (nucleic acid) component. **F.W. Ostwald** (1853–1932) Lativian-born German physical chemist proved that enzymes are catalysts. **Sir Arthur Harden** (1865–1940), British biochemist awarded along with **Euler-Chelpin** Nobel Prize for chemistry for their work on enzyme action in the fermentation of sugars. Harden demonstrated the existence of coenzymes and Euler-Chelpin determined the structure of *cozymase*. **Otto Heinrich Warburg** (1883–1970), German biochemist awarded the 1931 Nobel Prize for Physiology or Medicine for his work elucidating the chemistry of cell respiration.

In 1926, **James Batcheller Sumner** (1887–1955), US biochemist isolated the enzyme *urease* for the first time. He shared with **J.H. Northrop** and **W.M. Stanley** the 1946 Nobel Prize for Chemistry for his first crystallization of urease showing it was protein. **John Howard Northrop** (1891–1987). US biochemist who shared with **W.H. Stanley** and **J.B. Sumner** the 1946 Nobel Prize for Chemistry for first crystallization of the enzymes *chymotrypsin* (1935), *trypsin* (1936) and *pepsin* (1946) and also for first isolation of a *bacteriophage* (1938). **Wendell Meredith Stanley** (1904–1971) US biochemist who shared with *J. Northrop* and *J. Sumner* the 1946 Nobel Prize for Chemistry for his first crystallization of a virus (TMV or tobacco mosaic virus).

In 1932 **Warburg** and **Christian** discovered flavoprotein "yellow enzyme" (FAD). In 1933 **Krebs** and **Hansleit** discovered urea cycle. **Richard Kuhn** (1900–1967) German chemist awarded the 1938 Nobel Prize for Chemistry for his work on the carotenoids and vitamins. In 1937, **Sir Hans Adolf Krebs** (1900–1981) German-born British biochemist discovered the *citric acid cycle* or "Krebs cycle". He was awarded (with **Lipmann**) the 1953 Nobel Prize for Physiology or Medicine for his discovery. **Fritz Albert Lipmann** (1899–1986) German-born US biochemist who shared with *Sir H.A. Krebs* the 1953 Nobel Prize for Physiology or Medicine for his discovery in 1947 of *coenzyme A*. **George Wells Beadle** (1903–1989) US geneticist who shared part of the 1958 Nobel Prize for Physiology or Medicine with *E.L. Tatum* and *J.Lederberg* for work showing that individual genes controll the production of particular enzymes (1937–40). **Edward Lawrie Tatum** (1909–1975) US biochemist awarded the 1958 Nobel Prize for Physiology or Medicine with *Beadle* and *Lederberg* for work with Beadle showing that individual genes control production of particular enzymes (1937–40).

Joshua Lederberg (b. 1925) US geneticist awarded with *Beadle* and *Tatum* the 1958 Nobel Prize for Physiology or Medicine for his work on bacterial genetics.

Frederick Sanger (b. 1918), British biochemist awarded the 1958 Nobel Prize for Chemistry for his work on the proteins, particularly for first determining the complete structure of a protein, that of bovine **insulin**. In 1980, he became the fourth person to receive Nobel Prizes when he was given the Chemistry award jointly with *Berg* and *Gilbert* for their work on the chemistry of Nucleic acid. Sanger's contribution was to refine the techniques for determining the structure of more complex proteins. **Paul Berg** (b. 1926) US biochemist who shared 1980 Nobel Prize for Chemistry with *Sanger* and *Gilbert* for their work on the chemistry of Nucleic acid. In 1974 Berg was instrumental in calling a moratorium on Genetic Engineering experiments until the hazards could be evaluated. **Walter Gilbert** (b. 1932), US molecular biologist who shared the 1980 Nobel Prize for Chemistry for determining the nucleotide distribution of substances, including a particular kind of RNA found in yeast. **Sir Max Ferdinand Perutz** (1914–2002), Austrian-born British biochemist who shared with **Kendrew** the 1962 Nobel Prize for Chemistry for their research into the structure of **hemoglobin** and other globular proteins. **Sir John Cowdery Kendrew** (1917–1997). British biochemist awarded with **M.F. Perutz** the 1962 Nobel Prize for Chemistry for determining the structure of a globular protein (myoglobin). **Arthur Kornberg** (b. 1918) US biochemist who received the 1959 Nobel Prize for Physiology or Medicine for discovering an enzyme (**DNA polymerase**) that could produce from a mixture of nucleotides exact replicas of DNA molecule.

James Dewey Watson (b. 1928), US biochemist who shared with both *Crick* and *Wilkins* the 1962 Nobel Prize for Physiology or Medicine for his work with Crick establishing the "double helix" moelcular model of DNA (1953). His personalised account of research, *The Double Helix* (1968), became a bestseller. Nobel Laureate **Francis Harry Compton Crick** (1916–2004) is an English biochemist and biophyscicist who with *James Watson*, proposed the double helix model of DNA. Crick's subsequent work has been concerned with deciphering the functions of the individual codons in the genetic code. He formulated the '**Central dogma**', an important tenet of biology. Nobel Laureate **Maurice Hugh Frederick Wilkins** (b. 1916), British biophysicist shared his *X-ray diffraction* studies of DNA with Crick and Watson who determine DNA's molecular structure.

Feodor Lynen (1911–1979), German biochemist who shared with German-born American biochemist **K.E. Bloch** (1912–2000) the 1964 Nobel Prize for Physiology or Medicine for their independent work on the metabolism of cholesterol and the fatty acids. **Peter Dennis Mitchell** (1920–1992), British biochemist who received the 1978 Nobel Prize for Chemistry for his theory of how cells obtain energy from foodstuff and sunlight. **Cesar Milstein** (b. 1927) Argentinian biochemist based in the UK who developed *hybridoma cells* and *monoclonal antibodies* in 1975. He received the 1984 Nobel Prize for Physiology or Medicine jointly with **Niels Jerne** and **Georges Kohler** for discovering the principle for production of monoclonal antibodies (in 1975).

Stanley Cohen (b. 1922). US scientist who, with **Rita Levi-Montalcini**, received the 1986 Nobel Prize for Physiology or Medicine for discoveries of *neural growth* factors. **Susumu Tonegawa** (b. 1939) Japanese biologist based in the US who was awarded the 1987 Nobel Prize for Physiology or Medicine for his pioneering work in immunology, establishing that random grouping of genes on a chromosome achieves the necessary multiformity of antibodies against disease. **Johann Delbruck** (b. 1943). German biochemist who shared the 1988 Nobel Prize for Chemistry with **Robert Huber** and **Harmut Michel** for determining the precise arrangement of the more than 10,000 atoms comprising a bacterial protein complex that plays a central role in a simple type of photosynthesis. **Edwin G. Krebs** (b. 1918) US biochemist who shared with *E.H. Fischer* the 1992 Nobel Prize for Physiology or Medicine for their research into *reversible phosphorylation*, a mechanism governing the activity of cell proteins. **Edmond H. Fischer** (b. 1920). US biochemist, born in Shanghai to Swiss parents, who shared the 1992 Nobel Prize for Physiology or Medicine with *E.G. Krebs* for their discoveries on reversible phosphorylation. **Kary B. Mullis** (b. 1944) US biochemist who won the 1993 Nobel

Prize for Chemistry for his invention of the **PCR** (Polymerase chain reaction). **Alfred Gilman** and **Martin Rodbell** got 1994 Nobel Prize for Physiology or Medicine for investigating the structure and function for a GTP-binding protein called **G protein**. **Jens C. Skou**, **Paul Boyer** and **John Walker** awarded the 1997 Nobel Prize for Chemistry for the discovery of **sodium-potassium pump** (**Na^+/K^+–ATPase**).

John B. Fenn, **Koichi Tanaka** and **Kurt Wuthrich** awarded the 2002 Nobel Prize for Chemistry for improving the technique of MS (mass spectrometry). The 2004 Nobel Prize for Chemistry was awarded to **Aaron Ciechanover**, **Avram Hershko** and **Irwin Rose** for their work on ubiquitin protein and proteosomes. Both are employed inside cell for selective protein degradation. In 2006, **Andrew Z. Fire** and **Craig C. Mellow** were awarded Nobel Prize for discovering the role of RNA interference in the silencing of gene expression. **Martin Chalfie**, **Osamu Shimomura** and **Roger Tsien** awared 2008 Nobel Prize in for Chemistry for their isolation and development of **GFP** (= green florescent protein) from jellyfish and its use in revealing the movement of proteins within a living cell.

1.3 SCOPE OF BIOCHEMISTRY

Biochemistry is the foundation for understanding various biological processes. The knowledge and methods (Box 1.1) of biochemistry can be applied to different fields such as medicine, agriculture, biotechnology and other related fields to follow the complexity of the organisms. Advances made in biochemistry have caused significant contributions toward elucidation and understanding the mystry of life. Initially there was only two main branches of biochemistry: 1. *Descriptive biochemistry* and 2. *Dynamic biochemistry*. But recently various branches of biochemistry have emerged as follows: 1. *Enzymology*; 2. *Endocrinology*; 3. *Clinical biochemistry*; 4. *Physical biochemistry*; 5. *Neuro-chemistry*; 6. *Immunochemistry* and 7. *Molecular biochemistry*.

Box 1.1

A number of tools are used in biochemistry such as *ultracentrifuge*, *spectrophotometers*, and *amino acid analyzers*. Other tools and techniques used for biochemical studies include *labeling with radioactive isotopes* and separation techniques, such as *chromatography*, to analyze very small amounts of substances. *Electrophoresis* is used extensively in studying mixtures of proteins, nucleic acids, carbohydrates, enzymes, etc. Molecular structure may be determined by X-ray diffraction and so on.

For example, clinical biochemistry is an important area of biochemistry which aims at the diagnosis and treatment of diseases. Any change in normal physiology of an organism (*i.e.*, human being) which leads to a disease causes the changes in the biochemistry of the body. For instance, in most of the human diseases the changes appear in the chemistry of blood (Box 1.2). Such changes may be in plasma, serum or blood cells (RBCs, WBCs and platelets). There may be an increase or decrease in the concentration of blood components or appearance of some new substances in circulation. These changes signal the onset of some disease. By studying the biochemistry of blood and other body fluids (and urine, sputum) certain diseases can be diagnosed and treatment can be found out accordingly. Biochemistry and its various principles and techniques are quite significant in the growth of **pharmaceutical science**, **synthesis of drugs**, **hormones**, **antibodies**, etc.

Box 1.2

Study of biochemical details of blood and serum has important role in **clinical diagnosis**, *e.g.*, Hb-S of *sickle cell anaemic* person differs from normal Hb-A (=haemoglobin adult) in having glutamic acid in place of valine at 6th position of β-polypeptide chain. Similarly a person suffering from *haemolytic anaemia* was found to be deficient of Glucose-6-phosphate dehydrogenase enzyme. Both of these are genetical disease of human beings.

In the field of agriculture, biochemistry focuses on the development and innovation of new techniques for crop cultivation, pest control management and crop storage management. New varieties of plants which are resistant to disease or drought are being developed in collaboration with biotechnology. New improved varieties of plants and animals can be produced by applying the techniques of biochemistry in biotechnology.

Biochemical methods are widely used in different industries. Use of enzymes in food, detergents and personal care products is ever increasing. Biochemistry is ready to provide multiple solutions to problem of pollution. Biochemical methods for disposal of industrial and domestic wastes are the need of the hour.

Biochemistry as a subject holds lots of career prospects and job opportunities. There is a great demand of biochemistry in various research institutes and industries. Clinical biochemists are engaged in pathological labs. Pharmaceutical and agriculture research based industries require biochemists in good numbers.

1.4 WATER

The 70 or 90 per cent of living cells is water. In humans, about two-third of the body is formed of water and, of these, about 55 per cent (*i.e.*, 20–22 liters) is confined to cells as intracellular water. The remainder is found in extracellular fluids such as blood, tissue fluid and lymph.

Water is a Polar Molecule

Molecules that exhibit charge separation are called **polar molecules** because of their magnetic poles. Water is a polar molecule. In the water molecule, the oxygen atoms bear a partial negative charge (δ^-) and each hydrogen atom has a partial positive charge (δ^+) (Fig. 1.2A). In liquid water, the negatively charged oxygen atom of one molecule of water is attracted to the postively charged hydrogen atom of another molecule of water. The bond resulting from this attraction is called a **hydrogen bond** (Fig. 1.2B).

A hydrogen bond is a weak bond; it has about one-tenth (10%) of the strength of a covalent bond between a hydrogen atom and an oxygen atom. Weak hydrogen bonds are of great advantage in the biological systems. The reason is that the contact between the molecules becomes very brief. The molecules separate immediately after reacting with one another. Such advantage can be detected in the transmission of chemical signals to the receptor moelcules of brain. The signal molecules transmit the signal by using weak bonds. As a result, they can separate much before the initiation of response in the receiving cell. If there is covalent bonding between the signal molecule and the receptor molecule, the receiving cell will continue to respond long after the transmitting cell stops sending message.

Properties of Water

Water is the tiny polar molecule and passes readily through the membranes. It is already have discussed that in a water molecule the oxygen and hydrogen atoms with opposite polarity are bonded by hydrogen bond (Fig. 1.2B). Two electronegative atoms of oxygen of two water molecules share a hydrogen atom and become linked by hydrogen bond. In this manner, several water molecules are hydrogen bonded to form short-lived macromolecular aggregates. Such hydrogen bonding of several water molecules results in the formtion of a lattice structure. **Fluidity** of water is maintained by very rapid formation and dissociation of hydrogen bonds between water molecules.

Futher, water acts as a solvent in living organisms. To dissolve in water, a solute molecule must form hydrogen bonds with water molecules to join this lattice structure. Hydrophilic polar groups of solutes can form hydrogen bonds with water molecules. So, polar molecules join the lattice structure of water and dissolve in it. This solvent action of water enables the dissolved polar solutes such as glucose to diffuse uniformly in cytoplasm and to be transported in extra-cellular fluids. But hydrophobic non-polar groups cannot join the lattice structure of water. So non-polar molecules such as fats do not dissolve in water and have to be transported in water in association with polar-molecules.

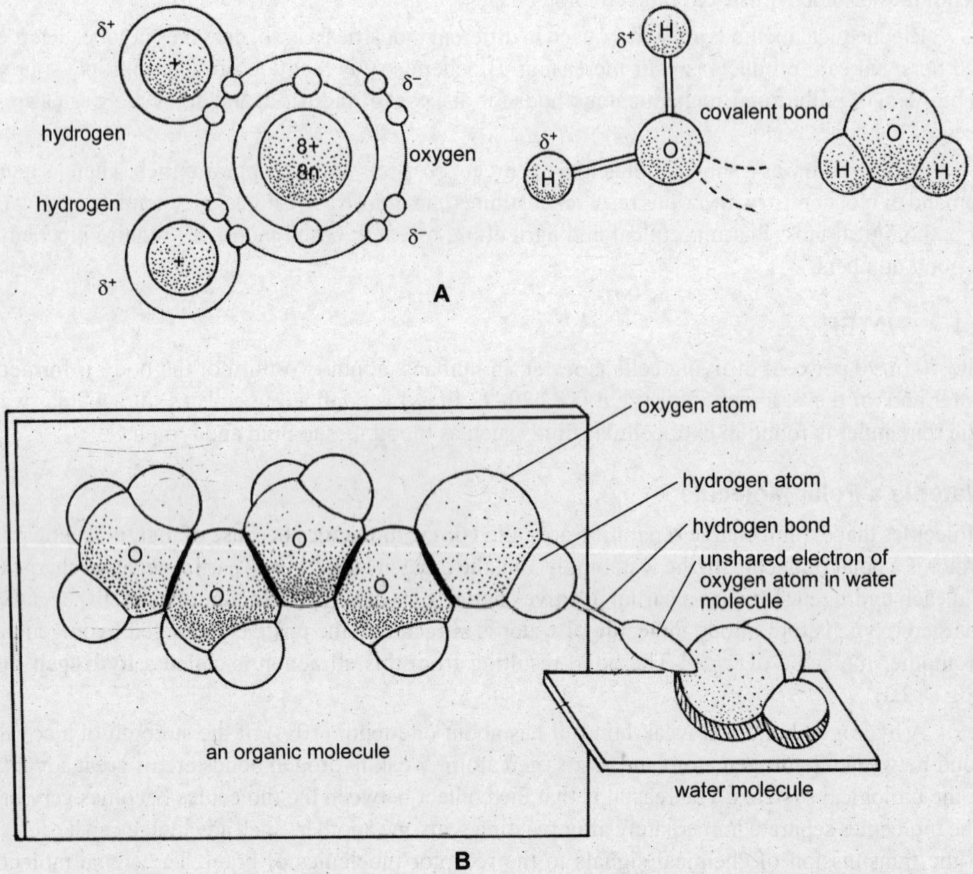

Fig. 1.2. Water molecule. A—Molecular structure of water; B—Hydrogen bond in water molecule.

Water maintains molecular conformations and stabilises structural organization of living organisms. In aqueous body fluids, proteins and nucleic acids become folded and coiled to assume specific 3-dimensional forms suitable for biological activity. Interaction between water and phospholipid molecules arranges the latter into the lipid bilayer of the membrane. Water ionizes to small but significant extent into H^+ and OH^-, depending on temperature. Proteins, nucleic acids and phospholipids attain specific ionic states by accepting or donating H^+ form or to water. Mineral salts also ionize in an aqueous medium of the cells. Owing to its solvent action, water forms an ideal medium for chemical reaction. The solute molecules and ions intermix and make intimate contact with each other. Water is a reagent and is a source of H^+ and OH^- for many biochemical reactions. During photosynthesis, water donates the electron to chlorophyll and is oxidised to molecular oxygen.

Water helps in maintaning the constancy of internal environment of an organism. Some substances, called **buffers**, are capable of neutralising acids or bases and remain in solution in the cytoplasm or extracellular fluids, *e.g.*, bicarbonate (HCO_3), carbonic acid, dibasic phosphate (HPO_4^{2-}) and monobasic phosphate ($H_2PO_4^-$). Acids and bases mix in the body fluids with these substances and are neutralised by them. Due to solvent action water aids in maintaining a constant pH. The high specific heat and mobility of water distribute heat uniformly throughout the body. Water also helps in heat loss from skin and plant surfaces. High latent heat of vapourisation of water causes elimination of excess heat through evaporation of sweat. This maintains a constant body temperature. Elimination of waste products through urine also helps in homeostasis.

Water is Nucleophilic

In addition to its physical properties (*e.g.*, viscocity, solvent power, osmotic pressure, etc.), the chemical properties of water are also important in biochemistry because water molecules can react with biological molecules. The electron-rich oxygen atoms determines much of water's **reactivity** in chemical reactions. Electron-rich oxygen atom determines much of water's **reactivity** in chemical reactions. Electron-rich chemicals are called **nucleophiles** (nucleus lovers) because they seek positively charged or electron deficient species, called **electrophiles** (electron lovers). Nucleophiles are either negatively charged or have unshared pairs of electrons. They attack electrophiles during substitution or addition reactions. The most common nucleophilic atoms in biology are oxygen, nitrogen, sulphur and carbon.

The oxygen atoms of water has two unshared pairs of electrons, which makes it nucleophilic. Although water is relatively weak nucleophile, its cellular concentration is so high that one might expect that many macromolecules would be easily degraded by nucleophilic attack by water, to release their monomeric units, amino acids (Fig. 1.3). The equilibrium for complete hydrolysis of a protein lies far in the direction of degradation; in other words, in the presence of water the ultimate fate of all proteins is destruction by hydrolysis!

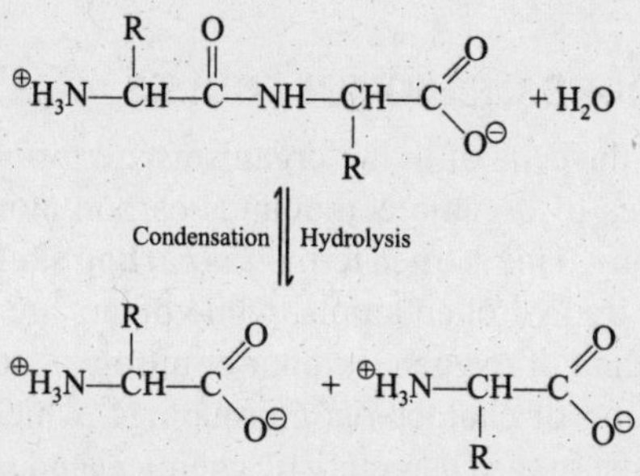

Fig. 1.3. Hydrolysis of peptide. In the presence of water the peptide bonds in proteins and peptides are hydrolyzed. Condensation, the reverse of hydrolysis, is not thermodynamically favoured (Horton *et al.*, 2006).

If there is so much water in cells, why are not all biopolymers rapidly degraded to their components? Similarly, if the equilibrium lies toward breakdown, how does biosynthesis occur in an aqueous environment? Cells circumwent these problems in several ways. The linkages (=bonds) between the monomeric units of macromolecules, such as the amide bonds in proteins and the ester linkages in DNA, are relatively stable in solution at cellular pH and temperature. In this case, the stability of linkages refers to their rate of hydrolysis in water and not their thermodynamic stability. The chemical properties of water, combined with its high concentration means that the Gibbs free energy change for hydrolysis (ΔG) is negative (Box 1.3). This means that the hydrolysis reaction is thermodynamically favourable (*i.e.*, spontaneous). However, the rate of the reaction inside the cell is so slow that macromolecules are not appreciably degraded by spontaneous hydrolysis during the average life time of a cell. It is important to keep in mind the distinction between the preferred direction of a reaction, as indicated by the Gibbs free energy change, and the rate of the reaction, as indicated by the rate constant.

Box 1.3

Gibbs free energy. The thermodynamic function $G = H - TS$, where H is enthalpy, T absolute temperature, and S entropy. Also known as free energy, free enthalpy, Gibbs function.

Enthalpy (or heat content). The sum of the internal energy of a system plus the product of the system's volume multiplied by the pressure exerted.

Entropy. Function of the state of a thermodynamic system whose change in any differential reversible process is equal to the heat absorbed by the system from its surroundings divided by the absolute temperature of the system.

Cells can synthesize macromolecules in an aqueous environment even though condensation reactions—the reverse hydrolysis—are thermodynamically unfavourable. They do this by using the chemical potential energy of ATP to overcome an unfavourable thermodynamic barrier. The enzymes that catalyze such reactions exclude water from the active site where synthetic reactions occur. Because synthetic reactions must proceed against very unfavourable equilibria, they follow two step chemical pathway that differ from the reversal of hydrolysis. The simple condensation pathway shown in Fig. 1.3, is not the pathway that is used in living cells because the presence of high concentration of water makes the direct condensation reaction extremely unfavourable. In the first synthetic step, which is thermodynamically uphill, the molecule to be transferred reacts with ATP to form a reactive intermediate. In the second step, the activated group is readily transferred to the attacking nucleophile. While studying the process of protein synthesis, we will see that the reactive intermediate in protein synthesis is an aminoacyl tRNA that is formed in a reaction that requires ATP. The net result of the biosynthesis reaction is to couple the condensation to the hydrolysis of ATP (**Horton** *et al.*, 2006).

1.5 FUNCTIONAL GROUPS OF BIOMOLECULES

Organic compounds found in the cells of living organisms are formed of carbon, hydrogen, oxygen and nitrogen. In the molecules of organic compounds carbon atoms are linked together forming straight or branched chain or ring. This is often termed as **carbon skeleton**. The majority of free bonds of carbon atoms are satisfied by hydrogen atoms, while others are linked with oxygen, sulphur or nitrogen. This additional bonding of oxygen, sulphur or nitrogen along with hydrogen in an organic molecule leads to the formation of characteristic groupings, which are called **functional groups** (Table 1.1). These attribute most of the characteristic chemical and physical properties to the organic compounds.

Table 1.1 Functional groups and the compounds in which these are found.

	Functional group	Name	Found in organic compound
1.	R—OH	Hydroxy	Alcohol
2.	R—C(=O)—H	Carbonyl (terminal)	Aldehyde
3.	R—C(=O)—R	Carbonyl (internal)	Ketone
4.	R—C(=O)—OH	Carboxy	Organic acid

	Functional group	Name	Found in organic compound
5.	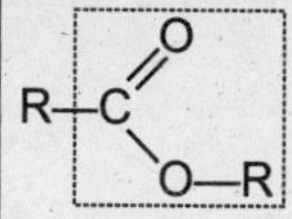	Ester	Ester
6.	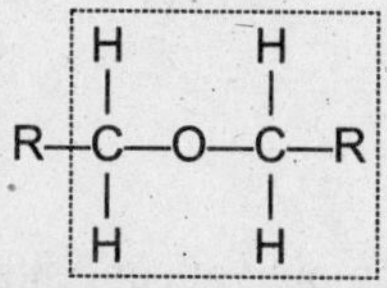	Ether	Ether
7.	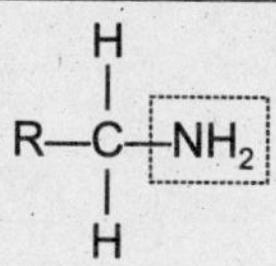	Amino	Amine
8.	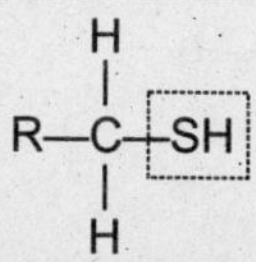	Sulphydryl	Thioalcohols

1. **Hydroxyl or Hydroxy Group (R–OH).** The **hydroxyl group** (–OH) is the functional group of alcohols. It exhibits following characteristic reactions:

   (*i*) **Reaction with phosphorus halides.** The hydroxyl group of alcohols is replaced by halogen group and halide is formed.

$$R-OH + PCl_5 \rightarrow RCl + POCl_3 + HCl$$

$$C_2H_5OH + PCl_5 \rightarrow \underset{\text{Ethyl chloride}}{C_2H_5Cl} + POCl_3 + HCl$$

$$3R-OH + PCl_3 \rightarrow 3RCl + H_3PO_3$$

$$3CH_3OH + PCl_3 \rightarrow 3CH_3Cl + H_3PO_3$$

   (*ii*) **Reaction with hydrogen halides (=halogen acids).** The hydroxyl group of alcohols is replaced by the halogen group from halogen acids and forms alkyl halide.

$$\underset{\text{Alcohol}}{ROH} + Hx \rightarrow \underset{\text{Alkyl halide}}{Rx} + H_2O$$

$$\underset{\text{Methyl alcohol}}{CH_3OH} + HBr \xrightarrow{H_2So_4} \underset{\text{Methyl bromide}}{CH_3Br} + H_2O$$

$$\underset{\text{Ethyl alcohol}}{CH_3CH_2OH} + HCl \xrightarrow{ZnCl_2} \underset{\text{Ethyl chloride}}{CH_3CH_2Cl} + H_2O$$

   (*iii*) **Reaction with aldehydes.** Aldehydes combine with acetal in presence of dry hydrogen chloride or $CaCl_2$ to form hemiacetal and then acetal.

$$R'OH + RCH=O \rightarrow \underset{\text{Hemiacetal}}{RCH\begin{matrix}OH\\OR'\end{matrix}} \xrightarrow{R'OH} \underset{\text{Acetal}}{RCH\begin{matrix}OR'\\OR'\end{matrix}} + H_2O$$

***(iv)* Reaction with ammonia.** When alcohol and ammonia are heated under pressure in presence of catalyst **alumina**, a mixture of three types of amines is formed.

(*a*) $ROH + HNH_2 \rightarrow \underset{\text{Primary amine}}{RNH_2} + H_2O$

(*b*) $RNH_2 + ROH \rightarrow \underset{\text{Secondary amine}}{R_2NH} + H_2O$

(*c*) $R_2NH + ROH \rightarrow \underset{\text{Tertiary amine}}{R_3N} + H_2O$

***(v)* Dehydration.** When an alcohol is treated with a large amount of concentrated H_2SO_4 at 170°C or with $ZnCl_2$, it dehydrates and form hydrocarbons.

$$\underset{\text{Ethyl alcohol}}{CH_3CH_2OH} \xrightarrow{H_2SO_4} \underset{\text{Ethylene}}{CH_2 = CH_2} + H_2O$$

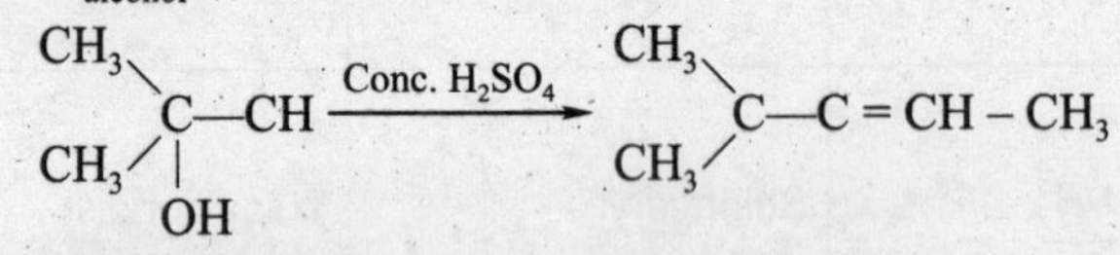

2-Methyl butanol-2 2-Methyl butane-2

2. Carbonyl Group (⟩C = O). Carbonyl group ⟩C = O is present in **aldehydes** and **ketones.** It is terminal in aldehydes but located in any position in the center of the molecule of ketones.

The **aldehydes**, therefore, have a general formula RCHO or 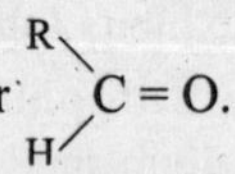

Ketones are represented by the general formula 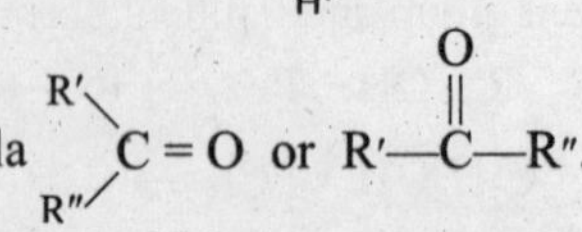

Carbonyl group exhibits following characters, which are common to both aldehydes and ketones.

A. Addition Reactions

1. Addition of hydrogen (reduction). Carbonyl group is reduced to alcoholic group by hydrogenation which can be brought about by the following methods:

***(i)* Catalytic hydrogenation.** This is accomplished with hydrogen and a catalyst such as Ni, Pt or Pd at room temperature:

(*a*) 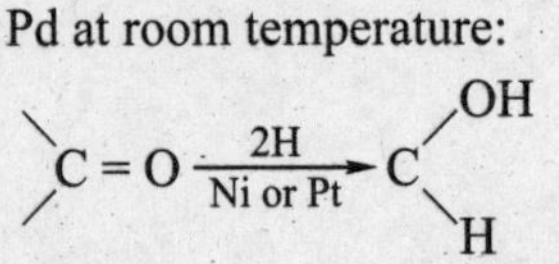

Carbonyl group Alcoholic group

(*b*) $\underset{\text{Acetaldehyde}}{CH_3CHO} \xrightarrow{2H} \underset{\text{Ethyl alcohol}}{CH_3CH_2OH}$

(*c*) 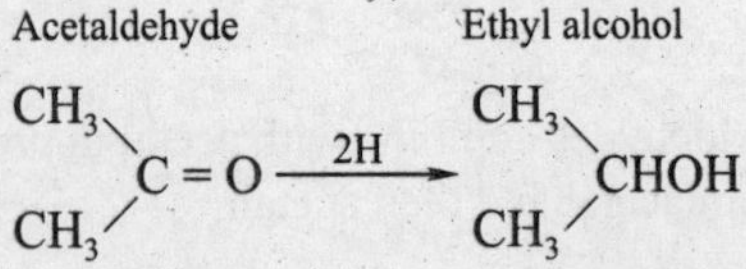

Acetone Isopropyl alcohol

***(ii)* Hydrogenation with active metals.** The carbonyl group is also reduced to alcoholic group when aqueous solution of carbonyl compound is treated with sodium amalgam.

$$C = O \xrightarrow[Na/Hg + H_2O]{2H} \rangle CHOH$$

(*iii*) **Clemmensen reduction.** Carbonyl compounds are reduced with amalgamated zinc and concentrated hydrochloric acid to yield hydrocarbon.

(*a*) 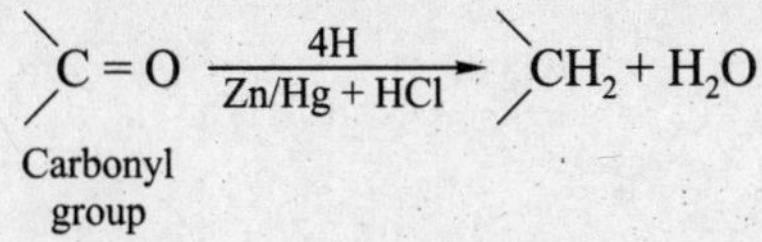

(*b*) 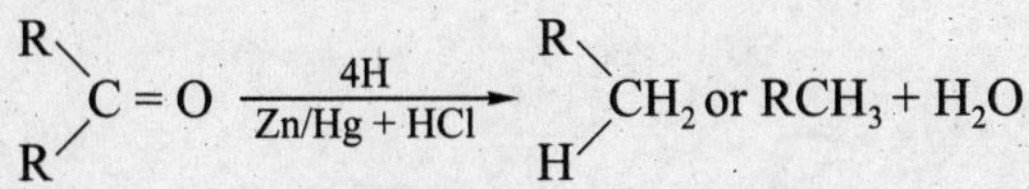

(*c*) 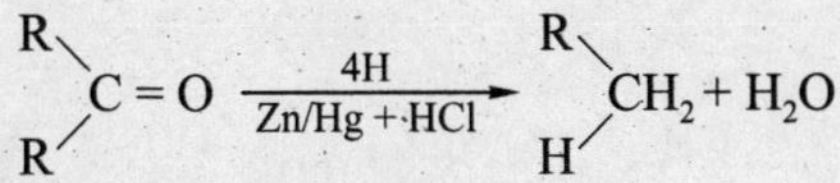

2. Addition of Hydrogen Cyanide. When carbonyl compounds react with HCN, they produce the following compounds:

(*i*) 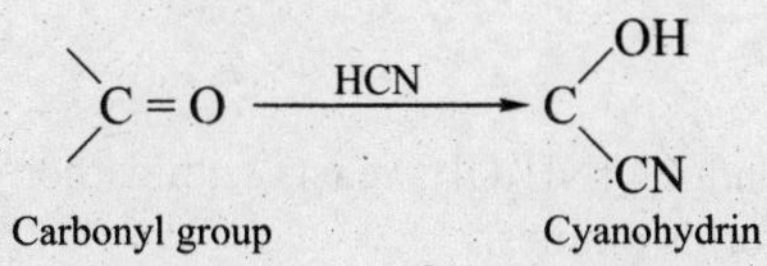

(*ii*) 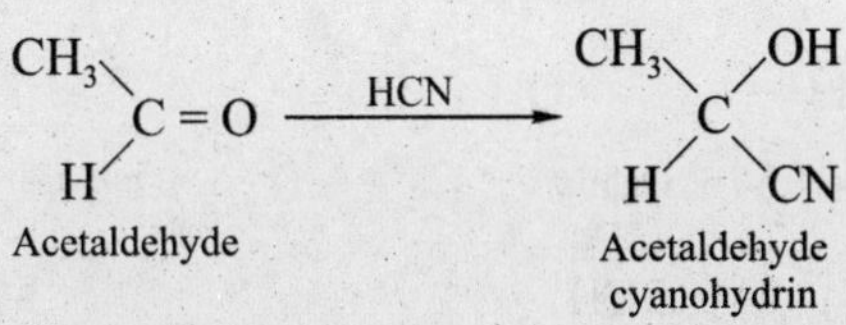

(*iii*) 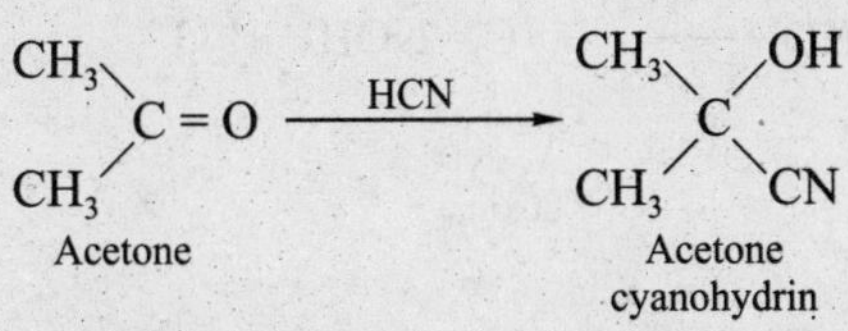

3. Addition of Sodium Bisulphite. Aldehydes and ketones react with a saturated solution of sodium bisulphite and form crystalline bisulphite compounds:

(*i*) 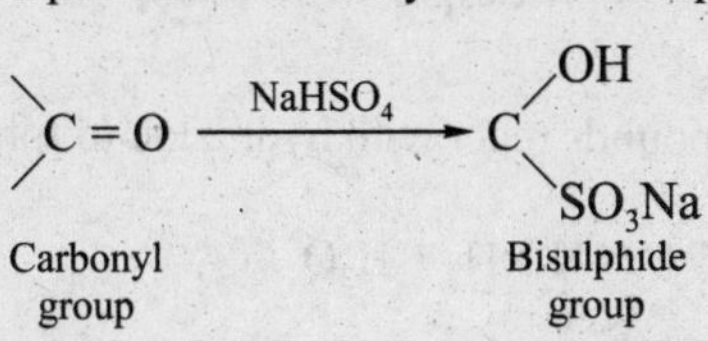

(*ii*) 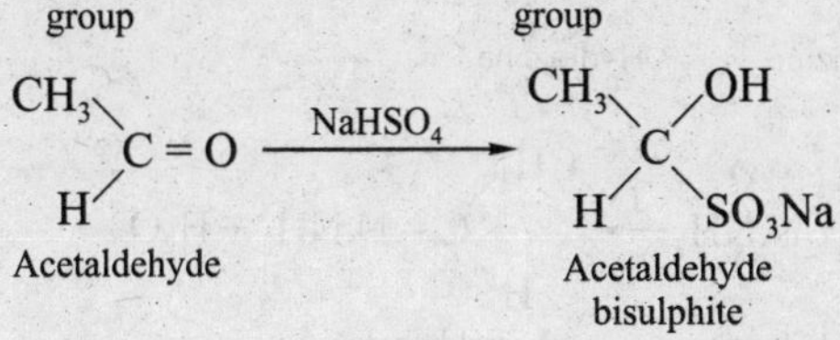

(*iii*) 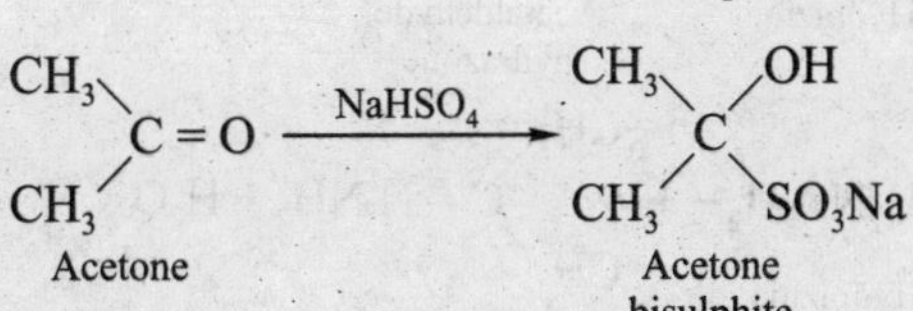

4. Reaction with Grignard Reagent. Aldehydes and ketones with Grignard reagent (RMgX) form a complex, which when decomposes with water forms secondary alcohol:

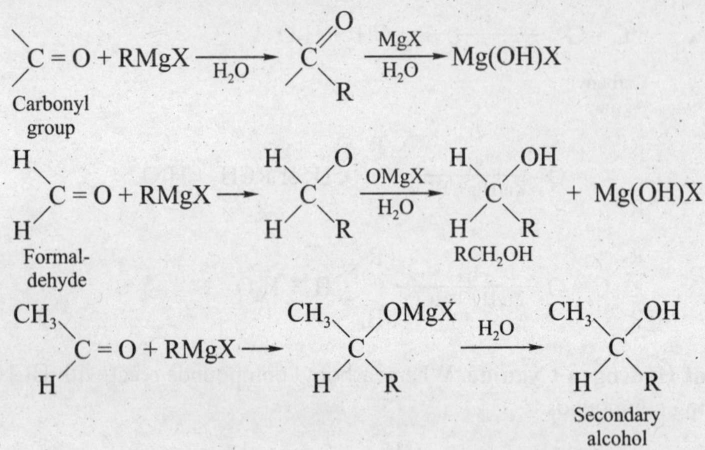

B. Replacement of Carbonyl Group Oxygen

(*i*) **Reaction with hydroxylamine.** Hydroxylamine (NH_2OH) reacts with carbonyl compounds in aqueous or alcohol-water medium and forms oximes:

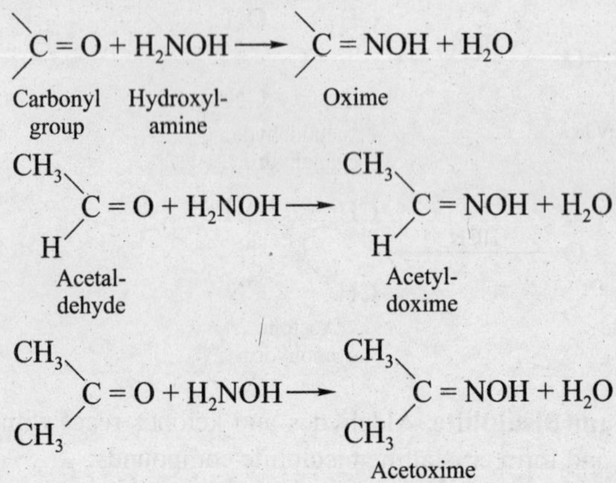

(*ii*) **Reaction with hydrazine.** Carbonyl compounds react with hydrazine to form hydrazones:

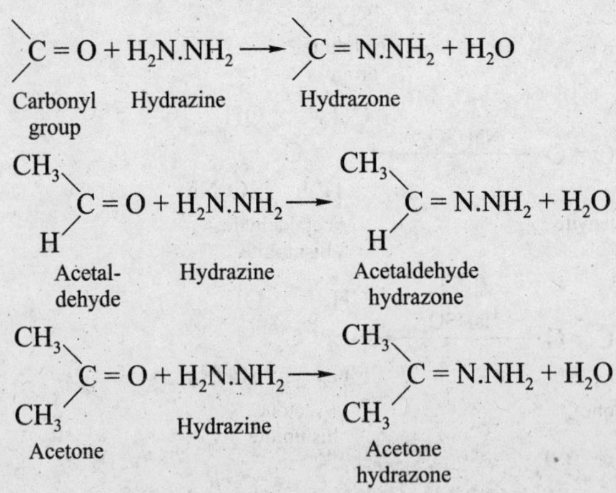

(*iii*) **Reaction with phenylhydrazine.** Reaction between carbonyl compounds and phenylhydrazine form phenylhydrazone:

$$>C=O + H_2N.NHC_6H_5 \longrightarrow C=N.NHC_6H_5 + H_2O$$

Phenylhydrazine → Phenylhydrazone

(*iv*) **Reaction with p-nitrophenyl hydrazine.** These form p-nitrophenyl hydrazones:

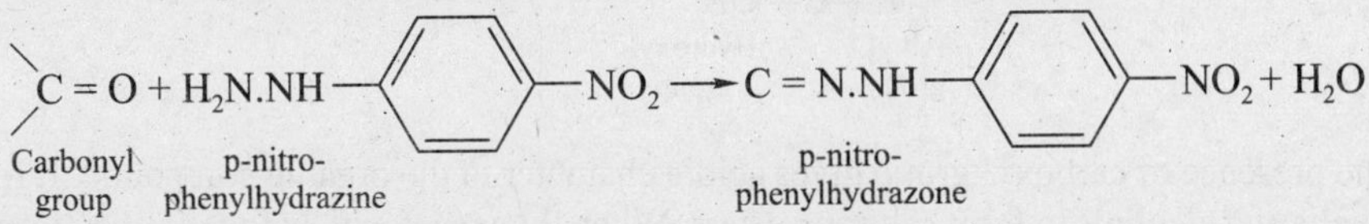

(*v*) **Reaction with 2, 4-dinitrophenyl hydrazine**

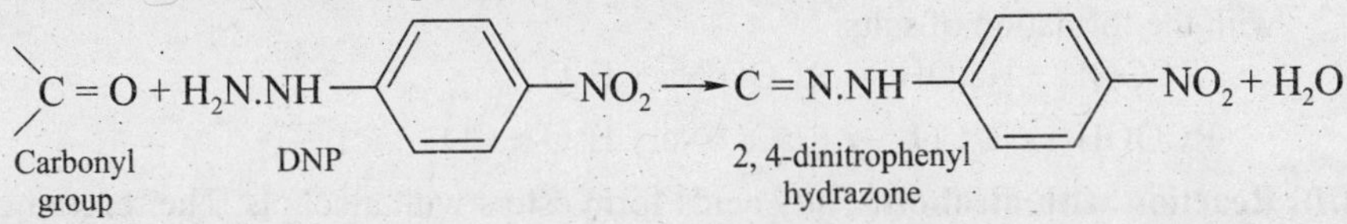

(*vi*) **Reaction with semi-carbazides.** When carbonyl group reacts with semi-carbazides, these form semicarbazones:

$$C=O + H_2N.NHCONH_2 \rightarrow C=N.NHCONH_2 + H_2O$$

Semicarbazide → Semicarbazone

(*vii*) **Reaction with phosphorus pentachloride.** The oxygen atom of carbonyl compounds is replaced by two chloride atoms from phosphorus pentachloride and alkylidene chloride or dichloride are formed:

$$>C=O + PCl_5 \longrightarrow C.Cl_2 + POCl_3$$

Carbonyl group → Alkylidene chloride

$$(CH_3)(H)C=O + PCl_5 \longrightarrow (CH_3)(H)C.Cl_2 + POCl_3$$

Ethylidine dichloride

$$(CH_3)_2C=O + PCl_5 \longrightarrow (CH_3)_2C.Cl_2 + POCl_3$$

Acetone → 2, 2-dichloro-propane

(*viii*) **Reaction with thioalcohols.** When carbonyl group reacts with thioalcohols (mercaptons), both condense to form mercaptal and mercaptols:

$$>C=O + 2RSH \longrightarrow C(SR)_2 + H_2O$$

Carbonyl group → Mercaptal

$$(CH_3)(H)C=:O: + 2RS:H: \longrightarrow (CH_3)(H)C(SR)_2 + H_2O$$

Acetaldehyde mercaptal

3. Carboxyl Group 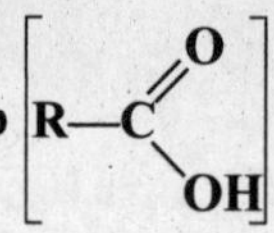

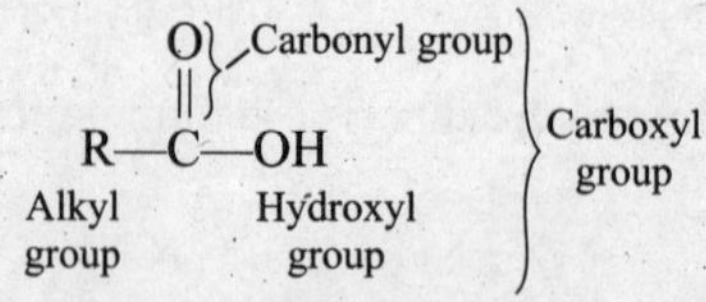

The presence of carboxyl group gives acidic character to the organic compound. It reacts with alkalis and alcohols to form salts and esters. When they react with halogens, they form halides.

(*i*) **Reaction of carboxyl group with bases.** Organic acids are neutralized by the alkaline with the formation of salts:

$$RCOOH + Na{:}OH \rightarrow RCOO.Na + H_2O$$

$$RCOOH + Na_2CO_3 \rightarrow RCOO.Na + H_2O + CO_2$$

(*ii*) **Reaction with alcohols.** Fatty acids form esters with alcohols. The reaction is carried out in the presence of sulphuric acid:

$$RCOOH + R'OH \xrightarrow{H_2SO_4} RCOOR' + H_2O$$

$$\underset{\text{Acetic acid}}{CH_3COOH} + \underset{\text{Ethyl alcohol}}{C_2H_5OH} \xrightarrow{H_2SO_4} \underset{\text{Etyl acetate}}{CH_3COO.C_2H_5} + H_2O$$

(*iii*) **Reaction with ammonia.** In presence of excess of ammonia carboxylic acids form ammonium, which on heating forms acid amides:

$$RCOOH + NH_3 \rightarrow RCOONH_4 \xrightarrow{\text{heated}} \underset{\text{Acid amides}}{RCONH_2}$$

$$CH_3COONH_2 \rightarrow \underset{\text{Acetamide}}{CH_3CONH_2} + H_2O$$

(*iv*) **Reaction with phosphorus halides.**

$$RCOOH + PCl_5 \rightarrow RCOCl + POCl_3 + HCl$$

$$\underset{\text{Acetic acid}}{CH_3COOH} + PCl_5 \rightarrow \underset{\text{Acetyl chloride}}{CH_3COCl} + POCl_3 + HCl$$

(*v*) **Decarboxylation.** When a sodium salt of fatty acid is heated with soda lime, alkanes are formed:

$$RCOONa \xrightarrow{H_2O} \underset{\text{Alkane}}{R.H} + Na_2CO_3$$

$$\underset{\text{Acetic acid}}{CH_3COOH} \xrightarrow[H_2O]{\text{Soda lime}} \underset{\text{Methane}}{CH_4} + Na_2CO_3$$

4. Amino Group 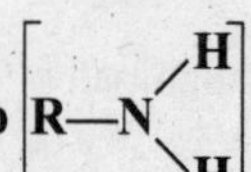

Amines are derivatives of ammonia obtained by replacing one or more hydrogen atoms from ammonia by hydrocarbon radicals.

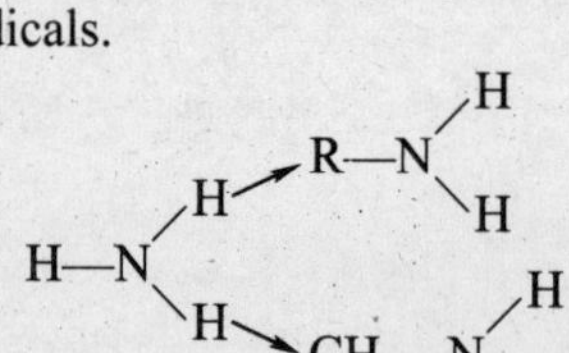

Methylamine

Amines are basic in nature. These exhibit following reactions:

(*i*) **Reaction of amino group with water.** Amines form substitution products with water which hydrolyses into hydroxyl ions and substitute ammonia ion:

$$RNH_2 + H_2O \rightarrow RNH_3OH$$

$$RNH_3OH \rightarrow [RNH_3]^+ + OH^-$$

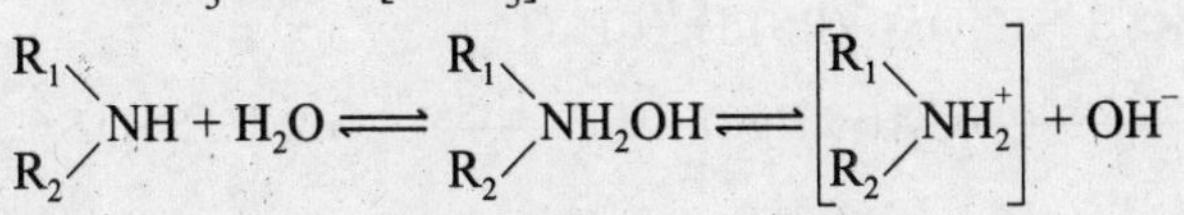

(*ii*) **Reaction of amino group with acids.** Amines combine with halogen acids to form salts:

$$RNH_2 + HCl \rightleftharpoons \underset{\text{Alkylamine hydrochloride}}{RNH_2HCl} \rightleftharpoons (RNH_3)^+$$

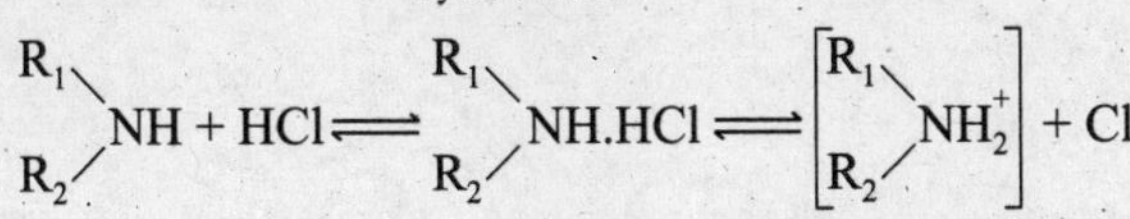

Dialkyl-ammonium chloride — dialkyl-ammonia

(*iii*) **Reaction with alkyl halides.** Amines react with alkyl halides to form alkyl-substituted ammonium halides.

$$C_2H_5NH_2 \xrightarrow[-HI]{C_2H_5I} (C_2H_5)_2\,NH \xrightarrow[-HI]{C_2H_5I} (C_2H_5)_3\,N$$

5. Sulphydril Group (–SH). Sulphydril group (–SH) is the functional radical of thio-alcohols or mercaptans. It is also known as thiol or mercapto group. This strongly resembles to –OH (hydroxyl) group of alcohols, but it is more acidic.

(*i*) **Reaction of –SH group with alkali metals.** When treated with alkali metals these evolve H_2 and form mercaptides.

$$2RSH + 2Na \rightarrow \underset{\text{Sodium mercaptide}}{2RSNa} + H_2$$

These mercaptides are salts and decompose into mercaptans in water

$$RSNa + H_2O \rightarrow RSH + NaOH$$

(*ii*) **Reaction with acids.** Thioalcohols react with acids forming thioesters.

$$R'SH + RCOOH \rightarrow \underset{\text{Thioester}}{RCOSR'} + H_2O$$

(*iii*) **Reaction with acid chlorides.** With acid chlorides –SH groups form thioesters:

$$\underset{\text{Acid}}{R'SH} + \underset{\text{Chloride}}{RCOCl} \rightarrow \underset{\text{Thioester}}{RCOSR'} + HCl$$

(*iv*) **Reaction with aldehyde and ketones.** The mercaptans unite with the carbonyl oxygen of aldehydes to form mercaptals (thiocetals) and with ketones form mercaptols:

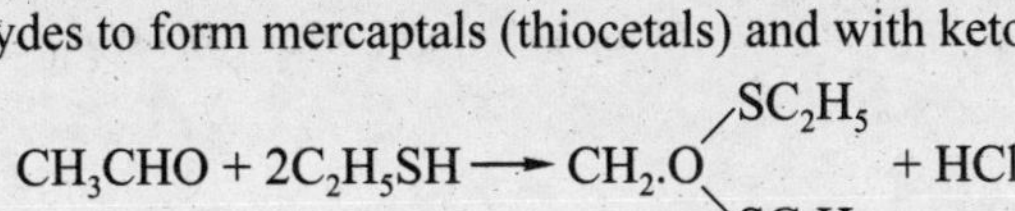

Acid — Mercaptal (Diethyl methyl mercaptal)

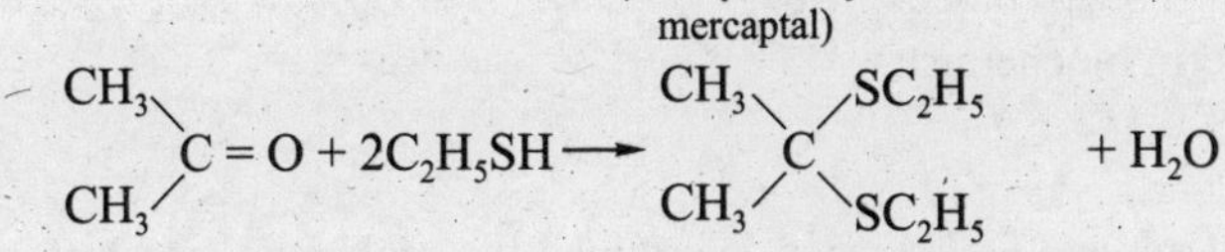

Acetone mercaptol (diethyl dimethyl merchaptol)

Acetone mercaptol on oxidation with potassium permangnate gives **sulphonal** which is an important sleep producing medicine.

(*v*) **Reaction with heavy metals.** Mercaptans form insoluble salts with heavy metals:

$$2RSH + \underset{\text{Lead acetate}}{(CH_3COO)_2\,Pb} \rightarrow \underset{\text{Lead mercaptide}}{(RS)_2\,Pb} + 2CH_3COOH$$

$$2RSH + HgO \rightarrow (RS)_2Hg + H_2O$$

$$C_2H_5SH \xrightarrow{HgCl_2} C_2H_5S.HgCl + HCl \xrightarrow{C_2H_5SH} C_2H_5.S.Hg.S.C_2H_5 + HCl$$

(*vi*) **Oxidation.**

(*a*) By mild oxidising agents, sulphydril group forms dialkyl disulphide:

$$2RSH \rightarrow \underset{\text{Dialkyl disulphide}}{R-S-S-R} + H_2O$$

(*b*) With strong oxidizing agents, it forms sulphonic acid:

$$RSH \rightarrow \underset{\substack{\text{Alkyl} \\ \text{sulphonic acid}}}{RSO_3H} \quad \text{or} \quad R-\overset{\overset{\displaystyle O}{\|}}{\underset{\underset{\displaystyle O}{\|}}{S}}-OH$$

(*vii*) **Formation of thioesters.** When vapours of mercapton are passed over a heated catalyst (AlO_3 + ZnS) at 300°C thiethers are obtained:

$$2RSH \xrightarrow[300°C]{Al_2O_3 + ZnS} \underset{\text{Thioether}}{R-S-R} + H_2S$$

QUESTIONS

Long Answer Questions

1. Why water is regarded as polar molecule? Describe nucleophilic nature of water. Describe properties of water in context with cell.
2. Describe growth of biochemistry in 19th century.
3. Write down six Nobel Prize winnig discoveries of 20th century.
4. Discuss the functional groups of biochemistry. (*Allahabad 2001*)
5. What are the important "Functional groups"? Give their characteristic chemical reactions.

Short Answer Questions

1. Differentiate between physiology and biochemistry. (*CCSU Meerut 2014*)
2. Write differences between the following:
   (*i*) Alcohol and Thioalcohols (*Gorakhpur 1999; Agra 2002*)
   (*ii*) Reactions of Aldehydes and Ketones
   (*iii*) Amines and Amino acids
3. Wirte short notes on the following:
   (*a*) Thioalcohols (*Bundelkhand 2000; Gorakhpur 2000; Avadh 2001*)
   (*b*) Water

Very Short Answer Questions

1. Define biochemistry. (*CCSU Meerut 2008*)
2. Name branches of modern biochemistry.
3. Who coined the term biochemistry.

Fill in the Blanks

1. The 1902 Nobel Laureate E.H. Fisher determined the molecular structure of ________ and ________.

2. In 1935, crystallization of enzyme ________ was first done by J.N. Northrop.
3. In 1905, ________ worked out β-oxidation of fatty acids.
4. Coenzyme A was discovered in 1947 by ________.
5. Urea cycle was discovered by ________ and ________ in 1933.
6. Electrophoresis was introduced in 1933 for separating proteins in solution by ________.
7. Central role of ATP in energy transfer cycle was postulated by ________.

Multiple Choice Questions

1. 80% of the cells is
   (*a*) minerals (*b*) water
   (*c*) proteins (*d*) carbohydrates
2. The first proteins to be sequenced was the insulin hormone. Who determined its amino acid sequence in 1953?
   (*a*) V. Ingram (*b*) F. Sanger
   (*c*) F. Lynen (*d*) E. Fischer

ANSWERS

Short Answer Questions

2. (*i*) Alcohols (R – OH) are those organic compounds in which a hydroxyl (–OH) group is attached to the carbon atom of the organic compound. It may be primary, secondary or tertiary carbon atoms.
When oxygen of the hydroxyl group in an alcohol is replaced by sulphur, the compound formed is known as **thioalcohol**. The SH group of a thioalcohol is known as the **thiol** group.

(*ii*) Chemical reactions of aldehyde and ketones are mainly due to the carbonyl group (having a double bond between carbon and oxygen atoms). In *aldehydes* the carbonyl group is terminal while in *ketones* it may be in any position in the center of the molecule. Both of them are **polar compounds**. The addition of polar reagents takes place due to electromeric effect. *Because of double bond these are very reactive*. These easily undergo hydrogenation into acids, producing alcohols or get oxidised forming salt with hydrogen cyanide, sodium bisulphite, alcohols, thioalcohols, etc. On hydrolysis these produce alcohols.

Very Short Answer Questions

1. Biochemistry is the study of the chemical changes within and produced by living organisms. In other words biochemistry is the discipline in which biological phenomena are analyzed in terms of chemistry.
2. 1. Descriptive biochemistry; and 2. Dynamic biochemistry.
3. German chemist **Carl Neuberg** in 1903.

Fill in the Blanks

1. Sugars, purines. **2.** Chymotrypsin. **3.** Knoop. **4.** F.A. Lipmann. **5.** H.A. Krebs, K. Hanseleit. **6.** Arne W.K. Tiselius. **7.** F.A. Lipmann.

Multiple Choice Questions

1. (*b*) **2.** (*b*)

2

Carbohydrates

Carbohydrates (L. *carbo* = coal; G. *hydro* = water) are compounds of carbon, hydrogen and oxygen, *e.g.*, glucose, fructose, sugar, starch, glycogen, cellulose, chitin, etc. They are also called **saccharides** (G. *sakcharon*—sugar) because they are sugars or polymers of sugars. On the basis of mass, carbohydrates are the most abundant class of biological molecules on Earth. Although all organisms can synthesize carbohydrate, much of it is produced by photosynthetic organisms, including bacteria, algae and plants. These organisms convert solar energy to chemical energy, which is then used to make carbohydrate from carbon dioxide. Carbohydrates play several crucial roles in living organisms. In animals and plants, carbohydrate polymers act as energy storage molecules (*e.g.*, starch in plants and glycogen in animals). Animals can ingest carbohydrates (*e.g.*, glucose) that can then be oxidized to yield energy for metabolic processes. Polymeric carbohydrates (*e.g.*, cellulose) are also found in cell walls and in the protective coatings (*e.g.*, chitin) of many organisms. Some other carbohydrate polymers are marker molecules (*e.g.*, antigens A, B, AB and O of RBCs) that allow one type of cell to recognize and interact with another type. Carbohydrate derivatives are found in a number of biological molecules, including some coenzymes and the nucleic acids DNA and RNA (*i.e.*, Deoxyribose and ribose sugars which form part of structural frame work of DNA and RNA, respectively).

2.1 CLASSIFICATION

Carbohydrates can be described by the number of monomeric units they contain: Monosaccharides, Oligosaccharides and Polysaccharides.

1. **Monosaccharides.** They are the smallest units of carbohydrate structure. The name carbohydrate, "hydrate of carbon", refers to their empirical formula $(CH_2O)n$, where n is 3 or greater (n is usually 5 or 6 but can be upto 9).
2. **Oligosaccharides.** They are polymers of 2 to about 20 monosaccharide residues. In them, monomers are joined together by **glucosidic bonds**. The most common oligosaccharides are disaccharides such as sucrose and maltose in plants and lactose in animals; they tend to have two linked monosaccharide residues.
3. **Polysaccharides.** They are polymers that contain many (usually more than 20) minosaccharide residues.

Monosaccharides and disaccharides are commonly known as **sugars** and are readily soluble in water. They are considered to be **micromolecules** whereas polysaccharides are considered to be **macromolecules**.

Oligosaccharides and polysaccharides do not have the empirical formula $(CH_2O)n$ because water is eliminated during polymer formation. The term **glycan** is more general term for carbohydrate polymers. It can refer to a polymer of identical sugars (called **homoglycan**) or of different sugars (called **heteroglycan**).

Glycoconjugates are carbohydrate derivatives in which one or more carbodydrate chains are linked covalently to a peptide, protein or lipid. These derivatives include **protoglycans**, **glycoproteins** and **glycolipids**.

2.2 MONOSACCHARIDES

Monosaccharides (G. *mono* = one) are water-soluble, white, crystalline solids that have a sweet taste. Examples include glucose and fructose. Chemically, most monosaccharides are **chiral** compounds (see Box 2.1). Monosaccharides are polyhydroxy aldehydes or **aldoses**, or polyhydroxy ketones or **ketoses** (Fig. 2.1). Monosaccharides are called monomers and form structural units of oligo- and polysaccharides. They are classified by their type of carboxyl group (Box 2.1) and their number of carbon atoms. As a rule, the suffix—*ose* is used in naming carbohydrates, although there are a number of exceptions. All monosaccharides contain at least three cabon atoms (Table 2.1). One of these is carbonyl carbon, and each of the remaining carbon atoms bear a hydroxyl group. In aldoses, the most oxidised carbon atom is designated C-1 and is drawn at the top of Fisher projection. In ketoses, the most oxidised carbon atom is usually C-2.

Box 2.1

Chiral atom. An atom with asymmetric substitutions that can exist in two different configurations. For example, carbon is a chiral atom since it has 4 different substituents, *i.e.*, in a complex organic compound, the tetrahedrally bonded carbon atom is attached to 4 different atoms or groups of atoms (see **Jain** 2012).

Carbonyl group.  A carbonyl group is a pair of atoms consisting of a carbon atom linked to an oxygen atom by a double bond (–C = O).

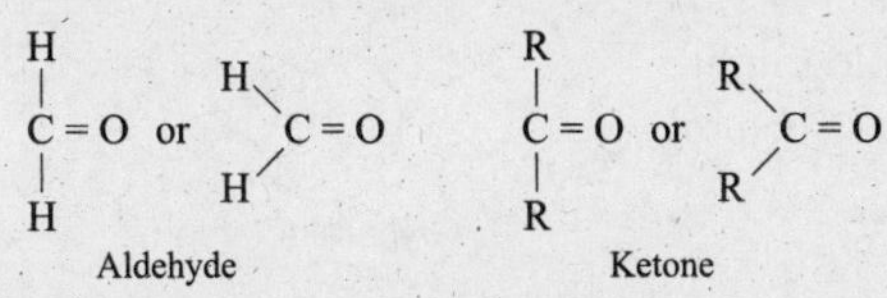

Fig. 2.1. General formula of aldehyde and ketone.

Table 2.1 Some monosaccharides.

	Carbohydrates	Number of carbon atoms	Aldoses	Ketoses
1.	Trioses ($C_3H_6O_3$)	3	Glycerose	Dihydroxyacetone
2.	Tetroses ($C_4H_8O_4$)	4	Erythrose and Threoses	Erythrulose
3.	Pentoses ($C_5H_{10}O_5$)	5	Ribose, Xylose, Lyxose, Arabinose	Ribulose, Xylulose
4.	Hexoses ($C_6H_{12}O_6$)	6	Glucose, Galactose, Mannose, Allose, Altrose Talose	Fructose (levulose), Sorbose, Psicose, Tagatose
5.	Heptoses ($C_7H_{14}O_7$)	7	Mannoheptulose	Sedoheptulose

The smallest monosaccharides are **trioses**, or three-carbon sugars. One- or two-carbon compounds having the general formula $(CH_2O)_n$ do not have properties typical of carbohydrates (such as sweet taste and the ability to crystalize). The aldehyde triose or aldotriose, is glyceraldehyde (Fig. 2.2A). Because of its central carbon, C-2 has four different groups attached to it, glyceraldehyde is chiral (Box 2.1). The ketonic triose, or ketotriose is dihydroxyacetone (Fig. 2.2B), which is achiral—it has no asymmetric carbon atom. All other monosaccharides, longer chain versions of these two sugars are chiral.

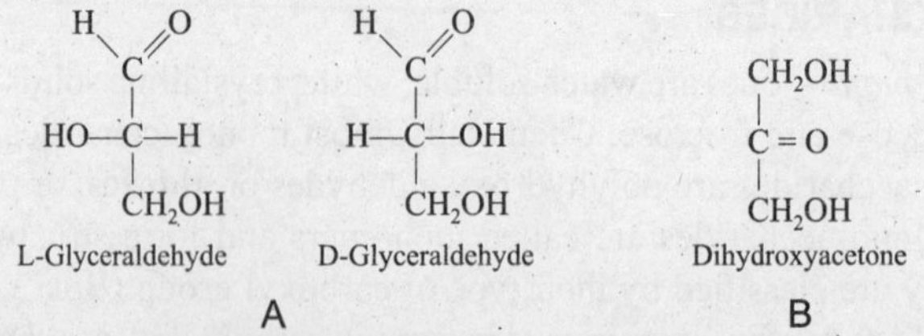

Fig. 2.2. Fischer projections of glyceraldehyde (A) and (B) dihydroxyacetone. The designation L (for left) and D (for right) for glyceraldehyde refers to the configuration of the hydroxyl group of the chiral carbon (C-2). Dihydroxyacetone is achiral.

In Fig. 2.2, the stereoisomers D- and L- glyceraldehyde are shown as ball- and -stick models. Chiral molecules are optically active; that is, they rotate the plane of polarised light. The convention for designating D and L isomers was originally based on the optical properties of glyceraldehyde. The form of glyceraldehyde that caused rotation to the right (dextrorotory) was designated D; the form that caused rotation to the left (levoratory) was designated L. X-ray crystallographic studies have confirmed these structural assignments.

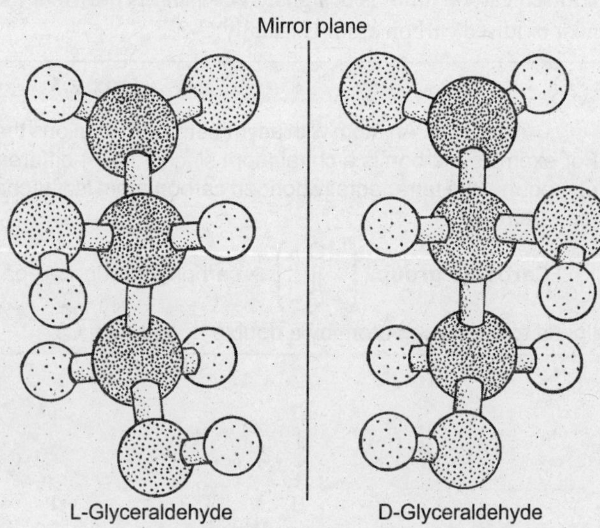

Fig. 2.3. View of L-glyceraldehyde (left) and D-glyceraldehyde (right).

Types of Monosaccharides

1. Trioses ($C_3H_6O_3$). Trioses are formed only during metabolic breakdown of hexoses. They do not occur in nature. They are formed of 3 carbon atom chains and are smallest carbohydrate molecules. *Glyceraldehyde* and *dihydroxyacetone* are two trioses formed during glycolysis of glucose.

2. Tetroses ($C_4H_8O_4$). They are four carbon compounds. Examples are **erythrose** and **threose**.

3. Pentoses ($C_5H_{10}O_5$). These are 5 carbon sugars. Examples of pentoses are **ribose**, **deoxyribose**, **ribulose**, **xylulose**, **arabinose**.

The pentose sugar, **ribose** is the important constituent molecule of the ribonucleic acid (RNA) and certain coenzymes as nicotinamide adenine dinucleotide (NAD), NAD phosphate (NADP), adenosine triphosphate (ATP) and coenzyme A (CoA). Another pentose sugar the **deoxyribose** is the important constituent of the deoxyribonucleic acid (DNA). The **ribulose** is a pentose sugar which is necessary for phtosynthetic mechanism.

4. Hexoses ($C_6H_{12}O_6$). These are compounds of 6-carbon atoms and are found as **glucose**, **fructose**, **galactose** and **mannose**. All of them are represented by the same formula $C_6H_{12}O_6$. These sugars are sweet, neutral in reaction, crystallizable, soluble in water and readily dialysable. They can undergo alcoholic fermentation. Hexoses form a major source of energy and also act as a raw material for other carbohydrates. All the carbohydrates taken into body (by feeding) are ultimately hydrolyzed into glucose which is absorbed in blood and transported to the cells. Only glucose and fructose occur in solution.

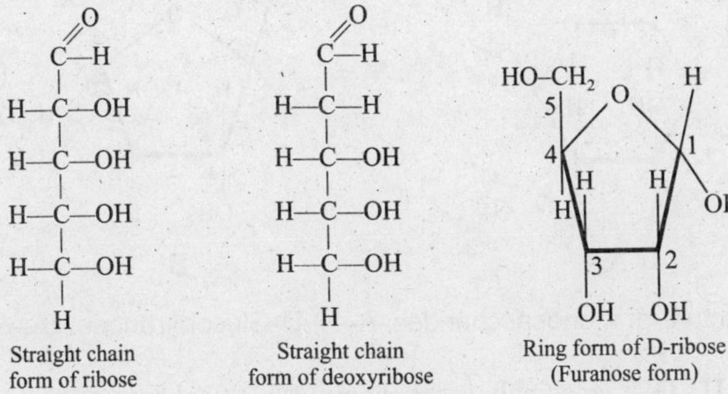

Fig. 2.4. Pentose sugars: ribose and deoxyribose.

Glucose, fructose and mannose differ only in their first and second carbon atoms. The carbonyl (C = O) group in glucose is attached to the first carbon atom and forms **aldehyde group**. But in fructose it is found with second carbon and forms **keto group**.

***(i)* Glucose** (G. *gleukos* = sweet new wine, sweetness). **Grape sugar** or **dextrose** occurs in fruits and honey along with fructose. In the body, glucose is present in the blood, in the intestine during digestion and in the urine of dibetics. It occurs in the sap of plants. Glucose is natural component of many disaccharides (*e.g.*, sucrose, lactose, maltose, cellobiose, trehalose) and polysaccharides (*e.g.*, starch, glycogen, cellulose). Glucose occurs in two isomeric forms: D- glucose and L- glucose. Glucose is a main metabolic fuel.

***(ii)* Fructose.** Fructose or fruit sugar is also known as **levulose.** It occurs in green plants, fruits and honey. Fructose has a composition similar to glucose but it is **keto-hexose** and is less readily absorbed by tissue cells. This is due to the difference in position of OH group. In liver and intestine it is changed to glucose.

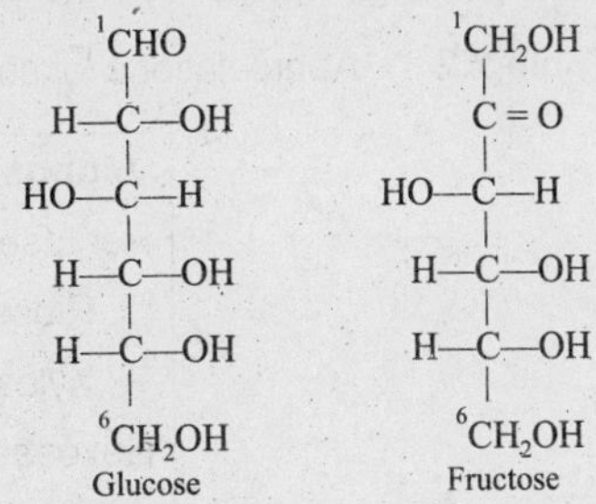

Fig. 2.5. Glucose and fructose.

Fructose is sweetest of all sugars and taste 1.73 times sweeter than sucrose (cane sugar), of which it is constituent. It is very soluble in water and crystallizes in large needles. Its some polysaccharide derivatives (such as **inulin,** which is made up of 30–35 β-D-fructose units per molecule) are energy stores of some plants.

***(iii)* Galactose** (G. *gala* = milk). It is found in milk-sugar or lactose along with glucose. Galactose is a simple aldohexose sugar and stereoisomeric with glucose. In the body it is changed into glucose to be utilised. Galactose does not occur freely in nature and is constituent of many plant polysaccharides (many gums such as gum arabic, mucilages, pectins), animal glycolipids (*i.e.*, occur in human nervous system) and glycoproteins.

Some human infants do not have the enzyme **uridyltransferase** to metabolise galactose of the milk sugar (of mother's milk), so they store galactose in their different tissues such as brain causing malnutrition and mental retardation. This inborn error of metabolism is called **galactosemia** (**Jain** 2012).

***(iv)* Mannose.** It is a 6-carbon aldo sugar that occurs naturally only in polymerised forms called **mannans**. It occurs in plants, yeasts and other fungi, bacteria. Mannose is a constituent of prosthetic polysaccharides of albumins, globulins, mucoproteins. It also occurs in glycoproteins. In animals, mannose changes to glucose before being used in metabolism. It is an epimer of glucose and serves as food energy store.

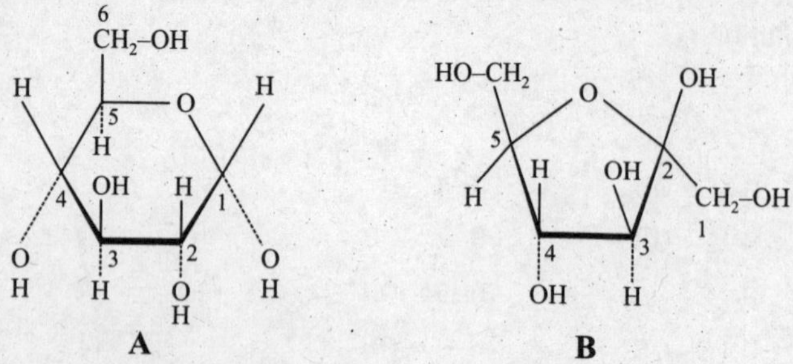

Fig. 2.6. Ring structure of monosaccharides. A—β-D-Glucopyranose; B—α-D-Fructofuranose.

5. Heptoses ($C_7H_{14}O_7$). They include sedoheptulose, glucoheptose, etc. **Sedoheptulose** is an amorphous ketose sugar that plays a role in carbohydrate metabolism but is not fermented by yeast and that is a laboratory source of D-altrose (a synthetic syrupy hexose sugar) and D-ribose.

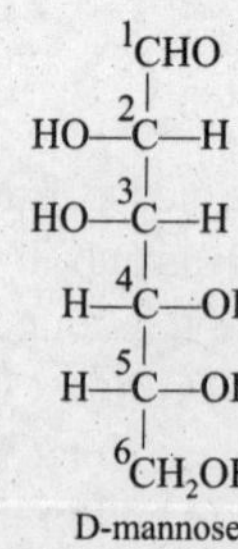

Fig. 2.7. D-mannose.

Derived Monosaccharides

There are many derivatives of the basic monosaccharide. These derivatives include polymerised monosaccharides, such as oligosaccharides and polysaccharides, as well as several classes of non-polymerised compounds (*i.e.*, Glycosides, sugar phosphates, deoxy-sugars, amino-sugars, sugar alcohols, sugar acids, ascorbic acids, Table 2.2).

Table 2.2 Abbreviations for some monosaccharides and their derivatives (after Horton *et al.*, 2006).

Monosaccharide derivative	Abbreviation
Pentoses	
Ribose	Rib
Xylose	Xyl
Hexoses	
Fructose	Fru
Galactose	Gal
Glucose	Glc
Mannose	Man
Deoxy sugars	
Abequose	Abe
Fucose	Fuc
Amino sugars	
Glucosamine	GlcN
Galactosamine	GalN
N-Acetylglucosamine	GlcNAc
N-Acetylgalactosamine	GalNAc
N-Acetylmuramic acid	MurNAc

Monosaccharide derivative	Abbreviation
Sugar acids	
Glucuronic acid	GlcUA
Iduronic acid	IdoA

1. Glycosides. The hydroxyl group on carbon 1 in the sugar molecule can be replaced by other radicals, forming compounds konwn as **glycosides**. Thus, if a *methyl group* replaces the hydroxyl group of glucose it results in the formation of **methyl glucosides**. There are two methyl glucosides possible, α and β, corresponding to α and β forms of glucose. Glucosides formed from *galactose* are known as **galactosides**. Among the complex glucosides occurring in nature are **digoxin** ($C_{41}H_{64}O_{14}$) which is obtained from the leaves of a foxglove plant (*Digitalis lanata*) and acts on the heart (a cardiotonic) and the antibiotic **streptomycin**.

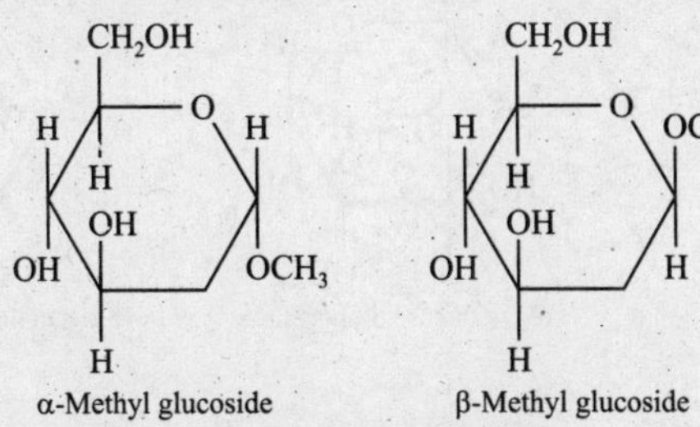

Fig. 2.8. Glycosides.

2. Sugar phosphates. Monosaccharides in metabolic pathways are often converted to phosphate esters. The triose phosphates, ribose 5-phosphate, and glucose 6-phosphate is a hemiacetal phosphate. The ability of UDP-glucose to act as a glycosyl donor is evidence of this reactivity.

Box 2.2

The aldehyde and ketone groups of carbohydrates with five and six carbon atoms react with alcohol group present in the neighbouring carbons to form **hemiacetal** and **hemiketals** respectively. This results in the formation of five to six membered ring (*i.e.*, pyranoses or furanoses).

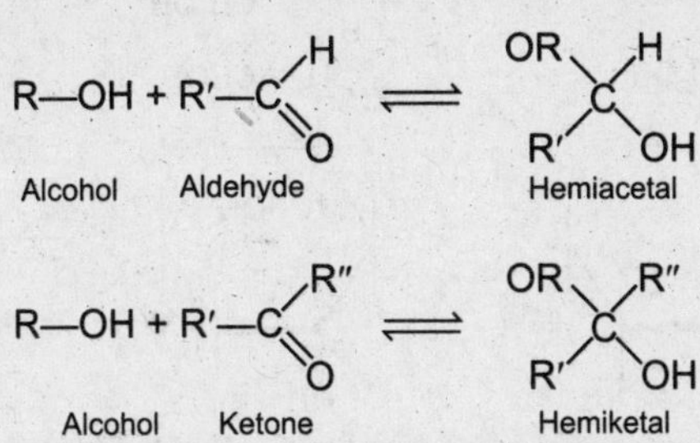

3. Deoxy sugars. In these derivatives, a hydrogen atom replaces one of the hydroxyl groups in the parent monosaccharide. 2 Deoxy-D-ribose is an important building block for DNA (Fig. 2.9). L-fucose (6-deoxy-L-galactose) is widely distributed in plants, animals and microorganisms. Fucose is derived metabolically from D-mannose.

4. Amino sugars. In a number of sugars, an amino group replaces one of the hydroxyl groups in the parent monosaccharide. Sometimes the amino group is acetylated. Amino sugars formed from glucose and galactose commonly occur in glycoconjugates (*e.g.*, α-D-Glucosamine and N-Acetyl-α-D-galactosamine). N-Acetylneuraminic acid (Neu NAC) is an acid formed from N-acetylmannosamine and pyruvate. When this compound cyclizes to form a pyranose, the carbonyl group at C-2 (from the pyruvate moiety) reacts with the hydroxyl group of C-6. NeuNAC is an important constituent of many glycoproteins and of a family of lipid called **gangliosides**. Neuraminic acid and its derivatives, including NeuNAC, are collectively known as **sialic acids**.

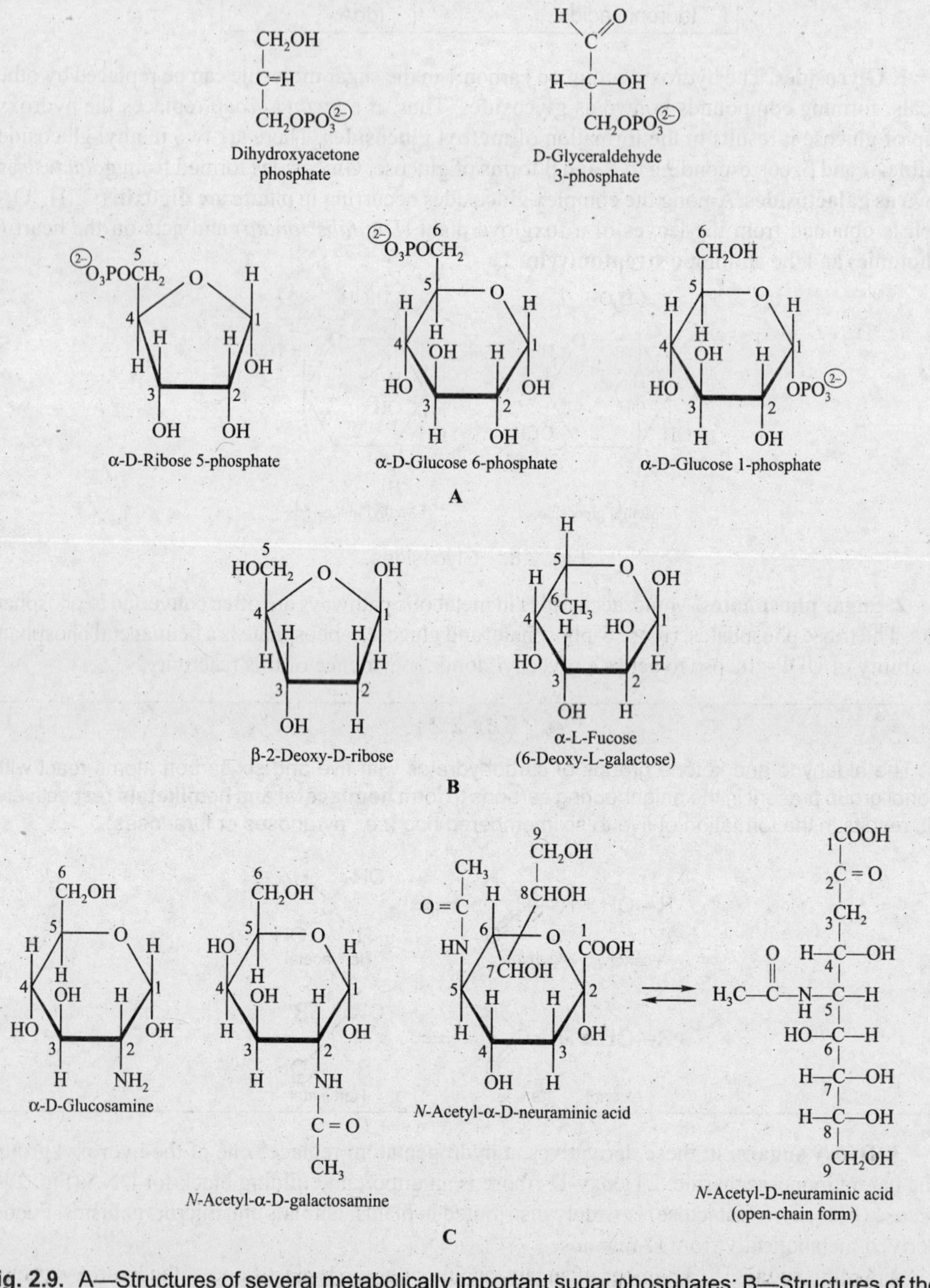

Fig. 2.9. A—Structures of several metabolically important sugar phosphates; B—Structures of the deoxy sugars 2-deoxy-D-ribose and L-fucose; C—Structures of several amino sugars.

5. Sugar alcohols. In a sugar alcohol, the carbonyl oxygen of the parent monosaccharide has been reduced, producing a polyhydroxy alcohol, *e.g.*, **glycerol**, **myo-Inositol** and **D-ribitol**

(Fig. 2.10). Glycerol and *myo*-inositol are important component of lipids. Ribitol is a component of flavin mononucleotide (FMN) and flavin adenine dinucleotide (FAD). In general, sugar alcohols are named by replacing the suffix *-ose* of the parent monosaccharides with *-itol*.

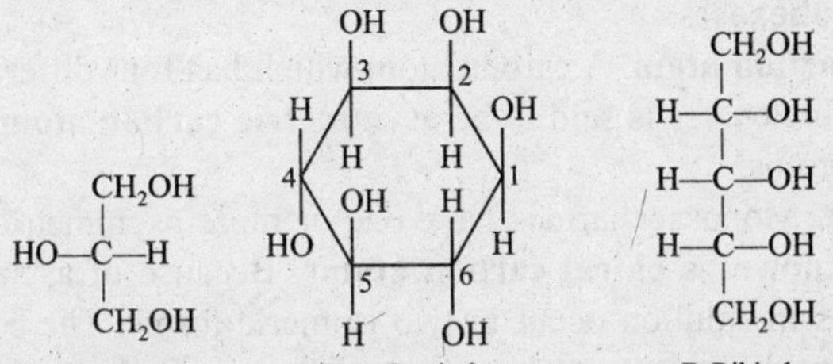

Fig. 2.10. Structure of some sugar alcohols. Glycerol (a reduced form of glyceraldehyde) and *myo*-inositol (metabolically derived from glucose) are important constituents of many lipids. Ribitol (a reduced form of ribose) is a constituent of the vitamin riboflavin and its coenzyme.

6. Sugar acids. Sugar acids are carboxylic acids derived from aldoses, either by oxidation of C-1 (the aldehydic carbon) to yield an aldonic acid or by oxidation of the highest-numbered carbon (the carbon bearing the primary alcohol) to yield an **alduronic acid**. The structures of the aldonic and alduronic derivatives of glucose-gluconate and glucuronate—are shown in Fig. 2.11. Aldonic acid exists in the open chain form in alkaline solution form lactones (intramolecular esters) on acidification. Alduronic acids can exist as pyranose, so they possess an anomeric carbon. N-Acetylneuraminic acid is an aldonic acid as well as an amino sugar. Sugar acids are important component of many polysaccharides.

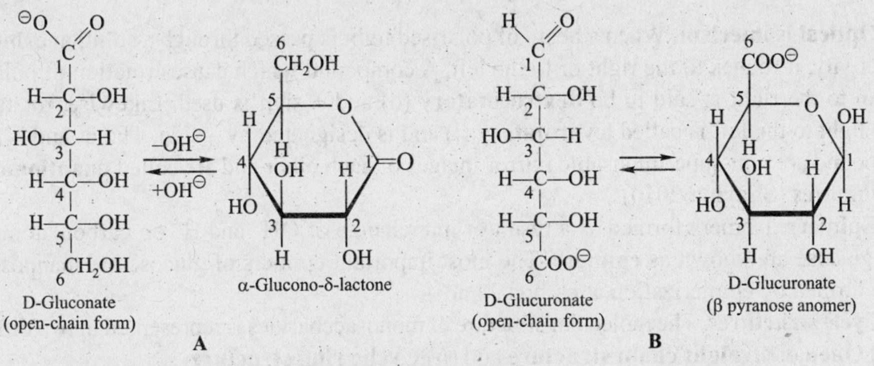

Fig. 2.11. Structures of sugar acids derived from D-glucose. A—Gluconate and its D-lactone. B—The open-chain and pyranose forms of glucuronate.

Many drugs, pesticides and environmental pollutants and hormones are coupled with glucuronic acid (*i.e.*, glucuronate) and excreted in urine or bile as glycuronides.

7. Ascorbic acid (Vitamin C). L-Ascorbic acid (Fig. 2.12) or vitamin C is an enediol of a lactone derived from D-glucuronate. Primates cannot convert-glucuronate to ascorbic acid and must therefore obtain ascorbic acid from the diet. Ascorbic acid is an essential cofactor for the enzymes that catalyze the hydroxylation of proline and lysine residues during collagen synthesis.

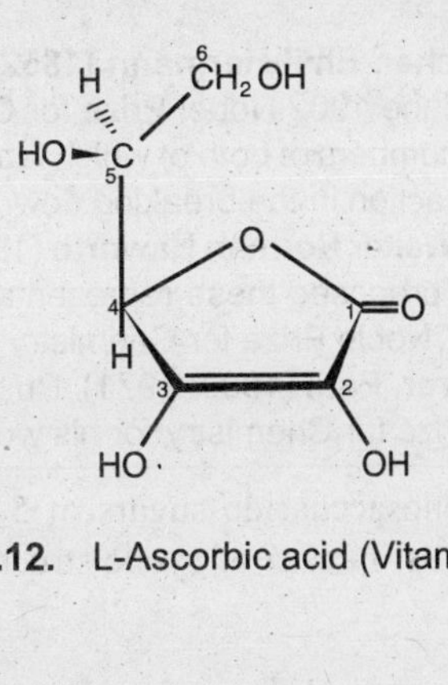

Fig. 2.12. L-Ascorbic acid (Vitamin C).

Structural Specialities of Monosaccharides

1. Aldose and ketose form. Monosaccharides or carbohydrate monomers are structural units of carbohydrates. They are mainly hexose sugars. They either occur as aldoses or ketoses and are called **aldohexoses** or **ketohexoses.**

2. Asymmetry of carbon atom. A carbon atom which has four different atoms or groups of atoms attached to its four valencies is said to be **asymmetric carbon atom**. All monosaccharides have asymmetric carbon atoms.

3. Stereoisomerism. Monosaccharides have one or more asymmetric carbon atoms. These asymmetric carbons are known as **chiral carbon atoms**. Because of asymmetric carbon atoms, monosaccharide molecules in solution occur in two isomeric forms. The compounds which have same structural formula but differ in spatial configuration are called **stereoisomers** or **geometric isomers**. (**Note.** *Configuration* means the spatial arrangement of atoms or groups in a molecule.) For example, glucose with four asymmetric carbon atoms have 16 isomeric forms.

4. D- and L- isomers. In monosaccharide molecules, when OH group on the second carbon, *i.e.*, adjacent to the carbon of terminal primary alcohol is on the right, the molecule is represented as **D-isomer**. When OH is on the left the monosaccharide molecule is **L-isomer**. The naturally occurring monosaccharides are mainly D-isomers.

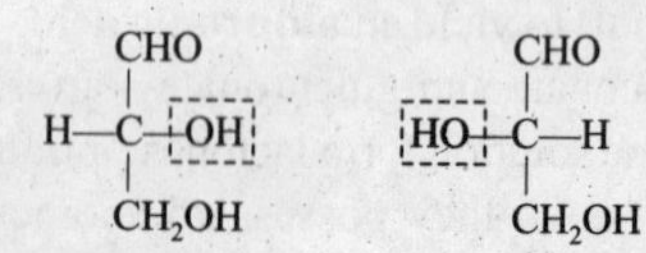

D-Glyceraldehyde L-Glyceraldehyde

Fig. 2.13. Two isomers of glyceraldehyde: D-isomer and L-isomer.

5. Optical isomerism. When a beam of polarised light is passed through a solution exhibiting optical activity, it rotates to the right or to the left. A compound which causes rotation of polarised light beam to the right is said to be **dextrorotatory** (d) and + sign is used. Likewise, rotation of polarised light to the left is called **levorotatory** (l) and is designated by – sign. The d- and l- forms of a compound are non-superimposable mirror images of each other and are called **enantiomorphs** or **enantiomores** (**Sharma** 2010).

6. Epimers. Isomers formed as a result of interchange of OH^- and H^+ on carbon atoms 2, 3 and 4 of glucose are known as **epimers**. The most important epimers of glucose are mannose and galactose formed by epimerization at carbon 2 and 4.

7. Cycle structures. The molecular structure of monosaccharides is represented in two different forms: (*i*) **Open** or **straight chain structure** and (*ii*) **Cyclic ring structure**.

The straight chain structural formula accounts for its aldohexose character. This was proposed by a **E.H. Fischer** (see Box 2.3) and is called **Fischer formula Fischer projection.**

Box 2.3

Fischer, Emil Hermann (1852–1919). German organic chemist and pioneer of biochemistry, awarded the 1902 Nobel Prize for Chemistry for his work on the structures of sugars and purines, simple members of both of which families he synthesized. He also synthesized peptides and studied enzyme action in the breaking down of proteins.

Sir Walter Norman Haworth (1883–1950) worked on the **cyclization reactions** of carbohydrates and first proposed these representations (*i.e.*, Haworth projections). He was awarded with **Karrer** the 1937 Noble Prize for Chemistry for his work on the structures of carbohydrates and vitamin C.

Karrer, Paul (1889–1971). Russian-born Swiss chemist awarded with W.N. Haworth the 1937 Nobel Prize for Chemistry for his work on the carotenoids and flavins, and on vitamins A and B.

Monosaccharide sugars of 5 or more carbon atoms have **cyclic structures** or **Haworth projections**. In case of cyclic structure, the carbonyl group (–C = O) forms a covalent bond with the oxygen of –OH group. In cyclic representation the molecule is viewed from the side and above the plane of the ring. By convention, the bonds nearst to the viewer are bold and thickened. X-ray diffraction analysis shows that the six-membered ring containing one oxygen atom is actually in the form of a **chair** or **boat** (**Murray** *et al.*, 2000; **Horton** *et al.*, 2006).

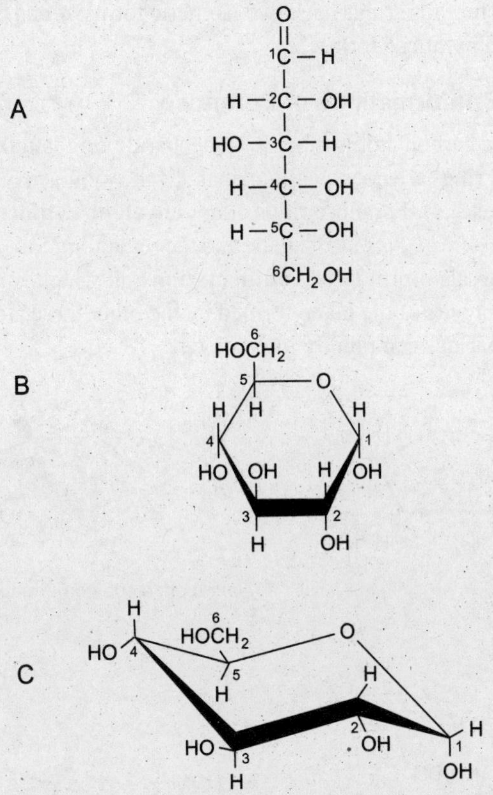

Fig. 2.14. Three types of molecule structures of D-glucose. A—Straight chain form; B—α-D-glucose-Haworth projection; C—α-D-glucose; chair form.

8. Mutarotation. The cyclic rings of monosaccharides occur in two stereoisomeric forms. These are named as α and β **anomers**. Interconversion of α and β forms in solution to maintain a balance in their concentration is called **mutarotation**. In mutarotation the hemiacetal ring opens and is reformed with change in the position of –H and –OH groups. Two anomers of D-glucose α and β forms occur naturally. Specific rotatin of α-D glucose is +112.2° while that of β-D glucose is 18.7°. In a solution D glucose α anomer is 36% and β anomer is 64%. The specific rotation of glucose solution is always maintained at 52.7%. When α and β D-glucose is dissolved in water, it gradually changes into another form till ratio is achieved.

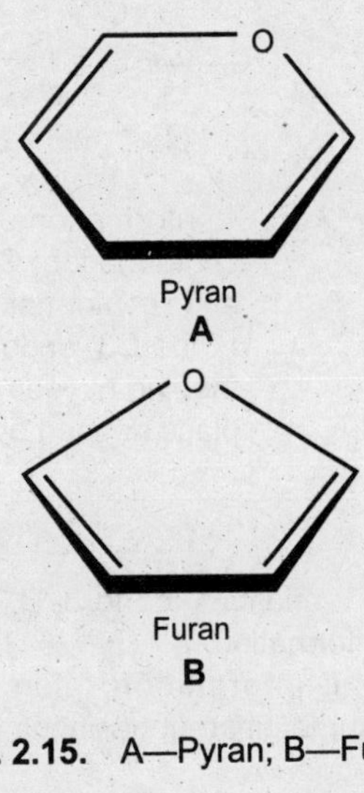

Fig. 2.15. A—Pyran; B—Furan.

9. Pyranose and furanose ring structure. The cyclic or ring like structure of 6 carbon sugars is described as a **pyranose ring**. A pyranose ring has pyran-like structure, *i.e.*, a 6-membered ring with a linkage between C-1 and C-5 carbons and 6th carbon remain outside the ring. The **furanose ring** of monosaccharide has a furan-like structure, *i.e.*, a 5-membered ring with 4 carbon atoms and one oxygen, the linkage is between C-1 and C-3 carbons and the 5th and 6th carbon remain outside. Thus pyranose ring is hexagonal while furanose ring is pentagonal.

Both pyranose and furanose rings have two isomeric forms: α and β. In a solution, the furanose forms are less stable than pyranose forms.

Envelope and Twist Conformations of Furanose

Since the geometry of the carbon atoms of a monosaccharide ring is **tetrahedral** (bond angles near 110°) so monosaccharide rings are not actually planar. Cyclic monosaccharides can exist in a variety of conformations (see Box 2.4). Furanose rings adopt **envelope conformation** in which one of the five ring atoms (either C-2 or C-3) is out-of-plane and the remaining four are approximately coplanar (Fig. 2.16). Furanoses can also form **twist conformations**, in which two of the five ring atoms are out-of-plane, one on either side of the plane formed by the other three atoms. Various conformers of unsubstituted monosaccharides can rapidly interconvert.

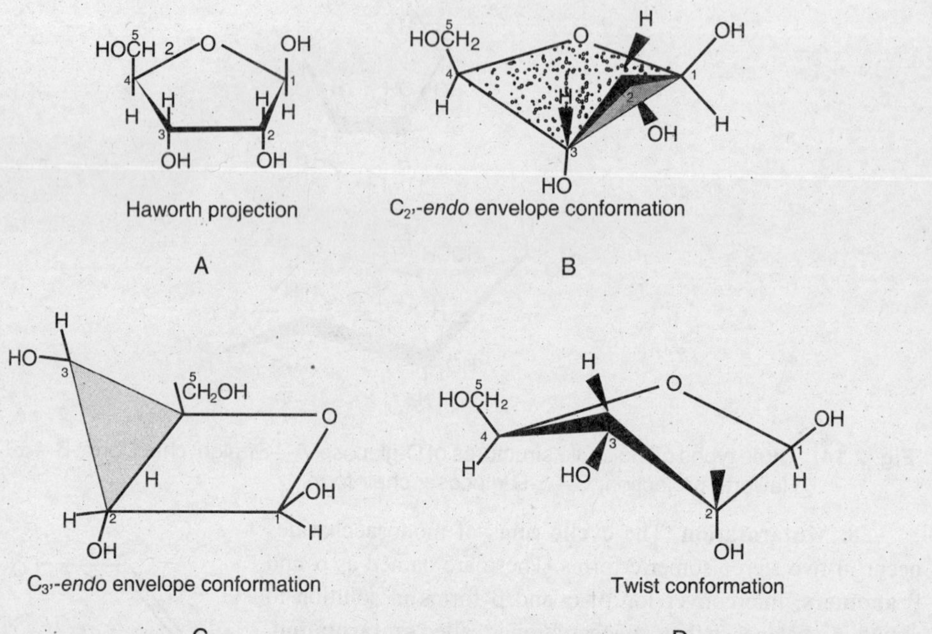

Fig. 2.16. Conformations of β-D-ribofuranose. A—Haworth projection; B—C2′-endo envelope conformation; C—C3′-endo envelope conformation; D—Twist conformation. In the C2′-endo conformation, C-2 lies above the plane defined by C-1, C-3, C-4, and the ring oxygen. In the C3′-endo conformation, C-3 lies above the plane defined by C-1, C-2, C-4, and the ring oxygen. In the twist conformation shown, C-3 lies above and C-2 lies below the place defined by C-1, C-4 and the ring oxygen (after Horton *et al.*, 2006).

Box 2.4 Conformation and Configuration

Variations in the 3-'D' structure of biomolecules are described in terms of configuration and conformation.

Conformation refers to the spatial arrangement of substituent groups that are free to assume different positions in space, without breaking any bonds, because of the freedom of bond rotation.

Configuration denotes the spatial arrangement of an organic molecule that is conferred by the presence of (a) double bonds, around which there is no freedom of rotation, or (b) chiral centers, around which substituent groups are arranged in a specific sequence.

Chair and Boat Conformations of Pyranose

Pyranose rings tend to assume one of the two conformations, the **chair conformations** or the **boat conformations** (Fig. 2.17). For each pyranose, there are two distinct conformers and six boat conformers. The chair conformations minimize steric repulsions among the ring substituents and are generally more stable than boat conformations. The –H, –OH and $-CH_2OH$ substituent of a pyranose ring in the chair conformation may occupy two different positions. In the axial position the substituent is above or below the plane of the ring, while in the equatorial position the substituent lies in the plane of the ring. In pyranoses, five substituents are axial and five are equatorial. Whether a group is axial or equatorial depends on which carbon atom (C-1 or C-4) extends above the plane of the ring

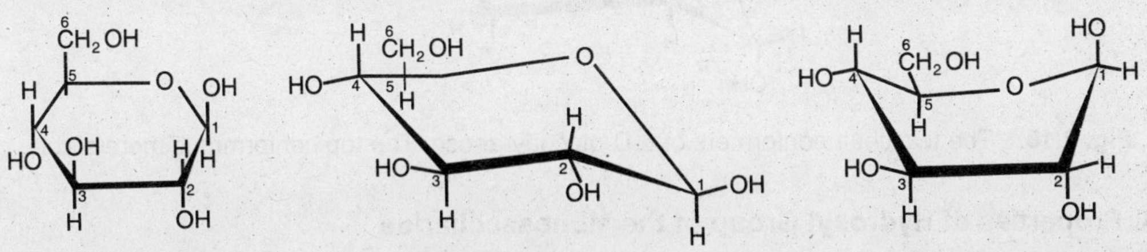

Fig. 2.17. Conformations of β-D-glucopyranose. Haworth projection, a chair conformation and a boat conformation.

when the ring is in the chair conformation. Figure 2.18 shows the two different chair conformers of β-D-glucopyranose. The more stable conformation is the one in which the bulkiest ring substituents are equatorial (Top structure). In fact, this conformation of β-D-glucose has the least strict strain of any aldohexose. Pyranose rings are occasionally forced to adopt slightly different conformations, such as the unstable half-chair adopted by a polysaccharide residue in the active site of lysozyme.

Properties of Monosaccharides

1. Physical Properties

(*i*) Monosaccharides are crystalline.

(*ii*) They are sweet and readily soluble in water.

(*iii*) In solution monosaccharides occur in two forms: in D-form and in L-form.

(*iv*) They exhibit stereo-isomerism.

(*v*) Monosaccharides occur in two forms: in the form of straight open chain and in ring form.

(*vi*) They occur in two types of rings: pyranose ring form and furanose ring form.

(*vii*) Monosaccharides cannot be hydrolysed into smaller molecules.

(*viii*) They can diffuse through semipermeable cell or plasma membrane.

(*ix*) Molecules of monosaccharides also exhibit the optic isomerism.

2. Chemical properties of monosaccharides. Chemically, monosaccharides are derivatives of **polyhydric alcohols.** Therefore, these exhibit properties of both hydroxyl and carbonyl groups. Since monosaccharides also contain either a free aldehyde or a ketone group, so exhibit reducing property. They are also called **reducing sugars.**

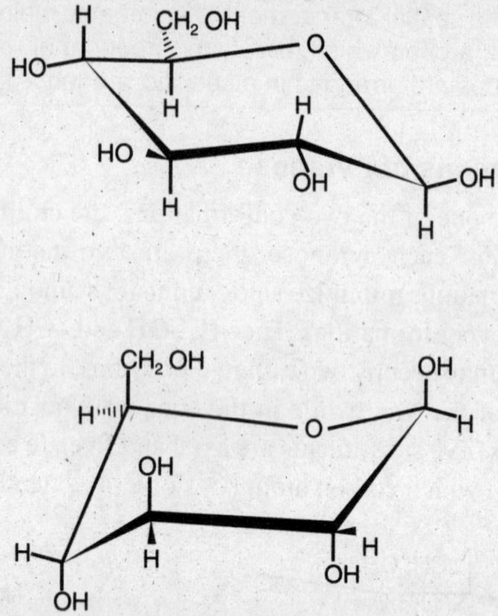

Fig. 2.18. The two chair conformers of β-D-glucopyranose. The top conformer is more stable.

A. Properties of Hydroxyl Group of the Monosaccharide

1. **Esterification (Formation of esters).** Due to the presence of hydroxyl group (–OH) monosaccharides form esters with organic and inorganic acids. Acetates, benzoates and phosphates are the common esters of monosaccharides. Hexose phosphates (glucose and fructose-phosphates) play important role in carbohydrate metabolism).

2. **Acetylation.** When monosaccharides are heated with acetic anhydride or acecyl chloride in presence of zinc chloride, their –OH groups are acetylated. The number of acetyl groups incorporated in a molecule of monosaccharide is equal to the number of –OH groups present in it.

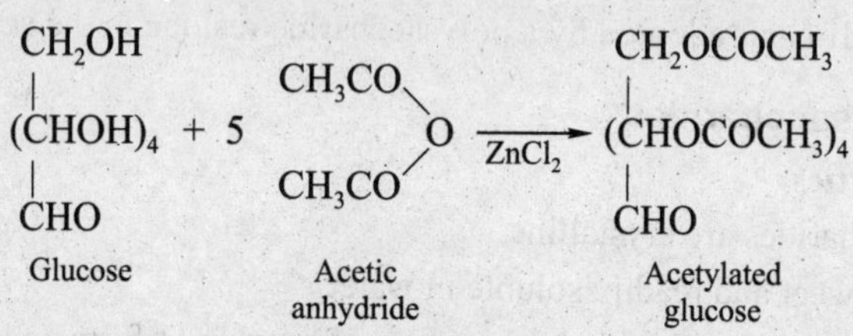

3. **Formation of methyl ether (Glucosides).** When monosaccharides react with methyl alcohol and dry HCl gas or methyl chloride or methyl iodide they form **ethers** or **glycosides**. The glycoside formed from glucose is **glucoside**, from galactose is **galactoside** and from fructose is **fructoside**. During this process the –OH group is replaced by methyl group.

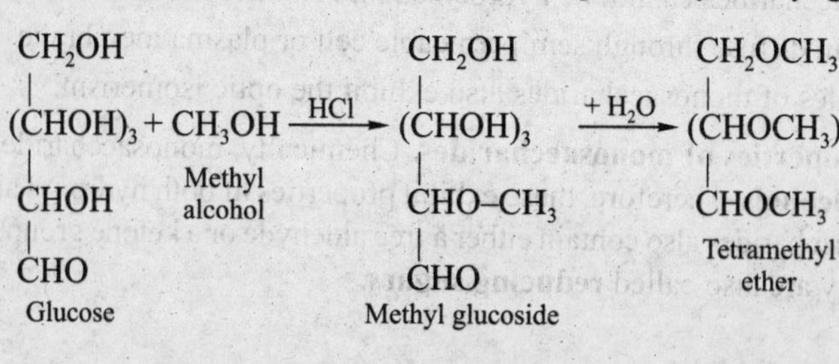

4. **Formation of glucosates or fructosates.** Hexoses form hexosates with certain metallic hydroxides:

$$\underset{\text{Glucose or Fructose}}{C_6H_{12}O_6} + \underset{\text{Calcium hydroxide}}{Ca(OH)_2} \rightarrow \underset{\text{Calcium glucosate or calcium fructosate}}{C_6H_{12}O_6CaO} + \underset{\text{Water}}{H_2O}$$

5. **Formation of hexosamine.** The hydroxyl group in hexose sugars can easily be replaced with amino groups and the end product is **hexosamine** or amino sugar. Hexosamine of glucose is **glucosamine** or of fructose is **fructosamine**.

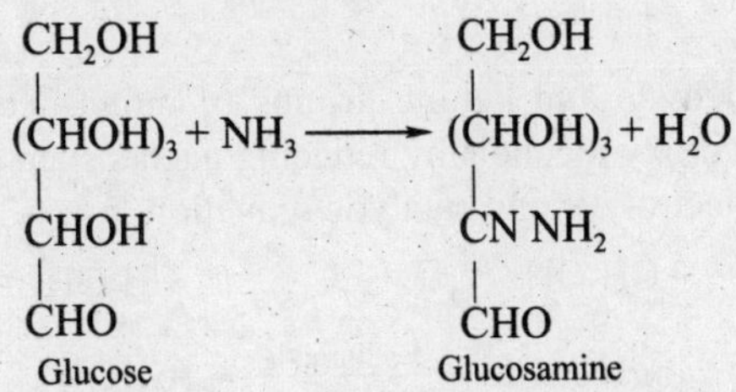

B. Reactions of Carboxyl Group

Monosaccharides contain a carboxyl group which may be an aldehyde or ketone and exhibit following reactions:

1. **Oxidation.** Aldoses can be oxidized with mild oxidizing agents such as Fehling's solution, Benedict's solution, bromine water, or Tollen's reagent. These form monocarboxylic hydroxy acids. If carbon atom one (C-1) is oxidised in a hexose, the oxidation product is **aldonic acid** (*e.g.*, glucose → **gluconic acid**), but if carbon atom six (C-6) is oxidised it is called **uronic acid** (glucose → **glucuronic acid**).

(*i*)

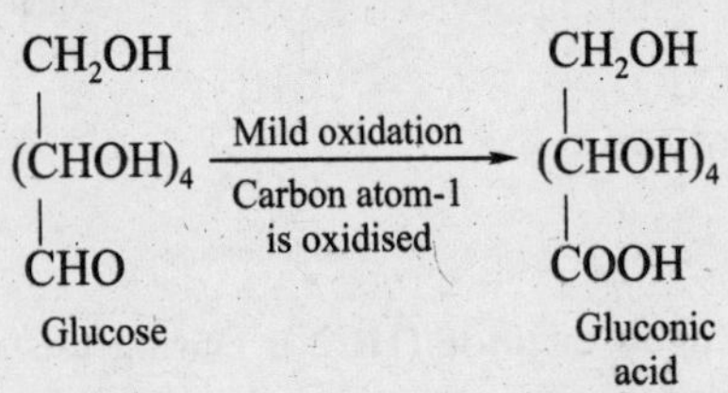

(*ii*)

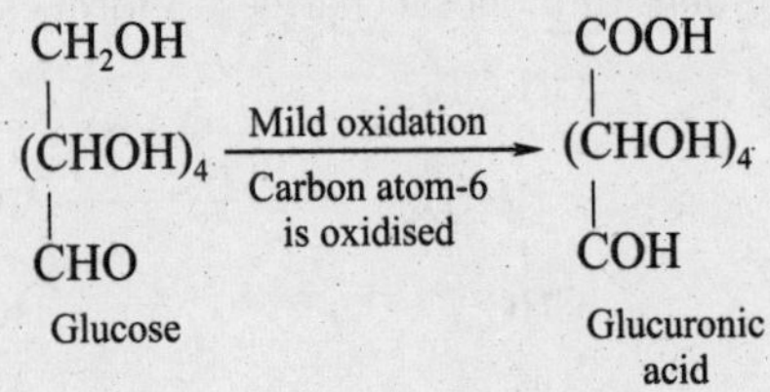

Aldoses, on being oxidised with dilute nitric acid produce dicarboxylic hydroxy acid or **aldaric acid** and with strong oxidising agent (glucose → glucaric acid or saccharic acid).

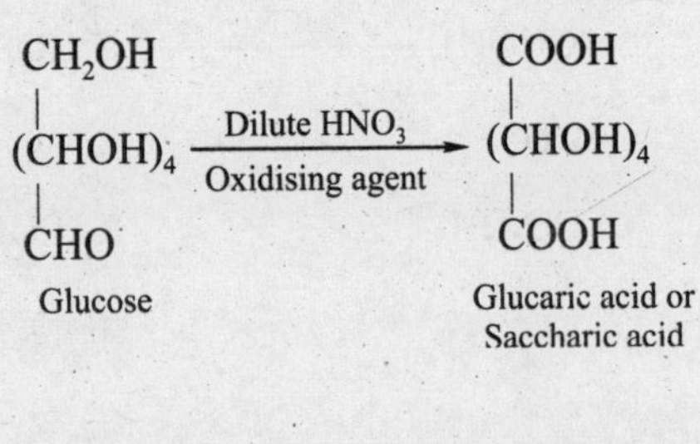

Ketoses when oxidised with nitric acid, undergoes cleavage at C = O group and form acids with a smaller number of carbon atoms.

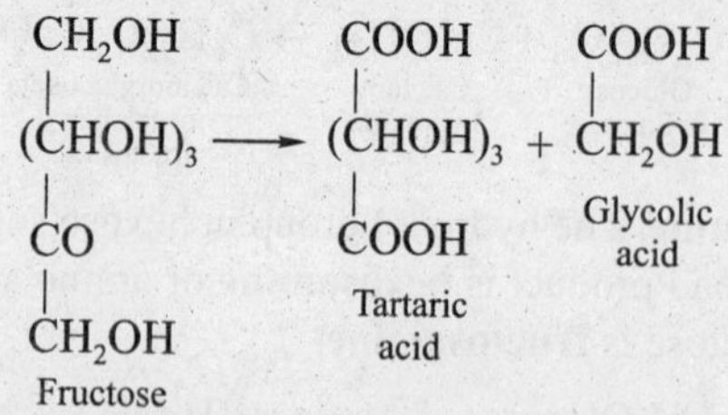

2. **Reduction.** Free aldehyde and ketone groups of monosaccharides are reduced to their corresponding polyhydroxy alcohols by reducing agents such as sodium amalgam (sodium-mercury amalgam), electrolytic and catalytic reduction.

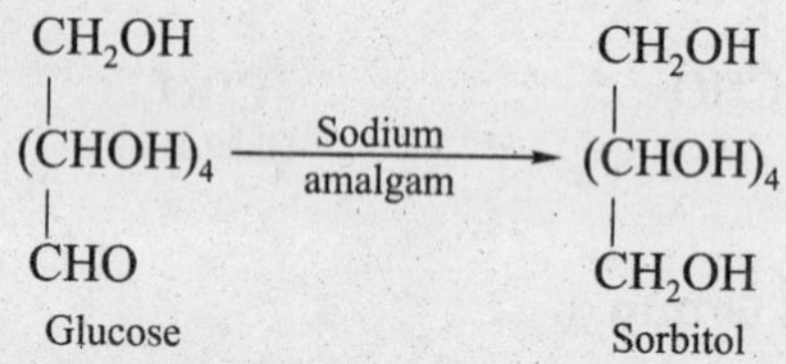

Thus, by reaction of reduction glucose yields **sorbitol**, galactose yields **gulcitol**, mannose yields **mannitol** and fructose produces mannitol and sorbitol.

3. **Iodo compounds.** When an aldose monosaccharide is heated with concentrated hydroiodic acid (H1) and red phosphorus at 100°C, it loses all its oxygen and is converted into iodo compound.

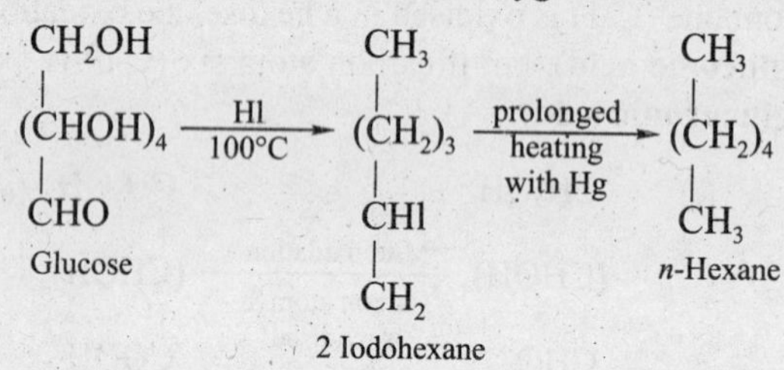

4. **Reaction with hydrogen cyanide (HCN).** During this reaction, the HCN is added to free aldehyde or ketone groups of monosaccharides to form cyanohydrins.

(*i*)

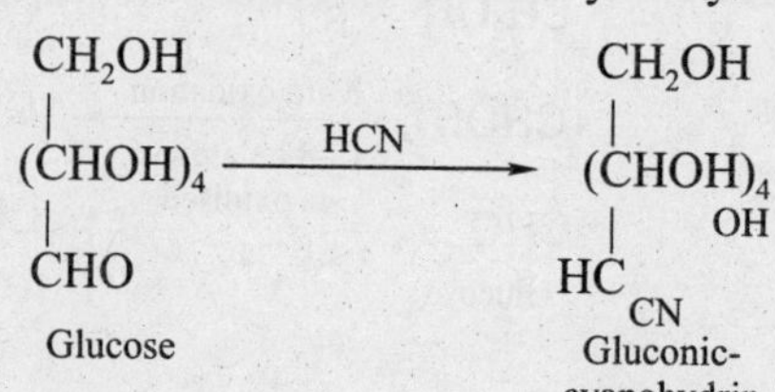

(*ii*)

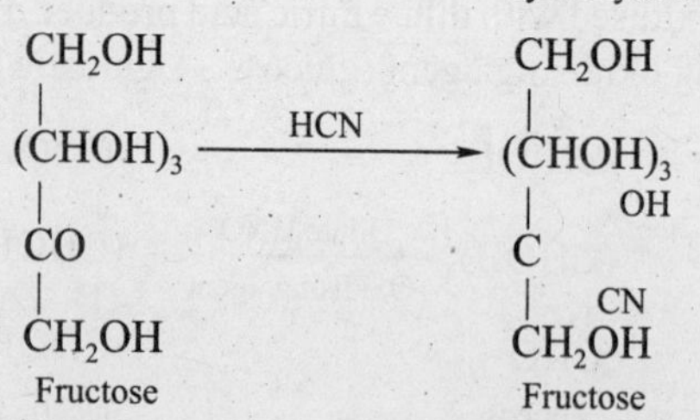

5. **Reaction with hydroxylamine.** Due to this reaction, saccharides condense to form oximes.

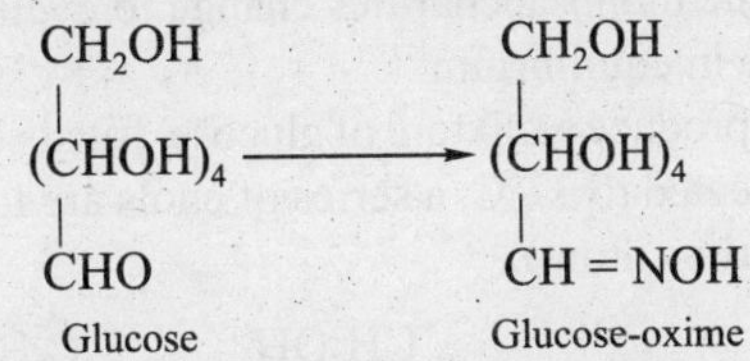

6. **Reaction with phenylhydrazin to form osazones.** All monosaccharides or reducing sugars in the slightly acidic medium and excess of hydrazine form **phenyl osazone** compounds. These have crystalline appearance. Phenylhydrazine ($C_6H_5NH.NH_2$) is a colourless toxic liquid.

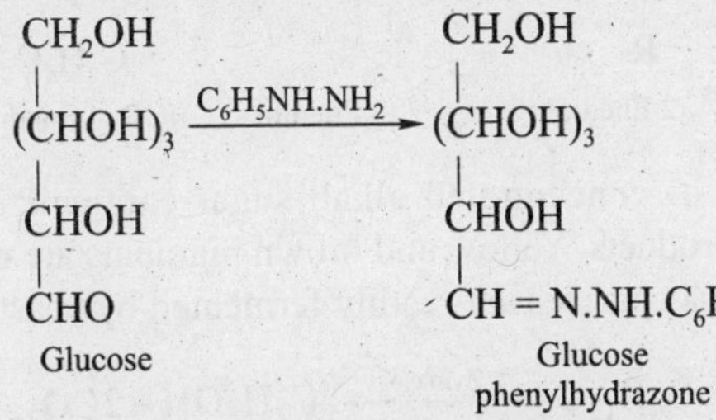

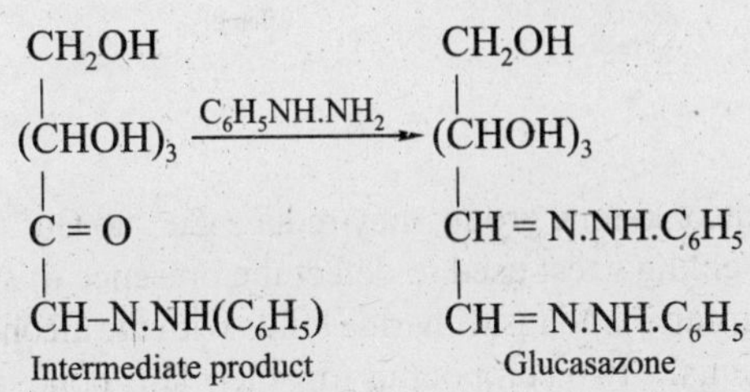

Similarly, fructose reacts with phenylhydrazin to form fructosazone. Reaction with phenylhydrazine involves only 2 carbon atoms *viz.*, the carbonyl (*i.e.*, the aldehyde or ketone group) and the adjacent one.

C. Other Reactions

1. **Reactions with concentrated mineral acids.** With concentrated HCl monosaccharides yield furfural (Box 2.5) or one of its derivatives, which react with variety of phenolic compounds to produce characteristic colours. Colour test of various sugars are based on such reaction.

Box 2.5

Furfural ($C_5H_4O_2$) is a heterocyclic aldehyde, similar to benzaldehyde, derived from furan (C_4H_4O). It is made by digesting corn cobs, oat and rice hulls, etc., with acid. It is also obtained on heating ribose. Furfural is a colourless liquid. It is used as a solvent and as an intermediate in the manufacture of plastics and other compounds.

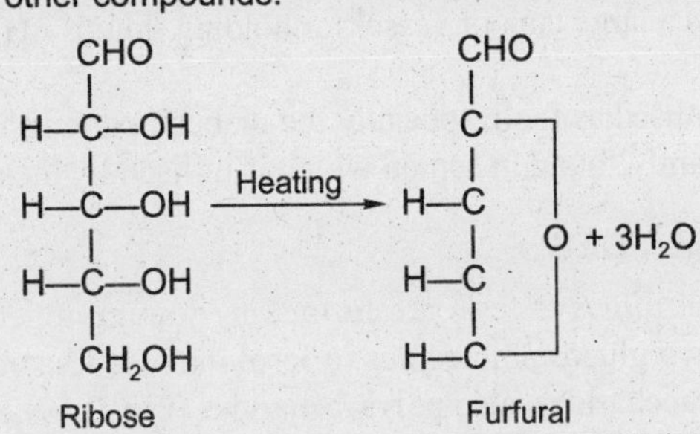

2. Reaction with alkali.

(*a*) In dilute alkalies the monosaccharides change to cyclic α and β structures and the two isomeric forms are in equilibrium.
On standing, these produce a mixture of glucose, fructose and mannose (inter-conversion).

(*b*) If this mixture is heated to 37°C, a series of enols are formed. (Here double bonds shifts from oxygen to carbon atom).

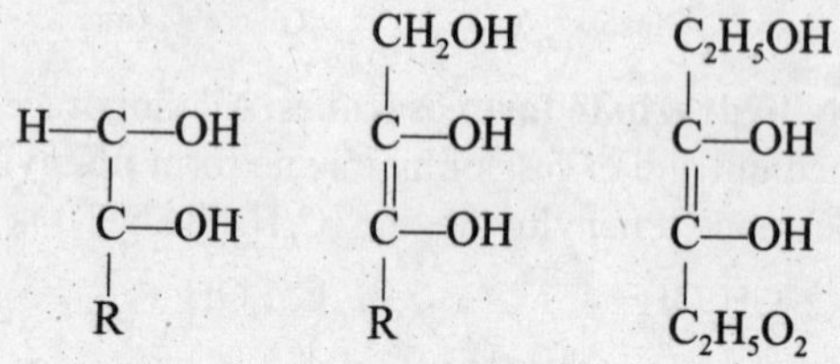

1, 2 Enediol   2, 3 Enediol   3, 4 Enediol

(*c*) In the presence of concentrated alkali sugar caramelizes and produces a series of decomposition products. Yellow and brown pigments are developed.

3. Fermentation. Monosaccharides are readily fermented by yeast to alcohols:

$$\underset{\text{Glucose or Fructose}}{C_6H_{12}O_6} \xrightarrow{\text{Zymase}} \underset{\text{Ethanol}}{2C_2H_5OH} + 2CO_2$$

Reducing Property

1. Due to a free aldehyde or ketone group, they reduce Cu^{++} to Cu^{+}. This property is the basis for Benedict's test and Fehling's test used to detect the presence of sugar (glucose in urine).
2. Aldehyde or ketone group of monosaccharide binds with the alcoholic group of another organic compound thus binding the two compounds togather. This bond is called **glycosidic bond**. On hydrolysis, the original compounds are formed.

Uses of Monosaccharides

1. Glyceraldehyde is fundamental in the process of glycolysis.
2. Deoxyribose and ribose (pentose) sugars are constituents of nucleic acids.
3. Ribose is a constituent of several coezymes.
4. Glucose is the source of energy and is the major product of photosynthesis.
5. Glucose is the easily accessible food for children and invalids.
6. Glucose is used in making sweets, candies, jellies, syrups and cakes.
7. Glucose is used for the manufacture of wines and glycerol.
8. Glucose forms raw material in the synthesis of vitamin C (ascorbic acid).
9. Dextroses (glucose in solution is in dextrorotary form) is frequently used in medical practice, *i.e.*, D-glucose solution is given intravenously to the patients.
10. Glucose is used as cheap reducing agent in the industry in silvering of mirrors and in vat dyeing with indigo. (*Vat* is a large tank or vessel for holding liquids in the process of dyeing, brewing or tanning.)
11. Fructose is chiefly used as food, specially the diabetic patients use it in place of cane sugar.
12. Fructose is abundantly found in semen which is utilised by the sperms for energy.

2.3 OLIGOSACCHARIDES

The oligosaccharides (G. *oligos* = few) are formed by joining together of 2–10 monosaccharide molecules. Linkage of two glucose molecules to form maltose illustrates the general pattern of the reactions forming oligosaccharides, also polysaccharides (Fig. 2.19).

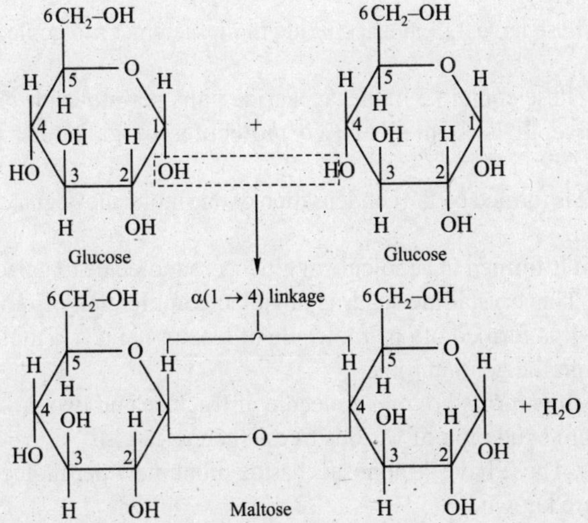

Fig. 2.19. Synthesis of the disaccharide maltose from two glucose molecules. The linkage formed by oxygen bridge is an alpha 1, 4 glycosidic.

Linkage (or bond) is established between the number 1 carbon of one glucose molecule and the number 4 carbon of the other glucose molecule by an oxygen bridge (C–O–C). The process involves loss of water and is called **condensation** or **dehydration**. The components of water, H^+ and OH^-, are split from the reacting chemical groups of the combining molecules. The chemical linkages joining the monosaccharide units are called **glycosidic bonds**. These bonds are designated as $1 \rightarrow 4$ or 1, 4 linkages, which may be alpha (α) or beta (β) depending on the orientation of the hydrogen at the number 1 carbon forming the bond. In a maltose molecule, the linkage is α $(1 \rightarrow 4)$ glycosidic because the hydrogen at the number 1 carbon of the bond is in the α position (see Box 2.6). The oligosaccharides are hydrolysed into component monosaccharides by an acid or enzyme.

Box 2.6

It may be noted that the linkages (*i.e.*, α-1-4 glycosidic bonds) are characteristic of energy providing carbohydrates, namely, sugars and starches, which can be digested by organisms. The structural carbohydrate cellulose has β-1, 4 glycosidic linkages between glucose units, and cannot be digested by most eukaryotic organisms.

Classification of Oligosaccharides

The oligosaccharides are classified according to the number of their monosaccharide units or monomers as follows (see Box 2.7):

Box 2.7

Disaccharides

- **Nonreducing** (or **Glycosylglycosides**)
  - C_1–C_1 *glycosidic linkage* *e.g.*, Trehalose
  - C_1–C_2 *glycosidic linkage* *e.g.*, Sucrose
- **Reducing** (or **Glycosylaldoses** or **Glycosylketoses**)
  - C_1–C_4 *glycosidic linkage* *e.g.*, Lactose, Maltose, Cellobiose
  - C_1–C_6 *glycosidic linkage* *e.g.*, Gentiobiose, Malibiose, Isomaltose

1. **Disaccharides.** These have 2 monosaccharide monomers per molecule, *e.g.*, maltose, sucrose, lactose, etc.
2. **Trisaccharides.** These contain 3 monosaccharide units per molecule. These include
   (*a*) **Mannotriose.** It is formed of two molecules of galactose and one molecule of glucose.
   (*b*) **Robinose.** It is formed by the condensation of one molecule of galactose and two molecules of rhamnose.
   (*c*) **Raffinose.** It is formed of a molcule of glucose, a molecule of galactose and one molecule of fructose. This crystalline sugar is obtained commercially from cottonseed meal.
   (*d*) **Gentinose.** It is formed of one molecule of fructose and two molecules of glucose. It is obtained from the gentian roots.
   (*e*) **Melezitose.** It is formed of one molecule of fructose and two molecules of glucose. It is obtained from exudation of various trees (such as larch).
3. **Tetrasaccharides.** These have 4 monosaccharide monomers per molecule. Only following to tetrasaccharides are known:
   (*a*) **Stachyose.** It is composed of D-glucose, fructose and two molecules of galatose. It is found in the haricot beans.
   (*b*) **Scorodose.** It is found in the bulbs of garlic and onion.
4. **Pentasaccharides.** These have 5 monosaccharide units per molecule.

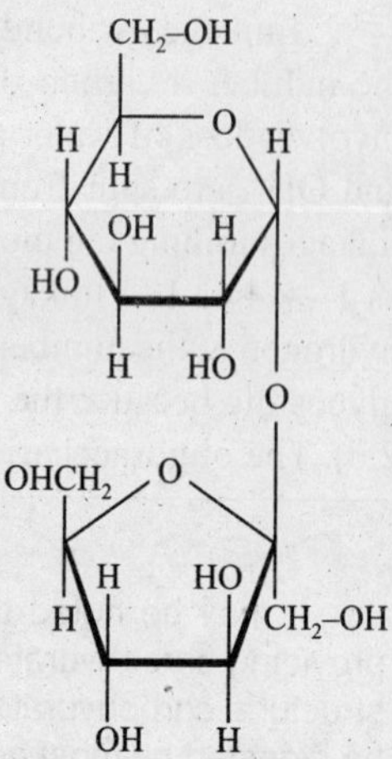

Fig. 2.20. Chemical formula of Sucrose.

Examples of Biologically Important Disaccharides

The disaccharides are the most abundant oligosaccharides in the cells. They, as already mentioned, have two monosaccharide monomers (or units) joined by glycosidic bond. They are also called **double sugars**. Biologically important disaccharides are sucrose, lactose, maltose, trehalose and cellobiose.

1. **Sucrose.** It is the common **cane sugar** or **table sugar**. Sucrose is common in higher plants and its commerical sources are sugarcane and sugar beet (*Beta vulgaris*). It is the storage product of photosynthesis in these plants. Sucrose is formed due to union of a glucose molecule with a fructose molecule (Fig. 2.20).
2. **Lactose.** It is called **milk sugar**. It is found in the milk of mammals. It is synthesized in mammary glands. Lactose is composed of one glucose molecule and one galactose molecule (Fig. 2.21). Souring of milk occurs when bacteria convert lactose in lactic acid.
3. **Maltose.** It is formed in the germinating starchy seeds and malt. It is called the **malt sugar**. Maltose is degradation product of larger polysaccharide molecule (starch). Maltose is composed of two glucose molecules (Fig. 2.22).

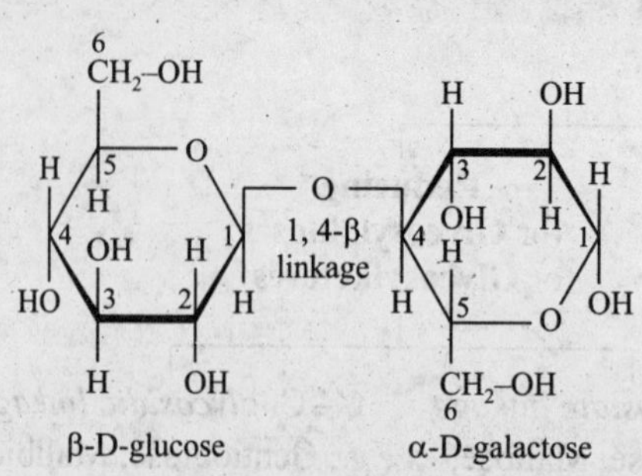

Fig. 2.21. Chemical formula of Lactose.

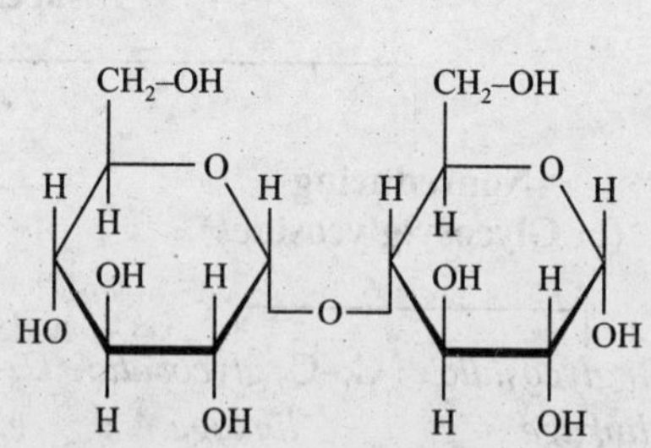

Fig. 2.22. Chemical formula of Maltose.

All the above disaccharides have an identical chemical formula $C_{12}H_{22}O_{11}$ derived as under:

$$\underset{\substack{\text{Glucose}\\\text{Fructose}\\\text{Galactose}}}{C_6H_{12}O_6 + C_6H_{12}O_6} \xrightarrow{\text{Condensation}} \underset{\substack{\text{Maltose}\\\text{Sucrose}\\\text{Lactose}}}{C_{12}H_{22}O_{11}} + \underset{\text{Water}}{H_2O}$$

4. Trehalose. It is the major sugar of insect haemolymph, in which it serves as an energy storage compound. Trehalose has no free aldehyde group and does not form an osazone. On hydrolysis, it yields glucose.

5. Cellobiose. It is formed during hydrolysis of cellulose. Cellobiose is composed of two molecules of D-glucose, and is identical to maltose except that it has β-1, 4 glycosidic linkage instead of α-1, 4 linkage in maltose.

Properties of Disaccharides

Like the monosaccharides, the disaccharides are also sweetish to taste and form a true solution in water. They, however, diffuse very slowly through the plasma membrane. They hydrolyze into free monosaccharide units when treated with a dilute acid or with an enzyme. Maltose and lactose are reducing disaccharides but sucrose is not.

Oligosaccharides may form unbranched or branched chains. Larger branched or unbranched oligosaccharide chains are found attached to plasma membrane.

Functions of Oligosaccharides

Some common functions of the oligosaccharides are the following:

1. Cellular interactions. The oligosaccharides attached to the plasma membrane help in recognizing cells of their own kind or species. Glycoproteins and glycolipids both are found on the outer cell surface and form a protective covering called **glycocalyx**.

2. Storage products. Sucrose is a reserve food in sugarcane and sugar beet.

3. Fuel. Disaccharides also serve as fuel to provide energy. **Lactose** or milk sugar is source of carbohydrates for infants and seriously ill persons.

2.4 POLYSACCHARIDES (Compound Sugars)

The polysaccharides are composed of ten to many thousands monosaccharides as the monomers in their macromolecules. Their empirical formula is $(C_6H_{10}O_6)n$. The molecules of the polysaccharides are of colloidal size having high molecular weights. The polysaccharides can be hydrolysed into simple sugars.

Naming of Polysaccharides

Polysaccharides are named after their monosaccharide monomers by changing the *–ose* ending of the monosaccharide to *–an*. Thus **cellulose**, a polysaccharide formed of *glucose* units, is called **glucan**. Fructose polymers are called **fructans** (previously fructosans). Likewise, galactose polysaccharides are called **galactans**; xylose polysaccharides are called **xylans**, pentose polymers are called **pentans** (formerly pentosans). Since the term glycose is used for any monosaccharide, the general name for any polysaccharide is **glycan**. If a polysaccharide contains more than one type of sugar unit the two sugars are named in alphabetical order. Thus a polysaccharide consisting of D-glucose and D-mannose units is termed D-gluco-D-mannoglycan. Some old names which have not been changed to conform with *–an* ending include starch, cellulose, pectin, amylopectin, inulin, chitin and heparin.

Classification of Polysaccharides

Polysaccharides have been classified as **homopolysaccharides (homoglycans)** and **heteropolysaccharides (heteroglycans)**. Homoglycans contain only one type of monosaccharide while heteroglycans contain at least two types of glycose monomers. Most polysaccharides contain one or two types of glycose units. Polysaccharides containing three types of glycose units are less common.

Some homopolysaccharides are formed of amino sugars such as *glucosamine* and *galactosamine* monosaccharides or of simple glucoronic acid and iduronic acid molecules.

Polysaccharides are formed by condensation of many molecules of monosaccharides with corresponding elimination of water molecules. The monosaccharide untis are joined by glycosidic bonds.

1. Homopolysaccharides

(*i*) **Glucans** (of glucose monomers), *e.g.*, starch, glycogen, cellulose, etc.

(*ii*) **Galactans** (of galactose monomers), *e.g.*, pectin, galactan from snails.

(*iii*) **Mannans** (of mannose monomers), *e.g.*, yeast mannan.

(*iv*) **Xylans** (of xylose monomers), *e.g.*, hemicellulose, xylan.

(*v*) **Fructans or fructosans** (of fructose monomers), *e.g.*, inulin.

(*a*) **Inulin** is the reserve carbohydrate of many plants of family compositae such as artichock, *Dahalia*, *Dendaleon*, garlic, onion, etc. It consists of 30 fructose units linked by $\beta(1 \rightarrow 2)$ bonds and has a molecular weight of 5000. Inulin is not metabolised in human body and is readily filtered through the kidney. It is therefore used in testing kidney function, especially glomerular filteration rate (GFR).

Paramylum. A reserve starch like unbranched nutrient homopolysaccharide of glucose found in various protozoans (*e.g.*, *Euglena*) and algae.

(*b*) **Starch.** Starch is a white soft amorphous powder and lacks sweetness. It is insoluble in water, alcohol and ether at ordinary temperature. Starch is a storage polysaccharide in green plants and the most important source of carbohydrate in our food. It is stored (see Box 2.8) in cereals, potatoes, legumes, sweet potatoes, tapioca and fruits (*e.g.*, banana). It is polymer of D-glycopyranose units linked by α-1-4 linkage or α-glycosidic bonds. Starch consists of a mixture of **amylose** and **amylopectin** in the proportion of 1 : 4 (see Box 2.9). Both are high molecular weight compounds. Amylose is linear while amylopectin is branched. Treating starch with hot water dissolves amylose, while amylopectin remain.

Box 2.8 Starch grains

Starch is stored as starch grains in chloroplasts or in special leucoplasts called **amyloplasts**. The diameter of starch grains (= granules) range from 3 to 100 μm. The starch grains are of two types: **simple** that occur singly and **compound** which occur in groups. A starch grain consists of several layers, the **shells**, arranged in concentric or eccentric around a point, the **hilum**. The microscopic form of the starch grains is characteristic of the source of starch. The starch grains from a particular source have a characteristic shape and can be identified by examination under the light microscope (Fig. 2.23). Starch gives blue colour with iodine solution.

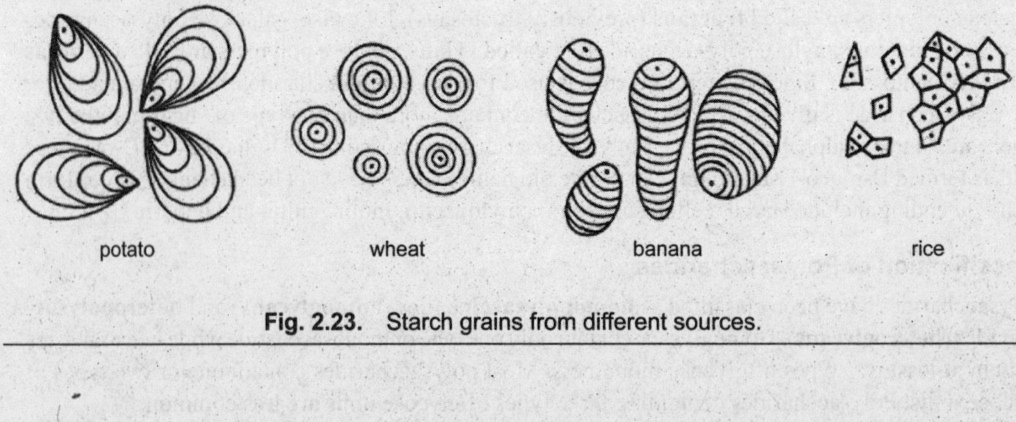

Fig. 2.23. Starch grains from different sources.

Box 2.9

Natural starches consists of two components: **amylose** (15–20%), a long unbranched straight chain component and **amylopectin** (80–85%), a branched chain polysaccharide. Potato and cereal starches are 20–30% amylose and 70–80% amylopectin. Starch from waxy corn is notable as it consists practically of amylopectin component, there being no amylose. However, starch of some varieties of pea having wrinkled surface may have as much as 98% of amylose.

***(i)* Amylose** (Fig. 2.24). Amylose is an unbranched polymer of about 100 to 1000 D-glucose residues connected by α-(1–4) glucosidic linkages, specifically termed **α(1 → 4) glycosidic bonds** because the anomeric carbons belong to glucose residues. Although it is not truely soluble in water, amylose forms hydrated micelles in water and can assume a helical structure under some conditions (Fig. 2.24). Amylose molecule is coiled like a watch spring. Each turn of amylose helix contains about six glucose units. Amylose gives blue colour with iodine.

***(ii)* Amylopectin.** Amylopectin is a branched version of amylose (Fig. 2.24). Branches, or polymeric side chains, are attached via α-(1 → 6) glucosidic bonds to linear chains of residues linked by α-(1 → 6) glucosidic bonds. Branching occurs, on average, once every 25 residues and the side chains contain about 15 to 25 glucose residues. Some side chains themselves are branched. Amylopectin molecules isolated from living cells may contain 300 to 6000 glucose residues (Horton *et al.*, 2006).

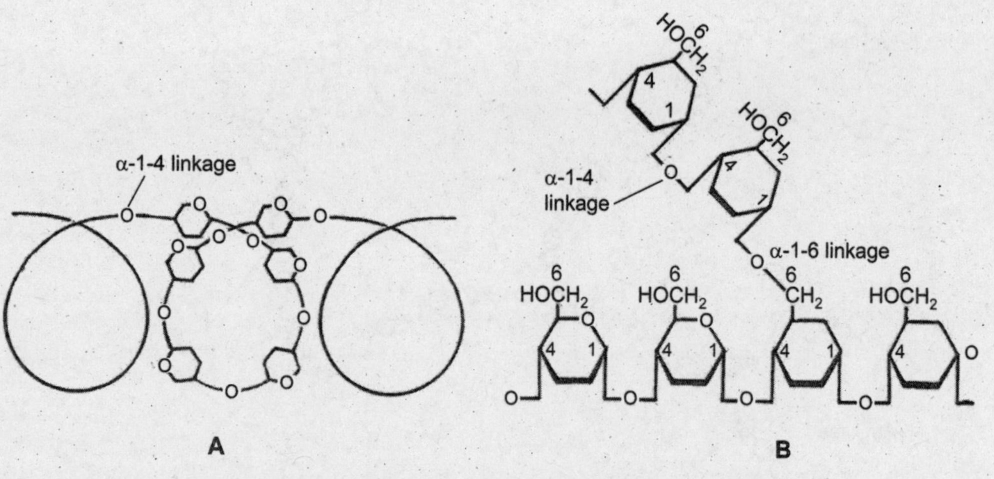

Fig. 2.24. Structure of starch. A—Amylose, showing helical coil structure (on suspension in water); B—Amylopectin, showing 1 → 6 branch (after Murray *et al.*, 2000).

Molecule of amylopectin is coiled concentrically. Because of its more complex structure, amylopectin is less soluble in water than amylose. It gives red-purple colour with iodine solution.

Dextrins and Limit Dextrins

Starch on hydrolysis yields lower molecular weight polysaccharides and finally maltose or glucose. Partial hydrolysis results in substances called **dextrins**. At first **amylodextrins** are formed which give rise violet colour with iodine. Then **erythrodextrins** are formed which gives red colour with iodine. Further hydrolysis results in the formation of **achrodextrins** which do not give any colour with iodine. Finally reducing sugars (maltose and glucose) appear. Dextrins are easily digested and are therefore used for feeding infants. The enzymes which bring about hydrolysis of starch are called **amylases**.

Dietary starch is degraded in the gastrointestinal tract by the actions of α-**amylase** and debranching enzyme. Enzyme α-amylase which is present in both animals and plants, is an **endoglycosidase** (it acts on internal glycosidic bonds). The enzyme catalyzes random hydrolysis of the α-(1 → 4) glucosidic bonds of amylose and amylopectin.

Another hydrolase, **β-amylase**, is found in seeds and tubers of some plants. Enzyme **β-amylase** is an **exoglycosidase** (it acts on terminal glucosidic bonds). It catalyzes sequential hydrolytic release of maltose from the free, nonreducing ends of amylopectin.

Despite their α and β designations, both types of amylases act only on α-(1 → 4)-D-glycosidic bonds. Figure 2.25, shows the action of α-amylase and β-amylase on amylopectin. The α-(1 → 6) linkages at branch points are not substrates for either α- or β-amylase. After amylase catalyzed hydrolysis of amylopectin, highly branched cores resistant to further hydrolysis, called **limit dextrins**, remain. Limit dextrins can be further degraded only after debranching enzymes have catalyzed hydrolysis of the α-(1 → 6) linkages at branch points.

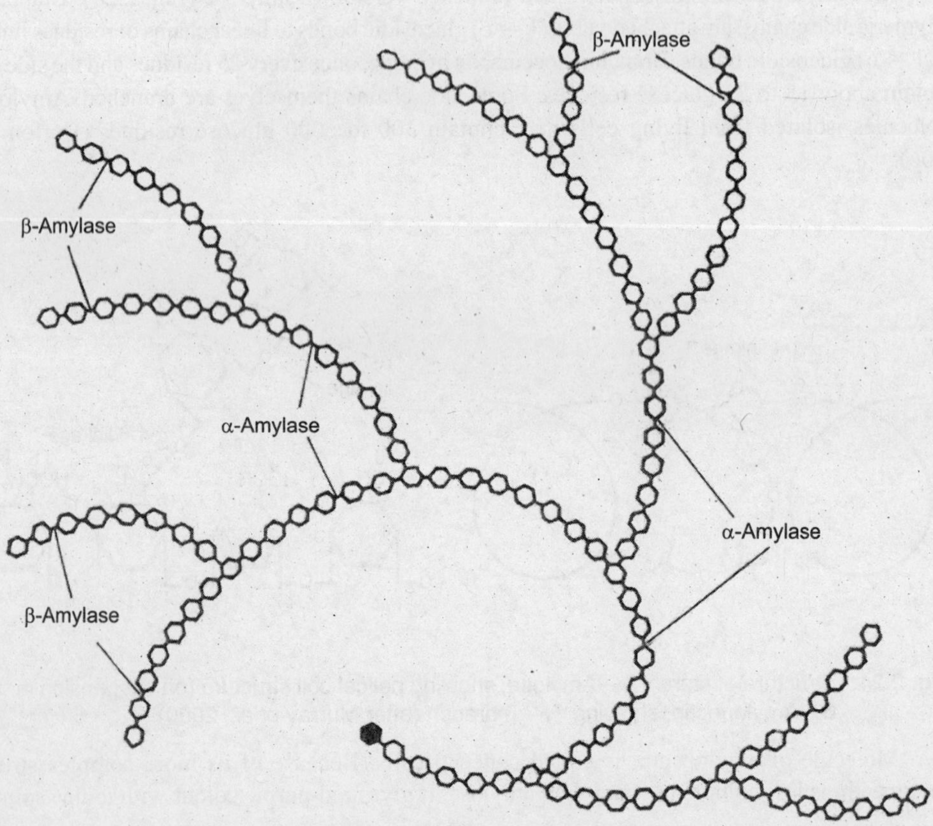

Fig. 2.25. Action of α-amylase and β-amylase on amylopectin. The α-amylase catalyzes random hydrolysis of internal α-(1 → 4) glycosidic bonds; β-amylase acts on the nonreducing ends. Each hexagon, represents a glucose residue; the single reducing end of the branched polymer is solid black (after Horton *et al.*, 2006).

The branched structures of amylopectin (and glycogen) possess only one reducing end but many nonreducing ends. It is at these nonreducing ends that enzymatic lengthening and degradation occur.

Table 2.3 Differences between amylose and amylopectin.

Amylose	Amylopectin
1. It forms 20% of the starch.	1. It forms 80% part of starch.
2. It is a straight chained polysaccharide. The chain is undranched.	2. It is a branched polysaccharide.
3. Each amylose molecule is formed of 100 to 1000 glucose residues.	3. An amylopectin molecule is formed of 3000 to 6000 glucose units.
4. Amylose is soluble in hot water.	4. Amylopectin is insoluble in water.
5. Its molecular weight ranges 50,000 to 60,000.	5. Its molecular weight is more than 500,000.
6. It gives blue colour with iodine reagent.	6. It gives pink purple (or red) colour with iodine.
7. Amylose is hydrolysed with β-amylase enzyme.	7. Amylopectin is hydrolysed into dextrin by enzyme β-amylase.

Glycogen. Glycogen is the main reserve food material of animal cells. It is known as **animal starch**. Glycogen is found mainly in liver, muscles and brain. It also occurs in cells of fungi and bacteria. The glycogen in the liver supplies glucose to all tissues through the blood. Muscle glycogen on the other hand is available during contraction of muscle.

Glycogen is a soluble polysaccharide. The molecule of glycogen is a **sphere** (Fig. 2.26) approximately 21 nm in diameter that can be visualised in electron microscope. It has a molecular mass of 10^7 Da and consists of polysaccharide chains each containing about 13 glucose residues. The chains are either branched or unbranched and are arranged in 12 concentric layers (in Fig. 2.26, only four layers are shown). The branch chains (each has two branches) are found in the inner layers and the unbranched chains in the outer layers.

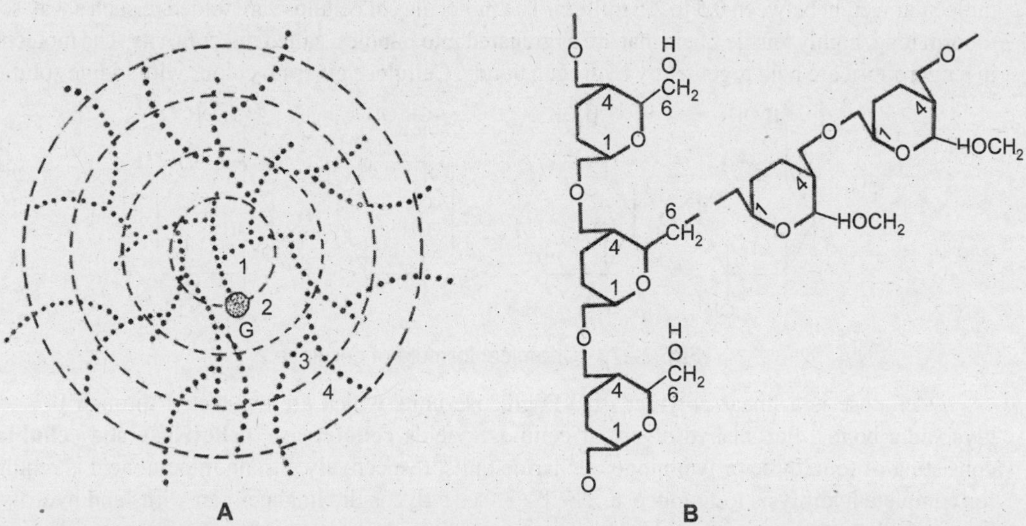

Fig. 2.26. The glycogen molecule. A—General structure; B—Enlargement of structure at a branch point. G=glycogenin, the primer molecule for glycogen sythesis (after Murray *et al.*, 2000).

Chemically glycogen is analogous to starch. Both are glucans consisting of glucose units linked by α-(1 → 4) glycosidic linkages, with branch chains formed at α-(1 → 6) glycosidic bonds. They differ mainly in their molecular weight and degree of branching. Glycogen's molecular weight is about 4,000,000. Each glycogen molecule contains about 50,000 glucose residues. The branching chains of glycogen average about 12 glucose units as compared to 25 or so for starch. Glycogen is non-reducing and gives a red colour with iodine.

Glycogen is produced from glucose in the mammalian liver and muscles when blood sugar levels are high, a process called **glycogenesis**, which is under the influence of insulin hormone. Glycogen in the liver can be broken down to glucose when blood sugar levels are low, a process called **glycogenolysis**, which is under the influence of **glucagon** hormone. Glycogen in the muscle, however, is broken down to lactic acid in glycolysis (**Jain** 2012).

Table 2.4 Differences in starch and glycogen.

Starch	Glycogen
1. It is a storage polysaccharide in the plants.	1. It is called animal starch. It is stored in animal cells such as liver and muscles.
2. It is formed of amylose and amylopectin. Both contain glucose units.	2. It is formed of glucose residues.
3. Amylose chains are unbranched while amylopectin chains are branched.	3. Glycogen molecules are branched similar to amylopectin.
4. The chains of amylopectin are branched after every 20–24 glucose units.	4. The glycogen chains are branched after every 8–10 glucose units.

Cellulose. It is most abundant carbohydrate, in fact, the most abundant organic compound on Earth. It is the main structural polysaccharide of the plants. Cellulose is totally absent in animals. Cellulose is common in the cell walls of most algae, all higher plants, certain fungi and some protists. It is especially abundant in wood (25–50%) and cotton fibers (90%). Cellulose molecule is a long unbranched chain of about 300 to more than 15000 glucose units (see **Horton** *et al.*, 2006) with molecular weight between 0.5 to 2.5 million. The molecules of cellulose are folded in such a way so as to form long, highly tensile fibers that are aggregated into bundles, called **microfibrils**. The molecules in a microfibril are held together by hydrogen bonds. Cellulose gives no colour with iodine solution.

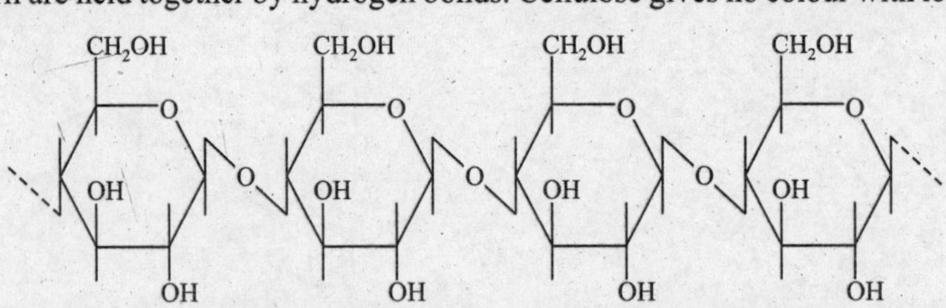

Fig. 2.27. Chemical formula of cellulose.

Cellulose is a linear polymer of β-D-glucose units which are connected through β-1 → 4 glycosidic bonds. Partial hydrolysis of cellulose yields **cellotetrose**, **cellotriose** and **cellobiose** consisting of four, three or two monosaccharide units, respectively. Strong mineral acid is required for complete hydrolysis to D-glucose. The β-(1 → 4) glycosidic linkages can withstand hydrolysis by glycosidases (enzymes) found in digestive tract of human beings and higher animals. Since, it is not broken down to glucose by animals, it is not useful as a food source. It simply forms "bulk" or "roughage" of the food necessary for proper functioning of the alimentary canal. Some animals, such as the farm animals ("ruminants" like cow and sheep) with the help of digestive system bacteria and the termites with the aid of intestinal flagellates can hydrolyse cellulose. These symbiotic organisms produce an enzyme **cellulase** for digesting cellulose. Some snails are said to produce the cellulase enzyme.

Uses of cellulose. Cellulose provides many useful materials. For example, cotton, linen, jute, rayon, cellulose acetate (used in fabrics, plastics and unbreakable glass), cellulose nitrate (used in propellant explosives), cellophane and carboxymethyl cellulose (added to icecreams, cosmetics and medicine). Cattle and buffaloes use cellulose as food because they can digest it by the help their symbiotic microorganisms. Cellulose serve as roughage in the human digestive tract. Cellulose rich wood is used in the manufacture of furniture, sports goods, tool handles and many more articles. Cellulose is also used as fuel (*e.g.*, fire wood) and in the production of paper.

Agar (Agar-Agar). It is the gelatinous extract from the cell wall of certain red algae (*e.g.*, *Gelidium*, etc.) and consists of the polysaccharide **agarose** (β-1, 3 linked D-galactose and 1, 4 linked anhydro-L-galactose) and **agropectin** (a sulphated galactan mixture) in a ratio of 7 : 3. Thus, this homopolysaccharide is composed of galactose monomer. Japan is the major producer of agar. Agar forms highly viscous gels. Once solidifies, agar gel does not melt below 100°C. Agar is an inert substance and hence provides no nutrients. Different nutrients may be added to make media suitable for different microorganisms.

Agar is used to solidify culture media on which bacteria are grown in the laboratory. Agar is also used as laxative and as food (deserts) in the East.

Heteropolysaccharides

These are composed of modified monosaccharide molecules of different types. Examples of heteropolysaccharides include mucopolysaccharides, glycoproteins, chitin, peptidoglycans, pectin, hemicellulose, etc.

I. Mucopolysaccharides. These are polymers of modified sugars (galactose and mannose) called sugar acids and amino sugars. Sugar acid has at least one hydroxyl group replaced by an acid group. Amino sugar has an amino group in place of one of the hydroxyl groups (Fig. 2.28).

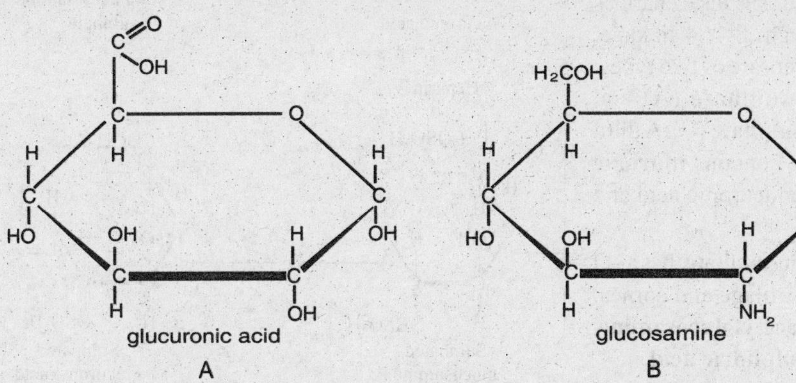

Fig. 2.28. Structure of sugar acid (A) and amino sugar (B).

The mucopolysaccharides are slimy compounds like that noticed in okra (bhindi) fruit. Important examples of mucopolysaccharides are hyaluronic acid, heparin, chondroitin sulphate and keratin sulphate.

Proteoglycans. The molecules of proteoglycans consist of much longer portion of polysaccharide and a small portion of protein. They are also called **mucoproteins**. The proteoglycans are amorphous and form gels which are able to hold large amounts of water.

The **cartilage proteoglycan** is found extracellularly in cartilage and bone. In its molecule, strands of protein, called **core protein**, extend radially from a long, central hyaluronic acid molecule. In each core protein strand, three carbohydrate bearing regions may be identified. The first region contains numerous oligosaccharides, the second region contains keratin sulphate chains and the third region contains chondroitin sulphate chains. This arrangement gives cartilage its resilience and tensile strength.

(*i*) **Hyaluronic acid.** It is polymer of D-glucuronic acid and N-acetyl D-glucosamine units arranged alternately (Fig. 2.29). Hyaluronic acid is present in the various lubricating fluids of the body such as synovial fluid of bone joints, cerebrospinal fluid of brain and spinal cord, viterous humour of the eye. It also occurs as a coat around the ovum and in the skin and between the cells.

(*ii*) **Heparin.** It is an anticoagulant and is found in the blood. It is used to prevent clotting of blood. Heparin is produced by the mast cells of the liver, intestinal muscosa, lung, spleen and kidney. Heparin is polymer of sulphated glucuronic acid and sulphated glucosamine units (Fig. 2.29).

(*iii*) **Chondroitin sulphate.** It is present in the ground substance of connective tissue and is predominant in cornea, cartilage, tendons, skin, heart valves and saliva. The repeating unit is a disaccharide consisting of **glucuronic acid** linked to sulphate ester of **N-acetyl galactosamine** through through β-1, 3 glucosidic bond. The disaccharides are linked through. β-1, 4 linkage. The sulphate esters are of two types: **chondroitin-4-sulphate** (A) and **chondroitin-6-sulphate** (C). A third type of sulphate contains **iduronic acid** instead of glucuronic acid and is labelled B.

(*iv*) **Keratin sulphate.** It is a–O component of cartilage and cornea. It consists of **N-acetylglucosamine, galactose** and **sulphuric acid**.

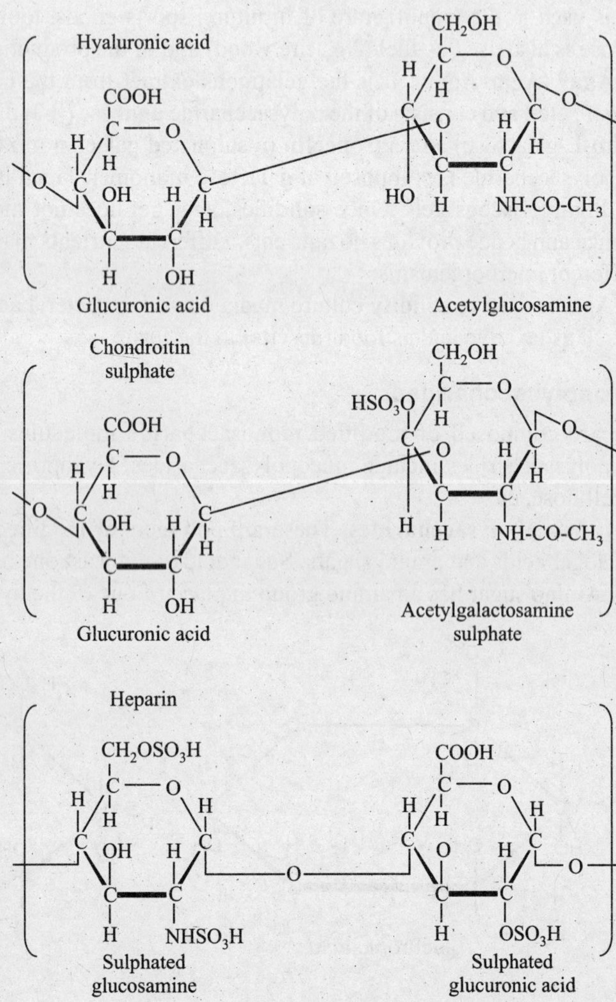

Fig. 2.29. Structure of some mucopolysaccharides.

II. Neutral Sugars

(*a*) **Hemicellulose.** It is found in association with cellulose in cell walls. The commonly found sugars in hemicellulose are **D-xylose, L-arabinose, D-galactose, L-rhamnose, D-mannose** and **D-glucuronic acid.**

(*b*) **Gums.** These are substances exuded by plants on mechanical injury or after bacterial, fungal or insect attack. The viscus substance seals the wound and thus protects the plant. Gums often contain polysaccharides, which are usually highly branched. The commercial Gum-Arabic occurs as a salt. Acidification of an aqueous solution of the gum yields various sugars such as *L-arabinose, L-rhamnose, 3-β-D-galactopyranosyl-L-arabinose, D-galactose, D-glucuronic acid* and *6-β-D-glucuropyranosyl-D-galactose.*

III. Glycoproteins. These are compounds of mucopolysaccharides and proteins. They contain acetylgucosamine, acetylgalactose-amine and hexose sugars (mannose or galactose) or their derivatives. Blood group substances A, B, and Rh antigens of the erythrocytes (see Box 2.10), luteinizing hormone (LH), mucus and plant mucilage all are glycoproteins. Mucilage in the husk of isabgol (*Plantago ovata*) and in the leaves of *Aloe* has medicinal value.

Box 2.10 ABO Blood Group

The ABO blood group was first discovered in 1901 by **Landsteiner**. Most primates display three different kinds of O- and N-linked oligosaccharides on their cell surfaces. The core structure of these oligosaccharides is called **H antigen**. It consists of various combinations of **galactose (Gal)**, **fucose (Fuc)**, **N-acetylglucoseamine (GlcNac)** and **N-acetyle-neuraminic acid** (sialic acid, NeuNAc). These monosaccharides are linked in various ways to form a short branched structure that exhibits considerable microheterogeneity. One of the most common **H antigen** structures is shown in following Fig. 2.30.

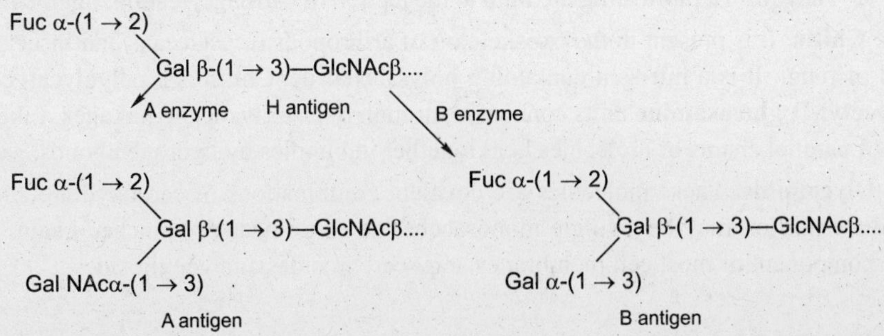

Fig. 2.30. Mode of conversion of H antigen into A antigen and B antigen (after Horton *et al.*, 2006).

The core structure (H antigen) can be modified in various ways. The addition of a GalNAc residue in α-(1 → 3) linkage forms A antigen. This reaction is catalyzed by A enzyme. The addition of Gal in α-(1 → 3) linkage is catalyzed by B enzyme.

If only A antigen is present a person will have **A blood group**. If only B antigen is present their blood type will be B (*i.e.*, B blood group). The **AB blood type** indicates that both A antigen and B antigen are present on cell surfaces. If neither GalNAc nor Gal have been added to the H antigen structure, then neither A antigen nor B antigen will be present and the blood type is O (*i.e.*, **O blood group**).

The ABO blood group is determined by a single gene on autosomal chromosome 9. Human (and other primate) populations contain many alleles of this gene. The original gene encoded A enzyme, which transfers GalNAc. Variants of this gene have altered specificity of the enzyme so that it no longer recognizes GalNAc but instead transfers Gal. These B enzymes differ by several amino acid residues from the allele that encodes the A enzyme. The structures of both types of **glycosyltransferase** enzymes have been solved and they reveal that only a single amino acid **substitution** is required to change the specificity from *N*-acetyl-aminogalactosyltransferase enzyme to galactosyltransferase enzyme.

The chromosome 9 locus can also contain several alleles that encode non-functional proteins one of the most common mutations is a single base pair deletion near the N-terminus of the coding region. This **deletion** shifts the reading frame for translation, making it impossible to synthesize a functional enzyme of either type. People who are homozygous for these non-functional O allele will not synthesize either A antigen or B antigen and their blood type will be O.

All cells contain some of the core oligosaccharide (H antigen) even if one's blood type is A, B, or AB. This is because not all of the H antigen structures are modified. Under normal circumstances human plasma will not contain antibodies against H antigen. However, O-type individuals will have antibodies against A antigen and B antigen because these structures are recognised as non-self. If

O-type individuals receive a blood transfusion from someone with A, B or AB blood, they will mount an immune response and reject it. Similarly, if one has A-type blood he will have anti-B antibodies and cannot receive a transfusion from someone with B or AB blood type.

The O allele (non-functional enzyme) is the most common allele in most human populations and the B allele is the rarest. Some Native American populations has type O blood type. Type O individuals are perfectly normal, indicating that the absence of the A and B oligosaccharide structures has no effect on normal growth and development (Horton *et al.*, 2006).

Glycoproteins are associated with following functions:

1. They form viscous solutions which function as lubricants and form a protective covering on the cell surface.
2. They provide support to the cells and bind them together.
3. They ensure exchange of substances between blood and tissue cells.
4. They form a lubricating medium to the juction of cartilage, tendon, ligaments and bones.

IV. Chitin. It is present in the exoskeleton of arthropods (crustaceans and insects) and in the cell wall of fungi. It is a nitrogen-containing polysaccharide. Chitin is a polyglycan consisting of both **N-acetyl-D-glucosamine** units connected through β-1, 4-glycosidic linkages. Like cellulose it consists of parallel chains of molecules held together in bundles by hydrogen bonds.

V. Glycolipids. These molecules are covalent combinations of carbohydrate and lipid. The carbohydrate portion may be a single monosaccharide or a linear or branched chain. Glycolipids form the component of most cell membranes, *e.g.*, cerebrosides and gangliosides.

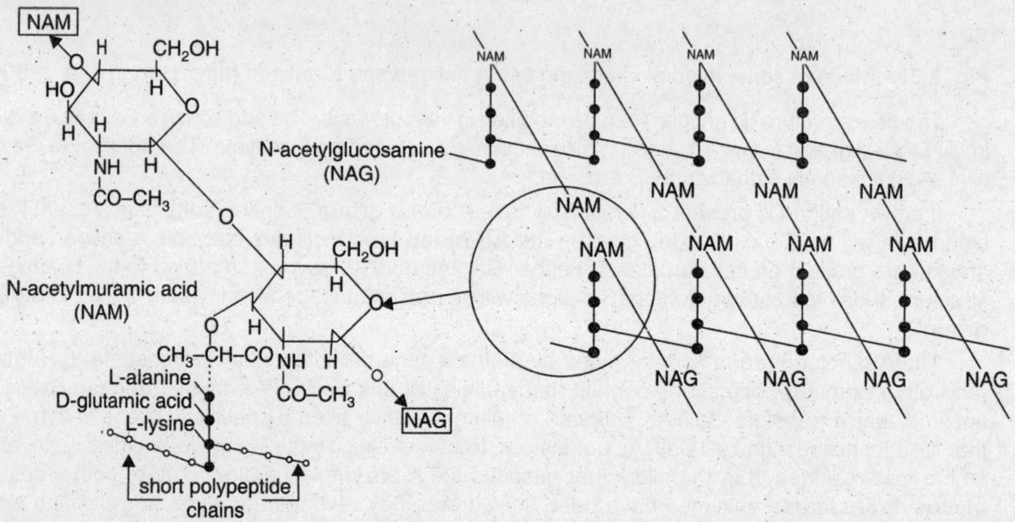

Fig. 2.31. Peptidoglycan sheet structure of bacterial cell wall.

VI. Peptidoglycans (Murein). Murein is a polysaccharide linked to short peptide chains. It is found in the cell walls of many bacteria. It consists of alternating units of **N-acetylglucosamine (NAG)** and **N-acetylmuramic acid (NAM)** linked by its carboxyl group to the peptide chain consisting of four amino acids (L-alanine, D-glutamic acid, L-lysine and D-alanine). Five glycine residues link peptide chain to its neighbour. The enzyme **lysozyme** hydrolyzes the chain into NAG-NAM disaccharides, resulting in rupture of cell wall.

Table 2.5 Difference between oligosaccharides and polysaccharides.

Oligosaccharides	Polysaccharides
1. They contain 2 – 10 monosaccharide residues.	1 Polysaccharides possess a large number of monosaccharide residues.
2. Oligosaccharides are soluble in water.	2. Polysaccharides are usually insoluble.
3. They are commonly sweet to taste.	3. They are tasteless.
4. Transport of carbohydrates (in plants) occurs in the oligosaccharide state.	4. Polysaccharides are not involved in transport.
5. Food storage in oligosaccharide state is less common.	5. Food storage occurs in the state of polysaccharide.
6. They form component of external surface of plasma membrane and cell coat.	6. They form component of cell wall.
Example: Sucrose, Maltose, Raffinose	Example: Starch, Glycogen, Cellulose

2.5 BIOLOGICAL FUNCTIONS OF CARBOHYDRATES

1. Storage substances of potential energy. The most important function of carbohydrates is to provide energy to the body. They are storage substances of potential energy in animals. About 60% of total energy requirement of human beings is provided by the breakdown of carbohydrates. One gram of carbohydrate on oxidation yields on average four Calories. Glucose supplies the immediate energy needed by tissues. Glucose is the only form of energy for the brain and other nervous tissues. Lack of glucose or of oxygen for the metabolism of brain leads to rapid damage to the brain. Carbohydrate is stored in the body in the form of glycogen. About 100 gm of glycogen is stored in the human liver and by its breakdown maintains the glucose level in the blood. Human muscles contain about 200 to 240 gm glycogen. This glycogen is, however, utilised only by the muscles themselves and it is not available for regulating the blood sugar level. In plants, storage carbohydrate is starch. In yeast and other fungi storage carbohydrate is glycogen.

2. Structural components. Carbohydrates are important components of some structural materials of living materials. In plant cells carbohydrate (cellulose) constitute the structural framework (cell wall). In bacteria, polysaccharides form the capsule.

Monosaccharides are important constituents of nucleic acids, coenzymes, flavoproteins and blood group substances (A, B antigens). *Ascorbic acid* (vitamin C) is related to sugars. *Immunopolysaccharides* play a vital part in resistance of infections. *Hyaluronic acid* is the viscous substance present in the matrix of connective tissue. *Heparin* prevents the clotting of blood. *Glucuronic acid*, which occurs in liver, acts as detoxifying agent by combining with toxic substances and bacterial byproducts. *Chondroitin sulphates* are present in cornea, cartilage, tendons, skin, heart valves and saliva. *Glycosides* are components of steroid hormones. *Galactolipins* are constituents of nervous tissue.

3. Regulation of fat metabolism. Some carbohydrates are essential for normal oxidation of fats. When carbohydrates are restricted in the diet there is more rapid metabolization of fats. This results in the accumulation of incompletely oxidized intermediate products leading to **ketosis**. This is common in uncontrolled diabetes mellitus.

4. Protein sparing function. Carbohydrate is preferentially metabolized in the body as a source of energy as long as it is present in the required quantity. This spares protein for building of tissue. When there is deficiency of calories in the diet, however, fat and then protein are utilized for supplying energy.

5. Role of carbohydrates in the normal functioning of gastrointestinal tract. Indigestible substances such as cellulose, hemicellulose and proteins provide the bulk of roughage.

6. Survival of Antarctic fish. The survival of Antarctic fish below –2° is attributed to the antifreeze glycoproteins. This glycoprotein consists of 50 repeating units of the tripeptide, alanine-alanine-threonine.

2.6 CLINICAL IMPORTANCE OF CARBOHYDRATES

1. **Glucose.** Glucose is the ultimate source of energy for all body cells. Its continuous supply is essential for following purposes:
   (*i*) As source of energy for the nervous system and for erythrocytes.
   (*ii*) As source of glyceride-glycerol in adipose tissue.
   (*iii*) For maintaning the level of intermediates of the citric acid cycle.
   (*iv*) As precursor of milk sugar, lactose, in the mammary gland.
   (*v*) As fuel for the supply of energy to the skeletal muscles especially under aerobic condition.

   Glycolysis and glyconeogenesis are two main metabolic activities that maintain an non-stop glucose supply to all body cells. Lowering the glucose supply below a critical level leads to brain dysfunction and may result in coma and death.
2. **Fructosurea.** Fructosurea is the appearance of fructose in urine. This clinical condition may occur under following circumstances:
   (*i*) Due to deficiency of **fructokinase** enzyme, fructose is not utilized completely and is excreted in the urine. This is a rare congenital disorder.
   (*ii*) When large quantities of fructose are ingested, some of it is excreted in urine. This is due to hepatic insufficiency.
   (*iii*) Due to deficiency of enzyme **aldolase**, fructose-1-phosphate is accumulated in the liver. This blocks pathways of fructose utilisation.
3. **Clinical significance of sorbitol.** Accumulation of sorbitol and dulicol in tissues may cause certain pathological conditions, *e.g.*, cataract, nephropathy.
4. **Clinical significance of glycogen.** Glycogen is animal starch. It is stored in the muscles and liver cells. **Muscle glycogen** provides hexose sugars to muscle cells for energy by glycolysis. **Liver glycogen** provides hexose sugars to all body cells for energy and maintains blood glucose level between meals. Liver glycogen is exhausted after 12 – 18 hours of fasting, whereas muscle glycogen is exhausted after prolonged vigorous exercise.

Inherited disorders result in **glycogen storage diseases**. They are caused due to inadequate mobilisation of glycogen and resultant deposition of glycogen in liver, heart, kidney and muscles. This condition is called **glycogenosis** and causes muscular weakness and may ultimately lead to death.

Types of Glycogen Storage Diseases

I. Von Gierke's disease (Type I glycogenosis). This disease is caused by the deficiency of enzyme **glucose-6-phosphatase**. In the absence of this enzyme glycogen is not broken down to glucose. Hence stored glycogen is not utilised, but the synthesis of glycogen from glucose-6-phosphate continues. This disease has clinical onset at birth or during infancy and has the following effects:

1. Enlargement of liver and kidney.
2. Children suffering from this disease develop **hypoglycemia** (low blood glucose level).
3. Hypoglycemia inhibits insulin secretion which in turn inhibit protein synthesis. This leads to retarded growth.
4. Hypoglycemia stimulates production of epinephrine. This causes breakdown of muscle glycogen forming lactate. The lactate interferes with the excretion of urate by the kidney raising blood urate level.

5. The persons suffering from this disease tend to use fat as energy source and develop **lipemia** (*i.e.*, the presence of an excess of fats or lipids in the blood); **acidemia** (*i.e.*, increased hydrogenion concentration is the blood); **ketosis** (*i.e.*, abnormal increase of ketone bodies in the body) and **nosebleeds**.
6. The liver shows fatty infiltration.

This disease is inherited as an autosomal recessive trait.

II. Pompe's disease (Type II glycogenosis). This fatal disease occurs due to deficiency of a lysosomal enzyme called **acid maltase**. Hence, this disease is also called **acid maltase deficiency**. Absence of acid maltase enzyme causes excessive accumulation of glycogen in the cells of liver, heart and muscles. The heart is enlarged. The patient shows extreamly weak muscles. Death occurs usually before the age of nine months.

III. Limit dextrinosis (Type III glycogenosis). This disease is caused by the deficiency of limit dextrin in liver and muscles. This condition is produced by the action of enzyme **phosphorylase** on glycogen.

IV. Amylopectinosis (Type IV glycogenosis). This disease is caused by the dificiency of branching enzyme in the liver. The disease is fatal. In the absence of this enzyme amylopectins are formed in the liver, heart, kidneys and muscles.

V. Mc Ardie disease (Type V glycogenosis). This glycogen storage disease is caused by the deficiency of **muscle phosphorylase** and affects the skeletal muscles. This leads to cramps in the muscles.

VI. Type VI glycogenosis. This disease is due to deficinecy of **liver phosphorylase** resulting in increased liver glycogen.

VII. Type VII glycogenosis. Deficiency of enzyme **phosphofructokinase** causes accumulation of glucose-6-phosphate and fructose-6-phosphate.

QUESTIONS

Long Answer Questions

1. What are carbohydrates? Give general account of chemical composition and molecular structure of carbohydrates. Write a note on their biological significance. (*Garhwal 1999; Kumaon 2004; Bundelkhand 2002; Poorvanchal 2003*)
2. What are monosaccharides, disaccharides and polysaccharides? Give examples of each and describe their biological significance.
3. What are carbohydrates? Classify carbohydrates used by animals. (*Avadh 2003*)
4. What are carbohydrates? Classify them and describe the characteristics of monosaccharides.
5. Classify carbohydrates on the basis of their composition. Write the structure of one typical example representing each class. (*Allahabad 2002*)
6. Classify carbohydrates into types giving nature and examples of each. (*Bundelkhand 2002*)
7. Describe monosaccharides and disaccharides.
8. Give an account of structure, chemical properties and uses of monosaccharides.
9. Explain properties of monosaccharides. (*Gorakhpur 2000*)
10. Give structural specialities of monosaccharides. (*Allahabad 2001; Kumaon 2002; Lucknow 2003*)
11. Describe structure, classification and function of polysaccharides.
12. How are polysaccharides formed? Name three biologically important homopolysaccharides and write a short note on each.
13. What are heteropolysaccharides? Discuss them briefly.
14. Describe in detail the clinical significance of glycogen storage.

15. What type of carbohydrates are the following?
Glucose, Maltose, Lactose, Glycogen, Cellulose, Heparin, Chitin, AB antigen of RBCs. Write a note on any one of them.
16. Give an account of properties and uses of monosaccharides. (*CCSU Meerut 2013*)

Short Answer Questions

1. Summarise clinical importance of glucose.
2. Describe clinical significance of glycogen.
3. What is fructosurea? Under what conditions does this occur?
4. In brief describe differences between reducing and non-reducing sugar.
5. Write down biological significance of carbohydrates.
6. What is mutarotation?
7. What is stereoisomerism? (*Kanpur 2004*)
8. What do you understand by dextro-rotatory and levo-rotatory monosaccharides?
9. Draw straight chain structural formula of glucose. Number the carbon atoms. Is this formula correct?
10. What is the difference between pyranose and furanose ring structure?
11. Differentiate between D-glucose and L-glucose.
12. Differentiate between amylose and amylopectin.
13. Give differences between glucose, fructose and galactose.
14. Give functions or biological role of monosaccharides. (*Gorakhpur 2002, 2003*)
15. Describe glycosidic bond.
16. Give functions of oligosaccharides.
17. Write short notes on the following:
    (*i*) Monosaccharides (*Gorkhpur 2002, 2003; Kanpur 2004; Garhwal 2004*)
    (*ii*) Sucrose (*Lucknow 2002*)
    (*iii*) Polysaccharides (*Gorakhpur 2002, 2003*)
    (*iv*) Carbohydrates (*Bundelkhand 2001*)
    (*v*) Starch (*Kanpur 2003*)
    (*vi*) Isomerism (*Kanpur 2004*)

Very Short Answer Questions

1. Name the bond by which monosaccharides are joined together to form a polysaccharide chain.
2. What is the natural source of maltose?
3. Which monosaccharides are found in lactose?
4. Which disaccharide is not a reducing sugar?
5. What is glycogenolysis?
6. Give two examples of storage polysaccharides.
7. What is animal starch?
8. What are mucopolysaccharides?
9. What is new name of mucopolysaccharides?
10. What is pectin?
11. What is economic use of pectin?
12. What are dextrins?
13. Where are glycosaminoglycans found?
14. Why sucrose which contains in its molecule glucose and fructose which are reducing sugars is itself a non-reducing sugar?
15. Name a disaccharide each of glucose only which is (*a*) reducing sugar; and (*b*) non-reducing sugar.

16. Name a disaccharide each of glucose and fructose; which is (*i*) a non-reducing sugar; and (*ii*) a reducing sugar.
17. What is L-fucose?
18. Name the compound that forms matrix of connective tissue and cartilage.
19. What is the function of hyaluronic acid?
20. Why are monosaccharides called reducing sugars?
21. Name any two amino sugars.
22. What is heparin?
23. Name two pentose sugars that are excreted in the urine in case of inherited disorder?
24. What is the cause of congenital disorder fructosauria?
25. What is galactosemia?
26. Define glycosauria. When do person suffers from glycosauria?
27. Differentiate between oligosaccharides and polysaccharides. (*CCSU Meerut 2014*)
28. Differentiate between ribose and deoxyribose sugars. (*CCSU Meerut 2014*)
29. Write a short note on hyaluronic acid. (*CCSU Meerut 2014*)
30. What are glycoproteins? (*CCSU Meerut 2013*)

True or False

Explain with reasons whether the following are true or false: Reducing sugars include

1. Glucose 2. Fructose 3. Sucrose 4. Galactose
5. Lactose

Multiple Choice Questions

1. Which of the following is not a disaccharide?
   (*a*) maltose (*b*) starch
   (*c*) sucrose (*d*) lactose
2. Which of the following is a reducing sugar?
   (*a*) maltose (*b*) sucrose
   (*c*) lactose (*d*) both (*a*) and (*c*)
3. Which one of the following is an example of glycoprotein?
   (*a*) haemoglobin (*b*) lecithin
   (*c*) mucin (*d*) casein
4. In which of the following group all are polysaccharides?
   (*a*) sucrose, glucose and fructose (*b*) maltose, lactose and fructose
   (*c*) glycogen, sucrose and maltose (*d*) glycogen, cellulose and starch
5. Glycosidic link is broken in digestion of
   (*a*) starch (*b*) protein
   (*c*) lipid (*d*) all of these
6. Glucose and galactose are
   (*a*) enantimers (*b*) D- and L- isomers
   (*c*) epimers (*d*) none of these
7. The polysaccharide with 200 – 1000 glucose units held by a $(1 \rightarrow 4)$ glycosidic linkage only is
   (*a*) starch (*b*) amylose
   (*c*) amylopectin (*d*) glycogen
8. Which of the following is also called fructosan?
   (*a*) starch (*b*) glycogen
   (*c*) inulin (*d*) cellulose

9. Hexoses and pentoses are
(a) monosaccharides (b) polysaccharides
(c) disaccharides (d) oligosaccharides

10. Most abundant water insoluble polysaccharide of plant cell wall is
(a) cellulose (b) hemicellulose
(c) pectin (d) lignin

11. Most common monomer of carbohydrate is
(a) glucose (b) fructose
(c) sucrose (d) maltose

12. Sweetest sugar is
(a) fructose (b) glucose
(c) mannose (d) lactose

13. Benedict's reagent test is conducted to confirm the presence of
(a) polysaccharide (b) lipids
(c) reducing sugars (d) proteins

ANSWERS

Short Answer Questions

6. A freshly prepared solution of glucose has a specific rotation of +112.2°. This goes on decreasing with time and becomes steady at a value of +52.7°. This change in specific rotation is called **mutarotation** (*muta* = change). The degree of mutarotation is speeded up by addition of alkali. A freshly prepared solution of glucose is predominantly alpha D glucose, which has a specific rotation of +112.2°.

   The alpha form in solution breaks and gets changed to beta form which has a lower specific rotation of +18.7°. Once an equilibrium is obtained for this conversion, there will be one part of alpha and three parts of beta with the net specific rotation of +52.7°.

9. Refer Fig. 2.5. This formula was given by **Fisher**. In this structural formula the numbering of carbon atoms begins with carbonyl group. This formula is not acceptable now because aldehyde group of glucose does behave like a typical aldehyde group. A cyclic structure was proposed for glucose molecule by **Haworth**.

Very Short Answer Questions

1. α-1, 4-glycosidic bond or β-1, 4-glycosidic bond.
2. Maltose is formed by the hydrolysis of starch, glycogen or dextrin.
3. Lactose is formed of α-D, glucose and β-D, galactose.
4. Sucrose.
5. Glycogenolysis is hydrolysis of glycogen into glucose molecule.
6. Starch and glycogen.
7. Glycogen.
8. The mucopolysaccharides are sticky compounds present in some cell secretions. These are formed of proteins and polysaccharides (glycoproteins).
9. Glycosaminoglycans.
10. Pectins are acidic polysaccharides. These are found in the matrix of cell wall of plant cells.
11. Pectin is extensively used in food industry to set jellies and jams. Commerical quantities of pectin are obtained from citrus and apple wastes after removal of juice.
12. Dextrins are white powders, soluble in water and have adhesive property. They are formed by partial digestion (=hydrolysis) of starch:

Starch → Dextrin → Maltose → Glucose

Dextrin is also formed when starch is heated dry.

13. Glycosaminoglycans are complex mucopolysaccharides. They are found in connective tissue, cartilage, tendon and wall of blood vessels. They form gel-like matrix or ground substance.
14. In the formation of a sucrose molecule the reducing aldehyde group of glucose and the ketonic group with adjacent hydroxyl group of fructose are changed with the formation of an oxygen bridge. Hence sucrose is a non-reducing sugar. But hydrolysis breaks the glycosidic oxygen bond releasing free glucose and fructose (**De** 1997).
15. (*a*) A reducing disaccharide of glucose-maltose; and (*b*) A non-reducing disaccharide of glucose-trehalose.
16. (*i*) A non-reducing sugar of glucose and fructose, *e.g.*, sucrose; and (*ii*) A reducing sugar of glucose and fructose, *e.g.*, Turanose.
17. L-fucose is an extraordinary carbohydrate (aldose sugar) in which $-CH_2OH$ group replaced by $-CH_2$ group. L-fucose occurs in the mucopolysaccharides of mammalian blood group, in human milk (as a polysaccharide) and in some brown algae.
18. The matrix of connective tissue and cartilage is formed of heteropolysaccharides, chondroitin sulphate.
19. Hyaluronic acid is glycosaminoglycan compound found at the joints of bones. It lubricates the joints and absorbs shock.
20. Monosaccharides have either aldehyde –CHO or keto (C = O) group. Both aldehyde and keto groups have reducing property.
21. (*i*) N-acetyl glucosamine; and (*ii*) Glucosaminoglycans or mucopolysaccharides.
22. Heparin is a complex mucopolysaccharide, found in blood and connective tissue. It help in blood clotting.
23. Pentose sugars *L-xylulose* and *L-arabitol* are excreted in urine in persons (*e.g.*, jews) suffering from a congenital disorder **pentosuria**.
24. Fructosauria is caused due to deficiency of enzyme **fructokinase** or **aldolase B**.
25. Galactosemia is an inherited disorder in which galactose concentration in the blood increases.
26. Glycosauria is a condition in which excess of glucose from blood is excreted in the urine. Whenever glucose concentration in venous blood exceed renal threshold value (*i.e.*, 160/180 mg/100 ml).

True or False

1. **True.** A sugar is said to be 'reducing' if it is capable of reducing alkaline cupric (Cu^{++}) sulphate to cuprous (Cu^+) sulphate (Fehling's test) or produces metallic silver from ammonical silver nitrate (silver mirror test). The reducing property is dependent upon the presence of an aldehyde group of carbon-1 when the sugar is in the open ring (straight chain) from glucose, an aldohexose. Glucose is therefore a reducing sugar.
2. **True.** Although fructose is a ketohexose it readily converts to glucose in weakly alkaline solutions and reacts positively as a reducing sugar.
3. **False.** Sucrose is not a reducing sugar although it is disaccharide consisting of glucose and fructose, both of which are themselves reducing sugars. The glycosidic bond is between carbon-1 of glucose and carbon-2 of fructose. These are the anomeric carbons of the two hexoses,

respectively, so that no ring opening is possible for either of them and no aldehydic reducing group is formed.

4. **True.** Galactose differs from glucose only by the configuration of the hydroxyl group at carbon-4. It has an aldehyde group at carbon-1 when in the straight chain form and is therefore a reducing sugar.
5. **True.** Lactose is a disaccharide consisting of galactose and glucose. It is a β-galactoside, linked 1, 4 to the glucose. The latter residue has potentially a free aldehyde group at carbon-1. It is therefore a reducing sugar.

Multiple Choice Questions

1. (*b*)	**2.** (*d*)	**3.** (*c*)	**4.** (*d*)	**5.** (*a*)	**6.** (*c*)
7. (*b*)	**8.** (*c*)	**9.** (*a*)	**10.** (*a*)	**11.** (*a*)	**12.** (*a*)
13. (*c*)					

3

Lipids

Lipids (Gr. *lipos* = fat) are oily, greasy and wax-like substances which are insoluble in water (a polar solvent) but soluble in non-polar organic solvents such as ether, chloroform, hot alcohol, benzene, petroleum or carbon disulphide. The lipids are a heterogenous group of organic compounds which include cooking oil, butter, ghee, waxes, cholesterol, essential oils, natural rubber, some plant pigments (carotene of carrots, lycopene of tomatoes), menthol (mint oil), vitamins A, E and K and eucalyptus oil. Lipids form major part of stored food material in animal body and a major structural component. The term **lipid** was used by German biochemist **Wilhelm Bloor** in the year 1943 to a group of fats and fat-like substances.

Percentage. The lipids form about 3.5 per cent of the cell content. In plants they are present in seeds, nuts and fruits (*e.g.*, olive), while in animals they are stored in adipose tissues, yellow bone marrow and nervous tissue.

3.1 SPECIAL FEATURES OF LIPIDS

1. Lipids are compounds of carbon, hydrogen and oxygen. In them the number of oxygen atoms is much less in comparison to carbon and hydrogen.
2. Chemically they are esters of fatty acids and polyhydric alcohols. Esters are formed when organic acids and alcohols are reacted just as inorganic acids and bases react to form salts. In one lipid molecule three molecules of fatty acids are joined to one glycerol molecule. The carboxyl groups (–COOH) of fatty acids form ester bonds with alcoholic groups (–OH) of glycerol molecule (Fig. 3.1). The characteristic of lipids depends on the nature of fatty acids.

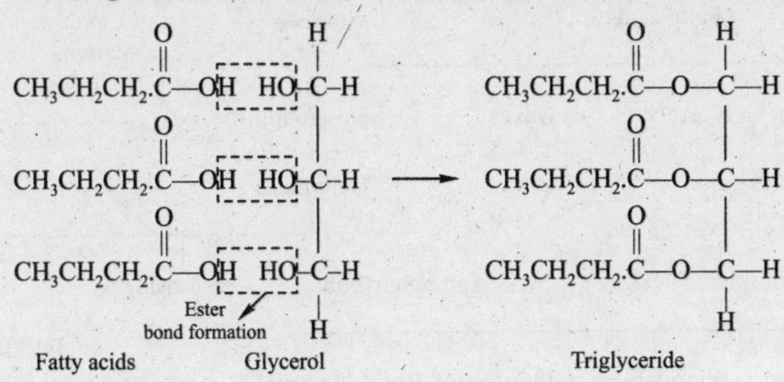

Fig. 3.1. Formation of a triglyceride or lipid.

3. Lipid molecules are **polar** and **hydrophobic**.
4. Generally the lipids do not polymerise to form macromolecules. However they may combine with carbohydrates and protein macromolecules.
5. Lipids can be hydrolysed by heating with alkalis. This process is called **saponification** (Fig. 3.2).

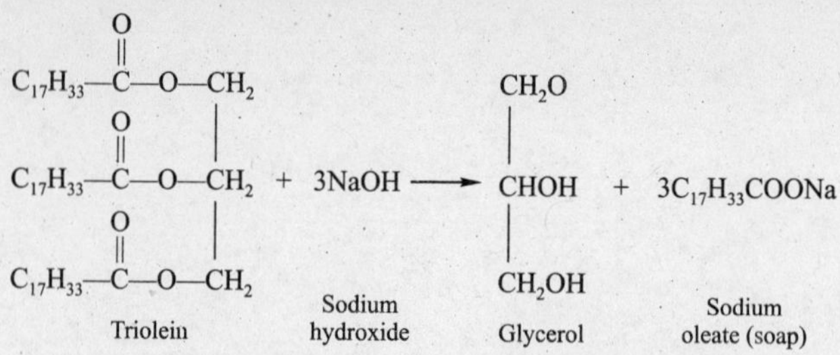

Fig. 3.2. Saponification of a neutral lipid.

3.2 CLASSIFICATION OF LIPIDS

Lipids are generally classified into the following groups:

I. Simple lipids
II. Compound lipids
III. Derived lipids

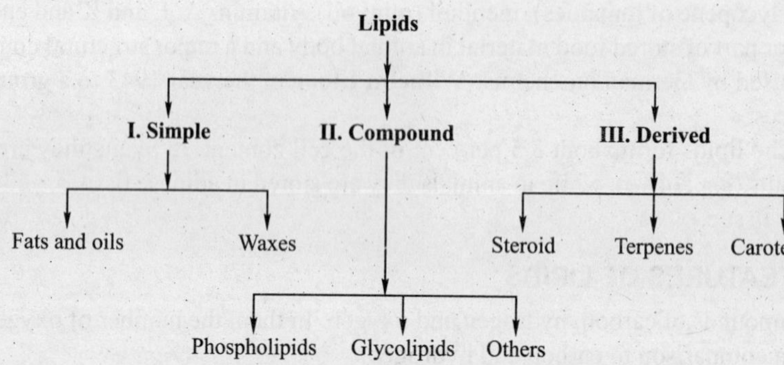

Fig. 3.3. A simple (classical) classification of lipids.

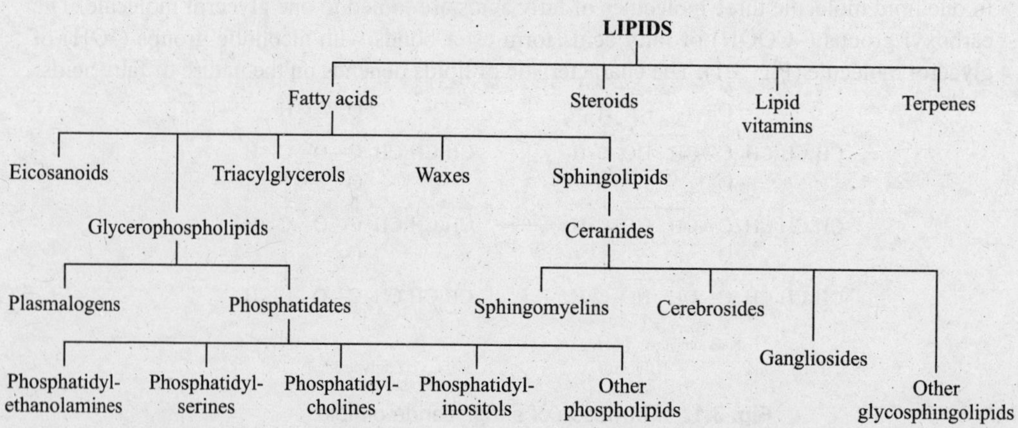

Fig. 3.4. Structural relationship of the major classes of lipids. Fatty-acids are the simplest lipids. Many other types of lipids either contain or are derived from fatty acids. Glycerophospholipids and sphingomyelins contain phosphate and are classified as phospholipids. Cerebrosides and gangliosides contain sphingosine and carbohydrate and are classified as glycosphingolipids. Steroids, lipid vitamins, and terpenes are called isoprenoids because they are related to the five-carbon molecule isoprene rather than to fatty acids.

I. Simple Lipids or Homolipids

They are esters of fatty acids with various alcohols. They are of two types:

1. **Neutral fats or true fats.** These are esters of fatty acids with glycerol.
2. **Waxes.** These are esters of fatty acids with alcohol other than glycerol (*i.e.*, higher molecular weight monohydric alcohols).

II. Compound or Conjugated Lipids or Heterolipids

Compound lipids are the esters of fatty acids containing groups other than and in addition to an alcohol and fatty acids.

1. **Phospholipids.** They have a nitrogen-containing base and a phosphate group in addition to fatty acids and glycerol, *e.g.*, lecithins, cephalins, sphinomyelins and plasmalogens.
2. **Glycolipids.** They have fatty acids, an amino alcohol and a carbohydrate, *e.g.*, phrenosin, kerasin, nervone and oxynervone.
3. **Lipoproteins.** They have lipds (mainly phospholipids) and proteins.
4. **Chromolipids**, *e.g.*, carotenoids and related pigments.

III. Derived Lipids

These are formed from the above lipids by hydrolysis. They include fatty acids, glycerol, vitamin D, cholesterol, bile acids, steroids, sterones, terpenes and prostaglandins.

IV. Unclassified Lipids

These include tocopherol (vitamin E) and vitamin K.

A. Neutral Fats or True Fats (Glycerides)

Neutral fats are also called **triglycerides** and are the main energy storing compounds in the body.

Composition. The neutral or true fats are composed of carbon, hydrogen and oxygen like the carbohydrates, but have far fewer oxygen atoms than carbon atoms unlike carbohydrates. Some lipids have phosphorus, nitrogen and sulphur also. A neutral fat molecule consists of two components: one molecule of alcohol called **glycerol** (or glycerine) and one or three molecules of the same or different **fatty acids**.

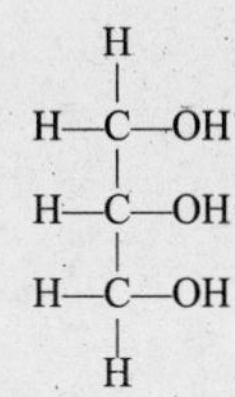

Fig. 3.5. Glycerol.

1. Glycerol. It is trihydroxy alcohol (with 3-OH groups). Glycerol has a molecular formula of $C_3H_5(OH)_3$, and it can form 3 ester bonds with 3 similar or dissimilar fatty acid molecules (Fig. 3.5). Glycerol is optically inactive. However, when it is esterified by two different fatty acids at 3 positions, the carbon 2 becomes asymmetric so the glycerol acts as a binder or carrier for fatty acids.

2. Fatty acids. These are organic molecules made up of long hydrocarbon chain which contain a terminal carboxyl, –COOH group, so fatty acids are always **monocarboxylic**. On the basis of nature of bond present, these fatty acids may be *saturated* or *unsaturated* (Table 3.1), but always have even number of carbon (C) atoms (varies from 2–88). These are generally straight chain organic compounds but may be cyclic ones or branched ones. Acetic acid (CH_3COOH) is the simplest fatty acid while stearic acid $CH_3(CH_2)_{16}COOH$ is a complex fatty acid.

Table 3.1 Differences between Saturated and Unsaturated fatty acids.

	Character	Saturated fatty acids	Unsturated fatty acids
1.	Occurrence	More common in **animal tissues**	More common in **plant tissues**
2.	Nature of chain	Straight chain compounds	Bending in their structure
3.	Nature of bonds	Single bonds only	One or more double or triple bonds

	Character	Saturated fatty acids	Unsturated fatty acids
4.	Physical nature	Solid at room temperature	Liquids at room temperature and called **oils**
5.	Melting point	Higher	Lower
6.	Reactivity	Less	More
7.	Examples	**Palmic acid** $CH_3–(CH_2)_{14}–COOH$ **Stearic acid** $CH_3–(CH_2)_{16}–COOH$	**Oleic acid** (18C)—$CH_3–(CH_2)_7–CH{=}CH–(CH_2)_7–COOH$

A fatty acid molecule has carboxyl group –COOH at one end and hydrogen at all or most other sites (Fig. 3.6). The carboxyl group gives the molecule its acidic properties. This group is **polar** as it can dissociate into a positively charged hydrogen ion (H^+) and negatively charged –COO group:

$$R–COOH = –R–COO^- + H^+$$

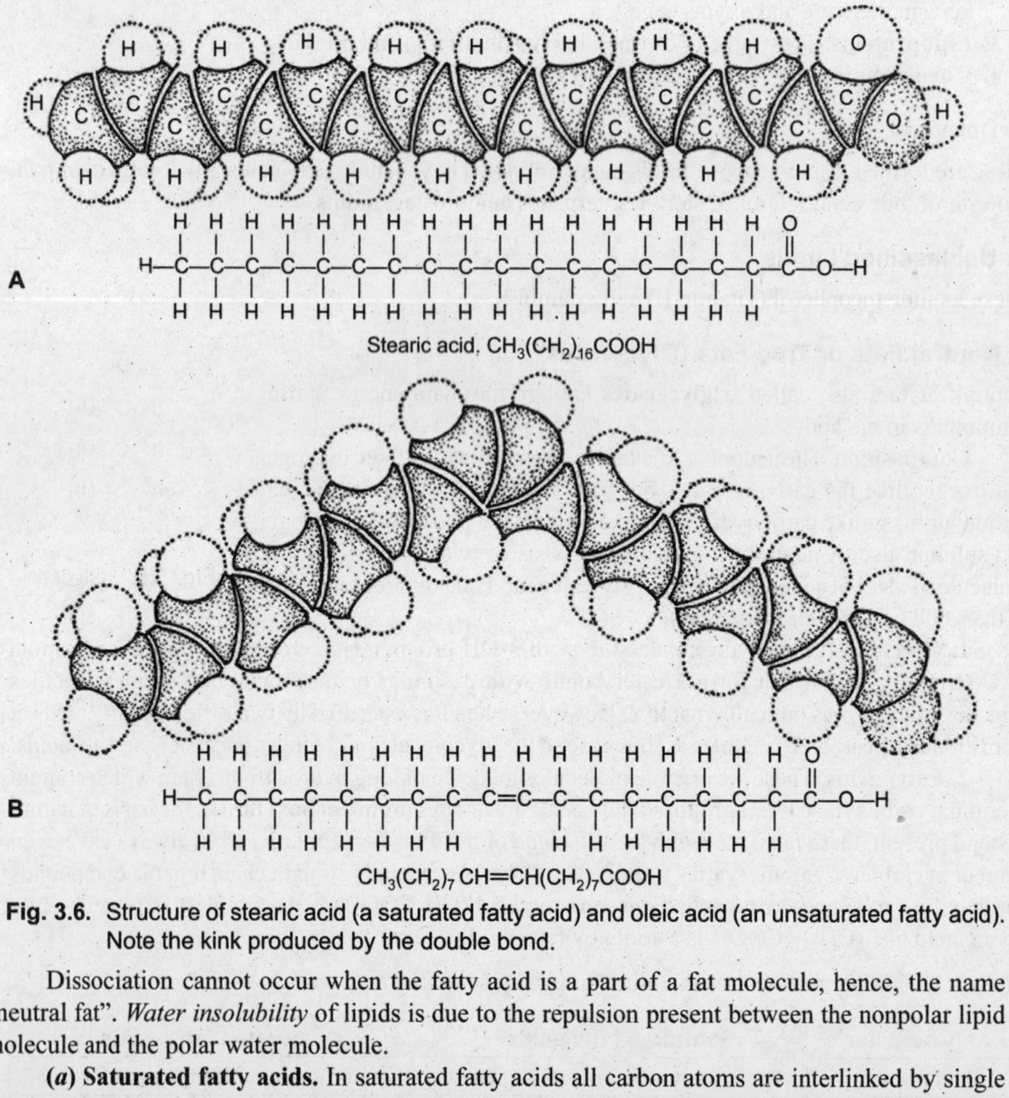

Fig. 3.6. Structure of stearic acid (a saturated fatty acid) and oleic acid (an unsaturated fatty acid). Note the kink produced by the double bond.

Dissociation cannot occur when the fatty acid is a part of a fat molecule, hence, the name "neutral fat". *Water insolubility* of lipids is due to the repulsion present between the nonpolar lipid molecule and the polar water molecule.

***(a)* Saturated fatty acids.** In saturated fatty acids all carbon atoms are interlinked by single bonds and each carbon atom carries a hydrogen atom, *e.g.*, Palmitic acid and stearic acid. The general chemical (molecular) formula for saturated fatty acids is $C_nH_{2n}O_2$. The chemical formula for stearic acid is $C_{18}H_{36}O_6$ and for palmitic acid is $C_{16}H_{32}O_2$ (Table 3.2).

General formula for saturated fatty acids is R–COOH where R represent CH_3–$(CH_2)_n$–(called **acyl radical**) and *n* varies from zero in acetic acid and 86 in mycolic acid.

Table 3.2 Important saturated fatty acids ($C_nH_{2n+1}COOH$).

	Name of fatty acid	Molecular formula	Number of C- atoms
1.	Acetic acid	CH_3–COOH	2
2.	Butyric acid	$CH_3(CH_2)_2COOH$	4
3.	Caproic acid	$CH_3(CH_2)_4COOH$	6
4.	Caprylic acid	$CH_3(CH_2)_6COOH$	8
5.	Capric acid	$CH_3(CH_2)_8COOH$	10
6.	Lauric acid	$CH_3(CH_2)_{10}COOH$	12
7.	Myristic acid	$CH_3(CH_2)_{12}COOH$	14
8.	Palmitic acid	$CH_3(CH_2)_{14}COOH$	16
9.	Stearic acid	$CH_3(CH_2)_{16}COOH$	18
10.	Arachidic acid	$CH_3(CH_2)_{18}COOH$	20
11.	Behenic acid	$CH_3(CH_2)_{20}COOH$	22
12.	Lignoceric acid	$CH_3(CH_2)_{22}COOH$	24

***(b)* Unsaturated fatty acids.** In unsaturated fatty acids, two adjacent carbon atoms at one or more places lack hydrogen atoms and are interlinked by double bonds. These include 18 carbon oleic, linoleic and linolenic acid (Table 3.3). They have 1, 2 and 3 double bonds respectively. A double bond in an unsaturated fatty acid may have a **cis** or **trans configuration** (Fig. 3.7). The double bond in most unsaturated fatty acids have the cis orientation, which introduces a bend (109° bond angle) in the hydrocarbon chain.

Table 3.3 Important unsaturated and special fatty acids.

I. Unsaturated acids		
A. Monoethenic, $C_nH_{2n-1}COOH$, one double bond		
(a) Palmitoleic Δ^9	$C_{15}H_{29}COOH$	Animal (Milk) and Vegetable (Sardine fats)
(b) Oleic acid Δ^9	$C_{17}H_{33}COOH$	
(c) Erucic acid Δ^{13}	$C_{21}H_{41}COOH$	
(d) Nervonic acid Δ^{15}	$C_{23}H_{45}COOH$	
B. Dienoic C_n–$H_{2n-3}COOH$, two double bonds		
Linoleic acid $\Delta^{9,12}$	$C_{17}H_{31}COOH$	Linseed oil, animal and plant fats, corn oil, soyabean oil, peanut oil etc.
C. Trienoic $C_nH_{2n-5}COOH$, three double bonds		
Linolenic acid $\Delta^{9,12,15}$	$C_{17}H_{29}COOH$	Linseed oil, rapeseed oil, soyabean oil, liver oils etc.
D. Polyenoic, $C_nH_{2n-7}COOH$, more than three double bonds		
Arachidonic acid $\Delta^{5,8,11,14}$	$C_{19}H_{31}COOH$	Fats, phospholipids etc.
II. Branched-chain acids		
(a) Isobutyric acid	C_3H_7COOH	Waxes
(b) Tuberculostearic acid	$C_{18}H_{37}COOH$	Wax of tubercle bacillus

III. Hydroxy acids		
(a) Cerebronic acid	$C_{23}H_{46}(OH)COOH$	Animal and vegetable fats
(b) Ricinoleic acid	$C_{17}H_{32}(OH)COOH$	
(c) Hydroxynervonic acid	$C_{23}H_{44}(OH)COOH$	
IV. Cyclic acids		
Chaulmoogric acid	$^*R—(CH_2)_{12}—COOH$	Seed oils
Hydnocarpic acid	$^*R—(CH_2)_{10}—COOH$	Seed oils

*where R is

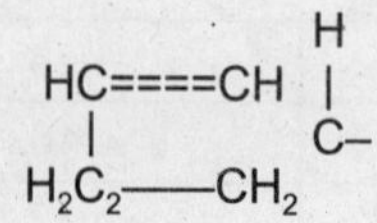

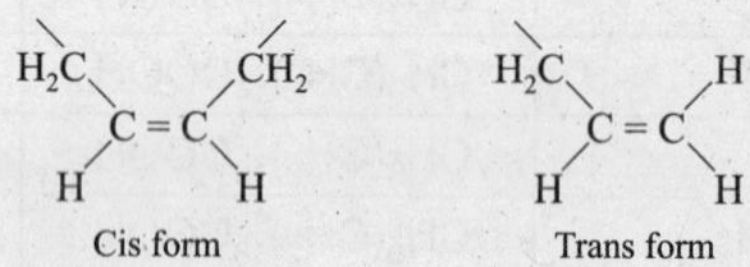

Fig. 3.7. Two configurations of double bond in an unsaturated fatty acid.

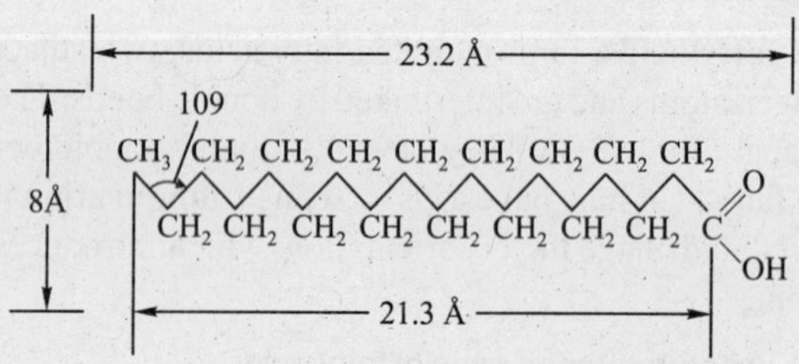

Fig. 3.8. Configuration of a saturated fatty acid.

Box 3.1 Trans Fatty Acids and Margarine

(**Trans.** (L. *trans* = through, across). *In organic chemistry*. In italics, a form of geometric isomerism in which the atoms attached to carbon atoms, joined by double bonds, are located on opposite sides of the molecule. *In Biochemistry*. A prefix to a group name in an enzyme name or reaction denoting transfer of that group from one compound to another, *e.g.*, transformylase.)

While the configuration of most double bonds in unsaturated fatty acids is *cis*, some fatty acids in the human diet have the *trans* configuration. *Trans* fatty acids can come from animal sources such as dairy products and ruminant meats. Most edible trans fatty acids, though are present as hydrogenated vegetable oils in some margarines or shortenings (*i.e.*, fat used for making pastry). Dietery *trans* monosaturated fatty acids can increase plasma levels of cholesterol and triglycerides, and thus their ingestion may increase the risk of cardiovascular disease.

Plant oils such as corn oil, sunflower oil, peanut oil, etc., can be converted to "spreadable" semisolid substances such as *margarines*. Margarines can be produced by the partial or complete hydrogenation of the double bonds in plant oils. The hydrogenation process itself not only saturates the carbon-carbon double bonds of fatty acid esters, but can also change the configuration of the remaining double bonds from *cis* to *trans*. The physical properties of these *trans* fatty acids are similar to those of saturated fatty acids. In order to reduce consumption of *trans* fatty acids, many margarines are now produced from plant oils without hydrogenation by adding other edible components such as skim milk powder (**Horton** *et al.*, 2006; **Jain** *et al.*, 2012).

The unsaturated fatty acids have lower melting points than the saturated fatty acids and are more abundant in the living organisms. Saturated fatty acids (SFA) have straight chains whereas unsaturated fatty acids (UFA) have a bend or kink at each double bond. The kinks affects the packing of UFA chains in cell organelles, making the chains more disordered and hence more fluid at ordinary temperatures. This property maintains the flexibility and fluidity of cell membranes. The general chemical formula for unsaturated fatty acids is $C_nH_{2n-2x}O_2$. The chemical formula of oleic acid with a single double bond is $C_{18}H_{34}O_2$ and of linoleic acid with two double bonds is $C_{18}H_{32}O_2$. Most plant fats contain unsaturated fatty acids and most animal fats have saturated fatty acids. Unsaturated fatty acids are of following types:

(*a*) Monounsaturated (or monoenoic) fatty acids. These fatty acids have only one unsaturation (*i.e.*, one double bond).

Examples.

(*i*) Palmitoleic acid (16C). It is derived from palmitic acid and has a double bond between 9th and 10th carbon (hence also called **9-hexadeconic**). It is found in milk fat and sardine oil. Its molecular formula is:

$$\underset{\text{Palmitoleic acid}}{CH_3-(CH_2)_5-CH=CH-(CH_3)_7-COOH}$$

(*ii*) Oleic acid (18C). It is a derivative of stearic acid and has a double bond between 9th and 10th carbon atom. It is also called **9-octadecenoic**. It is found in olive oil, pork fat. Oleic acid is most abundant fatty acid of human beings and is present in adipose tissue and milk. Its molecular formula is:

$$\underset{\text{Oleic acid}}{CH_3-(CH_2)_9-CH=CH-(CH_2)_7-COOH}$$

(*iii*) Euracic acid (22C). It is found in mustard oil and has one double bond between 13th and 14th carbon. Its molecular formula is:

$$\underset{\text{Euracic acid}}{CH_3(CH_2)_7CH=CH(CH_2)_{11}COOH}$$

(*b*) Polyunsaturated (Polyenoic) fatty acids. These fatty acids contain more than one double bond. Polyunsaturate oils are recommended by physicians for persons who have high blood **cholesterol** or **cardiovascular diseases**. There use lowers the blood cholesterol level. Polyunsaturated fatty acids include the following:

(*i*) Linoleic acid ($C_{18}H_{32}O_2$). It is derived from stearic acid and is found in many seed oils, *e.g.*, linseed oil, soyabean oil, ground nut oil, cotton seed oil, etc. It has two double bonds which are present between 9th and 10th carbon atoms and between 12th and 13th carbon atoms, so it is also called **dienoic fatty acid**. Its systematic name is **9, 12-octadecadienoic acid**. Linoleic acid has the following molecular formula:

$$\underset{\text{Linoleic acid}}{CH_3-(CH_2)_4-CH=CH-CH_2-CH=CH-(CH_2)_7-COOH}$$

Linoleic acid is an *essential fatty acid* and its deficiency in diet results in increased metabolic activity, failure in growth and even death.

(*ii*) Linolenic acid ($C_{18}H_{30}O_2$). This *essential fatty acid* is also derived from stearic acid and is present in linseed oil, soya bean oil and hempseed oil. It has three double bonds in its molecule; these are present between 9th/10th, 12th/13th and 15th/16th carbon atoms. Thus it is a **trienoic** and its systematic name is **9, 12, 15-octadecatrienoic**. Linolenic acid has the following molecular formula:

$$\underset{\text{Linolenic acid}}{CH_3-CH_2-CH=CH-CH_2-CH=CH-CH_2-CH=CH-(CH_2)_7-COOH}$$

(*iii*) Arachidonic acid (20C). It is an *essential fatty acid*, derived from arachidic acid and is found in mammalian tissues. It is a **tetraenoic** having four double bonds between 5th/6th, 8th/9th, 11th/12th and 14th/15th carbon atoms. Its systematic name is **5, 8, 11, 14-eicosatetraenoic acid** and biologic precursor of eicosanoids (*i.e.*, the prostaglandins, the thromboxanes, and the leukotrienes).

Thromboxanes or **Txs** is a class of molecules involved in platelet aggregation during blood clothing. **Leukotrienes** or **LTs** are of 6 types and secreted by white blood cells and lung tissues and involved in asthmatic attacks and stimulation of mucus secretion (**Jain** 2012). It has the following molecular formula:

$$CH_3-(CH_2)_4-CH=CH-CH_2-CH=CH-CH_2-CH=CH-CH_2-CH$$
$$=CH-(CH_2)_3-COOH$$

Essential and Non-essential Fatty Acids

On the basis of their synthesis, the fatty acids are of following two types:

1. **Essential fatty acid (EFAs).** These fatty acids are not synthesized in the body tissues of humans. So they must be present in human diet. These EFAs include **linoleic acid**, **linolenic acid** and **arachidonic acid**. These fatty acids are present in a number of vegetable cooking oils, *e.g.*, groundnut oil, sunflower oil and safflower oil.
2. **Non-essential fatty acids (NEFAs).** These fatty acids can be synthesized inside the human's body tissues. They may or may not be present in human food, *e.g.*, palmitic acid, stearic acid and oleic acid.

Formation of Fats

The three hydroxyl (–OH) groups of a glycerol molecule join with the carboxyl (–COOH) groups of three fatty acid molecules to form a fat molecule (Fig. 3.9). The chemical linkage between the glycerol and fatty acids is called the **ester bond**. It results from the elimination of 3 molecules of water, one for each linkage (condensation or dehydration). In a neutral fat the three fatty acids may be same or different. The fatty acid components of most of the animal fats are 16-carbon palmitic acid and 18-carbon stearic acid. The neutral fats formed by these fatty acids are named **tripalmitin** and **tristearin** respectively. Both are **pure neutral fats**. A fat having a molecule each of palmitic, oleic and stearic acid is called **palmito-oleio-stearin**. It is a **mixed neutral fat**. Butter is an example of the mixed fat.

The process that joins the glycerol and fatty acid molecules is called **esterification**.

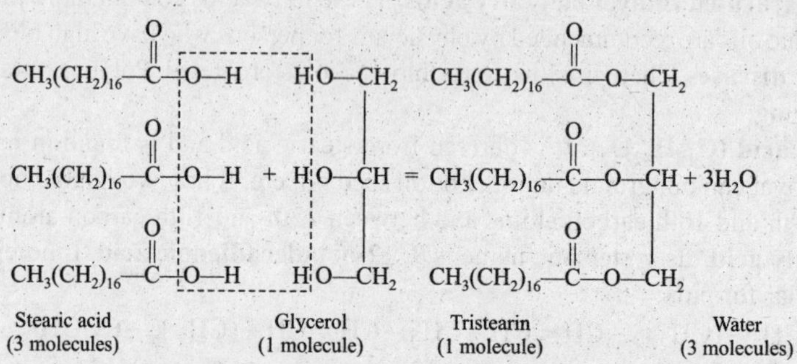

Fig. 3.9. Structural formula of tristearin.

Types of Fats

The neutral fats are of three types: monoglycerides, diglycerides and triglycerides. A true fat having 3 molecules of fatty acids is called **triglyceride**. Triglycerides are common in cells. Different species of animals usually have distinctive mixtures of fatty acids in their triglycerides, but these may change to some extent with the diet of the animal.

A neutral fat having only one or two fatty acid molecules is termed a **monoglyceride** or a **diglyceride**. These two types of fats occur as intermediates in certain biosynthetic reactions. Out of these three fats, triglycerides form about 98% of our dietary lipids while remaining 2% of lipids are formed of phospholipids and cholesterol. All the three types of neutral fats are given a common name **glyceride**.

Physical Nature of Fats

The physical nature of a neutral fat depends upon the kind of fatty acid present. Some fats have saturated fatty acids, and are solid at ordinary temperature, *e.g.*, butter, lard, beef tallow and bacon fat. These are called **hard fats**. Other neutral fats have unsaturated fatty acids and are liquid at ordinary temperature, *e.g.*, cod liver oil, shark liver oil, whale oil, groundnut oil, rape seed oil, mustard oil, sesame oil (or til oil), linseed oil, sunflower oil, safflower oil, corn oil, olive oil, castor oil, soya bean oil and cotton seed oil. These are termed **oils**. Most plant lipids (oils) have unsaturated fatty acids and most animal fats have saturated fatty acids. There is clinical evidence that saturated fats (animal fats) cause increased production of cholesterol and arteriosclerosis (hardening of arteries) in humans. This condition is dangerous, often fatal, for heart patients as it reduces the flow of blood through arteries.

Box. 3.2

Triglycerides having the saturated fatty acids are *solid* at room temperature because the aliphatic chains of fatty acids are arranged parallel to one another so that molecules become compact and need high amount of energy in order to melt them (Fig. 3.10A).

Fig. 3.10. Interactions between fatty acid chains. A—Saturated fatty acid chains. B—Unsaturated fatty acid chains.

Triglycerides having the unsaturated fatty acids are *liquid* at room temperature and are called *oils*. It is so as the unsaturated fatty acids are cis-isomers and cis double bonds interrupt the regular packing of their chains so these need less energy to melt (Fig. 3.10B). This disorganisation increases with the increase in the number of double bonds which lowers their melting point.

Properties of fats. True fats are insoluble in water but dissolve in organic solvents. They disperse in water as minute droplets, forming an emulsion. In cells they are dispersed through the cytoplasm. The have well defined melting points and solidifying points, the latter temperatures being only a few degrees lower. Fats have low specific gravity and float on water. Oils spread on water to form thin monomolecular layers. All fats are greasy. The neutral fats are hydrolysed in cells by the action of enzymes called **lipases**.

A rather unusual property of fats is that their glycerol end is hydrophilic (water loving) and the fatty acid end is lipophilic (fat loving). In other words, the glycerol end is polar and dissolves in water and the fatty acid end is nonpolar and dissolves in fat.

Fats are bad conductors of heat, so act as heat insulators. A layer of fats below skin provides an insulting blanket in all the warm blooded animals.

Chemical Properties of Fats

1. Hydrolysis. The fats are hydrolysed into glycerol and constituent fatty acids by superheated steam, alkalies and fat splitting enzymes. Water is utilised during this process. Fat splitting enzyme **lipase** acts at a slight alkaline pH (7.5–8.5) in a stepwise fashion.

$$\text{Triglyceride} + H_2O \xrightarrow{\text{Lipase}} \text{Diglyceride} + \text{Fatty acid}$$
$$\text{Diglyceride} + H_2O \xrightarrow{\text{Lipase}} \text{Monoglyceride} + \text{Fatty acid}$$
$$\text{Monoglyceride} + H_2O \xrightarrow{\text{Lipase}} \text{Glycerol} + \text{Fatty acid}$$

2. Saponification. The splitting of fat (triglyceride) by alkali is called **saponification**. The fatty acids formed as a result of hydrolysis of fats are neutralized by the alkali. These form the soap.

Triolein + Sodium hydroxide → Glycerol + Sodium oleate (soap)

3. Hydrogenation. When hydrogen is passed through unsaturated fats (or oils), these are converted into saturated fats. The process is completed rapidly in the presence of nickle (Ni) which acts as a catalyst. This process is called **hydrogenation** and is utilised for the manufacture of ghee (vanaspati) from vegetable oils.

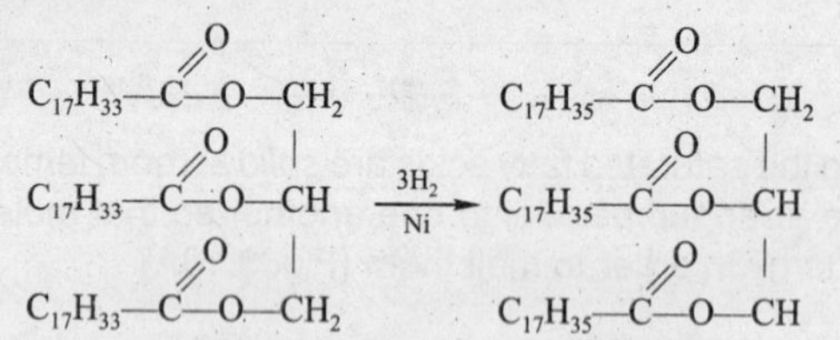

Triolein (oil) Tristearin (fat)

Fig. 3.11. Hydrogenation of lipid molecule.

4. Hydrogenolysis. When hydrogen is passed in excess the oils and fats break up in glycerol and fatty acids.

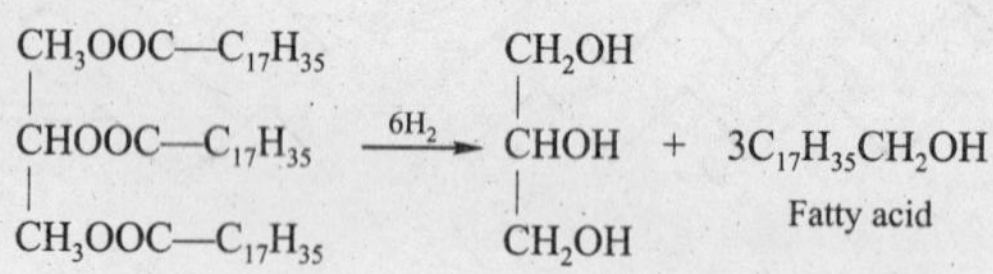

Fat Glycerol

Fig. 3.12. Hydrogenolysis of a fat molecule.

5. Rancidity. An exposure of air, fats develop unpleasant odour and flavour. This occurs due to oxidation of fat which releases volatile fatty acids having unpleasant odour. This is known as **rancidity**. It is mostly found in vegetable oils and unsaturated fats.

Rancidity of oils also takes place due to bacterial growth, moisture, heat and light. During rancidity, oxidation of fatty acids takes place. This is also known as **auto-oxidation** of fat. The incomplete hydrolysis of ester bonds in the fat produces fatty acid which also leads to rancidity.

The process of rancidity and auto-oxidation of some fats is prevented by certain substances such as vitamin E (which is an **antirancid agent**) which are naturally present in them. Examples of other antirancid agents include phenols, naphthols and quinones.

6. Fat analysis. For the analysis of fat the following chemical characters are used:

***(i)* Acetyl number.** Milligram (mg) of KOH required to neutralize the acetic acid obtained by saponification of one gram (g) of acetylated fat. This number indicates the number of hydroxyl groups in fatty acids.

***(ii)* Acid number.** The number of mg of KOH required to neutralize the free fatty acids present in 1 g fat or oil. High acid value indicates impurity.

***(iii)* Saponification number.** It is the number of milligrams of KOH required to saponify completely one gram of fat. If a fat is composed of fatty acid of higher molecular weight (such as palmitic, stearic and oleic acid), its saponification number will be low and will be high if the fat has shorter chain fatty acids.

***(iv)* Iodine number.** It is number of grams of iodine absorbed by 100 g of fat. It is used to determine the total amount of unsaturation in fat, but it does not indicate the number of double bonds.

***(v)* Reichert–Meissei number (R–M value).** The number of ml of 0.1 N alkali required to neutralise the volatile fatty acids contained in 5 g of fat. R–M value is of special significance in characterisation of fat and testing its purity.

Functions of Neutral Fats or Oils

True fats have the follwoing functions in plants and animals.

1. Reserve food materials. The neutral fats serve as food reserves in both plants and animals. They are stored in special regions—connective tissues (*e.g.*, adipose tissue) around the kidneys, under the dermis in mammals, in the fat bodies in frogs and lizards, and in the liver of fish. These fat deposits are not long term reserves of stored food used only in starvation, but are constantly being used up and reformed.

In plants fat is stored in the seeds to provide nourishment to the germinating embryo. The oil extracted from seeds of groundnut, mustard, soyabean, sunflower and coconut is used for cooking and other purposes (*e.g.*, castor-oil).

2. Fuel. The neutral fats form a concentrated fuel for the working of the animals and plants. They produce more than twice as much energy for gram as do the carbohydrates. They, thus, represent an economical food reserve in the body.

3. Insulation coats. The neutral fats form an insulating layer under the skin (the subcutaneous fat). Women tend to have thicker layer of adipose tissue than human beings have and are more tolerant to cold. The polar bear and whales have a very thick layer of adipose tissue, called **blubber**, under the skin for insulation.

4. Shock-absorbing cushions. The neutral fats serve as shock-absorbing cushion around the eyeballs, kidneys and gonads.

5. Improvement of physique. The subcutaneous fat in human beings keeps the skin firm and helps in rounding off the contour of the body. These features help in ethology of the animals.

B. Waxes

Waxes are those simple lipids in which one molecule of higher fatty acid (with longer side chains) is esterified with a higher monohydric aliphatic alcohol other than glycerol. The constituent fatty acid and alcohol usually have 24 to 36 carbon atoms. Waxes are chemically inert so are not digested by lipase enzymes. Molecules of waxes do not contain any double bond in their hydrocarbon chain and are insoluble in water, so are resistant to atmospheric oxidation. They have high melting points (60°C to 100°C).

Waxes are saponified with great difficulty (as they contain 31 to 55% unsaponified matter) than fats (which contain only 1 to 2% unsaponified matter).

Types of Waxes

1. Bee wax. It is a simple lipid formed by esterification of one molecule of **palmitic acid** ($C_{15}H_{31}COOH$) and one molecule of **mericyl alcohol** ($C_{30}H_{61}OH$) or *n-hexacosanol* ($C_{26}H_{53}OH$; Fig. 3.13). The hydrophobicity of **mericyl palmitate** (Fig. 3.14) makes bee wax very water-insoluble and its higher melting point (due to the long, saturate hydrocarbon chains) bees wax is hard and solid at typical outdoor temperatures.

$$\underbrace{CH_3(CH_2)_{14}CO}_{\text{Palmitic acid}}-\boxed{OH + H}-\underbrace{OCH_2(CH_2)_{24}-CH_3}_{\text{Hexacosanol}} \rightarrow$$

$$\underbrace{CH_3(CH_2)_{14}-\overset{\overset{\displaystyle O}{\|}}{C}-O-CH_2-(CH_2)_{24}CH_3}_{\text{Hexacosyl palmitate (Bee wax)}}$$

Fig. 3.13. Formation of bee wax.

$$H_3C-(CH_2)_{14}-\overset{\overset{\displaystyle O}{\|}}{C}-O-(CH_2)_{29}-CH_3$$

Fig. 3.14. Myricyl palmitate, a wax.

2. Ear wax. It is also known as **cerumen** (L. *cera* = wax). It is secreted by cells lining the auditory canal. Cerumen lubricates the canal and traps particles that could damage the eardrum. Ear wax is a complex mixture made up mostly of long chain fatty acids, cholesterol and ceramides. It also contains sequalene, triacylglycerols and true waxes (about 10% of weight) (**Horton** *et al.*, 2006).

3. Lanoline or wool fat. It is formed by esterification of one molecule of either palmitic acid ($C_{15}H_{31}COOH$) or stearic acid ($C_{17}H_{35}COOH$) or oleic acid ($C_{17}H_{33}COOH$) and one molecule of cholesterol. It is secreted by cutaneous glands and closely resembles the sebum.

4. Carnauba wax. It is formed by the esterification of cerotate acid (having 26C) with one molecule of mericyl alcohol (having 30C). It is extracted from a Brazilian palm tree.

5. Spermaceti (Fig. 3.15). It is palmitic ester of cetyl alcohol ($C_{16}H_{33}OH$). It is an oil extracted from the head of sperm whale.

$$CH_3(CH_2)_{14}-\overset{\overset{\displaystyle O}{\|}}{C}-O-(CH_2)_{15}-CH_3$$

Fig. 3.15. Cetyl palmitate (spermaceti).

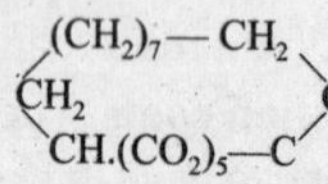

Fig. 3.16. Ambrettolide.

6. Ambrettolide (Fig. 3.16). It is extracted from the seeds of *Abelmoschus esculentus*. It is hydroxyl acid and gives a characteristic smell to the seeds.

Functions of Waxes

1. Waxes form a water-proof coating on plant parts such as leaves, etc., to reduce the rate of transpiration and also prevent their wetting.
2. Wax esters act as superior machine lubricant.
3. These form water resistant layers in insects, birds such as ducks and furred mammals.
4. Waxes form protective sheaths on fruits.
5. Waxes are used in polishing the furniture.
6. Lanolin fat is used in manufacture of cosmetic creams, ointments, etc.
7. Spermaceti is used in manufacture of polishes, ointments, candles, etc.
8. Ear wax or cerumen lubricates and protects the ear drum.
9. Sperm oil which is found along with spermaceti in head of sperm whale, acts as a valuable lubricant for instruments such as watches.
10. Waxy coating of fruits such as apple and citrous fruits prevent them from drying so these can be stored for long period of time.
11. Wax papers are commonly used for packing perishable materials.

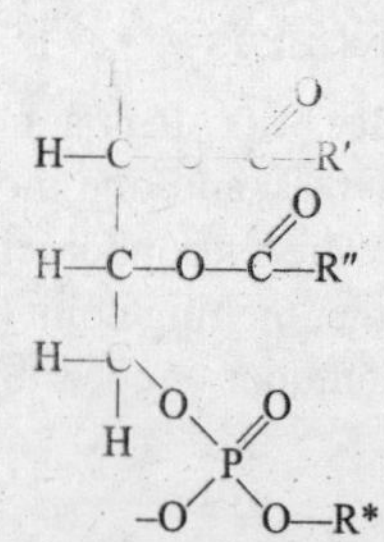

Fig. 3.17. Structure of a phospholipid molecule. R′ and R″ are fatty acid chains and R* a nitrogen containing base.

II. Compound Lipids

1. Phospholipids. Phospholipids are compounds of glycerol, fatty acids and phosphoric acid.

Structure. The phospholipids are composed of a molecule of glycerol or other alcohol having (*i*) a phosphate group joined to either of its outer –OH groups (never to the middle one), (*ii*) two fatty acid molecules linked to the other two –OH groups, and (*iii*) a base (alcohol) molecule usually containing one or more nitrogen atoms bound to the phosphate group (Fig. 3.17). Thus, in a phospholipid the phosphate group forms a linking bridge between the glycerol and the base. Both the fatty acid molecules of a phospholipid may be saturated or unsaturated, or one may be saturated and the other unsaturated (Fig. 3.18).

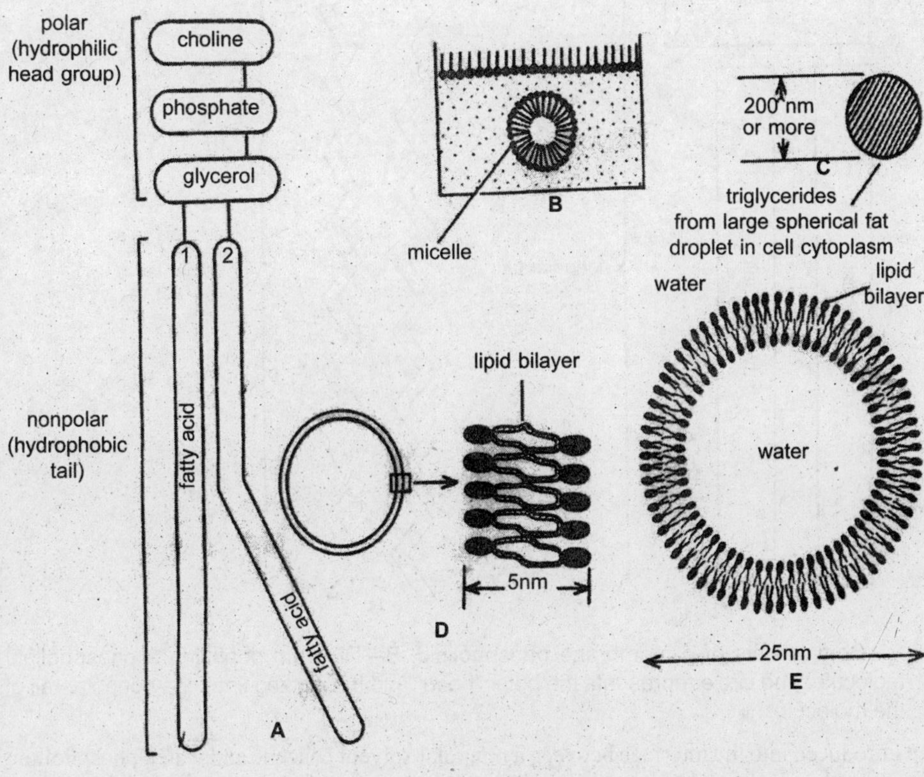

Fig. 3.18. Formation of various types of lipid aggregates. A—Schematic representation of a phospholipid molecule; B—Formation of micelle and monolayer film; C—Formation of a fat droplet by triglycerides; D—Formation of self-sealing lipid bilayer (*e.g.*, liposome); E—Cross-section of a liposome.

Special property. The phospholipids have a special property of dual solubility. A phospholipid molecule has a strongly nonpolar and hydrophobic (water hating or water insoluble) "tail" consisting of fatty acid chains, and a polar and hydrophilic (water loving or water soluble) "head" comprising a negatively charged phosphate group and a positively charged base (Fig. 3.19). The phospholipids are, thus, **amphipathic** (Gr. *amphi* – both + *pathos* = disease, meaning "tolerant to both") lipids, one end of a molecule being hydrophobic and the other hydrophilic. It is this dual nature of the phospholipids that makes them very important for the structure and functioning of the membranes of the cell.

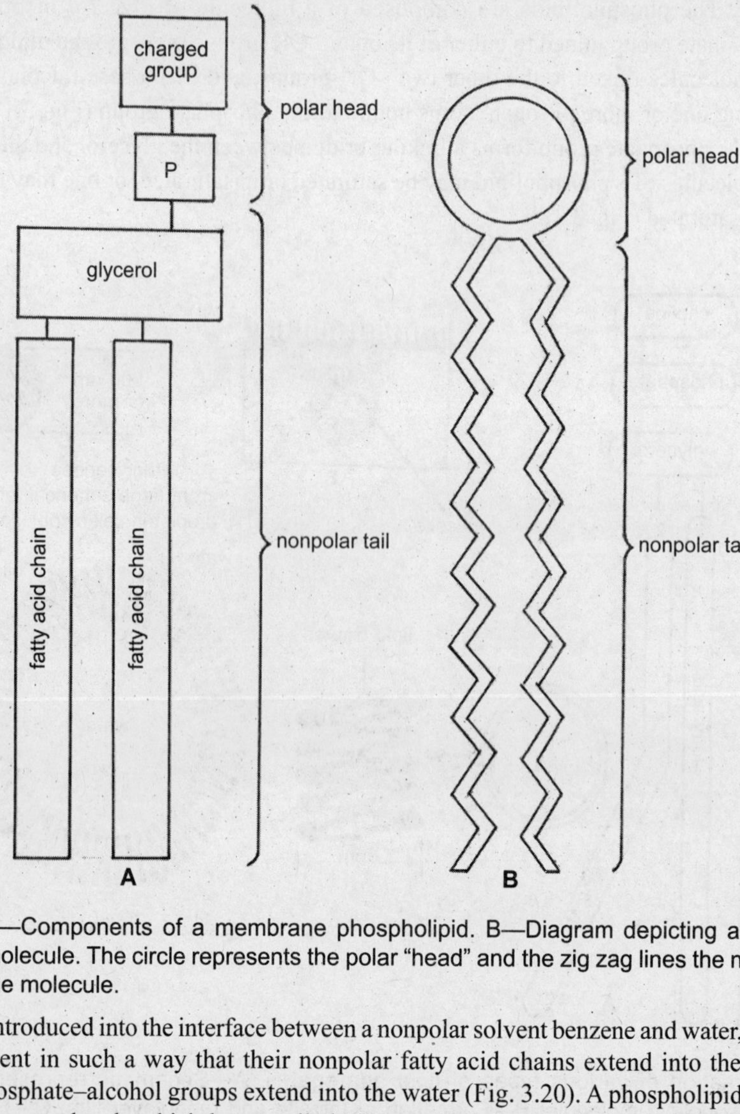

Fig. 3.19. A—Components of a membrane phospholipid. B—Diagram depicting a phospholipid molecule. The circle represents the polar "head" and the zig zag lines the nonpolar tail of the molecule.

When introduced into the interface between a nonpolar solvent benzene and water, phospholipid molecules orient in such a way that their nonpolar fatty acid chains extend into the benzene and their polar phosphate–alcohol groups extend into the water (Fig. 3.20). A phospholipid placed under water forms two molecules thick layer, called **bilayer**, in a characteristic way. The nonpolar fatty acid chains of the two layers associated together in the hydrophobic interior of the bilayer, and their polar phosphate-alcohol groups face the surrounding water molecule (Fig. 3.21). Bilayers of this type form the basic structural framework of the cell membranes. Phospholipid molecules placed in water may form liposomes and micelles also (Fig. 3.21). **Liposomes** are spherical bilayer structures with aqueous interior (Fig. 3.21). **Micelles** are spherical clusters having polar heads on the surface in contant with water and nonpolar fatty acid chains directed towards the center (Fig. 3.21).

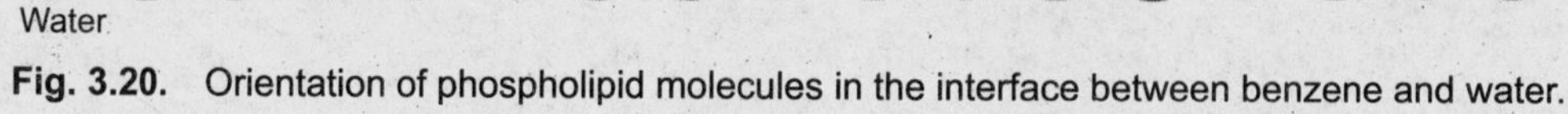

Fig. 3.20. Orientation of phospholipid molecules in the interface between benzene and water.

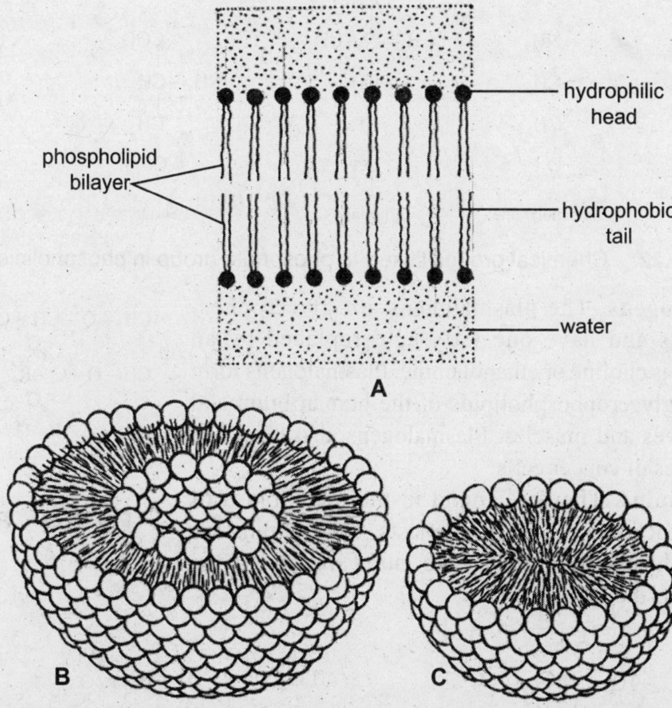

Fig. 3.21. A—Bilayer formed by phospholipid molecules placed under water. Nonpolar tails are associated together in the bilayer interior. Polar heads face the surrounding water molecules; B—liposome; C—micelle.

Liposomes. When aqueous suspensions of phospholipids are subjected to rapid agitation by using ultrasound (*i.e.*, insonation), the lipid disperses in the water and forms **liposomes** or lipid **vesicles**. Liposomes are small spherical bodies (25 nm to 1 μm in diameter) whose surface is formed by a bilayer of phospholipid molecules enclosing a small volume of the aqueous medium. They exhibit many of the permeability properties of natural membranes, *i.e.*, water soluble small molecules or ions can be enclosed by the liposomes and they can also traverse the lipid bilayer of latter. Recently, liposomes have been found to have great therapeutic promise, since, they can be used as vectors for the transfer of specific drugs, proteins, hormones, nucleic acids, ions or any other molecule into the specific types of animal cells. The contents of the liposomes can enter the target cells by two routes:

1. The liposomes can attach to the surface of target cells and may fuse with the plasma membrane, following which their contents are released into the cytosol or cytoplasmic matrix.
2. The entire liposomes may be endocytosed and degraded intracellularly.

Types of phospholipids. The phospholipids are classified on the basis of alcohol present in them. The more common types are glycerophosphatides phosphoinositides or phosphosphingosides.

(*i*) Glycerophosphatides. These phospholipids have the alcohol glycerol. They are of following types:

(*a*) Lecithins. The lecithins contain glycerol, fatty acids, phsophoric acid and a base choline (Fig. 3.22). They are especially abundant in the liver and egg yolk. They help in the emulsification of fats in the intestine where they keep cholesterol and its ester in the dissolved state.

Part of toxicity of snake venom is found to be related to the presence of enzyme named lecithinases, which hydrolyze the lecithins.

(*b*) Cephalins. The cephalins differ from the lecithins only in that the base is ethanolamine ($NH_2CH_2CH_2OH$) instead of choline (Fig. 3.22).

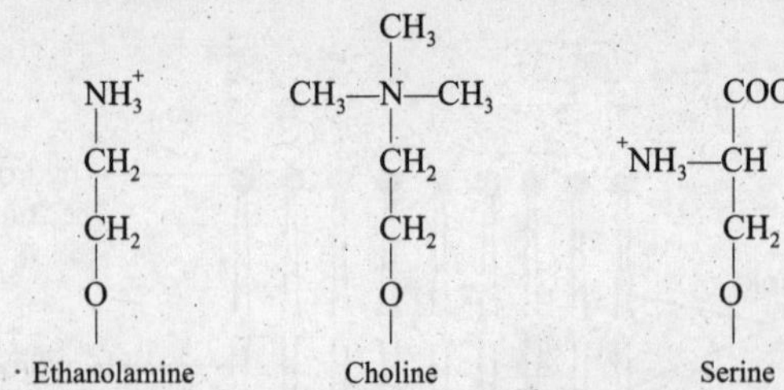

Fig. 3.22. Chemical groups linked to phosphate group in phospholipids.

(*c*) **Plasmalogens.** The plasmalogens are glycerophos-phosphospholipids and have one fatty acid replaced by an aldehyde. The base is choline or ethanolamine. Plasmalogens form about 23% of the glycerophospholipids of the human brain and membranes of nerves and muscles. Plasmalogens also form one of the characteristics of cancer cells.

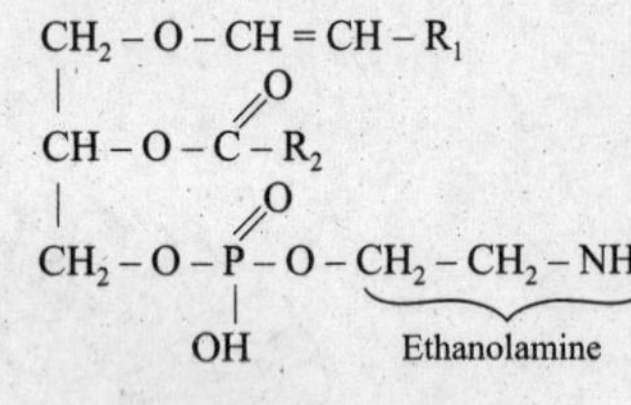

Fig. 3.23. Plasmalogen.

(*d*) **Cardiolipins.** They are found in the membranes of bacterial cells, mitochondria and chloroplasts. Cardiolipins are polymers of the phosphatidic acids. The latter are any of the phospholipids minus the base.

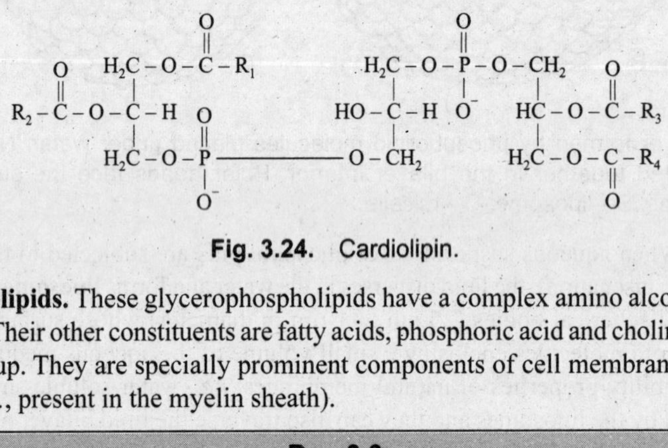

Fig. 3.24. Cardiolipin.

(*ii*) **Sphingolipids.** These glycerophospholipids have a complex amino alcohol sphingosine in place of glycerol. Their other constituents are fatty acids, phosphoric acid and choline. **Sphingmyelins** belong to this group. They are specially prominent components of cell membranes in the brain and nervous tissue (*i.e.*, present in the myelin sheath).

Box 3.3

Three major families of sphingolipids are the **sphingomyelins**, the **cerebrosides** and the **gangliosides.** Of these, only sphingomyelins contain phosphate and are classified as phospholipids. Cerebrosides and gangliosides contain carbohydrate residues and are classified as glycosphingolipids (**Horton** *et al.*, 2006).

(*iii*) **Phosphoinositides.** These phospholipids have the cyclic hexahydric alcohol named **inositol** in place of the base. Inositol $(CH_2O)_6$ exists in the inner membrane of mitochondria. It is a carbocyclic, hexahydric alcohol which is sweet in taste but is not sugar (carbohydrate). It is found in muscles (called **muscle sugar**), liver, kidneys, brain, cerebrospinal fluid (CSF), erythrocytes and tissues of eye. It is a fundamental ingradient of cell membranes. Among plants, inositol occurs in fruits, vegetables, whole grains and nuts as phytic acid esters. Milk and yeast too contain appreciable quantities of inositol.

Functions of phospholipids. Phospholipids serve the following functions:

1. Phospholipids are main components of plasma membrane and other cell membranes (*i.e.*, nuclear, mitochondrial, ER). They are thought to regulate the passage of materials into and out of the cell and cell organelles. Galactolipids and sulpholipids occur in plastid membranes.

2. Some phospholipids such as lecithins form an important component of yolk and protoplasm.
3. Small amount of phospholipids may occur in some storage fats.
4. Spingomyelins and cephalins insulate the nerve fibers (axons, *i.e.*, impulse) from its surrounding structures.
5. Phospholipids (*i.e.*, lipoproteins of serum) form important intermediates in the transport of lipids to and from the liver.
6. Lecithin also act as a component of acetylcholine—neurotransmitters in synaptic conduction of nerve impulses.

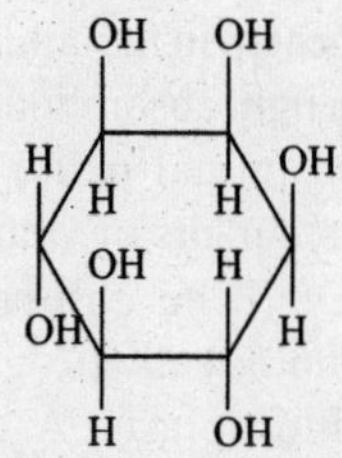

Fig. 3.25. Inositol.

7. Glycosphinolipids are the constituents of cell membranes so these are responsible for tissue immunity and cellular recognition.

2. Glycolipids. The glycolipids contain fatty acid, amino alcohol sphingosine (a high molecular weight fatty acid) and sugar. The sugar replaces one fatty acid molecule with galactose (Fig. 3.26). Their name refers to the presence of a carbohydrate in them. Glycolipids are of two types:

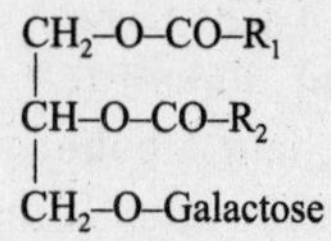

Fig. 3.26. Structure of a glycolipid molecule, R_1 and R_2 are fatty acids.

(*i*) Cerebrosides. These are found particularly in the myelin sheath of medullated nerve fibers and the white matter of the brain.

(*ii*) Gangliosides. These are rich in sialic acid (sugar) and ceramide (fatty acid) and found in ganglion cells of nervous tissue (forming 6% of gray matter). They are also found in parenchymatous tissues such as spleen and in RBCs. Gangliosides exist even in outer surface of plasma membrane. They have considerable physiological significance. Gangliosides are the receptors for certain bacterial protein toxins such as cholera toxins and certain pituitary glycoprotein hormones (**Sharma** 2010).

Other Compound Lipids

1. Sulpholipids. They are widely distributed in plants and are localised in chloroplasts. Sulpholipids are also found in the chromatophores of photosynthetic bacteria.

2. Lipoproteins. The lipoproteins contain lipids (mainly phospholipids) and proteins in their molecules. The lipid components consist of triacyl glycerol, phospholipids and cholesterol. The protein components of lipoproteins have a relatively high portion of non-polar amino acid residues that can participate in the binding of the lipids.

Phospholipids are found in membranes of mitochondria, endoplasmic reticulum and nuclei. The electron transport chain system of mitochondria appears to contain large amounts of lipoproteins. Lipoproteins are also found in mammalian blood plasma, milk and egg-yolk, etc. Some important examples of lipoproteins are: (*i*) **rhodopsin** from the retina; (*ii*) **chloroplasts** from leaves; (*iii*) **lipovitellin** and (*iv*) **lipovitellenin** from egg yolk.

Further, on the basis of centrifugational properties lipoproteins include following four types: (1) Chilomicrons; (2) Very low density lipoproteins (VLDL); (3) High density lipoproteins (HDL) and (4) Low density lipoproteins (LDL).

3. Chromolipids. This group of fats includes carotenoids and related pigments. The Kerotene $C_{40}H_{56}$ is a provitamin A found in deposits of fats of animals. The **carotenoids** are the compound lipids and they form the pigments of the animal and plant cells. There are about 70 carotenoids occurring in both types of cells. The important carotenoids of cells are the α, β and γ carotenes, retinene, xanthophylls, lactoflavin in milk, riboflavin (vitamin B_2), xanthocyanins, coenzyme Q, anthocyanins, flavones, flavonols and flavonones, etc. Chemically all carotenoids are long-chain isoprenoids having an alternating series of double bonds. They are synthesized by plant tissues and are located in the chloroplast lamellae to help in light absorption during photosynthesis. In animal cells, carotenoids serve as precursors of vitamin A.

The **chlorophylls** are essential photosynthetic green pigments of the chloroplasts. A chlorophyll molecule (Fig. 3.27) consists of a **head** and a **tail.** The head consists of a **porphyrin ring** or **tetrapyrrole nucleus** from which extends a hydrophobic tail which is made up of a 20-carbon grouping, called the **phytol**. Phytol ($C_{20}H_{39}$) is a long straight-chain alcohol containining a single double bond. It may be regarded as a hydrogenated carotene (vitamin A). The porphyrins (Gr., *porphyra* = purple) are complex cabron-nitrogen molecules that usually surround a metal, *i.e.*, it is formed from four pyrrol rings linked together by methane bridges and metal atom (Mg or Fe) is linked to pyrrol rings. In chlorophyll molecule, the porphyrin surrounds a magnesium ion, while in haeme of **haemoglobin**, it surrounds an iron ion (ferrous or Fe^{2+}; Fig. 3.28). Many other pigments of animal cells such as **myoglobin** and **cytochromes** have porphyrin rings in their molecules.

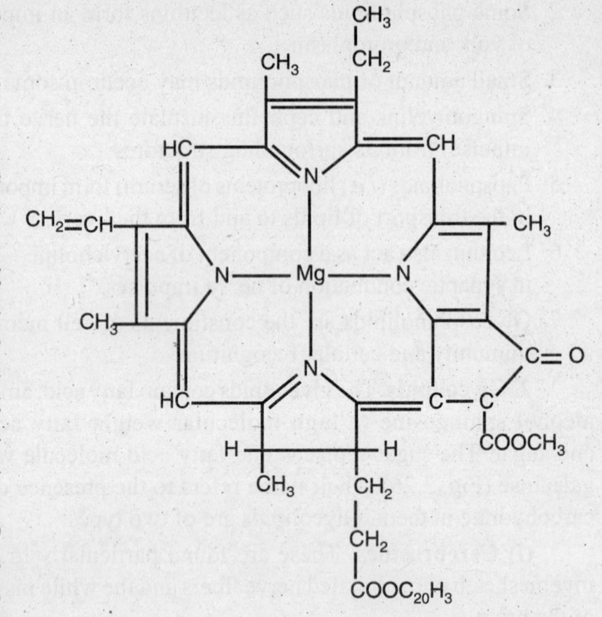

Fig. 3.27. Chemical formula of chlorophyll.

Derived Lipids

1. Steroids. The steroids (Gr., *stereos* = solid + *eidos* = form) do not contain fatty acids but are included in lipids because they have lipid-like properties. Unlike other lipids, they are not straight-chain compounds. They consist of a complex carbon framework of fused four rings, three of which have six carbon atoms each and the fourth have five (Fig. 3.29). The conventional method of naming the rings and numbering of seventeen positions in the rings have been shown in Fig. 3.29. The various steroids differ in the number and position of double bonds between the carbon atoms in the rings, and also in the side groups linked to the rings.

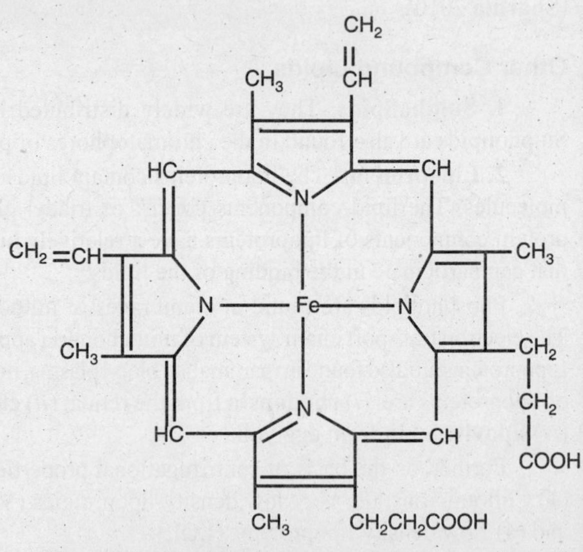

Fig. 3.28. Chemical formula of haeme portion of haemoglobin.

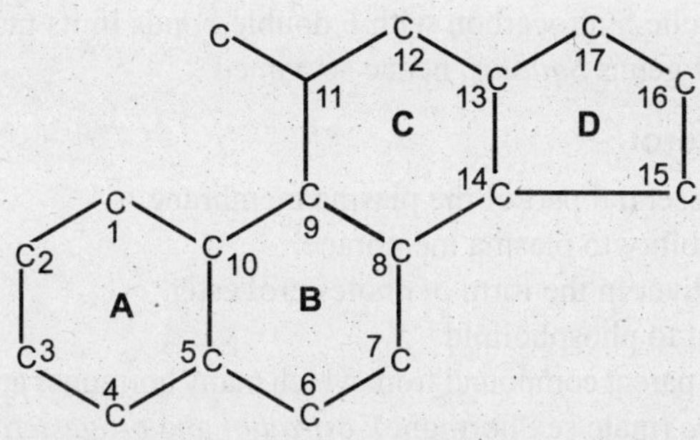

Fig. 3.29. Arrangement of four carbon rings in a steroid molecule.

Examples of steroids. The most abundant group of steroids includes the **sterols**. The sterols are complex alcohols (Gr., *stereos* = solid + *-ol* = alcohol), having a polar or hydrophilic hydroxyl (–OH) group linked at one end and a complex non-polar or hydrophobic carbon chain joined at the other end of the ring structure (Fig. 3.30). Thus, sterols are oriented in a way that they show their dual **solubility** properties. The sterols occur in both animal and plant cells. The common sterols include *cholesterol* and *ergosterol*. Other steroids are *bile acids*, vitamin D, male and female sex hormones and adrenal cortical hormones (such as cortisone, aldosterone) in vertebrates and *ecdysone* or moulting hormone of insects.

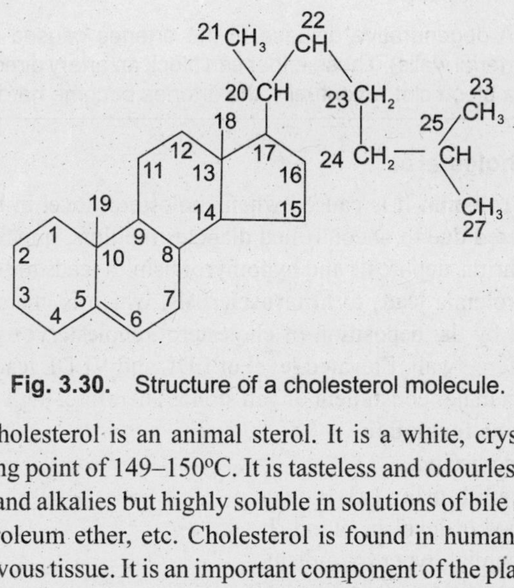

Fig. 3.30. Structure of a cholesterol molecule.

(*i*) **Cholesterol.** Cholesterol is an animal sterol. It is a white, crystalline and levo-rotatory substance having a melting point of 149–150°C. It is tasteless and odourless substance. Cholestrol is insoluble in water, acids and alkalies but highly soluble in solutions of bile salts and organic solvents such as chloroform, petroleum ether, etc. Cholesterol is found in human body in large quantities specially in brain and nervous tissue. It is an important component of the plasma membrane of animal cells (*e.g.*, erythrocytes). Various human organs such as adrenal glands, liver, kidney, spleen and skin also contain cholesterol. Egg yolk is rich in cholesterol. The normal concentration of cholesterol is 140–250 mg per 100 ml of blood. Its increase in the blood is harmful and may cause **atherosclerosis** (*i.e.*, narrowing of arteries) by deposition on their lining (such a deposition is called **atheroma**). Deposition occurs since both cholesterol and its esters are insoluble in water. Narrowing of cerebral arteries may lead to **stroke** while that of coronary arteries causes **heart attack**.

Cholesterol was first isolated from gall stones, hence its name which means 'a solid alcohol from bile' (Gr., *chole* = bile + *steros* = solid + ol = alcohol). Its formula is $C_{27}H_{45}OH$ (Fig. 3.30). It has a double bond between carbon 5 and 6. Cholesterol is synthesized in 10 steps. **Squalene** ($C_{30}H_{50}$) is one of the intermediates in its synthesis.

Squalene is an acyclic hydrocarbon with 6 double bonds in its molecule. It was first isolated from the liver of shark of genus *Squalus*, hence so named.

Functions of Cholesterol

1. Cholesterol is an essential part of the plasma membrane.
2. It provides permeability to plasma membrane.
3. It transports fat to liver in the form of cholesterol ester.
4. It acts as antagonist to phospholipid.
5. The cholesterol is a parent compound from which many hormones are synthesized in cells. These include *testosterone* (male sex hormone), *estradiol* and *progesterone* (female sex hormones), *aldosterone* and *cortisone* (adrenal cortex hormone) and *ecdysone* (insect moulting hormone).
6. It assists in the formation of bile acids and bile salts.
7. Cholesterol regulates the erythrocytes (RBCs) from being haemolized.
8. A cholesterol derivative forms vitamin D (calciferol) by the action of ultraviolet rays of sunlight.
9. Cholesterol is necessary for human life but the excess of it is known to cause cardiovascular problems such as arteriosclerosis, atherosclerosis, hypertension, heart attack, stroke, etc.

Box 3.4

Arteriosclerosis includes three pathological conditions: 1. an increase in the thickness of the arterial walls; 2. a reduction in elasticity of the vessel and 3. a constriction of diameter, which affects blood flow.

Atherosclerosis. A degenerative disease of the arteries caused by a atheroma (*i.e.*, fatty deposits or plaques in arterial walls). These either can block an artery directly or increase the chance of its being blocked by a blood clot. Eventually, the arteries become hardened and lose elasticity.

Clinical Effects of Cholesterol

1. Hypercholesterolemia. It is caused when cholesterol level in blood is raised beyond the threshold limit. It is caused due to uncontrolled diabetes mellitus, impairment of liver, obstructive jaundice, glomerulonephritis, nephrosis and hypothyroidism. It leads to following types of diseases:

(*i*) Hypercholesterolemia leads to **atherosclerosis**, which is characterized by the hardening of blood vessels caused by the deposition of cholesterol, cholesteryl ester and other lipids in the connective tissue of areterial wall. Elevated level of LDL and VLDL leads to atherosclerosis.

(*ii*) Cholesterol is a major constituent of gall stone. Therefore, high level of cholesterol always poses a threat of **gall stone formation**.

(*iii*) It also casuses obesity.

2. Hypocholesterolemia. (a) It is a condition of low cholesterol level in blood. It is found in cases of anaemia, hyperthyroidism, carcinoma, acute pancreatitis and in hepatic diseases. A decreased percentage of cholesterol ester is also known to cause hepatitis, cirrhosis and most types of liver damage.

(*b*) Ergosterol. It is a plant sterol. It is also found in the parasitic fungus, *Claviceps purpurea* (ergot) growing on the rye plant. It is present in yeast and vegetables. Its formula is $C_{28}H_{43}OH$ (Fig. 3.31). Ergosterol becomes vitamin D (calciferol) when activated by ultraviolet rays.

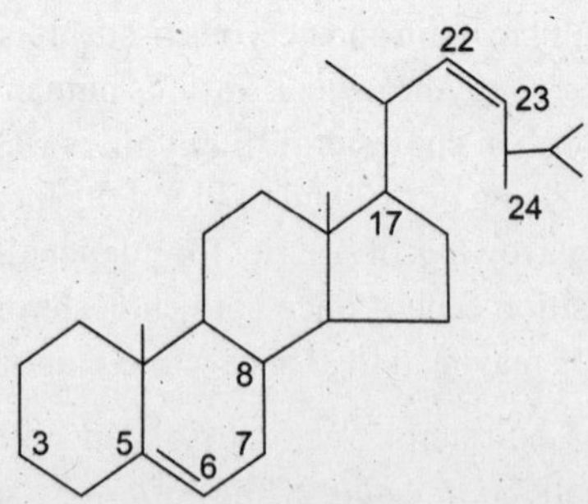

Fig. 3.31. Ergosterol.

(*c*) Bile acids and bile salts. The bile acids are steroids having 24 carbon atoms. These include cholic acid, deoxycholic acid, lithocholic acid, etc. The bile acids combine in liver with 'glycine' or 'taurine' to form 'glycocholic' or 'taurocholic' acid, which is then joined by sodium to form bile salt, namely, **sodium glycholate** or **sodium taurocholate**.

Functions.

1. Bile acids represent the only significant way of excretion of cholesterol. The bile acids and phospholipids solubilise cholesterol in the bile.
2. The bile salts lower the surface tension and thus helps in the emulsification of fats for their digestion in the intestine.
3. Bile acids increase the absorption of fat-soluble vitamins, particularly vitamin D, from the intenstine.

Steroid Hormones

Steroid hormones are fat soluble hormones having a sterol ring and are derived from the cholesterol. They can be classified according to the number of carbon atoms contained in their molecules:

1. **C_{18} hormones** which include **estrogens** such as esteradiole and esterone of ovary.
2. **C_{19} hormones** which include **androgens** such as testosterone of testes.
3. **C_{21} hormones** which include **progesterone** of corpus luteum and **mineralocorticoids** such as aldosterone and **glucocorticoids** such as cortisol and cortisone of adrenal cortex.

(A) Estrogens (C_{18} Hormone). Estrogens are produced by the follicular cells of membrana granulosa of Graafian follicle of ovary, corpus luteum of the ovary and by the placenta during the second and third trimesters of pregnancy. These include esteradiol, esterone and estriol. Out of these most important is **β-esteradiol**. It is found to act at a cellular level by binding to the nucleus of cells in its target organ, where it seems to stimulate the synthesis of RNA (transcription). Estradiol ($C_{18}H_{24}O_2$) promotes the development of endometrium (of uterus) and stimulate LH production in anterior lobe of pituitary gland. LH or **luteinizing hormone** in male humans stimulate interestial cells in testes to produce *testosterone*; in females, LH stimulates ovary to perform ovulation and causes ruptured ovarian follicle to develop into *progesterone*-secreting corpus luteum.

Functions of estrogens. Estrogens stimulate the growth, development and normal functioning of **female secondary sex organs** such as Fallopian tubes, uterus, vagina, etc. These hormones regulate the cyclic changes in uterus and ovaries during the menstrual cycle.

Estrogens also control the development of **female secondary sexual characters** such as breast-development, broadening of pelvis, growth of pubic and axillary hair and beginning of menstrual cycle. Exstrogens are also related with development of sexual desire in the female.

B. Androgens (C_{19} Hormone). Androgens are produced by the Leydig cells (interstitial cells), of the testes and the adrenal glands in both the sexes. The ovary also produces androgen in small amounts. **Testosterone** is the major androgen produced in the testes at the rate of 6–7 mg/day in adult male. Active form of testosterone is its metabolite called **5a-dihydrotestosterone (DHT)**. Other androgens are **androsterone, epiandrosterone** and **dehydro-epiandrosterone**.

Functions of testosterone. In humans and other mammals, testosterone controls the growth and development of **male secondary sex organs** such as epididymis, prostate gland, seminal vesicles and penis. It regulates the development of **male secondary sexual characters** such as growth of beared, moustaches, pubic hair, axial hair, deepening of voice, broadening of shoulders, increased height (due to elongation of bones and increased development of limbs). It also initiates sexual desire.

Testosterone also regulates spermatogenesis and maturation of sperms. It also increases the fertilizing power of the sperm.

C. Progesterone (C_{21} hormone). This steroid hormone is secreted by yellowish endocrine gland of ovary called **corpus luteum** which originates from ruptured ovarian follicle after the ovulation. Small amounts of progesteone is also secreted by adrenal cortex and placenta. Secretion of progesterone is stimulated by leuteinizing hormone (LH) of anterior lobe of pituitary gland.

Functions.

1. Progesterone stimulates the proliferation of the endometrium of uterus and prepare it for implantation.
2. It desensitizes the uterine muscles to the action of oxytocin.
3. It helps in implantation, placenta formation and normal development of foetus in the womb (uterus).
4. It suspends ovulation during pregnancy.
5. It stimulates the proliferation of mammary glands and enlargement of breasts.

D. Aldosterone (C_{21} hormone). It is the major **mineralocorticoid hormone** secreted by *zona glomerulosa* of adrenal cortex. Its secretion is stimulated by *decrease in sodium level* and *increase in potassium level* in the blood. Its secretion is controlled by ACTH of anterior lobe of pituitary.

Functions. Aldosterone maintains **hypernatraemia** (high Na^+ level in blood) and **hypokalaemia** (low K^+ level in blood) by increased sodium reabsorption from nephric filterate and decreased loss of sodium in sweat. This Na^+/K^+ balance helps in nerve impulse conduction, muscle contraction, maintenance of blood volume and blood pressure.

E. Cortisol (C_{21} hormone). It is a **glucocorticoid hormone** secreted by *zona fasciculata* of adrenal cortex. Its secretion is stimulated by low sugar level in the blood and is regulated through ACTH of anterior lobe of pituitary.

Functions. 1. It regulates **carbohydrate metabolism** and causes increase in sugar level by following ways: (*i*) increase rate of **gluconeogenesis** (conversion of proteins in liver into sugar); (*ii*) decrease in peripheral utilisation of glucose. 2. It has **anti-inflammatory** and **anti-allergic** processes.

2. Terpenes. The terpenes include certain **fat-soluble vitamins** (*e.g.*, vitamins A, E and K), **carotenoids** (*e.g.*, photosynthetic pigments of plants), and certain **coenzymes** (such as coenzyme Q or ubiquinone). All the terpenes are synthesized from various numbers of a five-carbon building block, called **isoprene unit**. The isoprene units are bonded together in a head-to-tail organization. Two isoprene units form a **monoterpene**, four form a **diterpene**, six form a **triterpene**, and so on. The monoterpenes are responsible for the characteristic odours and flavours of plants (*e.g.*, **geraniol** from geraniums, **menthol** from mint and **limoneme** from lemons). **Dolichol phosphate** is a **polyisoprenoid** (*i.e.*, long chain polymer of isoprene) and is used to carry activated sugars in the membrane-associated synthesis of glycoproteins and some polysaccharides.

3. Eicosanoids. Eicosanoids are **paracrine hormones**, which influence the nearby cells and are not circulated by blood to distant body parts. They are of following three types:

(*i*) Prostaglandins (PGs). The prostaglandins are recently discovered, hormone-like compounds. The name 'prostaglandin' was given by **U.S.V. Euler** in 1935. PGs are derivatives of 20-carbon polyunsaturated hydroxylated fatty acids such as **arachidonic acid**. They have a characteristic five-carbon ring in the middle of the chain and hydroxyl or ketone groups or certain carbons. Many different prostaglandins have been discovered.

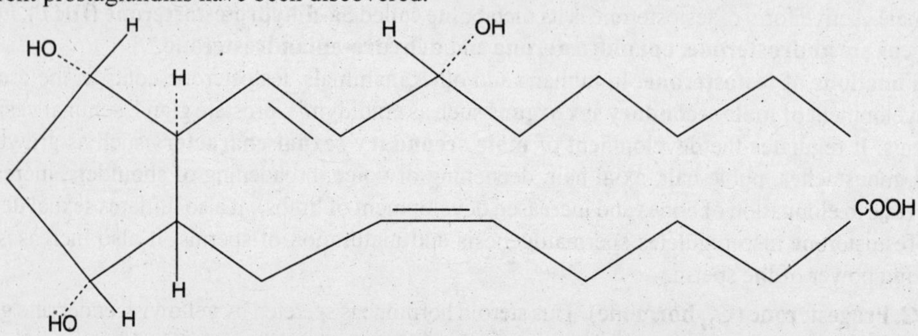

Fig. 3.32. Structure of one of the prostaglandins.

The prostaglandins were first found in the human semen, hence their name. It is now known that they are produced in many types of human tissues. They occur in lungs, thymus, kidneys, thyroid, brain, spinal cord, gastrointestinal mucosa, pancreas, endometrium, amniotic fluid, placenta and menstrual fluid. Many cells have minute amounts of prostaglandins in their membranes, but the richest sources are the secretion of seminal vesicles in males and menstrual fluid in females. Low-back pain in women during menstrual period may be due to prostaglandins released by disintegration of endometrial cells.

Prostaglandins are classified into 9 groups designated as A through I (*i.e.*, PGA through PGI). PGs have a half-life of 5 minute or less, hence they are destroyed very rapidly in the body. The short half-life of these tissue-hormones is thought to ensure their transient and limited response at the intermediate site of production (see **Jain** *et al.*, 2012).

Functions. The prostaglandins have a variety of effect on smooth visceral muscles, nervous system and blood pressure. They are thought to control many biological processes, perhaps by regulating the production cyclic AMP by adenyl cyclase. PGs increase uterine contractions. Hence, they are used to induce *parturition* and **abortion**. By increasing the activity of uterine muscles, the PGs may improve the chances of the union of sperm and egg. It has been found that the deficiency of prostaglandins in semen may cause male sterility. Prostaglandins also increase the motility of intestinal muscles, and may cause severe cramps, nausea, vomiting and diarrhoea. Prostaglandins inhibit gastric secretion, decrease arterial blood pressure, clear the nasal passages and relax the bronchioles. They have a sedative effect on many animals.

Clinical significance. From medical viewpoint, PGs are potentially most revolutionary therapeutic substances yet discovered. They have the following clinical applications:

1. PGs exert stimulatory effect on the contractions of human uterus. Therefore, PGE, PGE_2 and $PGE_2\alpha$ are given intravenously to induce labour.
2. PGE_1 is also given to asthmatic patients through inhalation for vasodilation.
3. PGE_1 is also an effective nasal decongestant.
4. PGA_1, PGE_1, PGE_2 and $PGE_2\alpha$ act as vasoconstrictor on blood vessels of nasal mucosa.
5. Prostaglanding are also used to control inflammation.
6. PGs lower down the blood pressure.

2. Leukotrienes (LTs). They are secreted by WBCs and are derived via lipoxygenase pathway. They have three conjugated double bonds. Leukotrienes stimulate mucus secretion and carry out vasoconstriction of bronchial muscles. Their excessive release causes **asthama**.

3. Thromboxanes (TXs). They are a class of molecules derived from arachidonate and are released by blood platelets or thrombocytes. They produce vasoconstriction and increase blood pressure. They cause release of serotonin and calcium ions (Ca^{2+}) for blood clotting.

QUESTIONS

Long Answer Questions

1. What are lipids? Classify lipids with suitable examples. Describe their importance for living beings. (*CCSU Meerut 2006, 2008*)
2. What are lipids? Discuss their structure. Classify lipids with suitable examples. (*CCSU Meerut 2009*)
3. What are lipids? Discuss their structure. Classify lipids with suitable examples and explain their hydrolysis. (*Avadh 2000, 2004; Garhwal 2002; Gorakhpur 2002*)
4. What are lipids? Give structure and classification of lipids. (*Garhwal 1998*)
5. What are lipids? Describe their importance for living beings. (*Allahabad 2000, 2004*)
6. Discuss the importance of lipids in human beings. (*Purvanchal 2003*)
7. What are lipids? Give examples of important lipids and lipid derivatives. Describe their chemical composition and role played by them in a living system.
8. What is significance of lipids in human body? Write about any two lipids present in plasma membrane. (*Allahabad 2000; Garhwal 2004*)
9. Describe derived lipids with examples. (*Gorakhpur 2001*)

10. Describe structure and properties of fats and oils.
11. What are fatty acids? Describe different types of fatty acids. (*Lucknow 2002*)
12. Discuss importance of prostaglandins.
13. What is ketosis? How are ketone bodies produced in the body?
14. Describe physiological or biological importance of cholesterol in the body. Summarise its clinical effect.
15. What are lipids? Describe significance of lipids in human body. (*CCSU Meerut 2014*)

Short Answer Questions

1. What are lipids? Classify lipids with suitable examples. (*CCSU Meerut 2010*)
2. Write a short note on hydrolysis of fats. (*Gorakhpur 2002, 03; Kanpur 2003*)
3. Describe biological significance of lipids. (*Gorakhpur 2002*)
4. Describe special properties of phospholipids. What is their significance or biological importance in living beings?
5. What is the reason that phospholipid molecules form a bilayer in living cells?
6. What are glycolipids? Give their functions.
7. Give functions of lipoproteins or proteiolipids.
8. Name the plasma lipoproteins found in blood plasma. Give their chemical composition.
9. What do you mean by iodine number of a fat?
10. What is rancidity?
11. Write short notes on the following:
    (*i*) Lipids (*Kanpur 2003*)
    (*ii*) Lipolysis (*Kanpur 2003*)
    (*iii*) Phospholipids (*Avadh 2002; Poorvanchal 2001*)
    (*iv*) Saturated fatty acids
    (*v*) Fatty acids
    (*vi*) Esterification
    (*vii*) Cholesterol (*Avadh 2001; Garhwal 2004*)
    (*viii*) Sterols (*Garhwal 2003*)
    (*ix*) Hydrolysis of fat (*Gorakhpur 2003*)
    (*x*) Glycoproteins (*Garhwal 2002*)
    (*xi*) Prostaglandins (*Garhwal 2002*)
12. Give differences between the following:
    (*a*) Choline and cholesterol.
    (*b*) Phospholipids and glycolipids. (*Gorakhpur 2002, 2005*)
    (*c*) Saturated and unsaturated fatty acids.
    (*d*) Fats and oils. (*CCSU Meerut 2013*)

Very Short Answer Questions

1. What are fatty acids?
2. Which fatty acids are called saturated?
3. What are monosaturated fatty acids?
4. Give two examples of polysaturated fatty acids?
5. What are carotenoids?
6. What is difference between wax and paraffin wax?
7. What are chilomicrons?
8. Name two groups of derived lipids?
9. Give full names of following: PUFA, HDL, VLDL, LDL. (*CCSU Meerut 2013*)
10. What is scientific name of lecitin?
11. Define saponification number.

12. If the saponification number of a fat is high, what does it indicate?
13. What types of lipids are present in the plasma membrane?
14. Which type of lipids are found in animal fats?
15. Who discovered essential and non-essential fatty acids?
16. Give empirical formula of fatty acids.
17. Name sources of essential fatty acids.
18. What is the number of fatty acid in a phospholipid volume?
19. What material is used for making the candles?
20. Which lipid forms essential part of our food?
21. Which fatty acid is most abundant in sunflower oil?
22. Name any one polyunsaturated fatty acid.
23. Who coin the term lipids? *(CCSU Meerut 2006, 07, 08)*
24. Which fatty acid causes sterility?
25. Which compound is being used as an antiferility agent?
26. How many molecules of glycerol and fatty acids form one fat molecule?
27. What is the effect of deficiency of essential fatty acids (EFA)?
28. Give clinical importance of LDL and VLDL.
29. What are prostaglandins?
30. What is meant by rancidity of fat?
31. How the fat can be protected from rancidity?
32. Give an example of following (with structure): A lipid containing glycerol, fatty acids, phosphate and ethanolamine.
33. What is causes of Norum's diseases?
34. What is causes of Tay-sach's disease?
35. What disease is caused by accumulations of sphingomyelins?
36. What is Gaucher's disease?

Multiple Choice Questions

1. Which of the following is incorrect about fatty acids?
   (*a*) they are straight chain compounds
   (*b*) all the carbon atoms are interlinked by single bonds
   (*c*) they have low melting point
   (*d*) they have general formula $C_nH_{2n}O_2$
2. The chemical linkage between glycerol and fatty acids is called
   (a) ester bond (*b*) glycosidic bond
   (*c*) phosphodiester bond (*d*) none of these
3. Atherosclerosis occurs due to high consumption of food rich in
   (*a*) proteins (*b*) carbohydrates (*c*) saturated fats (*d*) all of them
4. The most common lipids in the cell are
   (*a*) monoglycerides (*b*) diglycerides
   (*c*) triglycerides (*d*) polyglycerides
5. Which of the following represents the property of true fats?
   (*a*) they have high specific gravity (*b*) they have ill-defined melting points
   (*c*) they are insoluble in organic solvent (*d*) they disperse in water as minute droplets
6. Unusual property of fat is that
   (*a*) their glycerol end is lipophilic and fatty acid end is hydrophobic
   (*b*) their glycerol end is hydrophobic and fatty acid is hydrophilic
   (*c*) their glycerol end is hydrophilic and fatty acid end is lipophilic
   (*d*) their glycerol end is lipophilic and fatty acid end is hydrophilic

7. __________molecule has strongly nonpolar and hydrophobic tail consisting fatty acid chains and a polar hydrophilic head comprising a phosphate group
(*a*) phospholipid (*b*) glycolipid
(*c*) glyceride (*d*) tristearin
8. The nitrogenous base present in lecithinis
(*a*) choline (*b*) ethanolamine
(*c*) inositol (*d*) serine
9. One of the following is an amphipathic lipid
(*a*) phospholipids (*b*) fatty acid
(*c*) bile salts (*d*) all of these

ANSWERS

Very Short Answer Questions

1. Fatty acids are long-chain (4–30C atoms), aliphatic organic acids (Box 3.5) containing only C, H and O. They have a single carboxyl group attached to a long non-polar hydrocarbon tail which gives most lipids their hydrophobic and oily or greasy nature. They do not occur in free (*i.e.*, uncombined) state in cells or tissues but are present in covalently bound form in different classes of lipids.

Box 3.5

Aliphatic compounds. A major class of organic compounds with carbon atoms linked in straight or branched open chains. The other classes are: alicyclic, aromatic and heterocyclic compounds.

2. In saturated fatty acids all the carbon atoms in the hydrocarbon chain are saturated. It means all the carbon atoms in the chain are connected through a single bond, *e.g.*, palmitic acid.
3. Fatty acids with single double bond in their hydrocarbon chain are called monosaturated fatty acids, *e.g.*, oleic acid.
4. Linoleic acid and archidonic acid.
5. Carotenoids are fat soluble photosynthetic pigments. These are formed of **isoprene units**.
6. Waxes are esters of monohydro alcohols and fatty acids with long hydrocarbon chains. They have an odd number of carbon ranging from C_{25} to C_{35}.
Paraffin wax consists of a mixture of hydrocarbons of alkane series (*i.e.*, ethane series).
7. Chilomicrons are droplets of lipoproteins present in blood plasma. They are synthesized by smooth endoplasmic reticulum of epithelial cells lining the small intestine and serve to transport dietary lipids in the circulation.
Thus, chilomicrons are formed after digestion and absorption of fats. These are formed of triglycerols and cholesteryl esters and are surrounded by proteins, phospholipids and cholesterols.
8. 1. Terpens; and 2. Elcosanoids.
9. 1. PUFA: Polyunsaturated fatty acids; 2. HDL: High density lipids; and 3. VLDL: Very low density lipids.
10. Phosphatidyl choline.
11. Saponification number is the amount of KOH (in milligrams) which is used for the complete saponification of one gram of fat or oil.
12. High saponification number of fat indicates that it is formed of short hydrocarbon chains.
13. The lipids present in the plasma membrane are mostly unsaturated.
14. Animal fats are rich in saturated fatty acids.
15. **Evans** and **Burr**.
16. $C_8H_{2n}O_2$.

17. Vegetable oils and fish or fish liver oils.
18. Two fatty acids that form tail part of phospholipid.
19. Paraffin wax and stearic acid.
20. Linolenic acid.
21. Linolenic acid.
22. Arachidonic acid.
23. **Whilhelm Bloor**.
24. Deficiency of essential fatty acids causes sterility.
25. **Diosgenin** which is a crystalline steroid sapogenin $C_{27}H_{42}O_3$ obtained from yams (*Dioscorea* sp.) and used as a starting material for the synthesis of steroid hormones (as cortisone and progesterone).
26. In one fat molecule three molecules of fatty acids are joined to one glycerol molecule.
27. Deficiency of EFA causes swelling of mitochondrial membrane and reduction in the efficiency of oxidative phosphorylation.
28. Both LDL and VLDL are responsible for **atherosclerosis** and **coronary thrombosis**.
29. **Prostaglandins** are a group of cyclic fatty acids derived from essential fatty acids. They are present in many human's body cells and in semen and exhibit hormone like activity, *e.g.*, promotion of uterine contraction in human females.
30. The change in colour and odour of fats on standing, exposure to light, moisture and air is known as **rancidity**. It occurs due to action of enzyme lipase and changes in atmospheric moisture and temperature to bring partial hydrolysis of the fat. Rancidity is also accompanied with some degree of oxidation of the unsaturated fatty acids in the fat.
31. Rancidity of fats can be prevented in the presence of α-tocopherol (vitamin E) which acts as a radical chain breaker and thereby inhibits the destructive peroxidation of polysaturated fatty acids.
32. **Plasmalogen** is an example of a lipid containing glycerol, fatty acids and ethanolamine (Fig. 3.33).

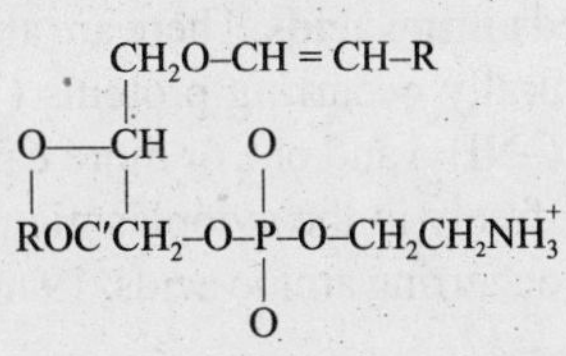

Fig. 3.33. Plasmalogen.

33. This disease is caused by the deficiency of *cholesterol acyl transferase* (*LCAT*).
34. Tay-Sach disease is caused due to accumulation of gangliosides in lysosomes of cells of the brain and nervous tissue. This lipid storage disease or lipidosis leads to progressive development of idiocy, blindness and paralysis. Ganglioside accumulation occurs because of absence of enzyme *N-acetylgalactosaminidase* (*hexosaminidase A*).
35. Accumulation of sphinogomyelins in brain, liver and spleen of children causes **Niemann-Pick disease.** The child suffering from this disease has enlarged liver, spleen, and enlarged abdomen and mental retardation.
36. Gaucher's disease is a hereditary disease characterised by the accumulation of glucocerebrosidase.

Multiple Choice Questions

1. (*c*)	**2.** (*a*)	**3.** (*c*)	**4.** (*c*)	**5.** (*d*)	6. (*c*)
7. (*a*)	**8.** (*a*)	**9.** (*d*)			

4

Amino Acids, Peptides and Proteins

Of all the macromolecules found in the cell, the ***proteins*** are chemically and physically more diverse. They are important constituents of the cell forming more than 50 percent of the cell's dry weight. The term protein was coined by Dutch chemist **G.J. Mulder** (1802–1880) and is derived from Greek word *proteios*, which means "of the first rank".

Proteins serve as the chief structural material of protoplasm and play numerous other essential roles in living systems. They form enzymes—globular proteins specialized to serve as catalysts in virtually all bio-chemical activities of the cells. Other proteins are antibodies (immunoglobulins), transport proteins, storage proteins, contractile proteins, and some hormones. In every living organism, there are thousands of different proteins, each fitted to perform a specific functional or structural role. Indeed, a single human cell may contain more than 10,000 different types of protein molecules. Chemically, proteins are polymers of amino acids.

Nobel Laureate **Emil Fischer** (1902) discovered that all proteins consist of chains (linear sequence) of smaller units that he named **amino acids**. There are about 20 different amino acids which occur regularly as constituents of naturally occurring proteins (Table 4.2). An organic compound containing one or more amino groups ($-NH_2$) and one or more carboxyl groups ($-COOH$) is known as amino acid. The amino acids occur freely in the cytoplasmic matrix and constitute the so called **amino acid pool**. Of the 20 commonly occurring amino acids, 19 may be represented by the following general formula (Fig. 4.1).

The sole exception is proline, where the amino group forms part of a ring structure. The central or **alpha** carbon atom of each amino acid is covalently bonded to four groups:

(*i*) a hydrogen atom,
(*ii*) an amino group ($-NH_2$),
(*iii*) an acid (or carboxyl) group, and
(*iv*) a side chain called an **R-group**. It is the particular chemical structure of the R-group that distinguishes one amino acid from another.

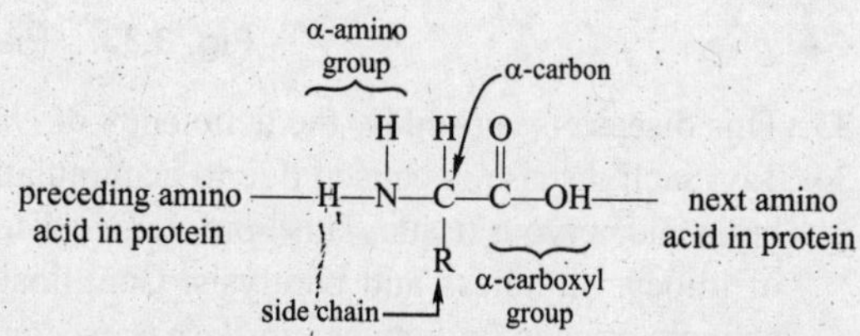

Fig. 4.1. Basic structure of an amino acid.

4.1 ACID DISSOCIATION CONSTANTS (K_a) OF WEAK ACIDS

Acids and bases which dissociate completely in water, such as hydrochloric acid and sodium hydroxide, are called **strong acids** and **strong bases**. Many other acids and bases, such as the amino acids from which proteins are made and the purines and pyrimidines from DNA and RNA, do not dissociate completely in water. These substances are known as **weak acids** and **weak bases**.

Acetic acid is the weak acid present in vinegar. The equilibrium reaction for the ionization of acetic acid is

$$\underset{\substack{\text{Acetic acid}\\ \text{(Weak acid)}}}{CH_3COOH} \xrightleftharpoons{K_2} \underset{\text{Proton}}{H^+} + \underset{\substack{\text{Acetate ion}\\ \text{(Conjugate base)}}}{CH_3COO^-} \quad ...(4.1)$$

The equilibrium constant for the dissociation of a proton from an acid in water is called the **acid dissociation constant, K_a**. When it reaches equilibrium, which happens very rapidly, the acid dissociation constant is equal to the concentration of the products divided by the concentration of the reactants. For reaction 4.1 the acid dissociation constant is

$$K_a = \frac{[H^+][CH_3COO^-]}{[CH_3COOH]} \quad ...(4.2)$$

The K_a for acetic acid at 25°C is 1.76×10^{-5} M. Because K_a values are numerically small and inconvienient in calculations, it is useful to place them on a logarithmic scale. The parameter pK_a is defined by analogy with pH

$$pK_a = -\log K_a = \log\frac{1}{K_a} \quad ...(4.3)$$

A pH value is a measure of the acidity of a solution and pK_a value is a measure of the strength of the base (ammonia, NH_3) in aqueous solution. The K_a values for some common substances are listed in Table 4.1.

Table 4.1 Dissociation constant [K_a (M)] and pK_a values of weak acids in aqueous solutions at 25°C (*Source:* Hurton *et al.*, 2006).

	Acid	K_a (M)	pK_a
1.	HCOOH (Formic acid)	1.77×10^{-4}	3.8
2.	CH_3COOH (Acetic acid)	1.76×10^{-5}	4.8
3.	$CH_3CHOHCOOH$ (Lactic acid)	1.36×10^{-4}	3.9
4.	H_3PO_4 (Phosphoric acid)	7.52×10^{-3}	2.2
5.	$H_2PO_4^-$ (Dihydrogen phosphate ion)	6.23×10^{-8}	7.2
6.	HPO_4^- (Monohydrogen phosphate ion)	2.20×10^{-13}	12.7
7.	H_2CO_3 (Carbonic acid)	4.30×10^{-7}	6.4
8.	HCO_3 (Bicarbonate ion)	5.61×10^{-11}	10.2
9.	NH_4^+ (Ammonium ion)	5.62×10^{-10}	9.2
10.	$CH_3NH_3^+$ (Methyl-ammonium ion)	2.70×10^{-11}	10.7

4.2 AMINO ACIDS

All organisms use the same 20 amino acids (Table 4.2) as building blocks for the assembly of protein molecules. These 20 amino acids are called the **common**, or **standard**, amino acids. Despite the limited number of amino acids an enormous variety of different polypeptides can be produced by connecting the 20 common amino acids in various combinations.

Table 4.2 The 20 amino acids which occur naturally.

Group of amino acid		Name of amino acid	Symbol of amino acid	Single letter code of amino acid
A.	**Aliphatic amino acids**			
1.	Simple amino acids or Mono-amino mono-carboxylic amino acids	1. Glycine	Gly	G
		2. Alanine	Ala	A
		3. Valine	Val	V
		4. Leucine	Leu	L
		5. Isoleucine	Ile	I
2.	Acidic amino acids or Monoaminomonocarboxylic acids	6. Aspartic acid	Asp	D
		7. Glutamic acid	Glm	E
3.	Basic amino acids or Diamino-monocarboxylic amino acids	8. Lysine	Lys	K
		9. Arginine	Arg	R
		10. Histidine	His	H
4.	Hydroxyl containing amino acids	11. Serine	Ser	S
		12. Threonine	Thr	T
5.	Sulphur-containing amino acids	13. Cysteine	Cys	C
		14. Methionine	Met	M
B.	**Aromatic amino acids**			
		15. Phenylalanine	Phe	F
		16. Tyrosine	Tyr	Y
		17. Tryptophan	Try	W
C.	**Secondary amino acids**			
		18. Proline	Pro	P
D.	**Amino acid amides**			
		19. Aspargine	Asn	N
		20. Glutamine	Glu	Q

1. Nomenclature of Amino Acids

Amino acids are called amino acids because they are amino derivatives of carboxylic acid. In the 20 common amino acids the amino group and carboxyl group are bonded to the same carbon atom: the α-carbon atom. Thus, all of the standard amino acids found in proteins are α-amino acids. Two other substituents are bound to the α-carbon—a hydrogen atom and a side chain (R group) that is distinctive for each amino acid. In the **chemical names** or **systematic names**, carbon atoms are identified by numbers, beginning with the carbon atom of the carboxyl group. [The correct chemical name follows rules established by **International Union of Pure and Applied Chemistry (IUPAC)** and the **International Union of Biochemistry** and **Molecular Biology (IUBMB)**]. If the R group is $-CH_3$, then the systematic name for that amino acid would be **2-aminopropanoic acid**. (Propanoic acid is CH_3-CH_2-COOH). The trivial name for $CH_3-CH(NH_2)-COOH$ is alanine. An alternate nomenclature uses Greek letters to identify the α-carbon atom and carbon atom of the side chain. The alternate nomenclature identifies the carbon atom relative to carboxyl group so the carbon atom of the carboxyl group is not specified, unlike the systematic nomenclature where this carbon atom is number 1 in the numbering system. Biochemists have traditionally used the old alternate nomenclature.

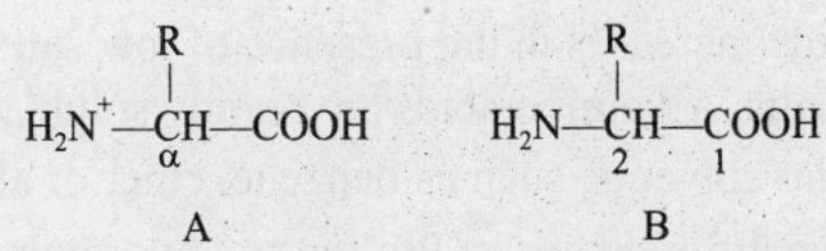

Fig. 4.2. Numbering conventions for amino acids. A—In traditional names, the carbon atoms adjacent to the carboxyl group are identified by the Greek letters α, β, γ, etc. B—In the official IUPAC/IUBMB chemical names or systematic names, the carbon atom in the carboxyl group is number 1 and the adjacent carbons are numbered sequentially. Thus, the α-carbon atom in traditional names is the carbon 2 atom in systematic names (after Horton *et al.*, 2006).

Figure 4.3A, shows the general structure of an amino acid in perspective. Figure 4.3B, shows a ball-and-stick model of representative amino acid, serine, whose side chain is $-CH_2OH$. The carbon atoms of a side chain are sequentially labelled β, γ, δ and ε, which refers to carbons 3, 4, 5 and 6, respectively. The systematic name for serine is 2-amino-3-hydroxypropanoic acid.

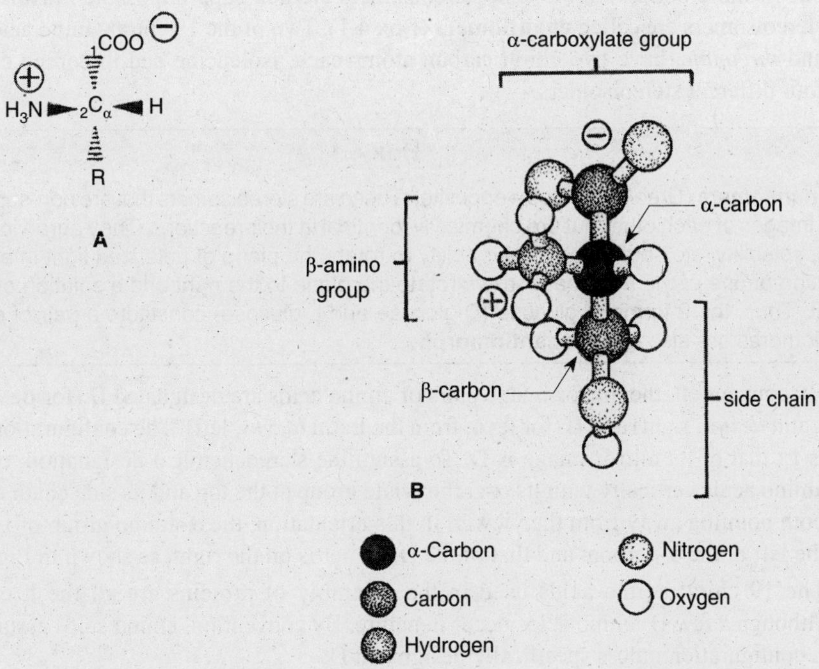

Fig. 4.3. A—General structure of an amino acid. An amino acid has a carboxylate group [whose carbon atom is designated C-1, an amino group, a hydrogen atom, and a side-chain (or R group), all attached to C-2 (the α-carbon). Solid wedges indicate bonds above the plane of the paper; dashed wedges indicate bonds below the plane of the water. B—Ball- and -stick model of serine (amino acid) whose R group $-CH_2OH$]. Note the alternate numbering and lettering system for the carbon (after Horton *et al.*, 2006).

2. Physical Properties of Amino Acids

(*i*) **Appearance.** Amino acids are colourless, crystalline compounds which appear as white power when in bulk. The crystals of amino acids have different appearance.

(*ii*) **Solubility.** Amino acids are soluble in polar solvents such as water but their solubility varies greatly. The solubility of amino acids is influenced by the presence of neutral salts in the solution.

Solubility of amino acids increases in the presence of low salt concentration, but decreases in high concentration of salts. All amino acids are readily soluble in dilute acids and alkalies but are insoluble in nonpolar solvents, such as benzene, ether or alcohol.

(*iii*) Taste. Amino acids have different tastes. Proline and hydroxyproline are sweet, valine is bitter sweet, tryptophan and isoleucine are bitter.

(*iv*) Melting point. The melting point of amino acids is above 200°C.

(*v*) Stereoisomerism. In 19 of the 20 amino acids used for the biosynthesis of proteins, the α-carbon atoms is **chiral**, or asymmetric, since it has four different groups bonded in it. The exception is glycine, whose R group is simply a hydrogen atom (the molecule is not chiral because the α-carbon atom is bonded to two identical hydrogen atoms). The 19 chiral amino acids can therefore exist as stereoisomers. **Stereoisomers** are compounds that have the same molecular formula but differ in the arrangement or configuration, of their atoms in space. The two stereoisomers are distinct molecules that cannot be easily converted from one form to other since a change in configuration requires the breaking of one or more bonds. Amino acid stereoisomers are non-superimposable mirror images. Such stereoisomers are called **enantiomers** (Box 4.1). Two of the 19 chiral amino acids, *isoleucine* and *threonine*, have two chiral carbon atoms each. Isoleucine and threonine can each form four different stereoisomers.

Box 4.1

Enantiomers (*Greek, enantios* = opposite). These are stereoisomers that are non-superimposable mirror images of each other but are chemically identical in their reactions. They agree in their melting points, solubility, etc., but differ in their ability to rotate the plane of polarized light in a polarimeter; a solution of one of the two enantiomers rotate the plane to the right and a solution of the other to the left. Thus, the 2 forms of glucose (D-glucose and L-glucose) constitute a pair of enantiomers. Enantiomores are also called **enantiomorphs**.

By convention, the mirror-image pairs of amino acids are designated **D** (for **dextro**, from the Latin *dexter*, right) and **L** (for **levo**, from the Latin *laevus*, left). The configuration in Fig. 4.3A is L; that of its mirror image is D. To assign the stereochemical designation, one draws the amino acids vertically with its α-carboxylate group at the top and its side chain at the bottom, both pointing away from the viewer. In this orientation, the α-amino group of L isomer is on the left of the α-carbon, and that of the D isomer is on the right, as shown in Fig. 4.4.

The 19 chiral amino acids used in the assembly of proteins are all the L configuration, although a few D-amino acids occur in nature. By convention, amino acids assumed to be the L configuration unless specifically designated D.

Evolutionary significance of levorotatory isomer of amino acids. The fact that all living organisms used the same standard amino acids in protein synthesis is evidence that all species on Earth are descended from a common ancestor. Like modern organisms, the last common ancestor (LCA) must have used L-amino acids and not D-amino acids. Mixtures of L- and D- amino acids are formed under conditions that mimic those present when life first arose on Earth 4 billion years ago, and both enantiomers are found in meteorites and in the vicinity of stars. It is not known how or why primitive life forms selected L-amino acids from the presumed mixture of the enantiomers present when life first arose. It is likely that the first proteins were composed of a small number of simple amino acids and selection of L-amino acids over D-amıno acids was a chance event. Modern living organisms do not select L-amino acids from a mixture since only L-amino acids are synthesized in sufficient quantities. Thus, the predominance of L-amino acids in modern species is due to the evolution of metabolic pathways that produce L-amino acids and not D-amino acids.

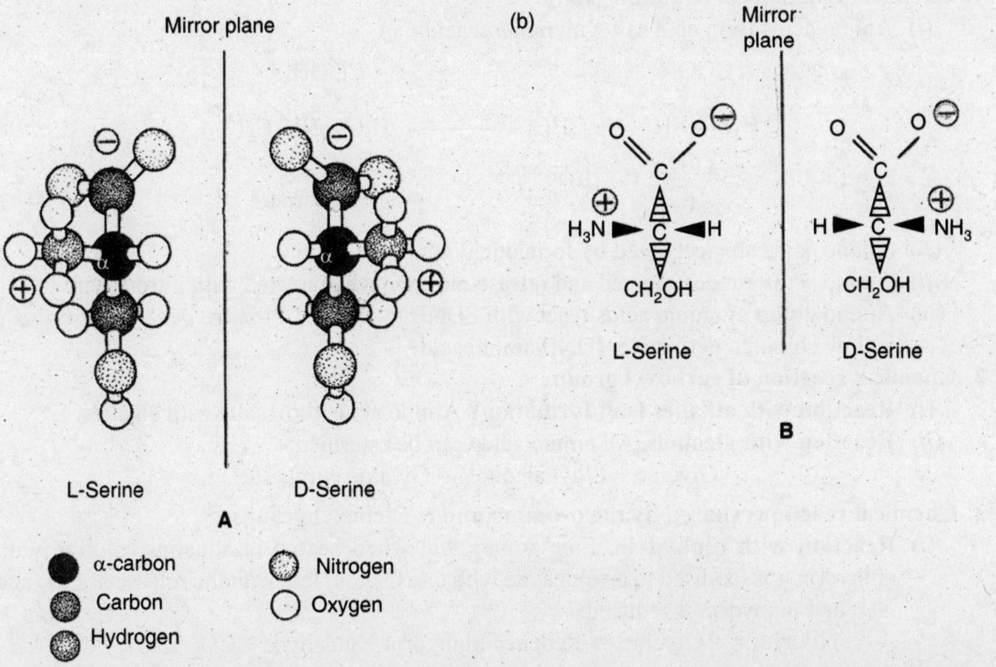

Fig. 4.4. Mirror-image pairs of amino acids. A—Ball-and-stick models of L-serine and D-serine. Note that the two molecules are not identical; they cannot be superimposed. B—L-serine and D-serine.

(*vi*) **Zwitter ions.** Inside a cell, under normal physiological conditions, the amino group is protonated ($-NH_3^+$) because pK_a of this group is close to 9. The carboxyl group is ionized ($-COO^-$) because the pK_a of the group is below 3 (Table 4.1). Thus, in the physiological pH range of 6.8 to 7.4 amino acids are **zwitter ions**, or **dipolar ions**, even though their net charges may be zero. Like a zwitter ion an amino acid has following characteristics:

(*a*) At its isoelectric pH, each amino acid has an equal number of plus (+) and minus (–) charges.

(*b*) In **acidic solution** an amino acid acts as a base and combines with H^+ ions of the solution.

(*c*) In **alkaline solution** amino acid acts as a acid and combines with OH^- ions of the solution.

(*d*) Zwitter ion is *ampholyte* because it acts as proton donor and proton acceptor.

(*e*) In a pure aqueous solution zwitter ions are not neutral because their acidic and basic strength are different.

3. Chemical Properties of Amino Acids

Amino acids are **amphoteric** having both amino and carboxyl groups (Box 4.2). Their chemical properties can be discussed under following three main headings:

Box 4.2 Amphoteric and Ampholytic Nature of Amino Acids

A compound is called *amphoteric* if it can react with both acids and bases. Both amino ($-NH_2$) and carboxyl (–COOH) groups of amino acids are ionizable and readily form salts either with acids or alkalies. Depending upon H^+ ions concentration of the medium, amino acids behave either negatively or positively charged.

1. **In acidic medium**, amino acids carry positive charge and act as base, yielding cations. In electric field, they migrate to cathode or positive pole.
2. **In alkaline medium**, amino acids carry negative charge and act as acid, yielding anions. In electric field, they migrate to anode or negative pole.

1. **Chemical reactions of α-amino group**

   (*i*) Amino acids form acid salt with mineral acids

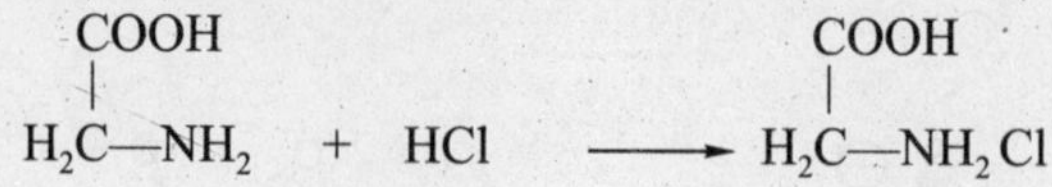

Glycine   Hydrochloric acid   Glycine hydrochloride

   (*ii*) Amino acids are acetylated by formaldehyde.

   (*iii*) Amino acids are deaminated and release nitrogen when treated with nitrous acid.

   (*iv*) Amino group of amino acids react with FDNB (1-fluro 2, 4-dinitro benzene) forming a yellow coloured derivative (DNB amino acid).

2. **Chemical reaction of carboxyl group**

   (*i*) **Reaction with alkalies (salt formation).** Amino acids form salts with alkalies.

   (*ii*) **Reaction with alcohols.** All amino acids can be esterified.

   Glycine + Ethyl alcohol → Glycine ethyl ester

3. **Chemical reactions shared by the α-amino and α-carboxyl groups**

   (*i*) **Reaction with ninhydrin.** Free amino acid when heated in aqueous solution with ninhydrin it is oxidised to an aldehyde with one C-atom less with the release of CO_2 and NH_3 and ninhydrin is reduced.

   Ninhydrin + Glycine → Reduced ninhydrin + aldehyde + CO_2 + NH_3

   (*ii*) **Peptide linkage.** Two amino acids join by the union of α-carboxyl group of one amino acid with the α-amino group of other amino acid forming a **dipeptide** and a molecule of water is eliminated.

   Alanine + Glycine → Anylglycine + H_2O

   (*iii*) **Chelation.** Amino acids also acts with ions of certain heavy metals by the process of *chelation* (*i.e.*, the binding of a metal ion to an organic molecule from which it can later be released). Such metallic ions are Cu^{2+}, Co^{2+} and Ca^{2+}.

4. Classification of Amino Acids

Examination of the structures reveals considerable variation in the side chains (*i.e.*, R groups) of the 20 amino acids. Some side chains are nonpolar and thus hydrophobic, whereas others are polar or ionized at neutral pH and therefore are hydrophilic.

The properties of the side chains greatly influence the overall three-dimensional shape, or conformation of a protein. For example, most of the hydrophobic side chains of a water-soluble protein fold into the interior, giving the protein a compact, globular shape.

Both the three-letter and one-letter abbreviations for each amino acid are shown in figures. The three-letter abbreviations is self-evident, but the simple one-letter abbreviation is less obvious. Several amino acids begin with the same letter so other letters of the alphabet have to be used in order to provide a unique label: for example, threonine = T, tyrosine = Y, and tryptophan = W (Table 4.2).

Standard amino acids are grouped by their general properties and the chemical structures of their side chains. The side chains or R groups fall into following chemical classes: 1. Aliphatic; 2. Aromatic; 3. Sulphur-containing; 4. Alcohols; 5. Basic; 6. Acidic and 7. Amides.

A. Aliphatic R Groups

Glycine (Gly, G) (Fig. 4.5) is the smallest amino acid since its R group is simply a hydrogen atom: consequently, the α-carbon of glycine is not chiral. The two hydrogen atoms of the α-carbon of glycine impart little hydrophobic character to the molecule. Glycine plays a unique role in the structure of many proteins becuase its side chain (R group) is small enough to fit into niches that can accommodate no other amino acid.

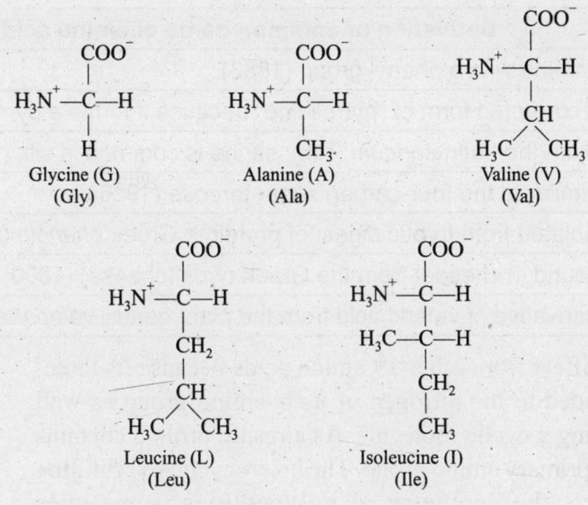

Fig. 4.5. Amino acids with aliphatic (R) groups.

Four amino acids, **alanine (Ala, A), valine (Val, V), leucine (Leu, L)**, and structural isomer of leucine, **isoleucine (Ile, I)**, have saturated aliphatic side chains. The side chain of alanine is a methyl group, whereas valine has a three-carbon branched side chain and leucine and isoleucine each contain a four-carbon branch side chain. Both α- and β-carbon atoms of isoleucine are asymmetric.

Box 4.3

Because isoleucine has two chiral centers, it has four possible stereoisomers. The stereoisomer used in proteins is called L-isoleucine, and the amino acid that differs at the β-carbon is called **L-alloisoleucine**. The other two stereoisomers are **D-isoleucine** and **D-alloleucine**.

Alanine, valine and leucine play an important role in establishing and maintaining the three-dimensional structures of proteins becuase of their tendency to cluster away from water. Valine, leucine and isoleucine are called **branched-chain amino acids** because their side chains (R groups) of carbon atoms contain branches. All three amino acids are highly hydrophobic.

Table 4.3 Common names of amino acids (*Source:* Horton *et al.*, 2006).

	Amino acid	Derivation of common name of amino acid and year
1.	Alanine	Probably from *a*ldehyde + "an" (for convenience) + am*ine* (1849).
2.	Arginine	Crystallizes as a silver salt, from Latin *argentum* (silver) (1886).
3.	Asparagine	First isolated from asparagus (1813).
4.	Aspartate	Similar to asparagine (1836).
5.	Glutamate	In the plant protein gluten (1866).
6.	Glutamine	Similar to glutamate (1866).
7.	Glycine	From the Greek *glykys* (sweet), tastes sweet (1848).
8.	Cysteine	From the Greek *kystis* (bladder), discovered in bladder stones (1882).
9.	Histidine	First isolated from sturgeon sperm, named for the Greek *histidin* (tissue) (1896).
10.	Isoleucine	Isomer of leucine.
11.	Lysine	Product of protein hydrolysis, from the Greek *lysis* (loosening) (1891).
12.	Methionine	Side chain is a sulphur (Greek *theion*) atom with a *me*thyl group (1928).

	Amino acid	Derivation of common name of amino acid and year
13.	Phenylalanine	Alanine with a phenyl group (1883).
14.	Proline	A corrupted form of "pyrrolidine" because it forms a pyrrolidine ring (1904).
15.	Serine	From the Latin *sericum* (silk), serine is common in silk (1865).
16.	Threonine	Similar to the four-carbon sugar threose (1936).
17.	Tryptophan	Isolated from tryptic digest of protein + Greek *phanein* (to appear) (1890).
18.	Tyrosine	Found in cheese, from the Greek *tyros* (cheese) (1890).
19.	Valine	Derivative of valeric acid from the plant genus *Valeriana* (1906).

Proline (**Pro**, **P**) differs from other 19 amino acids because its three-carbon side chain is bonded to the nitrogen of its α-amino group as well as to the α-carbon, creating a cyclic molecule. As a result, proline contains a secondary rather than a primary amino group. The heterocyclic **pyrrolidine ring** of proline restricts the geometry of polypeptides, sometimes introducing abrupt changes in the direction of the peptide chain. The cyclic structure of proline makes it much less hydrophobic than valine, leucine and isoleucine.

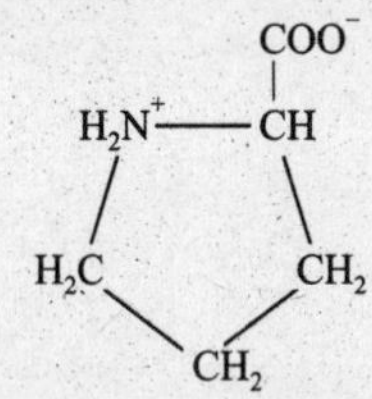

Fig. 4.6. Proline (P) (Pro).

B. Aromatic Groups

Phenylalanine (Phe, F), **tyrosine (Tyr, Y)**, and **tryptophan (Trp, W)**, have side chains with aromatic groups. Phenylalanine has a hydrophobic benzyl side chain. Tyrosine is structurally similar to phenylalanine; the *para* hydrogen of phenylalanine is replaced in tyrosine by a hydroxyl group (–OH) making tyrosine a phenol. The side chain of tryptophan contains a bicyclic indole group. Tyrosine and tryptophan are not as hydrophobic as phenyalanine because their side chains include polar groups.

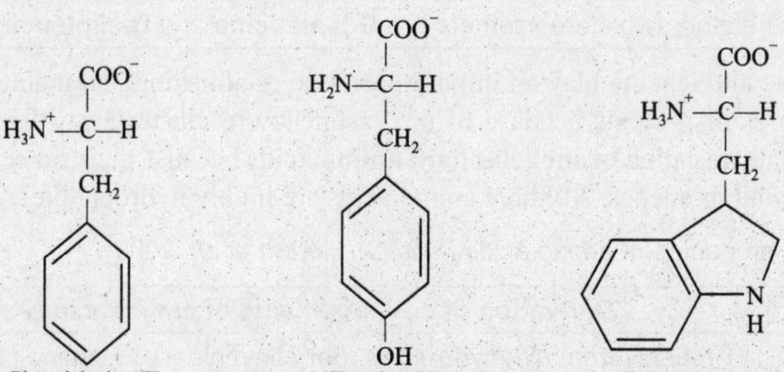

Phenylalanine (F) (Phe) — Tyrosine (Y) (Tyr) — Tryptophan (W) (Trp)

Fig. 4.7. Amino acids with aromatic R groups.

Box 4.4

All three aromatic amino acids absorb ultraviolet (UV) light. At neutral pH, both tryptophan and tyrosine absorb light at 24 nm. Such absorbance at 24 nm is routinely used to estimate the concentration of proteins in solution.

C. Sulphur-containing R Groups

Methionine (Met, M) and **cysteine (Cys, C)** are the two sulphur-containing amino acids. Methionine contains a nonpolar methyl thioether group in its side chain, and this makes it one of the most hydrophobic amino acids. Methionine plays a special role in protein synthesis since it is almost always the first amino acid in a polypeptide chain. The structure of cysteine resembles that of alanine with a hydrogen atom replaced by sulphydryl group (–SH).

Although the side chain of cysteine is somewhat hydrophobic, it is also highly reactive. Because the sulphur atom is polarizable, the sulphydryl group of cysteine can form weak hydrogen bonds with oxygen and nitrogen. Moreover, the sulphydryl group of cysteine is a weak acid, which allows it to lose its proton to become negatively charged thiolate ion.

A compound called **cystine** can be isolated when some proteins are hydrolyzed. Cystine is formed from two cysteine molecules that are linked by a **disulphide bonds** or **disulphide bridges**. Most proteins do not contain disulphide bonds because conditions inside the cell do not favour oxidation. However, many secreted or extracellular proteins contain disulphide bridges.

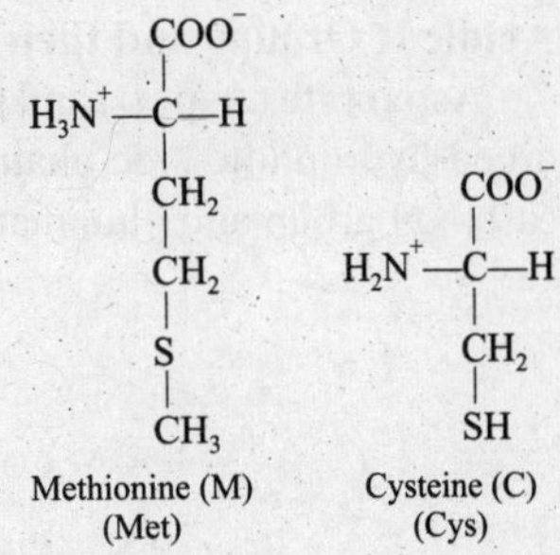

Fig. 4.8. Amino acids with sulphur containing R group.

D. Side Chains with Alcohol Groups

Serine (Ser, S) and **threonine (Thr, T)** have uncharged polar side chains containing β-hydroxyl groups. These alcohol groups give hydrophilic character to the aliphatic side chains. Unlike the more acidic phenolic side chain of tyrosine, the hydroxyl groups of serine and threonine have the weak ionization properties of primary and secondary alcohols. Threonine, like isoleucine, has two chiral centers, the α- and β-carbon atoms. L-Threonine is the only one of the four stereoisomers that commonly occurs in proteins. (The other stereoisomers are called D-threonine, L-allothreonine and D-allothreonine).

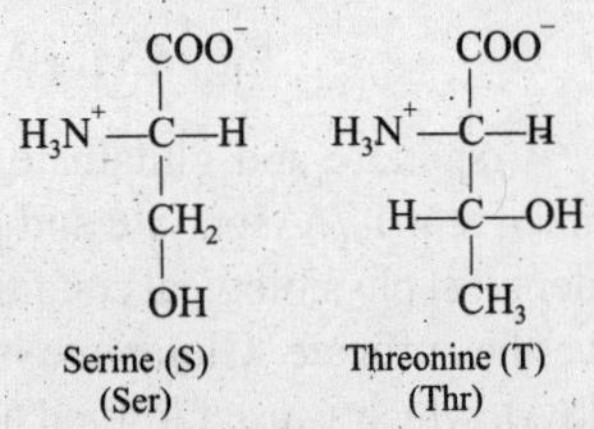

Fig. 4.9. Amino acids having side chains with alcohol groups.

E. Basic R Groups

Histidine (His, H), **lysine (Lys, K)** and **arginine (Arg, R)** have hydrophilic side chains that are nitrogenous bases and are positively charged at pH 7.

The side chains of histidine contains an **imidazole ring** substitute. The protonated form of this ring is called an **imidazolium ion**. Lysine is a diamino acid, having both α- and ε-amino groups. The ε-amino group exists as an alkylammonium ion ($-CH_2-NH_3^+$) at neutral pH and confers positive charge on proteins. Arginine is the most basic of the 20 amino acids because its side-chain guanidinium ion is protonated under all conditions normally found within a cell. Arginine side chains also contribute positive charges to proteins. (Note. Guanidinium ions means protonated form of arginine side chain).

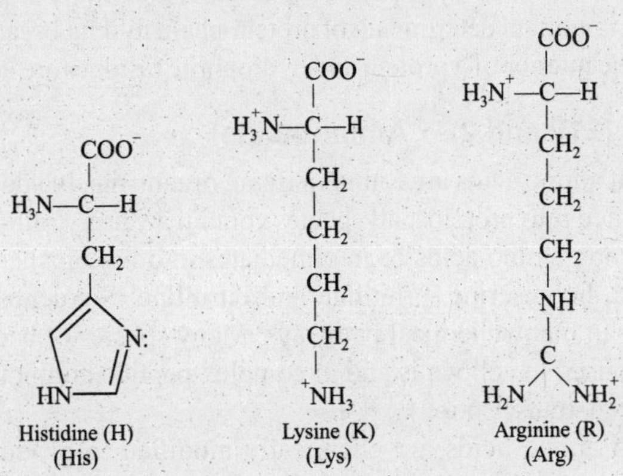

Fig. 4.10. Amino acids with basic R groups.

F. Acidic R Groups and their Amide Derivatives

Aspartate (Asp, D) and **glutamate (Glu, E)** are dicarboxylic amino acids and have negatively charged hydrophilic side chains at pH 7. In addition to α-carboxyl groups aspartate, possesses a β-carboxyl group and glutamate possesses a γ-carboxyl group.

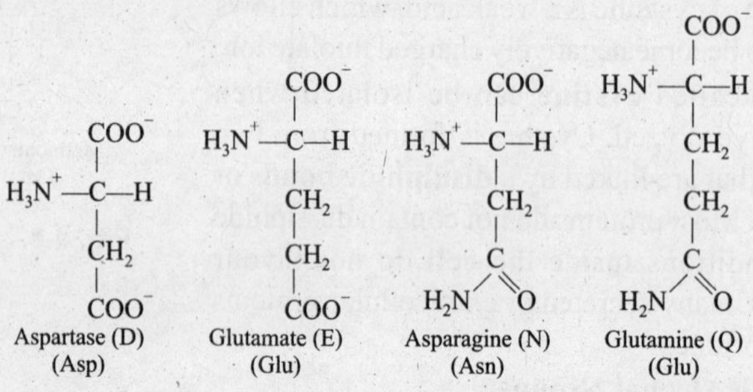

Fig. 4.11. Amino acids with acidic R groups and their amides.

Aspartate and glutamate confer negative charges on proteins because their side chains are ionized at pH 7. Aspartate and glutamate are sometimes called aspartic acid and glutamic acid, but under most physiological conditions they are found as the conjugate bases and, like other carboxylates, have the sufficate. Glutamate is probably familiar as its monosodium salt, monosodium glutamate (MSG), which is used in food as a flavor enhancer.

Asparagine (Asn, N) and **glutamine (Gln, Q)** are the *amides* of aspartic acid and glutamic acid, respectively. Although the side chains of asparagine and glutamine are uncharged, these amino acids are highly polar and often found on the surfaces of proteins, where they can interact with water molecules.

Hydropathy

Various side chains of amino acids range from highly *hydrophobic*, through weakly polar, to highly *hydrophilic*. The relative hydrophobicity or hydrophilicity of each amino acid is called its **hydropathy**. There are several ways of measuring hydropathy, but most of them rely on calculating the tendency of an amino acid to prefer a hydrophobic environment over a hydrophilic environment. Example of **highly hydrophobic** amino acids (hydropathy range: 3.1 to 1.1) are *five* amino acids such as isoleucine, phenylalanine, valine, leucine and methionine; **less hydrophobic** amino acids (Hydropathy range: 1.5 to –1.1) include *eight* amino acids such as tryptophan, alanine, glycine, cysteine, tyrosine, proline, threonine and serine. **Highly hydrophilic** amino acids (Hydropathy range: –1.7 to –7.5) include *seven* amino acids such as histidine, glutamate, asparagine, glutamine, aspartate, lysine and arginine.

Hydropathy is an important determinant of protein-chain folding because hydrophobic side chains tend to be clustered in the interior of a protein and hydrophilic residues are usually found on the surface.

Other Amino Acids (21st and 22nd Amino Acids)

More than 200 different amino acids are found in living organisms. In addition to the standard amino acids that are incorporated into proteins, all species contain a variety of L-amino acids that are either precursors of the common amino acids or intermediates in other biochemical pathways. Examples includes homocysteine, homoserine, ornithine and citrulline. S-Adenosylmethionine (SAM) is a common methyl donor in many chemical pathways. Many species of bacteria and fungi synthesize D-amino acids that are used in cell walls and in complex peptide antibiotics such as actinomycin D (*i.e.*, an inhibitor of DNA-transcription).

Several common amino acids are chemically modified to produce biologically important amines. These are synthesized by enzyme-catalyzed reactions that include decarboxylation and deamination. In the mammalian brain, for example, glutamate is converted to the neurotransmitter γ-amino butyrate (GABA). Mammals can also synthesize **histamine** from histidine. Histamine controls the constriction of certain blood vessels and also the secretion of hydrochloric acid by the stomach. In the adrenal medulla, tyrosine is metabolized to **epinephrine** or **adrenaline**. Epinephrine and its precursor, norepinephrine are hormones that help regulate metabolism in mammals. Tyrosine is also the precursor of the thyroid hormones such as **thyroxine** and **triiodothyronine**.

Some amino acids are chemically modified after they have been incorporated into polypeptides. For example, some proline residues in the protein collagen are oxidized to form **hydroxyproline** residues. Another common modification is the addition of complex carbohydrate chains, a process called **glycosylation**. Many proteins are **phosphorylated** by the addition of phosphoryl groups to the side chains of serine, threonine or tyrosine. The oxidation of pairs of cysteine residues to form cystine also occurs after a polypeptide has been synthesized. In bacteria, the first amino acid in a protein is usually methionine, which is modified by the addition of a formyl group to form *N*-formyl methionine.

An unexpected discovery was that a *21st amino acid*, **selenocysteine** (which contains selenium in place of the sulphur of cysteine), is incorporated into a few proteins in a wide variety of species. Selenocysteine is formed from *serine* during protein synthesis. The *22nd amino acid* is **pyrrolysine**, found in some species of archaebacteria. Pyrrolysine is a modified form of lysine that is synthesized before being added to a growing polypeptide chain by the translation machinery. Selenocysteine and pyrrolysine have their own codons, that is why they are considered additions to the standard reportoir of protein precursors.

5. Functions of Amino Acids

1. **Formation of proteins.** Amino acids are linked into long chains through their peptide linkage forming into polypeptides and proteins. Proteins form structural and functional components of the cell.
2. **Formation of other compounds.** Some amino acids give rise to other compounds, such as
   - (*i*) **Tyrosine** is part of hormone thyroxine and adrenaline.
   - (*ii*) **Glycine** participates in the synthesis of heme.
   - (*iii*) **Tryptophan** produces vitamin nicotinamide and plant hormone indole-3-acetic acid (IAA).
   - (*iv*) Amino acid **β-alanine** is needed in the synthesis of coenzyme-A and vitamin pantothenic acid.
3. **Formation of antibiotics.** Nonprotein amino acids are components of antibiotics.
4. **Formation of enkephalins.** These are small peptides (pentapeptides) formed in certain part of brain. These are associated with perception of pain and pleasure.
5. **Biological buffers.** Because of the amphoteric nature amino acids act as buffers in the body fluids. They maintain pH either by donating H^+ ions when pH increases or by accepting H^+ ions when pH falls.
6. **Synthesis of glucose.** When glucose level in blood comes down, glucose is synthesized from amino acids in the hepatic cells.
7. **Formation of histamine.** Amines in the body are derived from some amino acids by the loss of carboxyl group. For example histamine is derived from amino acid histidine. Histamine is an important vasodilator. It also causes constriction of bronchiole smooth muscles and stimulates gastric secretion.
8. **Prokaryotic cell wall.** Some peptides are components of **peptidoglycans** which form bacterial cell wall.

4.3 PEPTIDES

A peptide is a chain of amino acids linked by **peptide bond** or **amide bond**. The role of peptide bond in protein structure was understood by **Hofmeister** (1902). Peptide bond is a special type of amide bond formed between carboxyl (–COOH) group of one amino acid and amino (–NH_2) group of next amino acid with the removal of one molecule of water.

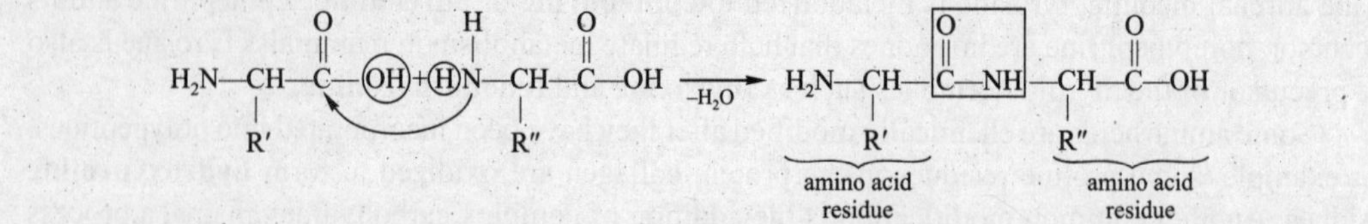

Fig. 4.12. Formation of a peptide bond or amide linkage to join two molecules of amino acids.

A peptide chain may have upto millions of amino acids. Each amino acid in the chain is called a **residue**. Depending on the number of amino acid residues, the peptides are called:

1. **Dipeptide.** A chain of two amino acids.
2. **Tripeptide.** A chain of three amino acids.
3. **Oligopeptide.** A chain of amino acid upto 10 amino acids.
4. **Polypeptide.** A peptide containing more than 10 amino acids.
5. **Macropeptide.** A polypeptide chain of more than 100 amino acids (Box 4.5).

Box 4.5

Strictly speaking, the proteins are polypeptides with more than 100 amino acids. The simplest peptide is glycylglicine. All naturally-occurring important peptides, however, possess a shorter individual name, such as tripeptide *glutathione* which consists of 3 amino acids—glutamic acid, cysteine and glycine. (Glutathione acts as sulphydryl buffer that maintains the cysteine residues of haemoglobin and other proteins of RBCs in the redued state. It also has a role in detoxifying H_2O_2 and organic peroxides, the harmful products of aerobic life.) The two pituitary hormones—ocytocin and vasopressin—are nonapeptides (having 9 amino acid residues). Peptides arise as partial hydrolysis products of proteins, *e.g.*, during digestion (**Jain** 2012).

1. Primary Structure of Peptides

The sequence of amino acids in a polypeptide chain is referred to as its **primary structure**. The two ends of peptide chains are marked, the **N-terminal** on the left and **C-terminal** at the right end of the polypeptide. The N-terminal has amino (–NH_2) group of left terminal amino acid and is basic, C-terminal has carboxyl (–COOH) group of the last amino acid and is acidic. These two terminal groups (one basic and one acidic) are ionisable.

The primary structure of a polypeptide is very important. The smallest change of even single amino acid in the primary structure often produces remarkable physiological effect. For example, substitution of a single amino acid in the sequence of 100 or more amino acids may change the biological activity having serious consequences.

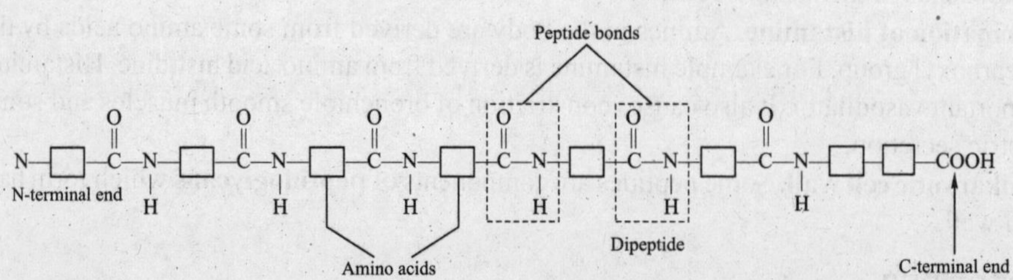

Fig. 4.13. A peptide chain showing peptide bonds between amino acids.

2. Biological Importance of Peptides

1. Peptides are intermediates in the formation of proteins.
2. Certain peptides serve as growth factors, *e.g.*, folic acid. **Streptogenins** is another peptide which serves as a growth factor for microorganisms.
3. Certain peptides act as horomes, *e.g.*, oxytocin (ocytocin), vasopressin, prolactin, etc.
4. Certain peptides such as **glutathione** participate in controlling oxidation-reduction potential.
5. Many peptides such as **penicilin G** have antibiotic activities.
6. Some peptides are constituents of alkaloids. For example, **ergotamine** is a peptide alkaloid from rye ergot (fungi).
7. Some peptides introduce psychiatric disturbances in human beings.

4.4 PROTEINS

I. Types of Proteins

Classification based on shape of proteins. According to the shape or conformation, two major types of proteins have been recognized:

(*a*) Fibrous proteins. Fibrous proteins are water-insoluble, thread-like proteins having greater length than their diameter. They contain secondary protein structure and occur in those cellular or extracellular structures, where strength, elasticity and rigidity are required, *e.g.*, collagen, elastin, keratin, fibrin (blood-clot proteins) and myosin (muscle contractile proteins).

(*b*) Globular proteins. Globular protein are water-soluble roughly spheroidal or ovoidal in shape. They readily go into colloidal suspension. They have tertiary protein structure and are usually functional proteins, *e.g.*, enzymes, hormones and immunoglobulins (antibodies). **Actin** of microfilaments and **tubulins** of microtubules are also globular proteins.

Classification based on solubility characteristics. According to this criterion proteins can be classified into two main types:

(*a*) Simple proteins. These proteins contain only amino acids in their molecules and they are of following types:

(*i*) Albumins. These are water soluble proteins found in all body cells and also in blood stream, *e.g.*, **lactalbumin**, found in milk and **serum albumin** found in blood.

(*ii*) Globulins. These are insoluble in water but are soluble in dilute salt solutions of strong acids and bases, *e.g.*, **lactoglobulin** found in milk and **ovoglobulin**.

(*iii*) Glutelins. These plant proteins are soluble in dilute acids and alkalis, *e.g.*, **glutenin** of wheat.

(*iv*) Prolamins. These plant proteins are soluble in 70 to 4 per cent alcohol, *e.g.*, gliadin of wheat and **zein** of corn.

(*v*) Scleroproteins. They are insoluble in all neutral solvents and in dilute alkalis and acids, *e.g.*, keratin and collagen.

(*vi*) Histones. These are water soluble proteins which are rich in basic amino acids such as arginine and lysine. In eukaryotes histones are associated with DNA of chromosomes to form nucleoproteins.

(*vii*) Protamines. These are water soluble, basic, light weight, **arginine** rich polypeptides. They are bound to DNA in spermatozoa of some fishes, *e.g.*, **salmine**, of salmon and **sturine** in sturgeons.

(*b*) Conjugated proteins. These proteins consist of simple proteins in combination with some non-protein components, called **prosthetic groups**. The prosthetic groups are permanently associated with the molecule, usually through covalent and/or non-covalent linkages with the side chains of certain amino acids. Conjugated proteins are of following types:

(*i*) Chromoproteins. Chromoproteins are a heterogeneous group of conjugated proteins which are in combination with a prosthetic group that is a pigment, *e.g.*, respiratory pigments such as **haemoglobin, myoglobin** and **haemocyanin**; **catalase, cytochromes, haemerythrins**; visual purple or **rhodopsin** of rods of retina of eye and yellow enzymes or **flavoproteins**.

(*ii*) **Glycoproteins.** Glycoproteins are proteins that contain various amounts (1 to 85 per cent) of carbohydrates. Of the known 100 monosaccharides, only nine are found to occur as regular constituents of glycoproteins (*e.g.*, glucose, galactose, mannose, fucose, acetylglucosamine, acetylgalactosamine, acetylneuraminic acid, arabinose and xylose). Glycoproteins are of two main types: **1. Intracellular glycoproteins** which are present in cell membranes and have an important role in membrane interaction and recognition. They also serve as antigenic determinants and receptor sites. **2. Secretory glycoproteins** are **plasma glycoproteins** secreted by the liver; **thyroglobulin**, secreted by the thyroid gland; **immunoglobulins** secreted by the plasma cells; **ovoalbumins** secreted by the cells of oviduct of hen; **ribonucleases** and **deoxyribonucleases. Mucus** and **synovial fluid** are also glycoproteins with lubricative properties.

(*iii*) **Lipoproteins.** Lipid containing proteins are called **lipoproteins**. Their lipid contents are 40 to 90 per cent of their molecular weight and this tends to affect the density of the molecule. There are four types of lipoproteins: **1. High density lipoproteins (HDL)** or α-lipoproteins; **2. Low density lipoproteins (LDL)** or β-lipoprotiens; **3. Very low density lipoproteins (VLDL)** or pre-β-lipoproteins; and **4. Chylomicrons.** Lipoproteins include some of the blood plasma proteins, various types of membrane proteins, lipovitellin of egg yolk and proteins of brain and nerve tissue.

(*iv*) **Nucleoproteins.** Nucleoproteins are proteins in combination with nucleic acids (DNA and RNA). However, these proteins are not true conjugated proteins since the nucleic acid involved cannot be regarded as prosthetic groups. Nucleoproteins are of two types: **1. Histones** which are quite similar in all plants and animals. Their highly basic nature accounts for the close associations histones form with the nucleic acids. Histones are involved in the tight packing of DNA molecules during the condensation of chromatin into chromosomes for the mitosis. **2. Nonhistones** have great heterogeneous amino acid composition and are acidic in nature. They have selective combination with certain stretches of nuclear DNA and, thus, are involved in the regulation of gene expression.

(*v*) **Metalloproteins.** Metalloproteins are proteins conjugated to metal ions which are not part of the prosthetic group, *e.g.*, **carbonic anhydrase** enzyme contains zinc ions and amino acids in its molecule; **caeruloplasmin**, an oxidase enzyme containing copper; and **siderophilin** contains iron.

(*vi*) **Phosphoproteins.** Phosphoproteins are proteins in combination with a phosphate group, e.g., casein of the milk and ovovitellin of eggs.

II. Formation of Proteins

Because a molecule of the amino acid contains both basic or amino ($-NH_2$) and acidic or carboxyl (–COOH) group, it can behave as an acid and base at a time. The molecules of such organic compounds which contain both acidic and basic properties are known as **amphoteric molecules**. Due to amphoteric molecules, the amino acids unite with one another to form complex and large protein molecules. When two molecules of amino acids are combined then the basic group ($-NH_2$) of one amino acid molecule combines with the carboxylic (–COOH) group of other amino acid and the loss of a water molecule takes place. This sort of condensation of two amino acid molecules by –NH–CO linkage or bond is known as **peptide linkage** or **peptide bond**. A combination of two amino acids by the peptide bond is known as **dipeptide** (Fig. 4.14). When three amino acids are united by two peptide bonds, they form **tripeptide**. Likewise, by condensation of few or many amino acids by the peptide bonds the **oligopeptides** and **polypeptides** are formed respectively (see Section 4.3). The various molecules of polypeptides unite to form the **peptones, proteases** and **proteins**. Thus, protein macromolecules are the polymers of many amino acid monomers. The size (molecular weight), shape, and function of proteins are determined by the number, type and distribution of the amino acids present in the molecule. Proteins occur in a wide spectrum of molecular sizes from small molecules such as the hormone ACTH (or adrenocorticotrophic hormone) which consists of only 39 amino acids and has a molecular weight of 4500, to extremely large proteins such as haemocyanin (an invertebrate blood pigment) which consists of 8200 amino acids and has a molecular weight greater than 900,000.

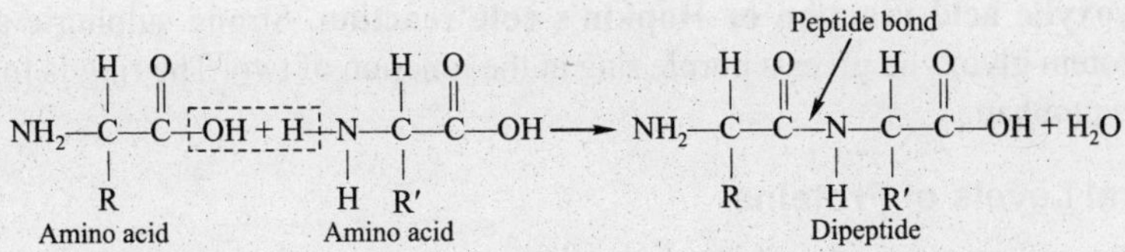

Fig. 4.14. A chemical reaction showing formation of a dipeptide.

III. General Properties of Proteins

Proteins exhibit following general properties:

1. Physical properties

***(i)* Colloid.** Proteins exist in colloidal state. Their molecules are of large size. Therefore, these are unable to diffuse through plasma membrane.

***(ii)* Solubility.** Because proteins are colloids of large-sized molecules, these form turbid solution in water. These are insoluble in alcohol. These are precipitated by acids in certain concentration.

2. Chemical properties

***(i)* Amphoteric nature.** Proteins are **amphoteric**. These behave as acid to alkaline solutions and alkaline to acidic solutions and form salts with them.

***(ii)* Coagulation.** On heating proteins are coagulated but temperature for coagulation differs for different proteins.

***(iii)* Optical properties.** It is already mentioned that amino acids are optically active except glycine. These are mostly **laevorotatory**. Some of the proteins, haemoglobin and nucleoproteins are dextrorotory.

***(iv)* Hydrolysis.** When proteins are boiled with dilute mineral acids in a reflux condenser, the protein molecules gradually break up into simple ones until these are reduced to amino acids. The process of hydrolysis of proteins involves the following steps:

$$\text{Proteins} \rightarrow \text{Proteoses} \rightarrow \text{Peptones} \rightarrow \text{Amino acids}$$

***(v)* Denaturation.** Denaturation of protein is the disruption of secondary, tertiary and quaternary structure of the functional protein resulting in the changes of its physical, chemical and biological characteristics. Various physical and chemical agents such as heat, UV light, ultrasonic waves, high pressure, acids, alkalies, urea, guanidine, ethanol and detergents can cause denaturation of proteins.

The denatured protein becomes insoluble and precipitate out. Their digestability by proteolytic enzymes increases, but their enzymatic or or hormonal activity is destroyed and antigenic antibody functions are altered.

***(vi)* Colour reactions**

***(a)* Biuret reaction.** All proteins give **biuret test** with caustic soda and a drop of dilute copper sulphate giving pinkish violet colour.

***(b)* Millon's reagent.** Mercury dissolved in strong nitric acid gives a white precipitate, which turns red on heating. The reaction occurs on account of amino acid tyrosine.

***(c)* Rosenheim's reaction.** Add a drop of formaldehyde solution to one volume of protein and shake thoroughly. Dilut with one volume of water. When concentrated sulphuric acid, is added slowly a beautiful purple colour appears at the junction. If a drop of ferric chloride solution is added to it the colour is accentuated.

***(d)* Xanthoproteic reaction.** Strong nitric acid (HNO_3) gives a white precipitate which turns yellow. The yellow colour changes to orange on addition of alkali. This is due to the presence of phenyl radical.

***(e)* Diazo reaction.** Proteins give a red colour on addition of diazobenzene sulphonic acid in a mild alkaline medium. The colour is due to the presence of histidine or tyrosine.

***(f)* Glyoxylic acid reaction or Hopkin's cole reaction.** Strong sulphuric acid added to a mixture of protein glyoxylic gives a purple ring at the junction of two. The ring is formed due to the reaction of tryptophan.

IV. Structural Levels of Proteins

The protein as synthesized on the ribosome is a linear sequence of amino acids, polymerized by the elimination of water between successive amino acids to from the peptide bond, and existing as a randomly coiled chain without specific shape and possessing no biological (*i.e.*, catalytic) acitivity. Within seconds of synthesis being completed, the protein folds into a specific three-dimensional form, which is the same for all molecules of the same type of protein and which now is capable of doing catalysis. According to their mode of folding the following four levels of protein organization have been recognized (Fig. 4.15):

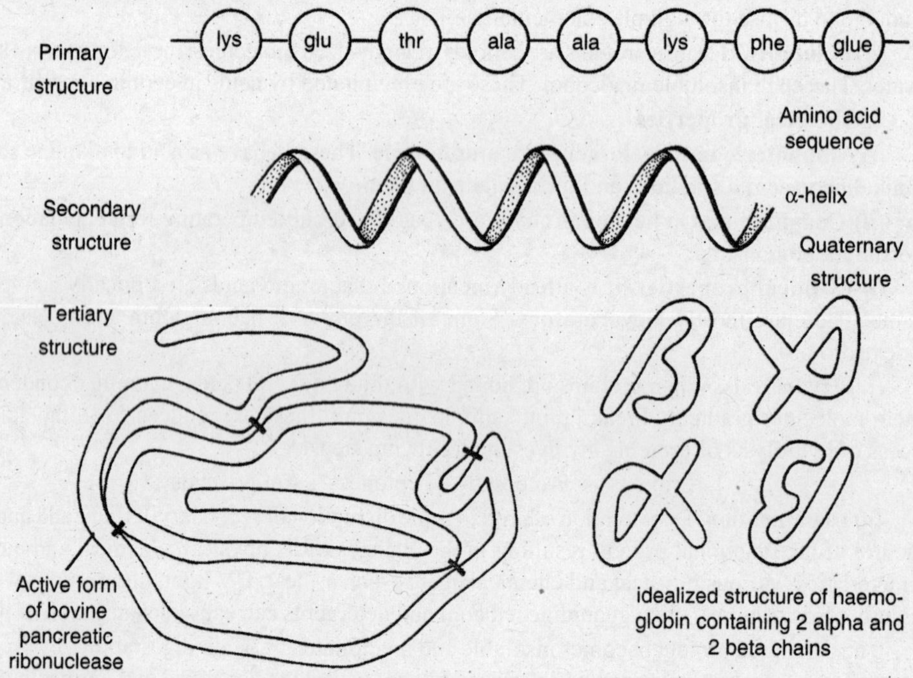

Fig. 4.15. The four levels of structural organization in protein molecules.

(*a*) **Primary structure of protein,** *e.g.*, insulin.
(*b*) **Secondary structure of protein,** *e.g.*, α-Keratin, collagen, globular proteins.
(*c*) **Tertiary structure of protein,** *e.g.*, myoglobin.
(*d*) **Quaternary structure of protein,** *e.g.*, haemoglobin.

(*a*) Primary protein structure. The primary protein structure is defined as the particular sequence of amino acids found in the protein. It is determined by the covalent peptide bondings between amino acids. Primary structure also includes other covalent linkages in proteins, for example the linkages that may exist between sulphur atoms of cysteine amino acids located in the chain of the protein insulin. The first protein to have its primary structure determined was of insulin, the pancreatic hormone that regulates glucose metabolism in mammals. Insulin has a molecular weight of 5,40 daltons and contains 51 amino acids. Insulin consists of two polypeptide chains of 21 and 30 amino acid residues, called the A and B chains, respectively (Fig. 4.16). (An amino acid residue is that which is left when the elements of water are split out during polymerization).

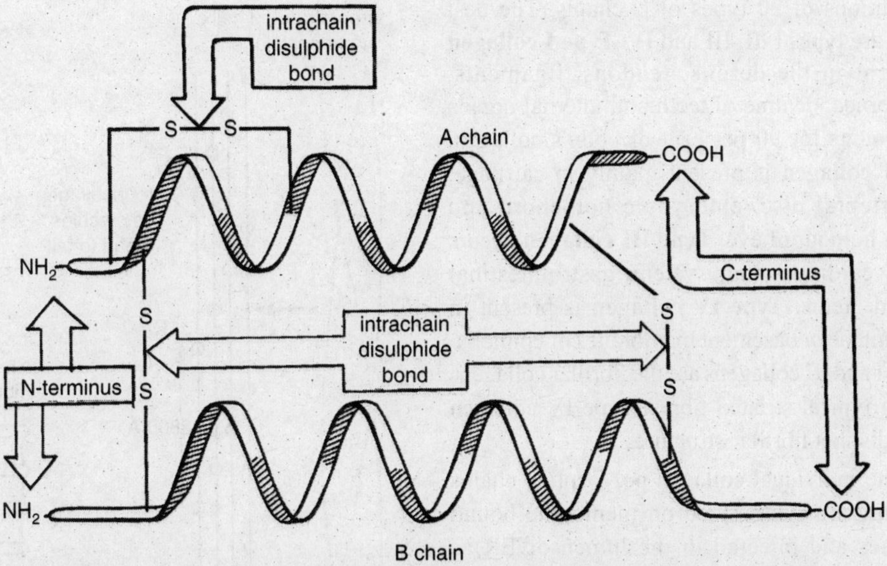

Fig. 4.16. Molecular structure of insulin (After Sheeler and Bianchi, 1987).

Since the elucidation of the primary structure of insulin in 1953 by **F. Sanger** (for which Sanger received a Nobel Prize), several hundred proteins have been fully sequenced. Among the fully sequenced proteins are ribonuclease and nearly 100 types of haemoglobin. For example; **Stein** and his coworkers established the amino acid sequence (*i.e.*, primary structure) of the enzyme **ribonuclease**. This enzyme is produced by the pancreas and secreted into the small intestine where it catalyzes the hydrolytic digestion of polyribonucleotide chains (RNA). The ribonuclease consists of a single 124 amino acid polypeptide chain having a molecular weight of about 12,000.

(*b*) Secondary protein structure. Secondary structure of the protein is any regular repeating organization of the polypeptide chain. There are three types of secondary protein structure:

1. Helical structure (*e.g.*, α-keratin and collagen);

2. Pleated sheet structure or β-structure (*e.g.*, fibroin of silk); and

3. Extended configuration (*e.g.*, stretched keratin). Most fibrous proteins have secondary structure. In globular protein, too, it is not uncommon for half of all the residues of each polypeptide to be organized into one or more specific secondary structures.

Collagen. The collagens (the source of leather, gelatin, glue, etc.) are a family of highly characteristic fibrous proteins found in all multicellular animals (*e.g.*, in connective tissues). They are secreted by the fibroblasts constituting most abundant (up to 25 per cent of total body's proteins) proteins of mammals. The characteristic feature of collagen (or **tropocollagen**) molecules is their stiff, triple-stranded helical structure (which was discovered by **Rich**, **Crick** and **Ramachandran**). Three collagen polypeptide chains are left-handed α-helices or alpha chains, each is about 1000 amino acid residues long. These chains are wound around one another in a regular superhelix to generate a rope-like collagen or tropocollagen molecule which is about 300 nm long and 1.5 nm is diameter (Fig. 4.17).

Collagens are exceptionally rich in **proline** (and **hydroxyproline**; both accounting for more than 20 percent of collagen's amino acids) and **glycine**. Other dominant amino acids of collagens are **lysine** and **alanine**.

So far, about 20 distinct collagen-chains have been identified, each encoded by a separate gene. About 10 types of triplet-stranded collagen molecules have been found to assemble from various combinations of 20 types of α-chains. The best defined are types I, II, III and IV. **Type I collagen** is present in the dermis, tendons, ligaments, bone, cornea, dentine of teeth and internal organs and accounts for 90 per cent of body's collagen. **Type II collagen** is present mainly in cartilage, intervertebral disc, embryonic notochord and vitreous humour of eye. **Type III collagen** occurs in skin, cardiovascular system, gastrointestinal tract and uterus. **Type IV collagen** is present in basal laminae or basement membranes of epithelia. Type I, II and III collagens are the fibrillar collagens showing typical striated fibres. Type IV collagen lacks a distinct fibrillar structure.

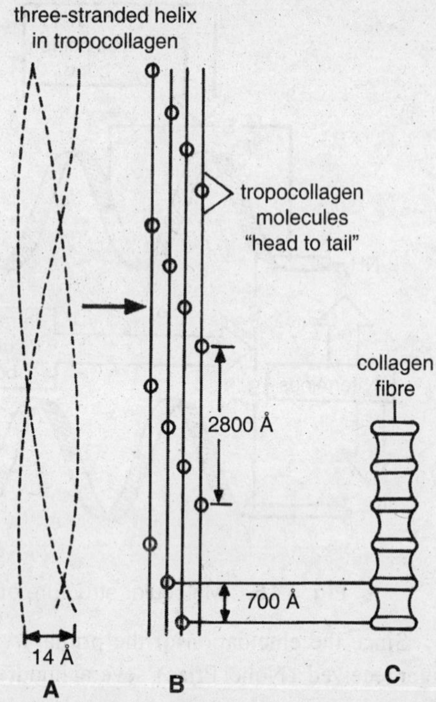

Fig. 4.17. A—A tropocollagen molecule (collagen or superhelix) with three intertwined left-handed α-helices; B—Staggered arrangement of tropocollagen molecules (super helices) in a collagen fibril; C—Striated appearance of a collagen fibril under the electron microscope (after Dyson, 1978).

The individual collagen polypeptide chains (α-chains) are synthesized on membrane bound ribosomes and injected in the lumen of ER as larger precursors, called pro-α-chains. These precursors have distinct polarity, containing terminal propeptides at their *N*-and *C*-terminus. Each pro-α-chain then combines with two others to form a hydrogen-bonded, triple stranded helical molecule, called **procollagen**. Such a procollagen of fibrillar collagens (I, II, and III) is serceted from the fibroblast in the extracellular space. Due to enzyme action its terminal propeptides are removed and the procollagen is converted into a **tropocollagen molecule**. Many tropocollagen molecules spontaneously assemble into the ordered arrays, called **collagen fibrils**. The collagen fibrils are thin (10 to 300 nm in diameter), cable-like structures, many micrometres long, exhibiting cross-striations every 67 nm and are clearly visible in the electron micrographs. The collagen fibrils often aggregate into larger bundles which can be seen in light microscope as **collagen fibres**. Type IV collagen molecules assemble to form a sheet-like **meshwork** that constitutes a major part of all basal laminae (**Martin** *et al.*, 1985, **Burgeson**, 1988).

(*c*) Tertiary protein structure. Tertiary protein structure refers to a more compact structure in which the helical and non-helical regions of a polypeptide chain are folded back on themselves. This structure is typical of globular protein structure, in which it is the non-helical region that permits the folding. The folding of a polypeptide chain is not random but occurs in a specific fashion, thereby imparting certain steric (*i.e.*, three-dimensional) properties to the protein. For example, in enzymes folding brings together **active amino acids**, which are otherwise scattered along the chain, and may form a distinctive cavity or cleft in which the substrate is bound.

The complete tertiary structure of a protein can only be deducted by a laborious analysis of X-ray scattering patterns from crystals. The first protein to have its secondary and tertiary structure determined was **myoglobin**, a 153-amino acid, oxygen-binding protein found primarily in red muscle and largely responsible for the colour of that tissue. The work was done at Cambridge under the direction of **J.C. Kendrew** (1961). Although at some points the polypeptide chain does have secondary structure (alpha-helical structure), the chain is mainly characterized by seemingly random loops and folds.

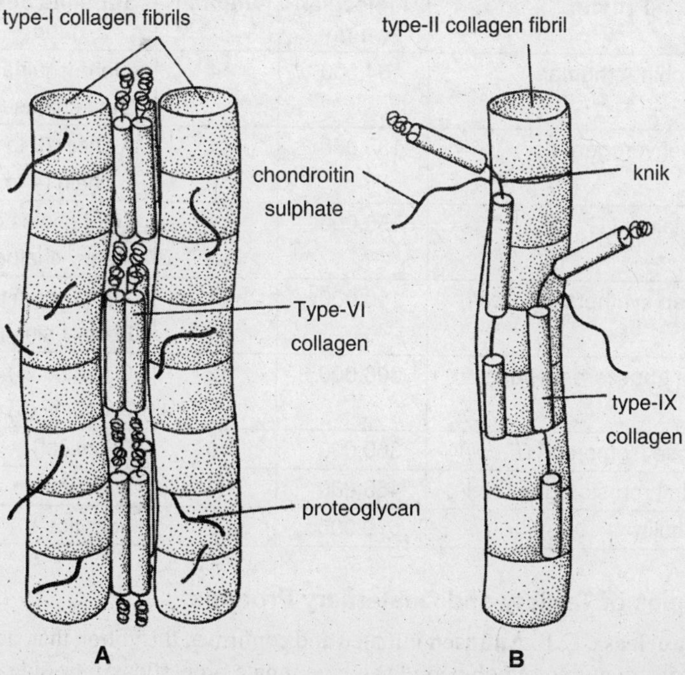

Fig. 4.18. Interactions of fibrous collagens with nonfibrous collagens. A—In tendons, type I collagen fibrils are associated with type-VI collagen. B—In cartilage, type II collagen fibrils are associated with type IX collagen (after Lodish *et al.*, 2004).

In a tertiary protein the polypeptide chain is held in position by weak secondary bonds which are of different types such as **ionic bonds** (or electrostatic bonds or salt or salt bridges); **hydrogen bonds**; **hydrophobic bonds** and **disulphide bonds**.

***(d)* Quaternary protein structure.** In proteins that are composed of two or more polypeptide chains, the quaternary structure refers to the specific orientation of these chains with respect to one another and the nature of the interactions that stabilize this orientation. The individual polypeptide chains of the protein are called **sub-units** and the active protein itself is called **multimer**. While multimeric proteins containing up to 32 subunits have been described, the most common multimers are **dimers**, **trimers**, **tetramers**, **pentamers** (*e.g.*, RNA polymerase) and **decamers** (*e.g.*, DNA polymerase III) (Table 4.4). If the protein consists of identical sub-units, it is called **homopolymers** and is said to have **homogeneous quaternary structure**, *e.g.*, the isozymes H_4 and M_4 of lactic dehydrogenase (LDH), enzyme phosphorylase and L-arabinose isomerase. The enzyme β-galactosidase, consists of four identical polypeptide chains. Lastly, when the sub-units of the protein are different, the protein is called **heteropolymer** and is said to have a **heterogeneous quaternary structure**, *e.g.*, haemoglobin and immunoglobulins. Quaternary proteins are usually joined by hydrophobic forces. Hydrogen bonds, ionic bonds and possibly disulphide bonds may also participate in forming quaternary structures.

Table 4.4 Subunits and molecular weight of some multimer proteins (*Source:* **Sheeler** and **Bianchi**, 1987).

	Protein	Molecular weight	Number	Subunits designation	Molecular weight
1.	Haemoglobin A (human)	64,500	4	Alpha chains (2)	15,700
				Beta chains (2)	16,500
2.	Lactate dehydrogenase	135,000	4	A chain (0 to 4)	33,600
				B chain (4 to 0)	33,600
3.	Immunoglobulin G	150,000	4	Light chains (2)	25,000
				Heavy chains (2)	50,000
4.	Tryptophan synthetase (*E.coli*)	150,000	4	Alpha chains (2)	29,500
				Beta chains (2)	45,000
5.	Aspartate transcarbamylase	306,000	12	C chains (6)	34,000
				R chains (6)	17,000
6.	L-arabinose isomerase (*E.coli*)	360,000	6	(identical)	60,000
7.	Apoferritin (iron storage protein)	456,000	24	(identical)	19,000
8.	Thyroglobulin	670,000	2	(identical)	3,35,000

Some Examples of Tertiary and Quaternary Proteins

***(i)* Ribonuclease. C.B. Anfinsen** initiated and confirmed the notion that, acting in concert, the specific primary structure of a polypeptide and the innate properties of the side chains of its amino acids cause the polypeptide to spontaneously assume its biologically active tertiary structure. In 1972, he got the Nobel Prize for this definitive work. **Anfinsen** identified four disulphide bridges in the ribonuclease protein, suggesting that the enzyme is highly folded. As is the case with almost all enzymes, the catalytic activity of ribonuclease depends on the maintenance of a particular three-dimensional shape. In concentrated solutions of β-mercaptoethanol and urea, the disulphide bridges of the enzyme are broken and the resultant unfolding of the polypeptide chain is accompanied by a loss of enzyme activity. The enzyme is said to be **denatured**. If the β-mercaptoethanol and urea are removed by dialysis and the denatured ribonuclease reacted with oxygen, the four disulphide bridges reform spontaneously, and essentially all the catalytic activity of the protein is restored. Similar observations have been made with other proteins, that is, they are capable of spontaneously reestablishing their biologically active tertiary (or even quaternary, *e.g.*, haemoglobin) structure after having undergone extensive molecular disorganization.

***(ii)* Haemoglobin.** Haemoglobin is one of the fully sequenced protein. Our present understanding of the structure and function of haemoglobin is the outcome of 50 years of research of **M.F. Perutz**. He got the Nobel Prize in 1962, along with **J.C. Kendrew**, for their studies of haemoglobin and myoglobin.

The haemoglobin is a conjugated globular protein, that is, it contains some non-protein part. In all but the lowest vertebrates, haemoglobin is a tetramer (a heteropolymer). In lampreys, however, haemoglobin is monomeric, that is, it contains a single globin chain like the myoglobin. In humans, most common type of haemoglobin is haemoglobin A (HbA), which consists of 574 amino acid residues and has a molecular weight of 64,500. Its secondary, tertiary and quaternary structure is typical of all higher vertebrate haemoglobins. The protein portion of the haemoglobin molecule, called **globin**, is composed, of four polypeptide chains, each of which is also globular in shape. The four globin chains consist of, two identical pairs: two **alpha chains** (141 amino acids each) and two **beta chains** (146 amino acids each). The non-protein portion of haemoglobin consists of four iron-containing haem groups, one associated with each of the four globin chains. Nineteen of the twenty biologically important amino acids are included in the globin of haemoglobin.

Haemoglobin molecule is highly symmetric; it can be divided into two identical halves, each consisting of an αβ-dimer. The complete tetramer is similar to a mildly flattened sphere having a maximum diameter of about 5.5 nm. The four polypeptide chains are arranged in such a way that unlike chains have numerous stabilizing interactions, whereas, like chains have few. A cavity about 2.5 nm long and varying in width from about 5 to 10 Å passes through the molecule along the axis. Each globin chain envelops its haem group in a deep cleft.

***(iii)* Immunoglobulins.** The ability of animals to resist infection by pathogens (viruses, bacteria and other unicellular parasites) and by multicellular endoparasites, is called **immunity**. Specific immune response function by recognizing particular chemical structures, known as **antigens**—on the surface of invading cells. An antigen can be a protein, lipid, carbohydrate or any other molecule. These antigens interact with protein molecules produced by the host, the **immunoglobulins**, which bind the antigen in much the same way as an enzyme binds its substrate. Specific immune responses involve many different types of cells. One type, the **B-lymphocyte** or **B-cell**, is capable of producing **free immunoglobulins**, called **antibodies**.

An immunoglobulin (Ig) molecule is a Y-shaped heteropolymer and is composed of two identical **H (heavy)** polypeptide chains and two smaller identical **L (Light)** polypeptide chains (Fig. 4.19). Heavy chains contain antigenic determinants in the "tail" (carboxyl) segments by which they can be classified as IgG, IgM, IgA, IgD or IgE. Light chains can likewise be typed as kappa or lambda. Within a H chain class or L chain type, these segments exhibit very little variation in primary structure from one individual to another and are called **constant regions (C)**. The amino ($-NH_2$) ends, however, are extremely diverse in primary structure, even within a class and are called **variable (V) regions**. The V_H and V_L regions together form two **antibody-combining sites** (called **antigen-binding sites**) for specific interactions with homologous antigen molecules. The C_H region consists of three or four similar segments, presumably derived evolutionarily by duplication of an ancestral gene and subsequent modification by mutations; the similar segments are called **domains** and are labelled CH_1, CH_2, CH_3, etc. A mature lymphocyte (plasma cell) produces antibodies with a single class H chain and a single type of L chain, hence, also a single antigen-binding specificity. The first antibodies produced by a developing plasma cells are usually of class IgM.

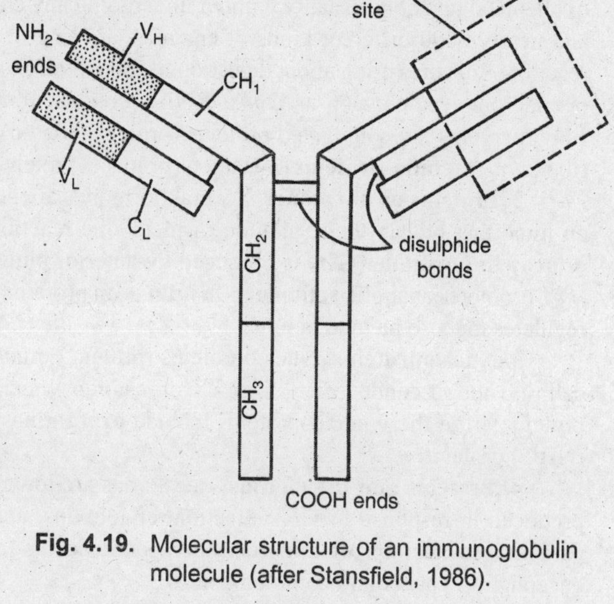

Fig. 4.19. Molecular structure of an immunoglobulin molecule (after Stansfield, 1986).

V. Chemical Bonds in Protein Structure

In a protein molecule amino acid residues are joined to each other by bonds forming chain. Amino acids of different chains of a protein are also joined together by different types of bonds or forces. These chemical bonds are of the following types:

1. Primary bonds. These are covalent **peptide bonds** between different amino acids of polypeptide chain. Peptide bonds are specialized amide bonds or amide linkages between $-COOH$ (carboxyl) and $-NH_2$ (amino) groups. These bonds form the backbone of peptide chain.

2. Secondary bonds. These bonds grant natural configuration to a polypeptide chain and help in maintaining the secondary structure of the proteins. These are of following types:

(*i*) Disulphide bonds (–S–S– bonds). These are covalent bonds which give some rigidity to the protein molecule. They are relatively more stable than the other types of bonds. A disulphide bond is formed between two *cysteine* residues having sulphydryl (–SH) groups. The enzyme **ribonuclease** consists of a single polypeptide chain of 124 residues and contains four disulphide bonds. **Insulin** and **oxytocin** are two protein hormones which also have disulphide bonds. The disulphide bonds are oxidised by **performic acid**, which is therefore used to denature proteins or to separate polypeptide chains held together by –S–S– bonds.

(*ii*) Hydrogen bonds (> CO–HN < bonds). They are weak bonds but since they are numerous they give considerable stability to the protein molecule. Hydrogen bonds can be formed by sharing *hydrogen* between *amide* (NH) *nitrogen* and *carbonyl* (*C=O*) *oxygen* of the peptide backbone. They are also formed between groups present in the side chain. The side chains of **aspartic** and **glutamic** acids, **serine** and **threonine** can all form hydrogen bonds.

Silk fibroin and wool's keratin have such hydrogen bonds.

(*iii*) Ionic, electrostatic bonds (or salt linkage/bridges). These bonds are formed when an acid and a basic amino acid are ionized and lie close together. Ionic bonds play an important role in binding a basic protein with an acidic macromolecule, *e.g.*, in the formation of nucleoproteins.

(*iv*) Nonpolar or hydrophibic bonds. These bonds are formed because of the tendency of the nonpolar side chains of neutral amino acids to closely associate with one another. When a protein molecule is submerged in an aqueous medium there is a tendency to expose a maximum number of its polar groups to the surrounding medium. Conversely a maximum number of nonpolar groups are directed internally. Although hydrophobic interactions do not form true bonds they are of great importance in the formation of tertiary structure.

VI. Biological Importance of Proteins

Proteins form a large part of the structure of cells and are present in all tissues. Many proteins also have special physiological functions:

1. Proteins as enzymes. Enzymes are special proteins produced within an organism which are capable of catalyzing specific catalytic reactions. They are **biocatalysts** which influence the rate of chemical reactions, usually without undergoing any change themselves. All enzymes are proteins and nearly 2000 different kinds of enzymes are known, each one catalyzing a different chemical reaction. Enzymes bring about digestion of food, hydrolysis of ATP, biosynthesis of cell components or macromolecules such as DNA, RNA, proteins, lipids, polysaccharides (carbohydrates), etc. **Flavoproteins** are conjugated proteins (protein + flavin); they are enzymes containing the vitamin riboflavin. **Metallo-proteins** (metal ions or ions + protein) are enzymes containing mineral elements.

2. Proteins as hormones. Several hormones are peptides and proteins (Table 4.5). They play an important role in the regulation of metabolic reactions. For example, **somatotropic hormone** or **growth hormone (GH)** is produced by anterior pituitary and regulates general somatic growth (*i.e.*, promotes anabolic activities). **Insulin** is an protein hormone which is secreted by pancreas and regulates glucose metabolism.

3. Nucleoproteins. Nucleoproteins (protein + nucleic acid) form chromatin material. During cell division, it condenses to form **chromosomes** which are carriers of hereditary characters. The protein part of these nucleoproteins include **protamines**, **histones** and **non-histone chromosomal (NHC)** proteins.

4. Proteins and interferons. Interferons are low molecular weight, regulatory glycoproteins produced in response to viral infection, endotoxins, antigenic stimuli, rickettesias and protozoan parasites. Interferons are important therapeutic agents for treating viral diseases, such as hepatitis, encephalitis, cancer and common cold.

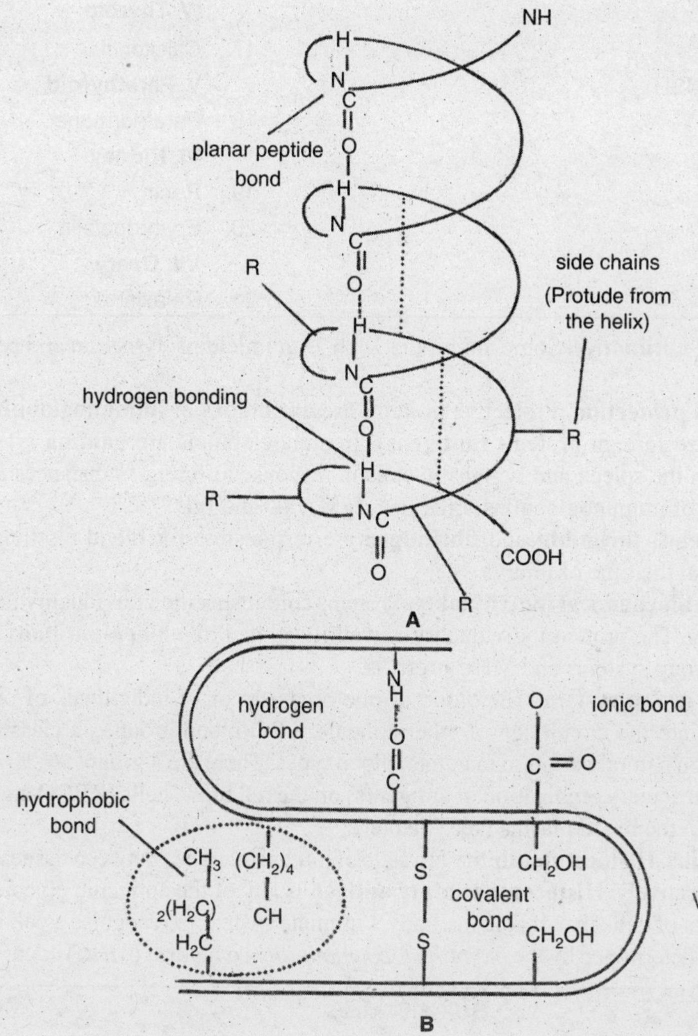

Fig. 4.20. Types of chemical bonds in protein structure: A—showing hydrogen bond and peptide bond; B—showing hydrophobic bond, ionic bond, covalent bond and Van der Waal's interactions. (**Note.** *Van der Waal's interaction*. Attractive forces that exist between atoms or molecules of all substances. Van der Waal's interactions are weak, non-specific, interatomic interactions that come into play when any two uncharged atoms are 3 to 4 A° apart. Since Van der Waals interactions are very weak, it is customary to call them as a force rather than a bond, **Jain** 2010).

Table 4.5 Peptide and protein hormones.

I. Pituitary	III. Alimentary canal
1. Somatotropin (growth hormone, GH)	12. Gastrin
2. Adrenocorticotropic hormone (ACTH)	13. Enterogasterone
3. Thyrotropic (thyroid stimulating hormone, TSH)	14. Cholecystokinin
4. Follicle stimulating hormone (FSH)	15. Secretin
5. Luteinizing hormone (LH)	16. Pancreozymin
6. Prolactin	

7. Oxytocin	**IV. Thyroid**
8. Vasopressin	17. Calcitonin
9. Melanotropin (MSH)	**V. Parathyroid**
II. Pancreas	18. Parathormone
10. Insulin	**VI. Kidney**
11. Glucagon	19. Renin
	20. Erythropoietin
	VII. Ovary
	21. Relaxin

5. Proteins and antibiotics. Some antibiotics such as gramicidins, tyrocidin and penicillin G are peptides.

6. Proteins and protection. Protective proteins are **antibodies** or **immunoglobulins.** These combine and neutralize foreign proteins (**antigens**). Immunoglobulins are **gamma (γ) globulins** which are produced in the spleen and lymphatic cells in response to foreign substances (antigens). There are five classes of immunoglobulins, IgG, IgA, IgM, IgD and IgE.

The blood proteins, **thrombin** and **fibrinogen** are responsible for blood clotting and thus prevent loss of blood at the time of injury.

7. Proteins and biochemical individuality. Proteins confer specific individuality to the living organisms and species. The protein individuality is well illustrated by **anigen-antibody reactions** such as ABO blood group systems and MHC proteins.

(*i*) ABO blood group system. The blood of one organism or of individuals of one species cannot be transfused into the circulation of other animals. ABO blood group is a classification of blood which is based on natural variation in human blood types. There are 4 groups: A, B, AB and O, each classified by a particular combination of antigen(s) on the red blood cells (RBCs) and naturally occurring antibodies in the blood plasma (see Chapter).

(*ii*) MHC proteins. Histocompatibility is a state of mutual tolerance between tissues that allow them to be grafted effectively. **Histocompatibility antigen** is any of the antigenic glycoproteins on the surface membranes of cells that enable the body's immune system to recognise a cell as a native or foreign and that are determined by the *major histocompatibility complexes* (*MHC*) *genes* (Box 4.6).

Box 4.6 MHC

During the first part of 20th century, clinical researches discovered that blood could be transfused from one person to another, as long as two individuals were compatiable for the ABO blood group system. The success of blood transfusion led to the proposal that skin might also be grafted between individuals. This idea was tested during World War II when skin grafts were attempted on piolets and other military personnel who had received serious burns. The grafts were rapidly and completely rejected. After the war, researchers set out to determine the basis for tissue rejection. It was discovered that skin could be grafted successfully between mice of same inbred strain, but that grafts between mice of different strains were rapidly rejected. Mice of the same inbred strain are like identical twins; they are *genetically indentical.* Subsequent studies revealed that the genes that governed this graft rejection were clustered in a region of the genome that was named **major histocompatibility complex (MHC).** Approximately 20 different MHC genes have been characterized, most of which are highly polymorphic: over 2000 different alleles of MHC genes have been identified, for more than any other loci in the human genome. *So, it is very unlikely that two individuals in a population have the same combination of MHC alleles.* This is the reason that transplanted organs are so likely to be rejected and why transplant patients are given drugs, such as **cyclosporin A**, to suppress the immune system following surgery. Cyclosporin A is a cyclic peptide produced by a soil fungus. Cyclosporin A inhibits a particular phosphatase enzyme in the signaling pathway leading to the production of cytokines required for T-cell activation (**Karp** 2010).

MHC proteins can be subdivided into two major groups: **MHC class I** and **MHC class II** molecules. MHC class I molecules consist of one polypeptide chain encoded by an MHC allele (known as heavy chain) associated noncovalently with a non-MHC polypeptide known as **β_2-microglobulin.** MHC class II molecules also consist of heterodimers but both subunits are encoded by MHC alleles. Both classes of MHC molecules as well as β_2-microglobulin contain Ig-like domains and are thus members of immunoglobulin superfamily (see **Karp** 2010).

8. Blood plasma proteins. Electrophoresis at pH 8.6 in barbital buffer separates six major blood's plasma components: **albumin**, **α_1-globulin**, **α_2-globulin**, **β-globulin**, **γ-globulin** and **fibrinogen**. Starch gel electrophoresis breaks up the bands into 20 zones containing 50 plasma proteins.

9. Proteins and transport. Some proteins bind and transport specific types of molecules via blood. For example

***(i)* Albumin** (MW 76,000) is synthesized in the liver and is the most abundant plasma protein. It maintains the colloid osmotic pressure of plasma and takes part in the transport of a variety of materials. It carries substances such as bilrubin and some drugs which are otherwise insoluble in plasma.

Serum albumin binds free fatty acid molecules and transport them between adipose tissue and other cells of the body.

***(ii)* Lipoproteins** (*e.g.*, HDL, LDL and VLDL) of blood plasma transport lipids (fats) between the intestine, liver and adipose tissue.

***(iii)* Haemoglobin** is a conjugated protein consisting of globin plus haeme (an Fe porphyrin prosthetic group). It consists of 4 polypeptide chains, 2 alpha chains and 2 beta chains. Each chain has a molecular weight of 16,750 and is attached to one atom of iron.

Haemoglobin in vertebrates and **haemocyanin** in some invertebrates transport oxygen from lungs to tissue cells (oxygen carrier of blood).

***(iv)* Myoglobin** (Fig. 4.21B) stores oxygen in muscles.

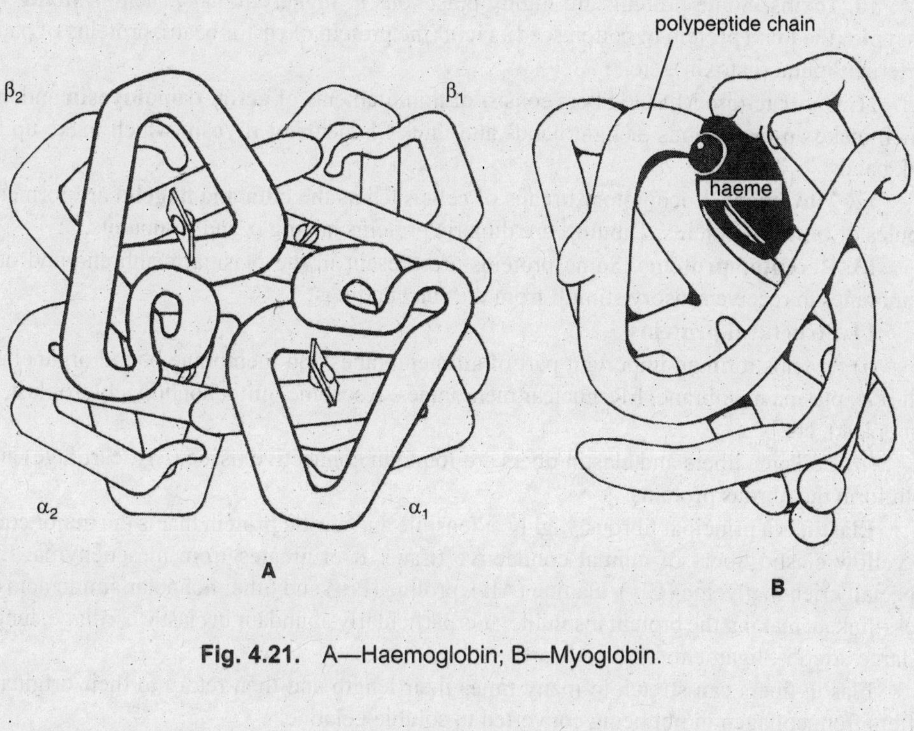

Fig. 4.21. A—Haemoglobin; B—Myoglobin.

***(v)* Transferrin** (MW 90,000) is β-globulin involved in the transfer of Fe to plasma.

***(vi)* Hepatoglobin** is an α-2-globulin. When RBCs are damaged, haemoglobin released into the plasma is bound by hepatoglobin and thus prevented from being excreted through the urine.

α_1-chymotrypsin (MW 45,000) inhibits the proteolytic action of protein-digesting enzymes such as trypsin, chymotrypsin and plasmin.

α_2-macroglobulin (MW 820,000) is an inhibitor protein of proteolytic enzymes.

***(vii)* Complement** is thermolabile group of proteins in normal blood serum and plasma that in combination with antibodies causes the destruction of particular antigens (as bacteria and foreign blood corpuscles). There are 9 components, C′1 to C′9.

***(viii)* Permeases.** These are transport proteins. They transport molecules (Na, K, Ca, glucose, etc.) through cell membrane.

(ix) Proteins help in the transport of organic food through phloem.

***(x)* Cytoplasmic motor proteins.** Proteins that utilize energy of ATP hydrolysis to generate mechanical forces that propel the proteins, as well as attached **cargo**, along one of the components of the cytoskeleton. Three families of motor proteins are known: **kinesins** and **dyneins** move along the microtubules and **myosins** move along the microfilaments (**Karp** 2010).

***(a)* Kinesin** is a plus end-directed motor protein that moves membranous vesicles and other organelles along microtubules through the cytoplasm.

***(b)* Dynein** is an exceptionally large, cargo-carrying multisubunit motor protein that moves along microtubules toward their minus ends. This family of proteins occurs as **cytoplasmic dyneins** and **ciliary** or **axonemal dyneins**.

***(c)* Myosin** is a large family of motor proteins that moves along actin-containing microfilaments. Most myosins are plus- end directed motors. **Conventional myosin** (*i.e.*, myosin II) is the protein that mediates muscle *contractility* as well as certain types of nonmuscle motility, such as **cytokinesis**. **Unconventional myosins** (*i.e.*, myosin I and III to XVIII) have many diverse roles including **organelle transport**.

10. Toxins. Some proteins are highly poisonous to higher animals, such as **snake venoms**, **gossypin** (the toxin protein of cotton seed), **ricin** (the protein of castor bean), proteins of pathogenic bacteria (diphtheria toxin), etc.

11. Contraction. Muscle fibers consist of thin filaments of **actin**, **tropomyosin** and **troponin** which make up the I bands or light bands and thick filaments of **myosin** which make up the A or dark bands.

12. Movement. Locomotory organs of cells such as the **cilia** and **flagella** are formed of fine tubules of **tubulin** proteins. Tubulins are dimeric proteins having α and β subunits.

13. Receptor proteins. Some proteins are present in the plasma membrane and other cell membranes to receive sensory stimuli from the surroundings.

14. Structural proteins.

(i) Proteins form an important part of all membranes and membrane bound organelles of the cell [*i.e.*, plasma membrane, ER, nuclear membranes, lysosome, mitochondria, chloroplast, vacuole (tonoplast), etc.].

(ii) Collagen fibers and elastin fibers are found in connective tissue (*e.g.*, cartilage) and bone; both form the fibrous proteins.

Elastin is a principal fibrous, highly extensible and elastic protein that is the major constituent of yellow elastic fibers of animal connective tissue. It originates from mesenchyma. Elastin is especially rich in glycine (Gly), alanine (Ala), proline (Pro) and other nonpolar amino acids and are cross-linked, making the protein insoluble. It is particularly abundant in elastic cartilage, lungs, walls of large arteries, ligaments and the heart.

Elastin fibers can stretch to many times their length and then return to their original size. It differs from collagen in not being converted to soluble gelatine.

Collagen is the most abundant fibrous (structural) protein of connective tissue, occuring in **white fibers**. It accounts for about 25 – 33% of total body protein of higher vertebrates, including human beings. Collagen found extensively in all connective tissue, *e.g.*, skin, cartilage, tendons, ligaments and bones. Collagen is also present in ECM (of epithelium). Collagen fibers are composed of masses of tropocolagen molecules, each a triple helix of collagen monomers; the three polypeptide chains are wound tightly around each other to give a **helix** (Fig. 4.17). It is especially rich in glycine (35%), alanine (11%) and proline + hydroxyproline (25%). Collagen is a rod-shaped molecule about 3,000 A° (= 300 nm) long and 15 A° (1.5 nm) in diameter. The helical motif of its three chains is entirely different from the α helix. Tropocollagen subunits are synthesized and released by fibroblasts.

Collagen fibers have **high tensile strength** (*i.e.*, resist stretching forces, as in tendon) and limited **elasticity** (unlike elastin). Collagen is essentially insoluble in water, but when boiled with water, the strands separate and undergo some hydrolysis, producing **gelatin**. Although a protein, collagen has little nutritional value because it lacks significant amount of many amino acids that are essential in the human diet (**Jain** 2012).

(*iii*) Protien **chondrin** and **ossein** form matrix of cartilage and bone respectively.

(*iv*) Scales, feathers, hair, fur wool, hoof, nail, horns, etc., formed of keratin sulphate, a protein derivative.

15. Special secretions. Spiders and silkworms secrete a thick solution of protein **fibroin** which solidifies into the thread of exceptional tensile strength and is used for weaving a web or cocoon.

16. Storage proteins. Proteins are stored as reserve food such as egg albumen or protein stored in various seeds (**glutellin** in rice), **zeins** in maize and **phaseolin** in pea.

17. Exotic proteins. These are proteins which are secreted by different animals or plants for specific purpose. For example:

(*i*) Some marine and aquatic animals secrete sticky proteinaceous substance for adhesion to the rock or substratum.

(*ii*) Fishes live in freezing cold water of Arctic and Antarctic oceans synthesize a anti-freeze protein to protect their blood from freezing.

(*iii*) Some african plants synthesize proteins **monellin** which is very sweet and is used as artificial sweetner for diabetic patients.

VII. Protein Purification Techniques

In order to study a particular protein in the laboratory, one must separate it from all other cell components, including other similar proteins. Purification steps usually exploit minor differences in the **solubilities, net charges, sizes** and **binding specificities** of proteins. Most purification techniques are performed at 0°C to 40°C to minimize temperature-dependent processes such as protein degradation and denaturation (unfolding). Protein purification method involves the following four main techniques:

1. Solution making and homogenization. The *first step* in protein purification is to prepare a solution of proteins. The source of a protein is often whole cells in which the target protein accounts far less than 0.1% of the total dry weight. **Isolation** of an intracellular protein requires that cells be suspended in a buffer solution and **homogenized**, or disrupted into cell fragments. Under these conditions, most proteins dissolve. (Major exceptions include membrane proteins, which require special purification procedures).

2. Fractionation. The *next step* in protein purification is often a crude separation, of **fractionation**, procedure that makes use of different solubilities of proteins in salt solutions. **Ammonium sulphate** is frequently used in such fractionations. Enough ammonium sulphate is mixed with the solution of proteins and other, more soluble proteins remain in the fluid, called the **supernatant fraction**. Next, more ammonium sulphate is added to the supernatant fraction until the desired protein is precipitated. The mixture is centrifuged, the fluid removed, and the precipitate is dissolved in a minimal volume of buffer solution. Typically, fractionation using ammonium sulphate gives a two to three fold purification, *i.e.*, one-half to two-thirds of the unwanted proteins have been removed from the resulting enriched protein fraction.

3. Dialysis. At this point, the solvent containing residual ammonium sulphate is exchanged by dialysis for a buffer solution suitable for chromatography. In dialysis, a protein solution is placed in a cylinder of **cellophane tubing** sealed at one end. The tubing is then sealed at the other end and suspended in a large volume of buffer. The cellophane membrane is semipermeable: high-molecular weight proteins are too large to pass through the pores of the membrane so proteins remain inside the tubing while low-molecular-weight solutes (including in this case, ammonium and sulphate ions) diffuse out and are replaced by solutes in this buffer.

4. Chromatography. Following four types of chromatographic techniques are applied to fractionate the mixture of proteins that remains after ammonium sulphate precipitation and dialysis: (*i*) column chromatography; (*ii*) ion-exchange chromatography; (*iii*) gel-filteration chromatography and (*iv*) affinity chromatography.

(*i*) Column chromatography. A cylindrical column is filled with an insoluble material such as substituated cellulose fibers or synthetic beads. The protein mixture is applied to the column and washed through the insoluble material by the addition of solvent. As solvent flows through the column, the **eluate** (the liquid emerging from the bottom of the column) is collected in many fractions, a few of which are represented in Fig. 4.22A. The rate at which proteins travel through the matrix depends on interactions between matrix and protein. For a given column, different proteins are eluated at different rates. The concentration of protein in each fraction can be determined by measuring the absorbance of the eluate at a wavelength of 24 nm (Fig. 4.22B) (*i.e.*, at neutral pH, tyrosine and tryptophan absorb UV light at 24 nm). To locate the target protein, the fractions containing protein must then be assayed, or tested, for biological activity or some other characteristic property.

Column chromatography may be performed under high pressure using small, tightly packed columns, with solvent flow controlled by a computer. This technique is called HPLC, for high-performance liquid chromatography.

(*ii*) Ion-exchange chromatography. In this case, matrix carries positive charges (anion-exchange resins) or negative charges (cation-exchange resins). Anion-exchange matrices bind negatively charged proteins, retaining them in the matrix for subsequent elution. Conversely, cation-exchange materials bind positively charged proteins. The bound proteins can be serially elute by gradually increasing the salt concetration in the solvent. As the salt concentration is increased, it eventually reaches a concentration where the salt ions compete proteins in binding to the matrix. At this concentration, the protein is released and collected in the eluate. Individual bound proteins are eluted at different salt concentrations, and this fractionation makes ion-exchange chromatography a powerful tool in protein purification.

(*iii*) Gel-filteration chromatography. It separates proteins on the basis of molecular size. The gel is a matrix of porous beads. Proteins that are smaller than the average pore size penetrate much of the internal volume of the beads and are therefore retarded by the matrix as the buffer solution flows through the column. The smaller is the protein, the later it elutes from the column. Fewer of the pores are accessible to larger protein molecules. Consequently, the largest proteins flow past the beads and elute first.

(*iv*) Affinity chromatography. It is most selective type of column chromatography. It relies on specific binding interactions between the target protein and some other molecules that is covalently bound to the matrix of the column. The molecule bound to the matrix may be a substance or a ligand that binds to a protein *in vivo*; an antibody that recognizes the target protein; or another protein that is known to interact with the target protein inside the cell. As a mixture of proteins passes through the column, only the target protein specifically binds to the matrix. The column is then washed with buffer several times to rid it of nonspecifically bound proteins.

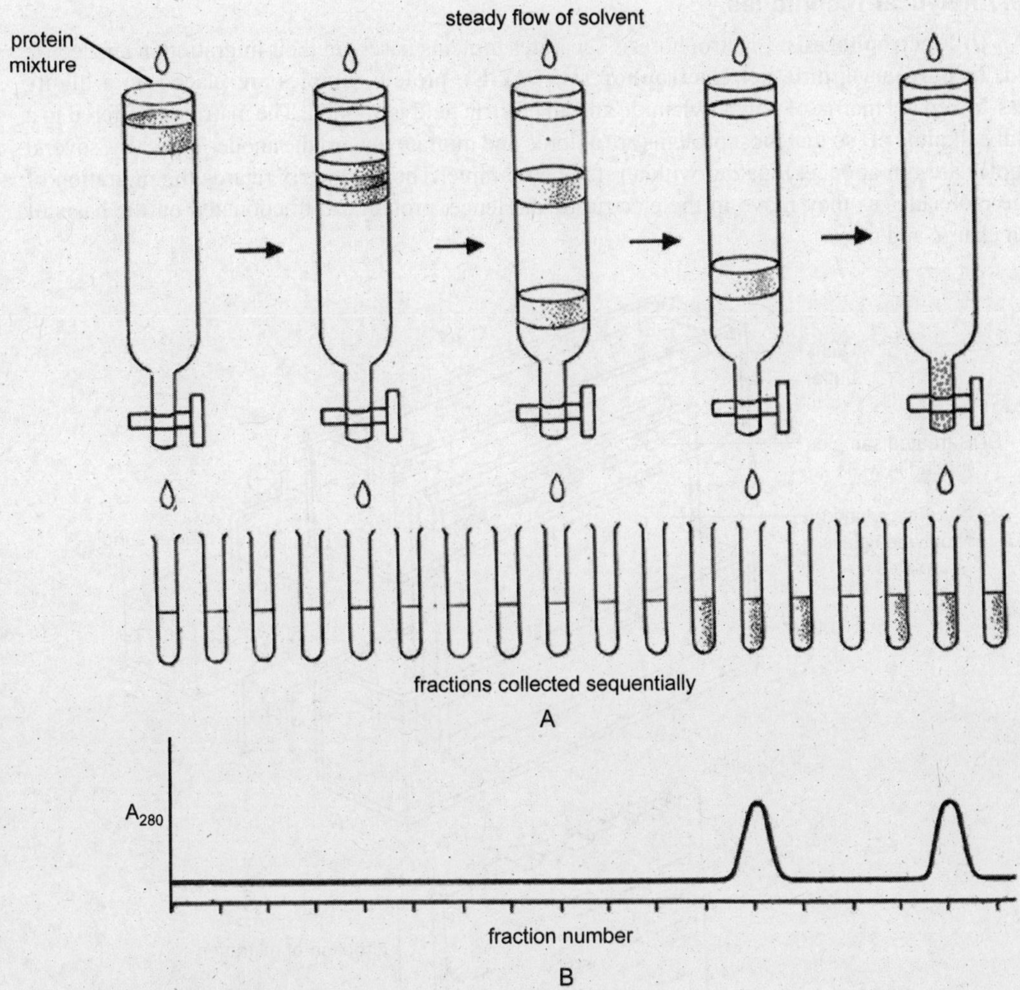

Fig. 4.22. Column chromatography. A—A mixture of proteins is added to a column containing a solid matrix. Solvent then flows into the column from a reservoir. Washed by solvent, different proteins (represented by o and x) travel through the column at different rates, depending on their interactions with the matrix. Eluate is collected in a series of fractions, a few of which are shown. B—The protein concentration of each fraction is determined by measuring the absorbance at 24 nm. The peaks correspond to the elution of the protein bands shown in A. The fractions are then tested for the presence of the target protein (after Horton *et al.*, 2006).

Finally, the target protein can be eluted by washing the column with a solvent containing a high concentration of salt that disrupts the interaction between the protein and column matrix. In some cases, bound protein can be selectively released from the affinity column by adding excess ligand to the eluation buffer. The target protein preferentially binds to the ligand in solution instead of lower concentration of ligand that is attached to the insoluble matrix of column. This method is most effective when the ligand is a small molecule. Affinity chromatography alone can sometimes purify a protein 1000 to 10,000 fold.

VIII. Analytical Techniques

(*i*) **Electrophoresis.** Electrophoresis separates proteins based on their migration in an electric field. In **polyacrylamide gel electrophoresis (PAGE)**, protein samples are placed on a highly cross-linked gel matrix of polyacrylamide and an electric field is applied. The matrix is buffered to a mildly alkaline pH so that most proteins are anionic and migrate toward the anode. Typically, several samples are run at once, togather with a reference sample. The gel matrix retards the migration of large molecules as they move in the electric field. Hence, protein are fractionated on the basis of both charge and mass.

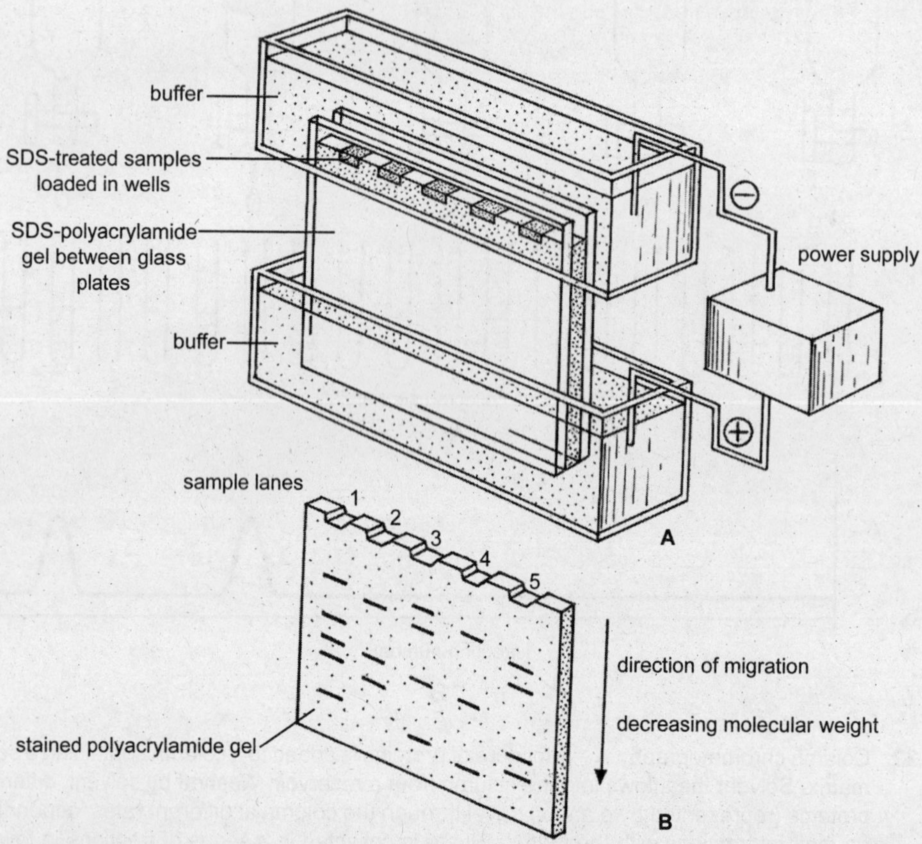

Fig. 4.23. SDS-PAGE. A—An electrophoresis apparatus includes as SDS-polyacrylamide gel between two glass plates and buffer in the upper and lower reservoirs. Samples are loaded into the wells of the gel, and voltage is applied. Since proteins complexed with SDS are negatively charged, they migrate toward the anode; B—The banding pattern of the proteins after electrophoresis can be visualized by staining. Since the smallest proteins migrate fastest, the proteins of lowest molecular weight are at the bottom at the gel (after Horton *et al.*, 2006).

A modification of the standard electrophoresis technique uses the negatively charged detergent sodium dodecyl sulphate (SDS) to overwhelm the native charge on proteins so that they are separated on the basis of mass only. **SDS-polyacrylamide gel electrophoresis (SDS-PAGE)** is used to assess the purity and to estimate the molecular weight of a protein. In SDS-PAGE, the detergent is added to the polyacrylamide gel as well as to protein samples. A reducing agent is also added to the samples to reduce any disulphide bonds. The dodecyl sulphate anion, which has a long hydrophobic tail [$CH_3(CH_2)_{11}OSO_3^-$] binds to hydrophobic side chains (R groups) of amino acid residues in the polypeptide chain. SDS binds at a ratio of approximately one molecule for every two residues of a typical protein. Since larger proteins bind approximately more SDS, the charge-to-mass ratios of all treated proteins are approximately the same. All the SDS-protein complexes are highly negatively charged and move toward the anode (Fig. 4.23). However, the rate of migration through the gel is inversely proportional to the logarithm of their mass, *i.e.*, larger proteins encounter more resistance and therefore migrate more slowly than smaller proteins. In SDS-PAGE, all molecules penetrate the pores of the gel, so the largest proteins travel most slowly. The protein bands that result from this differential migration (Fig. 4.24) can be visualized by staining. Molecular weights of unknown proteins can be estimated by comparing their migration to the migration of reference protein on the same gel.

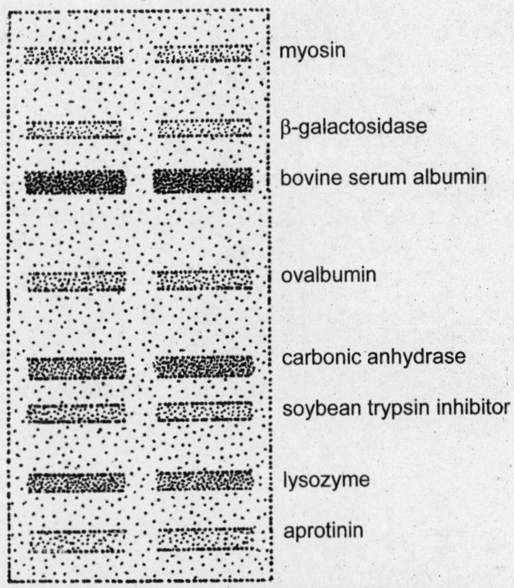

Fig. 4.24. Stained proteins separated on an SDS-polyacrylamide gel.

Although SDS-PAGE is primarily an analytical tool, it can be adapted for purifying proteins. Denatured proteins can be recovered from SDS-PAGE by cutting out the bands of a gel. The protein is then electroeluted by applying an electric current to allow the protein to migrate into a buffer solution. After concentration and the removal of salts, such protein preparations can be used for structural analysis, preparation of antibodies or other purposes.

***(ii)* Mass spectrometry.** Mass spectrometry is a technique that determines the mass of a molecule. The most basic type of mass spectrometer measures the time that it takes for charged gas phase molecule to travel from the point of injection to a sensitive detector. This time depends on the charge of a molecule and in mass and the result is reported as the mass/charge ratio. The technique of mass spectrometry has been used in chemistry for more than one hundred years but its application to proteins was limited because, until recently it was not possible to disperse charged protein molecules into a gaseous stream of particles.

This problem was solved in the late 194s with the development of two new types of mass spectrometry: electrospray mass spectrometry and matrix-assisted desorption ionization (MALDI).

(*a*) In **electropray mass spectrometry** the protein solution is pumped through a metal needle at high voltage to create tiny droplets. The liquid rapidly evaporates in a vacuum and the charged proteins are focused on a detector by a magnetic field.

(*b*) In MALDI technique the protein is mixed with a chemical matrix and the mixture is precipitated on a metal substrate. The matrix is a small organic molecule that absorbs light at a particular wavelength. A laser pulse at the absorption wavelength imparts energy to the protein molecules via the matrix. The proteins are instantly released from the substrate (desorbed) and directed to the detector (Fig. 4.25). When **time-of-flight (TOF)** is measured the technique is called **MALDI-TOF**.

Sophisticated computer programs analyze the data (Fig. 4.25) and calculate the correct masses of proteins. Mass spectrometry is very sensitive and highly accurate. Often the mass of a protein can be obtained from picomole quantities that are isolated from an SDS-PAGE gel. The correct mass can be determined with an accuracy of less than the mass of a single proton.

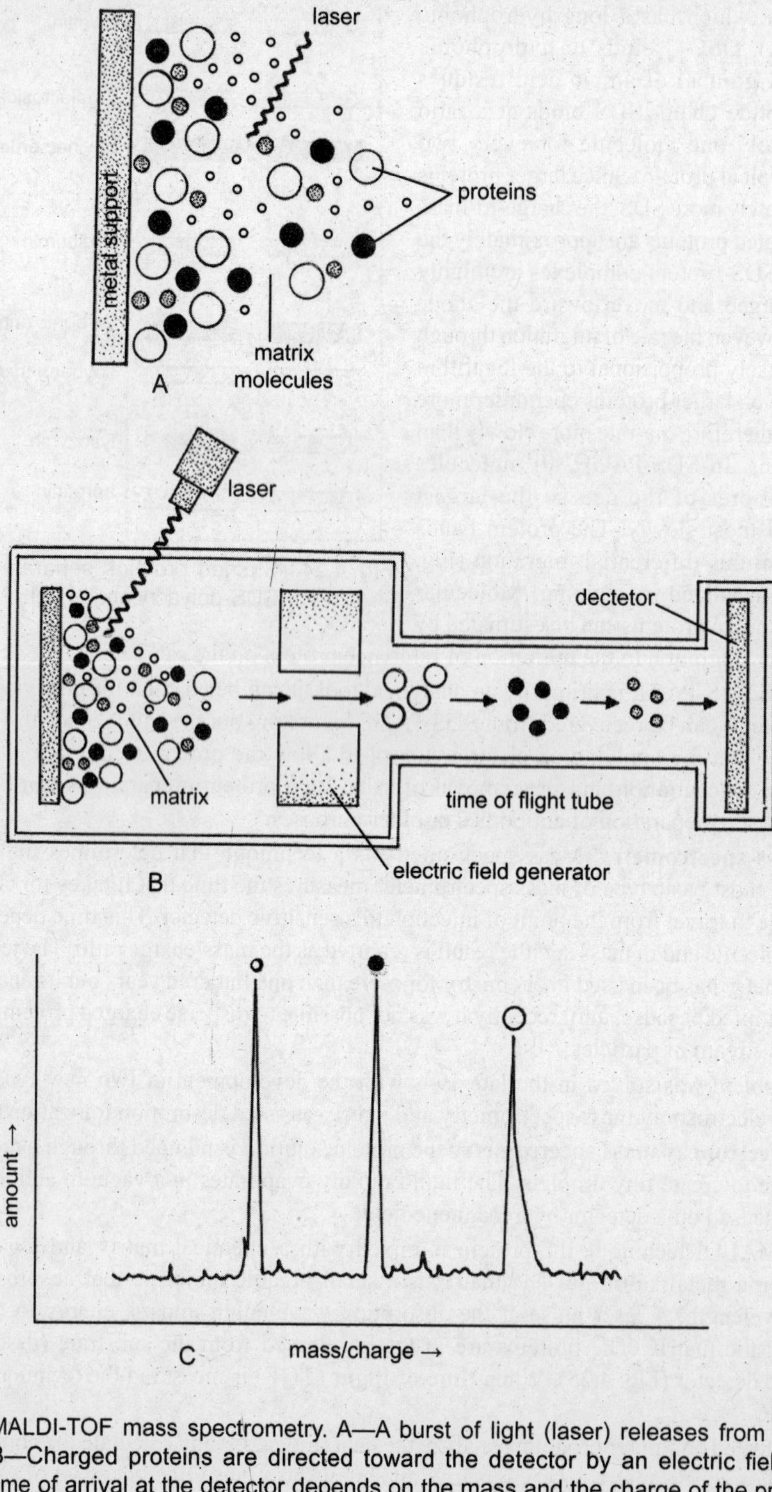

Fig. 4.25. MALDI-TOF mass spectrometry. A—A burst of light (laser) releases from the matrix; B—Charged proteins are directed toward the detector by an electric field; C—The time of arrival at the detector depends on the mass and the charge of the protein (after Horton *et al.*, 2006).

IX. Determination of Amino Acid Composition of Proteins

Once a protein has been isolated, its amino acid composition can be determined.

1. The peptide bonds of the protein are cleaved by acid hydrolysis, typically using 6M HCl.

2. The hydrolysed mixture, called **hydrolysate**, is subjected to a chromatographic procedure in which each of the amino acids is separated and quantitated, a process called **amino acid analysis**.

3. One method of amino acid analysis involves treatment of the protein hydrolysate with phenylisothiocyanate (PITC) at pH 9.0 to generate phenylthiocarbamoyl (PTC)-amino acid derivatives.

4. The PTC-amino acid mixture is then subjected to HPLC in a column of fine silica beads to which short hydrocarbon chains have been attached. The amino acids are separated by the hydrophobic properties of their side chains.

5. As each PTC-amino acid derivative is eluted, it is detected and its concentration is determined by measuring the absorbance of the eluate at 254 nm (the peak absorbance of the PTC moiety).

6. A plot of the absorbance of the eluate as a function of time is given in Fig. 4.26. Each peak corresponds to a specific PTC-amino acid representing a residue that was present in the original protein.

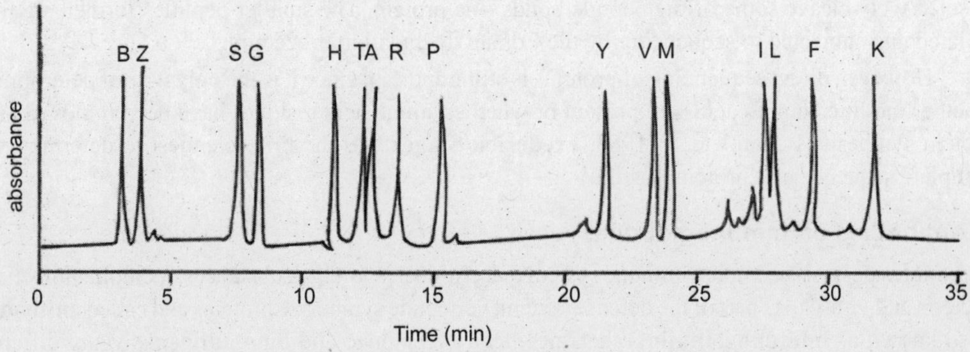

Fig. 4.26. Chromatogram obtained from HPLC separation of PTC-amino acids in the column eluate are detected by their absorbance at 254 nm. Peaks are labeled with one-letter abbreviations. The letters B and Z indicate totals of asparazine + aspartate, and glutamine + glutamate, respectively (Horton *et al.*, 2006).

Using various analytical techniques, the complete amino acid compositions of many proteins have been determined. The amino acid composition (and sequence) of proteins can also be determined from the sequence of its gene. In fact, these days it is often much easier to clone and sequence DNA than it is to purify and sequence a proteins.

In proteins of living organisms, the most common amino acids are leucine, alanine, and glycine, followed by serine, valine and glutamate. Tryptophan, cysteine and histidine are the least abundant amino acids in typical proteins.

X. Determining the Sequence of Amino Acid Residues

Amino acid analysis provides information on the composition of a protein but not its primary structure (sequence of residues). In 1950 **Pehr Edman** (1916–1977) developed a technique that permits removal and identification of one residue at a time from the N-terminus of protein. The **Edman degradation procedure** involves following steps:

1. Treating a protein at pH 9.0 with PITC, also known as the **Edman reagent**. PITC reacts with the free N-terminus of the chain to form a phenylthiocarbamoyl derivative, or PTC-peptide.

2. When the PTC-peptide is treated with an anhydrous acid, such as trifluoroacetic acid, the peptide bond of the N-terminal residue is selectively cleaved, releasing an anilinothiozolinone derivative of the residue.

3. This derivative can be extracted with an organic solvent, such as butyl chloride, leaving the remaining peptide in the aqueous phase.

4. The unstable anilinothiazolinone derivative is then treated with aqueous acid, which converts it to a stable phenylthiohydantion derivative of the amino acid that had been the N-terminal residue (PTH-amino acid).

5. The polypeptide chain in the aqueous phase, now one residue shorter (residue 2 of the original protein is now the N-terminus), can be adjusted back to pH 9.0 and treated again with PITC.

6. The entire procedure can be repeated serially using an automated instrument known as a **sequenator**. Each cycle yields a PTH-amino acid that can be identified chromatographically, usually by HPLC.

XI. Protein Sequencing Strategies

Most proteins contain too many residues to be completely sequenced by **Edman degradation proceeding** only from the N-terminus. Therefore, proteases (enzymes that catalyze the hydrolysis of peptide bonds in proteins) or certain chemical agents (such as cyanogen bromide, BrCN) are used to selectively cleave some of the peptide bonds of a protein. The smaller peptides formed are then isolated and subjected to sequencing by the Edman degradation procedure.

However, direct sequencing of proteins is still important since it is the only way of determining whether modified amino acids are present or whether amino acid residues have been removed after protein synthesis is complete. In 1953, **Frederick Singer** was the first scientiest to determine the complete sequence of a protein (insulin).

4. Antibodies or Immunoglobulins

Vertebrates possess a complex immune system that eliminates foreign substances, including infectious bacteria and viruses. As part of the defense system, vertebrate synthesize proteins and called **antibodies** (also known as **immunoglobulins**) that specifically recognize and bind **antigens**. Many different types of foreign compounds can serve as antigens that produce an immune response. Antibodies are synthesized by white blood cells called **lymphocytes**, each lymphocyte and its descendants synthesise the same antibody. Because animals are exposed to many foreign substances over their life-times, they develop a huge array of antibody-producing lymphocytes that persist at low levels for many years and can later respond to the immune system is the reason certain infections do not recur in an individual despite repeated exposure. **Vaccines** (i.e., inactivated) pathogens or analogs of toxins) administered to children are effective because immunity established in childhood lasts through adulthood.

When an antigen-either noval or previously encountered-binds to the surface of lymphocytes, these cells are stimulated to proliferate and produce soluble antibodies for secretion into the blood stream. The soluble antibodies bind to the foreign organism or substance, forming antibody antigen complex that precipitate and mark the antigen for destruction by a series of interacting proteases or by lymphocytes that engulf the antigen and digest it intracellulary.

Structure. 1. Immunoglobulins contain 4 polypeptide chains, linked by disulphide bonds. There are two identical low molecular weight polypeptide chains (called **light chains**) and two identical high molecular weight polypeptide chains (called **heavy chains**) (Fig. 4.27).

2. Both the heavy and light chains are divided into **constant** and **variable region**.

3. The amino-acid sequence in the constant regions of the different immunoglobulins are very similar. The constant regions are involved in the effector functions of antibodies.

4. The arrangement in the amino acids in the variable regions are distinct for each immunoglobulin. This variability allows for a vast number of unique antibodies each of which binds with specific affinities for different antigens.

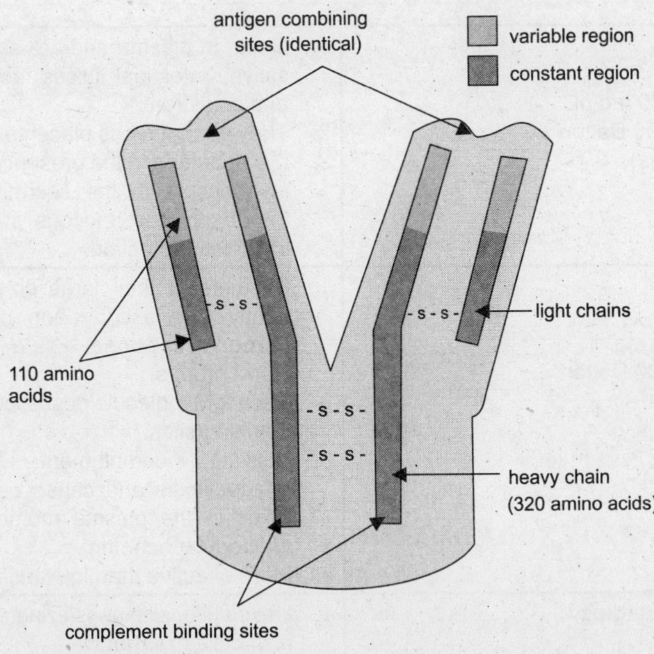

Fig. 4.27. Structure of an immunoglobulin molecule

Types of Immunoglobulins

Immunoglobulins can be divided into 5 classes based on differences in the 'constant' regions of their heavy chains. These are five types of heavy chains: λ (gamma), α (alpha), μ (mu), δ (delta) and ε (epsilon) and these give rise to the five classes of immunoglobulins, **IgG**, **IgA**, **IgM**, **IgD** and **IgE** respectively. Each of these classes had different effector functions (Table 4.6). These are two types of light chains, κ (kappa) and λ (lambda). Both light chains in any immunoglobulin is either κ or λ type.

Table 4.6 The major immunoglobulins (Igs) with their characteristic functions.

(i)	Type of Ig	Function
(ii) (iii) (iv) (v) (vi)	percentage of total Ig Normal plasma level Molecular weight (MW) Heavy chain Light chain	
1. (ii) (iii) (iv) (v) (vi)	**IgG** 71% 0.8 –1.6 gm/dL 1,60,000 Dalton γ (gamma) κ (kappa) or λ (lambda)	1. Antibodies of **secondary immune** response, because get stimulated by natural infection or artificial immunization with viruses or bacteria, therefore, produce major antiviral antibacterial and antitoxin activity in serum. 2. These antibodies are distributed equally between blood and tissue fluids. 3. They **cross the placenta** from mother to foetus by active transport, e.g., antibodies of Rh blood group system and found in milk, saliva, nasal and bronchial secretion. 4. They can serve as opsonin and promote characteristic activity of WBCs.

2. (ii) (iii) (iv) (v) (vi)	**IgA** 22% 140-420 mg/dL 1,70,000 Dalton α (alpha) κ or λ	1. Occur in plasma and are also secreted in tears, saliva, intestinal juices, respiratory secretions and colostrum. 2. They cannot cross placenta. 3. Lyse bacteria in the presence of lysozyme, which also occurs in the secretions containing IgA, thereby protect mucous surfaces, *i.e.*, provide localized protection.
3. (ii) (iii) (iv) (v) (vi)	**IgM** 7% 50-200 mg/dl 1,00,000 Dalton μ (mu) κ or λ	1. Because of their large size they are predominantly intravascular and produced in primary **immune response**, *e.g.*, α, β antibodies of ABO blood groups. 2. Each IgM molecule possesses at least 5 identical combing sites. 3. Activate complement system, promote phagocytosis and cause cell lysis by digesting holes in the plasma membrane at the site of antibody attachment. 4. More effective than IgG antibodies in lysing cells.
4. (ii) (iii) (iv) (v) (vi)	**IgE : Reagins** Traces 0.03 mg/dL 1,85,000 Dalton ε (eisilon) κ or λ	1. Heat labile, skin sensitizing antibodies, therefore, also called **reagins**. 2. When specific antigen is brought into contact with IgE-coated mast cells or blood basophils; these cells undergo degranulation with release of chemical mediators and E.C.F.–A (which is a chemical mediator of immediate hypersensitivity reaction). Thus play important role in allergies, parasite infections and anaphylactic type of *immediate hypersensitive disorders*.
5. (ii) (iii) (iv) (v) (vi)	**IgD** Traces 3 mg/dL 1,50,000 Dalton δ (delta) κ or λ	Present on the surface of B-lymphocytes with IgM, therefore involved in antigenic recognition.

Detailed structure of IgG immunoglobulin

The most abundant antibodies in the bloodstream are of the immunoglobulin G class (IgG)). These are Y-shaped oligomers composed of two identical **light chains** and two identical **heavy chains** connected by disulphide bonds (Fig. 4.28). Immunoglobulins are glycoproteins; they contain covalently bound carbohydrates attached to the heavy chain.

The N-termini of pairs of light and heavy chains are close together. Light chains contain two domains, and heavy chains contain four domains. All the domains consist of about 110 residues assembled into a common motif called the **immunoglobulin fold**. The characteristic feature of the immunoglobulin fold is a sandwich composed of two antiparallel β sheets. This domain structure is found in many other proteins of the immune system.

The N-terminal domains of antibodies, called the **variable domains** because of their sequence diversity, determine the specificity of antigen binding. X-ray crystallographic studies have shown that the antigen-binding site of a variable domain consists of three loops, called **hypervariable regions,** that differ widely in size and sequence. The loop from a light chain and a heavy chain combine to form a **barrel**, the upper surface of which is complementary to the shape and polarity of a specific antigen. The match between the antigen and antibody is so close that there is no space for water molecules between the two. The forces that stabilize the interaction of antigen with antibody are primarily hydrogen bonds and electrostatic interactions.

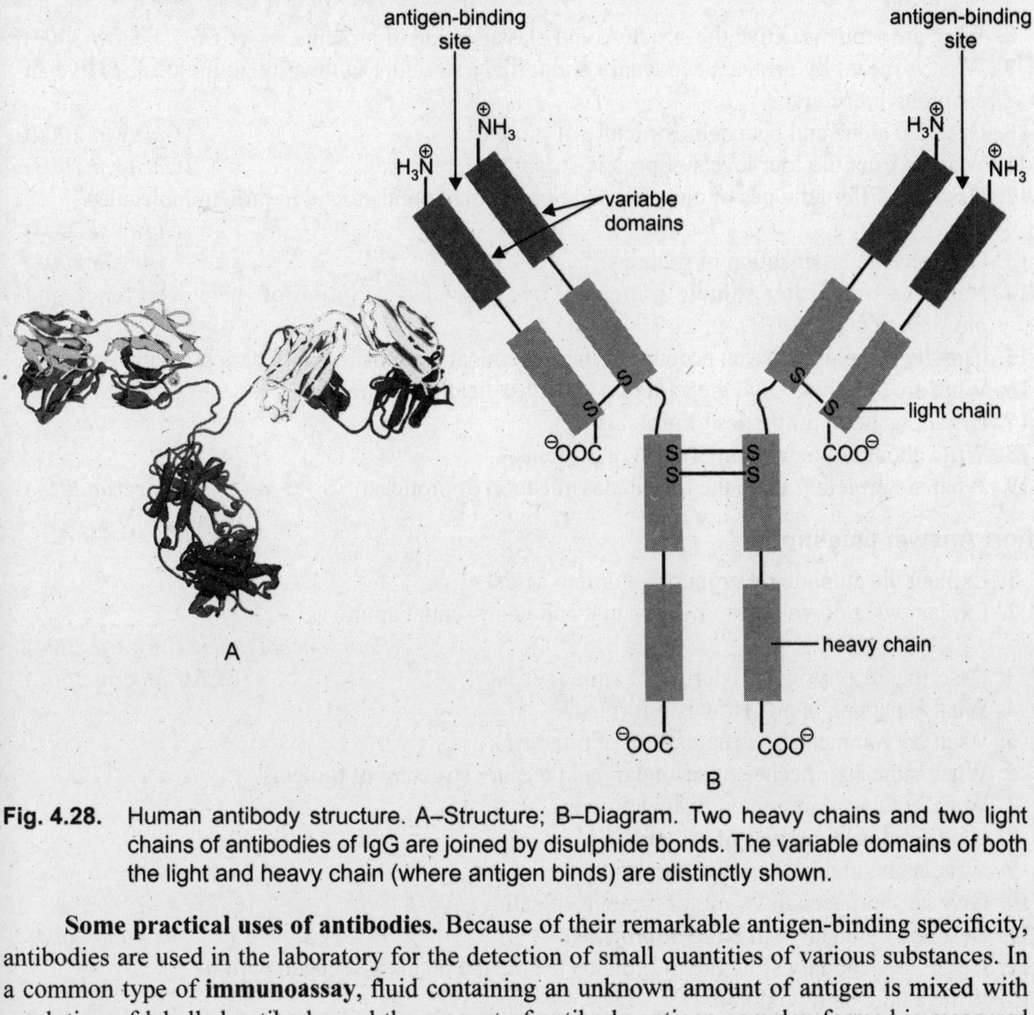

Fig. 4.28. Human antibody structure. A–Structure; B–Diagram. Two heavy chains and two light chains of antibodies of IgG are joined by disulphide bonds. The variable domains of both the light and heavy chain (where antigen binds) are distinctly shown.

Some practical uses of antibodies. Because of their remarkable antigen-binding specificity, antibodies are used in the laboratory for the detection of small quantities of various substances. In a common type of **immunoassay**, fluid containing an unknown amount of antigen is mixed with a solution of labelled antibody, and the amount of antibody-antigen complex formed is measured (**Horton** *et al.*, 2006).

QUESTIONS

Long Answer Questions

1. What are amino acids? Describe the physical and chemical properties of amino acids. (*CCSU Meerut 2006, 2010*)
2. What are amino acids? Describe briefly the physical and chemical properties of amino acids. (*Goraphpur 2003*)
3. What are amino acids? Classify various amino acids found in nature and give their properties. Explain how the amino acids are joined together to form a molecule? (*Avadh 2003, 04; Garhwal 2004*)
4. Describe chemical classification of amino acids. (*Gorakhpur 2000*)
5. What are amino acids? Which groups of amino acids take part in the formation of peptide bonds? What are functions of amino acids?

6. What are peptides? Describe the primary structure and biological importance of peptide chains.
7. Describe the composition, structure and biological significance of proteins in the body of living organisms.
8. What are protiens? Give the structure and classification of proteins. (*CCSU Meerut 2007*)
9. What is meant by primary, secondary and tertiary structure in the protein molecule? Give an example of each type.
10. Describe alpha and beta helix structure of proteins. (*Garhwal 2000*)
11. Write in brief the four levels of protein structure. (*Garhwal 2003*)
12. Describe differnet types of covalent and noncovalent bonds found in protein molecules. (*Garhwal 2004*)
13. Give recent classification of proteins. (*Avadh 2002*)
14. Name various water soluble proteins. Give a brief description of their occurrence and physiological role.
15. Classify proteins. Give an account of their chemical composition and biological importance.
16. What are proteins? Give the structure and classification of proteins. (*Garhwal 2004*)
17. Discuss general properties of proteins.
18. Write about biological importance of proteins.
19. What are proteins? Give the recent classification of protiens. (*CCSU Meerut 2014*)

Short Answer Questions

1. Explain the amphoteric property of amino acids.
2. Explain what do you mean by essential and nonessential amino acids. (*Poorvanchal 2003; Garhwal 2004*)
3. Describe the physical properties of amino acids. (*CCSU Meerut 2007*)
4. What is peptide bond? How is it formed?
5. What do you mean by denaturation of protein?
6. What is the significance of secondary and tertiary structure of proteins?
7. Write brief account of conjugated proteins.
8. Give any four functions of proteins.
9. Write name of any four transport proteins.
10. Give an short account of motor proteins of cell.
11. Describe secondary structure of proteins. (*Avadh 2004*)
12. Describe secondary structure of proteins. Name the bonds associated with it.
13. Write a note on myoglobin.
14. Write short notes on the following:
    (*i*) Amino acids (*Gorakhpur 2001; Avadh 2002*)
    (*ii*) Aromatic amino acids (*Kanpur 2003*)
    (*iii*) Zwitter ions (*Gorakhpur 2000*)
    (*iv*) Essential amino acids
    (*v*) Isoelectric point or isoelectric pH (*Gorakhpur 2001*)
    (*vi*) Peptide linkage or peptide bond (*Gorakhpur 2000, 03; Avadh 01; Poorvanchal 03*)
    (*vii*) Classification of aminod acids (*Garhwal 2001*)
    (*viii*) Conjugated proteins
    (*ix*) Primary structure of protein
    (*x*) Proteins
    (*xi*) Nucleoproteins (*Garhwal 2004*)
15. Write differences between the following:
    (*i*) D- and L- form of amino acids

(*ii*) Aromatic and Aliphatic amino acids
(*iii*) Polar and Nonpolar amino acids
(*iv*) Essential and non-essential amino acids
(*v*) Fibrous and globular proteins
(*vi*) Haemoglobin and Myoglobin (*Garhwal 2004; Avadh 2005*)
(*vii*) Secondary and Tertiary structure of proteins

Very Short Answer Questions

1. Name one amino acid which act as hormone.
2. Which amino acid forms part of vitamin nicotinamide?
3. What is peptide?
4. What do you mean by term amphoteric?
5. Name two polar charged amino acids.
6. What are heterocyclic amino acids?
7. Which chemical groups are involved in the formation of peptide bonds?
8. Name one fibrous and one globular proteins.
9. What is the role of myglobin?
10. What is an amino acid? (*CCSU Meerut 2014*)
11. What is peptide bond? (*CCSU Meerut 2013*)

Fill in the Blanks

1. Transplanted tissues and organs are rejected when the host's ________ attack antigens on the ________ of the transplant. This set of antigens, unique to each genotype, is known as the ________ complex.

Multiple Choice Questions

1. Which is following are essential amino acids?
   (*a*) lysine, linoleic acid (*b*) arachidonic acid, aspartic acid
   (*c*) leucine, valine (*d*) all of these
2. Non-essential amino acids combination is
   (*a*) alanine, valine, leucine, glycine
   (*b*) proline, glutamic acid, glycine, serine
   (*c*) glutamic acid, methionine, proline, aspartic acid
   (*d*) alanine, isoleucine, tyrosine, cysteine
3. A complete proteins is one which possesses
   (*a*) all non-essential amino acids (*b*) all essential amino acids
   (*c*) all the 20 amino acids (*d*) all the amino acids with complex structure
4. Which of the following can synthesize all of the amino acids?
   (*a*) animals (*b*) plants
   (*c*) both (*a*) and (*b*) (*d*) none of these
5. Which one is an imino acids?
   (*a*) pepsin (*b*) cysteine
   (*c*) proline (*d*) renin
6. Upon denaturation a protein loses its
   (*a*) primary structure (*b*) peptide bonds
   (*c*) sequence of amino acids (*d*) three dimensional shape
7. Proteins are formed by the condensation of
   (*a*) carbohydrate (*b*) fatty acids
   (*c*) amino acids (*d*) nucleic acids

8. A dipeptide contains
   (*a*) two amino acids (*b*) two peptide bonds
   (*c*) two amino groups but no carboxyl group (*d*) two carboxyl groups but no amino acid
9. The primary structure of protein is made by
   (*a*) hydrogen bonds (*b*) ionic bonds
   (*c*) peptide bonds (*d*) none of these
10. Which of the following is not conjugated protein?
   (*a*) peptide (*b*) phosphoprotein
   (*c*) lipoprotein (*d*) chromoprotein
11. A peptide chain assumes secondary structure through the formation of
   (*a*) interchain ionic bonds (*b*) intrachain hydrogen bonds
   (*c*) peptide bonds (*d*) intrachain disulphide linkages
12. Quaternary structure is present in molecule of
   (*a*) haemoglobin (*b*) histone
   (*c*) globulin (*d*) elastin
13. Primary structure of polypeptide is stabilised by
   (*a*) hydrogen bonds (*b*) hydrophobic interactions
   (*c*) ionic bonds (*d*) disulphide bonds
14. In β-pleated secondary protein structure there are
   (*a*) two or more polypeptide chains
   (*b*) hydrogen bonds between adjacent polypeptide chains
   (*c*) parallel or antiparallel polypeptides
   (*d*) all of the above
15. Most abundant protein in the human body is
   (*a*) haemoglobin (*b*) keratin
   (*c*) collagen (*d*) immunoglobulin
16. A globular protein is
   (*a*) elastin (*b*) keratin
   (*c*) albumin (*d*) collagen
17. The major fibrous proteins that provide external protection
   (*a*) collagen (*b*) elastin
   (*c*) keratin (*d*) none of these
18. Protein which play a significant role in aging is
   (*a*) elastin (*b*) collagen
   (*c*) actin (*d*) myosin
19. The protein present in wheat grain is
   (*a*) glutein (*b*) glycogen
   (*c*) zymase (*d*) albumin

ANSWERS

Very Short Answer Questions

1. Tyrosine.
2. Tryptophan.
3. Peptide is a organic compound containing two or more amino acids (*i.e.*, amino acid residues) linked through the amino group (–NH_2) of one acid and the carboxyl group (–COOH) of the other. The bond or linkage –NH–CO– is termed a peptide bond.
4. Compounds (*e.g.*, amino acids) having both acidic and alkaline groups are called amphoteric compounds.

5. Lysine, Arginine, Glutamic acid.
6. In heterocyclic amino acids, the ring side of chain has nitrogen as in proline and histidine.
7. A peptide bond is formed between carboxyl (–COOH) group of one amino acid and amino (–NH_2) group of succeeding amino acids.
8. Fibrous protein, *e.g.*, collagen and Globular protein, *e.g.*, Albumin.
9. Myoglobin stores oxygen.
10. Amino acid is an organic acid consisting of an α-carbon atom to which an amino group, a carboxylic group, a hydrogen atom, and a specific side chain (R group) are attached. Amino acids are the building blocks of proteins.
11. Peptide bond is the covalent secondary amide linkage that joins the carbonyl group of one amino acid residue to the amino nitrogen of another in peptides and proteins.

Fill in the Blanks

1. T cells, cell surfaces, major histocompatibility

Multiple Choice Questions

1. (*c*)	**2.** (*b*)	**3.** (*b*)	**4.** (*b*)	**5.** (*c*)	**6.** (*d*)
7. (*c*)	**8.** (*a*)	**9.** (*c*)	**10.** (*a*)	**11.** (*b*)	**12.** (*a*)
13. (*a*)	**14.** (*d*)	**15.** (*c*)	**16.** (*c*)	**17.** (*c*)	**18.** (*b*)
19. (*a*)					

5

Nucleic Acids

Nucleic Acids (DNA and **RNA)** represent the fourth major class of macromolecules. Like proteins and polysaccharides, they contain multiple similar monomeric units that are covalently joined to produce large polymers.

The nucleic acids (DNA and RNA) are the molecular repositories for genetic information and are jointly referred to as the "molecules of heredity." The structure of every protein, and ultimately of every cell constituent, is a product of information programmed into the nucleotide sequence of cell's nucleic acids.

The discovery of the substance that proved to be **deoxyribonucleic acid (DNA)** was made in 1869 by **Friedrich Miescher,** a young Swiss physician working in the laboratory of German physiological chemist **Felix Hoppe-Seyler.** Miescher treated white blood cells (which came from the pus on discarded surgical bandages) with hydrochloric acid to obtain nuclei for study. When the nuclei were subsequently treated with acid, a precipitate formed that contained *carbon, hydrogen, oxygen, nitrogen* and a high percentage of *phosphorus*. **Miescher** called the precipitate "nuclein" because it came from nuclei. Later, when it was found to be strongly acidic, its name was changed to **nucleic acid.** The term **nucleic acid** was introduced by **Altmann** in 1899. Although he did not know it, Miescher has discovered DNA. Soon afterward, **Hoppe-Seyler** isolated a similar substance from yeast cells; this substance is now known as **ribonucleic acid (RNA).** Both DNA and RNA are polymers of nucleotides, or **polynucleotides.**

In 1944, **Oswald Avery, Colin MacLeod** and **Maclyn McCarty** demonstrated that DNA is the molecule that carries the genetic information. Over the next few years, the structure of nucleotides were determined, and in 1953 **James D. Watson** and **Francis H.C. Crick** proposed their model of the structure of double-stranded DNA.

A living organism contains a set of instructions for every step required to construct a replica of itself. This information resides in the genetic material, or **genome**, of the organism. The genomes of all cells are composed of DNA. Some viral genomes are composed of RNA. A genome may consist of a single molecule of DNA, as in many species of bacteria. In eukaryotes, the genome is one complete set of DNA molecules found in the nucleus (*i.e.*, the haploid set of chromosomes in diploid organisms). By convention, the genome of a species does not include mitochondrial and chloroplast DNA. With rare exception, no two individuals in a species have exactly the same genome sequence.

In general, the information that specifies the primary structure of a protein is encoded in the sequence of nucleotides in DNA. This information is enzymatically copied during the synthesis of RNA, a process known as **transcription**. Some of the information contained in the transcribed RNA molecule is translated during the synthesis or polypeptide chain, which are then folded and assembled to form protein molecules. Thus, we can generalize that the biological information stored in a cell's DNA flows from DNA to RNA to protein.

5.1 COMPONENTS OF NUCLEIC ACIDS

Nucleic acids are polynucleotides, or polymers of nucleotides. Thus, nucleotides are the building blocks of nucleic acids. Nucleotides have three components: a five-carbon sugar, one or more phosphate groups and a weakly basic nitrogenous compound called **a base** (Fig. 5.1). The bases found in nucleotides are substituted pyrimidines and purines. The pentose is usually either ribose (D-ribofuranose) or 2-deoxyribose (2-deoxy-D-ribofuranose). The pyrimidine or purine N-glycosides of these sugars are called **nucleosides**. **Nucleotides** are the phosphate esters of nucleosides; the common *nucleotides* contain from one to three phosphoryl groups. Nucleotides containing ribose are called **ribonucleotides,** and nucleotides containing deoxyribose are called **deoxyribonucleotides.**

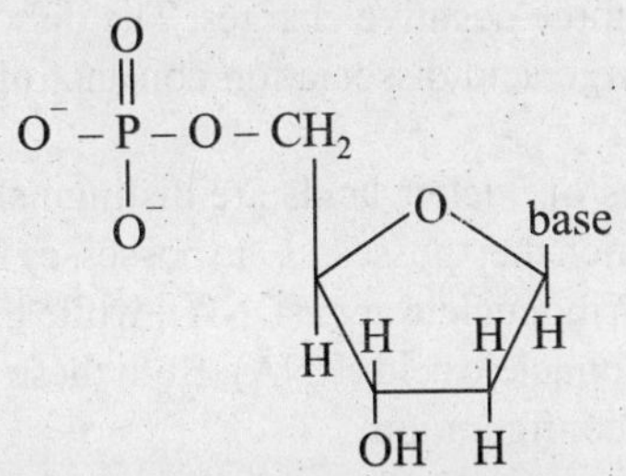

Fig. 5.1. Chemical structure of a nucleotide.

Upon hydrolysis under different sets of conditions, the two nucleic acids yield 3 components: phosphoric acid, a pentose sugar and nitrogenous bases (Fig. 5.2). Table 5.1 summarizes the structural components of the two types of nucleic acids, DNA and RNA.

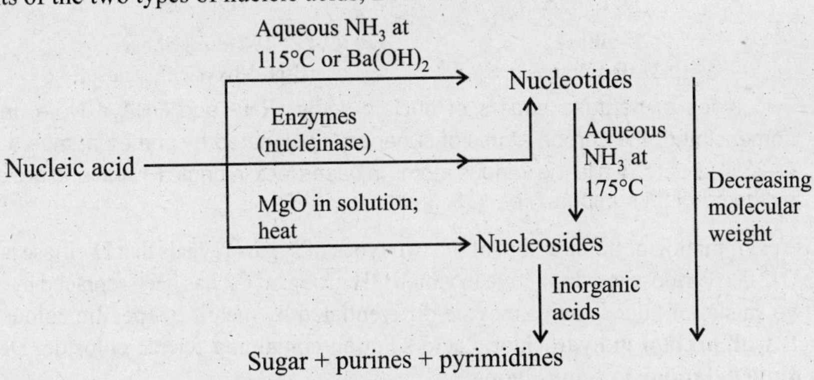

Fig. 5.2. Hydrolytic products of nucleic acid.

Table 5.1 Structural components of RNA and DNA.

Components	Ribonucleic acid	Deoxyribonucleic acid
1. Acid	Phosphoric acid	Phosphoric acid
2. Pentose sugar	D-ribose	D-2-deoxyribose
3. Nitrogenous bases		
(i) Purines	Adenine	Adenine
	Guanine	Guanine
(ii) Pyrimidines	Cytosine	Cytosine
	Uracil	Thymine

1. Phosphoric Acid. The molecular formula of phosphoric acid is H_3PO_4. It contains 3 monovalent hydroxyl groups and a divalent oxygen atom; all linked to the pentavalent phosphorus atom (Fig. 5.3).

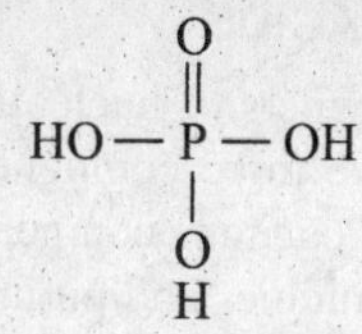

Fig. 5.3 Phosphoric acid

Phosphoric acid is a weak acid and a **polyprotic acid,** ***i.e.***, it possesses more than one dissociating group. It contains three different hydrogen atoms that can dissociate to form H^+ ions and corresponding conjugate bases with one, two or three negative charges. The dissociation of the first proton occurs readily and is associated with a large acid dissociation constant of 7.53×10^{-3} M and a pK_a of 2.2 in a aqueous solution.

2. Pentose sugar. Two types of nucleic acids are distinguished primarily on the basis of the 5′– carbon (pentose) aldo sugar which they possess. One possesses D-2-deoxyribose, hence the name deoxyribose nucleic acid or deoxyribonucleic acid (DNA), while the other contains D-ribose, hence the name ribose nucleic acid or ribonucleic acid (RNA). Both these sugars in nucleic acid are present in the furanose form and are of β-configuration.

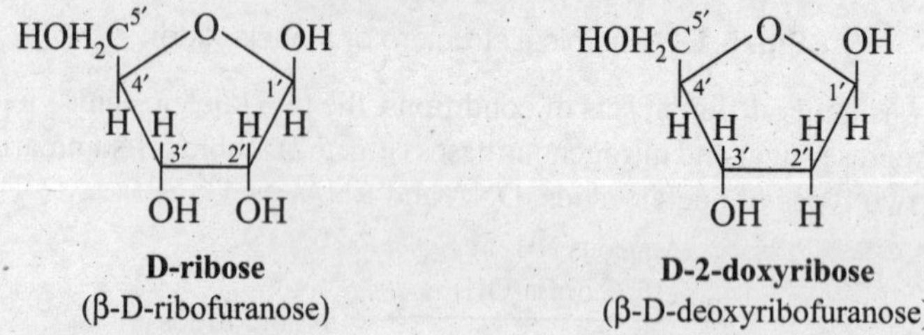

Fig. 5.4. Two types of pentose sugars of nucleic acids (RNA and DNA). (Note. In chemical nomenclature the carbon atoms of sugar are designated by primed numbers, *i.e.*, C–1′, C–2′, C–3′, etc., while the various atoms in bases lack in prime (′) sign and are designated by the cardinal numbers, *i.e.*, 1, 2, 3 etc.

A close examination of the structure of the two types of sugars reveals that D-ribose is the parent sugar while D-2 deoxyribose is a derivative in which OH group on C2 has been replaced by an H atom.

The two sugars of nucleic acids may be differentiated by means of specific colour reactions. Ribose reacts with **orcinol** in hydrochloric acid solution containing ferrich chloride. Deoxyribose reacts with **diphenylamine** in acid solution.

An important property of the pentoses is their capacity to form esters with phosphoric acid. In this reaction the OH groups of the pentose, especially those at C3 and C5, are involved forming a **3′, 5′–phosphodiester bond** between adjacent pentose residues. This bond, in fact, is an integral part of the structure of nucleic acids.

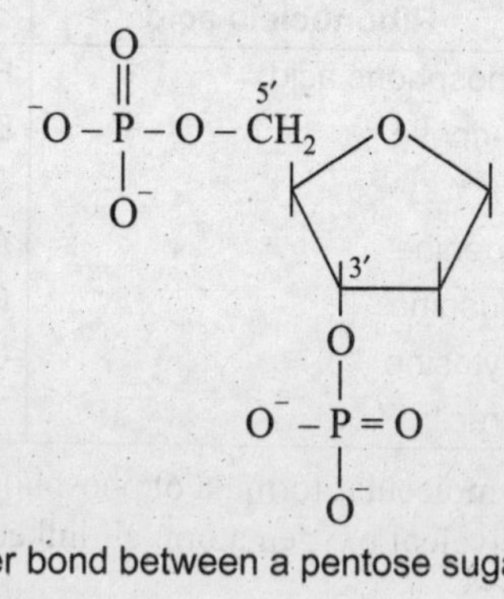

Fig. 5.5. 3′, 5′–phosphodiester bond between a pentose sugar and 2 phosphoric acids.

3. Nitrogenous bases. Two types of nitrogenous bases are found in all nucleic acids. The nitrogenous bases are derivatives of pyrimidine and purine. Owing to their π electron clouds, both the pyrimidine and purine bases are planar (*i.e.*, in the form of a plane or flat surface) molecules. (**Note:** π is a Greek letter *pi*. It represents the ratio of the circumference of a circle to its diameter; approximately 3.14159.)

A. Pyrimidine Derivatives

These are all derived from their parent heterocyclic compound pyrimidine (Fig. 5.6), which contains a six-membered ring with two nitrogen atoms and three double bonds. It has a meeting point of 22°C and boiling point of 123.5°C.

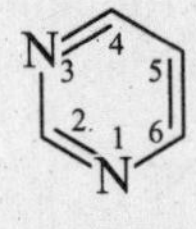

Fig. 5.6. Pyrimidine (Note: For pyrimidine, the Chemical Abstract use the numbering system, as depicted in formula of pyrimidine in the figure. This system is proposed by IUPAC; see J.L. Jain 2014).

The common derivatives of pyrimidine found in nucleic acids are *uracil, thymine*, and *cytosine* (Fig. 5.7).

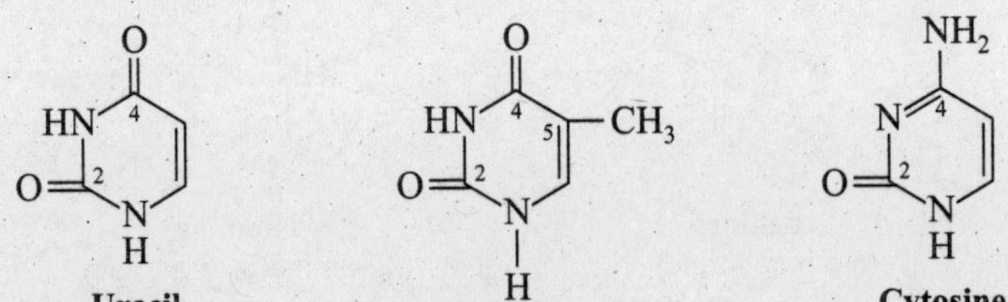

Uracil (2, 4-deoxypyrimidine or pyrimidinedione)

Thymine (5-methyl-2, 4-dioxypyrimidine or 5-methyluracil)

Cytosine (2-oxy-4-aminopyrimidine or aminopyrimidone)

Fig. 5.7. Major pyrimidine derivatives.

(*i*) **Uracil ($C_4H_4O_2N_2$).** It is found in RNA molecules only. It is a white crystalline pyrimidine base with MW = 112.10 daltons and a melting point 338°C.

(*ii*) **Thymine ($C_5H_6O_2N_2$).** It is found in DNA molecule only, has a M = 126.13 daltons. It was first isolated from thymus hence so named.

(*iii*) **Cytosine ($C_4H_5ON_3$).** It is found in both DNA and RNA. It is a white crystalline substance with MW = 111.12 daltons and melting point 320–325 °C.

B. Purine Derivatives

These are all derived from their parent compound purine, which contains a six-membered *pyrimidine ring* fused to the five-membered *imidazole ring* and is related to uric acid. It has a melting point of 216 °C. The common derivatives of purine found in nucleic acid are *adenine* and *guanine* (Fig. 5.9).

(*i*) **Adenine ($C_5H_5N_5$).** It is found both in RNA and DNA and is a white crystalline purine base. It has molecular weight 135.15 daltons and a melting point of 360-365 °C.

(*ii*) **Guanine ($C_5H_5ON_5$).** It is also found both in RNA and DNA and is a colourless, insoluble crystalline substance. It has molecular weight 151.15 daltons. It was first isolated from guano (bird manure), hence so named.

Minor or Modified Bases

The above 5 nitrogen bases (*i.e.*, Uracil, Thymine, Cytosine, Adenine and Guanine) were once believed to account for the total base composition of animal and plant nucleic acids and hence were designated as *major nitrogenous bases*. It is now known that other **minor bases** called **modified nitrogenous bases** also occur in polynucleotide structures. Examples of some naturally occurring forms of **modified pyrimidines** are **5, 6–dihydrouracil, pseudouracil, 4-thiouracil, 5-methyl-cytosine** and **5-hydroxymethyl cytosine**. Examples of some **modified purines** include **hypoxanthine** (6-oxypurine), **xanthine** (2, 6-dioxypurine), **uric acid** (2, 6, 8-trioxypurine), **6-methyl adenine** or **6-MeA** (6-methylaminopurine), (6-dimethyladednine or **6-DiMeA** (6-dimethylaminopurine), **6-N-isopentenyladenine** or **6-IPA, 1-methyl guanine** or **MeG** and **2-dimethylguanine** or **2-DiMeG.** Methylation is the most common form of purine modification.

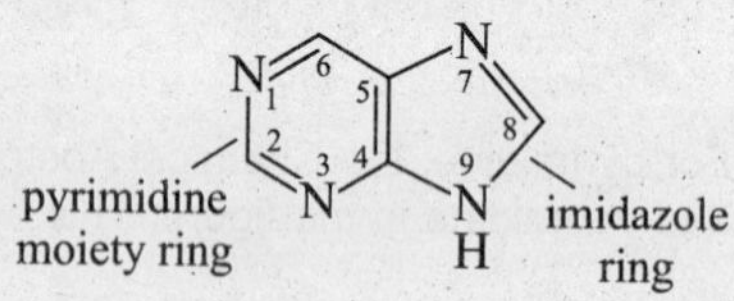

Fig. 5.8. Purine.

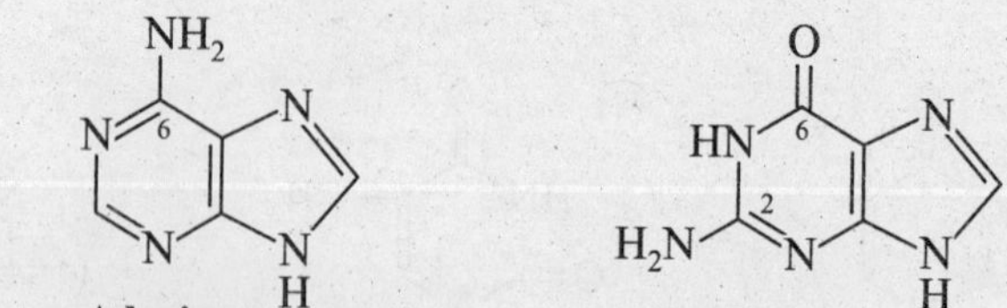

Fig. 5.9. Major purine derivatives.

5.2 NUCLEOSIDES

Nucleosides are composed of ribose or deoxyribose and heterocyclic base. In these compound, the nitrogenous bases are conjugated to the pentose sugars (ribose or deoxyribose) by a **β-glycosidic linkage.** This is the configuration in polymeric nucleic acids. The β-glycosidic linkage involves the C–1′ of sugar and the hydrogen atom of N-9 (in the case of purines) or N–1 (in the case of pyrimidines), thus eliminating a molecule of water. Therefore, the purine nucleosides are N–9 glycosides and the pyrimidine nucleosides are N–1 glycosides. Nucleosides are stable in alkali. Purine nucleosides are readily hydrolyzed by acids, whereas pyrimidine nucleosides are hydrolyzed only after prolonged treatment with concentrated acids.

The nucleosides are generally named for the particular purine or pyrimidine present. Nucleosides containing ribose are called **ribonucleosides**, while those possessing deoxyribose or **deoxyribo-nucleosides.** The single letter abbreviations for pyrimidines and purine bases are also commonly used to designate ribonucleosides: A, G, C and U (for adenosine, guanosine, cytidine and uridine respectively). The deoxyribonucleosides are abbreviated dA, dG, dC and dT when it is necessary to distinguish them from ribonucleosides.

Table 5.2 lists the trivial (= small) names of the purine and pyrimidine nucleosides which are related to the bases that occur in RNA and DNA. The table also includes the single-letter abbreviations for each of the nucleosides.

The structure of two nucleosides, one ribonucleoside (also called **riboside**) and another deoxyribonucleoside (or **deoxyriboside**) is given in Fig. 5.10.

Table 5.2 The nucleosides

Base	Sugar	Nucleoside	Trivial name	Single letter abbreviation
A. Ribo-nucleosides				
1. Adenine	Ribose	Adenine ribonucleo-side	Adenosine	A
2. Guanine	Ribose	Guanine ribonucleo-side	Guanosine	G
3. Cytosine	Ribose	Cytosine ribonucleo-side	Cytidine	C
4. Thymine	Ribose	Thymine ribonucleo-side	Thymidine	T
5. Uracil	Ribose	Uracil ribonucleo-side	Uridine	U
B. Deoxy-ribonucleoside				
1. Adenine	Deoxyribose	Adenine deoxyribo-nucleoside	Deoxyadenosine	dA
2. Guanine	Deoxyribose	Guanine deoxy-ribo-nucleoside	Deoxyguanosine	dG
3. Cytosine	Deoxyribose	Cytosine deoxy-rib-onucleoside	Deoxycytidine	dC
4. Thymine	Deoxyribose	Thymine deoxy-ribo-nucleoside	Deoxythymidine	dT
5. Uracil	Deoxyribose	Uracil deoxy-ribonu-cleoside	Deoxyuridine	dU

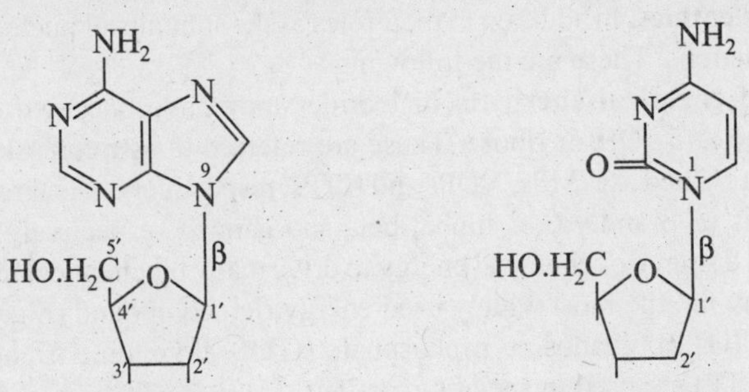

Adenosine
(1-β-D-ribofuranosyl adenine)

Deoxycytidine
(1-β-2′-deoxy-D-ribofuranosylcytosine)

Fig. 5.10. A ribo-and deoxyribonucleoside.

5.3 NUCLEOTIDES

Nucleotides are phosphorylated derivatives of nucleosides (*i.e.*, nucleotides are nucleoside phosphates, see **Tamarin** 2002). They are phosphoric acid esters of nucleosides. The ribose moiety of nucleosides contains three hydroxyl groups that can be phosphorylated (*i.e.*, C2′, C3′ and C5′);

the C1′ and C4′ are involved in the furanose ring formation of the ribose. On the contrary, the deoxyribose component of a 2′–deoxyribonucleoside contains only two such hydroxyl groups (C3′ and C5′); the Cl′ and C4′ are involved in the furanose ring of sugar and C2′ does not bear hydroxyl group). In naturally occurring nucleotides, the phosphoryl groups are usually attached to the oxygen atom of the 5′–hydroxyl group.

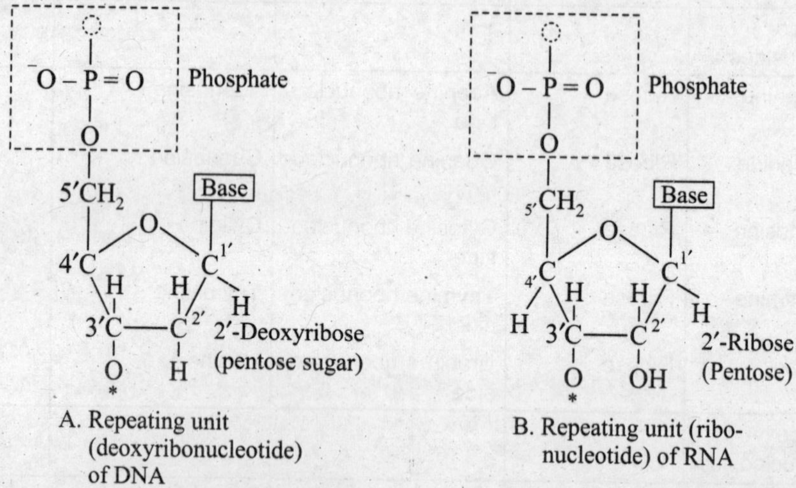

Fig. 5.11. The structure of nucleotides found in (A) DNA or deoxyribonucleic acid and (B) RNA or ribonucleic acid. In a DNA or RNA strand, the oxygen (*) on the 3′ carbon is linked to the phosphorus atom of phosphate in the adjacent nucleotide (see Fig. 5.5).

Accordingly, hydrolysis of the two nucleic acids, RNA and DNA, by various methods and under different set of conditions gives rise to isomeric nucleotides of 3 types and 2 types respectively. Table 5.3 lists some important ribonucleotides (or **ribotides**) and **deoxyribonucleotides** (or **deoxyribotides**). While the names of nucleosides and nucleotides are generally derived from the corresponding bases, there is one exception to this rule: the base corresponding to the nucleotide called **inosine** (and the derived nucleotide) is called **hypoxanthine.** The Fig. 5.11 shows the structure of nucleotides found in DNA and RNA both.

Functions of Nucleotides. In addition to their roles as the subunits of nucleic acids, nucleotides perform some other functions. These are the following :

1. As carriers of chemical energy. Nucleotides may have one, two or three phosphate groups covalently linked at 5′–OH of ribose. These are referred to as **nucleoside mono–, di–** and **tri-phosphates** and abbreviated as NMPs, NDPs and NTPs, respectively. The three phosphate groups are generally labelled as α, β and γ (*i.e.*, alpha, beta and gamma respectively), starting from the ribose. NTPs are used as a source of chemical energy to drive many biochemical reactions. Adenosine triphosphate (ATP) is, by far, the most widely used energy rich compound (Fig. 5.12). Others such as uridine triphosphate (UTP), guanosine triphosphate (GTP), thymidine triphosphate (TTP) and cytidine triphosphate (CTP) are used in specific reactions.

Table 5.3 The common nucleotides

Nucleotides	Trivial name	Abbreviations	
A. Ribonucleotides			
1. Adenosine-5′–monophosphate	Adenylic acid	AMP	Ado–5′–P
2. Guanosine–5′–monophosphate	Guanylic acid	GMP	Guo-5′–P

3. Cytidine–5′–monophosphate	Cytidylic acid	CMP	Cyd-5′–P
4. Uridine–5′–monophosphate	Uridylic acid	UMP	Urd-5′–P
B. 2′–Deoxyribonucleotides			
1. Deoxyadenosine–5′–monophosphate	Deoxyadenylic acid	dAMP	dAdo–5′–P
2. Deoxyguanosine 5′–monophosphate	Deoxyguanylic acid	dGMP	dGuo–5′–P
3. Deoxycytidine–5′–monophosphate	Deoxycytidylic acid	dCMP	dCyd–5′–P
4. Deoxythymidine–5′ monophosphate	Deoxythymidylic acid	dTMP	dUrd–5′–P

The hydrolysis of ATP and other nucleoside triphosphates is an **exergonic reaction.** The bond between the ribose and the α-phosphate is an ester linkage. The α-β and β–γ linkages are phosphoric acid anhydrides. Hydrolysis of the ester bond yields about 14 kJ/mol., whereas hydrolysis of each of the anhydride bond yields about 30 kJ/mol.

2. As the component of enzyme factor. Many enzyme cofactors and coenzymes (such as coenzyme A, NAD^+ and FAD) contain adenosine as part of their structure. They differ from each other except for the presence of adenosine. In these cofactors, adenosine does not participate directly, but removal of adenosine from these cofactors usually results in drastic reduction of their activities. For example, (removal of adenosine nucleotide from acetoacetyl-CoA reduces its reactivity as a substrate for β-ketoacyl-CoA transferase, an enzyme of lipid metabolism by a factor of 10^6.

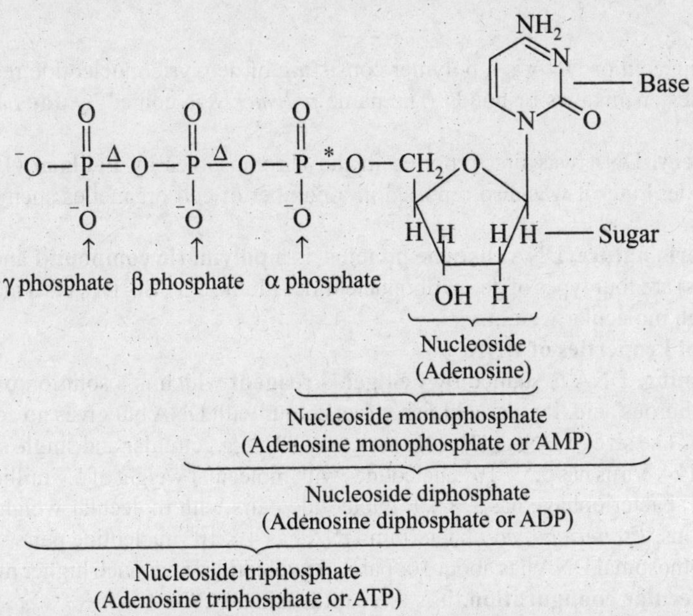

Fig. 5.12. The structure of a nucleoside and three nucleotides: AMP, ADP and ATP. * = phosphodiester bond; Δ = anhydride bonds. Hydrolysis of an anhydride bond yields more energy than hydrolysis of the ester.

3. As chemical messengers. The cells respond to their environment by taking cues from hormones or other chemical signals in the surrounding medium. The interactions of these chemical signals (called **first messengers**) with receptors on cell surface often leads to the formation of **second messengers** inside the cell, which in turn lead to adaptive changes inside the cell. Often the second messenger is a nucleotide.

One of the common second messenger is the nucleotide **adenosine 3′, 5′ cyclic monophosphate (cyclic AMP** or **cAMP)** formed from ATP in a reaction catalyzed by **adenylate cyclase,** associated with inner phase of the plasma membrane. Cyclic AMP serves regulatory functions in many types of cells. **Guanosine 3′–5′ cyclic monophosphate (cGMP)** occurs in many cells, and also has the regulatory function.

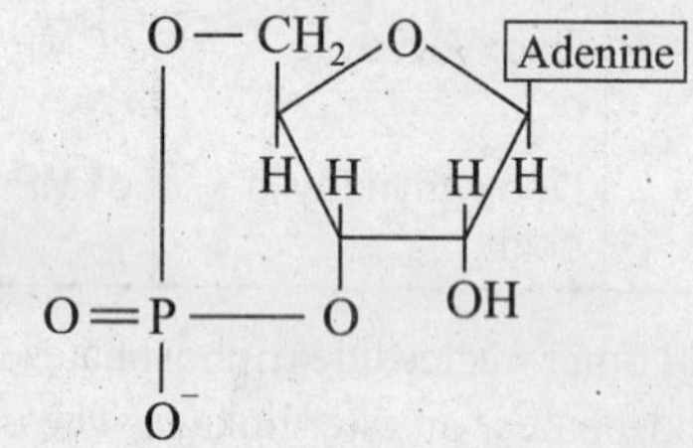

Fig. 5.13. Structure of cyclic AMP (adenosine 3′, 5′, - cyclic monophosphate).

5.4 TYPES OF NUCLEIC ACIDS

Beherns (1938) classified the nucleic acids into two classes : 1. **Deoxyribonucleic acid** or **DNA** and 2. **Ribonucleic acid** or **RNA.** All cells, prokaryotic and eukaryotic, contain two types of nucleic acids: DNA and RNA. Only viruses possess either one of the two types of nucleic acids, *i.e.*, either DNA or RNA. DNA serves as genetic material (*i.e.,* substance of a gene) of all cells and some viruses. Some viruses have genes which are made of RNA, called **genetic RNA**.

5.5 DNA

Deoxyribonucleic acid or DNA is a polymer consisting of deoxyribonucleotide residues joined by 3′–5′ phosphodiester linkages or bonds. The name *polymer* was coined by **Jons Jacob Berzelius** in 1887.

1. Discovery. DNA was first reported in the chromosomes by **Feulgen** (1924) by fuschin staining technique. Later it was also reported in a number of cell organelles such as mitochondria and chloroplasts.

2. Polymeric nature. DNA, like the proteins, is a **polymeric compound** and its monomeric units (or residues) are four types of deoxyribonucleotides (Table 5.3). So, DNA is a macrobiomolecule of extremely high molecular weight.

3. Physical Properties of DNA.

(*i*) **Staining.** DNA is stained by **Feulgen's reagent** which is a solution of acid fuschin in sulphurous acid. It gives a bluish violet colour with DNA but gives no colour with RNA.

(*ii*) **Size.** The size of DNA molecule varies widely, *e.g.*, circular and single stranded DNA of $\phi \times 174$. Virus has 5.5×10^3 nucleotides with molecular weight of 1.7 million daltons; DNA of T_2 bacteriophage has 2×10^5 nucleotide pairs with molecular weight of 130 million daltons. *Escherichia coli* bacterium DNA has 4×10^6 nucleotide pairs while eukaryotic chromosomal DNA has about 10^8 pairs of nucleotides and much higher molecular weight.

(*iii*) **Molecular configuration.** It is three dimensional (3D).

(*iv*) **As polyelectrolyte.** DNA behaves as a polyelectrolyte because the phosphate groups that lie on the outer surface of DNA, and impart an **anionic** character to it. DNA is thus *strongly acidic*. So DNA strongly bind divalent cations such as Mg^{2+} and Ca^{2+} and polycationic amines–**spermins** and **spermidine** which stabilises the DNA molecule and make it more flexible.

(*v*) **Optical rotation.** Double helix DNA shows a strong positive rotation (dextrorotatory).

(*vi*) **Viscosity.** DNA solutions are highly viscous due to rigidity of the double helix and immense length of DNA.

(*vii*) **Shape.** DNA molecule of organisms have two shape: *circular* and *linear*.

(a) Circular DNA.

1. It is found in prokaryotes (Monerans) such as bacteria, cyanobacteria; and also found in the mitochondria and plastids of eukaryotes.
2. It is circular-shaped double helix with no free ends.
3. It is not enveloped by basic proteins such as *histones*.
4. It is not surrounded by nuclear membrane so it is in direct contact with the cytoplasm.
5. It is irregularly folded to form a compact mass called **nucleoid** or **gonophore**. On one side, the nucleoid is close to the periphery of body and the DNA molecule is associated with plasma membrane.
6. Many bacteria also possess one or more small, extranucleoid circular DNA molecule called **plasmids**, found in cytoplasm outside the nucleoid.

(*b*) Linear DNA.

1. It is found in the chromosomes (or chromatin fiber) of eukaryotic nucleus.
2. It has free ends.
3. It is always covered by a proteinous coat of histones.
4. It is always surrounded by a double and porous nuclear membrane.

Chemical Composition of DNA

Linking in very large numbers, and in different linear sequences, only the four deoxyribo-nucleotides monomers (Table 5.3) polymerize to create innumerable different types of DNA molecules found in the diverse organisms.

Structure of DNA Molecules

1. Tetranucleotide hypothesis. The first attempt to elucidate the structure of nucleic acid molecules was made around 1910 by **Phoebus A. Levene.** Under his **tetranucleotide hypothesis, Levene** proposed that small tetranucleotide units, each composed of the four nucleotides in 1: 1: 1: 1 ratio form the basic structural units of the linear polynucleotides. His proposal obviously implied that each nucleic acid molecule contains equal amounts of the four component nucleotides. Until 1940. Levene's hypothesis was generally accepted.

2. Chargaff's equivalence rules. Between the years 1949 and 1953, **Erwin Chargaff** and his colleagues at Columbia analysed the **base composition** of DNA from various sources by refined quantitative chromatographic methods (Table 5.4) They made the following generalizations.

1. The sum of purines (Pu) is equal to the sum of pyrimidine (Py), *i.e.*, Pu/Py = 1. In other words, A + G = T + C.
2. The ratio of adenine to thymine is one, *i.e.*, A/T = 1.
3. The ratio of guanine to cytosine is also one, *i.e.*,, G/C = 1.
4. Bases with 6-amino groups are equal to bases with 6-keto (hydroxyl) groups, *i.e.*, A + C = G + T.
5. The ratio of A + T/G + C, known as **dissymmetry ratio,** varies greatly from one species of DNA to the other and is characteristic of that species. When the dissymmetry ratio exceeds one, such a DNA is called **AT type;** when the value is less than one, such a DNA is desinated as **GC type**.

Table 5.4 Percentage base composition of some DNAs (Source: Chargaff and Davidson 1955).

Species	Adenine	Thymine	Guanine	Cytosine
1. Human being (liver)	30.3	30.3	19.5	19.9
2. *Mycobacterium tuberculosis*	15.1	14.6	34.9	35.4
3. Sea urchin	32.8	32.1	17.7	18.4

These results indicated that **a specific relationship must exist between the two bases within each of the ratios.** These generalizations were collectively termed as **Chargaff's equivalence rules.** Later, the results of titration studies or analytical studies also suggested that the **long polynucleotide chains were held together by hydrogen bonding between the base residues.** Chargaff's observations disproved the tetra-nucleotide hypothesis; the four bases of DNA did not occur in a 1:1:1:1 ratio.

3. X-ray diffraction analysis. W.T. Astbury was the first biochemist to give any thought to the 3–'D' structure of DNA. By his X-ray cystallographic studies on DNA molecules, conducted in 1940s, he concluded that because DNA has a high density, so its polynucleotide was a stack of flat nucleotides, each of which was oriented perpendicularly to the long axis of the molecule and was placed 3.4 A° apart from each other, a concept given the name a **3.4 A° repeat.** The X-ray crystallographic studies of Astbury were continued in early 1950s by **Maurice Wilkins** and his student **Rosalind Franklin,** both of King's college, London, who obtained a superior X-ray diffraction photograph of a DNA fiber (Fig. 5.14). The molecules in a crystal are arranged in an orderly way, so that when a beam of X-rays is aimed at the crystal, the beam scatters in an orderly fashion. The scatter pattern can be recorded on photographic film or computer-controlled devices. The nature of this pattern depends on the structure of the crystal. The cross in the center of the photograph in Fig. 5.14 indicates that the molecule is a **helix;** the dark areas at the top and bottom come from the bases, stacked perpendicularly to the main axis of the molecule. This image of the DNA molecule stimulated **Watson** and **Cricks** understanding of its structure.

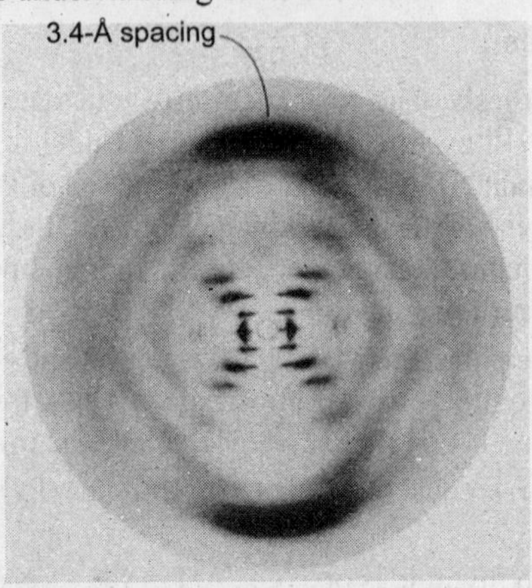

Fig. 5.14. X-ray diffraction photograph of a hydrated DNA fiber. The spots forming a cross in the center denote a helical structure. The strong arcs at the top and bottom correspond to the stack of base pairs which are 3.4 A° apart.

X-ray diffraction studies of Wilkins and Franklin confirmed Astbury's earlier inference of 3.4 A° internucleotide distance and suggested a **helical configuration** for DNA molecule. They also proposed a **34° repeat**, *i.e.*, the helix is folded into many turns and each turn causes a vertical rise of 34 A°. Furthermore their findings indicated a helical structure containing two or more strands of polydeoxyribonucleotides.

In February 1953, **Linus Pauling,** a Nobel laureate who had discovered helical structure of proteins, was investigating a **three-stranded structure** for the genetic material. **Pauling** suggested that the nitrogenous bases projected outward from a central core formed by the backbone phosphate group. But **Franklin** felt she had evidence that phosphate pointed toward outside and that the bases were in the center.

This was the situation around 1951 when **James Dewey Watson,** a 22 year old American post doctrol research fellow arrived in Cambridge and met **Francis Harry Compton Crick,** a physicist working for Ph.D. (degree in Biophysics. Crick had completed his Ph.D.) in 1954 and topic of Ph.D. thesis was X-ray diffraction: polypeptides and proteins." Although they were to work on different problems, they decided to collaborate on the study of DNA. Using intelligently all the available information, Watson and Crick, in April 1953, published an article proposing the double helical structure for DNA molecule in the same issue of journal 'Nature', in which **Wilkins** *et al.,* presented the X-ray evidence for that structure. Thus, we see that Watson and Crick structural model of DNA is the outcome of three types of studies *i.e.,* 1. the chemical nature of the components of DNA; 2. X-ray crystallography and 3. Chargaff's rule), admirably reasoned, interwoven and adjusted into a hypothetical double helical structure (Box 5.1). For this epoch-making discovery, **Watson** and **Crick** were awarded the Nobel Prize in Medicine and Physiology in 1962. They shared this Prize with **M.H.F. Wilkins.**

Box 5.1

1. About the discovery of structural model of DNA Watson once said, "actually it was a matter of 5 persons : Maurice Wilkins, Rosalind Franklin, Linus Pauling and F.H.C. Crick and me" (See J.L. Jain 2014).
2. In fact, their (Watson and Crick) discovery of the double helix model of DNA represents an elegant case of scientific sleuthing in the literature, thinking and master model building. This brilliant accomplishment ranks as one of the most significant in the history of biology because it led the way to an understanding of gene function in molecular terms.
3. Both Watson and Crick worked on the structure of DNA in their spare time. Neither Watson nor Crick performed experiments on DNA, concentrating instead on building structural models that theoretically should fit the known data about DNA, thus, **relying mainly on abstract thinking**. Using laboratory clamps and pieces of metal, they set about building a giant model of a section of DNA, which is now preserved and displayed in the London Science Museum.

Double Helical Structure of DNA

The salient features of the Watson-Crick model (Fig. 5.15) for the commonly found DNA, *i.e.,* the B-DNA are the following:

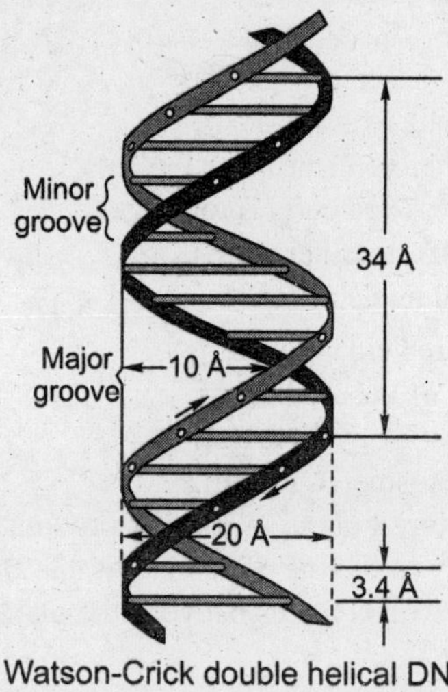

Fig. 5.15. Watson-Crick double helical DNA molecule.

1. Major and minor grooves of DNA. DNA molecule consists of two helical polynucleotide chains which are coiled around (or wrapped about) a common axis in the form of a right-handed double helix (Fig. 5.16). The two helices are wound in such a way as to produce two interchain spacings or grooves, a **major** or **wider groove** (width 12 A°, depth 8.5A°) and a **minor** or **narrow groove** (width 6Å, depth 7.5Å). The two grooves arise because the glycosidic bonds of a base pair are not diametrically opposite to each other. The minor groove contains the pyrimidine 0-2 and the purine N-3 of the base pair, and the major groove is an the opposite side of the pair. Each groove is lined by potential hydrogen bond donor and acceptor atoms. It is noteworthy that the major groove displays more distinctive features than does the minor groove. Also, the major groove is more accessible to proteins interacting with specific sequences of DNA.

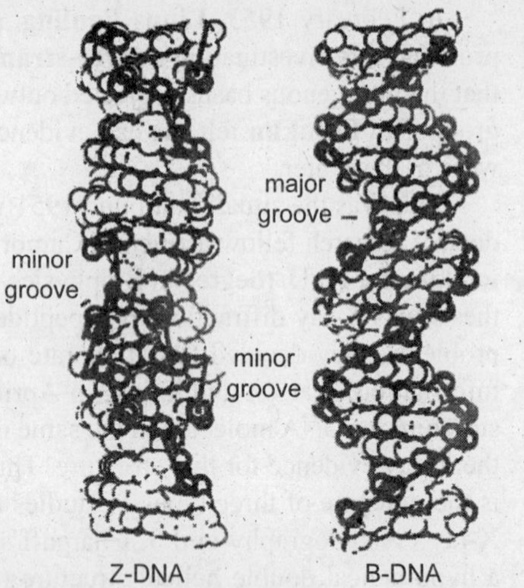

Fig. 5.16. Two molecular forms of DNA double helix, right-handed B-DNA and left-handed Z-DNA.

The two helices wind along the molecule parallel to the phosphodiester backbones. Such double helices cannot be pulled. They can be separated only by an unwinding process. They are called **plectonemic coils** *i.e.,* coils which are interlocked about the same axis. The two strands or chains run in opposite direction, *i.e.*, they are **antiparallel** which means that in one strand the ring oxygen of sugar moieties face upwards while in the outer strand, the ring oxygen faces downwards. Also, the two polymer chains are not identical but because of base pairing are **complementary** to each other.

Sometimes, one polymer chain of DNA is called the **Watson strand** or **W strand** and its complement the **Crick strand** or **C strand** as a gesture to the propounders of this model.

2. In DNA, the phosphate and deoxyribose units are found on the periphery of the helix whereas the purine and pyrimidine bases occur in the center (Fig. 5.17). The planes of the bases are perpendicular to the helix axis. The planes of sugars are almost at right angles to those of the bases.

3. The diameter of the DNA helix is 20Å. The bases are 3.4A° apart along the helix axis and are related by a rotation of 36 degree. Therefore, the helical structural repeats after 10 residues (nucleotides) on each chain, *i.e.*, at intervals of 34Å. In other words, each turn of the helix contains 10 nucleotides.

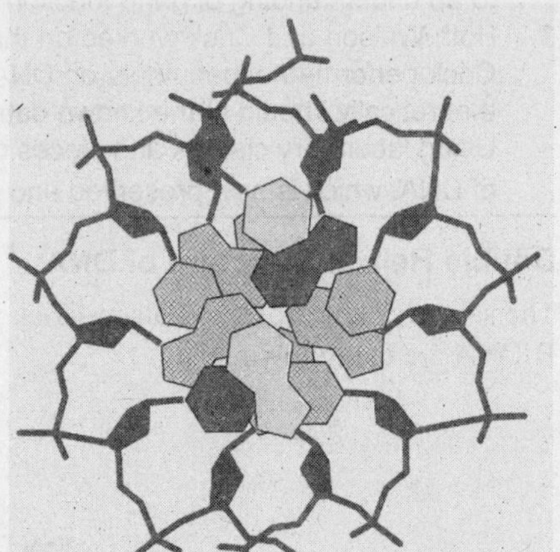

Fig. 5.17. Axial view of one of the strands of a DNA double helix (as viewed down the helix axis).

4. The two polymer chains are held together by **hydrogen bonds** between pairs of bases. Adenine (A) always pairs with thymine (T) by hydrogen bonds and guanine (G) with cytosine (C) with 3 hydrogen bonds are weak in nature but, as in case of proteins, a large number of them involved in DNA molecule confer stability to it. It is now thought that the stability of the DNA molecule is primarily an outcome of Van der Waals forces between the planes of stacked bases.

5. The sequence of bases along a polynucleotide chain is not restricted in any way. The precise sequence of bases carries the genetic information.

Antiparallelity of polynucleotide chains of DNA. The procedure of **nearest neighbour analysis** has shown that the two complementary chains do not run in the same direction with respect to their internucleotide linkages, but rather are **antiparallel**, *i.e.*, one strand runs in 5′→3′ direction and the other strand in the 3′→5′direction (Fig. 5.18). Thus, of the two chains, one ascends and the other descends. They have reverse polarity, that is to say the sequence of the atoms in the two chains is opposite to each other. This is analogous to 2 parallel streets each running one way but carrying traffic in opposite directions.

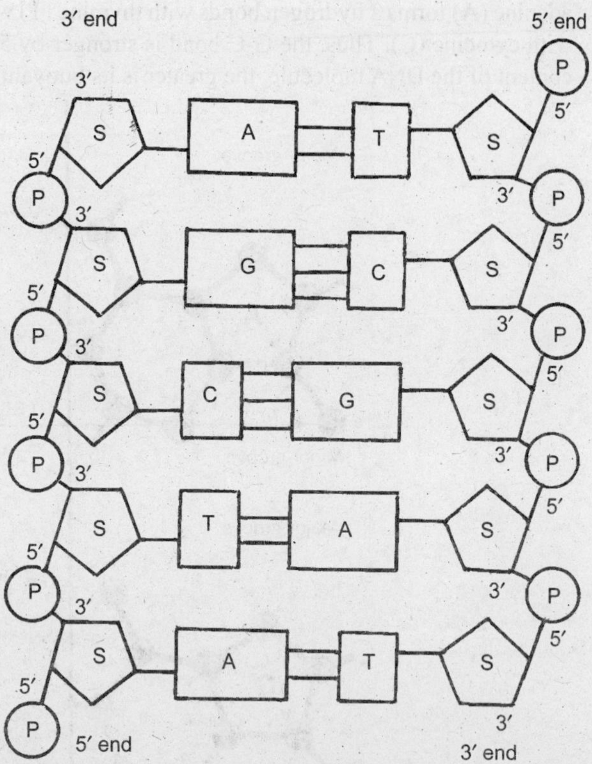

Fig. 5.18. A part of DNA molecule enlarged to show antiparallel, strands : 3′→5′ (left) and 5′→3′(right).

Base complementarity of the polynucleotide chains. An important feature of the double helix is the specificity of the pairing of bases. Pairing always occurs between adenine and thymine and between guanine and cytosine (Fig. 5.18.). This means that a 6-aminopurine will always bond with a 6-ketopyrimidine, and a 6-ketopurine bond with a 6-aminopyrimidine. Consequently, the two helices of a DNA molecule are complementary to each other and not identical. In fact, base pairing in DNA molecule is due to **steric** and hydrogen-bonding factors.

1. Steric factor. The steric restriction is imposed by the regular helical nature of the sugar phosphate backbone of each polynucleotide chain. The geometry of the base pair has some special consequences in that the distances between the glycosidic bonds are the same for both the base pairs and also because of the bond angle between the glycosidic bond direction and the line joining the C1 atoms is the same in each pair. With the result, the glycosidic bonds of all nucleotides are arranged in an identical manner in relation to the axis of the helix, despite the differences in the bases.

The glycosidic bonds that are attached to a base pair are always 10.85 A° apart. A purine-pyrimidine base pair fits perfectly in this space. In contrary to this, there is insufficient space for two purines, whereas there is more than enough space for two pyrimidines so that they would be far apart to form hydrogen bonds. Hence, one member of a base pair in a DNA helix must be invariably be a purine and the other a complementary pyrimidine because of steric reasons. Obviously, if the base order in one strand is known, the sequence of bases in the other strand can be predicted. For example, if one polymer chain of DNA has base sequence reading ACCGATC ....., then the corresponding region of the complementary chain will have TGGCTAG....

2. Hydrogen bonding factor. The base pairing is further restricted by hydrogen-bonding requirements. The hydrogen atoms in purine and pyrimidine bases have well-defined positions. Adenine (A) cannot pair with cytosine (C) because there would be two hydrogen atoms near one of the bonding positions and none at the other. Similarly, guanine (G) cannot pair with thymine (T). In contrast, adenine (A) forms 2 hydrogen bonds with thymine (T) whereas guanine (G) forms 3 hydrogen bonds with cytocine (C). Thus, the G-C bond is stronger by 50% than the A-T bond. The higher the G-C content of the DNA molecule, the greater is its buoyant density (Fig. 5.19).

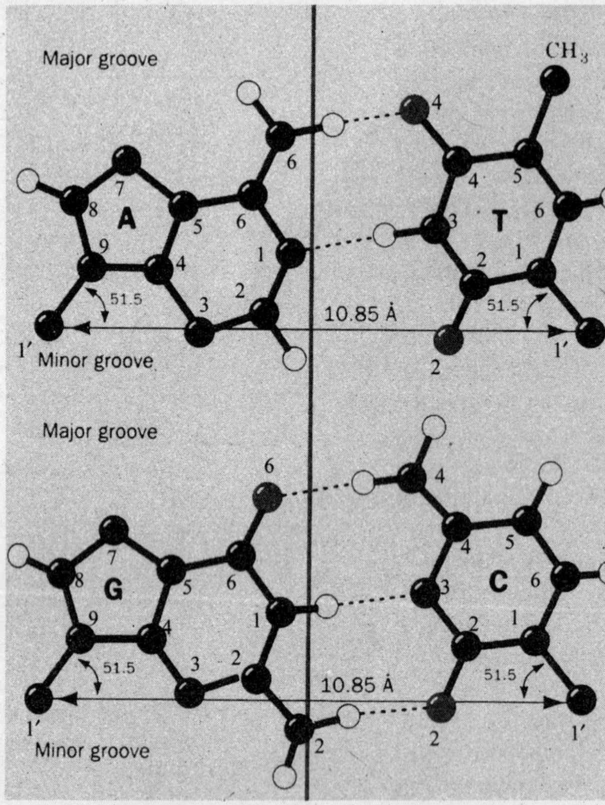

Fig. 5.19. Structure of the two Watson-Crick base pairs. The line joining the C1′ atoms in the same length in both base pairs and makes equal angles with the glycosidic bonds to the bases. This gives DNA a series of pseudo-twofold symmetry axes (often referred to as **dyad axes**) that pass through the center of each base pair (thick line) and are perpendicular to the helix axis. Note that A.T base pairs associate via two hydrogen bonds, whereas C.G base pairs are joined by three hydrogen bonds.

FORMS OF DNA

Under different conditions of isolation, purification and crystallization, six different forms of DNA double helix have been identified. Some of these forms are interconvertible. The difference in these DNA forms are based on: (i) the number of base pairs present in each turn of DNA helix. (ii) The pitch or angle between each base pair. (iii) The helical diameter of DNA molecule. (iv) The right or left-handedness of double helix.

1. B–DNA (Right handed DNA). This form of DNA occurs in all living beings under normal conditions, *i.e.*, under low salt concentration and high degree of hydration. Each coil or turn measures 34 A° or 3.4 nm. Each turn has 10 base pairs (bp) and each base pair occupies 3.4 A°. Tilt from the axis of helix (*y*) is 6.3°. The axial rise (*h*) is 3.37 A°. The two strands or polynucleotide chains of this DNA are wounded in right-handed turns.

The double helical structure described by Watson and Crick was B-DNA only.

2. A-DNA (Right handed DNA). A-DNA is the dehydrated form of DNA. It occurs under conditions of decreased hydration and increased concentration of Na^+, K^+ or Ca^{2+} ions. Like B-DNA, it also has right handed configuration, but A–DNA is more compact and more bulky. It has 11 base pairs per turn of the helix and a diameter 23A°. These base pairs are sufficiently tilted from the axis at helix (y = 20.2°). Axial rise (h) is about 2.56°. The orientation of bases is also different. These are tilted and displaced laterally in relation to the axis of helix. The appearance of major and minor grooves is also modified in comparison to B-DNA.

It is presumed that A-DNA is formed physiologically as a result of interaction with the hydrophobic molecules in changing cellular conditions.

3. C-DNA (Right-handed DNA). This form of DNA occurs under even greater dehydration conditions than A- and B-DNA. The C-DNA has only 9.3 base pairs per turn of helix and is only 19A° in diameter. It means C-DNA is less compact and narrower than other two right-handed DNAs. Its base pairs are negatively tilted as the axis of helix (y) is –7.8°. The axial rise (h) of C-DNA is 3.32 A°.

4. D-DNA and E-DNA (Right-handed DNAs). These two forms of DNA are also right-handed. These occur in helices which lack guanine in their base composition. These have 8 and 7½ nucleotides per turn, respectively. The base pairs are negatively tilted as the axis of the helix (y) by –16.7. The axial rise (h) is 3.03A°.

5. Z-DNA (Left-handed DNA). Z-DNA is a left-handed DNA. The phosphodiester backbone of its two polynucleotide strands follow a zig-zag course because of which this is called Z-DNA. The helix of Z-DNA is 18A° in diameter and contains 12 base pairs per turn. The value of y (tilt pairs from axis of helix) is 7°. The helix rise (h) per pair is 3.7A°. The major groove present in B-DNA is nearly absent in Z-DNA. Thus, Z-DNA is more compact. Because of this compactness the phosphate groups on opposite strands of Z-DNA are closely placed and create a greater electrostatic repulsion. Thus, Z-DNA is energetically less stable. Z-DNA is formed only when purines and pyrimidines are present alternately in the polynucleotide chains. Z-DNA also contains **repeating dinucleotide pair units** (poly G = C – C = G or poly A = T – C = G)$_n$.

Z-DNA was discovered by **Wang** and **Rich** in 1979. Z-DNA form appears when DNA is **brominated** or **methylated** and it can be stabilized by high salt concentration of specific cations. The existence of Z-DNA in *Drosophila* can be demonstrated by using antibodies that recognize and bind specifically to Z-DNA.

Table 5.5 Comparison of different forms of DNA

Characters	A-DNA	B-DNA	C-DNA	D-DNA	Z-DNA
1. Base pair per turn of the helix	11	10	9.33	8	12
2. Tilt of base pairs (y)	20.22°	6.3°	–7.8°	–16.7°	7°
3. Axial rise (h)	2.56Å	3.37Å	3.32Å	3.03Å	3.7Å
4. Helical diameter (Å)	23Å	20Å	19Å	–	18Å
5. Handedness of the double helix DNA	Right handed	Right handed	Right handed	Right handed	Left handed

Denaturation and Renaturation of DNA Helix

1. Denaturation of DNA is a loss of biological activity and is accompanied by cleavage of hydrogen bonds holding the complementary sequences of nucleotides together. This results in a separation of the double helix into the two constituent polynucleotide chains. In denaturation, the firm, helical, two-stranded native structure of DNA is converted to a flexible, single-stranded denatured state. The splitting of DNA molecule into its two strands or chains, during denaturation, is obvious because of the fact that the hydrogen bonds holding the bases are weaker than the bonds holding these bases to the sugar-phosphate groups. The transition from native to a denatured form is usually very abrupt and is accelerated by high temperature or by reagents such as urea and formide, which enhance the aqueous solubility of the purine and pyrimidine groups. Denaturation involves the following changes:

***(i)* Ultraviolet absorbency (hypochromatic and hyperchromatic effect).** DNA molecule absorbs UV region of spectrum due to heterocyclic aromatic nature of its purines and pyrimidines.

Although each base has a unique absorption spectrum, but overall DNA exhibits maximum absorbtion at 260 nm. However, native and, stranded DNA absorbs much less UV-light at 260 nm than would be expected from the summation of the light absorbed by its constituent mononucleotides. This is called **hypochromatic effect.** This is probably due to interactions between the electrons of neighbouring bases.

In a denatured DNA, there is a tremendous increase in the UV-absorbency (upto 40 per cent) at 260 nm. The total light absorption of fully denatured DNA is nearly equal to that of an equivalent number of free mononucleotides. The percentage increase in UV-absorbency is directly related to its number of free A-T base pairs. It is so as the stacking interactions between the bases are decreased in the random coils.

This phenomenon of increase in UV-absorbency at 200 nm is a denatured DNA is called **hyperchromatic effect.** This can be used to distinguish single or double stranded nature of DNA as single stranded DNA does not show hypochromatic effect.

***(ii)* Decrease in specific optical rotation.** Native DNA exhibits a strong positive rotation which is highly decreased upon denaturation. This change is analogous to the change in rotation observed when the proteins are denatured.

***(iii)* Decrease in viscocity.** The solutions of native DNA possess a high viscocity because of the relative rigid double helical structure and long, rod-like character of DNA. Disruption of the hydrogen bonds of DNA molecule causes a marked decrease in viscocity.

(A) Effect of pH on denaturation. Denaturation of DNA helix also occurs at **acidic** and **alkaline pH** values at which ionic changes of the substituents on the purine and pyrimidine bases can occur. (*Substituent* is a radical, atom or group, substituting or replacing another in a chemical reaction.) In acid solutions near pH2 to 3, at which amino groups bind protons, the DNA helix is disrupted. Similarly, in alkaline solutions near pH12, the enolic hydroxyl groups ionize, thus preventing the keto-amino group hydrogen bonding.

(B) Effect of temperature on denaturation. The DNA double helix, although stabilized by hydrogen bonding, can be denatured or melted by heat. The unwinding of the double helix is called **melting** because it occurs abruptly at a certain characteristic temperature called **denaturation temperature** on **melting temperature (T_m)**. The melting temperature is defined as the temperature at which half the helical structure is lost. The melting of DNA is readily monitored by measuring its absorbance of light at wavelength near 260 nm. As already stated, the unstacking of the base pairs results in increased absorbance a phenomenon called **hyperchromicity** (Fig. 5.20). T_m is analogous to the melting point of a crystal. It can be lowered by the addition of urea which disrupts hydrogen bonds. In 8 M urea, T_m is decreased by nearly 20 ºC. DNA can be completely denatured (*i.e.*, separated into a single-stranded structure) by 95% formamide at room temperature only.

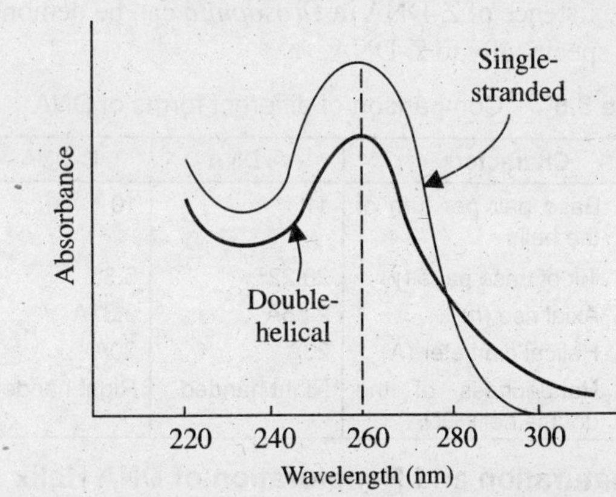

Fig. 5.20. The absorbance spectra of a DNA solution at 260 nm at pH 7.0 When mixtures of nucleotides are present, the wavelength at 260 nm (dashed vertical line) is used for measurement.

Since the G-C base pair has 3 hydrogen bonds as compared to 2 for A-T, it follows that DNA with high concentration of G and C might be more stable and have a higher T_m than those with high concentrations of A and T (Fig. 5.21).

In other words, the DNA molecules containing less G-C bond denature first as G-C bond has higher thermal stability. In fact, the T_m of DNA from many species varies linearly with G-C content, rising from 77 to 100 °C as the fraction of G-C pairs increases from 20 % to 78 % (Fig. 5.22). This measurement may, hence forth, be used as an **index of heterogeneity** of nucleic acid molecules. For instance, viral DNA has a much sharper thermal transition than that prepared from animal species.

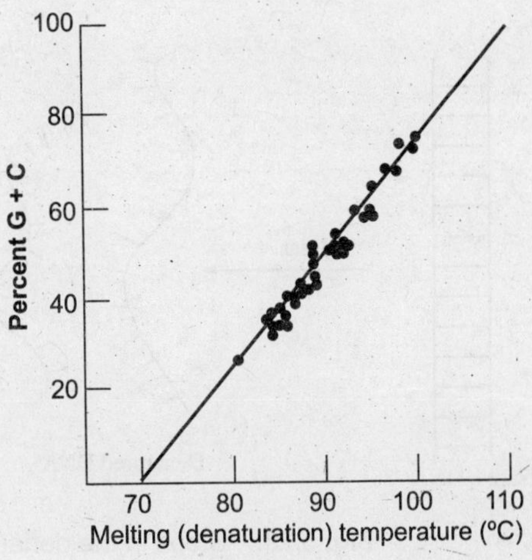

Fig. 5.21. Relationship of the number of hydrogen bonds (G-C content) and the thermal stability of DNA from different sources (after Tamarin 2002).

2. Renaturation and Annealing Complete rupture of the two-stranded helix by heating is not a readily reversible process. However, if a solution of denatured DNA, prepared by heating, is cooled slowly to room temperature, some amount of DNA is renatured. Maximum reversibility (50-60%) is usually attained by **annealing** (= slow cooling) the denatured DNA, *i.e.*, holding the solution at a temperature about 25 °C below T_m and above a concentration of 0.4 M Na^+ for several hours. Fast cooling will not reverse denaturation, but if the cooled solution is again heated and then cooled slowly, renaturation takes place. The restoration occurs because the complementary bases reunite by hydrogen bonds and double helix again forms.

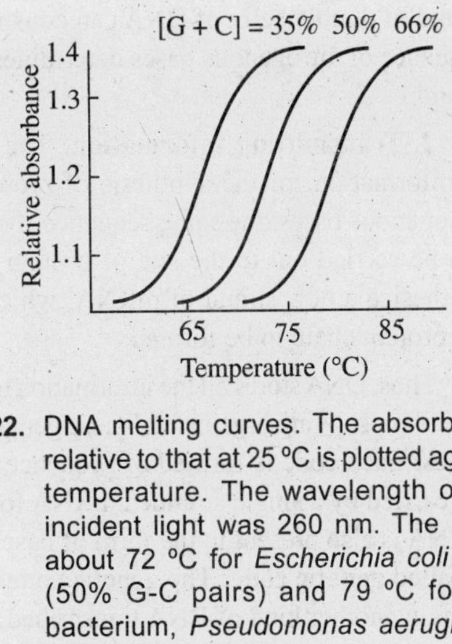

Fig. 5.22. DNA melting curves. The absorbance relative to that at 25 °C is plotted against temperature. The wavelength of the incident light was 260 nm. The T_m is about 72 °C for *Escherichia coli* DNA (50% G-C pairs) and 79 °C for the bacterium, *Pseudomonas aeruginosa* DNA (66% GC-pairs).

Renaturation of complementary single strands to produce fully double-stranded molecule requires following two reactions (Fig. 5.23).

(*i*) Nucleation reaction. In this reaction hydrogen bonds form between two complementary single strands; this is a bimolecular, second-order reaction.

(*ii*) Zippering reaction. In this reaction hydrogen bonds form between all the bases in the complementary strand; this is a unimolecular, first order reaction.

Significance of denaturation and renaturation. By denaturing two different DNA molecules and annealing them together, it is possible to form a **hybrid DNA duplex** formed of one strand from each DNA molecule. This technique is also used to produce DNA -RNA combinations.

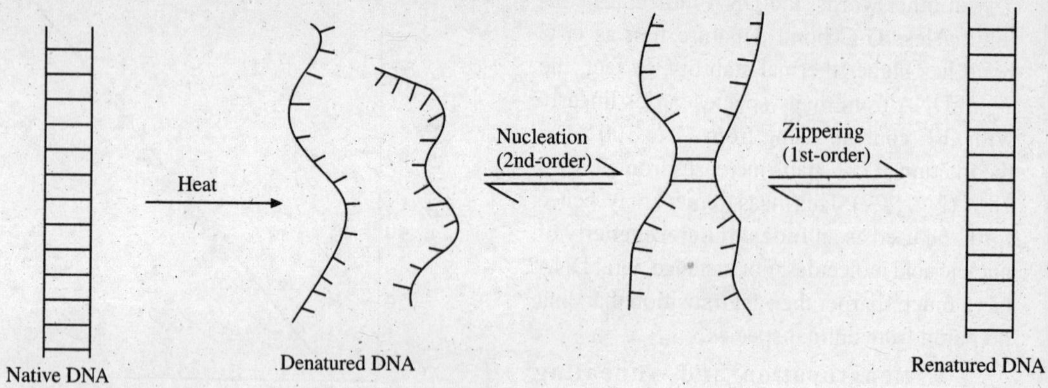

Fig. 5.23. Steps in the denaturation and renaturation of DNA fragments.

Biological Significance of Watson and Cricks Model of DNA

DNA as the hereditary or genetic material of cell is expected to possess following three separate special features:

1. Self duplication. The Watson and Crick's model of DNA provides the only possible device whereby a particular molecule with specific sequence of base pairs could be duplicated easily. Each strand of double helix of DNA can construct its complementary strand by acting as a template. The sequence of nitrogenous bases determines the laying down of bases in the second or complementary strand.

2. Transmitting information. The DNA structure (*i.e.*, double helix) explains transmitting of information for the synthesis of proteins. Since the information are coded in the sequence of nitrogenous bases, the same sequence if coded in some other molecule (messenger RNA or mRNA) can be carried out to the site of protein synthesis. One of the two strands of DNA molecule can synthesize a new strand of mRNA, which in turn controls the proper alignment of amino acids in the protein chain to be formed.

Thus, DNA stores all the information to control growth, development, differentiation, reproduction, etc., of most living organisms. These genetic informations of DNA are in the form of base triplet called **codes.** Each code is formed of sequence of three nucleotides of DNA. Thus informations of DNA are carried by a single stranded mRNA formed by the process of **transcription.** Genetic message or mRNA is also present in the form of base triplets called **codons.** The sequence of codons on mRNA is called **genetic code.** The genetic code is decoded on the ribosomes (*i.e.*, protein factories which contain another kind of RNA transcribed by DNA) to form proteins by linking specific number and types of amino acids in a specific sequence (of polypeptides and proteins). This process is called **translation.** Some proteins act as **structural proteins** (*e.g.*, cytoskeleton, collagen, muscles, etc.) while some proteins act as **enzymes** and regulate the functioning (metabolism) of the cell. As already has been mentioned earlier, DNA is also capable of its self-duplication or replication so that daughter cells have same blue print or genetic material as in the parent cell. This unidirectional flow of genetic information is called **central dogma** (Fig. 5.24).

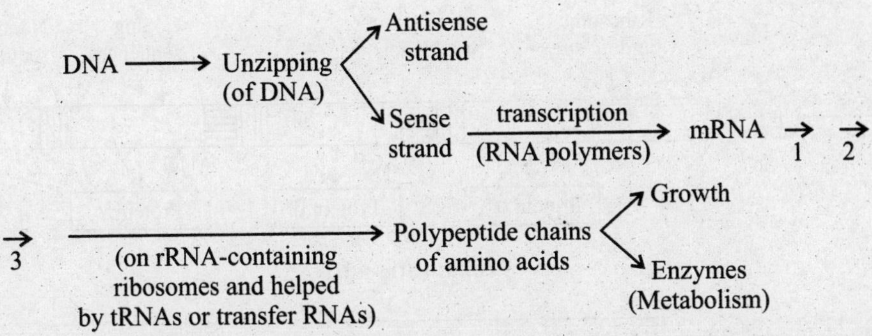

Fig. 5.24. Central dogma.

3. DNA repair. DNA molecule is relatively fragile and may get damaged from shear bending forces and by local changes in pH. It base sequence changes by chemical agents from environment and also by X-ray and ultraviolet radiations. Actually the double helical structure of DNA presents the most stable structure. It is most suited for the long term preservation of the coded informations. Because the base sequence in the two polynucleotide chains is complementary, the coded informations in a DNA molecule are preserved in duplicate. If sequence of bases on one chain is disturbed, the correct sequence of the other chain may help in correcting the errors that occur in the complementary strand and sends the correct message.

5.6 RNA

Ribonucleic acid (RNA) is a single-stranded nucleic acid composed of a polymeric chain of ribose-containing nucleotides.

Like DNA, RNA is a polymer consisting of thousands of ribonucleotide residues joined by 3′–5′ phosphodiester bonds forming linear sequence. The sugar moiety in RNA is ribose. Genetic information contained in DNA is transcribed in the synthesis of RNA. Some of which (mRNA) is translated in the synthesis of protein. RNA is generally found in the cytoplasm and the nucleolus. Besides this, it is also found in mitochondria and chloroplasts. As already mentioned, RNA is made up of long unbranched polynucleotide chain of single strand. The single strand coil itself to form a helix.

In RNA both types of nitrogenous bases-pyrimidines and purines are present. The purines are adenine and guanine as in DNA. The pyrimidine bases are cytosine and uracil. In RNA, unlike DNA, thymine is replaced by uracil. Thus, four types of ribonucleotides are present in RNA (Table 5.3).

Types of RNA

Following three types of RNA molecules are present in a cell:

1. Messenger RNA (mRNA)
2. Ribosomal RNA (rRNA), and
3. Transfer RNA (tRNA)

1. Messenger RNA (mRNA). Jacob and **Monod** (1961) proposed the term messenger RNA (mRNA) for the portion of cellular RNA. Carrying information for protein synthesis from DNA to the sites of protein synthesis. The process of protein synthesis can be divided into two steps :

1. The formation of mRNA from DNA (transcription) and
2. The formation of proteins from mRNA (translation).

Since proteins occur in different sizes, mRNA varies greatly in length and molecular weight. In prokaryotes a single mRNA may code one or more polypeptide chains. If mRNA codes for a single polypeptide chain, it is called **monocistronic** and if it codes for two or more different polypeptide chains, it is called **polycistronic** (Fig. 5.25).

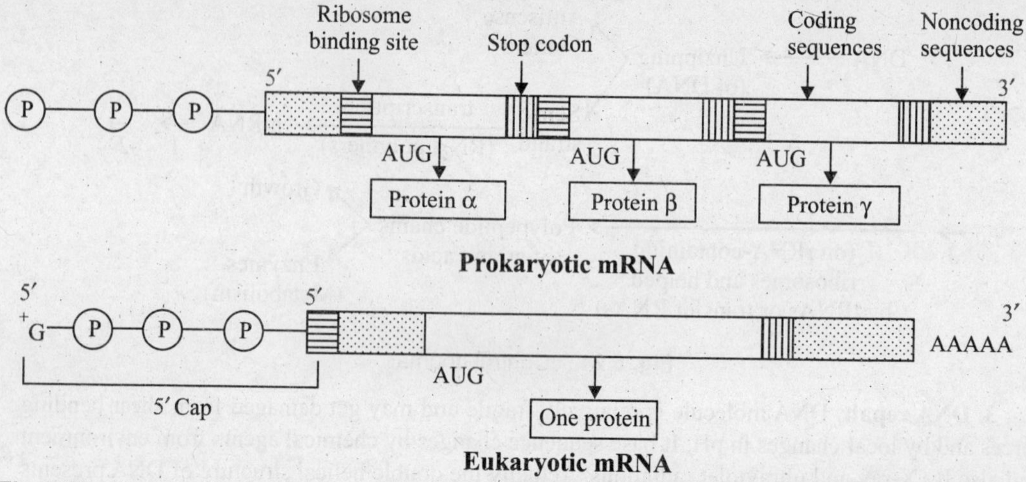

Fig. 5.25. A comparison of the structure of prokaryotic and eukaryotic messenger (mRNA) molecules.

The molecular weight of an average sized mRNA is about 500,000 Daltons and sedimentation coefficient is 85. It forms 5% to 10% of the total RNA present in the cell.

The mRNA are unstable in the bacterial systems with a half-life from a few seconds to about 2 minutes (*Half life* is the period of time after which half of the material is degraded). In mammalian systems, however, mRNA molecules are more stable with a half-life ranging from a few hours to one day.

The mRNA is formed as a complementary strand from one of the strands of DNA, the base sequence of each mRNA is complementary to that of DNA.

An eukaryotic mRNA has the following features:

1. Cap. A cap is found at the 5′ end of the mRNA in which methylated guanine is present. The rate of protein synthesis depends upon the presence of cap, without the cap the mRNA molecule binds very poorly to the ribosomes (the RNA containing sites of protein synthesis).

2. Noncoding region-I (NC-I). The cap is followed by a region of 10-100 nucleotides which is rich in A and C residues. This portion of mRNA does not translate into protein.

3. Initiation codon. In both prokaryotes and eukaryotes the initiation codon is AUG.

4. Coding region. This region translates protein.

5. Termination or stop region. Termination or stop region of translation on mRNA is carried by termination codon. Termination codons are UAA, UAG and UGA.

6. Non-coding region-2 (NC-2). This region consists of 50-150 nucleotides and does not translate protein.

7. Poly(A) sequence. At 3′ end poly A sequence is found. It initially consist of 200-250 nucleotides but becomes shorter with age.

2. Ribosomal RNA (rRNA). It is also called **insoluble RNA**. The ribosomal RNA is most stable form of RNA and is found in ribosomes. The rRNA molecules form 40-60% of the total weight of the ribosome. The rRNA forms 80% of the total RNA of the cell. The rRNA molecule may be short compact rod, a compact coil or extended strand depending upon ionic strength, temperature and pH. The rRNA consists of a single strand and the same strand is twisted upon itself in some regions and forms helix. In the helical region most of the base pairs are complementary and are joined with hydrogen bonds. In the unfolded single strand region the base pairs are not complementary. Hence rRNA does not show purines and pyrimidine equality. The rRNA differs in base contents from tRNA and mRNA. In rRNA the quantity of guanine and cytosine is relatively much more.

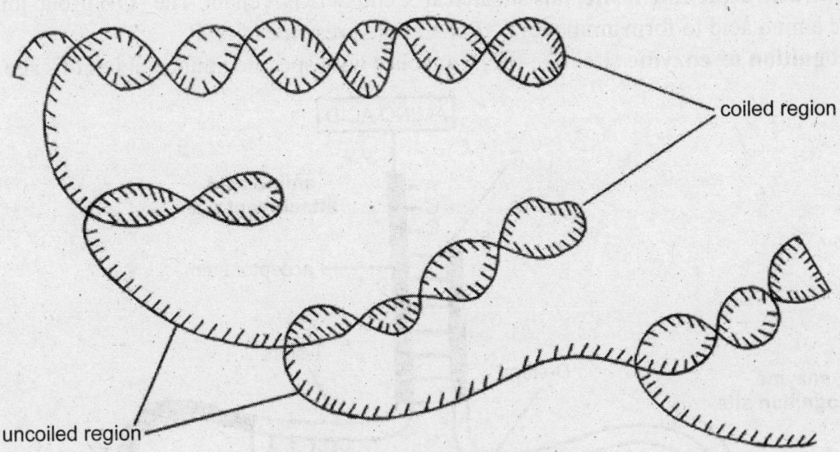

Fig. 5.26. Structure of rRNA.

The ribosomes of prokaryotic and eukaryotic cells possess different species of RNA (Table 5.6).

Table 5.6 Ribosomes and their RNA (*i.e.,* rRNAs).

Ribosomes	rRNAs
70S prokaryotic ribosomes	
30S subunit	16S
50S subunit	23S, 5S
80S eukaryotic ribosomes	
40S subunit	18S
60S subunit	28S, 5.8S, 5S

The ribosomes of prokaryotes and those found in the plastids and mitochondria of eukaryotes contain 3 different types of RNA: 23S rRNA (MW 1,200,000 daltons), 16S rRNA (MW 550,000 daltons) and 5S rRNA (MW 36000 daltons).

Eukaryotic (*e.g.*, mammalian, ribosomes has 4 types of rRNA. The 40S ribosomal subunit contains single 18S rRNA. The 60S ribosomal subunit contains 3 types of rRNAs: 5 S, 5.8 S and 28 S.

3. Transfer RNA (tRNA). It is also called **soluble RNA**. The tRNA is smallest polymeric form of RNA. The tRNA molecule is found in the cytoplasm. It comprises 10-15% of the total weight of RNA present in the cell. Its molecular weight is 25,000–30,000 dalton.

The transfer RNA plays a key role in protein synthesis. It picks up specific activated amino acid from the cytoplasm, carries it to the site of protein synthesis and attaches itself to ribosome in accord with the sequence specified by mRNA. Finally, it transmits the amino acid to the polypeptide chain which is being manufactured.

Accordingly, in the cytoplasm of the cell about 20 different types of tRNA molecules (one for each amino acid) are present which are quite similar to each other.

A tRNA molecule is made up of 75-80 nucleotides. Sedimentation constant of each tRNA molecule is 4S and molecular weight is 25000 daltons. Each molecule of tRNA coils itself in such a manner that its both arms are coiled around each other. Both ends of its polynucleotide chains show identical structure. At 3′ end of chain CCA base sequence is present and 5′ end terminates in G. At the turn of each tRNA molecule chain there is a definite sequence of three nitrogen bases which form **anticodon** which recognizes the codon on the mRNA (Fig. 5.27).

On the basis of nucleotide analysis of the RNA of yeast, **Robert Holey** (1965) proposed a **clover leaf model** of tRNA. According to this model tRNA has four arms out of which one arm has anticodon. On each tRNA molecule following four special sites are present.

1. Amino acid attachment site. It is situated at 3′ end of tRNA chain. The –OH group joins with any specific amino acid to form aminoacyl –tRNA in the presence of ATP.

2. Recognition or enzyme site. The tRNA attaches with specific amino acid at this site.

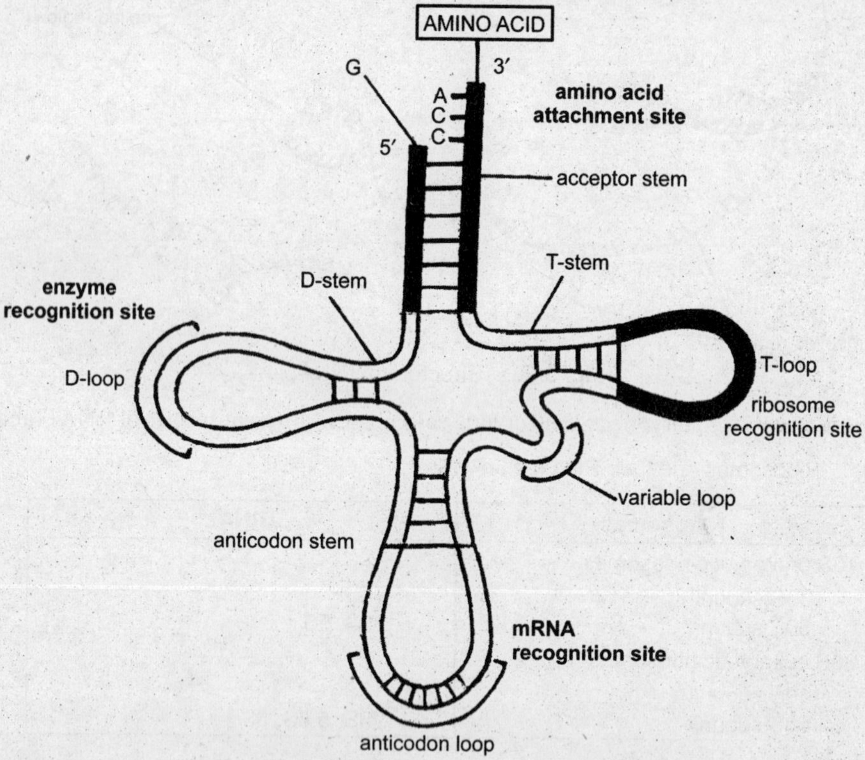

Fig. 5.27. Clover leaf model of alanine tRNA molecule.

3. Anticodon (codon recognition site). During protein synthesis, tRNA attaches with ribosome at this site.

In tRNA, besides normal nitrogenous bases—C, G, A and U, a few abnormal bases such as **pseudouridine, inosinic acid, methyl guanine, methylaminopurine,** etc., are also present. The presence of these bases does not create any hindrance of any type of pairing of tRNA with mRNA. The presence of these abnormal bases may possibly stop intramolecular base pairing of tRNA or they help in amino-acyl tRNA synthetase enzyme recognition.

In the cytoplasm, tRNA molecules occur in both active and inactive forms. The **inactive molecules** of tRNA lacks, the C-C-A sequence of nitrogenous bases at 3′ end of the chain either in full or part. By the addition of this nucleotide sequence an inactive tRNA molecule can be activated. The process is governed by the enzyme **cytidine triphosphate (CTP)** and ATP.

Table 5.7 Differences between DNA and RNA

Deoxyribonucleic acid (DNA)	Ribonucleic acid (RNA)
1. DNA is found in the chromosomes of the nucleus and is mainly concentrated in nucleus only.	1. RNA is chiefly concentrated in the cytoplasm although it also occurs in the nucleolus and to some extent in the nucleoplasm and associated with the chromosomes.
2. DNA is double stranded structure having a helical form. The two strands are coiled spirally in opposite directions.	2. RNA molecules are single stranded. The strand in some cases (in tRNA) is coiled on itself and may be connected by hydrogen bonds.

3. The sugar molecule is DNA is deoxyribose.	3. The sugar molecule in RNA is ribose.
4. The four nitrogenous bases found in DNA are: (*i*) Adenine } Purine bases (*ii*) Guanine (*ii*) Cytosine } Pyrimidine bases (*iv*) Thymine	4. The four nitrogenous bases found in RNA are: (*i*) Adenine } Purine bases (*ii*) Guanine (*iii*) Cytosine } Pyrimidine bases (*iv*) Uracil
5. The purines and pyrimidines in DNA are found in equal proportion, *i.e.*, A/T = G/C = 1.	5. It is not necessary that purines and pyrimidines occur in equal proportion.
6. DNA is the hereditary material. The informations for different cell activities and life processes are encoded and stored in DNA molecules.	6. RNA participation in protein synthesis.

QUESTIONS

Long Answer Questions

1. What is polynucleotide chain ? Explain how these chains contribute in the formation of genetic material? Elaborate the answer by giving structural formulae. (Avadh 2002)
2. Give an account of Watson and Crick's model of DNA. (Garhwal 2003; Lucknow 2004)
3. Describe the structure of DNA. (Rohilkhand 2004)
4. Give an account of the structure and biological significance of DNA. (Garhwal 2004)
5. Describe molecular structure of different forms of DNA.
6. Write an essay on DNA and RNA.
7. Describe chemical composition of RNA. What are different kinds of RNA? State their function. (Avadh 2003)
8. Name different types of RNAs found in animal cell. Discuss detailed structure and functions of tRNA. (Kanpur 2004)
9. Descrribe clover leaf model of transfer RNA. (Garhwal 2003)

Short Answer Questions

1. What facts you know about the molar ratio of nitrogenous bases in DNA? Who contributed to these facts?
2. Discuss biological significance of Watson and Crick's model of DNA.
3. Give differences between DNA and RNA.
4. Write differences between the following:
   (*i*) Purine and pyrimidine;
   (*ii*) Nucleoside and nucleotide;
   (*iii*) Polynucleotide chain and polypeptide chain;
   (*iv*) B-DNA and Z-DNA.
5. Write short notes on the following:
   (*a*) ribonucleic acid (RNA);
   (*b*) DNA;
   (*c*) Clover leaf model of Holley.

Very Short Answer Questions

1. Name various nitrogenous bases found in DNA.
2. What is Chargaff's rule about DNA structure ?

3. Who proposed working model of DNA?
4. What are monomers of deoxyribonucleic acid or DNA?
5. How the nucleotides are joined together in a polynucleotide chain of DNA?
6. What is the width of DNA helix?
7. What are genes made of?
8. What is transcription?
9. Define the following terms : anticodon, codon, translation.
10. Write about contributions of following scientists :
    (*i*) Hershey and Chase;
    (*ii*) Chargaff.
11. Give one word for the following :
    (*i*) Triplet bases on mRNA which code for an amino acid;
    (*ii*) Process of formation of mRNA on DNA template;
    (*iii*) Occurrence of more than one codon for the same amino acid.

Fill in the Blanks

1. Guanine + deoxyribose + phosphoric acid together form ______________.
2. Nucleic acids are macromolecular and polymers of ______________.
3. The nitrogenous bases of two polynucleotide chains of a DNA molecule are held together by __________.
4. The nucleotides of a polynucleotide chain are joined together by ________ bond, between __________ and phosphate.
5. The diameter of a double helix of DNA is __________ A°.
6. ____________ and ______________ got Nobel prize for double helical structure of DNA.
7. In tobacco mosaic virus (TMV) the hereditary material is __________. It was demonstrated by __________ and __________.

Multiple Choice Questions

1. DNA is mainly located inside:
   (a) nucleolus (b) nucleus
   (c) mitochondria (d) nucleus and mitochondria
2. In one of the following DNA is not found
   (a) mitochondria (b) chloroplast
   (c) bacteriophage (d) tobacco mosaic virus
3. Which one of the following molecule is genetic element in tobacco mosaic virus?
   (a) DNA (b) RNA
   (c) proteins (d) ATP
4. Name the discoverer of nuclein.
   (a) H. J. Muller (b) Friedrich Miesher
   (c) B. McClintok (d) Griffith
5. Which one of the following is specific for DNA
   (a) haemotoxylin (b) acetocarmine
   (c) Feulgen (d) fast green
6. Nucleic acids are polymers of
   (a) nucleoproteins (b) nucleosides
   (c) nucleotides (d) nitrogenous bases
7. DNA differs from RNA
   (a) in pyrimidines and sugar (b) in sugar
   (c) in purines (d) in purines and sugar

8. The nucleotides of two polynucleotide strands of DNA joined through hydrogen bonds are
(a) A = T, G ≡ C (b) A = C, G ≡ T
(c) A = G, T = C (d) A = U, G ≡ C

9. The nitrogenous bases found in RNA in place of thymine of DNA
(a) adenine (b) thymine
(c) uracil (d) guanine

10. Study of DNA by X-ray diffraction pattern was done by
(a) Watson and Crick (b) Avery, MacLeod, McCarty
(c) Wilkins and Franklin (d) Griffth

11. Synthesis of DNA molecule is called
(a) replication (b) transcription
(c) translation (d) amination

12. Transcription means the synthesis of
(a) DNA (b) lipids
(c) proteins (d) RNA

13. Transfer information from mRNA to a polypeptide chain is called
(a) amination (b) translation
(c) transamination (d) deamination

14. Pentose sugar present in DNA is
(a) ribose (b) deoxyribose
(c) sucrose (d) ribulose

15. Purines found in DNA are
(a) uracil and guanine (b) guanine and adenine
(c) adenine and cytosine (d) guanine and thymine

16. Double helical structure of DNA was proposed by
(a) Watson and Crick (b) Morgan
(c) Sutton (d) H.G. Khorana

17. Genetic information are encoded in
(a) nucleic acids (b) protein
(c) polysaccharide (d) lipids

18. DNA nucleotides in a polynucleotide chain are connected together by
(a) hydrogen bond (b) phospho-diester bond
(c) peptide bond (d) glycosidic bond

19. DNA acts as a template for
(a) DNA only (b) DNA and RNA
(c) protein (d) RNA only

ANSWERS

Very Short Answer Questions

1. Adenine, Guanine, Cytosine and Thymine.

2. According to Chargaff's rule:
(*i*) In a DNA molecule purines and pyrimidines occur in equal amounts.
(*ii*) Amount of Adenine is equivalent to Thymine and of Cytosine is equivalent to Guanine.

3. D.S. Watson and F.H.C. Crick.

4. Four types of nucleotides (deoxyribonucleotides).

5. By phosphodiester bond between deoxyribose molecules of two nucleotides through phosphate molecule.

6. 20 Å.

7. DNA (in almost all organisms) and RNA (*e.g.*, TMV).
8. Process of synthesis of mRNA from a DNA is called **transcription.**
9. **Anticodon** is a sequence of three nucleotides on a transfer RNA (or tRNA) molecule that recognises a complementary codon on mRNA during protein synthesis on the ribosome.

   Codon is a sequence of three nucleotides on mRNA. It codes for a particular amino acid in a polypeptide chain of protein molecule. It may also act for initiation or termination of translation.

   Translation is the process, whereby the sequence of codons in mRNA determines the sequence of amino acids in a polypeptide chain.
10. (*i*) DNA is a hereditary or genetic material of bacteriophage.

    (*ii*) Chargaff's rule for the molar ratio of nitrogenous bases in DNA.
11. (*i*) Codon.

    (*ii*) Transcription

    (*iii*) Degenerate code

Fill in the Blanks

1. Deoxyguanylic acid. **2.** Nucleotides, **3.** Hydrogen bonds, **4.** Phosphodiester, Deoxyribose, **5.** 20Å, **6.** Watson, Crick **7.** RNA, Fraenkel, Conrat, Stanley.

Multiple Choice Questions

1. (d)	**2.** (d)	**3.** (b)	**4.** (b)	**5.** (c)	**6.** (c)
7. (d)	**8.** (a)	**9.** (c)	**10.** (c)	**11.** (a)	**12.** (d)
13. (b)	**14.** (b)	**15.** (b)	**16.** (a)	**17.** (a)	**18.** (b)
19. (b)					

6

Enzymes

6.1 BIOENERGETICS

A living cell bustles with activity. Macromolecules of all types are assembled from raw materials, waste products are produced and excreted, genetic instructions flow from the nucleus to the cytoplasm, vesicles are moved from the Golgi apparatus to the plasma membrane, ions are pumped across cell membranes, and so forth. To maintain such a high level of activity, a cell must acquire and expand energy. The study of various types of energy transformations that occur in living organisms is referred to as **bioenergetics**.

1. Energy Transductions

Energy is defined as the capacity to do work, that is, the capacity to change or move some thing. Energy exists in two alternate states: potential and kinetic. A rock perched on the edge of a cliff possesses **potential energy** because it has the potential to perform work. Rock has this potential because it exists in a field of force, in this case, a gravitational field. If the rock is pushed over the edge, then gravity can act on it and cause it to fall, while falling, the rock has **kinetic energy** and can accomplish work, for example, by lifting another object (*i.e.*, fish in Fig. 6.1). Similarly, a resting nerve cell, has potential energy when it maintains a high concentration of sodium ions on the outer side of its plasma membrane and a low concentration of sodium ions on the inner side. Like the flow of water through a dam, the opening of specific channels in the plasma membrane allows sodium ions to flow across the membrane into the cell. The directed movement of sodium ions into the cell is a form of kinetic energy that can be used to accomplish work, as occurs when a nerve impulse passes along the membrane to the nerve cell.

In both of these cases just described–the falling rock and the directed movement of sodium ions–as in all cases of measurement of energy at work, two factors must be considered: a potential factor and a capacity factor (Table 6.1) The **potential factor** is proportional to the intensity of the field of force, whereas the **capacity factor** provides some measure of the "size" of the subject being considered. In the case of falling rock, the **potential factor** is the distance it will fall, and the capacity factor is the rock's mass. For the movement of charged ions, the potential factor is the voltage, and the **capacity factor** is the combined charge of the particles. The work, or energy released during these events, is a multiple of these two factors; as either one increases, so too does the amount of energy.

Fig. 6.1. Accomplishing work. The rock perched on top of the cliff has potential energy. The energy available for work is proportional to the mass of the rock and the distance it can fall. Once it leaves the edge of the cliff, the rock has kinetic energy that can be used to accomplish work, in this way by lifting both water and fish (after Karp, 2002).

Table 6.1 Potential and capacity factors in measuring energy (**source: Karp 2002**).

	Type of energy	Potential factor	Capacity factor
1.	**Mechanical**		
	Falling of rock	Height	Mass
	Compression	Pressure	Volume
	Stretching	Tension	Length
2.	**Electrical**		
	Movement of ions	Electric potential (voltage)	Charge
3.	**Osmotic**		
	Movement of water into a cell	Solute concentration gradient	Mass
4.	Energy available to do work	Temperature	Entropy

The Laws of Thermodynamics and the Concept of Entropy

Thermodynamics is the study of the changes in energy that accompany events in the universe.

The First Law of Thermodynamics. The first law of thermodynamics is the law of conservation of energy. It states that **energy** can neither be created nor destroyed. Energy can, however, be converted (**transduced**) from one form to another. Electric energy is converted to mechanical energy when we plug in a fan, and chemical energy is converted to thermal energy when fuel is burned in an oil heater. Cells are also capable of energy transduction. The chemical energy present in certain biological molecules, such as ATP, is converted to mechanical energy when organelles are moved from place to place in a cell, to thermal energy when heat is released during muscle cell contraction, or to electrical energy when ions flow across a membrane. *The most important energy transduction in the biological world is the conversion of sunlight into chemical energy—the process of photosynthesis*—which provides the fuel that directly or indirectly powers the activities of nearly all forms of life. A number of animals including fireflies and luminous fish are able to convert chemical energy back into light. Regardless of the transduction process, however, the total amount of energy in the universe remains constant.

To discuss energy transformations involving matter, we need to divide the universe into two parts: the **system** under study and the remainder of the universe, which we will refer to as the **surroundings**. A system can be defined in various ways: it may be a certain space in the universe or a certain amount of matter. For example, the system may be a living cell. The changes in a system's energy that occur during an event are manifested in two ways—as a change in heat content of a system and in the performance of work. Even though the system may gain or loss, energy, the first law of thermodynamics indicates that the loss or gain must be balanced by a corresponding gain or loss in the surroundings, so that the amount in the universe as a whole remains constant. The energy of the system is termed as **internal energy (E)**, and its change during a transformation in **ΔE**. One way to describe the first law of thermodynamics is that $\Delta E = Q - W$, where Q is the heat energy and W is the work energy.

Depending on the process, the internal energy of the system at the end can be greater than, equal to, or less than its internal energy at the start, depending on its relationship to its surroundings (Fig. 6.2). In other words ΔE (*i.e.*, change during the transformation) can be positive, zero, or negative.

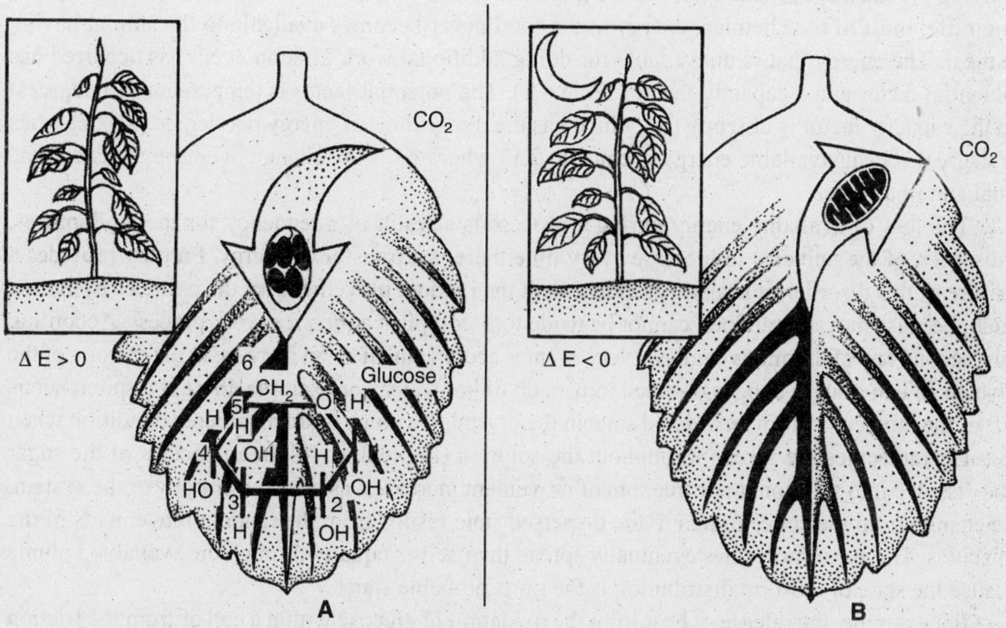

Fig. 6.2. A change in system's internal energy. In this example, the system will be defined as a particular leaf of a plant. A—During the day, sunlight is absorbed by photosynthetic pigments in the leaf's chloroplasts and used to convert CO_2 into carbohydrates, such as glucose molecule shown in the drawing (glucose is subsequently incorporated into sucrose or starch). As the cell continues to absorb light, its internal energy increases; the energy present in the remainder of the universe has to decrease. B—At night, the relationship between the cell and its surroundings is reversed as the carbohydrate produced during the day are oxidized to CO_2 in the mitochondria and the energy is used to run the cell's nocturnal activities (after Karp, 2002).

We can consider a system to be the contents of a reaction vessel. As long as there is no change in pressure or volume of the contents, there is no work being done by the system on its surroundings, or vice versa. In that case, the internal energy at the end of the transformation will be greater than at the beginning if heat is absorbed and less of heat is released. Reactions that lose heat are termed **exothermic**, and ones that gain heat are **endothermic**, and there are many reactions of both types.

The Second Law of Thermodynamics. The second law of thermodynamics expresses the concept that events in the universe have direction; they tend to proceed "downhill" from a state of higher energy to a state of lower energy. Thus, in any energy transformation, there is a decreasing availability of energy for doing additional work. Rocks fall off cliffs to the ground below, and once at the bottom, their ability to do additional work is reduced; it is very unlikely that they will lift themselves back to the top of the cliff. Similarly, opposite charges normally move together, not apart, and heat flows from a warmer to a cooler body not the reverse. Such events are said to be spontaneous, a term that indicates that they are thermodynamically favourable and *can occur without the input of external energy*.

The concept of the second law of thermodynamics was formulated originally for engines, and the law carried with the idea that it is thermodynamically impossible to construct a perpetual-motion machine. In other words, it is impossible for a machine to be 100 per cent efficient, which would be required if it were to continue functioning without the input of external energy. Some of the energy is inevitably lost as the machine carries out its activity. A similar relationship holds true for living organisms. For example, when a giraffe browses on the leaves of a tree or a lion preys on the giraffe, most of the chemical energy in the food never becomes available to the animal having the meal. The energy that is unavailable for doing additional work after an event has occurred has a potential factor and a capacity factor (Table 6.1). The potential factor is temperature (in degrees) and the capacity factor is **entropy (S)**, which has the dimensions of energy per degree (*e.g.*, calories per degree). The unavailable energy is equal to $T\Delta S$ where ΔS is the change in entropy between the initial and final states.

The loss of available energy during a process is a result of a tendency for the randomness, or disorder, of the universe to increase every time there is a transfer of energy. Entropy provides a measure of this disorder; it is also associated with the random movements of the particles of matter, which, because they are random, cannot be made to accomplish a *directed* work process. According to the second law of thermodynamics, every event is accompanied by an increase in the entropy of the universe. When a sugar cube is dropped into a cup of hot water, for example, there is a spontaneous shift of the molecules from an ordered state in the crystal to a much more disordered condition when the sugar molecules are spread throughout the solution (Fig. 6.3A). As the molecules of the sugar cube dissolve into solution, their freedom of movement increases, as does the entropy of the system. The change from the concentrated to the dispersed state results from the random movements of the molecules. The sugar molecules eventually spread themselves equally through the available volume because the state of uniform distribution is the most probable state.

For example, the release of heat from the oxidation of glucose within a cell or from the friction generated as blood flows through a vessel, is another example of an increase in entropy. The release of thermal energy by living organisms increases the rate of random movements of atoms and molecules; it cannot be redirected to accomplish additional work. Since the energy of molecular and atomic movements increases with temperature, so too does the entropy. It is only at absolute zero (0°K), when all movements ceases, that the entropy is zero.

As with other spontaneous events, one must distinguish between the system and its surroundings. The second law of thermodynamics indicate only that the total entropy in the universe must increase; the disorder within one part of the universe (the system) can decrease at the greater expense of its surroundings. The dissolved sugar can decrease in entropy it can be recrystallized by evaporating the water (Fig. 6.3B). The consequence of this process, however, is an increase in the entropy of the surroundings. The increased freedom of movement of the water molecules in the gaseous phase more than balances the decrease in freedom of the molecules of the sugar crystals.

Life operates on in a similar principle. Living organisms are able to decrease their own entropy by increasing the entropy of the environment. Entropy is decreased in an organism when relatively simple molecules, such as amino acids, are ordered into more complex molecules, such as the protein myoglobin in a muscle cell. For this to occur, however, the entropy of the environment must increase, which is accomplished as complex, ordered molecules such as glycogen stored in liver or muscle tissue are converted into heat and smaller less ordered compounds (such as CO_2 and H_2O) are released to the environment. It is this feature of metabolism that allows living organisms to maintain such a highly ordered and improbable state.

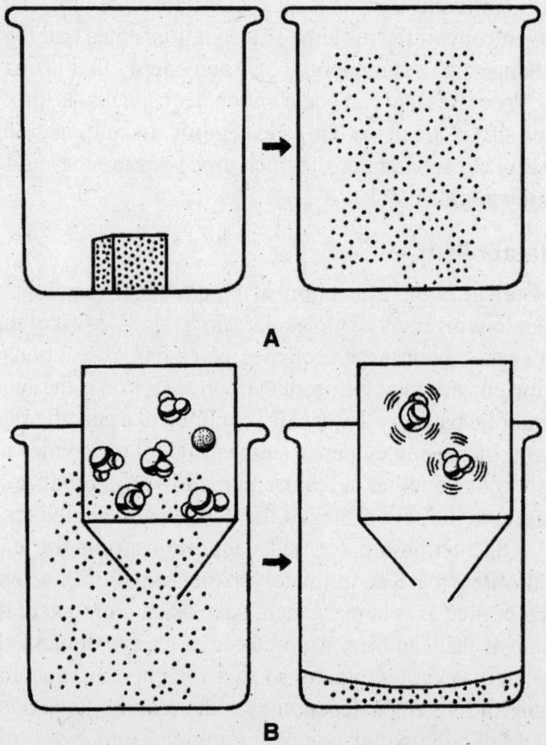

Fig. 6.3. Events are accompanied by an increase in the entropy of the universe. A—A sugar cube contains sucrose molecules in highly ordered arrangement in which the freedom of movement of the sucrose molecules is greatly increased, and their random movement causes them to become equally distributed throughout the available space. Once this occurs, there will be no further tendency for redistribution, and to entropy of the system is at a maximum. B—Sugar molecules spread randomly through a solution can be returned to a ordered state, but only if the entropy of the srroundings is increased, as occur when the more ordered water molecules of the liquid phase become disordered following evaporation (after Karp, 2002).

Another measure of the energy state of a living organism is provided by the information content of its macromolecules. **Information** is a subject that is difficult to define but easy to recognise. Information can be measured in terms of the ordered arrangement of a structure's subunits. For example, proteins and nucleic acids, in which the specific linear sequence of the subunits is highly ordered, are low in entropy and high in information content. Maintaining a state of high information content (low entropy) requires the input of energy. We can consider here just one molecule of DNA located in one cell of our liver. That cell has dozens of different proteins whose sole job is to patrol the DNA, looking for damage and repairing it. Nucleotide damage in an active cell can be so great that, without this expenditure of energy, the information content of DNA would rapidly deteriorate (See **Karp** 2002).

Free Energy. Together, the first and second laws of thermodynamics indicate that the energy of the universe is constant, but the entropy continues to increase toward a maximum. The concepts inherent in first two laws were put together by the American chemist, **Josiah Willard Gibbs** in 1878 into the expression $\Delta H = \Delta G + T\Delta S$, where ΔG is the change in **free energy**, that is, the change during a process in the energy available to do work; ΔH is the change in **enthalpy** or total energy content of the system (equivalent to ΔE), T is the absolute temperature (°K = °C + 273), and ΔS is the change in the entropy of the system. The equation states that the total energy change is equal to the sum of the changes in useful energy (ΔG) and energy that is unavailable to do further work ($T\Delta S$).

Processes that can occur spontaneously, that is, processes that are thermodynamically favoured (have a $-\Delta G$), are described as **exergonic**. In contrast, if the ΔG for a given process is positive, then it cannot occur spontaneously. Such processes are thermodynamically unfavourable and are described as **endergonic**.

2. Metabolism

Metabolism is the collection of biochemical reactions that occur within a cell, which include a tremendous diversity of molecular conversions. Most of these reactions can be grouped into **metabolic pathways** containing a sequence of chemical reactions in which each reaction is catalyzed by a specific enzyme, and the product of one reaction is the substrate of the next. The enzymes constituting a metabolic pathway are usually confined to a specific region of the cell, such as mitochondria or the cytosol. Increasing evidence suggests that the enzymes of a metabolic pathway are often physically linked to one another, a feature that allows the product of one enzyme to be delivered directly as a substrate to the active sites of the next enzyme in the reaction sequence.

The compound formed each step along the pathway are **metabolic intermediate** (or **metabolites**) that lead ultimately to the formation of an **end product**. End products are the molecules with a particular role in the cell, such as an amino acid that can be incorporated into a polypeptide, or a sugar that can be consumed for its energy content. The metabolic pathways of a cell are inter-connected at various points so that a compound generated by one pathway may be shuttled in a number of directions depending on the requirements of the cell at the time.

Metabolic pathways can be divided into two broad types. **Catabolic pathways** lead to the disassembly of complex molecules to form simple products. Catabolic pathways serve two functions: they make available the raw materials from which other molecules can be synthesized, and they provide chemical energy released by catabolic pathways stored temporarily in two forms: a high energy phosphate (primarily ATP) and high-energy electrons (primarily in NADPH). **Anabolic pathways** lead to the synthesis of more complex compounds from simple starting materials. Anabolic pathways are energy requiring (*i.e.*, endergonic) and utilize chemical energy released by the exergonic catabolic pathways.

Oxidation and Reduction

Both catabolic and anabolic pathways include key reactions in which electrons are transferred from one reactant to another. Reactions that involve a change in the electronic state of the reactants are called **oxidation-reduction** (or **redox**) **reactions**. Changes of this type involve the gain or loss of electrons, *e.g.*, the conversion of metallic iron (Fe^0) to the ferrous state (Fe^{2+}), by which iron atom loses a pair of electrons, thereby attaining a more positive state. When an atom loses one or more electrons, it is said to be **oxidized**. The reaction is reversible, meaning ferrous ions can be converted to metallic iron, a more negative state, by the acquisition of a pair of electrons. When an atom gains one or more electrons, it is said to be **reduced**.

For metallic iron to be oxidized, there must be some substance to accept the electrons that are released. Conversely, for ferrous ions to be reduced, there must be some substance to donate the necessary electrons. In other words, the oxidation of one reactant must be accompanied by the simultaneous reduction of some other reactant, and vice versa. One possible reaction involving iron might be

$$Fe^0 + Cu^{2+} \rightleftharpoons Fe^{2+} + Cu^0$$

The substance that is oxidized during an oxidation-reduction reaction, that is, the one that loses electrons, is called a **reducing agent**, and the one that is reduced, that is, the one that gains electrons, is called an **oxidizing agent**.

The oxidation or reduction of metals, such as iron or copper, involves the complete loss or gain of electrons. The same cannot occur with most organic compounds for the following reason; the oxidation and reduction of organic substrates during cellular metabolism involve carbon atoms that are covalently bonded to other atoms. When a pair of electrons are attracted more strongly to one of the two atoms of the polarized bond. In a C – H bond, the carbon atom has the strongest pull on the electrons; thus, it can be said that carbon atom is in a reduced state. In contrast, if a carbon atom is bonded to a more electronegative atom, as in a C – O or C – N bond, the electrons are pulled away from the carbon atom, which is thus is an oxidised state, since carbon has four outer-shell electrons it can share with other atoms, it can exist in a variety of oxidation states. This is illustrated by the carbon atom in a series of one carbon molecules (Fig. 6.4) ranging from the fully reduced state in methane (CH_4) to the fully oxidized state in carbon dioxide (CO_2). The relative oxidation state of an organic molecule can be roughly determined by counting the number of hydrogen versus oxygen and nitrogen atoms per carbon atom. The oxidation state of the carbon atoms in an organic molecule provides a measure of the molecule's free-energy content.

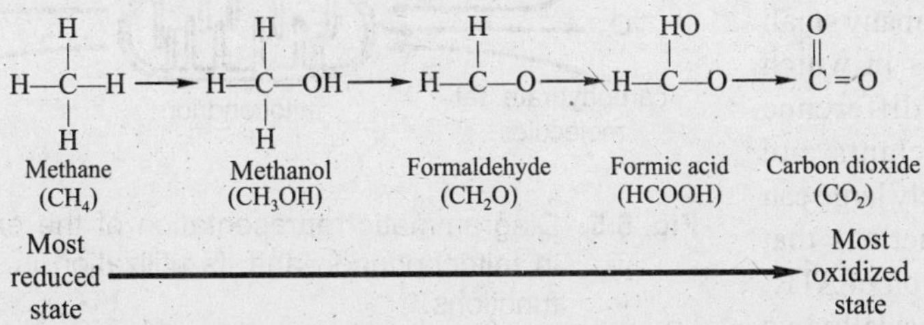

Fig. 6.4. The oxidation state of a carbon atom depends on the other atoms to which it is bonded. Each carbon atom can form a maximum of four bonds with other atoms. This series of simple, one-carbon molecules illustrates the various oxidation states in which the carbon atom can exist. In its most reduced state, the carbon is bonded to four hydrogens (forming methane); in its most oxidized state, the carbon atom is bonded to two oxygen (forming carbon dioxide) (after Karp 2002).

The Capture and Utilization of Energy

The compounds we use as chemical fuels to run automobiles are highly reduced organic compounds, such as natural gas (CH_4) and petroleum derivatives. Energy is released when these molecules are burned in the presence of oxygen, converting the carbon to more oxidized state, as in carbon dioxide and carbon monoxide gases. The degree of reduction of a compound is also a measure of its ability to perform chemical work within the cell. The more hydrogen atoms that can be stripped from a "fuel" molecule, the more ATP that ultimately can be produced. Carbohydrates are rich in chemical energy because they contain strings of (H–C–OH) units. Fats contain even greater energy per unit weight because they contain strings of more reduced (H–C–H) units.

As the sole building block of both starch and glycogen, glucose is a key molecule in the energy metabolism of plants and animals. The free energy released by the complete oxidation of glucose is very large:

$$C_6H_{12}O_6 + 6O_2 \rightarrow 6CO_2 + 6H_2O$$
$$\Delta G^{0'} = -686 \text{ kcal/mol}$$

By comparison, the free energy required to convert ADP to ATP is relatively small:

$$ADP + Pi \rightarrow ATP + H_2O$$
$$\Delta G^{0'} = 7.3 \text{ kcal/mol}$$

It is evident from these numbers that the complete oxidation of a molecule of glucose to CO_2 and H_2O can release enough energy to produce a large number of ATPs. As we have noted in Chapter 76 of Metabolism, upto 36 molecules of ATP are formed per molecule of glucose oxidized under conditions that exist in most cells. For this many ATP to be produced, the sugar molecule is disassembled in many small steps. Those steps in which the free-energy difference between the reactants and products is relatively large can be coupled to reactions that lead to the formation of ATP.

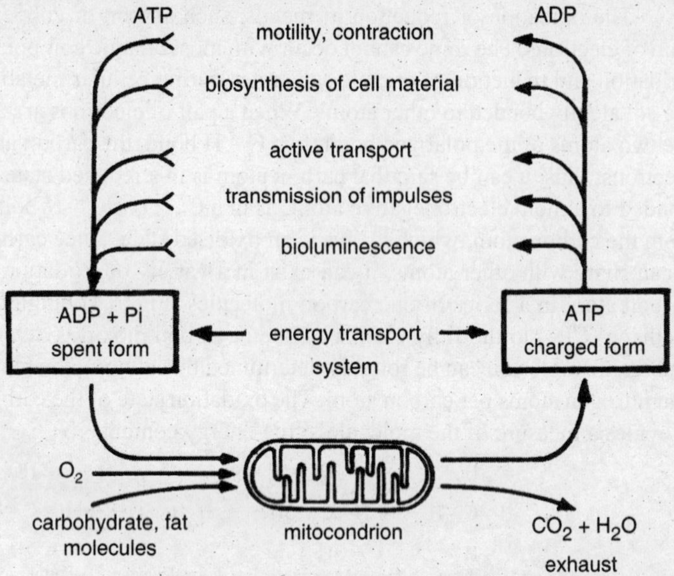

Fig. 6.5. Diagrammatic representation of the energy liberation in mitochondrion and its utilization in various cellular functions.

There are basically two stages in the catabolism of glucose, and they are virtually identical in all aerobic organisms. The first stage, **glycolysis**, occurs in the soluble phase of the cytoplasm (the cytosol) and leads to the formation of pyruvate. The second stage is the **tricarboxylic acid (TCA) cycle** or **Krebs cycle**, which occurs within the mitochondria of eukaryotic cells and the cytosol of prokaryotic cells and leads to the final oxidation of the carbon atoms to carbon dioxide. Most of the chemical energy of glucose is stored in the form of high-energy electrons, which are removed as substrate molecules are oxidized during both glycolysis and the TCA cycle. It is the energy of these electrons that is ultimately used to synthesize ATP.

Adenosine Triphosphate or ATP

The ATP consists of a purine base **adenine**, a pentose sugar **ribose** and three molecules of the **phosphoric acids**. The adenine and ribose sugar collectively constitute the nucleoside **adenosine** which by having one, two or three phosphate groups forms the adenosine monophosphate (AMP), adenosine diphosphate (ADP) and adenosine triphosphate (ATP) respectively. In ATP the last phosphate group is linked with ADP by a special bond known as "**energy rich bond**" because when the last phosphate group of the ATP is released the large amount of energy is released as shown by the following reaction:

$$\underset{\text{ATP}}{A{-}P{\sim}P{\sim}P} = \underset{\text{ADP}}{A{-}P{\sim}P} + \underset{\text{phosphate group}}{Pi} + \underset{\text{Energy}}{7300 \text{ calories}}$$

In the above reaction, we have seen that by the breaking of the energy rich bond about 7300 calories of energy are released, while the common chemical bond releases only 300 calories of energy. The chemical reactions which synthesize the energy rich bond or ~P bond require great amount of energy which is supplied by the oxidation of the food-stuffs in the mitochondria. The utility of energy rich phosphate bond (~P) of the ATP is that great amount of energy is kept stored in the ready state in a very limited space of the cell. The stored chemical energy is disposed of very quickly at the time of the need in various cellular functions such as respiratory cycle, protein and nucleic acid synthesis, nervous transmission, cell division, transportation and bioluminescence, etc. (Fig. 6.6).

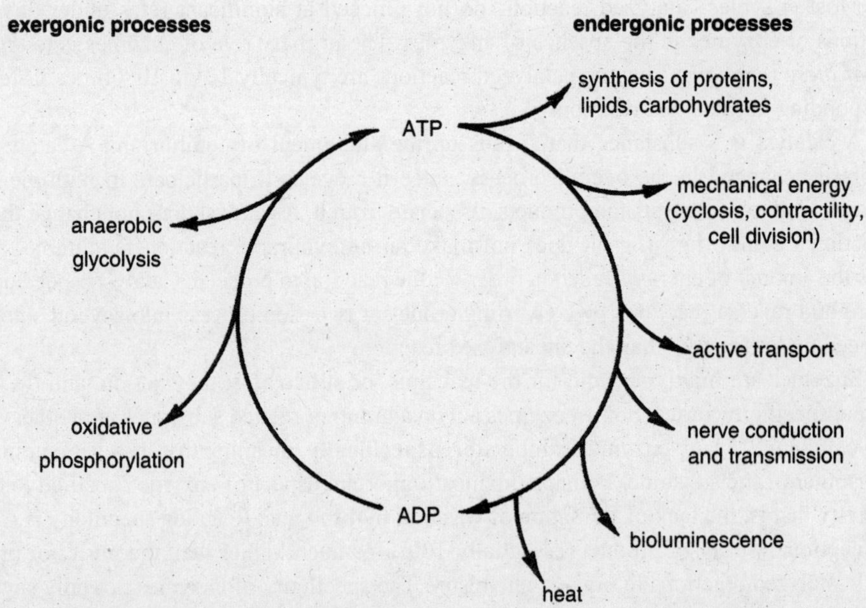

Fig. 6.6. Diagrammatic representation of uses and synthesis of ATP.

Because the terminal phosphate linkage in ATP is easily cleaved with release of free energy, ATP acts as an efficient phosphate donor in a large number of different phosphorylation reactions. In this way, ATP acts as a carrier molecule like the acetyl CoA and as coenzyme like the CoA or NAD.

Later on, besides ATP, certain other energy rich chemical compounds have been found to be active in the cellular metabolism. These are **cytosine triphosphate (CTP)**, **uridine triphosphate (UTP)** and **guanosine triphosphate (GTP)**. These compounds, however, derive the energy from the ATP by nucleoside diphosphokinases. The energy for the production of ATP or other energy rich molecules is produced during the breakdown of food molecules including carbohydrates, fats and proteins (**catabolic** and **exergonic activities**).

Energy transducers. The cell organelles, such as chloroplasts and mitochondria, which converts one form of energy to another are known as **energy transducers**. The plants and animals have not only evolved highly effective energy transducers but have also developed very efficient control systems to regulate the energy transformations and to enable the cells to adjust to variations in environmental conditions.

6.2 ENZYMES

Enzyme is a biological catalyst almost always a protein (Ribozyme is an RNA molecule with enzymatic activity). Some enzymes may require additional cofactors for activity. Virtually all biochemical reactions are catalyzed by specific enzymes.

1. Introduction

Enzymes are extraordinarily efficient selective biological catalysts. Every living cell has hundreds of different enzymes catalyzing the reactions essential for life. Even the simplest living organisms contain multiple copies of different enzymes. In multicellular organisms the complement of enzymes differentiates one cell type from another. Many enzymes catalyze the reactions of the central metabolic pathways necessary for the maintenance of life.

Most enzymes-catalyzed reactions do not proceed at significant rates under physiological conditions of enzymes in the absence of enzymes. The primary role of enzymes is to *enhance the rates of these reactions*. Enzyme-catalyzed reactions are typically 10^3 to 10^{20} times faster than the corresponding uncatalyzed reactions.

A catalyst is a substance that speeds up the attainment of equilibrium. A catalyst may be temporarily changed in the overall process, since it recycles to participate in multiple reactions. Reactants bind to a catalyst, and products dissociate from it. A catalyst does not change the position of reaction's equilibrium (*i.e.*, it does not make an unfavourable reaction favourable). Rather, it lowers the amount of energy needed in order for the reaction to proceed. Catalysts speed up both the forward and reverse reactions by converting a one- or two-step process into several smaller steps, each needing less energy than the uncatalyzed reaction.

Enzymes are highly specific for the reactants, or **substrates**, they act on, and the degree of substrate specificity varies. Some enzymes act on a group of related substrates, and others on only a single compound. Many enzymes exhibit **stereospecifically**, meaning that they act on only a single stereoisomer of the substrate. Perhaps the most important aspect of enzyme specificity is **reaction specificity** that is, the lack of formation of wasteful by-products. Reaction specificity is reflected in the exceptional purity of product (essentially 100%)—much higher than the purity of products of typical catalyzed reactions in organic chemistry. The specificity of enzymes not only saves energy for cells but also prevent the buildup of potentially toxic metabolic by products.

Enzymes can do more than simply increase the rate of a single highly specific reaction. Some can also combine, or couple, two reactions that would normally occur separately. This property allows the energy gained from one reaction to be used in a second reaction. **Coupled reactions** are a common feature of many enzymes; the hydrolysis of ATP, for example, is often coupled to less favourable metabolic reactions.

Some enzymatic reactions function as a control points in metabolism. Metabolism is regulated in a variety of ways, including alterations in the concentrations of enzymes, substrates, and enzyme inhibitors, and modulation of the activity levels of certain enzymes. Enzymes whose activity is regulated generally have a more complex structure than unregulated enzymes. With few exceptions, regulated enzymes are oligomeric molecules that have separate binding sites, for substrates and modulators, the compounds that act as regulatory signals. The fact that enzyme activity can be regulated is an important property that distinguishes biological catalysts from those encountered in a chemistry lab (*i.e.*, inorganic catalyst; Table 6.2)

Table 6.2 Difference between enzymes and chemical catalyst.

	Enzyme		Chemical catalyst
1.	Enzymes (biocatalysts) are always protein in nature [RNA enzymes (ribozymes) are notable exception].	1.	Inorganic or chemical catalysts are nonprotein in nature.
2.	They are highly specific and catalyses a specific reaction.	2.	They may catalyze different reactions.
3.	After completion of reaction, the enzymes do not always return to their original state.	3.	Catalyst always returns to its original state.
4.	Catalysis can occur through the active site present in enzyme.	4.	The catalyst takes part as a whole.
5.	Generally produced by living cells and acts independently of living cells.	5.	Generally, takes place outside the living cells.

2. History

The name enzyme is derived from a Greek and meaning "in yeast". It indicates that catalysts are present inside cells. In the late 1800s, scientists studied the formation of sugars by yeast cells. **Vitalists** (who maintained that organic compounds could be made only by living cells) said that intact cells were needed for fermentation. **Mechanists** claimed that enzymes in yeast cells catalyze the reactions of fermentation. The latter conclusion was supported by the observation that cell-free extracts of yeast can catalyst fermentation. This finding was soon followed by the identification of individual reactions and the enzymes that catalyze them.

In 1926 **James B. Sumner** (1887–1955), a US biochemist, 1946 Nobel laureate and "**Father of Modern Enzymology**", crystallised the first enzyme (**urease**) from jack bean (*Canavalia ensiformis*) and proved that it is protein. In next decade, five more enzymes were purified and also found to be proteins: **Pepsin**, **Trypsin**, **Chymotrypsin**, **Carboxypeptidase** and **Old Yellow Enzyme** (a flavoprotein NAD oxidase). Since then, almost all enzymes have been shown to be proteins plus cofactors. Certain RNA molecules also exhibit catalytic activity, but they are not usually referred to as enzymes.

3. Sources and Sites of Action of Enzymes

On the basis of sites of their action, the enzymes are classified as follows:

1. Endoenzymes. As most enzymes are macromolecules (proteins), they can not normally get into the cell through the plasma membranes, and therefore every cell must synthesize its own enzymes for carrying out the reactions goining within the cell. The enzymes which operate within the cells are called **endoenzymes**. These enzymes catalyze the metabolic reactions of the cell. Hence they are also referred to as **metabolic enzymes**. For example, cytochrome oxidase is such endoenzyme.

Cytochrome oxidase was discovered by **Warburg** in 1926. It is responsible for the catalysis of the oxidation of cytochrome by oxygen. During this reaction in the presence of free oxygen and cytochrome oxidase, cytochrome gives up its hydrogen to the oxygen with the formation of water.

2. Exoenzymes. These are normally extracellular, *i.e.*, they work outside the cells after they have been secreted (*i.e.*, synthesized and released) by them, *e.g.*, digestive enzymes.

3. Antienzymes. These are proteins or protein-like substances produced in the body to inhibit the catalytic action of enzymes, *e.g.*, **antipepsin**, **antitrypsin**, **antiamylase**, **antirennin**, **antiurease**, etc. These enzymes are secreted by parasites of intestine such as *Ascaris*, *Taenia*, etc.

4. Coagulative enzymes. These convert soluble calcium caseinogen into insoluble calcium caseinate, *e.g.*, rennin (or chymosin) and rannet.

4. Properties or Characteristics of Enzymes

Enzymes have the following common characteristics or properties:

1. Physical nature. The enzymes are generally colourless, but some are colouerd (*i.e.*, yellow, brown, red or green). Most of the enzymes are soluble in water, but some such as those located in the mitochondria are insoluble.

2. Protein nature. Except ribozymes which contain RNA, all enzymes are proteins. The protein structure provides weak associations due to side groups of the amino acids within it. The enzyme-substrate complex involves the association of the enzyme and substrate by weak associations rather than covalent bondings. Enzymes exhibit all typical properties of proteins, *viz.*, complex macromolecular organization, high molecular weights, colloidal behaviour, slow diffusion, inability to pass through most living membranes, and movement in response to an electric current.

3. Chemical activity. The enzymes may break a large molecule into two smaller molecules, or bring two small molecules together to form a large molecule.

4. Accelerate the rate of reaction. An enzyme can only change the rate of the biochemical reaction. Neither it can initiate a reaction nor it can change the direction of equilibrium. A reaction tends to proceed at a faster rate in presence of an enzyme, but the equilibrium between the substrate and product will remain constant.

5. Reversibility of enzyme action. Certain enzymes show reversibility of their action. This property is very helpful in metabolism. For example, *succinic dehydrogenase* catalyses reversibly the dehydrogenation of succinic acid. *Lipase* a fat splitting enzyme, hydrolyses ethyl butyrate to butyric acid and ethyl alcohol. If this process continues ethyl butyrate is formed again.

$$\underset{\text{Ethyl butyrate}}{C_3H_7COOC_2H_5} \xrightleftharpoons{\text{Lipase}} \underset{\text{Butyric acid}}{C_3H_7COOH} + \underset{\text{Ethyl alcohol}}{C_2H_5OH}$$

However, reversion of reaction may be checked by removing the products of enzyme reaction as quickly as they are formed.

6. Remain unaltered in the end. The enzymes combine temporarily with the substrate but do not undergo any permanent change during reaction they catalyse. The same enzymes can enter another biochemical reaction with another molecule of the same substrate.

7. High efficiency. Enzymes are reused and are therefore, required in small quantity by biological systems. Most enzymes have high **turnover number**. A turnover number of molecules of a substance acted upon by one molecule of enzyme per minute. A molecule of enzyme catalase from cattle liver decomposes 5,000,000 molecules of hydrogen peroxide in one minute at 0°C. The turnover number of catalase is, thus, 5,000,000 at 0°C. High turnover numbers of enzymes explain their remarkable effectiveness even though they occur in a cell in minute quantities.

8. Action specificity. Enzymes catalyse only one specific reaction or act upon only one kind of substrate. The specificity may be of any kind, *e.g.*, reaction specificity, substrate specificity, group specificity or optical specificity. For example, the enzymes lactase catalyses the hydrolysis of lactose and no other disaccharide.

(*i*) Reaction specificity. Most enzymes can catalyse the same type of reactions (*e.g.*, phosphate transfer, oxidation-reduction, etc.) with several structurally related substrates .

(*ii*) Substrate specificity. A particular enzyme will act only on a certain substrate (Box 6.1) For example, the enzyme **urease** is a specific enzyme, that acts on urea only. It would not exert any influence on the catabolism of protein or carbohydrate.

Box 6.1

Substrate specificity of enzyme is based on the "information" the enzyme molecule carries about its catalytic role. The "information" lies in the enzyme molecule in the form of its shape. By virtue of its specific shape, the enzyme molecule can "recognise" the substrate molecules with a given form and holds them for its action to the place. Molecules of other substrates fail to bind to the enzyme molecule, which, therefore, does not act on them. In short, the matching surface of the enzyme and substrate molecules form the basis of specificity.

(*iii*) Group specificity. A particular enzyme acts only on particular chemical groupings, *e.g.*, **glycosidase** on glycosides, **alcohol-dehydrogenase** on alcohols, **pepsin** and trypsin on peptide bonds and esterases on ester linkages. **Chymotrypsin** hydrolyses peptide bonds in which the carboxyl group is contributed by the aromatic amino acids: phenylalanine, tyrosine or tryptophan. **Carboxypeptidases** and **aminopeptidases** split amino acids one at a time from the carboxyl or amino terminal end of polypeptide chains, respectively.

(*iv*) Optical specificity (stereospecificity). With the exception of *epimerases* which interconvert optical isomers, enzymes generally show absolute optical specificity. For example, **maltase** catalyses the hydrolysis of α-glucosidic bond, while enzymes of the Embden-Meyerhof directs oxidative pathways and catalyse the interconversion of D-but not L-phosphosugars.

9. Denaturation. The protein enzymes become denatured by acids, high, salt concentration, alkaloid reagent or UV-light. The denaturation causes loss of **enzymatic activity**.

10. Temperature sensitivity. The enzymes function best at an optimum temperature (25°C – 40°C). Their activity decreases with decrease as well as increase in temperature and stops at 0°C and above 80°C.

11. pH sensitivity. The enzymes show maximum activity at an optimum pH (6 – 7.5). Their activity slows with decrease and increase in pH till it stops.

12. Activation. Most enzymes can be activated by addition of specific agents. In the absence of such factors enzymes becomes inactive and sluggish.

13. Inhibition. Enzyme activity is inhibited by the addition of specific agents, called **enzyme inhibitors**. For example, physostigmine (which is a salicylate-containing alkaloid which is obtained from Calabar beans) is a competitive inhibitor of acetylcholine.

14. Team work. The enzymes generally work in teams in the cell, the product of one enzyme controlled reaction serving as the substrate for the next. In germinating seeds, starch is changed into glucose, by two enzymes: amylase and maltase. Amylase splits the starch into the double sugar (disaccharide) maltose, which is then broken by maltase into the single sugar (monosaccharide) glucose. Eleven different enzymes work sequentially to convert glucose to lactic acid in animals as well as in plant cells (during glycolysis).

15. Destruction by poisons. The enzymes are destroyed by poisons such as cyanide and iodoacetic acid. **Cyanide poisoning** occurs due to the destruction of the respiratory cytochrome enzymes by the cyanide.

16. Lack of side reaction. In enzyme catalysed processes, there are virtually no side reactions, *i.e.*, the only products formed are the desired ones.

5. Nomenclature and Classification of Enzymes

Enzymes are named and classified by a system developed by a committee IUB or **International Union of Biochemists or IUBMB or International Union of Biochemistry and Molecular Biology**.

Nomenclature of Enzymes

Following five different criteria are followed in naming the enzymes:

1. According to substrate. Some enzymes are named by adding suffix '**ase**' to the name of substrate, on which they work. Thus, enzymes acting on sucrose, lipids and proteins are named **sucrase, lipase** and **proteinase**, respectively.

2. According to the type of reaction. Some enzymes are named according to the type of reactions they catalyze. For example, **alcohol dehydrogenase** catalyses the removal of hydrogen from alcohol (*i.e.*, oxidation of alcohols). Likewise the enzymes that removes CO_2 from the reactant are called **decarboxylases** and those add phosphate group to the reactant are **phosphates**.

3. According to history. A few enzymes such as **trypsin** and **amylase** are known by their historic names.

4. According to genes or characters. Many newly discovered enzymes are named after their genes or for some nondescriptive characteristics. For example, **RecA** is named after **recA** and **HSP70** is a heat shock protein; both enzymes catalyse the hydrolysis of ATP.

5. IUB/IUBMB system of naming enzymes. This is the most recent method of naming enzymes. The system assigns a **systematic name** to each enzyme. The systematic name of an enzyme has two parts. The first part represents the substrate and the second part represents the type of reaction catalysed. For example, enzyme **DNA polymerase** catalyse polymerisation or synthesis of DNA. Enzyme **3-phosphoglyceral dehydrogenase** removes hydrogen from substrate 3-phosphoglyceraldehyde.

IUBMB classification scheme assigns a unique number called the enzyme classification number, or **EC number**, to each enzyme.

Classification of Enzymes

To date over 2000 different enzymes are known (see **Jain** *et al.*, 2012). According IUB/IUBMB system of classification, enzymes are divided into following six main groups, based on the type of chemical reactions, they catalyse:

1. Oxireductases. These enzymes catalyse the oxidation-reduction reactions by transfer of hydrogen or electrons from one compound to another. A compound is **reduced** on accepting H or e^-, and **oxidized** on losing H^+ or e^-. Hence the enzyme involved is called *oxireductase*. Most of these enzymes (*i.e.*, oxireductases) are commonly referred to as **dehydrogenases**. Other enzymes in this class are called *oxidases*, *peroxidases*, *oxygenases*, or *reductases*.

An important example of oxireductase is *lactate dehydrogenase* (EC 1.1.1.27) also called *lactate: NAD oxireductase*. This enzyme catalyses the reversible conversion of L-lactate to pyruvate. The oxidation of L-lactate is coupled to the reduction of the coenzyme NAD^+ (nicotinamide adenine dinucleotide).

$$\text{L-Lactate} + NAD^+ \xrightleftharpoons{\text{Lactate dehydrogenase}} \text{Pyruvate} + NAD + H^+$$

Types and Examples of Oxyreductases

***(i)* Oxidases.** Oxidases bring about oxidation by addition of oxygen to a substrate or by removal of hydrogen or one or more electrons (Table 6.3). Oxidases or oxidising enzymes are intracellular enzymes (endoenzymes) which produce oxidative changes in the tissue cells during respiration and metabolism.

***(ii)* Reductases.** Reductases catalyse reduction by removing oxygen or by adding hydrogen or one or more electrons (Table 6.3).

Table 6.3 Summary of oxidation-reduction reactions.

	Oxidation (with respect to A)		Reduction (with respect to A)
(*i*)	Addition of oxygen: $A + BO \rightarrow AO + B$	(*i*)	Removal of oxygen: $AO + B \rightarrow A + BO$
(*ii*)	Removal of hydrogen: $AH + B \rightarrow A + BH$	(*ii*)	Addition of hydrogen: $A + BH \rightarrow AH + B$
(*iii*)	Removal of electrons: $A \rightarrow A^+ + e^-$	(*iii*)	Addition of electrons: $A + e^- \rightarrow A^-$
	All the above reactions release energy.		All the above reactions store energy.

Example:

(1) **Cytochrome oxidase** catalyses the transfer of hydrogen to oxygen, forming water in the last reaction of electron transport system in mitochondria:

$$NADH + H^+ + 1/2O_2 \xrightarrow{\text{Cytochrome oxidase}} NAD^+ + H_2O$$

(2) **3-Phosphoglyceraldehyde dehydrogenase** catalyses transfer of hydrogen from 3-phosphoglyceraldehyde (PGAL) to NAD^+ in glycolysis:

$$\text{3-PGAL} + NAD^+ + Pi^2 \xrightarrow{\text{3-phosphoglyceraldehyde dehydrogenase}} \text{1, 3-diphosphoglycerate} + NADH + H^+$$

(3) Reductases. These enzymes catalyse transfer of electrons from one coenzyme to the next in the electron transport system.

2. Transferases. These enzymes catalyse group-transfer reactions, and many require the presence of coenzymes. In group transfer reactions, a portion of the substrate molecule usually binds covalently to the enzyme or its coenzyme. This group includes **kinases**, enzymes that catalyse the transfer of a phosphoryl group from ATP.

Example:

(*i*) **Alanine transaminase** whose systematic name is *L-alanine: 2-oxyglutarate aminotransferase* (EC 2.6.12) is a typical example of transferase:

$$\text{L-Alanine} + \alpha - \text{Ketoglutarate} \xrightleftharpoons[]{\text{Alanine transaminase}} \text{Pyruvate} + \text{L-Glutamate}$$

(*ii*) **Hexokinase** transfers a phosphate group from ATP to glucose, forming glucose-6-phosphate in glycolysis.

(*iii*) **Glutamate pyruvate transaminase** transfers amino group from glutamate to pyruvate to form alanine.

$$\text{Glutamate} + \text{Pyruvate} \xrightarrow{\text{Glutamate pyruvate transaminase}} \text{Alanine} + \alpha\text{–Ketoglutarate}$$

3. Hydrolases. These enzymes catalyse hydrolysis. Hydrolases are special class of transferases, with water serving as the acceptor of the group transferred (**Horton** *et al.*, 2006).

Example: Pyrophosphatase is a simple example of a hydrolase. The systematic name of pyrophosphatase enzyme is *diphosphate phosphohydrolase* (EC 3.6.1.1).

$$\text{Pyrophosphate} + H_2O \xrightarrow{\text{Pyrophosphatase}} \text{Phosphate}$$

2. All digestive enzymes are hydrolases (Box 6.2). For example, pancreatic amylase hydrolyses starch and glycogen to maltose.

$$\text{Starch/Glycogen} \xrightarrow{\text{Pancreatic amylase}} \text{Maltose} + \text{Isomaltose} + \text{"Limit" Dextrins}$$

Box 6.2

Hydrolases are mostly exoenzymes; they act outside the cell and mainly help in digestion. Example of digestive enzymes include **amylases**, **saccharases**, **proteinases**, **lipases**, etc. Hydrolases also include certain endoenzymes that operate inside the cells and involve the use of water, *e.g.*, **deaminases**, **arginases**, **carbonic anhydrases**, etc.

Types of hydrolases. Hydrolases are divided into following 8 types of enzymes:

(*i*) **Proteolytic enzymes** which break down proteins into proteins and peptides, *e.g.*, pepsin, renin, trypsin and erepsin.

(*ii*) **Sucrolytic enzymes** that hydrolyse sugar molecules into simpler ones, *e.g.*, cellulase, amylase, lactase, fructase and galactase.

(*iii*) **Lipolytic enzymes** break down lipids or neutral fats into glycerol and fatty acids *e.g.*, lipase.

(*iv*) **Amylolytic enzymes** which bring about hydrolysis of starch into maltose.

(*v*) **Nucleases** break down nucleic acids (DNA and RNA) into nucleotides which are further hydrolysed by nucleotidases into nucleosides.

(*vi*) **Amidases** bring about breakages of urea, arginine, and purines, so accordingly are called urease, arginase, purinamidase, etc.

(*vii*) **Inverting enzymes** convert diasaccharides into monosaccharides, *e.g.*, invertase and maltase.

(*viii*) **Carbonic anhydrases** that split up carbonic acid into CO_2 and water.

4. Lyases or desmolases. Lyases catalyse breakage of specific covalent bonds and removal of some group without hydrolysis. Thus, lyases catalyse lysis of substrate, generating a double bond: these are **nonhydrolytic**, **nonoxidative**, **eliminative**, reactions. They play a role in respiration and fermentation.

Example:

(*i*) **Pyruvate decarboxylase** splits pyruvate into acetaldehyde and carbon dioxide. The systematic name of pyruvate decarboxylase is *2-oxoacid carboxylyase* (EC 4.1.1.1).

$$\text{Pyruvate} + H^+ \xrightarrow{\text{Pyruvate decarboxylase}} \text{Acetaldehyde} + CO_2$$

(*ii*) Enzyme **histidine decarboxylase** breaks histidine into CO_2 and histimine.

(*iii*) Enzyme **aldolase** splits fructose 1, 6-diposphate into dihydroxyacetone phosphate and 3-phosphoglyceraldehyde (3-PGAL) during glycolysis.

$$\text{Fructose 1, 6-diphosphate} \xrightarrow{\text{Aldolase}} \text{3-Phosphoglyceraldehyde (3-PGAL)} + \text{Dihydroxyacetone phosphate}$$

Synthase. In the reverse direction, lyases catalyse the addition of one substrate to a double bond of a second substrate. A lyase that catalyses an addition reaction in cells is often called a **synthase**.

5. Isomerases. Isomerases catalyse structural change within a single molecule (isomerization reactions). Thus, isomerases brings about the regrouping of atoms in the molecules (*i.e.*, intramolecular arrangement) and thus formation reactions have only one substrate and one product, they are among the simplest enzymatic reactions.

Example:

(*i*) *Alanine racemase* (EC 5.1.1.1) is an isomerase that catalyses the interconversion of L-alanine and D-alanine. Its common name is the same as the systematic name.

$$\text{L-Alanine} \xrightarrow{\text{Alanine racemase}} \text{D-Alanine}$$

(*ii*) During glycolysis *phosphohexose isomerase* enzyme changes glucos-6-phosphate to fructose-6-phosphate.

$$\text{Glucose-6-phosphate} \xrightarrow{\text{Isomerase}} \text{Fructose-6-phosphate}$$

6. Ligases or synthetases. These enzymes catalyse ligation or joining of two substrates. These reactions require the input of the chemical potential energy of a nucleotide triphosphate such as ATP.

Example:

(*i*) **Glutamine synthetase** or **L-glutamate**: *ammonia ligase* (ADP forming) (EC 6.3.1.2), uses the energy of ATP hydrolysis to join glutamate and ammonia to produce glutamine

$$\text{L-Glutamate} + \text{ATP} + NH_4^+ \xrightarrow{\text{Glutamine synthetase}} \text{L-Glutamine} + \text{ADP} + \underset{\text{(Inorganic phosphate)}}{\text{Pi}}$$

(*ii*) Enzyme **pyruvate carboxylase** catalyses joining of pyruvate and CO_2 to form oxaloacetate.

$$\text{Pyruvate} + CO_2 + \text{ATP} \xrightarrow{\text{Pyruvate carboxylase}} \text{Oxaloacetate} + \text{ADP} + \text{Pi}$$

(*iii*) Acetyl CoA synthesis is catalysed by enzyme **acetyl CoA synthetase**

$$\text{ATP} + \text{Acetate} + \text{CoA} \xrightarrow{\text{Acetyl CoA synthetase}} \text{Acetyl CoA} + \text{AMP} + \text{Pyrophosphate (PP)}$$

From the examples given above, we see that most enzymes have more than one substrate, although the second substrate may be only a molecule of water. Note also that although enzymes catalyse both forward and reverse reactions, one-way arrows are used when the equilibrium favours a great excess of product over substrate. At equilibrium, an enzyme catalyses both the forward and reverse reactions at the same rate.

6. Mechanism of Enzyme Action

An enzyme can speed up only those reactions which otherwise occur to some extent even in the absence of the enzyme. Several theories were put forward to explain how an enzyme speeds up the reaction.

1. Enzyme-substrate complex theory. This theory was put forward by **Heneri** and mathematically formulated by **Michaelis** (1875–1949) and **Menten** (1879–1960). According to this theory, an enzyme combines with its substrate to from an unstable **intermediate complex**, the **enzyme-substrate complex**. This intermediate complex immediately breaks down into the reaction products and the original enzyme. This can be represented in the following manner:

$$\underset{\text{Enzyme substrate}}{E+S} \underset{k_2}{\overset{k_1}{\rightleftharpoons}} E-S \text{ complex} \xrightarrow{k_3} \underset{\text{Product enzyme}}{E+P}$$

Where k_1 = Rate constant of formation of ES complex
k_2 = Rate constant of dissociation of ES complex
k_3 = Rate constant of formation of products from ES complex

Since the enzyme-substrate complex is a very transitory compound, some difficulty was faced in actually proving its existence. Direct evidence of the existence of enzyme substrate complex was obtained by **David Keilin** and **Britton Chance**. Chance isolated a brown coloured peroxidase enzyme from horse radishes (a cruciferous plant, *Armoracia rusticana*, with long lobed leaves) and when this was mixed with hydrogen peroxide, a green coloured enzyme substrate complex was formed. This was then changed to pale red enzyme substrate complex. The latter finally splitted up into enzyme and break down product.

The enzyme merely provides a "platform" or template on which certain molecules could react with each other. Such an enzyme platform brings reacting molecules in contact much faster than chance collisions at that temperature. The result is that the reactions are accelerated.

Enzymes reduce the energy of activation. The effect of enzymes is to lower the activation-energy requirements, there by promoting appreciable reaction rates at lower temperatures than would be possible otherwise. For example, decomposition of hydrogen peroxide without a catalase has energy activation of about 18,000 cal/mol, when catalase is added this energy of activation is reduced to 2000 cal/mol.

Enzymes accelerate reactions by reducing the free energy of activation. This principle may be interpreted in terms of the **transition state theory**. According to this concept reactants in order to react to form the product C, must pass through a transition state (A – B) in which the activated complex is formed and certain amount of activation energy (Ea) is required. When the proper enzyme is present, less energy is required for activation (Ea′). In other words, the combination of substrate with enzyme creates a new reaction pathway that has a transition state of lower energy than in absence of enzyme.

2. Lock and key theory (or rigid model of catalytic site). The larger enzyme molecule comes in contact with smaller substrate molecules to form an intermediate complex. Enzymes bear distinct cavities or **active sites** or **catalytic sites** or **active centers** in which specific types of substrate molecules can form intimate association by their specific areas called **reactive sites**. Because these active sites are specifically designed, only specific substrate molecules with proper geometric shape can fit correctly. This combination occurs by electrostatic hydrogen and/or hydrophobic bonds and produces a strain on the bond which is to be broken so that lesser amount of energy is required to break it. This **intermediate complex** is called **enzyme substrate complex (Michaelis complex)**. It soon breaks to release the enzyme and product of biochemical reaction. It is believed that when substrate molecule binds to the active site, its parts are held together in such a way to cause distortion of chemical bonds, *i.e.*, the bonds are weakened. These weakened bonds speed up the rate of reaction. The overall process can be summarised by following reaction:

$$\text{Enzyme} + \text{Substrate} \rightarrow \text{Enzyme substrate complex} \rightarrow \text{Enzyme} + \text{Product}$$

This simple model of enzyme activity indicates the **biphasic nature of enzyme mecnanism**. First step involves binding of substrate molecule with the active site of enzyme while the second step involves the chemical modification of the substrate.

Nature of active sites. Active sites contain special groups, *e.g.*, –OH group of hydroxyamino acids such as serine; $-NH_2$ of basic amino acids such as lysine, –COOH of acidic amino acids; –SH of cysteine amino acid for establishing contact with substrate molecules. Sometimes more than one side chains may form the active site. As a particular lock can be opened by a particular type of key so particular enzyme can act on particular substrate due to their specificity, the name for the mechanism has been given as **Lock and Key Mechanism (Fischer** 1989). He noticed that glycosidases which act on glycosides are highly specific with respect to their substrate which is due to close similarity in the shape of enzyme and substrate molecules.

3. Induced-fit model. Normally a large enzyme molecule is required to catalyse a reaction at the site which is only a tiny portion on its surface. It may be debated that if the active site represents only a small part of the enzyme molecule, what is the function of the rest of the enzyme molecule? In 1959, **Koshland Jr.**, suggested that the enzyme essentially consists of four categories of amino acids (Fig. 6.7):

(*a*) **Catalytic residues (catalytic sites).** These amino acids make and break chemical bonds and are required for catalytic activity.

(*b*) **Binding residues (binding sites).** These amino acids hold the substrates in place while the reaction (catalysis) is taking place.

(*c*) **Structural residues.** These amino acids hold the active site in the correct shape so that it can function properly.

(*d*) **Non-essential residues.** These amino acids have no specific function. They are often near the surface of the enzyme molecule and can be removed or replaced without loss of function.

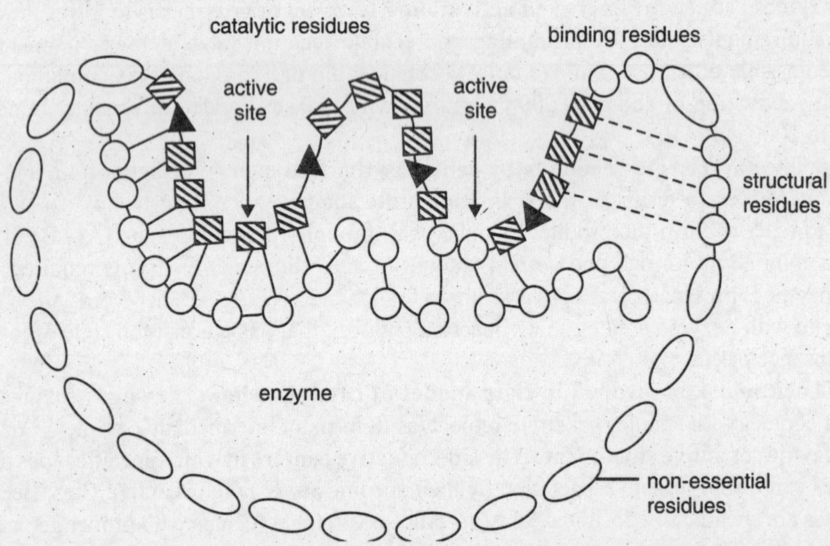

Fig. 6.7. Amino acid residues in an enzyme molecule. Active site is flexible to accommodate the substrate.

The idea of an "*enzyme wrapping around*" a substrate to form a more stable structure is called **induced-fit hypothesis**. According to the induced-fit model, the enzyme molecule on coming in contact with the substrate, does not retain its original shape and structure. Rather, the contact of the substrate **induces** some configurational or geometrical changes in the active site of the enzyme molecule (Fig. 6.8). Consequently, the enzyme molecule is **made to fit** completely the configuration and active centers of the substrates. At the same time, other amino acid residues may become buried in the interior of the molecule.

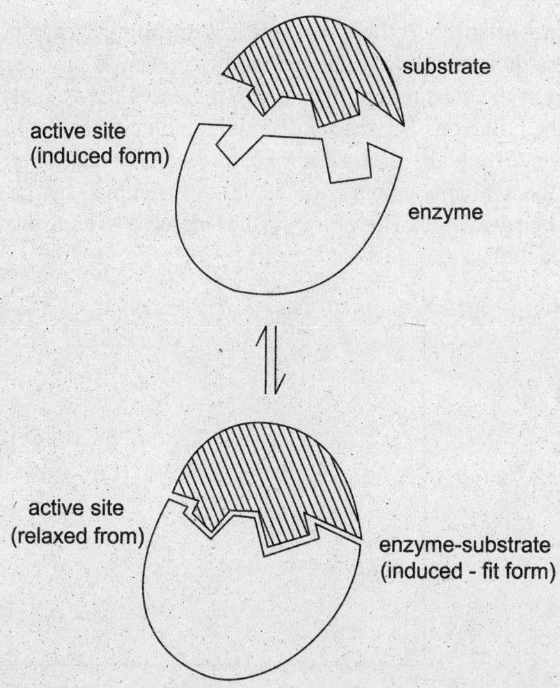

Fig. 6.8. Induced-fit model proposed by Koshland Jr.

7. Factors Affecting Enzyme Activity

Enzyme activity is affected by following nine important factors: 1. Enzyme concentration; 2. Substrate concentration; 3. Product concentration; 4. Temperature; 5. pH; 6. Poisons and radiations; 7. Activators; 8. Inhibitors; and 9. Enzyme-substrate complex.

1. Enzyme concentration. The rate of a chemical reaction is influenced by particular enzyme. In most cases it increases with increasing concentration of the enzyme, provided that the substrate is present abundantly and is not a limiting factors (Fig. 6.9).

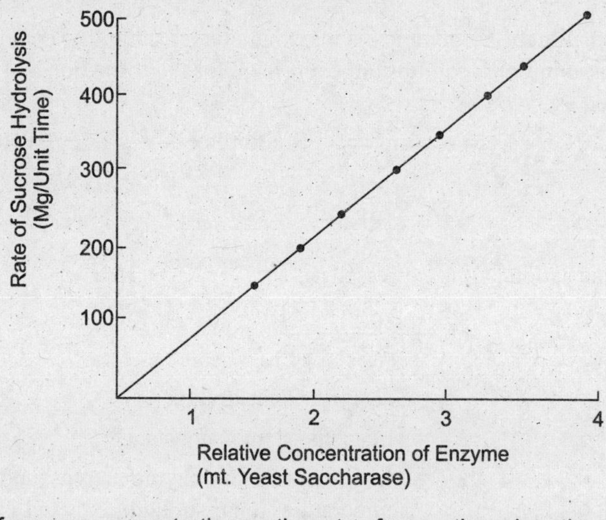

Fig. 6.9. Effect of enzyme concentration on the rate of a reaction when the substrate is available in excess.

2. Substrate concentration. If there are more enzyme molecules than substrate molecules, the two kinds of molecules do not collide frequently and the reaction proceeds slowly. In such a case, a progressive increase in the substrate molecule (S), increases the velocity (V) of their conversion to products because the collisions between the enzyme molecule and substrate molecules become more frequent. However, eventually the rate of reaction reaches a maximum (Fig. 6.10). At this state, the active sites of all the available enzyme molecules are occupied by the substrate molecules and the enzyme is said to be **saturated**. Therefore, further increase in concentration of substrate has no effect on the rate of reaction.

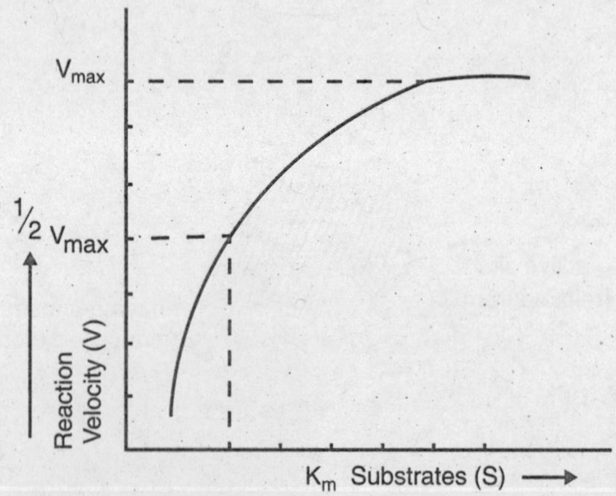

Fig. 6.10. Effect of substrate concentration on the velocity of enzyme action, S = substrates; V = reaction velocity, V_{max} = Maximum velocity; K_m = Michaelis-Menton constant.

Energy-kinetics of Enzymes (Michaelis-Menten Equation)

Michaelis and **Menten** (1913 proposed an equation which shows the inter-relationship between the enzyme activity and substrate concentration. This is called **Michaelis-Menten equation**. It is based on three basic assumptions:

1. ES complex is in a steady state, *i.e.*, concentration of ES complex remains constant.
2. Under saturating conditions, all enzyme molecules are involved to form ES complex and none is free.
3. Rate of formation of ES complex is maximum under saturating conditions.

Under these assumptions, equilibrium expression for the formation and breakdown of the ES complex is expressed as:

$$\text{E} + \text{S} \underset{k_2}{\overset{k_1}{\rightleftharpoons}} \text{ES complex} \xrightarrow{k_3} \text{E + Products}$$

So $$\text{k}_\text{m} = \frac{k_2 + k_3}{k_1}$$

Rate of enzyme action (V_i) is expressed as:

$$V_i = \frac{V_{max} + [\text{S}]}{k_m + [\text{S}]}$$

S = Substrate concentration

K_m = Michaelis-Menton constant

K_{max} = Maximum rate of enzyme activity under a given set of pH, temperature, ionic strength, etc. and is constant for specific enzyme

When substrate concentration [S] is equal to K_m, then,

$$V_i = \frac{V_{max} + [S]}{k_m + [S]}$$

$$V_i = \frac{V_{max} + [S]}{[S] + [S]}$$

$$V_i = \frac{V_{max} + [S]}{2[S]}$$

$$V_i = \frac{V_{max}\ [S]}{2[S]}$$

$$V_i = \frac{V_{max}}{2} = \frac{1}{2} V_{max}$$

This shows that when initial velocity is half-maximal, then the substrate concentration indicates Michaelis-Menton constant, or **Michaelis-Menten constant** *is equal to substrate concentration that gives half of the maximum velocity.*

It can be derived in another way: If the initial velocity is equal to half maximal, then

$$V_i = \frac{V_{max}\ [S]}{k_m\ [S]}$$

$$\frac{1}{2} V_{max} = \frac{V_{max}\ [S]}{k_m\ [S]}$$

$$k_m + [S] = \frac{2V_{max}\ [S]}{V_{max}}$$

$$k_m = [S]$$

Thus, K_m is equal to the substrate concentration at which the velocity of the reaction is half maximum.

The value of the Michaelis-Menton constant (K_m) is inversely proportional to the enzyme activity. A large value of K_m means that a high substrate concentration is needed to get half velocity of the maximum rate of reaction. In fact, it means that enzyme has lower affinity for the substrate.

When reciprocal values of enzyme activity and substrate concentration are plotted against each other, we get a straight line (Fig. 6.11). This double reciprocal plot is called the **Lineweaver-Burk plot**. From this plot, value of Km can be obtained by extending the line toward the abscissa. (Note. *Abcissa* is a mathematical term which means one of the coordinates of a two-dimensional coordinate system, usually the horizontal coordinate, denoted by *x*).

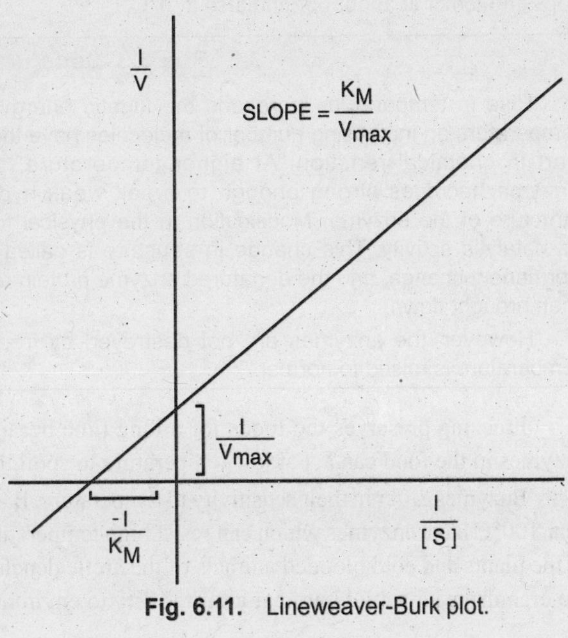

Fig. 6.11. Lineweaver-Burk plot.

3. Product concentration. Accumulation of the product of enzyme reaction lowers the enzyme activity. Enzyme molecules must be freed to combine with more substrate molecules. Normally the products are quickly removed from the site of formation and the reaction does not suffer.

4. Temperature. The optimum temperature for an enzyme usually corresponds to the body temperature of the organism. For example, the optimum temperature of enzyme of humans is 37°C which is also their normal body temperature. Most enzymes operate at their best in a temperature range of about 25°C to 40°C. Between a range of 0°C and 45°C the enzyme activity increases 2 – 3 times, with a rise of 10°C temperature. The increase in enzyme activity is due to two factors: (*i*) the influence of heat on the chemical reaction and (*ii*) the increased enzyme activity.

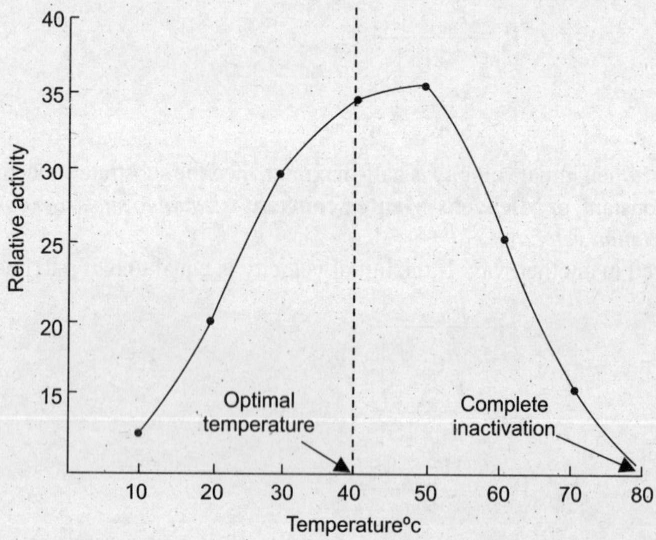

Fig. 6.12. Effect on increasing temperature on enzymatic activity.

In the range of 45° to 60°C thermal **denaturation** will destroy the catalytic activity of the enzyme (Box 6.3). This slows down the catalytic reaction and ultimately stops it. Enzyme's activity stops altogether at about 60° and also at 0°C.

Box 6.3 Denaturation

Rise in temperature increases the kinetic energy of the molecules. Therefore, at higher temperature an increasing number of molecules have the required activation energy and can take part in chemical reaction. At higher temperature, the kinetic activity of molecules in an enzyme becomes strong enough to break weak hydrogen bonds that maintain the tertiary structure of the enzyme. Modification in the physical form of the enzyme results in the loss of its catalytic activity. This change in structure is called **denaturation** of protein. This is a very permanent change, and the denatured enzyme protein remains inactive even if the temperature is then brought down.

However, the enzymes are not destroyed by freezing and regain their lost activity if the temperature is raised to normal.

Freezing preserves the foods for a long time because neither the microbial enzymes nor the enzymes in the food can act at low temperature to spoil them.

Enzymes differ in their sensitivity to temperature. Bacteria living in hot springs with temperature near 100°C have enzymes which can resist high temperature. On the other hand, the cellular proteins of the plants and cold blooded animals of the arctic denature readily at moderate temperature. Thus, the organisms have **biochemical adaptability** to environment.

5. Hydrogen ion concentration (pH). Some enzymes act best in an acid medium, others in an alkaline medium. For every enzyme, there is an optimum pH where its action is maximum. Most enzymes show maximum activity in a pH range of about 6.0 to 7.5 (Fig. 6.13A). A shift to alkaline or acid side rapidly decrease the enzyme activity and finally stop it altogether. This is due to denaturation of the enzyme molecule, *i.e.*, change in its physical structure. Some digestive enzymes have their optimum activity in the acidic or alkaline range, and drop in activity as the pH falls or rises (Fig. 6.13B, C). For example, **pepsin** of gastric juice has its optimum pH 2 (acidic) and **trypsin** of pancreatic juices show maximum activity at pH 8 (alkaline). Likewise, optimal range for activity of **amylases** is 5.2 and for **salivary amylase** between 6.7 and 6.8.

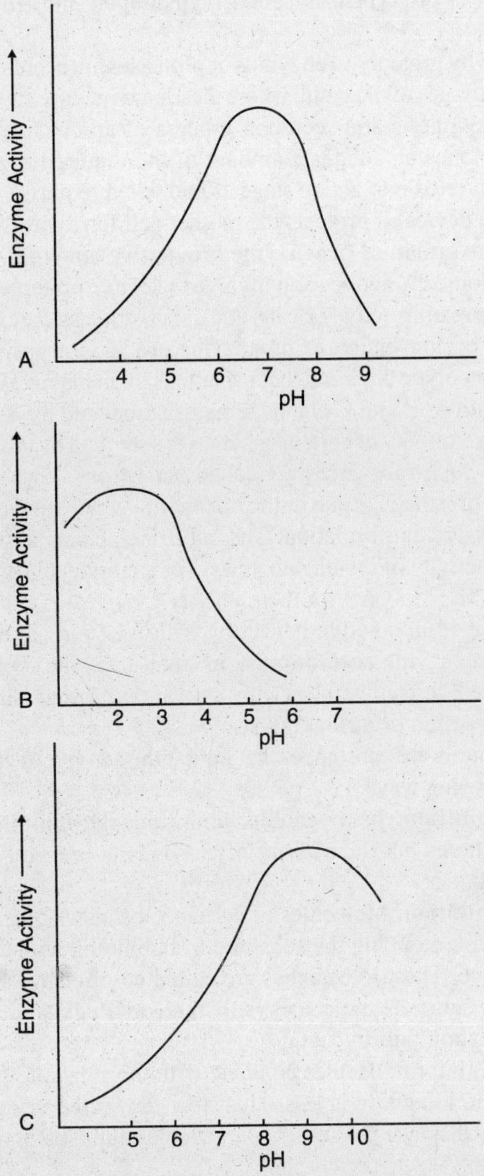

Fig. 6.13. Effect of pH changes on enzyme's activity. A—Enzyme with maximum activity at neutral pH. B—Enzyme with maximum activity at acid pH. C—Enzyme with maximum activity at alkaline pH.

6. Poisons and radiations. Poisons, such as cyanide, and radiations destroy the tertiary structure of the enzymes, making them in effective.

7. Activators. In living beings many proteins are manufactured and secreted in the form of inactive precursor protein. These are called **proprotein**. Conversion of a proprotion to the mature protein involves one or more successive "**proteolytic clips**". As a result of such selective proteolysis the inactive proprotein is converted into active mature protein.

Enzymes are proteins. Certain enzymes are produced by the living cells in the inactive (nonfunctional) form. These inactive or precursor enzymes are called **proenzymes** or **zymogens**. For example, all the proteolytic enzymesa are secreted in inactive form, such as (*i*) propepsin is called **pepsinogen**; (*ii*) protrypsin is called **trypsinogen** and (*iii*) prochymotrypsin is called **chymotrypsinogen**.

One basic reason why proteolytic enzymes or proteases are secreted in inactive form is to avoid the **autodigestion** of the tissues of of origin, where these enzymes are synthesized, stored and secreted.

Furthermore, the synthesis and secretion process of an enzyme may be slow relative to its physiological demand. Hence an adequate amount of the required enzyme is stored in inactive or precursor form to be converted into active stage as and when required. So, secretion of proteases as zymogens is a protective device. It protects the mother cell form **autolysis** or **self-digestion**.

Mechanism of conversion of proenzyme into active enzymes. Maturation of a proenzyme into functional or physiologically active form involves selective proteolysis. This introduces essential conformational changes creating active or catalytic or substrate site and the allosteric site if any. For example, in the process of conversion of proenzyme prochymotrypsin (pro CT) to chymotrypsin (α-CT) involves three proteolytic clips and the formations of an inactive intermediate π-chymotrypsin (π-CT). In α-chymotrypsin A, B and C chains remains associated by 2 interchain disulphide bonds.

Enzyme activators. Activity of certain enzymes is activated by the presence of some substances other than the substrate. Such substances are called **activators**. These may be components of the enzyme, a loosely bound prosthetic group or the metabolite which binds to specific site of an enzyme and changes it into an active conformation. The substrate acts as an **allosteric effector**. Specific ions are required for the activity of certain enzymes. For example, phosphorylating enzymes require divalent cations, such as Mg^{2+} or Ca^{2+} for their activity.

Pepsinogen is changed into active pepsin by hydrogen ions in the stomach. Trypsinogen is activated to trypsin by an enzyme **enterokinase** in small intestine. Further, once small amount of pepsin or trypsin is formed, it itself catalyses the activation of remaining proenzyme. This process is called **autocatalytic reaction** or **autocatalysis**.

8. Inhibitors. Inhibitors are substances that inhibit the activity of enzymes. Enzymes inhibition can be caused in the following ways:

(*i*) Reversible inhibition. In reversible inhibition, the inhibitors prevent enzymes from combining with the substrates but the activity of enzymes is restored when inhibitor is removed. This is of following two types:

1. Competitive inhibition. Molecules of certain substances other than the substrate, bind to the active site of the enzyme stopping the substrate from binding with the enzyme (Fig. 6.14). This makes the enzyme inactive. These substances are called **competitive inhibitors**. Their molecules are similar in shape to the substrate molecules (**substrate analogues**) but are different enough from them so that chemical reaction fails to occur.

The competitive inhibitor molecules compete with the substrate molecules. When enough of natural substrate is present, inhibition is less. This slows down the reaction. But when the inhibitor molecules are more, these displace the substrate molecules almost stopping the reaction.

Example:

1. Enzyme succinate dehydrogenase catalyses the oxidation of succinate to fumarate in Krebs cycle. In the presence of malonate [a salt of malonic acid, $CH_2(COOH)_2$] the reaction rate is slowed down. Malonate has a molecular structure very similar to that of succinate. It occupies the active center of the enzyme. However, effect of malonate can be reversed if enough succinate is added.

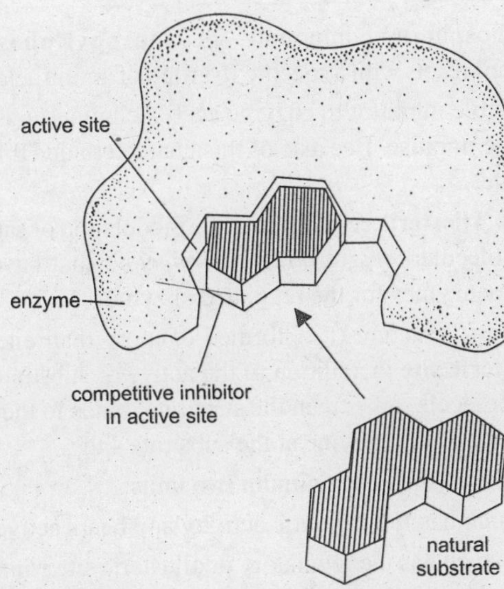

Fig. 6.14. Diagrammatic representation of competitive inhibition.

Uses of Competitive Inhibitor

1. The phenomenon of competitive inhibition can be compared to a lock which is jammed by a key similar the original one Competitive inhibitors are used as drugs in the control of bacterial pathogens. Sulpha drugs such as **prontosil** are used to combat bacterial infection. In molecular structure prontosil is similar to p-amino banzoic acid (PABA) which is used by bacteria to synthesize PABA, bacteria fails to synthesize folic acid and fail to grow and multiply.

2. Competitive inhibitors are used in treating some forms of cancer.

2. Non-competitive Inhibition. The non-competitive inhibitor does not compete with the substrate for binding at the active or substrate-binding site of the enzyme, because it does not bear any structural similarity with the substrate. It gets attached to the enzyme at a place other than the active site rendering it inactive.

In this case, the degree of inhibition cannot be reduced by increasing the number of substrate molecules. However, the enzyme activity can be restored because non-competitive inhibitors can be removed by dialysis.

Example:

1. Cyanide inhibits the mitochandrial enzyme cytochrome oxidase which is essential for cellular respiration. This kills the animals.

2. Ions of heavy metals such as mercury, silver and copper, Ag^{+}, Hg^{++}, and Cu^{++} combine with thiol (–SH) groups in the enzyme breaking the disulphide bridges. These bridges are important in maintaining tertiary structure. When these bridges are broken, the enzyme become denatured and inactive.

***(ii)* Non-reversible inhibition.** In non-reversible inhibition, the inhibitor molecule firmly (*i.e.*, permanently) attached to the R-group at the active site of enzyme. As a result the substrate molecule fails to bind to the enzyme molecule and the enzyme is rendered permanently inactive.

Example:

1. Many antibiotics are inhibitors of specific enzymes in bacteria. For instance, pencillin blocks the active site of an enzyme that many bacteria use to make their cell wall.

2. DDT and parathion are inhibitors of key enzymes in the nervous system.

3. Insecticide organophosphorus compound—**di isopropyl phosphofluoridate (DFP)** (marketed/trade name Malathion) reacts with a specific R-group of amino acid serine at the active site.

The same DFP is irreversible inhibitor to enzyme acetylocholinesterase which is essential for the orderly propagation of nerve impulse. Because of this characteristic DFP are classified as nerve gases and are used as insecticides.

3. Allosteric modulation. Allosteric regulation is the modulation of catalytic activity of certain regulatory enzymes by low molecular weight *allosteric effectors* that have little or no structural similarity to the substrates or coenzymes for the regulated enzyme.

Enzymes having quaternary structure (*i.e.*, formed of more than one polypeptide subunits) have a **regulatory site** or **allosteric site** in addition to the active or substrate site. **Allostery** means "different shape". An outside molecule, other than the substrate, binds to the regulatory site and can enhance or diminish the reactivity of the enzyme at the substrate site.

An allosteric hypothetical enzyme has minimum two units:

(*a*) **Catalytic subunit** which has full catalytic activity and bears active site or substrate site.

(*b*) **Regulatory subunit** which has the regulatory or allosteric site which binds to the inhibitor or activator.

The allosteric enzyme can exist in following two alternative forms:

(*i*) **The active form.** The active form of allosteric enzyme has full catalytic activity. Its active site on the catalytic subunit binds to the substrate and converts it into product or products. The regulatory or allosteric site of its regulatory subunit is deformed and cannot bind the inhibitor or activator.

(*ii*) **The inactive form.** The inactive form of allosteric enzyme is totally without enzyme activity. Its active site is distorted and unable to bind with the substrate. Its allosteric site can bind an effector (either inhibitor or activator).

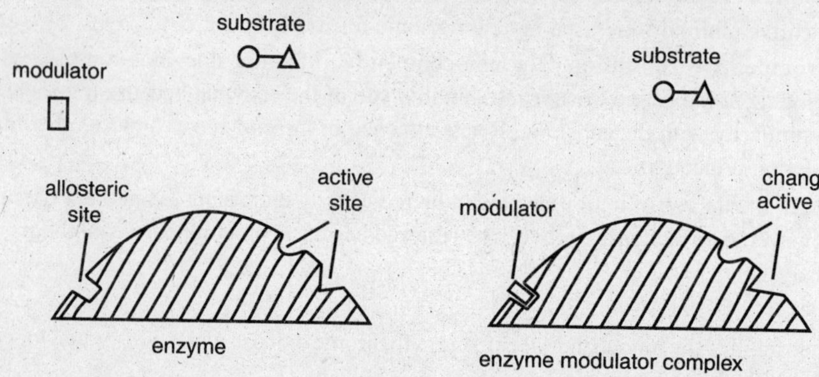

Fig. 6.15. Mechanism of allosteric regulation of enzyme action.

Mechanism of allosteric regulation. The active and inactive forms of allosteric enzyme occur in equilibrium.

When a substrate is added to the enzyme substrate solution, some of it bind to the active sites of the active enzyme molecules. The active molecules of enzyme that have entered enzyme-substrate complex are incapable of being converted to inactive form. However, to maintain equilibrium, the inactive form of enzyme is converted to active form. So there is an increase in the concentration of active form of enzyme. This increases reactivity of enzyme.

On the contrary, addition of allosteric inhibitor decreases the concentration of free enzyme and inhibits the reaction by binding to the allosteric site of inactive form and preventing its conversion to active form.

A specific case of allosteric enzyme inhibition has been cited. Glucose is changed to glucose 6-phosphate in glycolysis with the help of the enzyme hexokinase. Glucose 6-phosphate causes allosteric inhibition of hexokinase (Fig. 6.16). This is called **feedback allosteric inhibition**.

Inhibits

$$\text{Glusose} + \text{ATP} \xrightarrow{\text{Hexokinase}} \text{Glusose 6-phosphate} + \text{ADP}$$

Fig. 6.16. Feedback allosteric inhibition.

The mechanism of allosteric inhibition and promotion was proposed by **Monod** and his colleagues.

Table 6.4 Difference between competitive inhibition and allosteric inhibition.

	Competitive inhibition		Allosteric inhibition
1.	Inhibitor binds to active sites of enzymes.	1.	Inhibitor binds to allosteric sites of enzymes.
2.	Substrate fails to join enzyme as its active site is occupied by inhibitors.	2.	Substrate fails to join enzyme as its active site is deformed by inhibitors.
3.	Inhibitor closely resembles substrates in structure.	3.	Inhibitor does not resemble substrates in structure.
4.	Inhibitors is not an intermediate or a product of metabolic pathway catalyzed by the enzyme.	4.	Inhibitors may be an intermediate or a product of metabolic pathway catalyzed by the enzyme.
5.	It has non-regulatory function.	5.	It regulates the metabolic pathway and checks the excessive formation of the product.

9. Enzyme substrate complex. In bio-catalysts enzyme substrate complex is formed. The greater the affinity of the enzyme for its substrate, the greater is its catalytic activity.

Cooperativity. If an enzyme molecule comprising two or more subunits, the binding of one substrate molecule to the active site of one subunit causes all the subunits to assume their active conformation by induced fit mechanism (Fig. 6.17). An interaction of subunits of an enzyme in which a conformational change in one subunit results in a similar change in all other subunits is called **cooperativity**.

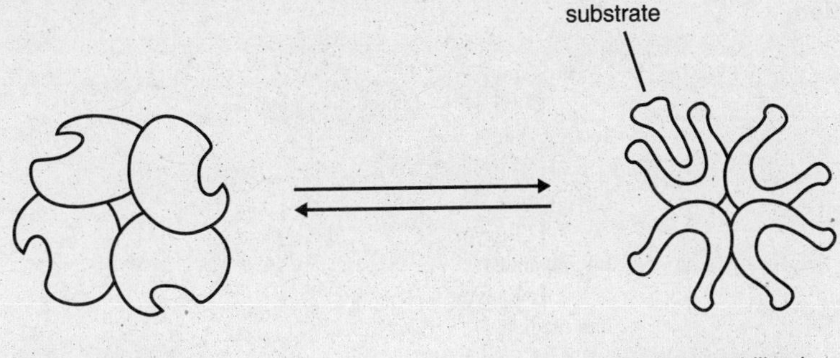

Fig. 6.17. Cooperativity. Joining of a substrate molecule to one subunit of an enzyme stabilizes the active form of all subunits.

8. Synthesis of Enzymes

Since all enzymes are proteins, their synthesis takes place on ribosomes like other proteins and is controlled by genes. There is generally one gene for each enzyme (*i.e.*, one gene-one enzyme theory of **Beadle** and **Tatum**). In case an enzyme is a complex of more than one protein subunit, more than one gene may control its synthesis.

Synthesis of specific enzymes in cells occurs at the right place and time and in right amount strictly according to the need. The presence of metabolite acts as a signal or stimulus for the synthesis of the desired enzyme or enzymes required for its metabolism.

9. Regulation of Gene Action

A large variety of chemical reactions occur in a cell simultaneously in a surprisingly coordinated manner, without interfering with one another. Each compound is produced when required and in the desired quantity. Such a precise regulation is achieved by two types of control mechanisms operating in the cells. One mechanism operates at the enzyme level and the other at the gene level.

1. Control at the enzyme level. In this type of gene regulation, the enzyme, substrate and product themselves regulate the chemical reaction. When the product of an enzyme reaction accumulates in the cell, it inhibits its own production by lowering the enzyme activity. This kind of control mechanism is called **feedback inhibition** or **negative feedback**.

Feedback inhibition. Many multienzyme systems possess the capability of self-regulation of their overall reaction rate. In such a system, the end product of the reaction sequence can inhibit the first enzyme, with the result that the rate of entire sequence is determined by the steady-state concentration of the product. The first enzyme that is inhibited by the end-product is called the regulatory or **allosteric enzyme**, the first step is called the **committed step**.

A colon bacterium *Escherichia coli* synthesizes the amino acid **isoleucine** from a substrate **threonine** by a series accumulates in amounts more than required, it stops its own production by inhibiting the activity of the enzyme **threonine deaminase** which catalyses the first reaction of the series (Fig. 6.19). The type of metabolic control, in which the first enzyme of a series is inhibited by the end product is known as **end-product inhibition.**

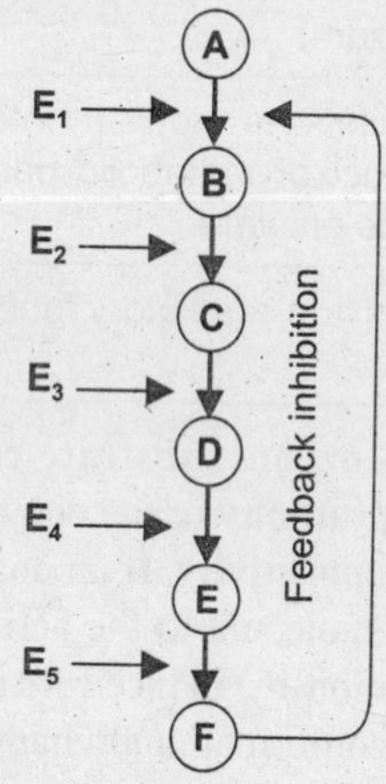

Fig. 6.18. Feed back inhibition.

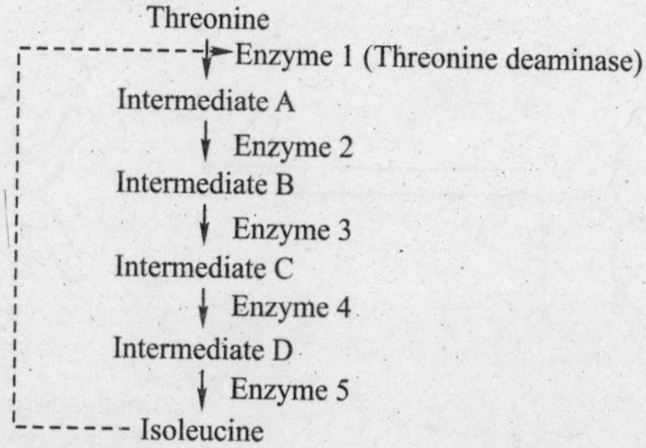

Fig. 6.19. Feedback inhibition of a pathway of *E. coli* involving five enzymes that catalyses successive steps. The end product inhibits the first enzyme.

2. Control at gene level. In this type of control mechanism, the gene regulates the production of enzymes. Proteins are the end products of gene (DNA) action. They are synthesized as summarised below:

$$\underset{1}{\text{DNA}} \xrightarrow{\text{Transcription}} \underset{2}{\text{mRNA}} \longrightarrow \underset{\substack{\uparrow \\ \text{Amino acid} \\ \text{carrying tRNAs}}}{\overset{}{\text{Ribosome}}\ {}^{3}} \xrightarrow{\text{Translation}} \underset{4}{\text{Protein (Enzyme)}}$$

The gene responsible for synthesis of enzyme is activated and inactivated by the substrate to be metabolised and end product accumulating in excess respectively.

The Operon concept. The phenomena of induction and repression were formulated by **Jacob** and **Monod** in 1961. The hypothesis propounded by them in context with *E.coli* is known as **Operon model. Operon** is defined as a complete unit of gene expression, which includes genes for the production of mRNA, a gene which regulates the synthesis and the site at which the control mechanism acts. The genome as such is composed of clusters of genes.

The **lac operon** of *E.coli* (Fig. 6.20) is the best known model which has distinct gene sites. The structural genes are three in number for the enzyme **β-galactosidase, permease** and **transacetylase**, represented by *z, y, a* respectively. Each structural gene transcribes a single mRNA to be used for the production of enzyme proteins.

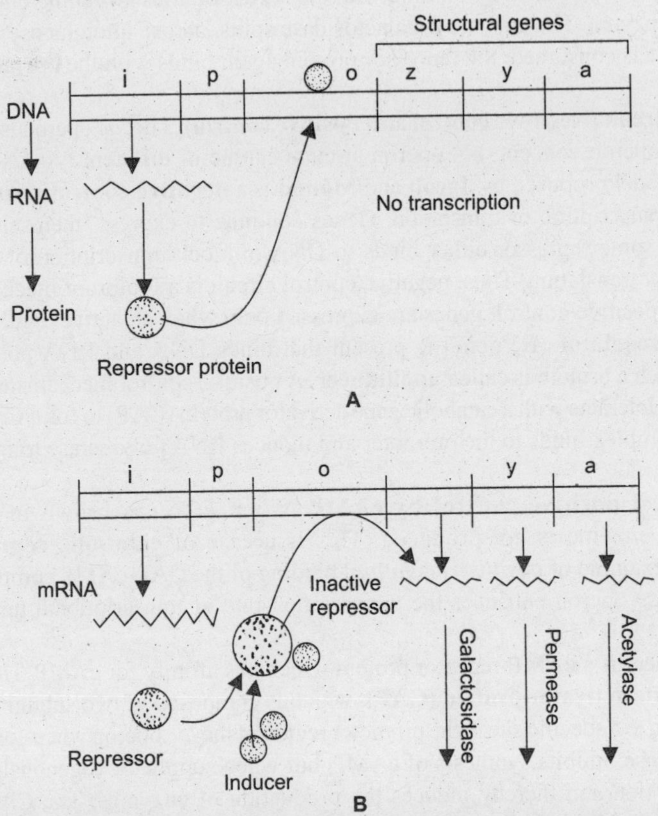

Fig. 6.20. Lac operon model as proposed by Jacob and Monod to show enzyme production in *E.coli*. *z*, *y* and *a* are structural genes, each of which produces an enzyme. Production of enzyme will stop when the operator (o) gene bind the repressor product of (i) gene. Here p stands for promoter gene and inducer stands for the lactose.

Another component of the *lac operon* is the *regulator gene* which is the key to the operation of the structural genes. The regulator is specified by the *i* gene, which codes for a metabolically important molecule which does not have enzyme activity and has to recognise and bind to a region of DNA to carry out its function. The *i* gene of *lac* operon produces a **repressor protein** which binds to the operator gene *O*. The operator is situated on the left of the structural genes. When the repressor protein binds the operator, the structural genes are switched off, preventing the synthesis of mRNA and ultimately the enzymes (Fig. 6.20A). Since the product of the *i* gene, the repressor, prevents the expression of the structural genes of the operon, the mechanism of control is said to be **negative**. Binding of the repressor to the operator is regulated by inducer substance (*e.g.* lactose). If the regulator gene undergoes mutation, it is unable to control the activity of the structural genes.

In the *lac* operon, there is a small segment of DNA having about 100 nucleotides lying in between the operator and the structural genes. This segment recognizes the DNA-dependent RNA polymerase (enzyme) to initiate transcription of structural genes and is called the **promoter gene** (*p*). There is a specific site on the promoter to bind cAMP.

Working of operon circuit is regulated by the regulator gene. The output of the *lac* circuit is a function of input received from the regulator. Transfer of information from the regulator to the structural genes is through *lac* repressor, which is the product of regulator gene. When the environment of *E.coli* lacks the disaccharide sugar lactose, repressor product is produced that binds tightly to the operator, preventing RNA polymerase to function. When lactose is present in the medium, it acts as **inducer** and puts on the circuit by binding with the repressor, thus switching on the structural genes to produce the enzyme. The enzyme β-galactosidase splits lactose into glucose and galactose units. When the lactose is consumed, the repressor protein again binds with the operator to switch off the structural genes.

Control circuit (Negative control and Positive control). The *lac* operon is the best understood example of the operon concept, but operon circuits can be of different types such as *positive* and *negative*. The model proposed by **Jacob** and **Monod** is a **negative control system**, which operates at the level of transcription or translation. Genes continue to express themselves unless they are switched off by some repressor either binds to DNA to block transcription of mRNA or binds to mRNA to prevent translation. Thus, negative control circuit is a foolproof mechanism.

In case of positive control, genes are expressed only when an active regulator protein [which is produced by **regulator** (**R**) **gene**] is present that binds DNA and RNA polymerase to initiate transcription. Such a protein is called **apoinducer**. A positive control mechanism has been provided by cAMP, which interacts with a catabolic gene activator protein (CAP) to form CAP-cAMP activator complex. The complex binds to the promoter and induces RNA polymerase to initiate transcription of the operator.

Example of positive control by cAMP. When *E.coli* is grown in glucose medium, β-galactosidase enzyme is not produced. This is a case of **catabolic repression**. Catabolic repression is an example of *positive regulation*: binding of the CAP-cAMP complex at the CAP site of promoter of *lac* operon enhances the transcription rate of transcriptional unit (see **Verma** and **Agarwal**, 2009).

In *E.coli* there is a cAMP-receptor protein which has affinity for cAMP. This receptor protein is called **catabolite activator protein (CAP)**, which is composed of two subunits, each of which is capable of binding to a specific site in the promoter region of the *lac* operon when combined with cAMP (Fig. 6.21). Glucose inhibits synthesis of cAMP, but when supplied exogenously to *E.coli* cells, it initiates transcription and thereby induces the production of enzymes. The CAP-cAMP complex binds the promoter of the *lac* operon, and consequently RNA polymerase recognises the promoter—activating positive control mechanism. The mechanism is responsible for the expression of most of inducible operons, which utilise sugars as maltose, galactose and arabinose. Thus, cAMP is required for the *E.coli* strains growing in changed environment.

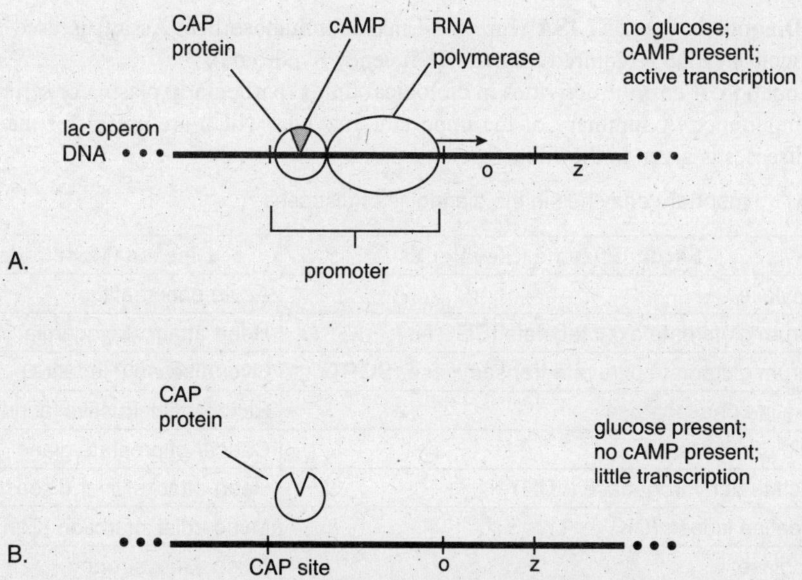

Fig. 6.21. Catabolite repression: When cAMP is present in the cell (no glucose is present), it binds with CAP protein and together they bind to the CAP site in various sugar metabolizing operons, such a *lac* operon shown here. The CAP-cAMP complex increases the transcription of the operon. When glucose is present, it inhibits the formation of cAMP. Thus no CAP-cAMP complex forms, and transcription of the same operon is reduced (after Tamarin, 2002).

10. Economic Importance of Enzymes

Enzymes have vital role in a variety of biological processes.

A. Biological uses of enzymes. A large number of chemical reactions occur in the living cells all the time. These reactions would normally occur outside the cells only at temperatures high enough to literally burn up the cells, or would occur so slowly as to be useful, being unable to sustain life. The enzymes makes the biochemical reactions go at ordinary (normal) temperature and virtually instantaneously. **G.R. Stark** has calculated that one molecule of the enzyme urease can break down 30,000 molecules of urea into carbon dioxide and ammonia in just one second. In the absence of that single enzyme molecule, the same reaction would take 3 million years. Thus, the enzymes serve as substitute for intensive heat, because, like heat, they bring the reactant molecules together for rapid interaction. The enzymes help in many biological processes:

$$\underbrace{\text{Urea + Water}}_{\text{Urine (substrate)}} \ \underbrace{\xrightarrow{\text{Urease}}}_{\text{Enzyme}} \ \underbrace{\text{Carbon dioxide + Ammonia}}_{\text{Product}}$$

1. Physiology. Enzymes digest food in alimentary canal of animals which will starve without the enzymes. The enzymes are also essential for other physiological activities such as respiration, blood clotting, muscle contraction, nerve impulse transmission and so on. A missing or defective enzyme can be very harmful. Lack of sucrase enzyme causes abdominal pain and diarrhoea in persons on eating sucrose.

In fact, the enzymes make the essential biochemical processes, *i.e.*, **metabolism**, to occur at body temperature. Thus, they make life possible.

2. Diagnostic uses. ELISA (Enzyme-Linked Immunosorbent Assay) is used for detecting diseases such as AIDS (Acquired Immuno-Deficiency Syndrome).

Estimation of enzyme activities in biological fluids (particularly plasma or serum) is of great clinical importance. A summary of the important enzymes which are useful for the diagnosis of specific diseases is given in the Table 6.5.

Table 6.5 Important enzymes in the diagnosis of diseases.

	Serum Enzyme (Elevated)	Disease (Most Important)
1.	Amylase	Acute pancreatitis
2.	Serum glutamate oxalo acetate (SGOT)	Heart attack (Myocardial infarction)
3.	Serum glutamate pyruvate transaminase (SGPT)	Liver disease (Hepatitis)
4.	Alkaline phosphatase	Rickets; Obstructive jaundine
5.	Acid phosphatase	Cancer of prostate gland
6.	Lactate dehydrogenase (LDH)	Heart attacks, liver diseases
7.	Creatine kinase (CK)	Myocardial infarction (Early marker)
8.	Aldolase	Muscular dystrophy
9.	5′–Nucleotidase	Hepatitis
10.	γ-Glutamyl transpeptidase	Alcoholism

3. Medical treatment. Enzyme **streptokinase** is used in dissolving blood clot formed inside blood vessels. This proteolytic enzyme is obtained from hemolytic streptococci.

4. Digestive aids. Enzyme **diastase** and others are used as digestive aid in patients with faulty digestive secretions.

5. Genetic engineering. Enzymes such as endonucleases, ligases, etc., are used in genetic engineering.

B. Industrial uses. Many enzymes have a great industrial value:

1. Alcoholic drinks. Enzyme **zymase** which is obtained from yeast is used in making alcoholic beverages by fermentation of sugar substances.

2. Cheese. Enzyme **rennin** produced from the calf's stomach is used in making cheese by coagulating milk protein caseinogen.

3. Detergents. Detergents for washing the clothes contain protease enzyme.

4. Biological detergents. Some enzymes serve as biological detergents such as:

(*a*) **Amylases** are used as detergents for machine dish washing to remove resistant starch residues.

(*b*) **Lipases** are used to assist in the removal of fatty and oily stains.

(*c*) **Cellulases** are used in biological fabric conditioner.

(*d*) **Catalase** is used in rubber industry to generate oxygen from peroxide to convert latex into foam rubber.

5. Cleaning for hides for leather industry. Enzyme **protease** is used for cleaning hides and softening of meat.

6. Clearing of juices. Enzyme **pectinase** is used for clearing fruit juices.

7. Baby food. Enzyme **trypsin** is added to predigest baby food partly.

8. Waste treatment. Bacteria in a box of waste matter produce the enzyme **urease** which catalyzes the breakdown of urea (from urine) into CO_2 and ammonia (NH_3).

Enzymes are also used for retting of fibers, degumming of silk, food processing and chemical synthesis.

6.3 COENZYMES

Many enzymes consist of large polypeptide or protein units (the protein portion of enzyme is called **apoenzyme**) plus some other type of substance known as **cofactor**. Without its cofactor, each of these enzymes are functionally inactive. The cofactor may be some simple divalent metalic ions such as Mg^{2+}, Ca^{2+}, Zn^{2+}, Co^{2+}, Mn^{2+}, etc., or sometimes a non-protein small molecular weight organic compound. If the organic cofactor is *loosely bound* with the protein (*i.e.*, apoenzyme), it is termed as **coenzyme** and if the cofactor is *firmly bound* with the apoenzyme it is called **prosthetic group.** The full intact enzyme sometimes referred to as **holoenzyme**.

$$\text{Cofactor} + \text{Apoenzyme} \rightarrow \text{Holoenzyme (or Enzyme)}$$

Nature of coenzyme. Coenzymes are nonprotein organic compounds. These usually have one of the vitamins as their active group.

1. Vitamins as coenzymes. The following vitamins and their derivatives act as coenzymes (Table 6.6):

Table 6.6 Derivatives of Vitamin B and AMP which act as coenzymes.

	Coenzyme	Abbreviation	Group transferred	Vitamin
	A. Oxido-reductase coenzymes (Hydrogen acceptor coenzyme)			
1.	Nicotin amide-adenine dinucleotide	NAD	Hydrogen	Nicotinamide
2.	Nicotin amide-adenine dinucle-otide phosphate	NADP	Hydrogen	Nicotinamide
3.	Flavin mononucleotide	FMN	Hydrogen	Riboflavin
4.	Flavin dinucleotide	FAD	Hydrogen	Riboflavin
5.	Cytochromes	Cyt	Electron acceptor	—
	B. Transferase coenzymes			
1.	Adenosine triphosphate	ATP	Phosphate	—
2.	Adenosyl methionine	—	Methyl group of methionine	—
3.	Biotin	—	Carboxyl	Biotin
4.	Thiamine pyrophosphate	TPP	C_2-aldehyde	Thiamine (Vitamin B_1)
5.	Pyrioxal phosphate	PALP	α-amino acid	Pyridoxine (Vitamin B_6)
6.	Coenzyme A	CoA	Acyl	Pantothenic acid
	C. Isomerase and Lyase coenzymes			
1.	B_{12} Coenzyme	—	Carboxyl displacement	Cobalamine
2.	Pyridoxal phosphate	PALP	Decarboxylation (in amino acid metabolism)	Pyridoxine (Vitamin B_6)
3.	Thiamine pyrophosphate	TPP	Decarboxylation	Thiamine (Vitamin B_1)
4.	Uridine diphosphate	UDP	Monosaccharide isomerization	Thiamine (Vitamin B_1)

	Coenzyme	Abbreviation	Group transferred	Vitamin
	D. Decarboxylation coenzymes			
1.	Thiamine pyrophosphate	TPP	Removes CO_2	Thiamine (Vitamin B_2)
2.	Pyroxal pyrophosphate	PALP	Removes CO_2	Pyridoxine (Vitamin B_6)

(*i*) **Thiamine or vitamin B_1 acts** as coenzyme in the form of thiamine pyrophosphate.

(*ii*) **Riboflavin or vitamin B_2 acts** as coenzyme in the form of following two compounds:
(*a*) **Flaving mononucleotide (FMN)** and
(*b*) **Flavin adenine dinucleotide (FAD)**; (Fig. 6.22).

(*iii*) **Nicotinic acid** acts as coenzyme in the form of following two compounds:
(*a*) *Nicotinamide adenine dinucleotide* (*NAD*; Fig. 6.23) and
(*b*) *Nicotinamide adenine dinucleotide phosphate* (*NADP*) or *coenzyme-II*.

(*iv*) **Pantothenic acid or vitamin-B_3.** It acts as a component of coenzyme A.

(*v*) **Pyridoxine or Vitamin-B_6.** The phosphate derivative of this vitamin pyridoxal-phosphate acts as a coenzyme.

(*vi*) **Folic acid or pteroylglutamic acid (PGA)** act as a coenzyme in DNA synthesis.

(*vii*) **Cyanocobalamin or vitamin-B_{12}.** Cobamide and methyl B_{12} act as coenzymes as methylcobalamin in methylation.

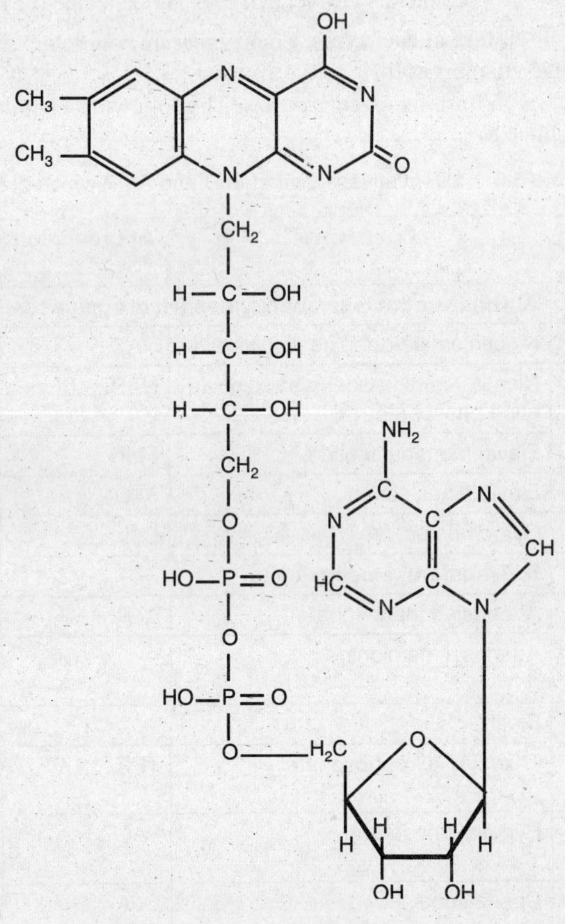

Fig. 6.22. Chemical formula of flavin adenine dinucleotide (FAD).

(*viii*) **Biotin or vitamin-H.** This vitamin acts as coenzyme for various carboxylating enzymes in CO_2 transfer reactions.

(*ix*) **Ascorbic acid or vitamin-C.** This vitamin acts as a coenzyme with the enzyme that changes p-hydroxy-phenyl puruvic acid into 2, 5-dihydroxy-phenyl pyruvic acid.

2. Other Organic Compounds as Coenzymes

(*i*) **Lipoic acid** acts as coenzyme in the decarboxylation of α-ketoacid.

(*ii*) **Haem peroxidase** acts as coenzyme with catalase and cytochrome-oxidase enzymes.

(*iii*) **Uridine diphosphate glucose (UDPG)** acts as a coenzyme with uridine diphospho-galactose-4-epimerase.

(*iv*) **Ubiquinone Q_1, Q_2, Q_6, Q_7, Q_8 and Q_{10}.** These organic compounds act as coenzymes in electron transport system.

3. Metals as cofactors/coenzymes. Some metals also act as coenzymes or cofactors: (*i*) **Zinc (Zn)** in carbonic anhydrase; (*ii*) **Copper (Cu)** in tyrosinase, phenolase and ascorbic oxidase; (*iii*) **Iron (Fe)** in cytochrome enzymes, catalase and peroxidase; (*iv*) **Calcium (Ca)** in lecithinase, ATPase and lipase; (*v*) **Cobalt (Co)** in peptidase; (*vi*) **Manganese (Mn)** in arginase and phosphoglucomutase and (*vii*) **Magnesium (Mg)** in phosphatases and hexokinases.

Mode of action of coenzymes. Coenzymes act as links between metabolic pathways. These can be considered as special intracellular enzyme-substrates. These help in oxido-reduction, group transfer and isomerization reactions and in reactions which form covalent bonds:

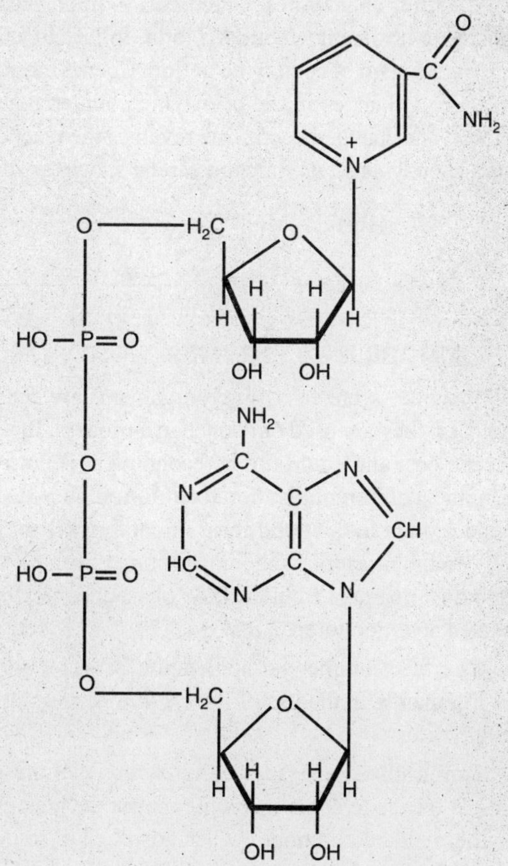

Fig. 6.23. Chemical formula of nicotinamide adenine dinucleotide (NAD^+).

A. Coenzymes acting as hydrogen acceptor in oxido-reduction reactions. These are NAD and NADP. Both these coenzymes act as **hydrogen acceptors**. For example:

(*a*) The enzyme *alcohol dehydrogenase* oxidises ethyl alcohol to acetaldehyde by removing 2H only in presence of coenzyme NAD.

(*b*) Coenzymes **FMN** and **FAD** act as prosthetic group and accept hydrogen atoms from the substrate or giving hydrogen atoms to the substrate in presence of specific apoenzyme.

(*c*) **Iron porphyrin** compounds act as prosthetic group and form integral part of cytochromes *a*, *b*, and *c* which are respiratory enzymes involved in oxidation and reduction in the cell.

B. Coenzymes acting with decarboxylases and transaminases. Pyridoxamine phosphate or pyridoxal (derivative of vitamin B_6) act as prosthetic group for a number of enzymes of amino acid metabolism (transaminases and decarboxylases).

C. Coenzymes helping in acetylation. Coenzyme A accepts acetyl group from one metabolite and donates it to another. It thus helps in metabolism of fat. Coenzyme A is formed of thioethanolamine, pantothenic acid (vitamin B) pyrophosphate and nucleotide adenylic acid.

D. Coenzyme for oxidative decarboxylation. For splitting of pyruvic acid into CO_2 and acetaldehyde in yeast cell enzyme carboxylase requires thiamine pyrophosphate as a coenzyme. In animal cells for oxidative carboxylation of pyruvic acid to yield CO_2 and acetly CoA, a coenzyme is required which contains lipoid acid and thiamine pyrophosphate.

E. Other coenzymes. Glucose-1, 6-diphosphate, glyceric acid 2, 3-diphosphate and glucose-1-phosphate-uridine nucleotide, biotin, folic acid and vitamin B12 act as coenzymes.

Mechanism of coenzyme action. Coenzyme acts as a second substrate or **cosubstrate** because the chemical changes in the coenzyme counterbalance the changes in the coenzyme occurring in substrate. The changed coenzyme reverses the reaction. Let us designate two metabolites as M_1 and M_2 and coenzyme A, its function can be shown as follows:

$$\text{Acetyl} - M_1 + \text{CoA} \xrightarrow{\text{Apoenzyme 1}} M_1 + \text{Acetyl CoA}$$

$$\text{Acetyl CoA} + M_2 \xrightarrow{\text{Apoenzyme 2}} \text{Acetyl} - M_2 + \text{CoA}$$

6.4 IMMOBILISED ENZYMES

Many enzymes secreted by microorganisms are available on a large scale and there is no effect on their cost if they are used only once in a process. In addition, many more enzymes are such that they affect the cost and could not be economical if not reused. Therefore, reuse of enzymes led to the development of **immobilisation techniques**. These techniques involve the conversion of water soluble enzyme protein into a solid form of catalyst (**Dubey** 2006). According to biotechnologist **Trevan** (1980), immobilisation is the imprisonment of an enzyme in a distinct phase that allows exchange with, but is separated from the bulk phase in which the substrate, effector or inhibitor molecules are dispersed and monitored.

The first commercial application of immobilised enzyme technology was accomplished in 1969 in Japan with the use of *Aspergillus oryzae* **amino ancyclase** for the industrial production of L-amino acids.

Immobilised enzymes are enzymes that are physically entrapped or covalently bonded by chemical means to some insoluble matrix such as glass-beads, polyacryl amide or cellulose. They are either rendered immobile when adsorbed to some solid surface or their movement is restricted.

Enzyme immobilization can be defined as confining the enzyme molecules to a distinct phase, different from the one in which substrates and products are present. Enzyme immobilization is achieved by fixing the enzyme molecules to or within some suitable material. The process of immobilisation does not literally mean that enzyme become immobile. In case of entrapment and membrane confinement, the enzyme molecules move freely within that phase. But in case of adsorption and covalent bonding, the enzyme becomes stationary.

Methods of Immobilisation

Following biotechnological methods are used for the immobilisation of enzymes:

1. Adsorption. In this case enzyme molecules adhere to the surface of carrier matrix by hydrophobic interactions and formation of salt links (or bonds). Here, bonds of low energy are involved, *e.g.*, ionic interactions, hydrogen bonds, van der waals forces, etc. Therefore binding is very strong. Commonly used matrices are porous carbon, clays, hydrous metal oxides (aluminium oxide), glasses, organic support (starch), polymeric resins and ion-exchange matrices.

2. Covalent bonding. In this binding, the enzyme molecules are attached to the carrier mataix by covalent bonds. The binding is very strong and there is no enzyme loss. The most commonly used matrices are agarose, cellulose and polyacryl amides, etc. **Sepharose**, an agarose, is a popularly used matrix, because it is highly hydrophilic, is inert to microbial growth and is commercially available as beads. It is activated by treating with chloroformates, carbomides, glutarldehyde or other compounds.

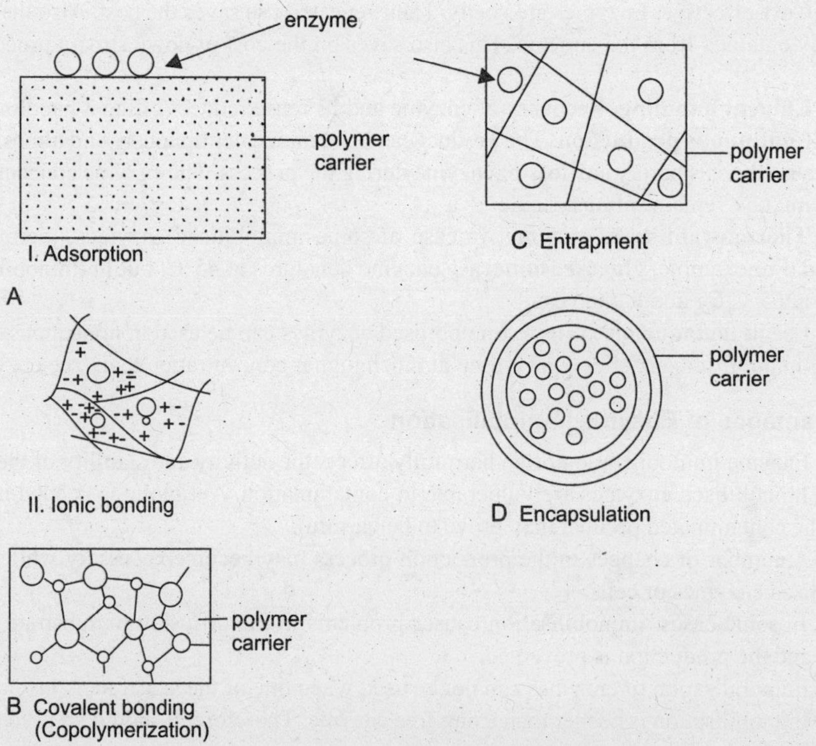

Fig. 6.24. Means of enzyme immobilisation.

3. Entrapment. The enzyme molecules are entrapped within suitable gels of fibers (*i.e.*, polyacrylamide, cellulose triacetate, agar, gelatin, carrageenan, alginate, etc.). There may or may not be covalent bond formation between the two. A noncovalent entrapment is somewhat like putting the enzyme molecule in a molecular cage. But for covalent entrapment the enzyme molecules are treated with a suitable reagent.

4. Membrane confinement (Encapsulation). In an aqueous solution, the enzyme molecules may be confined within a semipermeable membrane. This allows free movement of the substrate or products in either direction, but does not allow the enzyme molecules to escape. For this purpose following methods are adopted:

(*a*) The reaction vessel is partitioned into two chambers by a semipermeable membrane. One of these chambers contains the enzyme and other has the substrate or product.

(*b*) In this method the fibers of the matrix are submerged in the substrate solution. The enzyme is filled either in the lumen of hollow fibers of the membrane or in the hollow space between the fibers. This provides large surface area.

(*c*) Enzyme is packed in microcapsules formed by a polymerisation reaction by using 1, 6-diaminohexane. The enzyme or enzymes can be enclosed with liposomes.

Advantages of Enzyme Immobilization

Biotechnological techniques of enzyme immobilisation has the following advantages:

1. Recovery. Enzymes are not used during any reaction. Thus they can be used repeatedly, only if they can be recovered from the reaction mixture. Immobilisation of enzymes permits their repeated usage, since immobilized enzymes are separated from the reaction system, containing substrate and product.

2. Cost-effective. Enzymes are costly. Their repeated use saves the cost. Also the product can be readily obtained from the enzyme. This also saves on the cost of down stream processing of the product.

3. Effluent handling. Recovery of enzyme and its reuse reduces effluent handling problem.

4. Continuous production. The product can be obtained continuously and almost at the same rate, because there is hardly any loss of enzyme during the process. The enzyme concentration in the reaction mixture remains almost stable.

5. Thermostability of enzyme. In case of some immobilised enzymes, thermostability is increased. For example, **glucose isomerase** enzyme denatures at 45°C, but in immobilised state, it is stable at 65°C for about one year.

6. Use in nonaqueous system. Immobilised enzymes can be used in nonaquous system.

7. Immobilised enzymes can be used at much higher concentration than free enzymes.

Disadvantages of Enzyme Immobilisation

1. Enzyme immobilization often harmfully affects the activity and stability of the enzymes.
2. Immobilised enzymes are vulnerable to contamination. Antibiotic is needed in such cases or else the contaminated product may prove to be harmful.
3. A number of changes in the production process may become necessary while introducing immobilised enzymes or cells.
4. In some cases, immobilisation causes problems in the diffusion of substrate to reach the enzyme and the production is prevented.
5. Immobilisation of enzymes can not be used when one of the substrate is insoluble.
6. Immobilisation is costier than using free enzyme. Therefore, it should be used only when it is economical or safe or advantageous.

Economic Uses of Immobilised Enzymes

Immobilised enzymes have been used commercially in the production of following products:

1. High fructose corn syrup. Fructose is more sweet than glucose or sucrose. Therefore, fructose syrup is used for soft drinks. For this purpose, glucose syrup is treated with enzyme, **glucose isomerase**. This produces high fructose syrup. The enzyme glucose isomerase is obtained from *Actinoplanes missouriensis*, *Bacillus coagulans* and *Streptomyces* spp. The enzyme is used in the immobilised state and is cross-linked with glutarldehyde. Ions of Mg^{2+} and Co^{2+} are used as cofactors.

2. Immobilized lactase. Enzyme **lactase** is used to remove lactose (milk sugar) from milk and whey for getting lactose-free preparations, because some people are allergic to lactose. **Yeast lactase** is obtained from yeast *Kluyveromyces lactis*. It is immobilized in cellulose triacetate fibers. The immobilized enzyme is used for processing milk and sweet whey. Similarly, **fungal lactase** is obtained from *Aspergillus niger* and is immobilised on porous silica and is used for the treatment of whey.

3. Immobilised raffinase. Sugar raffinose is present in sugar beet juice. It is digested with immobilised enzyme **raffinase** by using immobilised cells of mould *Mortierella vinacea* variety *raffinoseutilizer*. The same enzyme is also used to remove raffinose and stachyose from soyabean milk, because these sugars cause flatulence.

4. Immobilised invertase. The enzyme *invertase* is obtained by the autolysis of yeast cells and is used to obtain **invertase** sugar. The invertase enzyme is immobilised by adsorption on to bone char. The purified enzyme-bone char mix is stabilized by cross-linking. The process is carried out in a continuous mode. One kg of granular invertase produces 16 tons of inverted sugar in packed bed reactor.

5. Acrylamide production. In the polymer industry, **acrylamide** is used to produce economically useful polymers in heavy chemical industry. It is obtained by the addition of a water molecule to acrylo-nitrite. The reaction is catalyzed by enzyme **nitrite hydratase**. This enzyme is obtained from bacterium *Rhodococcus rhodochrous*. These bacterial cells are used at 10°C in a semi-batchwise process.

6. Antibiotic production. For the mass-production of antibiotic enzymes **penicillin amidases** (also called **penicillin acylases**) are used. These enzymes are obtained from *E.coli* and are immobilised. The most popular way is by covalent linkage to sephadex 200, activated by cyanogen bromide. The enzyme can be used continuously for 100 days and produced about 2 tons of 6-amino-penicillonic acid per kg of immobilised enzyme.

QUESTIONS

Long Answer Questions

1. What are enzymes? How do they differ from catalysts? Describe the characteristics of enzymes. (*CCSU Meerut 2009*)
2. Define enzymes. Give their chemical nature and nomenclature of enzymes. (*Gorakhpur 2001*)
3. Give an account of chemical nature and nomenclature of enzymes. (*Gorakhpur 2002*)
4. What are enzymes? Describe their properties and mode of action. (*Bundelkhand 2003; Kanpur 2004*)
5. Define enzymes. Describe IUB system of nomenclature and classification of enzymes. (*Gorakhpur 2003; CCSU Meerut 2003*)
6. Give an account of chemical nature, classification and nomenclature of enzymes. (*Gorakhpur 2002, 04*)
7. What are enzymes? Give an account of various enzymes and their working. (*Rohilkhand 2002*)
8. What are enzymes? How does an enzyme work? Classify enzymes giving example.
9. Explain in brief the theories of mechanism of enzyme action. (*Garhwal 2003*)
10. Describe mechanism of action of enzyme. (*CCSU Meerut 2000, 03*)
11. What are enzymes? Explain lock and key mechanism of enzyme action.
12. What are enzymes? Describe factors affecting enzyme activity. (*Gorakhpur 2000; Kanpur 2003; CCSU Meerut 2013*)
13. Describe mode of enzyme action and also give factors affecting enzyme action.
14. What is enzyme? Explain how do the substrate concentration, pH and temperature affect the enzyme activity.
15. Write an essay on enzymes emphasizing their types, properties, mechanism of action and limiting factors.
16. Write an essay on coenzymes. (*Rohilkhand 2002*)
17. What are coenzymes? Describe the structure and functions of all types. (*Garhwal 2000*)
18. Enlist in a tabular form those derivatives of vitamin B and AMP which acts as coenzymes.
19. With the help Michaelis menton curve explain in brief the relationship between Km and V_{max}. (*Garhwal 2003*)
20. What do you mean by immobilized enzyme and immobilization of enzyme? Describe various methods of enzyme immobilization.

Short Answer Questions

1. Justify the statement: Almost all enzymes are proteins.
2. Define enzyme unit, specific activity and turnover number.
3. Give an example of competitive inhibition.
4. Why allosteric modulation is called feed back inhibition?
5. What is immobilisation of enzymes? Discuss advantages and disadvantages of enzyme immobilization.
6. What are endoenzymes? (*CCSU Meerut 2006*)
7. How do enzymes speed up biological reactions?

8. Explain turnover number of enzymes.
9. Discuss the effect of rise in temperature on the enzyme activity. What is denaturation of enzymes?
10. Write a note on IUB system of nomenclature. (*Kanpur 2004*)
11. Write a note on allosteric modulation.
12. Give the relationship between cofactor, coenzyme, apoenzyme and holoenzyme.
13. Write differences between the following:
    (*a*) Enzyme and Inorganic catalyst. (*Avadh 2004*)
    (*b*) Endoenzymes and Exoenzymes.
    (*c*) Pepsin and Trypsin.
14. Write short notes on the following:
    (*i*) Coenzymes. (*Bundelkhand 2000; CCSU Meerut 2000, 02; Poorvanchal 03*)
    (*ii*) Apoenzyme.
    (*iii*) Effect of pH and temperature on enzyme activity.
    (*iv*) Enzyme.
    (*v*) Enzyme action.
    (*vi*) Isoenzymes. (*Kanpur 2004*)
    (*vii*) Mechanism of enzyme action.
    (*viii*) Prosthetic group. (*Poorvanchal 2003*)
    (*ix*) International Union of Biochemistry (IUB) System. (*Kanpur 2003*)
    (*x*) Induced fit theory. (*CCSU Meerut 2000*)
    (*xi*) Proenzyme.
    (*xii*) Isozymes.
    (*xiii*) Zymogen. (*Garhwal 2000*)
    (*xiv*) Enzyme activators. (*Rohilkhand 2000*)
    (*xv*) Holoenzymes.

Very Short Answer Questions

1. Give two examples of coenzymes. (*CCSU Meerut 2009*)
2. Give name of the enzyme present in human saliva. (*CCSU Meerut 2006*)
3. Who discovered enzyme and when?
4. Who coined the term enzyme?
5. Name the scientist associated with lock and key theory of enzyme action.
6. Who proposed induced fit hypothesis for enzyme action?
7. How are enzymes named?
8. At what temperature enzymes act the best?
9. Give one example of reversible competitive inhibitor and one example of irreversible competitive inhibitor.
10. What is in active form of trypsin?
11. Name one vitamin which acts as coenzyme.
12. What is the pH at which enzyme salivary amylase in buccal cavity is active?
13. What is the term used for modified binding site?
14. Name the type of metabolic control on enzyme action in which end product inhibits the activity of first enzyme of the metabolic pathway.
15. Given on example of cofactor.
16. What is the name given to that part of enzyme where catalytic activity is carried out?
17. Why enzymes are required in very small amount?
18. Why enzymes are called biocatalysts?
19. What are proenzymes?
20. What happens to an enzyme when it is heated upto 100°C.
21. Name the class of enzymes which split larger molecules into smaller ones.

22. Which structural level enables the proteins to become functional enzyme?
23. Name the fastest acting enzyme.
24. What are ligases?
25. What is the fundamental chemical nature of enzymes?
26. Which proteins digesting enzyme functions in acidic medium?
27. Name the enzyme which is associated with transfer of electrons of oxidation reduction reaction.
28. Name the class of enzyme that catalyze the addition of a group of double bond without hydrolysis.
29. What is Prosthetic group?
30. The amino acids in a protein molecule are connected by a specific bond. What these bonds are called?
31. Give one example of isoenzyme.

True or False

The initial rate of an enzyme-catalysed reaction is

1. Normally directed proportional to substrate concentration.
2. Normally directly proportional to enzyme concentration.
3. Always maximum at neutral pH.
4. Always maximum when the substrate concentration is equal to the km (Michaelic constant).
5. Independent of temperature.

Fill in the Blanks

1. Trypsinogen is activated to trypsin by an enzyme called ________.
2. Cynide inhibits the mitochondrial enzyme ________ which is essential for cellular respiration.
3. The coenzyme pteroylglutamic acid or PGA represents ________.

Multiple Choice Questions

1. Who coined the term zymase for enzymes in yeast?
   (*a*) Kuhne (*b*) Louis Pasteur
   (*c*) Sumner (d) Eduard Buchner
2. Enzymes differ from ordinary catalysts, because enzyme consists of
   (*a*) vitamins (*b*) non-proteins
   (*c*) proteins (*d*) minerals
3. The temperature range, at which enzymes are maximum functional is
   (*a*) 30–55°C (*b*) 25–35°C
   (*c*) 15–25°C (*d*) 40–65°C
4. Which enzyme shows highest turn over number?
   (*a*) nuclear (*b*) phosphofructokinase
   (*c*) carbonic anhydrase (*d*) pepsin
5. An enzyme which increases the rate of permeability across the membrane is
   (*a*) permease (*b*) catalase
   (*c*) gelatinase (*d*) amylase
6. Feedback inhibition of enzyme action is affected by
   (*a*) enzyme (*b*) substrate
   (*c*) end product (*d*) none of these
7. Choose the correct substrate-enzyme pair
   (*a*) casein-rennin (*b*) protein-amylase
   (*c*) carbohydrate-lipase (*d*) lactose-maltose
8. Nomenclature of an enzyme consists of
   (*a*) first substrate name then reaction type (*b*) first reaction type and then product name

(*c*) only reaction type (*d*) only product type

9. Enzymes which possess an additional site other than active site are called
(*a*) apoenzyme (*b*) conjugated protein
(*c*) heloenzyme (*d*) allosteric enzymes

10. Substances which bring changes in allosteric sites are called
(*a*) promoters (*b*) activators
(*c*) modulators (*d*) inhibitors

11. "All enzymes are proteins". This statement is now modified because an apparent exception to this biological truth is
(*a*) arylsulphatase (*b*) dehydrogenase
(*c*) ribozyme (*d*) nitroreductase

12. A working combination of an aponenzyme and a coenzyme is termed as
(*a*) prosthetic group (*b*) enzyme substrate complex
(*c*) holoenzyme (*d*) enzyme product complex

13. The lock and key model of enzyme action illustrates that a particular enzyme molecule
(*a*) may be destroyed and resynthesized several times
(*b*) interacts with a specific type of substrate
(*c*) reacts at identical rates under all conditions
(*d*) forms a permanent enzyme-substrate complex

14. Michaelis-Menten constant (Km) of an enzyme is substrate concentration at which the reaction attains
(*a*) its maximum velocity (*b*) half its maximum velocity
(*c*) double its maximum velocity (*d*) its normal velocity

15. Glutamate pyruvate transaminase belongs to which of the following categories of enzymes?
(*a*) oxireductase (*b*) hydrolases
(*c*) lyases (*d*) transferase

16. An example of competitive inhibition of an enzyme is the inhibition of
(*a*) succinic dehydrogenase by melonic acid (*b*) cytochrome oxidase by cyanide
(*c*) hexokinase by glucose-6-phosphate (*d*) carbonic hydrase by carbon dioxide

17. The enzyme amylase belongs to the class
(*a*) hydrolases (*b*) isomerases
(*c*) transferases (*d*) lyases

18. The enzyme which is added to baby foods to partially predigest them is
(*a*) zymase (*b*) protease
(*c*) streptokinase (*d*) trypsin

19. A ribozyme is
(*a*) an enzyme associated with ribosome (*b*) a catalytic RNA
(*c*) an enzyme that helps in ribose synthesis (*d*) an enzyme that joins ribose with adenine

ANSWERS

Short Answer Questions

1. Almost all enzymes (with the exception of RNA-containing ribozymes) contain a protein part known as **apoenzyme**. Hence the enzymes show generally all properties which are identical to that of proteins. They are: (*i*) Water solubility. (*ii*) Precipitated by protein precipitating reagents. (*iii*) Non-dialysable. (*iv*) Amphoteric. (*v*) Have isoelectric point. (*vi*) Dependent on pH. (*vii*) Show UV absorption at 280 nm. (*viii*) Give the Biuret test. (*ix*) Have peptide bond. Hence except ribozymes which are made of ribonucleic acid (RNA), almost all enzymes are protein in nature.

2. **Enzyme unit.** It is defined as that amount causing transformation of 1 μ mole of substrate per min at 25°C under optimum conditions of measurement.
Specific activity. Most enzymes are protein in nature. The total enzyme unit present per mg of protein is known as **specific activity**. Specific activity is a measure of enzyme purity. It increases during purification of enzyme and becomes maximum and constant when the enzyme is in pure state.
Turnover number. It is defined as the number of substrate molecules transformed per unit time by a single enzyme molecule (or by single catalytic site) when the enzyme concentration alone is the rate limiting factor. For example, *carbonix anhydrase* (High concentration in RBC) has turnover number of 36 lakh per min per enzyme molecule.

$$CO_2 + H_2O \rightleftharpoons H_2CO_3$$

3. For competitive inhibition, the following may be mentioned as an example; succinic dehydrogenase.

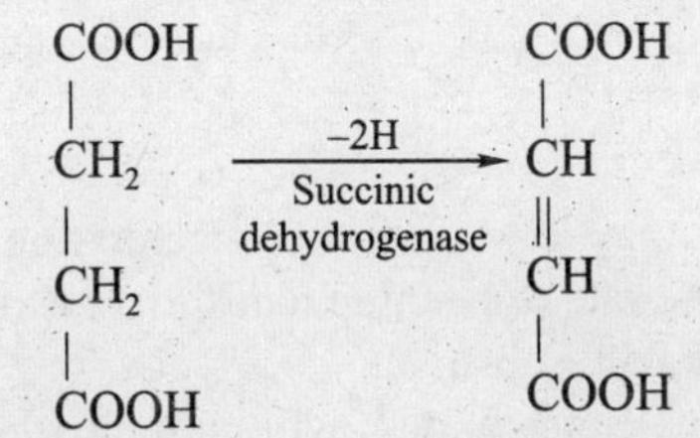

Succinic acid Fumaric acid

Here: Substrate → Succinic acid
Product → Fumaric acid
Enzyme → Succinic dehydrogenase
Competitive inhibitor → Malonate
Coenzyme → Vit B_2 (FAD)

4. Allosteric modulation is an enzyme regulatory mechanism where products or intermediates of a reaction function as temporary allosteric inhibitors if their concentration crosses a threshold value. It is, therefore, also called feedback inhibition, *e.g.*, inhibition of enzyme hexokinase by glucose-6-phosphate.

Very Short Answer Questions

1. Thiamine pyrophosphate and Flavin adenine dinucleotide (FAD).
2. Salivary amlase or ptyalin.
3. **Edward Buchner** (1897).
4. **Kuhn** (1878).
5. Fischer.
6. **Koshland**.
7. By adding *ase* to the word of substrate on which they act.
8. 5°C.
9. Sulpha drug PABA being used to kill bacteria, insecticide.
10. Trypsinogen.
11. Vitamin B_1 (Thiamine).
12. pH 7.4 (Slightly alkaline).
13. Allosteric site.
14. End product inhibition.
15. Cu, Fe or Zn.
16. Active site or substrate site.

17. Because enzymes are not used during reaction and enzymes have high turn over.
18. Because these are made of proteins and control metabolis reactions inside the organisms.
19. Inactive form of enzyme.
20. It gets denatured and functionless.
21. Hydrolases.
22. Tertiary structure.
23. Carbonic anhydrase.
24. Enzymes which brings about joining of two different molecules or substance.
25. Mostly proteins but sometimes RNA.
26. Pepsin.
27. Oxido-reductase.
28. Lyases.
29. Nonprotein molecule of conjugated enzyme.
30. Peptide bonds.
31. Lactate dehydrogenase.

True or False

1. **False.** This is only true at very low substrate concentration (relative to km). As the substrate concentration increases, the rate of reaction tends to a maximum value (Vmax). The enzyme is then said to be saturated with substrate.
2. **True.** Doubling the enzyme concentration will normally double the reaction rate, if all other conditions are kept constant and components are in excess.
3. **False.** The pH value at which activity is at a maximum (the optimum pH) depends upon the particular enzyme. Enzymes normally have an optimum pH suited to their particular environment.
4. **False.** The Km is the substrate concentration producing half the maximal velocity. Note that Km has the units of concentration.
5. **False.** An increase of 10°C normally causes an approximate doubling of the rate of a chemical reaction. Enzymes, though, being proteins, will generally denature and begin to lose activity if the temperature is increased much above 35–40°C. Thus the effect of temperature on an enzyme, catalysed reaction is not simple to predict.

Fill in the Blanks

1. Enterokinase. **2.** Cytochrome oxidase. **3.** Folic acid.

Multiple Choice Questions

1. (*d*)	**2.** (*c*)	**3.** (*b*)	**4.** (*c*)	**5.** (*a*)	**6.** (*c*)
7. (*a*)	**8.** (*a*)	**9.** (*d*)	**10.** (*c*)	**11.** (*c*)	**12.** (*c*)
13. (*b*)	**14.** (*b*)	**15.** (*d*)	**16.** (*a*)	**17.** (*a*)	**18.** (*d*)
19. (*b*)					

INDEX

A

B

C

D

E

F

G

M

N

O

P

T

U

V

W

X

Z

NOTES

NOTES